Atomic Masses of the Elements

Name	Symbol	Atomic Number	Atomic Mass[a]
Actinium	Ac	89	(227)[b]
Aluminum	Al	13	26.98
Americium	Am	95	(243)
Antimony	Sb	51	121.8
Argon	Ar	18	39.95
Arsenic	As	33	74.92
Astatine	At	85	(210)
Barium	Ba	56	137.3
Berkelium	Bk	97	(247)
Beryllium	Be	4	9.012
Bismuth	Bi	83	209.0
Bohrium	Bh	107	(264)
Boron	B	5	10.81
Bromine	Br	35	79.90
Cadmium	Cd	48	112.4
Calcium	Ca	20	40.08
Californium	Cf	98	(251)
Carbon	C	6	12.01
Cerium	Ce	58	140.1
Cesium	Cs	55	132.9
Chlorine	Cl	17	35.45
Chromium	Cr	24	52.00
Cobalt	Co	27	58.93
Copernicium	Cn	112	(285)
Copper	Cu	29	63.55
Curium	Cm	96	(247)
Darmstadtium	Ds	110	(271)
Dubnium	Db	105	(262)
Dysprosium	Dy	66	162.5
Einsteinium	Es	99	(252)
Erbium	Er	68	167.3
Europium	Eu	63	152.0
Fermium	Fm	100	(257)
Flerovium	Fl	114	(289)
Fluorine	F	9	19.00
Francium	Fr	87	(223)
Gadolinium	Gd	64	157.3
Gallium	Ga	31	69.72
Germanium	Ge	32	72.64
Gold	Au	79	197.0
Hafnium	Hf	72	178.5
Hassium	Hs	108	(265)
Helium	He	2	4.003
Holmium	Ho	67	164.9
Hydrogen	H	1	1.008
Indium	In	49	114.8
Iodine	I	53	126.9
Iridium	Ir	77	192.2
Iron	Fe	26	55.85
Krypton	Kr	36	83.80
Lanthanum	La	57	138.9
Lawrencium	Lr	103	(262)
Lead	Pb	82	207.2
Lithium	Li	3	6.941
Livermorium	Lv	116	(293)
Lutetium	Lu	71	175.0
Magnesium	Mg	12	24.31
Manganese	Mn	25	54.94
Meitnerium	Mt	109	(268)
Mendelevium	Md	101	(258)
Mercury	Hg	80	200.6
Molybdenum	Mo	42	95.94
Moscovium	Mc	115	(289)
Neodymium	Nd	60	144.2
Neon	Ne	10	20.18
Neptunium	Np	93	(237)
Nickel	Ni	28	58.69
Nihonium	Nh	113	(286)
Niobium	Nb	41	92.91
Nitrogen	N	7	14.01
Nobelium	No	102	(259)
Oganesson	Og	118	(294)
Osmium	Os	76	190.2
Oxygen	O	8	16.00
Palladium	Pd	46	106.4
Phosphorus	P	15	30.97
Platinum	Pt	78	195.1
Plutonium	Pu	94	(244)
Polonium	Po	84	(209)
Potassium	K	19	39.10
Praseodymium	Pr	59	140.9
Promethium	Pm	61	(145)
Protactinium	Pa	91	231.0
Radium	Ra	88	(226)
Radon	Rn	86	(222)
Rhenium	Re	75	186.2
Rhodium	Rh	45	102.9
Roentgenium	Rg	111	(272)
Rubidium	Rb	37	85.47
Ruthenium	Ru	44	101.1
Rutherfordium	Rf	104	(261)
Samarium	Sm	62	150.4
Scandium	Sc	21	44.96
Seaborgium	Sg	106	(266)
Selenium	Se	34	78.96
Silicon	Si	14	28.09
Silver	Ag	47	107.9
Sodium	Na	11	22.99
Strontium	Sr	38	87.62
Sulfur	S	16	32.07
Tantalum	Ta	73	180.9
Technetium	Tc	43	(99)
Tellurium	Te	52	127.6
Tennessine	Ts	117	(294)
Terbium	Tb	65	158.9
Thallium	Tl	81	204.4
Thorium	Th	90	232.0
Thulium	Tm	69	168.9
Tin	Sn	50	118.7
Titanium	Ti	22	47.87
Tungsten	W	74	183.8
Uranium	U	92	238.0
Vanadium	V	23	50.94
Xenon	Xe	54	131.3
Ytterbium	Yb	70	173.0
Yttrium	Y	39	88.91
Zinc	Zn	30	65.41
Zirconium	Zr	40	91.22

[a]Values for atomic masses are given to four significant figures.

[b]Values in parentheses are the mass number of an important radioactive isotope.

CHEMISTRY

An Introduction to General, Organic, and Biological Chemistry

CHEMISTRY

An Introduction to General, Organic, and Biological Chemistry

Thirteenth Edition

Karen Timberlake

Contributions by
MaryKay Orgill, Ph.D.
Professor of Chemistry,
University of Nevada, Las Vegas

330 Hudson Street, NY NY 10013

Courseware Portfolio Manager: Scott Dustan
Director, Courseware Portfolio Management: Jeanne Zalesky
Content Producer: Lizette Faraji
Managing Producer: Kristen Flathman
Courseware Analyst: Coleen Morrison
Courseware Director, Content Development: Jennifer Hart
Courseware Editorial Assistant: Fran Falk
Rich Media Content Producer: Jenny Moryan
Full Service Vendor: SPi Global
Full Service Project Manager: Karen Berry/Christian Arsenault
Copyeditor: Laura Patchkofsky
Design Manager: Mark Ong
Cover and Interior Designer: Tamara Newnam
Photo and Illustration Support: Stephanie Marquez, Imagineering Art
Rights and Permissions Project Manager: Kathleen Zander
Rights and Permissions Management: Ben Ferrini
Manufacturing Buyer: Maura Zaldivar-Garcia
Marketing Manager: Elizabeth Ellsworth Bell
Cover Image Credit: © Ralph Clevenger/Corbis Documentary/Getty Images

Credits and acknowledgments borrowed from other sources and reproduced, with permission, in this textbook appear on p. C-1.

Library of Congress Cataloging-in-Publication Data

Cataloging-in-Publication Data is on file with the Library of Congress.

3 17

ISBN-10: 0-134-42135-3
ISBN-13: 978-0-134-42135-3

www.pearsonhighered.com

Brief Contents

Table of Contents

1 Chemistry in Our Lives 1

2 Chemistry and Measurements 25

3 Matter and Energy 60

4 Atoms and Elements 98

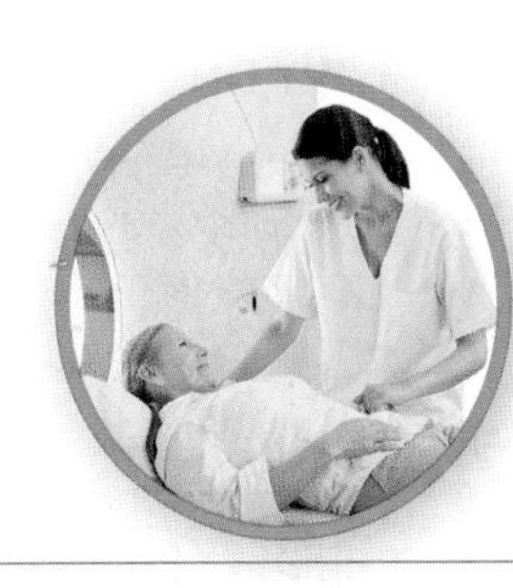

5 Nuclear Chemistry 136

6 Ionic and Molecular Compounds 168

10 Acids and Bases and Equilibrium 324

11 Introduction to Organic Chemistry: Hydrocarbons 363

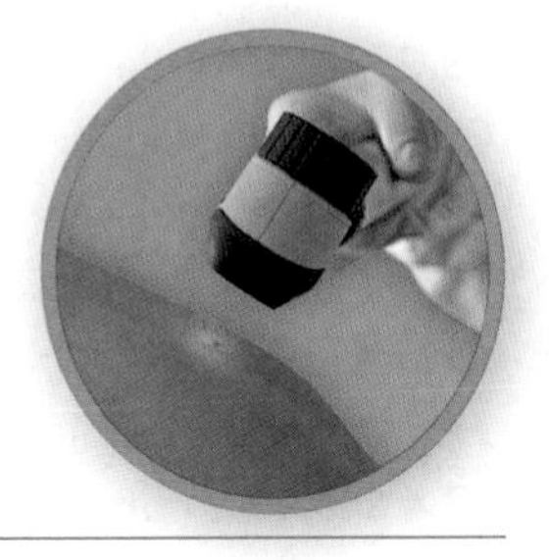

12 Alcohols, Thiols, Ethers, Aldehydes, and Ketones 398

17 Nucleic Acids and Protein Synthesis 584

18 Metabolic Pathways and ATP Production 619

Applications and Activities

KEY MATH SKILLS

CORE CHEMISTRY SKILLS

Interactive Videos

About the Author

KAREN TIMBERLAKE is Professor Emerita of chemistry at Los Angeles Valley College, where she taught chemistry for allied health and preparatory chemistry for 36 years. She received her bachelor's degree in chemistry from the University of Washington and her master's degree in biochemistry from the University of California at Los Angeles.

Professor Timberlake has been writing chemistry textbooks for 40 years. During that time, her name has become associated with the strategic use of pedagogical tools that promote student success in chemistry and the application of chemistry to real-life situations. More than one million students have learned chemistry using texts, laboratory manuals, and study guides written by Karen Timberlake. In addition to *An Introduction to General, Organic and Biological Chemistry*, thirteenth edition, she is also the author of *General, Organic, and Biological Chemistry*, fifth edition, with the accompanying *Study Guide and Selected Solutions Manual*, *Laboratory Manual* and *Essentials Laboratory Manual*, and *Basic Chemistry*, fifth edition, with the accompanying *Study Guide and Selected Solutions Manual.*

Professor Timberlake belongs to numerous scientific and educational organizations including the American Chemical Society (ACS) and the National Science Teachers Association (NSTA). She has been the Western Regional Winner of the Excellence in College Chemistry Teaching Award given by the Chemical Manufacturers Association. She received the McGuffey Award in Physical Sciences from the Textbook Authors Association for her textbook *Chemistry: An Introduction to General, Organic, and Biological Chemistry*, eighth edition, which has demonstrated her excellence over time. She received the "Texty" Textbook Excellence Award from the Textbook Authors Association for the first edition of *Basic Chemistry.* She has participated in education grants for science teaching including the Los Angeles Collaborative for Teaching Excellence (LACTE) and a Title III grant at her college. She speaks at conferences and educational meetings on the use of student-centered teaching methods in chemistry to promote the learning success of students.

When Professor Timberlake is not writing textbooks, she and her husband relax by playing tennis, ballroom dancing, traveling, trying new restaurants, cooking, and taking care of their grandchildren, Daniel and Emily.

DEDICATION

I dedicate this book to

- My husband, Bill, for his patience, loving support, and preparation of late meals
- My son, John, daughter-in-law, Cindy, grandson, Daniel, and granddaughter, Emily, for the precious things in life
- The wonderful students over many years whose hard work and commitment always motivated me and put purpose in my writing

FAVORITE QUOTES

The whole art of teaching is only the art of awakening the natural curiosity of young minds.

—Anatole France

One must learn by doing the thing; though you think you know it, you have no certainty until you try.

—Sophocles

Discovery consists of seeing what everybody has seen and thinking what nobody has thought.

—Albert Szent-Gyorgyi

I never teach my pupils; I only attempt to provide the conditions in which they can learn.

—Albert Einstein

Preface

Welcome to the thirteenth edition of *An Introduction to General, Organic, and Biological Chemistry*. This chemistry text was written and designed to help you prepare for a career in a health-related profession, such as nursing, dietetics, respiratory therapy, and environmental and agricultural science. This text assumes no prior knowledge of chemistry. My main objective in writing this text is to make the study of chemistry an engaging and a positive experience for you by relating the structure and behavior of matter to its role in health and the environment. This new edition introduces more problem-solving strategies, more problem-solving guides, new Analyze the Problem with Connect features, new Try It First and Engage features, conceptual and challenge problems, and new sets of combined problems.

It is my goal to help you become a critical thinker by understanding scientific concepts that will form a basis for making important decisions about issues concerning health and the environment. Thus, I have utilized materials that

- help you to learn and enjoy chemistry
- relate chemistry to careers that interest you
- develop problem-solving skills that lead to your success in chemistry
- promote learning and success in chemistry

New for the Thirteenth Edition

New and updated features have been added throughout this thirteenth edition, including the following:

- NEW AND UPDATED! **Chapter Openers** provide engaging clinical stories in the health profession and introduce the chemical concepts in each chapter.
- NEW! **Clinical Updates** added at the end of each Chapter continue the story of the chapter opener and describe the follow-up treatment.
- NEW! **Engage** feature in the margin asks students to think about the paragraph they are reading and to test their understanding by answering the Engage question, which is related to the topic.
- NEW! **Try It First** precedes the Solution section of each Sample Problem to encourage the student to work on the problem before reading the given Solution.
- NEW! **Connect** feature added to **Analyze** the **Problem** boxes indicates the relationships between *Given* and *Need.*
- NEW! **Clinical Applications** added to Practice Problems show the relevance between the chemistry content and medicine and health.
- NEW! **Strategies for Learning Chemistry** are added that utilize successful ways to study and learn chemistry.
- NEW! **TEST** feature added in the margin encourages students to solve related Practice Problems to practice retrieval of content for exams.
- NEW! **Interactive Videos** give students the experience of step-by-step problem solving for problems from the text.
- NEW! **Review** topics placed in the margin at the beginning of a Section list the Key Math Skills and Core Chemistry Skills from the previous chapters, which provide the foundation for learning new chemistry principles in the current chapter.
- UPDATED! **Solution Guides** are now included in selected Sample Problems.
- UPDATED! **Key Math Skills** review basic math relevant to the chemistry the students are learning throughout the text. A **Key Math Skill Review** at the end of each chapter summarizes and gives additional examples.
- UPDATED! **Core Chemistry Skills** identify the key chemical principles in each chapter that are required for successfully learning chemistry. A **Core Chemistry Skill Review** at the end of each chapter helps reinforce the material and gives additional examples.
- UPDATED! **Analyze the Problem** features included in the Solutions of the Sample Problems strengthen critical-thinking skills and illustrate the breakdown of a word problem into the components required to solve it.
- UPDATED! **Practice Problems**, **Sample Problems**, and **art** demonstrate the connection between the chemistry being discussed and how these skills will be needed in professional experience.
- UPDATED! **Combining Ideas** features offer sets of integrated problems that test students' understanding and develop critical thinking by integrating topics from two or more previous chapters.

Chapter Organization of the Thirteenth Edition

In each textbook I write, I consider it essential to relate every chemical concept to real-life issues. Because a chemistry course may be taught in different time frames, it may be difficult to cover all the chapters in this text. However, each chapter is a complete package, which allows some chapters to be skipped or the order of presentation to be changed.

Chapter 1, Chemistry in Our Lives, discusses the Scientific Method in everyday terms, guides students in developing a study plan for learning chemistry, with a section of Key Math

Skills that reviews the basic math, including scientific notation, needed in chemistry calculations.

- The Chapter Opener tells the story of a murder and features the work and career of forensic scientists.
- A new Clinical Update feature describes the forensic evidence that helps to solve the murder and includes Clinical Applications.
- "Scientific Method: Thinking Like a Scientist" is expanded to include *law* and *theory*.
- Writing Numbers in Scientific Notation is now a new Section.
- An updated Section titled Studying and Learning Chemistry expands the discussion of strategies that improve learning and understanding of content.
- Key Math Skills are: Identifying Place Values, Using Positive and Negative Numbers in Calculations, Calculating Percentages, Solving Equations, Interpreting Graphs, and Writing Numbers in Scientific Notation.

Chapter 2, Chemistry and Measurements, looks at measurement and emphasizes the need to understand numerical relationships of the metric system. Significant figures are discussed in the determination of final answers. Prefixes from the metric system are used to write equalities and conversion factors for problem-solving strategies. Density is discussed and used as a conversion factor.

- The Chapter Opener tells the story of a patient with high blood pressure and features the work and career of a registered nurse.
- A new Clinical Update describes the patient's status and follow-up visit with his doctor.
- New photos, including an endoscope, propranolol tablets, cough syrup, people exercising, a urine dipstick, and a pint of blood, are added to improve visual introduction to clinical applications of chemistry. Previous art is updated to improve clarity.
- Sample Problems relate problem solving to health-related topics such as the measurements of blood volume, omega-3 fatty acids, radiological imaging, body fat, cholesterol, and medication orders.
- New Clinical Applications feature questions about measurements, daily values for minerals and vitamins, equalities and conversion factors for medications.
- New material illustrates how to count significant figures in equalities and in conversion factors used in a problem setup.
- A new Key Math Skill, Rounding Off, has been added.
- Core Chemistry Skills are: Counting Significant Figures, Using Significant Figures in Calculations, Using Prefixes, Writing Conversion Factors from Equalities, Using Conversion Factors, and Using Density as a Conversion Factor.

Chapter 3, Matter and Energy, classifies matter and states of matter, describes temperature measurement, and discusses energy, specific heat, energy in nutrition, and changes of state. Physical and chemical properties and physical and chemical changes are discussed.

- The chapter opener describes diet and exercise for an overweight adolescent at risk for type 2 diabetes and features the work and career of a dietitian.
- A new Clinical Update describes the new diet prepared with a dietitian for weight loss.
- Practice Problems and Sample Problems include high temperatures used in cancer treatment, the energy produced by a high-energy shock output of a defibrillator, body temperature lowering using a cooling cap, ice bag therapy for muscle injury, and energy values for food.
- Core Chemistry Skills are: Identifying Physical and Chemical Changes, Converting between Temperature Scales, Using Energy Units, Using the Heat Equation, and Calculating Heat for Change of State.
- The interchapter problem set, Combining Ideas from Chapters 1 to 3, completes the chapter.

Chapter 4, Atoms and Elements, introduces elements and atoms and the periodic table. The names and symbols for the newest elements 113, Nihonium, Nh, 115, Moscovium, Mc, 117, Tennessine, Ts, and 118, Oganesson, Og, are added to the periodic table. Electron arrangements are written for atoms and the trends in periodic properties are described. Atomic numbers and mass numbers are determined for isotopes. The most abundant isotope of an element is determined by its atomic mass.

- The Chapter Opener and Follow Up feature the work and career of a farmer.
- A new Clinical Update describes the improvement in crop production by the farmer.
- Atomic number and mass number are used to calculate the number of protons and neutrons in an atom.
- The number of protons and neutrons are used to calculate the mass number and to write the atomic symbol for an isotope.
- The trends in periodic properties are described for valence electrons, atomic size, ionization energy, and metallic character.
- Core Chemistry Skills are: Counting Protons and Neutrons, Writing Atomic Symbols for Isotopes, Writing Electron Arrangements, Identifying Trends in Periodic Properties, and Drawing Lewis Symbols.

Chapter 5, Nuclear Chemistry, looks at the types of radiation emitted from the nuclei of radioactive atoms. Nuclear equations are written and balanced for both naturally occurring radioactivity and artificially produced radioactivity. The half-lives of radioisotopes are discussed, and the amount of time for a sample to decay is calculated. Radioisotopes important in the field of nuclear medicine are described. Fission and fusion and their role in energy production are discussed.

- The new chapter opener describes a patient with possible coronary heart disease who undergoes a nuclear stress test and features the work and career of a radiation technologist.
- A new Clinical Update discusses the results of cardiac imaging using the radioisotope Tl-201.
- Sample Problems and Practice Problems use nursing and medical examples, including phosphorus-32 for the treatment of leukemia, titanium seeds containing a radioactive isotope implanted in the body to treat cancer, yttrium injections for arthritis pain, and millicuries in a dose of phosphorus-32.
- Core Chemistry Skills are: Writing Nuclear Equations and Using Half-Lives.

Chapter 6, Ionic and Molecular Compounds, describes the formation of ionic and covalent bonds. Chemical formulas are written, and ionic compounds—including those with polyatomic ions—and molecular compounds are named.

- The chapter opener describes aspirin as a molecular compound and features the work and career of a pharmacy technician.
- A new Clinical Update describes several types of compounds at a pharmacy and includes Clinical Applications.
- Section 6.6 is now titled "Lewis Structures for Molecules," 6.7 is "Electronegativity and Bond Polarity," 6.8 is "Shapes of Molecules," and 6.9 is "Polarity of Molecules and Intermolecular Forces."
- The term Lewis structure has replaced the term electron-dot formula.
- Updated material on polyatomic ions compares the names of *ate* ions and *ite* ions, the charge of carbonate and hydrogen carbonate, and the formulas and charges of halogen polyatomic ions with oxygen.
- A new art comparing the particles and bonding of ionic compounds and molecular compounds has been added.
- A new flowchart for naming chemical compounds in Section 6.5 shows naming patterns for ionic and molecular compounds.
- Core Chemistry Skills are: Writing Positive and Negative Ions, Writing Ionic Formulas, Naming Ionic Compounds, Writing the Names and Formulas for Molecular Compounds, Drawing Lewis Structures, Using Electronegativity, Predicting Shape, and Identifying Polarity of Molecules and Intermolecular Forces.
- The interchapter problem set, Combining Ideas from Chapters 4 to 6, completes the chapter.

Chapter 7, Chemical Quantities and Reactions, discusses Avogadro's number, the mole, and molar masses of compounds, which are used in calculations to determine the mass or number of particles in a given quantity of an element or a substance. Students learn to balance chemical equations and to recognize the types of chemical reactions: combination, decomposition, single replacement, double replacement, and combustion. Chapter discussion includes Oxidation–Reduction Reactions using real-life examples, including biological reactions, Mole Relationships in Chemical Equations, Mass Calculations for Chemical Reactions, and Energy in Chemical Reactions, which discusses activation energy and energy changes in exothermic and endothermic reactions.

- The chapter opener describes the symptoms of pulmonary emphysema and discusses the career of an exercise physiologist.
- A new Clinical Update explains the treatment for interstitial lung disease.
- Sample Problems and Challenge Problems use nursing and medical examples.
- New expanded art shows visible evidence of a chemical reaction.
- Core Chemistry Skills are: Converting Particles to Moles, Calculating Molar Mass, Using Molar Mass as a Conversion Factor, Balancing a Chemical Equation, Classifying Types of Chemical Reactions, Identifying Oxidized and Reduced Substances, Using Mole–Mole Factors, and Converting Grams to Grams.

Chapter 8, Gases, discusses the properties of gases and calculates changes in gases using the gas laws: Boyle's, Charles's, Gay-Lussac's, Avogadro's, and Dalton's. Problem-solving strategies enhance the discussion and calculations with gas laws.

- The chapter opener features the work and career of a respiratory therapist.
- New Clinical Update describes exercise to prevent exercise-induced asthma. Clinical Applications are related to lung volume and gas laws.
- Sample Problems and Challenge Problems use nursing and medical examples, including, calculating the volume of oxygen gas delivered through a face mask during oxygen therapy, preparing a heliox breathing mixture for a scuba diver, and home oxygen tanks.
- Core Chemistry Skills are: Using the Gas Laws and Calculating Partial Pressure.

Chapter 9, Solutions, describes solutions, electrolytes, saturation and solubility, insoluble salts, concentrations, and osmosis. The concentrations of solutions are used to determine volume or mass of solute. The volumes and molarities of solutions are used in calculations of dilutions and titrations. Properties of solutions, osmosis in the body, and dialysis are discussed.

- The chapter opener describes a patient with kidney failure and dialysis treatment and features the work and career of a dialysis nurse.

- A new Clinical Update explains dialysis treatment and electrolyte levels in dialysate fluid.
- Art updates include gout and intravenous solutions.
- Table 9.6 on electrolytes in intravenous solutions is expanded.
- Core Chemistry Skills are: Using Solubility Rules, Calculating Concentration, and Using Concentration as a Conversion Factor.
- The interchapter problem set, Combining Ideas from Chapters 7 to 9, completes the chapter.

Chapter 10, Acids and Bases and Equilibrium, discusses acids and bases and conjugate acid–base pairs. The dissociation of strong and weak acids and bases is related to their strengths as acids or bases. The dissociation of water leads to the water dissociation expression, K_w, the pH scale, and the calculation of pH. The reactions of acids and bases with metals, carbonates, and bicarbonates are discussed. Chemical equations for acids in reactions are balanced and titration of an acid is illustrated. Buffers are discussed along with their role in the blood.

- The chapter opener describes an accident victim with respiratory acidosis and the work and career of a clinical laboratory technician.
- A Clinical Update discusses the symptoms and treatment for acid reflux disease.
- The section "Acid–Base Equilibrium" includes Le Châtelier's principle.
- Clinical Applications include calculating $[OH^-]$ or $[H_3O^+]$ of body fluids, foods, blood plasma, and the pH of body fluids.
- Key Math Skills are: Calculating pH from $[H_3O^+]$ and Calculating $[H_3O^+]$ from pH.
- New Core Chemistry Skills are: Identifying Conjugate Acid–Base Pairs, Using Le Chatelier's Principle, Calculating $[H_3O^+]$ and $[OH^-]$ in Solutions, Writing Equations for Reactions of Acids and Bases, and Calculating Molarity or Volume of an Acid or Base in a Titration.

Chapter 11, Introduction to Organic Chemistry: Hydrocarbons, compares inorganic and organic compounds, and describes the structures and naming of alkanes, alkenes including cis–trans isomers, alkynes, and aromatic compounds.

- The chapter opener describes a fire victim and the search for traces of accelerants and fuel at the arson scene and features the work and career of a firefighter/emergency medical technician.
- A new Clinical Update describes the treatment of burns in the hospital and the types of fuels identified in the fire.
- Wedge–dash models have been added to the representations of methane and ethane.
- Line-angle formulas are now included in Table 11.2 IUPAC Names and Formulas of the First Ten Alkanes.
- Core Chemistry Skills are: Naming and Drawing Alkanes and Writing Equations for Hydrogenation and Hydration.

Chapter 12, Alcohols, Thiols, Ethers, Aldehydes, and Ketones, describes the functional groups and names of alcohols, thiols, ethers, aldehydes, and ketones. The solubility of alcohols, phenols, aldehydes, and ketones in water is discussed.

- A new chapter opener describes the risk factors for melanoma and discusses work and career of a dermatology nurse.
- A new Clinical Update discusses melanoma, skin protection, and functional groups of sunscreens.
- A table Solubility of Selected Aldehydes and Ketones has been updated.
- New material on antiseptics is added.
- The oxidation of methanol in the body is included in the Chemistry Link to Health "Oxidation of Alcohol in the Body."
- Core Chemistry Skills are: Identifying Functional Groups, Naming Alcohols and Phenols, Naming Aldehydes and Ketones, Writing Equations for the Dehydration of Alcohols, and Writing Equations for the Oxidation of Alcohols.
- The interchapter problem set, Combining Ideas from Chapters 10 to 12, completes the chapter.

Chapter 13, Carbohydrates, describes the carbohydrate molecules monosaccharides, disaccharides, and polysaccharides and their formation by photosynthesis. Monosaccharides are classified as aldo or keto pentoses or hexoses. Chiral molecules are discussed along with Fischer projections and D and L notations. Chiral objects are modeled using gumdrops and toothpicks. Carbohydrates used as sweeteners are described and carbohydrates used in blood typing are discussed. The formation of glycosidic bonds in disaccharides and polysaccharides is described.

- A chapter opener describes a diabetes patient and her diet and features the work and career of a diabetes nurse.
- A new Clinical Update describes a diet to lower blood glucose.
- Chiral molecules are discussed and Fischer projections are drawn.
- A new Sample Problem identifies chiral carbons in glycerol and ibuprofen.
- New art shows that insulin needed for the metabolism of glucose is produced in the pancreas.
- Examples of chiral molecules in nature are included to Chemistry Link to Health, "Enantiomers in Biological Systems."
- New Clinical Applications include psicose in foods, lyxose in bacterial glycolipids, xylose in absorption tests, and tagatose in fruit.

- New art shows the rotation of groups on carbon 5 for the Haworth structures of glucose and galactose.
- Drawing Haworth Structures is updated.
- The Chemistry Link to Health "Blood Types and Carbohydrates" has updated structures of the saccharides that determine each blood type.
- Core Chemistry Skills are: Identifying Chiral Molecules, Identifying D and L Fischer Projections, and Drawing Haworth Structures.

Chapter 14, Carboxylic Acids, Esters, Amines, and Amides, discusses the functional groups and naming of carboxylic acids, esters, amines, and amides. Chemical reactions include esterification, amidation, and acid and base hydrolysis of esters and amides.

- A chapter opener describes pesticides and pharmaceuticals used on a ranch and discusses the career of an environmental health practitioner.
- A new Clinical Update describes an insecticide used to spray animals.
- Line-angle structures for carboxylic acids are added to Table 14.1.
- Core Chemistry Skills are: Naming Carboxylic Acids, Hydrolyzing Esters, and Forming Amides.

Chapter 15, Lipids, discusses fatty acids and the formation of ester bonds in triacylglycerols and glycerophospholipids. Chemical properties of fatty acids and their melting points along with the hydrogenation of unsaturated triacylglycerols are discussed. Steroids, such as cholesterol and bile salts, are described. Chemistry Links to Health include "Converting Unsaturated Fats to Saturated Fats: Hydrogenation." The role of phospholipids in the lipid bilayer of cell membranes is discussed as well as the lipids that function as steroid hormones.

- A new chapter opener describes a patient with symptoms of familial hypercholesterolemia and features the work and career of a clinical lipid specialist.
- A new Clinical Update describes a program to lower cholesterol.
- New notation for number of carbon atoms and double bonds in a fatty acid is added.
- New art of unsaturated fatty acids with cis and trans double bonds is added.
- New art of normal and damaged myelin sheath shows deterioration in multiple sclerosis.
- New art of the gallbladder and the bile duct where gallstones pass causing obstruction and pain.
- Core Chemistry Skills are: Identifying Fatty Acids, Drawing Structures for Triacylglycerols, Drawing the Products for the Hydrogenation, Hydrolysis, and Saponification of a Triacylglycerol, and Identifying the Steroid Nucleus.
- The interchapter problem set, Combining Ideas from Chapters 13 to 15, completes the chapter.

Chapter 16, Amino Acids, Proteins, and Enzymes, discusses amino acids, formation of peptide bonds and proteins, structural levels of proteins, enzymes, and enzyme action. The structures of amino acids are drawn at physiological pH. Enzymes are discussed as biological catalysts, along with the impact of inhibitors and denaturation on enzyme action.

- A new chapter opener discusses the symptoms of sickle-cell anemia in a child, the mutation in amino acids that causes the crescent shape of abnormal red blood cells, and the career of a physician assistant.
- The use of electrophoresis to diagnose sickle-cell anemia was added to Chemistry Link to Health "Sickle-Cell Anemia."
- Abbreviations for amino acid names use three letters as well as one letter.
- New ribbon models of beta-amyloid proteins in normal brain and an Alzheimer's brain are added to Chemistry Link to Health "Protein Secondary Structures and Alzheimer's Disease".
- Diagrams illustrate enzyme action and the effect of competitive and noncompetitive inhibitors on enzyme structure.
- Core Chemistry Skills are: Drawing the Structure for an Amino Acid at Physiological pH, Identifying the Primary, Secondary, Tertiary, and Quaternary Structures of Proteins, and Describing Enzyme Action.

Chapter 17, Nucleic Acids and Protein Synthesis, describes the nucleic acids and their importance as biomolecules that store and direct information for the synthesis of cellular components. The role of complementary base pairing is discussed in both DNA replication and the formation of mRNA during protein synthesis. The role of RNA is discussed in the relationship of the genetic code to the sequence of amino acids in a protein. Mutations describe ways in which the nucleotide sequences are altered in genetic diseases.

- A new chapter opener describes a patient's diagnosis and treatment of breast cancer and discusses the work and career of a histology technician.
- A new Clinical Update describes estrogen-positive tumors, the impact of the altered genes BRCA1 and BRCA2 on the estrogen receptor, and medications to suppress tumor growth.
- A new Section discusses recombinant DNA, polymerase chain reaction, and DNA fingerprinting.
- New art illustrates point mutation, deletion mutation, and insertion mutation.
- Core Chemistry Skills are: Writing the Complementary DNA Strand, Writing the mRNA Segment for a DNA Template, and Writing the Amino Acid for an mRNA Codon.

Chapter 18, Metabolic Pathways and ATP Production, describes the metabolic pathways of biomolecules from the digestion of foodstuffs to the synthesis of ATP. The stages of catabolism and the digestion of carbohydrates along with the coenzymes required in metabolic pathways are described. The breakdown of glucose to pyruvate is described using glycolysis, which is followed by the decarboxylation of pyruvate to acetyl CoA and the entry of acetyl CoA into the citric acid cycle. Electron transport, oxidative phosphorylation, and the synthesis of ATP is described. The oxidation of lipids and the degradation of amino acids are also discussed.

- A new chapter opener describes elevated levels of liver enzymes for a patient with chromic hepatitis C infection and discusses the career of a public health nurse.
- A new Clinical Update describes interferon and ribavirin therapy for hepatitis C.
- Updated art for glycolysis, the citric acid cycle, and electron transport is added.
- The values of ATP produced from the metabolism of glucose, fatty acids, and amino acids is calculated using the updated values of 2.5 ATP for NADH and 1.5 ATP for $FADH_2$.
- Core Chemistry Skills are: Identifying the Compounds in Glycolysis, Describing the Reactions in the Citric Acid Cycle, Calculating the ATP Produced from Glucose, and Calculating the ATP from Fatty Acid Oxidation (β Oxidation).
- The interchapter problem set, Combining Ideas from Chapters 16 to 18, completes the chapter.

Acknowledgments

The preparation of a new text is a continuous effort of many people. I am thankful for the support, encouragement, and dedication of many people who put in hours of tireless effort to produce a high-quality book that provides an outstanding learning package. I am thankful for the outstanding contributions of Professor MaryKay Orgill whose updates and clarifications enhanced the content of the biochemistry chapters 16 to 18. The editorial team at Pearson has done an exceptional job. I want to thank Jeanne Zalesky, Director, Courseware Portfolio Management, and Scott Dustan, Courseware Portfolio Manager, who supported our vision of this thirteenth edition.

I appreciate all the wonderful work of Lizette Faraji, Content Producer, who skillfully brought together reviews, art, web site materials, and all the things it takes to prepare a book for production. I appreciate the work of Karen Berry and Christian Arsenault at SPi Global, who brilliantly coordinated all phases of the manuscript to the final pages of a beautiful book. Thanks to Mark Quirie, manuscript and accuracy reviewer, and Laura Patchkofsky and Linda Smith, who precisely analyzed and edited the initial and final manuscripts and pages to make sure the words and problems were correct to help students learn chemistry. Their keen eyes and thoughtful comments were extremely helpful in the development of this text.

I am especially proud of the art program in this text, which lends beauty and understanding to chemistry. I would like to thank Wynne Au Yeung and Stephanie Marquez, art specialists; Mark Ong and Tamara Newnam, interior and cover designers, whose creative ideas provided the outstanding design for the cover and pages of the book. Eric Schrader, photo researcher, was outstanding in researching and selecting vivid photos for the text so that students can see the beauty of chemistry. Thanks also to *Bio-Rad Laboratories* for their courtesy and use of *KnowItAll ChemWindows,* drawing software that helped us produce chemical structures for the manuscript. The macro-to-micro illustrations designed by Production Solutions and Precision Graphics give students visual impressions of the atomic and molecular organization of everyday things and are a fantastic learning tool. I also appreciate all the hard work in the field put in by the marketing team and Elizabeth Ellsworth, marketing manager.

I am extremely grateful to an incredible group of peers for their careful assessment of all the new ideas for the text; for their suggested additions, corrections, changes, and deletions; and for providing an incredible amount of feedback about improvements for the book. I admire and appreciate every one of you.

If you would like to share your experience with chemistry, or have questions and comments about this text, I would appreciate hearing from you.

Karen Timberlake
Email: khemist@aol.com

Instructor and Student Supplements

Chemistry: An Introduction to General, Organic, and Biological Chemistry, thirteenth edition, provides an integrated teaching and learning package of support material for both students and professors.

Name of Supplement	Available in Print	Available Online	Instructor or Student Supplement	Description
Study Guide and Selected Solutions Manual (9780134553986)	✓		Supplement for Students	The *Study Guide and Selected Solutions Manual*, by Karen Timberlake and Mark Quirie, promotes active learning through a variety of exercises with answers as well as practice tests that are connected directly to the learning goals of the textbook. Complete solutions to odd-numbered problems are included.
MasteringChemistry™ (www.masteringchemistry.com) (9780134551272)		✓	Supplement for Students and Instructors	This product includes all of the resources of MasteringChemistry™ plus the now fully mobile eText 2.0. eText 2.0 mobile app offers offline access and can be downloaded for most iOS and Android phones/tablets from the Apple App Store or Google Play. Added integration brings videos and other rich media to the student's reading experience. MasteringChemistry™ from Pearson is the leading online homework, tutorial, and assessment system, designed to improve results by engaging students with powerful content. Instructors ensure students arrive ready to learn by assigning educationally effective content and encourage critical thinking and retention with in-class resources such as Learning Catalytics™. Students can further master concepts through traditional and adaptive homework assignments that provide hints and answer specific feedback. The Mastering™ gradebook records scores for all automatically-graded assignments in one place, while diagnostic tools give instructors access to rich data to assess student understanding and misconceptions. http://www.masteringchemistry.com.
Pearson eText enhanced with media (stand-alone: ISBN 9780134545684; within MasteringChemistry™: 9780134552170)		✓	Supplement for Students	The thirteenth edition of *Chemistry: An Introduction to General, Organic, and Biological Chemistry* features a Pearson eText enhanced with media within Mastering. In conjunction with Mastering assessment capabilities, new **Interactive Videos and 3D animations** will improve student engagement and knowledge retention. Each chapter contains a balance of interactive animations, videos, sample calculations, and self-assessments / quizzes embedded directly in the eText. Additionally, the Pearson eText offers students the power to create notes, highlight text in different colors, create bookmarks, zoom, and view single or multiple pages. Icons in the margins throughout the text signify that there is a new **Interactive Video** or animation located within MasteringChemistry™ for *Chemistry: An Introduction to General, Organic, and Biological Chemistry*, thirteenth edition.
Laboratory Manual by Karen Timberlake (9780321811851)	✓		Supplement for Students	This best-selling lab manual coordinates 35 experiments with the topics in *Chemistry: An Introduction to General, Organic, and Biological Chemistry*, thirteenth edition, uses laboratory investigations to explore chemical concepts, develop skills of manipulating equipment, reporting data, solving problems, making calculations, and drawing conclusions.
Instructor's Solutions Manual (9780134564661)		✓	Supplement for Instructors	Prepared by Mark Quirie, the Instructor's Solutions Manual highlights chapter topics, and includes answers and solutions for all Practice Problems in the text.
Instructor Resource Materials–Download Only (9780134552262)		✓	Supplement for Instructors	Includes all the art, photos, and tables from the book in JPEG format for use in classroom projection or when creating study materials and tests. In addition, the instructors can access modifiable PowerPoint™ lecture outlines. Also available are downloadable files of the Instructor's Solutions Manual and a set of "clicker questions" designed for use with classroom-response systems. Also visit the Pearson Education catalog page for Timberlake's *Chemistry: An Introduction to General, Organic, Biological Chemistry*, thirteenth edition, at www.pearsonhighered.com to download available instructor supplements.
TestGen Test Bank-Download Only (9780134564678)		✓	Supplement for Instructors	Prepared by William Timberlake, this resource includes more than 1600 questions in multiple-choice, matching, true/false, and short-answer format.
Online Instructor Manual for Laboratory Manual (9780321812858)		✓	Supplement for Instructors	This manual contains answers to report sheet pages for the *Laboratory Manual* and a list of the materials needed for each experiment with amounts given for 20 students working in pairs, available for download at www.pearsonhighered.com.

Career Focus Engages Students

Best-selling author Karen Timberlake connects chemistry to real-world and career applications like no one else. The 13th edition of *Chemistry: An Introduction to General, Organic, and Biological Chemistry* engages students by helping them to see the connections between chemistry, the world around them, and future careers.

3 Matter and Energy

CHARLES IS 13 YEARS OLD AND OVERWEIGHT. His doctor is worried that Charles is at risk for type 2 diabetes and advises his mother to make an appointment with a dietitian. Daniel, a dietitian, explains to them that choosing the appropriate foods is important to living a healthy lifestyle, losing weight, and preventing or managing diabetes.

Daniel also explains that food contains potential or stored energy and different foods contain different amounts of potential energy. For instance, carbohydrates contain 4 kcal/g (17 kJ/g), whereas fats contain 9 kcal/g (38 kJ/g). He then explains that diets high in fat require more exercise to burn the fats, as they contain more energy. When Daniel looks at Charles's typical daily diet, he calculates that Charles obtains 2500 kcal in one day. The American Heart Association recommends 1800 kcal for boys 9 to 13 years of age. Daniel encourages Charles and his mother to include whole grains, fruits, and vegetables in their diet instead of foods high in fat. They also discuss food labels and the fact that smaller serving sizes of healthy foods are necessary to lose weight. Daniel also recommends that Charles exercises at least 60 minutes every day. Before leaving, Charles and his mother make an appointment for the following week to look at a weight loss plan.

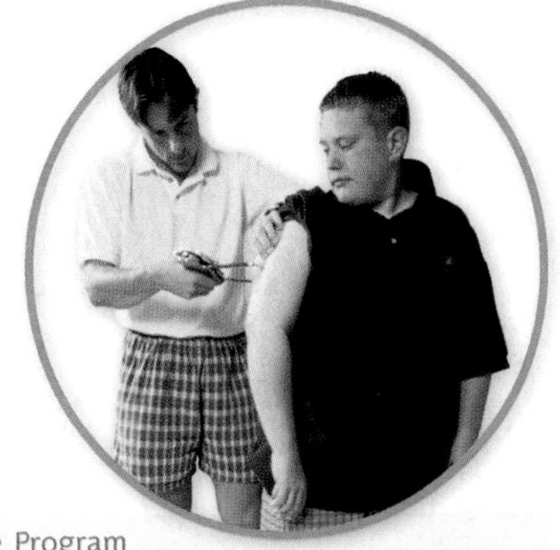

CAREER Dietitian

Dietitians specialize in helping individuals learn about good nutrition and the need for a balanced diet. This requires them to understand biochemical processes, the importance of vitamins and food labels, as well as the differences between carbohydrates, fats, and proteins in terms of their energy value and how they are metabolized. Dietitians work in a variety of environments, including hospitals, nursing homes, school cafeterias, and public health clinics. In these roles, they create specialized diets for individuals diagnosed with a specific disease or create meal plans for those in a nursing home.

CLINICAL UPDATE A Diet and Exercise Program

When Daniel sees Charles and his mother, they discuss a menu for weight loss. Charles is going to record his food intake and return to discuss his diet with Daniel. You can view the results in the **CLINICAL UPDATE A Diet and Exercise Program** on page 87, and calculate the kilocalories that Charles consumes in one day, and also the weight that Charles has lost.

Chapter Openers emphasize clinical connections by showing students relevant, engaging, topical examples of how health professionals use chemistry everyday. Clinical Updates at the end of each chapter relate the chemistry the student learns in the chapter to expand the clinical content in the Chapter Opener and include clinical applications.

Chemistry Links to Health, woven throughout each chapter, apply chemical concepts to topics in health and medicine such as weight loss and weight gain, alcohol abuse, blood buffers, and kidney dialysis, illustrating the importance of understanding chemistry in real-life situations.

CHEMISTRY LINK TO HEALTH

Breathing Mixtures

The air we breathe is composed mostly of the gases oxygen (21%) and nitrogen (79%). The homogeneous breathing mixtures used by scuba divers differ from the air we breathe depending on the depth of the dive. Nitrox is a mixture of oxygen and nitrogen, but with more oxygen gas (up to 32%) and less nitrogen gas (68%) than air. A breathing mixture with less nitrogen gas decreases the risk of *nitrogen narcosis* associated with breathing regular air while diving. Heliox contains oxygen and helium, which is typically used for diving to more than 200 ft. By replacing nitrogen with helium, nitrogen narcosis does not occur. However, at dive depths over 300 ft, helium is associated with severe shaking and a drop in body temperature.

A breathing mixture used for dives over 400 ft is trimix, which contains oxygen, helium, and some nitrogen. The addition of some nitrogen lessens the problem of shaking that comes with breathing high levels of helium. Heliox and trimix are used only by professional, military, or other highly trained divers.

In hospitals, heliox may be used as a treatment for respiratory disorders and lung constriction in adults and premature infants. Heliox is less dense than air, which reduces the effort of breathing and helps distribute the oxygen gas to the tissues.

A nitrox mixture is used to fill scuba tanks.

Builds Students' Critical-Thinking and Problem-Solving Skills

One of Karen Timberlake's goals is to help students to become critical thinkers. Color-coded tips found throughout each chapter are designed to provide guidance and to encourage students to really think about what they are reading, helping to develop important critical-thinking skills.

3.3 Temperature

LEARNING GOAL Given a temperature, calculate the corresponding temperature on another scale.

REVIEW
Using Positive and Negative Numbers in Calculations (1.4)
Solving Equations (1.4)
Counting Significant Figures (2.2)

Temperatures in science are measured and reported in *Celsius* (°C) units. On the Celsius scale, the reference points are the freezing point of water, defined as 0 °C, and the boiling point, 100 °C. In the United States, everyday temperatures are commonly reported in *Fahrenheit* (°F) units. On the Fahrenheit scale, water freezes at 32 °F and boils at 212 °F. A typical room temperature of 22 °C would be the same as 72 °F. Normal human body temperature is 37.0 °C, which is the same temperature as 98.6 °F.

On the Celsius and Fahrenheit temperature scales, the temperature difference between freezing and boiling is divided into smaller units called *degrees*. On the Celsius scale, there are 100 degrees Celsius between the freezing and boiling points of water, whereas the Fahrenheit scale has 180 degrees Fahrenheit between the freezing and boiling points of water. That makes a degree Celsius almost twice the size of a degree Fahrenheit: 1 °C = 1.8 °F (see **FIGURE 3.4**).

$$180 \text{ degrees Fahrenheit} = 100 \text{ degrees Celsius}$$

$$\frac{180 \text{ degrees Fahrenheit}}{100 \text{ degrees Celsius}} = \frac{1.8\ ^\circ\text{F}}{1\ ^\circ\text{C}}$$

We can write a temperature equation that relates a Fahrenheit temperature and its corresponding Celsius temperature.

ENGAGE
Why is a degree Celsius a larger unit of temperature than a degree Fahrenheit?

$$T_F = 1.8(T_C) + 32$$ Temperature equation to obtain degrees Fahrenheit

Changes °C to °F; Adjusts freezing point

In the equation, the Celsius temperature is multiplied by 1.8 to change °C to °F; then 32 is added to adjust the freezing point from 0 °C to the Fahrenheit freezing point, 32 °F. The values, 1.8 and 32, used in the temperature equation are exact numbers and are not used to determine significant figures in the answer.

CORE CHEMISTRY SKILL
Converting between Temperature Scales

To convert from degrees Fahrenheit to degrees Celsius, the temperature equation is rearranged to solve for T_C. First, we subtract 32 from both sides since we must apply the same operation to both sides of the equation.

$$T_F - 32 = 1.8(T_C) + \cancel{32} - \cancel{32}$$

$$T_F - 32 = 1.8(T_C)$$

NEW! Review Feature lists the core chemistry skills and key math skills from previous chapters which provide the foundation for learning the new chemistry principles in the current chapter.

NEW! Engage Feature asks students to think about the paragraph they are reading and immediately test their understanding by answering the Engage question, which is related to the topic. Students connect new concepts to prior knowledge to increase retrieval of content.

UPDATED! Core Chemistry Skills found throughout the chapter identify the fundamental chemistry concepts that students need to understand in the current chapter.

Four NEW problem solving features enhance Karen Timberlake's unmatched problem-solving strategies and help students deepen their understanding of content while improving their problem-solving skills.

NEW! Try It First precedes the Solution section of each Sample Problem to encourage the student to work on the problem before reading the given Solution.

NEW! Connect Feature added to Analyze the Problem boxes indicates the relationships between Given and Need.

NEW! Solution Guide provides **STEPS** for successful Problem Solving within the Sample Problem.

SAMPLE PROBLEM 3.7 **Using Specific Heat**

TRY IT FIRST

During surgery or when a patient has suffered a cardiac arrest or stroke, lowering the body temperature will reduce the amount of oxygen needed by the body. Some methods used to lower body temperature include cooled saline solution, cool water blankets, or cooling caps worn on the head. How many kilojoules are lost when the body temperature of a surgery patient with a blood volume of 5500 mL is cooled from 38.5 °C to 33.2 °C? (Assume that the specific heat and density of blood are the same as for water.)

A cooling cap lowers the body temperature to reduce the oxygen required by the tissues.

SOLUTION GUIDE

STEP 1 State the given and needed quantities.

	Given	Need	Connect
ANALYZE THE PROBLEM	5500 mL of blood = 5500 g of blood, cooled from 38.5 °C to 33.2 °C	kilojoules removed	heat equation, specific heat of water

STEP 2 Calculate the temperature change (ΔT).

$\Delta T = 38.5\ °\text{C} - 33.2\ °\text{C} = 5.3\ °\text{C}$

STEP 3 Write the heat equation and needed conversion factors.

$$\text{Heat} = m \times \Delta T \times SH$$

$$SH_{\text{water}} = \frac{4.184\ \text{J}}{\text{g}\ °\text{C}} \qquad 1\ \text{kJ} = 1000\ \text{J}$$

$$\frac{4.184\ \text{J}}{\text{g}\ °\text{C}} \text{ and } \frac{\text{g}\ °\text{C}}{4.184\ \text{J}} \qquad \frac{1000\ \text{J}}{1\ \text{kJ}} \text{ and } \frac{1\ \text{kJ}}{1000\ \text{J}}$$

STEP 4 Substitute in the given values and calculate the heat, making sure units cancel.

$$\text{Heat} = \underset{\text{Two SFs}}{5500\ \text{g}} \times \underset{\text{Two SFs}}{5.3\ \cancel{°\text{C}}} \times \overset{\text{Four SFs}}{\underset{\text{Exact}}{\frac{4.184\ \cancel{\text{J}}}{\cancel{\text{g}}\ \cancel{°\text{C}}}}} \times \overset{\text{Exact}}{\underset{\text{Exact}}{\frac{1\ \text{kJ}}{1000\ \cancel{\text{J}}}}} = \underset{\text{Two SFs}}{120\ \text{kJ}}$$

The copper on a pan conducts heat rapidly to the food in the pan.

STUDY CHECK 3.7

Some cooking pans have a layer of copper on the bottom. How many kilojoules are needed to raise the temperature of 125 g of copper from 22 °C to 325 °C (see Table 3.11)?

ANSWER

14.6 kJ

TEST

Try Practice Problems 3.39 to 3.42

NEW! Test Feature added in the margin encourages students to solve related Practice Problems to practice retrieval of content for exams.

Continuous Learning Before, During, and After Class

BEFORE CLASS

Dynamic Study Modules

NEW! 66 Dynamic Study Modules, specific to GOB Chemistry, help students study effectively on their own by continuously assessing their activity and performance in real time.

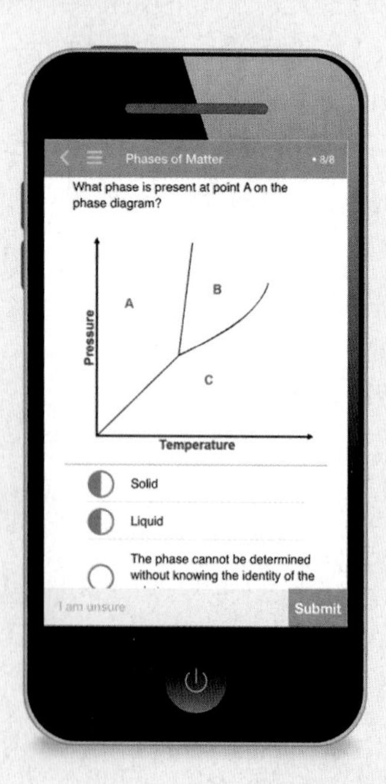

Students complete a set of questions with a unique answer format that also asks them to indicate their confidence level. Questions repeat until the student can answer them all correctly and confidently. Once completed, Dynamic Study Modules explain the concept. These are available as graded assignments prior to class, and accessible on smartphones, tablets, and computers.

Chemistry Primer

NEW! Chemistry Primer is a series of tutorials focused on remediating students taking their first college chemistry course. Topics include math in the context of chemistry, chemical skills and literacy, as well as some basics of balancing chemical equations, mole–mole factors, and mass–mass calculations—all of which were chosen based on extensive surveys of chemistry professors across the country.

The main body of each item in the primer offers diagnostic questions designed to help students recognize that they need help. If they struggle, the primer offers extensive formative help in the hint structure via wrong answer feedback, instructional videos, and step-wise worked examples that provide scaffolding to build up students' understanding as needed. The primer is offered as a pre-built assignment that is automatically generated with all chemistry courses.

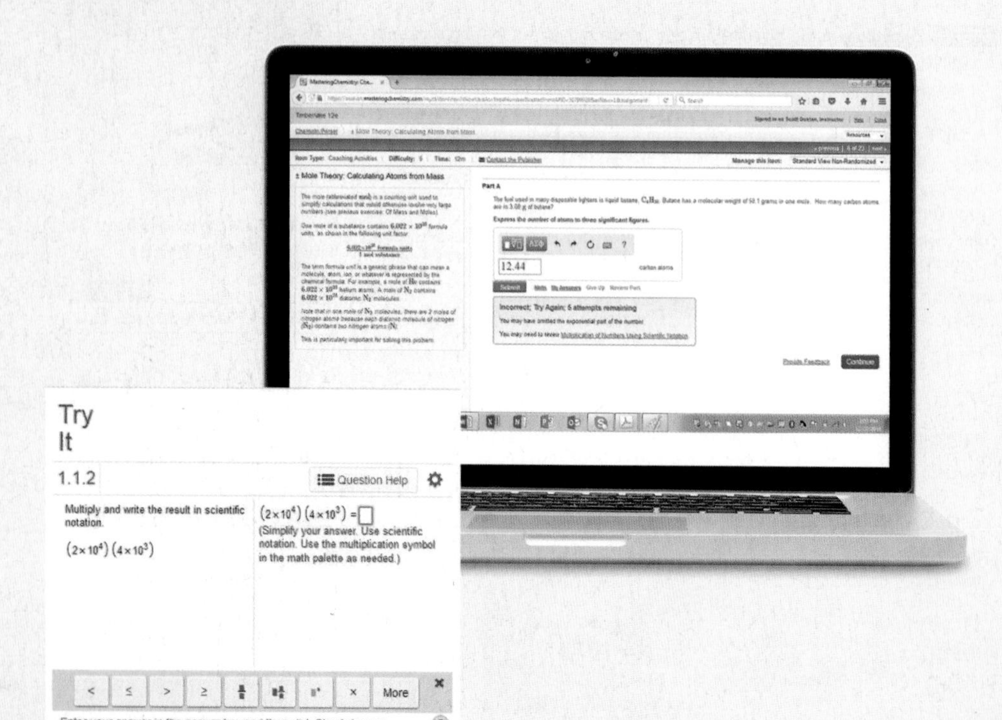

with MasteringChemistry™

DURING CLASS

Learning Catalytics

Learning Catalytics generates class discussion, guides your lecture, and promotes peer-to-peer learning with real-time analytics. MasteringChemistry with eText now provides Learning Catalytics—an interactive student response tool that uses students' smartphones, tablets, or laptops to engage them in more sophisticated tasks and thinking. Instructors can:

- **NEW!** Upload a full PowerPoint® deck for easy creation of slide questions.
- Help students develop critical thinking skills.
- Monitor responses to find out where students are struggling.
- Rely on real-time data to adjust teaching strategies.
- Automatically group students for discussion, teamwork, and peer-to-peer learning.

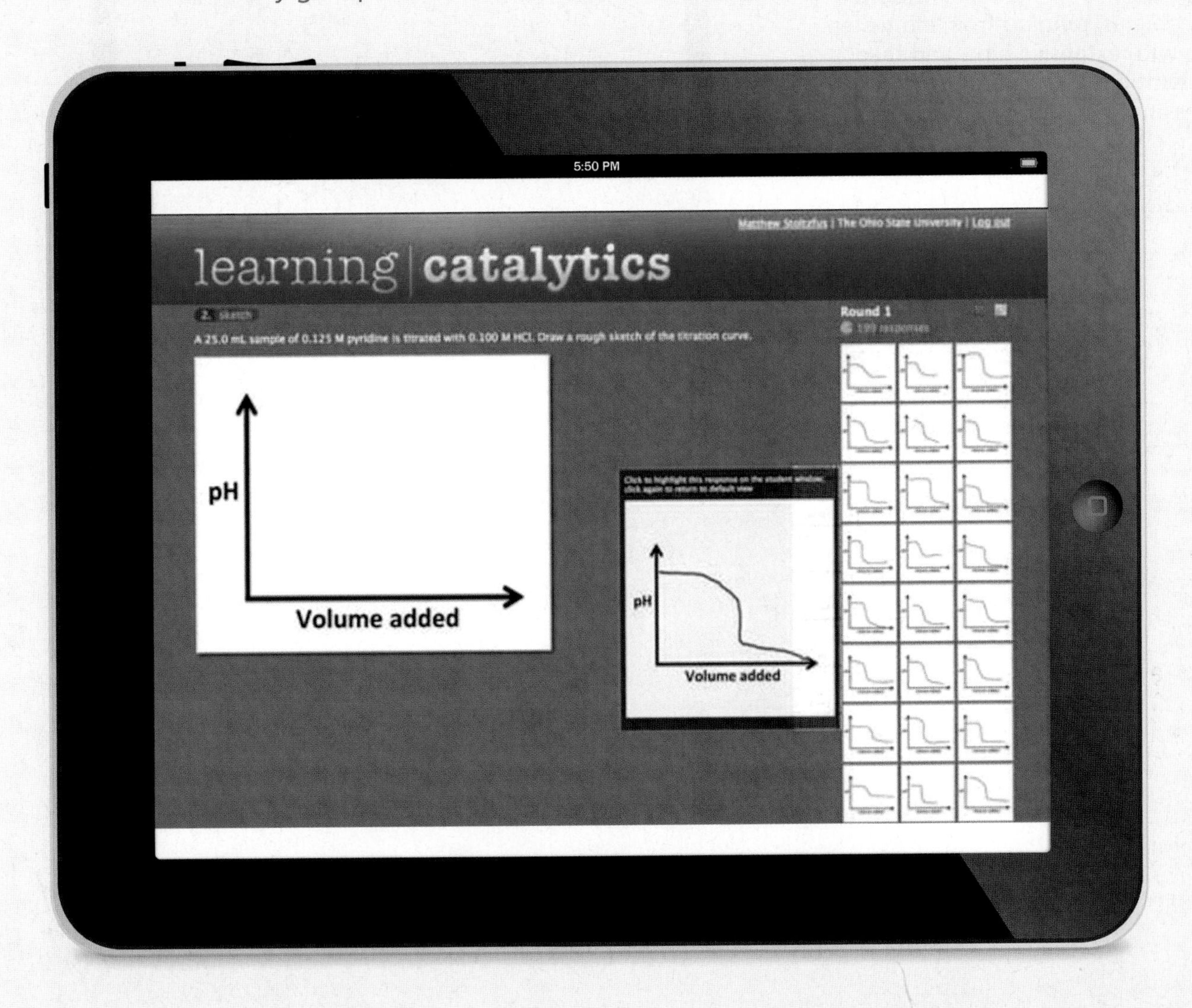

MasteringChemistry™

AFTER CLASS

NEW! Interactive Videos clarify and reinforce important concepts such as solving equations, conversion factors, solutions, and more. Sample Calculations now correspond to a key concept/topic in most chapters, giving students an opportunity to reinforce what they just learned by showing how chemistry works in real life and introducing a bit of humor into chemical problem solving and demonstrations.

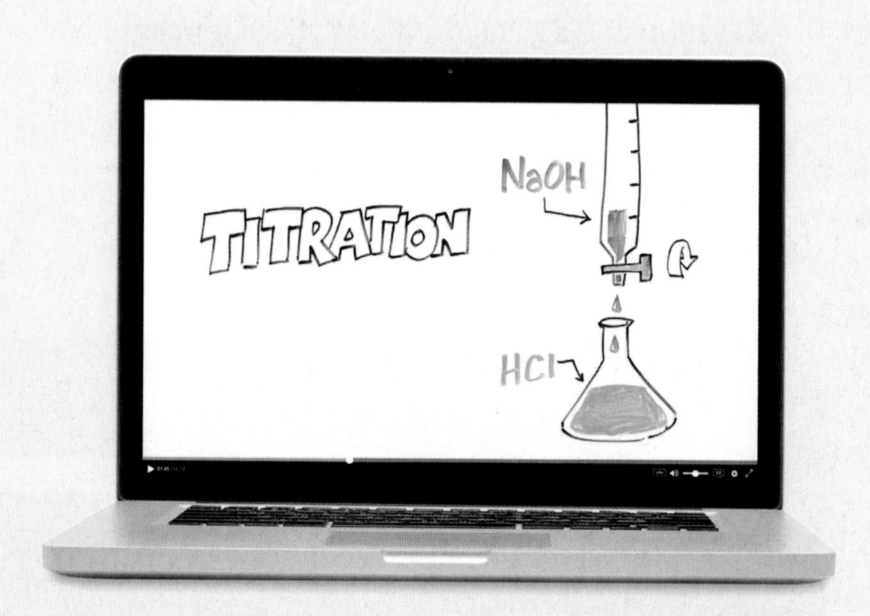

MasteringChemistry™ offers a wide variety of problems, ranging from multi-step tutorials with extensive hints and feedback to multiple-choice End-of-Chapter Problems and Test Bank questions.

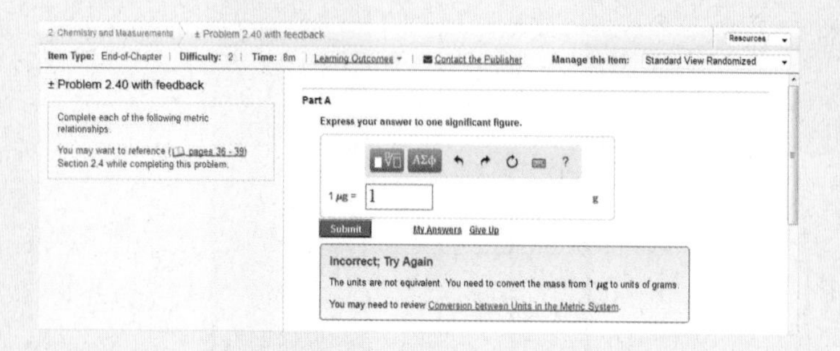

To provide additional scaffolding for students moving from Tutorial Problems to End-of-Chapter Problems we created **New!** Enhanced End-of-Chapter problems that now contain specific wrong-answer feedback.

eText 2.0

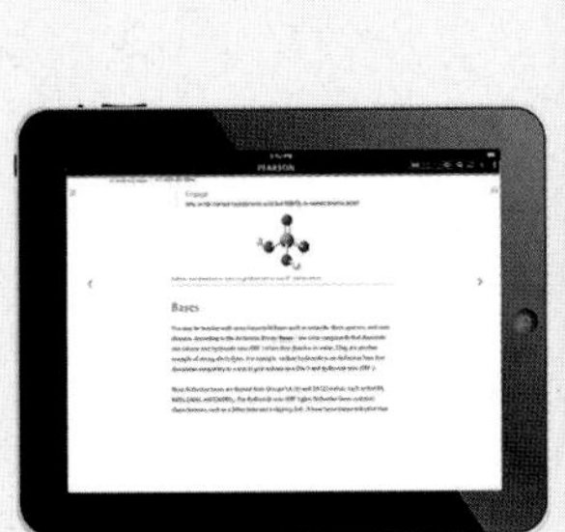

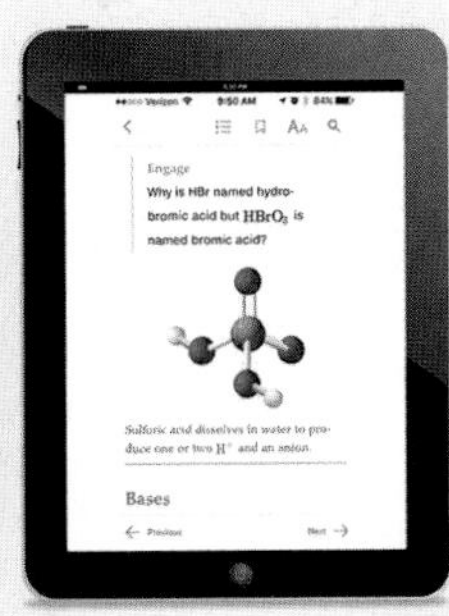

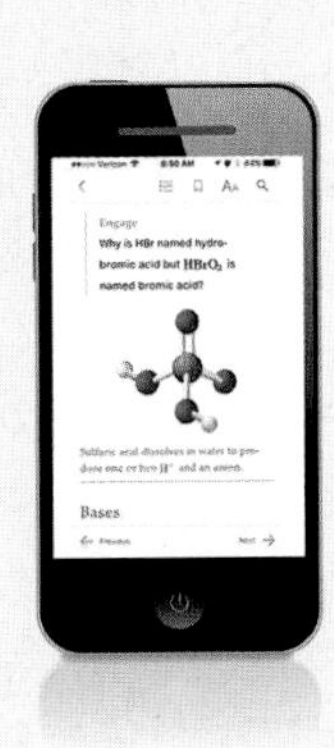

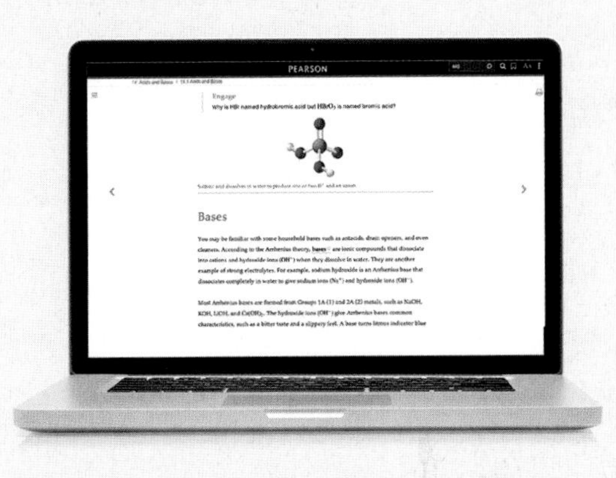

eText 2.0

- Full eReader functionality includes page navigation, search, glossary, highlighting, note taking, annotations, and more.
- A responsive design allows the eText to reflow/resize to a device or screen. eText 2.0 now works on supported smartphones, tablets, and laptop/desktop computers.
- In-context glossary offers students instant access to definitions by simply hovering over key terms.
- Seamlessly integrated videos and activities allow students to watch and practice key concepts within the eText learning experience.
- Accessible (screen-reader ready).
- Configurable reading settings, including resizable type and night reading mode.
- Study Check Questions allow students to interact in eText 2.0 with the questions which follow each Sample Problem. With one click, these activities are brought to life, allowing students to study on their own and test their understanding in real-time. These interactives help students extinguish misconceptions and enhance their problem-solving skills.

1 Chemistry in Our Lives

A CALL CAME IN TO 911 FROM A MAN WHO arrived home from work to find his wife lying on the floor of their home. When the police arrived, they prounouced the woman dead. The victim's body was lying on the floor of the living room. There was no blood at the scene, but the police did find a glass on the side table that contained a small amount of liquid. In an adjacent laundry room, the police found a half-empty bottle of antifreeze, which contains the toxic compound ethylene glycol. The bottle, glass, and liquid were bagged and sent to the forensic laboratory.

In another 911 call, a man was found lying on the grass outside his home. Blood was present on his body, and some bullet casings were found on the grass. Inside the victim's home, a weapon was recovered. The bullet casings and the weapon were bagged and sent to the forensic laboratory.

Sarah and Mark, forensic scientists, use scientific procedures and chemical tests to examine the evidence from law enforcement agencies. Sarah analyzes blood, stomach contents, and the unknown liquid from the first victim's home. She will look for the presence of drugs, poisons, and alcohol. Her lab partner, Mark, analyzes the fingerprints on the glass. He will also match the characteristics of the bullet casings to the weapon that was found at the second crime scene.

CAREER Forensic Scientist

Most forensic scientists work in crime laboratories that are part of city or county legal systems where they analyze bodily fluids and tissue samples collected by crime scene investigators. In analyzing these samples, forensic scientists identify the presence or absence of specific chemicals within the body to help solve the criminal case. Some of the chemicals they look for include alcohol, illegal or prescription drugs, poisons, arson debris, metals, and various gases such as carbon monoxide. In order to identify these substances, a variety of chemical instruments and highly specific methodologies are used. Forensic scientists analyze samples from criminal suspects, athletes, and potential employees. They also work on cases involving environmental contamination and animal samples for wildlife crimes. Forensic scientists usually have a bachelor's degree that includes courses in math, chemistry, and biology.

CLINICAL UPDATE Forensic Evidence Helps Solve the Crime

In the forensic laboratory, Sarah analyzes the victim's stomach contents and blood for toxic compounds. You can view the results of the tests on the forensic evidence in the CLINICAL UPDATE **Forensic Evidence Helps Solve the Crime**, page 19, and determine if the victim ingested a toxic level of ethylene glycol (antifreeze).

LOOKING AHEAD

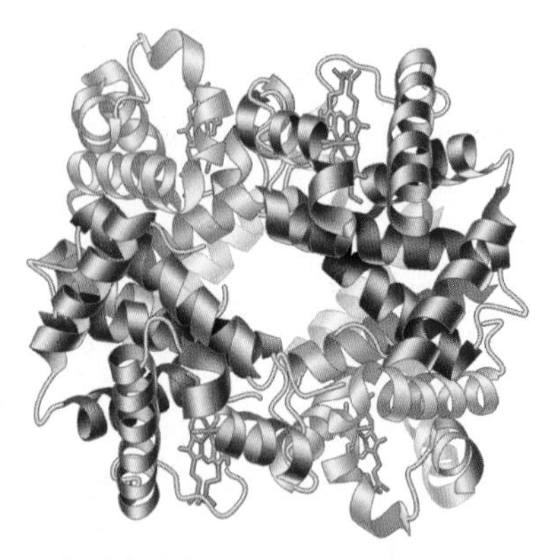

In the blood, hemoglobin transports oxygen to the tissues and carbon dioxide to the lungs.

Antacid tablets undergo a chemical reaction when dropped into water.

ENGAGE

Why is water a chemical?

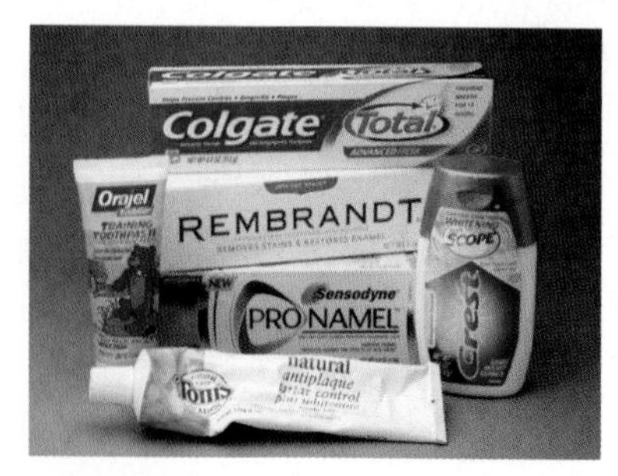

Toothpaste is a combination of many chemicals.

TEST

Try Practice Problems 1.1 to 1.6

1.1 Chemistry and Chemicals

LEARNING GOAL Define the term chemistry and identify substances as chemicals.

Now that you are in a chemistry class, you may be wondering what you will be learning. What questions in science have you been curious about? Perhaps you are interested in what hemoglobin does in the blood or how aspirin relieves a headache. Just like you, chemists are curious about the world we live in.

What does hemoglobin do in the body? Hemoglobin consists of four polypeptide chains, each containing a heme group with an iron atom that binds to oxygen (O_2) in the lungs. From the lungs, hemoglobin transports oxygen to the tissues of the body where it is used to provide energy. Once the oxygen is released, hemoglobin binds to carbon dioxide (CO_2) for transport to the lungs where it is released.

Why does aspirin relieve a headache? When a part of the body is injured, substances called prostaglandins are produced, which cause inflammation and pain. Aspirin acts to block the production of prostaglandins, reducing inflammation and pain. Chemists in the medical field develop new treatments for diabetes, genetic defects, cancer, AIDS, and other diseases. For the chemist in the forensic laboratory, the nurse in the dialysis unit, the dietitian, the chemical engineer, or the agricultural scientist, chemistry plays a central role in understanding problems and assessing possible solutions.

Chemistry

Chemistry is the study of the composition, structure, properties, and reactions of matter. *Matter* is another word for all the substances that make up our world. Perhaps you imagine that chemistry takes place only in a laboratory where a chemist is working in a white coat and goggles. Actually, chemistry happens all around you every day and has an impact on everything you use and do. You are doing chemistry when you cook food, add bleach to your laundry, or start your car. A chemical reaction has taken place when silver tarnishes or an antacid tablet fizzes when dropped into water. Plants grow because chemical reactions convert carbon dioxide, water, and energy to carbohydrates. Chemical reactions take place when you digest food and break it down into substances that you need for energy and health.

Chemicals

A **chemical** is a substance that always has the same composition and properties wherever it is found. All the things you see around you are composed of one or more chemicals. Chemical processes take place in chemistry laboratories, manufacturing plants, and pharmaceutical labs as well as every day in nature and in our bodies. Often the terms *chemical* and *substance* are used interchangeably to describe a specific type of matter.

Every day, you use products containing substances that were developed and prepared by chemists. Soaps and shampoos contain chemicals that remove oils on your skin and scalp. In cosmetics and lotions, chemicals are used to moisturize, prevent deterioration of the product, fight bacteria, and thicken the product. Perhaps you wear a ring or watch made of gold, silver, or platinum. Your breakfast cereal is probably fortified with iron, calcium, and phosphorus, whereas the milk you drink is enriched with vitamins A and D. When you brush your teeth, the substances in toothpaste clean your teeth, prevent plaque formation, and stop tooth decay. Some of the chemicals used to make toothpaste are listed in **TABLE 1.1**.

TABLE 1.1 Chemicals Commonly Used in Toothpaste

Chemical	Function
Calcium carbonate	Used as an abrasive to remove plaque
Sorbitol	Prevents loss of water and hardening of toothpaste
Sodium lauryl sulfate	Used to loosen plaque
Titanium dioxide	Makes toothpaste white and opaque
Sodium fluorophosphate	Prevents formation of cavities by strengthening tooth enamel with fluoride
Methyl salicylate	Gives toothpaste a pleasant wintergreen flavor

PRACTICE PROBLEMS

1.1 Chemistry and Chemicals

LEARNING GOAL Define the term chemistry and identify substances as chemicals.

In every chapter, odd-numbered exercises in the *Practice Problems* are paired with even-numbered exercises. The answers for the magenta, odd-numbered *Practice Problems* are given at the end of each chapter. The complete solutions to the odd-numbered *Practice Problems* are in the *Study Guide and Student Solutions Manual*.

1.1 Write a one-sentence definition for each of the following:
a. chemistry **b.** chemical

1.2 Ask two of your friends (not in this class) to define the terms in problem 1.1. Do their answers agree with the definitions you provided?

Clinical Applications

1.3 Obtain a bottle of multivitamins and read the list of ingredients. What are four chemicals from the list?

1.4 Obtain a box of breakfast cereal and read the list of ingredients. What are four chemicals from the list?

1.5 Read the labels on some items found in your medicine cabinet. What are the names of some chemicals contained in those items?

1.6 Read the labels on products used to wash your dishes. What are the names of some chemicals contained in those products?

1.2 Scientific Method: Thinking Like a Scientist

LEARNING GOAL Describe the activities that are part of the scientific method.

When you were very young, you explored the things around you by touching and tasting. As you grew, you asked questions about the world in which you live. What is lightning? Where does a rainbow come from? Why is the sky blue? As an adult, you may have wondered how antibiotics work or why vitamins are important to your health. Every day, you ask questions and seek answers to organize and make sense of the world around you.

When the late Nobel Laureate Linus Pauling described his student life in Oregon, he recalled that he read many books on chemistry, mineralogy, and physics. "I mulled over the properties of materials: why are some substances colored and others not, why are some minerals or inorganic compounds hard and others soft?" He said, "I was building up this tremendous background of empirical knowledge and at the same time asking a great number of questions." Linus Pauling won two Nobel Prizes: the first, in 1954, was in chemistry for his work on the nature of chemical bonds and the determination of the structures of complex substances; the second, in 1962, was the Peace Prize.

Linus Pauling won the Nobel Prize in Chemistry in 1954.

The Scientific Method

The process of trying to understand nature is unique to each scientist. However, the **scientific method** is a process that scientists use to make observations in nature, gather data, and explain natural phenomena.

1. **Observations** The first step in the scientific method is to make observations about nature and ask questions about what you observe. When an observation always seems to be true, it may be stated as a *law* that predicts that behavior and is often measurable. However, a law does not explain that observation. For example, we can use the *Law of Gravity* to predict that if we drop our chemistry book it would fall on the table or the floor but this law does not explain why our book falls.
2. **Hypothesis** A scientist forms a hypothesis, which gives a possible explanation of an observation or a law. The hypothesis must be stated in such a way that it can be tested by experiments.
3. **Experiments** To determine if a hypothesis is true or false, experiments are done to find a relationship between the hypothesis and the observations. The results of the experiments may confirm the hypothesis. However, if the experiments do not confirm the hypothesis, it is modified or discarded. Then new experiments will be designed to test the hypothesis.
4. **Conclusion/Theory** When the results of the experiments are analyzed, a conclusion is made as to whether the hypothesis is *true* or *false*. When experiments give consistent results, the hypothesis may be stated to be true. Even then, the hypothesis continues

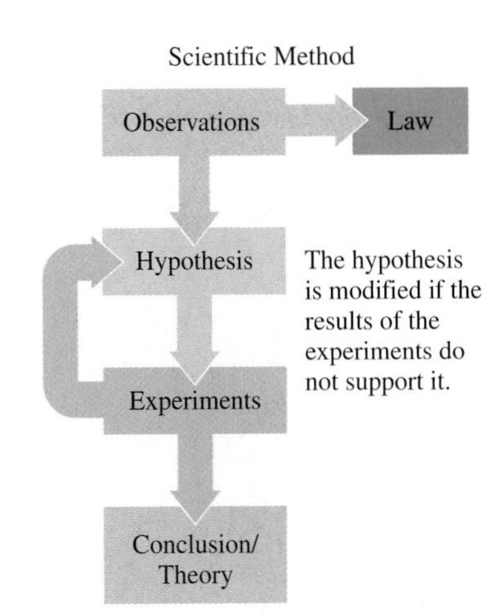

The scientific method develops a conclusion or theory about nature using observations, hypotheses, and experiments.

to be tested and, based on new experimental results, may need to be modified or replaced. If many additional experiments by a group of scientists continue to support the hypothesis, it may become a *scientific theory*, which gives an explanation for the initial observations.

CHEMISTRY LINK TO HEALTH

Early Chemist: Paracelsus

For many centuries, chemistry has been the study of changes in matter. From the time of the ancient Greeks to the sixteenth century, alchemists described matter in terms of four components of nature: earth, air, fire, and water. By the eighth century, alchemists believed that they could change metals such as copper and lead into gold and silver. Although these efforts failed, the alchemists provided information on the chemical reactions involved in the extraction of metals from ores. The alchemists also designed some of the first laboratory equipment and developed early laboratory procedures. These early efforts were some of the first observations and experiments using the scientific method.

Paracelsus (1493–1541) was a physician and an alchemist who thought that alchemy should be about preparing new medicines. Using observation and experimentation, he proposed that a healthy body was regulated by a series of chemical processes that could be unbalanced by certain chemical compounds and rebalanced by using minerals and medicines. For example, he determined that inhaled dust caused lung disease in miners. He also thought that goiter was a problem caused by contaminated water, and he treated syphilis with compounds of mercury. His opinion of medicines was that the right dose makes the difference between a poison and a cure. Paracelsus changed alchemy in ways that helped establish modern medicine and chemistry.

Swiss physician and alchemist Paracelsus (1493–1541) believed that chemicals and minerals could be used as medicines.

Through observation you may think that you are allergic to cats.

ENGAGE

Why would the following statement "Today I placed two tomato seedlings in the garden, and two more in a closet. I will give all the plants the same amount of water and fertilizer." be considered an experiment?

Using the Scientific Method in Everyday Life

You may be surprised to realize that you use the scientific method in your everyday life. Suppose you visit a friend in her home. Soon after you arrive, your eyes start to itch and you begin to sneeze. Then you observe that your friend has a new cat. Perhaps you form the hypothesis that you are allergic to cats. To test your hypothesis, you leave your friend's home. If the sneezing stops, perhaps your hypothesis is correct. You test your hypothesis further by visiting another friend who also has a cat. If you start to sneeze again, your experimental results support your hypothesis and you come to the conclusion that you are allergic to cats. However, if you continue sneezing after you leave your friend's home, your hypothesis is not supported. Now you need to form a new hypothesis, which could be that you have a cold.

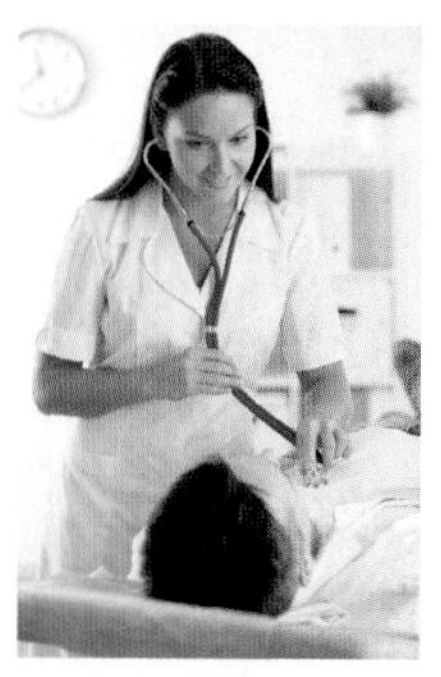

Nurses make observations in the hospital.

SAMPLE PROBLEM 1.1 Scientific Method

TRY IT FIRST

Identify each of the following as an observation, a hypothesis, an experiment, or a conclusion:

a. During an assessment in the emergency room, a nurse writes that the patient has a resting pulse of 30 beats/min.

b. Repeated studies show that lowering sodium in the diet leads to a decrease in blood pressure.

c. A nurse thinks that an incision from a recent surgery that is red and swollen is infected.

SOLUTION

a. observation **b.** conclusion **c.** hypothesis

STUDY CHECK 1.1

Identify each of the following as an observation, a hypothesis, an experiment, or a conclusion:

a. Drinking coffee at night keeps me awake.
b. I will try drinking coffee only in the morning.
c. If I stop drinking coffee in the afternoon, I will be able to sleep at night.

ANSWER

a. observation **b.** experiment **c.** hypothesis

TEST

Try Practice Problems 1.7 to 1.10

PRACTICE PROBLEMS

1.2 Scientific Method: Thinking Like a Scientist

LEARNING GOAL Describe the activities that are part of the scientific method.

1.7 Identify each activity, **a** to **f**, as an observation, a hypothesis, an experiment, or a conclusion.

At a popular restaurant, where Chang is the head chef, the following occurred:

Customers rated the sesame seed dressing as the best.

a. Chang determined that sales of the house salad had dropped.
b. Chang decided that the house salad needed a new dressing.
c. In a taste test, Chang prepared four bowls of lettuce, each with a new dressing: sesame seed, olive oil and balsamic vinegar, creamy Italian, and blue cheese.
d. Tasters rated the sesame seed salad dressing as the favorite.
e. After two weeks, Chang noted that the orders for the house salad with the new sesame seed dressing had doubled.
f. Chang decided that the sesame seed dressing improved the sales of the house salad because the sesame seed dressing enhanced the taste.

1.8 Identify each activity, **a** to **f**, as an observation, a hypothesis, an experiment, or a conclusion.

Lucia wants to develop a process for dyeing shirts so that the color will not fade when the shirt is washed. She proceeds with the following activities:

a. Lucia notices that the dye in a design fades when the shirt is washed.
b. Lucia decides that the dye needs something to help it combine with the fabric.
c. She places a spot of dye on each of four shirts and then places each one separately in water, salt water, vinegar, and baking soda and water.
d. After one hour, all the shirts are removed and washed with a detergent.
e. Lucia notices that the dye has faded on the shirts in water, salt water, and baking soda, whereas the dye did not fade on the shirt soaked in vinegar.
f. Lucia thinks that the vinegar binds with the dye so it does not fade when the shirt is washed.

Clinical Applications

1.9 Identify each of the following as an observation, a hypothesis, an experiment, or a conclusion:

a. One hour after drinking a glass of regular milk, Jim experienced stomach cramps.
b. Jim thinks he may be lactose intolerant.
c. Jim drinks a glass of lactose-free milk and does not have any stomach cramps.
d. Jim drinks a glass of regular milk to which he has added lactase, an enzyme that breaks down lactose, and has no stomach cramps.

1.10 Identify each of the following as an observation, a hypothesis, an experiment, or a conclusion:

a. Sally thinks she may be allergic to shrimp.
b. Yesterday, one hour after Sally ate a shrimp salad, she broke out in hives.
c. Today, Sally had some soup that contained shrimp, but she did not break out in hives.
d. Sally realizes that she does not have an allergy to shrimp.

1.3 Studying and Learning Chemistry

LEARNING GOAL Identify strategies that are effective for learning. Develop a study plan for learning chemistry.

Here you are taking chemistry, perhaps for the first time. Whatever your reasons for choosing to study chemistry, you can look forward to learning many new and exciting ideas.

Strategies to Improve Learning and Understanding

Success in chemistry utilizes good study habits, connecting new information with your knowledge base, rechecking what you have learned and what you have forgotten, and retrieving what you have learned for an exam. Let's take a look at ways that can help you study and learn chemistry. Suppose you were asked to indicate if you think each of the following common study habits is helpful or not helpful:

	Helpful	Not helpful
Highlighting		
Underlining		
Reading the chapter many times		
Memorizing the key words		
Testing practice		
Cramming		
Studying different ideas at the same time		
Retesting a few days later		

Learning something requires us to place new information in our long-term memory, which allows us to remember those ideas for an exam, a process called retrieval. Thus, our evaluation of study habits depends on their value in helping us to recall knowledge. The study habits that are not very helpful in retrieval include highlighting, underlining, reading the chapter many times, memorizing key words, and cramming. If we want to recall new information, we need to connect it with prior knowledge that we can retrieve. This can be accomplished by developing study habits that involve a lot of practice testing ourselves on how to retrieve new information. We can determine how much we have learned by going back a few days later and retesting. Another useful learning strategy is to study different ideas at the same time, which allows us to connect those ideas and how to differentiate them. Although these study habits may take more time and seem more difficult, they help us find the gaps in our knowledge and connect new information with what we already know. In the long run, you retain and retrieve more information, making your study for exams less stressful.

Tips for Using New Study Habits for Successful Learning

1. **Do not keep rereading text or notes.** Reading the same material over and over will make that material seem familiar but does not mean that you have learned it. You need to test yourself to find out what you do and do not know.
2. **Ask yourself questions as you read.** Asking yourself questions as you read requires you to interact continually with new material. For example, you might ask yourself how the new material is related to previous material, which helps you make connections. By linking new material with long-term knowledge, you make pathways for retrieving new material.
3. **Self-test by giving yourself quizzes.** Using problems in the text or sample exams, practice taking tests frequently.
4. **Study at a regular pace rather than cramming.** Once you have tested yourself, go back in a few days and practice testing and retrieving information again. We do not recall all the information when we first read it. By frequent quizzing and retesting, we identify what we still need to learn. Sleep is also important for strengthening the associations between newly learned information. Lack of sleep may interfere with retrieval of information as well. So staying up all night to cram for your chemistry exam is not a good idea. Success in chemistry is a combined effort to learn new information and then to retrieve that information when you need it for an exam.
5. **Study different topics in a chapter and relate the new concepts to concepts you know.** We learn material more efficiently by relating it to information we already know. By increasing connections between concepts, we can retrieve information when we need it.

Helpful	**Not helpful**
Testing practice	Highlighting
Studying different ideas at the same time	Underlining
Retesting a few days later	Reading the chapter many times
	Memorizing the key words
	Cramming

ENGAGE

Why is self-testing helpful for learning new concepts?

SAMPLE PROBLEM 1.2 Strategies for Learning Chemistry

TRY IT FIRST

Predict which student will obtain the best exam score.

a. A student who reads the chapter four times.
b. A student who reads the chapter two times and works all the problems at the end of each Section.
c. A student who reads the chapter the night before the exam.

SOLUTION

b. A student who reads the chapter two times and works all the problems at the end of each Section has interacted with the content in the chapter using self-testing to make connections between concepts and practicing retrieving information learned previously.

STUDY CHECK 1.2

What is another way that student **b** in Sample Problem 1.2 could improve his or her retrieval of information?

ANSWER

Student **b** in Sample Problem 1.2 could also wait two or three days and practice working the problems in each Section again to determine how much he or she has learned. Retesting strengthens connections between new and previously learned information for longer lasting memory and more efficient retrieval.

Features in This Text That Help You Study and Learn Chemistry

This text has been designed with study features to complement your individual learning style. On the inside of the front cover is a periodic table of the elements. On the inside of the back cover are tables that summarize useful information needed throughout your study of chemistry. Each chapter begins with *Looking Ahead*, which outlines the topics in the chapter. *Key Terms* are bolded when they first appear in the text, and are summarized at the end of each chapter. They are also listed and defined in the comprehensive *Glossary and Index*, which appears at the end of the text. *Key Math Skills* and *Core Chemistry Skills* that are critical to learning chemistry are indicated by icons in the margin, and summarized at the end of each chapter.

KEY MATH SKILL

CORE CHEMISTRY SKILL

REVIEW

Before you begin reading, obtain an overview of a chapter by reviewing the topics in Looking Ahead. As you prepare to read a Section of the chapter, look at the Section title and turn it into a question. Asking yourself questions about new topics builds new connections to material you have already learned. For example, for Section 1.1, "Chemistry and Chemicals," you could ask, "What is chemistry?" or "What are chemicals?" At the beginning of each Section, a *Learning Goal* states what you need to understand and a *Review* box lists the Key Math Skills and Core Chemistry Skills from previous chapters that relate to new material in the chapter. As you read the text, you will see *Engage* features in the margin, which remind you to pause your reading and test yourself with a question related to the material.

ENGAGE

What is the purpose of an Engage question?

Several *Sample Problems* are included in each Chapter. The *Try It First* feature reminds you to work the problem before you look at the Solution. The *Analyze the Problem* feature includes *Given*, the information you have; *Need*, what you have to accomplish; and *Connect*, how you proceed. It is helpful to try to work a problem first because it helps you link what you know to what you need to learn. This process will help you develop successful problem-solving techniques. Many Sample Problems include a *Solution Guide* that shows the steps you can use for problem solving. Work the associated *Study Check* and compare your answer to the one provided.

TRY IT FIRST

ANALYZE THE PROBLEM	Given	Need	Connect
	165 lb	kilograms	conversion factor

At the end of each chapter Section, you will find a set of *Practice Problems* that allows you to apply problem solving immediately to the new concepts. Throughout each

TEST

Section, *Test* suggestions remind you to solve the indicated Practice Problems as you study. The *Clinical Applications* in the Practice Problems relate the content to health and medicine. The problems are paired, which means that each of the odd-numbered problems is matched to the following even-numbered problem. At the end of each chapter, the answers to all the odd-numbered problems are provided. If the answers match yours, you most likely understand the topic; if not, you need to study the Section again.

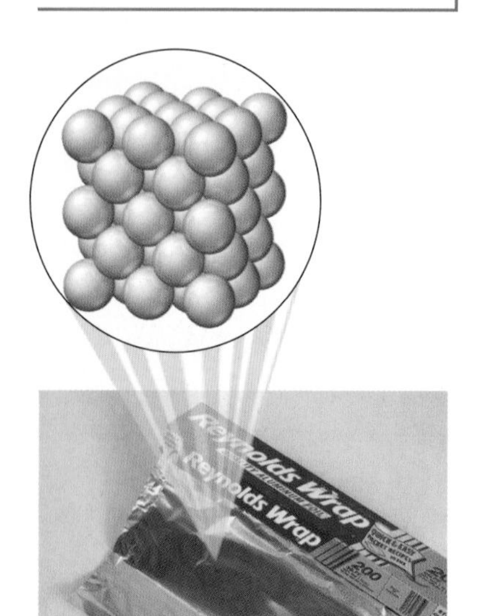

Illustrating the atoms of aluminum in aluminum foil is an example of macro-to-micro art.

Throughout each chapter, boxes titled *Chemistry Link to Health* and *Chemistry Link to the Environment* help you relate the chemical concepts you are learning to real-life situations. Many of the figures and diagrams use macro-to-micro illustrations to depict the atomic level of organization of ordinary objects, such as the atoms in aluminum foil. These visual models illustrate the concepts described in the text and allow you to "see" the world in a microscopic way. *Interactive Video* suggestions illustrate content as well as problem solving.

At the end of each chapter, you will find several study aids that complete the chapter. *Chapter Reviews* provide a summary in easy-to-read bullet points and *Concept Maps* visually show the connections between important topics. *Understanding the Concepts* are problems that use art and models to help you visualize concepts and connect them to your background knowledge. *Additional Practice Problems* and *Challenge Problems* provide additional exercises to test your understanding of the topics in the chapter. *Answers* to all of the odd-numbered problems complete the chapter allowing you to compare your answers to the ones provided.

After some chapters, problem sets called *Combining Ideas* test your ability to solve problems containing material from more than one chapter.

INTERACTIVE VIDEO

Many students find that studying with a group can be beneficial to learning. In a group, students motivate each other to study, fill in gaps, and correct misunderstandings by teaching and learning together. Studying alone does not allow the process of peer correction. In a group, you can cover the ideas more thoroughly as you discuss the reading and problem solve with other students.

Making a Study Plan

As you embark on your journey into the world of chemistry, think about your approach to studying and learning chemistry. You might consider some of the ideas in the following list. Check those ideas that will help you successfully learn chemistry. Commit to them now. *Your* success depends on *you*.

My study plan for learning chemistry will include the following:

________ reading the chapter before class
________ going to class
________ reviewing the *Learning Goals*
________ keeping a problem notebook
________ reading the text
________ working the *Test* problems as I read each Section
________ answering the *Engage* questions
________ trying to work the *Sample Problem* before looking at the *Solution*
________ working the *Practice Problems* at the end of each Section and checking answers
________ studying different topics at the same time
________ organizing a study group
________ seeing the professor during office hours
________ reviewing *Key Math Skills* and *Core Chemistry Skills*
________ attending review sessions
________ studying as often as I can

Studying in a group can be beneficial to learning.

SAMPLE PROBLEM 1.3 A Study Plan for Learning Chemistry

TRY IT FIRST

Which of the following activities should you include in your study plan for learning chemistry successfully?

a. reading the chapter over and over until you think you understand it
b. going to the professor's office hours
c. self-testing during and after reading each Section
d. waiting to study until the night before the exam
e. trying to work the Sample Problem before looking at the Solution
f. retesting on new information a few days later

SOLUTION

Your success in chemistry can be improved by:

b. going to the professor's office hours
c. self-testing during and after reading each Section
e. trying to work the Sample Problem before looking at the Solution
f. retesting on new information a few days later

STUDY CHECK 1.3

Which of the following will help you learn chemistry?

a. skipping review sessions
b. working problems as you read a Section
c. staying up all night before an exam
d. reading the assignment before class

ANSWER

b and **d**

TEST

Try Practice Problems 1.11 to 1.14

PRACTICE PROBLEMS

1.3 Studying and Learning Chemistry

LEARNING GOAL Identify strategies that are effective for learning. Develop a study plan for learning chemistry.

1.11 What are four things you can do to help yourself to succeed in chemistry?

1.12 What are four things that would make it difficult for you to learn chemistry?

1.13 A student in your class asks you for advice on learning chemistry. Which of the following might you suggest?
a. forming a study group
b. skipping class
c. asking yourself questions while reading the text
d. waiting until the night before an exam to study
e. answering the Engage questions

1.14 A student in your class asks you for advice on learning chemistry. Which of the following might you suggest?
a. studying different topics at the same time
b. not reading the text; it's never on the test
c. attending review sessions
d. working the problems again after a few days
e. keeping a problem notebook

1.4 Key Math Skills for Chemistry

LEARNING GOAL Review math concepts used in chemistry: place values, positive and negative numbers, percentages, solving equations, and interpreting graphs.

During your study of chemistry, you will work many problems that involve numbers. You will need various math skills and operations. We will review some of the key math skills that are particularly important for chemistry. As we move through the chapters, we will also reference the key math skills as they apply.

KEY MATH SKILL

Identifying Place Values

Identifying Place Values

For any number, we can identify the *place value* for each of the digits in that number. These place values have names such as the ones place (first place to the left of the decimal point) or the tens place (second place to the left of the decimal point). A premature baby has a mass of 2518 g. We can indicate the place values for the number 2518 as follows:

Digit	Place Value
2	thousands
5	hundreds
1	tens
8	ones

ENGAGE

In the number 8.034, how do you know the 0 is in the tenths place?

We also identify place values such as the tenths place (first place to the right of the decimal point) and the hundredths place (second place to the right of the decimal point). A silver coin has a mass of 6.407 g. We can indicate the place values for the number 6.407 as follows:

Digit	Place Value
6	ones
4	ten**ths**
0	hundred**ths**
7	thousand**ths**

Note that place values ending with the suffix *ths* refer to the decimal places to the right of the decimal point.

SAMPLE PROBLEM 1.4 Identifying Place Values

TRY IT FIRST

A bullet found at a crime scene has a mass of 15.24 g. What are the place values for each of the digits in the mass of the bullet?

SOLUTION

Digit	Place Value
1	tens
5	ones
2	tenths
4	hundredths

STUDY CHECK 1.4

A bullet found at a crime scene contains 0.925 g of lead. What are the place values for each of the digits in the mass of the lead?

ANSWER

Digit	Place Value
9	tenths
2	hundredths
5	thousandths

TEST

Try Practice Problems 1.15 and 1.16

Using Positive and Negative Numbers in Calculations

A *positive number* is any number that is greater than zero and has a positive sign (+). Often the positive sign is understood and not written in front of the number. For example, the number +8 can also be written as 8. A *negative number* is any number that is less than zero and is written with a negative sign (−). For example, a negative eight is written as −8.

KEY MATH SKILL

Using Positive and Negative Numbers in Calculations

Multiplication and Division of Positive and Negative Numbers

When two positive numbers or two negative numbers are multiplied, the answer is positive (+).

$$2 \times 3 = +6$$
$$(-2) \times (-3) = +6$$

When a positive number and a negative number are multiplied, the answer is negative (−).

$$2 \times (-3) = -6$$
$$(-2) \times 3 = -6$$

The rules for the division of positive and negative numbers are the same as the rules for multiplication. When two positive numbers or two negative numbers are divided, the answer is positive (+).

$$\frac{6}{3} = 2 \qquad \frac{-6}{-3} = 2$$

When a positive number and a negative number are divided, the answer is negative (−).

$$\frac{-6}{3} = -2 \qquad \frac{6}{-3} = -2$$

Addition of Positive and Negative Numbers

When positive numbers are added, the sign of the answer is positive.

$3 + 4 = 7$ The + sign (+7) is understood.

When negative numbers are added, the sign of the answer is negative.

$$(-3) + (-4) = -7$$

When a positive number and a negative number are added, the smaller number is subtracted from the larger number, and the result has the same sign as the larger number.

$$12 + (-15) = -3$$

ENGAGE

Why does (−5) + 4 = −1, whereas (−5) + (−4) = −9?

Subtraction of Positive and Negative Numbers

When two numbers are subtracted, change the sign of the number to be subtracted and follow the rules for addition shown above.

$$12 - (\mathbf{+5}) = 12 \mathbf{- 5} = 7$$
$$12 - (\mathbf{-5}) = 12 \mathbf{+ 5} = 17$$
$$-12 - (\mathbf{-5}) = -12 \mathbf{+ 5} = -7$$
$$-12 - (\mathbf{+5}) = -12 \mathbf{- 5} = -17$$

TEST

Try Practice Problems 1.17 and 1.18

Calculator Operations

On your calculator, there are four keys that are used for basic mathematical operations. The change sign [+/−] key is used to change the sign of a number.

To practice these basic calculations on the calculator, work through the problem going from the left to the right doing the operations in the order they occur. If your calculator has a change sign [+/−] key, a negative number is entered by pressing the number and then pressing the change sign [+/−] key. At the end, press the equals [=] key or ANS or ENTER.

Multiplication
Division
Subtraction
Equals
Change sign
Addition

Addition and Subtraction

Example 1: $15 - 8 + 2 =$

Solution: $15 \boxed{-} 8 \boxed{+} 2 \boxed{=} 9$

Example 2: $4 + (-10) - 5 =$

Solution: $4 \boxed{+} 10 \boxed{+/-} \boxed{-} 5 \boxed{=} -11$

Multiplication and Division

Example 3: $2 \times (-3) =$

Solution: $2 \boxed{\times} 3 \boxed{+/-} \boxed{=} -6$

Example 4: $\dfrac{8 \times 3}{4} =$

Solution: $8 \boxed{\times} 3 \boxed{\div} 4 \boxed{=} 6$

KEY MATH SKILL
Calculating Percentages

Calculating Percentages

To determine a percentage, divide the parts by the total (whole) and multiply by 100%. For example, if an aspirin tablet contains 325 mg of aspirin (active ingredient) and the tablet has a mass of 545 mg, what is the percentage of aspirin in the tablet?

$$\frac{325 \text{ mg aspirin}}{545 \text{ mg tablet}} \times 100\% = 59.6\% \text{ aspirin}$$

ENGAGE
Why is the value of 100% used in the calculation of a percentage?

When a value is described as a percentage (%), it represents the number of parts of an item in 100 of those items. If the percentage of red balls is 5, it means there are 5 red balls in every 100 balls. If the percentage of green balls is 50, there are 50 green balls in every 100 balls.

$$5\% \text{ red balls} = \frac{5 \text{ red balls}}{100 \text{ balls}} \qquad 50\% \text{ green balls} = \frac{50 \text{ green balls}}{100 \text{ balls}}$$

A bullet casing at a crime scene is marked as evidence.

SAMPLE PROBLEM 1.5 Calculating a Percentage

TRY IT FIRST

A bullet found at a crime scene may be used as evidence in a trial if the percentage of metals is a match to the composition of metals in a bullet from the suspect's ammunition. If a bullet found at a crime scene contains 13.9 g of lead, 0.3 g of tin, and 0.9 g of antimony, what is the percentage of each metal in the bullet? Express your answers to the ones place.

SOLUTION

Total mass = 13.9 g + 0.3 g + 0.9 g = 15.1 g

Percentage of lead

$$\frac{13.9 \text{ g}}{15.1 \text{ g}} \times 100\% = 92\% \text{ lead}$$

Percentage of tin

$$\frac{0.3 \text{ g}}{15.1 \text{ g}} \times 100\% = 2\% \text{ tin}$$

Percentage of antimony

$$\frac{0.9 \text{ g}}{15.1 \text{ g}} \times 100\% = 6\% \text{ antimony}$$

STUDY CHECK 1.5

A bullet seized from the suspect's ammunition has a composition of lead 11.6 g, tin 0.5 g, and antimony 0.4 g.

a. What is the percentage of each metal in the bullet? Express your answers to the ones place.

b. Could the bullet removed from the suspect's ammunition be considered as evidence that the suspect was at the crime scene mentioned in Sample Problem 1.5?

ANSWER

a. The bullet from the suspect's ammunition is lead 93%, tin 4%, and antimony 3%.

b. The composition of this bullet does not match the bullet from the crime scene and cannot be used as evidence.

TEST

Try Practice Problems 1.19 and 1.20

Solving Equations

KEY MATH SKILL

Solving Equations

In chemistry, we use equations that express the relationship between certain variables. Let's look at how we would solve for x in the following equation:

$$2x + 8 = 14$$

Our overall goal is to rearrange the items in the equation to obtain x on one side.

ENGAGE

Why is the number 8 subtracted from both sides of this equation?

1. *Place all like terms on one side.* The numbers 8 and 14 are like terms. To remove the 8 from the left side of the equation, we subtract 8. To keep a balance, we need to subtract 8 from the 14 on the other side.

$$2x + \cancel{8} - \cancel{8} = 14 - 8$$
$$2x = 6$$

2. *Isolate the variable you need to solve for.* In this problem, we obtain x by dividing both sides of the equation by 2. The value of x is the result when 6 is divided by 2.

$$\frac{\cancel{2}x}{\cancel{2}} = \frac{6}{2}$$
$$x = 3$$

3. *Check your answer.* Check your answer by substituting your value for x back into the original equation.

$$2(3) + 8 = 14$$
$$6 + 8 = 14$$
$$14 = 14 \quad \text{Your answer } x = 3 \text{ is correct.}$$

Summary: To solve an equation for a particular variable, be sure you perform the same mathematical operations on *both* sides of the equation.

If you eliminate a symbol or number by subtracting, you need to subtract that same symbol or number on the opposite side.

If you eliminate a symbol or number by adding, you need to add that same symbol or number on the opposite side.

If you cancel a symbol or number by dividing, you need to divide both sides by that same symbol or number.

If you cancel a symbol or number by multiplying, you need to multiply both sides by that same symbol or number.

When we work with temperature, we may need to convert between degrees Celsius and degrees Fahrenheit using the following equation:

$$T_F = 1.8(T_C) + 32$$

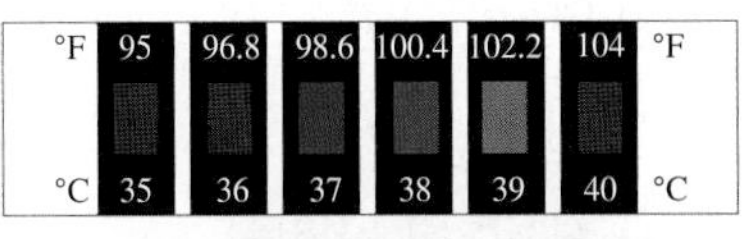

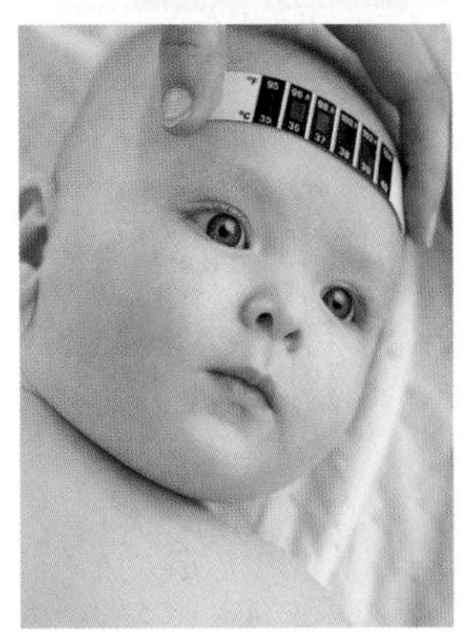

A plastic strip thermometer changes color to indicate body temperature.

To obtain the equation for converting degrees Fahrenheit to degrees Celsius, we subtract 32 from both sides.

$$T_F = 1.8(T_C) + 32$$
$$T_F - 32 = 1.8(T_C) + \cancel{32} - \cancel{32}$$
$$T_F - 32 = 1.8(T_C)$$

To obtain T_C by itself, we divide both sides by 1.8.

$$\frac{T_F - 32}{1.8} = \frac{\cancel{1.8}(T_C)}{\cancel{1.8}} = T_C$$

SAMPLE PROBLEM 1.6 Solving Equations

TRY IT FIRST

Solve the following equation for V_2:

$P_1V_1 = P_2V_2$

INTERACTIVE VIDEO

Solving Equations

SOLUTION

$P_1V_1 = P_2V_2$

ENGAGE

Why is the numerator divided by P_2 on both sides of the equation?

To solve for V_2, divide both sides by the symbol P_2.

$$\frac{P_1V_1}{P_2} = \frac{\cancel{P_2}V_2}{\cancel{P_2}}$$

$$V_2 = \frac{P_1V_1}{P_2}$$

STUDY CHECK 1.6

Solve the following equation for m:

heat $= m \times \Delta T \times SH$

TEST

Try Practice Problems 1.21 and 1.22

ANSWER

$$m = \frac{\text{heat}}{\Delta T \times SH}$$

KEY MATH SKILL

Interpreting Graphs

Interpreting Graphs

A graph represents the relationship between two variables. These quantities are plotted along two perpendicular axes, which are the x axis (horizontal) and y axis (vertical).

Example

In the graph Volume of a Balloon Versus Temperature, the volume of a gas in a balloon is plotted against its temperature.

Title

Look at the title. What does it tell us about the graph? The title indicates that the volume of a balloon was measured at different temperatures.

Vertical Axis

Look at the label and the numbers on the vertical (y) axis. The label indicates that the volume of the balloon was measured in liters (L). The numbers, which are chosen to include the low and high measurements of the volume of the gas, are evenly spaced from 22.0 L to 30.0 L.

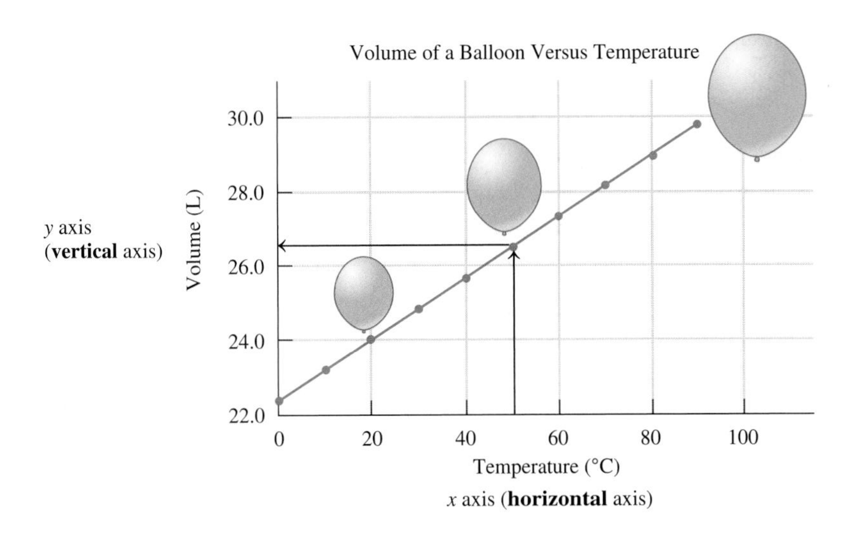

ENGAGE

Why are the numbers on the vertical and horizontal axes placed at regular intervals?

Horizontal Axis

The label on the horizontal (x) axis indicates that the temperature of the balloon was measured in degrees Celsius (°C). The numbers are measurements of the Celsius temperature, which are evenly spaced from 0 °C to 100 °C.

Points on the Graph

Each point on the graph represents a volume in liters that was measured at a specific temperature. When these points are connected, a line is obtained.

Interpreting the Graph

From the graph, we see that the volume of the gas increases as the temperature of the gas increases. This is called a *direct relationship*. Now we use the graph to determine the volume at various temperatures. For example, suppose we want to know the volume of the gas at 50 °C. We would start by finding 50 °C on the x axis and then drawing a line up to the plotted line. From there, we would draw a horizontal line that intersects the y axis and read the volume value where the line crosses the y axis as shown on the graph above.

SAMPLE PROBLEM 1.7 Interpreting a Graph

TRY IT FIRST

A nurse administers Tylenol to lower a child's fever. The graph shows the body temperature of the child plotted against time.

a. What is measured on the vertical axis?
b. What is the range of values on the vertical axis?
c. What is measured on the horizontal axis?
d. What is the range of values on the horizontal axis?

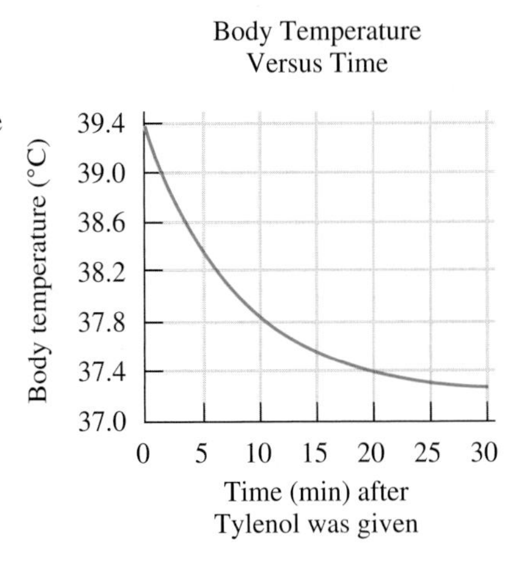

SOLUTION

a. body temperature, in degrees Celsius
b. 37.0 °C to 39.4 °C
c. time, in minutes, after Tylenol was given
d. 0 min to 30 min

STUDY CHECK 1.7

a. Using the graph in Sample Problem 1.7, what was the child's temperature 15 min after Tylenol was given?
b. How many minutes elapsed before the temperature decreased to 38.0 °C?

ANSWER

a. 37.6 °C **b.** 8 min

TEST

Try Practice Problems 1.23 and 1.24

PRACTICE PROBLEMS

1.4 Key Math Skills for Chemistry

LEARNING GOAL Review math concepts used in chemistry: place values, positive and negative numbers, percentages, solving equations, and interpreting graphs.

1.15 What is the place value for the bold digit?
- **a.** 7.32**8**8
- **b.** 1**6**.1234
- **c.** 4**6**75.99

1.16 What is the place value for the bold digit?
- **a.** 97.**5**689
- **b.** 375.88
- **c.** 46.1**0**00

1.17 Evaluate each of the following:
- **a.** $15 - (-8) =$ _____
- **b.** $-8 + (-22) =$ _____
- **c.** $4 \times (-2) + 6 =$ _____

1.18 Evaluate each of the following:
- **a.** $-11 - (-9) =$ _____
- **b.** $34 + (-55) =$ _____
- **c.** $\frac{-56}{8} =$ _____

Clinical Applications

1.19 **a.** A clinic had 25 patients on Friday morning. If 21 patients were given flu shots, what percentage of the patients received flu shots? Express your answer to the ones place.
b. An alloy contains 56 g of pure silver and 22 g of pure copper. What is the percentage of silver in the alloy? Express your answer to the ones place.
c. A collection of coins contains 11 nickels, 5 quarters, and 7 dimes. What is the percentage of dimes in the collection? Express your answer to the ones place.

1.20 **a.** At a local hospital, 35 babies were born. If 22 were boys, what percentage of the newborns were boys? Express your answer to the ones place.
b. An alloy contains 67 g of pure gold and 35 g of pure zinc. What is the percentage of zinc in the alloy? Express your answer to the ones place.
c. A collection of coins contains 15 pennies, 14 dimes, and 6 quarters. What is the percentage of pennies in the collection? Express your answer to the ones place.

1.21 Solve each of the following for a:
- **a.** $4a + 4 = 40$
- **b.** $\frac{a}{6} = 7$

1.22 Solve each of the following for b:
- **a.** $2b + 7 = b + 10$
- **b.** $3b - 4 = 24 - b$

Use the following graph for problems 1.23 and 1.24:

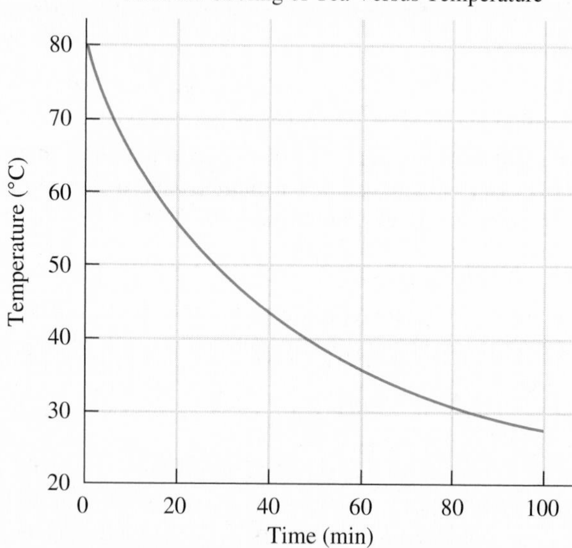

1.23 **a.** What does the title indicate about the graph?
b. What is measured on the vertical axis?
c. What is the range of values on the vertical axis?
d. Does the temperature increase or decrease with an increase in time?

1.24 **a.** What is measured on the horizontal axis?
b. What is the range of values on the horizontal axis?
c. What is the temperature of the tea after 20 min?
d. How many minutes were needed to reach a temperature of 45 °C?

1.5 Writing Numbers in Scientific Notation

LEARNING GOAL Write a number in scientific notation.

In chemistry, we often work with numbers that are very large and very small. We might measure something as tiny as the width of a human hair, which is about 0.000 008 m. Or perhaps we want to count the number of hairs on the average human scalp, which is about 100 000 hairs. In this text, we add spaces between sets of three digits when it helps make the places easier to count. However, we will see that it is more convenient to write large and small numbers in *scientific notation*.

A number written in **scientific notation** has two parts: a coefficient and a power of 10. For example, the number 2400 is written in scientific notation as 2.4×10^3. The coefficient, 2.4,

Humans have an average of 1×10^5 hairs on their scalps. Each hair is about 8×10^{-6} m wide.

Standard Number	Scientific Notation
0.000 008 m	8×10^{-6} m
100 000 hairs	1×10^5 hairs

is obtained by moving the decimal point to the left to give a number that is at least 1 but less than 10. Because we moved the decimal point three places to the left, the power of 10 is a positive 3, which is written as 10^3. When a number greater than 1 is converted to scientific notation, the power of 10 is positive.

KEY MATH SKILL
Writing Numbers in Scientific Notation

Standard Number		**Scientific Notation**		
2400.	=	2.4	×	10^3
← 3 places		Coefficient		Power of 10

ENGAGE
Why is 530 000 written as 5.3×10^5 in scientific notation?

In another example, 0.000 86 is written in scientific notation as 8.6×10^{-4}. The coefficient, 8.6, is obtained by moving the decimal point to the right. Because the decimal point is moved four places to the right, the power of 10 is a negative 4, written as 10^{-4}. When a number less than 1 is written in scientific notation, the power of 10 is negative.

Standard Number		**Scientific Notation**		
0.00086	=	8.6	×	10^{-4}
4 places →		Coefficient		Power of 10

ENGAGE
Why is 0.000 053 written as 5.3×10^{-5} in scientific notation?

TABLE 1.2 gives some examples of numbers written as positive and negative powers of 10. The powers of 10 are a way of keeping track of the decimal point in the number. **TABLE 1.3** gives several examples of writing measurements in scientific notation.

TABLE 1.2 Some Powers of 10

Standard Number	Multiples of 10	Scientific Notation	
10 000	$10 \times 10 \times 10 \times 10$	1×10^4	Some positive powers of 10
1 000	$10 \times 10 \times 10$	1×10^3	
100	10×10	1×10^2	
10	10	1×10^1	
1	0	1×10^0	
0.1	$\frac{1}{10}$	1×10^{-1}	Some negative powers of 10
0.01	$\frac{1}{10} \times \frac{1}{10} = \frac{1}{100}$	1×10^{-2}	
0.001	$\frac{1}{10} \times \frac{1}{10} \times \frac{1}{10} = \frac{1}{1\,000}$	1×10^{-3}	
0.0001	$\frac{1}{10} \times \frac{1}{10} \times \frac{1}{10} \times \frac{1}{10} = \frac{1}{10\,000}$	1×10^{-4}	

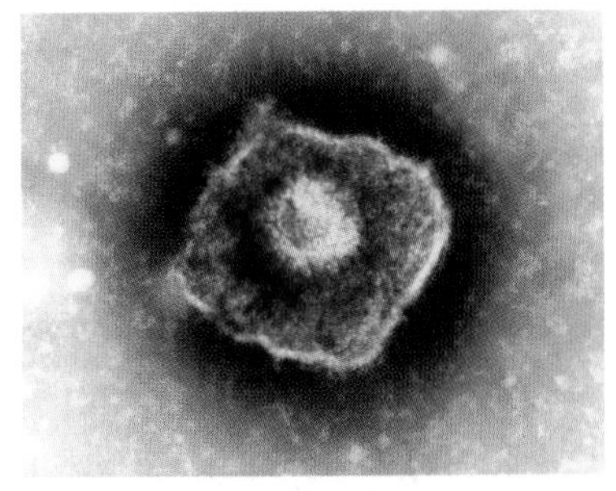
A chickenpox virus has a diameter of 3×10^{-7} m.

TABLE 1.3 Some Measurements Written as Standard Numbers and in Scientific Notation

Measured Quantity	Standard Number	Scientific Notation
Volume of gasoline used in the United States each year	550 000 000 000 L	5.5×10^{11} L
Diameter of Earth	12 800 000 m	1.28×10^{7} m
Average volume of blood pumped in 1 day	8500 L	8.5×10^{3} L
Time for light to travel from the Sun to Earth	500 s	5×10^{2} s
Mass of a typical human	68 kg	6.8×10^{1} kg
Mass of stirrup bone in ear	0.003 g	3×10^{-3} g
Diameter of a chickenpox (*Varicella zoster*) virus	0.000 000 3 m	3×10^{-7} m
Mass of bacterium (mycoplasma)	0.000 000 000 000 000 000 1 kg	1×10^{-19} kg

SAMPLE PROBLEM 1.8 Writing a Number in Scientific Notation

TRY IT FIRST

Write each of the following in scientific notation:

a. 3500 **b.** 0.000 016

SOLUTION GUIDE

	Given	Need	Connect
ANALYZE THE PROBLEM	standard number	scientific notation	coefficient is at least 1 but less than 10

a. 3500

STEP 1 Move the decimal point to obtain a coefficient that is at least 1 but less than 10. For a number greater than 1, the decimal point is moved to the left three places to give a coefficient of 3.5.

STEP 2 Express the number of places moved as a power of 10. Moving the decimal point three places to the left gives a power of 3, written as 10^3.

STEP 3 Write the product of the coefficient multiplied by the power of 10. 3.5×10^3

b. 0.000 016

STEP 1 Move the decimal point to obtain a coefficient that is at least 1 but less than 10. For a number less than 1, the decimal point is moved to the right five places to give a coefficient of 1.6.

STEP 2 Express the number of places moved as a power of 10. Moving the decimal point five places to the right gives a power of negative 5, written as 10^{-5}.

STEP 3 Write the product of the coefficient multiplied by the power of 10. 1.6×10^{-5}

STUDY CHECK 1.8

Write each of the following in scientific notation:

a. 425 000 **b.** 0.000 000 86

ANSWER

a. 4.25×10^5 **b.** 8.6×10^{-7}

TEST

Try Practice Problems 1.25 to 1.28

Scientific Notation and Calculators

You can enter a number in scientific notation on many calculators using the [EE or EXP] key. After you enter the coefficient, press the [EE or EXP] key and enter the power 10. To enter a negative power of 10, press the [+/−] key or the [−] key, depending on your calculator.

Number to Enter	Procedure	Calculator Display
4×10^6	4 [EE or EXP] 6	4 06 or 4^{06} or 4E06
2.5×10^{-4}	2.5 [EE or EXP] [+/−] 4	2.5−04 or 2.5^{-04} or 2.5E−04

When a calculator answer appears in scientific notation, the coefficient is shown as a number that is at least 1 but less than 10, followed by a space or E and the power of 10. To express this display in scientific notation, write the coefficient value, write × 10, and use the power of 10 as an exponent.

ENGAGE

Describe how you enter a number in scientific notation on your calculator.

Calculator Display	Expressed in Scientific Notation
7.52 04 or 7.52^{04} or 7.52E04	7.52×10^4
5.8−02 or 5.8^{-02} or 5.8E−02	5.8×10^{-2}

On many calculators, a number is converted into scientific notation using the appropriate keys. For example, the number 0.000 52 is entered, followed by pressing the 2nd or 3rd function key and the SCI key. The scientific notation appears in the calculator display as a coefficient and the power of 10.

0.000 52 [2nd or 3rd function key] [SCI] = 5.2−04 or 5.2^{-04} or 5.2E−04 = 5.2×10^{-4}

Calculator display

PRACTICE PROBLEMS

1.5 Writing Numbers in Scientific Notation

LEARNING GOAL Write a number in scientific notation.

1.25 Write each of the following in scientific notation:
a. 55 000 **b.** 480 **c.** 0.000 005
d. 0.000 14 **e.** 0.0072 **f.** 670 000

1.26 Write each of the following in scientific notation:
a. 180 000 000 **b.** 0.000 06 **c.** 750
d. 0.15 **e.** 0.024 **f.** 1500

1.27 Which number in each of the following pairs is larger?
a. 7.2×10^3 or 8.2×10^2
b. 4.5×10^{-4} or 3.2×10^{-2}
c. 1×10^4 or 1×10^{-4}
d. 0.000 52 or 6.8×10^{-2}

1.28 Which number in each of the following pairs is smaller?
a. 4.9×10^{-3} or 5.5×10^{-9}
b. 1250 or 3.4×10^2
c. 0.000 000 4 or 5.0×10^2
d. 2.50×10^2 or 4×10^5

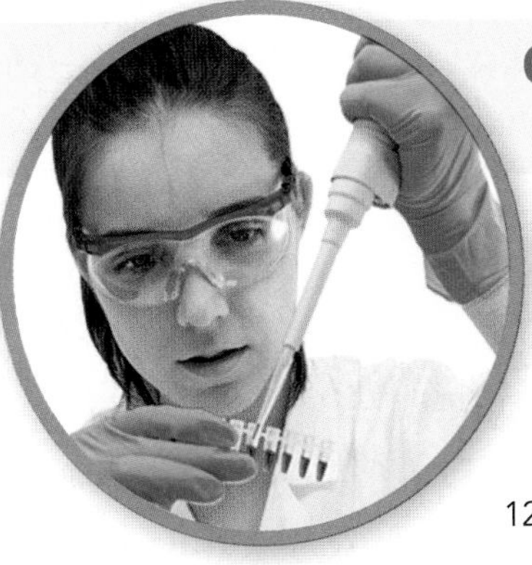

CLINICAL UPDATE Forensic Evidence Helps Solve the Crime

Using a variety of laboratory tests, Sarah finds ethylene glycol in the victim's blood. The quantitative tests indicate that the victim had ingested 125 g of ethylene glycol.

Sarah determines that the liquid in a glass found at the crime scene was ethylene glycol that had been added to an alcoholic beverage. Ethylene glycol is a clear, sweet-tasting, thick liquid that is odorless and mixes with water. It is easy to obtain since it is used as antifreeze in automobiles and in brake fluid. Because the initial symptoms of ethylene glycol poisoning are similar to being intoxicated, the victim is often unaware of its presence.

If ingestion of ethylene glycol occurs, it can cause depression of the central nervous system, cardiovascular damage, and kidney failure. If discovered quickly, hemodialysis may be used to remove ethylene glycol from the blood. A toxic amount of ethylene glycol is 1.5 g of ethylene glycol/kg of body mass. Thus, 75 g could be fatal for a 50-kg (110-lb) person.

Mark determines that fingerprints on the glass containing the ethylene glycol were those of the victim's husband. This evidence along with the container of antifreeze found in the home led to the arrest and conviction of the husband for poisoning his wife.

Clinical Applications

1.29 A container was found in the home of the victim that contained 120 g of ethylene glycol in 450 g of liquid. What was the percentage of ethylene glycol? Express your answer to the ones place.

1.30 If the toxic quantity is 1.5 g of ethylene glycol per 1000 g of body mass, what percentage of ethylene glycol is fatal?

CONCEPT MAP

CHEMISTRY IN OUR LIVES

- deals with **Substances** called **Chemicals**
- uses the **Scientific Method** starting with **Observations** that lead to **Hypothesis**, **Experiments**, **Conclusion/Theory**
- is learned by **Reading the Text**, **Practicing Problem Solving**, **Self-Testing**, **Working with a Group**, **Engaging**, **Trying It First**
- uses key math skills **Identifying Place Values**, **Using Positive and Negative Numbers**, **Calculating Percentages**, **Solving Equations**, **Interpreting Graphs**, **Writing Numbers in Scientific Notation**

CHAPTER REVIEW

1.1 Chemistry and Chemicals

LEARNING GOAL Define the term chemistry and identify substances as chemicals.

- Chemistry is the study of the composition, structure, properties, and reactions of matter.
- A chemical is any substance that always has the same composition and properties wherever it is found.

1.2 Scientific Method: Thinking Like a Scientist

LEARNING GOAL Describe the activities that are part of the scientific method.

- The scientific method is a process of explaining natural phenomena beginning with making observations, forming a hypothesis, and performing experiments.
- After repeated successful experiments, a hypothesis may become a theory.

1.3 Studying and Learning Chemistry

LEARNING GOAL Identify strategies that are effective for learning. Develop a study plan for learning chemistry.

- A plan for learning chemistry utilizes the features in the text that help develop a successful approach to learning chemistry.
- By using the *Learning Goals*, *Reviews*, *Analyze the Problems*, and *Try It First* in the chapter and working the *Sample Problems*, *Study Checks*, and the *Practice Problems* at the end of each Section, you can successfully learn the concepts of chemistry.

1.4 Key Math Skills for Chemistry

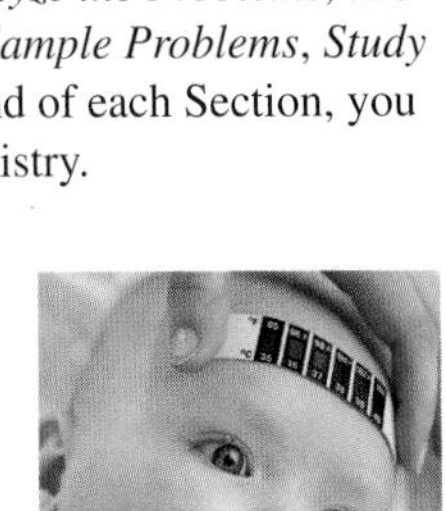

LEARNING GOAL Review math concepts used in chemistry: place values, positive and negative numbers, percentages, solving equations, and interpreting graphs.

- Solving chemistry problems involves a number of math skills: identifying place values, using positive and negative numbers, calculating percentages, solving equations, and interpreting graphs.

1.5 Writing Numbers in Scientific Notation

LEARNING GOAL Write a number in scientific notation.

- A number written in scientific notation has two parts, a coefficient and a power of 10.
- When a number greater than 1 is converted to scientific notation, the power of 10 is positive.
- When a number less than 1 is written in scientific notation, the power of 10 is negative.

KEY TERMS

chemical A substance that has the same composition and properties wherever it is found.

chemistry The study of the composition, structure, properties, and reactions of matter.

conclusion An explanation of an observation that has been validated by repeated experiments that support a hypothesis.

experiment A procedure that tests the validity of a hypothesis.

hypothesis An unverified explanation of a natural phenomenon.

observation Information determined by noting and recording a natural phenomenon.

scientific method The process of making observations, proposing a hypothesis, and testing the hypothesis; after repeated experiments validate the hypothesis, it may become a theory.

scientific notation A form of writing large and small numbers using a coefficient that is at least 1 but less than 10, followed by a power of 10.

theory An explanation for an observation supported by additional experiments that confirm the hypothesis.

KEY MATH SKILLS

The chapter Section containing each Key Math Skill is shown in parentheses at the end of each heading.

Identifying Place Values (1.4)

- The place value identifies the numerical value of each digit in a number.

Example: Identify the place value for each of the digits in the number 456.78.

Answer:

Digit	Place Value
4	hundreds
5	tens
6	ones
7	tenths
8	hundredths

Using Positive and Negative Numbers in Calculations (1.4)

- A *positive number* is any number that is greater than zero and has a positive sign (+). A *negative number* is any number that is less than zero and is written with a negative sign (−).
- When two positive numbers are added, multiplied, or divided, the answer is positive.
- When two negative numbers are multiplied or divided, the answer is positive. When two negative numbers are added, the answer is negative.
- When a positive and a negative number are multiplied or divided, the answer is negative.
- When a positive and a negative number are added, the smaller number is subtracted from the larger number and the result has the same sign as the larger number.
- When two numbers are subtracted, change the sign of the number to be subtracted then follow the rules for addition.

Example: Evaluate each of the following:

a. $-8 - 14 =$ _____ **b.** $6 \times (-3) =$ _____

Answer: **a.** -22 **b.** -18

Calculating Percentages (1.4)

- A percentage is the part divided by the total (whole) multiplied by 100%.

Example: A drawer contains 6 white socks and 18 black socks. What is the percentage of white socks?

Answer: $\dfrac{6 \text{ white socks}}{24 \text{ total socks}} \times 100\% = 25\%$ white socks

Solving Equations (1.4)

An equation in chemistry often contains an unknown. To rearrange an equation to obtain the unknown factor by itself, you keep it balanced by performing matching mathematical operations on both sides of the equation.

- If you eliminate a number or symbol by subtracting, subtract that same number or symbol on the opposite side.
- If you eliminate a number or symbol by adding, add that same number or symbol on the opposite side.
- If you cancel a number or symbol by dividing, divide both sides by that same number or symbol.
- If you cancel a number or symbol by multiplying, multiply both sides by that same number or symbol.

Example: Solve the equation for a: $3a - 8 = 28$

Answer: *Add 8 to both sides* $3a - \cancel{8} + \cancel{8} = 28 + 8$

$3a = 36$

Divide both sides by 3 $\dfrac{\cancel{3}a}{\cancel{3}} = \dfrac{36}{3}$

$a = 12$

Check: $3(12) - 8 = 28$

$36 - 8 = 28$

$28 = 28$

Your answer $a = 12$ is correct.

Interpreting Graphs (1.4)

- A graph represents the relationship between two variables.
- The quantities are plotted along two perpendicular axes, which are the x axis (horizontal) and y axis (vertical).
- The title indicates the components of the x and y axes.
- Numbers on the x and y axes show the range of values of the variables.
- The graph shows the relationship between the component on the y axis and that on the x axis.

Example:

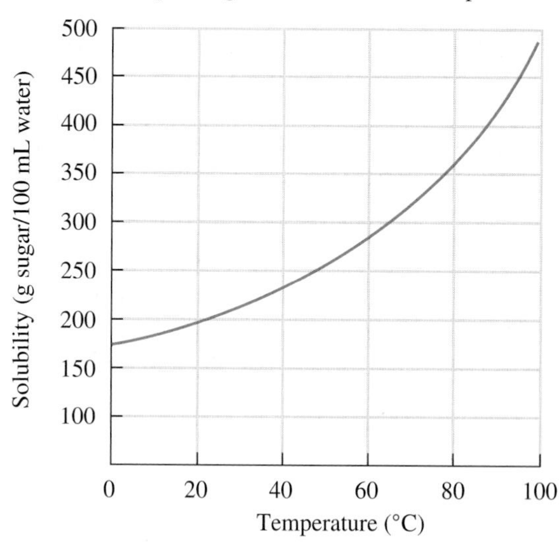

a. Does the amount of sugar that dissolves in 100 mL of water increase or decrease when the temperature increases?

b. How many grams of sugar dissolve in 100 mL of water at 70 °C?

c. At what temperature (°C) will 275 g of sugar dissolve in 100 mL of water?

Answer: **a.** increase
b. 320 g
c. 55 °C

Writing Numbers in Scientific Notation (1.5)

- A number written in scientific notation consists of a coefficient and a power of 10.

A number is written in scientific notation by:

- Moving the decimal point to obtain a coefficient that is at least 1 but less than 10.
- Expressing the number of places moved as a power of 10. The power of 10 is positive if the decimal point is moved to the left, negative if the decimal point is moved to the right.

Example: Write the number 28 000 in scientific notation.

Answer: Moving the decimal point four places to the left gives a coefficient of 2.8 and a positive power of 10, 10^4. The number 28 000 written in scientific notation is 2.8×10^4.

UNDERSTANDING THE CONCEPTS

The chapter Sections to review are shown in parentheses at the end of each problem.

1.31 A "chemical-free" shampoo includes the following ingredients: water, cocamide, glycerin, and citric acid. Is the shampoo truly "chemical-free"? (1.1)

1.32 A "chemical-free" sunscreen includes the following ingredients: titanium dioxide, vitamin E, and vitamin C. Is the sunscreen truly "chemical-free"? (1.1)

1.33 According to Sherlock Holmes, "One must follow the rules of scientific inquiry, gathering, observing, and testing data, then formulating, modifying, and rejecting hypotheses, until only one remains." Did Holmes use the scientific method? Why or why not? (1.2)

1.34 In *A Scandal in Bohemia*, Sherlock Holmes receives a mysterious note. He states, "I have no data yet. It is a capital mistake to theorize before one has data. Insensibly one begins to twist facts to suit theories, instead of theories to suit facts." What do you think Holmes meant? (1.2)

Sherlock Holmes is a fictional detective in novels written by Arthur Conan Doyle.

1.35 For each of the following, indicate if the answer has a positive or negative sign: (1.4)

a. Two negative numbers are added.

b. A positive and negative number are multiplied.

1.36 For each of the following, indicate if the answer has a positive or negative sign: (1.4)

a. A negative number is subtracted from a positive number.

b. Two negative numbers are divided.

Clinical Applications

1.37 Classify each of the following statements as an observation or a hypothesis: (1.2)

a. A patient breaks out in hives after receiving penicillin.

b. Dinosaurs became extinct when a large meteorite struck Earth and caused a huge dust cloud that severely decreased the amount of light reaching Earth.

c. The 100-yd dash was run in 9.8 s.

1.38 Classify each of the following statements as an observation or a hypothesis: (1.2)

a. Analysis of 10 ceramic dishes showed that four dishes contained lead levels that exceeded federal safety standards.

b. Marble statues undergo corrosion in acid rain.

c. A child with a high fever and a rash may have chickenpox.

ADDITIONAL PRACTICE PROBLEMS

1.39 Select the correct phrase(s) to complete the following statement: If experimental results do not support your hypothesis, you should: (1.2)

a. pretend that the experimental results support your hypothesis

b. modify your hypothesis

c. do more experiments

1.40 Select the correct phrase(s) to complete the following statement: A hypothesis is confirmed when: (1.2)

a. one experiment proves the hypothesis

b. many experiments validate the hypothesis

c. you think your hypothesis is correct

1.41 Which of the following will help you develop a successful study plan? (1.3)

a. skipping class and just reading the text

b. working the *Sample Problems* as you go through a chapter

c. self-testing

d. reading through the chapter, but working the problems later

1.42 Which of the following will help you develop a successful study plan? (1.3)

a. studying all night before the exam

b. forming a study group and discussing the problems together

c. working problems in a notebook for easy reference

d. highlighting important ideas in the text

1.43 Evaluate each of the following: (1.4)

a. $4 \times (-8) =$ _____ **b.** $-12 - 48 =$ _____

c. $\dfrac{-168}{-4} =$ _____

1.44 Evaluate each of the following: (1.4)

a. $-95 - (-11) =$ _____ **b.** $\dfrac{152}{-19} =$ _____

c. $4 - 56 =$ _____

1.45 A bag of gumdrops contains 16 orange gumdrops, 8 yellow gumdrops, and 16 black gumdrops. (1.4)

a. What is the percentage of yellow gumdrops? Express your answer to the ones place.

b. What is the percentage of black gumdrops? Express your answer to the ones place.

1.46 On the first chemistry test, 12 students got As, 18 students got Bs, and 20 students got Cs. (1.4)

a. What is the percentage of students who received Bs? Express your answer to the ones place.

b. What is the percentage of students who received Cs? Express your answer to the ones place.

1.47 Write each of the following in scientific notation: (1.5)

a. 120 000 **b.** 0.000 000 34

c. 0.066 **d.** 2700

1.48 Write each of the following in scientific notation: (1.5)

a. 0.0042 **b.** 310

c. 890 000 000 **d.** 0.000 000 056

Clinical Applications

1.49 Identify each of the following as an observation, a hypothesis, an experiment, or a conclusion: (1.2)

a. A patient has a high fever and a rash on her back.

b. A nurse tells a patient that her baby who gets sick after drinking milk may be lactose intolerant.

c. Numerous studies have shown that omega-3 fatty acids lower triglyceride levels.

1.50 Identify each of the following as an observation, a hypothesis, an experiment, or a conclusion: (1.2)

a. Every spring, you have congestion and a runny nose.

b. An overweight patient decides to exercise more to lose weight.

c. Many research studies have linked obesity to heart disease.

CHALLENGE PROBLEMS

The following problems are related to the topics in this chapter. However, they do not all follow the chapter order, and they require you to combine concepts and skills from several Sections. These problems will help you increase your critical thinking skills and prepare for your next exam.

1.51 Classify each of the following as an observation, a hypothesis, an experiment, or a conclusion: (1.2)

a. The bicycle tire is flat.

b. If I add air to the bicycle tire, it will expand to the proper size.

c. When I added air to the bicycle tire, it was still flat.

d. The bicycle tire has a leak in it.

1.52 Classify each of the following as an observation, a hypothesis, an experiment, or a conclusion: (1.2)

a. A big log in the fire does not burn well.

b. If I chop the log into smaller wood pieces, it will burn better.

c. The small wood pieces burn brighter and make a hotter fire.

d. The small wood pieces are used up faster than burning the big log.

1.53 Solve each of the following for x: (1.4)

a. $2x + 5 = 41$ b. $\frac{5x}{3} = 40$

1.54 Solve each of the following for z: (1.4)

a. $3z - (-6) = 12$ b. $\frac{4z}{-12} = -8$

Use the following graph for problems 1.55 and 1.56:

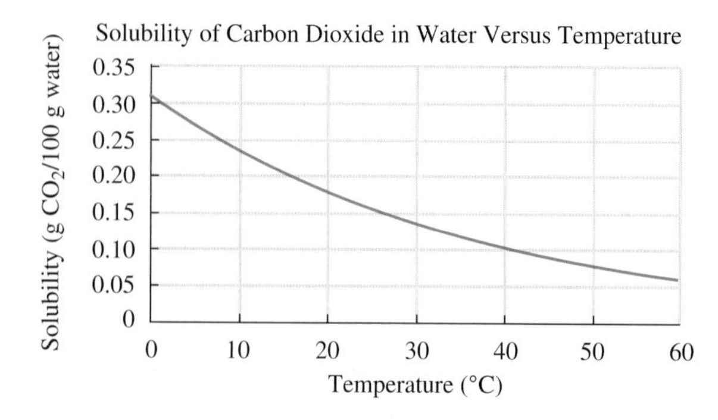

1.55 **a.** What does the title indicate about the graph? (1.4)
b. What is measured on the vertical axis?
c. What is the range of values on the vertical axis?
d. Does the solubility of carbon dioxide increase or decrease with an increase in temperature?

1.56 **a.** What is measured on the horizontal axis? (1.4)
b. What is the range of values on the horizontal axis?
c. What is the solubility of carbon dioxide in water at 25 °C?
d. At what temperature does carbon dioxide have a solubility of 0.20 g/100 g water?

ANSWERS

Answers to Selected Practice Problems

1.1 **a.** Chemistry is the study of the composition, structure, properties, and reactions of matter.
b. A chemical is a substance that has the same composition and properties wherever it is found.

1.3 Many chemicals are listed on a vitamin bottle such as vitamin A, vitamin B_3, vitamin B_{12}, vitamin C, and folic acid.

1.5 Typical items found in a medicine cabinet and some of the chemicals they contain are as follows:

Antacid tablets: calcium carbonate, cellulose, starch, stearic acid, silicon dioxide

Mouthwash: water, alcohol, thymol, glycerol, sodium benzoate, benzoic acid

Cough suppressant: menthol, beta-carotene, sucrose, glucose

1.7 **a.** observation **b.** hypothesis **c.** experiment **d.** observation **e.** observation **f.** conclusion

1.9 **a.** observation **b.** hypothesis **c.** experiment **d.** experiment

1.11 There are several things you can do that will help you successfully learn chemistry: forming a study group, retesting, doing *Try It First* before reading the Solution, checking *Review*, working *Sample Problems* and *Study Checks*, working *Practice Problems* and checking *Answers*, reading the assignment ahead of class, and keeping a problem notebook.

1.13 a, c, and e

1.15 **a.** thousandths **b.** ones **c.** hundreds

1.17 **a.** 23 **b.** −30 **c.** −2

1.19 **a.** 84% **b.** 72% **c.** 30%

1.21 **a.** 9 **b.** 42

1.23 **a.** The graph shows the relationship between the temperature of a cup of tea and time.
b. temperature, in °C
c. 20 °C to 80 °C
d. decrease

1.25 **a.** 5.5×10^4 **b.** 4.8×10^2 **c.** 5×10^{-6} **d.** 1.4×10^{-4} **e.** 7.2×10^{-3} **f.** 6.7×10^5

1.27 **a.** 7.2×10^3 **b.** 3.2×10^{-2} **c.** 1×10^4 **d.** 6.8×10^{-2}

1.29 27% ethylene glycol

1.31 No. All of the ingredients are chemicals.

1.33 Yes. Sherlock's investigation includes making observations (gathering data), formulating a hypothesis, testing the hypothesis, and modifying it until one of the hypotheses is validated.

1.35 **a.** negative **b.** negative

1.37 **a.** observation **b.** hypothesis **c.** observation

1.39 b and c

1.41 b and c

1.43 **a.** −32 **b.** −60 **c.** 42

1.45 **a.** 20% **b.** 40%

1.47 **a.** 1.2×10^5 **b.** 3.4×10^{-7} **c.** 6.6×10^{-2} **d.** 2.7×10^3

1.49 **a.** observation **b.** hypothesis **c.** conclusion

1.51 **a.** observation **b.** hypothesis **c.** experiment **d.** conclusion

1.53 **a.** 18 **b.** 24

1.55 **a.** The graph shows the relationship between the solubility of carbon dioxide in water and temperature.
b. solubility of carbon dioxide (g CO_2/100 g water)
c. 0 to 0.35 g of CO_2/100 g of water
d. decrease

2 Chemistry and Measurements

DURING THE PAST FEW MONTHS, GREG experienced an increased number of headaches, dizzy spells, and nausea. He goes to his doctor's office where Sandra, the registered nurse, completes the initial part of his exam by recording several measurements: weight 74.5 kg, height 171 cm, temperature 37.2 °C, and blood pressure 155/95. Normal blood pressure is 120/80 or lower.

When Greg sees his doctor, he is diagnosed with high blood pressure (hypertension). The doctor prescribes 80 mg of Inderal (propranolol), which is available in 40.-mg tablets. Inderal is a beta blocker, which relaxes the muscles of the heart. It is used to treat hypertension, angina (chest pain), arrhythmia, and migraine headaches.

Two weeks later, Greg visits his doctor again, who determines that Greg's blood pressure is now 152/90. The doctor increases the dosage of Inderal to 160 mg. The registered nurse, Sandra, informs Greg that he needs to increase his daily dosage from two tablets to four tablets.

CAREER Registered Nurse

In addition to assisting physicians, registered nurses work to promote patient health and prevent and treat disease. They provide patient care and help patients cope with illness. They take measurements such as a patient's weight, height, temperature, and blood pressure; make conversions; and calculate drug dosage rates. Registered nurses also maintain detailed medical records of patient symptoms and prescribed medications.

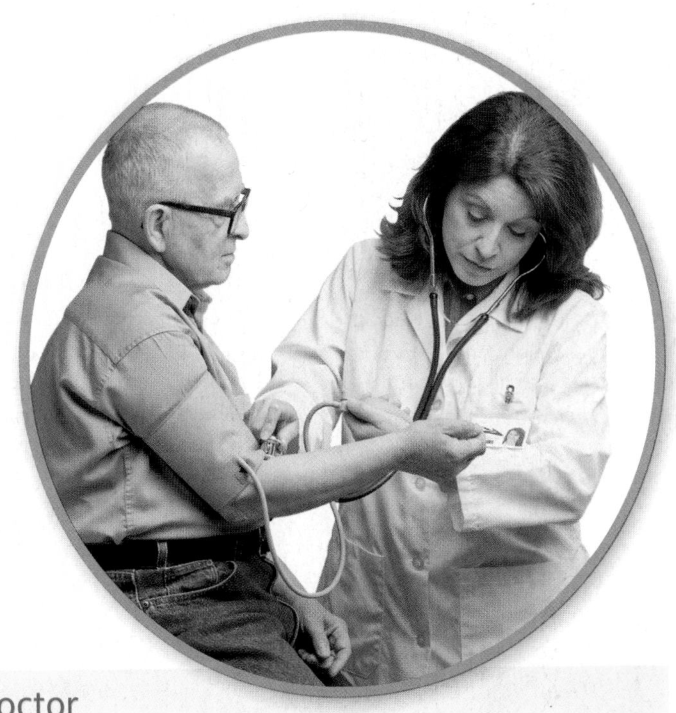

CLINICAL UPDATE Greg's Visit with His Doctor

A few weeks later, Greg complained to his doctor that he was feeling tired. He had a blood test to determine if his iron level was low. You can see the results of Greg's blood serum iron level in the **CLINICAL UPDATE Greg's Visit with His Doctor**, page 52, and determine if Greg should be given an iron supplement.

2.1 Units of Measurement

LEARNING GOAL Write the names and abbreviations for the metric or SI units used in measurements of volume, length, mass, temperature, and time.

Think about your day. You probably took some measurements. Perhaps you checked your weight by stepping on a bathroom scale. If you made rice for dinner, you added two cups of water to one cup of rice. If you did not feel well, you may have taken your temperature. Whenever you take a measurement, you use a measuring device such as a scale, a measuring cup, or a thermometer.

Scientists and health professionals throughout the world use the **metric system** of measurement. It is also the common measuring system in all but a few countries in the world. The **International System of Units (SI)**, or Système International, is the official system of measurement throughout the world except for the United States. In chemistry, we use metric units and SI units for volume, length, mass, temperature, and time, as listed in TABLE 2.1.

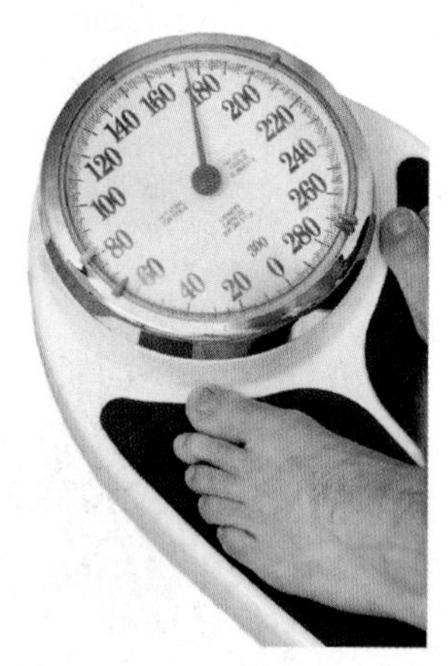

Your weight on a bathroom scale is a measurement.

TABLE 2.1 Units of Measurement and Their Abbreviations

Measurement	Metric	SI
Volume	liter (L)	cubic meter (m^3)
Length	meter (m)	meter (m)
Mass	gram (g)	kilogram (kg)
Temperature	degree Celsius (°C)	kelvin (K)
Time	second (s)	second (s)

Suppose you walked 1.3 mi to campus today, carrying a backpack that weighs 26 lb. The temperature was 72 °F. Perhaps you weigh 128 lb and your height is 65 in. These measurements and units may seem familiar to you because they are stated in the U.S. system of measurement. However, in chemistry, we use the *metric system* in making our measurements. Using the metric system, you walked 2.1 km to campus, carrying a backpack that has a mass of 12 kg, when the temperature was 22 °C. You have a mass of 58.0 kg and a height of 1.7 m.

There are many measurements in everyday life.

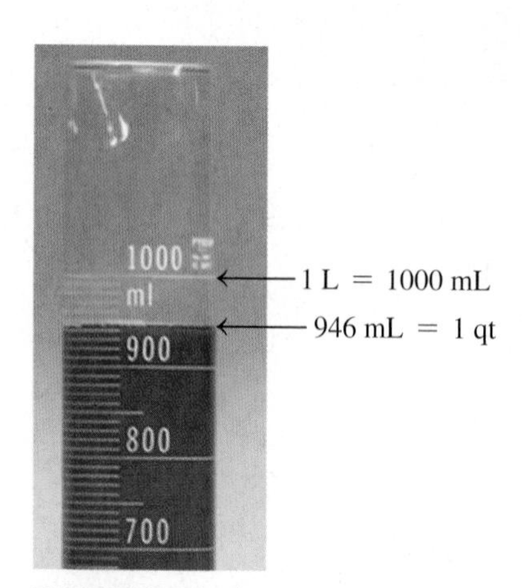

FIGURE 2.1 In the metric system, volume is based on the liter.

Q How many milliliters are in 1 quart?

Volume

Volume (*V*) is the amount of space a substance occupies. The metric unit for volume is the **liter (L)**, which is slightly larger than a quart (qt). In a laboratory or a hospital, chemists work with metric units of volume that are smaller and more convenient, such as the **milliliter (mL)**. There are 1000 mL in 1 L. (See FIGURE 2.1.) Some relationships between units for volume are

1 L = 1000 mL   1 L = 1.06 qt   946 mL = 1 qt

Meterstick
1 m = 39.4 in.
Yardstick
1 ft 2 ft 3 ft
Centimeters
1 in. = 2.54 cm
Inches

FIGURE 2.2 Length in the metric (SI) system is based on the meter, which is slightly longer than a yard.

Q How many centimeters are in a length of 1 inch?

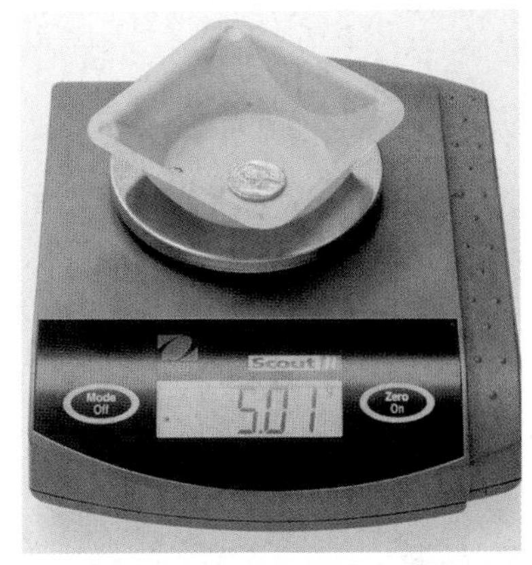

FIGURE 2.3 On an electronic balance, the digital readout gives the mass of a nickel, which is 5.01 g.

Q What is the mass of 10 nickels?

Length

The metric and SI unit of length is the **meter (m)**. The **centimeter (cm)**, a smaller unit of length, is commonly used in chemistry and is about equal to the width of your little finger (see **FIGURE 2.2**). Some relationships between units for length are

1 m = 100 cm 1 m = 39.4 in. 1 m = 1.09 yd 2.54 cm = 1 in.

Mass

The **mass** of an object is a measure of the quantity of material it contains. The SI unit of mass, the **kilogram (kg)**, is used for larger masses such as body mass. In the metric system, the unit for mass is the **gram (g)**, which is used for smaller masses. There are 1000 g in 1 kg. One pound (lb) is equal to 454 g. Some relationships between units for mass are

1 kg = 1000 g 1 kg = 2.20 lb 454 g = 1 lb

You may be more familiar with the term *weight* than with mass. Weight is a measure of the gravitational pull on an object. On Earth, an astronaut with a mass of 75.0 kg has a weight of 165 lb. On the Moon where the gravitational pull is one-sixth that of Earth, the astronaut has a weight of 27.5 lb. However, the mass of the astronaut is the same as on Earth, 75.0 kg. Scientists measure mass rather than weight because mass does not depend on gravity.

In a chemistry laboratory, an electronic balance is used to measure the mass in grams of a substance (see **FIGURE 2.3**).

FIGURE 2.4 A thermometer is used to determine temperature.

Q What kinds of temperature readings have you made today?

Temperature

Temperature tells us how hot something is, how cold it is outside, or helps us determine if we have a fever (see **FIGURE 2.4**). In the metric system, temperature is measured using Celsius temperature. On the **Celsius (°C) temperature scale**, water freezes at 0 °C and boils at 100 °C, whereas on the Fahrenheit (°F) scale, water freezes at 32 °F and boils at 212 °F. In the SI system, temperature is measured using the **Kelvin (K) temperature scale** on which the lowest possible temperature is 0 K. A unit on the Kelvin scale is called a kelvin (K) and is not written with a degree sign.

Time

We typically measure time in units such as years (yr), days, hours (h), minutes (min), or seconds (s). Of these, the SI and metric unit of time is the **second (s)**. The standard now used to determine a second is an atomic clock. Some relationships between units for time are

1 day = 24 h 1 h = 60 min 1 min = 60 s

A stopwatch is used to measure the time of a race.

SAMPLE PROBLEM 2.1 Units of Measurement

TRY IT FIRST

On a typical day, a nurse encounters several situations involving measurement. State the name and type of measurement indicated by the units in each of the following:

a. A patient has a temperature of 38.5 °C.
b. A physician orders 1.5 g of cefuroxime for injection.
c. A physician orders 1 L of a sodium chloride solution to be given intravenously.
d. A medication is to be given to a patient every 4 h.

SOLUTION

a. A degree Celsius is a unit of temperature.
b. A gram is a unit of mass.
c. A liter is a unit of volume.
d. An hour is a unit of time.

STUDY CHECK 2.1

State the name and type of measurement indicated by an infant that is 32.5 cm long.

ANSWER

A centimeter is a unit of length.

TEST

Try Practice Problems 2.1 to 2.8

PRACTICE PROBLEMS

2.1 Units of Measurement

LEARNING GOAL Write the names and abbreviations for the metric or SI units used in measurements of volume, length, mass, temperature, and time.

2.1 Write the abbreviation for each of the following:
a. gram **b.** degree Celsius
c. liter **d.** pound
e. second

2.2 Write the abbreviation for each of the following:
a. kilogram **b.** kelvin
c. quart **d.** meter
e. centimeter

2.3 State the type of measurement in each of the following statements:
a. I put 12 L of gasoline in my gas tank.
b. My friend is 170 cm tall.
c. Earth is 385 000 km away from the Moon.
d. The horse won the race by 1.2 s.

2.4 State the type of measurement in each of the following statements:
a. I rode my bicycle 15 km today.
b. My dog weighs 12 kg.
c. It is hot today. It is 30 °C.
d. I added 2 L of water to my fish tank.

2.5 State the name of the unit and the type of measurement indicated for each of the following quantities:
a. 4.8 m **b.** 325 g **c.** 1.5 mL
d. 4.8×10^2 s **e.** 28 °C

2.6 State the name of the unit and the type of measurement indicated for each of the following quantities:
a. 0.8 L **b.** 3.6 cm **c.** 4 kg
d. 3.5 h **e.** 373 K

Clinical Applications

2.7 On a typical day, medical personnel may encounter several situations involving measurement. State the name and type of measurement indicated by the units in each of the following:
a. The clotting time for a blood sample is 12 s.
b. A premature baby weighs 2.0 kg.
c. An antacid tablet contains 1.0 g of calcium carbonate.
d. An infant has a temperature of 39.2 °C.

2.8 On a typical day, medical personnel may encounter several situations involving measurement. State the name and type of measurement indicated by the units in each of the following:
a. During open-heart surgery, the temperature of a patient is lowered to 29 °C.
b. The circulation time of a red blood cell through the body is 20 s.
c. A patient with a persistent cough is given 10. mL of cough syrup.
d. The amount of iron in the red blood cells of the body is 2.5 g.

2.2 Measured Numbers and Significant Figures

LEARNING GOAL Identify a number as measured or exact; determine the number of significant figures in a measured number.

When you make a measurement, you use some type of measuring device. For example, you may use a meterstick to measure your height, a scale to check your weight, or a thermometer to take your temperature.

Measured Numbers

Measured numbers are the numbers you obtain when you measure a quantity such as your height, weight, or temperature. Suppose you are going to measure the lengths of the objects in **FIGURE 2.5**. To report the length of the object, you observe the numerical values of the marked lines at the end of the object. Then you can *estimate* by visually dividing the space between the marked lines. This estimated value is the final digit in a measured number.

For example, in Figure 2.5a, the end of the object is between the marks of 4 cm and 5 cm, which means that the length is more than 4 cm but less than 5 cm. If you estimate that the end of the object is halfway between 4 cm and 5 cm, you would report its length as 4.5 cm. Another student might report the length of the same object as 4.4 cm because people do not estimate in the same way.

The metric ruler shown in Figure 2.5b is marked at every 0.1 cm. Now you can determine that the end of the object is between 4.5 cm and 4.6 cm. Perhaps you report its length as 4.55 cm, whereas another student reports its length as 4.56 cm. Both results are acceptable.

In Figure 2.5c, the end of the object appears to line up with the 3-cm mark. Because the end of the object is on the 3-cm mark, the estimated digit is 0, which means the measurement is reported as 3.0 cm.

0 cm 1 2 3 4 5

(a)

0 cm 1 2 3 4 5

(b)

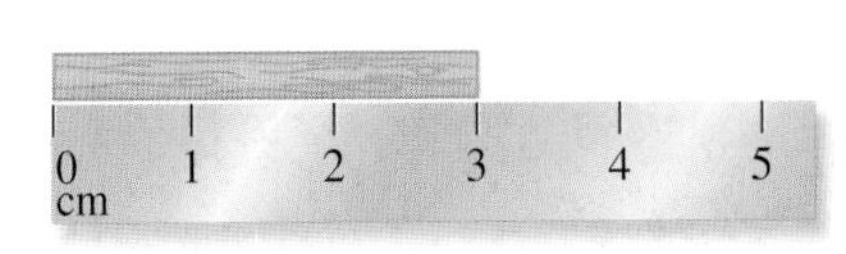

(c)

FIGURE 2.5 The lengths of the rectangular objects are measured as **(a)** 4.5 cm and **(b)** 4.55 cm.

Q Why is the length of the object in **(c)** reported as 3.0 cm not 3 cm?

Significant Figures

CORE CHEMISTRY SKILL
Counting Significant Figures

In a measured number, the **significant figures (SFs)** are all the digits including the estimated digit. Nonzero numbers are always counted as significant figures. However, a zero may or may not be a significant figure depending on its position in a number. **TABLE 2.2** gives the rules and examples of counting significant figures.

TABLE 2.2 Significant Figures in Measured Numbers

Rule	Measured Number	Number of Significant Figures
1. A number is a *significant figure* if it is		
a. not a zero	4.5 g 122.35 m	2 5
b. a zero between nonzero digits	205 °C 5.008 kg	3 4
c. a zero at the end of a decimal number	50. L 16.00 mL	2 4
d. in the coefficient of a number written in scientific notation	4.8×10^5 m 5.70×10^{-3} g	2 3
2. A zero is *not significant* if it is		
a. at the beginning of a decimal number	0.0004 s 0.075 cm	1 2
b. used as a placeholder in a large number without a decimal point	850 000 m 1 250 000 g	2 3

TEST
Try Practice Problems 2.9 to 2.12

ENGAGE

Why is the zero in the coefficient of 3.20×10^4 cm a significant figure?

TEST

Try Practice Problems 2.13 to 2.16

Significant Zeros and Scientific Notation

In this text, we will place a decimal point after a significant zero at the end of a number. For example, if a measurement is written as 500. g, the decimal point after the second zero indicates that *both zeros* are significant. To show this more clearly, we can write it as 5.00×10^2 g. When the first zero in the measurement 300 m is a significant zero, but the second zero is not, the measurement is written as 3.0×10^2 m. We will assume that all zeros at the end of large standard numbers without a decimal point are not significant. Therefore, we write 400 000 g as 4×10^5 g, which has only one significant figure.

Exact Numbers

Exact numbers *are those numbers obtained by counting items or using a definition that compares two units in the same measuring system.* Suppose a friend asks you how many classes you are taking. You would answer by counting the number of classes in your schedule. Suppose you are asked to state the number of seconds in one minute. Without using any measuring device, you would give the definition: There are 60 s in 1 min. *Exact numbers are not measured, do not have a limited number of significant figures, and do not affect the number of significant figures in a calculated answer.* For more examples of exact numbers, see **TABLE 2.3**.

The number of baseballs is counted, which means 2 is an exact number.

TABLE 2.3 Examples of Some Exact Numbers

Counted Numbers	Defined Equalities	
Items	Metric System	U.S. System
8 doughnuts	1 L = 1000 mL	1 ft = 12 in.
2 baseballs	1 m = 100 cm	1 qt = 4 cups
5 capsules	1 kg = 1000 g	1 lb = 16 oz

For example, a mass of 42.2 g and a length of 5.0×10^{-3} cm are measured numbers because they are obtained using measuring tools. There are three SFs in 42.2 g because all nonzero digits are always significant. There are two SFs in 5.0×10^{-3} cm because all the digits in the coefficient of a number written in scientific notation are significant. However, a quantity of three eggs is an exact number that is obtained by counting. In the equality 1 kg = 1000 g, the masses of 1 kg and 1000 g are both exact numbers because this equality is a definition in the metric system.

SAMPLE PROBLEM 2.2 **Measured and Exact Numbers**

TRY IT FIRST

Identify each of the following numbers as measured or exact and give the number of significant figures (SFs) in each of the measured numbers:

a. 0.170 L **b.** 4 knives
c. 6.3×10^{-6} s **d.** 1 m = 100 cm

SOLUTION

a. measured; three SFs **b.** exact
c. measured; two SFs **d.** exact

STUDY CHECK 2.2

Identify each of the following numbers as measured or exact and give the number of significant figures (SFs) in each of the measured numbers:

a. 0.020 80 kg **b.** 5.06×10^4 h **c.** 4 chemistry books

ANSWER

a. measured; four SFs **b.** measured; three SFs **c.** exact

TEST

Try Practice Problems 2.17 to 2.22

PRACTICE PROBLEMS

2.2 Measured Numbers and Significant Figures

LEARNING GOAL Identify a number as measured or exact; determine the number of significant figures in a measured number.

2.9 How many significant figures are in each of the following?
a. 11.005 g
b. 0.000 32 m
c. 36 000 000 km
d. 1.80×10^4 kg
e. 0.8250 L
f. 30.0 °C

2.10 How many significant figures are in each of the following?
a. 20.60 mL
b. 1036.48 kg
c. 4.00 m
d. 20.8 °C
e. 60 800 000 g
f. 5.0×10^{-3} L

2.11 In which of the following pairs do both numbers contain the same number of significant figures?
a. 11.0 m and 11.00 m
b. 0.0250 m and 0.205 m
c. 0.000 12 s and 12 000 s
d. 250.0 L and 2.5×10^{-2} L

2.12 In which of the following pairs do both numbers contain the same number of significant figures?
a. 0.005 75 g and 5.75×10^{-3} g
b. 405 K and 405.0 K
c. 150 000 s and 1.50×10^4 s
d. 3.8×10^{-2} L and 3.0×10^5 L

2.13 Indicate if the zeros are significant in each of the following measurements:
a. 0.0038 m
b. 5.04 cm
c. 800. L
d. 3.0×10^{-3} kg
e. 85 000 g

2.14 Indicate if the zeros are significant in each of the following measurements:
a. 20.05 °C
b. 5.00 m
c. 0.000 02 g
d. 120 000 yr
e. 8.05×10^2 L

2.15 Write each of the following in scientific notation with two significant figures:
a. 5000 L
b. 30 000 g
c. 100 000 m
d. 0.000 25 cm

2.16 Write each of the following in scientific notation with two significant figures:
a. 5 100 000 g
b. 26 000 s
c. 40 000 m
d. 0.000 820 kg

2.17 Identify the numbers in each of the following statements as measured or exact:
a. A patient has a mass of 67.5 kg.
b. A patient is given 2 tablets of medication.
c. In the metric system, 1 L is equal to 1000 mL.
d. The distance from Denver, Colorado, to Houston, Texas, is 1720 km.

2.18 Identify the numbers in each of the following statements as measured or exact:
a. There are 31 students in the laboratory.
b. The oldest known flower lived 1.20×10^8 yr ago.
c. The largest gem ever found, an aquamarine, has a mass of 104 kg.
d. A laboratory test shows a blood cholesterol level of 184 mg/dL.

2.19 Identify the measured number(s), if any, in each of the following pairs of numbers:
a. 3 hamburgers and 6 oz of hamburger
b. 1 table and 4 chairs
c. 0.75 lb of grapes and 350 g of butter
d. 60 s = 1 min

2.20 Identify the exact number(s), if any, in each of the following pairs of numbers:
a. 5 pizzas and 50.0 g of cheese
b. 6 nickels and 16 g of nickel
c. 3 onions and 3 lb of onions
d. 5 miles and 5 cars

Clinical Applications

2.21 Identify each of the following as measured or exact and give the number of significant figures (SFs) in each measured number:
a. The mass of a neonate is 1.607 kg.
b. The Daily Value (DV) for iodine for an infant is 130 mcg.
c. There are 4.02×10^6 red blood cells in a blood sample.
d. In November, 23 babies were born in a hospital.

2.22 Identify each of the following as measured or exact and give the number of significant figures (SFs) in each measured number:
a. An adult with the flu has a temperature of 103.5 °F.
b. A blister (push-through) pack of prednisone contains 21 tablets.
c. The time for a nerve impulse to travel from the feet to the brain is 0.46 s.
d. A brain contains 1.20×10^{10} neurons.

2.3 Significant Figures in Calculations

LEARNING GOAL Adjust calculated answers to give the correct number of significant figures.

REVIEW
Identifying Place Values (1.4)
Using Positive and Negative Numbers in Calculations (1.4)

In the sciences, we measure many things: the length of a bacterium, the volume of a gas sample, the temperature of a reaction mixture, or the mass of iron in a sample. The number of significant figures in measured numbers determines the number of significant figures in the calculated answer.

Using a calculator will help you perform calculations faster. However, calculators cannot think for you. It is up to you to enter the numbers correctly, press the correct function keys, and give the answer with the correct number of significant figures.

KEY MATH SKILL

Rounding Off

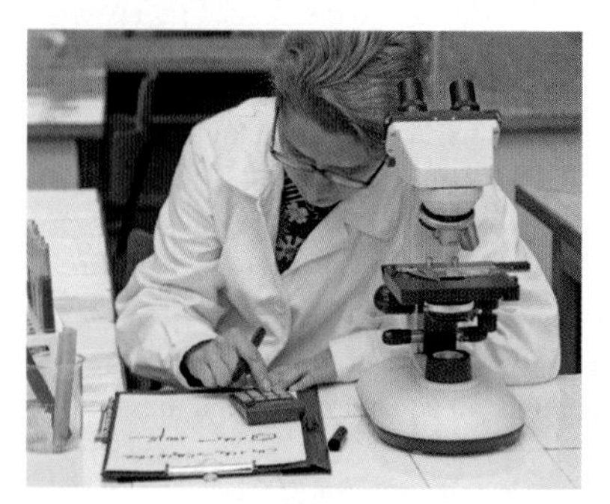

A technician uses a calculator in the laboratory.

Rounding Off

Suppose you decide to buy carpeting for a room that has a length of 5.52 m and a width of 3.58 m. To determine how much carpeting you need, you would calculate the area of the room by multiplying 5.52 times 3.58 on your calculator. The calculator shows the number 19.7616 in its display. Because each of the original measurements has only three significant figures, the calculator display (19.7616) is *rounded off* to three significant figures, 19.8.

5.52	×	3.58	=	19.7616	=	19.8 m^2
Three SFs		Three SFs		Calculator display		Final answer, rounded off to three SFs

Therefore, you can order carpeting that will cover an area of 19.8 m^2.

Each time you use a calculator, it is important to look at the original measurements and determine the number of significant figures that can be used for the answer. You can use the following rules to round off the numbers shown in a calculator display.

Rules for Rounding Off

1. If the first digit to be dropped is *4 or less*, then it and all following digits are simply dropped from the number.
2. If the first digit to be dropped is *5 or greater*, then the last retained digit of the number is increased by 1.

Number to Round Off	Three Significant Figures	Two Significant Figures
8.4234	8.42 (drop 34)	8.4 (drop 234)
14.780	14.8 (drop 80, increase the last retained digit by 1)	15 (drop 780, increase the last retained digit by 1)
3256	3260* (drop 6, increase the last retained digit by 1, add 0) (3.26×10^3)	3300* (drop 56, increase the last retained digit by 1, add 00) (3.3×10^3)

*The value of a large number is retained by using placeholder zeros to replace dropped digits.

ENGAGE

Why is 10.07208 rounded off to three significant figures equal to 10.1?

SAMPLE PROBLEM 2.3 Rounding Off

TRY IT FIRST

Round off each of the following numbers to three significant figures:

a. 35.7823 m **b.** 0.002 621 7 L **c.** 3.8268×10^3 g

SOLUTION

a. 35.8 m **b.** 0.002 62 L **c.** 3.83×10^3 g

STUDY CHECK 2.3

Round off each of the numbers in Sample Problem 2.3 to two significant figures.

ANSWER

a. 36 m **b.** 0.0026 L **c.** 3.8×10^3 g

TEST

Try Practice Problems 2.23 to 2.26

Multiplication and Division with Measured Numbers

CORE CHEMISTRY SKILL

Using Significant Figures in Calculations

In multiplication or division, the final answer is written so that it has the same number of significant figures (SFs) as the measurement with the fewest SFs. An example of rounding off a calculator display follows:

Perform the following operations with measured numbers:

$$\frac{2.8 \times 67.40}{34.8} =$$

When the problem has multiple steps, the numbers in the numerator are multiplied and then divided by each of the numbers in the denominator.

2.8	×	67.40	÷	34.8	=	5.422988506	=	5.4
Two SFs		Four SFs		Three SFs		Calculator display		Answer, rounded off to two SFs

Because the calculator display has more digits than the significant figures in the measured numbers allow, we need to round it off. Using the measured number that has the smallest number (two) of significant figures, 2.8, we round off the calculator display to an answer with two SFs.

A calculator is helpful in working problems and doing calculations faster.

Adding Significant Zeros

Sometimes, a calculator display gives a small whole number. For example, suppose the calculator display is 4, but you used measurements that have three significant numbers. Then two significant zeros are *added* to give 4.00 as the correct answer.

$$\frac{8.00}{2.00} = 4. = 4.00$$

Three SFs (8.00); Three SFs (2.00); Calculator display (4.); Final answer, two zeros added to give three SFs (4.00)

SAMPLE PROBLEM 2.4 Significant Figures in Multiplication and Division

TRY IT FIRST

Perform the following calculations with measured numbers. Write each answer with the correct number of significant figures.

a. 56.8×0.37 **b.** $\dfrac{(2.075)\,(0.585)}{(8.42)\,(0.0245)}$ **c.** $\dfrac{25.0}{5.00}$

SOLUTION GUIDE

ANALYZE THE PROBLEM	Given	Need	Connect
	multiplication and division	answer with SFs	rules for rounding off, adding zeros

STEP 1 Determine the number of significant figures in each measured number.

a. 56.8×0.37 (Three SFs, Two SFs)

b. $\dfrac{(2.075)\,(0.585)}{(8.42)\,(0.0245)}$ (Four SFs, Three SFs over Three SFs, Three SFs)

c. $\dfrac{25.0}{5.00}$ (Three SFs over Three SFs)

STEP 2 Perform the indicated calculation.

a. 21.016 (Calculator display) **b.** 5.884313345 (Calculator display) **c.** 5. (Calculator display)

STEP 3 Round off (or add zeros) to give the same number of significant figures as the measurement having the fewest significant figures.

a. 21 **b.** 5.88 **c.** 5.00

STUDY CHECK 2.4

Perform the following calculations with measured numbers and give the answers with the correct number of significant figures:

a. 45.26×0.01088 **b.** $2.6 \div 324$ **c.** $\dfrac{4.0 \times 8.00}{16}$

ANSWER

a. 0.4924 **b.** 0.0080 or 8.0×10^{-3} **c.** 2.0

ENGAGE

Why is the answer for the multiplication of 0.3×52.6 written with one significant figure?

TEST

Try Practice Problems 2.27 and 2.28

Addition and Subtraction with Measured Numbers

In addition or subtraction, the final answer is written so that it has the same number of decimal places as the measurement having the fewest decimal places.

2.045	Thousandths place
+ 34.1	Tenths place
36.145	Calculator display
36.1	Answer, rounded off to the tenths place

When numbers are added or subtracted to give an answer ending in zero, the zero does not appear after the decimal point in the calculator display. For example, 14.5 g − 2.5 g = 12.0 g. However, if you do the subtraction on your calculator, the display shows 12. To write the correct answer, a significant zero is written after the decimal point.

ENGAGE

Why is the answer for the addition of 55.2 and 2.506 written with one decimal place?

SAMPLE PROBLEM 2.5 Decimal Places in Addition and Subtraction

TRY IT FIRST

Perform the following calculations and give each answer with the correct number of decimal places:

a. 104.45 mL + 0.838 mL + 46 mL **b.** 153.247 g − 14.82 g

SOLUTION GUIDE

	Given	Need	Connect
ANALYZE THE PROBLEM	addition and subtraction	correct number of decimal places	rules for rounding off

STEP 1 Determine the number of decimal places in each measured number.

a.

104.45 mL	Hundredths place
0.838 mL	Thousandths place
+ 46 mL	Ones place

b.

153.247 g	Thousandths place
− 14.82 g	Hundredths place

STEP 2 Perform the indicated calculation.

a. 151.288 Calculator display **b.** 138.427 Calculator display

STEP 3 Round off the answer to give the same number of decimal places as the measurement having the fewest decimal places.

a. 151 mL **b.** 138.43 g

STUDY CHECK 2.5

Perform the following calculations and give each answer with the correct number of decimal places:

a. 82.45 mg + 1.245 mg + 0.000 56 mg **b.** 4.259 L − 3.8 L

ANSWER

a. 83.70 mg **b.** 0.5 L

TEST

Try Practice Problems 2.29 and 2.30

PRACTICE PROBLEMS

2.3 Significant Figures in Calculations

LEARNING GOAL Adjust calculated answers to give the correct number of significant figures.

2.23 Round off each of the following calculator answers to three significant figures:
a. 1.854 kg
b. 88.2038 L
c. 0.004 738 265 cm
d. 8807 m
e. 1.832×10^5 s

2.24 Round off each of the calculator answers in problem 2.23 to two significant figures.

2.25 Round off or add zeros to each of the following to three significant figures:
a. 56.855 m
b. 0.002 282 g
c. 11 527 s
d. 8.1 L

2.26 Round off or add zeros to each of the following to two significant figures:
a. 3.2805 m
b. 1.855×10^2 g
c. 0.002 341 mL
d. 2 L

2.27 Perform each of the following calculations, and give an answer with the correct number of significant figures:
a. 45.7×0.034
b. $0.002\,78 \times 5$
c. $\frac{34.56}{1.25}$
d. $\frac{(0.2465)(25)}{1.78}$
e. $(2.8 \times 10^4)(5.05 \times 10^{-6})$
f. $\frac{(3.45 \times 10^{-2})(1.8 \times 10^5)}{(8 \times 10^3)}$

2.28 Perform each of the following calculations, and give an answer with the correct number of significant figures:
a. 400×185
b. $\frac{2.40}{(4)(125)}$
c. $0.825 \times 3.6 \times 5.1$
d. $\frac{(3.5)(0.261)}{(8.24)(20.0)}$
e. $\frac{(5 \times 10^{-5})(1.05 \times 10^4)}{(8.24 \times 10^{-8})}$
f. $\frac{(4.25 \times 10^2)(2.56 \times 10^{-3})}{(2.245 \times 10^{-3})(56.5)}$

2.29 Perform each of the following calculations, and give an answer with the correct number of decimal places:
a. 45.48 cm + 8.057 cm
b. 23.45 g + 104.1 g + 0.025 g
c. 145.675 mL − 24.2 mL
d. 1.08 L − 0.585 L

2.30 Perform each of the following calculations, and give an answer with the correct number of decimal places:
a. 5.08 g + 25.1 g
b. 85.66 cm + 104.10 cm + 0.025 cm
c. 24.568 mL − 14.25 mL
d. 0.2654 L − 0.2585 L

2.4 Prefixes and Equalities

LEARNING GOAL Use the numerical values of prefixes to write a metric equality.

REVIEW
Writing Numbers in Scientific Notation (1.5)

The special feature of the metric system is that a **prefix** can be placed in front of any unit to increase or decrease its size by some factor of 10. For example, the prefixes *milli* and *micro* are used to make the smaller units, milligram (mg) and microgram (μg).

The U.S. Food and Drug Administration has determined the Daily Values (DV) for nutrients for adults and children age 4 or older. Examples of these recommended Daily Values, some of which use prefixes, are listed in TABLE 2.4.

The prefix *centi* is like cents in a dollar. One cent would be a "centidollar" or 0.01 of a dollar. That also means that one dollar is the same as 100 cents. The prefix *deci* is like dimes in a dollar. One dime would be a "decidollar" or 0.1 of a dollar. That also means that one dollar is the same as 10 dimes. TABLE 2.5 lists some of the metric prefixes, their symbols, and their numerical values.

The relationship of a prefix to a unit can be expressed by replacing the prefix with its numerical value. For example, when the prefix *kilo* in kilometer is replaced with its value of 1000, we find that a kilometer is equal to 1000 m. Other examples follow:

1 **kilo**meter (1 km) = **1000** meters (1000 m = 10^3 m)
1 **kilo**liter (1 kL) = **1000** liters (1000 L = 10^3 L)
1 **kilo**gram (1 kg) = **1000** grams (1000 g = 10^3 g)

TABLE 2.4 Daily Values for Selected Nutrients

Nutrient	Amount Recommended
Calcium	1.0 g
Copper	2 mg
Iodine	150 μg (150 mcg)
Iron	18 mg
Magnesium	400 mg
Niacin	20 mg
Phosphorus	800 mg
Potassium	3.5 g
Selenium	70. μg (70. mcg)
Sodium	2.4 g
Zinc	15 mg

CORE CHEMISTRY SKILL
Using Prefixes

TABLE 2.5 Metric and SI Prefixes

Prefix	Symbol	Numerical Value	Scientific Notation	Equality
Prefixes That Increase the Size of the Unit				
tera	T	1 000 000 000 000	10^{12}	$1\ \text{Ts} = 1 \times 10^{12}\ \text{s}$ $1\ \text{s} = 1 \times 10^{-12}\ \text{Ts}$
giga	G	1 000 000 000	10^{9}	$1\ \text{Gm} = 1 \times 10^{9}\ \text{m}$ $1\ \text{m} = 1 \times 10^{-9}\ \text{Gm}$
mega	M	1 000 000	10^{6}	$1\ \text{Mg} = 1 \times 10^{6}\ \text{g}$ $1\ \text{g} = 1 \times 10^{-6}\ \text{Mg}$
kilo	k	1 000	10^{3}	$1\ \text{km} = 1 \times 10^{3}\ \text{m}$ $1\ \text{m} = 1 \times 10^{-3}\ \text{km}$
Prefixes That Decrease the Size of the Unit				
deci	d	0.1	10^{-1}	$1\ \text{dL} = 1 \times 10^{-1}\ \text{L}$ $1\ \text{L} = 10\ \text{dL}$
centi	c	0.01	10^{-2}	$1\ \text{cm} = 1 \times 10^{-2}\ \text{m}$ $1\ \text{m} = 100\ \text{cm}$
milli	m	0.001	10^{-3}	$1\ \text{ms} = 1 \times 10^{-3}\ \text{s}$ $1\ \text{s} = 1 \times 10^{3}\ \text{ms}$
micro	μ*	0.000 001	10^{-6}	$1\ \mu\text{g} = 1 \times 10^{-6}\ \text{g}$ $1\ \text{g} = 1 \times 10^{6}\ \mu\text{g}$
nano	n	0.000 000 001	10^{-9}	$1\ \text{nm} = 1 \times 10^{-9}\ \text{m}$ $1\ \text{m} = 1 \times 10^{9}\ \text{nm}$
pico	p	0.000 000 000 001	10^{-12}	$1\ \text{ps} = 1 \times 10^{-12}\ \text{s}$ $1\ \text{s} = 1 \times 10^{12}\ \text{ps}$

*In medicine, the abbreviation *mc* for the prefix *micro* is used because the symbol μ may be misread, which could result in a medication error. Thus, 1 μg would be written as 1 mcg.

ENGAGE
Why is 60. mg of vitamin C the same as 0.060 g of vitamin C?

SAMPLE PROBLEM 2.6 Prefixes and Equalities

TRY IT FIRST

An endoscopic camera has a width of 1 mm. Complete each of the following equalities involving millimeters:

a. 1 m = ______ mm **b.** 1 cm = ______ mm

SOLUTION

a. 1 m = 1000 mm **b.** 1 cm = 10 mm

STUDY CHECK 2.6

What is the relationship between millimeters and micrometers?

ANSWER

1 mm = 1000 μm (mcm)

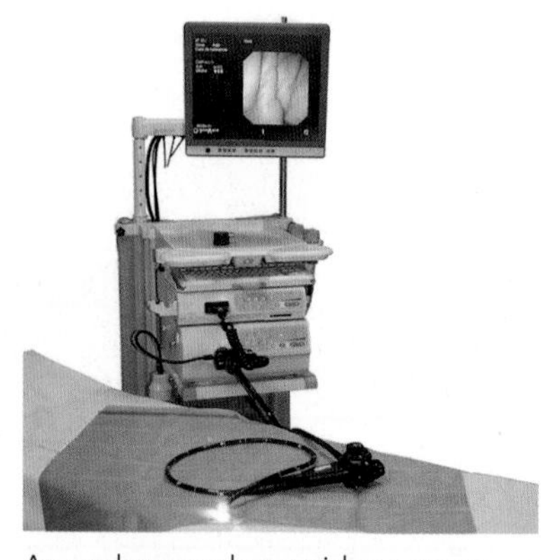

An endoscope has a video camera with a width of 1 mm attached to the end of a thin cable.

Using a retinal camera, an ophthalmologist photographs the retina of an eye.

Measuring Length

An ophthalmologist may measure the diameter of the retina of an eye in centimeters (cm), whereas a surgeon may need to know the length of a nerve in millimeters (mm). When the prefix *centi* is used with the unit meter, it becomes *centimeter*, a length that is one-hundredth of a meter (0.01 m). When the prefix *milli* is used with the unit meter, it becomes *millimeter*, a length that is one-thousandth of a meter (0.001 m). There are 100 cm and 1000 mm in a meter.

If we compare the lengths of a millimeter and a centimeter, we find that 1 mm is 0.1 cm; there are 10 mm in 1 cm. These comparisons are examples of **equalities**, which show the relationship between two units that measure the same quantity. Examples of equalities between different metric units of length follow:

1 m = 100 cm = 1×10^2 cm
1 m = 1000 mm = 1×10^3 mm
1 cm = 10 mm = 1×10^1 mm

Some metric units for length are compared in **FIGURE 2.6**.

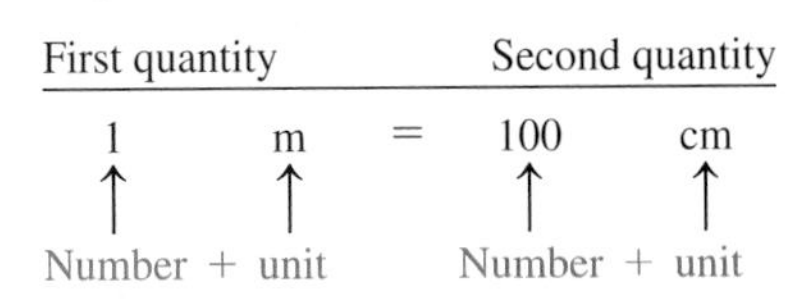

This example of an equality shows the relationship between meters and centimeters.

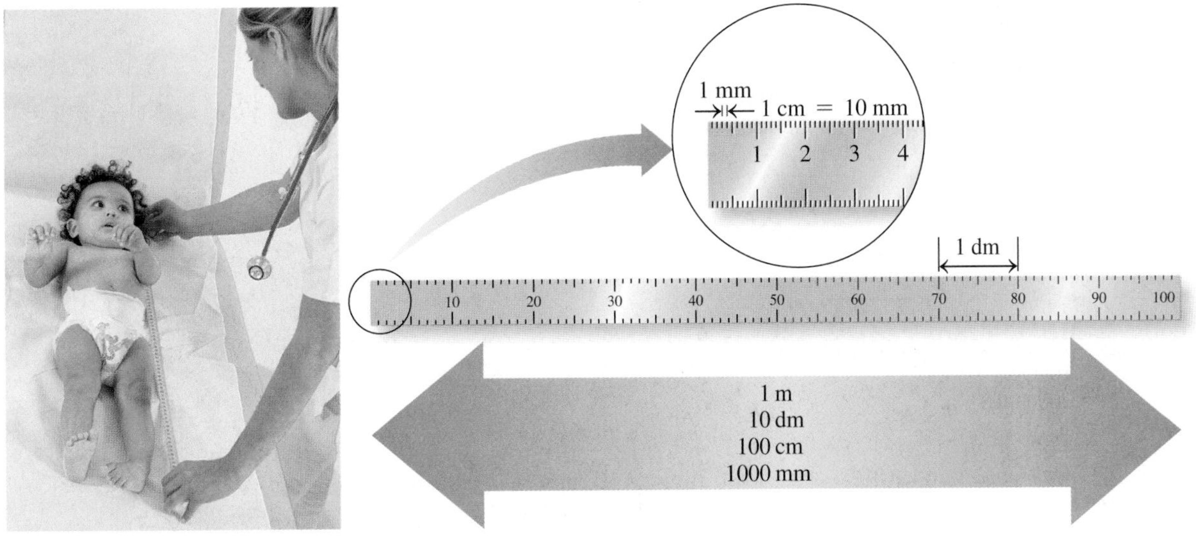

FIGURE 2.6 The metric length of 1 m is the same length as 10 dm, 100 cm, or 1000 mm.

Q How many millimeters (mm) are in 1 centimeter (cm)?

Measuring Volume

Volumes of 1 L or smaller are common in the health sciences. When a liter is divided into 10 equal portions, each portion is a deciliter (dL). There are 10 dL in 1 L. Laboratory results for bloodwork are often reported in mass per deciliter. **TABLE 2.6** lists normal laboratory test values for some substances in the blood.

TABLE 2.6 Some Normal Laboratory Test Values

Substance in Blood	Normal Range
Albumin	3.5–5.4 g/dL
Ammonia	20–70 μg/dL (mcg/dL)
Calcium	8.5–10.5 mg/dL
Cholesterol	105–250 mg/dL
Iron (male)	80–160 μg/dL (mcg/dL)
Protein (total)	6.0–8.5 g/dL

TEST

Try Practice Problems 2.31 to 2.38

When a liter is divided into a thousand parts, each of the smaller volumes is a milliliter (mL). In a 1-L container of physiological saline, there are 1000 mL of solution (see **FIGURE 2.7**). Examples of equalities between different metric units of volume follow:

1 L = 10 dL = 1×10^1 dL
1 L = 1000 mL = 1×10^3 mL
1 dL = 100 mL = 1×10^2 mL
1 mL = 1000 μL (mcL) = 1×10^3 μL (mcL)

The **cubic centimeter** (abbreviated as **cm^3** or **cc**) is the volume of a cube whose dimensions are 1 cm on each side. A cubic centimeter has the same volume as a milliliter, and the units are often used interchangeably.

1 cm^3 = 1 cc = 1 mL

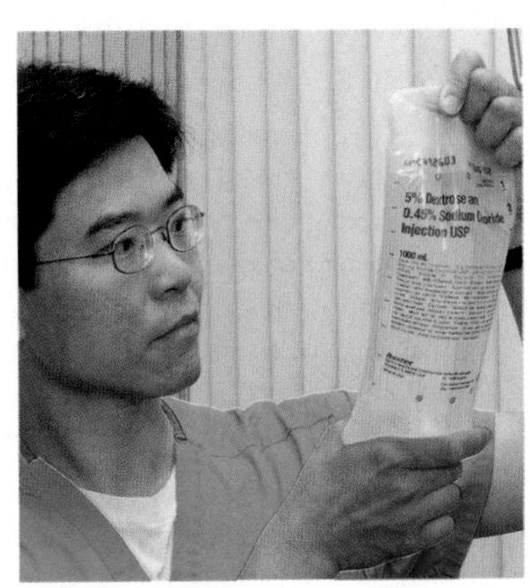

FIGURE 2.7 A plastic intravenous fluid container contains 1000 mL.

Q How many liters of solution are in the intravenous fluid container?

When you see *1 cm*, you are reading about length; when you see *1 cm^3* or *1 cc* or *1 mL*, you are reading about volume. A comparison of units of volume is illustrated in **FIGURE 2.8**.

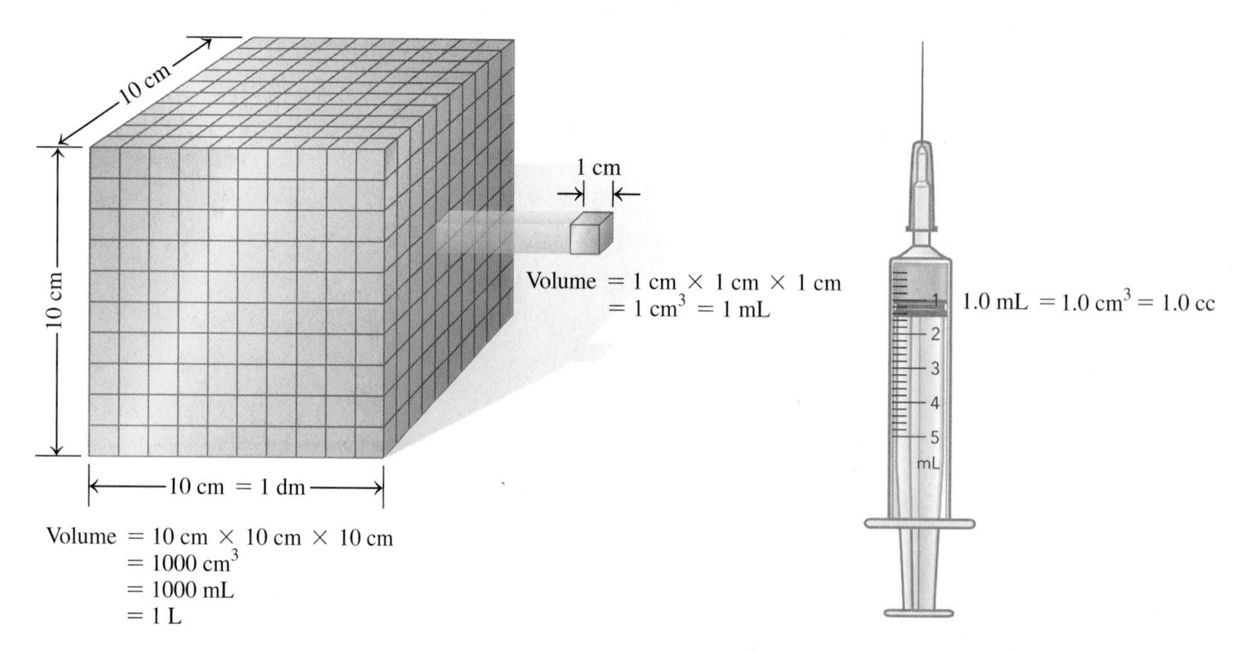

FIGURE 2.8 A cube measuring 10 cm on each side has a volume of 1000 cm^3, or 1 L; a cube measuring 1 cm on each side has a volume of 1 cm^3 (cc) or 1 mL.

Q What is the relationship between a milliliter (mL) and a cubic centimeter (cm^3)?

Measuring Mass

When you go to the doctor for a physical examination, your mass is recorded in kilograms, whereas the results of your laboratory tests are reported in grams, milligrams (mg), or micrograms (μg or mcg). A kilogram is equal to 1000 g. One gram represents the same mass as 1000 mg, and one mg equals 1000 μg (or 1000 mcg). Examples of equalities between different metric units of mass follow:

TEST

Try Practice Problems 2.39 to 2.42

$$1 \text{ kg} = 1000 \text{ g} = 1 \times 10^3 \text{ g}$$
$$1 \text{ g} = 1000 \text{ mg} = 1 \times 10^3 \text{ mg}$$
$$1 \text{ mg} = 1000\ \mu\text{g (mcg)} = 1 \times 10^3\ \mu\text{g (mcg)}$$

PRACTICE PROBLEMS

2.4 Prefixes and Equalities

LEARNING GOAL Use the numerical values of prefixes to write a metric equality.

2.31 Write the abbreviation for each of the following units:
a. milligram **b.** deciliter
c. kilometer **d.** picogram

2.32 Write the abbreviation for each of the following units:
a. gigagram **b.** megameter
c. microliter **d.** nanosecond

2.33 Write the complete name for each of the following units:
a. cL **b.** kg **c.** ms **d.** Gm

2.34 Write the complete name for each of the following units:
a. dL **b.** Ts **c.** mcg **d.** pm

2.35 Write the numerical value for each of the following prefixes:
a. centi **b.** tera **c.** milli **d.** deci

2.36 Write the numerical value for each of the following prefixes:
a. giga **b.** micro **c.** mega **d.** nano

2.37 Use a prefix to write the name for each of the following:
a. 0.1 g **b.** 10^{-6} g **c.** 1000 g **d.** 0.01 g

2.38 Use a prefix to write the name for each of the following:
a. 10^9 m **b.** 10^6 m **c.** 0.001 m **d.** 10^{-12} m

2.39 Complete each of the following metric relationships:
a. 1 m = ________ cm **b.** 1 m = ________ nm
c. 1 mm = ________ m **d.** 1 L = ________ mL

2.40 Complete each of the following metric relationships:
a. 1 Mg = ________ g **b.** 1 mL = ________ μL
c. 1 g = ________ kg **d.** 1 g = ________ mg

2.41 For each of the following pairs, which is the larger unit?
a. milligram or kilogram **b.** milliliter or microliter
c. m or km **d.** kL or dL
e. nanometer or picometer

2.42 For each of the following pairs, which is the smaller unit?
a. mg or g **b.** centimeter or nanometer
c. millimeter or micrometer **d.** mL or dL
e. centigram or megagram

2.5 Writing Conversion Factors

LEARNING GOAL Write a conversion factor for two units that describe the same quantity.

REVIEW
Calculating Percentages (1.4)

CORE CHEMISTRY SKILL
Writing Conversion Factors from Equalities

Many problems in chemistry and the health sciences require you to change from one unit to another unit. Suppose you worked 2.0 h on your homework, and someone asked you how many minutes that was. You would answer 120 min. You must have multiplied 2.0 h × 60 min/h because you knew the equality (1 h = 60 min) that related the two units. When you expressed 2.0 h as 120 min, you did not change the amount of time you spent studying. You changed only the unit of measurement used to express the time. *Any equality can be written as fractions called* **conversion factors** *with one of the quantities in the numerator and the other quantity in the denominator*. Two conversion factors are always possible from any equality. Be sure to include the units when you write the conversion factors.

TEST
Try Practice Problems 2.43 and 2.44

Two Conversion Factors for the Equality: 1 h = 60 min

$$\frac{\text{Numerator}}{\text{Denominator}} \longrightarrow \frac{60\ \text{min}}{1\ \text{h}} \quad \text{and} \quad \frac{1\ \text{h}}{60\ \text{min}}$$

These factors are read as "60 minutes per 1 hour" and "1 hour per 60 minutes." The term *per* means "divide." Some common relationships are given in **TABLE 2.7**.

TABLE 2.7 Some Common Equalities

Quantity	Metric (SI)	U.S.	Metric–U.S.
Length	1 km = 1000 m	1 ft = 12 in.	2.54 cm = 1 in. (exact)
	1 m = 1000 mm	1 yd = 3 ft	1 m = 39.4 in.
	1 cm = 10 mm	1 mi = 5280 ft	1 km = 0.621 mi
Volume	1 L = 1000 mL	1 qt = 4 cups	946 mL = 1 qt
	1 dL = 100 mL	1 qt = 2 pt	1 L = 1.06 qt
	1 mL = 1 cm^3	1 gal = 4 qt	473 mL = 1 pt
	1 mL = 1 cc*		5 mL = 1 t (tsp)*
			15 mL = 1 T (tbsp)*
Mass	1 kg = 1000 g	1 lb = 16 oz	1 kg = 2.20 lb
	1 g = 1000 mg		454 g = 1 lb
	1 mg = 1000 mcg*		
Time	1 h = 60 min	1 h = 60 min	
	1 min = 60 s	1 min = 60 s	

*Used in medicine.

The numbers in any equality between two metric units or between two U.S. system units are definitions. Because numbers in a definition are exact, they are not used to determine significant figures. For example, the equality of 1 g = 1000 mg is a definition, which means that both of the numbers 1 and 1000 are exact.

When an equality consists of a metric unit and a U.S. unit, one of the numbers in the equality is obtained by measurement and counts toward the significant figures in the answer. For example, the equality of 1 lb = 454 g is obtained by measuring the grams in exactly 1 lb. In this equality, the measured quantity 454 g has three significant figures, whereas the 1 is exact. An exception is the relationship of 1 in. = 2.54 cm, which has been defined as exact.

TEST
Try Practice Problems 2.45 and 2.46

Metric Conversion Factors

We can write two metric conversion factors for any of the metric relationships. For example, from the equality for meters and centimeters, we can write the following factors:

Metric Equality	Conversion Factors
1 m = 100 cm	$\frac{100\ \text{cm}}{1\ \text{m}}$ and $\frac{1\ \text{m}}{100\ \text{cm}}$

ENGAGE
Why does the equality 1 day = 24 h have two conversion factors?

Both are proper conversion factors for the relationship; one is just the inverse of the other. *The usefulness of conversion factors is enhanced by the fact that we can turn a conversion factor over and use its inverse.* The numbers 100 and 1 in this equality and its conversion factors are both *exact* numbers.

FIGURE 2.9 In the United States, the contents of many packaged foods are listed in both U.S. and metric units.

Q What are some advantages of using the metric system?

Metric–U.S. System Conversion Factors

Suppose you need to convert from pounds, a unit in the U.S. system, to kilograms in the metric system. A relationship you could use is

$$1 \text{ kg} = 2.20 \text{ lb}$$

The corresponding conversion factors would be

$$\frac{2.20 \text{ lb}}{1 \text{ kg}} \quad \text{and} \quad \frac{1 \text{ kg}}{2.20 \text{ lb}}$$

FIGURE 2.9 illustrates the contents of some packaged foods in both U.S. and metric units.

Equalities and Conversion Factors Stated Within a Problem

An equality may also be stated within a problem that applies only to that problem. For example, the speed of a car in kilometers per hour or the milligrams of vitamin C in a tablet would be specific relationships for that problem only. From each of the following statements, we can write an equality and two conversion factors, and identify each number as exact or give the number of significant figures.

TEST

Try Practice Problems 2.47 to 2.52

The car was traveling at a speed of 85 km/h.

Equality	Conversion Factors	Significant Figures or Exact
85 km = 1 h	$\frac{85 \text{ km}}{1 \text{ h}}$ and $\frac{1 \text{ h}}{85 \text{ km}}$	The 85 km is measured: It has two significant figures. The 1 h is exact.

One tablet contains 500 mg of vitamin C.

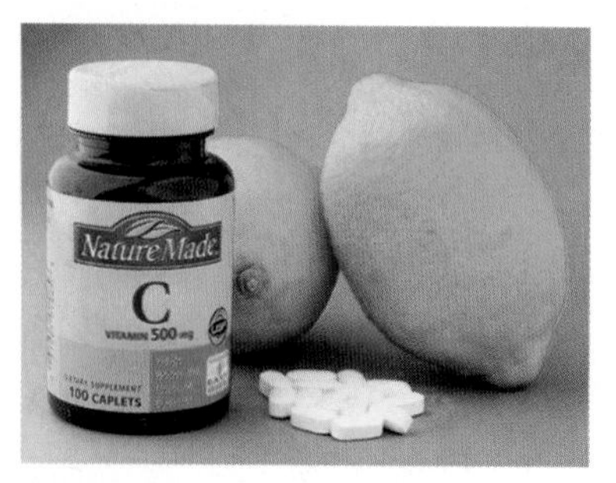

Vitamin C, an antioxidant needed by the body, is found in fruits such as lemons.

Equality	Conversion Factors	Significant Figures or Exact
1 tablet = 500 mg of vitamin C	$\frac{500 \text{ mg vitamin C}}{1 \text{ tablet}}$ and $\frac{1 \text{ tablet}}{500 \text{ mg vitamin C}}$	The 500 mg is measured: It has one significant figure. The 1 tablet is exact.

Conversion Factors from a Percentage

A percentage (%) is written as a conversion factor by choosing a unit and expressing the numerical relationship of the parts of this unit to 100 parts of the whole. For example, a person might have 18% body fat by mass. The percentage quantity can be written as 18 mass units of body fat in every 100 mass units of body mass. Different mass units such as grams (g), kilograms (kg), or pounds (lb) can be used, but both units used for the factor must be the same.

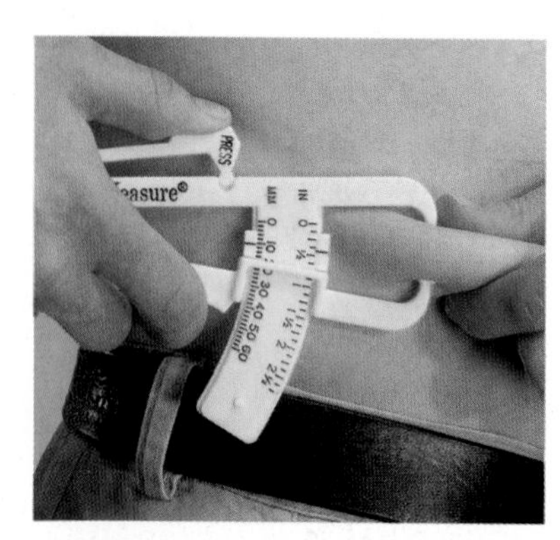

The thickness of the skin-fold at the abdomen is used to determine the percentage of body fat.

Equality	Conversion Factors	Significant Figures or Exact
18 kg of body fat = 100 kg of body mass	$\frac{18 \text{ kg body fat}}{100 \text{ kg body mass}}$ and $\frac{100 \text{ kg body mass}}{18 \text{ kg body fat}}$	The 18 kg is measured: It has two significant figures. The 100 kg is exact.

Conversion Factors from Dosage Problems

Equalities stated within dosage problems for medications can also be written as conversion factors. For example, Keflex (cephalexin), an antibiotic used for respiratory and ear infections, is available in 250-mg capsules. The quantity of Keflex in a capsule can be written as an equality from which two conversion factors are possible.

Equality	Conversion Factors	Significant Figures or Exact
1 capsule = 250 mg of Keflex	$\frac{250 \text{ mg Keflex}}{1 \text{ capsule}}$ and $\frac{1 \text{ capsule}}{250 \text{ mg Keflex}}$	The 250 mg is measured: It has two significant figures. The 1 capsule is exact.

Keflex (cephalexin), used to treat respiratory infections, is available in 250-mg capsules.

SAMPLE PROBLEM 2.7 Equalities and Conversion Factors in a Problem

TRY IT FIRST

Write the equality and two conversion factors, and identify each number as exact or give the number of significant figures for each of the following:

a. The medication that Greg takes for his high blood pressure contains 40. mg of propranolol in 1 tablet.

b. Cold-water fish such as salmon contains 1.9% omega-3 fatty acids by mass.

SOLUTION

a. There are 40. mg of propranolol in 1 tablet.

Equality	Conversion Factors	Significant Figures or Exact
1 tablet = 40. mg of propranolol	$\frac{40. \text{ mg propranolol}}{1 \text{ tablet}}$ and $\frac{1 \text{ tablet}}{40. \text{ mg propranolol}}$	The 40. mg is measured: It has two significant figures. The 1 tablet is exact.

Propranolol is used to lower high blood pressure.

b. Cold-water fish such as salmon contains 1.9% omega-3 fatty acids by mass.

Equality	Conversion Factors	Significant Figures or Exact
1.9 g of omega-3 fatty acids = 100 g of salmon	$\frac{1.9 \text{ g omega-3 fatty acids}}{100 \text{ g salmon}}$ and $\frac{100 \text{ g salmon}}{1.9 \text{ g omega-3 fatty acids}}$	The 1.9 g is measured: It has two significant figures. The 100 g is exact.

Salmon contains high levels of omega-3 fatty acids.

STUDY CHECK 2.7

Levsin (hyoscyamine), used to treat stomach and bladder problems, is available as drops with 0.125 mg Levsin per 1 mL of solution. Write the equality and two conversion factors, and identify each number as exact or give the number of significant figures.

ANSWER

0.125 mg of Levsin = 1 mL of solution

$$\frac{0.125 \text{ mg Levsin}}{1 \text{ mL solution}} \quad \text{and} \quad \frac{1 \text{ mL solution}}{0.125 \text{ mg Levsin}}$$

The 0.125 mg is measured: It has three SFs. The 1 mL is exact.

ENGAGE

How is a percentage used to write an equality and two conversion factors?

TEST

Try Practice Problems 2.53 and 2.54

PRACTICE PROBLEMS

2.5 Writing Conversion Factors

LEARNING GOAL Write a conversion factor for two units that describe the same quantity.

2.43 Why can two conversion factors be written for an equality such as 1 m = 100 cm?

2.44 How can you check that you have written the correct conversion factors for an equality?

2.45 Write the equality and two conversion factors for each of the following pairs of units:
a. centimeters and meters **b.** nanograms and grams
c. liters and kiloliters **d.** seconds and milliseconds

2.46 Write the equality and two conversion factors for each of the following pairs of units:
a. centimeters and inches **b.** kilometers and miles
c. pounds and grams **d.** quarts and liters

2.47 Write the equality and two conversion factors, and identify the numbers as exact or give the number of significant figures for each of the following:
a. One yard is 3 ft.
b. One kilogram is 2.20 lb.
c. A car goes 27 mi on 1 gal of gas.
d. Sterling silver is 93% silver by mass.

2.48 Write the equality and two conversion factors, and identify the numbers as exact or give the number of significant figures for each of the following:
a. One liter is 1.06 qt.
b. At the store, oranges are $1.29 per lb.
c. One deciliter contains 100 mL.
d. An 18-carat gold ring contains 75% gold by mass.

2.49 Write the equality and two conversion factors, and identify the numbers as exact or give the number of significant figures for each of the following:
a. A bee flies at an average speed of 3.5 m per second.
b. The Daily Value (DV) for potassium is 3.5 g.
c. An automobile traveled 26.0 km on 1 L of gasoline.
d. Silicon makes up 28.2% by mass of Earth's crust.

2.50 Write the equality and two conversion factors, and identify the numbers as exact or give the number of significant figures for each of the following:
a. The Daily Value (DV) for iodine is 150 mcg.
b. Gold jewelry contains 58% gold by mass.
c. The price of a liter of milk is $1.65.
d. A metric ton is 1000 kg.

Clinical Applications

2.51 Write the equality and two conversion factors, and identify the numbers as exact or give the number of significant figures for each of the following:
a. A calcium supplement contains 630 mg of calcium per tablet.
b. The Daily Value (DV) for vitamin C is 60 mg.
c. The label on a bottle reads 50 mg of atenolol per tablet.
d. A low-dose aspirin contains 81 mg of aspirin per tablet.

2.52 Write the equality and two conversion factors, and identify the numbers as exact or give the number of significant figures for each of the following:
a. The label on a bottle reads 10 mg of furosemide per 1 mL.
b. The Daily Value (DV) for selenium is 70. mcg.
c. An IV of normal saline solution has a flow rate of 85 mL per hour.
d. One capsule of fish oil contains 360 mg of omega-3 fatty acids.

2.53 Write an equality and two conversion factors for each of the following medications:
a. 10 mg of Atarax per 5 mL of Atarax syrup
b. 0.25 g of Lanoxin per 1 tablet of Lanoxin
c. 300 mg of Motrin per 1 tablet of Motrin

2.54 Write an equality and two conversion factors for each of the following medications:
a. 2.5 mg of Coumadin per 1 tablet of Coumadin
b. 100 mg of Clozapine per 1 tablet of Clozapine
c. 1.5 g of Cefuroxime per 1 mL of Cefuroxime

2.6 Problem Solving Using Unit Conversion

LEARNING GOAL Use conversion factors to change from one unit to another.

The process of problem solving in chemistry often requires one or more conversion factors to change a given unit to the needed unit. For the problem, the unit of the given and the unit of the needed are identified. From there, the problem is set up with one or more conversion factors used to convert the given unit to the needed unit as seen in Sample Problem 2.8.

Given unit × one or more conversion factors = needed unit

SAMPLE PROBLEM 2.8 Using Conversion Factors

TRY IT FIRST

Greg's doctor has ordered a PET scan of his heart. In radiological imaging, dosages of pharmaceuticals are based on body mass. If Greg weighs 164 lb, what is his body mass in kilograms?

SOLUTION GUIDE

STEP 1 State the given and needed quantities.

ANALYZE THE PROBLEM	Given	Need	Connect
	164 lb	kilograms	conversion factor (kg/lb)

STEP 2 Write a plan to convert the given unit to the needed unit.

pounds → U.S.–Metric factor → kilograms

STEP 3 State the equalities and conversion factors.

$$1\text{ kg} = 2.20\text{ lb}$$

$$\frac{2.20\text{ lb}}{1\text{ kg}} \text{ and } \frac{1\text{ kg}}{2.20\text{ lb}}$$

STEP 4 Set up the problem to cancel units and calculate the answer. Write the given, 164 lb, and multiply by the conversion factor that has lb in the denominator (bottom number) to cancel lb in the given.

Unit for answer goes here

$$\underset{\text{Given}}{164\,\cancel{\text{lb}}} \times \underset{\text{Conversion factor}}{\frac{1\text{ kg}}{2.20\,\cancel{\text{lb}}}} = \underset{\text{Answer}}{74.5\text{ kg}}$$

The given unit lb cancels out and the needed unit kg is in the numerator. *The unit you want in the final answer is the one that remains after all the other units have canceled out.* This is a helpful way to check that you set up a problem properly.

$$\cancel{\text{lb}} \times \frac{\text{kg}}{\cancel{\text{lb}}} = \text{kg} \quad \text{Unit needed for answer}$$

The calculator display gives the numerical answer, which is adjusted to give a final answer with the proper number of significant figures (SFs).

$$\underset{\text{Three SFs}}{164} \times \underset{\text{Three SFs}}{\overset{\text{Exact}}{\frac{1}{2.20}}} = 164 \div 2.20 = \underset{\text{Calculator display}}{74.54545454} = \underset{\text{Three SFs (rounded off)}}{74.5}$$

The value of 74.5 combined with the unit, kg, gives the final answer of 74.5 kg. With few exceptions, answers to numerical problems contain a number and a unit.

ENGAGE

When you convert one unit to another, how do you know which unit of the conversion factor to place in the denominator?

STUDY CHECK 2.8

A total of 2500 mL of a boric acid antiseptic solution is prepared from boric acid concentrate. How many quarts of boric acid have been prepared?

ANSWER

2.6 qt

INTERACTIVE VIDEO

Conversion Factors

TEST

Try Practice Problems 2.55 and 2.56

Using Two or More Conversion Factors

In problem solving, two or more conversion factors are often needed to complete the change of units. In setting up these problems, one factor follows the other. Each factor is arranged to cancel the preceding unit until the needed unit is obtained. Once the problem is set up to cancel units properly, the calculations can be done without writing intermediate results.

CORE CHEMISTRY SKILL

Using Conversion Factors

The process is worth practicing until you understand unit cancelation, the steps on the calculator, and rounding off to give a final answer. In this text, when two or more conversion factors are required, the final answer will be based on obtaining a final calculator display and rounding off (or adding zeros) to give the correct number of significant figures as shown in Sample Problem 2.9.

SAMPLE PROBLEM 2.9 Using Two Conversion Factors

TRY IT FIRST

Greg has been diagnosed with diminished thyroid function. His doctor prescribes a dosage of 0.150 mg of Synthroid to be taken once a day. If tablets in stock contain 75 mcg of Synthroid, how many tablets are required to provide the prescribed medication?

SOLUTION GUIDE

STEP 1 State the given and needed quantities.

	Given	Need	Connect
ANALYZE THE PROBLEM	0.150 mg of Synthroid	number of tablets	1 tablet = 75 mcg of Synthroid

STEP 2 Write a plan to convert the given unit to the needed unit.

milligrams → Metric factor → micrograms → Clinical factor → number of tablets

STEP 3 State the equalities and conversion factors.

$$1\text{ mg} = 1000\text{ mcg}$$
$$\frac{1000\text{ mcg}}{1\text{ mg}} \text{ and } \frac{1\text{ mg}}{1000\text{ mcg}}$$

$$1\text{ tablet} = 75\text{ mcg of Synthroid}$$
$$\frac{75\text{ mcg Synthroid}}{1\text{ tablet}} \text{ and } \frac{1\text{ tablet}}{75\text{ mcg Synthroid}}$$

STEP 4 Set up the problem to cancel units and calculate the answer. The problem can be set up using the metric factor to cancel milligrams, and then the clinical factor to obtain the number of tablets as the final unit.

$$\underset{\text{Three SFs}}{0.150\ \cancel{\text{mg Synthroid}}} \times \underset{\text{Exact}}{\overset{\text{Exact}}{\frac{1000\ \cancel{\text{mcg}}}{1\ \cancel{\text{mg}}}}} \times \underset{\text{Two SFs}}{\overset{\text{Exact}}{\frac{1\text{ tablet}}{75\ \cancel{\text{mcg Synthroid}}}}} = 2\text{ tablets}$$

One teaspoon of cough syrup is measured for a patient.

STUDY CHECK 2.9

A bottle contains 120 mL of cough syrup. If one teaspoon (5 mL) is given four times a day, how many days will elapse before a refill is needed?

ANSWER

6 days

TEST

Try Practice Problems 2.57 to 2.60

SAMPLE PROBLEM 2.10 Using a Percentage as a Conversion Factor

TRY IT FIRST

A person who exercises regularly has 16% body fat by mass. If this person weighs 155 lb, what is the mass, in kilograms, of body fat?

SOLUTION GUIDE

STEP 1 State the given and needed quantities.

ANALYZE THE PROBLEM	Given	Need	Connect
	155 lb body weight	kilograms of body fat	conversion factors (kg/lb, percent body fat)

STEP 2 Write a plan to convert the given unit to the needed unit.

pounds of body weight → U.S.–Metric factor → kilograms of body mass → Percentage factor → kilograms of body fat

STEP 3 State the equalities and conversion factors.

1 kg of body mass = 2.20 lb of body weight

$$\frac{2.20 \text{ lb body weight}}{1 \text{ kg body mass}} \text{ and } \frac{1 \text{ kg body mass}}{2.20 \text{ lb body weight}}$$

16 kg of body fat = 100 kg of body mass

$$\frac{16 \text{ kg body fat}}{100 \text{ kg body mass}} \text{ and } \frac{100 \text{ kg body mass}}{16 \text{ kg body fat}}$$

STEP 4 Set up the problem to cancel units and calculate the answer.

$$155 \cancel{\text{lb body weight}} \times \frac{1 \cancel{\text{kg body mass}}}{2.20 \cancel{\text{lb body weight}}} \times \frac{16 \text{ kg body fat}}{100 \cancel{\text{kg body mass}}} = 11 \text{ kg of body fat}$$

Three SFs; Exact / Three SFs; Two SFs / Exact; Two SFs

Exercising regularly helps reduce body fat.

STUDY CHECK 2.10

A package contains 1.33 lb of ground round. If it contains 29% fat, how many grams of fat are in the ground round?

ANSWER

180 g of fat

TEST

Try Practice Problems 2.61 to 2.66

PRACTICE PROBLEMS

2.6 Problem Solving Using Unit Conversion

LEARNING GOAL Use conversion factors to change from one unit to another.

2.55 Perform each of the following conversions using metric conversion factors:
- **a.** 44.2 mL to liters
- **b.** 8.65 m to nanometers
- **c.** 5.2×10^8 g to megagrams
- **d.** 0.72 ks to milliseconds

2.56 Perform each of the following conversions using metric conversion factors:
- **a.** 4.82×10^{-5} L to picoliters
- **b.** 575.2 dm to kilometers
- **c.** 5×10^{-4} kg to micrograms
- **d.** 6.4×10^{10} ps to seconds

2.57 Perform each of the following conversions using metric and U.S. conversion factors:
- **a.** 3.428 lb to kilograms
- **b.** 1.6 m to inches
- **c.** 4.2 L to quarts
- **d.** 0.672 ft to millimeters

2.58 Perform each of the following conversions using metric and U.S. conversion factors:
- **a.** 0.21 lb to grams
- **b.** 11.6 in. to centimeters
- **c.** 0.15 qt to milliliters
- **d.** 35.41 kg to pounds

2.59 Use metric conversion factors to solve each of the following problems:
- **a.** If a student is 175 cm tall, how tall is the student in meters?
- **b.** A cooler has a volume of 5000 mL. What is the capacity of the cooler in liters?
- **c.** A hummingbird has a mass of 0.0055 kg. What is the mass, in grams, of the hummingbird?
- **d.** A balloon has a volume of 3500 cm^3. What is the volume in liters?

2.60 Use metric conversion factors to solve each of the following problems:
- **a.** The Daily Value (DV) for phosphorus is 800 mg. How many grams of phosphorus are recommended?
- **b.** A glass of orange juice contains 3.2 dL of juice. How many milliliters of orange juice are in the glass?
- **c.** A package of chocolate instant pudding contains 2840 mg of sodium. How many grams of sodium are in the pudding?
- **d.** A jar contains 0.29 kg of olives. How many grams of olives are in the jar?

2.61 Solve each of the following problems using one or more conversion factors:

a. A container holds 0.500 qt of liquid. How many milliliters of lemonade will it hold?

b. What is the mass, in kilograms, of a person who weighs 175 lb?

c. An athlete has 15% body fat by mass. What is the weight of fat, in pounds, of a 74-kg athlete?

d. A plant fertilizer contains 15% nitrogen (N) by mass. In a container of soluble plant food, there are 10.0 oz of fertilizer. How many grams of nitrogen are in the container?

Agricultural fertilizers applied to a field provide nitrogen for plant growth.

2.62 Solve each of the following problems using one or more conversion factors:

a. Wine is 12% alcohol by volume. How many milliliters of alcohol are in a 0.750-L bottle of wine?

b. Blueberry high-fiber muffins contain 51% dietary fiber by mass. If a package with a net weight of 12 oz contains six muffins, how many grams of fiber are in each muffin?

c. A jar of crunchy peanut butter contains 1.43 kg of peanut butter. If you use 8.0% of the peanut butter for a sandwich, how many ounces of peanut butter did you take out of the container?

d. In a candy factory, the nutty chocolate bars contain 22.0% pecans by mass. If 5.0 kg of pecans were used for candy last Tuesday, how many pounds of nutty chocolate bars were made?

Clinical Applications

2.63 Using conversion factors, solve each of the following clinical problems:

a. You have used 250 L of distilled water for a dialysis patient. How many gallons of water is that?

b. A patient needs 0.024 g of a sulfa drug. There are 8-mg tablets in stock. How many tablets should be given?

c. The daily dose of ampicillin for the treatment of an ear infection is 115 mg/kg of body weight. What is the daily dose for a 34-lb child?

d. You need 4.0 oz of a steroid ointment. How many grams of ointment does the pharmacist need to prepare?

2.64 Using conversion factors, solve each of the following clinical problems:

a. The physician has ordered 1.0 g of tetracycline to be given every six hours to a patient. If your stock on hand is 500-mg tablets, how many will you need for one day's treatment?

b. An intramuscular medication is given at 5.00 mg/kg of body weight. What is the dose for a 180-lb patient?

c. A physician has ordered 0.50 mg of atropine, intramuscularly. If atropine were available as 0.10 mg/mL of solution, how many milliliters would you need to give?

d. During surgery, a patient receives 5.0 pt of plasma. How many milliliters of plasma were given?

2.65 Using conversion factors, solve each of the following clinical problems:

a. A nurse practitioner prepares 500. mL of an IV of normal saline solution to be delivered at a rate of 80. mL/h. What is the infusion time, in hours, to deliver 500. mL?

b. A nurse practitioner orders Medrol to be given 1.5 mg/kg of body weight. Medrol is an anti-inflammatory administered as an intramuscular injection. If a child weighs 72.6 lb and the available stock of Medrol is 20. mg/mL, how many milliliters does the nurse administer to the child?

2.66 Using conversion factors, solve each of the following clinical problems:

a. A nurse practitioner prepares an injection of promethazine, an antihistamine used to treat allergic rhinitis. If the stock bottle is labeled 25 mg/mL and the order is a dose of 12.5 mg, how many milliliters will the nurse draw up in the syringe?

b. You are to give ampicillin 25 mg/kg to a child with a mass of 67 lb. If stock on hand is 250 mg/capsule, how many capsules should be given?

2.7 Density

LEARNING GOAL Calculate the density of a substance; use the density to calculate the mass or volume of a substance.

The mass and volume of any object can be measured. If we compare the mass of the object to its volume, we obtain a relationship called **density**.

$$\text{Density} = \frac{\text{mass of substance}}{\text{volume of substance}}$$

Every substance has a unique density, which distinguishes it from other substances. For example, lead has a density of 11.3 g/mL, whereas cork has a density of 0.26 g/mL. From these densities, we can predict if these substances will sink or float in water. *If an object is less dense than a liquid, the object floats when placed in the liquid.* If a substance, such as cork, is less dense than water, it will float. However, a lead object sinks because its density is greater than that of water (see **FIGURE 2.10**).

ENGAGE

If a piece of iron sinks in water, how does its density compare to that of water?

Density is used in chemistry in many ways. If we calculate the density of a pure metal as 10.5 g/mL, then we could identify it as silver, but not gold or aluminum. Metals such as

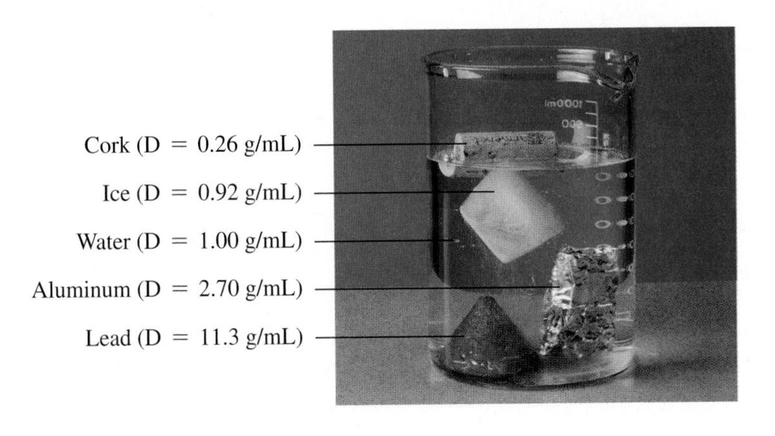

FIGURE 2.10 Objects that sink in water are more dense than water; objects that float are less dense.

Why does an ice cube float and a piece of aluminum sink?

gold and silver have higher densities, whereas gases have low densities. In the metric system, the densities of solids and liquids are usually expressed as grams per cubic centimeter (g/cm^3) or grams per milliliter (g/mL). The densities of gases are usually stated as grams per liter (g/L). **TABLE 2.8** gives the densities of some common substances.

TABLE 2.8 Densities of Some Common Substances

Solids (at 25 °C)	Density (g/mL)	Liquids (at 25 °C)	Density (g/mL)	Gases (at 0 °C)	Density (g/L)
Cork	0.26	Gasoline	0.74	Hydrogen	0.090
Body fat	0.909	Ethanol	0.79	Helium	0.179
Ice (at 0 °C)	0.92	Olive oil	0.92	Methane	0.714
Muscle	1.06	Water (at 4 °C)	1.00	Neon	0.902
Sugar	1.59	Urine	1.003–1.030	Nitrogen	1.25
Bone	1.80	Plasma (blood)	1.03	Air (dry)	1.29
Salt (NaCl)	2.16	Milk	1.04	Oxygen	1.43
Aluminum	2.70	Blood	1.06	Carbon dioxide	1.96
Iron	7.86	Mercury	13.6		
Copper	8.92				
Silver	10.5				
Lead	11.3				
Gold	19.3				

Calculating Density

We can calculate the density of a substance from its mass and volume as shown in Sample Problem 2.11.

SAMPLE PROBLEM 2.11 Calculating Density

TRY IT FIRST

High-density lipoprotein (HDL) is a type of cholesterol, sometimes called "good cholesterol," that is measured in a routine blood test. If a 0.258-g sample of HDL has a volume of 0.215 mL what is the density, in grams per milliliter, of the HDL sample?

SOLUTION GUIDE

STEP 1 State the given and needed quantities.

	Given	Need	Connect
ANALYZE THE PROBLEM	0.258 g of HDL, 0.215 mL	density (g/mL) of HDL	density expression

STEP 2 Write the density expression.

$$\text{Density} = \frac{\text{mass of substance}}{\text{volume of substance}}$$

STEP 3 Express mass in grams and volume in milliliters.

Mass of HDL sample = 0.258 g

Volume of HDL sample = 0.215 mL

STEP 4 Substitute mass and volume into the density expression and calculate the density.

$$\text{Density} = \frac{\overset{\text{Three SFs}}{0.258\text{ g}}}{\underset{\text{Three SFs}}{0.215\text{ mL}}} = \frac{1.20\text{ g}}{1\text{ mL}} = \underset{\text{Three SFs}}{1.20\text{ g/mL}}$$

STUDY CHECK 2.11

Low-density lipoprotein (LDL), sometimes called "bad cholesterol," is also measured in a routine blood test. If a 0.380-g sample of LDL has a volume of 0.362 mL, what is the density, in grams per milliliter, of the LDL sample?

ANSWER

1.05 g/mL

TEST

Try Practice Problems 2.67 to 2.70

Density of Solids Using Volume Displacement

The volume of a solid can be determined by volume displacement. When a solid is completely submerged in water, it displaces a volume that is equal to the volume of the solid. In **FIGURE 2.11**, the water level rises from 35.5 mL to 45.0 mL after the zinc object is added. This means that 9.5 mL of water is displaced and that the volume of the object is 9.5 mL. The density of the zinc is calculated using volume displacement as follows:

$$\text{Density} = \frac{\overset{\text{Four SFs}}{68.60\text{ g Zn}}}{\underset{\text{Two SFs}}{9.5\text{ mL}}} = \underset{\text{Two SFs}}{7.2\text{ g/mL}}$$

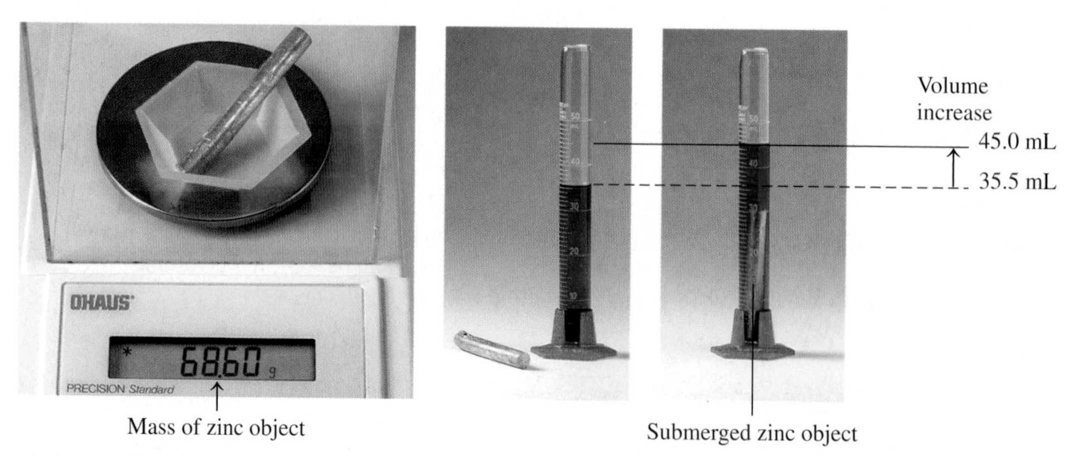

FIGURE 2.11 The density of a solid can be determined by volume displacement because a submerged object displaces a volume of water equal to its own volume.

Q How is the volume of the zinc object determined?

CHEMISTRY LINK TO HEALTH

Bone Density

The density of our bones is a measure of their health and strength. Our bones are constantly gaining and losing calcium, magnesium, and phosphate. In childhood, bones form at a faster rate than they break down. As we age, the breakdown of bone occurs more rapidly than new bone forms. As the loss of bone increases, bones begin to thin, causing a decrease in mass and density. Thinner bones lack strength, which increases the risk of fracture. Hormonal changes, disease, and certain medications can also contribute to the thinning of bone. Eventually, a condition of severe thinning of bone known as *osteoporosis* may occur. *Scanning electron micrographs* (SEMs) show **(a)** normal bone and **(b)** bone with osteoporosis due to loss of bone minerals.

Bone density is often determined by passing low-dose X-rays through the narrow part at the top of the femur (hip) and the spine **(c)**. These locations are where fractures are more likely to occur, especially as we age. Bones with high density will block more of the X-rays compared to bones that are less dense. The results of a bone density test are compared to a healthy young adult as well as to other people of the same age.

Recommendations to improve bone strength include supplements of calcium and vitamin D. Weight-bearing exercise such as walking and lifting weights can also improve muscle strength, which in turn increases bone strength.

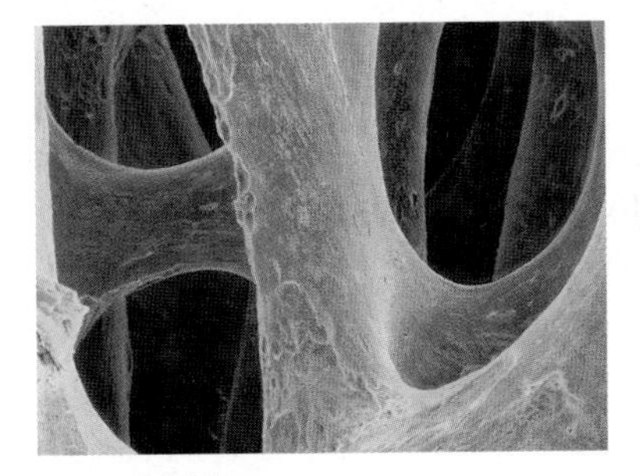

(a) Normal bone

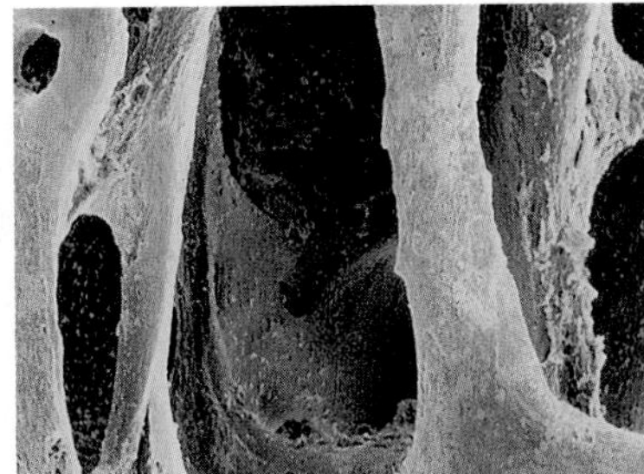

(b) Bone with osteoporosis

(c) Viewing a low-dose X-ray of the spine

Problem Solving Using Density

CORE CHEMISTRY SKILL
Using Density as a Conversion Factor

Density can be used as a conversion factor. For example, if the volume and the density of a sample are known, the mass in grams of the sample can be calculated as shown in Sample Problem 2.12.

SAMPLE PROBLEM 2.12 Problem Solving Using Density

TRY IT FIRST

Greg has a blood volume of 5.9 qt. If the density of blood is 1.06 g/mL, what is the mass, in grams, of Greg's blood?

SOLUTION GUIDE

STEP 1 State the given and needed quantities.

	Given	Need	Connect
ANALYZE THE PROBLEM	5.9 qt of blood	grams of blood	conversion factors (qt/mL, density of blood)

STEP 2 Write a plan to calculate the needed quantity.

quarts → U.S.–Metric factor → milliliters → Density factor → grams

STEP 3 Write the equalities and their conversion factors including density.

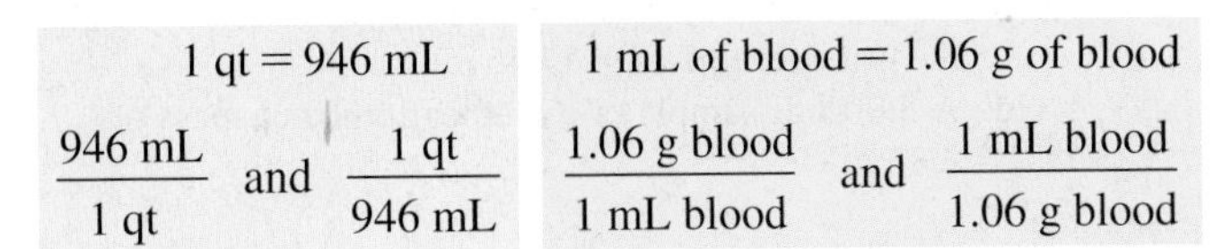

$$1 \text{ qt} = 946 \text{ mL} \qquad 1 \text{ mL of blood} = 1.06 \text{ g of blood}$$

$$\frac{946 \text{ mL}}{1 \text{ qt}} \text{ and } \frac{1 \text{ qt}}{946 \text{ mL}} \qquad \frac{1.06 \text{ g blood}}{1 \text{ mL blood}} \text{ and } \frac{1 \text{ mL blood}}{1.06 \text{ g blood}}$$

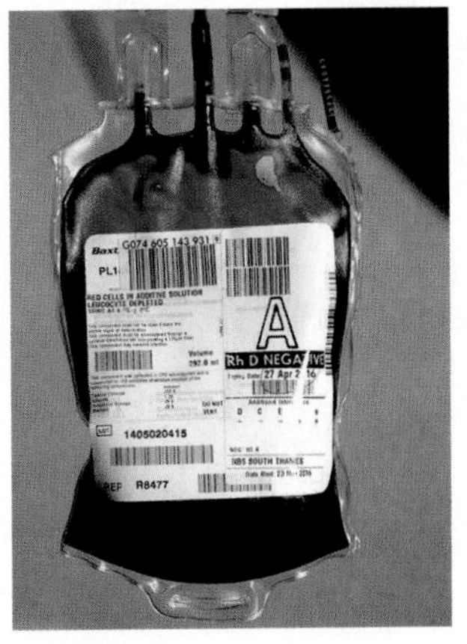

1 pt of blood contains 473 mL.

STEP 4 Set up the problem to calculate the needed quantity.

$$\underset{\text{Two SFs}}{7.5\ \cancel{\text{qt blood}}} \times \underset{\text{Exact}}{\overset{\text{Three SFs}}{\frac{946\ \cancel{\text{mL}}}{1\ \cancel{\text{qt}}}}} \times \underset{\text{Exact}}{\overset{\text{Three SFs}}{\frac{1.06\ \text{g blood}}{1\ \cancel{\text{mL blood}}}}} = \underset{\text{Two SFs}}{7500\ \text{g of blood}}$$

STUDY CHECK 2.12

During surgery, a patient receives 3.0 pt of blood. How many kilograms of blood (density = 1.06 g/mL) were needed for the transfusion?

ANSWER

1.5 kg of blood

TEST

Try Practice Problems 2.71 to 2.76

Specific Gravity

Specific gravity (sp gr) is a relationship between the density of a substance and the density of water. Specific gravity is calculated by dividing the density of a sample by the density of water, which is 1.00 g/mL at 4 °C. A substance with a specific gravity of 1.00 has the same density as water (1.00 g/mL).

$$\text{Specific gravity} = \frac{\text{density of sample}}{\text{density of water}}$$

TEST

Try Practice Problems 2.77 and 2.78

Specific gravity is one of the few unitless values you will encounter in chemistry. The specific gravity of urine helps evaluate the water balance in the body and the substances in the urine. In **FIGURE 2.12**, a hydrometer is used to measure the specific gravity of urine. The normal range of specific gravity for urine is 1.003 to 1.030. The specific gravity can decrease with *type 2 diabetes* and kidney disease. Increased specific gravity may occur with dehydration, kidney infection, and liver disease. In a clinic or hospital, a dipstick containing chemical pads is used to evaluate specific gravity.

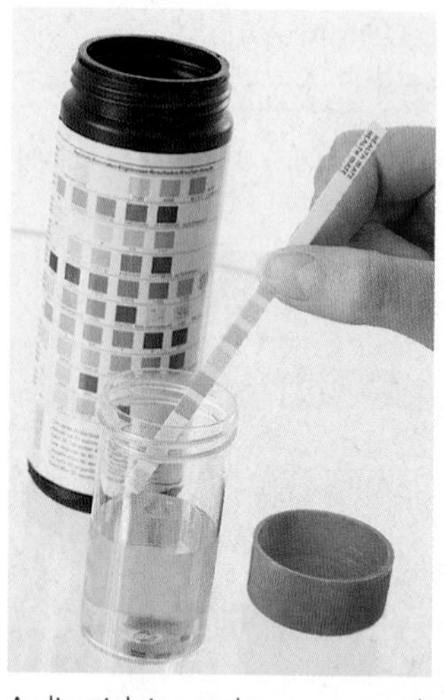

A dipstick is used to measure the specific gravity of a urine sample.

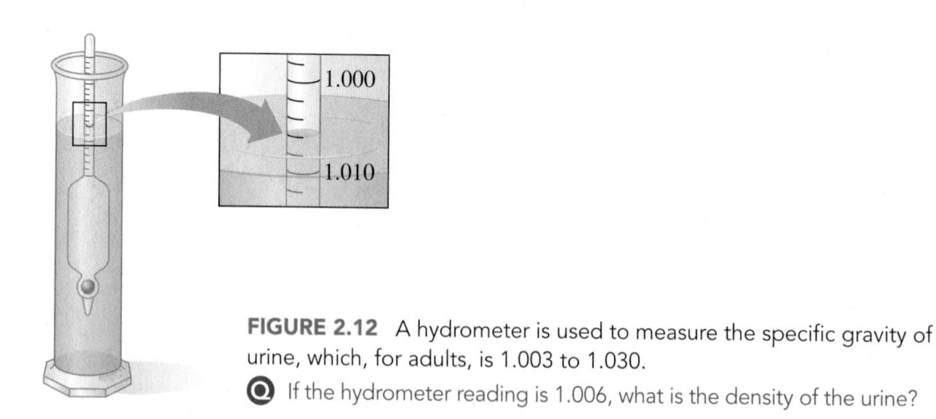

FIGURE 2.12 A hydrometer is used to measure the specific gravity of urine, which, for adults, is 1.003 to 1.030.

Q If the hydrometer reading is 1.006, what is the density of the urine?

PRACTICE PROBLEMS

2.7 Density

LEARNING GOAL Calculate the density of a substance; use the density to calculate the mass or volume of a substance.

2.67 Determine the density (g/mL) for each of the following:
- **a.** A 20.0-mL sample of a salt solution has a mass of 24.0 g.
- **b.** A cube of butter weighs 0.250 lb and has a volume of 130.3 mL.
- **c.** A gem has a mass of 4.50 g. When the gem is placed in a graduated cylinder containing 12.00 mL of water, the water level rises to 13.45 mL.
- **d.** A 3.00-mL sample of a medication has a mass of 3.85 g.

2.68 Determine the density (g/mL) for each of the following:
a. The fluid in a car battery has a volume of 125 mL and a mass of 155 g.
b. A plastic material weighs 2.68 lb and has a volume of 3.5 L.
c. A 4.000-mL urine sample from a person suffering from diabetes mellitus has a mass of 4.004 g.
d. A solid object has a mass of 1.65 lb and a volume of 170 mL.

2.69 What is the density (g/mL) of each of the following samples?
a. A lightweight head on a golf club is made of titanium. The volume of a sample of titanium is 114 cm^3 and the mass is 514.1 g.

Lightweight heads on golf clubs are made of titanium.

b. A syrup is added to an empty container with a mass of 115.25 g. When 0.100 pt of syrup is added, the total mass of the container and syrup is 182.48 g.

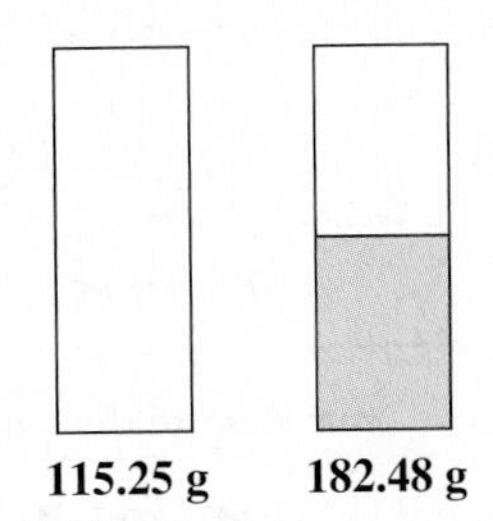

c. A block of aluminum metal has a volume of 3.15 L and a mass of 8.51 kg.

2.70 What is the density (g/mL) of each of the following samples?
a. An ebony carving has a mass of 275 g and a volume of 207 cm^3.
b. A 14.3-cm^3 sample of tin has a mass of 0.104 kg.
c. A bottle of acetone (fingernail polish remover) contains 55.0 mL of acetone with a mass of 43.5 g.

2.71 Use the density values in Table 2.8 to solve each of the following problems:
a. How many liters of ethanol contain 1.50 kg of ethanol?
b. How many grams of mercury are present in a barometer that holds 6.5 mL of mercury?
c. A sculptor has prepared a mold for casting a silver figure. The figure has a volume of 225 cm^3. How many ounces of silver are needed in the preparation of the silver figure?

2.72 Use the density values in Table 2.8 to solve each of the following problems:
a. A graduated cylinder contains 18.0 mL of water. What is the new water level, in milliliters, after 35.6 g of silver metal is submerged in the water?
b. A thermometer containing 8.3 g of mercury has broken. What volume, in milliliters, of mercury spilled?
c. A fish tank holds 35 gal of water. How many kilograms of water are in the fish tank?

2.73 Use the density values in Table 2.8 to solve each of the following problems:
a. What is the mass, in grams, of a cube of copper that has a volume of 74.1 cm^3?
b. How many kilograms of gasoline fill a 12.0-gal gas tank?
c. What is the volume, in cubic centimeters, of an ice cube that has a mass of 27 g?

2.74 Use the density values in Table 2.8 to solve each of the following problems:
a. If a bottle of olive oil contains 1.2 kg of olive oil, what is the volume, in milliliters, of the olive oil?
b. A cannon ball made of iron has a volume of 115 cm^3. What is the mass, in kilograms, of the cannon ball?
c. A balloon filled with helium has a volume of 7.3 L. What is the mass, in grams, of helium in the balloon?

2.75 In an old trunk, you find a piece of metal that you think may be aluminum, silver, or lead. You take it to a lab, where you find it has a mass of 217 g and a volume of 19.2 cm^3. Using Table 2.8, what is the metal you found?

2.76 Suppose you have two 100-mL graduated cylinders. In each cylinder, there is 40.0 mL of water. You also have two cubes: one is lead, and the other is aluminum. Each cube measures 2.0 cm on each side. After you carefully lower each cube into the water of its own cylinder, what will the new water level be in each of the cylinders? Use Table 2.8 for density values.

Clinical Applications

2.77 Solve each of the following problems:
a. A urine sample has a density of 1.030 g/mL. What is the specific gravity of the sample?
b. A 20.0-mL sample of a glucose IV solution has a mass of 20.6 g. What is the density of the glucose solution?
c. The specific gravity of a vegetable oil is 0.92. What is the mass, in grams, of 750 mL of vegetable oil?
d. A bottle containing 325 g of cleaning solution is used to clean hospital equipment. If the cleaning solution has a specific gravity of 0.850, what volume, in milliliters, of solution was used?

2.78 Solve each of the following problems:
a. A glucose solution has a density of 1.02 g/mL. What is its specific gravity?
b. A 0.200-mL sample of very-low-density lipoprotein (VLDL) has a mass of 190 mg. What is the density of the VLDL?
c. Butter has a specific gravity of 0.86. What is the mass, in grams, of 2.15 L of butter?
d. A 5.000-mL urine sample has a mass of 5.025 g. If the normal range for the specific gravity of urine is 1.003 to 1.030, would the specific gravity of this urine sample indicate that the patient could have type 2 diabetes?

CLINICAL UPDATE Greg's Visit with His Doctor

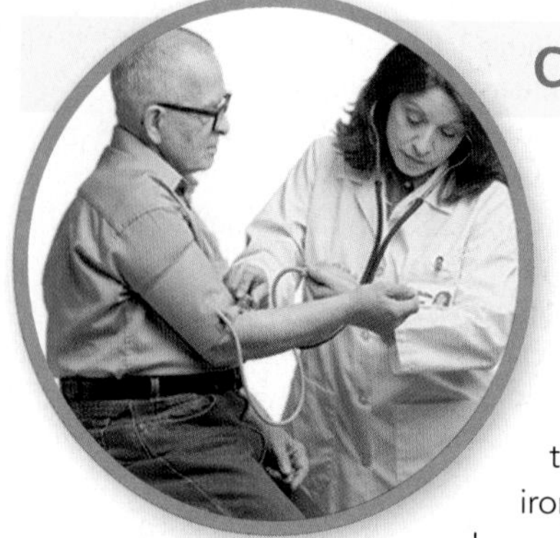

On Greg's last visit to his doctor, he complained of feeling tired. Sandra, the registered nurse, withdrew 8.0 mL of blood, which was sent to the lab and tested for iron. When the iron level is low, a person may have fatigue and decreased immunity.

The normal range for serum iron in men is 80 to 160 mcg/dL. Greg's iron test showed a blood serum iron level of 42 mcg/dL, which indicates that Greg has *iron-deficiency anemia*. His doctor orders an iron supplement to be taken twice daily. One tablet of the iron supplement contains 65 mg of iron.

Clinical Applications

2.79 **a.** Write an equality and two conversion factors for Greg's serum iron level.

b. How many micrograms of iron were in the 8.0-mL sample of Greg's blood?

2.80 **a.** Write an equality and two conversion factors for one tablet of the iron supplement.

b. How many grams of iron will Greg consume in one week?

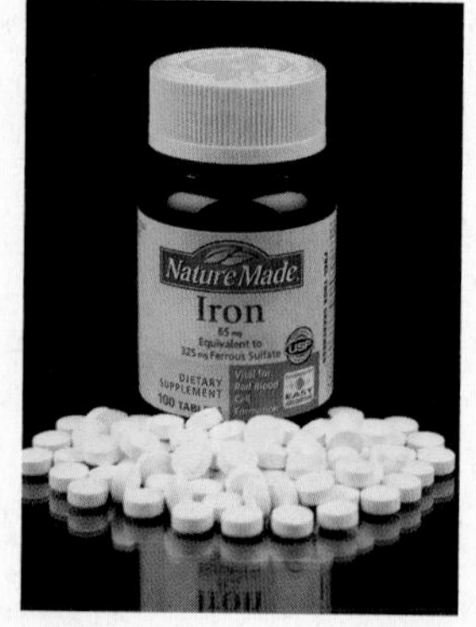

Each tablet contains 65 mg of iron, which is given for iron supplementation.

CONCEPT MAP

CHEMISTRY AND MEASUREMENTS

Measurements in chemistry involve

- **Metric Units** for measuring
  - **Length (m)**
  - **Mass (g)**
  - **Volume (L)**
  - **Temperature (°C)**
  - **Time (s)**
  - Mass and Volume give **Density** and **Specific Gravity**
- **Measured Numbers** have **Significant Figures** that require **Rounding Off Answers** or **Adding Zeros**
- **Prefixes** that change the size of **Metric Units** to give **Equalities** used for **Conversion Factors** to change units in **Problem Solving**

CHAPTER REVIEW

2.1 Units of Measurement

LEARNING GOAL Write the names and abbreviations for the metric or SI units used in measurements of volume, length, mass, temperature, and time.

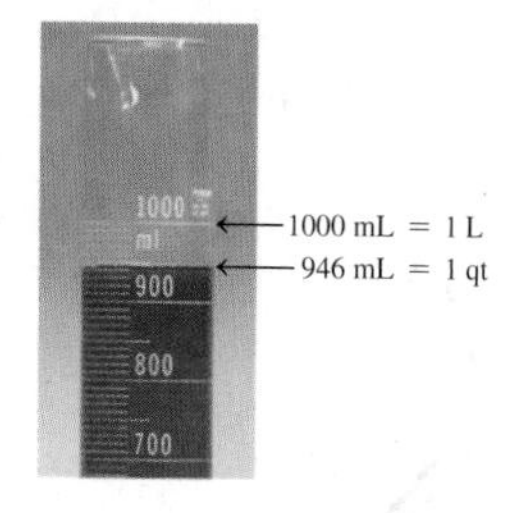

- In science, physical quantities are described in units of the metric or International System of Units (SI).
- Some important units are liter (L) for volume, meter (m) for length, gram (g) and kilogram (kg) for mass, degree Celsius (°C) and kelvin (K) for temperature, and second (s) for time.

2.2 Measured Numbers and Significant Figures

LEARNING GOAL Identify a number as measured or exact; determine the number of significant figures in a measured number.

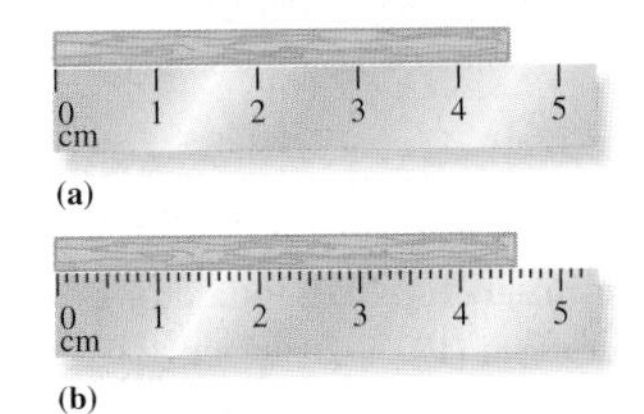

- A measured number is any number obtained by using a measuring device.
- An exact number is obtained by counting items or from a definition; no measuring device is needed.
- Significant figures are the numbers reported in a measurement including the estimated digit.
- Zeros in front of a decimal number or at the end of a nondecimal number are not significant.

2.3 Significant Figures in Calculations

LEARNING GOAL Adjust calculated answers to give the correct number of significant figures.

- In multiplication and division, the final answer is written so that it has the same number of significant figures as the measurement with the fewest significant figures.
- In addition and subtraction, the final answer is written so that it has the same number of decimal places as the measurement with the fewest decimal places.

2.4 Prefixes and Equalities

LEARNING GOAL Use the numerical values of prefixes to write a metric equality.

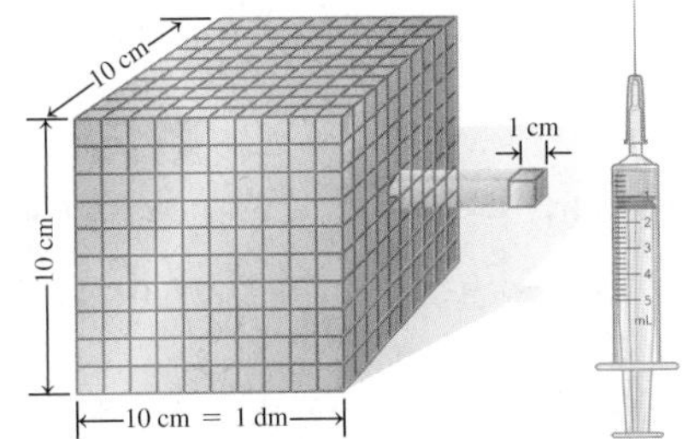

- A prefix placed in front of a metric or SI unit changes the size of the unit by factors of 10.
- Prefixes such as *centi*, *milli*, and *micro* provide smaller units; prefixes such as *kilo*, *mega*, and *tera* provide larger units.
- An equality shows the relationship between two units that measure the same quantity of volume, length, mass, or time.
- Examples of metric equalities are 1 L = 1000 mL, 1 m = 100 cm, 1 kg = 1000 g, and 1 min = 60 s.

2.5 Writing Conversion Factors

LEARNING GOAL Write a conversion factor for two units that describe the same quantity.

- Conversion factors are used to express a relationship in the form of a fraction.
- Two conversion factors can be written for any relationship in the metric or U.S. system.
- A percentage is written as a conversion factor by expressing matching units as the parts in 100 parts of the whole.

2.6 Problem Solving Using Unit Conversion

LEARNING GOAL Use conversion factors to change from one unit to another.

- Conversion factors are useful when changing a quantity expressed in one unit to a quantity expressed in another unit.
- In the problem-solving process, a given unit is multiplied by one or more conversion factors that cancel units until the needed answer is obtained.

2.7 Density

LEARNING GOAL Calculate the density of a substance; use the density to calculate the mass or volume of a substance.

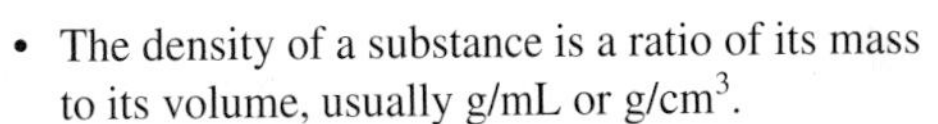
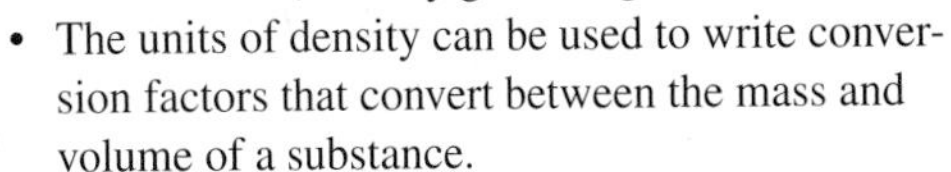
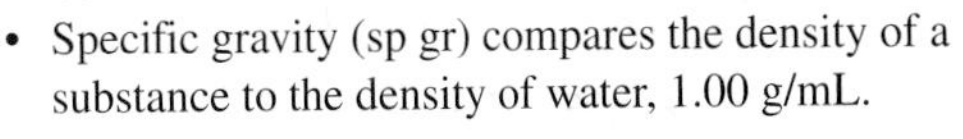

- The density of a substance is a ratio of its mass to its volume, usually g/mL or g/cm^3.
- The units of density can be used to write conversion factors that convert between the mass and volume of a substance.
- Specific gravity (sp gr) compares the density of a substance to the density of water, 1.00 g/mL.

KEY TERMS

Celsius (°C) temperature scale A temperature scale on which water has a freezing point of 0 °C and a boiling point of 100 °C.

centimeter (cm) A unit of length in the metric system; there are 2.54 cm in 1 in.

conversion factor A ratio in which the numerator and denominator are quantities from an equality or given relationship. For example, the two conversion factors for the equality 1 kg = 2.20 lb are written as

$$\frac{2.20 \text{ lb}}{1 \text{ kg}} \quad \text{and} \quad \frac{1 \text{ kg}}{2.20 \text{ lb}}$$

cubic centimeter (cm^3, cc) The volume of a cube that has 1-cm sides; 1 cm^3 is equal to 1 mL.

density The relationship of the mass of an object to its volume expressed as grams per cubic centimeter (g/cm^3), grams per milliliter (g/mL), or grams per liter (g/L).

equality A relationship between two units that measure the same quantity.

exact number A number obtained by counting or by definition.

gram (g) The metric unit used in measurements of mass.

International System of Units (SI) The official system of measurement throughout the world except for the United States that modifies the metric system.

Kelvin (K) temperature scale A temperature scale on which the lowest possible temperature is 0 K.

kilogram (kg) A metric mass of 1000 g, equal to 2.20 lb. The kilogram is the SI standard unit of mass.

liter (L) The metric unit for volume that is slightly larger than a quart.

mass A measure of the quantity of material in an object.

measured number A number obtained when a quantity is determined by using a measuring device.

meter (m) The metric unit for length that is slightly longer than a yard. The meter is the SI standard unit of length.

metric system A system of measurement used by scientists and in most countries of the world.

milliliter (mL) A metric unit of volume equal to one-thousandth of a liter (0.001 L).

prefix The part of the name of a metric unit that precedes the base unit and specifies the size of the measurement. All prefixes are related on a decimal scale.

second (s) A unit of time used in both the SI and metric systems.

SI See International System of Units (SI).

significant figures (SFs) The numbers recorded in a measurement.

specific gravity (sp gr) A relationship between the density of a substance and the density of water:

$$\text{sp gr} = \frac{\text{density of sample}}{\text{density of water}}$$

temperature An indicator of the hotness or coldness of an object.

volume (*V*) The amount of space occupied by a substance.

KEY MATH SKILL

The chapter Section containing each Key Math Skill is shown in parentheses at the end of each heading.

Rounding Off (2.3)

Calculator displays are rounded off to give the correct number of significant figures.

- If the first digit to be dropped is *4 or less*, then it and all following digits are simply dropped from the number.
- If the first digit to be dropped is *5 or greater*, then the last retained digit of the number is increased by 1.

One or more significant zeros are added when the calculator display has fewer digits than the needed number of significant figures.

Example: Round off each of the following to three significant figures:

a. 3.608 92 L
b. 0.003 870 298 m
c. 6 g

Answer: **a.** 3.61 L
b. 0.003 87 m
c. 6.00 g

CORE CHEMISTRY SKILLS

The chapter Section containing each Core Chemistry Skill is shown in parentheses at the end of each heading.

Counting Significant Figures (2.2)

The significant figures (SFs) are all the measured numbers including the last, estimated digit:

- All nonzero digits
- Zeros between nonzero digits
- Zeros within a decimal number
- All digits in a coefficient of a number written in scientific notation

An *exact* number is obtained from counting or a definition and has no effect on the number of significant figures in the final answer.

Example: State the number of significant figures in each of the following:

a. 0.003 045 mm
b. 15 000 m
c. 45.067 kg
d. 5.30×10^3 g
e. 2 cans of soda

Answer: **a.** four SFs
b. two SFs
c. five SFs
d. three SFs
e. exact

Using Significant Figures in Calculations (2.3)

- In multiplication or division, the final answer is written so that it has the same number of significant figures as the measurement with the fewest SFs.
- In addition or subtraction, the final answer is written so that it has the same number of decimal places as the measurement having the fewest decimal places.

Example: Perform the following calculations using measured numbers and give answers with the correct number of SFs:

a. $4.05 \text{ m} \times 0.6078 \text{ m}$
b. $\dfrac{4.50 \text{ g}}{3.27 \text{ mL}}$
c. $0.758 \text{ g} + 3.10 \text{ g}$
d. $13.538 \text{ km} - 8.6 \text{ km}$

Answer: **a.** 2.46 m^2
b. 1.38 g/mL
c. 3.86 g
d. 4.9 km

Using Prefixes (2.4)

- In the metric and SI systems of units, a prefix attached to any unit increases or decreases its size by some factor of 10.
- When the prefix *centi* is used with the unit meter, it becomes centimeter, a length that is one-hundredth of a meter (0.01 m).
- When the prefix *milli* is used with the unit meter, it becomes millimeter, a length that is one-thousandth of a meter (0.001 m).

Example: Complete each of the following metric relationships:

a. 1000 m = 1 ____m
b. 0.01 g = 1 ____g

Answer: **a.** 1000 m = 1 km
b. 0.01 g = 1 cg

Writing Conversion Factors from Equalities (2.5)

- A conversion factor allows you to change from one unit to another.
- Two conversion factors can be written for any equality in the metric, U.S., or metric–U.S. systems of measurement.
- Two conversion factors can be written for a relationship stated within a problem.

Example: Write two conversion factors for the equality: 1 L = 1000 mL.

Answer: $\dfrac{1000 \text{ mL}}{1 \text{ L}}$ and $\dfrac{1 \text{ L}}{1000 \text{ mL}}$

Using Conversion Factors (2.6)

In problem solving, conversion factors are used to cancel the given unit and to provide the needed unit for the answer.

- State the given and needed quantities.
- Write a plan to convert the given unit to the needed unit.
- State the equalities and conversion factors.
- Set up the problem to cancel units and calculate the answer.

Example: A computer chip has a width of 0.75 in. What is that distance in millimeters?

Answer: $0.75\ \cancel{\text{in.}} \times \dfrac{2.54\ \cancel{\text{cm}}}{1\ \cancel{\text{in.}}} \times \dfrac{10\ \text{mm}}{1\ \cancel{\text{cm}}} = 19\ \text{mm}$

Using Density as a Conversion Factor (2.7)

Density is an equality of mass and volume for a substance, which is written as the *density expression.*

$$\text{Density} = \frac{\text{mass of substance}}{\text{volume of substance}}$$

Density is useful as a conversion factor to convert between mass and volume.

Example: The element tungsten used in light bulb filaments has a density of 19.3 g/cm^3. What is the volume, in cubic centimeters, of 250 g of tungsten?

Answer: $250\ \cancel{\text{g}} \times \dfrac{1\ \text{cm}^3}{19.3\ \cancel{\text{g}}} = 13\ \text{cm}^3$

UNDERSTANDING THE CONCEPTS

The chapter Sections to review are shown in parentheses at the end of each problem.

2.81 In which of the following pairs do both numbers contain the same number of significant figures? (2.2)

a. 2.0500 m and 0.0205 m
b. 600.0 K and 60 K
c. 0.000 75 s and 75 000 s
d. 6.240 L and 6.240×10^{-2} L

2.82 In which of the following pairs do both numbers contain the same number of significant figures? (2.2)

a. 3.44×10^{-3} g and 0.0344 g
b. 0.0098 s and 9.8×10^{4} s
c. 6.8×10^{3} m and 68 000 m
d. 258.000 g and 2.58×10^{-2} g

2.83 Indicate if each of the following is answered with an exact number or a measured number: (2.2)

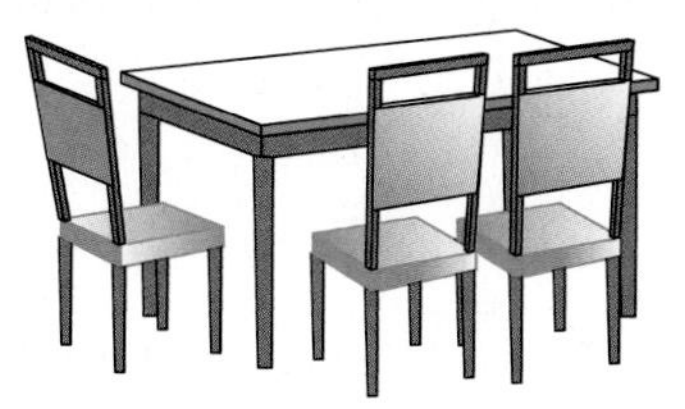

a. number of legs
b. height of table
c. number of chairs at the table
d. area of tabletop

2.84 Measure the length of each of the objects in diagrams **(a)**, **(b)**, and **(c)** using the metric ruler in the figure. Indicate the number of significant figures for each and the estimated digit for each. (2.2)

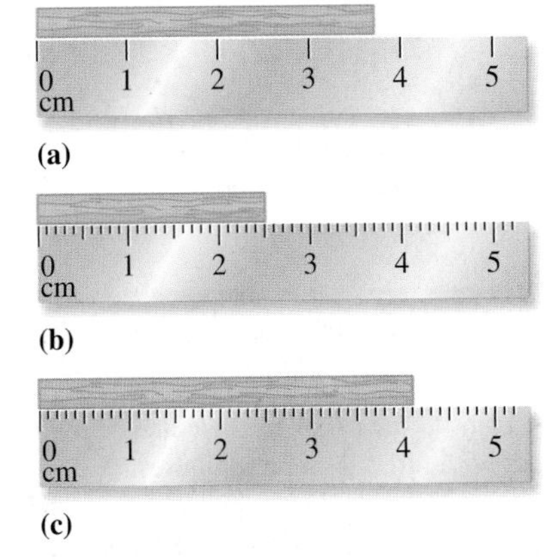

2.85 State the temperature on the Celsius thermometer to the correct number of significant figures: (2.3)

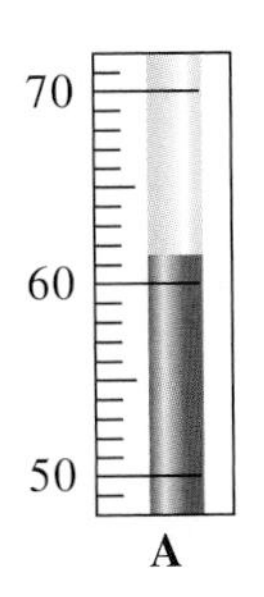

A

2.86 State the temperature on the Celsius thermometer to the correct number of significant figures: (2.3)

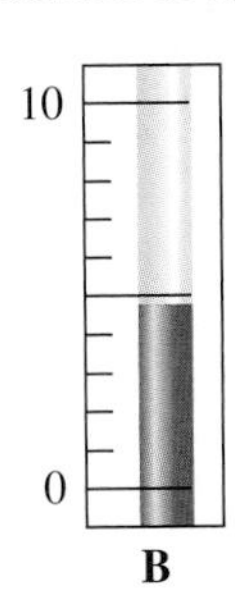

B

2.87 The length of this rug is 38.4 in. and the width is 24.2 in. (2.3, 2.6)

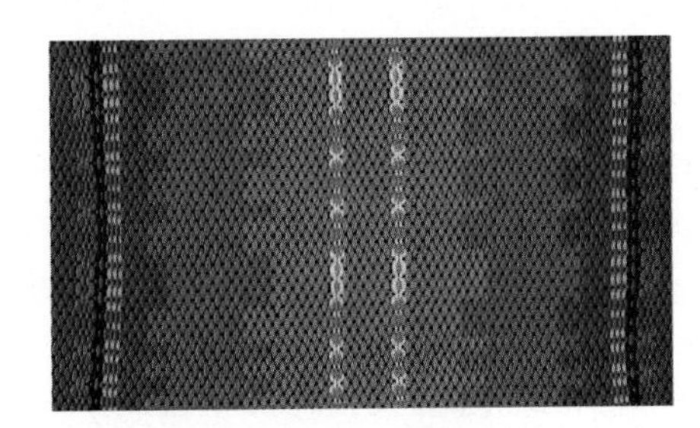

a. What is the length of this rug, in centimeters?
b. What is the width of this rug, in centimeters?
c. How many significant figures are in the length measurement?
d. Calculate the area of the rug, in square centimeters, to the correct number of significant figures. (Area = Length $\times$ Width)

2.88 A shipping box has a length of 7.00 in., a width of 6.00 in., and a height of 4.00 in. (2.3, 2.6)

a. What is the length of the box, in centimeters?
b. What is the width of the box, in centimeters?
c. How many significant figures are in the width measurement?
d. Calculate the volume of the box, in cubic centimeters, to the correct number of significant figures.
(Volume = Length × Width × Height)

2.89 Each of the following diagrams represents a container of water and a cube. Some cubes float while others sink. Match diagrams **1**, **2**, **3**, or **4** with one of the following descriptions and explain your choices: (2.7)

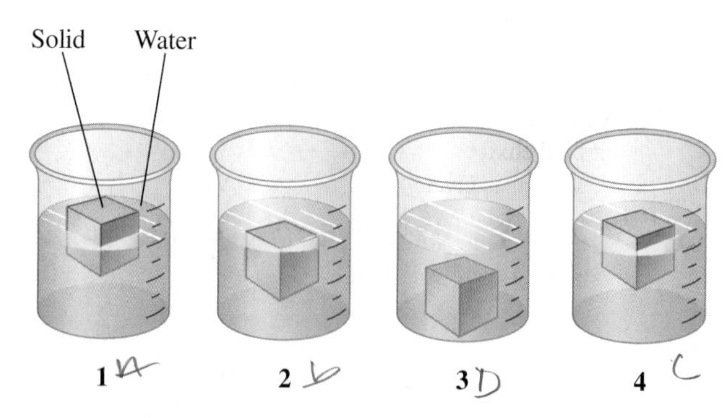

a. The cube has a greater density than water.
b. The cube has a density that is 0.80 g/mL.
c. The cube has a density that is one-half the density of water.
d. The cube has the same density as water.

2.90 What is the density of the solid object that is weighed and submerged in water? (2.7)

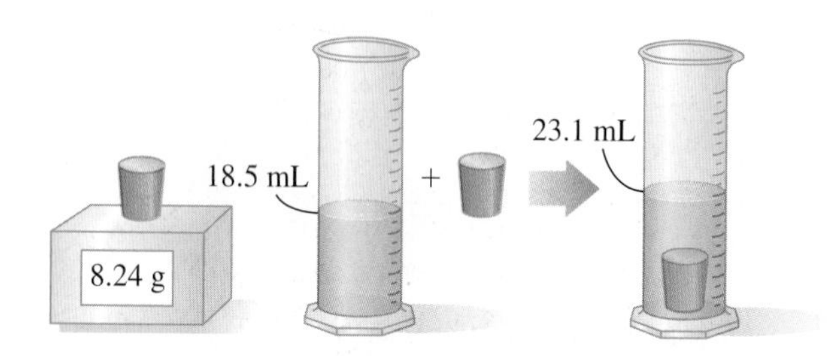

2.91 Consider the following solids. The solids **A**, **B**, and **C** represent aluminum (D = 2.70 g/mL), gold (D = 19.3 g/mL), and silver (D = 10.5 g/mL). If each has a mass of 10.0 g, what is the identity of each solid? (2.7)

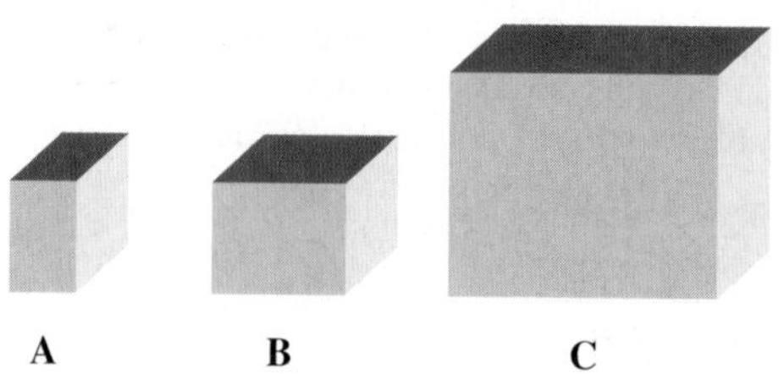

2.92 A graduated cylinder contains three liquids **A**, **B**, and **C**, which have different densities and do not mix: mercury (D = 13.6 g/mL), vegetable oil (D = 0.92 g/mL), and water (D = 1.00 g/mL). Identify the liquids **A**, **B**, and **C** in the cylinder. (2.7)

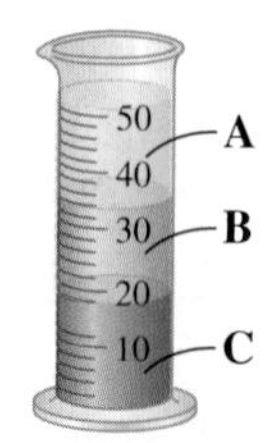

2.93 The gray cube has a density of 4.5 g/cm^3. Is the density of the green cube the same, lower than, or higher than that of the gray cube? (2.7)

2.94 The gray cube has a density of 4.5 g/cm^3. Is the density of the green cube the same, lower than, or higher than that of the gray cube? (2.7)

ADDITIONAL PRACTICE PROBLEMS

2.95 Round off or add zeros to the following calculated answers to give a final answer with three significant figures: (2.2)
a. 0.000 012 58 L
b. 3.528×10^2 kg
c. 125 111 m
d. 34.9673 s

2.96 Round off or add zeros to the following calculated answers to give a final answer with three significant figures: (2.2)
a. 58.703 mL
b. 3×10^{-3} s
c. 0.010 826 g
d. 1.7484×10^3 ms

2.97 A dessert contains 137.25 g of vanilla ice cream, 84 g of fudge sauce, and 43.7 g of nuts. (2.3, 2.6)
 a. What is the total mass, in grams, of the dessert?
 b. What is the total weight, in pounds, of the dessert?

2.98 A fish company delivers 22 kg of salmon, 5.5 kg of crab, and 3.48 kg of oysters to your seafood restaurant. (2.3, 2.6)
 a. What is the total mass, in kilograms, of the seafood?
 b. What is the total number of pounds?

2.99 In France, grapes are 1.95 euros per kilogram. What is the cost of grapes, in dollars per pound, if the exchange rate is 1.14 dollars/euro? (2.6)

2.100 In Mexico, avocados are 48 pesos per kilogram. What is the cost, in cents, of an avocado that weighs 0.45 lb if the exchange rate is 18 pesos to the dollar? (2.6)

2.101 Bill's recipe for onion soup calls for 4.0 lb of thinly sliced onions. If an onion has an average mass of 115 g, how many onions does Bill need? (2.6)

2.102 The price of 1 lb of potatoes is $1.75. If all the potatoes sold today at the store bring in $1420, how many kilograms of potatoes did grocery shoppers buy? (2.6)

2.103 During a workout at the gym, you set the treadmill at a pace of 55.0 m/min. How many minutes will you walk if you cover a distance of 7500 ft? (2.6)

2.104 The distance between two cities is 1700 km. How long will it take, in hours, to drive from one city to the other if your average speed is 63 mi/h? (2.6)

2.105 The water level in a graduated cylinder initially at 215 mL rises to 285 mL after a piece of lead is submerged. What is the mass, in grams, of the lead (see Table 2.8)? (2.7)

2.106 A graduated cylinder contains 155 mL of water. A 15.0-g piece of iron and a 20.0-g piece of lead are added. What is the new water level, in milliliters, in the cylinder (see Table 2.8)? (2.7)

2.107 How many milliliters of gasoline have a mass of 1.2 kg (see Table 2.8)? (2.7)

2.108 What is the volume, in quarts, of 3.40 kg of ethanol (see Table 2.8)? (2.7)

Clinical Applications

2.109 The following nutrition information is listed on a box of crackers: (2.6)

Serving size 0.50 oz (6 crackers)
Fat 4 g per serving; Sodium 140 mg per serving

 a. If the box has a net weight (contents only) of 8.0 oz, about how many crackers are in the box?
 b. If you ate 10 crackers, how many ounces of fat did you consume?
 c. How many servings of crackers in part **a** would it take to obtain the Daily Value (DV) for sodium, which is 2.4 g?

2.110 A dialysis unit requires 75 000 mL of distilled water. How many gallons of water are needed? (2.6)

2.111 To treat a bacterial infection, a doctor orders 4 tablets of amoxicillin per day for 10 days. If each tablet contains 250 mg of amoxicillin, how many ounces of the medication are given in 10 days? (2.6)

2.112 Celeste's diet restricts her intake of protein to 24 g per day. If she eats 1.2 oz of protein, has she exceeded her protein limit for the day? (2.6)

2.113 A doctor orders 5.0 mL of phenobarbital elixir. If the phenobarbital elixir is available as 30. mg per 7.5 mL, how many milligrams is given to the patient? (2.6)

2.114 A doctor orders 2.0 mg of morphine. The vial of morphine on hand is 10. mg/mL. How many milliliters of morphine should you administer to the patient? (2.6)

CHALLENGE PROBLEMS

The following problems are related to the topics in this chapter. However, they do not all follow the chapter order, and they require you to combine concepts and skills from several Sections. These problems will help you increase your critical thinking skills and prepare for your next exam.

2.115 A balance measures mass to 0.001 g. If you determine the mass of an object that weighs about 31 g, would you record the mass as 31 g, 31.1 g, 31.08 g, 31.075 g, or 31.0750? Explain your choice by writing two to three complete sentences that describe your thinking. (2.3)

2.116 When three students use the same meterstick to measure the length of a paper clip, they obtain results of 5.8 cm, 5.75 cm, and 5.76 cm. If the meterstick has millimeter markings, what are some reasons for the different values? (2.3)

2.117 A car travels at 55 mi/h and gets 11 km/L of gasoline. How many gallons of gasoline are needed for a 3.0-h trip? (2.6)

2.118 A sunscreen preparation contains 2.50% benzyl salicylate by mass. If a tube contains 4.0 oz of sunscreen, how many kilograms of benzyl salicylate are needed to manufacture 325 tubes of sunscreen? (2.6)

2.119 How many milliliters of olive oil have the same mass as 1.50 L of gasoline (see Table 2.8)? (2.7)

2.120 A 50.0-g silver object and a 50.0-g gold object are both added to 75.5 mL of water contained in a graduated cylinder. What is the new water level, in milliliters, in the cylinder (see Table 2.8)? (2.7)

Clinical Applications

2.121 **a.** An athlete with a body mass of 65 kg has 3.0% body fat. How many pounds of body fat does that person have? (2.6)
 b. In liposuction, a doctor removes fat deposits from a person's body. If body fat has a density of 0.909 g/mL and 3.0 L of fat is removed, how many pounds of fat were removed from the patient?

2.122 A mouthwash is 21.6% ethanol by mass. If each bottle contains 0.358 pt of mouthwash with a density of 0.876 g/mL, how many kilograms of ethanol are in 180 bottles of the mouthwash? (2.6, 2.7)

A mouthwash may contain over 20% ethanol.

ANSWERS

Answers to Selected Practice Problems

2.1 **a.** g **b.** L **c.** °C **d.** lb **e.** s

2.3 **a.** volume **b.** length **c.** length **d.** time

2.5 **a.** meter, length **b.** gram, mass **c.** milliliter, volume **d.** second, time **e.** degree Celsius, temperature

2.7 **a.** second, time **b.** kilogram, mass **c.** gram, mass **d.** degree Celsius, temperature

2.9 **a.** five SFs **b.** two SFs **c.** two SFs **d.** three SFs **e.** four SFs **f.** three SFs

2.11 b and c

2.13 **a.** not significant **b.** significant **c.** significant **d.** significant **e.** not significant

2.15 **a.** 5.0×10^3 L **b.** 3.0×10^4 g **c.** 1.0×10^5 m **d.** 2.5×10^{-4} cm

2.17 **a.** measured **b.** exact **c.** exact **d.** measured

2.19 **a.** 6 oz **b.** none **c.** 0.75 lb, 350 g **d.** none (definitions are exact)

2.21 **a.** measured, four SFs **b.** measured, two SFs **c.** measured, three SFs **d.** exact

2.23 **a.** 1.85 kg **b.** 88.2 L **c.** 0.004 74 cm **d.** 8810 m **e.** 1.83×10^5 s

2.25 **a.** 56.9 m **b.** 0.002 28 g **c.** 11500 s (1.15×10^4 s) **d.** 8.10 L

2.27 **a.** 1.6 **b.** 0.01 **c.** 27.6 **d.** 3.5 **e.** 0.14 (1.4×10^{-1}) **f.** 0.8 (8×10^{-1})

2.29 **a.** 53.54 cm **b.** 127.6 g **c.** 121.5 mL **d.** 0.50 L

2.31 **a.** mg **b.** dL **c.** km **d.** pg

2.33 **a.** centiliter **b.** kilogram **c.** millisecond **d.** gigameter

2.35 **a.** 0.01 **b.** 10^{12} **c.** 0.001 **d.** 0.1

2.37 **a.** decigram **b.** microgram **c.** kilogram **d.** centigram

2.39 **a.** 100 cm **b.** 1×10^9 nm **c.** 0.001 m **d.** 1000 mL

2.41 **a.** kilogram **b.** milliliter **c.** km **d.** kL **e.** nanometer

2.43 A conversion factor can be inverted to give a second conversion factor.

2.45 **a.** 1 m = 100 cm; $\frac{100\text{ cm}}{1\text{ m}}$ and $\frac{1\text{ m}}{100\text{ cm}}$

b. 1 g = 1×10^9 ng; $\frac{1 \times 10^9\text{ ng}}{1\text{ g}}$ and $\frac{1\text{ g}}{1 \times 10^9\text{ ng}}$

c. 1 kL = 1000 L; $\frac{1000\text{ L}}{1\text{ kL}}$ and $\frac{1\text{ kL}}{1000\text{ L}}$

d. 1 s = 1000 ms; $\frac{1000\text{ ms}}{1\text{ s}}$ and $\frac{1\text{ s}}{1000\text{ ms}}$

2.47 **a.** 1 yd = 3 ft; $\frac{3\text{ ft}}{1\text{ yd}}$ and $\frac{1\text{ yd}}{3\text{ ft}}$
The 1 yd and 3 ft are both exact.

b. 1 kg = 2.20 lb; $\frac{2.20\text{ lb}}{1\text{ kg}}$ and $\frac{1\text{ kg}}{2.20\text{ lb}}$
The 2.20 lb is measured: It has three SFs. The 1 kg is exact.

c. 1 gal = 27 mi; $\frac{27\text{ mi}}{1\text{ gal}}$ and $\frac{1\text{ gal}}{27\text{ mi}}$
The 27 mi is measured: It has two SFs. The 1 gal is exact.

d. 93 g of silver = 100 g of sterling;
$\frac{93\text{ g silver}}{100\text{ g sterling}}$ and $\frac{100\text{ g sterling}}{93\text{ g silver}}$
The 93 g is measured: It has two SFs. The 100 g is exact.

2.49 **a.** 3.5 m = 1 s; $\frac{3.5\text{ m}}{1\text{ s}}$ and $\frac{1\text{ s}}{3.5\text{ m}}$
The 3.5 m is measured: It has two SFs. The 1 s is exact.

b. 3.5 g of potassium = 1 day;
$\frac{3.5\text{ g potassium}}{1\text{ day}}$ and $\frac{1\text{ day}}{3.5\text{ g potassium}}$
The 3.5 g is measured: It has two SFs. The 1 day is exact.

c. 26.0 km = 1 L; $\frac{26.0\text{ km}}{1\text{ L}}$ and $\frac{1\text{ L}}{26.0\text{ km}}$
The 26.0 km is measured: It has three SFs. The 1 L is exact.

d. 28.2 g of silicon = 100 g of crust;
$\frac{28.2\text{ g silicon}}{100\text{ g crust}}$ and $\frac{100\text{ g crust}}{28.2\text{ g silicon}}$
The 28.2 g is measured: It has three SFs. The 100 g is exact.

2.51 **a.** 1 tablet = 630 mg of calcium;
$\frac{630\text{ mg calcium}}{1\text{ tablet}}$ and $\frac{1\text{ tablet}}{630\text{ mg calcium}}$
The 630 mg is measured: It has two SFs. The 1 tablet is exact.

b. 60 mg of vitamin C = 1 day;
$\frac{60\text{ mg vitamin C}}{1\text{ day}}$ and $\frac{1\text{ day}}{60\text{ mg vitamin C}}$
The 60 mg is measured: It has one SF. The 1 day is exact.

c. 1 tablet = 50 mg of atenolol;
$\frac{50\text{ mg atenolol}}{1\text{ tablet}}$ and $\frac{1\text{ tablet}}{50\text{ mg atenolol}}$
The 50 mg is measured: It has one SF. The 1 tablet is exact.

d. 1 tablet = 81 mg of aspirin;
$\frac{81\text{ mg aspirin}}{1\text{ tablet}}$ and $\frac{1\text{ tablet}}{81\text{ mg aspirin}}$
The 81 mg is measured: It has two SFs. The 1 tablet is exact.

2.53 **a.** 5 mL of syrup = 10 mg of Atarax;

$\frac{10 \text{ mg Atarax}}{5 \text{ mL syrup}}$ and $\frac{5 \text{ mL syrup}}{10 \text{ mg Atarax}}$

b. 1 tablet = 0.25 g of Lanoxin;

$\frac{0.25 \text{ g Lanoxin}}{1 \text{ tablet}}$ and $\frac{1 \text{ tablet}}{0.25 \text{ g Lanoxin}}$

c. 1 tablet = 300 mg of Motrin;

$\frac{300 \text{ mg Motrin}}{1 \text{ tablet}}$ and $\frac{1 \text{ tablet}}{300 \text{ mg Motrin}}$

2.55 **a.** 0.0442 L **b.** 8.65×10^9 nm **c.** 5.2×10^2 Mg **d.** 7.2×10^5 ms

2.57 **a.** 1.56 kg **b.** 63 in. **c.** 4.5 qt **d.** 205 mm

2.59 **a.** 1.75 m **b.** 5 L **c.** 5.5 g **d.** 3.5 L

2.61 **a.** 473 mL **b.** 79.5 kg **c.** 24 lb **d.** 43 g

2.63 **a.** 66 gal **b.** 3 tablets **c.** 1800 mg **d.** 110 g

2.65 **a.** 6.3 h **b.** 2.5 mL

2.67 **a.** 1.20 g/mL **b.** 0.871 g/mL **c.** 3.10 g/mL **d.** 1.28 g/mL

2.69 **a.** 4.51 g/mL **b.** 1.42 g/mL **c.** 2.70 g/mL

2.71 **a.** 1.9 L of ethanol **b.** 88 g of mercury **c.** 83.3 oz of silver

2.73 **a.** 661 g **b.** 34 kg **c.** 29 cm^3

2.75 Since we calculate the density to be 11.3 g/cm^3, we identify the metal as lead.

2.77 **a.** 1.03 **b.** 1.03 g/mL **c.** 690 g **d.** 382 mL

2.79 **a.** 42 mcg of iron = 1 dL of blood;

$\frac{42 \text{ mcg iron}}{1 \text{ dL blood}}$ and $\frac{1 \text{ dL blood}}{42 \text{ mcg iron}}$

b. 3.4 mcg of iron

2.81 c and **d**

2.83 **a.** exact **b.** measured **c.** exact **d.** measured

2.85 61.5 °C

2.87 **a.** 97.5 cm **b.** 61.5 cm **c.** three SFs **d.** 6.00×10^3 cm^2

2.89 **a.** Diagram 3; a cube that has a greater density than the water will sink to the bottom.

b. Diagram 4; a cube with a density of 0.80 g/mL will be about two-thirds submerged in the water.

c. Diagram 1; a cube with a density that is one-half the density of water will be one-half submerged in the water.

d. Diagram 2; a cube with the same density as water will float just at the surface of the water.

2.91 **A** would be gold; it has the highest density (19.3 g/mL) and the smallest volume.

B would be silver; its density is intermediate (10.5 g/mL) and the volume is intermediate.

C would be aluminum; it has the lowest density (2.70 g/mL) and the largest volume.

2.93 The green cube has the same volume as the gray cube. However, the green cube has a larger mass on the scale, which means that its mass/volume ratio is larger. Thus, the density of the green cube is higher than the density of the gray cube.

2.95 **a.** 0.000 012 6 L (1.26×10^{-5} L)
b. 353 kg (3.53×10^2 kg)
c. 125 000 m (1.25×10^5 m)
d. 35.0 s

2.97 **a.** 265 g **b.** 0.584 lb

2.99 $1.01 per lb

2.101 16 onions

2.103 42 min

2.105 790 g

2.107 1600 mL (1.6×10^3 mL)

2.109 **a.** 96 crackers **b.** 0.2 oz of fat **c.** 17 servings

2.111 0.35 oz

2.113 20. mg

2.115 You would record the mass as 31.075 g. Since the balance will weigh to the nearest 0.001 g, the mass value would be reported to 0.001 g.

2.117 6.4 gal

2.119 1200 mL

2.121 **a.** 4.3 lb of body fat **b.** 6.0 lb

3 Matter and Energy

CHARLES IS 13 YEARS OLD AND OVERWEIGHT. His doctor is worried that Charles is at risk for type 2 diabetes and advises his mother to make an appointment with a dietitian. Daniel, a dietitian, explains to them that choosing the appropriate foods is important to living a healthy lifestyle, losing weight, and preventing or managing diabetes.

Daniel also explains that food contains potential or stored energy and different foods contain different amounts of potential energy. For instance, carbohydrates contain 4 kcal/g (17 kJ/g), whereas fats contain 9 kcal/g (38 kJ/g). He then explains that diets high in fat require more exercise to burn the fats, as they contain more energy. When Daniel looks at Charles's typical daily diet, he calculates that Charles obtains 2500 kcal in one day. The American Heart Association recommends 1800 kcal for boys 9 to 13 years of age. Daniel encourages Charles and his mother to include whole grains, fruits, and vegetables in their diet instead of foods high in fat. They also discuss food labels and the fact that smaller serving sizes of healthy foods are necessary to lose weight. Daniel also recommends that Charles exercises at least 60 minutes every day. Before leaving, Charles and his mother make an appointment for the following week to look at a weight loss plan.

CAREER Dietitian

Dietitians specialize in helping individuals learn about good nutrition and the need for a balanced diet. This requires them to understand biochemical processes, the importance of vitamins and food labels, as well as the differences between carbohydrates, fats, and proteins in terms of their energy value and how they are metabolized. Dietitians work in a variety of environments, including hospitals, nursing homes, school cafeterias, and public health clinics. In these roles, they create specialized diets for individuals diagnosed with a specific disease or create meal plans for those in a nursing home.

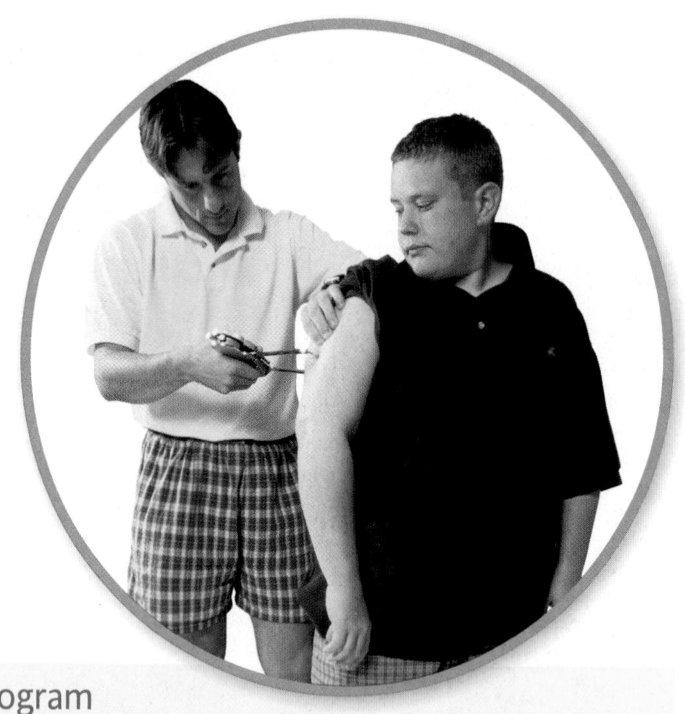

CLINICAL UPDATE A Diet and Exercise Program

When Daniel sees Charles and his mother, they discuss a menu for weight loss. Charles is going to record his food intake and return to discuss his diet with Daniel. You can view the results in the **CLINICAL UPDATE A Diet and Exercise Program,** page 87, and calculate the kilocalories that Charles consumes in one day, and also the weight that Charles has lost.

3.1 Classification of Matter

LEARNING GOAL Classify examples of matter as pure substances or mixtures.

Matter is anything that has mass and occupies space. Matter is everywhere around us: the orange juice we had for breakfast, the water we put in the coffee maker, the plastic bag we put our sandwich in, our toothbrush and toothpaste, the oxygen we inhale, and the carbon dioxide we exhale. To a scientist, all of this material is matter. The different types of matter are classified by their composition.

LOOKING AHEAD

Pure Substances: Elements and Compounds

A **pure substance** is matter that has a fixed or definite composition. There are two kinds of pure substances: *elements* and *compounds*. An **element**, the simplest type of a pure substance, is composed of only one type of material such as silver, iron, or aluminum. Every element is composed of *atoms*, which are extremely tiny particles that make up each type of matter. Silver is composed of silver atoms, iron of iron atoms, and aluminum of aluminum atoms. A full list of the elements is found on the inside front cover of this text.

A **compound** is also a pure substance, but it consists of atoms of two or more elements always chemically combined in the same proportion. For example, in the compound water, there are two hydrogen atoms for every one oxygen atom, which is represented by the formula H_2O. This means that water always has the same composition of H_2O. Another compound that consists of a chemical combination of hydrogen and oxygen is hydrogen peroxide. It has two hydrogen atoms for every two oxygen atoms and is represented by the formula H_2O_2. Thus, water (H_2O) and hydrogen peroxide (H_2O_2) are different compounds that have different properties even though they contain the same elements, hydrogen and oxygen.

Pure substances that are compounds can be broken down by chemical processes into their elements. They cannot be broken down through physical methods such as boiling or sifting. For example, ordinary table salt consists of the compound NaCl, which can be separated by chemical processes into sodium metal and chlorine gas, as seen in **FIGURE 3.1**. Elements cannot be broken down further.

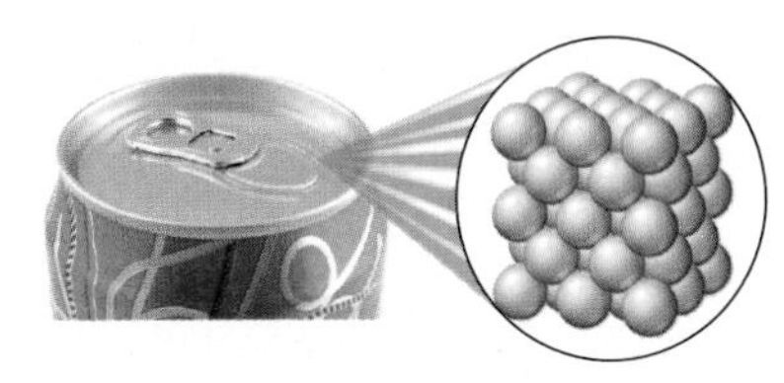

An aluminum can consists of many atoms of aluminum.

ENGAGE

Why are elements and compounds both pure substances?

Water, H_2O, consists of two atoms of hydrogen (white) for one atom of oxygen (red).

TEST

Try Practice Problems 3.1 and 3.2

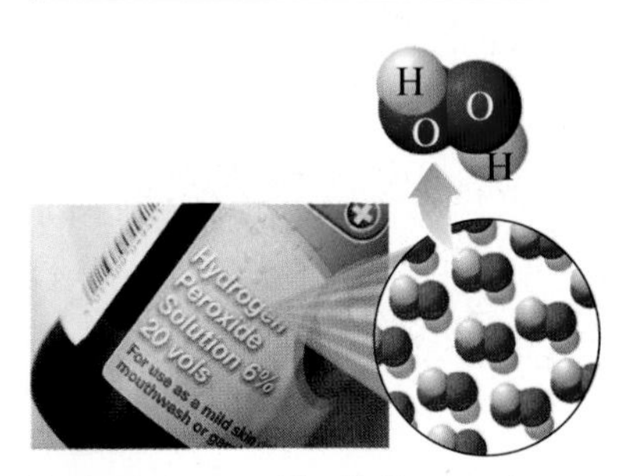

Hydrogen peroxide, H_2O_2, consists of two atoms of hydrogen (white) for every two atoms of oxygen (red).

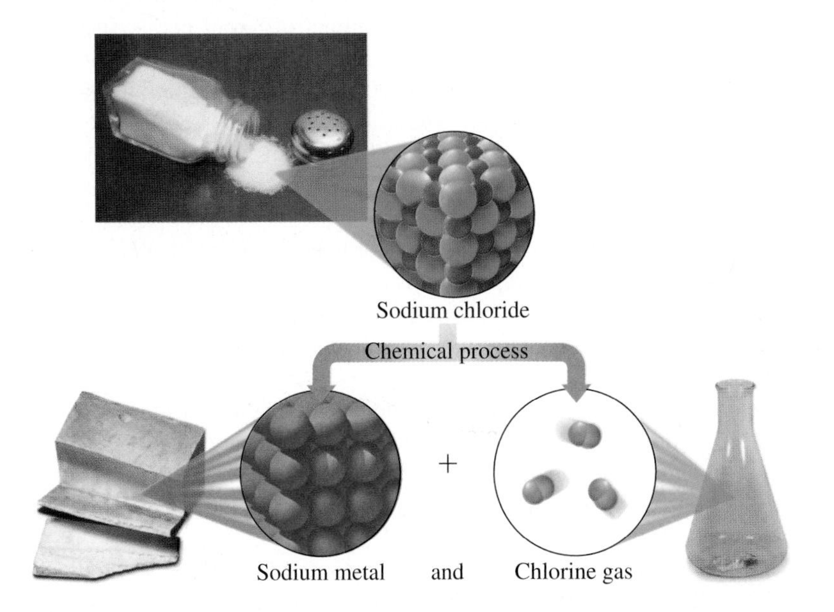

FIGURE 3.1 The decomposition of salt, NaCl, produces the elements sodium and chlorine.

Q How do elements and compounds differ?

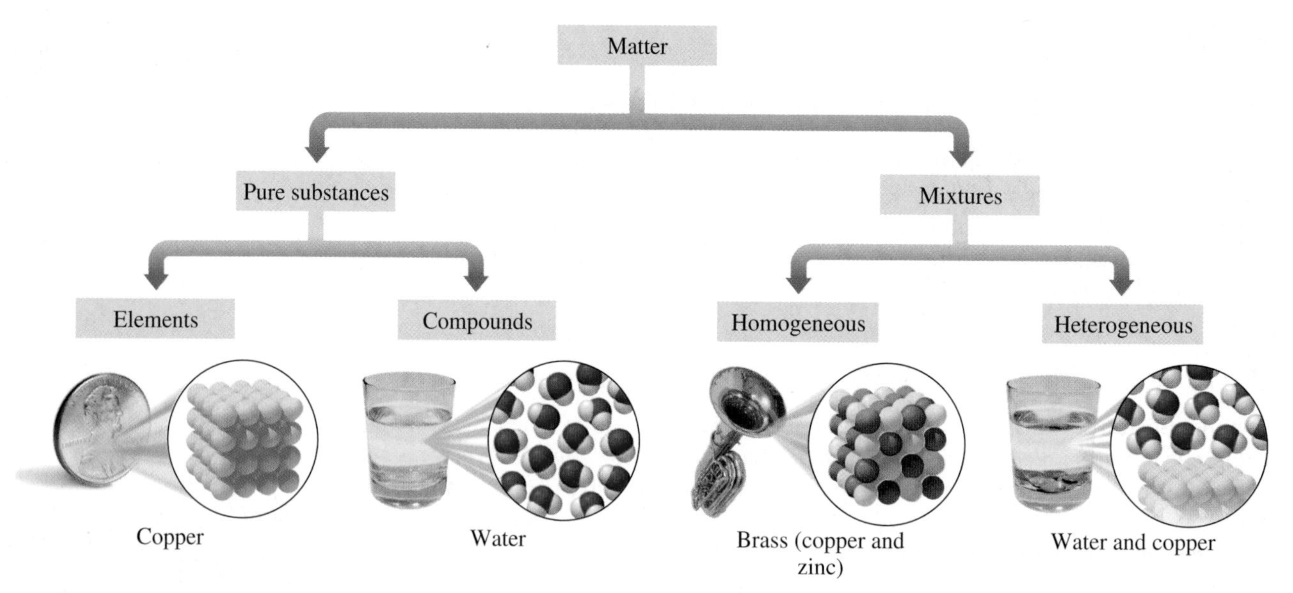

FIGURE 3.2 Matter is organized by its components: elements, compounds, and mixtures.

Why are copper and water pure substances, but brass is a mixture?

FIGURE 3.3 A mixture of spaghetti and water is separated using a strainer, a physical method of separation.

Why can physical methods be used to separate mixtures but not compounds?

Mixtures

In a **mixture**, two or more different substances are physically mixed, but not chemically combined. Much of the matter in our everyday lives consists of mixtures. The air we breathe is a mixture of mostly oxygen and nitrogen gases. The steel in buildings and railroad tracks is a mixture of iron, nickel, carbon, and chromium. The brass in doorknobs and musical instruments is a mixture of zinc and copper (see **FIGURE 3.2**). Tea, coffee, and ocean water are mixtures too. Unlike compounds, the proportions of substances in a mixture are not consistent but can vary. For example, two sugar–water mixtures may look the same, but the one with the higher ratio of sugar to water would taste sweeter.

Physical processes can be used to separate mixtures because there are no chemical interactions between the components. For example, different coins, such as nickels, dimes, and quarters, can be separated by size; iron particles mixed with sand can be picked up with a magnet; and water is separated from cooked spaghetti by using a strainer (see **FIGURE 3.3**).

Types of Mixtures

Mixtures are classified further as homogeneous or heterogeneous. In a *homogeneous mixture*, also called a *solution*, the composition is uniform throughout the sample. We cannot see the individual components, which appear as one state. Familiar examples of homogeneous mixtures are air, which contains oxygen and nitrogen gases, and seawater, a solution of salt and water.

In a *heterogeneous mixture*, the components do not have a uniform composition throughout the sample. The components appear as two separate regions. For example, a mixture of oil and water is heterogeneous because the oil floats on the surface of the water. Other examples of heterogeneous mixtures are a cookie with raisins and orange juice with pulp.

In the chemistry laboratory, mixtures are separated by various methods. Solids are separated from liquids by *filtration*, which involves pouring a mixture through a filter paper set in a funnel. The solid (residue) remains in the filter paper and the filtered liquid (filtrate) moves through. In *chromatography*, different components of a liquid mixture separate as they move at different rates up the surface of a piece of chromatography paper.

ENGAGE

Why is a pizza heterogeneous, whereas vinegar is a homogeneous mixture?

(a)

(b)

(a) A mixture of a liquid and a solid is separated by filtration. **(b)** Different substances in ink are separated as they travel at different rates up the surface of chromatography paper.

CHEMISTRY LINK TO HEALTH

Breathing Mixtures

The air we breathe is composed mostly of the gases oxygen (21%) and nitrogen (79%). The homogeneous breathing mixtures used by scuba divers differ from the air we breathe depending on the depth of the dive. Nitrox is a mixture of oxygen and nitrogen, but with more oxygen gas (up to 32%) and less nitrogen gas (68%) than air. A breathing mixture with less nitrogen gas decreases the risk of *nitrogen narcosis* associated with breathing regular air while diving. Heliox contains oxygen and helium, which is typically used for diving to more than 200 ft. By replacing nitrogen with helium, nitrogen narcosis does not occur. However, at dive depths over 300 ft, helium is associated with severe shaking and a drop in body temperature.

A breathing mixture used for dives over 400 ft is trimix, which contains oxygen, helium, and some nitrogen. The addition of some nitrogen lessens the problem of shaking that comes with breathing high levels of helium. Heliox and trimix are used only by professional, military, or other highly trained divers.

In hospitals, heliox may be used as a treatment for respiratory disorders and lung constriction in adults and premature infants. Heliox is less dense than air, which reduces the effort of breathing and helps distribute the oxygen gas to the tissues.

A nitrox mixture is used to fill scuba tanks.

SAMPLE PROBLEM 3.1 Classifying Mixtures

TRY IT FIRST

Classify each of the following as a pure substance (element or compound) or a mixture (homogeneous or heterogeneous):

a. copper in wire
b. a chocolate-chip cookie
c. nitrox, a combination of oxygen and nitrogen used to fill scuba tanks

SOLUTION

a. Copper is an element, which is a pure substance.
b. A chocolate-chip cookie does not have a uniform composition, which makes it a heterogeneous mixture.
c. The gases oxygen and nitrogen have a uniform composition in nitrox, which makes it a homogeneous mixture.

STUDY CHECK 3.1

A salad dressing is prepared with oil, vinegar, and chunks of blue cheese. Is this a homogeneous or heterogeneous mixture?

ANSWER

heterogeneous mixture

TEST

Try Practice Problems 3.3 to 3.6

PRACTICE PROBLEMS

3.1 Classification of Matter

LEARNING GOAL Classify examples of matter as pure substances or mixtures.

3.1 Classify each of the following pure substances as an element or a compound:
a. a silicon (Si) chip **b.** hydrogen peroxide (H_2O_2)
c. oxygen gas (O_2) **d.** rust (Fe_2O_3)
e. methane (CH_4) in natural gas

3.2 Classify each of the following pure substances as an element or a compound:
a. helium gas (He) **b.** sulfur (S)
c. sugar ($C_{12}H_{22}O_{11}$) **d.** mercury (Hg) in a thermometer
e. lye (NaOH)

3.3 Classify each of the following as a pure substance or a mixture:
a. baking soda ($NaHCO_3$) **b.** a blueberry muffin
c. ice (H_2O) **d.** zinc (Zn)
e. trimix (oxygen, nitrogen, and helium) in a scuba tank

3.4 Classify each of the following as a pure substance or a mixture:
a. a soft drink **b.** propane (C_3H_8)
c. a cheese sandwich **d.** an iron (Fe) nail
e. salt substitute (KCl)

Clinical Applications

3.5 A dietitian includes one of the following mixtures in the lunch menu. Classify each as homogeneous or heterogeneous.
a. vegetable soup **b.** tea
c. fruit salad **d.** tea with ice and lemon slices

3.6 A dietitian includes one of the following mixtures in the lunch menu. Classify each as homogeneous or heterogeneous.
a. nonfat milk **b.** chocolate-chip ice cream
c. peanut butter sandwich **d.** cranberry juice

3.2 States and Properties of Matter

LEARNING GOAL Identify the states and the physical and chemical properties of matter.

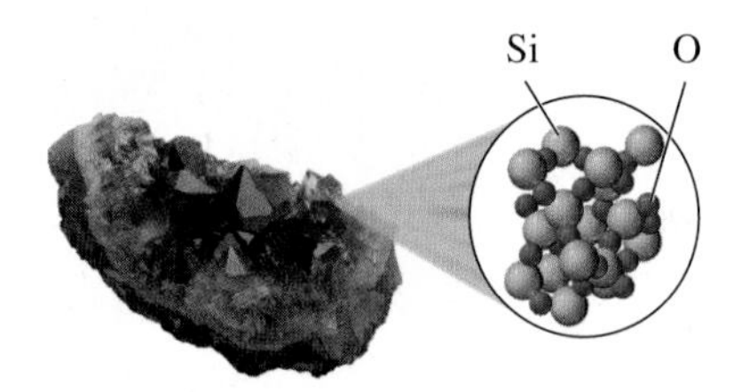

Amethyst, a solid, is a purple form of quartz that contains atoms of Si and O.

On Earth, matter exists in one of three *physical forms* called the **states of matter**: *solids*, *liquids*, and *gases*. Water is a familiar substance that we routinely observe in all three states. In the solid state, water can be an ice cube or a snowflake. It is a liquid when it comes out of a faucet or fills a pool. Water forms a gas, or vapor, when it evaporates from wet clothes or boils in a pan. A **solid**, such as a pebble or a baseball, has a definite shape and volume. You can probably recognize several solids within your reach right now such as books, pencils, or a computer mouse. In a solid, strong attractive forces hold the particles close together. The particles in a solid are arranged in such a rigid pattern, their only movement is to vibrate slowly in fixed positions. For many solids, this rigid structure produces a crystal such as that seen in amethyst.

A **liquid** has a definite volume, but not a definite shape. In a liquid, the particles move in random directions but are sufficiently attracted to each other to maintain a definite volume, although not a rigid structure. Thus, when water, oil, or vinegar is poured from one container to another, the liquid maintains its own volume but takes the shape of the new container.

A **gas** does not have a definite shape or volume. In a gas, the particles are far apart, have little attraction to each other, and move at high speeds, taking the shape and volume of their container. When you inflate a bicycle tire, the air, which is a gas, fills the entire volume of the tire. The propane gas in a tank fills the entire volume of the tank. **TABLE 3.1** compares the three states of matter.

ENGAGE

Why does a gas take both the shape and volume of its container?

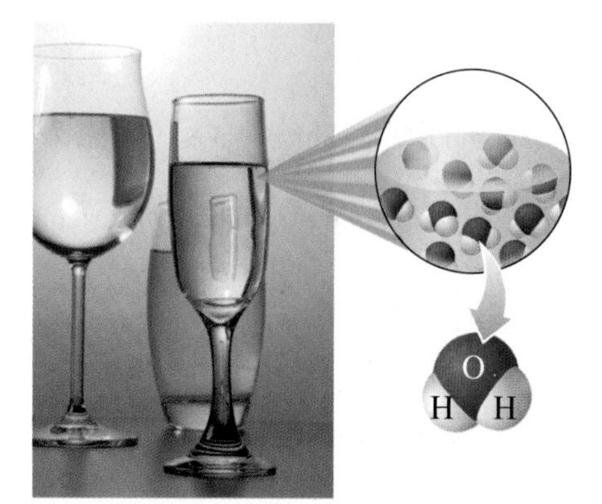

Water as a liquid takes the shape of its container.

A gas takes the shape and volume of its container.

TEST

Try Practice Problems 3.7 and 3.8

TABLE 3.1 A Comparison of Solids, Liquids, and Gases

Characteristic	Solid	Liquid	Gas
Shape	Has a definite shape	Takes the shape of the container	Takes the shape of the container
Volume	Has a definite volume	Has a definite volume	Fills the volume of the container
Arrangement of Particles	Fixed, very close	Random, close	Random, far apart
Interaction between Particles	Very strong	Strong	Essentially none
Movement of Particles	Very slow	Moderate	Very fast
Examples	Ice, salt, iron	Water, oil, vinegar	Water vapor, helium, air

Physical Properties and Physical Changes

One way to describe matter is to observe its properties. For example, if you were asked to describe yourself, you might list characteristics such as your height and weight, the color of your eyes and skin, or the length, color, and texture of your hair.

Physical properties are those characteristics that can be observed or measured without affecting the identity of a substance. In chemistry, typical physical properties include the shape, color, melting point, boiling point, and physical state of a substance. For example, some of the physical properties of a penny include its round shape, orange-red color (from copper), solid state, and shiny luster. **TABLE 3.2** gives more examples of physical properties of copper found in pennies, electrical wiring, and copper pans.

Copper, used in cookware, is a good conductor of heat.

TABLE 3.2 Some Physical Properties of Copper

State at 25 °C	Solid
Color	Orange-red
Odor	Odorless
Melting Point	1083 °C
Boiling Point	2567 °C
Luster	Shiny
Conduction of Electricity	Excellent
Conduction of Heat	Excellent

Water is a substance that is commonly found in all three states: solid, liquid, and gas. When matter undergoes a **physical change**, its state, size, or appearance will change, but its composition remains the same. The solid state of water, snow or ice, has a different appearance than its liquid or gaseous state, but all three states are water.

The physical appearance of a substance can change in other ways too. Suppose that you dissolve some salt in water. The appearance of the salt changes, but you could re-form the salt crystals by evaporating the water. Thus, in a physical change of state, no new substances are produced.

Chemical Properties and Chemical Changes

Chemical properties are those that describe the ability of a substance to change into a new substance. When a **chemical change** takes place, the original substance is converted into one or more new substances, which have new physical and chemical properties. For example, the rusting or corrosion of a metal, such as iron, is a chemical property. In the rain, an iron (Fe) nail undergoes a chemical change when it reacts with oxygen (O_2) to form rust (Fe_2O_3). A chemical change has taken place: Rust is a new substance with new physical and chemical properties. **TABLE 3.3** gives examples of some physical and chemical changes. **TABLE 3.4** summarizes physical and chemical properties and changes.

ENGAGE

Why is bending an iron nail a physical change, whereas the formation of rust on an iron nail is a chemical change?

TABLE 3.3 Examples of Some Physical and Chemical Changes

Physical Changes	Chemical Changes
Water boils to form water vapor.	Shiny, silver metal reacts in air to give a black, grainy coating.
Copper is drawn into thin copper wires.	A piece of wood burns with a bright flame and produces heat, ashes, carbon dioxide, and water vapor.
Sugar dissolves in water to form a solution.	Heating sugar forms a smooth, caramel-colored substance.
Paper is cut into tiny pieces of confetti.	Iron, which is gray and shiny, combines with oxygen to form orange-red rust.

A chemical change occurs when sugar is heated, forming a caramelized topping for flan.

CORE CHEMISTRY SKILL

Identifying Physical and Chemical Changes

TABLE 3.4 Summary of Physical and Chemical Properties and Changes

	Physical	Chemical
Property	A characteristic of a substance: color, shape, odor, luster, size, melting point, or density.	A characteristic that indicates the ability of a substance to form another substance: paper can burn, iron can rust, silver can tarnish.
Change	A change in a physical property that retains the identity of the substance: a change of state, a change in size, or a change in shape.	A change in which the original substance is converted to one or more new substances: paper burns, iron rusts, silver tarnishes.

A gold ingot is hammered to form gold leaf.

INTERACTIVE VIDEO

Chemical vs. Physical Changes

TEST

Try Practice Problems 3.9 to 3.14

SAMPLE PROBLEM 3.2 Physical and Chemical Changes

TRY IT FIRST

Classify each of the following as a physical or chemical change:

a. A gold ingot is hammered to form gold leaf.
b. Gasoline burns in air.
c. Garlic is chopped into small pieces.

SOLUTION

a. A physical change occurs when the gold ingot changes shape.
b. A chemical change occurs when gasoline burns and forms different substances with new properties.
c. A physical change occurs when the size of the garlic pieces changes.

STUDY CHECK 3.2

Classify each of the following as a physical or chemical change:

a. Water freezes on a pond.
b. Gas bubbles form when baking powder is placed in vinegar.
c. A log is cut for firewood.

ANSWER

a. physical change **b.** chemical change **c.** physical change

PRACTICE PROBLEMS

3.2 States and Properties of Matter

LEARNING GOAL Identify the states and the physical and chemical properties of matter.

3.7 Indicate whether each of the following describes a gas, a liquid, or a solid:
a. The breathing mixture in a scuba tank has no definite volume or shape.
b. The neon atoms in a lighting display do not interact with each other.
c. The particles in an ice cube are held in a rigid structure.

3.8 Indicate whether each of the following describes a gas, a liquid, or a solid:
a. Lemonade has a definite volume but takes the shape of its container.
b. The particles in a tank of oxygen are very far apart.
c. Helium occupies the entire volume of a balloon.

3.9 Describe each of the following as a physical or chemical property:
a. Chromium is a steel-gray solid.
b. Hydrogen reacts readily with oxygen.
c. A patient has a temperature of 40.2 °C.
d. Milk will sour when left in a warm room.
e. Butane gas in an igniter burns in oxygen.

3.10 Describe each of the following as a physical or chemical property:
a. Neon is a colorless gas at room temperature.
b. Apple slices turn brown when they are exposed to air.
c. Phosphorus will ignite when exposed to air.
d. At room temperature, mercury is a liquid.
e. Propane gas is compressed to a liquid for placement in a small cylinder.

3.11 What type of change, physical or chemical, takes place in each of the following?
a. Water vapor condenses to form rain.
b. Cesium metal reacts explosively with water.
c. Gold melts at 1064 °C.
d. A puzzle is cut into 1000 pieces.
e. Cheese is grated.

3.12 What type of change, physical or chemical, takes place in each of the following?
a. Pie dough is rolled into thin pieces for a crust.
b. A silver pin tarnishes in the air.
c. A tree is cut into boards at a saw mill.
d. Food is digested.
e. A chocolate bar melts.

3.13 Describe each property of the element fluorine as physical or chemical.
a. is highly reactive
b. is a gas at room temperature
c. has a pale, yellow color
d. will explode in the presence of hydrogen
e. has a melting point of −220 °C

3.14 Describe each property of the element zirconium as physical or chemical.
a. melts at 1852 °C
b. is resistant to corrosion
c. has a grayish white color
d. ignites spontaneously in air when finely divided
e. is a shiny metal

3.3 Temperature

LEARNING GOAL Given a temperature, calculate the corresponding temperature on another scale.

REVIEW
Using Positive and Negative Numbers in Calculations (1.4)
Solving Equations (1.4)
Counting Significant Figures (2.2)

Temperatures in science are measured and reported in *Celsius* (°C) units. On the Celsius scale, the reference points are the freezing point of water, defined as 0 °C, and the boiling point, 100 °C. In the United States, everyday temperatures are commonly reported in *Fahrenheit* (°F) units. On the Fahrenheit scale, water freezes at 32 °F and boils at 212 °F. A typical room temperature of 22 °C would be the same as 72 °F. Normal human body temperature is 37.0 °C, which is the same temperature as 98.6 °F.

On the Celsius and Fahrenheit temperature scales, the temperature difference between freezing and boiling is divided into smaller units called *degrees*. On the Celsius scale, there are 100 degrees Celsius between the freezing and boiling points of water, whereas the Fahrenheit scale has 180 degrees Fahrenheit between the freezing and boiling points of water. That makes a degree Celsius almost twice the size of a degree Fahrenheit: 1 °C = 1.8 °F (see **FIGURE 3.4**).

$$180 \text{ degrees Fahrenheit} = 100 \text{ degrees Celsius}$$

$$\frac{180 \text{ degrees Fahrenheit}}{100 \text{ degrees Celsius}} = \frac{1.8\ °\text{F}}{1\ °\text{C}}$$

We can write a temperature equation that relates a Fahrenheit temperature and its corresponding Celsius temperature.

ENGAGE
Why is a degree Celsius a larger unit of temperature than a degree Fahrenheit?

$$T_F = 1.8(T_C) + 32$$ Temperature equation to obtain degrees Fahrenheit

Changes °C to °F — Adjusts freezing point

In the equation, the Celsius temperature is multiplied by 1.8 to change °C to °F; then 32 is added to adjust the freezing point from 0 °C to the Fahrenheit freezing point, 32 °F. The values, 1.8 and 32, used in the temperature equation are exact numbers and are not used to determine significant figures in the answer.

CORE CHEMISTRY SKILL
Converting between Temperature Scales

To convert from degrees Fahrenheit to degrees Celsius, the temperature equation is rearranged to solve for T_C. First, we subtract 32 from both sides since we must apply the same operation to both sides of the equation.

$$T_F - 32 = 1.8(T_C) + \cancel{32} - \cancel{32}$$

$$T_F - 32 = 1.8(T_C)$$

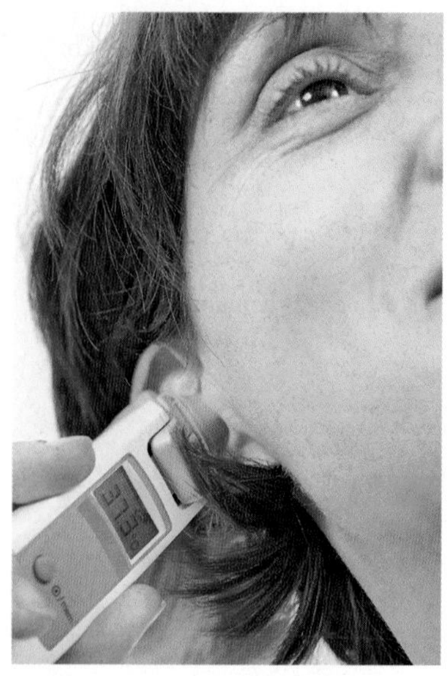

A digital ear thermometer is used to measure body temperature.

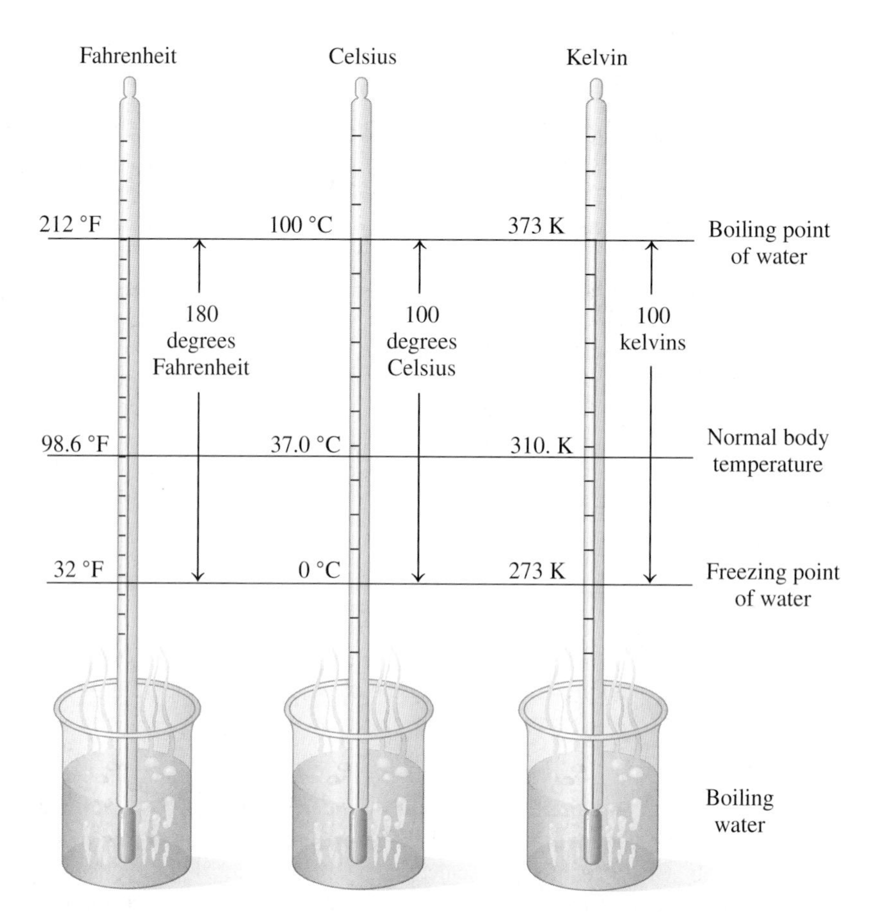

FIGURE 3.4 A comparison of the Fahrenheit, Celsius, and Kelvin temperature scales between the freezing and boiling points of water.

Q What is the difference in the freezing points of water on the Celsius and Fahrenheit temperature scales?

Second, we solve the equation for T_C by dividing both sides by 1.8.

$$\frac{T_F - 32}{1.8} = \frac{\cancel{1.8}(T_C)}{\cancel{1.8}}$$

$$\frac{T_F - 32}{1.8} = T_C \quad \text{Temperature equation to obtain degrees Celsius}$$

Scientists have learned that the coldest temperature possible is −273 °C (more precisely, −273.15 °C). On the *Kelvin* scale, this temperature, called *absolute zero*, has the value of 0 K. Units on the Kelvin scale are called kelvins (K); *no degree symbol is used*. Because there are no lower temperatures, the Kelvin scale has no negative temperature values. Between the freezing point of water, 273 K, and the boiling point, 373 K, there are 100 kelvins, which makes a kelvin equal in size to a degree Celsius.

$$1\ \text{K} = 1\ °\text{C}$$

We can write an equation that relates a Celsius temperature to its corresponding Kelvin temperature by adding 273 to the Celsius temperature. **TABLE 3.5** gives a comparison of some temperatures on the three scales.

$$T_K = T_C + 273 \quad \text{Temperature equation to obtain kelvins}$$

An antifreeze mixture in a car radiator will not freeze until the temperature drops to −37 °C. We can calculate the temperature of the antifreeze mixture in kelvins by adding 273 to the temperature in degrees Celsius.

$$T_K = -37\ °\text{C} + 273 = 236\ \text{K}$$

TABLE 3.5 A Comparison of Temperatures

Example	Fahrenheit (°F)	Celsius (°C)	Kelvin (K)
Sun	9937	5503	5776
A hot oven	450	232	505
Water boils	212	100	373
A high fever	104	40	313
Normal body temperature	98.6	37.0	310
Room temperature	70	21	294
Water freezes	32	0	273
A northern winter	−66	−54	219
Nitrogen liquefies	−346	−210	63
Absolute zero	−459	−273	0

SAMPLE PROBLEM 3.3 Calculating Temperature

TRY IT FIRST

A dermatologist uses cryogenic nitrogen at −196 °C to remove skin lesions and some skin cancers. What is the temperature, in degrees Fahrenheit, of the nitrogen?

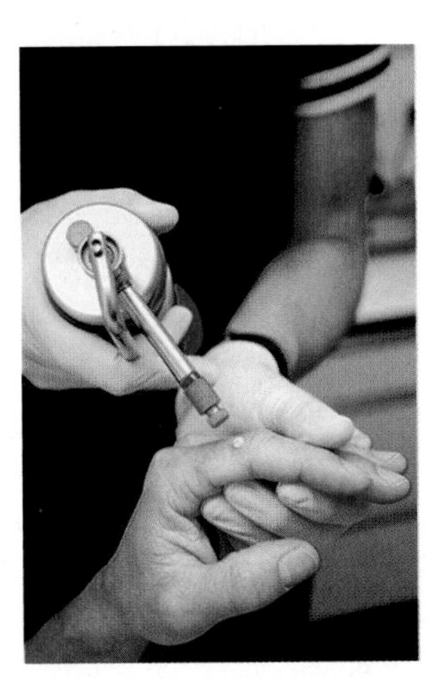

The low temperature of cryogenic nitrogen is used to destroy skin lesions.

SOLUTION GUIDE

STEP 1 State the given and needed quantities.

ANALYZE THE PROBLEM	Given	Need	Connect
	−196 °C	T in degrees Fahrenheit	temperature equation

STEP 2 Write a temperature equation.

$$T_F = 1.8(T_C) + 32$$

STEP 3 Substitute in the known values and calculate the new temperature.

$$T_F = 1.8(-196) + 32 \quad \text{1.8 is exact; 32 is exact}$$

$$= -353 + 32$$

$$= -321\ °F \quad \text{Answer to the ones place}$$

In the equation, *the values of 1.8 and 32 are exact numbers.*

STUDY CHECK 3.3

In the process of making ice cream, rock salt is added to crushed ice to chill the ice cream mixture. If the temperature drops to −11 °C, what is it in degrees Fahrenheit?

ANSWER

12 °F

ENGAGE

Show that −40. °C is the same temperature as −40. °F.

TEST

Try Practice Problems 3.15 to 3.18

SAMPLE PROBLEM 3.4 Calculating Degrees Celsius and Kelvins

TRY IT FIRST

In a type of cancer treatment called *thermotherapy*, temperatures as high as 113 °F are used to destroy cancer cells or make them more sensitive to radiation. What is that temperature in degrees Celsius? In kelvins?

SOLUTION GUIDE

STEP 1 State the given and needed quantities.

ANALYZE THE PROBLEM	Given	Need	Connect
	113 °F	T in degrees Celsius, kelvins	temperature equations

STEP 2 Write a temperature equation.

$$T_C = \frac{T_F - 32}{1.8} \qquad T_K = T_C + 273$$

STEP 3 Substitute in the known values and calculate the new temperature.

$$T_C = \frac{(113 - 32)}{1.8}$$ 32 is exact; 1.8 is exact

$$= \frac{\overset{\text{Two SFs}}{81}}{\underset{\text{Exact}}{1.8}} = \underset{\text{Two SFs}}{45\ °C}$$

Using the equation that converts degrees Celsius to kelvins, we substitute in degrees Celsius.

$$T_K = \underset{\text{Ones place}}{45} + \underset{\text{Ones place}}{273} = \underset{\text{Ones place}}{318\ K}$$

STUDY CHECK 3.4

A child has a temperature of 103.6 °F. What is this temperature on a Celsius thermometer?

ANSWER

39.8 °C

TEST

Try Practice Problems 3.19 and 3.20

PRACTICE PROBLEMS

3.3 Temperature

LEARNING GOAL Given a temperature, calculate the corresponding temperature on another scale.

3.15 Your friend who is visiting from Canada just took her temperature. When she reads 99.8 °F, she becomes concerned that she is quite ill. How would you explain this temperature to your friend?

3.16 You have a friend who is using a recipe for flan from a Mexican cookbook. You notice that he set your oven temperature at 175 °F. What would you advise him to do?

3.17 Calculate the unknown temperature in each of the following:

a. 37.0 °C = _______ °F **b.** 65.3 °F = _______ °C
c. −27 °C = _______ K **d.** 62 °C = _______ K
e. 114 °F = _______ °C

3.18 Calculate the unknown temperature in each of the following:

a. 25 °C = _______ °F **b.** 155 °C = _______ °F
c. −25 °F = _______ °C **d.** 224 K = _______ °C
e. 145 °C = _______ K

Clinical Applications

3.19 **a.** A patient with hyperthermia has a temperature of 106 °F. What does this read on a Celsius thermometer?

b. Because high fevers can cause convulsions in children, the doctor needs to be called if the child's temperature goes over 40.0 °C. Should the doctor be called if a child has a temperature of 103 °F?

3.20 **a.** Water is heated to 145 °F. What is the temperature of the hot water in degrees Celsius?

b. During extreme hypothermia, a child's temperature dropped to 20.6 °C. What was his temperature in degrees Fahrenheit?

CHEMISTRY LINK TO HEALTH

Variation in Body Temperature

Normal body temperature is considered to be 37.0 °C, although it varies throughout the day and from person to person. Oral temperatures of 36.1 °C are common in the morning and climb to a high of 37.2 °C between 6 P.M. and 10 P.M. Temperatures above 37.2 °C for a person at rest are usually an indication of illness. Individuals who are involved in prolonged exercise may also experience elevated temperatures. Body temperatures of marathon runners can range from 39 °C to 41 °C as heat production during exercise exceeds the body's ability to lose heat.

Changes of more than 3.5 °C from the normal body temperature begin to interfere with bodily functions. Body temperatures above 41 °C, *hyperthermia*, can lead to convulsions, particularly in children, which may cause permanent brain damage. Heatstroke occurs above 41.1 °C. Sweat production stops, and the skin becomes hot and dry. The pulse rate is elevated, and respiration becomes weak and rapid. The person can become lethargic and lapse into a coma. Damage to internal organs is a major concern, and treatment, which must be immediate, may include immersing the person in an ice-water bath.

At the low temperature extreme of *hypothermia*, body temperature can drop as low as 28.5 °C. The person may appear cold and pale and have an irregular heartbeat. Unconsciousness can occur if the body temperature drops below 26.7 °C. Respiration becomes slow and shallow, and oxygenation of the tissues decreases. Treatment involves providing oxygen and increasing blood volume with glucose and saline fluids. Injecting warm fluids (37.0 °C) into the peritoneal cavity may restore the internal temperature.

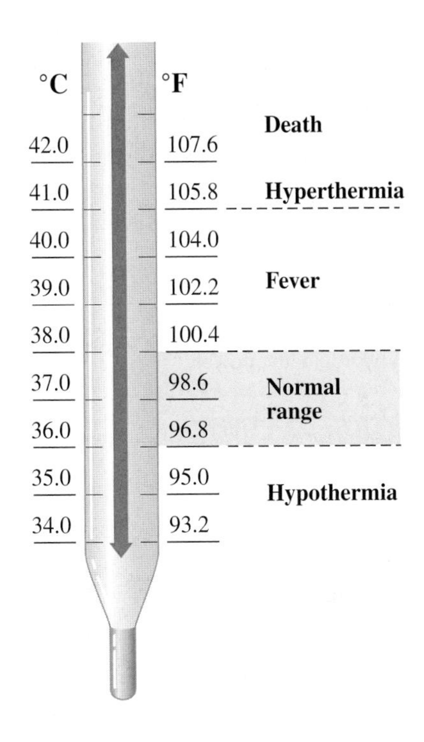

3.4 Energy

LEARNING GOAL Identify energy as potential or kinetic; convert between units of energy.

REVIEW
Rounding Off (2.3)
Using Significant Figures in Calculations (2.3)
Writing Conversion Factors from Equalities (2.5)
Using Conversion Factors (2.6)

Almost everything you do involves energy. When you are running, walking, dancing, or thinking, you are using energy to do *work*, any activity that requires energy. In fact, **energy** is defined as the ability to do work. Suppose you are climbing a steep hill and you become too tired to go on. At that moment, you do not have the energy to do any more work. Now suppose you sit down and have lunch. In a while, you will have obtained some energy from the food, and you will be able to do more work and complete the climb.

Kinetic and Potential Energy

Energy can be classified as kinetic energy or potential energy. **Kinetic energy** is the energy of motion. Any object that is moving has kinetic energy. **Potential energy** is determined by the position of an object or by the chemical composition of a substance. A boulder resting on top of a mountain has potential energy because of its location. If the boulder rolls down the mountain, the potential energy becomes kinetic energy. Water stored in a reservoir has potential energy. When the water goes over the dam and falls to the stream below, its potential energy is converted to kinetic energy. Foods and fossil fuels have potential energy. When you digest food or burn gasoline in your car, potential energy is converted to kinetic energy to do work.

ENGAGE
Why does a book have more potential energy when it is on the top of a high table than when it is on the floor?

TEST
Try Practice Problems 3.21 to 3.24

Heat and Energy

Heat is the energy associated with the motion of particles. An ice cube feels cold because heat flows from your hand into the ice cube. The faster the particles move, the greater the heat or thermal energy of the substance. In the ice cube, the particles are moving very slowly. As heat is added, the motion of the particles in the ice cube increases. Eventually, the particles have enough energy to make the ice cube melt as it changes from a solid to a liquid.

Water at the top of the dam stores potential energy. When the water flows over the dam, potential energy is converted to hydroelectric power.

Units of Energy

The SI unit of energy and work is the **joule (J)** (pronounced "jewel"). The joule is a small amount of energy, so scientists often use the kilojoule (kJ), 1000 joules. To heat water for one cup of tea, you need about 75 000 J or 75 kJ of heat. TABLE 3.6 shows a comparison of energy in joules for several energy sources or uses.

You may be more familiar with the unit **calorie (cal)**, from the Latin *caloric*, meaning "heat." The calorie was originally defined as the amount of energy (heat) needed to raise the temperature of 1 g of water by 1 °C. Now, one calorie is defined as exactly 4.184 J. This equality can be written as two conversion factors:

$$1 \text{ cal} = 4.184 \text{ J (exact)} \qquad \frac{4.184 \text{ J}}{1 \text{ cal}} \text{ and } \frac{1 \text{ cal}}{4.184 \text{ J}}$$

One *kilocalorie* (kcal) is equal to 1000 calories, and one *kilojoule* (kJ) is equal to 1000 joules. The equalities and conversion factors follow:

CORE CHEMISTRY SKILL
Using Energy Units

$$1 \text{ kcal} = 1000 \text{ cal} \qquad \frac{1000 \text{ cal}}{1 \text{ kcal}} \text{ and } \frac{1 \text{ kcal}}{1000 \text{ cal}}$$

$$1 \text{ kJ} = 1000 \text{ J} \qquad \frac{1000 \text{ J}}{1 \text{ kJ}} \text{ and } \frac{1 \text{ kJ}}{1000 \text{ J}}$$

TABLE 3.6 A Comparison of Energy for Various Resources and Uses

Energy in Joules	
10^{27}	
	Energy radiated by the Sun in 1 s (10^{26})
10^{24}	
	World reserves of fossil fuel (10^{23})
10^{21}	
	Energy consumption for 1 yr in the United States (10^{20})
10^{18}	
	Solar energy reaching Earth in 1 s (10^{17})
10^{15}	
10^{12}	
	Energy use per person in 1 yr in the United States (10^{11})
10^{9}	
	Energy from 1 gal of gasoline (10^{8})
10^{6}	Energy from one serving of pasta, a doughnut, or needed to bicycle for 1 h (10^{6})
	Energy used to sleep for 1 h (10^{5})
10^{3}	
10^{0}	

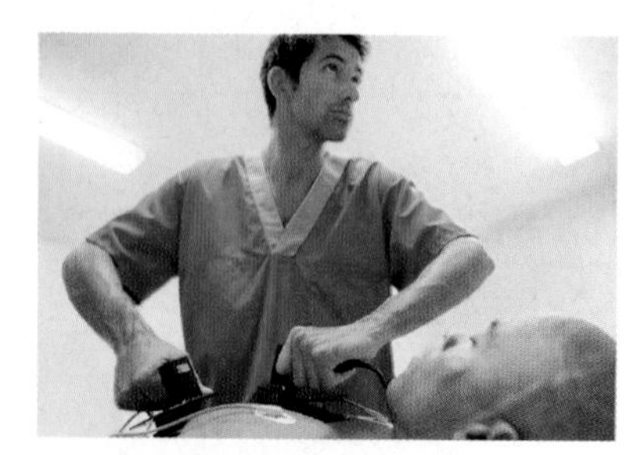

A defibrillator provides electrical energy to heart muscle to re-establish normal rhythm.

SAMPLE PROBLEM 3.5 Energy Units

TRY IT FIRST

A defibrillator gives a high-energy-shock output of 360 J. What is this quantity of energy in calories?

SOLUTION GUIDE

STEP 1 State the given and needed quantities.

ANALYZE THE PROBLEM	Given	Need	Connect
	360 J	calories	energy factor

STEP 2 Write a plan to convert the given unit to the needed unit.

joules → Energy factor → calories

STEP 3 State the equalities and conversion factors.

$$1 \text{ cal} = 4.184 \text{ J}$$

$$\frac{4.184 \text{ J}}{1 \text{ cal}} \quad \text{and} \quad \frac{1 \text{ cal}}{4.184 \text{ J}}$$

STEP 4 Set up the problem to calculate the needed quantity.

$$\underset{\text{Two SFs}}{360 \text{ J}} \times \overset{\text{Exact}}{\underset{\text{Exact}}{\frac{1 \text{ cal}}{4.184 \text{ J}}}} = \underset{\text{Two SFs}}{86 \text{ cal}}$$

STUDY CHECK 3.5

When 1.0 g of glucose is metabolized in the body, it produces 3.9 kcal. How many kilojoules are produced?

ANSWER

16 kJ

TEST

Try Practice Problems 3.25 to 3.28

PRACTICE PROBLEMS

3.4 Energy

LEARNING GOAL Identify energy as potential or kinetic; convert between units of energy.

3.21 Discuss the changes in the potential and kinetic energy of a roller-coaster ride as the roller-coaster car climbs to the top and goes down the other side.

3.22 Discuss the changes in the potential and kinetic energy of a ski jumper taking the elevator to the top of the jump and going down the ramp.

3.23 Indicate whether each of the following statements describes potential or kinetic energy:
a. water at the top of a waterfall **b.** kicking a ball
c. the energy in a lump of coal **d.** a skier at the top of a hill

3.24 Indicate whether each of the following statements describes potential or kinetic energy:
a. the energy in your food **b.** a tightly wound spring
c. a car speeding down the freeway **d.** an earthquake

3.25 Convert each of the following energy units:
a. 3500 cal to kcal **b.** 415 J to cal
c. 28 cal to J **d.** 4.5 kJ to cal

3.26 Convert each of the following energy units:
a. 8.1 kcal to cal **b.** 325 J to kJ
c. 2550 cal to kJ **d.** 2.50 kcal to J

Clinical Applications

3.27 The energy needed to keep a 75-watt light bulb burning for 1.0 h is 270 kJ. Calculate the energy required to keep the light bulb burning for 3.0 h in each of the following energy units:
a. joules **b.** kilocalories

3.28 A person uses 750 kcal on a long walk. Calculate the energy used for the walk in each of the following energy units:
a. joules **b.** kilojoules

CHEMISTRY LINK TO THE ENVIRONMENT

Carbon Dioxide and Climate Change

The Earth's climate is a product of interactions between sunlight, the atmosphere, and the oceans. The Sun provides us with energy in the form of solar radiation. Some of this radiation is reflected back into space. The rest is absorbed by the clouds, atmospheric gases including carbon dioxide (CO_2), and Earth's surface. For millions of years, concentrations of carbon dioxide have fluctuated. However in the last 100 years, the amount of CO_2 gas in our atmosphere has increased significantly. From the years 1000 to 1800, the atmospheric carbon dioxide level averaged 280 ppm. The ppm indicates the parts per million by volume, which for gases is the same as mL of CO_2 per kL of air.

g of the Industrial Revolution in 1800 to 2016, ric carbon dioxide has risen from about 280 ppm 40% increase.
ieric CO_2 level increases, more solar radiation is es the temperature at the surface of Earth. Some scien- ed that if the carbon dioxide level doubles from its level before the strial Revolution, the average global temperature could increase by 2.0 to 4.4 °C. Although this seems to be a small temperature change, it could have dramatic impact worldwide. Even now, glaciers and snow cover in much of the world have diminished. Ice sheets in Antarctica and Greenland are melting faster and breaking apart. Although no one knows for sure how rapidly the ice in the polar regions is melting, this accelerating change will contribute to a rise in sea level. In the twentieth century, the sea level rose by 15 to 23 cm. Some scientists predict the sea level will rise by 1 m in this century. Such an increase will have a major impact on coastal areas.

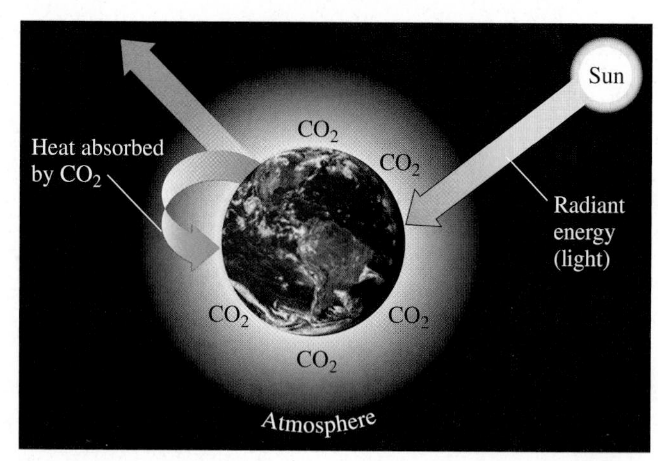

Heat from the Sun is trapped by the CO_2 layer in the atmosphere.

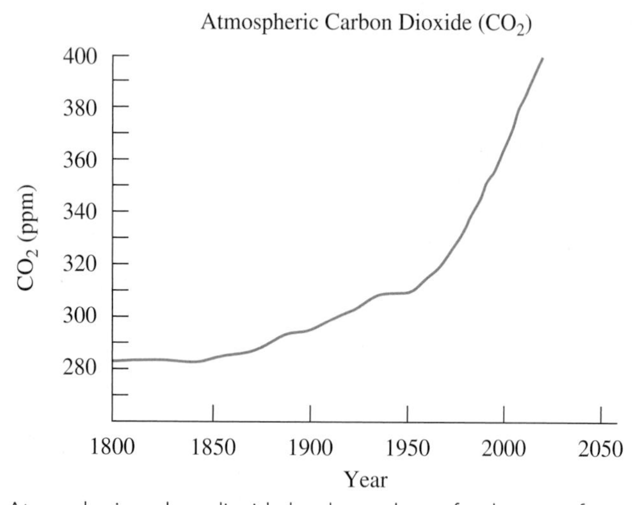

Atmospheric carbon dioxide levels are shown for the years from 1800 C.E. to 2016 C.E.

Until recently, the carbon dioxide level was maintained as algae in the oceans and the trees in the forests utilized the carbon dioxide. However, the ability of plant life to absorb carbon dioxide is not keeping up with the increase in carbon dioxide. Most scientists agree that the primary source of the increase of carbon dioxide is the burning of fossil fuels such as gasoline, coal, and natural gas. The cutting and burning of trees in the rainforests (deforestation) also reduces the amount of carbon dioxide removed from the atmosphere.

Worldwide efforts are being made to reduce the carbon dioxide produced by burning fossil fuels that heat our homes, run our cars, and provide energy for industries. Scientists are exploring ways to provide alternative energy sources and to reduce deforestation. Meanwhile, we can reduce energy use in our homes by using appliances that are more energy efficient such as replacing incandescent light bulbs with light-emitting diodes (LEDs). Such an effort worldwide will reduce the possible impact of climate change and at the same time save our fuel resources.

3.5 Energy and Nutrition

LEARNING GOAL Use the energy values to calculate the kilocalories (kcal) or kilojoules (kJ) for a food.

The food we eat provides energy to do work in the body, which includes the growth and repair of cells. Carbohydrates are the primary fuel for the body, but if the carbohydrate reserves are exhausted, fats and then proteins are used for energy.

For many years in the field of nutrition, the energy from food was measured as Calories or kilocalories. The nutritional unit *Calorie*, *Cal* (with an uppercase C), is the same as 1000 cal, or 1 kcal. The international unit, kilojoule (kJ), is becoming more prevalent. For example, a baked potato has an energy content of 100 Calories, which is 100 kcal or 440 kJ. A typical diet that provides 2100 Cal (kcal) is the same as an 8800 kJ diet.

TEST

Try Practice Problems 3.29 and 3.30

$$1 \text{ Cal} = 1 \text{ kcal} = 1000 \text{ cal}$$
$$1 \text{ Cal} = 4.184 \text{ kJ} = 4184 \text{ J}$$

In the nutrition laboratory, foods are burned in a *calorimeter* to determine their *energy value* (kcal/g or kJ/g) (see **FIGURE 3.5**). A sample of food is placed in a steel container called a calorimeter filled with oxygen. A measured amount of water is added to fill the

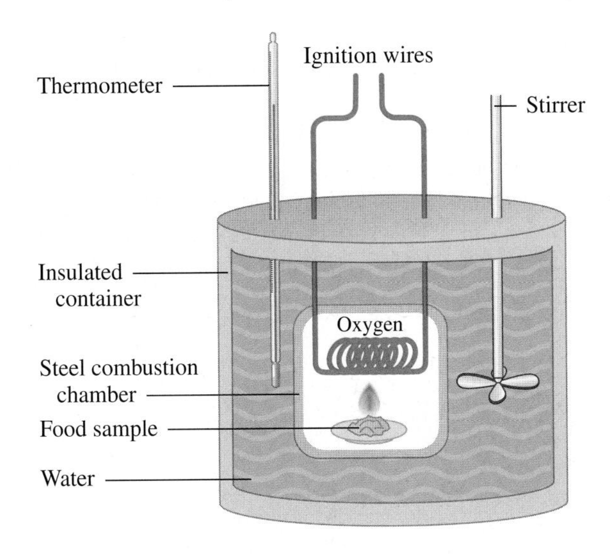

FIGURE 3.5 Heat released from burning a food sample in a calorimeter is used to determine the energy value for the food.

What happens to the temperature of water in a calorimeter during the combustion of a food sample?

area surrounding the combustion chamber. The food sample is burned, releasing heat that increases the temperature of the water. From the known mass of the food and water as well as the measured temperature increase, the energy value for the food is calculated. We assume that the energy absorbed by the calorimeter is negligible.

Energy Values for Foods

The **energy values** for food are the kilocalories or kilojoules obtained from burning 1 g of carbohydrate, fat, or protein, which are listed in TABLE 3.7. Using these energy values, we can calculate the total energy for a food if the mass of each food type is known.

$$\text{kilocalories} = \cancel{g} \times \frac{\text{kcal}}{\cancel{g}} \qquad \text{kilojoules} = \cancel{g} \times \frac{\text{kJ}}{\cancel{g}}$$

On packaged food, the energy content is listed on the Nutrition Facts label, usually in terms of the number of Calories or kilojoules for one serving. The composition and energy content for some foods are given in TABLE 3.8. The total energy in kilocalories for each food type was calculated using energy values in kilocalories. Total energy in kilojoules was calculated using energy values in kilojoules. The energy for each food type was rounded off to the tens place.

TABLE 3.7 Typical Energy Values for the Three Food Types

Food Type	kcal/g	kJ/g
Carbohydrate	4	17
Fat	9	38
Protein	4	17

ENGAGE

What type of food provides the most energy per gram?

TABLE 3.8 Composition and Energy Content for Some Foods

Food	Carbohydrate (g)	Fat (g)	Protein (g)	Energy
Apple, 1 medium	15	0	0	60 kcal (260 kJ)
Banana, 1 medium	26	0	1	110 kcal (460 kJ)
Beef, ground, 3 oz	0	14	22	220 kcal (900 kJ)
Broccoli, 3 oz	4	0	3	30 kcal (120 kJ)
Carrots, 1 cup	11	0	2	50 kcal (220 kJ)
Chicken, no skin, 3 oz	0	3	20	110 kcal (450 kJ)
Egg, 1 large	0	6	6	70 kcal (330 kJ)
Milk, nonfat, 1 cup	12	0	9	90 kcal (350 kJ)
Potato, baked	23	0	3	100 kcal (440 kJ)
Salmon, 3 oz	0	5	16	110 kcal (460 kJ)
Steak, 3 oz	0	27	19	320 kcal (1350 kJ)

Snack Crackers

Nutrition Facts
Serving Size 14 crackers (31g)
Servings Per Container About 7

Amount Per Serving	
Calories 130	Calories from Fat 40
Kilojoules 500	kJ from Fat 150
	% Daily Value*
Total Fat 4g	6%
Saturated Fat 0.5g	3%
Trans Fat 0g	
Polyunsaturated Fat 0.5%	
Monounsaturated Fat 1.5g	
Cholesterol 0mg	0%
Sodium 310mg	13%
Total Carbohydrate 19g	6%
Dietary Fiber Less than 1g	4%
Sugars 2g	
Proteins 2g	

The Nutrition Facts include the total Calories and kilojoules, and the grams of carbohydrate, fat, and protein per serving.

SAMPLE PROBLEM 3.6 Calculating the Energy from a Food

TRY IT FIRST

While working on his diet log, Charles observed that the Nutrition Facts label for crackers states that one serving contains 19 g of carbohydrate, 4 g of fat, and 2 g of protein. If Charles eats one serving of Snack Crackers, what is the energy, in kilocalories, from each food type and the total kilocalories? Round off the kilocalories for each food type to the tens place.

SOLUTION GUIDE

STEP 1 State the given and needed quantities.

ANALYZE THE PROBLEM	Given	Need	Connect
	19 g of carbohydrate, 4 g of fat, 2 g of protein	total number of kilocalories	energy values

STEP 2 Use the energy value for each food type to calculate the kilocalories, rounded off to the tens place.

Food Type	Mass		Energy Value		Energy
Carbohydrate	$19\ \cancel{g}$	$\times$	$\frac{4\ \text{kcal}}{1\ \cancel{g}}$	=	80 kcal
Fat	$4\ \cancel{g}$	$\times$	$\frac{9\ \text{kcal}}{1\ \cancel{g}}$	=	40 kcal
Protein	$2\ \cancel{g}$	$\times$	$\frac{4\ \text{kcal}}{1\ \cancel{g}}$	=	10 kcal

STEP 3 Add the energy for each food type to give the total energy from the food.

Total energy = 80 kcal + 40 kcal + 10 kcal = 130 kcal

STUDY CHECK 3.6

a. For one serving of crackers in Sample Problem 3.6, calculate the energy, in kilojoules, for each food type. Round off the kilojoules for each food type to the tens place.

b. What is the total energy, in kilojoules, for one serving of crackers?

ANSWER

a. carbohydrate, 320 kJ; fat, 150 kJ; protein, 30 kJ

b. 500 kJ

TEST

Try Practice Problems 3.31 to 3.36

CHEMISTRY LINK TO HEALTH

Losing and Gaining Weight

The number of kilocalories or kilojoules needed in the daily diet of an adult depends on gender, age, and level of physical activity. Some typical levels of energy needs are given in **TABLE 3.9**.

A person gains weight when food intake exceeds energy output. The amount of food a person eats is regulated by the hunger center in the hypothalamus, which is located in the brain. Food intake is normally proportional to the nutrient stores in the body. If these nutrient stores are low, you feel hungry; if they are high, you do not feel like eating.

TABLE 3.9 Typical Energy Requirements for Adults

Gender	Age	Moderately Active kcal (kJ)	Highly Active kcal (kJ)
Female	19–30	2100 (8800)	2400 (10 000)
	31–50	2000 (8400)	2200 (9200)
Male	19–30	2700 (11 300)	3000 (12 600)
	31–50	2500 (10 500)	2900 (12 100)

A person loses weight when food intake is less than energy output. Many diet products contain cellulose, which has no nutritive value but provides bulk and makes you feel full. Some diet drugs depress the hunger center and must be used with caution because they excite the nervous system and can elevate blood pressure. Because muscular exercise is an important way to expend energy, an increase in daily exercise aids weight loss. **TABLE 3.10** lists some activities and the amount of energy they require.

One hour of swimming uses 2100 kJ of energy.

TABLE 3.10 Energy Expended by a 70.0-kg (154-lb) Adult

Activity	Energy (kcal/h)	Energy (kJ/h)
Sleeping	60	250
Sitting	100	420
Walking	200	840
Swimming	500	2100
Running	750	3100

PRACTICE PROBLEMS

3.5 Energy and Nutrition

LEARNING GOAL Use the energy values to calculate the kilocalories (kcal) or kilojoules (kJ) for a food.

3.29 Calculate the kilocalories for each of the following:
  a. one stalk of celery that produces 125 kJ when burned in a calorimeter
  b. a waffle that produces 870. kJ when burned in a calorimeter

3.30 Calculate the kilocalories for each of the following:
  a. one cup of popcorn that produces 131 kJ when burned in a calorimeter
  b. a sample of butter that produces 23.4 kJ when burned in a calorimeter

3.31 Using the energy values for foods (see Table 3.7), determine each of the following (round off the answer for each food type to the tens place):
  a. the total kilojoules for one cup of orange juice that contains 26 g of carbohydrate, no fat, and 2 g of protein
  b. the grams of carbohydrate in one apple if the apple has no fat and no protein and provides 72 kcal of energy
  c. the kilocalories in one tablespoon of vegetable oil, which contains 14 g of fat and no carbohydrate or protein
  d. the grams of fat in one avocado that has 405 kcal, 13 g of carbohydrate, and 5 g of protein

3.32 Using the energy values for foods (see Table 3.7), determine each of the following (round off the answer for each food type to the tens place):
  a. the total kilojoules in two tablespoons of crunchy peanut butter that contains 6 g of carbohydrate, 16 g of fat, and 7 g of protein
  b. the grams of protein in one cup of soup that has 110 kcal with 9 g of carbohydrate and 6 g of fat
  c. the kilocalories in one can of cola if it has 40. g of carbohydrate and no fat or protein
  d. the total kilocalories for a diet that consists of 68 g of carbohydrate, 9 g of fat, and 150 g of protein

Clinical Applications

3.33 For dinner, Charles had one cup of clam chowder, which contains 16 g of carbohydrate, 12 g of fat, and 9 g of protein. How much energy, in kilocalories and kilojoules, is in the clam chowder? (Round off the answer for each food type to the tens place.)

3.34 For lunch, Charles consumed 3 oz of skinless chicken, 3 oz of broccoli, 1 medium apple, and 1 cup of nonfat milk (see Table 3.8). How many kilocalories did Charles obtain from the lunch?

3.35 A patient receives 3.2 L of intravenous (IV) glucose solution. If 100. mL of the solution contains 5.0 g of glucose (carbohydrate), how many kilocalories did the patient obtain from the glucose solution?

3.36 A high-protein diet contains 70.0 g of carbohydrate, 5.0 g of fat, and 150 g of protein. How much energy, in kilocalories and kilojoules, does this diet provide? (Round off the answer for each food type to the tens place.)

3.6 Specific Heat

LEARNING GOAL Use specific heat to calculate heat loss or gain.

Every substance has its own characteristic ability to absorb heat. When you bake a potato, you place it in a hot oven. If you are cooking pasta, you add the pasta to boiling water. You already know that adding heat to water increases its temperature until it boils. Certain substances must absorb more heat than others to reach a certain temperature.

TABLE 3.11 Specific Heats for Some Substances

Substance	cal/g °C	J/g °C
Elements		
Aluminum, Al(*s*)	0.214	0.897
Copper, Cu(*s*)	0.0920	0.385
Gold, Au(*s*)	0.0308	0.129
Iron, Fe(*s*)	0.108	0.452
Silver, Ag(*s*)	0.0562	0.235
Titanium, Ti(*s*)	0.125	0.523
Compounds		
Ammonia, $NH_3(g)$	0.488	2.04
Ethanol, $C_2H_6O(l)$	0.588	2.46
Sodium chloride, NaCl(*s*)	0.207	0.864
Water, $H_2O(l)$	1.00	4.184
Water, $H_2O(s)$	0.485	2.03

The energy requirements for different substances are described in terms of a physical property called *specific heat* (see **TABLE 3.11**). The **specific heat (*SH*)** for a substance is defined as the amount of heat needed to raise the temperature of exactly 1 g of a substance by exactly 1 °C. This temperature change is written as ΔT (*delta T*), where the delta symbol means "change in."

$$\text{Specific heat } (SH) = \frac{\text{heat}}{\text{mass } \Delta T} = \frac{\text{cal (or J)}}{\text{g °C}}$$

The specific heat for water is written using our definition of the calorie and joule.

$$SH \text{ for } H_2O(l) = \frac{1.00 \text{ cal}}{\text{g °C}} = \frac{4.184 \text{ J}}{\text{g °C}}$$

TEST

Try Practice Problems 3.37 and 3.38

The specific heat of water is about five times larger than the specific heat of aluminum. Aluminum has a specific heat that is about twice that of copper. Adding the same amount of heat (1.00 cal or 4.184 J) will raise the temperature of 1 g of aluminum by about 5 °C and 1 g of copper by about 10 °C. The low specific heats of aluminum and copper mean they transfer heat efficiently, which makes them useful in cookware.

The high specific heat of water has a major impact on the temperatures in a coastal city compared to an inland city. A large mass of water near a coastal city can absorb or release five times the energy absorbed or released by the same mass of rock near an inland city. This means that in the summer a body of water absorbs large quantities of heat, which cools a coastal city, and then in the winter that same body of water releases large quantities of heat, which provides warmer temperatures. A similar effect happens with our bodies, which contain 70% water by mass. Water in the body absorbs or releases large quantities of heat to maintain body temperature.

The high specific heat of water keeps temperatures more moderate in summer and winter.

Calculations Using Specific Heat

When we know the specific heat of a substance, we can calculate the heat lost or gained by measuring the mass of the substance and the initial and final temperature. We can substitute these measurements into the specific heat expression that is rearranged to solve for heat, which we call the *heat equation.*

CORE CHEMISTRY SKILL

Using the Heat Equation

$$\text{Heat} = \text{mass} \times \text{temperature change} \times \text{specific heat}$$

$$\text{Heat} = m \times \Delta T \times SH$$

$$\text{cal} = \cancel{\text{g}} \times \cancel{°\text{C}} \times \frac{\text{cal}}{\cancel{\text{g}}\,\cancel{°\text{C}}}$$

$$\text{J} = \cancel{\text{g}} \times \cancel{°\text{C}} \times \frac{\text{J}}{\cancel{\text{g}}\,\cancel{°\text{C}}}$$

SAMPLE PROBLEM 3.7 Using Specific Heat

TRY IT FIRST

During surgery or when a patient has suffered a cardiac arrest or stroke, lowering the body temperature will reduce the amount of oxygen needed by the body. Some methods used to lower body temperature include cooled saline solution, cool water blankets, or cooling caps worn on the head. How many kilojoules are lost when the body temperature of a surgery patient with a blood volume of 5500 mL is cooled from 38.5 °C to 33.2 °C? (Assume that the specific heat and density of blood are the same as for water.)

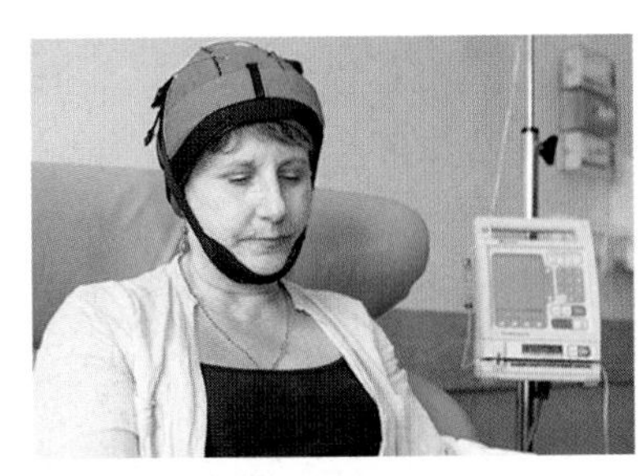

A cooling cap lowers the body temperature to reduce the oxygen required by the tissues.

SOLUTION GUIDE

STEP 1 State the given and needed quantities.

	Given	Need	Connect
ANALYZE THE PROBLEM	5500 mL of blood = 5500 g of blood, cooled from 38.5 °C to 33.2 °C	kilojoules removed	heat equation, specific heat of water

STEP 2 Calculate the temperature change (ΔT).

$$\Delta T = 38.5\ °C - 33.2\ °C = 5.3\ °C$$

STEP 3 Write the heat equation and needed conversion factors.

$$\text{Heat} = m \times \Delta T \times SH$$

$$SH_{\text{water}} = \frac{4.184\ \text{J}}{\text{g}\ °C} \qquad \frac{4.184\ \text{J}}{\text{g}\ °C} \text{ and } \frac{\text{g}\ °C}{4.184\ \text{J}}$$

$$1\ \text{kJ} = 1000\ \text{J} \qquad \frac{1000\ \text{J}}{1\ \text{kJ}} \text{ and } \frac{1\ \text{kJ}}{1000\ \text{J}}$$

STEP 4 Substitute in the given values and calculate the heat, making sure units cancel.

$$\text{Heat} = \underset{\text{Two SFs}}{5500\ \cancel{\text{g}}} \times \underset{\text{Two SFs}}{5.3\ \cancel{°C}} \times \overset{\text{Four SFs}}{\underset{\text{Exact}}{\frac{4.184\ \cancel{\text{J}}}{\cancel{\text{g}}\ \cancel{°C}}}} \times \overset{\text{Exact}}{\underset{\text{Exact}}{\frac{1\ \text{kJ}}{1000\ \cancel{\text{J}}}}} = \underset{\text{Two SFs}}{120\ \text{kJ}}$$

STUDY CHECK 3.7

Some cooking pans have a layer of copper on the bottom. How many kilojoules are needed to raise the temperature of 125 g of copper from 22 °C to 325 °C (see Table 3.11)?

The copper on a pan conducts heat rapidly to the food in the pan.

ANSWER

14.6 kJ

TEST

Try Practice Problems 3.39 to 3.42

PRACTICE PROBLEMS

3.6 Specific Heat

LEARNING GOAL Use specific heat to calculate heat loss or gain.

3.37 If the same amount of heat is supplied to samples of 10.0 g each of aluminum, iron, and copper all at 15.0 °C, which sample would reach the highest temperature (see Table 3.11)?

3.38 Substances A and B are the same mass and at the same initial temperature. When they are heated, the final temperature of A is 55 °C higher than the temperature of B. What does this tell you about the specific heats of A and B?

3.39 Use the heat equation to calculate the energy for each of the following (see Table 3.11):
a. calories to heat 8.5 g of water from 15 °C to 36 °C
b. joules lost when 25 g of water cools from 86 °C to 61 °C
c. kilocalories to heat 150 g of water from 15 °C to 77 °C
d. kilojoules to heat 175 g of copper from 28 °C to 188 °C

3.40 Use the heat equation to calculate the energy for each of the following (see Table 3.11):
a. calories lost when 85 g of water cools from 45 °C to 25 °C
b. joules to heat 75 g of water from 22 °C to 66 °C
c. kilocalories to heat 5.0 kg of water from 22 °C to 28 °C
d. kilojoules to heat 224 g of gold from 18 °C to 185 °C

3.41 Use the heat equation to calculate the energy, in joules and calories, for each of the following (see Table 3.11):
a. to heat 25.0 g of water from 12.5 °C to 25.7 °C
b. to heat 38.0 g of copper from 122 °C to 246 °C
c. lost when 15.0 g of ethanol, C_2H_6O, cools from 60.5 °C to −42.0 °C
d. lost when 125 g of iron cools from 118 °C to 55 °C

3.42 Use the heat equation to calculate the energy, in joules and calories, for each of the following (see Table 3.11):
a. to heat 5.25 g of water from 5.5 °C to 64.8 °C
b. lost when 75.0 g of water cools from 86.4 °C to 2.1 °C
c. to heat 10.0 g of silver from 112 °C to 275 °C
d. lost when 18.0 g of gold cools from 224 °C to 118 °C

REVIEW
Interpreting Graphs (1.4)

3.7 Changes of State

LEARNING GOAL Describe the changes of state between solids, liquids, and gases; calculate the energy released or absorbed.

Matter undergoes a **change of state** when it is converted from one state to another state (see **FIGURE 3.6**).

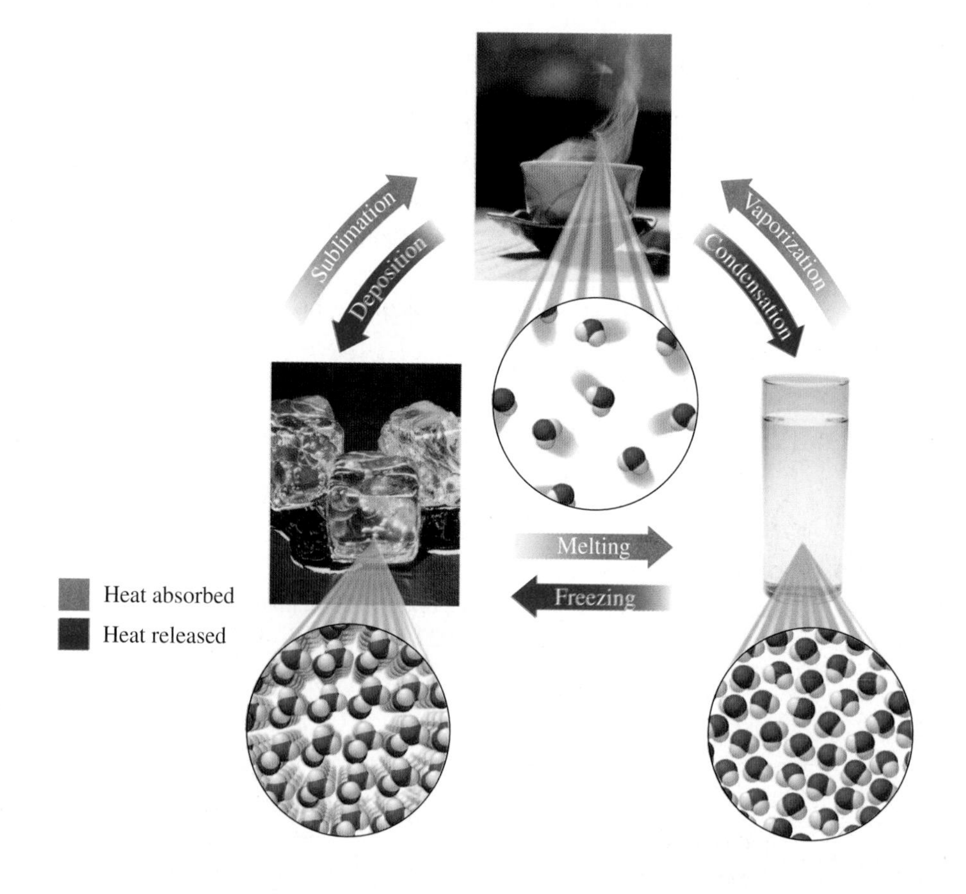

FIGURE 3.6 Changes of state include melting and freezing, vaporization and condensation, sublimation and deposition.

Q Is heat added or released when liquid water freezes?

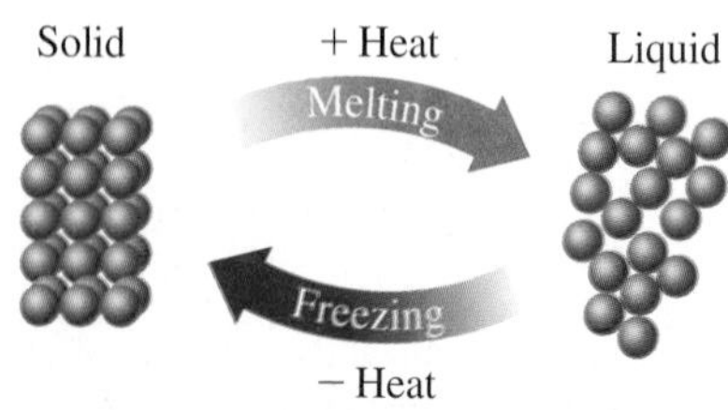

Melting and freezing are reversible processes.

Melting and Freezing

When heat is added to a solid, the particles move faster. At a temperature called the **melting point (mp)**, the particles of a solid gain sufficient energy to overcome the attractive forces that hold them together. The particles in the solid separate and move about in random patterns. The substance is **melting**, changing from a solid to a liquid.

If the temperature of a liquid is lowered, the reverse process takes place. Kinetic energy is lost, the particles slow down, and attractive forces pull the particles close together. The substance is **freezing**. A liquid changes to a solid at the **freezing point (fp)**, which is the same temperature as its melting point. Every substance has its own freezing (melting) point: Solid water (ice) melts at 0 °C when heat is added, and liquid water freezes at 0 °C when heat is removed.

ENGAGE

Why is the freezing point of water the same as its melting point?

Heat of Fusion

During melting, the **heat of fusion** is the energy that must be added to convert exactly 1 g of solid to liquid at the melting point. For example, 80. cal (334 J) of heat is needed to melt exactly 1 g of ice at its melting point (0 °C).

$$H_2O(s) + 80.\ \text{cal/g (or 334 J/g)} \longrightarrow H_2O(l)$$

Heat of Fusion for Water

$$\frac{80.\ \text{cal}}{1\ \text{g}\ H_2O} \text{ and } \frac{1\ \text{g}\ H_2O}{80.\ \text{cal}} \qquad \frac{334\ \text{J}}{1\ \text{g}\ H_2O} \text{ and } \frac{1\ \text{g}\ H_2O}{334\ \text{J}}$$

The heat of fusion (80. cal/g or 334 J/g) is also the quantity of heat that must be removed to freeze exactly 1 g of water at its freezing point (0 °C).

$$H_2O(l) \longrightarrow H_2O(s) + 80.\ \text{cal/g (or 334 J/g)}$$

The heat of fusion can be used as a conversion factor in calculations. For example, to determine the heat needed to melt a sample of ice, the mass of ice, in grams, is multiplied by the heat of fusion. Because the temperature remains constant as long as the ice is melting, there is no temperature change given in the calculation, as shown in Sample Problem 3.8.

Calculating Heat to Melt (or Freeze) Water

CORE CHEMISTRY SKILL

Calculating Heat for Change of State

$$\text{Heat} = \text{mass} \times \text{heat of fusion}$$

$$\text{cal} = \cancel{\text{g}} \times \frac{80.\ \text{cal}}{\cancel{\text{g}}} \qquad \text{J} = \cancel{\text{g}} \times \frac{334\ \text{J}}{\cancel{\text{g}}}$$

SAMPLE PROBLEM 3.8 Using a Heat Conversion Factor

TRY IT FIRST

Ice bags are used by sports trainers to treat muscle injuries. If 260. g of ice is placed in an ice bag, how much heat, in joules, will be released when all the ice melts at 0 °C?

An ice bag is used to treat a sports injury.

SOLUTION GUIDE

STEP 1 State the given and needed quantities.

	Given	Need	Connect
ANALYZE THE PROBLEM	260. g of ice at 0 °C	joules released when ice melts at 0 °C	heat of fusion

STEP 2 Write a plan to convert the given quantity to the needed quantity.

grams of ice → Heat of fusion → joules

STEP 3 Write the heat conversion factor and any metric factor.

$$1 \text{ g of } H_2O\ (s \rightarrow l) = 334 \text{ J}$$
$$\frac{334 \text{ J}}{1 \text{ g } H_2O} \text{ and } \frac{1 \text{ g } H_2O}{334 \text{ J}}$$

STEP 4 Set up the problem and calculate the needed quantity.

$$\underset{\text{Three SFs}}{260.\ \text{g } H_2O} \times \overset{\text{Three SFs}}{\underset{\text{Exact}}{\frac{334 \text{ J}}{1 \text{ g } H_2O}}} = \underset{\text{Three SFs}}{86\,800 \text{ J } (8.68 \times 10^4 \text{ J})}$$

STUDY CHECK 3.8

In a freezer, 125 g of water at 0 °C is placed in an ice cube tray. How much heat, in kilojoules, must be removed to form ice cubes at 0 °C?

ANSWER

41.8 kJ

TEST

Try Practice Problems 3.45 and 3.46

Sublimation and Deposition

In a process called **sublimation**, the particles on the surface of a solid change directly to a gas without going through the liquid state. In the reverse process called **deposition**, gas particles change directly to a solid. For example, dry ice, which is solid carbon dioxide, sublimes at −78 °C. It is called "dry" because it does not form a liquid as it warms. In extremely cold areas, snow does not melt but sublimes directly to water vapor.

When frozen foods are left in the freezer for a long time, so much water sublimes that foods, especially meats, become dry and shrunken, a condition called *freezer burn*. Deposition occurs in a freezer when water vapor forms ice crystals on the surface of freezer bags and frozen food.

Freeze-dried foods prepared by sublimation are convenient for long-term storage and for camping and hiking. A frozen food is placed in a vacuum chamber where it dries as the ice sublimes. The dried food retains all of its nutritional value and needs only water to be edible. Freeze-dried foods do not need refrigeration because bacteria cannot grow without moisture.

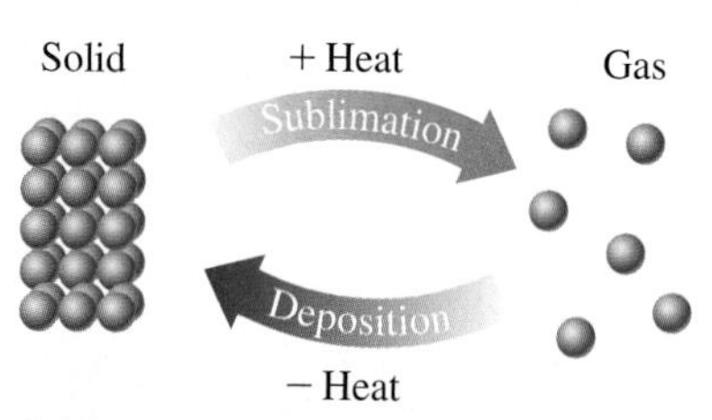

Sublimation and deposition are reversible processes.

TEST

Try Practice Problems 3.43 and 3.44

Evaporation, Boiling, and Condensation

Water in a mud puddle disappears, unwrapped food dries out, and clothes hung on a line dry. **Evaporation** is taking place as water particles with sufficient energy escape from the liquid surface and enter the gas phase (see **FIGURE 3.7A**). The loss of the "hot" water particles removes heat, which cools the liquid water. As heat is added, more and more water particles evaporate. At the **boiling point (bp)**, the particles within a liquid have enough energy to overcome their attractive forces and become a gas. We observe the **boiling** of a liquid as gas bubbles form, rise to the surface, and escape (see **FIGURE 3.7B**).

When heat is removed, a reverse process takes place. In **condensation**, water vapor is converted back to liquid as the water particles lose kinetic energy and slow down. Condensation occurs at the same temperature as boiling. You may have noticed that condensation occurs when you take a hot shower and the water vapor forms water droplets on a mirror.

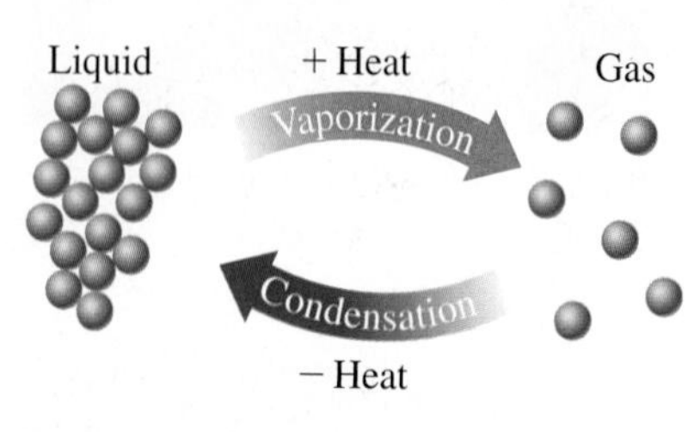

Vaporization and condensation are reversible processes.

TEST

Try Practice Problems 3.47 and 3.48

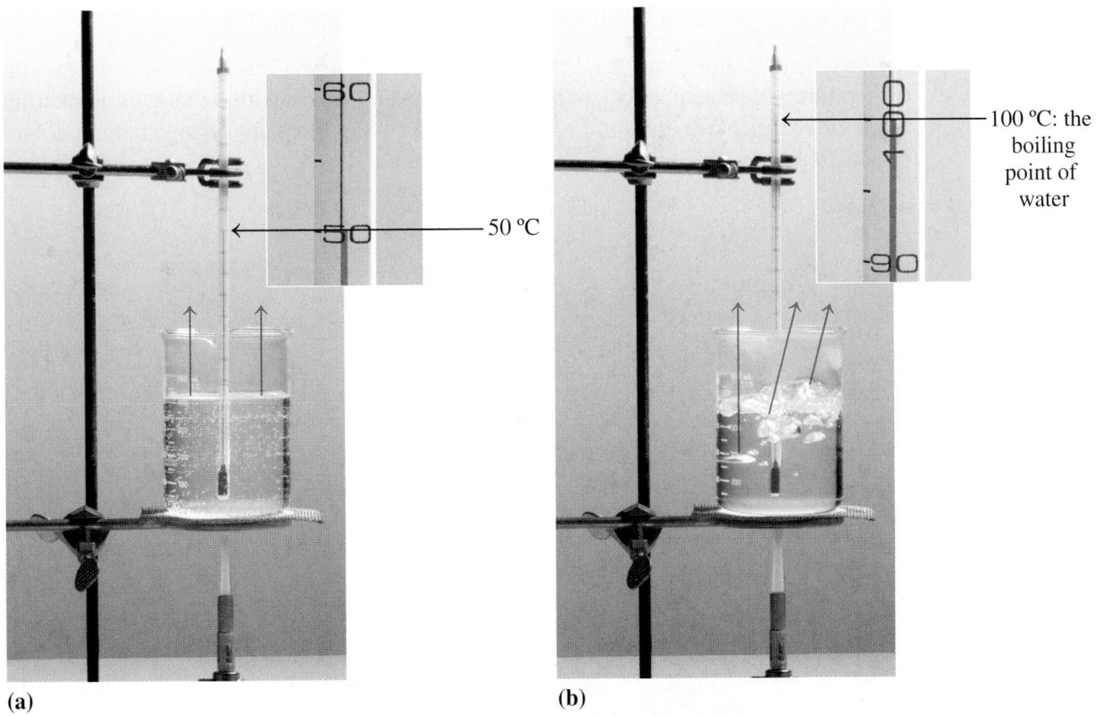

FIGURE 3.7 **(a)** Evaporation occurs at the surface of a liquid. **(b)** Boiling occurs as bubbles of gas form throughout the liquid.

Q Why does water evaporate faster at 85 °C than at 15 °C?

Heat of Vaporization and Condensation

The **heat of vaporization** is the energy that must be added to convert exactly 1 g of liquid to gas at its boiling point. For water, 540 cal (2260 J) is needed to convert 1 g of water to vapor at 100 °C.

$$H_2O(l) + 540 \text{ cal/g (or 2260 J/g)} \longrightarrow H_2O(g)$$

This same amount of heat is released when 1 g of water vapor (gas) changes to liquid at 100 °C.

$$H_2O(g) \longrightarrow H_2O(l) + 540 \text{ cal/g (or 2260 J/g)}$$

Therefore, 540 cal/g or 2260 J/g is also the *heat of condensation* of water.

Heat of Vaporization (Condensation) for Water

$$\frac{540 \text{ cal}}{1 \text{ g } H_2O} \text{ and } \frac{1 \text{ g } H_2O}{540 \text{ cal}} \qquad \frac{2260 \text{ J}}{1 \text{ g } H_2O} \text{ and } \frac{1 \text{ g } H_2O}{2260 \text{ J}}$$

To determine the heat needed to boil a sample of water, the mass, in grams, is multiplied by the heat of vaporization. Because the temperature remains constant as long as the water is boiling, there is no temperature change in the calculation.

Calculating Heat to Vaporize (or Condense) Water

$$\text{Heat} = \text{mass} \times \text{heat of vaporization}$$

$$\text{cal} = \cancel{g} \times \frac{540 \text{ cal}}{\cancel{g}} \qquad \text{J} = \cancel{g} \times \frac{2260 \text{ J}}{\cancel{g}}$$

TEST

Try Practice Problems 3.49 and 3.50

Heating and Cooling Curves

All the changes of state during the heating or cooling of a substance can be illustrated visually. On a **heating curve** or **cooling curve**, the temperature is shown on the vertical axis and the loss or gain of heat is shown on the horizontal axis.

Steps on a Heating Curve

The first diagonal line indicates the warming of a solid as heat is added. When the melting temperature is reached, a horizontal line, or plateau, indicates that the solid is melting. As melting takes place, the solid is changing to liquid without any change in temperature (see **FIGURE 3.8**).

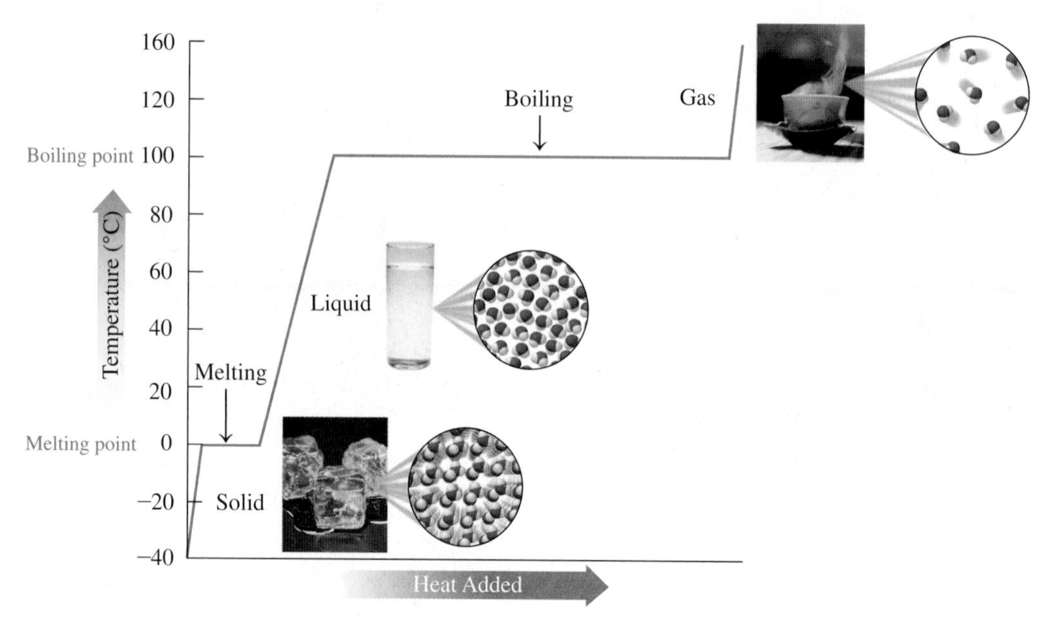

FIGURE 3.8 A heating curve diagrams the temperature increases and changes of state as heat is added.

Q What does the plateau at 100 °C represent on the heating curve for water?

TEST

Try Practice Problems 3.51 and 3.52

When all the particles are in the liquid state, adding more heat will increase the temperature of the liquid, which is shown as a diagonal line. Once the liquid reaches its boiling point, a horizontal line indicates that the liquid changes to gas at constant temperature. Because the heat of vaporization is greater than the heat of fusion, the horizontal line at the boiling point is longer than the horizontal line at the melting point. When all the liquid becomes gas, adding more heat increases the temperature of the gas.

Steps on a Cooling Curve

A cooling curve is a diagram in which the temperature decreases as heat is removed. Initially, a diagonal line is drawn to show that heat is removed from a gas until it begins to condense. At the condensation point, a horizontal line indicates a change of state as gas condenses to form a liquid. When all the gas has changed to liquid, further cooling lowers the temperature. The decrease in temperature is shown as a diagonal line from the condensation point to the freezing point, where another horizontal line indicates that liquid is changing to solid. When all the substance is frozen, the removal of more heat decreases the temperature of the solid below its freezing point, which is shown as a diagonal line.

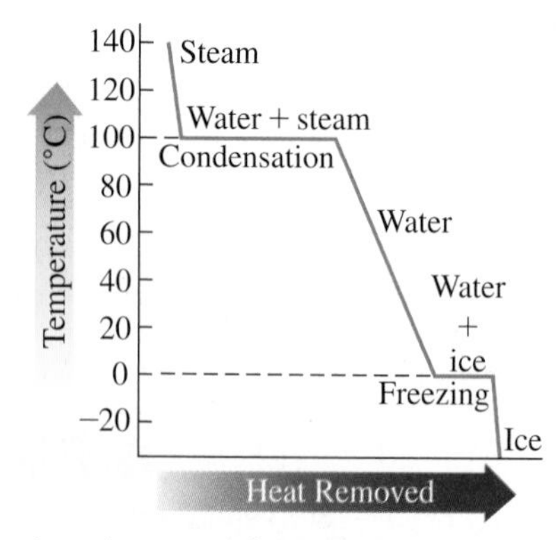

A cooling curve for water illustrates the change in temperature and changes of state as heat is removed.

Combining Energy Calculations

Up to now, we have calculated one step in a heating or cooling curve. However, some problems require a combination of steps that include a temperature change as well as a change of state. The heat is calculated for each step separately and then added together to find the total energy, as seen in Sample Problem 3.9.

SAMPLE PROBLEM 3.9 Combining Heat Calculations

TRY IT FIRST

Charles has increased his activity by doing more exercise. After a session of using weights, he has a sore arm. An ice bag is filled with 125 g of ice at 0.0 °C. How much heat, in kilojoules, is absorbed to melt the ice, and raise the temperature of the water to body temperature, 37.0 °C?

SOLUTION GUIDE

STEP 1 State the given and needed quantities.

	Given	Need	Connect
ANALYZE THE PROBLEM	125 g of ice at 0.0 °C	total kilojoules to melt ice at 0.0 °C and to raise temperature of water to 37.0 °C	combine heat from change of state (heat of fusion) and temperature change (specific heat of water)

STEP 2 Write a plan to convert the given quantity to the needed quantity.

Total heat = kilojoules needed to melt the ice at 0.0 °C and heat the water from 0.0 °C (freezing point) to 37.0 °C

STEP 3 Write the heat conversion factors and any metric factor.

$$1 \text{ g of } H_2O\ (s \longrightarrow l) = 334 \text{ J}$$

$$\frac{334 \text{ J}}{1 \text{ g } H_2O} \text{ and } \frac{1 \text{ g } H_2O}{334 \text{ J}}$$

$$SH_{\text{water}} = \frac{4.184 \text{ J}}{\text{g °C}}$$

$$\frac{4.184 \text{ J}}{\text{g °C}} \text{ and } \frac{\text{g °C}}{4.184 \text{ J}}$$

$$1 \text{ kJ} = 1000 \text{ J}$$

$$\frac{1000 \text{ J}}{1 \text{ kJ}} \text{ and } \frac{1 \text{ kJ}}{1000 \text{ J}}$$

STEP 4 Set up the problem and calculate the needed quantity.

$\Delta T = 37.0 \text{ °C} - 0.0 \text{ °C} = 37.0 \text{ °C}$

Heat needed to change ice (solid) to water (liquid) at 0.0 °C:

$$\underset{\text{Three SFs}}{125 \text{ g ice}} \times \underset{\text{Exact}}{\overset{\text{Three SFs}}{\frac{334 \text{ J}}{1 \text{ g ice}}}} \times \underset{\text{Exact}}{\overset{\text{Exact}}{\frac{1 \text{ kJ}}{1000 \text{ J}}}} = \underset{\text{Three SFs}}{41.8 \text{ kJ}}$$

Heat needed to warm water (liquid) from 0.0 °C to water (liquid) at 37.0 °C:

$$\underset{\text{Three SFs}}{125 \text{ g}} \times \underset{\text{Three SFs}}{37.0 \text{ °C}} \times \underset{\text{Exact}}{\overset{\text{Four SFs}}{\frac{4.184 \text{ J}}{\text{g °C}}}} \times \underset{\text{Exact}}{\overset{\text{Exact}}{\frac{1 \text{ kJ}}{1000 \text{ J}}}} = \underset{\text{Three SFs}}{19.4 \text{ kJ}}$$

Calculate the total heat:

Melting ice at 0.0 °C	41.8 kJ
Heating water (0.0 °C to 37.0 °C)	19.4 kJ
Total heat needed	61.2 kJ

STUDY CHECK 3.9

How many kilojoules are released when 75.0 g of steam at 100 °C condenses, cools to 0 °C, and freezes at 0 °C? (*Hint*: The solution will require three energy calculations.)

ANSWER

226 kJ

TEST

Try Practice Problems 3.53 to 3.56

CHEMISTRY LINK TO HEALTH

Steam Burns

Hot water at 100 °C will cause burns and damage to the skin. However, getting steam on the skin is even more dangerous. If 25 g of hot water at 100 °C falls on a person's skin, the temperature of the water will drop to body temperature, 37 °C. The heat released during cooling can cause severe burns. The amount of heat can be calculated from the mass, the temperature change, 100 °C − 37 °C = 63 °C, and the specific heat of water, 4.184 J/g °C.

$$25\ \cancel{g} \times 63\ \cancel{°C} \times \frac{4.184\ \text{J}}{\cancel{g}\ \cancel{°C}} = 6600\ \text{J of heat released when water cools}$$

The condensation of the same quantity of steam to liquid at 100 °C releases much more heat. The heat released when steam condenses can be calculated using the heat of vaporization, which is 2260 J/g for water at 100 °C.

$$25\ \cancel{g} \times \frac{2260\ \text{J}}{1\ \cancel{g}} = 57\,000\ \text{J released when water (gas) condenses to water (liquid) at 100 °C}$$

The total heat released is calculated by combining the heat from the condensation at 100 °C and the heat from cooling of the steam from 100 °C to 37 °C (body temperature). We can see that most of the heat is from the condensation of steam.

Condensation (100 °C)	= 57 000 J
Cooling (100 °C to 37 °C)	= 6 600 J
Heat released	= 64 000 J (rounded off)

The amount of heat released from steam is almost ten times greater than the heat from the same amount of hot water. This large amount of heat released on the skin is what causes damage from steam burns.

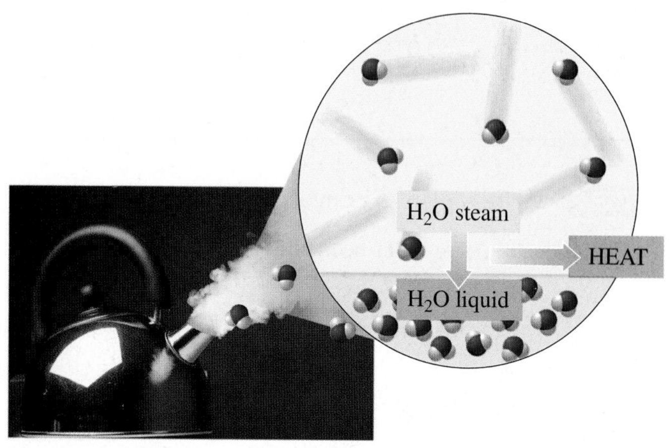

When 1 g of steam condenses, 2260 J or 540 cal is released.

PRACTICE PROBLEMS

3.7 Changes of State

LEARNING GOAL Describe the changes of state between solids, liquids, and gases; calculate the energy released or absorbed.

3.43 Identify each of the following changes of state as melting, freezing, sublimation, or deposition:
- **a.** The solid structure of a substance breaks down as liquid forms.
- **b.** Coffee is freeze-dried.
- **c.** Water on the street turns to ice during a cold wintry night.
- **d.** Ice crystals form on a package of frozen corn.

3.44 Identify each of the following changes of state as melting, freezing, sublimation, or deposition:
- **a.** Dry ice in an ice-cream cart disappears.
- **b.** Snow on the ground turns to liquid water.
- **c.** Heat is removed from 125 g of liquid water at 0 °C.
- **d.** Frost (ice) forms on the walls of a freezer unit of a refrigerator.

3.45 Calculate the heat change at 0 °C for each of the following, and indicate whether heat was absorbed/released:
- **a.** calories to melt 65 g of ice
- **b.** joules to melt 17.0 g of ice
- **c.** kilocalories to freeze 225 g of water
- **d.** kilojoules to freeze 50.0 g of water

3.46 Calculate the heat change at 0 °C for each of the following, and indicate whether heat was absorbed/released:
- **a.** calories to freeze 35 g of water
- **b.** joules to freeze 275 g of water
- **c.** kilocalories to melt 140 g of ice
- **d.** kilojoules to melt 5.00 g of ice

3.47 Identify each of the following changes of state as evaporation, boiling, or condensation:
- **a.** The water vapor in the clouds changes to rain.
- **b.** Wet clothes dry on a clothesline.
- **c.** Lava flows into the ocean and steam forms.
- **d.** After a hot shower, your bathroom mirror is covered with water.

3.48 Identify each of the following changes of state as evaporation, boiling, or condensation:
- **a.** At 100 °C, the water in a pan changes to steam.
- **b.** On a cool morning, the windows in your car fog up.
- **c.** A shallow pond dries up in the summer.
- **d.** Your teakettle whistles when the water is ready for tea.

3.49 Calculate the heat change at 100 °C for each of the following, and indicate whether heat was absorbed/released:
- **a.** calories to vaporize 10.0 g of water
- **b.** joules to vaporize 5.00 g of water
- **c.** kilocalories to condense 8.0 kg of steam
- **d.** kilojoules to condense 175 g of steam

3.50 Calculate the heat change at 100 °C for each of the following, and indicate whether heat was absorbed/released:
a. calories to condense 10.0 g of steam
b. joules to condense 7.60 g of steam
c. kilocalories to vaporize 44 g of water
d. kilojoules to vaporize 5.00 kg of water

3.51 Draw a heating curve for a sample of ice that is heated from −20 °C to 150 °C. Indicate the segment of the graph that corresponds to each of the following:
a. solid **b.** melting **c.** liquid
d. boiling **e.** gas

3.52 Draw a cooling curve for a sample of steam that cools from 110 °C to −10 °C. Indicate the segment of the graph that corresponds to each of the following:
a. solid **b.** freezing **c.** liquid
d. condensation **e.** gas

3.53 Using the values for the heat of fusion, specific heat of water, and/or heat of vaporization, calculate the amount of heat energy in each of the following:
a. joules needed to melt 50.0 g of ice at 0 °C and to warm the liquid to 65.0 °C
b. kilocalories released when 15.0 g of steam condenses at 100 °C and the liquid cools to 0 °C
c. kilojoules needed to melt 24.0 g of ice at 0 °C, warm the liquid to 100 °C, and change it to steam at 100 °C

3.54 Using the values for the heat of fusion, specific heat of water, and/or heat of vaporization, calculate the amount of heat energy in each of the following:
a. joules released when 125 g of steam at 100 °C condenses and cools to liquid at 15.0 °C
b. kilocalories needed to melt a 525-g ice sculpture at 0 °C and to warm the liquid to 15.0 °C
c. kilojoules released when 85.0 g of steam condenses at 100 °C, cools, and freezes at 0 °C

Clinical Applications

3.55 A patient arrives in the emergency room with a burn caused by steam. Calculate the heat, in kilocalories, that is released when 18.0 g of steam at 100. °C hits the skin, condenses, and cools to body temperature of 37.0 °C.

3.56 A sports trainer applies an ice bag to the back of an injured athlete. Calculate the heat, in kilocalories, that is absorbed if 145 g of ice at 0.0 °C is placed in an ice bag, melts, and warms to body temperature of 37.0 °C.

CLINICAL UPDATE A Diet and Exercise Program

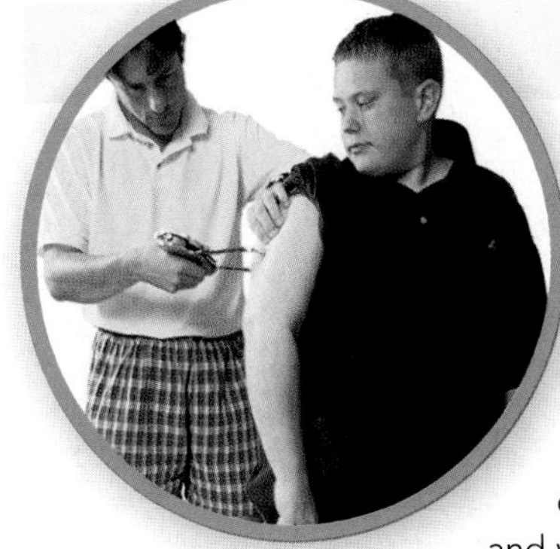

After Charles has been on his diet and exercise plan for a month, he and his mother meet again with the dietitian. Daniel looks at the diary of food intake and exercise, and weighs Charles to evaluate how well the diet is working. The following is what Charles ate in one day:

Breakfast
1 banana, 1 cup of nonfat milk, 1 egg

Lunch
1 cup of carrots, 3 oz of ground beef, 1 apple, 1 cup of nonfat milk

Dinner
6 oz of skinless chicken, 1 baked potato, 3 oz of broccoli, 1 cup of nonfat milk

Clinical Applications

3.57 Using energy values from Table 3.8, determine each of the following:
a. the total kilocalories for each meal
b. the total kilocalories for one day
c. If Charles consumes 1800 kcal per day, he will maintain his weight. Would he lose weight on his new diet?
d. If expending 3500 kcal is equal to a loss of 1.0 lb, how many days will it take Charles to lose 5.0 lb?

3.58 **a.** During one week, Charles swam for a total of 2.5 h and walked for a total of 8.0 h. If Charles expends 340 kcal/h swimming and 160 kcal/h walking, how many total kilocalories did he expend for one week?
b. For the amount of exercise that Charles did for one week in part **a**, if expending 3500 kcal is equal to a loss of 1.0 lb, how many pounds did he lose?
c. How many hours would Charles have to walk to lose 1.0 lb?
d. How many hours would Charles have to swim to lose 1.0 lb?

CONCEPT MAP

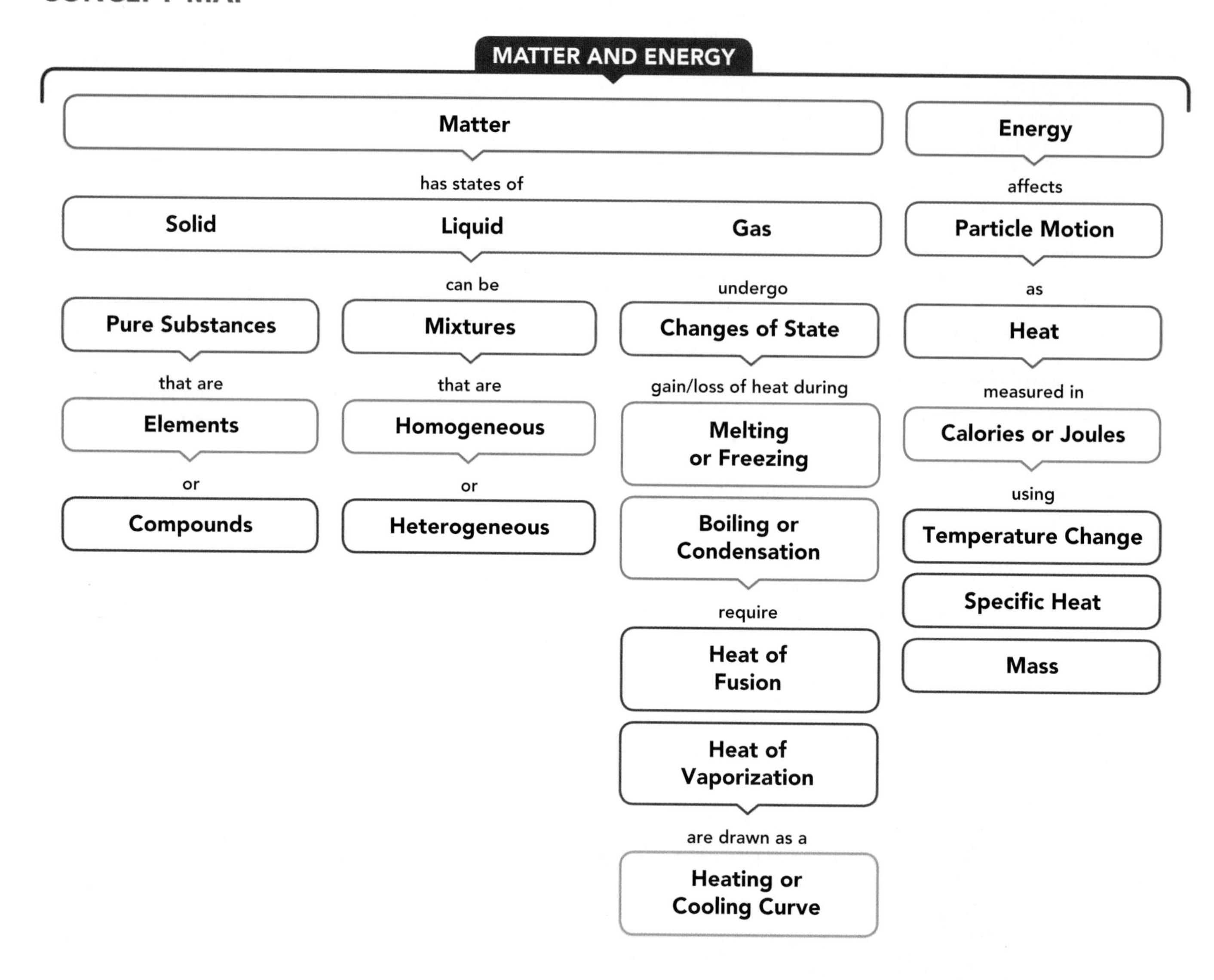

CHAPTER REVIEW

3.1 Classification of Matter

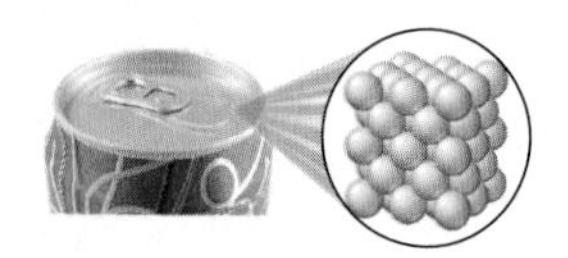

LEARNING GOAL Classify examples of matter as pure substances or mixtures.

- Matter is anything that has mass and occupies space.
- Matter is classified as pure substances or mixtures.
- Pure substances, which are elements or compounds, have fixed compositions, and mixtures have variable compositions.
- The substances in mixtures can be separated using physical methods.

3.2 States and Properties of Matter

LEARNING GOAL Identify the states and the physical and chemical properties of matter.

- The three states of matter are solid, liquid, and gas.
- A physical property is a characteristic of a substance in that it can be observed or measured without affecting the identity of the substance.
- A physical change occurs when physical properties change, but not the composition of the substance.

- A chemical property indicates the ability of a substance to change into another substance.
- A chemical change occurs when one or more substances react to form a substance with new physical and chemical properties.

3.3 Temperature

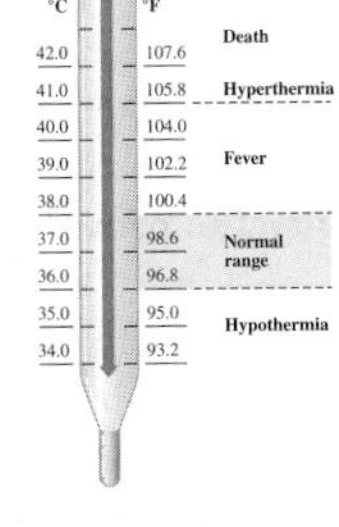

LEARNING GOAL Given a temperature, calculate the corresponding temperature on another scale.

- In science, temperature is measured in degrees Celsius (°C) or kelvins (K).
- On the Celsius scale, there are 100 units between the freezing point (0 °C) and the boiling point (100 °C) of water.
- On the Fahrenheit scale, there are 180 units between the freezing point (32 °F) and the boiling point (212 °F) of water. A Fahrenheit temperature is related to its Celsius temperature by the equation $T_F = 1.8(T_C) + 32$.
- The SI unit, kelvin, is related to the Celsius temperature by the equation $T_K = T_C + 273$.

3.4 Energy

LEARNING GOAL Identify energy as potential or kinetic; convert between units of energy.

- Energy is the ability to do work.
- Potential energy is stored energy; kinetic energy is the energy of motion.
- Common units of energy are the calorie (cal), kilocalorie (kcal), joule (J), and kilojoule (kJ).
- One calorie is equal to 4.184 J.

3.5 Energy and Nutrition

Snack Crackers

Nutrition Facts
Serving Size 14 crackers (31g)
Servings Per Container About 7
Amount Per Serving
Calories 130 Calories from Fat 40
Kilojoules 500 kJ from Fat 150

LEARNING GOAL Use the energy values to calculate the kilocalories (kcal) or kilojoules (kJ) for a food.

- The nutritional Calorie is the same amount of energy as 1 kcal or 1000 calories.
- The energy of a food is the sum of kilocalories or kilojoules from carbohydrate, fat, and protein.

3.6 Specific Heat

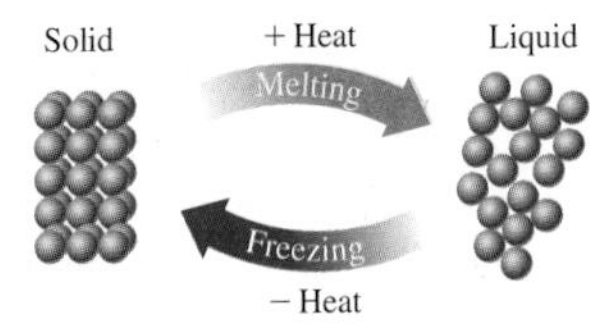

LEARNING GOAL Use specific heat to calculate heat loss or gain.

- Specific heat is the amount of energy required to raise the temperature of exactly 1 g of a substance by exactly 1 °C.
- The heat lost or gained by a substance is determined by multiplying its mass, the temperature change, and its specific heat.

3.7 Changes of State

Solid
+ Heat
Melting
Liquid
Freezing
− Heat

LEARNING GOAL Describe the changes of state between solids, liquids, and gases; calculate the energy released or absorbed.

- Melting occurs when the particles in a solid absorb enough energy to break apart and form a liquid.
- The amount of energy required to convert exactly 1 g of solid to liquid is called the heat of fusion.
- For water, 80. cal (334 J) is needed to melt 1 g of ice or must be removed to freeze 1 g of water.
- Evaporation occurs when particles in a liquid state absorb enough energy to break apart and form gaseous particles.
- Boiling is the vaporization of liquid at its boiling point. The heat of vaporization is the amount of heat needed to convert exactly 1 g of liquid to vapor.
- For water, 540 cal (2260 J) is needed to vaporize 1 g of water or must be removed to condense 1 g of steam.
- Sublimation is a process whereby a solid changes directly to a gas.
- A heating or cooling curve illustrates the changes in temperature and state as heat is added to or removed from a substance. Plateaus on the graph indicate changes of state.
- The total heat absorbed or removed from a substance undergoing temperature changes and changes of state is the sum of energy calculations for change(s) of state and change(s) in temperature.

KEY TERMS

boiling The formation of bubbles of gas throughout a liquid.

boiling point (bp) The temperature at which a liquid changes to gas (boils) and gas changes to liquid (condenses).

calorie (cal) The amount of heat energy that raises the temperature of exactly 1 g of water by exactly 1 °C.

change of state The transformation of one state of matter to another; for example, solid to liquid, liquid to solid, liquid to gas.

chemical change A change during which the original substance is converted into a new substance that has a different composition and new physical and chemical properties.

chemical properties The properties that indicate the ability of a substance to change into a new substance.

compound A pure substance consisting of two or more elements, with a definite composition, that can be broken down into simpler substances only by chemical methods.

condensation The change of state from a gas to a liquid.

cooling curve A diagram that illustrates temperature changes and changes of state for a substance as heat is removed.

deposition The change of a gas directly into a solid; the reverse of sublimation.

element A pure substance containing only one type of matter, which cannot be broken down by chemical methods.

energy The ability to do work.

energy value The kilocalories (or kilojoules) obtained per gram of the food types: carbohydrate, fat, and protein.

evaporation The formation of a gas (vapor) by the escape of high-energy particles from the surface of a liquid.
freezing The change of state from liquid to solid.
freezing point (fp) The temperature at which a liquid changes to a solid (freezes) and a solid changes to a liquid (melts).
gas A state of matter that does not have a definite shape or volume.
heat The energy associated with the motion of particles in a substance.
heat of fusion The energy required to melt exactly 1 g of a substance at its melting point. For water, 80. cal (334 J) is needed to melt 1 g of ice; 80. cal (334 J) is released when 1 g of water freezes.
heat of vaporization The energy required to vaporize exactly 1 g of a substance at its boiling point. For water, 540 cal (2260 J) is needed to vaporize 1 g of liquid; 1 g of steam gives off 540 cal (2260 J) when it condenses.
heating curve A diagram that illustrates the temperature changes and changes of state of a substance as it is heated.
joule (J) The SI unit of heat energy; 4.184 J = 1 cal.
kinetic energy The energy of moving particles.
liquid A state of matter that takes the shape of its container but has a definite volume.
matter The material that makes up a substance and has mass and occupies space.
melting The change of state from a solid to a liquid.
melting point (mp) The temperature at which a solid becomes a liquid (melts). It is the same temperature as the freezing point.
mixture The physical combination of two or more substances that does not change the identities of the mixed substances.
physical change A change in which the physical properties of a substance change but its identity stays the same.
physical properties The properties that can be observed or measured without affecting the identity of a substance.
potential energy A type of energy related to position or composition of a substance.
pure substance A type of matter that has a definite composition.
solid A state of matter that has its own shape and volume.
specific heat (*SH*) A quantity of heat that changes the temperature of exactly 1 g of a substance by exactly 1 °C.
states of matter Three forms of matter: solid, liquid, and gas.
sublimation The change of state in which a solid is transformed directly to a gas without forming a liquid first.

CORE CHEMISTRY SKILLS

The chapter Section containing each Core Chemistry Skill is shown in parentheses at the end of each heading.

Identifying Physical and Chemical Changes (3.2)

- Physical properties can be observed or measured without changing the identity of a substance.
- Chemical properties describe the ability of a substance to change into a new substance.
- When matter undergoes a physical change, its state or its appearance changes, but its composition remains the same.
- When a chemical change takes place, the original substance is converted into a new substance, which has different physical and chemical properties.

Example: Classify each of the following as a physical or chemical property:

a. Helium in a balloon is a gas.
b. Methane, in natural gas, burns.
c. Hydrogen sulfide smells like rotten eggs.

Answer: **a.** A gas is a state of matter, which makes it a physical property.
b. When methane burns, it changes to different substances with new properties, which is a chemical property.
c. The odor of hydrogen sulfide is a physical property.

Converting between Temperature Scales (3.3)

- The temperature equation $T_F = 1.8(T_C) + 32$ is used to convert from Celsius to Fahrenheit and can be rearranged to convert from Fahrenheit to Celsius.
- The temperature equation $T_K = T_C + 273$ is used to convert from Celsius to Kelvin and can be rearranged to convert from Kelvin to Celsius.

Example: Convert 75.0 °C to degrees Fahrenheit.

Answer: $T_F = 1.8(T_C) + 32$

$T_F = 1.8(75.0) + 32 = 135 + 32$

$= 167\ °F$

Example: Convert 355 K to degrees Celsius.

Answer: $T_K = T_C + 273$

To solve the equation for T_C, subtract 273 from both sides.

$T_K - 273 = T_C + \cancel{273} - \cancel{273}$

$T_C = T_K - 273$

$T_C = 355 - 273$

$= 82\ °C$

Using Energy Units (3.4)

- Equalities for energy units include 1 cal = 4.184 J, 1 kcal = 1000 cal, and 1 kJ = 1000 J.
- Each equality for energy units can be written as two conversion factors:

$$\frac{4.184\text{ J}}{1\text{ cal}} \text{ and } \frac{1\text{ cal}}{4.184\text{ J}} \quad \frac{1000\text{ cal}}{1\text{ kcal}} \text{ and } \frac{1\text{ kcal}}{1000\text{ cal}} \quad \frac{1000\text{ J}}{1\text{ kJ}} \text{ and } \frac{1\text{ kJ}}{1000\text{ J}}$$

- The energy unit conversion factors are used to cancel given units of energy and to obtain the needed unit of energy.

Example: Convert 45 000 J to kilocalories.

Answer: Using the conversion factors above, we start with the given 45 000 J and convert it to kilocalories.

$$45\,000\ \cancel{\text{J}} \times \frac{1\ \cancel{\text{cal}}}{4.184\ \cancel{\text{J}}} \times \frac{1\text{ kcal}}{1000\ \cancel{\text{cal}}} = 11\text{ kcal}$$

Using the Heat Equation (3.6)

- The quantity of heat absorbed or lost by a substance is calculated using the heat equation.

Heat = $m \times \Delta T \times SH$

- Heat, in calories, is obtained when the specific heat of a substance in cal/g °C is used.
- Heat, in joules, is obtained when the specific heat of a substance in J/g °C is used.
- To cancel, the unit grams is used for mass, and the unit °C is used for temperature change.

Example: How many joules are required to heat 5.25 g of titanium from 85.5 °C to 132.5 °C?

Answer: $m = 5.25$ g, $\Delta T = 132.5\ °C - 85.5\ °C = 47.0\ °C$
SH for titanium = 0.523 J/g °C

The known values are substituted into the heat equation making sure units cancel.

$$\text{Heat} = m \times \Delta T \times SH = 5.25\ \cancel{g} \times 47.0\ \cancel{°C} \times \frac{0.523\ \text{J}}{\cancel{g}\ \cancel{°C}}$$

$$= 129\ \text{J}$$

Calculating Heat for Change of State (3.7)

- At the melting/freezing point, the heat of fusion is absorbed/released to convert 1 g of a solid to a liquid or 1 g of liquid to a solid.
- For example, 334 J of heat is needed to melt (freeze) exactly 1 g of ice at its melting (freezing) point (0 °C).
- At the boiling/condensation point, the heat of vaporization is absorbed/released to convert exactly 1 g of liquid to gas or 1 g of gas to liquid.
- For example, 2260 J of heat is needed to boil (condense) exactly 1 g of water/steam at its boiling (condensation) point, 100 °C.

Example: What is the quantity of heat, in kilojoules, released when 45.8 g of steam (water) condenses at its boiling (condensation) point?

Answer: $45.8\ \cancel{\text{g steam}} \times \frac{2260\ \cancel{\text{J}}}{1\ \cancel{\text{g steam}}} \times \frac{1\ \text{kJ}}{1000\ \cancel{\text{J}}} = 104\ \text{kJ}$

UNDERSTANDING THE CONCEPTS

The chapter Sections to review are shown in parentheses at the end of each problem.

3.59 Identify each of the following as an element, a compound, or a mixture. Explain your choice. (3.1)

a.

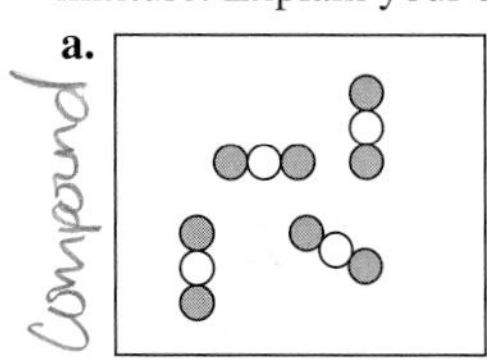

b.

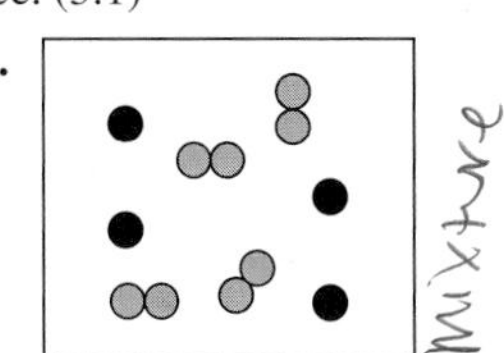

c.

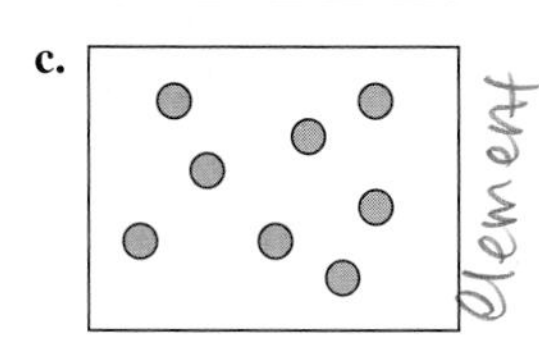

3.60 Identify each of the following as a homogeneous or heterogeneous mixture. Explain your choice. (3.1)

a.

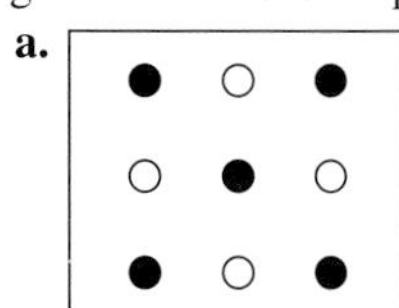

b.

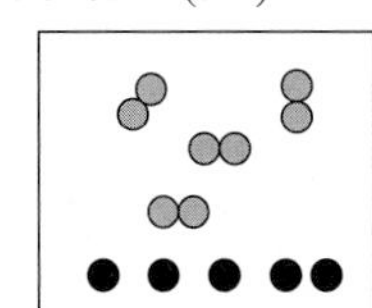

c.

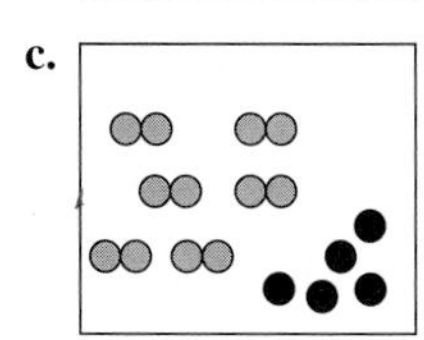

3.61 Classify each of the following as a homogeneous or heterogeneous mixture: (3.1)

a. lemon-flavored water
b. stuffed mushrooms
c. eye drops

3.62 Classify each of the following as a homogeneous or heterogeneous mixture: (3.1)

a. ketchup **b.** hard-boiled egg **c.** tortilla soup

3.63 State the temperature on the Celsius thermometer and convert to Fahrenheit. (3.3)

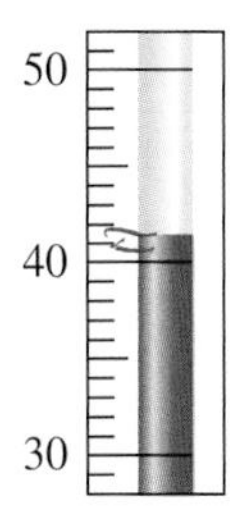

3.64 State the temperature on the Celsius thermometer and convert to Fahrenheit. (3.3)

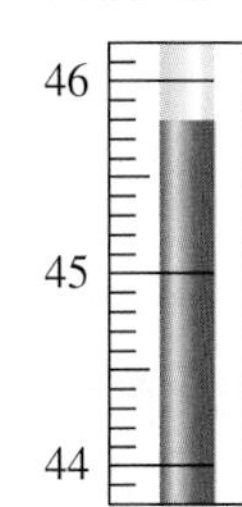

3.65 Compost can be made at home from grass clippings, kitchen scraps, and dry leaves. As microbes break down organic matter, heat is generated and the compost can reach a temperature of 155 °F, which kills most pathogens. What is this temperature in degrees Celsius? In kelvins? (3.3)

Compost produced from decayed plant material is used to enrich the soil.

3.66 After a week, biochemical reactions in compost slow, and the temperature drops to 45 °C. The dark brown organic-rich mixture is ready for use in the garden. What is this temperature in degrees Fahrenheit? In kelvins? (3.3)

3.67 Calculate the energy to heat two cubes (gold and aluminum) each with a volume of 10.0 cm^3 from 15 °C to 25 °C. Refer to Tables 2.8 and 3.11. (3.6)

3.68 Calculate the energy to heat two cubes (silver and copper), each with a volume of 10.0 cm^3 from 15 °C to 25 °C. Refer to Tables 2.8 and 3.11. (3.6)

Clinical Applications

3.69 A 70.0-kg person had a quarter-pound cheeseburger, french fries, and a chocolate shake. (3.5)

Item	Carbohydrate (g)	Fat (g)	Protein (g)
Cheeseburger	46	40.	47
French fries	47	16	4
Chocolate shake	76	10.	10.

a. Using Table 3.7, calculate the total kilocalories for each food type in this meal (round off the kilocalories to the tens place).
b. Determine the total kilocalories for the meal (round off to the tens place).
c. Using Table 3.10, determine the number of hours of sleep needed to burn off the kilocalories in this meal.
d. Using Table 3.10, determine the number of hours of running needed to burn off the kilocalories in this meal.

3.70 Your friend, who has a mass of 70.0 kg, has a slice of pizza, a cola soft drink, and ice cream. (3.5)

Item	Carbohydrate (g)	Fat (g)	Protein (g)
Pizza	29	10.	13
Cola	51	0	0
Ice cream	44	28	8

a. Using Table 3.7, calculate the total kilocalories for each food type in this meal (round off the kilocalories to the tens place).
b. Determine the total kilocalories for the meal (round off to the tens place).
c. Using Table 3.10, determine the number of hours of sitting needed to burn off the kilocalories in this meal.
d. Using Table 3.10, determine the number of hours of swimming needed to burn off the kilocalories in this meal.

ADDITIONAL PRACTICE PROBLEMS

3.71 Classify each of the following as an element, a compound, or a mixture: (3.1)
a. carbon in pencils
b. carbon monoxide (CO) in automobile exhaust
c. orange juice

3.72 Classify each of the following as an element, a compound, or a mixture: (3.1)
a. neon gas in lights
b. a salad dressing of oil and vinegar
c. sodium hypochlorite (NaClO) in bleach

3.73 Classify each of the following mixtures as homogeneous or heterogeneous: (3.1)
a. hot fudge sundae **b.** herbal tea **c.** vegetable oil

3.74 Classify each of the following mixtures as homogeneous or heterogeneous: (3.1)
a. water and sand **b.** mustard **c.** blue ink

3.75 Identify each of the following as solid, liquid, or gas: (3.2)
a. vitamin tablets in a bottle **b.** helium in a balloon
c. milk in a bottle **d.** the air you breathe
e. charcoal briquettes on a barbecue

3.76 Identify each of the following as solid, liquid, or gas: (3.2)
a. popcorn in a bag **b.** water in a garden hose
c. a computer mouse **d.** air in a tire
e. hot tea in a teacup

3.77 Identify each of the following as a physical or chemical property: (3.2)
a. Gold is shiny.
b. Gold melts at 1064 °C.
c. Gold is a good conductor of electricity.
d. When gold reacts with sulfur, a black sulfide compound forms.

3.78 Identify each of the following as a physical or chemical property: (3.2)
a. A candle is 10 in. high and 2 in. in diameter.
b. A candle burns.
c. The wax of a candle softens on a hot day.
d. A candle is blue.

3.79 Identify each of the following as a physical or chemical change: (3.2)
a. A plant grows a new leaf.
b. Chocolate is melted for a dessert.
c. Wood is chopped for the fireplace.
d. Wood burns in a woodstove.

3.80 Identify each of the following as a physical or chemical change: (3.2)
a. Aspirin tablets are broken in half.
b. Carrots are grated for use in a salad.
c. Malt undergoes fermentation to make beer.
d. A copper pipe reacts with air and turns green.

3.81 Calculate each of the following temperatures in degrees Celsius and kelvins: (3.3)

a. The highest recorded temperature in the continental United States was 134 °F in Death Valley, California, on July 10, 1913.

b. The lowest recorded temperature in the continental United States was −69.7 °F in Rodgers Pass, Montana, on January 20, 1954.

3.82 Calculate each of the following temperatures in kelvins and degrees Fahrenheit: (3.3)

a. The highest recorded temperature in the world was 58.0 °C in El Azizia, Libya, on September 13, 1922.

b. The lowest recorded temperature in the world was −89.2 °C in Vostok, Antarctica, on July 21, 1983.

3.83 What is −15 °F in degrees Celsius and in kelvins? (3.3)

3.84 What is 56 °F in degrees Celsius and in kelvins? (3.3)

3.85 A 0.50-g sample of vegetable oil is placed in a calorimeter. When the sample is burned, 18.9 kJ is given off. What is the energy value (kcal/g) for the oil? (3.5)

3.86 A 1.3-g sample of rice is placed in a calorimeter. When the sample is burned, 22 kJ is given off. What is the energy value (kcal/g) for the rice? (3.5)

3.87 On a hot day, the beach sand gets hot but the water stays cool. Would you predict that the specific heat of sand is higher or lower than that of water? Explain. (3.6)

On a sunny day, the sand gets hot but the water stays cool.

3.88 On a hot sunny day, you get out of the swimming pool and sit in a metal chair, which is very hot. Would you predict that the specific heat of the metal is higher or lower than that of water? Explain. (3.6)

3.89 The following graph is a heating curve for chloroform, a solvent for fats, oils, and waxes: (3.7)

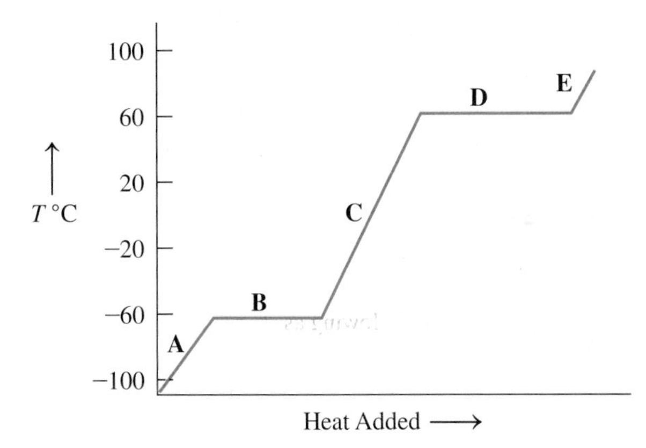

a. What is the approximate melting point of chloroform?

b. What is the approximate boiling point of chloroform?

c. On the heating curve, identify the segments **A**, **B**, **C**, **D**, and **E** as solid, liquid, gas, melting, or boiling.

d. At the following temperatures, is chloroform a solid, liquid, or gas?

−80 °C; −40 °C; 25 °C; 80 °C

3.90 Associate the contents of the beakers (**1** to **5**) with segments (**A** to **E**) on the following heating curve for water: (3.7)

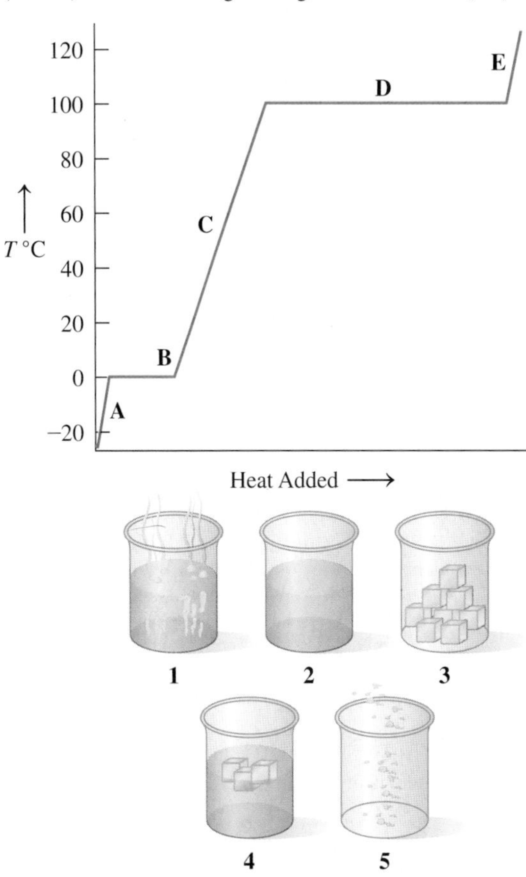

3.91 The melting point of dibromomethane is −53 °C and its boiling point is 97 °C. Sketch a heating curve for dibromomethane from −100 °C to 120 °C. (3.7)

a. What is the state of dibromomethane at −75 °C?

b. What happens on the curve at −53 °C?

c. What is the state of dibromomethane at −18 °C?

d. What is the state of dibromomethane at 110 °C?

e. At what temperature will both solid and liquid be present?

3.92 The melting point of benzene is 5.5 °C and its boiling point is 80.1 °C. Sketch a heating curve for benzene from 0 °C to 100 °C. (3.7)

a. What is the state of benzene at 15 °C?

b. What happens on the curve at 5.5 °C?

c. What is the state of benzene at 63 °C?

d. What is the state of benzene at 98 °C?

e. At what temperature will both liquid and gas be present?

Clinical Applications

3.93 If you want to lose 1 lb of "body fat," which is 15% water, how many kilocalories do you need to expend? (3.5)

3.94 A young patient drinks whole milk as part of her diet. Calculate the total kilocalories if the glass of milk contains 12 g of carbohydrate, 9 g of fat, and 9 g of protein. (Round off answers for each food type to the tens place.) (3.5)

3.95 A hot-water bottle for a patient contains 725 g of water at 65 °C. If the water cools to body temperature (37 °C), how many kilojoules of heat could be transferred to sore muscles? (3.6)

3.96 The highest recorded body temperature that a person has survived is 46.5 °C. Calculate that temperature in degrees Fahrenheit and in kelvins. (3.3)

CHALLENGE PROBLEMS

The following problems are related to the topics in this chapter. However, they do not all follow the chapter order, and they require you to combine concepts and skills from several Sections. These problems will help you increase your critical thinking skills and prepare for your next exam.

3.97 When a 0.66-g sample of olive oil is burned in a calorimeter, the heat released increases the temperature of 370 g of water from 22.7 °C to 38.8 °C. What is the energy value for the olive oil in kcal/g? (3.5, 3.6)

3.98 A 45-g piece of ice at 0.0 °C is added to a sample of water at 8.0 °C. All of the ice melts and the temperature of the water decreases to 0.0 °C. How many grams of water were in the sample? (3.6, 3.7)

3.99 In a large building, oil is used in a steam boiler heating system. The combustion of 1.0 lb of oil provides 2.4×10^7 J. (3.4, 3.6)

a. How many kilograms of oil are needed to heat 150 kg of water from 22 °C to 100 °C?

b. How many kilograms of oil are needed to change 150 kg of water to steam at 100 °C?

3.100 When 1.0 g of gasoline burns, it releases 11 kcal. The density of gasoline is 0.74 g/mL. (3.4, 3.6)

a. How many megajoules are released when 1.0 gal of gasoline burns?

b. If a television requires 150 kJ/h to run, how many hours can the television run on the energy provided by 1.0 gal of gasoline?

3.101 An ice bag containing 275 g of ice at 0.0 °C was used to treat sore muscles. When the bag was removed, the ice had melted and the liquid water had a temperature of 24.0 °C. How many kilojoules of heat were absorbed? (3.6, 3.7)

3.102 A 115-g sample of steam at 100 °C is emitted from a volcano. It condenses, cools, and falls as snow at 0.0 °C. How many kilojoules were released? (3.6, 3.7)

3.103 A 70.0-g piece of copper metal at 54.0 °C is placed in 50.0 g of water at 26.0 °C. If the final temperature of the water and metal is 29.2 °C, what is the specific heat (J/g °C) of copper? (3.6)

3.104 A 125-g piece of metal is heated to 288 °C and dropped into 85.0 g of water at 12.0 °C. The metal and water come to the same temperature of 24.0 °C. What is the specific heat, in J/g °C, of the metal? (3.6)

3.105 A metal is thought to be titanium or aluminum. When 4.7 g of the metal absorbs 11 J, its temperature rises by 4.5 °C. (3.6)

a. What is the specific heat (J/g °C) of the metal?

b. Would you identify the metal as titanium or aluminum (see Table 3.11)?

3.106 A metal is thought to be copper or gold. When 18 g of the metal absorbs 58 cal, its temperature rises by 35 °C. (3.6)

a. What is the specific heat, in cal/g °C, of the metal?

b. Would you identify the metal as copper or gold (see Table 3.11)?

ANSWERS

Answers to Selected Practice Problems

3.1 **a.** element **b.** compound **c.** element **d.** compound **e.** compound

3.3 **a.** pure substance **b.** mixture **c.** pure substance **d.** pure substance **e.** mixture

3.5 **a.** heterogeneous **b.** homogeneous **c.** heterogeneous **d.** heterogeneous

3.7 **a.** gas **b.** gas **c.** solid

3.9 **a.** physical **b.** chemical **c.** physical **d.** chemical **e.** chemical

3.11 **a.** physical **b.** chemical **c.** physical **d.** physical **e.** physical

3.13 **a.** chemical **b.** physical **c.** physical **d.** chemical **e.** physical

3.15 In the United States, we still use the Fahrenheit temperature scale. In °F, normal body temperature is 98.6. On the Celsius scale, her temperature would be 37.7 °C, a mild fever.

3.17 **a.** 98.6 °F **b.** 18.5 °C **c.** 246 K **d.** 335 K **e.** 46 °C

3.19 **a.** 41 °C **b.** No. The temperature is equivalent to 39 °C.

3.21 When the roller-coaster car is at the top of the ramp, it has its maximum potential energy. As it descends, potential energy changes to kinetic energy. At the bottom, all the energy is kinetic.

3.23 **a.** potential **b.** kinetic **c.** potential **d.** potential

3.25 **a.** 3.5 kcal **b.** 99.2 cal **c.** 120 J **d.** 1100 cal

3.27 **a.** 8.1×10^5 J **b.** 190 kcal

3.29 **a.** 29.9 kcal **b.** 208 kcal

3.31 **a.** 470 kJ **b.** 18 g **c.** 130 kcal **d.** 37 g

3.33 210 kcal, 880 kJ

3.35 640 kcal

3.37 Copper has the lowest specific heat of the samples and will reach the highest temperature.

3.39 **a.** 180 cal **b.** 2600 J **c.** 9.3 kcal **d.** 10.8 kJ

3.41 **a.** 1380 J, 330. cal **b.** 1810 J, 434 cal **c.** 3780 J, 904 cal **d.** 3600 J, 850 cal

3.43 **a.** melting **b.** sublimation **c.** freezing **d.** deposition

3.45 **a.** 5200 cal absorbed **b.** 5680 J absorbed
c. 18 kcal released **d.** 16.7 kJ released

3.47 **a.** condensation **b.** evaporation
c. boiling **d.** condensation

3.49 **a.** 5400 cal absorbed **b.** 11 300 J absorbed
c. 4300 kcal released **d.** 396 kJ released

3.51

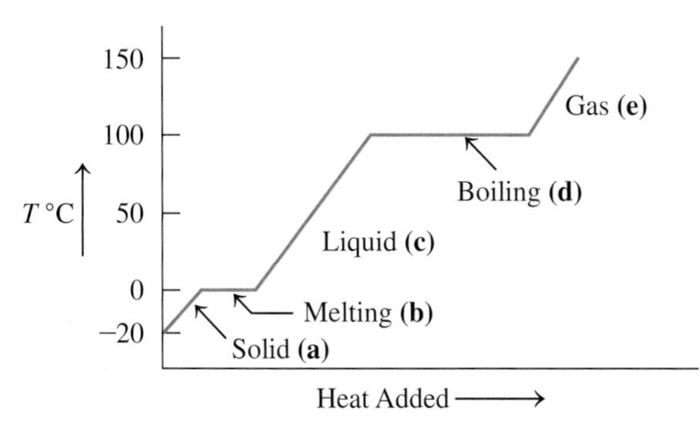

3.53 **a.** 30 300 J **b.** 9.6 kcal
c. 72.2 kJ

3.55 10.8 kcal

3.57 **a.** Breakfast, 270 kcal; Lunch, 420 kcal; Dinner, 440 kcal
b. 1130 kcal total
c. Yes. Charles should be losing weight.
d. 26 days

3.59 **a.** compound, the particles have a 2:1 ratio of atoms
b. mixture, has two different kinds of particles
c. element, has a single kind of atom

3.61 **a.** homogeneous **b.** heterogeneous
c. homogeneous

3.63 41.5 °C, 107 °F

3.65 68.3 °C, 341 K

3.67 gold, 250 J or 60. cal; aluminum, 240 J or 58 cal

3.69 **a.** carbohydrate, 680 kcal; fat, 590 kcal; protein, 240 kcal
b. 1510 kcal
c. 25 h
d. 2.0 h

3.71 **a.** element **b.** compound
c. mixture

3.73 **a.** heterogeneous **b.** homogeneous
c. homogeneous

3.75 **a.** solid **b.** gas
c. liquid **d.** gas
e. solid

3.77 **a.** physical **b.** physical
c. physical **d.** chemical

3.79 **a.** chemical **b.** physical
c. physical **d.** chemical

3.81 **a.** 56.7 °C, 330. K **b.** −56.5 °C, 217 K

3.83 −26 °C, 247 K

3.85 9.0 kcal/g

3.87 The same amount of heat causes a greater temperature change in the sand than in the water; thus the sand must have a lower specific heat than that of water.

3.89 **a.** about −60 °C
b. about 60 °C
c. The diagonal line A represents the solid state as temperature increases. The horizontal line B represents the change from solid to liquid or melting of the substance. The diagonal line C represents the liquid state as temperature increases. The horizontal line D represents the change from liquid to gas or boiling of the liquid. The diagonal line E represents the gas state as temperature increases.
d. At −80 °C, solid; at −40 °C, liquid; at 25 °C, liquid; at 80 °C, gas

3.91

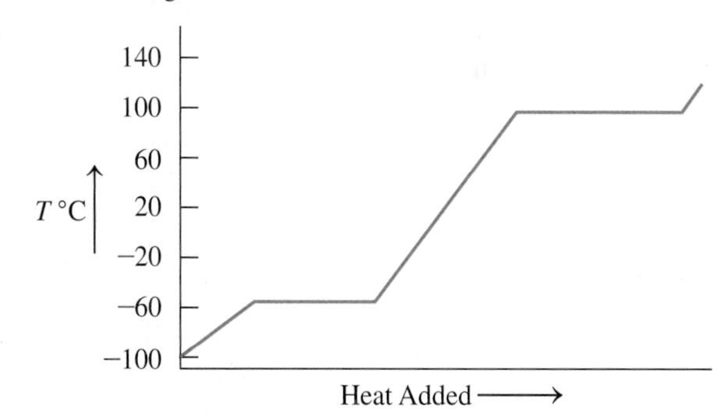

a. solid **b.** solid dibromomethane melts
c. liquid **d.** gas
e. −53 °C

3.93 3500 kcal

3.95 85 kJ

3.97 9.0 kcal/g

3.99 **a.** 0.93 kg **b.** 6.4 kg

3.101 119.5 kJ

3.103 Specific heat = 0.385 J/g °C

3.105 **a.** 0.52 J/g °C **b.** titanium

COMBINING IDEAS from Chapters 1 to 3

CI.1 Gold, one of the most sought-after metals in the world, has a density of 19.3 g/cm^3, a melting point of 1064 °C, and a specific heat of 0.129 J/g °C. A gold nugget found in Alaska in 1998 weighed 20.17 lb. (2.4, 2.6, 2.7, 3.3, 3.5)

Gold nuggets, also called native gold, can be found in streams and mines.

a. How many significant figures are in the measurement of the weight of the nugget?
b. Which is the mass of the nugget in kilograms?
c. If the nugget were pure gold, what would its volume be in cubic centimeters?
d. What is the melting point of gold in degrees Fahrenheit and kelvins?
e. How many kilocalories are required to raise the temperature of the nugget from 500. °C to 1064 °C?
f. If the price of gold is \$42.06 per gram, what is the nugget worth in dollars?

CI.2 The mileage for a motorcycle with a fuel-tank capacity of 22 L is 35 mi/gal. (2.5, 2.6, 2.7, 3.4)

When 1.00 g of gasoline burns, 47 kJ of energy is released.

a. How long a trip, in kilometers, can be made on one full tank of gasoline?
b. If the price of gasoline is \$2.82 per gallon, what would be the cost of fuel for the trip?
c. If the average speed during the trip is 44 mi/h, how many hours will it take to reach the destination?
d. If the density of gasoline is 0.74 g/mL, what is the mass, in grams, of the fuel in the tank?
e. When 1.00 g of gasoline burns, 47 kJ of energy is released. How many kilojoules are produced when the fuel in one full tank is burned?

CI.3 Answer the following for the water samples **A** and **B** shown in the diagrams: (3.1, 3.2, 3.3, 3.5)

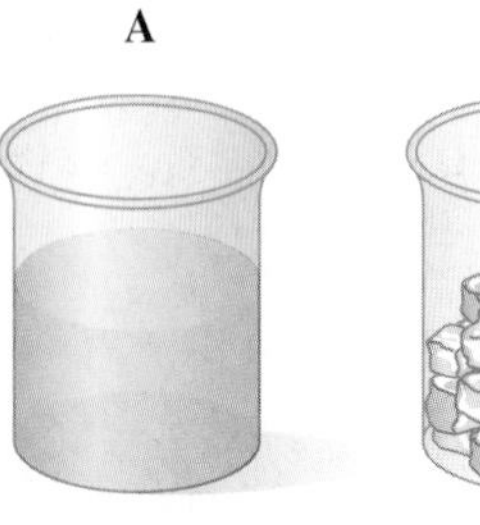

a. In which sample (**A** or **B**) does the water have its own shape?
b. Which diagram (**1** or **2** or **3**) represents the arrangement of particles in water sample **A**?
c. Which diagram (**1** or **2** or **3**) represents the arrangement of particles in water sample **B**?

Answer the following for diagrams **1**, **2**, and **3**:

d. The state of matter indicated in diagram **1** is a _____ ; in diagram **2**, it is a _____ ; and in diagram **3**, it is a _____.
e. The motion of the particles is slowest in diagram _____.
f. The arrangement of particles is farthest apart in diagram _____.
g. The particles fill the volume of the container in diagram _____.
h. If the water in diagram **2** has a mass of 19 g and a temperature of 45 °C, how much heat, in kilojoules, is removed to cool the liquid to 0 °C?

CI.4 The label of an energy bar with a mass of 68 g lists the nutrition facts as 39 g of carbohydrate, 5 g of fat, and 10. g of protein. (2.5, 2.6, 3.4, 3.6)

An energy bar contains carbohydrate, fat, and protein.

a. Using the energy values for carbohydrates, fats, and proteins (see Table 3.7), what are the total kilocalories listed for the energy bar? (Round off answers for each food type to the tens place.)
b. What are the kilojoules for the energy bar? (Round off answers for each food type to the tens place.)
c. If you obtain 160 kJ, how many grams of the energy bar did you eat?
d. If you are walking and using energy at a rate of 840 kJ/h, how many minutes will you need to walk to expend the energy from two energy bars?

CI.5 In one box of nails, there are 75 iron nails weighing 0.250 lb. The density of iron is 7.86 g/cm^3. The specific heat of iron is 0.452 J/g °C. The melting point of iron is 1535 °C. (2.5, 2.6, 2.7, 3.4, 3.5)

Nails made of iron have a density of 7.86 g/cm^3.

a. What is the volume, in cubic centimeters, of the iron nails in the box?
b. If 30 nails are added to a graduated cylinder containing 17.6 mL of water, what is the new level of water, in milliliters, in the cylinder?
c. How much heat, in joules, must be added to the nails in the box to raise their temperature from 16 °C to 125 °C?
d. How much heat, in joules, is required to heat one nail from 25 °C to its melting point?

CI.6 A hot tub is filled with 450 gal of water. (2.5, 2.6, 2.7, 3.3, 3.4, 3.5)

A hot tub filled with water is heated to 105 °F.

a. What is the volume of water, in liters, in the tub?
b. What is the mass, in kilograms, of water in the tub?
c. How many kilocalories are needed to heat the water from 62 °F to 105 °F?
d. If the hot-tub heater provides 5900 kJ/min, how long, in minutes, will it take to heat the water in the hot tub from 62 °F to 105 °F?

ANSWERS

CI.1 **a.** four SFs
b. 9.17 kg
c. 475 cm^3
d. 1947 °F; 1337 K
e. 159 kcal
f. $386 000

CI.3 **a.** **B**
b. **A** is represented by diagram **2**.
c. **B** is represented by diagram **1**.
d. solid; liquid; gas
e. diagram **1**
f. diagram **3**
g. diagram **3**
h. 3.6 kJ

CI.5 **a.** 14.4 cm^3
b. 23.4 mL
c. 5590 J
d. 1030 J

4 Atoms and Elements

JOHN PREPARES FOR THE NEXT GROWING season by deciding how much of each crop should be planted and their location on his farm. Part of this decision is determined by the quality of the soil, including the pH, the amount of moisture, and the nutrient content in the soil. He begins by sampling the soil and performing a few chemical tests on the soil. John determines that several of his fields need additional fertilizer before the crops can be planted. John considers several different types of fertilizers as each supplies different nutrients to the soil to help increase crop production. Plants need three basic elements for growth: potassium, nitrogen, and phosphorus. Potassium (K on the periodic table) is a metal, whereas nitrogen (N) and phosphorus (P) are nonmetals. Fertilizers may also contain several other elements including calcium (Ca), magnesium (Mg), and sulfur (S). John applies a fertilizer containing a mixture of all of these elements to his soil and plans to re-check the soil nutrient content in a few days.

CAREER Farmer

Farming involves much more than growing crops and raising animals. Farmers must understand how to perform chemical tests and how to apply fertilizer to soil and pesticides or herbicides to crops. Pesticides are chemicals used to kill insects that could destroy the crop, whereas herbicides are chemicals used to kill weeds that would compete with the crops for water and nutrients. This requires knowledge of how these chemicals work and their safety, effectiveness, and storage. In using this information, farmers are able to grow crops that produce a higher yield, greater nutritional value, and better taste.

CLINICAL UPDATE Improving Crop Production

Last year, John noticed that the potatoes from one of his fields had brown spots and were undersized. Last week, he obtained soil samples from that field to check the nutrient levels. You can see how John improved the soil and crop production in the **CLINICAL UPDATE Improving Crop Production**, page 127, and learn what kind of fertilizer John used and the quantity applied.

LOOKING AHEAD

4.1 Elements and Symbols

LEARNING GOAL Given the name of an element, write its correct symbol; from the symbol, write the correct name.

All matter is composed of *elements*, of which there are 118 different kinds. Of these, 88 elements occur naturally and make up all the substances in our world. Many elements are already familiar to you. Perhaps you use aluminum in the form of foil or drink soft drinks from aluminum cans. You may have a ring or necklace made of gold, silver, or perhaps platinum. If you play tennis or golf, then you may have noticed that your racket or clubs may be made from the elements titanium or carbon. In our bodies, calcium and phosphorus form the structure of bones and teeth, iron and copper are needed in the formation of red blood cells, and iodine is required for the proper functioning of the thyroid.

Elements are pure substances from which all other things are built. Elements cannot be broken down into simpler substances. Over the centuries, elements have been named for planets, mythological figures, colors, minerals, geographic locations, and famous people. Some sources of names of elements are listed in **TABLE 4.1**. A complete list of all the elements and their symbols are found on the inside front cover of this text.

TABLE 4.1 Some Elements, Symbols, and Source of Names

Element	Symbol	Source of Name
Uranium	U	The planet Uranus
Titanium	Ti	Titans (mythology)
Chlorine	Cl	*Chloros*: "greenish yellow" (Greek)
Iodine	I	*Ioeides*: "violet" (Greek)
Magnesium	Mg	Magnesia, a mineral
Tennessine	Ts	Tennessee
Curium	Cm	Marie and Pierre Curie
Copernicium	Cn	Nicolaus Copernicus

Chemical Symbols

Chemical symbols are one- or two-letter abbreviations for the names of the elements. Only the first letter of an element's symbol is capitalized. If the symbol has a second letter, it is lowercase so that we know when a different element is indicated. If two letters are capitalized, they represent the symbols of two different elements. For example, the element cobalt has the symbol Co. However, the two capital letters CO specify two elements, carbon (C) and oxygen (O).

One-Letter Symbols		Two-Letter Symbols	
C	Carbon	Co	Cobalt
S	Sulfur	Si	Silicon
N	Nitrogen	Ne	Neon
H	Hydrogen	He	Helium

ENGAGE

What are the names and symbols of some elements that you encounter every day?

TABLE 4.2 Names and Symbols of Some Common Elements

Name*	Symbol	Name*	Symbol	Name*	Symbol
Aluminum	Al	Gallium	Ga	Oxygen	O
Argon	Ar	Gold (*aurum*)	Au	Phosphorus	P
Arsenic	As	Helium	He	Platinum	Pt
Barium	Ba	Hydrogen	H	Potassium (*kalium*)	K
Boron	B	Iodine	I	Radium	Ra
Bromine	Br	Iron (*ferrum*)	Fe	Silicon	Si
Cadmium	Cd	Lead (*plumbum*)	Pb	Silver (*argentum*)	Ag
Calcium	Ca	Lithium	Li	Sodium (*natrium*)	Na
Carbon	C	Magnesium	Mg	Strontium	Sr
Chlorine	Cl	Manganese	Mn	Sulfur	S
Chromium	Cr	Mercury (*hydrargyrum*)	Hg	Tin (*stannum*)	Sn
Cobalt	Co	Neon	Ne	Titanium	Ti
Copper (*cuprum*)	Cu	Nickel	Ni	Uranium	U
Fluorine	F	Nitrogen	N	Zinc	Zn

*Names given in parentheses are ancient Latin or Greek words from which the symbols are derived.

Aluminum

Carbon

Gold

Silver

Sulfur

Although most of the symbols use letters from the current names, some are derived from their ancient names. For example, Na, the symbol for sodium, comes from the Latin word *natrium*. The symbol for iron, Fe, is derived from the Latin name *ferrum*. **TABLE 4.2** lists the names and symbols of some common elements. Learning their names and symbols will greatly help your learning of chemistry.

SAMPLE PROBLEM 4.1 Names and Symbols of Chemical Elements

TRY IT FIRST

Complete the following table with the correct name or symbol for each element:

Name	Symbol
nickel	____
nitrogen	____
____	Zn
____	K
iron	____

SOLUTION

Name	Symbol
nickel	Ni
nitrogen	N
zinc	Zn
potassium	K
iron	Fe

STUDY CHECK 4.1

Write the chemical symbols for the elements silicon, sulfur, and silver.

ANSWER

Si, S, and Ag

TEST

Try Practice Problems 4.1 to 4.6

CHEMISTRY LINK TO HEALTH

Toxicity of Mercury

Mercury (Hg) is a silvery, shiny element that is a liquid at room temperature. Mercury can enter the body through inhaled mercury vapor, contact with the skin, or ingestion of foods or water contaminated with mercury. In the body, mercury destroys proteins and disrupts cell function. Long-term exposure to mercury can damage the brain and kidneys, cause mental retardation, and decrease physical development. Blood, urine, and hair samples are used to test for mercury.

In both freshwater and seawater, bacteria convert mercury into toxic methylmercury, which attacks the central nervous system. Because fish absorb methylmercury, we are exposed to mercury by consuming mercury-contaminated fish. As levels of mercury ingested from fish became a concern, the Food and Drug Administration set a maximum level of 1 mg of mercury in every kilogram of seafood. Fish higher in the food chain, such as swordfish and shark, can have such high levels of mercury that the U.S. Environmental Protection Agency (EPA) recommends they be consumed no more than once a week.

One of the worst incidents of mercury poisoning occurred in Minamata and Niigata, Japan, in 1965. At that time, the ocean was polluted with high levels of mercury from industrial wastes. Because fish were a major food in the diet, more than 2000 people were affected with mercury poisoning and died or developed neural damage. In the United States between 1988 and 1997, the use of mercury decreased by 75% when the use of mercury was banned in paint and pesticides, and regulated in batteries and other products. Certain batteries and compact fluorescent light bulbs (CFLs) contain mercury, and instructions for their safe disposal should be followed.

The element mercury is a silvery, shiny liquid at room temperature.

PRACTICE PROBLEMS

4.1 Elements and Symbols

LEARNING GOAL Given the name of an element, write its correct symbol; from the symbol, write the correct name.

4.1 Write the symbols for the following elements:
a. copper **b.** platinum **c.** calcium **d.** manganese
e. iron **f.** barium **g.** lead **h.** strontium

4.2 Write the symbols for the following elements:
a. oxygen **b.** lithium **c.** uranium **d.** titanium
e. hydrogen **f.** chromium **g.** tin **h.** gold

Clinical Applications

4.3 Write the name for the symbol of each of the following elements found in the body:
a. C **b.** Cl **c.** I **d.** Se
e. N **f.** S **g.** Zn **h.** Co

4.4 Write the name for the symbol of each of the following elements found in the body:
a. V **b.** P **c.** Na **d.** As
e. Ca **f.** Mo **g.** Mg **h.** Si

4.5 Write the names for the elements in each of the following formulas of compounds used in medicine:
a. table salt, NaCl
b. plaster casts, $CaSO_4$
c. Demerol, $C_{15}H_{22}ClNO_2$
d. treatment of bipolar disorder, Li_2CO_3

4.6 Write the names for the elements in each of the following formulas of compounds used in medicine:
a. salt substitute, KCl
b. dental cement, $Zn_3(PO_4)_2$
c. antacid, $Mg(OH)_2$
d. contrast agent for X-ray, $BaSO_4$

4.2 The Periodic Table

LEARNING GOAL Use the periodic table to identify the group and the period of an element; identify the element as a metal, a nonmetal, or a metalloid.

By the late 1800s, scientists recognized that certain elements looked alike and behaved the same way. In 1869, a Russian chemist, Dmitri Mendeleev, arranged the 60 elements known at that time into groups with similar properties and placed them in order of increasing atomic masses. Today, this arrangement of 118 elements is known as the **periodic table** (see FIGURE 4.1).

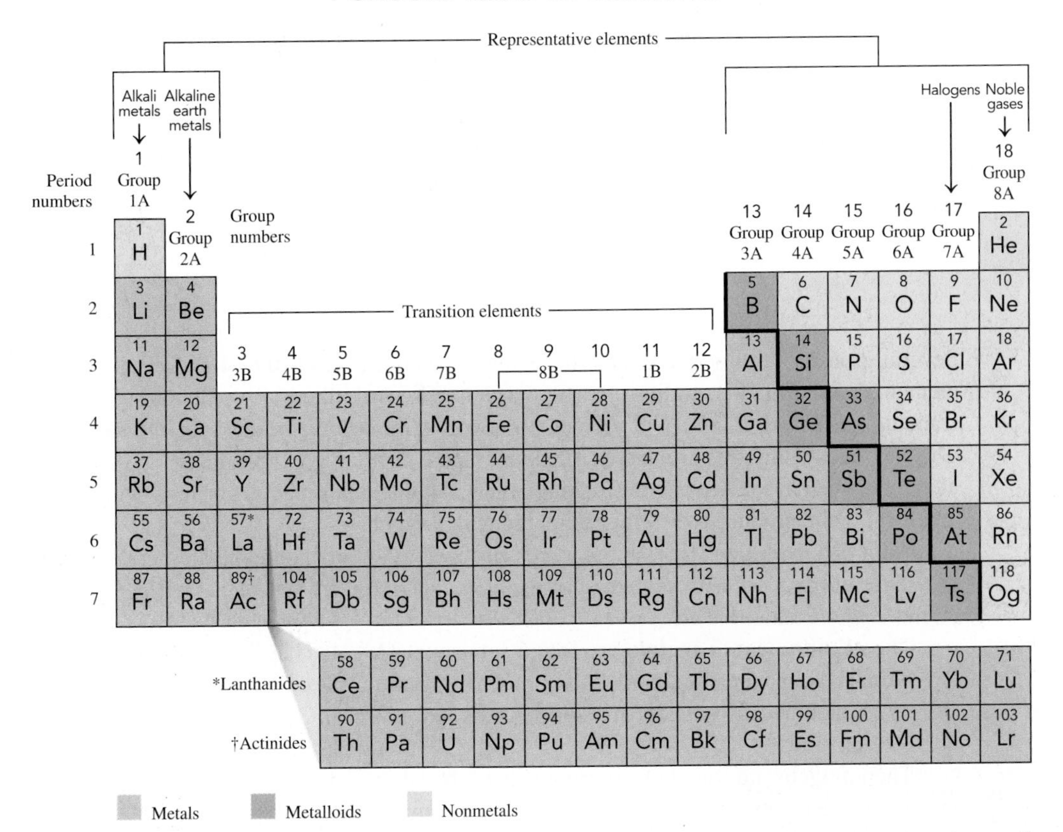

FIGURE 4.1 On the periodic table, groups are the elements arranged as vertical columns, and periods are the elements in each horizontal row.

Q What is the symbol of the alkali metal in Period 3?

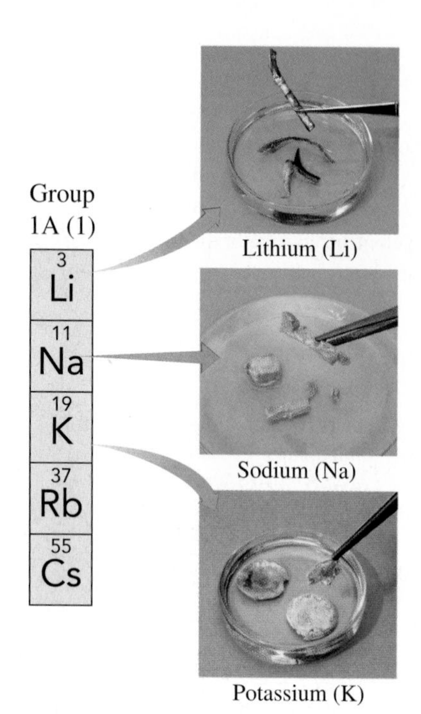

FIGURE 4.3 Lithium (Li), sodium (Na), and potassium (K) are alkali metals from Group 1A (1).

Q What other properties do these alkali metals have in common?

Groups and Periods

Each vertical column on the periodic table contains a **group** (or family) of elements that have similar properties. A **group number** is written at the top of each vertical column (group) in the periodic table. For many years, the **representative elements** have had group numbers 1A to 8A. In the center of the periodic table is a block of elements known as the **transition elements**, which have numbers followed by the letter "B." A newer system assigns numbers 1 to 18 to the groups going left to right across the periodic table. Because both systems are in use, they are shown on the periodic table and are included in our discussions of elements and group numbers. The two rows of 14 elements called the *lanthanides* and *actinides* (or the inner transition elements) are placed at the bottom of the periodic table to allow them to fit on a page.

Each horizontal row in the periodic table is a **period**. The periods are counted down from the top of the table as Periods 1 to 7. The first period contains two elements: hydrogen (H) and helium (He). The second period contains eight elements: lithium (Li), beryllium (Be), boron (B), carbon (C), nitrogen (N), oxygen (O), fluorine (F), and neon (Ne). The third period also contains eight elements beginning with sodium (Na) and ending with argon (Ar). The fourth period, which begins with potassium (K), and the fifth period, which begins with rubidium (Rb), have 18 elements each. The sixth period, which begins with cesium (Cs), has 32 elements. The seventh period, contains 32 elements, for a total of 118 elements.

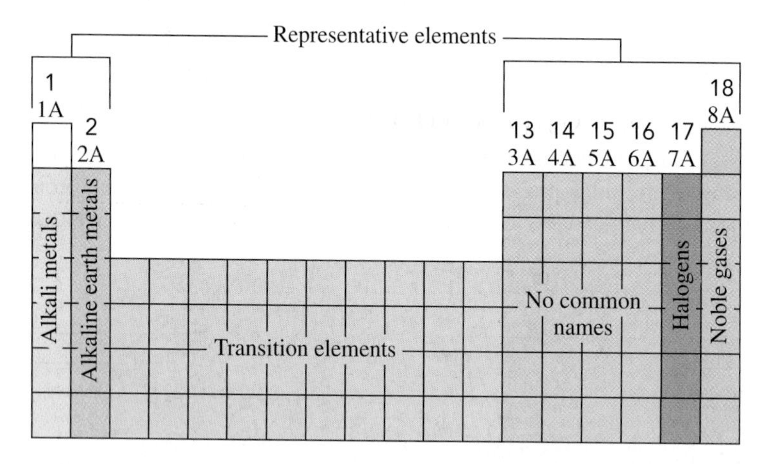

FIGURE 4.2 Certain groups on the periodic table have common names.

Q What is the common name for the group of elements that includes helium and argon?

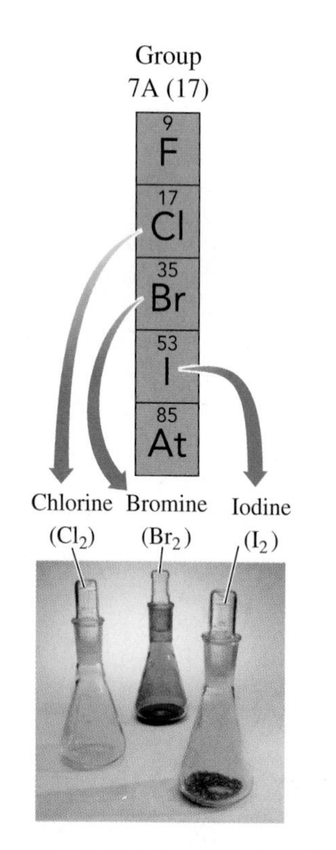

FIGURE 4.4 Chlorine (Cl_2), bromine (Br_2), and iodine (I_2) are halogens from Group 7A (17).

Q What other elements are in the halogen group?

Names of Groups

Several groups in the periodic table have special names (see **FIGURE 4.2**). Group 1A (1) elements—lithium (Li), sodium (Na), potassium (K), rubidium (Rb), cesium (Cs), and francium (Fr)—are a family of elements known as the **alkali metals** (see **FIGURE 4.3**). The elements within this group are soft, shiny metals that are good conductors of heat and electricity and have relatively low melting points. Alkali metals react vigorously with water and form white products when they combine with oxygen.

Although hydrogen (H) is at the top of Group 1A (1), it is not an alkali metal and has very different properties than the rest of the elements in this group. Thus, hydrogen is not included in the alkali metals.

The **alkaline earth metals** are found in Group 2A (2). They include the elements beryllium (Be), magnesium (Mg), calcium (Ca), strontium (Sr), barium (Ba), and radium (Ra). The alkaline earth metals are shiny metals like those in Group 1A (1), but they are not as reactive.

The **halogens** are found on the right side of the periodic table in Group 7A (17). They include the elements fluorine (F), chlorine (Cl), bromine (Br), iodine (I), and astatine (At) (see **FIGURE 4.4**). The halogens, especially fluorine and chlorine, are highly reactive and form compounds with most of the elements.

The **noble gases** are found in Group 8A (18). They include helium (He), neon (Ne), argon (Ar), krypton (Kr), xenon (Xe), radon (Rn), and oganesson (Og). They are quite unreactive and are seldom found in combination with other elements.

Metals, Nonmetals, and Metalloids

Another feature of the periodic table is the heavy zigzag line that separates the elements into the *metals* and the *nonmetals*. *Except for hydrogen*, the metals are to the left of the line with the nonmetals to the right.

In general, most **metals** are shiny solids, such as copper (Cu), gold (Au), and silver (Ag). Metals can be shaped into wires (ductile) or hammered into a flat sheet (malleable). Metals are good conductors of heat and electricity. They usually melt at higher temperatures than nonmetals. All the metals are solids at room temperature, except for mercury (Hg), which is a liquid.

Nonmetals are not especially shiny, ductile, or malleable, and they are often poor conductors of heat and electricity. They typically have low melting points and low densities. Some examples of nonmetals are hydrogen (H), carbon (C), nitrogen (N), oxygen (O), chlorine (Cl), and sulfur (S).

Except for aluminum, the elements located along the heavy line are **metalloids**: B, Si, Ge, As, Sb, Te, Po, At, and Ts. Metalloids are elements that exhibit some properties that are typical of the metals and other properties that are characteristic of the nonmetals. For example, they are better conductors of heat and electricity than the nonmetals, but not as good as the metals. The metalloids are *semiconductors* because they can be modified to function as conductors or insulators. TABLE 4.3 compares some characteristics of silver, a metal, with those of antimony, a metalloid, and sulfur, a nonmetal.

ENGAGE

If element 119 were produced, in what group and period would it be placed?

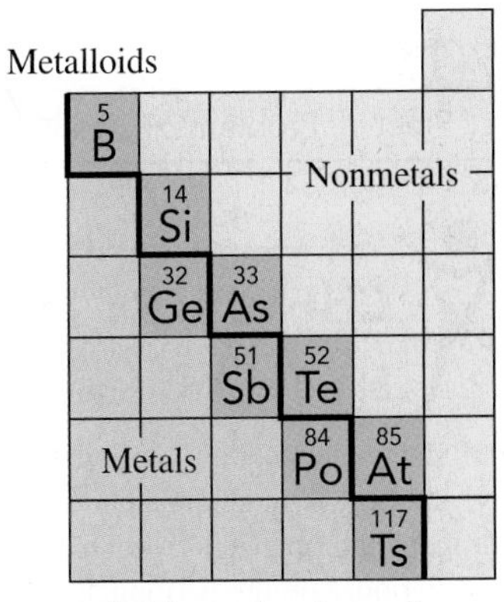

The metalloids on the zigzag line exhibit characteristics of both metals and nonmetals.

TABLE 4.3 Some Characteristics of a Metal, a Metalloid, and a Nonmetal

Silver (Ag)	Antimony (Sb)	Sulfur (S)
Metal	Metalloid	Nonmetal
Shiny	Blue-gray, shiny	Dull, yellow
Extremely ductile	Brittle	Brittle
Can be hammered into sheets (malleable)	Shatters when hammered	Shatters when hammered
Good conductor of heat and electricity	Poor conductor of heat and electricity	Poor conductor of heat and electricity, good insulator
Used in coins, jewelry, tableware	Used to harden lead, color glass and plastics	Used in gunpowder, rubber, fungicides
Density 10.5 g/mL	Density 6.7 g/mL	Density 2.1 g/mL
Melting point 962 °C	Melting point 630 °C	Melting point 113 °C

Silver is a metal, antimony is a metalloid, and sulfur is a nonmetal.

SAMPLE PROBLEM 4.2 Metals, Nonmetals, and Metalloids

TRY IT FIRST

Use the periodic table to classify each of the following elements by its group and period, group name (if any), and as a metal, a nonmetal, or a metalloid:

a. Na, important in nerve impulses, regulates blood pressure
b. I, needed to produce thyroid hormones
c. Si, needed for tendons and ligaments

SOLUTION

a. Na (sodium), Group 1A (1), Period 3, is an alkali metal.
b. I (iodine), Group 7A (17), Period 5, halogen, is a nonmetal.
c. Si (silicon), Group 4A (14), Period 3, is a metalloid.

Strontium provides the red color in fireworks.

TEST

Try Practice Problems 4.7 to 4.16

STUDY CHECK 4.2

Strontium is an element that gives a brilliant red color to fireworks.

a. In what group is strontium found?
b. What is the name of this chemical family?
c. In what period is strontium found?
d. Is strontium a metal, a nonmetal, or a metalloid?

ANSWER

a. Group 2A (2)
b. alkaline earth metals
c. Period 5
d. metal

CHEMISTRY LINK TO HEALTH

Elements Essential to Health

Of all the elements, only about 20 are essential for the well-being and survival of the human body. Of those, four elements—oxygen, carbon, hydrogen, and nitrogen—which are representative elements in Period 1 and Period 2 on the periodic table, make up 96% of our body mass. Most of the food in our daily diet provides these elements to maintain a healthy body. These elements are found in carbohydrates, fats, and proteins. Most of the hydrogen and oxygen is found in water, which makes up 55 to 60% of our body mass.

The *macrominerals*—Ca, P, K, Cl, S, Na, and Mg—are located in Period 3 and Period 4 of the periodic table. They are involved in the formation of bones and teeth, maintenance of heart and blood vessels, muscle contraction, nerve impulses, acid–base balance of body fluids, and regulation of cellular metabolism. The macrominerals are present in lower amounts than the major elements, so that smaller amounts are required in our daily diets.

The other essential elements, called *microminerals* or *trace elements*, are mostly transition elements in Period 4 along with Si in Period 3 and Mo and I in Period 5. They are present in the human body in very small amounts, some less than 100 mg. In recent years, the detection of such small amounts has improved so that researchers can more easily identify the roles of trace elements. Some trace elements such as arsenic, chromium, and selenium are toxic at high levels in the body but are still required by the body. Other elements, such as tin and nickel, are thought to be essential, but their metabolic role has not yet been determined. Some examples and the amounts present in a 60.-kg person are listed in **TABLE 4.4**.

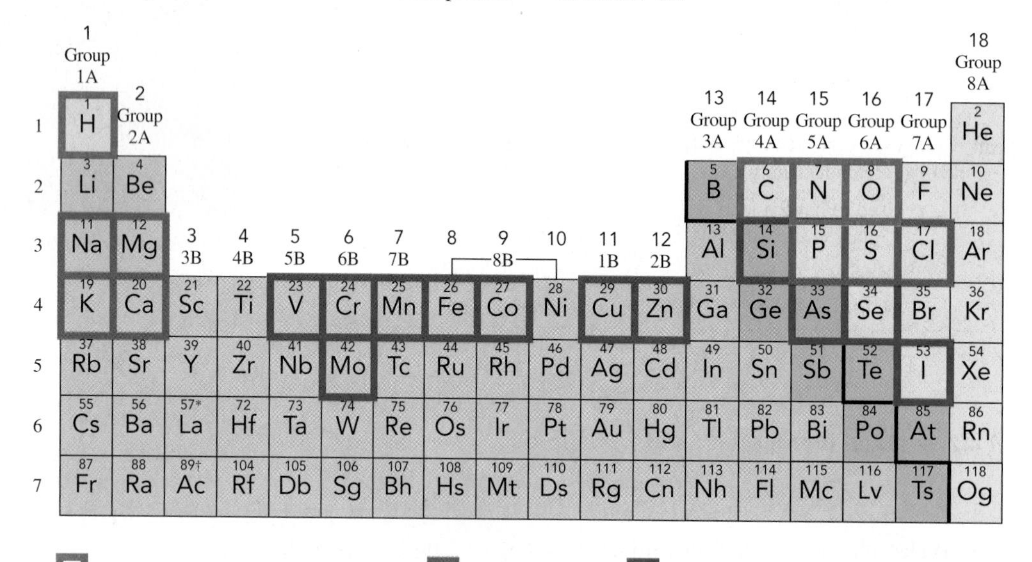

Major elements in the human body | Macrominerals | Microminerals (trace elements)

TABLE 4.4 Typical Amounts of Essential Elements in a 60.-kg Adult

Element	Quantity	Function
Major Elements		
Oxygen (O)	39 kg	Building block of biomolecules and water (H_2O)
Carbon (C)	11 kg	Building block of organic and biomolecules
Hydrogen (H)	6 kg	Component of biomolecules, water (H_2O), regulates pH of body fluids, stomach acid (HCl)
Nitrogen (N)	2 kg	Component of proteins and nucleic acids

(continued)

TABLE 4.4 Typical Amounts of Essential Elements in a 60.-kg Adult (*Continued*)

Element	Quantity	Function
Macrominerals		
Calcium (Ca)	1000 g	Needed for bones and teeth, muscle contraction, nerve impulses
Phosphorus (P)	600 g	Needed for bones and teeth, nucleic acids
Potassium (K)	120 g	Most common positive ion (K^+) in cells, muscle contraction, nerve impulses
Chlorine (Cl)	100 g	Most common negative ion (Cl^-) in fluids outside cells, stomach acid (HCl)
Sulfur (S)	86 g	Component of proteins, liver, vitamin B_1, insulin
Sodium (Na)	60 g	Most common positive ion (Na^+) in fluids outside cells, water balance, muscle contraction, nerve impulses
Magnesium (Mg)	36 g	Component of bones, required for metabolic reactions
Microminerals (Trace Elements)		
Iron (Fe)	3600 mg	Component of oxygen carrier hemoglobin
Silicon (Si)	3000 mg	Needed for growth and maintenance of bones and teeth, tendons and ligaments, hair and skin
Zinc (Zn)	2000 mg	Needed for metabolic reactions in cells, DNA synthesis, growth of bones, teeth, connective tissue, immune system
Copper (Cu)	240 mg	Needed for blood vessels, blood pressure, immune system
Manganese (Mn)	60 mg	Needed for growth of bones, blood clotting, metabolic reactions
Iodine (I)	20 mg	Needed for proper thyroid function
Molybdenum (Mo)	12 mg	Needed to process Fe and N from food
Arsenic (As)	3 mg	Needed for growth and reproduction
Chromium (Cr)	3 mg	Needed for maintenance of blood sugar levels, synthesis of biomolecules
Cobalt (Co)	3 mg	Component of vitamin B_{12}, red blood cells
Selenium (Se)	2 mg	Used in the immune system, health of heart and pancreas
Vanadium (V)	2 mg	Needed in the formation of bones and teeth, energy from food

PRACTICE PROBLEMS

4.2 The Periodic Table

LEARNING GOAL Use the periodic table to identify the group and the period of an element; identify the element as a metal, a nonmetal, or a metalloid.

4.7 Identify the group or period number described by each of the following:
a. contains C, N, and O
b. begins with helium
c. contains the alkali metals
d. ends with neon

4.8 Identify the group or period number described by each of the following:
a. contains Na, K, and Rb
b. begins with Be
c. contains the noble gases
d. contains B, N, and F

4.9 Give the symbol of the element described by each of the following:
a. Group 4A (14), Period 2
b. the noble gas in Period 1
c. the alkali metal in Period 3
d. Group 2A (2), Period 4
e. Group 3A (13), Period 3

4.10 Give the symbol of the element described by each of the following:
a. the alkaline earth metal in Period 2
b. Group 5A (15), Period 3
c. the noble gas in Period 4
d. the halogen in Period 5
e. Group 4A (14), Period 4

4.11 Identify each of the following elements as a metal, a nonmetal, or a metalloid:
a. calcium
b. sulfur
c. a shiny element
d. an element that is a gas at room temperature
e. located in Group 8A (18)
f. bromine
g. boron
h. silver

4.12 Identify each of the following elements as a metal, a nonmetal, or a metalloid:
a. located in Group 2A (2)
b. a good conductor of electricity
c. chlorine
d. arsenic
e. an element that is not shiny
f. oxygen
g. nitrogen
h. tin

Clinical Applications

4.13 Using Table 4.4, identify the function of each of the following in the body and classify each as an alkali metal, an alkaline earth metal, a transition element, or a halogen:
a. Ca
b. Fe
c. K
d. Cl

4.14 Using Table 4.4, identify the function of each of the following in the body and classify each as an alkali metal, an alkaline earth metal, a transition element, or a halogen:
a. Mg **b.** Cu **c.** I **d.** Na

4.15 Using the Chemistry Link to Health: Elements Essential to Health, answer each of the following:
a. What is a macromineral?
b. What is the role of sulfur in the human body?
c. How many grams of sulfur would be a typical amount in a 60.-kg adult?

4.16 Using the Chemistry Link to Health: Elements Essential to Health, answer each of the following:
a. What is a micromineral?
b. What is the role of iodine in the human body?
c. How many milligrams of iodine would be a typical amount in a 60.-kg adult?

4.3 The Atom

LEARNING GOAL Describe the electrical charge and location in an atom for a proton, a neutron, and an electron.

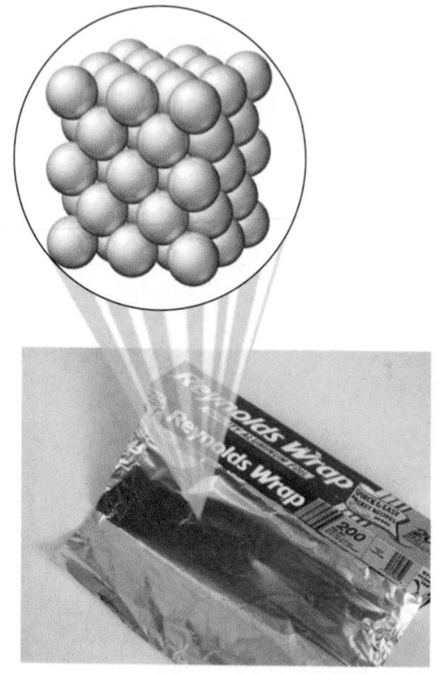

Aluminum foil consists of atoms of aluminum.

All the elements listed on the periodic table are made up of atoms. An **atom** is the smallest particle of an element that retains the characteristics of that element. Imagine that you are tearing a piece of aluminum foil into smaller and smaller pieces. Now imagine that you have a microscopic piece so small that it cannot be divided any further. Then you would have a single atom of aluminum.

The concept of the atom is relatively recent. Although the Greek philosophers in 500 B.C.E. reasoned that everything must contain minute particles they called *atomos*, the idea of atoms did not become a scientific theory until 1808. Then, John Dalton (1766–1844) developed an atomic theory that proposed that atoms were responsible for the combinations of elements found in compounds.

Dalton's Atomic Theory

1. All matter is made up of tiny particles called atoms.
2. All atoms of a given element are the same and different from atoms of other elements.
3. Atoms of two or more different elements combine to form compounds. A particular compound is always made up of the same kinds of atoms and always has the same number of each kind of atom.
4. A chemical reaction involves the rearrangement, separation, or combination of atoms. Atoms are neither created nor destroyed during a chemical reaction.

ENGAGE
What does Dalton's atomic theory say about the atoms in a compound?

Dalton's atomic theory formed the basis of current atomic theory, although we have modified some of Dalton's statements. We now know that atoms of the same element are not completely identical to each other and consist of even smaller particles. However, an atom is still the smallest particle that retains the properties of an element.

Although atoms are the building blocks of everything we see around us, we cannot see an atom or even a billion atoms with the naked eye. However, when billions and billions of atoms are packed together, the characteristics of each atom are added to those of the next until we can see the characteristics we associate with the element. For example, a small piece of the element gold consists of many, many gold atoms. A special kind of microscope called a *scanning tunneling microscope* (STM) produces images of individual atoms (see **FIGURE 4.5**).

FIGURE 4.5 Images of gold atoms are produced when magnified 16 million times by a scanning microscope.
Why is a microscope with extremely high magnification needed to see these atoms?

Electrical Charges in an Atom

By the end of the 1800s, experiments with electricity showed that atoms were not solid spheres but were composed of smaller bits of matter called **subatomic particles**, three of which are the *proton*, *neutron*, and *electron*. Two of these subatomic particles were discovered because they have electrical charges.

An electrical charge can be positive or negative. Experiments show that like charges repel or push away from each other. When you brush your hair on a dry day, electrical charges that are alike build up on the brush and in your hair. As a result, your hair flies away from the brush. However, opposite or unlike charges attract. The crackle of clothes taken from the clothes dryer indicates the presence of electrical charges. The clinginess of the clothing results from the attraction of opposite, unlike charges (see FIGURE 4.6).

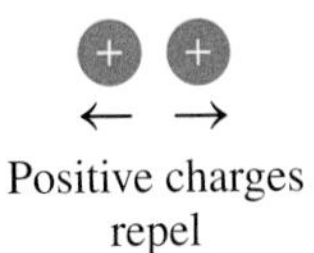

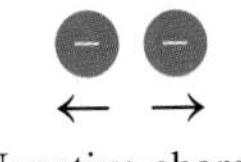

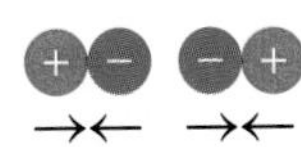

FIGURE 4.6 Like charges repel and unlike charges attract.

Q Why are the electrons attracted to the protons in the nucleus of an atom?

Structure of the Atom

In 1897, J. J. Thomson, an English physicist, applied electricity to electrodes sealed in a glass tube, which produced streams of small particles called *cathode rays*. Because these rays were attracted to a positively charged electrode, Thomson realized that the particles in the rays must be negatively charged. In further experiments, these particles called **electrons** were found to be much smaller than the atom and to have extremely small masses. Because atoms are neutral, scientists soon discovered that atoms contained positively charged particles called **protons** that were much heavier than the electrons.

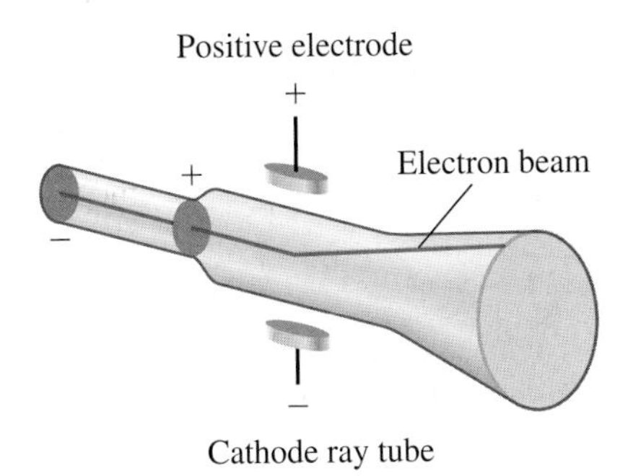

Negatively charged cathode rays (electrons) are attracted to the positive electrode.

Thomson proposed a "plum-pudding" model for the atom in which the electrons and protons were randomly distributed in a positively charged cloud like plums in a pudding. In 1911, Ernest Rutherford worked with Thomson to test this model. In Rutherford's experiment, positively charged particles were aimed at a thin sheet of gold foil (see FIGURE 4.7). If the Thomson model were correct, the particles would travel in straight paths through the gold foil. Rutherford was greatly surprised to find that some of the particles were deflected as they passed through the gold foil, and a few particles were deflected so much that they went back in the opposite direction. According to Rutherford, it was as though he had shot a cannonball at a piece of tissue paper, and it bounced back at him.

From his gold-foil experiments, Rutherford realized that the protons must be contained in a small, positively charged region at the center of the atom, which he called the **nucleus**. He proposed that the electrons in the atom occupy the space surrounding the nucleus through which most of the particles traveled undisturbed. Only the particles that

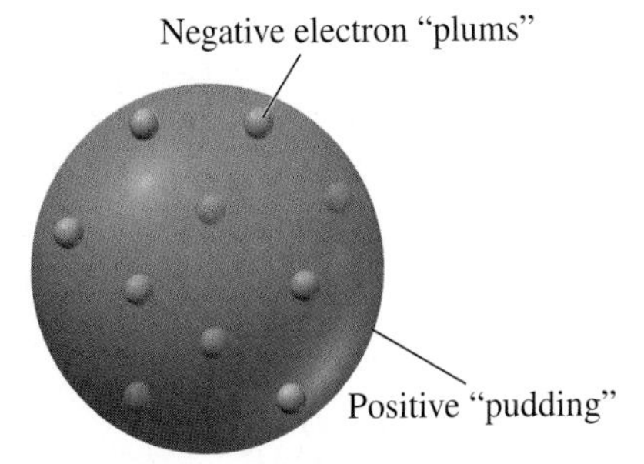

Thomson's "plum-pudding" model had protons and electrons scattered throughout the atom.

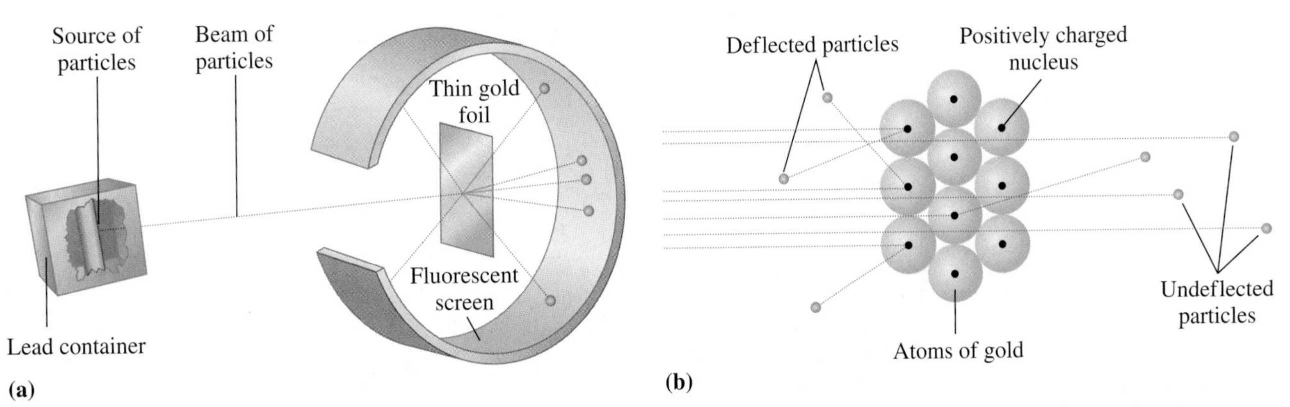

FIGURE 4.7 **(a)** Positive particles are aimed at a piece of gold foil. **(b)** Particles that come close to the atomic nuclei are deflected from their straight path.

Q Why are some particles deflected whereas most pass through the gold foil undeflected?

INTERACTIVE VIDEO

Rutherford's Gold-Foil Experiment

came near this dense, positive center were deflected. If an atom were the size of a football stadium, the nucleus would be about the size of a golf ball placed in the center of the field.

Scientists knew that the nucleus was heavier than the mass of the protons, so they looked for another subatomic particle. Eventually, they discovered that the nucleus also contained a particle that is neutral, which they called a **neutron**. Thus, the masses of the protons and neutrons in the nucleus determine the mass of an atom (see **FIGURE 4.8**).

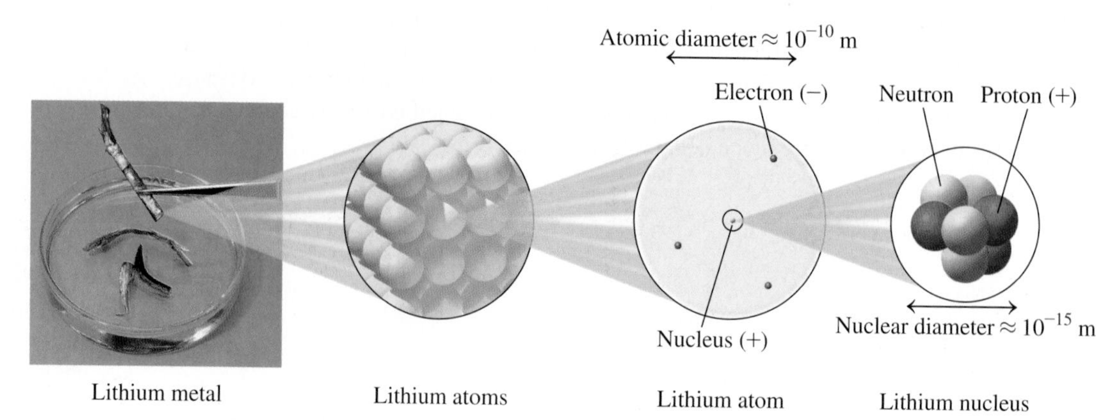

FIGURE 4.8 In an atom, the protons and neutrons that make up almost all the mass are packed into the tiny volume of the nucleus. The rapidly moving electrons (negative charge) surround the nucleus and account for the large volume of the atom.

Q Why can we say that the atom is mostly empty space?

Mass of the Atom

All the subatomic particles are extremely small compared with the things you see around you. One proton has a mass of 1.67×10^{-24} g, and the neutron is about the same. However, the electron has a mass 9.11×10^{-28} g, which is much less than the mass of either a proton or neutron. Because the masses of subatomic particles are so small, chemists use a very small unit of mass called an **atomic mass unit (amu)**. An amu is defined as one-twelfth of the mass of a carbon atom, which has a nucleus containing six protons and six neutrons. In biology, the atomic mass unit is called a *Dalton* (Da) in honor of John Dalton. On the amu scale, the proton and neutron each have a mass of about 1 amu. Because the electron mass is so small, it is usually ignored in atomic mass calculations. **TABLE 4.5** summarizes some information about the subatomic particles in an atom.

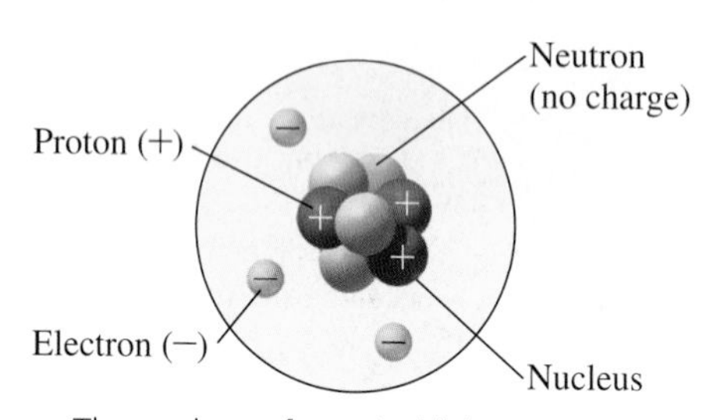

The nucleus of a typical lithium atom contains three protons and four neutrons.

TABLE 4.5 Subatomic Particles in the Atom

Particle	Symbol	Charge	Mass (amu)	Location in Atom
Proton	p or p^+	1+	1.007	Nucleus
Neutron	n or n^0	0	1.008	Nucleus
Electron	e^-	1−	0.000 55	Outside nucleus

SAMPLE PROBLEM 4.3 Subatomic Particles

TRY IT FIRST

Indicate whether each of the following is *true* or *false*:

a. A proton is heavier than an electron.
b. An electron is attracted to a neutron.
c. The nucleus contains all the protons and neutrons of an atom.

SOLUTION

a. True
b. False; an electron is attracted to a proton.
c. True

STUDY CHECK 4.3

Is the following statement *true* or *false*?
The nucleus occupies a large volume in an atom.

ANSWER

False, most of the volume of the atom is outside the nucleus.

TEST

Try Practice Problems 4.17 to 4.24

PRACTICE PROBLEMS

4.3 The Atom

LEARNING GOAL Describe the electrical charge and location in an atom for a proton, a neutron, and an electron.

4.17 Identify each of the following as describing either a proton, a neutron, or an electron:
a. has the smallest mass
b. has a 1+ charge
c. is found outside the nucleus
d. is electrically neutral

4.18 Identify each of the following as describing either a proton, a neutron, or an electron:
a. has a mass about the same as a proton
b. is found in the nucleus
c. is attracted to the protons
d. has a 1− charge

4.19 What did Rutherford determine about the structure of the atom from his gold-foil experiment?

4.20 How did Thomson determine that the electrons have a negative charge?

4.21 Is each of the following statements *true* or *false*?
a. A proton and an electron have opposite charges.
b. The nucleus contains most of the mass of an atom.
c. Electrons repel each other.
d. A proton is attracted to a neutron.

4.22 Is each of the following statements *true* or *false*?
a. A proton is attracted to an electron.
b. A neutron has twice the mass of a proton.
c. Neutrons repel each other.
d. Electrons and neutrons have opposite charges.

4.23 On a dry day, your hair flies apart when you brush it. How would you explain this?

4.24 Sometimes clothes cling together when removed from a dryer. What kinds of charges are on the clothes?

4.4 Atomic Number and Mass Number

REVIEW

Using Positive and Negative Numbers in Calculations (1.4)

LEARNING GOAL Given the atomic number and the mass number of an atom, state the number of protons, neutrons, and electrons.

All the atoms of the same element always have the same number of protons. This feature distinguishes atoms of one element from atoms of all the other elements.

Atomic Number

CORE CHEMISTRY SKILL

Counting Protons and Neutrons

The **atomic number** of an element is equal to the number of protons in every atom of that element. The atomic number is the whole number that appears above the symbol of each element on the periodic table.

Atomic number = number of protons in an atom

The periodic table on the inside front cover of this text shows the elements in order of atomic number from 1 to 118. We can use an atomic number to identify the number of protons in an atom of any element. For example, a lithium atom, with atomic number 3, has 3 protons. Any atom with 3 protons is always a lithium atom. In the same way, we determine that a carbon atom, with atomic number 6, has 6 protons. Any atom with 6 protons is carbon.

An atom is electrically neutral. That means that the number of protons in an atom is equal to the number of electrons, which gives every atom an overall charge of zero. Thus, the atomic number also gives the number of electrons.

ENGAGE

Why does every atom of barium have 56 protons and 56 electrons?

Mass Number

We now know that the protons and neutrons determine the mass of the nucleus. Thus, for a single atom, we assign a **mass number**, which is the total number of protons and neutrons in its nucleus. However, the mass number does not appear on the periodic table because it applies to single atoms only.

ENGAGE

Which subatomic particles in an atom determine its mass number?

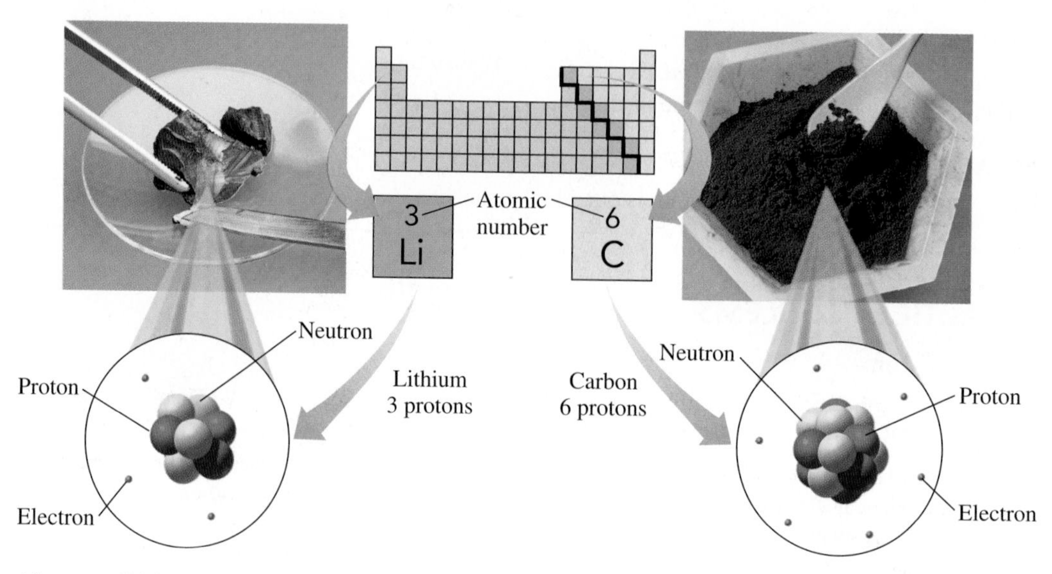

All atoms of lithium (left) contain three protons and three electrons, and all atoms of carbon (right) contain six protons and six electrons.

$$\text{Mass number} = \text{number of protons} + \text{number of neutrons}$$

For example, the nucleus of an oxygen atom that contains 8 protons and 8 neutrons has a mass number of 16. An atom of iron that contains 26 protons and 32 neutrons has a mass number of 58.

If we are given the mass number of an atom and its atomic number, we can calculate the number of neutrons in its nucleus.

ENGAGE

How many neutrons are in an atom of tin that has a mass number of 102?

$$\text{Number of neutrons in a nucleus} = \text{mass number} - \text{number of protons}$$

For example, if we are given a mass number of 37 for an atom of chlorine (atomic number 17), we can calculate the number of neutrons in its nucleus.

$$\text{Number of neutrons} = 37\ (\text{mass number}) - 17\ (\text{protons}) = 20\ \text{neutrons}$$

TABLE 4.6 illustrates the relationships between atomic number, mass number, and the number of protons, neutrons, and electrons in examples of single atoms for different elements.

TABLE 4.6 Composition of Some Atoms of Different Elements

Element	Symbol	Atomic Number	Mass Number	Number of Protons	Number of Neutrons	Number of Electrons
Hydrogen	H	1	1	1	0	1
Nitrogen	N	7	14	7	7	7
Oxygen	O	8	16	8	8	8
Chlorine	Cl	17	37	17	20	17
Iron	Fe	26	58	26	32	26
Gold	Au	79	197	79	118	79

TEST

Try Practice Problems 4.25 to 4.28

SAMPLE PROBLEM 4.4 Calculating Numbers of Protons, Neutrons, and Electrons

TRY IT FIRST

Zinc, a micronutrient, is needed for metabolic reactions in cells, DNA synthesis, the growth of bone, teeth, and connective tissue, and the proper functioning of the immune system. For an atom of zinc that has a mass number of 68, determine the number of:

a. protons
b. neutrons
c. electrons

SOLUTION

ANALYZE THE PROBLEM	Given	Need	Connect
	zinc (Zn), mass number 68	number of protons, neutrons, electrons	periodic table, atomic number

a. Zinc (Zn), with an atomic number of 30, has 30 protons.

b. The number of neutrons in this atom is found by subtracting the number of protons (atomic number) from the mass number.

Mass number	−	atomic number	=	number of neutrons
68	−	30	=	38

c. Because a zinc atom is neutral, the number of electrons is equal to the number of protons. A zinc atom has 30 electrons.

STUDY CHECK 4.4

How many neutrons are in the nucleus of a bromine atom that has a mass number of 80?

ANSWER

45

TEST

Try Practice Problems 4.29 to 4.32

CHEMISTRY LINK TO THE ENVIRONMENT

Many Forms of Carbon

Carbon has the symbol C. However, its atoms can be arranged in different ways to give several different substances. Two forms of carbon—diamond and graphite—have been known since prehistoric times. A diamond is transparent and harder than any other substance, whereas graphite is black and soft. In diamond, carbon atoms are arranged in a rigid structure. In graphite, carbon atoms are arranged in flat sheets that slide over each other. Graphite is used as pencil lead and as a lubricant.

Two other forms of carbon have been discovered more recently. In the form called *Buckminsterfullerene* or *buckyball* (named after R. Buckminster Fuller, who popularized the geodesic dome), 60 carbon atoms are arranged as rings of five and six atoms to give a spherical, cage-like structure. When a fullerene structure is stretched out, it produces a cylinder with a diameter of only a few nanometers called a *nanotube*. Practical uses for buckyballs and nanotubes are not yet developed, but they are expected to find use in lightweight structural materials, heat conductors, computer parts, and medicine. Recent research has shown that carbon nanotubes (CNT) can carry many drug molecules that can be released once the CNT enter the targeted cells.

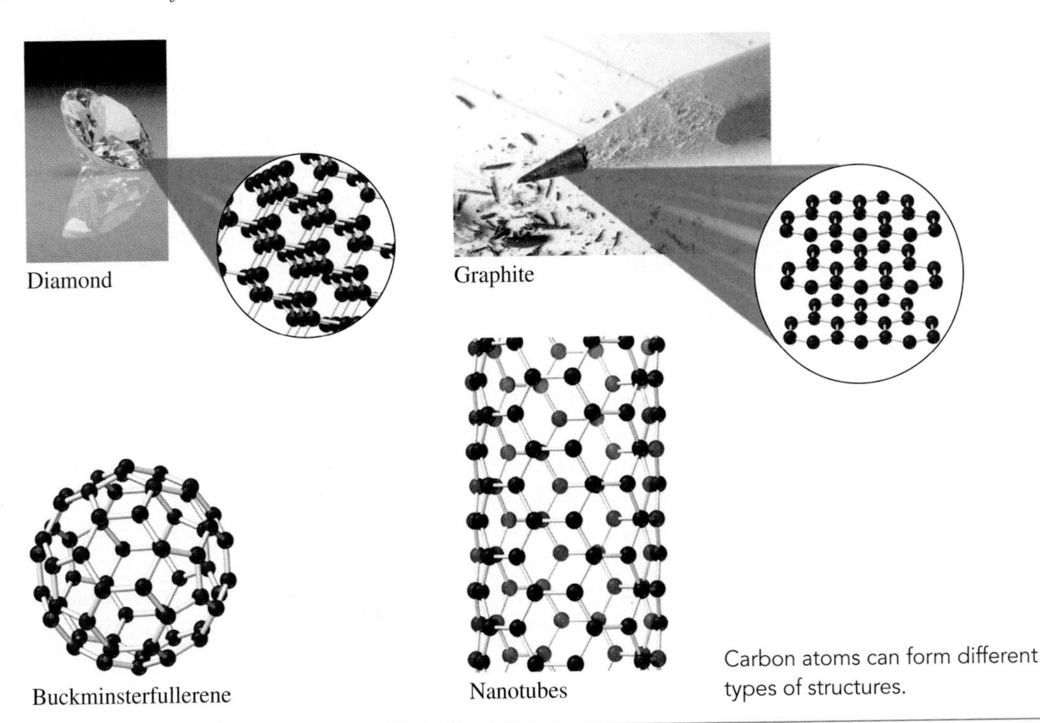

Carbon atoms can form different types of structures.

PRACTICE PROBLEMS

4.4 Atomic Number and Mass Number

LEARNING GOAL Given the atomic number and the mass number of an atom, state the number of protons, neutrons, and electrons.

4.25 Would you use the atomic number, mass number, or both to determine each of the following?
a. number of protons in an atom
b. number of neutrons in an atom
c. number of particles in the nucleus
d. number of electrons in a neutral atom

4.26 Identify the type of subatomic particles described by each of the following:
a. atomic number
b. mass number
c. mass number − atomic number
d. mass number + atomic number

4.27 Write the names and symbols for the elements with the following atomic numbers:
a. 3 **b.** 9 **c.** 20 **d.** 30
e. 10 **f.** 14 **g.** 53 **h.** 8

4.28 Write the names and symbols for the elements with the following atomic numbers:
a. 1 **b.** 11 **c.** 19 **d.** 82
e. 35 **f.** 47 **g.** 15 **h.** 2

4.29 How many protons and electrons are there in a neutral atom of each of the following elements?
a. argon **b.** manganese
c. iodine **d.** cadmium

4.30 How many protons and electrons are there in a neutral atom of each of the following elements?
a. carbon **b.** fluorine
c. tin **d.** nickel

Clinical Applications

4.31 Complete the following table for atoms of essential elements in the body:

Name of the Element	Symbol	Atomic Number	Mass Number	Number of Protons	Number of Neutrons	Number of Electrons
	Zn		66			
		12			12	
Potassium					20	
				16	15	
			56			26

4.32 Complete the following table for atoms of essential elements in the body:

Name of the Element	Symbol	Atomic Number	Mass Number	Number of Protons	Number of Neutrons	Number of Electrons
	N		15			
Calcium			42			
				53	72	
		14			16	
		29	65			

4.5 Isotopes and Atomic Mass

LEARNING GOAL Determine the number of protons, neutrons, and electrons in one or more of the isotopes of an element; identify the most abundant isotope of an element.

We have seen that all atoms of the same element have the same number of protons and electrons. However, the atoms are not entirely identical because the atoms of most elements have different numbers of neutrons. When a sample of an element consists of two or more atoms with differing numbers of neutrons, those atoms are called *isotopes*.

Atoms and Isotopes

CORE CHEMISTRY SKILL
Writing Atomic Symbols for Isotopes

Isotopes are atoms of the same element that have the same atomic number but different numbers of neutrons. For example, all atoms of the element magnesium (Mg) have an atomic number of 12. Thus, every magnesium atom has 12 protons. However, some naturally occurring magnesium atoms have 12 neutrons, others have 13 neutrons, and still others have 14 neutrons. The different numbers of neutrons give the magnesium atoms different mass numbers but do not change their chemical behavior.

To distinguish between the different isotopes of an element, we write an **atomic symbol** that indicates the mass number in the upper left corner and the atomic number in the lower left corner.

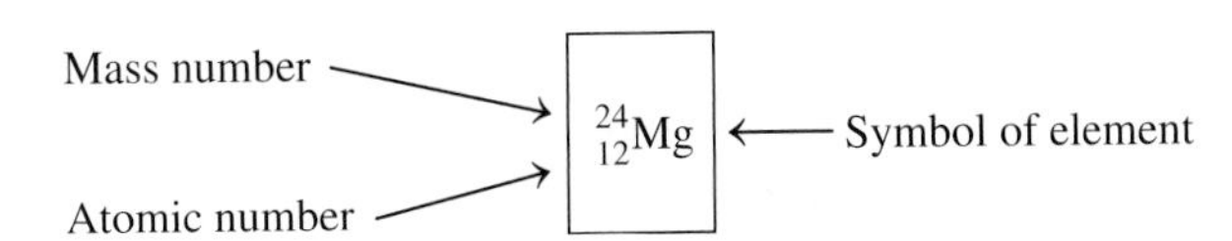

Atomic symbol for an isotope of magnesium, Mg-24.

An isotope may be referred to by its name or symbol, followed by its mass number, such as magnesium-24 or Mg-24. Magnesium has three naturally occurring isotopes, as shown in **TABLE 4.7**. In a large sample of naturally occurring magnesium atoms, each type of isotope can be present as a low percentage or a high percentage. For example, the Mg-24 isotope makes up almost 80% of the total sample, whereas Mg-25 and Mg-26 each make up only about 10% of the total number of magnesium atoms.

TABLE 4.7 Isotopes of Magnesium

Atomic Symbol	$^{24}_{12}Mg$	$^{25}_{12}Mg$	$^{26}_{12}Mg$
Name	Mg-24	Mg-25	Mg-26
Number of Protons	12	12	12
Number of Electrons	12	12	12
Mass Number	24	25	26
Number of Neutrons	12	13	14
Percent	78.70	10.13	11.17

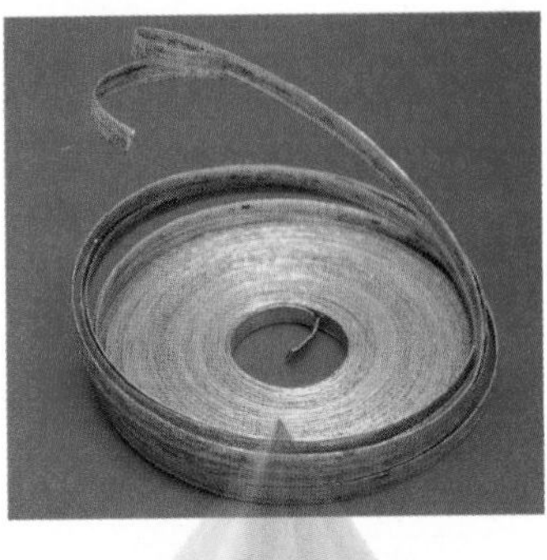

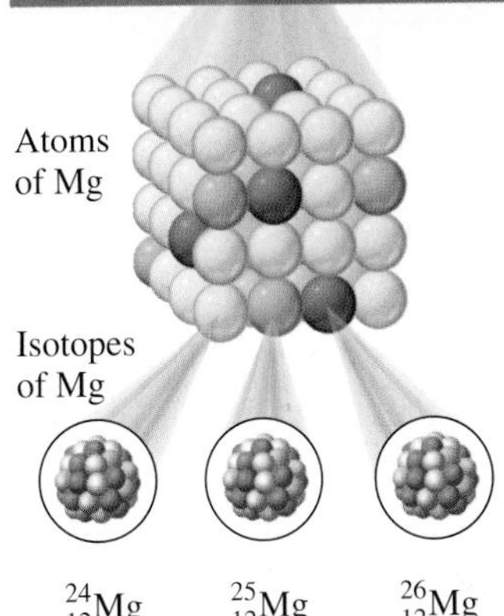

The nuclei of three naturally occurring magnesium isotopes have the same number of protons but different numbers of neutrons.

SAMPLE PROBLEM 4.5 Identifying Protons and Neutrons in Isotopes

TRY IT FIRST

Chromium, a micronutrient needed for maintenance of blood sugar levels, has four naturally occurring isotopes. Calculate the number of protons and number of neutrons in each of the following isotopes:

a. $^{50}_{24}Cr$ **b.** $^{52}_{24}Cr$ **c.** $^{53}_{24}Cr$ **d.** $^{54}_{24}Cr$

SOLUTION

ANALYZE THE PROBLEM	Given	Need	Connect
	atomic symbols for Cr isotopes	number of protons, number of neutrons	periodic table, atomic number

In the atomic symbol, the mass number is shown in the upper left corner of the symbol, and the atomic number is shown in the lower left corner of the symbol. Thus, each isotope of Cr, atomic number 24, has 24 protons. The number of neutrons is found by subtracting the number of protons (24) from the mass number of each isotope.

Atomic Symbol	Atomic Number	Mass Number	Number of Protons	Number of Neutrons
a. $^{50}_{24}Cr$	24	50	24	26 (50 − 24)
b. $^{52}_{24}Cr$	24	52	24	28 (52 − 24)
c. $^{53}_{24}Cr$	24	53	24	29 (53 − 24)
d. $^{54}_{24}Cr$	24	54	24	30 (54 − 24)

STUDY CHECK 4.5

Vanadium is a micronutrient needed in the formation of bones and teeth. Vanadium has two naturally occurring isotopes, V-50 and V-51. Write the atomic symbols for these isotopes of vanadium.

ANSWER

$^{50}_{23}V$ $^{51}_{23}V$

ENGAGE

What is different and what is the same for an atom of Sn-105 and an atom of Sn-132?

TEST

Try Practice Problems 4.33 to 4.36

ENGAGE

What is the difference between mass number and atomic mass?

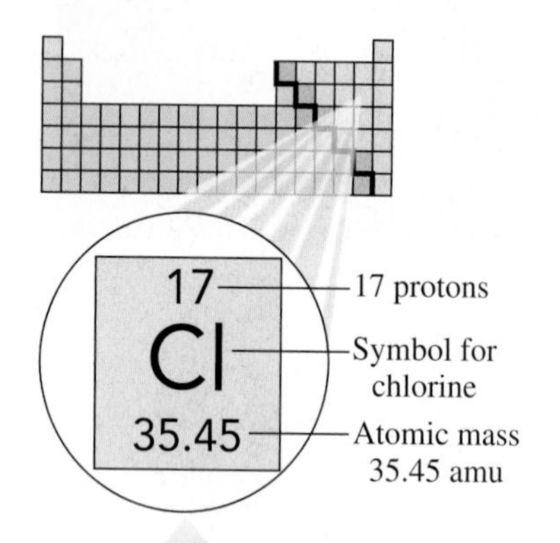

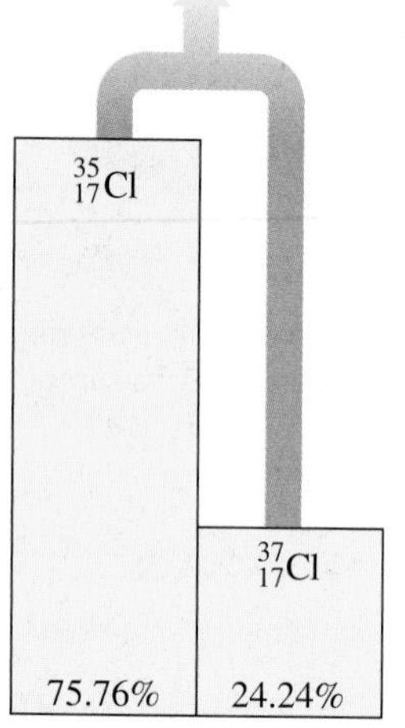

Chlorine, with two naturally occurring isotopes, has an atomic mass of 35.45 amu.

TEST

Try Practice Problems 4.37 to 4.44

Atomic Mass

In laboratory work, a chemist generally uses samples that contain all the different isotopes of an element. Because each isotope has a different mass, chemists have calculated an **atomic mass** for an "average atom," which is the *average* of the masses of the naturally occurring isotopes of that element. On the periodic table, the atomic mass is the number including decimal places that is given below the symbol of each element. Most elements consist of two or more isotopes, which is one reason that the atomic masses on the periodic table are seldom whole numbers.

For example, the atomic mass of 35.45 amu for chlorine that we see on the periodic table is the average mass of a sample of Cl atoms, although no individual Cl atom actually has this mass. By experiment, we know that chlorine consists of two naturally occurring isotopes, Cl-35 and Cl-37. The atomic mass of 35.45 is closer to the mass number of Cl-35, which indicates there is a higher percentage of $^{35}_{17}Cl$ atoms in the chlorine sample. In fact, there are about three atoms of $^{35}_{17}Cl$ for every one atom of $^{37}_{17}Cl$ in a sample of chlorine atoms.

TABLE 4.8 lists the naturally occurring isotopes of some selected elements and their atomic masses along with their most abundant isotopes.

TABLE 4.8 The Atomic Mass of Some Elements

Element	Atomic Symbols	Atomic Mass	Most Abundant Isotope
Lithium	$^{6}_{3}Li$, $^{7}_{3}Li$	6.941 amu	$^{7}_{3}Li$
Carbon	$^{12}_{6}C$, $^{13}_{6}C$, $^{14}_{6}C$	12.01 amu	$^{12}_{6}C$
Oxygen	$^{16}_{8}O$, $^{17}_{8}O$, $^{18}_{8}O$	16.00 amu	$^{16}_{8}O$
Fluorine	$^{19}_{9}F$	19.00 amu	$^{19}_{9}F$
Sulfur	$^{32}_{16}S$, $^{33}_{16}S$, $^{34}_{16}S$, $^{36}_{16}S$	32.07 amu	$^{32}_{16}S$
Potassium	$^{39}_{19}K$, $^{40}_{19}K$, $^{41}_{19}K$	39.10 amu	$^{39}_{19}K$
Copper	$^{63}_{29}Cu$, $^{65}_{29}Cu$	63.55 amu	$^{63}_{29}Cu$

We can illustrate how the most abundant isotope is determined for a sample containing two or more isotopes. For example, the naturally occurring isotopes of magnesium are Mg-24, Mg-25, and Mg-26. On the periodic table, the atomic mass of magnesium, shown below the element symbol, is 24.31. When we compare the atomic mass with the masses of the three isotopes of magnesium, it is closest to the mass of Mg-24. Thus, magnesium-24 is the most abundant isotope of magnesium (see **FIGURE 4.9**).

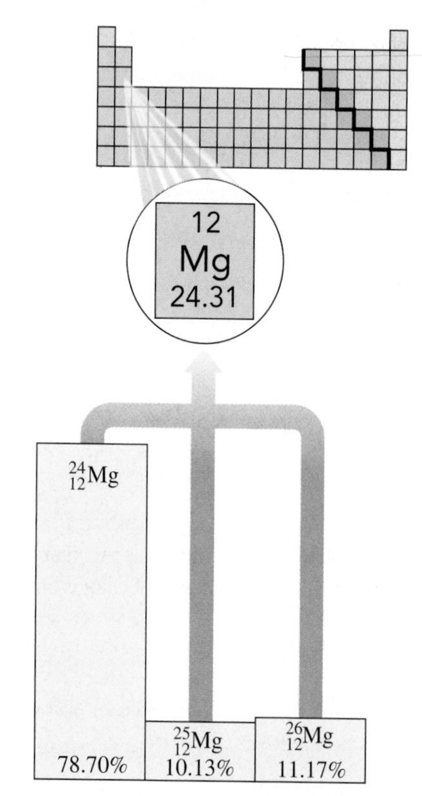

FIGURE 4.9 Magnesium, with three naturally occurring isotopes, has an atomic mass of 24.31 amu.

Q Why is the atomic mass of magnesium closest to that of Mg-24?

PRACTICE PROBLEMS

4.5 Isotopes and Atomic Mass

LEARNING GOAL Determine the number of protons, neutrons, and electrons in one or more of the isotopes of an element; identify the most abundant isotope of an element.

4.33 What are the number of protons, neutrons, and electrons in the following isotopes?

a. $^{89}_{38}Sr$ **b.** $^{52}_{24}Cr$
c. $^{34}_{16}S$ **d.** $^{81}_{35}Br$

4.34 What are the number of protons, neutrons, and electrons in the following isotopes?

a. $^{2}_{1}H$ **b.** $^{14}_{7}N$
c. $^{26}_{14}Si$ **d.** $^{70}_{30}Zn$

4.35 Write the atomic symbol for the isotope with each of the following characteristics:

a. 15 protons and 16 neutrons
b. 35 protons and 45 neutrons
c. 50 electrons and 72 neutrons
d. a chlorine atom with 18 neutrons
e. a mercury atom with 122 neutrons

4.36 Write the atomic symbol for the isotope with each of the following characteristics:

a. an oxygen atom with 10 neutrons
b. 4 protons and 5 neutrons
c. 25 electrons and 28 neutrons
d. a mass number of 24 and 13 neutrons
e. a nickel atom with 32 neutrons

4.37 Argon has three naturally occurring isotopes, with mass numbers 36, 38, and 40.

a. Write the atomic symbol for each of these atoms.
b. How are these isotopes alike?
c. How are they different?
d. Why is the atomic mass of argon listed on the periodic table not a whole number?
e. Which isotope is the most abundant in a sample of argon?

4.38 Strontium has four naturally occurring isotopes, with mass numbers 84, 86, 87, and 88.

a. Write the atomic symbol for each of these atoms.
b. How are these isotopes alike?
c. How are they different?
d. Why is the atomic mass of strontium listed on the periodic table not a whole number?
e. Which isotope is the most abundant in a sample of strontium?

4.39 Two isotopes of gallium occur naturally, $^{69}_{31}Ga$ and $^{71}_{31}Ga$. Which isotope of gallium is more abundant?

4.40 Two isotopes of rubidium occur naturally, $^{85}_{37}Rb$ and $^{87}_{37}Rb$. Which isotope of rubidium is more abundant?

4.41 Copper consists of two isotopes, $^{63}_{29}Cu$ and $^{65}_{29}Cu$. If the atomic mass for copper on the periodic table is 63.55, are there more atoms of $^{63}_{29}Cu$ and $^{65}_{29}Cu$ in a sample of copper?

4.42 Indium consists of two isotopes, $^{113}_{49}In$ and $^{115}_{49}In$. If the atomic mass for indium on the periodic table is 114.8, are there more atoms of $^{113}_{49}In$ or $^{115}_{49}In$ in a sample of indium?

4.43 There are two naturally occurring isotopes of thallium: $^{203}_{81}Tl$ and $^{205}_{81}Tl$. Use the atomic mass of thallium listed on the periodic table to identify the more abundant isotope.

4.44 There are five naturally occurring isotopes of zinc: $^{64}_{30}Zn$, $^{66}_{30}Zn$, $^{67}_{30}Zn$, $^{68}_{30}Zn$, and $^{70}_{30}Zn$. None of these isotopes has the atomic mass of 65.41 listed for zinc on the periodic table. Explain.

4.6 Electron Energy Levels

LEARNING GOAL Given the name or symbol of one of the first 20 elements in the periodic table, write the electron arrangement.

When we listen to a radio, use a microwave oven, turn on a light, see the colors of a rainbow, or have an X-ray taken, we are experiencing various forms of *electromagnetic radiation.* Light and other electromagnetic radiation consist of energy particles that move as waves of energy. In a wave, just like the waves in an ocean, the distance between the peaks is called the *wavelength.* All forms of electromagnetic radiation travel in space at the speed of light, 3.0×10^8 m/s, but differ in energy and wavelength. High-energy radiation has short wavelengths compared to low-energy radiation, which has longer wavelengths. The *electromagnetic spectrum* shows the arrangement of different types of electromagnetic radiation in order of increasing energy (see **FIGURE 4.10**).

When the light from the Sun passes through a prism, the light separates into a continuous color spectrum, which consists of the colors we see in a rainbow. In contrast, when light from a heated element passes through a prism, it separates into distinct lines of color separated by dark areas called an *atomic spectrum.* Each element has its own unique atomic spectrum.

A rainbow forms when light passes through water droplets.

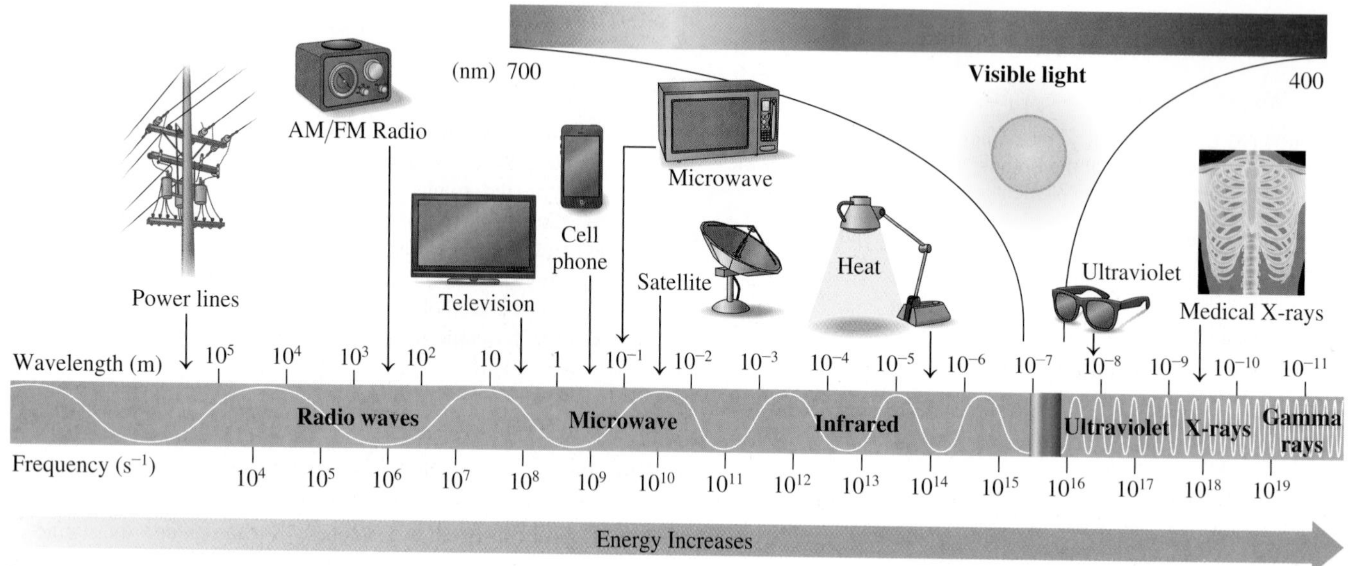

FIGURE 4.10 The electromagnetic spectrum shows the arrangement of wavelengths of electromagnetic radiation. The visible portion consists of wavelengths from 700 nm to 400 nm.

Q How do the energy and wavelength of ultraviolet light compare to that of a microwave?

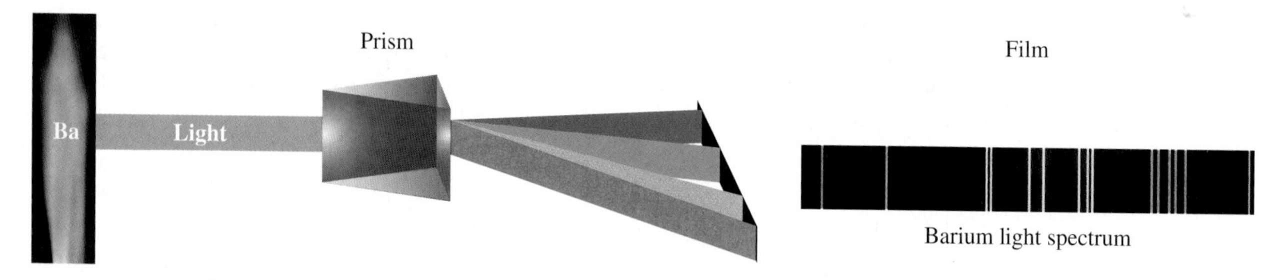

In an atomic spectrum, light from a heated element separates into distinct lines.

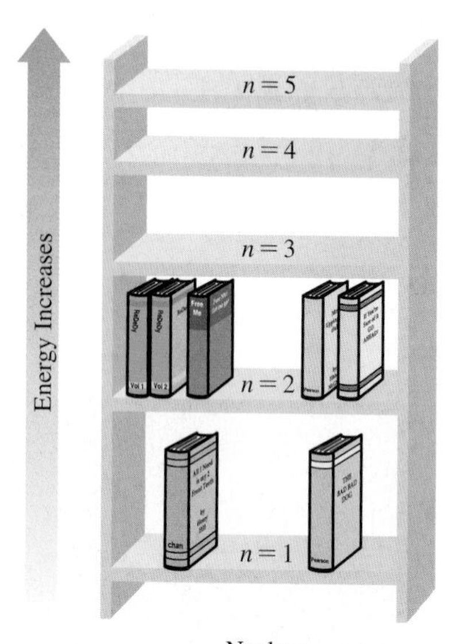

An electron can have only the energy of one of the energy levels in an atom.

Electron Energy Levels

Scientists have determined that the lines in the atomic spectra of elements are associated with changes in the energies of the electrons. In an atom, each electron has a specific energy known as its **energy level**, which is assigned values called *principal quantum numbers* (n), ($n = 1, n = 2, \ldots$). Generally, electrons in the lower energy levels are closer to the nucleus, while electrons in the higher energy levels are farther away. The energy of an electron is *quantized*, which means that the energy of an electron can only have specific energy values, but cannot have values between them.

Principal Quantum Number (n)

$1 < 2 < 3 < 4 < 5 < 6 < 7$

Lowest energy ⟶ Highest energy

All the electrons with the same energy are grouped in the same energy level. As an analogy, we can think of the energy levels of an atom as similar to the shelves in a bookcase. The first shelf is the lowest energy level; the second shelf is the second energy level, and so on. If we are arranging books on the shelves, it would take less energy to fill the bottom shelf first, and then the second shelf, and so on. However, we could never get any book to stay in the space between any of the shelves. Similarly, the energy of an electron must be at specific energy levels, and not between.

Unlike standard bookcases, however, there is a large difference between the energy of the first and second levels, but then the higher energy levels are closer together. Another difference is that the lower electron energy levels hold fewer electrons than the higher energy levels.

Changes in Electron Energy Level

An electron can change from one energy level to a higher level only if it absorbs the energy equal to the difference in energy levels. When an electron changes to a lower energy level, it emits energy equal to the difference between the two levels (see **FIGURE 4.11**). If the energy emitted is in the visible range, we see one of the colors of visible light. The yellow color of sodium streetlights and the red color of neon lights are examples of electrons emitting energy in the visible color range.

TEST

Try Practice Problems 4.45 to 4.48

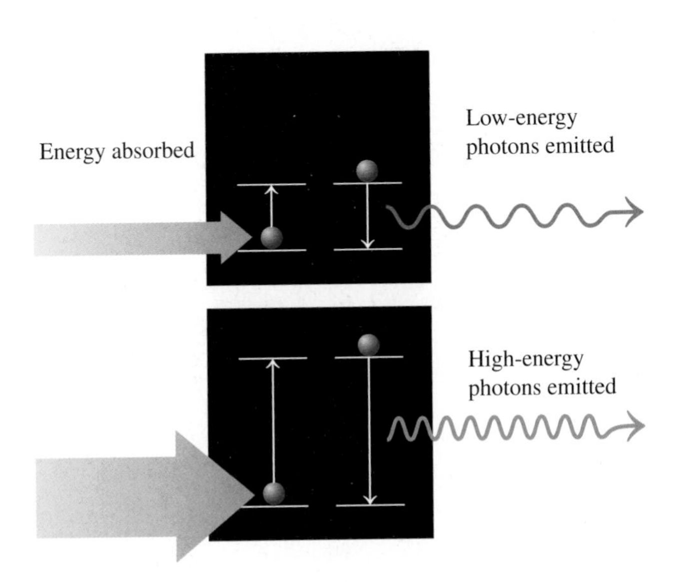

FIGURE 4.11 Electrons absorb a specific amount of energy to move to a higher energy level. When electrons lose energy, a specific quantity of energy is emitted.

What causes electrons to move to higher energy levels?

Electron Arrangements for the First 20 Elements

CORE CHEMISTRY SKILL

Writing Electron Arrangements

The **electron arrangement** of an atom gives the number of electrons in each energy level. We can write the electron arrangements for the first 20 elements by placing electrons in energy levels beginning with the lowest. There is a limit to the number of electrons allowed in each energy level. Only a few electrons can occupy the lower energy levels, while more electrons can be accommodated in higher energy levels. We can now look at the numbers of electrons in the first four energy levels for the first 20 elements as shown in **TABLE 4.9**.

Period 1

The single electron of hydrogen goes into energy level 1, and the two electrons of helium fill energy level 1. Thus, energy level 1 can hold just two electrons. The electron arrangements for H and He are shown in the margin.

Hydrogen	1
Helium	2

Period 2

For the elements of the second period (lithium, Li, to neon, Ne), we fill the first energy level with two electrons, and place the remaining electrons in the second energy level. For example, lithium has three electrons. Two of those electrons fill energy level 1. The remaining electron goes into the second energy level. We can write this electron arrangement as 2,1. Going across Period 2, more electrons are added to the second energy level. For example, an atom of carbon, with a total of six electrons, fills energy level 1, which leaves the four remaining electrons in the second energy level. The electron arrangement for carbon can be written 2,4. For neon, the last element in Period 2, both the first and second energy levels are filled to give an electron arrangement of 2,8.

Lithium	2,1
Carbon	2,4
Neon	2,8

Period 3

Sodium	2,8,1
Sulfur	2,8,6
Argon	2,8,8

For sodium, the first element in Period 3, 10 electrons fill the first and second energy levels, which means that the remaining electron must go into the third energy level. The electron arrangement for sodium can be written 2,8,1. The elements that follow sodium in the third period add electrons, one at a time, until the third level is complete. For example, a sulfur atom with 16 electrons fills the first and second energy levels, which leaves six electrons in the third level. The electron arrangement for sulfur can be written 2,8,6. At the end of Period 3, argon has eight electrons in the third level.

ENGAGE

How many electrons are in the $n = 2$ energy level of nitrogen?

TABLE 4.9 Electron Arrangements for the First 20 Elements

			Number of Electrons in Energy Level			
Element	Symbol	Atomic Number	1	2	3	4
Hydrogen	H	1	1			
Helium	He	2	2			
Lithium	Li	3	2	1		
Beryllium	Be	4	2	2		
Boron	B	5	2	3		
Carbon	C	6	2	4		
Nitrogen	N	7	2	5		
Oxygen	O	8	2	6		
Fluorine	F	9	2	7		
Neon	Ne	10	2	8		
Sodium	Na	11	2	8	1	
Magnesium	Mg	12	2	8	2	
Aluminum	Al	13	2	8	3	
Silicon	Si	14	2	8	4	
Phosphorus	P	15	2	8	5	
Sulfur	S	16	2	8	6	
Chlorine	Cl	17	2	8	7	
Argon	Ar	18	2	8	8	
Potassium	K	19	2	8	8	1
Calcium	Ca	20	2	8	8	2

Period 4

Potassium	2,8,8,1
Calcium	2,8,8,2

For potassium, the first element in Period 4, electrons fill the first and second energy levels, and eight electrons are in the third energy level. The remaining electron of potassium enters energy level 4, which gives the electron arrangement of 2,8,8,1. For calcium, the process is similar, except that calcium has two electrons in the fourth energy level. The electron arrangement for calcium can be written 2,8,8,2.

Elements Beyond 20

Following calcium, the next 10 electrons (scandium to zinc) are added to the third energy level, which already has 8 electrons, until it is complete with 18 electrons. The higher energy levels 5, 6, and 7 can theoretically accommodate up to 50, 72, and 98 electrons, but they are not completely filled. Beyond the first 20 elements, the electron arrangements become more complicated and will not be covered in this text.

Energy Level (n)	1	2	3	4	5	6	7
Number of Electrons	2	8	18	32	32	18	8

SAMPLE PROBLEM 4.6 Writing Electron Arrangements

TRY IT FIRST

Write the electron arrangement for each of the following:

a. oxygen **b.** chlorine

SOLUTION

a. Oxygen with atomic number 8 has eight electrons, which are arranged with two electrons in energy level 1 and six electrons in energy level 2.

2,6

b. Chlorine with atomic number 17 has 17 electrons, which are arranged with two electrons in energy level 1, eight electrons in energy level 2, and seven electrons in energy level 3.

2,8,7

STUDY CHECK 4.6

What element has an electron arrangement of 2,8,5?

ANSWER

phosphorus

TEST

Try Practice Problems 4.49 to 4.52

CHEMISTRY LINK TO HEALTH

Biological Reactions to UV Light

Our everyday life depends on sunlight, but exposure to sunlight can have damaging effects on living cells, and too much exposure can even cause their death. Light energy, especially ultraviolet (UV), excites electrons and may lead to unwanted chemical reactions. The list of damaging effects of sunlight includes sunburn; wrinkling; premature aging of the skin; changes in the DNA of the cells, which can lead to skin cancers; inflammation of the eyes; and perhaps cataracts. Some drugs, like the acne medications Accutane and Retin-A, as well as antibiotics, diuretics, sulfonamides, and estrogen, make the skin extremely sensitive to light.

In a disorder called *seasonal affective disorder* or SAD, people experience mood swings and depression during the winter. Some research suggests that SAD is the result of a decrease in serotonin, or an increase in melatonin, when there are fewer hours of sunlight. One treatment for SAD is therapy using bright light provided by a lamp called a light box. A daily exposure to blue light (460 nm) for 30 to 60 min seems to reduce symptoms of SAD.

Phototherapy uses light to treat certain skin conditions, including psoriasis, eczema, and dermatitis. In the treatment of psoriasis, for example, oral drugs are given to make the skin more photosensitive; then exposure to UV radiation follows. Low-energy radiation (blue light) with wavelengths from 390 to 470 nm is used to treat babies with neonatal jaundice, which converts high levels of bilirubin to water-soluble compounds that can be excreted from the body. Sunlight is also a factor in stimulating the immune system.

A light box is used to provide light, which reduces symptoms of SAD.

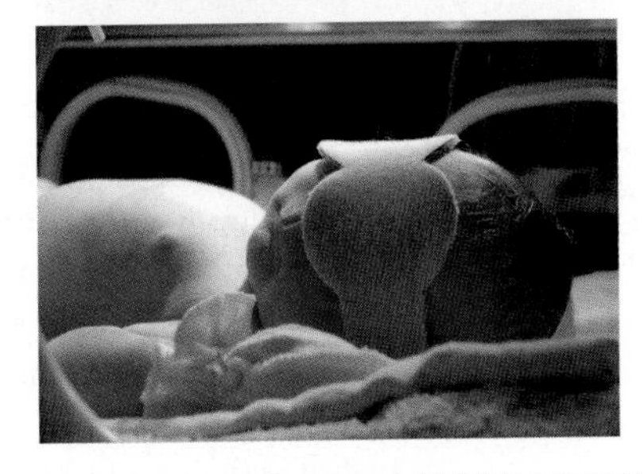

Babies with neonatal jaundice are treated with UV light.

PRACTICE PROBLEMS

4.6 Electron Energy Levels

LEARNING GOAL Given the name or symbol of one of the first 20 elements in the periodic table, write the electron arrangement.

4.45 Electrons can move to higher energy levels when they _______ (absorb/emit) energy.

4.46 Electrons drop to lower energy levels when they _______ (absorb/emit) energy.

4.47 Identify the form of electromagnetic radiation in each pair that has the greater energy:
a. green light or yellow light
b. microwaves or blue light

4.48 Identify the form of electromagnetic radiation in each pair that has the greater energy:
a. radio waves or violet light
b. infrared light or ultraviolet light

4.49 Write the electron arrangement for each of the following elements: (*Example*: sodium 2,8,1)
a. carbon **b.** argon **c.** potassium
d. silicon **e.** helium **f.** nitrogen

4.50 Write the electron arrangement for each of the following elements: (*Example*: sodium 2,8,1)
a. phosphorus **b.** neon **c.** sulfur
d. magnesium **e.** aluminum **f.** fluorine

4.51 Identify the elements that have the following electron arrangements:

	Energy Level			
	1	2	3	4
a.	2	1		
b.	2	8	2	
c.	1			
d.	2	8	7	
e.	2	6		

4.52 Identify the elements that have the following electron arrangements:

	Energy Level			
	1	2	3	4
a.	2	5		
b.	2	8	6	
c.	2	4		
d.	2	8	8	1
e.	2	8	3	

4.7 Trends in Periodic Properties

LEARNING GOAL Use the electron arrangement of elements to explain the trends in periodic properties.

SPRING SUMMER AUTUMN WINTER

The change in temperature with the seasons is a periodic property.

The electron arrangements of atoms is an important factor in the physical and chemical properties of the elements. Now we look at the *valence electrons* in atoms, the trends in *atomic size*, *ionization energy*, and *metallic character*. Known as *periodic properties*, each property increases or decreases across a period, and then the trend is repeated in each successive period. We can use the seasonal changes in temperatures as an analogy for periodic properties. In the winter, temperatures are cold and become warmer in the spring. By summer, the outdoor temperatures are hot but begin to cool in the fall. By winter, we expect cold temperatures again as the pattern of decreasing and increasing temperatures repeats for another year.

Group Number and Valence Electrons

CORE CHEMISTRY SKILL
Identifying Trends in Periodic Properties

The chemical properties of representative elements in Groups 1A (1) to 8A (18) are mostly due to the **valence electrons**, which are the electrons in the outermost energy level. The group number gives the number of valence electrons for each group of representative elements. For example, all the elements in Group 1A (1) have one valence electron. All the elements in Group 2A (2) have two valence electrons. The halogens in Group 7A (17) all have seven valence electrons. **TABLE 4.10** shows how the number of valence electrons for common representative elements is consistent with the group number.

TABLE 4.10 Comparison of Electron Arrangements, by Group, for Some Representative Elements

			Number of Electrons in Energy Level			
Group Number	**Element**	**Symbol**	**1**	**2**	**3**	**4**
1A (1)	Lithium	Li	2	**1**		
	Sodium	Na	2	8	**1**	
	Potassium	K	2	8	8	**1**
2A (2)	Beryllium	Be	2	**2**		
	Magnesium	Mg	2	8	**2**	
	Calcium	Ca	2	8	8	**2**
3A (13)	Boron	B	2	**3**		
	Aluminum	Al	2	8	**3**	
	Gallium	Ga	2	8	18	**3**
4A (14)	Carbon	C	2	**4**		
	Silicon	Si	2	8	**4**	
	Germanium	Ge	2	8	18	**4**
5A (15)	Nitrogen	N	2	**5**		
	Phosphorus	P	2	8	**5**	
	Arsenic	As	2	8	18	**5**
6A (16)	Oxygen	O	2	**6**		
	Sulfur	S	2	8	**6**	
	Selenium	Se	2	8	18	**6**
7A (17)	Fluorine	F	2	**7**		
	Chlorine	Cl	2	8	**7**	
	Bromine	Br	2	8	18	**7**
8A (18)	Helium	He	**2**			
	Neon	Ne	2	**8**		
	Argon	Ar	2	8	**8**	
	Krypton	Kr	2	8	18	**8**

SAMPLE PROBLEM 4.7 Using Group Numbers

TRY IT FIRST

Using the periodic table, write the group number and the number of valence electrons for each of the following elements:

a. cesium **b.** iodine

SOLUTION

a. Cesium (Cs) is in Group 1A (1); cesium has one valence electron.
b. Iodine (I) is in Group 7A (17); iodine has seven valence electrons.

STUDY CHECK 4.7

What is the group number of elements with atoms that have five valence electrons?

ANSWER

Group 5A (15)

TEST

Try Practice Problems 4.53 and 4.54

Lewis Symbols

CORE CHEMISTRY SKILL
Drawing Lewis Symbols

A **Lewis symbol** is a convenient way to represent the valence electrons as dots, which are placed on the sides, top, or bottom of the symbol for the element. One to four valence electrons are arranged as single dots. When an atom has five to eight valence electrons, one or more electrons are paired. Any of the following would be an acceptable Lewis symbol for magnesium, which has two valence electrons:

Lewis Symbols for Magnesium

Mg· Mg ·Mg ·Mg· Mg· ·Mg

Lewis symbols for selected elements are given in **TABLE 4.11**.

TABLE 4.11 Lewis Symbols for Selected Elements in Periods 1 to 4

	Group Number							
	1A (1)	2A (2)	3A (13)	4A (14)	5A (15)	6A (16)	7A (17)	8A (18)
Number of Valence Electrons	1	2	3	4	5	6	7	8
	Number of Valence Electrons Increases →							
Lewis Symbol	H·							He:*
	Li·	Be·	·B·	·C·	·N·	·O:	·F:	:Ne:
	Na·	Mg·	·Al·	·Si·	·P·	·S:	·Cl:	:Ar:
	K·	Ca·	·Ga·	·Ge·	·As·	·Se:	·Br:	:Kr:

*Helium (He) is stable with two valence electrons.

SAMPLE PROBLEM 4.8 Drawing Lewis Symbols

TRY IT FIRST

Draw the Lewis symbol for each of the following:

a. bromine **b.** aluminum

SOLUTION

a. Because the group number for bromine is 7A (17), bromine has seven valence electrons, which are drawn as seven dots, three pairs and one single dot, around the symbol Br.

·Br:

b. Aluminum, in Group 3A (13), has three valence electrons, which are shown as three single dots around the symbol Al.

·Al·

STUDY CHECK 4.8

What is the Lewis symbol for phosphorus?

ANSWER

·P·

TEST
Try Practice Problems 4.55 and 4.56

Atomic Size

The atomic size is determined by the distance of the valence electrons from the nucleus of the atom. For each group of representative elements, the atomic size *increases* going from the top

Atomic Size Decreases

Atomic Size Increases

Period	1A (1)	2A (2)	3A (13)	4A (14)	5A (15)	6A (16)	7A (17)	8A (18)
1	H							He
2	Li	Be	B	C	N	O	F	Ne
3	Na	Mg	Al	Si	P	S	Cl	Ar
4	K	Ca	Ga	Ge	As	Se	Br	Kr
5	Rb	Sr	In	Sn	Sb	Te	I	Xe

FIGURE 4.12 For representative elements, the atomic size increases going down a group but decreases going from left to right across a period.

Q Why does the atomic size increase going down a group of representative elements?

ENGAGE

Why is a phosphorus atom larger than a nitrogen atom but smaller than an aluminum atom?

to the bottom because the outermost electrons in each energy level are farther from the nucleus. For example, in Group 1A (1), Li has a valence electron in energy level 2; Na has a valence electron in energy level 3; and K has a valence electron in energy level 4. This means that a K atom is larger than a Na atom, and a Na atom is larger than a Li atom (see **FIGURE 4.12**).

For the elements in a period, an increase in the number of protons in the nucleus increases the attraction for the outermost electrons. As a result, the outer electrons are pulled closer to the nucleus, which means that the size of representative elements *decreases* going from left to right across a period.

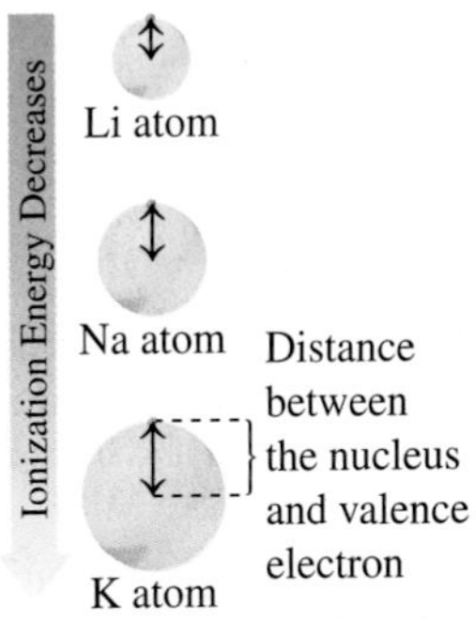

Atomic size increases going down each group as the distance from the nucleus to the valence electrons increases.

SAMPLE PROBLEM 4.9 Size of Atoms

TRY IT FIRST

Identify the smaller atom in each of the following pairs and explain your choice:

a. N or F **b.** K or Kr **c.** Ca and Sr

SOLUTION

a. The F atom has a greater positive charge on the nucleus, which pulls electrons closer, and makes the F atom smaller than the N atom. Atomic size decreases going from left to right across a period.

b. The Kr atom has a greater positive charge on the nucleus, which pulls electrons closer, and makes the Kr atom smaller than the K atom. Atomic size decreases going from left to right across a period.

c. The outer electrons in the Ca atom are closer to the nucleus than in the Sr atom, which makes the Ca atom smaller than the Sr atom. Atomic size increases going down a group.

STUDY CHECK 4.9

Which atom has the largest atomic size, Mg, Ca, or Cl?

ANSWER

Ca

TEST

Try Practice Problems 4.57 to 4.60

Ionization Energy

In an atom, negatively charged electrons are attracted to the positive charge of the protons in the nucleus. Thus, a quantity of energy known as the **ionization energy** is required to remove one of the outermost electrons. When an electron is removed from a neutral atom, a positive particle called a *cation* with a 1+ charge is formed.

$$Na(g) + \text{energy (ionization)} \longrightarrow Na^+(g) + e^-$$

The ionization energy *decreases* going down a group. Less energy is needed to remove an electron because nuclear attraction decreases when electrons are farther from the nucleus. Going across a period from left to right, the ionization energy *increases.* As the positive charge of the nucleus increases, more energy is needed to remove an electron.

In Period 1, the valence electrons are close to the nucleus and strongly held. H and He have high ionization energies because a large amount of energy is required to remove an electron. The ionization energy for He is the highest of any element because He has a full, stable, energy level which is disrupted by removing an electron. The high ionization energies of the noble gases indicate that their electron arrangements are especially stable. In general, the ionization energy is low for the metals and high for the nonmetals (see **FIGURE 4.13**).

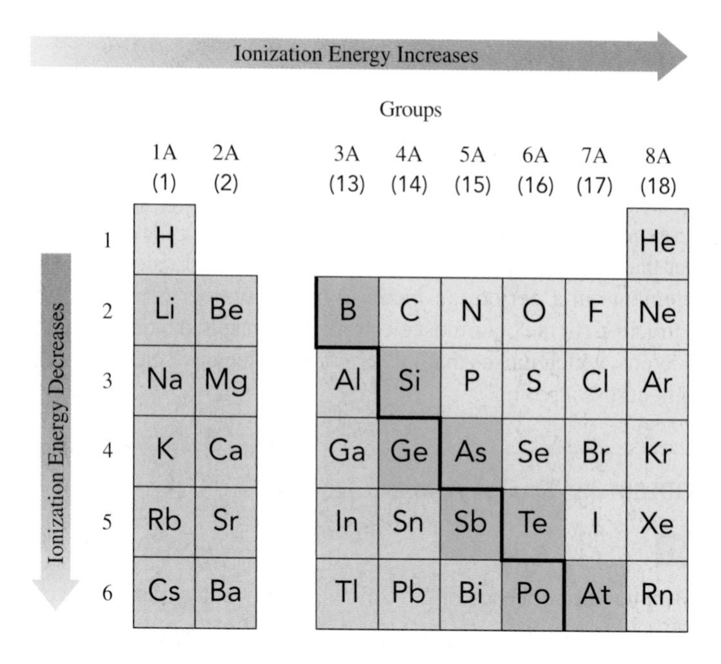

FIGURE 4.13 Ionization energies for the representative elements decrease going down a group and increase going across a period.

Why is the ionization energy for F greater than that for Cl?

SAMPLE PROBLEM 4.10 Ionization Energy

TRY IT FIRST

Indicate the element in each group that has the higher ionization energy and explain your choice.

a. K or Na **b.** Mg or Cl **c.** F, N, or C

SOLUTION

a. Na. In Na, an electron is removed from an energy level closer to the nucleus, which requires a higher ionization energy for Na compared to K.

b. Cl. The increased nuclear charge of Cl increases the attraction for the valence electrons, which requires a higher ionization energy for Cl compared to Mg.

c. F. The increased nuclear charge of F increases the attraction for the valence electrons, which requires a higher ionization energy for F compared to C or N.

STUDY CHECK 4.10

Arrange Sn, Sr, and I in order of increasing ionization energy.

ANSWER

Ionization energy increases going from left to right across a period: Sr, Sn, I.

TEST

Try Practice Problems 4.61 to 4.64

Metallic Character

An element that has **metallic character** is an element that loses valence electrons easily. Metallic character is more common in the elements (metals) on the left side of the periodic table and decreases going from left to right across a period. The elements (nonmetals) on the right side of the periodic table do not easily lose electrons, which means they are the least metallic. Most of the metalloids between the metals and nonmetals tend to lose electrons, but not as easily as the metals. Thus, in Period 3, sodium, which loses electrons most easily, would be the most metallic. Going across from left to right in Period 3, metallic character decreases to argon, which has the least metallic character.

For elements in the same group of representative elements, metallic character increases going from top to bottom. Atoms at the bottom of any group have more electron levels, which makes it easier to lose electrons. Thus, the elements at the bottom of a group on the periodic table have lower ionization energy and are more metallic compared to the elements at the top (see **FIGURE 4.14**).

ENGAGE

When a magnesium atom ionizes to form a magnesium ion Mg^{2+}, which electrons are lost?

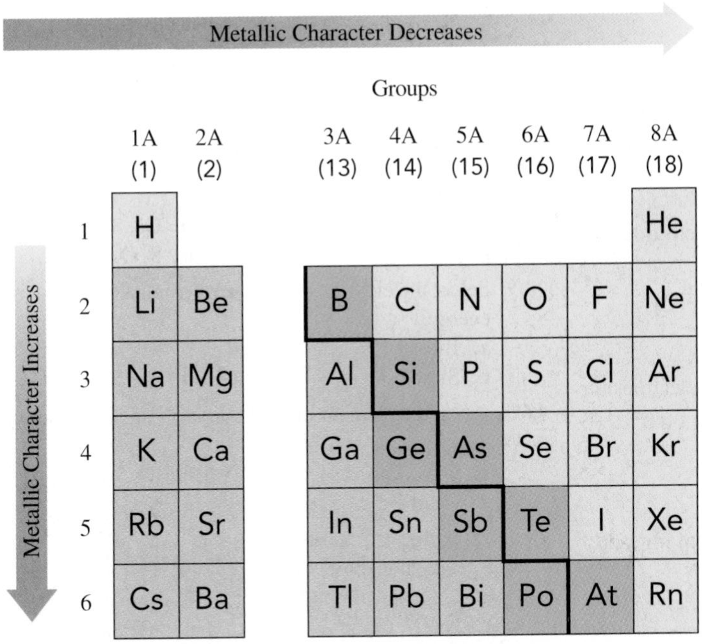

Metallic Character Decreases

Groups

Period	1A (1)	2A (2)	3A (13)	4A (14)	5A (15)	6A (16)	7A (17)	8A (18)
1	H							He
2	Li	Be	B	C	N	O	F	Ne
3	Na	Mg	Al	Si	P	S	Cl	Ar
4	K	Ca	Ga	Ge	As	Se	Br	Kr
5	Rb	Sr	In	Sn	Sb	Te	I	Xe
6	Cs	Ba	Tl	Pb	Bi	Po	At	Rn

Metallic Character Increases

FIGURE 4.14 Metallic character of the representative elements increases going down a group and decreases going from left to right across a period.

Why is the metallic character greater for Rb than for Li?

A summary of the trends in periodic properties we have discussed is given in **TABLE 4.12**.

TABLE 4.12 Summary of Trends in Periodic Properties of Representative Elements

Periodic Property	Top to Bottom within a Group	Left to Right across a Period
Valence Electrons	Remains the same	Increases
Atomic Size	Increases due to the increase in number of energy levels	Decreases due to the increase of protons in the nucleus that pull electrons closer
Ionization Energy	Decreases because valence electrons are easier to remove when they are farther from the nucleus	Increases as the attraction of the protons for outer electrons requires more energy to remove an electron
Metallic Character	Increases because valence electrons are easier to remove when they are farther from the nucleus	Decreases as the attraction of the protons makes it more difficult to remove an electron

SAMPLE PROBLEM 4.11 Periodic Properties

TRY IT FIRST

For Na, P, S, Cl, and F, write the symbol for the element that

a. has five valence electrons
b. is in Group 6A (16)
c. has the smallest atomic size
d. has the lowest ionization energy
e. has the most metallic character

SOLUTION

a. P **b.** S **c.** F
d. Na **e.** Na

STUDY CHECK 4.11

Of the elements Na, K, and S, which has the most metallic character?

ANSWER

K

TEST

Try Practice Problems 4.65 to 4.74

PRACTICE PROBLEMS

4.7 Trends in Periodic Properties

LEARNING GOAL Use the electron arrangement of elements to explain the trends in periodic properties.

4.53 What is the group number and number of valence electrons for each of the following elements?
a. magnesium **b.** iodine
c. oxygen **d.** phosphorus
e. tin **f.** boron

4.54 What is the group number and number of valence electrons for each of the following elements?
a. potassium **b.** silicon
c. neon **d.** aluminum
e. barium **f.** bromine

4.55 Write the group number and draw the Lewis symbol for each of the following elements:
a. sulfur **b.** nitrogen
c. calcium **d.** sodium
e. gallium

4.56 Write the group number and draw the Lewis symbol for each of the following elements:
a. carbon **b.** oxygen
c. argon **d.** lithium
e. chlorine

4.57 Select the larger atom in each pair.
a. Na or Cl **b.** Na or Rb
c. Na or Mg **d.** Rb or I

4.58 Select the larger atom in each pair.
a. S or Ar **b.** S or O
c. S or K **d.** S or Mg

4.59 Place the elements in each set in order of decreasing atomic size.
a. Al, Si, Mg **b.** Cl, I, Br
c. Sb, Sr, I **d.** P, Si, Na

4.60 Place the elements in each set in order of decreasing atomic size.
a. Cl, S, P **b.** Ge, Si, C
c. Ba, Ca, Sr **d.** S, O, Se

4.61 Select the element in each pair with the higher ionization energy.
a. Br or I **b.** Mg or Sr
c. Si or P **d.** I or Xe

4.62 Select the element in each pair with the higher ionization energy.
a. O or Ne **b.** K or Br
c. Ca or Ba **d.** Ne or N

4.63 Place the elements in each set in order of increasing ionization energy.
a. F, Cl, Br **b.** Na, Cl, Al
c. Na, K, Cs **d.** As, Ca, Br

4.64 Place the elements in each set in order of increasing ionization energy.
a. O, N, C **b.** S, P, Cl
c. As, P, N **d.** Al, Si, P

4.65 Fill in the following blanks using *larger* or *smaller*, *more metallic* or *less metallic*. Na has a _______ atomic size and is _______ than P.

4.66 Fill in the following blanks using *larger* or *smaller*, *higher* or *lower*. Mg has a _______ atomic size and a _______ ionization energy than Cs.

4.67 Place the following in order of decreasing metallic character:
Br, Ge, Ca, Ga

4.68 Place the following in order of increasing metallic character:
Na, P, Al, Ar

4.69 Fill in the following blanks using *higher* or *lower*, *more* or *less*:

Sr has a _______ ionization energy and _______ metallic character than Sb.

4.70 Fill in the following blanks using *higher* or *lower*, *more* or *less*:

N has a _______ ionization energy and _______ metallic character than As.

4.71 Complete each of the statements **a** to **d** using **1**, **2**, or **3**:

1. decreases **2.** increases
3. remains the same

Going down Group 6A (16),

a. the ionization energy _______
b. the atomic size _______
c. the metallic character _______
d. the number of valence electrons _______

4.72 Complete each of the statements **a** to **d** using **1**, **2**, or **3**:

1. decreases **2.** increases
3. remains the same

Going from left to right across Period 4,

a. the ionization energy _______
b. the atomic size _______
c. the metallic character _______
d. the number of valence electrons _______

4.73 Which statements completed with **a** to **e** will be *true* and which will be *false*?

An atom of N compared to an atom of Li has a larger (greater)

a. atomic size **b.** ionization energy
c. number of protons **d.** metallic character
e. number of valence electrons

4.74 Which statements completed with **a** to **e** will be *true* and which will be *false*?

An atom of C compared to an atom of Sn has a larger (greater)

a. atomic size **b.** ionization energy
c. number of protons **d.** metallic character
e. number of valence electrons

CLINICAL UPDATE Improving Crop Production

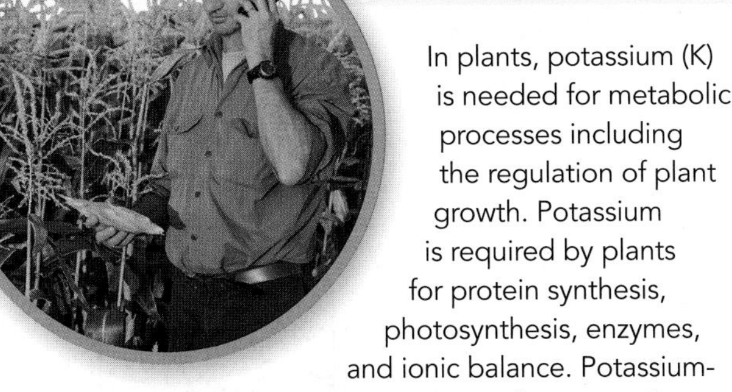

In plants, potassium (K) is needed for metabolic processes including the regulation of plant growth. Potassium is required by plants for protein synthesis, photosynthesis, enzymes, and ionic balance. Potassium-deficient potato plants may show purple or brown spots and plant, root, and seed growth is reduced. John has noticed that the leaves of his recent crop of potatoes had brown spots, the potatoes were undersized, and the crop yield was low.

Tests on soil samples indicated that the potassium levels were below 100 ppm, which indicated that supplemental potassium was needed. John applied a fertilizer containing potassium chloride (KCl). To apply the correct amount of potassium, John needs to apply 170 kg of fertilizer per hectare.

A mineral deficiency of potassium causes brown spots on potato leaves.

Clinical Applications

4.75 **a.** What is the group number and name of the group that contains potassium?
b. Is potassium a metal, a nonmetal, or a metalloid?
c. How many protons are in an atom of potassium?
d. Potassium has three naturally occurring isotopes. They are K-39, K-40, and K-41. Using the atomic mass of potassium, determine which isotope of potassium is the most abundant.

4.76 **a.** How many neutrons are in K-41?
b. Write the electron arrangement for potassium.
c. Which is the larger atom, K or Cs?
d. Which is the smallest atom, K, As, or Br?
e. If John's potato field has an area of 34.5 hectares, how many pounds of fertilizer does John need to use?

CONCEPT MAP

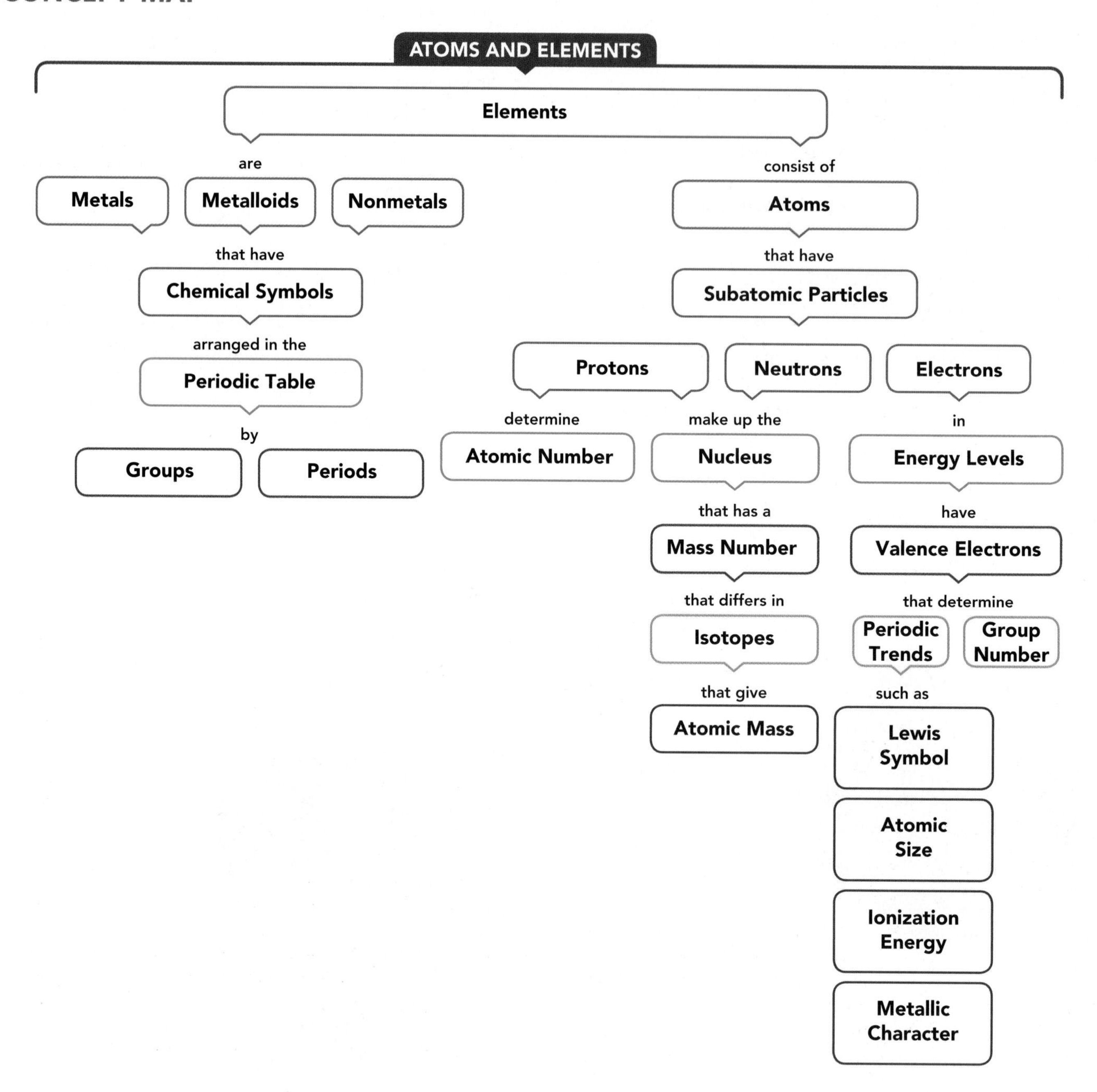

CHAPTER REVIEW

4.1 Elements and Symbols

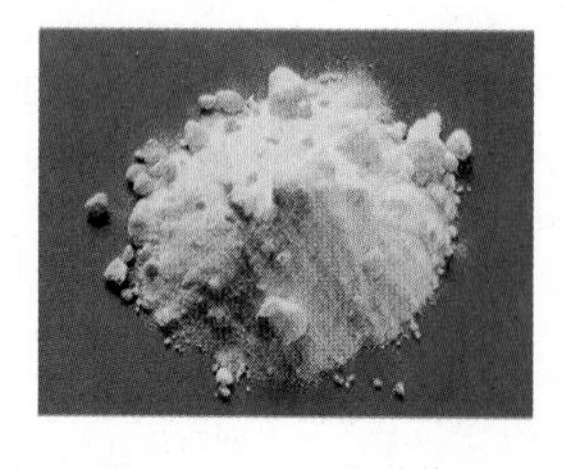

LEARNING GOAL Given the name of an element, write its correct symbol; from the symbol, write the correct name.

- Elements are the primary substances of matter.
- Chemical symbols are one- or two-letter abbreviations of the names of the elements.

4.2 The Periodic Table

LEARNING GOAL Use the periodic table to identify the group and the period of an element; identify the element as a metal, a nonmetal, or a metalloid.

- The periodic table is an arrangement of the elements by increasing atomic number.

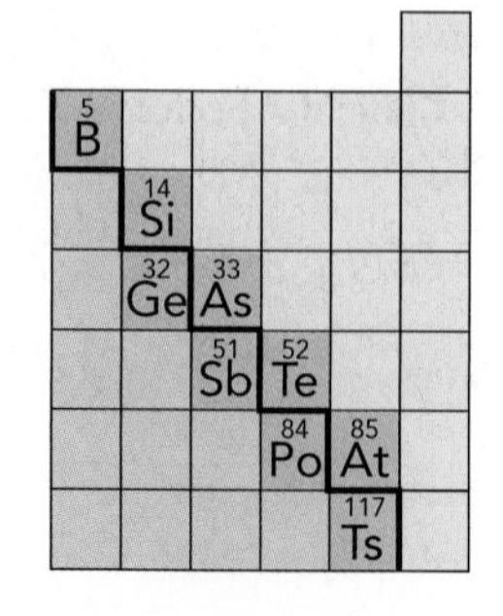

- A horizontal row is called a *period.*
- A vertical column on the periodic table containing elements with similar properties is called a *group.*
- Elements in Group 1A (1) are called the *alkali metals*; Group 2A (2), the *alkaline earth metals*; Group 7A (17), the *halogens*; and Group 8A (18), the *noble gases.*
- On the periodic table, *metals* are located on the left of the heavy zigzag line, and *nonmetals* are to the right of the heavy zigzag line.
- Except for aluminum, elements located along the heavy zigzag line are called *metalloids.*

4.3 The Atom

LEARNING GOAL Describe the electrical charge and location in an atom for a proton, a neutron, and an electron.

Electrons (−)

Nucleus (+)

- An atom is the smallest particle that retains the characteristics of an element.
- Atoms are composed of three types of subatomic particles.
- Protons have a positive charge (+), electrons carry a negative charge (−), and neutrons are electrically neutral.
- The protons and neutrons are found in the tiny, dense nucleus; electrons are located outside the nucleus.

4.4 Atomic Number and Mass Number

LEARNING GOAL Given the atomic number and the mass number of an atom, state the number of protons, neutrons, and electrons.

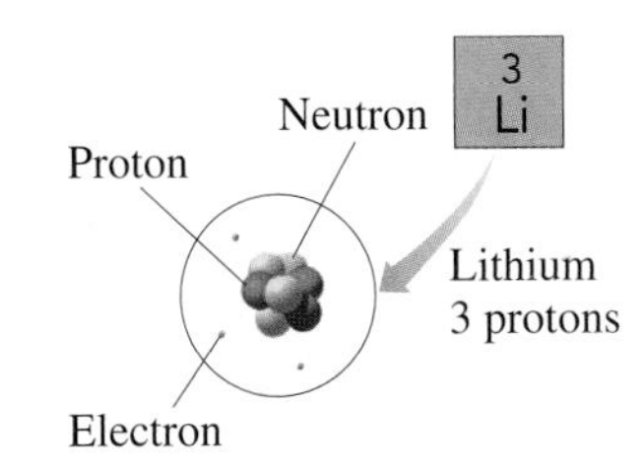

- The atomic number gives the number of protons in all the atoms of the same element.
- In a neutral atom, the number of protons and electrons is equal.
- The mass number is the total number of protons and neutrons in an atom.

4.5 Isotopes and Atomic Mass

LEARNING GOAL Determine the number of protons, neutrons, and electrons in one or more of the isotopes of an element; identify the most abundant isotope of an element.

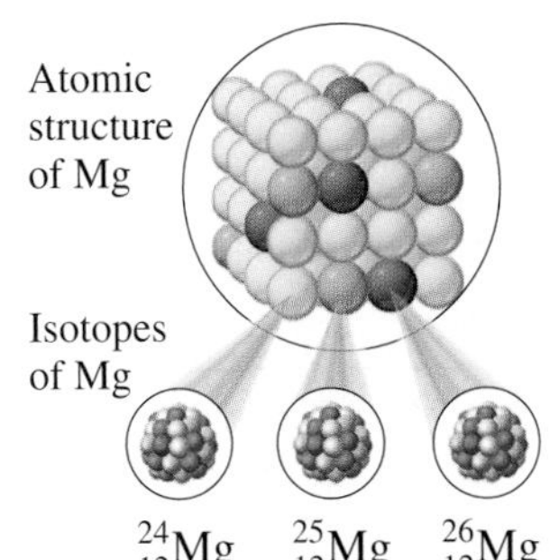

- Atoms that have the same number of protons but different numbers of neutrons are called *isotopes.*
- The atomic mass of an element is the average mass of all the isotopes in a naturally occurring sample of that element.

4.6 Electron Energy Levels

LEARNING GOAL Given the name or symbol of one of the first 20 elements in the periodic table, write the electron arrangement.

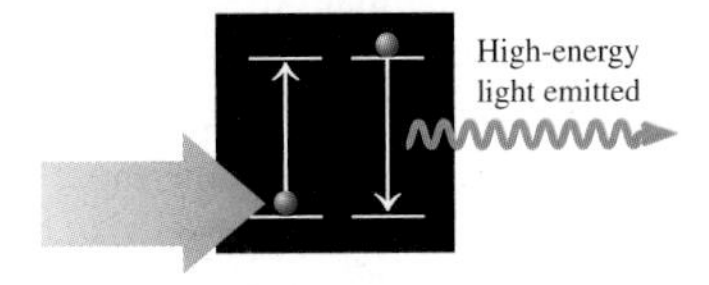

- Every electron has a specific amount of energy.
- In an atom, the electrons of similar energy are grouped in specific energy levels.
- The first level nearest the nucleus can hold 2 electrons, the second level can hold 8 electrons, and the third level will take up to 18 electrons.
- The electron arrangement is written by placing the number of electrons in that atom in order from the lowest energy levels and filling to higher levels.

4.7 Trends in Periodic Properties

LEARNING GOAL Use the electron arrangement of elements to explain the trends in periodic properties.

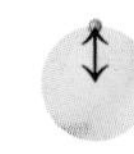
Li atom

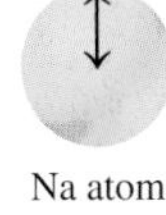
Na atom

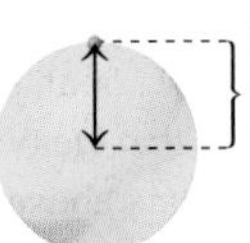

K atom

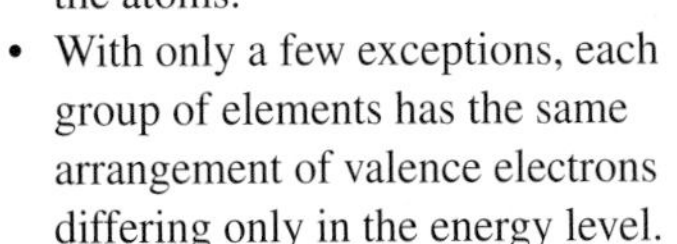
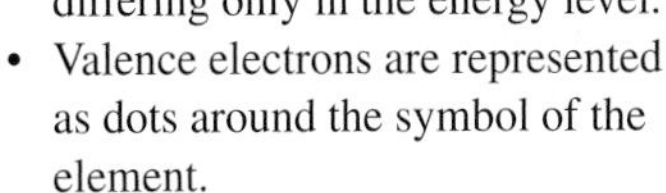

- The properties of elements are related to the valence electrons of the atoms.
- With only a few exceptions, each group of elements has the same arrangement of valence electrons differing only in the energy level.
- Valence electrons are represented as dots around the symbol of the element.
- The size of an atom increases going down a group and decreases going from left to right across a period.
- The energy required to remove a valence electron is the ionization energy, which decreases going down a group and increases going from left to right across a period.
- The metallic character of an element increases going down a group and decreases going from left to right across a period.

KEY TERMS

alkali metals Elements of Group 1A (1) except hydrogen; these are soft, shiny metals with one valence electron.

alkaline earth metals Group 2A (2) elements, which have two valence electrons.

atom The smallest particle of an element that retains the characteristics of the element.

atomic mass The average mass of all the naturally occurring isotopes of an element.

atomic mass unit (amu) A small mass unit used to describe the mass of very small particles such as atoms and subatomic particles; 1 amu is equal to one-twelfth the mass of a $^{12}_{6}C$ atom.

atomic number A number that is equal to the number of protons in an atom.

atomic size The distance between the outermost electrons and the nucleus.

atomic symbol An abbreviation used to indicate the mass number and atomic number of an isotope.

chemical symbol An abbreviation that represents the name of an element.

electron A negatively charged subatomic particle having a very small mass that is usually ignored in calculations; its symbol is e^-.
electron arrangement A list of the number of electrons in each energy level arranged by increasing energy levels.
energy level A group of electrons with similar energy.
group A vertical column in the periodic table that contains elements having similar physical and chemical properties.
group number A number that appears at the top of each vertical column (group) in the periodic table and indicates the number of electrons in the outermost energy level.
halogen Elements in Group 7A (17), which have seven valence electrons.
ionization energy The energy needed to remove an electron from the outermost energy level of an atom.
isotope An atom that differs only in mass number from another atom of the same element. Isotopes have the same atomic number (number of protons), but different numbers of neutrons.
Lewis symbol The representation of an atom that shows valence electrons as dots around the symbol of the element.
mass number The total number of protons and neutrons in the nucleus of an atom.
metal An element that is shiny, malleable, ductile, and a good conductor of heat and electricity. The metals are located to the left of the heavy zigzag line on the periodic table.
metallic character A measure of how easily an element loses a valence electron.
metalloid Elements with properties of both metals and nonmetals, located along the heavy zigzag line on the periodic table.
neutron A neutral subatomic particle having a mass of about 1 amu and found in the nucleus of an atom; its symbol is n or n^0.
noble gas An element in Group 8A (18) of the periodic table.
nonmetal An element with little or no luster that is a poor conductor of heat and electricity. The nonmetals are located to the right of the heavy zigzag line on the periodic table.
nucleus The compact, very dense center of an atom, containing the protons and neutrons of the atom.
period A horizontal row of elements in the periodic table.
periodic table An arrangement of elements by increasing atomic number such that elements having similar chemical behavior are grouped in vertical columns.
proton A positively charged subatomic particle having a mass of about 1 amu and found in the nucleus of an atom; its symbol is p or p^+.
representative elements Elements found in Groups 1A (1) through 8A (18) excluding B groups (3–12) of the periodic table.
subatomic particle A particle within an atom; protons, neutrons, and electrons.
transition elements Elements located between Groups 2A (2) and 3A (13) on the periodic table.
valence electrons Electrons in the outermost energy level of an atom.

CORE CHEMISTRY SKILLS

The chapter Section containing each Core Chemistry Skill is shown in parentheses at the end of each heading.

Counting Protons and Neutrons (4.4)

- The atomic number of an element is equal to the number of protons in every atom of that element. The atomic number is the whole number that appears above the symbol of each element on the periodic table.

Atomic number = number of protons in an atom

- Because atoms are neutral, the number of electrons is equal to the number of protons. Thus, the atomic number gives the number of electrons.
- The mass number is the total number of protons and neutrons in the nucleus of an atom.

Mass number = number of protons + number of neutrons

- The number of neutrons is calculated from the mass number and atomic number.

Number of neutrons = mass number − number of protons

Example: Calculate the number of protons, neutrons, and electrons in a krypton atom with a mass number of 80.

Answer:

Element	Atomic Number	Mass Number	Number of Protons	Number of Neutrons	Number of Electrons
Kr	36	80	equal to atomic number 36	equal to mass number − number of protons 80 − 36 = 44	equal to number of protons 36

Writing Atomic Symbols for Isotopes (4.5)

- Isotopes are atoms of the same element that have the same atomic number but different numbers of neutrons.
- An atomic symbol is written for a particular isotope, with its mass number (protons and neutrons) shown in the upper left corner and its atomic number (protons) shown in the lower left corner.

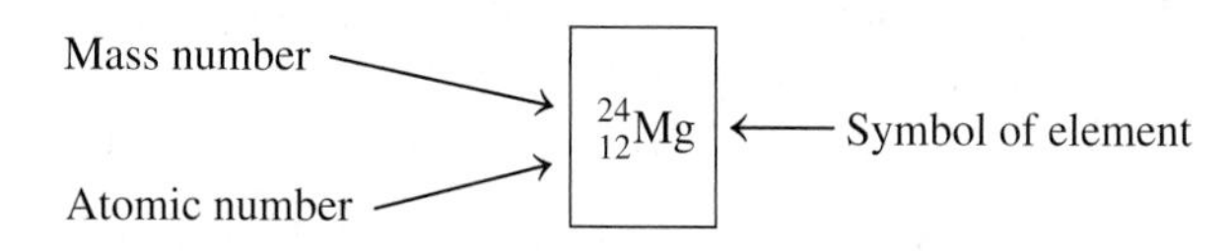

Example: Calculate the number of protons and neutrons in the cadmium isotope $^{112}_{48}Cd$.

Answer:

Atomic Symbol	Atomic Number	Mass Number	Number of Protons	Number of Neutrons
$^{112}_{48}Cd$	number in lower left corner 48	number in upper left corner 112	equal to atomic number 48	equal to mass number − number of protons 112 − 48 = 64

Writing Electron Arrangements (4.6)

- The electron arrangement for an atom specifies the energy levels occupied by the electrons of an atom.
- An electron arrangement is written starting with the lowest energy level, followed by the next lowest energy level.

Example: Write the electron arrangement for phosphorus.

Answer: Phosphorus has atomic number 15, which means it has 15 protons and 15 electrons.

2,8,5

Identifying Trends in Periodic Properties (4.7)

- The number of valence electrons remains the same going down a group and increases going from left to right across a period.
- The size of an atom increases going down a group and decreases going from left to right across a period.
- The ionization energy decreases going down a group and increases going from left to right across a period.
- The metallic character of an element increases going down a group and decreases going from left to right across a period.

Example: For Mg, P, and Cl, identify which has the

a. largest atomic size
b. highest ionization energy
c. most metallic character

Answer: **a.** Mg **b.** Cl **c.** Mg

Drawing Lewis Symbols (4.7)

- The valence electrons are the electrons in the outermost energy level.
- The number of valence electrons is the same as the group number for the representative elements.
- A Lewis symbol represents the number of valence electrons shown as dots placed around the symbol for the element.

Example: Give the group number and number of valence electrons, and draw the Lewis symbol for each of the following:

a. Rb **b.** Se **c.** Xe

Answer: **a.** Group 1A (1), one valence electron, Rb•
b. Group 6A (16), six valence electrons, •Se:
c. Group 8A (18), eight valence electrons, :Xe:

UNDERSTANDING THE CONCEPTS

The chapter Sections to review are shown in parentheses at the end of each problem.

4.77 According to Dalton's atomic theory, which of the following are *true* or *false*? If false, correct the statement to make it true. (4.3)
a. Atoms of an element are identical to atoms of other elements.
b. Every element is made of atoms.
c. Atoms of different elements combine to form compounds.
d. In a chemical reaction, some atoms disappear and new atoms appear.

4.78 Use Rutherford's gold-foil experiment to answer each of the following: (4.3)
a. What did Rutherford expect to happen when he aimed particles at the gold foil?
b. How did the results differ from what he expected?
c. How did he use the results to propose a model of the atom?

4.79 Match the subatomic particles (**1** to **3**) to each of the descriptions below: (4.4)
1. protons **2.** neutrons **3.** electrons
a. atomic mass
b. atomic number
c. positive charge
d. negative charge
e. mass number – atomic number

4.80 Match the subatomic particles (**1** to **3**) to each of the descriptions below: (4.4)
1. protons **2.** neutrons **3.** electrons
a. mass number
b. surround the nucleus
c. in the nucleus
d. charge of 0
e. equal to number of electrons

4.81 Consider the following atoms in which X represents the chemical symbol of the element: (4.4, 4.5)

$^{16}_{8}X$ $^{16}_{9}X$ $^{18}_{10}X$ $^{17}_{8}X$ $^{18}_{8}X$

a. What atoms have the same number of protons?
b. Which atoms are isotopes? Of what element?
c. Which atoms have the same mass number?
d. What atoms have the same number of neutrons?

4.82 For each of the following, write the symbol and name for X and the number of protons and neutrons. Which are isotopes of each other? (4.4, 4.5)
a. $^{124}_{47}X$ **b.** $^{116}_{49}X$ **c.** $^{116}_{50}X$
d. $^{124}_{50}X$ **e.** $^{116}_{48}X$

4.83 Indicate if the atoms in each pair have the same number of protons, neutrons, or electrons. (4.4, 4.5)
a. $^{37}_{17}Cl$, $^{38}_{18}Ar$ **b.** $^{36}_{16}S$, $^{34}_{16}S$ **c.** $^{40}_{18}Ar$, $^{39}_{17}Cl$

4.84 Complete the following table for the three naturally occurring isotopes of silicon, the major component in computer chips: (4.4, 4.5)

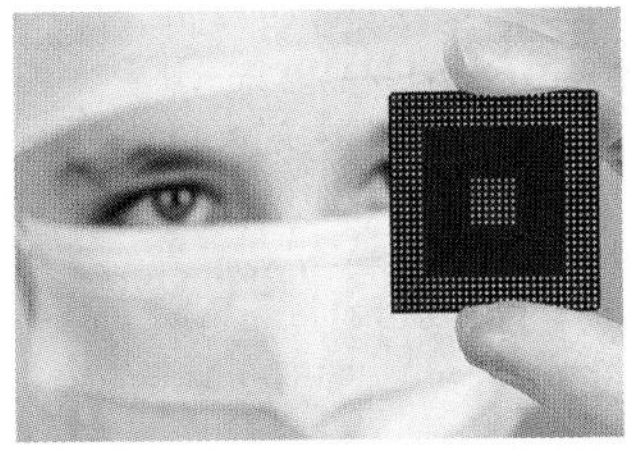

Computer chips consist primarily of the element silicon.

	Isotope		
	$^{28}_{14}Si$	$^{29}_{14}Si$	$^{30}_{14}Si$
Atomic Number			
Mass Number			
Number of Protons			
Number of Neutrons			
Number of Electrons			

4.85 For each representation of a nucleus **A** through **E**, write the atomic symbol and identify which are isotopes. (4.4, 4.5)

Proton ●
Neutron ●

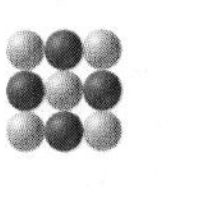
A

B

C

D

E

4.86 Identify the element represented by each nucleus **A** through **E** in problem 4.85 as a metal, a nonmetal, or a metalloid. (4.3)

4.87 Match the spheres **A** through **D** with atoms of Li, Na, K, and Rb. (4.7)

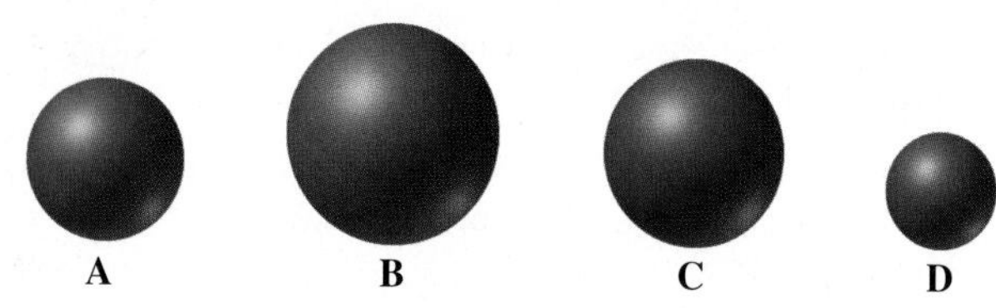

4.88 Match the spheres **A** through **D** with atoms of K, Ge, Ca, and Kr. (4.7)

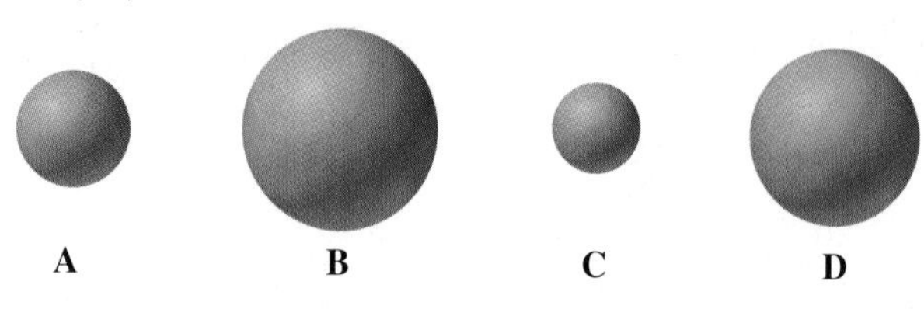

4.89 Of the elements Na, Mg, Si, S, Cl, and Ar, identify one that fits each of the following: (4.2, 4.6, 4.7)

a. largest atomic size
b. a halogen
c. electron arrangement 2,8,4
d. highest ionization energy
e. in Group 6A (16)
f. most metallic character
g. two valence electrons

4.90 Of the elements Sn, Xe, Te, Sr, I, and Rb, identify one that fits each of the following: (4.2, 4.6, 4.7)

a. smallest atomic size
b. an alkaline earth metal
c. a metalloid
d. lowest ionization energy
e. in Group 4A (14)
f. least metallic character
g. seven valence electrons

ADDITIONAL PRACTICE PROBLEMS

4.91 Give the group and period numbers for each of the following elements: (4.2)

a. bromine **b.** argon
c. potassium **d.** radium

4.92 Give the group and period numbers for each of the following elements: (4.2)

a. radon **b.** arsenic
c. carbon **d.** neon

4.93 Indicate if each of the following statements is *true* or *false*: (4.3)

a. The proton is a negatively charged particle.
b. The neutron is 2000 times as heavy as a proton.
c. The atomic mass unit is based on a carbon atom with six protons and six neutrons.
d. The nucleus is the largest part of the atom.
e. The electrons are located outside the nucleus.

4.94 Indicate if each of the following statements is *true* or *false*: (4.3)

a. The neutron is electrically neutral.
b. Most of the mass of an atom is due to the protons and neutrons.
c. The charge of an electron is equal, but opposite, to the charge of a neutron.
d. The proton and the electron have about the same mass.
e. The mass number is the number of protons.

4.95 Complete the following statements: (4.2, 4.3)

a. The atomic number gives the number of _______ in the nucleus.
b. In an atom, the number of electrons is equal to the number of _______.
c. Sodium and potassium are examples of elements called _______.

4.96 Complete the following statements: (4.2, 4.3)

a. The number of protons and neutrons in an atom is also the _______ number.
b. The elements in Group 7A (17) are called the _______.
c. Elements that are shiny and conduct heat are called _______.

4.97 Write the name and symbol of the element with the following atomic number: (4.1, 4.2)

a. 28 **b.** 56 **c.** 88
d. 33 **e.** 50 **f.** 55
g. 79 **h.** 80

4.98 Write the name and symbol of the element with the following atomic number: (4.1, 4.2)

a. 36 **b.** 22 **c.** 48
d. 26 **e.** 54 **f.** 78
g. 83 **h.** 92

4.99 For the following atoms, determine the number of protons, neutrons, and electrons: (4.4, 4.5)

a. $^{114}_{48}Cd$ **b.** $^{98}_{43}Tc$ **c.** $^{199}_{79}Au$
d. $^{222}_{86}Rn$ **e.** $^{136}_{54}Xe$

4.100 For the following atoms, determine the number of protons, neutrons, and electrons: (4.4, 4.5)

a. $^{202}_{80}Hg$ **b.** $^{127}_{53}I$ **c.** $^{75}_{35}Br$
d. $^{133}_{55}Cs$ **e.** $^{195}_{78}Pt$

4.101 Complete the following table: (4.4, 4.5)

Name	Atomic Symbol	Number of Protons	Number of Neutrons	Number of Electrons
	$^{34}_{16}S$			
		28	34	
Magnesium			14	
	$^{220}_{86}Rn$			

4.102 Complete the following table: (4.4, 4.5)

Name	Atomic Symbol	Number of Protons	Number of Neutrons	Number of Electrons
Potassium			22	
	$^{51}_{23}V$			
		48	64	
Barium			82	

4.103 Write the atomic symbol for each of the following: (4.4, 4.5)
- **a.** an atom with four protons and five neutrons
- **b.** an atom with 12 protons and 14 neutrons
- **c.** a calcium atom with a mass number of 46
- **d.** an atom with 30 electrons and 40 neutrons

4.104 Write the atomic symbol for each of the following: (4.4, 4.5)
- **a.** an aluminum atom with 14 neutrons
- **b.** an atom with atomic number 26 and 32 neutrons
- **c.** a strontium atom with 50 neutrons
- **d.** an atom with a mass number of 72 and atomic number 33

4.105 The most abundant isotope of lead is ${}^{208}_{82}Pb$. (4.4, 4.5)
- **a.** How many protons, neutrons, and electrons are in ${}^{208}_{82}Pb$?
- **b.** What is the atomic symbol of another isotope of lead with 132 neutrons?
- **c.** What is the atomic symbol and name of an atom with the same mass number as in part **b** and 131 neutrons?

4.106 The most abundant isotope of silver is ${}^{107}_{47}Ag$. (4.4, 4.5)
- **a.** How many protons, neutrons, and electrons are in ${}^{107}_{47}Ag$?
- **b.** What is the atomic symbol of another isotope of silver with 62 neutrons?
- **c.** What is the atomic symbol and name of an atom with the same mass number as in part **b** and 61 neutrons?

4.107 Write the group number and the electron arrangement for each of the following: (4.6)
- **a.** oxygen **b.** sodium **c.** neon **d.** boron

4.108 Write the group number and the electron arrangement for each of the following: (4.6)
- **a.** magnesium **b.** chlorine **c.** beryllium **d.** argon

4.109 Why is the ionization energy of Ca higher than K, but lower than that of Mg? (4.7)

4.110 Why is the ionization energy of Cl lower than F, but higher than that of S? (4.7)

4.111 Of the elements Li, Be, N, and F, which (4.7)
- **a.** is an alkaline earth metal?
- **b.** has the largest atomic size?
- **c.** has the highest ionization energy?
- **d.** is found in Group 5A (15)?
- **e.** has the most metallic character?

4.112 Of the elements F, Br, Cl, and I, which (4.7)
- **a.** has the largest atomic size?
- **b.** has the smallest atomic size?
- **c.** has the lowest ionization energy?
- **d.** requires the most energy to remove an electron?
- **e.** is found in Period 4?

CHALLENGE PROBLEMS

The following problems are related to the topics in this chapter. However, they do not all follow the chapter order, and they require you to combine concepts and skills from several Sections. These problems will help you increase your critical thinking skills and prepare for your next exam.

4.113 There are four naturally occurring isotopes of strontium: ${}^{84}_{38}Sr$, ${}^{86}_{38}Sr$, ${}^{87}_{38}Sr$, and ${}^{88}_{38}Sr$. (4.4, 4.5)
- **a.** How many protons, neutrons, and electrons are in Sr-87?
- **b.** What is the most abundant isotope in a strontium sample?
- **c.** How many neutrons are in Sr-84?
- **d.** Why don't any of the isotopes of strontium have the atomic mass of 87.62 amu listed on the periodic table?

4.114 There are four naturally occurring isotopes of iron: ${}^{54}_{26}Fe$, ${}^{56}_{26}Fe$, ${}^{57}_{26}Fe$, and ${}^{58}_{26}Fe$. (4.4, 4.5)
- **a.** How many protons, neutrons, and electrons are in ${}^{58}_{26}Fe$?
- **b.** What is the most abundant isotope in an iron sample?
- **c.** How many neutrons are in ${}^{57}_{26}Fe$?
- **d.** Why don't any of the isotopes of iron have the atomic mass of 55.85 amu listed on the periodic table?

4.115 Give the symbol of the element that has the (4.7)
- **a.** smallest atomic size in Group 6A (16)
- **b.** smallest atomic size in Period 3
- **c.** highest ionization energy in Group 5A (15)
- **d.** lowest ionization energy in Period 3
- **e.** most metallic character in Group 2A (2)

4.116 Give the symbol of the element that has the (4.7)
- **a.** largest atomic size in Group 1A (1)
- **b.** largest atomic size in Period 4
- **c.** highest ionization energy in Group 2A (2)
- **d.** lowest ionization energy in Group 7A (17)
- **e.** least metallic character in Group 4A (14)

ANSWERS

Answers to Selected Practice Problems

4.1 **a.** Cu **b.** Pt **c.** Ca **d.** Mn **e.** Fe **f.** Ba **g.** Pb **h.** Sr

4.3 **a.** carbon **b.** chlorine **c.** iodine **d.** selenium **e.** nitrogen **f.** sulfur **g.** zinc **h.** cobalt

4.5
- **a.** sodium, chlorine
- **b.** calcium, sulfur, oxygen
- **c.** carbon, hydrogen, chlorine, nitrogen, oxygen
- **d.** lithium, carbon, oxygen

4.7 **a.** Period 2 **b.** Group 8A (18) **c.** Group 1A (1) **d.** Period 2

4.9 **a.** C **b.** He **c.** Na **d.** Ca **e.** Al

4.11 **a.** metal **b.** nonmetal **c.** metal **d.** nonmetal **e.** nonmetal **f.** nonmetal **g.** metalloid **h.** metal

4.13
- **a.** needed for bones and teeth, muscle contraction, nerve impulses; alkaline earth metal
- **b.** component of hemoglobin; transition element
- **c.** muscle contraction, nerve impulses; alkali metal
- **d.** found in fluids outside cells; halogen

4.15
- **a.** A macromineral is an element essential to health, which is present in the body in amounts from 5 to 1000 g.
- **b.** Sulfur is a component of proteins, liver, vitamin B_1, and insulin.
- **c.** 86 g

4.17 **a.** electron **b.** proton
c. electron **d.** neutron

4.19 Rutherford determined that an atom contains a small, compact nucleus that is positively charged.

4.21 **a.** True **b.** True
c. True **d.** False; a proton is attracted to an electron

4.23 In the process of brushing hair, strands of hair become charged with like charges that repel each other.

4.25 **a.** atomic number **b.** both
c. mass number **d.** atomic number

4.27 **a.** lithium, Li **b.** fluorine, F
c. calcium, Ca **d.** zinc, Zn
e. neon, Ne **f.** silicon, Si
g. iodine, I **h.** oxygen, O

4.29 **a.** 18 protons and 18 electrons
b. 25 protons and 25 electrons
c. 53 protons and 53 electrons
d. 48 protons and 48 electrons

4.31

Name of the Element	Symbol	Atomic Number	Mass Number	Number of Protons	Number of Neutrons	Number of Electrons
Zinc	Zn	30	66	30	36	30
Magnesium	Mg	12	24	12	12	12
Potassium	K	19	39	19	20	19
Sulfur	S	16	31	16	15	16
Iron	Fe	26	56	26	30	26

4.33 **a.** 38 protons, 51 neutrons, 38 electrons
b. 24 protons, 28 neutrons, 24 electrons
c. 16 protons, 18 neutrons, 16 electrons
d. 35 protons, 46 neutrons, 35 electrons

4.35 **a.** $^{31}_{15}\text{P}$ **b.** $^{80}_{35}\text{Br}$ **c.** $^{122}_{50}\text{Sn}$
d. $^{35}_{17}\text{Cl}$ **e.** $^{202}_{80}\text{Hg}$

4.37 **a.** $^{36}_{18}\text{Ar}$ $^{38}_{18}\text{Ar}$ $^{40}_{18}\text{Ar}$
b. They all have the same number of protons and electrons.
c. They have different numbers of neutrons, which gives them different mass numbers.
d. The atomic mass of Ar listed on the periodic table is the average atomic mass of all the naturally occurring isotopes.
e. The isotope Ar-40 is most abundant because its mass is closest to the atomic mass of Ar on the periodic table.

4.39 Ga-69

4.41 Since the atomic mass of copper is closer to 63 amu, there are more atoms of $^{63}_{29}\text{Cu}$.

4.43 Since the atomic mass of thallium is 204.4 amu, the most abundant isotope is $^{205}_{81}\text{Tl}$.

4.45 absorb

4.47 **a.** green light **b.** blue light

4.49 **a.** 2,4 **b.** 2,8,8 **c.** 2,8,8,1
d. 2,8,4 **e.** 2 **f.** 2,5

4.51 **a.** Li **b.** Mg **c.** H
d. Cl **e.** O

4.53 **a.** Group 2A (2), 2 e^- **b.** Group 7A (17), 7 e^-
c. Group 6A (16), 6 e^- **d.** Group 5A (15), 5 e^-
e. Group 4A (14), 4 e^- **f.** Group 3A (13), 3 e^-

4.55 **a.** Group 6A (16)
·S: (with six dots)
b. Group 5A (15)
·N· (with five dots)
c. Group 2A (2)
Ca· (with two dots)
d. Group 1A (1)
Na·
e. Group 3A (13)
·Ga· (with three dots)

4.57 **a.** Na **b.** Rb **c.** Na **d.** Rb

4.59 **a.** Mg, Al, Si **b.** I, Br, Cl **c.** Sr, Sb, I **d.** Na, Si, P

4.61 **a.** Br **b.** Mg **c.** P **d.** Xe

4.63 **a.** Br, Cl, F **b.** Na, Al, Cl **c.** Cs, K, Na **d.** Ca, As, Br

4.65 larger, more metallic

4.67 Ca, Ga, Ge, Br

4.69 lower, more

4.71 **a.** decreases **b.** increases
c. increases **d.** remains the same

4.73 **b**, **c**, and **e** are true; **a** and **d** are false.

4.75 **a.** Group 1A (1), alkali metals **b.** metal
c. 19 protons **d.** K-39

4.77 **a.** False; All atoms of a given element are different from atoms of other elements.
b. True
c. True
d. False; In a chemical reaction atoms are never created or destroyed.

4.79 **a.** 1 + 2 **b.** 1 **c.** 1
d. 3 **e.** 2

4.81 **a.** $^{16}_{8}\text{X}$, $^{17}_{8}\text{X}$, $^{18}_{8}\text{X}$ all have eight protons.
b. $^{16}_{8}\text{X}$, $^{17}_{8}\text{X}$, $^{18}_{8}\text{X}$ are all isotopes of oxygen.
c. $^{16}_{8}\text{X}$ and $^{16}_{9}\text{X}$ and have mass numbers of 16, and $^{18}_{10}\text{X}$ and $^{18}_{8}\text{X}$ have mass numbers of 18.
d. $^{16}_{8}\text{X}$ and $^{18}_{10}\text{X}$ have eight neutrons each.

4.83 **a.** Both have 20 neutrons.
b. Both have 16 protons and 16 electrons.
c. Both have 22 neutrons.

4.85 **a.** $^{9}_{4}\text{Be}$ **b.** $^{11}_{5}\text{B}$ **c.** $^{13}_{6}\text{C}$ **d.** $^{10}_{5}\text{B}$ **e.** $^{12}_{6}\text{C}$
Representations **B** and **D** are isotopes of boron; **C** and **E** are isotopes of carbon.

4.87 Li is **D**, Na is **A**, K is **C**, and Rb is **B**.

4.89 **a.** Na **b.** Cl **c.** Si **d.** Ar
e. S **f.** Na **g.** Mg

4.91 **a.** Group 7A (17), Period 4 **b.** Group 8A (18), Period 3
c. Group 1A (1), Period 4 **d.** Group 2A (2), Period 7

4.93 **a.** False **b.** False **c.** True **d.** False **e.** True

4.95 **a.** protons **b.** protons **c.** alkali metals

4.97 **a.** nickel, Ni **b.** barium, Ba
c. radium, Ra **d.** arsenic, As
e. tin, Sn **f.** cesium, Cs
g. gold, Au **h.** mercury, Hg

4.99 **a.** 48 protons, 66 neutrons, 48 electrons
b. 43 protons, 55 neutrons, 43 electrons
c. 79 protons, 120 neutrons, 79 electrons
d. 86 protons, 136 neutrons, 86 electrons
e. 54 protons, 82 neutrons, 54 electrons

4.101

Name	Atomic Symbol	Number of Protons	Number of Neutrons	Number of Electrons
Sulfur	$^{34}_{16}S$	16	18	16
Nickel	$^{62}_{28}Ni$	28	34	28
Magnesium	$^{26}_{12}Mg$	12	14	12
Radon	$^{220}_{86}Rn$	86	134	86

4.103 **a.** $^{9}_{4}Be$ **b.** $^{26}_{12}Mg$ **c.** $^{46}_{20}Ca$ **d.** $^{70}_{30}Zn$

4.105 **a.** 82 protons, 126 neutrons, 82 electrons
b. $^{214}_{82}Pb$
c. $^{214}_{83}Bi$, bismuth

4.107 **a.** O; Group 6A (16); 2,6 **b.** Na; Group 1A (1); 2,8,1
c. Ne; Group 8A (18); 2,8 **d.** B; Group 3A (13); 2,3

4.109 Calcium has a greater number of protons than K. The increase in positive charge increases the attraction for electrons, which means that more energy is required to remove the valence electrons. The valence electrons are farther from the nucleus in Ca than in Mg and less energy is needed to remove them.

4.111 **a.** Be **b.** Li **c.** F **d.** N **e.** Li

4.113 **a.** 38 protons, 49 neutrons, 38 electrons
b. $^{88}_{38}Sr$
c. 46 neutrons
d. The atomic mass on the periodic table is the average of all the naturally occurring isotopes.

4.115 **a.** O **b.** Ar **c.** N **d.** Na **e.** Ra

5 Nuclear Chemistry

SIMONE'S DOCTOR IS CONCERNED ABOUT her elevated cholesterol, which could lead to coronary heart disease and a heart attack. He sends her to a nuclear medicine center to undergo a cardiac stress test. Pauline, the radiation technologist, explains to Simone that a stress test measures the blood flow to her heart muscle at rest and then during stress. The test is performed in a similar way as a routine exercise stress test, but images are produced that show areas of low blood flow through the heart and areas of damaged heart muscle. It involves taking two sets of images of her heart, one when the heart is at rest, and one when she is walking on a treadmill.

Pauline tells Simone that she will inject thallium-201 into her bloodstream. She explains that Tl-201 is a radioactive isotope that has a half-life of 3.0 days. Simone is curious about the term "half-life." Pauline explains that a half-life is the amount of time it takes for one-half of a radioactive sample to break down. She assures Simone that after four half-lives, the radiation emitted will be almost zero. Pauline tells Simone that Tl-201 decays to Hg-201 and emits energy similar to X-rays. When the Tl-201 reaches any areas within her heart with restricted blood supply, smaller amounts of the radioisotope will accumulate.

CAREER Radiation Technologist

A radiation technologist works in a hospital or imaging center where nuclear medicine is used to diagnose and treat a variety of medical conditions. In a diagnostic test, a radiation technologist uses a scanner, which converts radiation into images. The images are evaluated to determine any abnormalities in the body. A radiation technologist operates the instrumentation and computers associated with nuclear medicine such as computed tomography (CT), magnetic resonance imaging (MRI), and positron emission tomography (PET). A radiation technologist must know how to handle radioisotopes safely, use the necessary type of shielding, and give radioactive isotopes to patients. In addition, they must physically and mentally prepare patients for imaging. A patient may be given radioactive tracers such as technetium-99m, iodine-131, gallium-67, and thallium-201 that emit gamma radiation, which is detected and used to develop an image of the kidneys or thyroid or to follow the blood flow in the heart muscle.

CLINICAL UPDATE Cardiac Imaging Using a Radioisotope

When Simone arrives at the clinic for her nuclear stress test, a radioactive dye is injected that will show images of her heart muscle. You can read about Simone's scans and the results of her nuclear stress test in the **CLINICAL UPDATE Cardiac Imaging Using a Radioisotope**, page 161.

LOOKING AHEAD

REVIEW

Writing Atomic Symbols for Isotopes (4.5)

5.1 Natural Radioactivity

LEARNING GOAL Describe alpha, beta, positron, and gamma radiation.

Most naturally occurring isotopes of elements up to atomic number 19 have stable nuclei. Elements with atomic numbers 20 and higher usually have one or more isotopes that have unstable nuclei in which the nuclear forces cannot offset the repulsions between the protons. An unstable nucleus is *radioactive*, which means that it spontaneously emits small particles of energy called **radiation** to become more stable. Radiation may take the form of alpha (α) and beta (β) particles, positrons (β^+), or pure energy such as gamma (γ) rays. An isotope of an element that emits radiation is called a **radioisotope**. For most types of radiation, there is a change in the number of protons in the nucleus, which means that an atom is converted into an atom of a different element. This kind of nuclear change was not evident to Dalton when he made his predictions about atoms. Elements with atomic numbers of 93 and higher are produced artificially in nuclear laboratories and consist only of radioactive isotopes.

Symbols for Radioisotopes

The atomic symbols for the different isotopes are written with the mass number in the upper left corner and the atomic number in the lower left corner. The mass number is the sum of the number of protons and neutrons in the nucleus, and the atomic number is equal to the number of protons. For example, a radioactive isotope of carbon used for archaeological dating has a mass number of 14 and an atomic number of 6.

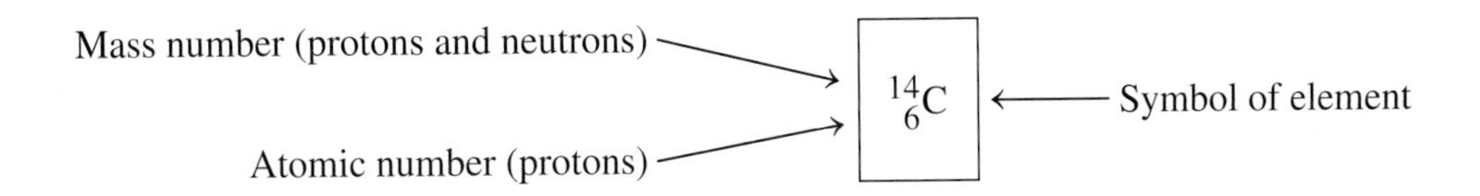

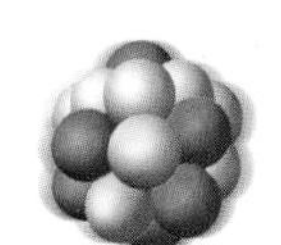

$^{14}_{6}C$ carbon-14

6 protons (red)

8 neutrons (white)

Radioactive isotopes are identified by writing the mass number after the element's name or symbol. Thus, in this example, the isotope is called carbon-14 or C-14. **TABLE 5.1** compares some stable, nonradioactive isotopes with some radioactive isotopes.

TABLE 5.1 Stable and Radioactive Isotopes of Some Elements

Magnesium	Iodine	Uranium
Stable Isotopes		
$^{24}_{12}Mg$	$^{127}_{53}I$	None
Magnesium-24	Iodine-127	
Radioactive Isotopes		
$^{23}_{12}Mg$	$^{125}_{53}I$	$^{235}_{92}U$
Magnesium-23	Iodine-125	Uranium-235
$^{27}_{12}Mg$	$^{131}_{53}I$	$^{238}_{92}U$
Magnesium-27	Iodine-131	Uranium-238

Types of Radiation

By emitting radiation, an unstable nucleus forms a more stable, lower energy nucleus. One type of radiation consists of *alpha particles*. An **alpha particle** is identical to a helium (He) nucleus, which has two protons and two neutrons. An alpha particle has a mass number of 4, an atomic number of 2, and a charge of 2+. The symbol for an alpha particle is the Greek letter alpha (α) or the symbol of a helium nucleus except that the 2+ charge is omitted.

Alpha particle $^{4}_{2}He$ or α

Another type of radiation occurs when a radioisotope emits a *beta particle*. A **beta particle**, which is a high-energy electron, has a charge of 1−, and because its mass is so much less than the mass of a proton, it has a mass number of 0. It is represented by the

Beta particle $^{0}_{-1}e$ or β

Greek letter beta (β) or by the symbol for the electron including the mass number and the charge ($_{-1}^{0}e$).

Positron $_{+1}^{0}e$ or β^+

A **positron**, similar to a beta particle, has a positive (1+) charge with a mass number of 0. It is represented by the Greek letter beta with a 1+ charge, β^+, or by the symbol of an electron, which includes the mass number and the charge ($_{+1}^{0}e$). A positron is an example of *antimatter*, a term physicists use to describe a particle that is the opposite of another particle, in this case, an electron.

Gamma ray $_{0}^{0}\gamma$ or γ

Gamma rays are high-energy radiation, released when an unstable nucleus undergoes a rearrangement of its particles to give a more stable, lower energy nucleus. Gamma rays are often emitted along with other types of radiation. A gamma ray is written as the Greek letter gamma γ. Because gamma rays are energy only, zeros are used to show that a gamma ray has no mass or charge ($_{0}^{0}\gamma$).

TABLE 5.2 summarizes the types of radiation we use in nuclear equations.

ENGAGE

What is the charge and the mass number of an alpha particle emitted by a radioactive atom?

TABLE 5.2 Some Forms of Radiation

Type of Radiation	Symbol		Mass Number	Charge
Alpha Particle	$_{2}^{4}He$	α	4	2+
Beta Particle	$_{-1}^{0}e$	β	0	1−
Positron	$_{+1}^{0}e$	β^+	0	1+
Gamma Ray	$_{0}^{0}\gamma$	γ	0	0
Proton	$_{1}^{1}H$	p	1	1+
Neutron	$_{0}^{1}n$	n	1	0

SAMPLE PROBLEM 5.1 **Radiation Particles**

TRY IT FIRST

Identify and write the symbol for each of the following types of radiation:

a. contains two protons and two neutrons
b. has a mass number of 0 and a 1− charge

SOLUTION

a. An alpha particle, $_{2}^{4}He$ or α, has two protons and two neutrons.
b. A beta particle, $_{-1}^{0}e$ or β, has a mass number of 0 and a 1− charge.

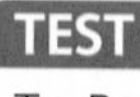
TEST

Try Practice Problems 5.1 to 5.10

STUDY CHECK 5.1

Identify and write the symbol for the type of radiation that has a mass number of zero and a 1+ charge.

ANSWER

A positron, $_{+1}^{0}e$ has a mass number of 0 and a 1+ charge.

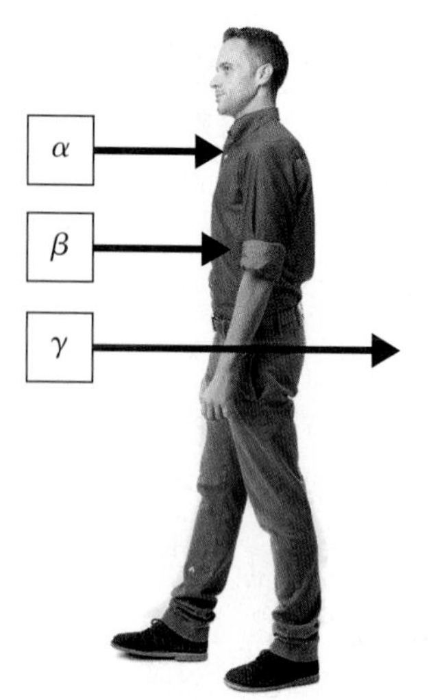

Different types of radiation penetrate the body to different depths.

Biological Effects of Radiation

When radiation strikes molecules in its path, electrons may be knocked away, forming unstable ions. If this *ionizing radiation* passes through the human body, it may interact with water molecules, removing electrons and producing H_2O^+, which can cause undesirable chemical reactions.

The cells most sensitive to radiation are the ones undergoing rapid division—those of the bone marrow, skin, reproductive organs, and intestinal lining, as well as all cells of growing children. Damaged cells may lose their ability to produce necessary materials. For example, if radiation damages cells of the bone marrow, red blood cells may no longer be produced. If sperm cells, ova, or the cells of a fetus are damaged, birth defects may result. In contrast, cells of the nerves, muscles, liver, and adult bones are much less sensitive to radiation because they undergo little or no cellular division.

Cancer cells are another example of rapidly dividing cells. Because cancer cells are highly sensitive to radiation, large doses of radiation are used to destroy them. The normal tissue that surrounds cancer cells divides at a slower rate and suffers less damage from radiation. However, radiation may cause malignant tumors, leukemia, anemia, and genetic mutations.

Radiation Protection

Nuclear medicine technologists, chemists, doctors, and nurses who work with radioactive isotopes must use proper radiation protection. Proper *shielding* is necessary to prevent exposure. Alpha particles, which have the largest mass and charge of the radiation particles, travel only a few centimeters in the air before they collide with air molecules, acquire electrons, and become helium atoms. A piece of paper, clothing, and our skin are protection against alpha particles. Lab coats and gloves will also provide sufficient shielding. However, if alpha emitters are ingested or inhaled, the alpha particles they give off can cause serious internal damage.

Beta particles have a very small mass and move much faster and farther than alpha particles, traveling as much as several meters through air. They can pass through paper and penetrate as far as 4 to 5 mm into body tissue. External exposure to beta particles can burn the surface of the skin, but they do not travel far enough to reach the internal organs. Heavy clothing such as lab coats and gloves are needed to protect the skin from beta particles.

Gamma rays travel great distances through the air and pass through many materials, including body tissues. Because gamma rays penetrate so deeply, exposure to gamma rays can be extremely hazardous. Only very dense shielding, such as lead or concrete, will stop them. Syringes used for injections of radioactive materials use shielding made of lead or heavyweight materials such as tungsten and plastic composites.

When working with radioactive materials, medical personnel wear protective clothing and gloves and stand behind a shield (see **FIGURE 5.1**). Long tongs may be used to pick up vials of radioactive material, keeping them away from the hands and body. **TABLE 5.3** summarizes the shielding materials required for the various types of radiation.

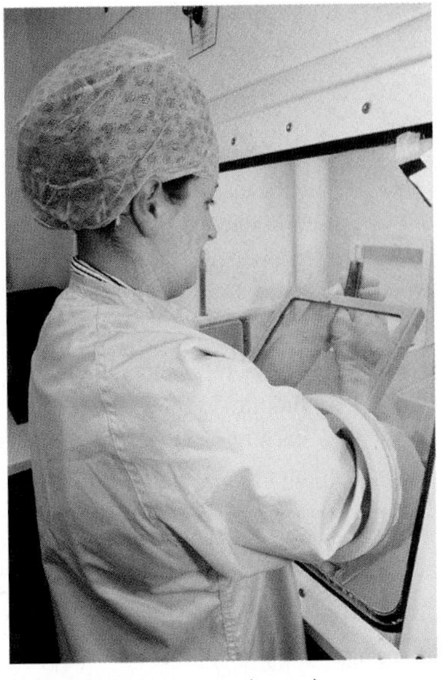

FIGURE 5.1 In a nuclear pharmacy, a person working with radioisotopes wears protective clothing and gloves and uses a lead glass shield on a syringe.

What types of radiation does a lead shield block?

TABLE 5.3 Properties of Radiation and Shielding Required

Property	Alpha (α) Particle	Beta (β) Particle	Gamma (γ) Ray
Travel Distance in Air	2 to 4 cm	200 to 300 cm	500 m
Tissue Depth	0.05 mm	4 to 5 mm	50 cm or more
Shielding	Paper, clothing	Heavy clothing, lab coats, gloves	Lead, thick concrete
Typical Source	Radium-226	Carbon-14	Technetium-99m

If you work in an environment where radioactive materials are present, such as a nuclear medicine facility, try to keep the time you spend in a radioactive area to a minimum. Remaining in a radioactive area twice as long exposes you to twice as much radiation.

Keep your distance! The greater the distance from the radioactive source, the lower the intensity of radiation received. By doubling your distance from the radiation source, the intensity of radiation drops to $(\frac{1}{2})^2$, or one-fourth of its previous value.

TEST

Try Practice Problems 5.11 and 5.12

PRACTICE PROBLEMS

5.1 Natural Radioactivity

LEARNING GOAL Describe alpha, beta, positron, and gamma radiation.

5.1 Identify the type of particle or radiation for each of the following:
a. $^{4}_{2}He$ **b.** $^{0}_{+1}e$ **c.** $^{0}_{0}\gamma$

5.2 Identify the type of particle or radiation for each of the following:
a. $^{0}_{-1}e$ **b.** $^{1}_{1}H$ **c.** $^{1}_{0}n$

5.3 Naturally occurring potassium consists of three isotopes: potassium-39, potassium-40, and potassium-41.
a. Write the atomic symbol for each isotope.
b. In what ways are the isotopes similar, and in what ways do they differ?

5.4 Naturally occurring iodine is iodine-127. Medically, radioactive isotopes of iodine-125 and iodine-131 are used.
a. Write the atomic symbol for each isotope.
b. In what ways are the isotopes similar, and in what ways do they differ?

5.5 Identify each of the following:
a. ${}^{0}_{-1}X$ **b.** ${}^{4}_{2}X$ **c.** ${}^{1}_{0}X$
d. ${}^{38}_{18}X$ **e.** ${}^{14}_{6}X$

5.6 Identify each of the following:
a. ${}^{1}_{1}X$ **b.** ${}^{81}_{35}X$ **c.** ${}^{0}_{0}X$
d. ${}^{59}_{26}X$ **e.** ${}^{0}_{+1}X$

Clinical Applications

5.7 Write the atomic symbol for each of the following isotopes used in nuclear medicine:
a. copper-64 **b.** selenium-75
c. sodium-24 **d.** nitrogen-15

5.8 Write the atomic symbol for each of the following isotopes used in nuclear medicine:
a. indium-111 **b.** palladium-103
c. barium-131 **d.** rubidium-82

5.9 Supply the missing information in the following table:

Medical Use	Atomic Symbol	Mass Number	Number of Protons	Number of Neutrons
Heart imaging	${}^{201}_{81}Tl$			
Radiation therapy		60	27	
Abdominal scan			31	36
Hyperthyroidism	${}^{131}_{53}I$			
Leukemia treatment		32		17

5.10 Supply the missing information in the following table:

Medical Use	Atomic Symbol	Mass Number	Number of Protons	Number of Neutrons
Cancer treatment	${}^{131}_{55}Cs$			
Brain scan			43	56
Blood flow		141	58	
Bone scan		85		47
Lung function	${}^{133}_{54}Xe$			

5.11 Match the type of radiation (**1** to **3**) with each of the following statements:
1. alpha particle
2. beta particle
3. gamma radiation

a. does not penetrate skin
b. shielding protection includes lead or thick concrete
c. can be very harmful if ingested

5.12 Match the type of radiation (**1** to **3**) with each of the following statements:
1. alpha particle
2. beta particle
3. gamma radiation

a. penetrates farthest into skin and body tissues
b. shielding protection includes lab coats and gloves
c. travels only a short distance in air

REVIEW

Using Positive and Negative Numbers in Calculations (1.4)
Solving Equations (1.4)
Counting Protons and Neutrons (4.4)

5.2 Nuclear Reactions

LEARNING GOAL Write a balanced nuclear equation for radioactive decay, showing mass numbers and atomic numbers.

CORE CHEMISTRY SKILL

Writing Nuclear Equations

In a process called **radioactive decay**, a nucleus spontaneously breaks down by emitting radiation. This process is shown by writing a *nuclear equation* with the atomic symbols of the original radioactive nucleus on the left, an arrow, and the new nucleus and the type of radiation emitted on the right.

$$\text{Radioactive nucleus} \longrightarrow \text{new nucleus} + \text{radiation } (\alpha, \beta, \beta^{+}, \gamma)$$

In a nuclear equation, the sum of the mass numbers and the sum of the atomic numbers on one side of the arrow must equal the sum of the mass numbers and the sum of the atomic numbers on the other side.

Alpha Decay

An unstable nucleus may emit an alpha particle, which consists of two protons and two neutrons. Thus, the mass number of the radioactive nucleus decreases by 4, and its atomic number decreases by 2. For example, when uranium-238 emits an alpha particle, the new nucleus that forms has a mass number of 234. Compared to uranium with 92 protons, the new nucleus has 90 protons, which is thorium.

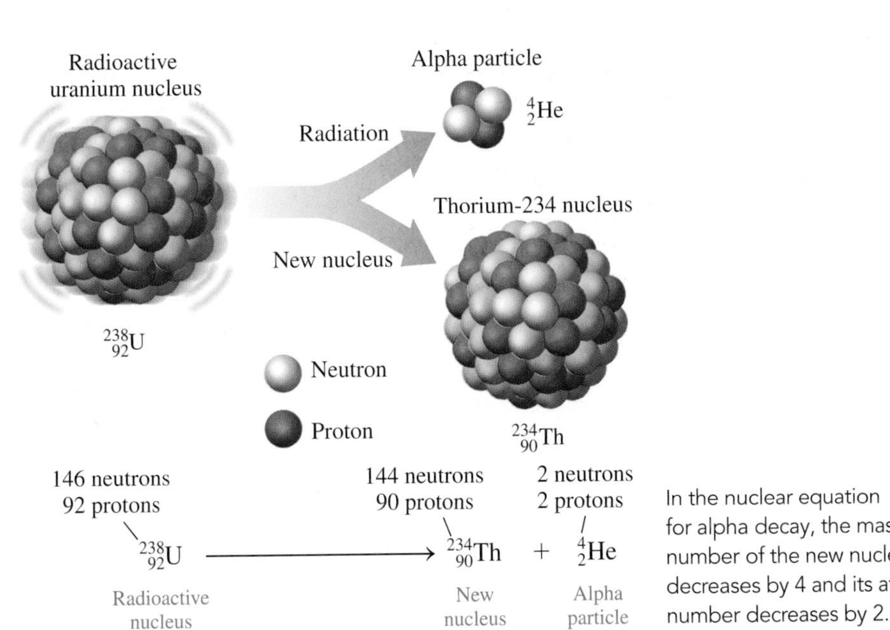

In the nuclear equation for alpha decay, the mass number of the new nucleus decreases by 4 and its atomic number decreases by 2.

ENGAGE

What happens to the U-238 nucleus when an alpha particle is emitted?

We can look at writing a balanced nuclear equation for americium-241, which undergoes alpha decay as shown in Sample Problem 5.2.

SAMPLE PROBLEM 5.2 Writing a Nuclear Equation for Alpha Decay

TRY IT FIRST

Smoke detectors that are used in homes and apartments contain americium-241, which undergoes alpha decay. When alpha particles collide with air molecules, charged particles are produced that generate an electrical current. If smoke particles enter the detector, they interfere with the formation of charged particles in the air, and the electrical current is interrupted. This causes the alarm to sound and warns the occupants of the danger of fire. Write the balanced nuclear equation for the alpha decay of americium-241.

A smoke detector sounds an alarm when smoke enters its ionization chamber.

SOLUTION GUIDE

	Given	Need	Connect
ANALYZE THE PROBLEM	Am-241, alpha decay	balanced nuclear equation	mass number, atomic number of new nucleus

STEP 1 Write the incomplete nuclear equation.

$$^{241}_{95}Am \longrightarrow ? + \ ^{4}_{2}He$$

STEP 2 Determine the missing mass number. In the equation, the mass number, 241, is equal to the sum of the mass numbers of the new nucleus and the alpha particle.

$241 = ? + 4$
$241 - 4 = ?$
$241 - 4 = 237$ (mass number of new nucleus)

STEP 3 Determine the missing atomic number. The atomic number, 95, must equal the sum of the atomic numbers of the new nucleus and the alpha particle.

$95 = ? + 2$
$95 - 2 = ?$
$95 - 2 = 93$ (atomic number of new nucleus)

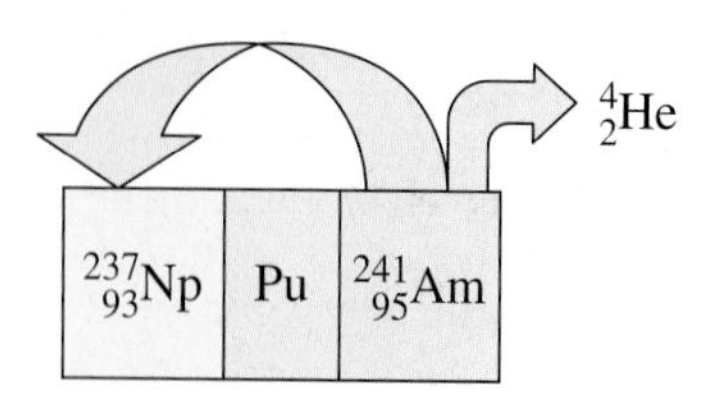

STEP 4 Determine the symbol of the new nucleus. On the periodic table, the element that has atomic number 93 is neptunium, Np. The atomic symbol for this isotope of Np is written $^{237}_{93}Np$.

STEP 5 Complete the nuclear equation.

$$^{241}_{95}Am \longrightarrow ^{237}_{93}Np + ^{4}_{2}He$$

STUDY CHECK 5.2

Write the balanced nuclear equation for the alpha decay of Po-214.

ANSWER

$$^{214}_{84}Po \longrightarrow ^{210}_{82}Pb + ^{4}_{2}He$$

TEST

Try Practice Problems 5.13 and 5.14

CHEMISTRY LINK TO HEALTH

Radon in Our Homes

The presence of radon gas has become a much publicized environmental and health issue because of the radiation danger it poses. Radioactive isotopes such as radium-226 are naturally present in many types of rocks and soils. Radium-226 emits an alpha particle and is converted into radon gas, which diffuses out of the rocks and soil.

$$^{226}_{88}Ra \longrightarrow ^{222}_{86}Rn + ^{4}_{2}He$$

Outdoors, radon gas poses little danger because it disperses in the air. However, if the radioactive source is under a house or building, the radon gas can enter the house through cracks in the foundation or other openings. Those who live or work there may inhale the radon. In the lungs, radon-222 emits alpha particles to form polonium-218, which is known to cause lung cancer.

$$^{222}_{86}Rn \longrightarrow ^{218}_{84}Po + ^{4}_{2}He$$

The U.S. Environmental Protection Agency (EPA) estimates that radon causes about 20 000 lung cancer deaths in one year. The EPA recommends that the maximum level of radon not exceed 4 picocuries (pCi) per liter of air in a home. One picocurie (pCi) is equal to 10^{-12} curies (Ci); curies are described in Section 5.3. The EPA estimates that more than 6 million homes have radon levels that exceed this maximum.

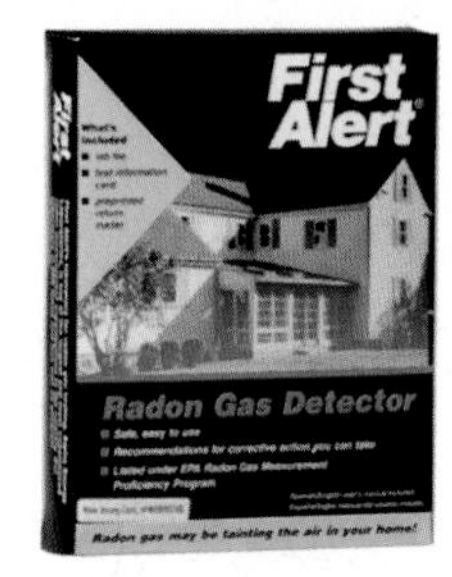

A radon gas detector is used to determine radon levels in buildings.

Beta Decay

The formation of a beta particle is the result of the breakdown of a neutron into a proton and an electron (beta particle). Because the proton remains in the nucleus, the number of protons increases by one, whereas the number of neutrons decreases by one. Thus, in a nuclear equation for beta decay, the mass number of the radioactive nucleus and the mass number of the new nucleus are the same. However, the atomic number of the new nucleus increases by one, which makes it a nucleus of a different element (*transmutation*). For example, the beta decay of a carbon-14 nucleus produces a nitrogen-14 nucleus.

ENGAGE

What happens to a C-14 nucleus when a beta particle is emitted?

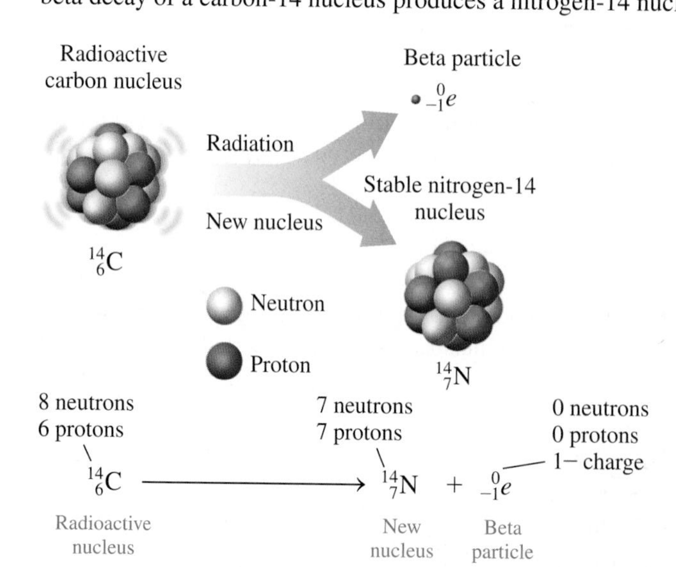

In the nuclear equation for beta decay, the mass number of the new nucleus remains the same and its atomic number increases by 1.

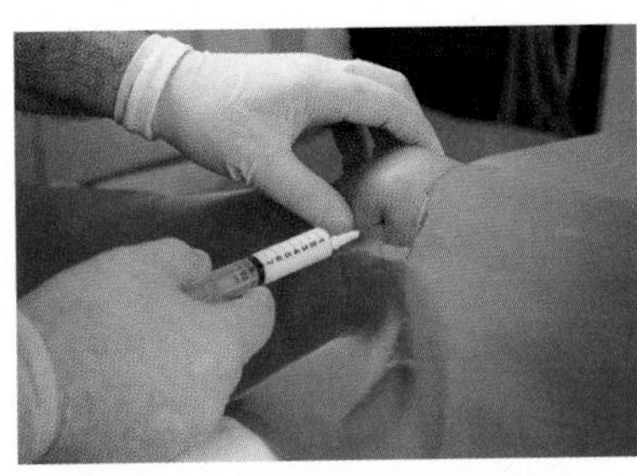

A radioisotope is injected into the joint to relieve the pain caused by arthritis.

SAMPLE PROBLEM 5.3 Writing a Nuclear Equation for Beta Decay

TRY IT FIRST

The radioactive isotope yttrium-90, a beta emitter, is used in cancer treatment and as a colloidal injection into large joints to relieve the pain caused by arthritis. Write the balanced nuclear equation for the beta decay of yttrium-90.

SOLUTION GUIDE

	Given	Need	Connect
ANALYZE THE PROBLEM	Y-90, beta decay	balanced nuclear equation	mass number, atomic number of new nucleus

STEP 1 Write the incomplete nuclear equation.

$$^{90}_{39}\text{Y} \longrightarrow ? + {}^{0}_{-1}e$$

STEP 2 **Determine the missing mass number.** In the equation, the mass number, 90, is equal to the sum of the mass numbers of the new nucleus and the beta particle.

$$90 = ? + 0$$
$$90 - 0 = ?$$
$$90 - 0 = 90 \text{ (mass number of new nucleus)}$$

STEP 3 **Determine the missing atomic number.** The atomic number, 39, must equal the sum of the atomic numbers of the new nucleus and the beta particle.

$$39 = ? - 1$$
$$39 + 1 = ?$$
$$39 + 1 = 40 \text{ (atomic number of new nucleus)}$$

STEP 4 **Determine the symbol of the new nucleus.** On the periodic table, the element that has atomic number 40 is zirconium, Zr. The atomic symbol for this isotope of Zr is written $^{90}_{40}\text{Zr}$.

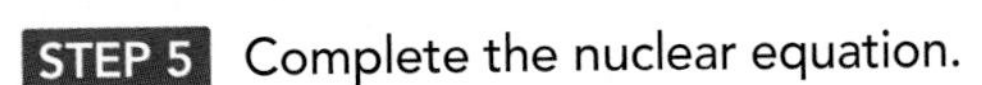

STEP 5 Complete the nuclear equation.

$$^{90}_{39}\text{Y} \longrightarrow {}^{90}_{40}\text{Zr} + {}^{0}_{-1}e$$

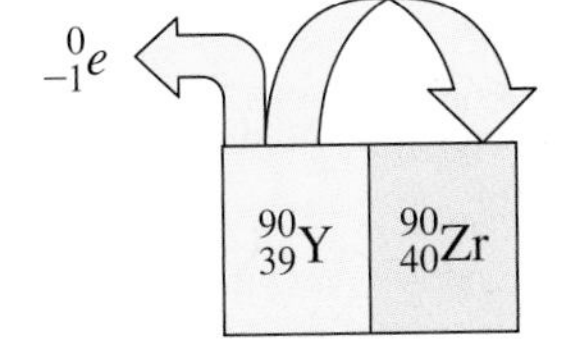

STUDY CHECK 5.3

Write the balanced nuclear equation for the beta decay of chromium-51.

ANSWER

$$^{51}_{24}\text{Cr} \longrightarrow {}^{51}_{25}\text{Mn} + {}^{0}_{-1}e$$

Positron Emission

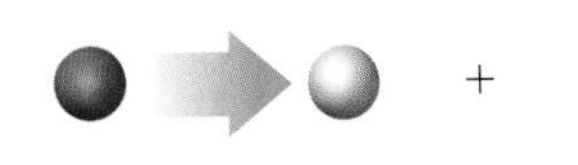

$$^{24}_{13}\text{Al} \longrightarrow {}^{24}_{12}\text{Mg} + {}^{0}_{+1}e$$

Positron

In positron emission, a proton in an unstable nucleus is converted to a neutron and a positron. The neutron remains in the nucleus, but the positron is emitted from the nucleus. In a nuclear equation for positron emission, the mass number of the radioactive nucleus and the mass number of the new nucleus are the same. However, the atomic number of the new nucleus decreases by one, indicating a change of one element into another. For example, an aluminum-24 nucleus undergoes positron emission to produce a magnesium-24 nucleus. The atomic number of magnesium (12) and the charge of the positron (1+) give the atomic number of aluminum (13).

ENGAGE

Why does the atomic number but not the mass number change in positron emission?

SAMPLE PROBLEM 5.4 Writing a Nuclear Equation for Positron Emission

TRY IT FIRST

Write the balanced nuclear equation for manganese-49, which decays by emitting a positron.

SOLUTION GUIDE

	Given	Need	Connect
ANALYZE THE PROBLEM	Mn-49, positron emission	balanced nuclear equation	mass number, atomic number of new nucleus

STEP 1 Write the incomplete nuclear equation.

$${}^{49}_{25}\text{Mn} \longrightarrow ? + {}^{0}_{+1}e$$

STEP 2 Determine the missing mass number. In the equation, the mass number, 49, is equal to the sum of the mass numbers of the new nucleus and the positron.

$$49 = ? + 0$$
$$49 - 0 = ?$$
$$49 - 0 = 49 \text{ (mass number of new nucleus)}$$

STEP 3 Determine the missing atomic number. The atomic number, 25, must equal the sum of the atomic numbers of the new nucleus and the positron.

$$25 = ? + 1$$
$$25 - 1 = ?$$
$$25 - 1 = 24 \text{ (atomic number of new nucleus)}$$

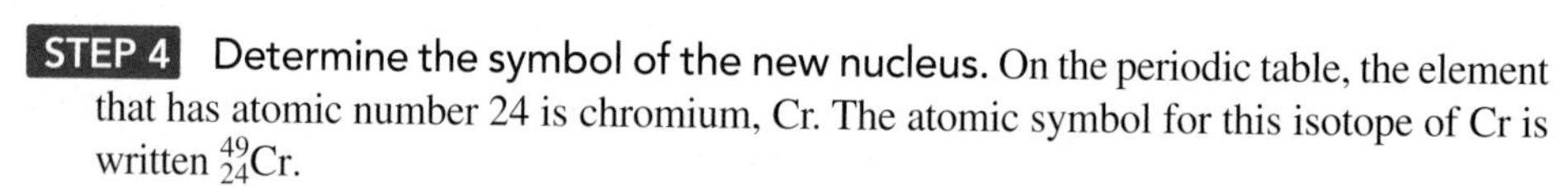

STEP 4 Determine the symbol of the new nucleus. On the periodic table, the element that has atomic number 24 is chromium, Cr. The atomic symbol for this isotope of Cr is written ${}^{49}_{24}\text{Cr}$.

STEP 5 Complete the nuclear equation.

$${}^{49}_{25}\text{Mn} \longrightarrow {}^{49}_{24}\text{Cr} + {}^{0}_{+1}e$$

STUDY CHECK 5.4

Write the balanced nuclear equation for xenon-118, which undergoes positron emission.

ANSWER

$${}^{118}_{54}\text{Xe} \longrightarrow {}^{118}_{53}\text{I} + {}^{0}_{+1}e$$

TEST

Try Practice Problems 5.15 to 5.18

Gamma Emission

Pure gamma emitters are rare, although gamma radiation accompanies most alpha and beta radiation. In radiology, one of the most commonly used gamma emitters is technetium (Tc). The unstable isotope of technetium is written as the *metastable* (symbol m) isotope technetium-99m, Tc-99m, or ${}^{99m}_{43}\text{Tc}$. By emitting energy in the form of gamma rays, the nucleus becomes more stable.

$${}^{99m}_{43}\text{Tc} \longrightarrow {}^{99}_{43}\text{Tc} + {}^{0}_{0}\gamma$$

FIGURE 5.2 summarizes the changes in the nucleus for alpha, beta, positron, and gamma radiation.

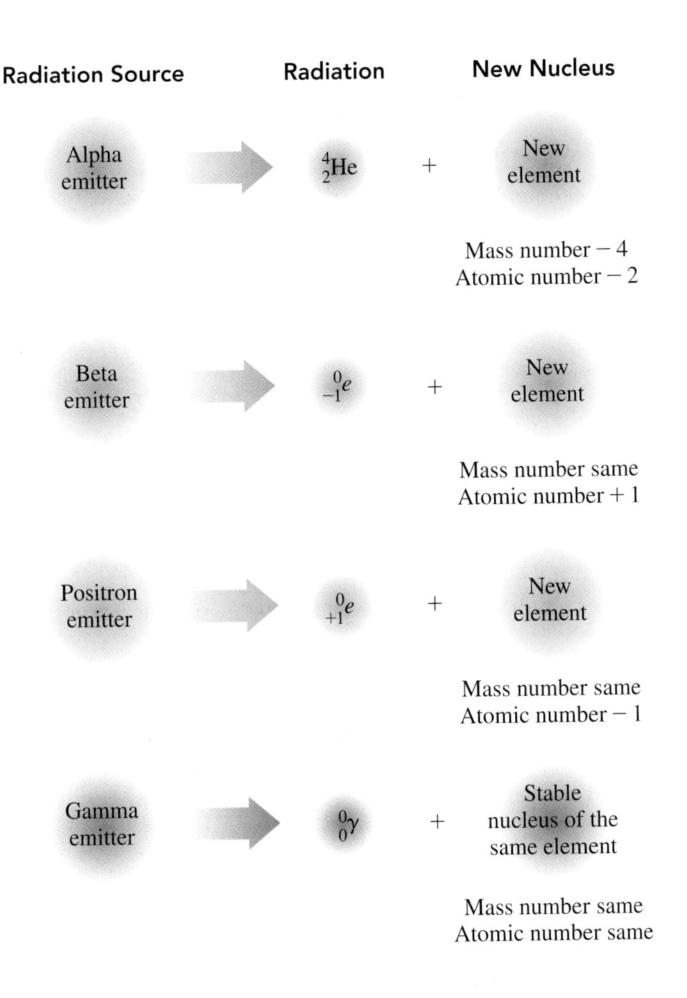

FIGURE 5.2 When the nuclei of alpha, beta, positron, and gamma emitters emit radiation, new, more stable nuclei are produced.

Q What changes occur in the number of protons and neutrons when an alpha emitter gives off radiation?

Producing Radioactive Isotopes

Today, many radioisotopes are produced in small amounts by bombarding stable, nonradioactive isotopes with high-speed particles such as alpha particles, protons, neutrons, and small nuclei. When one of these particles is absorbed, the stable nucleus is converted to a radioactive isotope and usually some type of radiation particle.

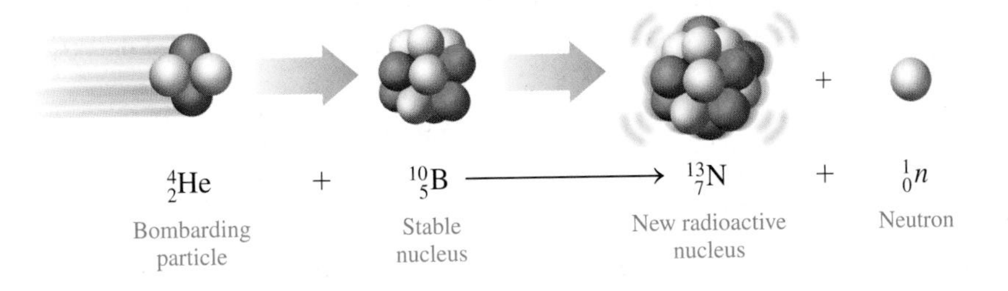

When nonradioactive B-10 is bombarded by an alpha particle, the products are radioactive N-13 and a neutron.

All elements that have an atomic number greater than 92 have been produced by bombardment. Most have been produced in only small amounts and exist for only a short time, making it difficult to study their properties. For example, when californium-249 is bombarded with nitrogen-15, the radioactive element dubnium-260 and four neutrons are produced.

$$^{15}_{7}N + ^{249}_{98}Cf \longrightarrow ^{260}_{105}Db + 4^{1}_{0}n$$

INTERACTIVE VIDEO
Writing Equations for an Isotope Produced by Bombardment

Technetium-99m is a radioisotope used in nuclear medicine for several diagnostic procedures, including the detection of brain tumors and examinations of the liver and spleen. The source of technetium-99m is molybdenum-99, which is produced in a nuclear reactor by neutron bombardment of molybdenum-98.

$$^{1}_{0}n + ^{98}_{42}Mo \longrightarrow ^{99}_{42}Mo$$

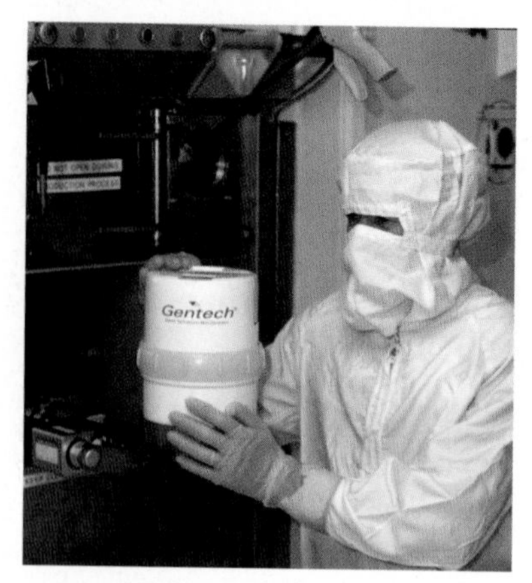

A generator is used to prepare technetium-99m.

Many radiology laboratories have small generators containing molybdenum-99, which decays to the technetium-99m radioisotope.

$$^{99}_{42}\text{Mo} \longrightarrow {}^{99m}_{43}\text{Tc} + {}^{0}_{-1}e$$

The technetium-99m radioisotope decays by emitting gamma rays. Gamma emission is desirable for diagnostic work because the gamma rays pass through the body to the detection equipment.

$$^{99m}_{43}\text{Tc} \longrightarrow {}^{99}_{43}\text{Tc} + {}^{0}_{0}\gamma$$

SAMPLE PROBLEM 5.5 Writing a Nuclear Equation for an Isotope Produced by Bombardment

TRY IT FIRST

Write the balanced nuclear equation for the bombardment of nickel-58 by a proton, $^{1}_{1}\text{H}$, which produces a radioactive isotope and an alpha particle.

SOLUTION GUIDE

	Given	Need	Connect
ANALYZE THE PROBLEM	Ni-58, proton bombardment	balanced nuclear equation	mass number, atomic number of new nucleus

STEP 1 Write the incomplete nuclear equation.

$$^{1}_{1}\text{H} + {}^{58}_{28}\text{Ni} \longrightarrow ? + {}^{4}_{2}\text{He}$$

STEP 2 Determine the missing mass number. In the equation, the sum of the mass numbers of the proton, 1, and the nickel, 58, must equal the sum of the mass numbers of the new nucleus and the alpha particle.

$1 + 58 = ? + 4$
$59 - 4 = ?$
$59 - 4 = 55$ (mass number of new nucleus)

STEP 3 Determine the missing atomic number. The sum of the atomic numbers of the proton, 1, and nickel, 28, must equal the sum of the atomic numbers of the new nucleus and the alpha particle.

$1 + 28 = ? + 2$
$29 - 2 = ?$
$29 - 2 = 27$ (atomic number of new nucleus)

STEP 4 Determine the symbol of the new nucleus. On the periodic table, the element that has atomic number 27 is cobalt, Co. The atomic symbol for this isotope of Co is written $^{55}_{27}\text{Co}$.

STEP 5 Complete the nuclear equation.

$$^{1}_{1}\text{H} + {}^{58}_{28}\text{Ni} \longrightarrow {}^{55}_{27}\text{Co} + {}^{4}_{2}\text{He}$$

ENGAGE

If the bombardment of N-14 by an alpha particle produces a proton, how do you know the other product is O-17?

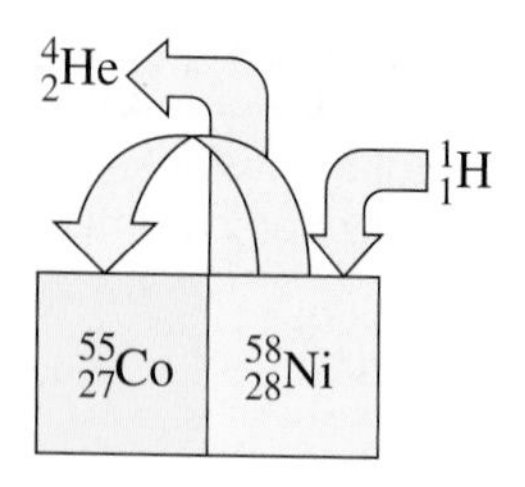

STUDY CHECK 5.5

The first radioactive isotope was produced in 1934 by the bombardment of aluminum-27 by an alpha particle to produce a radioactive isotope and one neutron. Write the balanced nuclear equation for this bombardment.

ANSWER

$$^{4}_{2}\text{He} + {}^{27}_{13}\text{Al} \longrightarrow {}^{30}_{15}\text{P} + {}^{1}_{0}n$$

TEST

Try Practice Problems 5.19 to 5.22

PRACTICE PROBLEMS

5.2 Nuclear Reactions

LEARNING GOAL Write a balanced nuclear equation for radioactive decay, showing mass numbers and atomic numbers.

5.13 Write a balanced nuclear equation for the alpha decay of each of the following radioactive isotopes:
a. $^{208}_{84}Po$ **b.** $^{232}_{90}Th$ **c.** $^{251}_{102}No$ **d.** radon-220

5.14 Write a balanced nuclear equation for the alpha decay of each of the following radioactive isotopes:
a. curium-243 **b.** $^{252}_{99}Es$
c. $^{251}_{98}Cf$ **d.** $^{261}_{107}Bh$

5.15 Write a balanced nuclear equation for the beta decay of each of the following radioactive isotopes:
a. $^{25}_{11}Na$ **b.** $^{20}_{8}O$
c. strontium-92 **d.** iron-60

5.16 Write a balanced nuclear equation for the beta decay of each of the following radioactive isotopes:
a. $^{44}_{19}K$ **b.** iron-59
c. potassium-42 **d.** $^{141}_{56}Ba$

5.17 Write a balanced nuclear equation for the positron emission of each of the following radioactive isotopes:
a. silicon-26 **b.** cobalt-54 **c.** $^{77}_{37}Rb$ **d.** $^{93}_{45}Rh$

5.18 Write a balanced nuclear equation for the positron emission of each of the following radioactive isotopes:
a. boron-8 **b.** $^{15}_{8}O$
c. $^{40}_{19}K$ **d.** nitrogen-13

5.19 Complete each of the following nuclear equations and describe the type of radiation:
a. $^{28}_{13}Al \longrightarrow ? + ^{0}_{-1}e$ **b.** $^{180m}_{73}Ta \longrightarrow ^{180}_{73}Ta + ?$
c. $^{66}_{29}Cu \longrightarrow ^{66}_{30}Zn + ?$ **d.** $? \longrightarrow ^{234}_{90}Th + ^{4}_{2}He$
e. $^{188}_{80}Hg \longrightarrow ? + ^{0}_{+1}e$

5.20 Complete each of the following nuclear equations and describe the type of radiation:
a. $^{11}_{6}C \longrightarrow ^{11}_{5}B + ?$ **b.** $^{35}_{16}S \longrightarrow ? + ^{0}_{-1}e$
c. $? \longrightarrow ^{90}_{39}Y + ^{0}_{-1}e$ **d.** $^{210}_{83}Bi \longrightarrow ? + ^{4}_{2}He$
e. $? \longrightarrow ^{89}_{39}Y + ^{0}_{+1}e$

5.21 Complete each of the following bombardment reactions:
a. $^{1}_{0}n + ^{9}_{4}Be \longrightarrow ?$ **b.** $^{1}_{0}n + ^{131}_{52}Te \longrightarrow ? + ^{0}_{-1}e$
c. $^{1}_{0}n + ? \longrightarrow ^{24}_{11}Na + ^{4}_{2}He$ **d.** $^{4}_{2}He + ^{14}_{7}N \longrightarrow ? + ^{1}_{1}H$

5.22 Complete each of the following bombardment reactions:
a. $? + ^{40}_{18}Ar \longrightarrow ^{43}_{19}K + ^{1}_{1}H$ **b.** $^{1}_{0}n + ^{238}_{92}U \longrightarrow ?$
c. $^{1}_{0}n + ? \longrightarrow ^{14}_{6}C + ^{1}_{1}H$ **d.** $? + ^{64}_{28}Ni \longrightarrow ^{272}_{111}Rg + ^{1}_{0}n$

5.3 Radiation Measurement

LEARNING GOAL Describe the detection and measurement of radiation.

One of the most common instruments for detecting beta and gamma radiation is the Geiger counter. It consists of a metal tube filled with a gas such as argon. When radiation enters a window on the end of the tube, it forms charged particles in the gas, which produce an electrical current. Each burst of current is amplified to give a click and a reading on a meter.

$$Ar + \text{radiation} \longrightarrow Ar^{+} + e^{-}$$

Measuring Radiation

Radiation is measured in several different ways. When a radiology laboratory obtains a radioisotope, the *activity* of the sample is measured in terms of the number of nuclear disintegrations per second. The **curie (Ci)**, the original unit of activity, was defined as the number of disintegrations that occur in 1 s for 1 g of radium, which is equal to 3.7×10^{10} disintegrations/s. The unit was named for Polish scientist Marie Curie, who along with her husband, Pierre, discovered the radioactive elements radium and polonium. The SI unit of radiation activity is the **becquerel (Bq)**, which is 1 disintegration/s.

A radiation counter is used to determine radiation levels in workers at the Fukushima Daiichi nuclear power plant.

ENGAGE

What is the difference between becquerels and rems?

The **rad (radiation absorbed dose)** is a unit that measures the amount of radiation absorbed by a gram of material such as body tissue. The SI unit for absorbed dose is the **gray (Gy)**, which is defined as the joules of energy absorbed by 1 kg of body tissue. The gray is equal to 100 rad.

The **rem (radiation equivalent in humans)** is a unit that measures the biological effects of different kinds of radiation. Although alpha particles do not penetrate the skin, if they should enter the body by some other route, they can cause extensive damage within a short distance in tissue. High-energy radiation, such as beta particles, high-energy protons, and neutrons that travel into tissue, causes more damage. Gamma rays are damaging because they travel a long way through body tissue.

To determine the **equivalent dose** or rem dose, the absorbed dose (rad) is multiplied by a factor that adjusts for biological damage caused by a particular form of radiation. For beta and gamma radiation the factor is 1, so the biological damage in rems is the same as the absorbed radiation (rad). For high-energy protons and neutrons, the factor is about 10, and for alpha particles it is 20.

$$\text{Biological damage (rem)} = \text{Absorbed dose (rad)} \times \text{Factor}$$

Often, the measurement for an equivalent dose will be in units of millirems (mrem). One rem is equal to 1000 mrem. The SI unit is the **sievert (Sv)**. One sievert is equal to 100 rem. **TABLE 5.4** summarizes the units used to measure radiation.

TABLE 5.4 Units of Radiation Measurement

Measurement	Common Unit	SI Unit	Relationship
Activity	curie (Ci) 1 Ci = 3.7×10^{10} disintegrations/s	becquerel (Bq) 1 Bq = 1 disintegration/s	1 Ci = 3.7×10^{10} Bq
Absorbed Dose	rad	gray (Gy) 1 Gy = 1 J/kg of tissue	1 Gy = 100 rad
Biological Damage	rem	sievert (Sv)	1 Sv = 100 rem

CHEMISTRY LINK TO HEALTH

Radiation and Food

Foodborne illnesses caused by pathogenic bacteria such as *Salmonella*, *Listeria*, and *Escherichia coli* have become a major health concern in the United States. *E. coli* has been responsible for outbreaks of illness from contaminated ground beef, fruit juices, lettuce, and alfalfa sprouts.

The U.S. Food and Drug Administration (FDA) has approved the use of 0.3 kGy to 1 kGy of radiation produced by cobalt-60 or cesium-137 for the treatment of foods. The irradiation technology is much like that used to sterilize medical supplies. Cobalt pellets are placed in stainless steel tubes, which are arranged in racks. When food moves through the series of racks, the gamma rays pass through the food and kill the bacteria.

It is important for consumers to understand that when food is irradiated, it never comes in contact with the radioactive source. The gamma rays pass through the food to kill bacteria, but that does not make the food radioactive. The radiation kills bacteria because it stops their ability to divide and grow. We cook or heat food thoroughly for the same purpose. Radiation, as well as heat, has little effect on the food itself because its cells are no longer dividing or growing. Thus irradiated food is not harmed, although a small amount of vitamins B_1 and C may be lost.

Currently, tomatoes, blueberries, strawberries, and mushrooms are being irradiated to allow them to be harvested when completely ripe and extend their shelf life (see **FIGURE 5.3**). The FDA has also approved the irradiation of pork, poultry, and beef to decrease potential infections and to extend shelf life. Currently, irradiated vegetable and meat products are available in retail markets in more than 40 countries. In the United States, irradiated foods such as tropical fruits, spinach, and ground meats are found in some stores. *Apollo 17* astronauts ate irradiated foods on the Moon, and some U.S. hospitals and nursing homes now use irradiated poultry to reduce the possibility of salmonella infections among residents. The extended shelf life of irradiated food also makes it useful for campers and military personnel. Soon, consumers concerned about food safety will have a choice of irradiated meats, fruits, and vegetables at the market.

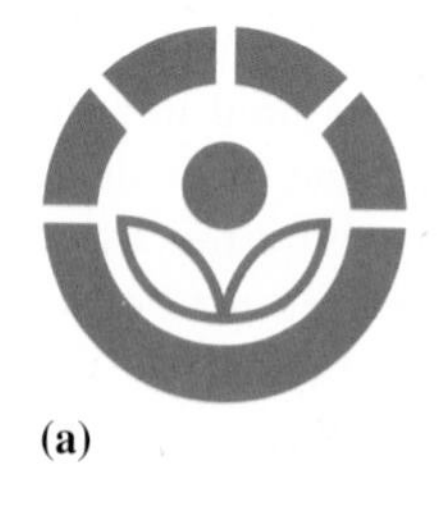
(a)

(b)

FIGURE 5.3 **(a)** The FDA requires this symbol to appear on irradiated retail foods. **(b)** After two weeks, the irradiated strawberries on the right show no spoilage. Mold is growing on the nonirradiated ones on the left.

Q Why are irradiated foods used on spaceships and in nursing homes?

People who work in radiology laboratories wear dosimeters attached to their clothing to determine any exposure to radiation such as X-rays, gamma rays, or beta particles. A dosimeter can be thermoluminescent (TLD), optically stimulated luminescence (OSL), or electronic personal (EPD). Dosimeters provide real-time radiation levels measured by monitors in the work area.

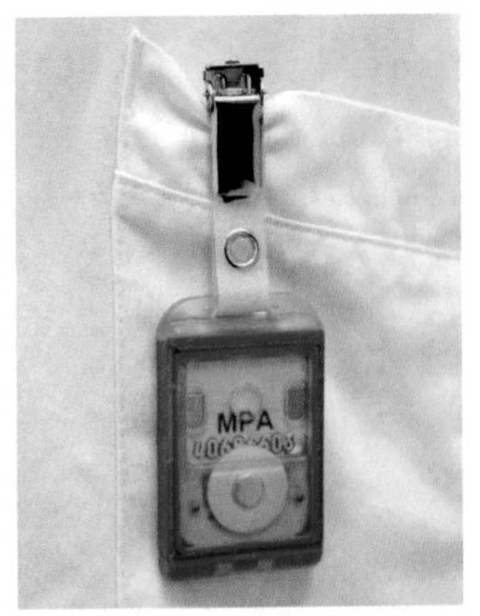

A dosimeter measures radiation exposure.

SAMPLE PROBLEM 5.6 Radiation Measurement

TRY IT FIRST

One treatment for bone pain involves intravenous administration of the radioisotope phosphorus-32, which is incorporated into bone. A typical dose of 7 mCi can produce up to 450 rad in the bone. What is the difference between the units of mCi and rad?

SOLUTION

The millicuries (mCi) indicate the activity of the P-32 in terms of nuclei that break down in 1 s. The radiation absorbed dose (rad) is a measure of amount of radiation absorbed by the bone.

STUDY CHECK 5.6

For Sample Problem 5.6, what is the absorbed dose of radiation in grays (Gy)?

ANSWER

4.5 Gy

TEST

Try Practice Problems 5.23 to 5.28

Exposure to Radiation

Every day, we are exposed to low levels of radiation from naturally occurring radioactive isotopes in the buildings where we live and work, in our food and water, and in the air we breathe. For example, potassium-40, a naturally occurring radioactive isotope, is present in any potassium-containing food. Other naturally occurring radioisotopes in air and food are carbon-14, radon-222, strontium-90, and iodine-131. The average person in the United States is exposed to about 3.6 mSv of radiation annually. Medical sources of radiation, including dental, hip, spine, and chest X-rays and mammograms, add to our radiation exposure. **TABLE 5.5** lists some common sources of radiation.

Another source of background radiation is cosmic radiation produced in space by the Sun. People who live at high altitudes or travel by airplane receive a greater amount of cosmic radiation because there are fewer molecules in the atmosphere to absorb the radiation. For example, a person living in Denver receives about twice the cosmic radiation as a person living in Los Angeles. A person living close to a nuclear power plant normally does not receive much additional radiation, perhaps 0.001 mSv in 1 yr. (One Sv equals 1000 mSv.) However, in the accident at the Chernobyl nuclear power plant in 1986 in Ukraine, it is estimated that people in a nearby town received as much as 0.01 Sv/h.

TABLE 5.5 Average Annual Radiation Received by a Person in the United States

Source	Dose (mSv)
Natural	
Ground	0.2
Air, water, food	0.3
Cosmic rays	0.4
Wood, concrete, brick	0.5
Medical	
Chest X-ray	0.2
Dental X-ray	0.2
Mammogram	0.4
Hip X-ray	0.6
Lumbar spine X-ray	0.7
Upper gastrointestinal tract X-ray	2
Other	
Nuclear power plants	0.001
Television	0.2
Air travel	0.1
Radon	2*

*Varies widely.

Radiation Sickness

The larger the dose of radiation received at one time, the greater the effect on the body. Exposure to radiation of less than 0.25 Sv usually cannot be detected. Whole-body exposure of 1 Sv produces a temporary decrease in the number of white blood cells. If the exposure to radiation is greater than 1 Sv, a person may suffer the symptoms of radiation sickness: nausea, vomiting, fatigue, and a reduction in white-cell count. A whole-body dosage greater than 3 Sv can decrease the white-cell count to zero. The person suffers diarrhea, hair loss, and infection. Exposure to radiation of 5 Sv is expected to cause death in 50% of the people receiving that dose. This amount of radiation to the whole body is called the *lethal dose for one-half the population*, or the LD_{50}. The LD_{50} varies for different life forms, as **TABLE 5.6** shows. Whole-body radiation of 6 Sv or greater would be fatal to all humans within a few weeks.

TABLE 5.6 Lethal Doses of Radiation for Some Life Forms

Life Form	LD_{50} (Sv)
Insect	1000
Bacterium	500
Rat	8
Human	5
Dog	3

PRACTICE PROBLEMS

5.3 Radiation Measurement

LEARNING GOAL Describe the detection and measurement of radiation.

5.23 Match each property (**1** to **3**) with its unit of measurement.
1. activity
2. absorbed dose
3. biological damage

a. rad
b. mrem
c. mCi
d. Gy

5.24 Match each property (**1** to **3**) with its unit of measurement.
1. activity
2. absorbed dose
3. biological damage

a. mrad
b. gray
c. becquerel
d. Sv

Clinical Applications

5.25 Two technicians in a nuclear laboratory were accidentally exposed to radiation. If one was exposed to 8 mGy and the other to 5 rad, which technician received more radiation?

5.26 Two samples of a radioisotope were spilled in a nuclear laboratory. The activity of one sample was 8 kBq and the other 15 mCi. Which sample produced the higher amount of radiation?

5.27 **a.** The recommended dosage of iodine-131 is 4.20 μCi/kg of body mass. How many microcuries of iodine-131 are needed for a 70.0-kg person with hyperthyroidism?
b. A person receives 50 rad of gamma radiation. What is that amount in grays?

5.28 **a.** The dosage of technetium-99m for a lung scan is 20. μCi/kg of body mass. How many millicuries of technetium-99m should be given to a 50.0-kg person (1 mCi = 1000 μCi)?
b. Suppose a person absorbed 50 mrad of alpha radiation. What would be the equivalent dose in millisieverts?

REVIEW

Interpreting Graphs (1.4)
Using Conversion Factors (2.6)

5.4 Half-Life of a Radioisotope

LEARNING GOAL Given the half-life of a radioisotope, calculate the amount of radioisotope remaining after one or more half-lives.

The **half-life** of a radioisotope is the amount of time it takes for one-half of a sample to decay. For example, $^{131}_{53}\text{I}$ has a half-life of 8.0 days. As $^{131}_{53}\text{I}$ decays, it produces the nonradioactive isotope $^{131}_{54}\text{Xe}$ and a beta particle.

$$^{131}_{53}\text{I} \longrightarrow {}^{131}_{54}\text{Xe} + {}^{0}_{-1}e$$

ENGAGE

If a 24-mg sample of Tc-99m has a half-life of 6.0 h, why are only 3 mg of Tc-99m radioactive after 18 h?

Suppose we have a sample that initially contains 20. mg of $^{131}_{53}\text{I}$. In 8.0 days, one-half (10. mg) of all the I-131 nuclei in the sample will decay, which leaves 10. mg of I-131. After 16 days (two half-lives), 5.0 mg of the remaining I-131 decays, which leaves 5.0 mg of I-131. After 24 days (three half-lives), 2.5 mg of the remaining I-131 decays, which leaves 2.5 mg of I-131 nuclei still capable of producing radiation.

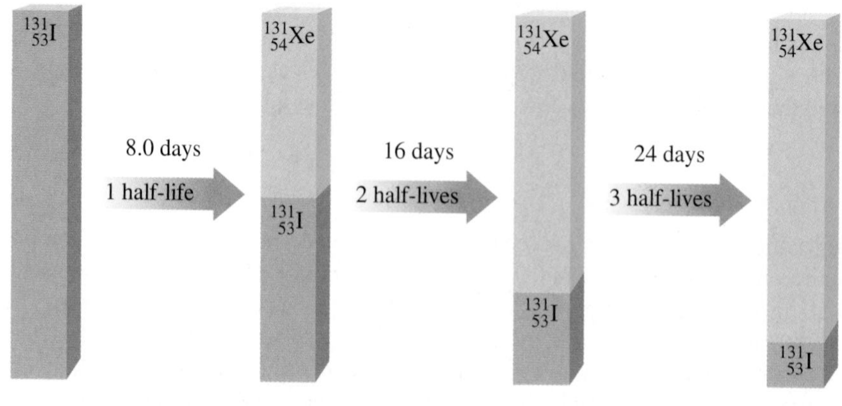

$$20.\text{ mg of } ^{131}_{53}\text{I} \xrightarrow{\text{1 half-life}} 10.\text{ mg of } ^{131}_{53}\text{I} \xrightarrow{\text{2 half-lives}} 5.0\text{ mg of } ^{131}_{53}\text{I} \xrightarrow{\text{3 half-lives}} 2.5\text{ mg of } ^{131}_{53}\text{I}$$

In one half-life, the activity of an isotope decreases by half.

A **decay curve** is a diagram of the decay of a radioactive isotope. **FIGURE 5.4** shows such a curve for the $^{131}_{53}\text{I}$ we have discussed.

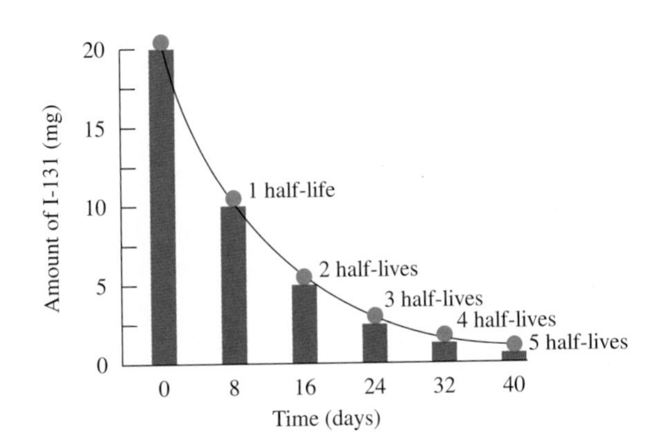

FIGURE 5.4 The decay curve for iodine-131 shows that one-half of the radioactive sample decays and one-half remains radioactive after each half-life of 8.0 days.

Q How many milligrams of the 20.-mg sample remain radioactive after two half-lives?

CORE CHEMISTRY SKILL

Using Half-Lives

SAMPLE PROBLEM 5.7 Using Half-Lives of a Radioisotope

TRY IT FIRST

Phosphorus-32, a radioisotope used in the treatment of leukemia, has a half-life of 14.3 days. If a sample contains 8.0 mg of phosphorus-32, how many milligrams of phosphorus-32 remain after 42.9 days?

SOLUTION GUIDE

STEP 1 State the given and needed quantities.

ANALYZE THE PROBLEM	Given	Need	Connect
	8.0 mg of P-32, 42.9 days elapsed, half-life = 14.3 days	milligrams of P-32 remaining	number of half-lives

STEP 2 Write a plan to calculate the unknown quantity.

days → Half-life → number of half-lives

milligrams of $^{32}_{15}P$ → Number of half-lives → milligrams of $^{32}_{15}P$ remaining

STEP 3 Write the half-life equality and conversion factors.

1 half-life = 14.3 days

$$\frac{14.3 \text{ days}}{1 \text{ half-life}} \quad \text{and} \quad \frac{1 \text{ half-life}}{14.3 \text{ days}}$$

STEP 4 Set up the problem to calculate the needed quantity. First, we determine the number of half-lives in the amount of time that has elapsed.

$$\text{number of half-lives} = 42.9 \cancel{\text{days}} \times \frac{1 \text{ half-life}}{14.3 \cancel{\text{days}}} = 3.00 \text{ half-lives}$$

Now we can determine how much of the sample decays in three half-lives and how many milligrams of the phosphorus remain.

$$8.0 \text{ mg of } ^{32}_{15}P \xrightarrow{\text{1 half-life}} 4.0 \text{ mg of } ^{32}_{15}P \xrightarrow{\text{2 half-lives}} 2.0 \text{ mg of } ^{32}_{15}P \xrightarrow{\text{3 half-lives}} 1.0 \text{ mg of } ^{32}_{15}P$$

STUDY CHECK 5.7

Iron-59 has a half-life of 44 days. If a nuclear laboratory receives a sample of 8.0 μg of iron-59, how many micrograms of iron-59 are still active after 176 days?

ANSWER

0.50 μg of iron-59

INTERACTIVE VIDEO

Half-Lives

TABLE 5.7 Half-Lives of Some Radioisotopes

Element	Radioisotope	Half-Life	Type of Radiation
Naturally Occurring Radioisotopes			
Carbon-14	${}^{14}_{6}C$	5730 yr	Beta
Potassium-40	${}^{40}_{19}K$	1.3×10^9 yr	Beta, gamma
Radium-226	${}^{226}_{88}Ra$	1600 yr	Alpha
Strontium-90	${}^{90}_{38}Sr$	38.1 yr	Alpha
Uranium-238	${}^{238}_{92}U$	4.5×10^9 yr	Alpha
Some Medical Radioisotopes			
Carbon-11	${}^{11}_{6}C$	20. min	Positron
Chromium-51	${}^{51}_{24}Cr$	28 days	Gamma
Iodine-131	${}^{131}_{53}I$	8.0 days	Gamma
Oxygen-15	${}^{15}_{8}O$	2.0 min	Positron
Iron-59	${}^{59}_{26}Fe$	44 days	Beta, gamma
Radon-222	${}^{222}_{86}Rn$	3.8 days	Alpha
Technetium-99m	${}^{99m}_{43}Tc$	6.0 h	Beta, gamma

TEST

Try Practice Problems 5.29 to 5.34

ENGAGE

Why do radioisotopes used in nuclear medicine have shorter half-lives than naturally occurring radioisotopes?

Naturally occurring isotopes of the elements usually have long half-lives, as shown in TABLE 5.7. They disintegrate slowly and produce radiation over a long period of time, even hundreds or millions of years. In contrast, the radioisotopes used in nuclear medicine have much shorter half-lives. They disintegrate rapidly and produce almost all their radiation in a short period of time. For example, technetium-99m emits half of its radiation in the first 6 h. This means that a small amount of the radioisotope given to a patient is essentially gone within two days. The decay products of technetium-99m are totally eliminated by the body.

CHEMISTRY LINK TO THE ENVIRONMENT

Dating Ancient Objects

Radiological dating is a technique used by geologists, archaeologists, and historians to determine the age of ancient objects. The age of an object derived from plants or animals (such as wood, fiber, natural pigments, bone, and cotton or woolen clothing) is determined by measuring the amount of carbon-14, a naturally occurring radioactive form of carbon. In 1960, Willard Libby received the Nobel Prize for the work he did developing carbon-14 dating techniques during the 1940s. Carbon-14 is produced in the upper atmosphere by the bombardment of ${}^{14}_{7}N$ by high-energy neutrons from cosmic rays.

$${}^{1}_{0}n + {}^{14}_{7}N \longrightarrow {}^{14}_{6}C + {}^{1}_{1}H$$

Neutron from cosmic rays; Nitrogen in atmosphere; Radioactive carbon-14; Proton

The carbon-14 reacts with oxygen to form radioactive carbon dioxide, ${}^{14}_{6}CO_2$. Living plants continuously absorb carbon dioxide, which incorporates carbon-14 into the plant material. The uptake of carbon-14 stops when the plant dies.

$${}^{14}_{6}C \longrightarrow {}^{14}_{7}N + {}^{0}_{-1}e$$

As the carbon-14 decays, the amount of radioactive carbon-14 in the plant material steadily decreases. In a process called *carbon dating*, scientists use the half-life of carbon-14 (5730 yr) to calculate the length of time since the plant died. For example, a wooden beam found in an ancient dwelling might have one-half of the carbon-14 found in living plants today. Because one half-life of carbon-14 is 5730 yr, the dwelling was constructed about 5730 yr ago. Carbon-14 dating was used to determine that the Dead Sea Scrolls are about 2000 yr old.

The age of the Dead Sea Scrolls was determined using carbon-14.

A radiological dating method used for determining the age of much older items is based on the radioisotope uranium-238, which decays through a series of reactions to lead-206. The uranium-238 isotope has an incredibly long half-life, about 4×10^9 (4 billion) yr. Measurements of the amounts of uranium-238 and lead-206 enable geologists to determine the age of rock samples. The older rocks will have a higher percentage of lead-206 because more of the uranium-238 has decayed. The age of rocks brought back from the Moon by the *Apollo* missions, for example, was determined using uranium-238. They were found to be about 4×10^9 yr old, approximately the same age calculated for Earth.

SAMPLE PROBLEM 5.8 Dating Using Half-Lives

TRY IT FIRST

The bones of humans and animals assimilate carbon until death. Using radiocarbon dating, the number of half-lives of carbon-14 from a bone sample determines the age of the bone. Suppose a sample is obtained from a prehistoric animal and used for radiocarbon dating. We can calculate the age of the bone or the years elapsed since the animal died by using the half-life of carbon-14, which is 5730 yr. A bone sample from the skeleton of a prehistoric animal has 25% of the activity of C-14 found in a living animal. How many years ago did the prehistoric animal die?

The age of a bone sample from a skeleton can be determined by carbon dating.

SOLUTION GUIDE

STEP 1 State the given and needed quantities.

	Given	Need	Connect
ANALYZE THE PROBLEM	1 half-life of C-14 = 5730 yr, 25% of initial C-14 activity	years elapsed	number of half-lives

STEP 2 Write a plan to calculate the unknown quantity.

$$\text{Activity:}\quad \underset{\text{(initial)}}{100\%} \xrightarrow{\text{1.0 half-life}} 50\% \xrightarrow{\text{2.0 half-lives}} 25\%$$

STEP 3 Write the half-life equality and conversion factors.

$$1\text{ half-life} = 5730\text{ yr}$$
$$\frac{5730\text{ yr}}{1\text{ half-life}} \quad\text{and}\quad \frac{1\text{ half-life}}{5730\text{ yr}}$$

STEP 4 Set up the problem to calculate the needed quantity.

$$\text{Years elapsed} = 2.0\ \cancel{\text{half-lives}} \times \frac{5730\text{ yr}}{1\ \cancel{\text{half-life}}} = 11\,000\text{ yr}$$

We would estimate that the animal died 11 000 yr ago.

STUDY CHECK 5.8

Suppose that a piece of wood found in a cave had one-eighth of its original carbon-14 activity. About how many years ago was the wood part of a living tree?

ANSWER

17 000 yr

PRACTICE PROBLEMS

5.4 Half-Life of a Radioisotope

LEARNING GOAL Given the half-life of a radioisotope, calculate the amount of radioisotope remaining after one or more half-lives.

5.29 For each of the following, indicate if the number of half-lives elapsed is:
1. one half-life
2. two half-lives
3. three half-lives

a. a sample of Pd-103 with a half-life of 17 days after 34 days
b. a sample of C-11 with a half-life of 20 min after 20 min
c. a sample of At-211 with a half-life of 7 h after 21 h

5.30 For each of the following, indicate if the number of half-lives elapsed is:
1. one half-life
2. two half-lives
3. three half-lives

a. a sample of Ce-141 with a half-life of 32.5 days after 32.5 days
b. a sample of F-18 with a half-life of 110 min after 330 min
c. a sample of Au-198 with a half-life of 2.7 days after 5.4 days

Clinical Applications

5.31 Technetium-99m is an ideal radioisotope for scanning organs because it has a half-life of 6.0 h and is a pure gamma emitter. Suppose that 80.0 mg were prepared in the technetium generator this morning. How many milligrams of technetium-99m would remain after each of the following intervals?
a. one half-life **b.** two half-lives
c. 18 h **d.** 1.5 days

5.32 A sample of sodium-24 with an activity of 12 mCi is used to study the rate of blood flow in the circulatory system. If sodium-24 has a half-life of 15 h, what is the activity after each of the following intervals?
a. one half-life **b.** 30 h
c. three half-lives **d.** 2.5 days

5.33 Strontium-85, used for bone scans, has a half-life of 65 days.
a. How long will it take for the radiation level of strontium-85 to drop to one-fourth of its original level?
b. How long will it take for the radiation level of strontium-85 to drop to one-eighth of its original level?

5.34 Fluorine-18, which has a half-life of 110 min, is used in PET scans.
a. If 100. mg of fluorine-18 is shipped at 8:00 A.M., how many milligrams of the radioisotope are still active after 110 min?
b. If 100. mg of fluorine-18 is shipped at 8:00 A.M., how many milligrams of the radioisotope are still active when the sample arrives at the radiology laboratory at 1:30 P.M.?

5.5 Medical Applications Using Radioactivity

LEARNING GOAL Describe the use of radioisotopes in medicine.

The first radioactive isotope was used to treat a person with leukemia at the University of California at Berkeley. In 1946, radioactive iodine was successfully used to diagnose thyroid function and to treat hyperthyroidism and thyroid cancer. Radioactive isotopes are now used to produce images of organs including the liver, spleen, thyroid, kidneys, brain, and heart.

To determine the condition of an organ in the body, a nuclear medicine technologist may use a radioisotope that concentrates in that organ. The cells in the body do not differentiate between a nonradioactive atom and a radioactive one, so these radioisotopes are easily incorporated. Then the radioactive atoms are detected because they emit radiation. Some radioisotopes used in nuclear medicine are listed in **TABLE 5.8**.

TABLE 5.8 Medical Applications of Radioisotopes

Isotope	Half-Life	Radiation	Medical Application
Au-198	2.7 days	Beta	Liver imaging; treatment of abdominal carcinoma
Ce-141	32.5 days	Beta	Gastrointestinal tract diagnosis; measuring blood flow to the heart
Cs-131	9.7 days	Gamma	Prostate brachytherapy
F-18	110 min	Positron	Positron emission tomography (PET)
Ga-67	78 h	Gamma	Abdominal imaging; tumor detection
Ga-68	68 min	Gamma	Detection of pancreatic cancer
I-123	13.2 h	Gamma	Treatment of thyroid, brain, and prostate cancer
I-131	8.0 days	Beta	Treatment of Graves' disease, goiter, hyperthyroidism, thyroid and prostate cancer
Ir-192	74 days	Gamma	Treatment of breast and prostate cancer
P-32	14.3 days	Beta	Treatment of leukemia, excess red blood cells, and pancreatic cancer
Pd-103	17 days	Gamma	Prostate brachytherapy
Sm-153	46 h	Beta	Treatment of bone cancer
Sr-85	65 days	Gamma	Detection of bone lesions; brain scans
Tc-99m	6.0 h	Gamma	Imaging of skeleton and heart muscle, brain, liver, heart, lungs, bone, spleen, kidney, and thyroid; most widely used radioisotope in nuclear medicine
Xe-133	5.2 days	Beta	Pulmonary function diagnosis
Y-90	2.7 days	Beta	Treatment of liver cancer

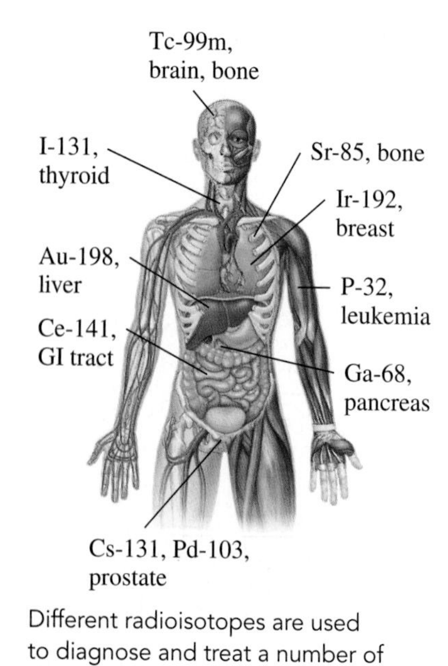

Different radioisotopes are used to diagnose and treat a number of diseases.

SAMPLE PROBLEM 5.9 Using Half-Lives in Medicine

TRY IT FIRST

In the treatment of abdominal carcinoma, a person is treated with "seeds" of gold-198, which is a beta emitter. Write the balanced nuclear equation for the beta decay of gold-198.

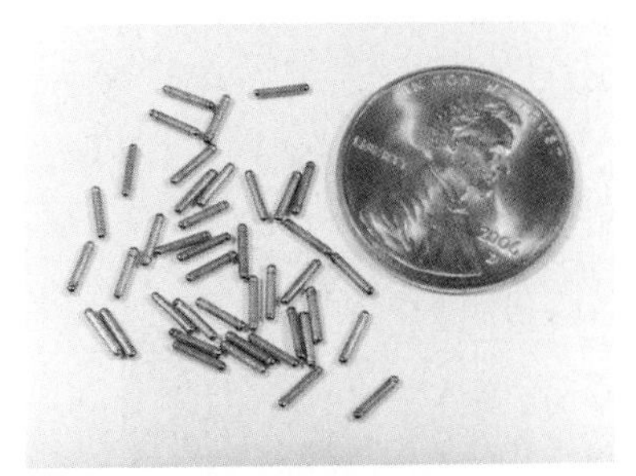

Titanium "seeds" filled with a radioactive isotope are implanted in the body to treat cancer.

SOLUTION GUIDE

	Given	Need	Connect
ANALYZE THE PROBLEM	gold-198, beta decay	balanced nuclear equation	mass number, atomic number of new nucleus

STEP 1 Write the incomplete nuclear equation.

$${}^{198}_{79}\text{Au} \longrightarrow ? + {}^{0}_{-1}e$$

STEP 2 Determine the missing mass number. In the equation, the mass number, 198, is equal to the sum of the mass numbers of the new nucleus and the beta particle.

$198 = ? + 0$
$198 - 0 = ?$
$198 - 0 = 198$ (mass number of new nucleus)

STEP 3 Determine the missing atomic number. The atomic number, 79, must equal the sum of the atomic numbers of the new nucleus and the beta particle.

$79 = ? - 1$
$79 + 1 = ?$
$79 + 1 = 80$ (mass number of new nucleus)

STEP 4 Determine the symbol of the new nucleus. On the periodic table, the element that has atomic number 80 is mercury, Hg. The atomic symbol for this isotope of Hg is written ${}^{198}_{80}\text{Hg}$.

STEP 5 Complete the nuclear equation.

$${}^{198}_{79}\text{Au} \longrightarrow {}^{198}_{80}\text{Hg} + {}^{0}_{-1}e$$

STUDY CHECK 5.9

In an experimental treatment, a person is given boron-10, which is taken up by malignant tumors. When bombarded with neutrons, boron-10 decays by emitting alpha particles that destroy the surrounding tumor cells. Write the balanced nuclear equation for the reaction for this experimental procedure.

ANSWER

$${}^{1}_{0}n + {}^{10}_{5}\text{B} \longrightarrow {}^{7}_{3}\text{Li} + {}^{4}_{2}\text{He}$$

ENGAGE

If Y-90 used to treat liver cancer is a beta emitter, why is the product Zr-90?

TEST

Try Practice Problems 5.35 to 5.40

Scans with Radioisotopes

After a person receives a radioisotope, the radiation technologist determines the level and location of radioactivity emitted by the radioisotope. An apparatus called a *scanner* is used to produce an image of the organ. The scanner moves slowly across the body above the region where the organ containing the radioisotope is located. The gamma rays emitted from the radioisotope in the organ can be used to expose a photographic plate, producing a *scan* of the organ. On a scan, an area of decreased or increased radiation can indicate conditions such as a disease of the organ, a tumor, a blood clot, or edema.

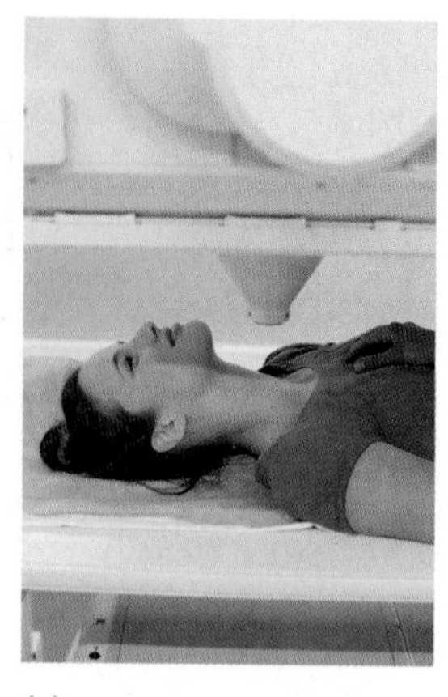

(a)

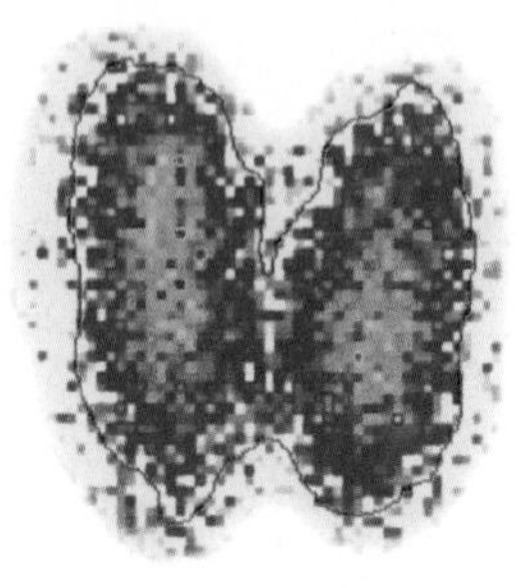

(b)

FIGURE 5.5 **(a)** A scanner detects radiation from a radioisotope in an organ. **(b)** A scan shows radioactive iodine-131 in the thyroid.

Q What type of radiation would move through body tissues to create a scan?

A common method of determining thyroid function is the use of *radioactive iodine uptake*. Taken orally, the radioisotope iodine-131 mixes with the iodine already present in the thyroid. Twenty-four hours later, the amount of iodine taken up by the thyroid is determined. A detection tube held up to the area of the thyroid gland detects the radiation coming from the iodine-131 that has located there (see **FIGURE 5.5**).

A person with a hyperactive thyroid will have a higher than normal level of radioactive iodine, whereas a person with a hypoactive thyroid will have lower values. If a person has hyperthyroidism, treatment is begun to lower the activity of the thyroid. One treatment involves giving a therapeutic dosage of radioactive iodine, which has a higher radiation level than the diagnostic dose. The radioactive iodine goes to the thyroid where its radiation destroys some of the thyroid cells. The thyroid produces less thyroid hormone, bringing the hyperthyroid condition under control.

Positron Emission Tomography

Positron emitters with short half-lives such as carbon-11, oxygen-15, nitrogen-13, and fluorine-18 are used in an imaging method called *positron emission tomography* (PET). A positron-emitting isotope such as fluorine-18 combined with substances in the body such as glucose is used to study brain function, metabolism, and blood flow.

$$^{18}_{9}\mathrm{F} \longrightarrow {}^{18}_{8}\mathrm{O} + {}^{0}_{+1}e$$

As positrons are emitted, they combine with electrons to produce gamma rays that are detected by computerized equipment to create a three-dimensional image of the organ (see **FIGURE 5.6**).

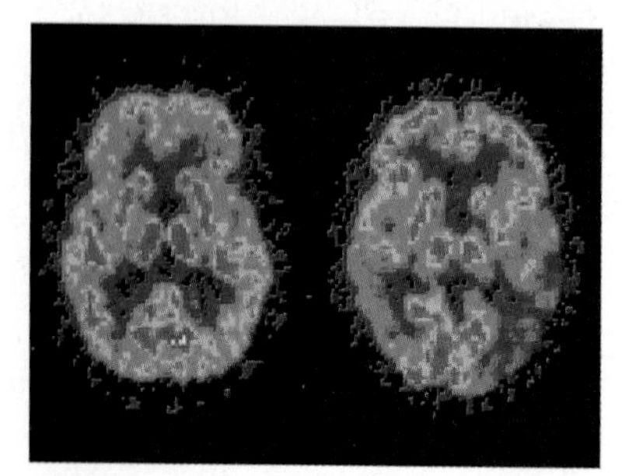

FIGURE 5.6 These PET scans of the brain show a normal brain on the left and a brain affected by Alzheimer's disease on the right.

Q When positrons collide with electrons, what type of radiation is produced that gives an image of an organ?

Nonradioactive Imaging

Computed Tomography

Another imaging method used to scan organs such as the brain, lungs, and heart is *computed tomography* (CT). A computer monitors the absorption of 30 000 X-ray beams directed at successive layers of the target organ. Based on the densities of the tissues and fluids in the organ, the differences in absorption of the X-rays provide a series of images of the organ. This technique is successful in the identification of hemorrhages, tumors, and atrophy.

Magnetic Resonance Imaging

Magnetic resonance imaging (MRI) is a powerful imaging technique that does not involve X-ray radiation. It is the least invasive imaging method available. MRI is based on the absorption of energy when the protons in hydrogen atoms are excited by a strong

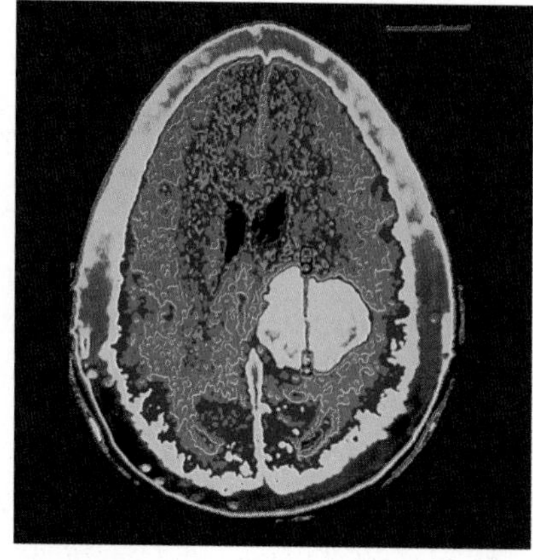

A CT scan shows a tumor (yellow) in the brain.

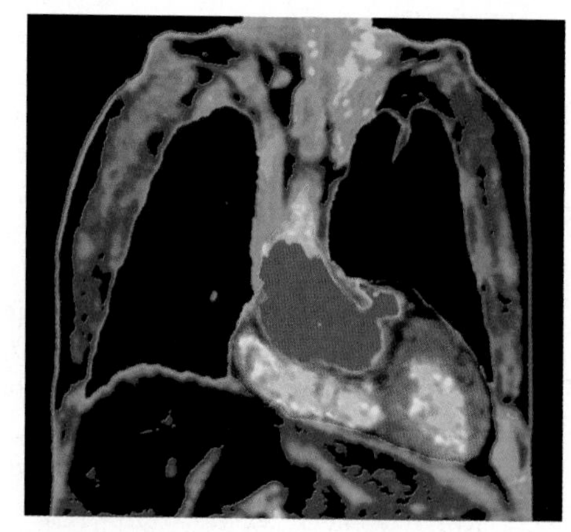

An MRI scan provides images of the heart and lungs.

magnetic field. The difference in energy between the two states is released, which produces the electromagnetic signal that the scanner detects. These signals are sent to a computer system, where a color image of the body is generated. MRI is particularly useful in obtaining images of soft tissues, which con in large amounts of hydrogen atoms in the form of water.

CHEMISTRY LINK TO HEALTH

Brachytherapy

The process called *brachytherapy*, or seed implantation, is an internal form of radiation therapy. The prefix *brachy* is from the Greek word for short distance. With internal radiation, a high dose of radiation is delivered to a cancerous area, whereas normal tissue sustains minimal damage. Because higher doses are used, fewer treatments of shorter duration are needed. Conventional external treatment delivers a lower dose per treatment, but requires six to eight weeks of treatment.

Permanent Brachytherapy

One of the most common forms of cancer in males is prostate cancer. In addition to surgery and chemotherapy, one treatment option is to place 40 or more titanium capsules, or "seeds," in the malignant area. Each seed, which is the size of a grain of rice, contains radioactive iodine-125, palladium-103, or cesium-131, which decay by gamma emission. The radiation from the seeds destroys the cancer by interfering with the reproduction of cancer cells with minimal damage to adjacent normal tissues. Ninety percent (90%) of the radioisotopes decay within a few months because they have short half-lives.

Isotope	I-125	Pd-103	Cs-131
Radiation	Gamma	Gamma	Gamma
Half-Life	60 days	17 days	10 days
Time Required to Deliver 90% of Radiation	7 months	2 months	1 month

Almost no radiation passes out of the patient's body. The amount of radiation received by a family member is no greater than that received on a long plane flight. Because the radioisotopes decay to products that are not radioactive, the inert titanium capsules can be left in the body.

Temporary Brachytherapy

In another type of treatment for prostate cancer, long needles containing iridium-192 are placed in the tumor. However, the needles are removed after 5 to 10 min, depending on the activity of the iridium isotope. Compared to permanent brachytherapy, temporary brachytherapy can deliver a higher dose of radiation over a shorter time. The procedure may be repeated in a few days.

Brachytherapy is also used following breast cancer lumpectomy. An iridium-192 isotope is inserted into the catheter implanted in the space left by the removal of the tumor. Radiation is delivered primarily to the tissue surrounding the cavity that contained the tumor and where the cancer is most likely to recur. The procedure is repeated twice a day for five days to give an absorbed dose of 34 Gy (3400 rad). The catheter is removed, and no radioactive material remains in the body.

In conventional external beam therapy for breast cancer, a patient is given 2 Gy once a day for six to seven weeks, which gives a total absorbed dose of about 80 Gy or 8000 rad. The external beam therapy irradiates the entire breast, including the tumor cavity.

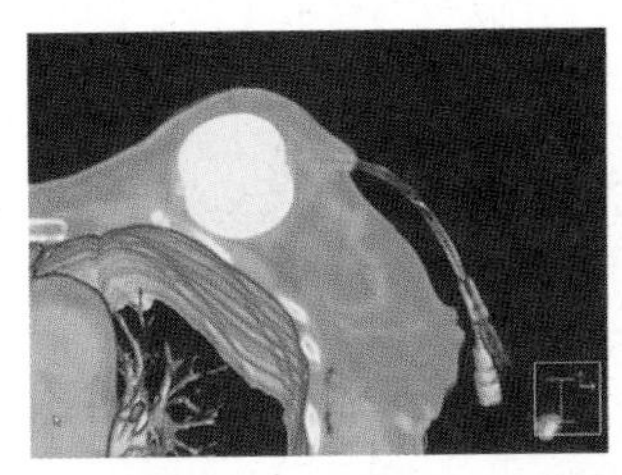
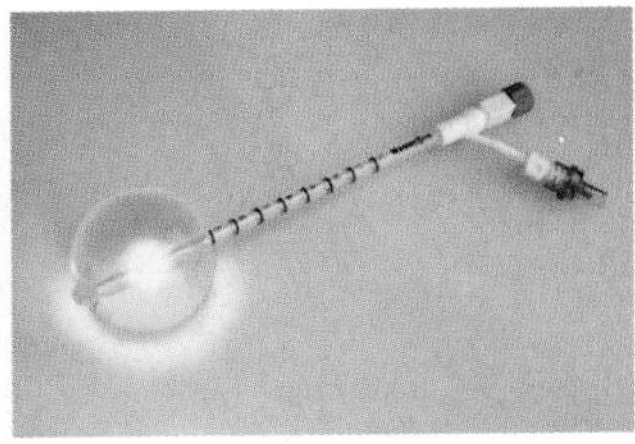

A catheter placed temporarily in the breast for radiation from Ir-192.

PRACTICE PROBLEMS

5.5 Medical Applications Using Radioactivity

LEARNING GOAL Describe the use of radioisotopes in medicine.

Clinical Applications

5.35 Bone and bony structures contain calcium and phosphorus.
  a. Why would the radioisotopes calcium-47 and phosphorus-32 be used in the diagnosis and treatment of bone diseases?
  b. During nuclear tests, scientists were concerned that strontium-85, a radioactive product, would be harmful to the growth of bone in children. Explain.

5.36 **a.** Technetium-99m emits only gamma radiation. Why would this type of radiation be used in diagnostic imaging rather than an isotope that also emits beta or alpha radiation?
  b. A person with *polycythemia vera* (excess production of red blood cells) receives radioactive phosphorus-32. Why would this treatment reduce the production of red blood cells in the bone marrow of the patient?

5.37 In a diagnostic test for leukemia, a person receives 4.0 mL of a solution containing selenium-75. If the activity of the selenium-75 is 45 μCi/mL, what dose, in microcuries, does the patient receive?

5.38 A vial contains radioactive iodine-131 with an activity of 2.0 mCi/mL. If a thyroid test requires 3.0 mCi in an "atomic cocktail," how many milliliters are used to prepare the iodine-131 solution?

5.39 Gallium-68 is taken up by tumors; the emission of positrons allows the tumors to be located.

a. Write an equation for the positron emission of Ga-68.

b. If the half-life is 68 min, how much of a 64-mcg sample is active after 136 min?

5.40 Xenon-133 is used to test lung function; it decays by emitting a beta particle.

a. Write an equation for the beta decay of Xe-133.

b. If the half-life of Xe-133 is 5.2 h, how much of a 20.-mCi sample is still active after 15.6 h?

5.6 Nuclear Fission and Fusion

LEARNING GOAL Describe the processes of nuclear fission and fusion.

During the 1930s, scientists bombarding uranium-235 with neutrons discovered that the U-235 nucleus splits into two smaller nuclei and produces a great amount of energy. This was the discovery of nuclear **fission**. The energy generated by splitting the atom was called *atomic energy*. A typical equation for nuclear fission is:

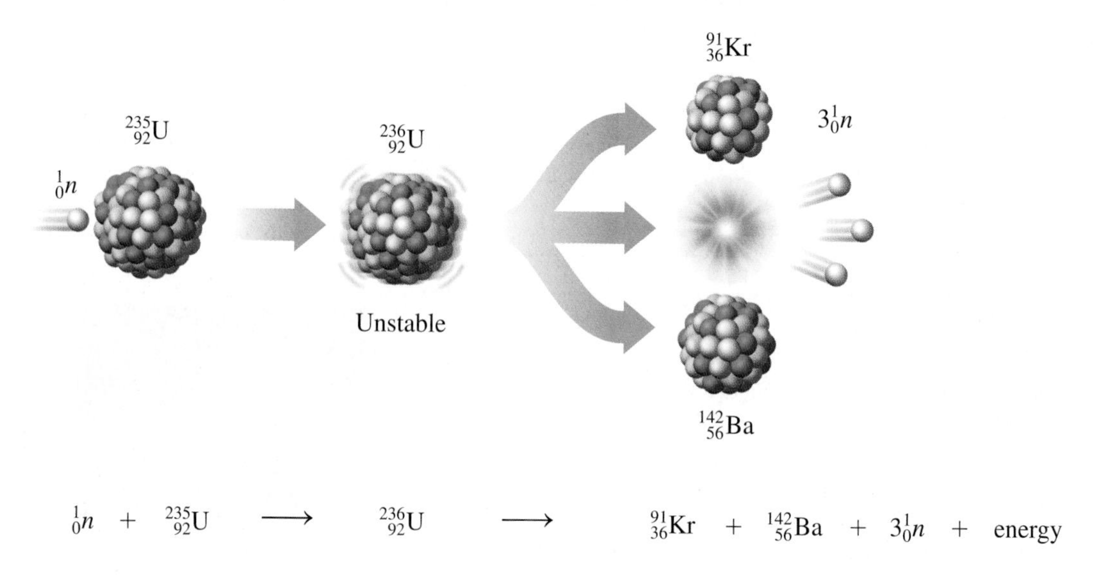

$$^{1}_{0}n + ^{235}_{92}\text{U} \longrightarrow ^{236}_{92}\text{U} \longrightarrow ^{91}_{36}\text{Kr} + ^{142}_{56}\text{Ba} + 3^{1}_{0}n + \text{energy}$$

If we could determine the mass of the products krypton, barium, and three neutrons with great accuracy, we would find that their total mass is slightly less than the mass of the starting materials. The missing mass has been converted into an enormous amount of energy, consistent with the famous equation derived by Albert Einstein:

$$E = mc^2$$

where E is the energy released, m is the mass lost, and c is the speed of light, 3×10^8 m/s. Even though the mass loss is very small, when it is multiplied by the speed of light squared, the result is a large value for the energy released. The fission of 1 g of uranium-235 produces about as much energy as the burning of 3 tons of coal.

Chain Reaction

Fission begins when a neutron collides with the nucleus of a uranium atom. The resulting nucleus is unstable and splits into smaller nuclei. This fission process also releases several neutrons and large amounts of gamma radiation and energy. The neutrons emitted have high energies and bombard other uranium-235 nuclei. In a **chain reaction**, there is a rapid increase in the number of high-energy neutrons available to react with more uranium. To sustain a nuclear chain reaction, sufficient quantities of uranium-235 must be brought together to provide a *critical mass* in which almost all the neutrons immediately collide with more uranium-235 nuclei. So much heat and energy build up that an atomic explosion can occur (see **FIGURE 5.7**).

ENGAGE

Why is a critical mass of uranium-235 necessary to sustain a nuclear chain reaction?

TEST

Try Practice Problems 5.41 to 5.44

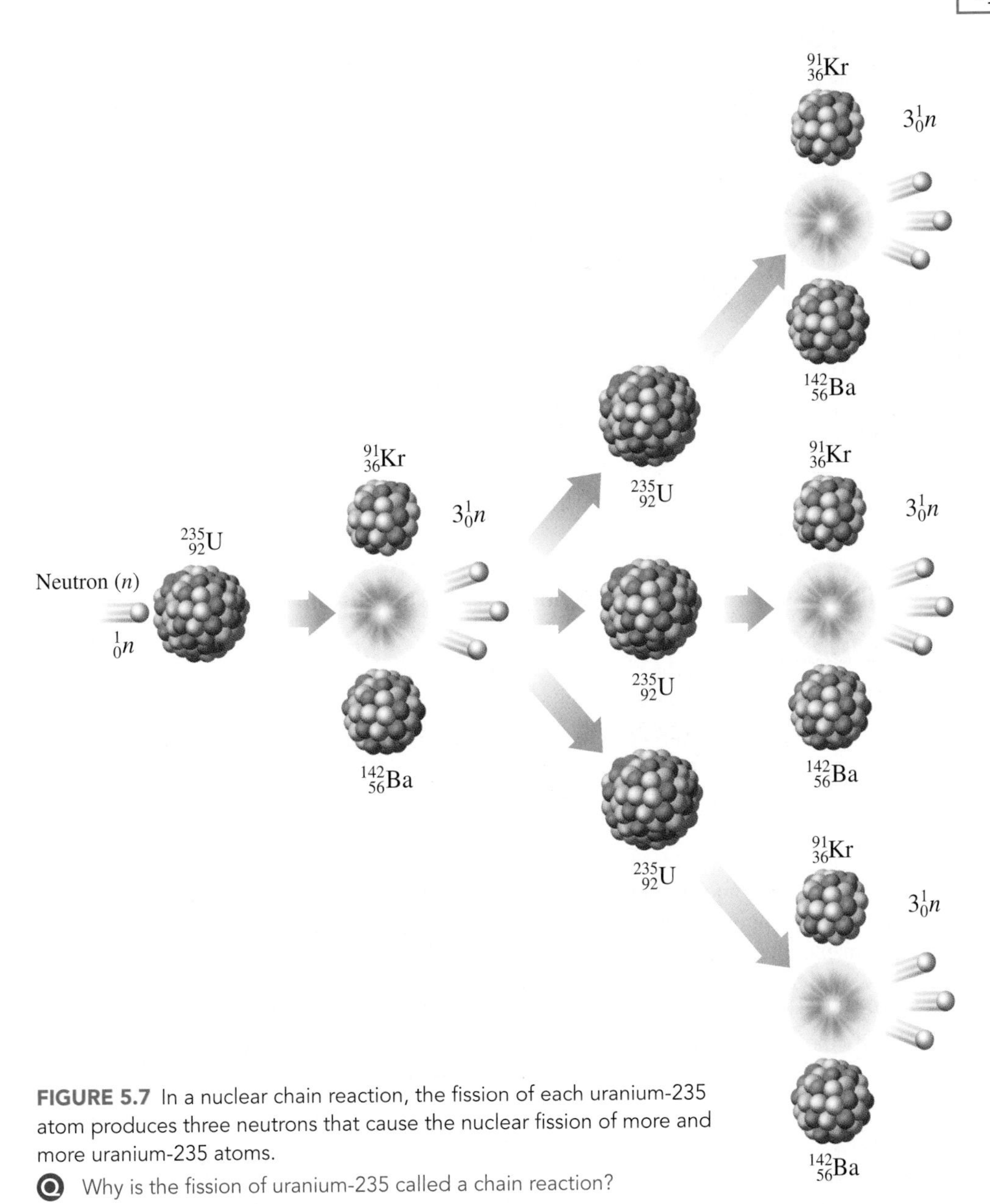

FIGURE 5.7 In a nuclear chain reaction, the fission of each uranium-235 atom produces three neutrons that cause the nuclear fission of more and more uranium-235 atoms.

Why is the fission of uranium-235 called a chain reaction?

Nuclear Fusion

In **fusion**, two small nuclei combine to form a larger nucleus. Mass is lost, and a tremendous amount of energy is released, even more than the energy released from nuclear fission. However, a fusion reaction requires a temperature of 100 000 000 °C to overcome the repulsion of the hydrogen nuclei and cause them to undergo fusion. Fusion reactions occur continuously in the Sun and other stars, providing us with heat and light. The huge amounts of energy produced by our sun come from the fusion of 6×10^{11} kg of hydrogen every second. In a fusion reaction, isotopes of hydrogen combine to form helium and large amounts of energy.

Scientists expect less radioactive waste with shorter half-lives from fusion reactors. However, fusion is still in the experimental stage because the extremely high temperatures needed have been difficult to reach and even more difficult to maintain. Research groups around the world are attempting to develop the technology needed to make the harnessing of the fusion reaction for energy a reality in our lifetime.

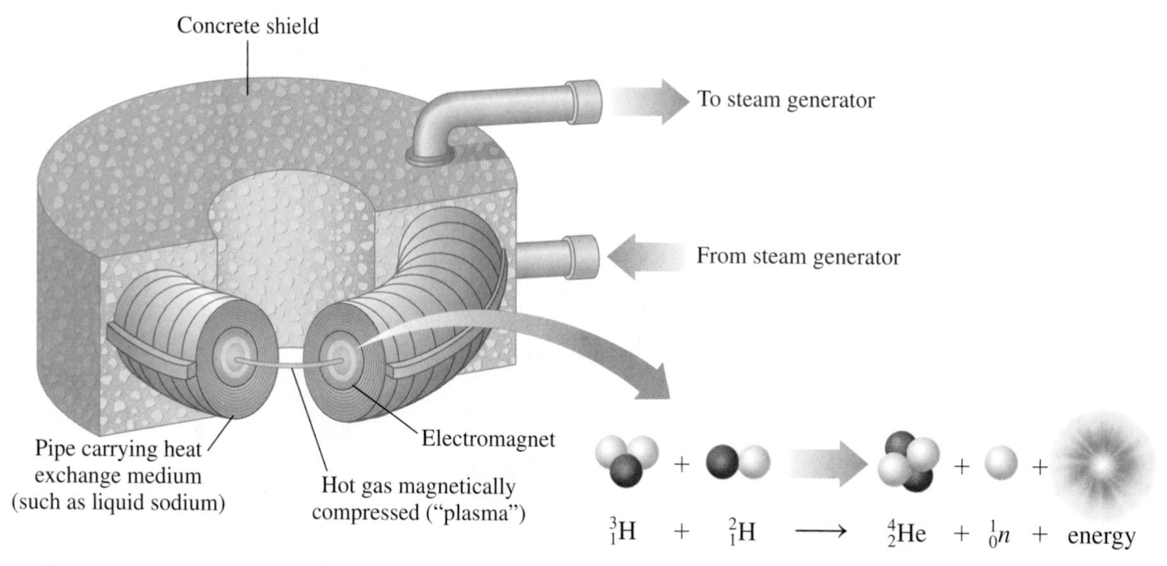

In a fusion reactor, high temperatures are needed to combine hydrogen atoms.

SAMPLE PROBLEM 5.10 Identifying Fission and Fusion

TRY IT FIRST

Classify the following as pertaining to fission, fusion, or both:

a. A large nucleus breaks apart to produce smaller nuclei.
b. Large amounts of energy are released.
c. Extremely high temperatures are needed for reaction.

SOLUTION

a. When a large nucleus breaks apart to produce smaller nuclei, the process is fission.
b. Large amounts of energy are generated in both the fusion and fission processes.
c. An extremely high temperature is required for fusion.

STUDY CHECK 5.10

Classify the following nuclear equation as pertaining to fission, fusion, or both:

$$^{3}_{1}H + ^{2}_{1}H \longrightarrow ^{4}_{2}He + ^{1}_{0}n + \text{energy}$$

ANSWER

When small nuclei combine to release energy, the process is fusion.

TEST

Try Practice Problems 5.45 and 5.46

PRACTICE PROBLEMS

5.6 Nuclear Fission and Fusion

LEARNING GOAL Describe the processes of nuclear fission and fusion.

5.41 What is nuclear fission?

5.42 How does a chain reaction occur in nuclear fission?

5.43 Complete the following fission reaction:

$$^{1}_{0}n + ^{235}_{92}U \longrightarrow ^{131}_{50}Sn + ? + 2^{1}_{0}n + \text{energy}$$

5.44 In another fission reaction, uranium-235 bombarded with a neutron produces strontium-94, another small nucleus, and three neutrons. Write the balanced nuclear equation for the fission reaction.

5.45 Indicate whether each of the following is characteristic of the fission or fusion process, or both:
- **a.** Neutrons bombard a nucleus.
- **b.** The nuclear process occurs in the Sun.
- **c.** A large nucleus splits into smaller nuclei.
- **d.** Small nuclei combine to form larger nuclei.

5.46 Indicate whether each of the following is characteristic of the fission or fusion process, or both:
- **a.** Very high temperatures are required to initiate the reaction.
- **b.** Less radioactive waste is produced.
- **c.** Hydrogen nuclei are the reactants.
- **d.** Large amounts of energy are released when the nuclear reaction occurs.

CHEMISTRY LINK TO THE ENVIRONMENT

Nuclear Power Plants

In a nuclear power plant, the quantity of uranium-235 is held below a critical mass, so it cannot sustain a chain reaction. The fission reactions are slowed by placing control rods, which absorb some of the fast-moving neutrons, among the uranium samples. In this way, less fission occurs, and there is a slower, controlled production of energy. The heat from the controlled fission is used to produce steam. The steam drives a generator, which produces electricity. Approximately 20% of the electrical energy produced in the United States is generated in nuclear power plants.

Although nuclear power plants help meet some of our energy needs, there are some problems associated with nuclear power. One of the most serious problems is the production of radioactive byproducts that have very long half-lives, such as plutonium-239 with a half-life of 24 000 yr. It is essential that these waste products be stored safely in a place where they do not contaminate the environment.

Nuclear power plants supply about 20% of electricity in the United States.

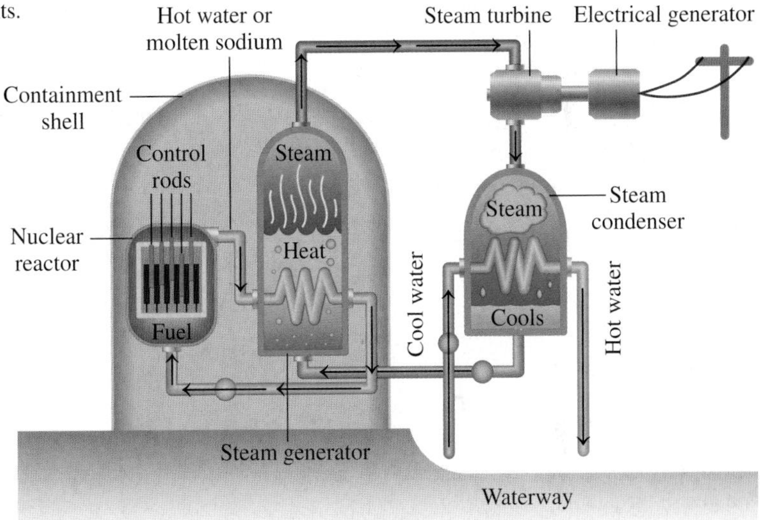

Heat from nuclear fission is used to generate electricity.

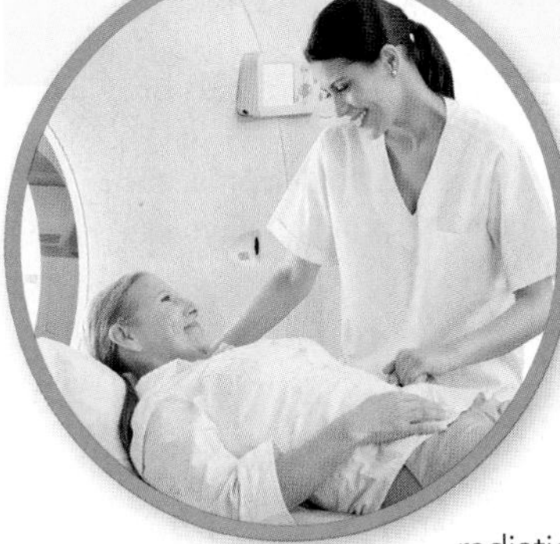

CLINICAL UPDATE Cardiac Imaging Using a Radioisotope

As part of her nuclear stress test, Simone starts to walk on the treadmill. When she reaches the maximum level, Pauline injects a radioactive dye containing Tl-201 with an activity of 74 MBq. The radiation emitted from areas of the heart is detected by a scanner and produces images of her heart muscle. A thallium stress test can determine how effectively coronary arteries provide blood to the heart. If Simone has any damage to her coronary arteries, reduced blood flow during stress would show a narrowing of an artery or a blockage. After Simone rests for 3 h, Pauline injects more dye with Tl-201 and she is placed under the scanner again. A second set of images of her heart muscle at rest are taken. When Simone's doctor reviews her scans, he assures her that she had normal blood flow to her heart muscle, both at rest and under stress.

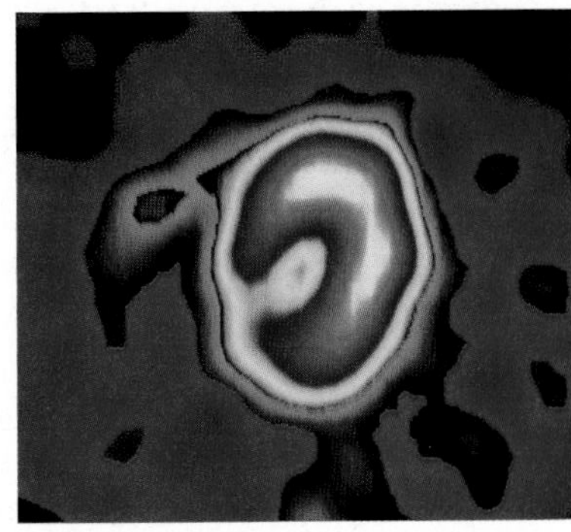

A thallium stress test can show narrowing of an artery during stress.

Clinical Applications

5.47 What is the activity of the radioactive dye injection for Simone
a. in curies? b. in millicuries?

5.48 If the half-life of Tl-201 is 3.0 days, what is its activity, in megabecquerels
a. after 3.0 days? b. after 6.0 days?

5.49 How many days will it take until the activity of the Tl-201 in Simone's body is one-eighth of the initial activity?

5.50 Radiation from Tl-201 to Simone's kidneys can be 24 mGy. What is this amount of radiation in rads?

CONCEPT MAP

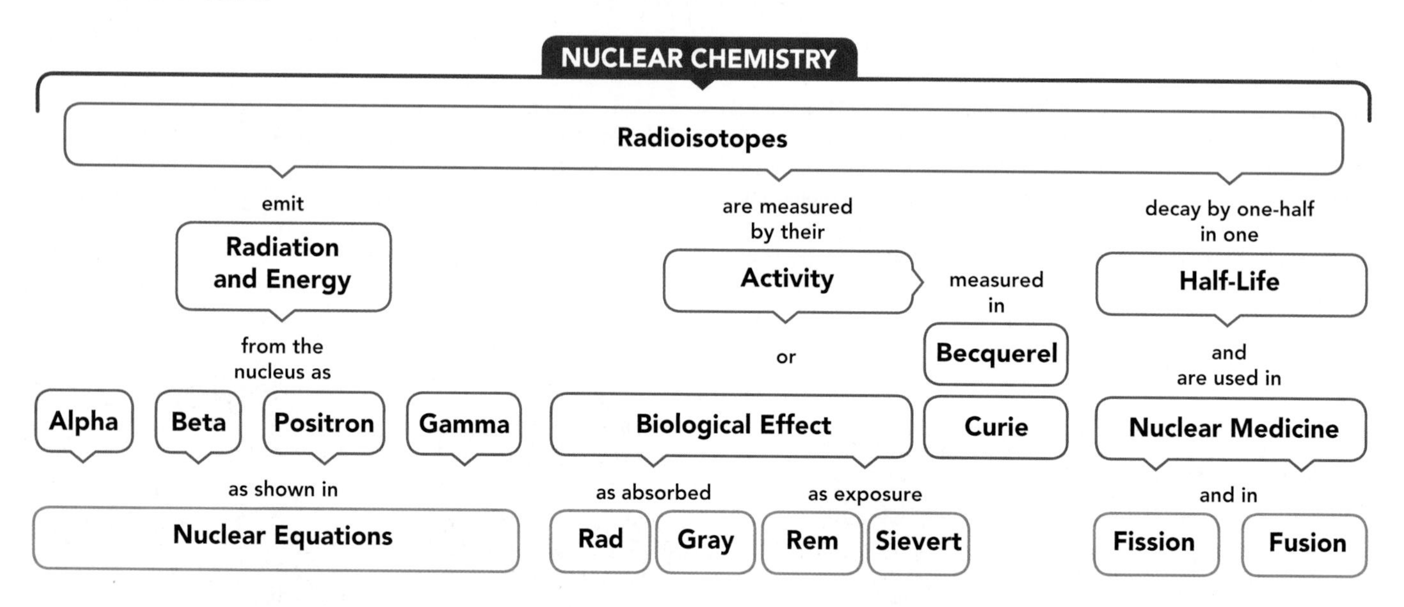

CHAPTER REVIEW

5.1 Natural Radioactivity

$^{4}_{2}He$ or α
Alpha particle

LEARNING GOAL Describe alpha, beta, positron, and gamma radiation.

- Radioactive isotopes have unstable nuclei that break down (decay), spontaneously emitting alpha (α), beta (β), positron (β^+), and gamma (γ) radiation.
- Because radiation can damage the cells in the body, proper protection must be used: shielding, limiting the time of exposure, and distance.

5.2 Nuclear Reactions

LEARNING GOAL Write a balanced nuclear equation for radioactive decay, showing mass numbers and atomic numbers.

- A balanced nuclear equation is used to represent the changes that take place in the nuclei of the reactants and products.
- The new isotopes and the type of radiation emitted can be determined from the symbols that show the mass numbers and atomic numbers of the isotopes in the nuclear equation.
- A radioisotope is produced artificially when a nonradioactive isotope is bombarded by a small particle.

5.3 Radiation Measurement

LEARNING GOAL Describe the detection and measurement of radiation.

- In a Geiger counter, radiation produces charged particles in the gas contained in a tube, which generates an electrical current.
- The curie (Ci) and the becquerel (Bq) measure the activity, which is the number of nuclear transformations per second.
- The amount of radiation absorbed by a substance is measured in the rad or the gray (Gy).
- The rem and the sievert (Sv) are units used to determine the biological damage from the different types of radiation.

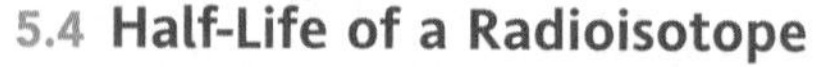

5.4 Half-Life of a Radioisotope

LEARNING GOAL Given the half-life of a radioisotope, calculate the amount of radioisotope remaining after one or more half-lives.

- Every radioisotope has its own rate of emitting radiation.
- The time it takes for one-half of a radioactive sample to decay is called its half-life.

- For many medical radioisotopes, such as Tc-99m and I-131, half-lives are short.
- For other isotopes, usually naturally occurring ones such as C-14, Ra-226, and U-238, half-lives are extremely long.

5.5 Medical Applications Using Radioactivity

LEARNING GOAL Describe the use of radioisotopes in medicine.

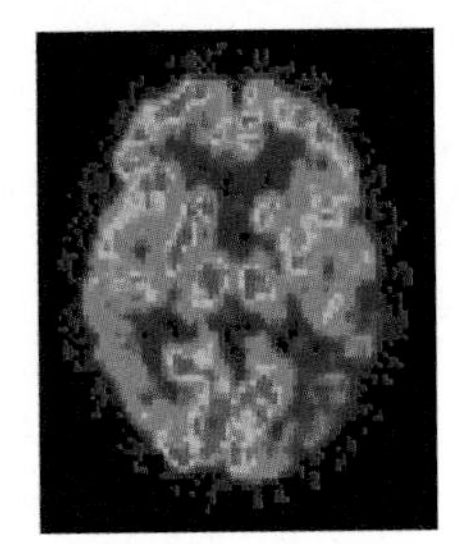

- In nuclear medicine, radioisotopes that go to specific sites in the body are given to the patient.
- By detecting the radiation they emit, an evaluation can be made about the location and extent of an injury, disease, tumor, or the level of function of a particular organ.
- Higher levels of radiation are used to treat or destroy tumors.

5.6 Nuclear Fission and Fusion

LEARNING GOAL
Describe the processes of nuclear fission and fusion.

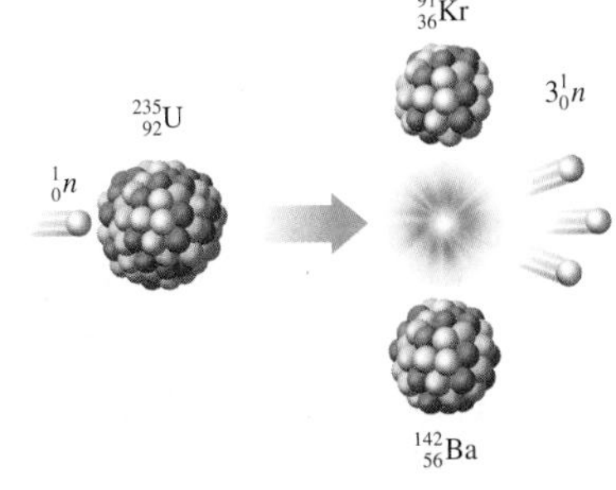

- In fission, the bombardment of a large nucleus breaks it apart into smaller nuclei, releasing one or more types of radiation and a great amount of energy.
- In fusion, small nuclei combine to form larger nuclei while great amounts of energy are released.

KEY TERMS

alpha particle A nuclear particle identical to a helium nucleus, symbol α or $^{4}_{2}He$.

becquerel (Bq) A unit of activity of a radioactive sample equal to one disintegration per second.

beta particle A particle identical to an electron, symbol $^{0}_{-1}e$ or β, that forms in the nucleus when a neutron changes to a proton and an electron.

chain reaction A fission reaction that will continue once it has been initiated by a high-energy neutron bombarding a heavy nucleus such as uranium-235.

curie (Ci) A unit of activity of a radioactive sample equal to 3.7×10^{10} disintegrations/s.

decay curve A diagram of the decay of a radioactive element.

equivalent dose The measure of biological damage from an absorbed dose that has been adjusted for the type of radiation.

fission A process in which large nuclei are split into smaller pieces, releasing large amounts of energy.

fusion A reaction in which large amounts of energy are released when small nuclei combine to form larger nuclei.

gamma ray High-energy radiation, symbol $^{0}_{0}\gamma$, emitted by an unstable nucleus.

gray (Gy) A unit of absorbed dose equal to 100 rad.

half-life The length of time it takes for one-half of a radioactive sample to decay.

positron A particle of radiation with no mass and a positive charge, symbol β^{+} or $^{0}_{+1}e$, produced when a proton is transformed into a neutron and a positron.

rad (radiation absorbed dose) A measure of an amount of radiation absorbed by the body.

radiation Energy or particles released by radioactive atoms.

radioactive decay The process by which an unstable nucleus breaks down with the release of high-energy radiation.

radioisotope A radioactive atom of an element.

rem (radiation equivalent in humans) A measure of the biological damage caused by the various kinds of radiation (rad $\times$ radiation biological factor).

sievert (Sv) A unit of biological damage (equivalent dose) equal to 100 rem.

CORE CHEMISTRY SKILLS

The chapter Section containing each Core Chemistry Skill is shown in parentheses at the end of each heading.

Writing Nuclear Equations (5.2)

- A nuclear equation is written with the atomic symbols of the original radioactive nucleus on the left, an arrow, and the new nucleus and the type of radiation emitted on the right.
- The sum of the mass numbers and the sum of the atomic numbers on one side of the arrow must equal the sum of the mass numbers and the sum of the atomic numbers on the other side.
- When an alpha particle is emitted, the mass number of the new nucleus decreases by 4, and its atomic number decreases by 2.
- When a beta particle is emitted, there is no change in the mass number of the new nucleus, but its atomic number increases by one.
- When a positron is emitted, there is no change in the mass number of the new nucleus, but its atomic number decreases by one.
- In gamma emission, there is no change in the mass number or the atomic number of the new nucleus.

Example: **a.** Write a balanced nuclear equation for the alpha decay of Po-210.

b. Write a balanced nuclear equation for the beta decay of Co-60.

Answer: **a.** When an alpha particle is emitted, we calculate the decrease of 4 in the mass number (210) of the polonium, and a decrease of 2 in its atomic number.

$$^{210}_{84}Po \longrightarrow {}^{206}_{82}? + {}^{4}_{2}He$$

Because lead has atomic number 82, the new nucleus must be an isotope of lead.

$$^{210}_{84}Po \longrightarrow {}^{206}_{82}Pb + {}^{4}_{2}He$$

b. When a beta particle is emitted, there is no change in the mass number (60) of the cobalt, but there is an increase of 1 in its atomic number.

$$^{60}_{27}Co \longrightarrow {}^{60}_{28}? + {}^{0}_{-1}e$$

Because nickel has atomic number 28, the new nucleus must be an isotope of nickel.

$$^{60}_{27}\text{Co} \longrightarrow {}^{60}_{28}\text{Ni} + {}^{0}_{-1}e$$

Using Half-Lives (5.4)

- The half-life of a radioisotope is the amount of time it takes for one-half of a sample to decay.
- The remaining amount of a radioisotope is calculated by dividing its quantity or activity by one-half for each half-life that has elapsed.

Example: Co-60 has a half-life of 5.3 yr. If the initial sample of Co-60 has an activity of 1200 Ci, what is its activity after 15.9 yr?

Answer:

$$\text{number of half-lives} = 15.9\ \cancel{\text{yr}} \times \frac{1\ \text{half-life}}{5.3\ \cancel{\text{yr}}} = 3.0\ \text{half-lives}$$

In 15.9 yr, three half-lives have passed. Thus, the activity was reduced from 1200 Ci to 150 Ci.

$$1200\,\text{Ci} \xrightarrow{\text{1 half-life}} 600\,\text{Ci} \xrightarrow{\text{2 half-lives}} 300\,\text{Ci} \xrightarrow{\text{3 half-lives}} 150\,\text{Ci}$$

UNDERSTANDING THE CONCEPTS

The chapter Sections to review are shown in parentheses at the end of each problem.

In problems 5.51 to 5.54, a nucleus is shown with protons and neutrons.

 proton  neutron

5.51 Draw the new nucleus when this isotope emits a positron to complete the following: (5.2)

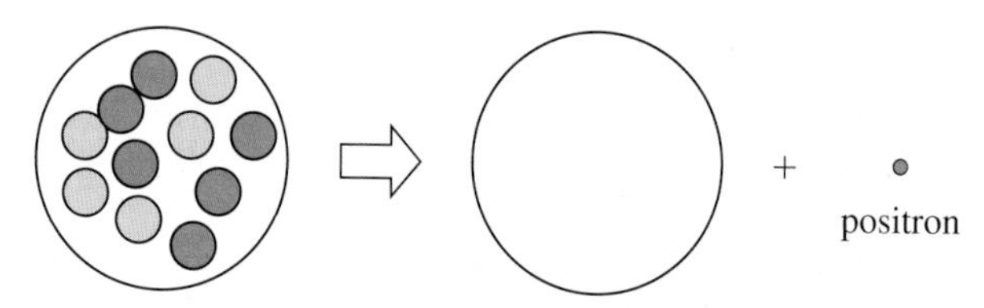

5.52 Draw the nucleus that emits a beta particle to complete the following: (5.2)

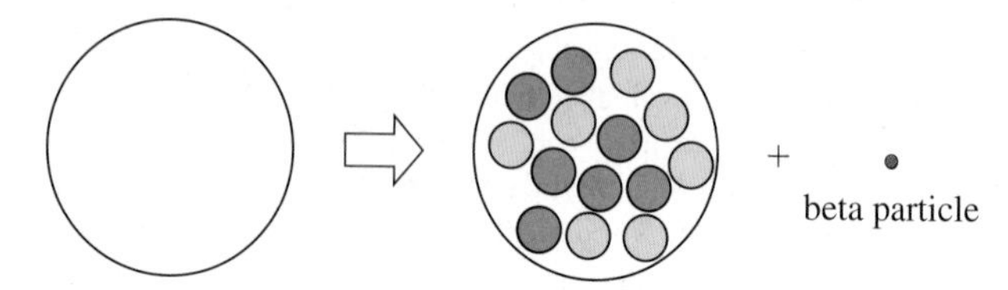

5.53 Draw the nucleus of the isotope that is bombarded in the following: (5.2)

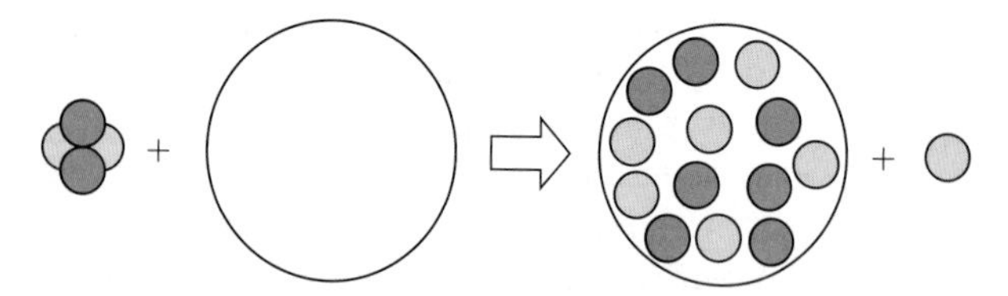

5.54 Complete the following bombardment reaction by drawing the nucleus of the new isotope that is produced in the following: (5.2)

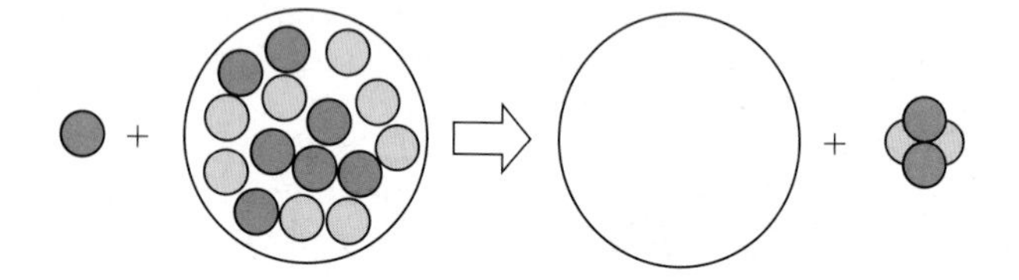

5.55 Carbon dating of small bits of charcoal used in cave paintings has determined that some of the paintings are from 10 000 to 30 000 yr old. Carbon-14 has a half-life of 5730 yr. In a 1-μg sample of carbon from a live tree, the activity of carbon-14 is 6.4 μCi. If researchers determine that 1 μg of charcoal from a prehistoric cave painting in France has an activity of 0.80 μCi, what is the age of the painting? (5.4)

The technique of carbon dating is used to determine the age of ancient cave paintings.

5.56 Use the following decay curve for iodine-131 to answer problems **a** to **c**: (5.4)

a. Complete the values for the mass of radioactive iodine-131 on the vertical axis.

b. Complete the number of days on the horizontal axis.

c. What is the half-life, in days, of iodine-131?

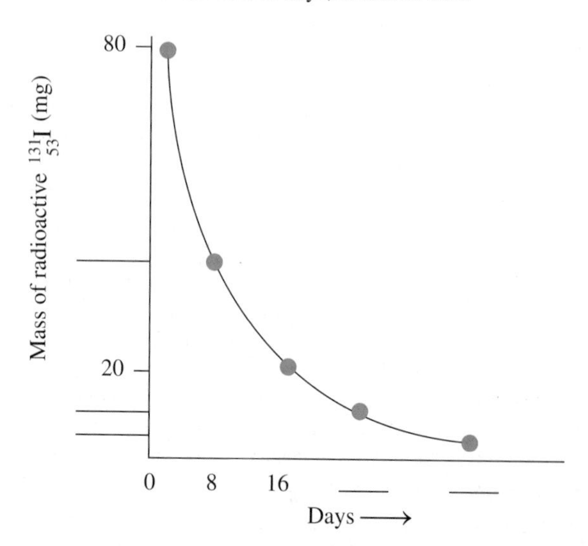

ADDITIONAL PRACTICE PROBLEMS

5.57 Determine the number of protons and number of neutrons in the nucleus of each of the following: (5.1)
a. sodium-25 **b.** nickel-61
c. rubidium-84 **d.** silver-110

5.58 Determine the number of protons and number of neutrons in the nucleus of each of the following: (5.1)
a. boron-10 **b.** zinc-72
c. iron-59 **d.** gold-198

5.59 Identify each of the following as alpha decay, beta decay, positron emission, or gamma emission: (5.1, 5.2)
a. ${}^{27m}_{13}\text{Al} \longrightarrow {}^{27}_{13}\text{Al} + {}^{0}_{0}\gamma$
b. ${}^{8}_{5}\text{B} \longrightarrow {}^{8}_{4}\text{Be} + {}^{0}_{+1}e$
c. ${}^{220}_{86}\text{Rn} \longrightarrow {}^{216}_{84}\text{Po} + {}^{4}_{2}\text{He}$

5.60 Identify each of the following as alpha decay, beta decay, positron emission, or gamma emission: (5.1, 5.2)
a. ${}^{127}_{55}\text{Cs} \longrightarrow {}^{127}_{54}\text{Xe} + {}^{0}_{+1}e$
b. ${}^{90}_{38}\text{Sr} \longrightarrow {}^{90}_{39}\text{Y} + {}^{0}_{-1}e$
c. ${}^{218}_{85}\text{At} \longrightarrow {}^{214}_{83}\text{Bi} + {}^{4}_{2}\text{He}$

5.61 Write the balanced nuclear equation for each of the following: (5.1, 5.2)
a. Th-225 (α decay) **b.** Bi-210 (α decay)
c. cesium-137 (β decay) **d.** tin-126 (β decay)
e. F-18 (β^+ emission)

5.62 Write the balanced nuclear equation for each of the following: (5.1, 5.2)
a. potassium-40 (β decay) **b.** sulfur-35 (β decay)
c. platinum-190 (α decay) **d.** Ra-210 (α decay)
e. In-113m (γ emission)

5.63 Complete each of the following nuclear equations: (5.2)
a. ${}^{4}_{2}\text{He} + {}^{14}_{7}\text{N} \longrightarrow ? + {}^{1}_{1}\text{H}$
b. ${}^{4}_{2}\text{He} + {}^{27}_{13}\text{Al} \longrightarrow {}^{30}_{14}\text{Si} + ?$
c. ${}^{1}_{0}n + {}^{235}_{92}\text{U} \longrightarrow {}^{90}_{38}\text{Sr} + 3{}^{1}_{0}n + ?$
d. ${}^{23m}_{12}\text{Mg} \longrightarrow ? + {}^{0}_{0}\gamma$

5.64 Complete each of the following nuclear equations: (5.2)
a. $? + {}^{59}_{27}\text{Co} \longrightarrow {}^{56}_{25}\text{Mn} + {}^{4}_{2}\text{He}$
b. $? \longrightarrow {}^{14}_{7}\text{N} + {}^{0}_{-1}e$
c. ${}^{0}_{-1}e + {}^{76}_{36}\text{Kr} \longrightarrow ?$
d. ${}^{4}_{2}\text{He} + {}^{241}_{95}\text{Am} \longrightarrow ? + 2{}^{1}_{0}n$

5.65 Write the balanced nuclear equation for each of the following: (5.2)
a. When two oxygen-16 atoms collide, one of the products is an alpha particle.
b. When californium-249 is bombarded by oxygen-18, a new element, seaborgium-263, and four neutrons are produced.
c. Radon-222 undergoes alpha decay.
d. An atom of strontium-80 emits a positron.

5.66 Write the balanced nuclear equation for each of the following: (5.2)
a. Polonium-210 decays to give lead-206.
b. Bismuth-211 emits an alpha particle.
c. A radioisotope emits a positron to form titanium-48.
d. An atom of germanium-69 emits a positron.

5.67 A 120-mg sample of technetium-99m is used for a diagnostic test. If technetium-99m has a half-life of 6.0 h, how many milligrams of the technetium-99m sample remains active 24 h after the test? (5.4)

5.68 The half-life of oxygen-15 is 124 s. If a sample of oxygen-15 has an activity of 4000 Bq, how many minutes will elapse before it has an activity of 500 Bq? (5.4)

5.69 What is the difference between fission and fusion? (5.6)

5.70 **a.** What are the products in the fission of uranium-235 that make possible a nuclear chain reaction? (5.6)
b. What is the purpose of placing control rods among uranium samples in a nuclear reactor?

5.71 Where does fusion occur naturally? (5.6)

5.72 Why are scientists continuing to try to build a fusion reactor even though the very high temperatures it requires have been difficult to reach and maintain? (5.6)

Clinical Applications

5.73 The activity of K-40 in a 70.-kg human body is estimated to be 120 nCi. What is this activity in becquerels? (5.3)

5.74 The activity of C-14 in a 70.-kg human body is estimated to be 3.7 kBq. What is this activity in microcuries? (5.3)

5.75 If the amount of radioactive phosphorus-32, used to treat leukemia, in a sample decreases from 1.2 mg to 0.30 mg in 28.6 days, what is the half-life of phosphorus-32? (5.4)

5.76 If the amount of radioactive iodine-123, used to treat thyroid cancer, in a sample decreases from 0.4 mg to 0.1 mg in 26.4 h, what is the half-life of iodine-123? (5.4)

5.77 Calcium-47, used to evaluate bone metabolism, has a half-life of 4.5 days. (5.2, 5.4)
a. Write the balanced nuclear equation for the beta decay of calcium-47.
b. How many milligrams of a 16-mg sample of calcium-47 remain after 18 days?
c. How many days have passed if 4.8 mg of calcium-47 decayed to 1.2 mg of calcium-47?

5.78 Cesium-137, used in cancer treatment, has a half-life of 30 yr. (5.2, 5.4)
a. Write the balanced nuclear equation for the beta decay of cesium-137.
b. How many milligrams of a 16-mg sample of cesium-137 remain after 90 yr?
c. How many years are required for 28 mg of cesium-137 to decay to 3.5 mg of cesium-137?

CHALLENGE PROBLEMS

The following problems are related to the topics in this chapter. However, they do not all follow the chapter order, and they require you to combine concepts and skills from several Sections. These problems will help you increase your critical thinking skills and prepare for your next exam.

5.79 Write the balanced nuclear equation for each of the following radioactive emissions: (5.2)
a. an alpha particle from Hg-180
b. a beta particle from Au-198
c. a positron from Rb-82

5.80 Write the balanced nuclear equation for each of the following radioactive emissions: (5.2)

a. an alpha particle from Gd-148

b. a beta particle from Sr-90

c. a positron from Al-25

5.81 All the elements beyond uranium, the transuranium elements, have been prepared by bombardment and are not naturally occurring elements. The first transuranium element neptunium, Np, was prepared by bombarding U-238 with neutrons to form a neptunium atom and a beta particle. Complete the following equation: (5.2)

$$^{1}_{0}n + ^{238}_{92}\text{U} \longrightarrow ? + ?$$

5.82 One of the most recent transuranium elements, oganesson-294 (Og-294), atomic number 118, was prepared by bombarding californium-249 with another isotope. Complete the following equation for the preparation of this new element: (5.2)

$$? + ^{249}_{98}\text{Cf} \longrightarrow ^{294}_{118}\text{Og} + 3^{1}_{0}n$$

5.83 A 64-μCi sample of Tl-201 decays to 4.0 μCi in 12 days. What is the half-life, in days, of Tl-201? (5.3, 5.4)

5.84 A wooden object from the site of an ancient temple has a carbon-14 activity of 10 counts/min compared with a reference piece of wood cut today that has an activity of 40 counts/min. If the half-life for carbon-14 is 5730 yr, what is the age of the ancient wood object? (5.3, 5.4)

5.85 The half-life for the radioactive decay of calcium-47 is 4.5 days. If a sample has an activity of 1.0 μCi after 27 days, what was the initial activity, in microcuries, of the sample? (5.3, 5.4)

5.86 The half-life for the radioactive decay of Ce-141 is 32.5 days. If a sample has an activity of 4.0 μCi after 130 days have elapsed, what was the initial activity, in microcuries, of the sample? (5.3, 5.4)

5.87 Element 114 was recently named flerovium, symbol Fl. The reaction for its synthesis involves bombarding Pu-244 with Ca-48. Write the balanced nuclear equation for the synthesis of flerovium. (5.2)

5.88 Element 116 was recently named livermorium, symbol Lv. The reaction for its synthesis involves bombarding Cm-248 with Ca-48. Write the balanced nuclear equation for the synthesis of livermorium. (5.2)

Clinical Applications

5.89 A nuclear technician was accidentally exposed to potassium-42 while doing brain scans for possible tumors. The error was not discovered until 36 h later when the activity of the potassium-42 sample was 2.0 μCi. If potassium-42 has a half-life of 12 h, what was the activity of the sample at the time the technician was exposed? (5.3, 5.4)

5.90 The radioisotope sodium-24 is used to determine the levels of electrolytes in the body. A 16-μg sample of sodium-24 decays to 2.0 μg in 45 h. What is the half-life, in hours, of sodium-24? (5.4)

ANSWERS

Answers to Selected Practice Problems

5.1 **a.** alpha particle **b.** positron
c. gamma radiation

5.3 **a.** $^{39}_{19}\text{K}$, $^{40}_{19}\text{K}$, $^{41}_{19}\text{K}$
b. They all have 19 protons and 19 electrons, but they differ in the number of neutrons.

5.5 **a.** $^{0}_{-1}e$ or β **b.** $^{4}_{2}\text{He}$ or α
c. $^{1}_{0}n$ or n **d.** $^{38}_{18}\text{Ar}$
e. $^{14}_{6}\text{C}$

5.7 **a.** $^{64}_{29}\text{Cu}$ **b.** $^{75}_{34}\text{Se}$
c. $^{24}_{11}\text{Na}$ **d.** $^{15}_{7}\text{N}$

5.9

Medical Use	Atomic Symbol	Mass Number	Number of Protons	Number of Neutrons
Heart imaging	$^{201}_{81}\text{Tl}$	201	81	120
Radiation therapy	$^{60}_{27}\text{Co}$	60	27	33
Abdominal scan	$^{67}_{31}\text{Ga}$	67	31	36
Hyperthyroidism	$^{131}_{53}\text{I}$	131	53	78
Leukemia treatment	$^{32}_{15}\text{P}$	32	15	17

5.11 **a.** 1, alpha particle **b.** 3, gamma radiation
c. 1, alpha particle

5.13 **a.** $^{208}_{84}\text{Po} \longrightarrow ^{204}_{82}\text{Pb} + ^{4}_{2}\text{He}$
b. $^{232}_{90}\text{Th} \longrightarrow ^{228}_{88}\text{Ra} + ^{4}_{2}\text{He}$
c. $^{251}_{102}\text{No} \longrightarrow ^{247}_{100}\text{Fm} + ^{4}_{2}\text{He}$
d. $^{220}_{86}\text{Rn} \longrightarrow ^{216}_{84}\text{Po} + ^{4}_{2}\text{He}$

5.15 **a.** $^{25}_{11}\text{Na} \longrightarrow ^{25}_{12}\text{Mg} + ^{0}_{-1}e$
b. $^{20}_{8}\text{O} \longrightarrow ^{20}_{9}\text{F} + ^{0}_{-1}e$
c. $^{92}_{38}\text{Sr} \longrightarrow ^{92}_{39}\text{Y} + ^{0}_{-1}e$
d. $^{60}_{26}\text{Fe} \longrightarrow ^{60}_{27}\text{Co} + ^{0}_{-1}e$

5.17 **a.** $^{26}_{14}\text{Si} \longrightarrow ^{26}_{13}\text{Al} + ^{0}_{+1}e$
b. $^{54}_{27}\text{Co} \longrightarrow ^{54}_{26}\text{Fe} + ^{0}_{+1}e$
c. $^{77}_{37}\text{Rb} \longrightarrow ^{77}_{36}\text{Kr} + ^{0}_{+1}e$
d. $^{93}_{45}\text{Rh} \longrightarrow ^{93}_{44}\text{Ru} + ^{0}_{+1}e$

5.19 **a.** $^{28}_{14}\text{Si}$, beta decay
b. $^{0}_{0}\gamma$, gamma emission
c. $^{0}_{-1}e$, beta decay
d. $^{238}_{92}\text{U}$, alpha decay
e. $^{188}_{79}\text{Au}$, positron emission

5.21 **a.** $^{10}_{4}\text{Be}$ **b.** $^{132}_{53}\text{I}$
c. $^{27}_{13}\text{Al}$ **d.** $^{17}_{8}\text{O}$

5.23 **a.** 2, absorbed dose **b.** 3, biological damage
c. 1, activity **d.** 2, absorbed dose

5.25 The technician exposed to 5 rad received the higher amount of radiation.

5.27 **a.** 294 μCi **b.** 0.5 Gy

5.29 **a.** two half-lives **b.** one half-life
c. three half-lives

5.31 **a.** 40.0 mg **b.** 20.0 mg
c. 10.0 mg **d.** 1.25 mg

5.33 **a.** 130 days **b.** 195 days

5.35 **a.** Because the elements Ca and P are part of the bone, the radioactive isotopes of Ca and P will become part of the bony structures of the body, where their radiation can be used to diagnose or treat bone diseases.
b. Strontium (Sr) acts much like calcium (Ca) because both are Group 2A (2) elements. The body will accumulate radioactive strontium in bones in the same way that it incorporates calcium. Radioactive strontium is harmful to children because the radiation it produces causes more damage in cells that are dividing rapidly.

5.37 180 μCi

5.39 **a.** $^{68}_{31}Ga \longrightarrow ^{68}_{30}Mn + ^{0}_{+1}e$ **b.** 16 mcg

5.41 Nuclear fission is the splitting of a large atom into smaller fragments with the release of large amounts of energy.

5.43 $^{103}_{42}Mo$

5.45 **a.** fission **b.** fusion
c. fission **d.** fusion

5.47 **a.** 2.0×10^{-3} Ci **b.** 2.0 mCi

5.49 9.0 days

5.51

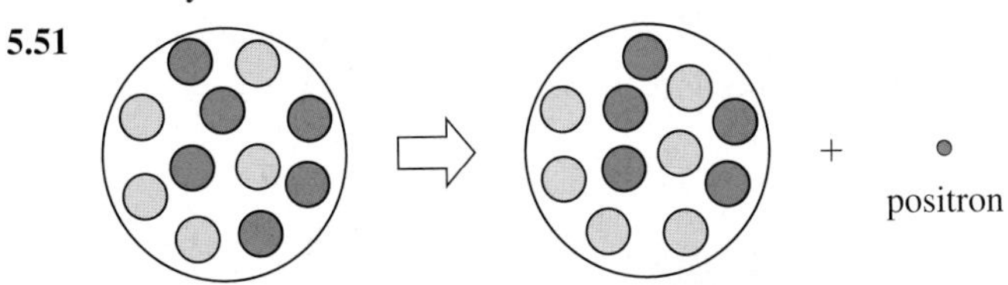

5.53

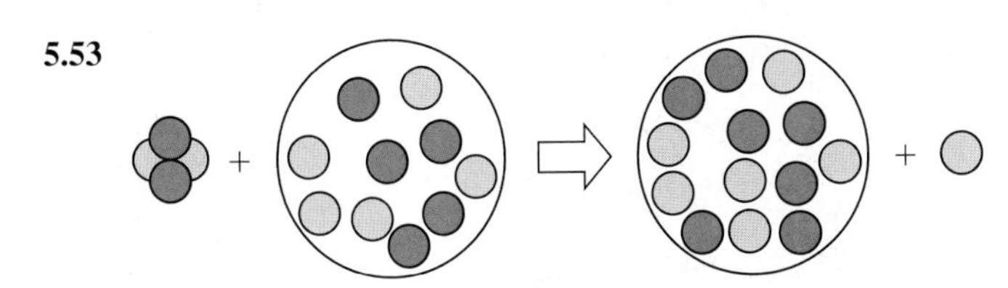

5.55 17 000 yr old

5.57 **a.** 11 protons and 14 neutrons
b. 28 protons and 33 neutrons
c. 37 protons and 47 neutrons
d. 47 protons and 63 neutrons

5.59 **a.** gamma emission
b. positron emission
c. alpha decay

5.61 **a.** $^{225}_{90}Th \longrightarrow ^{221}_{88}Ra + ^{4}_{2}He$
b. $^{210}_{83}Bi \longrightarrow ^{206}_{81}Tl + ^{4}_{2}He$
c. $^{137}_{55}Cs \longrightarrow ^{137}_{56}Ba + ^{0}_{-1}e$
d. $^{126}_{50}Sn \longrightarrow ^{126}_{51}Sb + ^{0}_{-1}e$
e. $^{18}_{9}F \longrightarrow ^{18}_{8}O + ^{0}_{+1}e$

5.63 **a.** $^{17}_{8}O$ **b.** $^{1}_{1}H$
c. $^{143}_{54}Xe$ **d.** $^{23}_{12}Mg$

5.65 **a.** $^{16}_{8}O + ^{16}_{8}O \longrightarrow ^{28}_{14}Si + ^{4}_{2}He$
b. $^{18}_{8}O + ^{249}_{98}Cf \longrightarrow ^{263}_{106}Sg + 4^{1}_{0}n$
c. $^{222}_{86}Rn \longrightarrow ^{218}_{84}Po + ^{4}_{2}He$
d. $^{80}_{38}Sr \longrightarrow ^{80}_{37}Rb + ^{0}_{+1}e$

5.67 7.5 mg of Tc-99m

5.69 In the fission process, an atom splits into smaller nuclei. In fusion, small nuclei combine (fuse) to form a larger nucleus.

5.71 Fusion occurs naturally in the Sun and other stars.

5.73 4.4×10^{3} Bq

5.75 14.3 days

5.77 **a.** $^{47}_{20}Ca \longrightarrow ^{47}_{21}Sc + ^{0}_{-1}e$
b. 1.0 mg of Ca-47
c. 9.0 days

5.79 **a.** $^{180}_{80}Hg \longrightarrow ^{176}_{78}Pt + ^{4}_{2}He$
b. $^{198}_{79}Au \longrightarrow ^{198}_{80}Hg + ^{0}_{-1}e$
c. $^{82}_{37}Rb \longrightarrow ^{82}_{36}Kr + ^{0}_{+1}e$

5.81 $^{1}_{0}n + ^{238}_{92}U \longrightarrow ^{239}_{93}Np + ^{0}_{-1}e$

5.83 3.0 days

5.85 64 μCi

5.87 $^{48}_{20}Ca + ^{244}_{94}Pu \longrightarrow ^{292}_{114}Fl$

5.89 16 μCi

6 Ionic and Molecular Compounds

RICHARD'S DOCTOR HAS RECOMMENDED THAT he take a low-dose aspirin (81 mg) every day to prevent a heart attack or stroke. Richard is concerned about taking aspirin and asks Sarah, a pharmacist working at a local pharmacy, about the effects of aspirin. Sarah explains to Richard that aspirin is acetylsalicylic acid and has the chemical formula $C_9H_8O_4$. Aspirin is a molecular compound, often referred to as an organic molecule because it contains the nonmetal carbon (C) combined with the nonmetals hydrogen (H) and oxygen (O). Sarah explains to Richard that aspirin is used to relieve minor pains, to reduce inflammation and fever, and to slow blood clotting. Aspirin is one of several nonsteroidal anti-inflammatory drugs (NSAIDs) that reduce pain and fever by blocking the formation of prostaglandins, which are chemical messengers that transmit pain signals to the brain and cause fever. Some potential side effects of aspirin may include heartburn, upset stomach, nausea, and an increased risk of a stomach ulcer.

CAREER Pharmacy Technician

Pharmacy technicians work in hospitals, pharmacies, clinics, and long-term care facilities where they are responsible for the preparation and distribution of pharmaceutical medications based on a doctor's orders. They obtain the proper medication, and also calculate, measure, and label the patient's medication. Pharmacy technicians advise clients and health care practitioners on the selection of both prescription and over-the-counter drugs, proper dosages, and geriatric considerations, as well as possible side effects and interactions. They may also administer vaccinations; prepare sterile intravenous solutions; and advise clients about health, diet, and home medical equipment. Pharmacy technicians also prepare insurance claims and create and maintain patient profiles.

CLINICAL UPDATE Compounds at the Pharmacy

Two weeks later, Richard returns to the pharmacy. Using Sarah's recommendations, he purchases Epsom salts for a sore toe, an antacid for an upset stomach, and an iron supplement. You can see the chemical formulas of these medications in the **CLINICAL UPDATE Compounds at the Pharmacy**, page 203.

6.1 Ions: Transfer of Electrons

LEARNING GOAL Write the symbols for the simple ions of the representative elements.

Most of the elements, except the noble gases, are found in nature combined as compounds. The noble gases are so stable that they form compounds only under extreme conditions. One explanation for the stability of noble gases is that they have a filled valence electron energy level.

Compounds form when electrons are transferred or shared between atoms to give stable electron arrangements. In the formation of either an *ionic bond* or a *covalent bond*, atoms lose, gain, or share valence electrons to acquire an **octet** of eight valence electrons. This tendency of atoms to attain a stable electron arrangement is known as the **octet rule** and provides a key to our understanding of the ways in which atoms bond and form compounds. A few elements achieve the stability of helium with two valence electrons. However, the octet rule is not used with transition elements.

Ionic bonds occur when the valence electrons of atoms of a metal are transferred to atoms of nonmetals. For example, sodium atoms lose electrons and chlorine atoms gain electrons to form the ionic compound NaCl. *Covalent bonds* form when atoms of nonmetals share valence electrons. In the molecular compounds H_2O and C_3H_8, atoms share electrons (see **TABLE 6.1**).

LOOKING AHEAD

TABLE 6.1 Types of Particles and Bonds in Compounds

Type	Ionic Compounds	Molecular Compounds	
Particles	Ions	Molecules	
Bonds	Ionic	Covalent	
Examples	Na^+ Cl^- ions	H_2O molecules	C_3H_8 molecules

REVIEW

Using Positive and Negative Numbers in Calculations (1.4)
Writing Electron Arrangements (4.6)

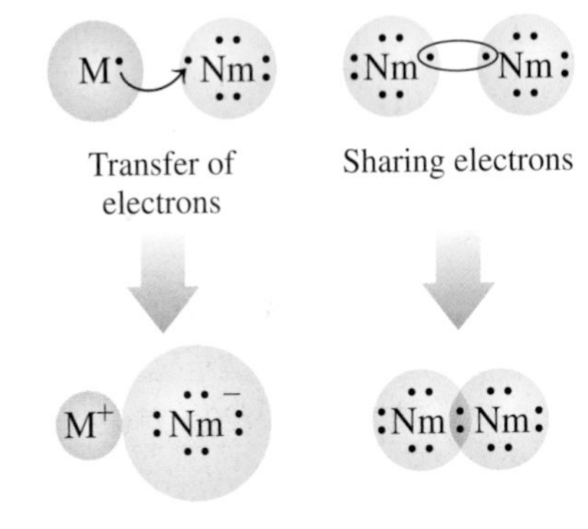

Positive Ions: Loss of Electrons

In ionic bonding, **ions**, which have electrical charges, form when atoms lose or gain electrons to form a stable electron arrangement. Because the ionization energies of metals of Groups 1A (1), 2A (2), and 3A (13) are low, metal atoms readily lose their valence electrons. In doing so, they form ions with positive charges. A metal atom obtains the same electron arrangement as its nearest noble gas (usually eight valence electrons). For example, when a sodium atom loses its single valence electron, the remaining electrons have a stable arrangement. By losing an electron, sodium has 10 negatively charged electrons instead of 11. Because there

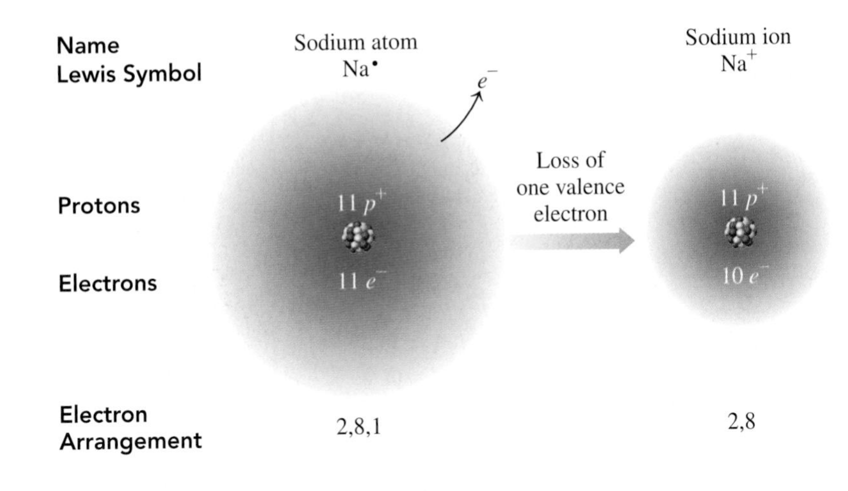

ENGAGE

A cesium atom has 55 protons and 55 electrons. A cesium ion has 55 protons and 54 electrons. Why does the cesium ion have a 1+ charge?

are still 11 positively charged protons in its nucleus, the atom is no longer neutral. It is now a sodium ion with a positive electrical charge, called an **ionic charge**, of 1+. In the Lewis symbol for the sodium ion, the ionic charge of 1+ is written in the upper right corner, Na^+, where the 1 is understood. The sodium ion is smaller than the sodium atom because the ion has lost its outermost electron from the third energy level. A positively charged ion of a metal is called a **cation** (pronounced *cat-eye-un*) and uses the name of the element.

$$\begin{aligned} \text{Ionic charge} &= \text{Charge of protons} + \text{Charge of electrons} \\ 1+ &= (11+) \qquad\qquad + (10-) \end{aligned}$$

Magnesium, a metal in Group 2A (2), obtains a stable electron arrangement by losing two valence electrons to form a magnesium ion with a 2+ ionic charge, Mg^{2+}. The magnesium ion is smaller than the magnesium atom because the outermost electrons in the third energy level were removed. The octet in the magnesium ion is made up of electrons that fill its second energy level.

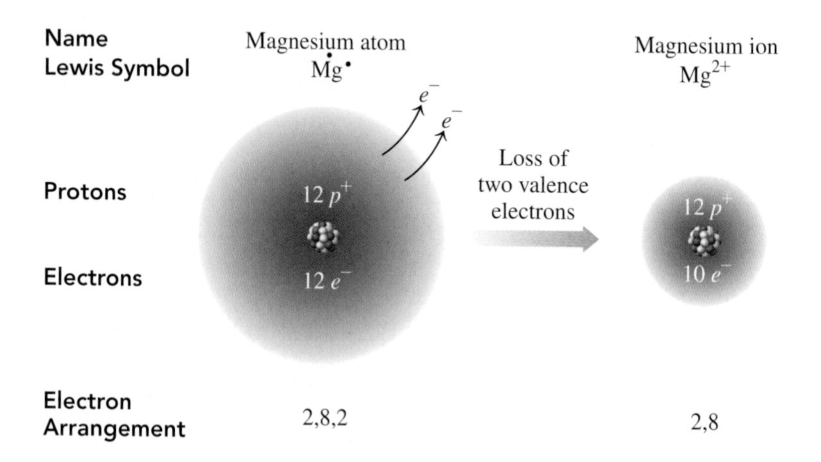

Negative Ions: Gain of Electrons

CORE CHEMISTRY SKILL

Writing Positive and Negative Ions

The ionization energy of a nonmetal atom in Groups 5A (15), 6A (16), or 7A (17) is high. In an ionic compound, a nonmetal atom gains one or more valence electrons to obtain a stable electron arrangement. By gaining electrons, a nonmetal atom forms a negatively charged ion. For example, an atom of chlorine with seven valence electrons gains one electron to form an octet. Because it now has 18 electrons and 17 protons in its nucleus, the chlorine atom is no longer neutral. It is a *chloride ion* with an ionic charge of 1−, which is written as Cl^-, with the 1 understood. A negatively charged ion, called an **anion** (pronounced *an-eye-un*), is named by using the first syllable of its element name followed by *ide*. The chloride ion is larger than the chlorine atom because the ion has an additional electron, which completes its outermost energy level.

ENGAGE

Why does Li form a positive ion, Li^+, whereas Br forms a negative ion, Br^-?

$$\begin{aligned} \text{Ionic charge} &= \text{Charge of protons} + \text{Charge of electrons} \\ 1- &= (17+) \qquad\qquad + (18-) \end{aligned}$$

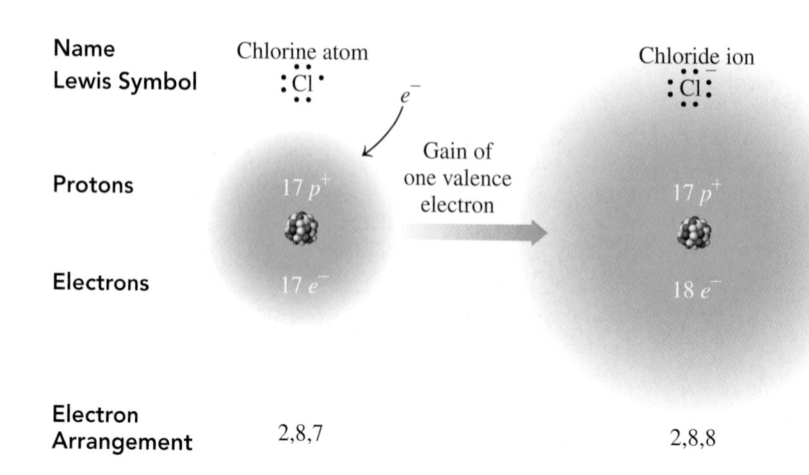

TABLE 6.2 lists the names of some important metal and nonmetal ions.

TABLE 6.2 Formulas and Names of Some Common Ions

Metals			Nonmetals		
Group Number	Cation	Name of Cation	Group Number	Anion	Name of Anion
1A (1)	Li^+	Lithium	5A (15)	N^{3-}	Nitride
	Na^+	Sodium		P^{3-}	Phosphide
	K^+	Potassium	6A (16)	O^{2-}	Oxide
2A (2)	Mg^{2+}	Magnesium		S^{2-}	Sulfide
	Ca^{2+}	Calcium	7A (17)	F^-	Fluoride
	Ba^{2+}	Barium		Cl^-	Chloride
3A (13)	Al^{3+}	Aluminum		Br^-	Bromide
				I^-	Iodide

SAMPLE PROBLEM 6.1 Ions

TRY IT FIRST

a. Write the symbol and name for the ion that has 7 protons and 10 electrons.
b. Write the symbol and name for the ion that has 20 protons and 18 electrons.

SOLUTION

a. The element with 7 protons is nitrogen. In an ion of nitrogen with 10 electrons, the ionic charge would be 3−, $[(7+) + (10-) = 3-]$. The ion, written as N^{3-}, is the *nitride* ion.

b. The element with 20 protons is calcium. In an ion of calcium with 18 electrons, the ionic charge would be 2+, $[(20+) + (18-) = 2+]$. The ion, written as Ca^{2+}, is the *calcium* ion.

STUDY CHECK 6.1

How many protons and electrons are in each of the following ions?

a. Sr^{2+} **b.** Cl^-

ANSWER

a. 38 protons, 36 electrons **b.** 17 protons, 18 electrons

TEST

Try Practice Problems 6.1 to 6.8

Ionic Charges from Group Numbers

In ionic compounds, representative elements usually lose or gain electrons to give eight valence electrons like their nearest noble gas (or two for helium). We can use the group numbers in the periodic table to determine the charges for the ions of the representative elements. The elements in Group 1A (1) lose one electron to form ions with a 1+ charge. The elements in Group 2A (2) lose two electrons to form ions with a 2+ charge. The elements in Group 3A (13) lose three electrons to form ions with a 3+ charge. In this text, we do not use the group numbers of the transition elements to determine their ionic charges.

In ionic compounds, the elements in Group 7A (17) gain one electron to form ions with a 1− charge. The elements in Group 6A (16) gain two electrons to form ions with a 2− charge. The elements in Group 5A (15) gain three electrons to form ions with a 3− charge.

The nonmetals of Group 4A (14) do not typically form ions. However, the metals Sn and Pb in Group 4A (14) lose electrons to form positive ions. **TABLE 6.3** lists the ionic charges for some common monatomic ions of representative elements.

ENGAGE

Why do all the atoms in Group 2A (2) form ions with 2+ charges?

TABLE 6.3 Examples of Monatomic Ions and Their Nearest Noble Gases

	Metals Lose Valence Electrons			Nonmetals Gain Valence Electrons		
Noble Gas	1A (1)	2A (2)	3A (13)	5A (15)	6A (16)	7A (17)
He	Li^+					
Ne	Na^+	Mg^{2+}	Al^{3+}	N^{3-}	O^{2-}	F^-
Ar	K^+	Ca^{2+}		P^{3-}	S^{2-}	Cl^-
Kr	Rb^+	Sr^{2+}			Se^{2-}	Br^-
Xe	Cs^+	Ba^{2+}				I^-

SAMPLE PROBLEM 6.2 Writing Symbols for Ions

TRY IT FIRST

Consider the elements aluminum and oxygen.

a. Identify each as a metal or a nonmetal.
b. State the number of valence electrons for each.
c. State the number of electrons that must be lost or gained for each to achieve an octet.
d. Write the symbol, including its ionic charge, and the name for each resulting ion.

SOLUTION

Aluminum	Oxygen
a. metal	nonmetal
b. 3 valence electrons	6 valence electrons
c. loses 3 e^-	gains 2 e^-
d. Al^{3+}, [(13+) + (10−) = 3+], aluminum ion	O^{2-}, [(8+) + (10−) = 2−], oxide ion

STUDY CHECK 6.2

Write the symbols for the ions formed by potassium and sulfur.

ANSWER

K^+ and S^{2-}

TEST

Try Practice Problems 6.9 to 6.14

CHEMISTRY LINK TO HEALTH

Some Important Ions in the Body

Several ions in body fluids have important physiological and metabolic functions. Some are listed in **TABLE 6.4**.

Foods such as bananas, milk, cheese, and potatoes provide the body with ions that are important in regulating body functions.

Milk, cheese, bananas, cereal, and potatoes provide ions for the body.

TABLE 6.4 Ions in the Body

Ion	Occurrence	Function	Source	Result of Too Little	Result of Too Much
Na^+	Principal cation outside the cell	Regulation and control of body fluids	Salt, cheese, pickles	Hyponatremia, anxiety, diarrhea, circulatory failure, decrease in body fluid	Hypernatremia, little urine, thirst, edema
K^+	Principal cation inside the cell	Regulation of body fluids and cellular functions	Bananas, orange juice, milk, prunes, potatoes	Hypokalemia (hypopotassemia), lethargy, muscle weakness, failure of neurological impulses	Hyperkalemia (hyperpotassemia), irritability, nausea, little urine, cardiac arrest
Ca^{2+}	Cation outside the cell; 90% of calcium in the body in bones	Major cation of bones; needed for muscle contraction	Milk, yogurt, cheese, greens, spinach	Hypocalcemia, tingling fingertips, muscle cramps, osteoporosis	Hypercalcemia, relaxed muscles, kidney stones, deep bone pain
Mg^{2+}	Cation outside the cell; 50% of magnesium in the body in bones	Essential for certain enzymes, muscles, nerve control	Widely distributed (part of chlorophyll of all green plants), nuts, whole grains	Disorientation, hypertension, tremors, slow pulse	Drowsiness
Cl^-	Principal anion outside the cell	Gastric juice, regulation of body fluids	Salt	Same as for Na^+	Same as for Na^+

PRACTICE PROBLEMS

6.1 Ions: Transfer of Electrons

LEARNING GOAL Write the symbols for the simple ions of the representative elements.

6.1 State the number of electrons that must be lost by atoms of each of the following to achieve a stable electron arrangement:
a. Li **b.** Ca **c.** Ga **d.** Cs **e.** Ba

6.2 State the number of electrons that must be gained by atoms of each of the following to achieve a stable electron arrangement:
a. Cl **b.** Se **c.** N **d.** I **e.** S

6.3 State the number of electrons lost or gained when the following elements form ions:
a. Sr **b.** P **c.** Group 7A (17)
d. Na **e.** Br

6.4 State the number of electrons lost or gained when the following elements form ions:
a. O **b.** Group 2A (2) **c.** F
d. K **e.** Rb

6.5 Write the symbols for the ions with the following number of protons and electrons:
a. 3 protons, 2 electrons **b.** 9 protons, 10 electrons
c. 12 protons, 10 electrons **d.** 26 protons, 23 electrons

6.6 Write the symbols for the ions with the following number of protons and electrons:
a. 8 protons, 10 electrons **b.** 19 protons, 18 electrons
c. 35 protons, 36 electrons **d.** 50 protons, 46 electrons

6.7 State the number of protons and electrons in each of the following:
a. Cu^{2+} **b.** Se^{2-}
c. Br^- **d.** Fe^{3+}

6.8 State the number of protons and electrons in each of the following:
a. S^{2-} **b.** Ni^{2+}
c. Au^{3+} **d.** Ag^+

6.9 Write the symbol for the ion of each of the following:
a. chlorine **b.** cesium
c. nitrogen **d.** radium

6.10 Write the symbol for the ion of each of the following:
a. fluorine **b.** barium
c. sodium **d.** iodine

6.11 Write the names for each of the following ions:
a. Li^+ **b.** Ca^{2+}
c. Ga^{3+} **d.** P^{3-}

6.12 Write the names for each of the following ions:
a. Rb^+ **b.** Sr^{2+}
c. S^{2-} **d.** F^-

Clinical Applications

6.13 State the number of protons and electrons in each of the following ions:
a. O^{2-}, used to build biomolecules and water
b. K^+, most prevalent positive ion in cells; needed for muscle contraction, nerve impulses
c. I^-, needed for thyroid function
d. Na^+, most prevalent positive ion in extracellular fluid

6.14 State the number of protons and electrons in each of the following ions:
a. P^{3-}, needed for bones and teeth
b. F^-, used to strengthen tooth enamel
c. Mg^{2+}, needed for bones and teeth
d. Ca^{2+}, needed for bones and teeth

Rubies and sapphires are the ionic compound aluminum oxide, with chromium ions in rubies, and titanium and iron ions in sapphires.

6.2 Ionic Compounds

LEARNING GOAL Using charge balance, write the correct formula for an ionic compound.

We utilize ionic compounds such as salt, NaCl, and baking soda, $NaHCO_3$, every day. Milk of magnesia, $Mg(OH)_2$, or calcium carbonate, $CaCO_3$, may be taken to settle an upset stomach. In a mineral supplement, iron may be present as iron(II) sulfate, $FeSO_4$, iodine as potassium iodide, KI, and manganese as manganese(II) sulfate, $MnSO_4$. Some sunscreens contain zinc oxide, ZnO, while tin(II) fluoride, SnF_2, in toothpaste provides fluoride to help prevent tooth decay. Gemstones are ionic compounds that are cut and polished to make jewelry. For example, sapphires and rubies are aluminum oxide, Al_2O_3. Impurities of chromium ions make rubies red, and iron and titanium ions make sapphires blue.

Properties of Ionic Compounds

In an **ionic compound**, one or more electrons are transferred from metals to nonmetals, which form positive and negative ions. The attraction between these ions is called an *ionic bond.*

The physical and chemical properties of an ionic compound such as NaCl are very different from those of the original elements. For example, the original elements of NaCl were sodium, which is a soft, shiny metal, and chlorine, which is a yellow-green poisonous gas. However, when they react and form positive and negative ions, they produce NaCl, which is ordinary table salt, a hard, white, crystalline substance that is important in our diet.

In a crystal of NaCl, the larger Cl^- ions are arranged in a three-dimensional structure in which the smaller Na^+ ions occupy the spaces between the Cl^- ions (see **FIGURE 6.1**). In this crystal, every Na^+ ion is surrounded by six Cl^- ions, and every Cl^- ion is surrounded by six Na^+ ions. Thus, there are many strong attractions between the positive and negative ions, which account for the high melting points of ionic compounds. For example, the melting point of NaCl is 801 °C. At room temperature, ionic compounds are solids.

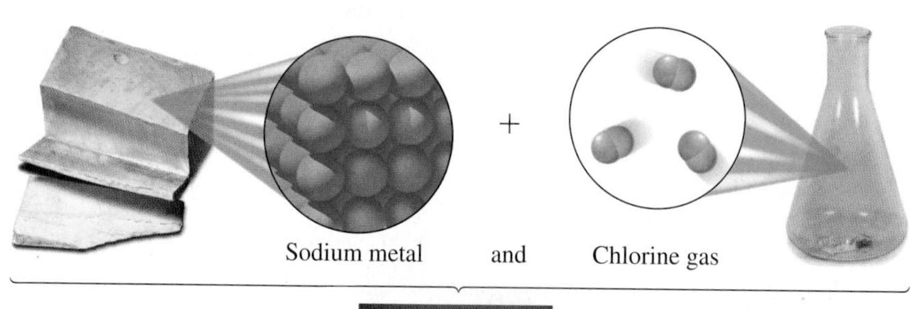

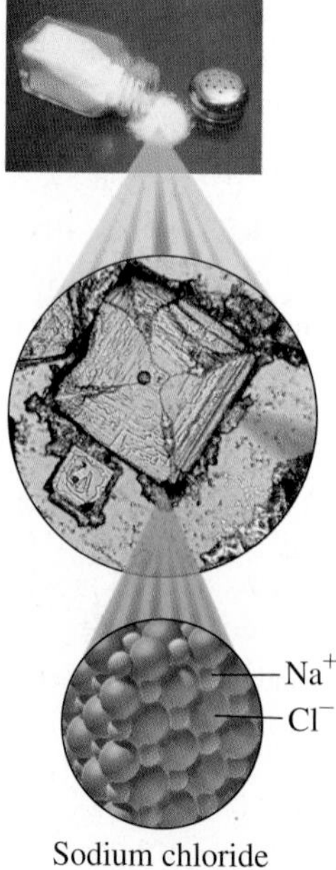

FIGURE 6.1 The elements sodium and chlorine react to form the ionic compound sodium chloride, the compound that makes up table salt. The magnification of NaCl crystals shows the arrangement of Na^+ and Cl^- ions.

Q What is the type of bonding between Na^+ and Cl^- ions in NaCl?

Chemical Formulas of Ionic Compounds

The **chemical formula** of a compound represents the symbols and subscripts in the lowest whole-number ratio of the atoms or ions. In the formula of an ionic compound, the sum of the ionic charges in the formula is always zero. *Thus, the total amount of positive charge is equal to the total amount of negative charge*. For example, to achieve a stable electron arrangement, one Na atom (metal) loses its one valence electron to form Na^+, and one Cl atom (nonmetal) gains one electron to form a Cl^- ion. The formula NaCl indicates that the compound has charge balance because there is one sodium ion, Na^+, for every chloride ion, Cl^-. Although the ions are positively or negatively charged, they are not shown in the formula of the compound.

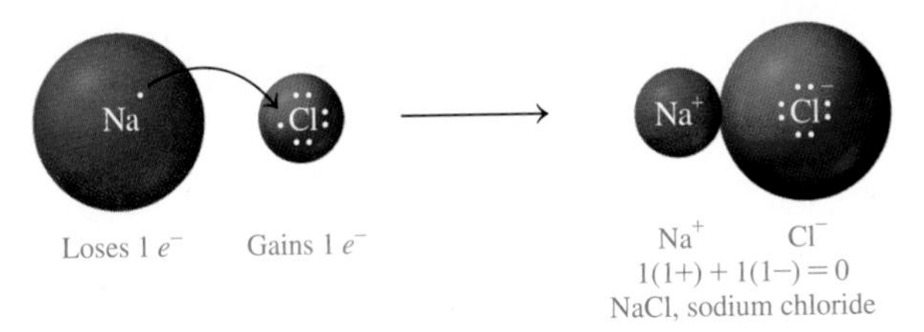

Subscripts in Formulas

Consider a compound of magnesium and chlorine. To achieve a stable electron arrangement, one Mg atom (metal) loses its two valence electrons to form Mg^{2+}. Two Cl atoms (nonmetals) each gain one electron to form two Cl^- ions. The two Cl^- ions are needed to balance the positive charge of Mg^{2+}. This gives the formula $MgCl_2$, magnesium chloride, in which the subscript 2 shows that two Cl^- ions are needed for charge balance.

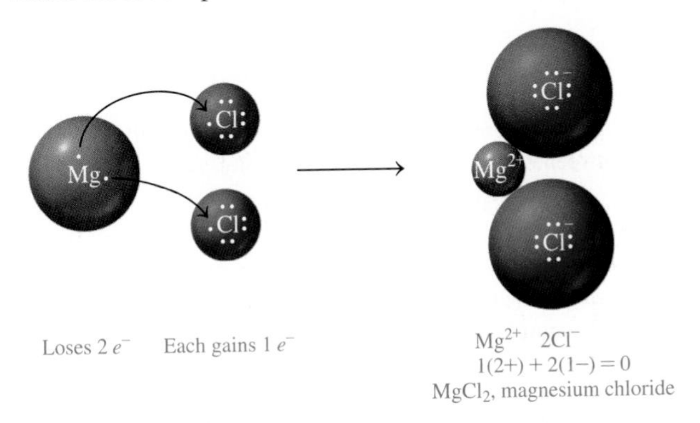

Writing Ionic Formulas from Ionic Charges

CORE CHEMISTRY SKILL
Writing Ionic Formulas

The subscripts in the formula of an ionic compound represent the number of positive and negative ions that give an overall charge of zero. Thus, we can now write a formula directly from the ionic charges of the positive and negative ions. Suppose we wish to write the formula for the ionic compound containing Na^+ and S^{2-} ions. To balance the ionic charge of the S^{2-} ion, we will need to place two Na^+ ions in the formula. This gives the formula Na_2S, which has an overall charge of zero. In the formula of an ionic compound, the cation is written first followed by the anion. Appropriate subscripts are used to show the number of each of the ions. This formula is the lowest ratio of ions in the ionic compound. Since ionic compounds do not exist as molecules, this lowest ratio of ions is called a *formula unit*.

Each loses 1 e^- Gains 2 e^-

$2Na^+$ S^{2-}

$2(1+) + 1(2-) = 0$

Na_2S, sodium sulfide

ENGAGE

How are the charges of ions used to write an ionic formula?

SAMPLE PROBLEM 6.3 Writing Formulas from Ionic Charges

TRY IT FIRST

Write the symbols for the ions, and the correct formula for the ionic compound formed when lithium and nitrogen react.

SOLUTION

Lithium, which is a metal in Group 1A (1), forms Li^+; nitrogen, which is a nonmetal in Group 5A (15), forms N^{3-}. The charge of 3− is balanced by three Li^+ ions.

$$3(1+) + 1(3-) = 0$$

Writing the cation (positive ion) first and the anion (negative ion) second gives the formula Li_3N.

STUDY CHECK 6.3

Write the symbols for the ions, and the correct formula for the ionic compound that would form when calcium and oxygen react.

ANSWER

Ca^{2+}, O^{2-}, CaO

TEST

Try Practice Problems 6.15 to 6.20

PRACTICE PROBLEMS

6.2 Ionic Compounds

LEARNING GOAL Using charge balance, write the correct formula for an ionic compound.

6.15 Which of the following pairs of elements are likely to form an ionic compound?
- **a.** lithium and chlorine
- **b.** oxygen and bromine
- **c.** potassium and oxygen
- **d.** sodium and neon
- **e.** cesium and magnesium
- **f.** nitrogen and fluorine

6.16 Which of the following pairs of elements are likely to form an ionic compound?
- **a.** helium and oxygen
- **b.** magnesium and chlorine
- **c.** chlorine and bromine
- **d.** potassium and sulfur
- **e.** sodium and potassium
- **f.** nitrogen and iodine

6.17 Write the correct ionic formula for the compound formed between each of the following pairs of ions:
- **a.** Na^+ and O^{2-}
- **b.** Al^{3+} and Br^-
- **c.** Ba^{2+} and N^{3-}
- **d.** Mg^{2+} and F^-
- **e.** Al^{3+} and S^{2-}

6.18 Write the correct ionic formula for the compound formed between each of the following pairs of ions:
- **a.** Al^{3+} and Cl^-
- **b.** Ca^{2+} and S^{2-}
- **c.** Li^+ and S^{2-}
- **d.** Rb^+ and P^{3-}
- **e.** Cs^+ and I^-

6.19 Write the symbols for the ions, and the correct formula for the ionic compound formed by each of the following:
- **a.** potassium and sulfur
- **b.** sodium and nitrogen
- **c.** aluminum and iodine
- **d.** gallium and oxygen

6.20 Write the symbols for the ions, and the correct formula for the ionic compound formed by each of the following:
- **a.** calcium and chlorine
- **b.** rubidium and bromine
- **c.** sodium and phosphorus
- **d.** magnesium and oxygen

REVIEW

Solving Equations (1.4)

6.3 Naming and Writing Ionic Formulas

LEARNING GOAL Given the formula of an ionic compound, write the correct name; given the name of an ionic compound, write the correct formula.

CORE CHEMISTRY SKILL

Naming Ionic Compounds

In the name of an ionic compound made up of two elements, the name of the metal ion, which is written first, is the same as its element name. The name of the nonmetal ion is obtained by using the first syllable of its element name followed by *ide*. In the name of any ionic compound, a space separates the name of the cation from the name of the anion. Subscripts are not used; they are understood because of the charge balance of the ions in the compound (see **TABLE 6.5**).

TABLE 6.5 Names of Some Ionic Compounds

Compound	Metal Ion	Nonmetal Ion	Name
KI	K^+ Potassium	I^- Iodide	Potassium iodide
$MgBr_2$	Mg^{2+} Magnesium	Br^- Bromide	Magnesium bromide
Al_2O_3	Al^{3+} Aluminum	O^{2-} Oxide	Aluminum oxide

Iodized salt contains KI to prevent iodine deficiency.

SAMPLE PROBLEM 6.4 **Naming Ionic Compounds**

TRY IT FIRST

Write the name for the ionic compound Mg_3N_2.

SOLUTION GUIDE

	Given	Need	Connect
ANALYZE THE PROBLEM	Mg_3N_2	name	cation, anion

STEP 1 Identify the cation and anion. The cation is Mg^{2+} and the anion is N^{3-}.

STEP 2 Name the cation by its element name. The cation Mg^{2+} is magnesium.

STEP 3 Name the anion by using the first syllable of its element name followed by *ide*. The anion N^{3-} is nitride.

STEP 4 Write the name for the cation first and the name for the anion second.
magnesium nitride

STUDY CHECK 6.4

Name the compound Ga_2S_3.

ANSWER

gallium sulfide

TEST

Try Practice Problems 6.21 and 6.22

Metals with Variable Charges

We have seen that the charge of an ion of a representative element can be obtained from its group number. However, we cannot determine the charge of a transition element because it typically forms two or more positive ions. The transition elements lose electrons, but they are lost from the highest energy level and sometimes from a lower energy level as well. This is also true for metals of representative elements in Groups 4A (14) and 5A (15), such as Pb, Sn, and Bi.

In some ionic compounds, iron is in the Fe^{2+} form, but in other compounds, it has the Fe^{3+} form. Copper also forms two different ions, Cu^+ and Cu^{2+}. When a metal can form two or more types of ions, it has *variable charge*. Then we cannot predict the ionic charge from the group number.

For metals that form two or more ions, a naming system is used to identify the particular cation. To do this, a Roman numeral that is equal to the ionic charge is placed in parentheses immediately after the name of the metal. For example, Fe^{2+} is iron(II), and Fe^{3+} is iron(III). **TABLE 6.6** lists the ions of some metals that produce more than one ion.

ENGAGE

Why is a Roman numeral placed after the name of the cations of most transition elements?

TABLE 6.6 Some Metals That Form More Than One Positive Ion

Element	Possible Ions	Name of Ion
Bismuth	Bi^{3+} Bi^{5+}	Bismuth(III) Bismuth(V)
Chromium	Cr^{2+} Cr^{3+}	Chromium(II) Chromium(III)
Cobalt	Co^{2+} Co^{3+}	Cobalt(II) Cobalt(III)
Copper	Cu^{+} Cu^{2+}	Copper(I) Copper(II)
Gold	Au^{+} Au^{3+}	Gold(I) Gold(III)
Iron	Fe^{2+} Fe^{3+}	Iron(II) Iron(III)
Lead	Pb^{2+} Pb^{4+}	Lead(II) Lead(IV)
Manganese	Mn^{2+} Mn^{3+}	Manganese(II) Manganese(III)
Mercury	Hg_2^{2+} Hg^{2+}	Mercury(I)* Mercury(II)
Nickel	Ni^{2+} Ni^{3+}	Nickel(II) Nickel(III)
Tin	Sn^{2+} Sn^{4+}	Tin(II) Tin(IV)

*Mercury(I) ions form an ion pair with a 2+ charge.

The transition elements form more than one positive ion except for zinc (Zn^{2+}), cadmium (Cd^{2+}), and silver (Ag^{+}), which form only one ion. Thus, no Roman numerals are used with zinc, cadmium, and silver when naming their cations in ionic compounds. Metals in Groups 4A (14) and 5A (15) also form more than one type of positive ion. For example, lead and tin in Group 4A (14) form cations with charges of 2+ and 4+, and bismuth in Group 5A (15) forms cations with charges of 3+ and 5+.

Determination of Variable Charge

When you name an ionic compound, you need to determine if the metal is a representative element or a transition element. If it is a transition element, except for zinc, cadmium, or silver, you will need to use its ionic charge as a Roman numeral as part of its name. The calculation of ionic charge depends on the negative charge of the anions in the formula. For example, we use charge balance to determine the charge of a copper cation in the ionic compound $CuCl_2$. Because there are two chloride ions, each with a 1− charge, the total negative charge is 2−. To balance this 2− charge, the copper ion must have a charge of 2+, or Cu^{2+}:

$CuCl_2$

$Cu^{?}$ Cl^{-}

Cl^{-}

$$1(?) + 2(1-) = 0$$

$$? = 2+$$

To indicate the 2+ charge for the copper ion Cu^{2+}, we place the Roman numeral (II) immediately after copper when naming this compound: copper(II) chloride. Some ions and their location on the periodic table are seen in **FIGURE 6.2**.

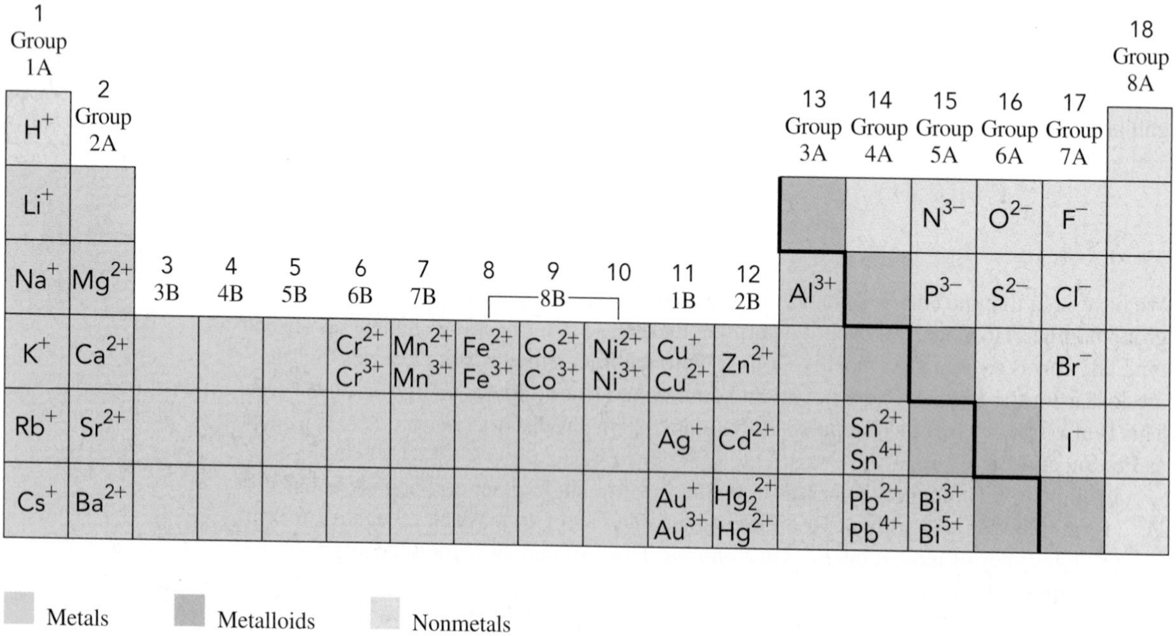

1 Group 1A	2 Group 2A	3 3B	4 4B	5 5B	6 6B	7 7B	8 8B	9 8B	10 8B	11 1B	12 2B	13 Group 3A	14 Group 4A	15 Group 5A	16 Group 6A	17 Group 7A	18 Group 8A
H^{+}																	
Li^{+}														N^{3-}	O^{2-}	F^{-}	
Na^{+}	Mg^{2+}											Al^{3+}		P^{3-}	S^{2-}	Cl^{-}	
K^{+}	Ca^{2+}				Cr^{2+} Cr^{3+}	Mn^{2+} Mn^{3+}	Fe^{2+} Fe^{3+}	Co^{2+} Co^{3+}	Ni^{2+} Ni^{3+}	Cu^{+} Cu^{2+}	Zn^{2+}					Br^{-}	
Rb^{+}	Sr^{2+}									Ag^{+}	Cd^{2+}		Sn^{2+} Sn^{4+}			I^{-}	
Cs^{+}	Ba^{2+}									Au^{+} Au^{3+}	Hg_2^{2+} Hg^{2+}		Pb^{2+} Pb^{4+}	Bi^{3+} Bi^{5+}			

Metals Metalloids Nonmetals

FIGURE 6.2 In ionic compounds, metals form positive ions, and nonmetals form negative ions.

What are the typical ions of calcium, copper, and oxygen in ionic compounds?

TABLE 6.7 lists the names of some ionic compounds in which the transition elements and metals from Groups 4A (14) and 5A (15) have more than one positive ion.

TABLE 6.7 Some Ionic Compounds of Metals That Form Two compounds

Compound	Systematic Name
$FeCl_2$	Iron(II) chloride
Fe_2O_3	Iron(III) oxide
Cu_3P	Copper(I) phosphide
$CrBr_2$	Chromium(II) bromide
$SnCl_2$	Tin(II) chloride
PbS_2	Lead(IV) sulfide
BiF_3	Bismuth(III) fluoride

SAMPLE PROBLEM 6.5 Naming Ionic Compounds with Variable Charge Metal Ions

TRY IT FIRST

Antifouling paint contains Cu_2O, which prevents the growth of barnacles and algae on the bottoms of boats. What is the name of Cu_2O?

The growth of barnacles is prevented by using a paint with Cu_2O on the bottom of a boat.

SOLUTION GUIDE

ANALYZE THE PROBLEM	Given	Need	Connect
	Cu_2O	name	cation, anion, charge balance

STEP 1 Determine the charge of the cation from the anion.

	Metal	Nonmetal
Element	copper (Cu)	oxygen (O)
Group	transition element	6A (16)
Ion	$Cu^?$	O^{2-}
Charge Balance	$2Cu^? +$	$2- = 0$
	$\frac{2Cu^?}{2} = \frac{2+}{2} = 1+$	
Ion	Cu^+	O^{2-}

STEP 2 Name the cation by its element name and use a Roman numeral in parentheses for the charge. copper(I)

STEP 3 Name the anion by using the first syllable of its element name followed by *ide*. oxide

STEP 4 Write the name for the cation first and the name for the anion second. copper(I) oxide

STUDY CHECK 6.5

Write the name for the compound with the formula Mn_2S_3.

ANSWER

manganese(III) sulfide

TEST

Try Practice Problems 6.23 to 6.28

Writing Formulas from the Name of an Ionic Compound

The formula for an ionic compound is written from the first part of the name that describes the metal ion, including its charge, and the second part of the name that specifies the nonmetal ion. Subscripts are added, as needed, to balance the charge. The steps for writing a formula from the name of an ionic compound are shown in Sample Problem 6.6.

SAMPLE PROBLEM 6.6 Writing Formulas for Ionic Compounds

TRY IT FIRST

Write the correct formula for iron(III) chloride.

SOLUTION GUIDE

ANALYZE THE PROBLEM	Given	Need	Connect
	iron(III) chloride	formula	cation, anion, charge balance

STEP 1 Identify the cation and anion.

Type of Ion	Cation	Anion
Name	iron(III)	chloride
Group	transition element	7A (17)
Symbol of Ion	Fe^{3+}	Cl^-

The pigment green chrome oxide contains chromium(III) oxide.

STEP 2 Balance the charges. The charge of 3+ is balanced by three Cl^- ions.

$$\mathbf{1}(3+) + \mathbf{3}(1-) = 0$$

STEP 3 Write the formula, cation first, using subscripts from the charge balance.

$FeCl_3$

STUDY CHECK 6.6

Write the correct formula for chromium(III) oxide used in green chrome oxide pigment.

ANSWER

Cr_2O_3

TEST

Try Practice Problems 6.29 to 6.34

PRACTICE PROBLEMS

6.3 Naming and Writing Ionic Formulas

LEARNING GOAL Given the formula of an ionic compound, write the correct name; given the name of an ionic compound, write the correct formula.

6.21 Write the name for each of the following ionic compounds:
a. AlF_3 **b.** $CaCl_2$ **c.** Na_2O
d. Mg_3P_2 **e.** KI **f.** BaF_2

6.22 Write the name for each of the following ionic compounds:
a. $MgCl_2$ **b.** K_3P **c.** Li_2S
d. CsF **e.** MgO **f.** $SrBr_2$

6.23 Write the name for each of the following ions (include the Roman numeral when necessary):
a. Fe^{2+} **b.** Cu^{2+} **c.** Zn^{2+}
d. Pb^{4+} **e.** Cr^{3+} **f.** Mn^{2+}

6.24 Write the name for each of the following ions (include the Roman numeral when necessary):
a. Ag^+ **b.** Cu^+ **c.** Bi^{3+}
d. Sn^{2+} **e.** Au^{3+} **f.** Ni^{2+}

6.25 Write the name for each of the following ionic compounds:
a. $SnCl_2$ **b.** FeO **c.** Cu_2S
d. CuS **e.** $CdBr_2$ **f.** $HgCl_2$

6.26 Write the name for each of the following ionic compounds:
a. Ag_3P **b.** PbS **c.** SnO_2
d. $MnCl_3$ **e.** Bi_2O_3 **f.** $CoCl_2$

6.27 Write the symbol for the cation in each of the following ionic compounds:
a. $AuCl_3$ **b.** Fe_2O_3 **c.** PbI_4 **d.** AlP

6.28 Write the symbol for the cation in each of the following ionic compounds:
a. $FeCl_2$ **b.** CrO **c.** Ni_2S_3 **d.** $SnCl_2$

6.29 Write the formula for each of the following ionic compounds:
a. magnesium chloride **b.** sodium sulfide
c. copper(I) oxide **d.** zinc phosphide
e. gold(III) nitride **f.** cobalt(III) fluoride

6.30 Write the formula for each of the following ionic compounds:
a. nickel(III) oxide **b.** barium fluoride
c. tin(IV) chloride **d.** silver sulfide
e. bismuth(V) chloride **f.** potassium nitride

6.31 Write the formula for each of the following ionic compounds:
a. cobalt(III) chloride **b.** lead(IV) oxide
c. silver iodide **d.** calcium nitride
e. copper(I) phosphide **f.** chromium(II) chloride

6.32 Write the formula for each of the following ionic compounds:
a. zinc bromide **b.** iron(III) sulfide
c. manganese(IV) oxide **d.** chromium(III) iodide
e. lithium nitride **f.** gold(I) oxide

Clinical Applications

6.33 The following compounds contain ions that are required in small amounts by the body. Write the formula for each.
a. potassium phosphide **b.** copper(II) chloride
c. iron(III) bromide **d.** magnesium oxide

6.34 The following compounds contain ions that are required in small amounts by the body. Write the formula for each.
a. calcium chloride **b.** nickel(II) iodide
c. manganese(II) oxide **d.** zinc nitride

6.4 Polyatomic Ions

LEARNING GOAL Write the name and formula for an ionic compound containing a polyatomic ion.

An ionic compound may also contain a *polyatomic ion* as one of its cations or anions. A **polyatomic ion** is a group of covalently bonded atoms that has an overall ionic charge. Most polyatomic ions consist of a nonmetal such as phosphorus, sulfur, carbon, or nitrogen covalently bonded to oxygen atoms.

Almost all the polyatomic ions are anions with charges 1−, 2−, or 3−. Only one common polyatomic ion, NH_4^+, has a positive charge. Some models of common polyatomic ions are shown in **FIGURE 6.3**.

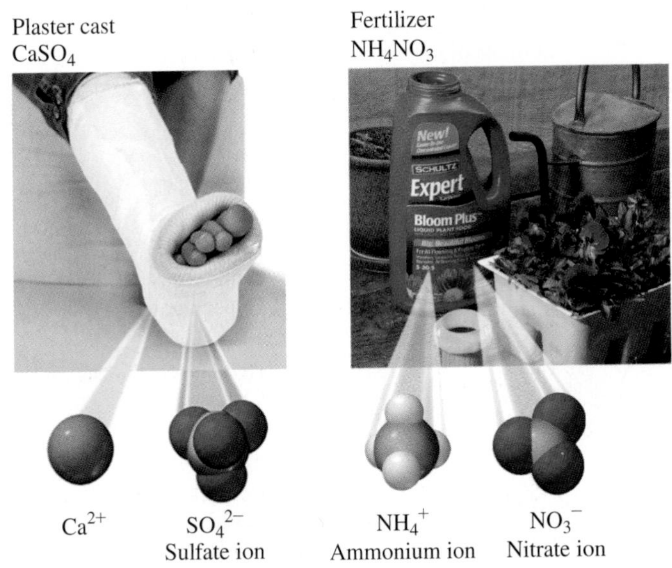

FIGURE 6.3 Many products contain polyatomic ions, which are groups of atoms that have an ionic charge.

Q What is the charge of a sulfate ion?

Names of Polyatomic Ions

The names of the most common polyatomic ions, which end in *ate*, are shown in bold in **TABLE 6.8**. When a related ion has one less O atom, the *ite* ending is used. For the same nonmetal, the *ate* ion and the *ite* ion have the same charge. For example, the sulfate ion is SO_4^{2-}, and the sulfite ion, which has one less oxygen atom, is SO_3^{2-}.

Formula	Charge	Name
SO_4^{2-}	2−	sulf**ate**
SO_3^{2-}	2−	sulf**ite**

Phosphate and phosphite ions each have a 3− charge.

Formula	Charge	Name
PO_4^{3-}	3−	phosph**ate**
PO_3^{3-}	3−	phosph**ite**

The formula of hydrogen carbonate, or *bicarbonate*, is written by placing a hydrogen in front of the polyatomic ion formula for carbonate (CO_3^{2-}), and the charge is decreased from 2− to 1− to give HCO_3^-.

Formula	Charge	Name
CO_3^{2-}	2−	carbon**ate**
HCO_3^-	1−	hydrogen carbon**ate**

The halogens form four different polyatomic ions with oxygen. The prefix *per* is used for one more O than the *ate* ion and the prefix *hypo* for one less O than the *ite* ion.

Formula	Charge	Name
ClO_4^-	1−	**per**chlor**ate**
ClO_3^-	1−	chlor**ate**
ClO_2^-	1−	chlor**ite**
ClO^-	1−	**hypo**chlor**ite**

Recognizing these prefixes and endings will help you identify polyatomic ions in the name of a compound. The hydroxide ion (OH^-) and cyanide ion (CN^-) are exceptions to this naming pattern.

TABLE 6.8 Names and Formulas of Some Common Polyatomic Ions

Nonmetal	Formula of Ion*	Name of Ion
Hydrogen	OH^-	Hydroxide
Nitrogen	NH_4^+	Ammonium
	$\mathbf{NO_3^-}$	**Nitrate**
	NO_2^-	Nitrite
Chlorine	ClO_4^-	Perchlorate
	$\mathbf{ClO_3^-}$	**Chlorate**
	ClO_2^-	Chlorite
	ClO^-	Hypochlorite
Carbon	$\mathbf{CO_3^{2-}}$	**Carbonate**
	HCO_3^-	Hydrogen carbonate (or bicarbonate)
	CN^-	Cyanide
	$C_2H_3O_2^-$	Acetate
Sulfur	$\mathbf{SO_4^{2-}}$	**Sulfate**
	HSO_4^-	Hydrogen sulfate (or bisulfate)
	SO_3^{2-}	Sulfite
	HSO_3^-	Hydrogen sulfite (or bisulfite)
Phosphorus	$\mathbf{PO_4^{3-}}$	**Phosphate**
	HPO_4^{2-}	Hydrogen phosphate
	$H_2PO_4^-$	Dihydrogen phosphate
	PO_3^{3-}	Phosphite

*Formulas and names in bold type indicate the most common polyatomic ion for that element.

TEST

Try Practice Problems 6.35 to 6.38

Sodium chlorite is used in the processing and bleaching of pulp from wood fibers and recycled cardboard.

Writing Formulas for Compounds Containing Polyatomic Ions

No polyatomic ion exists by itself. Like any ion, a polyatomic ion must be associated with ions of opposite charge. The bonding between polyatomic ions and other ions is one of electrical attraction. For example, the compound sodium chlorite consists of sodium ions (Na^+) and chlorite ions (ClO_2^-) held together by ionic bonds.

To write correct formulas for compounds containing polyatomic ions, we follow the same rules of charge balance that we used for writing the formulas for simple ionic compounds. The total negative and positive charges must equal zero. For example, consider the formula for a compound containing sodium ions and chlorite ions. The ions are written as

Na^+	ClO_2^-
Sodium ion	Chlorite ion

$$(1+) + (1-) = 0$$

Because one ion of each balances the charge, the formula is written as

$NaClO_2$

Sodium chlorite

When more than one polyatomic ion is needed for charge balance, parentheses are used to enclose the formula of the ion. A subscript is written outside the right parenthesis of the polyatomic ion to indicate the number needed for charge balance. Consider the formula for magnesium nitrate. The ions in this compound are the magnesium ion and the nitrate ion, a polyatomic ion.

Mg^{2+}	NO_3^-
Magnesium ion	Nitrate ion

To balance the positive charge of 2+ on the magnesium ion, two nitrate ions are needed. In the formula of the compound, parentheses are placed around the nitrate ion, and the subscript 2 is written outside the right parenthesis.

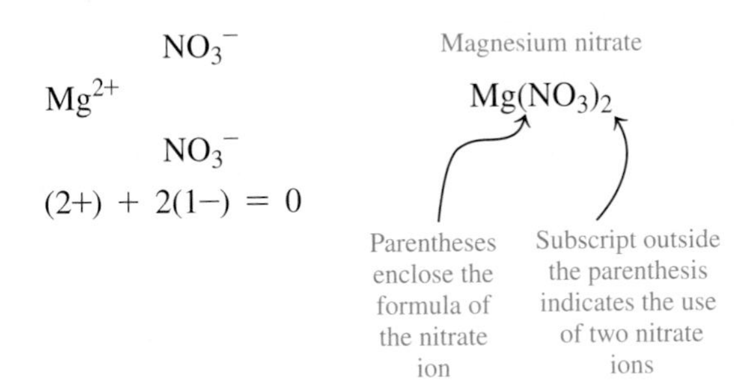

ENGAGE

Why does the formula for magnesium nitrate contain two NO_3^- ions?

SAMPLE PROBLEM 6.7 Writing Formulas Containing Polyatomic Ions

TRY IT FIRST

An antacid called Amphojel contains aluminum hydroxide, which treats acid indigestion and heartburn. Write the formula for aluminum hydroxide.

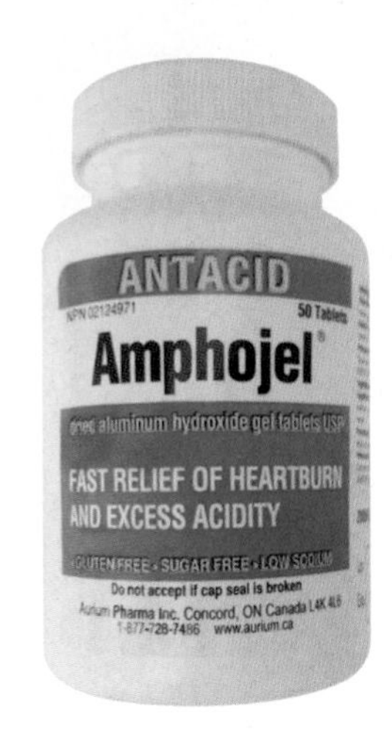

Aluminum hydroxide is an antacid used to treat acid indigestion.

SOLUTION GUIDE

	Given	Need	Connect
ANALYZE THE PROBLEM	aluminum hydroxide	formula	cation, polyatomic anion, charge balance

STEP 1 Identify the cation and polyatomic ion (anion).

Cation	Polyatomic Ion (anion)
aluminum	hydroxide
Al^{3+}	OH^-

STEP 2 Balance the charges. The charge of 3+ is balanced by three OH^- ions.

$$\mathbf{1}(3+) + \mathbf{3}(1-) = 0$$

STEP 3 Write the formula, cation first, using the subscripts from charge balance. The formula for the compound is written by enclosing the formula of the hydroxide ion, OH^-, in parentheses and writing the subscript 3 outside the right parenthesis.

$Al(OH)_3$

STUDY CHECK 6.7

Write the formula for a compound containing ammonium ions and phosphate ions.

ANSWER

$(NH_4)_3PO_4$

Naming Ionic Compounds Containing Polyatomic Ions

When naming ionic compounds containing polyatomic ions, we first write the positive ion, usually a metal, and then we write the name for the polyatomic ion. It is important that you learn to recognize the polyatomic ion in the formula and name it correctly. As with other ionic compounds, no prefixes are used.

Na_2SO_4	$FePO_4$	$Al_2(CO_3)_3$
$Na_2\boxed{SO_4}$	$Fe\boxed{PO_4}$	$Al_2(\boxed{CO_3})_3$
Sodium sulfate	Iron(III) phosphate	Aluminum carbonate

TABLE 6.9 lists the formulas and names of some ionic compounds that include polyatomic ions and also gives their uses in medicine and industry.

A solution of Epsom salts, magnesium sulfate, $MgSO_4$, may be used to soothe sore muscles.

TABLE 6.9 Some Ionic Compounds That Contain Polyatomic Ions

Formula	Name	Medical Use
$AlPO_4$	Aluminum phosphate	Antacid
$Al_2(SO_4)_3$	Aluminum sulfate	Antiperspirant, anti-infective
$BaSO_4$	Barium sulfate	Contrast medium for X-rays
$CaCO_3$	Calcium carbonate	Antacid, calcium supplement
$Ca_3(PO_4)_2$	Calcium phosphate	Calcium dietary supplement
$CaSO_4$	Calcium sulfate	Plaster casts
$MgSO_4$	Magnesium sulfate	Cathartic, Epsom salts
K_2CO_3	Potassium carbonate	Alkalizer, diuretic
$AgNO_3$	Silver nitrate	Topical anti-infective
$NaHCO_3$	Sodium bicarbonate or sodium hydrogen carbonate	Antacid
$Zn_3(PO_4)_2$	Zinc phosphate	Dental cement

SAMPLE PROBLEM 6.8 Naming Compounds Containing Polyatomic Ions

TRY IT FIRST

Name the following ionic compounds:

a. $Cu(NO_2)_2$ **b.** $KClO_3$

SOLUTION

ANALYZE THE PROBLEM	Given	Need	Connect
	formula	name	cation, polyatomic ion

Formula	Cation	Anion	Name of Cation	Name of Anion	Name of Compound
a. $Cu(NO_2)_2$	Cu^{2+}	NO_2^-	Copper(II) ion	Nitrite ion	Copper(II) nitrite
b. $KClO_3$	K^+	ClO_3^-	Potassium ion	Chlorate ion	Potassium chlorate

STUDY CHECK 6.8

What is the name of $Co_3(PO_4)_2$?

ANSWER

cobalt(II) phosphate

TEST

Try Practice Problems 6.39 to 6.46

PRACTICE PROBLEMS

6.4 Polyatomic Ions

LEARNING GOAL Write the name and formula for an ionic compound containing a polyatomic ion.

6.35 Write the formula including the charge for each of the following polyatomic ions:
a. hydrogen carbonate (bicarbonate)
b. ammonium
c. phosphite
d. chlorate

6.36 Write the formula including the charge for each of the following polyatomic ions:
a. nitrite b. sulfite
c. hydroxide d. acetate

6.37 Name the following polyatomic ions:
a. SO_4^{2-} b. CO_3^{2-}
c. HSO_3^- d. NO_3^-

6.38 Name the following polyatomic ions:
a. OH^- b. PO_4^{3-}
c. CN^- d. NO_2^-

6.39 Complete the following table with the formula and name of the compound that forms between each pair of ions:

	NO_2^-	CO_3^{2-}	HSO_4^-	PO_4^{3-}
Li^+				
Cu^{2+}				
Ba^{2+}				

6.40 Complete the following table with the formula and name of the compound that forms between each pair of ions:

	NO_3^-	HCO_3^-	SO_3^{2-}	HPO_4^{2-}
NH_4^+				
Al^{3+}				
Pb^{4+}				

6.41 Write the correct formula for the following ionic compounds:
a. barium hydroxide b. sodium hydrogen sulfate
c. iron(II) nitrite d. zinc phosphate
e. iron(III) carbonate

6.42 Write the correct formula for the following ionic compounds:
a. aluminum chlorate b. ammonium oxide
c. magnesium bicarbonate d. sodium nitrite
e. copper(I) sulfate

6.43 Write the formula for the polyatomic ion and name each of the following compounds:
a. Na_2CO_3 b. $(NH_4)_2S$
c. $Ca(OH)_2$ d. $Sn(NO_2)_2$

6.44 Write the formula for the polyatomic ion and name each of the following compounds:
a. $MnCO_3$ b. Au_2SO_4
c. $Mg_3(PO_4)_2$ d. $Fe(HCO_3)_3$

Clinical Applications

6.45 Name each of the following ionic compounds:
a. $Zn(C_2H_3O_2)_2$, cold remedy
b. $Mg_3(PO_4)_2$, antacid
c. NH_4Cl, expectorant
d. $NaHCO_3$, corrects pH imbalance
e. $NaNO_2$, meat preservative

6.46 Name each of the following ionic compounds:
a. Li_2CO_3, antidepressant
b. $MgSO_4$, Epsom salts
c. $NaClO$, disinfectant
d. Na_3PO_4, laxative
e. $Ba(OH)_2$, component of antacids

6.5 Molecular Compounds: Sharing Electrons

LEARNING GOAL Given the formula of a molecular compound, write its correct name; given the name of a molecular compound, write its formula.

A **molecular compound** consists of atoms of two or more nonmetals that share one or more valence electrons. The shared atoms are held together by **covalent bonds** that form a **molecule**. There are many more molecular compounds than there are ionic ones. For example, water (H_2O) and carbon dioxide (CO_2) are both molecular compounds. Molecular compounds consist of molecules, which are discrete groups of atoms in a definite proportion. A molecule of water (H_2O) consists of two atoms of hydrogen and one atom of oxygen. When you have iced tea, perhaps you add molecules of sugar ($C_{12}H_{22}O_{11}$), which is a molecular compound. Other familiar molecular compounds include propane (C_3H_8), alcohol (C_2H_6O), the antibiotic amoxicillin ($C_{16}H_{19}N_3O_5S$), and the antidepressant Prozac ($C_{17}H_{18}F_3NO$).

CORE CHEMISTRY SKILL

Writing the Names and Formulas for Molecular Compounds

Names and Formulas of Molecular Compounds

When naming a molecular compound, the first nonmetal in the formula is named by its element name; the second nonmetal is named using the first syllable of its element name, followed by *ide*. When a subscript indicates two or more atoms of an element, a prefix is shown in front of its name. **TABLE 6.10** lists prefixes used in naming molecular compounds.

The names of molecular compounds need prefixes because several different compounds can be formed from the same two nonmetals. For example, carbon and oxygen can form two different compounds, carbon monoxide, CO, and carbon dioxide, CO_2, in which the number of atoms of oxygen in each compound is indicated by the prefixes *mono* or *di* in their names.

When the vowels *o* and *o* or *a* and *o* appear together, the first vowel is omitted, as in carbon monoxide. In the name of a molecular compound, the prefix *mono* is usually omitted, as in NO, nitrogen oxide. Traditionally, however, CO is named carbon monoxide. **TABLE 6.11** lists the formulas, names, and commercial uses of some molecular compounds.

TABLE 6.10 Prefixes Used in Naming Molecular Compounds

1 mono	6 hexa
2 di	7 hepta
3 tri	8 octa
4 tetra	9 nona
5 penta	10 deca

TABLE 6.11 Some Common Molecular Compounds

Formula	Name	Commercial Uses
CO_2	Carbon dioxide	Fire extinguishers, dry ice, propellant in aerosols, carbonation of beverages
CS_2	Carbon disulfide	Manufacture of rayon
N_2O	Dinitrogen oxide	Inhalation anesthetic, "laughing gas"
NO	Nitrogen oxide	Stabilizer, biochemical messenger in cells
SO_2	Sulfur dioxide	Preserving fruits, vegetables; disinfectant in breweries; bleaching textiles
SF_6	Sulfur hexafluoride	Electrical circuits
SO_3	Sulfur trioxide	Manufacture of explosives

ENGAGE

Why are prefixes used to name molecular compounds?

SAMPLE PROBLEM 6.9 Naming Molecular Compounds

TRY IT FIRST

Name the molecular compound NCl_3.

SOLUTION GUIDE

ANALYZE THE PROBLEM	Given	Need	Connect
	NCl_3	name	prefixes

STEP 1 **Name the first nonmetal by its element name.** In NCl_3, the first nonmetal (N) is nitrogen.

STEP 2 **Name the second nonmetal by using the first syllable of its element name followed by *ide*.** The second nonmetal (Cl) is named chloride.

STEP 3 **Add prefixes to indicate the number of atoms (subscripts).** Because there is one nitrogen atom, no prefix is needed. The subscript 3 for the Cl atoms is shown as the prefix *tri*. The name of NCl_3 is nitrogen trichloride.

STUDY CHECK 6.9

Name each of the following molecular compounds:

a. $SiBr_4$ **b.** Br_2O

ANSWER

a. silicon tetrabromide **b.** dibromine oxide

TEST

Try Practice Problems 6.47 to 6.50

Writing Formulas from the Names of Molecular Compounds

In the name of a molecular compound, the names of two nonmetals are given along with prefixes for the number of atoms of each. To write the formula from the name, we use the symbol for each element and a subscript if a prefix indicates two or more atoms.

SAMPLE PROBLEM 6.10 Writing Formulas for Molecular Compounds

TRY IT FIRST

Write the formula for the molecular compound diboron trioxide.

SOLUTION GUIDE

ANALYZE THE PROBLEM	Given	Need	Connect
	diboron trioxide	formula	subscripts from prefixes

STEP 1 Write the symbols in the order of the elements in the name.

Name of Element	Boron	Oxygen
Symbol of Element	B	O
Subscript	2 (di)	3 (tri)

STEP 2 Write any prefixes as subscripts. The prefix *di* in *di*boron indicates that there are two atoms of boron, shown as a subscript 2 in the formula. The prefix *tri* in *tri*oxide indicates that there are three atoms of oxygen, shown as a subscript 3 in the formula.

B_2O_3

STUDY CHECK 6.10

Write the formula for the molecular compound iodine pentafluoride.

ANSWER

IF_5

TEST

Try Practice Problems 6.51 to 6.54

Summary of Naming Ionic and Molecular Compounds

We have now examined strategies for naming ionic and molecular compounds. In general, compounds having two elements are named by stating the first element name followed by the name of the second element with an *ide* ending. If the first element is a metal, the compound is usually ionic; if the first element is a nonmetal, the compound is usually molecular. For ionic compounds, it is necessary to determine whether the metal can form more than one type of positive ion; if so, a Roman numeral following the name of the metal indicates the particular ionic charge. One exception is the ammonium ion, NH_4^+, which is also written first as a positively charged polyatomic ion. Ionic compounds having three or more elements include some type of polyatomic ion. They are named by ionic rules but have an *ate* or *ite* ending when the polyatomic ion has a negative charge.

In naming molecular compounds having two elements, prefixes are necessary to indicate two or more atoms of each nonmetal as shown in that particular formula (see **FIGURE 6.4**).

SAMPLE PROBLEM 6.11 Naming Ionic and Molecular Compounds

TRY IT FIRST

Identify each of the following compounds as ionic or molecular and give its name:

a. K_3P **b.** $NiSO_4$ **c.** SO_3

SOLUTION

a. K_3P, consisting of a metal and a nonmetal, is an ionic compound. As a representative element in Group 1A (1), K forms the potassium ion, K^+. Phosphorus, as a representative element in Group 5A (15), forms a phosphide ion, P^{3-}. Writing the name of the cation followed by the name of the anion gives the name potassium phosphide.

b. $NiSO_4$, consisting of a cation of a transition element and a polyatomic ion SO_4^{2-}, is an ionic compound. As a transition element, Ni forms more than one type of ion. In this formula, the 2− charge of SO_4^{2-} is balanced by one nickel ion, Ni^{2+}. In the name, a Roman numeral written after the metal name, nickel(II), specifies the 2+ charge. The anion SO_4^{2-} is a polyatomic ion named sulfate. The compound is named nickel(II) sulfate.

c. SO_3 consists of two nonmetals, which indicates that it is a molecular compound. The first element S is sulfur (no prefix is needed). The second element O, oxide, has subscript 3, which requires a prefix *tri* in the name. The compound is named sulfur trioxide.

ENGAGE

Why is sodium phosphate an ionic compound and diphosphorus pentoxide a molecular compound?

STUDY CHECK 6.11

What is the name of $Fe(NO_3)_3$?

ANSWER

iron(III) nitrate

TEST

Try Practice Problems 6.55 and 6.56

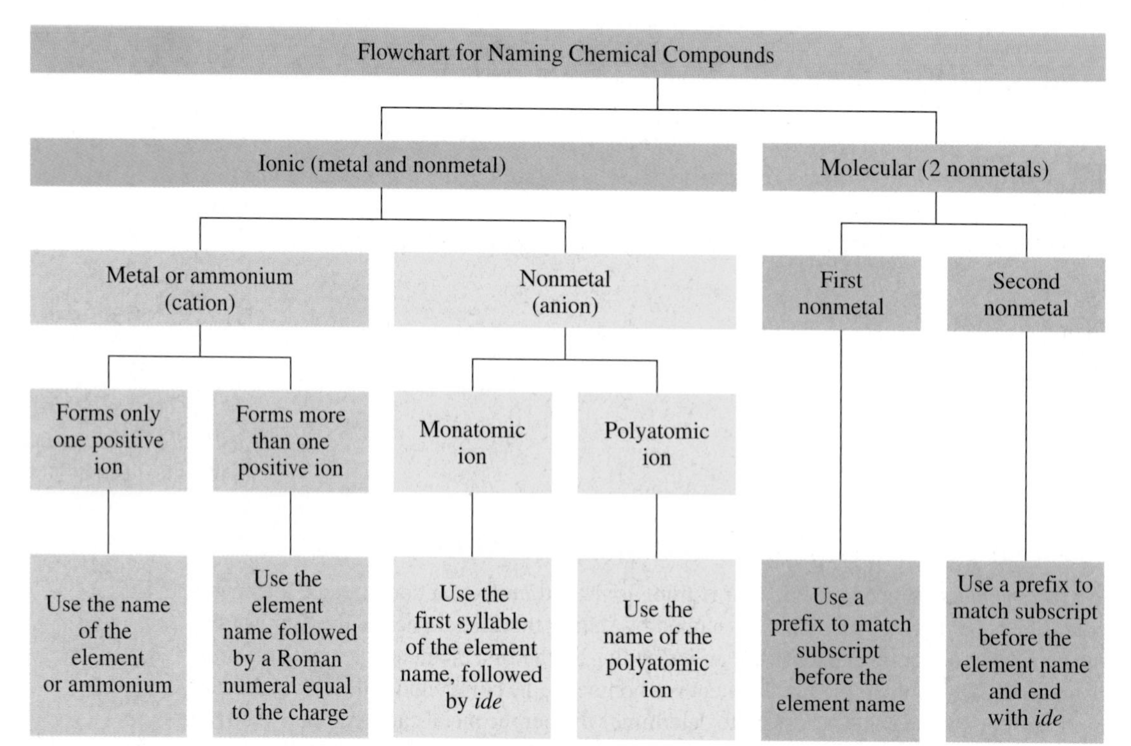

FIGURE 6.4 A flowchart illustrates naming for ionic and molecular compounds.

Q Why are the names of some metal ions followed by a Roman numeral in the name of a compound?

PRACTICE PROBLEMS

6.5 Molecular Compounds: Sharing Electrons

LEARNING GOAL Given the formula of a molecular compound, write its correct name; given the name of a molecular compound, write its formula.

6.47 Name each of the following molecular compounds:
a. PBr_3 **b.** Cl_2O **c.** CBr_4 **d.** HF **e.** NF_3

6.48 Name each of the following molecular compounds:
a. CS_2 **b.** P_2O_5 **c.** SiO_2 **d.** PCl_3 **e.** CO

6.49 Name each of the following molecular compounds:
a. N_2O_3 **b.** Si_2Br_6 **c.** P_4S_3 **d.** PCl_5 **e.** SeF_6

6.50 Name each of the following molecular compounds:
a. SiF_4 **b.** IBr_3 **c.** CO_2 **d.** N_2F_2 **e.** N_2S_3

6.51 Write the formula for each of the following molecular compounds:
a. carbon tetrachloride **b.** carbon monoxide
c. phosphorus trifluoride **d.** dinitrogen tetroxide

6.52 Write the formula for each of the following molecular compounds:
a. sulfur dioxide **b.** silicon tetrachloride
c. iodine trifluoride **d.** dinitrogen oxide

6.53 Write the formula for each of the following molecular compounds:
a. oxygen difluoride
b. boron trichloride
c. dinitrogen trioxide
d. sulfur hexafluoride

6.54 Write the formula for each of the following molecular compounds:
a. sulfur dibromide
b. carbon disulfide
c. tetraphosphorus hexoxide
d. dinitrogen pentoxide

Clinical Applications

6.55 Name each of the following ionic or molecular compounds:
a. $Al_2(SO_4)_3$, antiperspirant
b. $CaCO_3$, antacid
c. N_2O, "laughing gas," inhaled anesthetic
d. $Mg(OH)_2$, laxative

6.56 Name each of the following ionic or molecular compounds:
a. $Al(OH)_3$, antacid
b. $FeSO_4$, iron supplement in vitamins
c. NO, vasodilator
d. $Cu(OH)_2$, fungicide

6.6 Lewis Structures for Molecules

REVIEW
Drawing Lewis Symbols (4.7)

LEARNING GOAL Draw the Lewis structures for molecular compounds with single and multiple bonds.

Now we can investigate more complex chemical bonds and how they contribute to the structure of a molecule. *Lewis structures* use Lewis symbols to diagram the sharing of valence electrons for single and multiple bonds in molecules.

Lewis Structure for the Hydrogen Molecule

The simplest molecule is hydrogen, H_2. When two H atoms are far apart, there is no attraction between them. As the H atoms move closer, the positive charge of each nucleus attracts the electron of the other atom. This attraction, which is greater than the repulsion between the valence electrons, pulls the H atoms closer until they share a pair of valence electrons (see **FIGURE 6.5**). The result is called a *covalent bond*, in which the shared electrons give the stable electron arrangement of He to *each* of the H atoms. When the H atoms form H_2, they are more stable than two individual H atoms.

H· ·H
Far apart; no attractions
H· ·H
Attractions pull atoms closer
H : H
H_2 molecule
Energy Increases
Distance between Nuclei Decreases

FIGURE 6.5 A covalent bond forms as H atoms move close together to share electrons.
Q What determines the attraction between two H atoms?

Lewis Structures for Molecular Compounds

A molecule is represented by a **Lewis structure** in which the valence electrons of all the atoms are arranged to give octets, except for hydrogen, which has two electrons. The shared electrons, or *bonding pairs*, are shown as two dots or a single line between atoms. The nonbonding pairs of electrons, or *lone pairs*, are placed on the outside. For example, a fluorine molecule, F_2, consists of two fluorine atoms, which are in Group 7A (17), each with seven

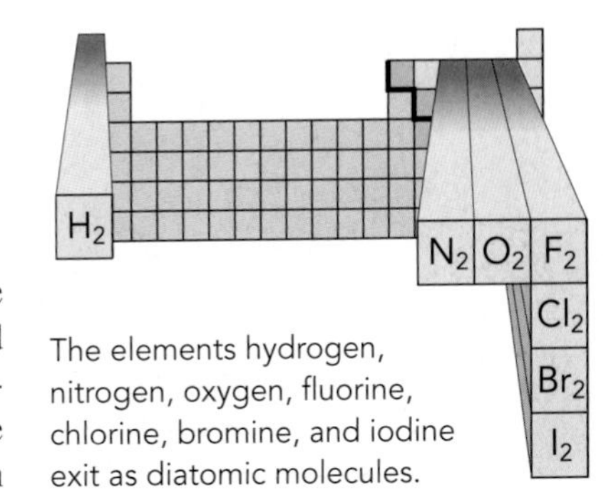

The elements hydrogen, nitrogen, oxygen, fluorine, chlorine, bromine, and iodine exit as diatomic molecules.

valence electrons. In the Lewis structure for the F_2 molecule, each F atom achieves an octet by sharing its unpaired valence electron.

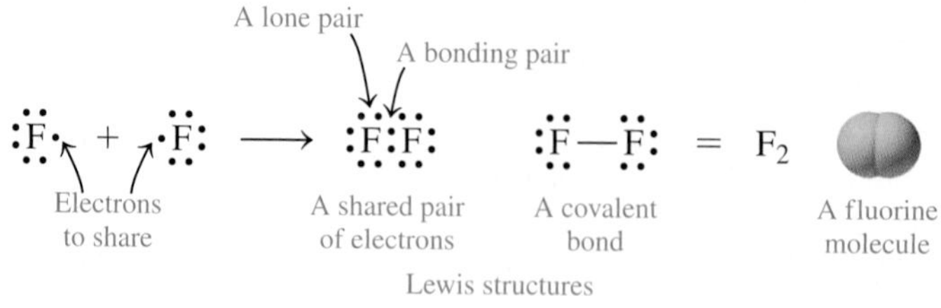

A lone pair
A bonding pair
$:\ddot{\underset{..}{F}}\cdot + \cdot\ddot{\underset{..}{F}}: \longrightarrow :\ddot{\underset{..}{F}}:\ddot{\underset{..}{F}}: \quad :\ddot{\underset{..}{F}}-\ddot{\underset{..}{F}}: = F_2$
Electrons to share — A shared pair of electrons — A covalent bond — A fluorine molecule
Lewis structures

TABLE 6.12 Elements That Exist as Diatomic Molecules

Diatomic Molecule	Name
H_2	Hydrogen
N_2	Nitrogen
O_2	Oxygen
F_2	Fluorine
Cl_2	Chlorine
Br_2	Bromine
I_2	Iodine

Hydrogen (H_2) and fluorine (F_2) are examples of nonmetal elements whose natural state is diatomic; that is, they contain two like atoms. The elements that exist as diatomic molecules are listed in **TABLE 6.12**.

Drawing Lewis Structures

The number of electrons that a nonmetal atom shares and the number of covalent bonds it forms are usually equal to the number of electrons it needs to achieve a stable electron arrangement.

To draw the Lewis structure for CH_4, we first draw the Lewis symbols for carbon and hydrogen.

$\cdot\dot{\underset{\cdot}{C}}\cdot \qquad \cdot H$

Then we determine the number of valence electrons needed for carbon and hydrogen. When a carbon atom shares its four electrons with four hydrogen atoms, carbon obtains an octet and each hydrogen atom is complete with two shared electrons. The Lewis structure is drawn with the carbon atom as the central atom, with the hydrogen atoms on each of the sides. The bonding pairs of electrons, which are single covalent bonds, may also be shown as single lines between the carbon atom and each of the hydrogen atoms.

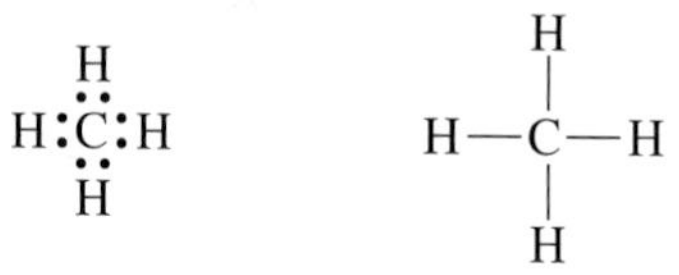

$$\begin{matrix} & H & \\ H & :\ddot{\underset{..}{C}}: & H \\ & H & \end{matrix} \qquad \begin{matrix} & H & \\ & | & \\ H- & C & -H \\ & | & \\ & H & \end{matrix}$$

TABLE 6.13 gives examples of Lewis structures and molecular models for some molecules.

TABLE 6.13 Lewis Structures for Some Molecular Compounds

CH_4	NH_3	H_2O
Lewis Structures		
H:C:H with H above and below (dot pairs)	H:N:H with lone pair above, H below	:O:H with lone pairs, H below
H—C—H with H above and below (lines)	H—N—H with lone pair, H below	:O—H with lone pairs, H below
Molecular Models		
Methane molecule	Ammonia molecule	Water molecule

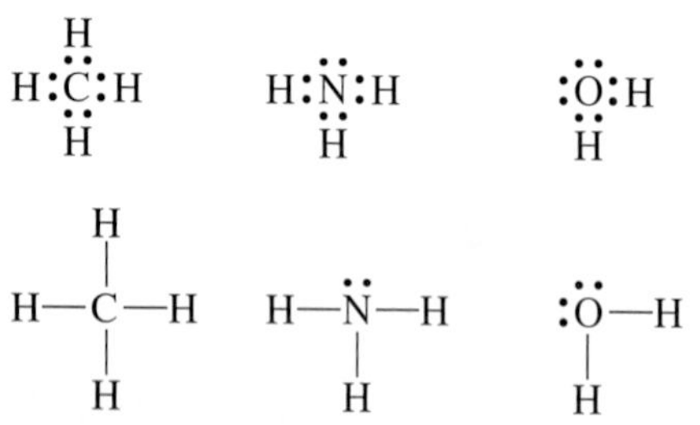

CORE CHEMISTRY SKILL
Drawing Lewis Structures

SAMPLE PROBLEM 6.12 Drawing Lewis Structures

TRY IT FIRST

Draw the Lewis structure for PCl_3, phosphorus trichloride, used commercially to prepare insecticides and flame retardants.

SOLUTION GUIDE

	Given	Need	Connect
ANALYZE THE PROBLEM	PCl_3	Lewis structure	total valence electrons

STEP 1 **Determine the arrangement of atoms.** In PCl_3, the central atom is P because there is only one P atom.

Cl P Cl
Cl

STEP 2 **Determine the total number of valence electrons.** We use the group number to determine the number of valence electrons for each of the atoms in the molecule.

Element	Group	Atoms		Valence Electrons	=	Total
P	5A (15)	1 P	×	5 e^-	=	5 e^-
Cl	7A (17)	3 Cl	×	7 e^-	=	21 e^-
				Total valence electrons for PCl_3	=	26 e^-

STEP 3 **Attach each bonded atom to the central atom with a pair of electrons.** Each bonding pair can also be represented by a bond line.

Cl:P:Cl (with :Cl below P) or Cl—P—Cl (with Cl bonded below P)

Six electrons (3 × 2 e^-) are used to bond the central P atom to three Cl atoms. Twenty valence electrons are left.

26 valence e^- − 6 bonding e^- = 20 e^- remaining

STEP 4 **Use the remaining electrons to complete octets.** We use the remaining 20 electrons as lone pairs, which are placed around the outer Cl atoms and on the P atom, such that all the atoms have octets.

:C̤̈l:P̤̈:C̤̈l: (with :C̤̈l: below P) or :C̤̈l—P̈—C̤̈l: (with :C̤̈l: bonded below P)

STUDY CHECK 6.12

Draw the Lewis structure for Cl_2O.

ANSWER

:C̤̈l:Ö̤:C̤̈l: or :C̤̈l—Ö̤—C̤̈l:

TEST

Try Practice Problems 6.57 and 6.58

Double and Triple Bonds

Up to now, we have looked at bonding in molecules having only single bonds. In many molecular compounds, atoms share two or three pairs of electrons to complete their octets. Double and triple bonds form when the number of valence electrons is not enough to complete the octets of all the atoms in the molecule. Then one or more lone pairs of electrons from the atoms attached to the central atom are shared with the central atom. A **double bond** occurs when two pairs of electrons are shared; in a **triple bond**, three pairs of electrons are shared. Atoms of carbon, oxygen, nitrogen, and sulfur are most likely to form multiple bonds.

Atoms of hydrogen and the halogens do not form double or triple bonds. The process of drawing a Lewis structure with multiple bonds is shown in Sample Problem 6.13.

SAMPLE PROBLEM 6.13 Drawing Lewis Structures with Multiple Bonds

TRY IT FIRST

Draw the Lewis structure for carbon dioxide, CO_2, in which the central atom is C.

SOLUTION GUIDE

ANALYZE THE PROBLEM	Given	Need	Connect
	CO_2	Lewis structure	total valence electrons

STEP 1 **Determine the arrangement of atoms.** O C O

STEP 2 **Determine the total number of valence electrons.** We use the group number to determine the number of valence electrons for each of the atoms in the molecule.

Element	Group	Atoms		Valence Electrons	=	Total
C	4A (14)	1 C	×	4 e^-	=	4 e^-
O	6A (16)	2 O	×	6 e^-	=	12 e^-
				Total valence electrons for CO_2	=	16 e^-

STEP 3 **Attach each bonded atom to the central atom with a pair of electrons.**

O:C:O or O—C—O

We use four valence electrons to attach the central C atom to two O atoms.

STEP 4 **Use the remaining electrons to complete octets, using multiple bonds if needed.**

The 12 remaining electrons are placed as six lone pairs of electrons on the outside O atoms. However, this does not complete the octet of the C atom.

:Ö:C:Ö: or :Ö—C—Ö:

To obtain an octet, the C atom must share pairs of electrons from each of the O atoms. When two bonding pairs occur between atoms, it is known as a double bond.

ENGAGE

How do you know when you need to add one or more multiple bonds to complete a Lewis structure?

Lone pairs converted to bonding pairs

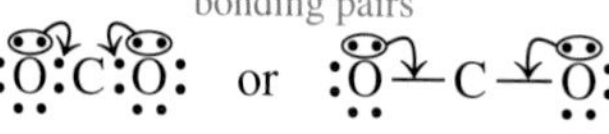

Double bonds Double bonds

:O::C::O: or :O=C=O:

Molecule of carbon dioxide

STUDY CHECK 6.13

Draw the Lewis structure for HCN, which has a triple bond.

ANSWER

H:C:::N: or H—C≡N:

TEST

Try Practice Problems 6.59 and 6.60

Exceptions to the Octet Rule

Although the octet rule is useful for bonding in many compounds, there are exceptions. We have already seen that a hydrogen (H_2) molecule requires just two electrons or a single bond. Usually the nonmetals form octets. However, in BCl_3, the B atom has only three valence electrons to share. Boron compounds typically have six valence electrons on the central B atoms and form just three bonds. Although we will generally see compounds of P, S, Cl, Br, and I with octets, they can form molecules in which they share more of their valence electrons. This expands their valence electrons to 10, 12, or even 14 electrons. For example, we have seen that the P atom in PCl_3 has an octet, but in PCl_5, the P atom has five bonds with 10 valence electrons. In H_2S, the S atom has an octet, but in SF_6, there are six bonds to sulfur with 12 valence electrons.

PRACTICE PROBLEMS

6.6 Lewis Structures for Molecules

LEARNING GOAL Draw the Lewis structures for molecular compounds with single and multiple bonds.

6.57 Determine the total number of valence electrons for each of the following:
a. H_2S **b.** I_2 **c.** CCl_4

6.58 Determine the total number of valence electrons for each of the following:
a. SBr_2 **b.** NBr_3 **c.** CH_3OH

6.59 Draw the Lewis structure for each of the following molecules:
a. HF **b.** SF_2 **c.** NBr_3
d. $ClNO_2$ (N is the central atom)

6.60 Draw the Lewis structure for each of the following molecules:
a. HCCH **b.** CS_2
c. H_2CO (C is the central atom) **d.** SiF_4

6.7 Electronegativity and Bond Polarity

LEARNING GOAL Use electronegativity to determine the polarity of a bond.

We can learn more about the chemistry of compounds by looking at how bonding electrons are shared between atoms. The bonding electrons are shared equally in a bond between identical nonmetal atoms. However, when a bond is between atoms of different elements, the electron pairs are usually shared unequally. Then the shared pairs of electrons are attracted to one atom in the bond more than the other.

The **electronegativity** of an atom is its ability to attract the shared electrons in a chemical bond. Nonmetals have higher electronegativities than do metals because nonmetals have a greater attraction for electrons than metals. On the electronegativity scale, the nonmetal fluorine located in the upper right corner of the periodic table, was assigned the highest value of 4.0, and the electronegativities for all other elements were determined relative to the attraction of fluorine for shared electrons. The metal cesium located in the lower left corner of the periodic table, has the lowest electronegativity value of 0.7. The electronegativities for the representative elements are shown in FIGURE 6.6. Note that there are no electronegativity values for the noble gases because they do not typically form bonds. The electronegativity values for transition elements are also low, but we have not included them in our discussion.

CORE CHEMISTRY SKILL
Using Electronegativity

ENGAGE
Why is the electronegativity of a chlorine atom greater than that of an iodine atom?

TEST
Try Practice Problems 6.61 to 6.64

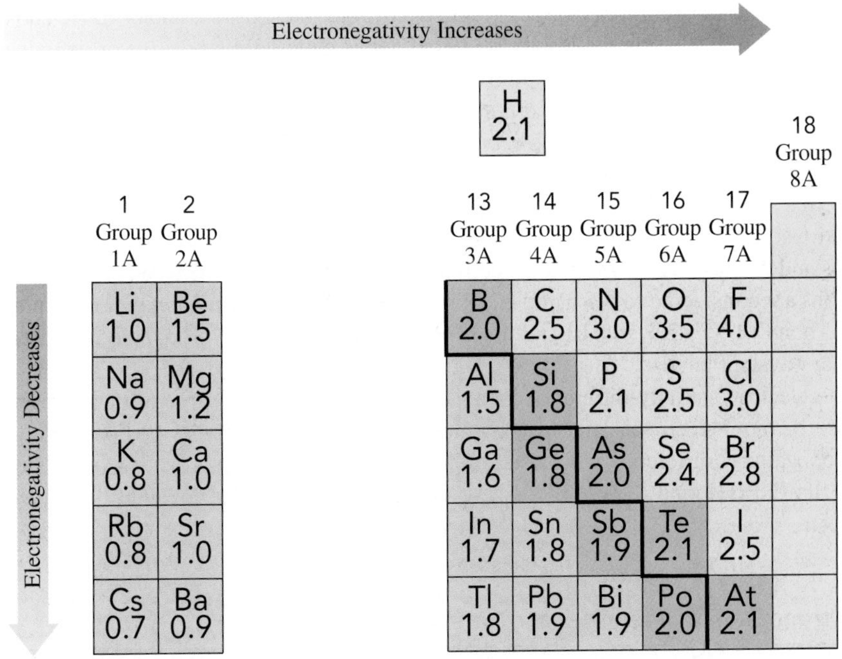

Electronegativity Increases →

Electronegativity Decreases ↓

1 Group 1A	2 Group 2A	13 Group 3A	14 Group 4A	15 Group 5A	16 Group 6A	17 Group 7A	18 Group 8A
H 2.1							
Li 1.0	Be 1.5	B 2.0	C 2.5	N 3.0	O 3.5	F 4.0	
Na 0.9	Mg 1.2	Al 1.5	Si 1.8	P 2.1	S 2.5	Cl 3.0	
K 0.8	Ca 1.0	Ga 1.6	Ge 1.8	As 2.0	Se 2.4	Br 2.8	
Rb 0.8	Sr 1.0	In 1.7	Sn 1.8	Sb 1.9	Te 2.1	I 2.5	
Cs 0.7	Ba 0.9	Tl 1.8	Pb 1.9	Bi 1.9	Po 2.0	At 2.1	

FIGURE 6.6 The electronegativity values of representative elements in Group 1A (1) to Group 7A (17), which indicate the ability of atoms to attract shared electrons, increase going across a period from left to right and decrease going down a group.

Q What element on the periodic table has the strongest attraction for shared electrons?

Polarity of Bonds

The difference in the electronegativity values of two atoms can be used to predict the type of chemical bond, ionic or covalent, that forms. For the H—H bond, the electronegativity difference is zero ($2.1 - 2.1 = 0$), which means the bonding electrons are shared equally. We illustrate this by drawing a symmetrical electron cloud around the H atoms. A bond between atoms with identical or very similar electronegativity values is a **nonpolar covalent bond**. However, when covalent bonds are between atoms with different electronegativity values, the electrons are shared unequally; the bond is a **polar covalent bond**. The electron cloud for a polar covalent bond is unsymmetrical. For the H—Cl bond, there is an electronegativity difference of 3.0 (Cl) − 2.1 (H) = 0.9, which means that the H—Cl bond is polar covalent (see FIGURE 6.7). When finding the electronegativity difference, the smaller electronegativity is always subtracted from the larger; thus, the difference is always a positive number.

The **polarity** of a bond depends on the difference in the electronegativity values of its atoms. In a polar covalent bond, the shared electrons are attracted to the more electronegative

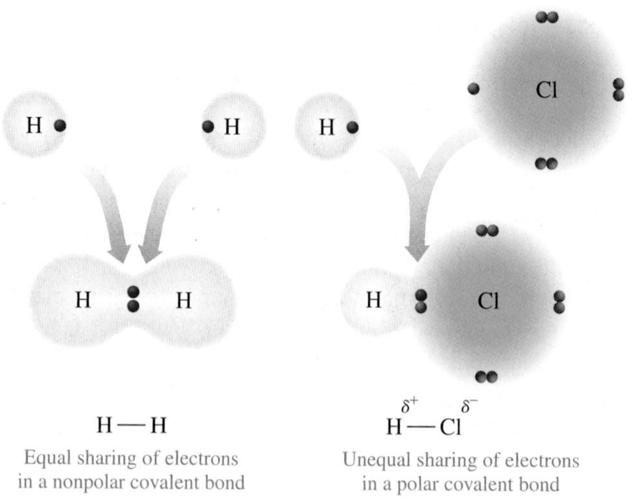

H • • H H • • Cl

H : H H : Cl

H—H
Equal sharing of electrons in a nonpolar covalent bond

δ^+ δ^-
H—Cl
Unequal sharing of electrons in a polar covalent bond

FIGURE 6.7 In the nonpolar covalent bond of H_2, electrons are shared equally. In the polar covalent bond of HCl, electrons are shared unequally.

Q H_2 has a nonpolar covalent bond, but HCl has a polar covalent bond. Explain.

Examples of Dipoles in Polar Covalent Bonds

$\overset{\delta^+}{C}-\overset{\delta^-}{O}$ $\overset{\delta^+}{N}-\overset{\delta^-}{O}$ $\overset{\delta^+}{Cl}-\overset{\delta^-}{F}$

$\longmapsto$ $\longmapsto$ $\longmapsto$

atom, which makes it partially negative, because of the negatively charged electrons around that atom. At the other end of the bond, the atom with the lower electronegativity becomes partially positive because of the lack of electrons at that atom.

A bond becomes more *polar* as the electronegativity difference increases. A polar covalent bond that has a separation of charges is called a **dipole**. The positive and negative ends of the dipole are indicated by the lowercase Greek letter delta with a positive or negative sign, δ^+ and δ^-. Sometimes we use an arrow that points from the positive charge to the negative charge $\longmapsto$ to indicate the dipole.

Variations in Bonding

The variations in bonding are continuous; there is no definite point at which one type of bond stops and the next starts. When the electronegativity difference is between 0.0 and 0.4, the electrons are considered to be shared equally in a *nonpolar covalent bond*. For example, the C—C bond (2.5 − 2.5 = 0.0) and the C—H bond (2.5 − 2.1 = 0.4) are classified as nonpolar covalent bonds.

ENGAGE

Use electronegativity differences to explain why a Si—S bond is polar covalent and a Si—P bond is nonpolar covalent.

As the electronegativity difference increases, the shared electrons are attracted more strongly to the more electronegative atom, which increases the polarity of the bond. When the electronegativity difference is from 0.5 to 1.8, the bond is a *polar covalent bond*. For example, the O—H bond (3.5 − 2.1 = 1.4) is classified as a polar covalent bond (see **TABLE 6.14**).

TABLE 6.14 Electronegativity Differences and Types of Bonds

Electronegativity Difference	0.0 to 0.4	0.5 to 1.8	1.9 to 3.3
Bond Type	Nonpolar covalent	Polar covalent	Ionic
Electron Bonding	Electrons shared equally	Electrons shared unequally	Electrons transferred
		δ^+ δ^-	+ −

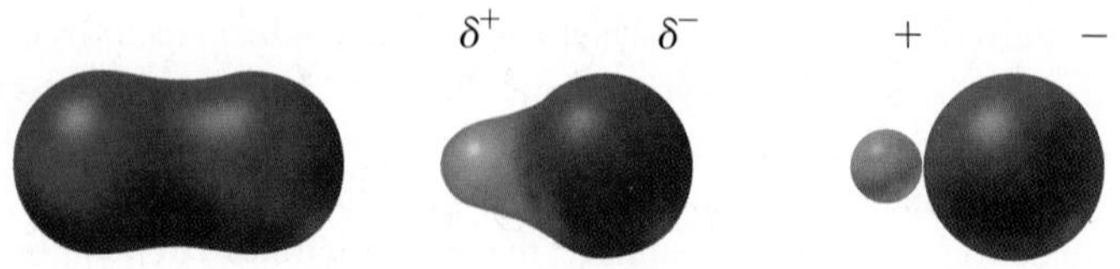

When the electronegativity difference is greater than 1.8, electrons are transferred from one atom to another, which results in an *ionic bond*. For example, the electronegativity difference for the ionic compound NaCl is 3.0 − 0.9 = 2.1. Thus, for large differences in electronegativity, we would predict an ionic bond (see **TABLE 6.15**).

TABLE 6.15 Predicting Bond Type from Electronegativity Differences

Molecule	Bond	Type of Electron Sharing	Electronegativity Difference*	Type of Bond	Reason
H_2	H—H	Shared equally	2.1 − 2.1 = 0.0	Nonpolar covalent	From 0.0 to 0.4
BrCl	Br—Cl	Shared about equally	3.0 − 2.8 = 0.2	Nonpolar covalent	From 0.0 to 0.4
HBr	$\overset{\delta^+}{H}—\overset{\delta^-}{Br}$	Shared unequally	2.8 − 2.1 = 0.7	Polar covalent	From 0.5 to 1.8
HCl	$\overset{\delta^+}{H}—\overset{\delta^-}{Cl}$	Shared unequally	3.0 − 2.1 = 0.9	Polar covalent	From 0.5 to 1.8
NaCl	Na^+Cl^-	Electron transfer	3.0 − 0.9 = 2.1	Ionic	From 1.9 to 3.3
MgO	$Mg^{2+}O^{2-}$	Electron transfer	3.5 − 1.2 = 2.3	Ionic	From 1.9 to 3.3

*Values are taken from Figure 6.6.

SAMPLE PROBLEM 6.14 Bond Polarity

TRY IT FIRST

Using electronegativity values, classify each of the following bonds as nonpolar covalent, polar covalent, or ionic and label any polar covalent bond with δ^+ and δ^- and show the direction of the dipole:

a. K and O **b.** As and Cl **c.** N and N **d.** P and Br

SOLUTION

ANALYZE THE PROBLEM	Given	Need	Connect
	bonds	type of bonds	electronegativity values

For each bond, we obtain the electronegativity values and calculate the difference in electronegativity.

Bond	Electronegativity Difference	Type of Bond	Dipole
a. K and O	3.5 − 0.8 = 2.7	Ionic	
b. As and Cl	3.0 − 2.0 = 1.0	Polar covalent	$\overset{\delta^+}{As}—\overset{\delta^-}{Cl}$ $\longmapsto$
c. N and N	3.0 − 3.0 = 0.0	Nonpolar covalent	
d. P and Br	2.8 − 2.1 = 0.7	Polar covalent	$\overset{\delta^+}{P}—\overset{\delta^-}{Br}$ $\longmapsto$

STUDY CHECK 6.14

Classify each of the following bonds as nonpolar covalent, polar covalent, or ionic and label any polar covalent bond with δ^+ and δ^- and show the direction of the dipole:

a. P and Cl **b.** Br and Br **c.** Na and O

ANSWER

a. polar covalent (0.9) $\overset{\delta^+}{P}—\overset{\delta^-}{Cl}$ $\longmapsto$ **b.** nonpolar covalent (0.0) **c.** ionic (2.6)

TEST

Try Practice Problems 6.65 to 6.70

PRACTICE PROBLEMS

6.7 Electronegativity and Bond Polarity

LEARNING GOAL Use electronegativity to determine the polarity of a bond.

6.61 Describe the trend in electronegativity as *increases* or *decreases* for each of the following:
a. from B to F **b.** from Mg to Ba **c.** from F to I

6.62 Describe the trend in electronegativity as *increases* or *decreases* for each of the following:
a. from Al to Cl **b.** from Br to K **c.** from Li to Cs

6.63 Using the periodic table, arrange the atoms in each of the following sets in order of increasing electronegativity:
a. Li, Na, K **b.** Na, Cl, P **c.** Se, Ca, O

6.64 Using the periodic table, arrange the atoms in each of the following sets in order of increasing electronegativity:
a. Cl, F, Br **b.** B, O, N **c.** Mg, F, S

6.65 Which electronegativity difference (**a**, **b**, or **c**) would you expect for a nonpolar covalent bond?
a. from 0.0 to 0.4 **b.** from 0.5 to 1.8 **c.** from 1.9 to 3.3

6.66 Which electronegativity difference (**a**, **b**, or **c**) would you expect for a polar covalent bond?
a. from 0.0 to 0.4 **b.** from 0.5 to 1.8 **c.** from 1.9 to 3.3

6.67 Predict whether each of the following bonds is nonpolar covalent, polar covalent, or ionic:
a. Si and Br **b.** Li and F **c.** Br and F
d. I and I **e.** N and P **f.** C and P

6.68 Predict whether each of the following bonds is nonpolar covalent, polar covalent, or ionic:
a. Si and O **b.** K and Cl **c.** S and F
d. P and Br **e.** Li and O **f.** N and S

6.69 For each of the following bonds, indicate the positive end with δ^+ and the negative end with δ^-. Draw an arrow to show the dipole for each.
a. N and F **b.** Si and Br **c.** C and O
d. P and Br **e.** N and P

6.70 For each of the following bonds, indicate the positive end with δ^+ and the negative end with δ^-. Draw an arrow to show the dipole for each.
a. P and Cl **b.** Se and F **c.** Br and F
d. N and H **e.** B and Cl

6.8 Shapes of Molecules

LEARNING GOAL Predict the three-dimensional structure of a molecule.

Using the Lewis structures, we can predict the three-dimensional shapes of many molecules. The shape is important in our understanding of how molecules interact with enzymes or certain antibiotics or produce our sense of taste and smell. The three-dimensional shape of a molecule is determined by drawing its Lewis structure and counting the number of electron groups around the central atom. Each of the following is counted as *one* electron group: a lone pair, a single bond, a double bond, or a triple bond. In the **valence shell electron-pair repulsion (VSEPR) theory**, the electron groups are arranged as far apart as possible around the central atom to minimize the repulsion between their negative charges. Once we have counted the number of electron groups surrounding the central atom, we can determine its specific shape from the number of atoms bonded to the central atom.

Central Atoms with Two Electron Groups

CORE CHEMISTRY SKILL
Predicting Shape

In the Lewis structure for CO_2, there are two electron groups (two double bonds) attached to the central atom. According to VSEPR theory, minimal repulsion occurs when two electron groups are on opposite sides of the central C atom. This gives the CO_2 molecule a *linear* electron-group geometry and a shape that is **linear** with a bond angle of 180°.

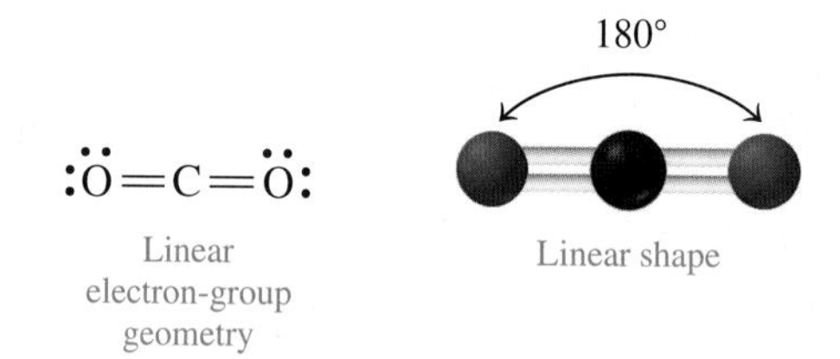

Central Atoms with Three Electron Groups

In the Lewis structure for formaldehyde, H_2CO, the central atom C is attached to two H atoms by single bonds and to the O atom by a double bond. Minimal repulsion occurs when three electron groups are as far apart as possible around the central C atom, which gives

:O: ‖ H—C—H

Lewis structure

:O: ‖ C / \ H H

Trigonal planar electron-group geometry

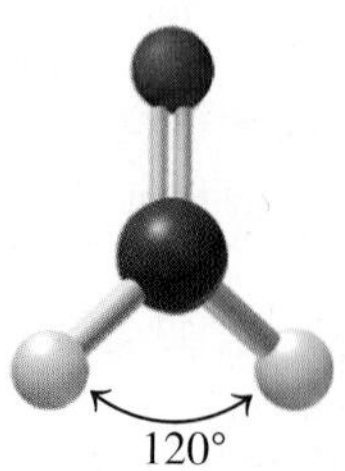

Trigonal planar shape

120° bond angles. This type of electron-group geometry is *trigonal planar* and gives a shape for H_2CO called **trigonal planar**. When the central atom has the same number of electron groups as bonded atoms, the shape and the electron-group geometry have the same name.

In the Lewis structure for SO_2, there are also three electron groups around the central S atom: a single bond to an O atom, a double bond to another O atom, and a lone pair of electrons. As in H_2CO, three electron groups have minimal repulsion when they form trigonal planar electron-group geometry. However, in SO_2 one of the electron groups is a lone pair of electrons. Therefore, the shape of the SO_2 molecule is determined by the two O atoms bonded to the central S atom, which gives the SO_2 molecule a shape that is **bent** with a bond angle of 120°. When the central atom has more electron groups than bonded atoms, the shape and the electron-group geometry have different names.

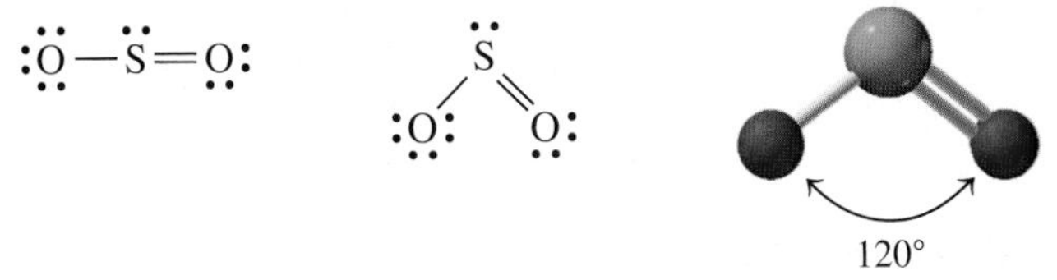

Lewis structure | Trigonal planar electron-group geometry | 120° Bent shape

Central Atom with Four Electron Groups

In a molecule of methane, CH_4, the central C atom is bonded to four H atoms. From the Lewis structure, you may think that CH_4 is planar with 90° bond angles. However, the best geometry for minimal repulsion is *tetrahedral*, giving bond angles of 109°. When there are four atoms attached to four electron groups, the shape of the molecule is **tetrahedral**.

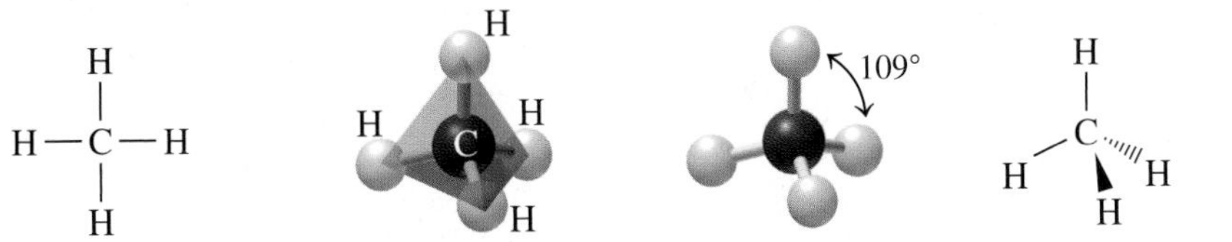

Lewis structure | Tetrahedral electron-group geometry | 109° Tetrahedral shape | Tetrahedral wedge–dash notation

A way to represent the three-dimensional structure of methane is to use the *wedge–dash notation*. In this representation, the two bonds connecting carbon to hydrogen by solid lines are in the plane of the paper. The wedge represents a carbon-to-hydrogen bond coming out of the page toward us, whereas the dash represents a carbon-to-hydrogen bond going into the page away from us.

Now we can look at molecules that have four electron groups, of which one or more are lone pairs of electrons. Then the central atom is attached to only two or three atoms. For example, in the Lewis structure for ammonia, NH_3, four electron groups have a tetrahedral electron-group geometry. However, in NH_3 one of the electron groups is a lone pair of electrons. Therefore, the shape of NH_3 is determined by the three H atoms bonded to the central N atom. Therefore, the shape of the NH_3 molecule is **trigonal pyramidal**, with a bond angle of 109°. The wedge–dash notation can also represent this three-dimensional structure of ammonia with one N—H bond in the plane, one N—H bond coming toward us, and one N—H bond going away from us.

ENGAGE

If the four electron groups in a PH_3 molecule have a tetrahedral geometry, why does a PH_3 molecule have a trigonal pyramidal shape and not a tetrahedral shape?

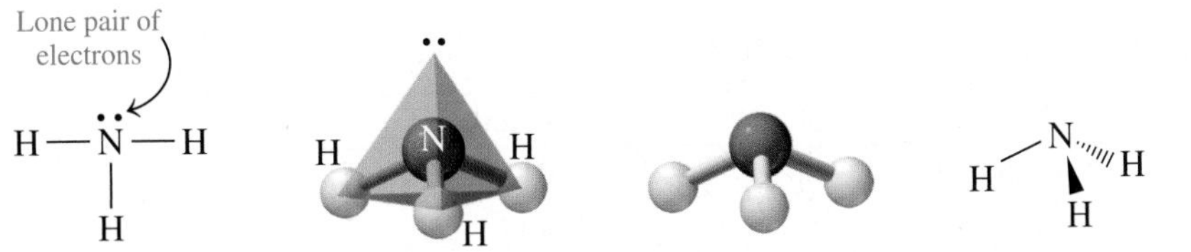

Lone pair of electrons

Lewis structure | Tetrahedral electron-group geometry | Trigonal pyramidal shape | Trigonal pyramidal wedge–dash notation

In the Lewis structure for water, H_2O, there are also four electron groups, which have minimal repulsion when the electron-group geometry is tetrahedral. However, in H_2O, two of the electron groups are lone pairs of electrons. Because the shape of H_2O is determined by the two H atoms bonded to the central O atom, the shape of the H_2O molecule is **bent** with a bond angle of 109°. **TABLE 6.16** gives the shapes of molecules with two, three, or four bonded atoms.

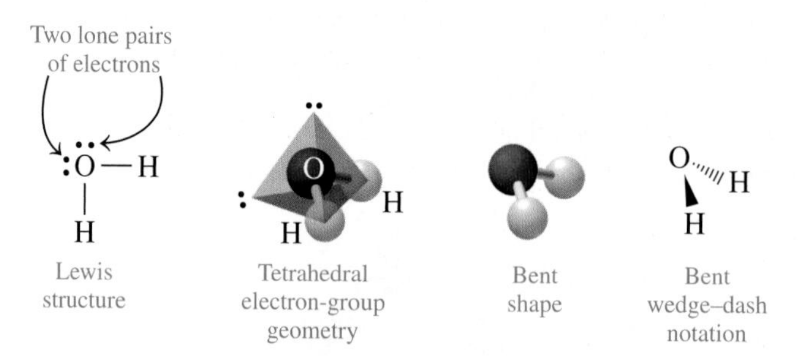

TABLE 6.16 Molecular Shapes for a Central Atom with Two, Three, and Four Bonded Atoms

Electron Groups	Electron-Group Geometry	Bonded Atoms	Lone Pairs	Bond Angle*	Molecular Shape	Example	Three-Dimensional Model
2	Linear	2	0	180°	Linear	CO_2	
3	Trigonal planar	3	0	120°	Trigonal planar	H_2CO	
3	Trigonal planar	2	1	120°	Bent	SO_2	
4	Tetrahedral	4	0	109°	Tetrahedral	CH_4	
4	Tetrahedral	3	1	109°	Trigonal pyramidal	NH_3	
4	Tetrahedral	2	2	109°	Bent	H_2O	

*The bond angles in actual molecules may vary slightly.

SAMPLE PROBLEM 6.15 Predicting Shapes of Molecules

TRY IT FIRST

Use VSEPR theory to predict the shape of the molecule $SiCl_4$.

SOLUTION GUIDE

	Given	Need	Connect
ANALYZE THE PROBLEM	$SiCl_4$	shape	Lewis structure, electron groups, bonded atoms

STEP 1 Draw the Lewis structure.

Name of Element	Silicon		Chlorine
Symbol of Element	Si		Cl
Atoms of Element	1		4
Valence Electrons	$4\,e^-$		$7\,e^-$
Total Electrons	$1(4\,e^-)$	+	$4(7\,e^-) = 32\,e^-$

Using 32 e^-, we draw the bonds and lone pairs for the Lewis structure of $SiCl_4$.

```
      ..
     :Cl:
  ..  ..  ..
 :Cl:Si:Cl:
  ..  ..  ..
     :Cl:
      ..
```

STEP 2 **Arrange the electron groups around the central atom to minimize repulsion.** Each single bond around silicon counts as one electron group. Four electron groups arranged around a central atom have a tetrahedral geometry.

STEP 3 **Use the atoms bonded to the central atom to determine the shape.** Because the central Si atom is bonded to four atoms, the $SiCl_4$ molecule has a tetrahedral shape.

STUDY CHECK 6.15

Use VSEPR theory to predict the shape of SCl_2.

ANSWER

The central atom S has four electron groups: two bonded atoms and two lone pairs of electrons. The shape of SCl_2 is bent, 109°.

TEST

Try Practice Problems 6.71 to 6.78

PRACTICE PROBLEMS

6.8 Shapes of Molecules

LEARNING GOAL Predict the three-dimensional structure of a molecule.

6.71 Choose the shape (**1** to **6**) that matches each of the following descriptions (**a** to **c**):

1. linear **2.** bent (109°)
3. trigonal planar **4.** bent (120°)
5. trigonal pyramidal **6.** tetrahedral

a. a molecule with a central atom that has four electron groups and four bonded atoms
b. a molecule with a central atom that has four electron groups and three bonded atoms
c. a molecule with a central atom that has three electron groups and three bonded atoms

6.72 Choose the shape (**1** to **6**) that matches each of the following descriptions (**a** to **c**):

1. linear **2.** bent (109°)
3. trigonal planar **4.** bent (120°)
5. trigonal pyramidal **6.** tetrahedral

a. a molecule with a central atom that has four electron groups and two bonded atoms
b. a molecule with a central atom that has two electron groups and two bonded atoms
c. a molecule with a central atom that has three electron groups and two bonded atoms

6.73 Complete each of the following statements for a molecule of SeO_3:

a. There are ________ electron groups around the central Se atom.
b. The electron-group geometry is ________.
c. The number of atoms attached to the central Se atom is ________.
d. The shape of the molecule is ________.

6.74 Complete each of the following statements for a molecule of H_2S:

a. There are ________ electron groups around the central S atom.
b. The electron-group geometry is ________.
c. The number of atoms attached to the central S atom is ________.
d. The shape of the molecule is ________.

6.75 Compare the Lewis structures of CF_4 and NF_3. Why do these molecules have different shapes?

6.76 Compare the Lewis structures of CH_4 and H_2O. Why do these molecules have similar bond angles but different molecular shapes?

6.77 Use VSEPR theory to predict the shape of each of the following:

a. GaH_3 **b.** OF_2
c. HCN **d.** CCl_4

6.78 Use VSEPR theory to predict the shape of each of the following:

a. CI_4 **b.** NCl_3
c. $SeBr_2$ **d.** CS_2

6.9 Polarity of Molecules and Intermolecular Forces

LEARNING GOAL Use the three-dimensional structure of a molecule to classify it as polar or nonpolar. Describe the intermolecular forces between ions, polar covalent molecules, and nonpolar covalent molecules.

CORE CHEMISTRY SKILL

Identifying Polarity of Molecules and Intermolecular Forces

We have seen that covalent bonds in molecules can be polar or nonpolar. Now we look at how the bonds in a molecule and its shape determine whether that molecule is classified as polar or nonpolar.

Nonpolar Molecules

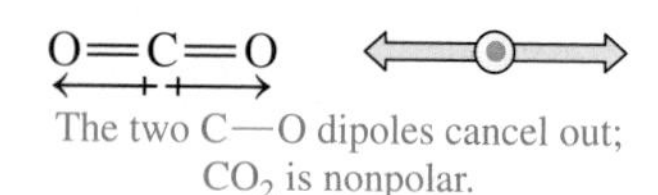

The two C—O dipoles cancel out; CO_2 is nonpolar.

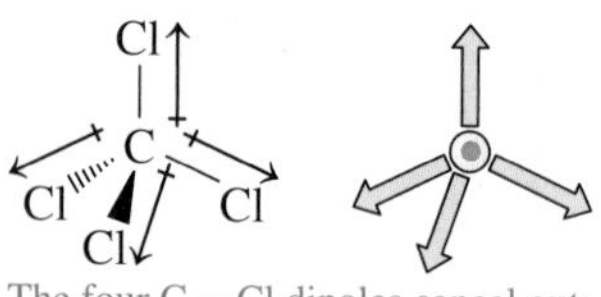

The four C—Cl dipoles cancel out; CCl_4 is nonpolar.

In a **nonpolar molecule**, all the bonds are nonpolar or the polar bonds cancel each other out. Molecules such as H_2, Cl_2, and CH_4 are nonpolar because they contain only nonpolar covalent bonds. A *nonpolar molecule* also occurs when polar bonds (dipoles) cancel each other because they are in a symmetrical arrangement. For example, CO_2, a linear molecule, contains two equal polar covalent bonds whose dipoles point in opposite directions. As a result, the dipoles cancel out, making a CO_2 molecule nonpolar.

Another example of a nonpolar molecule is the CCl_4 molecule, which has four polar bonds symmetrically arranged around the central C atom. Each of the C—Cl bonds has the same polarity, but because they have a tetrahedral arrangement, their opposing dipoles cancel out. As a result, a molecule of CCl_4 is nonpolar.

Polar Molecules

A single dipole does not cancel out; HCl is polar.

In a **polar molecule**, one end of the molecule is more negatively charged than the other end. Polarity in a molecule occurs when the dipoles from the individual polar bonds do not cancel each other. For example, HCl is a polar molecule because it has one covalent bond that is polar.

In molecules with two or more electron groups, the shape, such as bent or trigonal pyramidal, determines whether the dipoles cancel. For example, we have seen that H_2O has a bent shape. Thus, a water molecule is polar because the individual dipoles do not cancel.

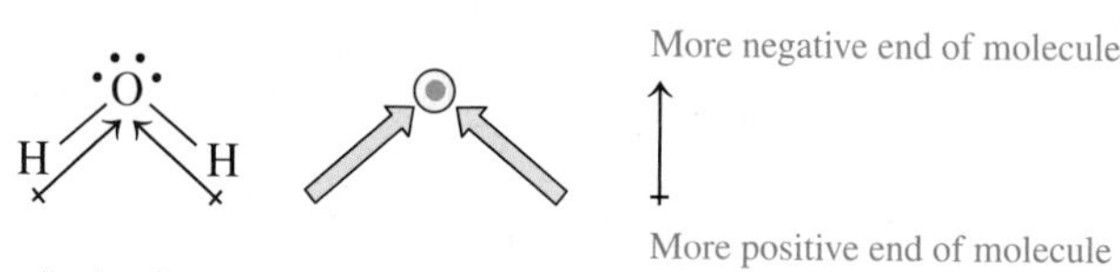

The dipoles do not cancel out; H_2O is polar.

The NH_3 molecule has a tetrahedral electron-group geometry with three bonded atoms, which gives it a trigonal pyramidal shape. Thus, the NH_3 molecule is polar because the individual N—H dipoles do not cancel.

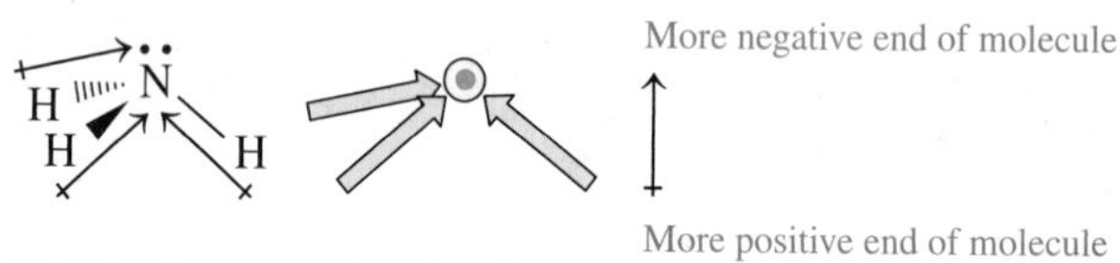

The dipoles do not cancel out; NH_3 is polar.

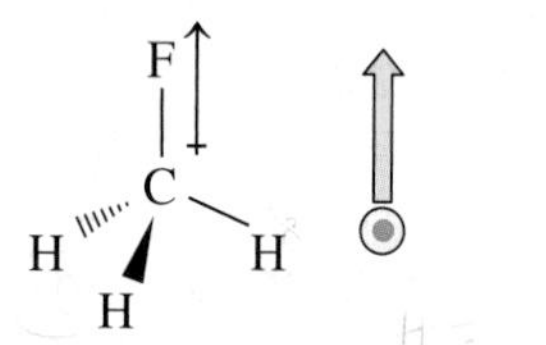

The dipole does not cancel out; CH_3F is polar.

In the molecule CH_3F, the C—F bond is polar covalent, but the three C—H bonds are nonpolar covalent. Because there is only one dipole in CH_3F, CH_3F is a polar molecule.

SAMPLE PROBLEM 6.16 Polarity of Molecules

TRY IT FIRST

Determine whether a molecule of OF_2 is polar or nonpolar.

SOLUTION GUIDE

ANALYZE THE PROBLEM	Given	Need	Connect
	OF_2	polarity	Lewis structure, bond polarity

STEP 1 **Determine if the bonds are polar covalent or nonpolar covalent.** From Figure 6.6, F and O have an electronegativity difference of 0.5 (4.0 − 3.5 = 0.5), which makes each of the O—F bonds polar covalent.

STEP 2 **If the bonds are polar covalent, draw the Lewis structure and determine if the dipoles cancel.** The Lewis structure for OF_2 has four electron groups and two bonded atoms. The molecule has a bent shape in which the dipoles of the O—F bonds do not cancel. The OF_2 molecule would be polar.

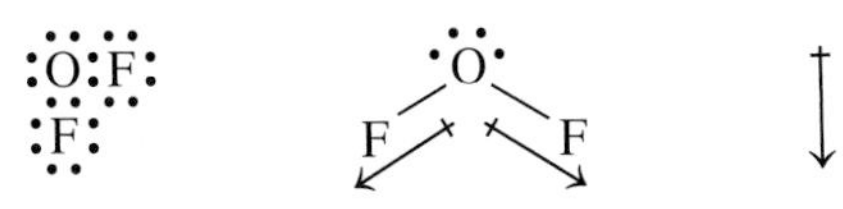

The dipoles do not cancel out; OF_2 is polar.

STUDY CHECK 6.16

Would a PCl_3 molecule be polar or nonpolar?

ANSWER

polar

TEST

Try Practice Problems 6.79 to 6.84

Intermolecular Forces

In gases, the interactions between particles are minimal, which allows gas molecules to move far apart from each other. In solids and liquids, there are sufficient interactions between the particles to hold them close together. Such differences in properties are explained by looking at the various kinds of *intermolecular* forces between particles including *dipole–dipole attractions*, *hydrogen bonding*, *dispersion forces*, and *ionic bonds*.

Ionic Bonds

Ionic bonds are the strongest of the attractive forces found in compounds. Thus, most ionic compounds are solids at room temperature. The ionic compound sodium chloride, NaCl, melts at 801 °C. Large amounts of energy are needed to overcome the strong attractive forces between positive and negative ions and change solid sodium chloride to a liquid.

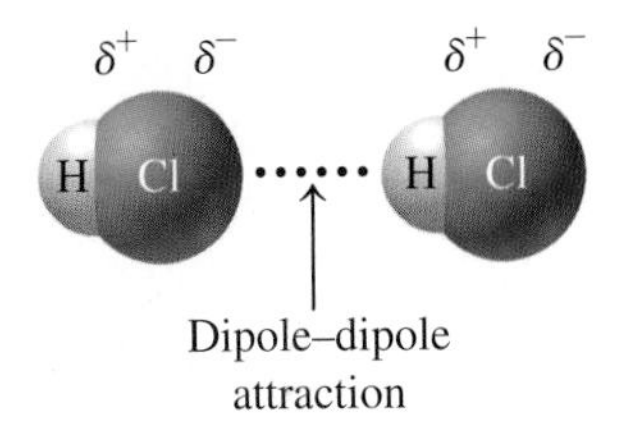

Dipole–Dipole Attractions

All polar molecules are attracted to each other by **dipole–dipole attractions**. Because polar molecules have dipoles, the positively charged end of the dipole in one molecule is attracted to the negatively charged end of the dipole in another molecule.

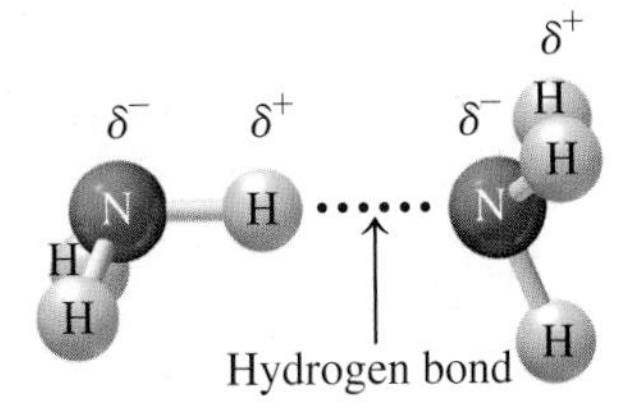

Hydrogen Bonds

Polar molecules containing hydrogen atoms bonded to highly electronegative atoms of nitrogen, oxygen, or fluorine form especially strong dipole–dipole attractions. This type of attraction, called a **hydrogen bond**, occurs between the partially positive hydrogen atom in one molecule and the partially negative nitrogen, oxygen, or fluorine atom in another molecule. Hydrogen bonds are the strongest type of attractive forces between polar covalent molecules. They are attractions between polar molecules and not bonds that hold molecules together.

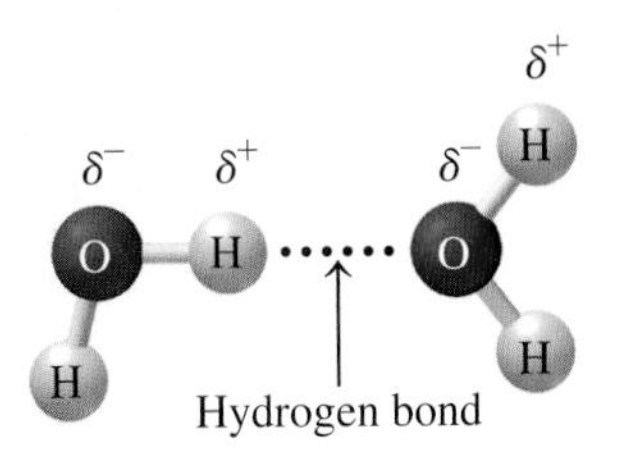

Dispersion Forces

Very weak attractions called **dispersion forces** are the only intermolecular forces that occur between nonpolar molecules. Usually, the electrons in a nonpolar covalent molecule are distributed symmetrically. However, the movement of the electrons may place more of them in one part of the molecule than another, which forms a *temporary dipole*. These momentary dipoles align the molecules so that the positive end of one molecule is attracted to the negative end of another molecule. Although dispersion forces are very weak, they make it possible for nonpolar molecules to form liquids and solids.

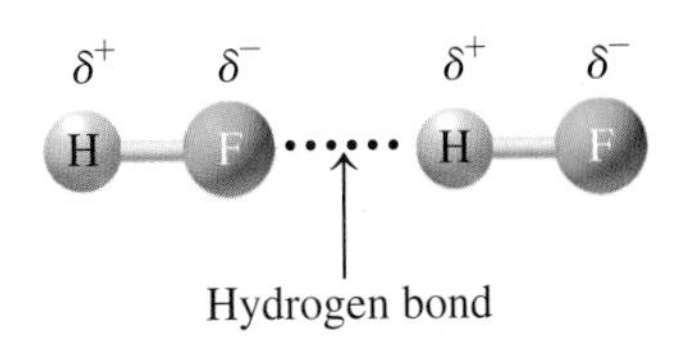

TABLE 6.17 Melting Points of Selected Substances

Substance	Melting Point (°C)
Ionic Bonds	
MgF_2	1248
NaCl	801
Hydrogen Bonds	
H_2O	0
NH_3	−78
Dipole–Dipole Attractions	
HI	−51
HBr	−89
HCl	−115
Dispersion Forces	
Br_2	−7
Cl_2	−101
F_2	−220

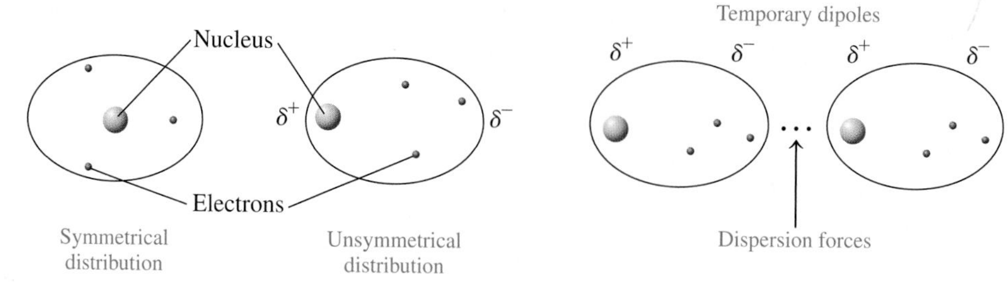

Nonpolar covalent molecules have weak attractions when they form temporary dipoles.

Attractive Forces and Melting Points

The melting point of a substance is related to the strength of the attractive forces between its particles. A compound with weak attractive forces, such as dispersion forces, has a low melting point because only a small amount of energy is needed to separate the molecules and form a liquid. A compound with dipole–dipole attractions requires more energy to break the attractive forces between the molecules. A compound that can form hydrogen bonds requires even more energy to overcome the attractive forces that exist between its molecules. Larger amounts of energy are needed to overcome the strong attractive forces between positive and negative ions and to melt an ionic solid. For example, the ionic solid MgF_2 melts at 1248 °C. **TABLE 6.17** compares the melting points of some substances with various kinds of attractive forces. The various types of attractions between particles in solids and liquids are summarized in **TABLE 6.18**.

TABLE 6.18 Comparison of Bonding and Attractive Forces

Type of Force	Particle Arrangement	Example	Strength
Between Atoms or Ions			Strong
Ionic bond	+ − + / − + −	Na^+Cl^-	
Covalent bond (X = nonmetal)	X:X	Cl—Cl	
Between Molecules			
Hydrogen bond (X = N, O, or F)	δ^+ δ^- H X ··· δ^+ δ^- H X	$H^{\delta^+}—F^{\delta^-} \cdots H^{\delta^+}—F^{\delta^-}$	
Dipole–dipole attractions (X and Y = nonmetals)	δ^+ δ^- Y X ··· δ^+ δ^- Y X	$H^{\delta^+}—Cl^{\delta^-} \cdots H^{\delta^+}—Cl^{\delta^-}$	
Dispersion forces (temporary shift of electrons in nonpolar bonds)	δ^+ δ^- X:X ··· δ^+ δ^- X:X (temporary dipoles)	$F^{\delta^+}—F^{\delta^-} \cdots F^{\delta^+}—F^{\delta^-}$	Weak

SAMPLE PROBLEM 6.17 Intermolecular Forces between Particles

TRY IT FIRST

Indicate the major type of intermolecular forces—dipole–dipole attractions, hydrogen bonding, or dispersion forces—expected of each of the following:

a. HF **b.** Br_2 **c.** PCl_3

SOLUTION

a. HF is a polar molecule that has a highly electronegative fluorine atom bonded to hydrogen. Hydrogen bonding is the major type of intermolecular force between HF molecules.

b. Br_2 is nonpolar; dispersion forces are the major type of intermolecular forces between Br_2 molecules.

c. PCl_3 is a polar molecule. Dipole–dipole attractions are the major type of intermolecular forces between PCl_3 molecules.

STUDY CHECK 6.17

Indicate the major type of intermolecular forces in H_2S and H_2O.

ANSWER

The intermolecular forces between H_2S molecules are dipole–dipole attractions, whereas the intermolecular forces between H_2O molecules are hydrogen bonds.

TEST

Try Practice Problems 6.85 to 6.88

PRACTICE PROBLEMS

6.9 Polarity of Molecules and Intermolecular Forces

LEARNING GOAL Use the three-dimensional structure of a molecule to classify it as polar or nonpolar. Describe the intermolecular forces between ions, polar covalent molecules, and nonpolar covalent molecules.

6.79 Why is F_2 a nonpolar molecule, but HF is a polar molecule?

6.80 Why is CCl_4 a nonpolar molecule, but PCl_3 is a polar molecule?

6.81 Identify each of the following molecules as polar or nonpolar:
a. CS_2 **b.** NF_3 **c.** CHF_3 **d.** SO_3

6.82 Identify each of the following molecules as polar or nonpolar:
a. SeF_2 **b.** PBr_3 **c.** SiF_4 **d.** SO_2

6.83 The molecule CO_2 is nonpolar, but CO is a polar molecule. Explain.

6.84 The molecules CH_4 and CH_3Cl both have tetrahedral shapes. Why is CH_4 nonpolar whereas CH_3Cl is polar?

6.85 Identify the major type of intermolecular forces between the particles of each of the following:
a. BrF **b.** KCl **c.** NF_3 **d.** Cl_2

6.86 Identify the major type of intermolecular forces between the particles of each of the following:
a. HCl **b.** MgF_2 **c.** PBr_3 **d.** NH_3

6.87 Identify the strongest intermolecular forces between the particles of each of the following:
a. CH_3OH **b.** CO **c.** CF_4 **d.** CH_3CH_3

6.88 Identify the strongest intermolecular forces between the particles of each of the following:
a. O_2 **b.** SiH_4 **c.** CH_3Cl **d.** H_2O_2

CLINICAL UPDATE Compounds at the Pharmacy

A few days ago, Richard went back to the pharmacy to pick up aspirin, $C_9H_8O_4$, and acetaminophen, $C_8H_9NO_2$. He also wanted to talk to Sarah about a way to treat his sore toe. Sarah recommended soaking his foot in a solution of Epsom salts, which is magnesium sulfate. Richard also asked Sarah to recommend an antacid for his upset stomach and an iron supplement. Sarah suggested an antacid that contains calcium carbonate and aluminum hydroxide, and iron(II) sulfate as an iron supplement. Richard also picked up toothpaste containing tin(II) fluoride, and carbonated water, which contains carbon dioxide.

Clinical Applications

6.89 Write the chemical formula for each of the following:
a. magnesium sulfate
b. tin(II) fluoride
c. aluminum hydroxide

6.90 Write the chemical formula for each of the following:
a. calcium carbonate
b. carbon dioxide
c. iron(II) sulfate

6.91 Identify each of the compounds in problem 6.89 as ionic or molecular.

6.92 Identify each of the compounds in problem 6.90 as ionic or molecular.

CONCEPT MAP

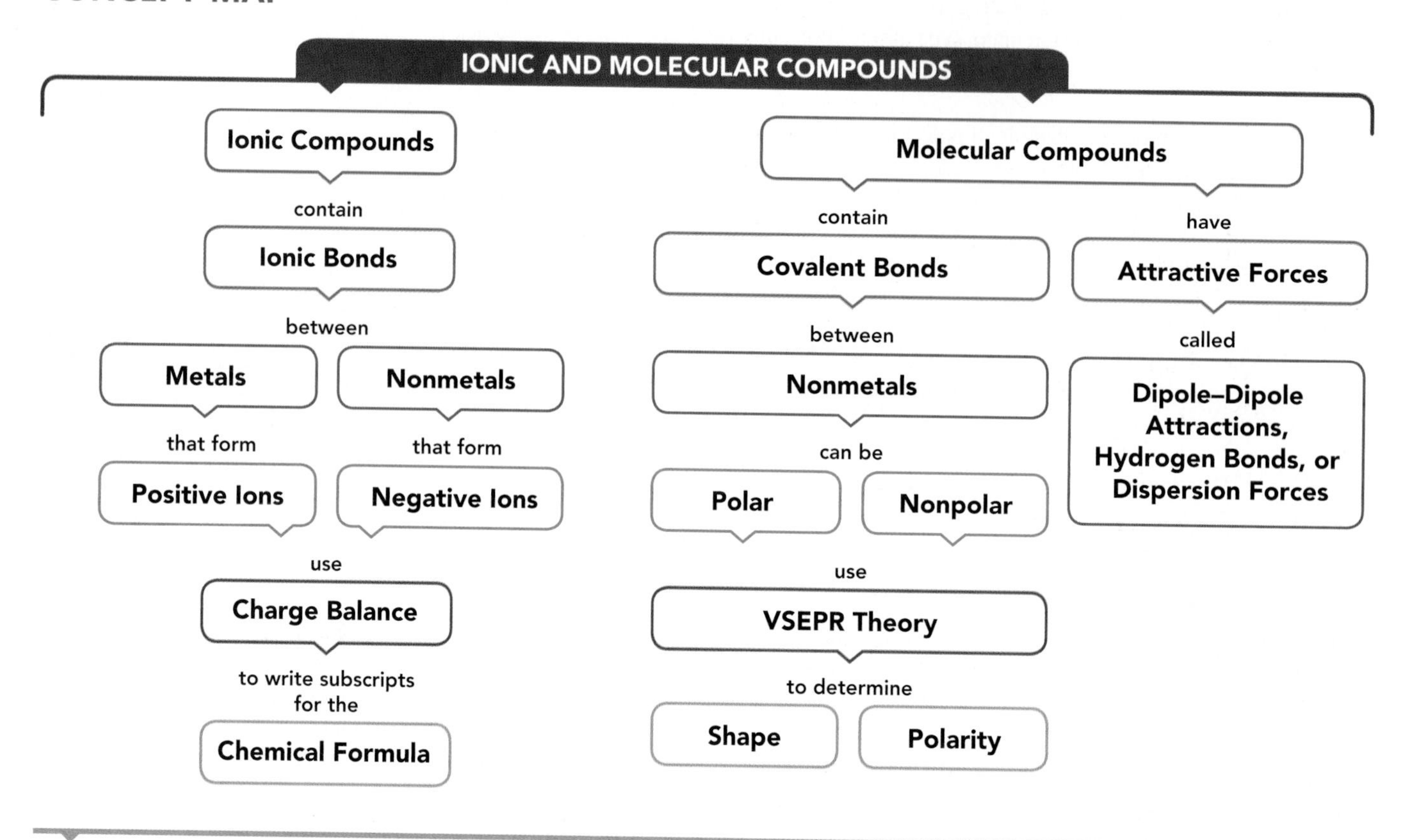

CHAPTER REVIEW

6.1 Ions: Transfer of Electrons

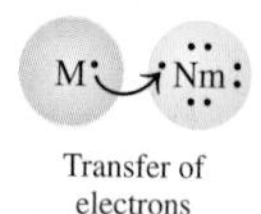

Transfer of electrons

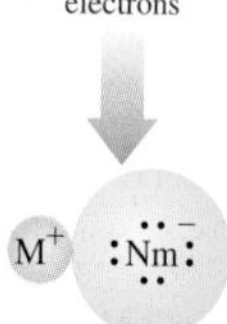

Ionic bond

LEARNING GOAL Write the symbols for the simple ions of the representative elements.

- The stability of the noble gases is associated with a stable electron arrangement in the outermost energy level.
- With the exception of helium, which has two electrons, noble gases have eight valence electrons, which is an octet.
- Atoms of elements in Groups 1A to 7A (1, 2, 13 to 17) achieve stability by losing, gaining, or sharing their valence electrons in the formation of compounds.
- Metals of the representative elements lose valence electrons to form positively charged ions (cations): Group 1A (1), 1+, Group 2A (2), 2+, and Group 3A (13), 3+.
- When reacting with metals, nonmetals gain electrons to form octets and form negatively charged ions (anions): Groups 5A (15), 3−, 6A (16), 2−, and 7A (17), 1−.

6.2 Ionic Compounds

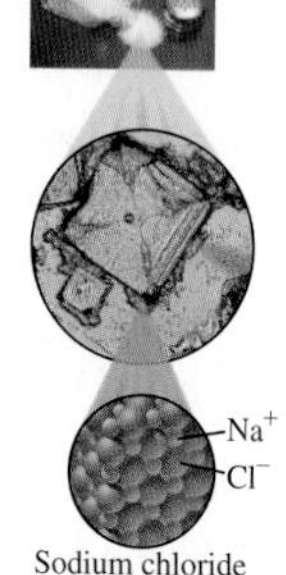

Sodium chloride

LEARNING GOAL Using charge balance, write the correct formula for an ionic compound.

- The total positive and negative ionic charge is balanced in the formula of an ionic compound.
- Charge balance in a formula is achieved by using subscripts after each symbol so that the overall charge is zero.

6.3 Naming and Writing Ionic Formulas

LEARNING GOAL Given the formula of an ionic compound, write the correct name; given the name of an ionic compound, write the correct formula.

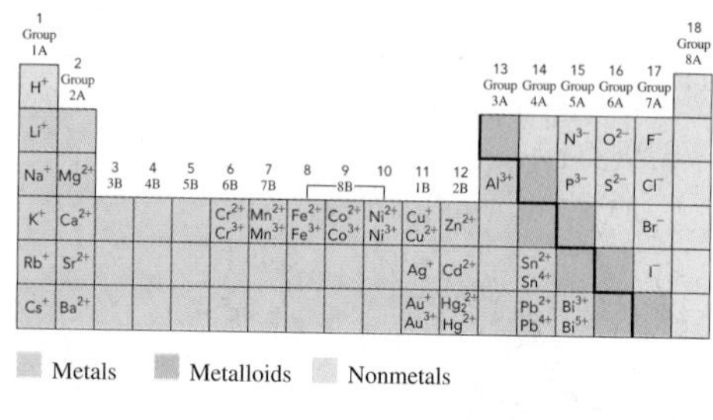

- In naming ionic compounds, the positive ion is given first followed by the name of the negative ion.
- The names of ionic compounds containing two elements end with *ide*.
- Except for Ag, Cd, and Zn, transition elements form cations with two or more ionic charges.
- The charge of the cation is determined from the total negative charge in the formula and included as a Roman numeral immediately following the name of the metal that has a variable charge.

6.4 Polyatomic Ions

LEARNING GOAL Write the name and formula for an ionic compound containing a polyatomic ion.

- A polyatomic ion is a covalently bonded group of atoms with an electrical charge; for example, the carbonate ion has the formula CO_3^{2-}.

- Most polyatomic ions have names that end with *ate* or *ite*.
- Most polyatomic ions contain a nonmetal and one or more oxygen atoms.
- The ammonium ion, NH_4^+, is a positive polyatomic ion.
- When more than one polyatomic ion is used for charge balance, parentheses enclose the formula of the polyatomic ion.

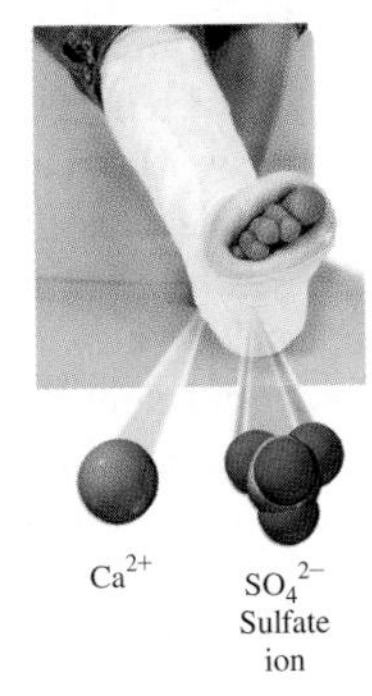

6.5 Molecular Compounds: Sharing Electrons

LEARNING GOAL Given the formula of a molecular compound, write its correct name; given the name of a molecular compound, write its formula.

1	mono
2	di
3	tri
4	tetra
5	penta

- In a covalent bond, atoms of nonmetals share valence electrons such that each atom has a stable electron arrangement.
- The first nonmetal in a molecular compound uses its element name; the second nonmetal uses the first syllable of its element name followed by *ide*.
- The name of a molecular compound with two different atoms uses prefixes to indicate the subscripts in the formula.

6.6 Lewis Structures for Molecules

LEARNING GOAL Draw the Lewis structures for molecular compounds with single and multiple bonds.

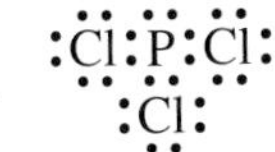

- The total number of valence electrons is determined for all the atoms in the molecule.
- In the Lewis structure, a bonding pair of electrons is placed between the central atom and each of the attached atoms.
- Any remaining valence electrons are used as lone pairs to complete the octets of the surrounding atoms and then the central atom.
- If octets are not completed, one or more lone pairs of electrons are placed as bonding pairs forming double or triple bonds.

6.7 Electronegativity and Bond Polarity

LEARNING GOAL Use electronegativity to determine the polarity of a bond.

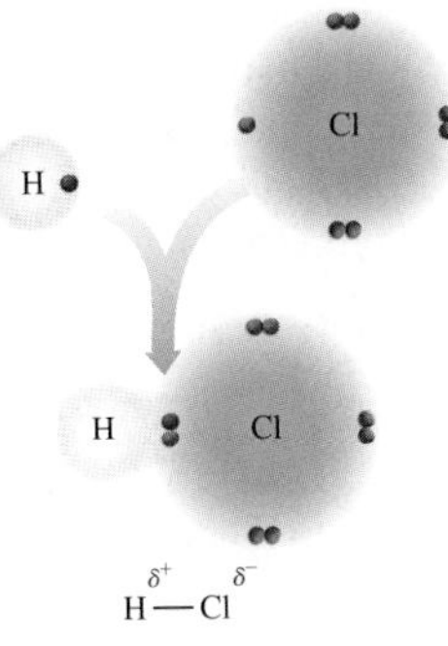

Unequal sharing of electrons in a polar covalent bond

- Electronegativity is the ability of an atom to attract the electrons it shares with another atom. In general, the electronegativities of metals are low, whereas nonmetals have high electronegativities.
- In a nonpolar covalent bond, atoms share electrons equally.
- In a polar covalent bond, the electrons are unequally shared because they are attracted to the more electronegative atom.
- The atom in a polar bond with the lower electronegativity is partially positive (δ^+), and the atom with the higher electronegativity is partially negative (δ^-).
- Atoms that form ionic bonds have large differences in electronegativities.

6.8 Shapes of Molecules

LEARNING GOAL Predict the three-dimensional structure of a molecule.

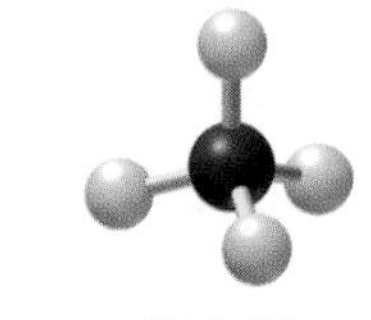
Tetrahedral shape

- The shape of a molecule is determined from the Lewis structure, the electron-group geometry, and the number of bonded atoms.
- The electron-group geometry around a central atom with two electron groups is linear; with three electron groups, the geometry is trigonal planar; and with four electron groups, the geometry is tetrahedral.
- When all the electron groups are bonded to atoms, the shape has the same name as the electron arrangement.
- A central atom with three electron groups and two bonded atoms has a bent shape, 120°.
- A central atom with four electron groups and three bonded atoms has a trigonal pyramidal shape.
- A central atom with four electron groups and two bonded atoms has a bent shape, 109°.

6.9 Polarity of Molecules and Intermolecular Forces

LEARNING GOAL Use the three-dimensional structure of a molecule to classify it as polar or nonpolar. Describe the intermolecular forces between ions, polar covalent molecules, and nonpolar covalent molecules.

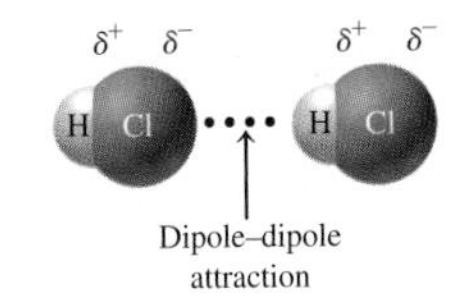

- Nonpolar molecules contain nonpolar covalent bonds or have an arrangement of bonded atoms that causes the dipoles to cancel out.
- In polar molecules, the dipoles do not cancel.
- In ionic solids, oppositely charged ions are held in a rigid structure by ionic bonds.
- Intermolecular forces called dipole–dipole attractions and hydrogen bonds hold the solid and liquid states of polar molecular compounds together.
- Nonpolar compounds form solids and liquids by weak attractions between temporary dipoles called dispersion forces.

KEY TERMS

anion A negatively charged ion such as Cl^-, O^{2-}, or SO_4^{2-}.

bent The shape of a molecule with two bonded atoms and one lone pair or two lone pairs.

cation A positively charged ion such as Na^+, Mg^{2+}, Al^{3+}, or NH_4^+.

chemical formula The group of symbols and subscripts that represents the atoms or ions in a compound.

covalent bond A sharing of valence electrons by atoms.

dipole The separation of positive and negative charge in a polar bond indicated by an arrow that is drawn from the more positive atom to the more negative atom.

dipole–dipole attractions Attractive forces between oppositely charged ends of polar molecules.

dispersion forces Weak dipole bonding that results from a momentary polarization of nonpolar molecules.
double bond A sharing of two pairs of electrons by two atoms.
electronegativity The relative ability of an element to attract electrons in a bond.
hydrogen bond The attraction between a partially positive H atom and a strongly electronegative atom of N, O, or F.
ion An atom or group of atoms having an electrical charge because of a loss or gain of electrons.
ionic charge The difference between the number of protons (positive) and the number of electrons (negative) written in the upper right corner of the symbol for the element or polyatomic ion.
ionic compound A compound of positive and negative ions held together by ionic bonds.
Lewis structure A structure drawn in which the valence electrons of all the atoms are arranged to give octets except two electrons for hydrogen.
linear The shape of a molecule that has two bonded atoms and no lone pairs.
molecular compound A combination of atoms in which stable electron arrangements are attained by sharing electrons.
molecule The smallest unit of two or more atoms held together by covalent bonds.
nonpolar covalent bond A covalent bond in which the electrons are shared equally between atoms.
nonpolar molecule A molecule that has only nonpolar bonds or in which the bond dipoles cancel.
octet A set of eight valence electrons.
octet rule Elements in Groups 1A to 7A (1, 2, 13 to 17) react with other elements by forming ionic or covalent bonds to produce a stable electron arrangement, usually eight electrons in the outer shell.
polar covalent bond A covalent bond in which the electrons are shared unequally between atoms.
polar molecule A molecule containing bond dipoles that do not cancel.
polarity A measure of the unequal sharing of electrons, indicated by the difference in electronegativities.
polyatomic ion A group of covalently bonded nonmetal atoms that has an overall electrical charge.
tetrahedral The shape of a molecule with four bonded atoms.
trigonal planar The shape of a molecule with three bonded atoms and no lone pairs.
trigonal pyramidal The shape of a molecule that has three bonded atoms and one lone pair.
triple bond A sharing of three pairs of electrons by two atoms.
valence shell electron-pair repulsion (VSEPR) theory A theory that predicts the shape of a molecule by moving the electron groups on a central atom as far apart as possible to minimize the mutual repulsion of the electrons.

CORE CHEMISTRY SKILLS

The chapter Section containing each Core Chemistry Skill is shown in parentheses at the end of each heading.

Writing Positive and Negative Ions (6.1)

- In the formation of an ionic bond, atoms of a metal lose and atoms of a nonmetal gain valence electrons to acquire a stable electron arrangement, usually eight valence electrons.
- This tendency of atoms to attain a stable electron arrangement is known as the octet rule.

Example: State the number of electrons lost or gained by atoms and the ion formed for each of the following to obtain a stable electron arrangement:

a. Br **b.** Ca **c.** S

Answer: **a.** Br atoms gain one electron to achieve a stable electron arrangement, Br^-.

b. Ca atoms lose two electrons to achieve a stable electron arrangement, Ca^{2+}.

c. S atoms gain two electrons to achieve a stable electron arrangement, S^{2-}.

Writing Ionic Formulas (6.2)

- The chemical formula of a compound represents the lowest whole-number ratio of the atoms or ions.
- In the chemical formula of an ionic compound, the sum of the positive and negative charges is always zero.
- Thus, in a chemical formula of an ionic compound, the total positive charge is equal to the total negative charge.

Example: Write the formula for magnesium phosphide.

Answer: Magnesium phosphide is an ionic compound that contains the ions Mg^{2+} and P^{3-}.

Using charge balance, we determine the number(s) of each type of ion.

$$3(2+) + 2(3-) = 0$$

$3Mg^{2+}$ and $2P^{3-}$ give the formula Mg_3P_2.

Naming Ionic Compounds (6.3)

- In the name of an ionic compound made up of two elements, the name of the metal ion, which is written first, is the same as its element name.
- For metals that form two or more ions, a Roman numeral that is equal to the ionic charge is placed in parentheses immediately after the name of the metal.
- The name of a nonmetal ion is obtained by using the first syllable of its element name followed by *ide*.

Example: What is the name of PbS?

Answer: This compound contains the S^{2-} ion, which has a 2− charge.

For charge balance, the positive ion must have a charge of 2+.

$$Pb? + (2-) = 0; \quad Pb = 2+$$

Because lead can form two different positive ions, a Roman numeral (II) is used in the name of the compound: lead(II) sulfide.

Writing the Names and Formulas for Molecular Compounds (6.5)

- When naming a molecular compound, the first nonmetal in the formula is named by its element name; the second nonmetal is named using the first syllable of its element name followed by *ide*.
- When a subscript indicates two or more atoms of an element, a prefix is shown in front of its name.

Example: Name the molecular compound BrF_5.

Answer: Two nonmetals share electrons and form a molecular compound. Br (first nonmetal) is bromine; F (second nonmetal) is fluoride. In the name for a molecular compound, prefixes indicate the subscripts in the formulas. The subscript 1 is understood for Br. The subscript 5 for fluoride is written with the prefix *penta*. The name is bromine pentafluoride.

Drawing Lewis Structures (6.6)

- The Lewis structure for a molecule shows the sequence of atoms, the bonding pairs of electrons shared between atoms, and the nonbonding or *lone pairs* of electrons.
- Double or triple bonds result when a second or third electron pair is shared between the same atoms to complete octets.

Example: Draw the Lewis structure for CS_2.

Answer: The central atom is C.

S C S

Determine the total number of valence electrons.

$$2\ S \times 6\ e^- = 12\ e^-$$
$$1\ C \times 4\ e^- = 4\ e^-$$
$$\text{Total} = 16\ e^-$$

Attach each bonded atom to the central atom using a pair of electrons. Two bonding pairs use four electrons.

S:C:S

Place the 12 remaining electrons as lone pairs around the S atoms.

:S̤̈:C:S̤̈:

To complete the octet for C, a lone pair of electrons from each of the S atoms is shared with C, which forms two double bonds.

:S̈::C::S̈: or :S̈=C=S̈:

Using Electronegativity (6.7)

- The electronegativity values indicate the ability of atoms to attract shared electrons.
- Electronegativity values increase going across a period from left to right, and decrease going down a group.
- A nonpolar covalent bond occurs between atoms with identical or very similar electronegativity values such that the electronegativity difference is 0.0 to 0.4.
- A polar covalent bond typically occurs when electrons are shared unequally between atoms with electronegativity differences from 0.5 to 1.8.
- An ionic bond typically occurs when the difference in electronegativity for two atoms is greater than 1.8.

Example: Use electronegativity values to classify each of the following bonds as nonpolar covalent, polar covalent, or ionic:

a. Sr and Cl **b.** C and S **c.** O and Br

Answer:
a. An electronegativity difference of 2.0 (Cl 3.0 − Sr 1.0) makes this an ionic bond.
b. An electronegativity difference of 0.0 (C 2.5 − S 2.5) makes this a nonpolar covalent bond.
c. An electronegativity difference of 0.7 (O 3.5 − Br 2.8) makes this a polar covalent bond.

Predicting Shape (6.8)

- The three-dimensional shape of a molecule is determined by drawing a Lewis structure and identifying the number of electron groups (one or more electron pairs) around the central atom and the number of bonded atoms.
- In the valence shell electron-pair repulsion (VSEPR) theory, the electron groups are arranged as far apart as possible around a central atom to minimize the repulsion.
- A central atom with two electron groups bonded to two atoms is linear. A central atom with three electron groups bonded to three atoms is trigonal planar, and to two atoms is bent (120°). A central atom with four electron groups bonded to four atoms is tetrahedral, to three atoms is trigonal pyramidal, and to two atoms is bent (109°).

Example: Predict the shape of $AsCl_3$.

Answer: From its Lewis structure, we see that $AsCl_3$ has four electron groups with three bonded atoms.

:C̤̈l:Ä̤s:C̤̈l:
:C̤̈l:

The electron-group geometry is tetrahedral, but with the central atom bonded to three Cl atoms, the shape is trigonal pyramidal.

Identifying Polarity of Molecules and Intermolecular Forces (6.9)

- A molecule is nonpolar if all of its bonds are nonpolar or it has polar bonds that cancel out. CCl_4 is a nonpolar molecule that consists of four polar bonds that cancel out.

Cl
C
Cl Cl
Cl

- A molecule is polar if it contains polar bonds that do not cancel out. H_2O is a polar molecule that consists of polar bonds that do not cancel out.

Ö
H H

Example: Predict whether $AsCl_3$ is polar or nonpolar.

Answer: From its Lewis structure, we see that $AsCl_3$ has four electron groups with three bonded atoms.

:C̤̈l:Ä̤s:C̤̈l:
:C̤̈l:

The shape of a molecule of $AsCl_3$ would be trigonal pyramidal with three polar bonds (As—Cl = 3.0 − 2.0 = 1.0) that do not cancel out. Thus, it is a polar molecule.

As
Cl Cl Cl

- Dipole–dipole attractions occur between the dipoles in polar compounds because the positively charged end of one molecule is attracted to the negatively charged end of another molecule.

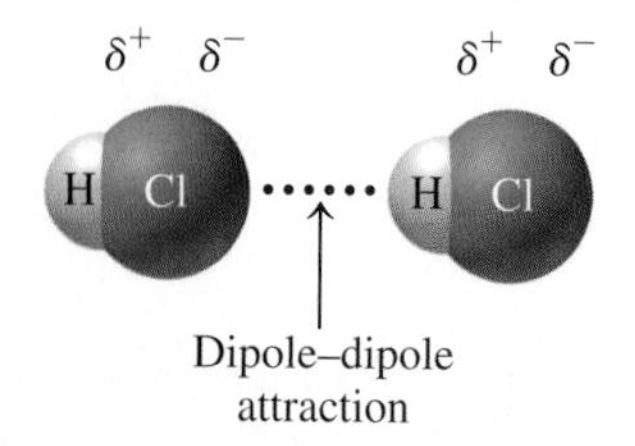

- Strong dipole–dipole attractions called hydrogen bonds occur in compounds in which H is bonded to N, O, or F. The partially positive H atom in one molecule has a strong attraction to the partially negative N, O, or F in another molecule.
- Dispersion forces are very weak attractive forces between nonpolar molecules that occur when *temporary dipoles* form as electrons are unsymmetrically distributed.

Example: Identify the strongest type of intermolecular forces in each of the following:

a. HF **b.** F_2 **c.** NF_3

Answer: **a.** HF molecules, which are polar with H bonded to F, have hydrogen bonding.
b. Nonpolar F_2 molecules have only dispersion forces.
c. NF_3 molecules, which are polar, have dipole–dipole attractions.

UNDERSTANDING THE CONCEPTS

The chapter Sections to review are shown in parentheses at the end of each problem.

6.93 **a.** How does the octet rule explain the formation of a magnesium ion? (6.1)
b. What noble gas has the same electron arrangement as the magnesium ion?
c. Why are Group 1A (1) and Group 2A (2) elements found in many compounds, but not Group 8A (18) elements?

6.94 **a.** How does the octet rule explain the formation of a chloride ion? (6.1)
b. What noble gas has the same electron arrangement as the chloride ion?
c. Why are Group 7A (17) elements found in many compounds, but not Group 8A (18) elements?

6.95 Identify each of the following atoms or ions: (6.1)

A	B	C	D
$18\ e^-$, $15\ p^+$, $16\ n$	$8\ e^-$, $8\ p^+$, $8\ n$	$28\ e^-$, $30\ p^+$, $35\ n$	$23\ e^-$, $26\ p^+$, $28\ n$

6.96 Identify each of the following atoms or ions: (6.1)

A	B	C	D
$2\ e^-$, $3\ p^+$, $4\ n$	$0\ e^-$, $1\ p^+$	$3\ e^-$, $3\ p^+$, $4\ n$	$10\ e^-$, $7\ p^+$, $8\ n$

6.97 Consider the following Lewis symbols for elements X and Y: (6.1, 6.2, 6.5)

X· ·Ÿ· (Y with five dots)

a. What are the group numbers of X and Y?
b. Will a compound of X and Y be ionic or molecular?
c. What ions would be formed by X and Y?
d. What would be the formula of a compound of X and Y?
e. What would be the formula of a compound of X and sulfur?
f. What would be the formula of a compound of Y and chlorine?
g. Is the compound in part **f** ionic or molecular?

6.98 Consider the following Lewis symbols for elements X and Y: (6.1, 6.2, 6.5)

Ẋ· ·Ÿ· (X with two dots, Y with five dots)

a. What are the group numbers of X and Y?
b. Will a compound of X and Y be ionic or molecular?
c. What ions would be formed by X and Y?
d. What would be the formula of a compound of X and Y?
e. What would be the formula of a compound of X and sulfur?
f. What would be the formula of a compound of Y and chlorine?
g. Is the compound in part **f** ionic or molecular?

6.99 Using each of the following electron arrangements, give the formulas for the cation and anion that form, the formula for the compound they form, and its name. (6.2, 6.3)

Electron Arrangements		Cation	Anion	Formula of Compound	Name of Compound
2,8,2	2,5				
2,8,8,1	2,6				
2,8,3	2,8,7				

6.100 Using each of the following electron arrangements, give the formulas for the cation and anion that form, the formula for the compound they form, and its name. (6.2, 6.3)

Electron Arrangements		Cation	Anion	Formula of Compound	Name of Compound
2,8,1	2,7				
2,8,8,2	2,8,6				
2,1	2,8,5				

6.101 State the number of valence electrons, bonding pairs, and lone pairs in each of the following Lewis structures: (6.6)

a. H:H **b.** H:B̤̈r: **c.** :B̤̈r:B̤̈r:

6.102 State the number of valence electrons, bonding pairs, and lone pairs in each of the following Lewis structures: (6.6)

a. H:Ö: (with H above O) **b.** H:N̈:H (with H below N) **c.** :B̤̈r:Ö:B̤̈r:

6.103 Match each of the Lewis structures (**a** to **c**) with the correct diagram (**1** to **3**) of its shape, and name the shape; indicate if each molecule is polar or nonpolar. Assume X and Y are nonmetals and all bonds are polar covalent. (6.6, 6.8, 6.9)

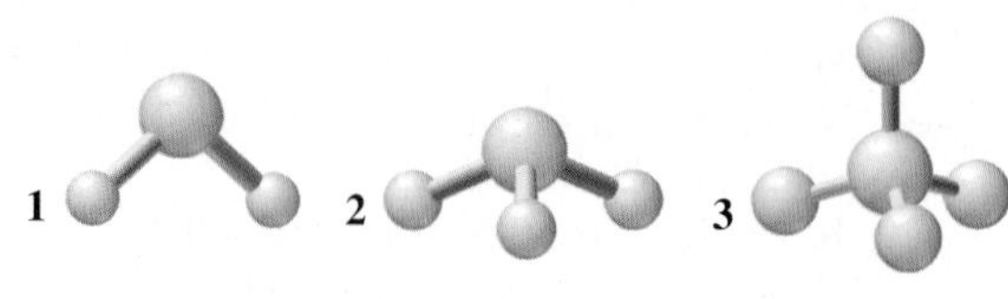

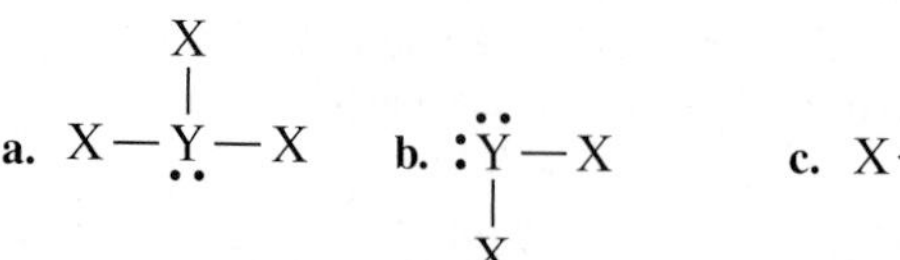

6.104 Match each of the formulas (**a** to **c**) with the correct diagram (**1** to **3**) of its shape, and name the shape; indicate if each molecule is polar or nonpolar. (6.6, 6.8, 6.9)

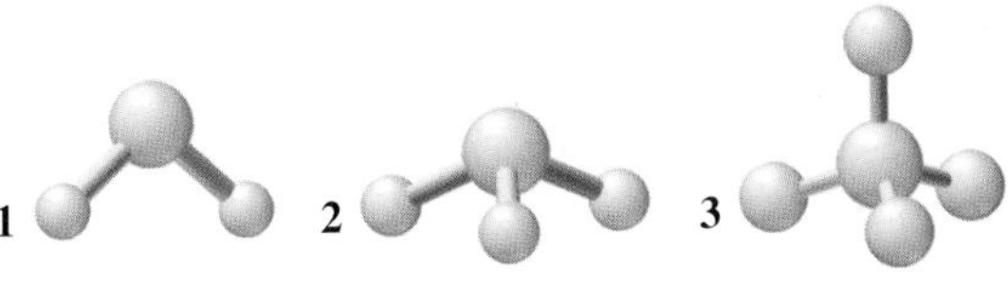

a. PBr_3 **b.** $SiCl_4$ **c.** OF_2

6.105 Consider the following bonds: Ca and O, C and O, K and O, O and O, and N and O. (6.7)
a. Which bonds are polar covalent?
b. Which bonds are nonpolar covalent?
c. Which bonds are ionic?
d. Arrange the covalent bonds in order of decreasing polarity.

6.106 Consider the following bonds: F and Cl, Cl and Cl, Cs and Cl, O and Cl, and Ca and Cl. (6.7)
a. Which bonds are polar covalent?
b. Which bonds are nonpolar covalent?
c. Which bonds are ionic?
d. Arrange the covalent bonds in order of decreasing polarity.

6.107 Identify the major intermolecular forces between each of the following atoms or molecules: (6.9)
a. PH_3 **b.** NO_2
c. CH_3NH_2 **d.** Ar

6.108 Identify the major intermolecular forces between each of the following atoms or molecules: (6.9)
a. He **b.** HBr
c. SnH_4 **d.** $CH_3—CH_2—CH_2—OH$

ADDITIONAL PRACTICE PROBLEMS

6.109 Write the name for the following: (6.1)
a. N^{3-} **b.** Mg^{2+} **c.** O^{2-} **d.** Al^{3+}

6.110 Write the name for the following: (6.1)
a. K^+ **b.** Na^+ **c.** Ba^{2+} **d.** Cl^-

6.111 Consider an ion with the symbol X^{2+} formed from a representative element. (6.1, 6.2, 6.3)
a. What is the group number of the element?
b. What is the Lewis symbol of the element?
c. If X is in Period 3, what is the element?
d. What is the formula of the compound formed from X and the nitride ion?

6.112 Consider an ion with the symbol Y^{3-} formed from a representative element. (6.1, 6.2, 6.3)
a. What is the group number of the element?
b. What is the Lewis symbol of the element?
c. If Y is in Period 3, what is the element?
d. What is the formula of the compound formed from the barium ion and Y?

6.113 One of the ions of tin is tin(IV). (6.1, 6.2, 6.3, 6.4)
a. What is the symbol for this ion?
b. How many protons and electrons are in the ion?
c. What is the formula of tin(IV) oxide?
d. What is the formula of tin(IV) phosphate?

6.114 One of the ions of gold is gold(III). (6.1, 6.2, 6.3, 6.4)
a. What is the symbol for this ion?
b. How many protons and electrons are in the ion?
c. What is the formula of gold(III) sulfate?
d. What is the formula of gold(III) nitrate?

6.115 Write the formula for each of the following ionic compounds: (6.2, 6.3)
a. tin(II) sulfide **b.** lead(IV) oxide
c. silver chloride **d.** calcium nitride
e. copper(I) phosphide **f.** chromium(II) bromide

6.116 Write the formula for each of the following ionic compounds: (6.2, 6.3)
a. nickel(III) oxide **b.** iron(III) sulfide
c. lead(II) sulfate **d.** chromium(III) iodide
e. lithium nitride **f.** gold(I) oxide

6.117 Name each of the following molecular compounds: (6.5)
a. NCl_3 **b.** N_2S_3 **c.** N_2O
d. IF **e.** BF_3 **f.** P_2O_5

6.118 Name each of the following molecular compounds: (6.5)
a. CF_4 **b.** SF_6 **c.** BrCl
d. N_2O_4 **e.** SO_2 **f.** CS_2

6.119 Write the formula for each of the following molecular compounds: (6.5)
a. carbon sulfide **b.** diphosphorus pentoxide
c. dihydrogen sulfide **d.** sulfur dichloride

6.120 Write the formula for each of the following molecular compounds: (6.5)
a. silicon dioxide **b.** carbon tetrabromide
c. diphosphorus tetraiodide **d.** dinitrogen trioxide

6.121 Classify each of the following as ionic or molecular, and give its name: (6.3, 6.5)
a. $FeCl_3$ **b.** Na_2SO_4 **c.** NO_2
d. Rb_2S **e.** PF_5 **f.** CF_4

6.122 Classify each of the following as ionic or molecular, and give its name: (6.3, 6.5)
a. $Al_2(CO_3)_3$ **b.** ClF_5 **c.** BCl_3
d. Mg_3N_2 **e.** ClO_2 **f.** $CrPO_4$

6.123 Write the formula for each of the following: (6.3, 6.4, 6.5)
a. tin(II) carbonate **b.** lithium phosphide
c. silicon tetrachloride **d.** manganese(III) oxide
e. tetraphosphorus triselenide **f.** calcium bromide

6.124 Write the formula for each of the following: (6.3, 6.4, 6.5)
a. sodium carbonate **b.** nitrogen dioxide
c. aluminum nitrate **d.** copper(I) nitride
e. potassium phosphate **f.** cobalt(III) sulfate

6.125 Determine the total number of valence electrons in each of the following: (6.6)
a. HNO_2 **b.** CH_3CHO **c.** CH_3NH_2

6.126 Determine the total number of valence electrons in each of the following: (6.6)
a. $COCl_2$ **b.** N_2O **c.** $SeCl_2$

6.127 Draw the Lewis structure for each of the following: (6.6)
a. Cl_2O
b. H_2NOH (N is the central atom)
c. H_2CCCl_2

6.128 Draw the Lewis structure for each of the following: (6.6)
a. H_3COCH_3 (the atoms are in the order C O C)
b. HNO_2 (the atoms are in the order HONO)
c. OBr_2

6.129 Use the periodic table to arrange the following atoms in order of increasing electronegativity: (6.7)
a. I, F, Cl
b. Li, K, S, Cl
c. Mg, Sr, Ba, Be

6.130 Use the periodic table to arrange the following atoms in order of increasing electronegativity: (6.7)
a. Cl, Br, Se
b. Na, Cs, O, S
c. O, F, B, Li

6.131 Select the more polar bond in each of the following pairs: (6.7)
a. C—N or C—O
b. N—F or N—Br
c. Br—Cl or S—Cl
d. Br—Cl or Br—I
e. N—F or N—O

6.132 Select the more polar bond in each of the following pairs: (6.7)
a. C—C or C—O
b. P—Cl or P—Br
c. Si—S or Si—Cl
d. F—Cl or F—Br
e. P—O or P—S

6.133 Show the dipole arrow for each of the following bonds: (6.7)
a. Si—Cl
b. C—N
c. F—Cl
d. C—F
e. N—O

6.134 Show the dipole arrow for each of the following bonds: (6.7)
a. P—O
b. N—F
c. O—Cl
d. S—Cl
e. P—F

6.135 Calculate the electronegativity difference and classify each of the following bonds as nonpolar covalent, polar covalent, or ionic: (6.7)
a. Si and Cl
b. C and C
c. Na and Cl
d. C and H
e. F and F

6.136 Calculate the electronegativity difference and classify each of the following bonds as nonpolar covalent, polar covalent, or ionic: (6.7)
a. C and N
b. Cl and Cl
c. K and Br
d. H and H
e. N and F

6.137 For each of the following, draw the Lewis structure and determine the shape: (6.6, 6.8)
a. NF_3
b. $SiBr_4$
c. CSe_2

6.138 For each of the following, draw the Lewis structure and determine the shape: (6.6, 6.8)
a. $COCl_2$ (C is the central atom)
b. HCCH
c. SO_2

6.139 Use the Lewis structure to determine the shape for each of the following: (6.6, 6.8)
a. CBr_4
b. H_2O

6.140 Use the Lewis structure to determine the shape for each of the following: (6.6, 6.8)
a. PH_3
b. HCN

6.141 Predict the shape and polarity of each of the following molecules, which have polar covalent bonds: (6.8, 6.9)
a. A central atom with three identical bonded atoms and one lone pair.
b. A central atom with two bonded atoms and two lone pairs.

6.142 Predict the shape and polarity of each of the following molecules, which have polar covalent bonds: (6.8, 6.9)
a. A central atom with four identical bonded atoms and no lone pairs.
b. A central atom with four bonded atoms that are not identical and no lone pairs.

6.143 Classify each of the following molecules as polar or nonpolar: (6.7, 6.9)
a. HBr
b. SiO_2
c. NCl_3
d. CH_3Cl
e. NI_3
f. H_2O

6.144 Classify each of the following molecules as polar or nonpolar: (6.8, 6.9)
a. GeH_4
b. I_2
c. CF_3Cl
d. PCl_3
e. BCl_3
f. SCl_2

6.145 Indicate the major type of intermolecular forces—(1) ionic bonds, (2) dipole–dipole attractions, (3) hydrogen bonds, (4) dispersion forces—that occurs between particles of the following: (6.9)
a. NF_3
b. ClF
c. Br_2
d. Cs_2O
e. C_4H_{10}
f. CH_3OH

6.146 Indicate the major type of intermolecular forces—(1) ionic bonds, (2) dipole–dipole attractions, (3) hydrogen bonds, (4) dispersion forces—that occurs between particles of the following: (6.9)
a. $CHCl_3$
b. H_2O
c. LiCl
d. OBr_2
e. HBr
f. IBr

CHALLENGE PROBLEMS

The following problems are related to the topics in this chapter. However, they do not all follow the chapter order, and they require you to combine concepts and skills from several Sections. These problems will help you increase your critical thinking skills and prepare for your next exam.

6.147 Complete the following table for atoms or ions: (6.1)

Atom or Ion	Number of Protons	Number of Electrons	Electrons Lost/Gained
K^+			
	$12\ p^+$	$10\ e^-$	
	$8\ p^+$		$2\ e^-$ gained
		$10\ e^-$	$3\ e^-$ lost

6.148 Complete the following table for atoms or ions: (6.1)

Atom or Ion	Number of Protons	Number of Electrons	Electrons Lost/Gained
	$30\ p^+$		$2\ e^-$ lost
	$36\ p^+$	$36\ e^-$	
	$16\ p^+$		$2\ e^-$ gained
		$46\ e^-$	$4\ e^-$ lost

6.149 Identify the group number in the periodic table of X, a representative element, in each of the following ionic compounds: (6.2)
a. XCl_3
b. Al_2X_3
c. XCO_3

6.150 Identify the group number in the periodic table of X, a representative element, in each of the following ionic compounds: (6.2)

a. X_2O_3 b. X_2SO_3 c. Na_3X

6.151 Classify each of the following as ionic or molecular, and name each: (6.2, 6.3, 6.4, 6.5)

a. Li_2HPO_4 b. ClF_3 c. $Mg(ClO_2)_2$
d. NF_3 e. $Ca(HSO_4)_2$ f. $KClO_4$
g. $Au_2(SO_3)_3$

6.152 Classify each of the following as ionic or molecular, and name each: (6.2, 6.3, 6.4, 6.5)

a. $FePO_3$ b. Cl_2O_7 c. $Ca_3(PO_4)_2$
d. PCl_3 e. $Al(ClO_2)_3$ f. $Pb(C_2H_3O_2)_2$
g. $MgCO_3$

6.153 Complete the Lewis structure for each of the following: (6.6)

```
      H   O
      |   |
a. H—N—C—H

      H
      |
b. Cl—C—C—N
      |
      H
```

```
c. H—N—N—H

        O     H
        |     |
d. Cl—C—O—C—H
              |
              H
```

6.154 Identify the errors in each of the following Lewis structures and draw the correct formula: (6.6)

```
    ..     ..
a. :Cl=O=Cl:

      ..
     :O:
      |
b. H—C—H

       ..  ..
c. H—N=O—H
      |
      H
```

6.155 Predict the shape of each of the following molecules: (6.8)

a. NH_2Cl (N is the central atom) b. TeO_2

6.156 Classify each of the following molecules as polar or nonpolar: (6.7, 6.8, 6.9)

a. N_2 b. NH_2Cl (N is the central atom)

ANSWERS

Answers to Selected Practice Problems

6.1 a. 1 b. 2 c. 3 d. 1 e. 2

6.3 a. 2 e^- lost b. 3 e^- gained c. 1 e^- gained
d. 1 e^- lost e. 1 e^- gained

6.5 a. Li^+ b. F^- c. Mg^{2+} d. Fe^{3+}

6.7 a. 29 protons, 27 electrons b. 34 protons, 36 electrons
c. 35 protons, 36 electrons d. 26 protons, 23 electrons

6.9 a. Cl^- b. Cs^+ c. N^{3-} d. Ra^{2+}

6.11 a. lithium b. calcium c. gallium d. phosphide

6.13 a. 8 protons, 10 electrons b. 19 protons, 18 electrons
c. 53 protons, 54 electrons d. 11 protons, 10 electrons

6.15 a and c

6.17 a. Na_2O b. $AlBr_3$ c. Ba_3N_2
d. MgF_2 e. Al_2S_3

6.19 a. K^+ and S^{2-}, K_2S b. Na^+ and N^{3-}, Na_3N
c. Al^{3+} and I^-, AlI_3 d. Ga^{3+} and O^{2-}, Ga_2O_3

6.21 a. aluminum fluoride b. calcium chloride
c. sodium oxide d. magnesium phosphide
e. potassium iodide f. barium fluoride

6.23 a. iron(II) b. copper(II) c. zinc
d. lead(IV) e. chromium(III) f. manganese(II)

6.25 a. tin(II) chloride b. iron(II) oxide
c. copper(I) sulfide d. copper(II) sulfide
e. cadmium bromide f. mercury(II) chloride

6.27 a. Au^{3+} b. Fe^{3+} c. Pb^{4+} d. Al^{3+}

6.29 a. $MgCl_2$ b. Na_2S c. Cu_2O d. Zn_3P_2
e. AuN f. CoF_3

6.31 a. $CoCl_3$ b. PbO_2 c. AgI d. Ca_3N_2
e. Cu_3P f. $CrCl_2$

6.33 a. K_3P b. $CuCl_2$ c. $FeBr_3$ d. MgO

6.35 a. HCO_3^- b. NH_4^+ c. PO_3^{3-} d. ClO_3^-

6.37 a. sulfate b. carbonate
c. hydrogen sulfite (bisulfite) d. nitrate

6.39

	NO_2^-	CO_3^{2-}	HSO_4^-	PO_4^{3-}
Li^+	$LiNO_2$ Lithium nitrite	Li_2CO_3 Lithium carbonate	$LiHSO_4$ Lithium hydrogen sulfate	Li_3PO_4 Lithium phosphate
Cu^{2+}	$Cu(NO_2)_2$ Copper(II) nitrite	$CuCO_3$ Copper(II) carbonate	$Cu(HSO_4)_2$ Copper(II) hydrogen sulfate	$Cu_3(PO_4)_2$ Copper(II) phosphate
Ba^{2+}	$Ba(NO_2)_2$ Barium nitrite	$BaCO_3$ Barium carbonate	$Ba(HSO_4)_2$ Barium hydrogen sulfate	$Ba_3(PO_4)_2$ Barium phosphate

6.41 a. $Ba(OH)_2$ b. $NaHSO_4$ c. $Fe(NO_2)_2$
d. $Zn_3(PO_4)_2$ e. $Fe_2(CO_3)_3$

6.43 a. CO_3^{2-}, sodium carbonate b. NH_4^+, ammonium sulfide
c. OH^-, calcium hydroxide d. NO_2^-, tin(II) nitrite

6.45 a. zinc acetate b. magnesium phosphate
c. ammonium chloride
d. sodium bicarbonate or sodium hydrogen carbonate
e. sodium nitrite

6.47 a. phosphorus tribromide b. dichlorine oxide
c. carbon tetrabromide d. hydrogen fluoride
e. nitrogen trifluoride

6.49 a. dinitrogen trioxide b. disilicon hexabromide
c. tetraphosphorus trisulfide d. phosphorus pentachloride
e. selenium hexafluoride

6.51 a. CCl_4 b. CO c. PF_3 d. N_2O_4

6.53 **a.** OF_2 **b.** BCl_3 **c.** N_2O_3 **d.** SF_6

6.55 **a.** aluminum sulfate **b.** calcium carbonate
c. dinitrogen oxide **d.** magnesium hydroxide

6.57 **a.** 8 valence electrons **b.** 14 valence electrons
c. 32 valence electrons

6.59 **a.** HF (8 e^-)
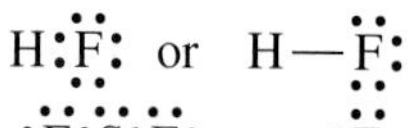

b. SF_2 (20 e^-)
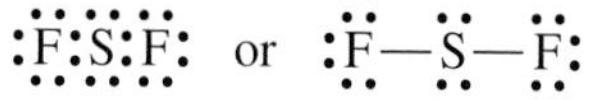

c. NBr_3 (26 e^-)
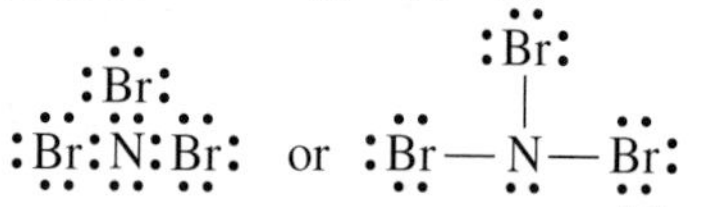

d. $ClNO_2$ (24 e^-)
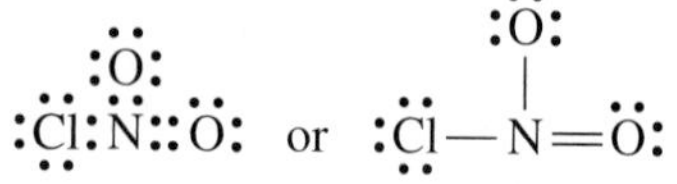

6.61 **a.** increases **b.** decreases **c.** decreases

6.63 **a.** K, Na, Li **b.** Na, P, Cl **c.** Ca, Se, O

6.65 **a.** from 0.0 and 0.4

6.67 **a.** polar covalent **b.** ionic
c. polar covalent **d.** nonpolar covalent
e. polar covalent **f.** nonpolar covalent

6.69 **a.** $\overset{\delta^+}{\mathrm{N}}-\overset{\delta^-}{\mathrm{F}}$ ($+\!\!\longrightarrow$) **b.** $\overset{\delta^+}{\mathrm{Si}}-\overset{\delta^-}{\mathrm{Br}}$ ($+\!\!\longrightarrow$)
c. $\overset{\delta^+}{\mathrm{C}}-\overset{\delta^-}{\mathrm{O}}$ ($+\!\!\longrightarrow$) **d.** $\overset{\delta^+}{\mathrm{P}}-\overset{\delta^-}{\mathrm{Br}}$ ($+\!\!\longrightarrow$)
e. $\overset{\delta^-}{\mathrm{N}}-\overset{\delta^+}{\mathrm{P}}$ ($\longleftarrow\!\!+$)

6.71 **a.** 6, tetrahedral **b.** 5, trigonal pyramidal
c. 3, trigonal planar

6.73 **a.** three **b.** trigonal planar
c. three **d.** trigonal planar

6.75 In CF_4, the central atom C has four bonded atoms and no lone pairs of electrons, which gives it a tetrahedral shape. In NF_3, the central atom N has three bonded atoms and one lone pair of electrons, which gives NF_3 a trigonal pyramidal shape.

6.77 **a.** trigonal planar **b.** bent (109°)
c. linear **d.** tetrahedral

6.79 Electrons are shared equally between two identical atoms and unequally between nonidentical atoms.

6.81 **a.** nonpolar **b.** polar **c.** polar **d.** nonpolar

6.83 In the molecule CO_2, the two C—O dipoles cancel out; in CO, there is only one dipole.

6.85 **a.** dipole–dipole attractions **b.** ionic bonds
c. dipole–dipole attractions **d.** dispersion forces

6.87 **a.** hydrogen bonds **b.** dipole–dipole attractions
c. dispersion forces **d.** dispersion forces

6.89 **a.** $MgSO_4$ **b.** SnF_2 **c.** $Al(OH)_3$

6.91 **a.** ionic **b.** ionic **c.** ionic

6.93 **a.** By losing two valence electrons from the third energy level, magnesium achieves an octet in the second energy level.
b. The magnesium ion Mg^{2+} has the same electron arrangement as Ne (2,8).
c. Group 1A (1) and 2A (2) elements achieve octets by losing electrons to form compounds. Group 8A (18) elements are stable with octets (or two electrons for helium).

6.95 **a.** P^{3-} ion **b.** O atom **c.** Zn^{2+} ion **d.** Fe^{3+} ion

6.97 **a.** X = Group 1A (1), Y = Group 6A (16)
b. ionic **c.** X^+ and Y^{2-} **d.** X_2Y
e. X_2S **f.** YCl_2 **g.** molecular

6.99

Electron Arrangements		Cation	Anion	Formula of Compound	Name of Compound
2,8,2	2,5	Mg^{2+}	N^{3-}	Mg_3N_2	Magnesium nitride
2,8,8,1	2,6	K^+	O^{2-}	K_2O	Potassium oxide
2,8,3	2,8,7	Al^{3+}	Cl^-	$AlCl_3$	Aluminum chloride

6.101 **a.** two valence electrons, one bonding pair, no lone pairs
b. eight valence electrons, one bonding pair, three lone pairs
c. 14 valence electrons, one bonding pair, six lone pairs

6.103 **a.** 2, trigonal pyramidal, polar
b. 1, bent (109°), polar
c. 3, tetrahedral, nonpolar

6.105 **a.** C—O and N—O
b. O—O
c. Ca—O and K—O
d. C—O, N—O, O—O

6.107 **a.** dispersion forces **b.** dipole–dipole attractions
c. hydrogen bonds **d.** dispersion forces

6.109 **a.** nitride **b.** magnesium
c. oxide **d.** aluminum

6.111 **a.** 2A (2) **b.** $\dot{\mathrm{X}}\cdot$ **c.** Mg **d.** X_3N_2

6.113 **a.** Sn^{4+} **b.** 50 protons and 46 electrons
c. SnO_2 **d.** $Sn_3(PO_4)_4$

6.115 **a.** SnS **b.** PbO_2 **c.** AgCl
d. Ca_3N_2 **e.** Cu_3P **f.** $CrBr_2$

6.117 **a.** nitrogen trichloride **b.** dinitrogen trisulfide
c. dinitrogen oxide **d.** iodine fluoride
e. boron trifluoride **f.** diphosphorus pentoxide

6.119 **a.** CS **b.** P_2O_5 **c.** H_2S **d.** SCl_2

6.121 **a.** ionic, iron(III) chloride
b. ionic, sodium sulfate
c. molecular, nitrogen dioxide
d. ionic, rubidium sulfide
e. molecular, phosphorus pentafluoride
f. molecular, carbon tetrafluoride

6.123 **a.** $SnCO_3$ **b.** Li_3P **c.** $SiCl_4$
d. Mn_2O_3 **e.** P_4Se_3 **f.** $CaBr_2$

6.125 **a.** 1 + 5 + 2(6) = 18 valence electrons
b. 2(4) + 4(1) + 6 = 18 valence electrons
c. 4 + 5(1) + 5 = 14 valence electrons

6.127 **a.** Cl_2O (20 e^-)
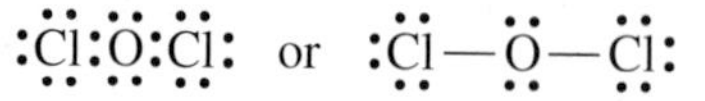

b. H_2NOH (14 e^-)
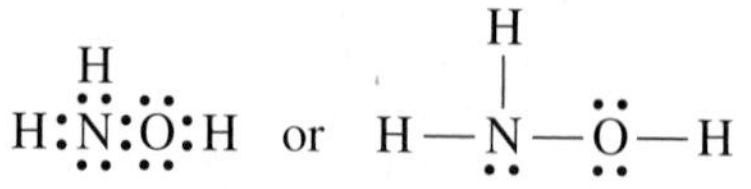

c. H_2CCCl_2 (24 e^-)
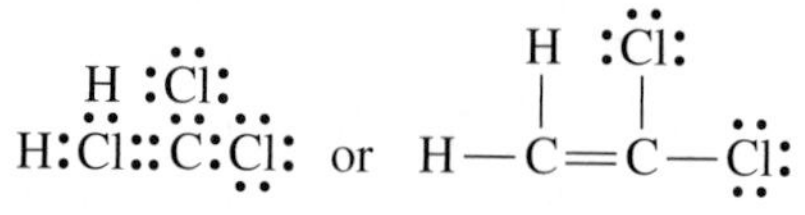

6.129 **a.** I, Cl, F **b.** K, Li, S, Cl **c.** Ba, Sr, Mg, Be

6.131 a. C—O **b.** N—F **c.** S—Cl
d. Br—I **e.** N—F

6.133 a. Si—Cl (⟼) **b.** C—N (⟼)
c. F—Cl (⟻) **d.** C—F (⟼)
e. N—O (⟼)

6.135 a. polar covalent **b.** nonpolar covalent
c. ionic **d.** nonpolar covalent
e. nonpolar covalent

6.137 a. NF_3 (26 e^-) :F̤̈—N̈—F̤̈: with :F̤: below N — trigonal pyramidal

b. $SiBr_4$ (32 e^-) Si bonded to four :Br̤̈: — tetrahedral

c. CSe_2 (16 e^-) :S̈e=C=S̈e: linear

6.139 a. tetrahedral **b.** bent (109°)

6.141 a. trigonal pyramidal, polar
b. bent (109°), polar

6.143 a. polar **b.** nonpolar
c. nonpolar **d.** polar
e. polar **f.** polar

6.145 a. (2) dipole–dipole attractions
b. (2) dipole–dipole attractions
c. (4) dispersion forces
d. (1) ionic bonds
e. (4) dispersion forces
f. (3) hydrogen bonds

6.147

Atom or Ion	Number of Protons	Number of Electrons	Electrons Lost/ Gained
K^+	19 p^+	18 e^-	1 e^- lost
Mg^{2+}	12 p^+	10 e^-	2 e^- lost
O^{2-}	8 p^+	10 e^-	2 e^- gained
Al^{3+}	13 p^+	10 e^-	3 e^- lost

6.149 a. Group 3A (13) **b.** Group 6A (16)
c. Group 2A (2)

6.151 a. ionic, lithium hydrogen phosphate
b. molecular, chlorine trifluoride
c. ionic, magnesium chlorite
d. molecular, nitrogen trifluoride
e. ionic, calcium bisulfate or calcium hydrogen sulfate
f. ionic, potassium perchlorate
g. ionic, gold(III) sulfite

6.153 a. (18 e^-) H—N̤(—H)—C(=Ö:)—H

b. (22 e^-) :C̤̈l—C(H)(H)—C≡N:

c. (12 e^-) H—N̈=N̈—H

d. (30 e^-) :C̤̈l—C(=Ö:)—Ö̤—C(H)(H)—H

6.155 a. trigonal pyramidal **b.** bent, 120°

COMBINING IDEAS from Chapters 4 to 6

CI.7 For parts **a** to **f**, consider the loss of electrons by atoms of the element X, and a gain of electrons by atoms of the element Y, if X is in Group 2A (2), Period 3, and Y is in Group 7A (17), Period 3. (4.6, 6.1, 6.2, 6.3, 6.8)

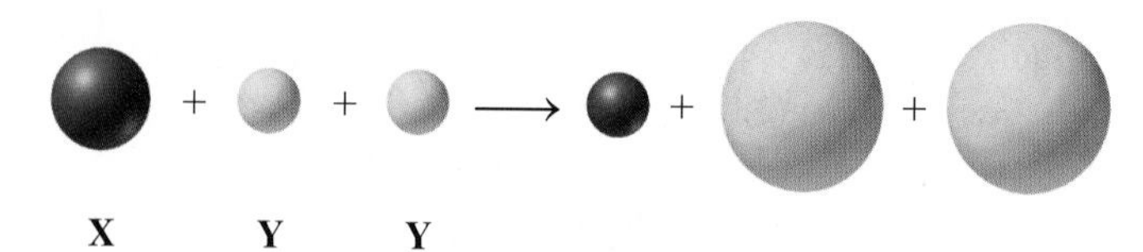

a. Which reactant has the higher electronegativity?
b. What are the ionic charges of **X** and **Y** in the product?
c. Write the electron arrangements for the atoms **X** and **Y**.
d. Write the electron arrangements for the ions of **X** and **Y**.
e. Give the names for the noble gases with the same electron arrangements as each of these ions.
f. Write the formula and name for the ionic compound formed by the ions of **X** and **Y**.

CI.8 A sterling silver bracelet, which is 92.5% silver by mass, has a volume of 25.6 cm^3 and a density of 10.2 g/cm^3. (2.10, 4.5, 5.2, 5.4)

Sterling silver is 92.5% silver by mass.

a. What is the mass, in kilograms, of the bracelet?
b. Determine the number of protons and neutrons in each of the two stable isotopes of silver:

$^{107}_{47}Ag$ $\quad$ $^{109}_{47}Ag$

c. Ag-112 decays by beta emission. Write the balanced nuclear equation for the decay of Ag-112.
d. A 64.0-μCi sample of Ag-112 decays to 8.00 μCi in 9.3 h. What is the half-life of Ag-112?

CI.9 The naturally occurring isotopes of silicon are listed in the following table: (4.5, 4.6, 5.2, 5.4, 6.6, 6.7)

a. Complete the table with the number of protons, neutrons, and electrons for each isotope listed:

Isotope	Number of Protons	Number of Neutrons	Number of Electrons
$^{28}_{14}Si$			
$^{29}_{14}Si$			
$^{30}_{14}Si$			

b. What is the electron arrangement of silicon?
c. Which isotope is the most abundant, if the atomic mass for Si is 28.09?
d. Write the balanced nuclear equation for the beta decay of the radioactive isotope Si-31.
e. Draw the Lewis structure and predict the shape of $SiCl_4$.
f. How many hours are needed for a sample of Si-31 with an activity of 16 μCi to decay to 2.0 μCi?

CI.10 The iceman known as Ötzi was discovered in a high mountain pass on the Austrian–Italian border. Samples of his hair and bones had carbon-14 activity that was 50% of that present in new hair or bone. Carbon-14 undergoes beta decay and has a half-life of 5730 yr. (5.2, 5.4)

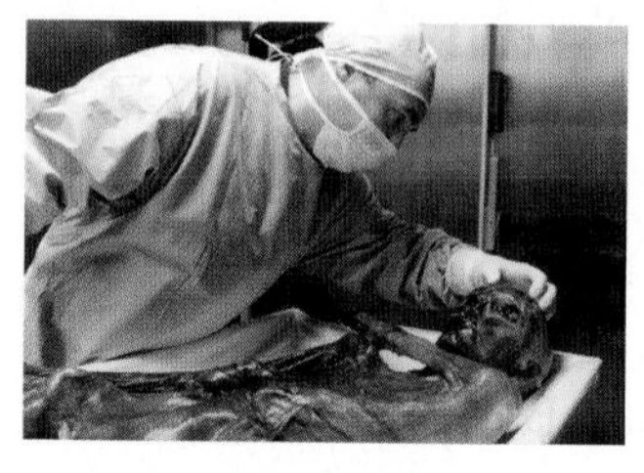
The mummified remains of Ötzi were discovered in 1991.

a. How long ago did Ötzi live?
b. Write a balanced nuclear equation for the decay of carbon-14.

CI.11 K^+ is an electrolyte required by the human body and found in many foods as well as salt substitutes. One of the isotopes of potassium is potassium-40, which has a natural abundance of 0.012% and a half-life of 1.30×10^9 yr. The isotope potassium-40 decays to calcium-40 or to argon-40. A typical activity for potassium-40 is 7.0 μCi per gram. (5.2, 5.3, 5.4)

Potassium chloride is used as a salt substitute.

a. Write a balanced nuclear equation for each type of decay and identify the particle emitted.
b. A shaker of salt substitute contains 1.6 oz of K. What is the activity, in millicuries and becquerels, of the potassium in the shaker?

CI.12 Of much concern to environmentalists is radon-222, which is a radioactive noble gas that can seep from the ground into basements of homes and buildings. Radon-222 is a product of the decay of radium-226 that occurs naturally in rocks and soil in much of the United States. Radon-222, which has a half-life of 3.8 days, decays by emitting an alpha particle. Radon-222, which is a gas, can be inhaled into the lungs where it is strongly associated with lung cancer. Radon levels in a home can be measured with a home radon-detection kit. Environmental agencies have set the maximum level of radon-222 in a home at 4 picocuries per liter (pCi/L) of air. (5.2, 5.3, 5.4)

a. Write the balanced nuclear equation for the decay of Ra-226.

b. Write the balanced nuclear equation for the decay of Rn-222.
c. If a room contains 24 000 atoms of radon-222, how many atoms of radon-222 remain after 15.2 days?
d. Suppose a room has a volume of 72 000 L (7.2×10^4 L). If the radon level is the maximum allowed (4 pCi/L), how many alpha particles are emitted from Rn-222 in 1 day? (1 Ci = 3.7×10^{10} disintegrations/s)

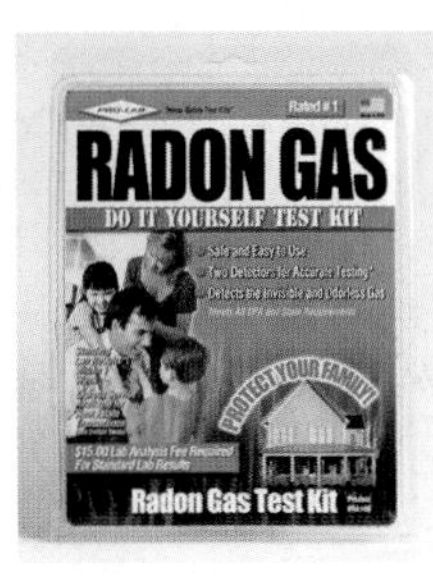

A home detection kit is used to measure the level of radon-222.

ANSWERS

CI.7 **a.** Y has the higher electronegativity.
b. X^{2+}, Y^-
c. X = 2,8,2 Y = 2,8,7
d. X^{2+} = 2,8 Y^- = 2,8,8
e. X^{2+} has the same electron arrangement as Ne. Y^- has the same electron arrangement as Ar.
f. $MgCl_2$, magnesium chloride

CI.9 **a.**

Isotope	Number of Protons	Number of Neutrons	Number of Electrons
$^{28}_{14}Si$	14	14	14
$^{29}_{14}Si$	14	15	14
$^{30}_{14}Si$	14	16	14

b. 2,8,4
c. Si-28
d. $^{31}_{14}Si \longrightarrow ^{31}_{15}P + ^{0}_{-1}e$
e. Lewis structure: Si bonded to four :Cl: atoms (each Cl with three lone pairs) tetrahedral
f. 7.8 h

CI.11 **a.** $^{40}_{19}K \longrightarrow ^{40}_{20}Ca + ^{0}_{-1}e$ beta particle
$^{40}_{19}K \longrightarrow ^{40}_{18}Ar + ^{0}_{+1}e$ positron
b. 3.8×10^{-5} mCi; 1.4×10^3 Bq

7 Chemical Quantities and Reactions

NATALIE WAS RECENTLY DIAGNOSED WITH MILD pulmonary emphysema due to secondhand cigarette smoke. She was referred to Angela, an exercise physiologist, who begins to assess Natalie's condition by connecting her to an electrocardiogram (ECG or EKG), a pulse oximeter, and a blood pressure cuff. The ECG records the electrical activity of Natalie's heart, which is used to measure the rate and rhythm of her heartbeat and detect the possible presence of heart damage. The pulse oximeter measures her pulse and the saturation level of oxygen in her arterial blood (the percentage of hemoglobin that is saturated with O_2). The blood pressure cuff determines the pressure exerted by the heart in pumping her blood.

To determine possible heart disease, Natalie has an exercise stress test on a treadmill to measure how her heart rate and blood pressure respond to exertion by walking faster as the slope of the treadmill is increased. Electrical leads are attached to measure the heart rate and blood pressure first at rest and then on the treadmill. Additional equipment using a face mask collects expired air and measures Natalie's maximal volume of oxygen uptake, or $V_{O_2\,max}$.

CAREER Exercise Physiologist

Exercise physiologists work with athletes as well as patients who have been diagnosed with diabetes, heart disease, pulmonary disease, or other chronic disabilities or diseases. Patients who have been diagnosed with one of these diseases are often prescribed exercise as a form of treatment, and they are referred to an exercise physiologist. The exercise physiologist evaluates the patient's overall health and then creates a customized exercise program for that individual. The program for an athlete might focus on reducing the number of injuries, whereas a program for a cardiac patient would focus on strengthening the heart muscles. The exercise physiologist also monitors the patient for improvement and determines if the exercise is helping to reduce or reverse the progression of the disease.

CLINICAL UPDATE Improving Natalie's Overall Fitness

Natalie's test results indicate that her blood oxygen level was below normal. You can view Natalie's test results and diagnosis of lung function in the **CLINICAL UPDATE Improving Natalie's Overall Fitness**, page 249. After reviewing the test results, Angela teaches Natalie how to improve her respiration and her overall fitness.

7.1 The Mole

LEARNING GOAL Use Avogadro's number to determine the number of particles in a given number of moles.

At the grocery store, you buy eggs by the dozen or soda by the case. In an office-supply store, pencils are ordered by the gross and paper by the ream. Common terms *dozen*, *case*, *gross*, and *ream* are used to count the number of items present. For example, when you buy a dozen eggs, you know you will get 12 eggs in the carton.

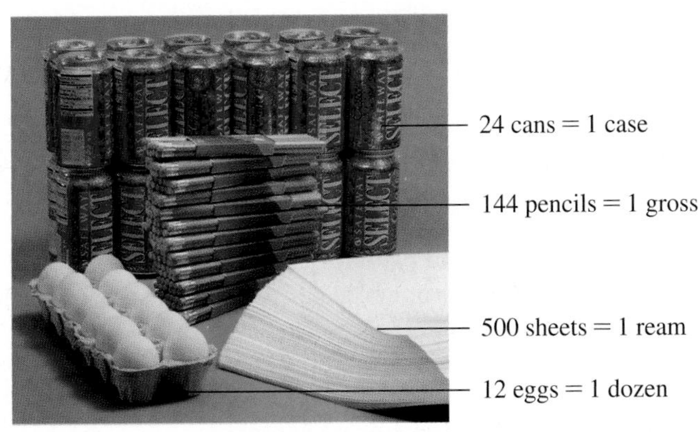

Collections of items include dozen, gross, ream, and mole.

Avogadro's Number

In chemistry, particles such as atoms, molecules, and ions are counted by the **mole**, which contains 6.02×10^{23} items. This value, known as **Avogadro's number**, is a very big number because atoms are so small that it takes an extremely large number of atoms to provide a sufficient amount to weigh and use in chemical reactions. Avogadro's number is named for Amedeo Avogadro (1776–1856), an Italian physicist.

Avogadro's number

$$6.02 \times 10^{23} = 602\,000\,000\,000\,000\,000\,000\,000$$

One mole of any element always contains Avogadro's number of atoms. For example, 1 mole of carbon contains 6.02×10^{23} carbon atoms; 1 mole of aluminum contains 6.02×10^{23} aluminum atoms; 1 mole of sulfur contains 6.02×10^{23} sulfur atoms.

$$1 \text{ mole of an element} = 6.02 \times 10^{23} \text{ atoms of that element}$$

One mole of a molecular compound contains Avogadro's number of molecules. For example, 1 mole of CO_2 contains 6.02×10^{23} molecules of CO_2. One mole of an ionic compound contains Avogadro's number of **formula units**, which are the groups of ions represented by the formula of an ionic compound. One mole of NaCl contains 6.02×10^{23} formula units of NaCl (Na^+, Cl^-). **TABLE 7.1** gives examples of the number of particles in some 1-mole quantities.

TABLE 7.1 Number of Particles in 1-Mole Quantities

Substance	Number and Type of Particles
1 mole of Al	6.02×10^{23} atoms of Al
1 mole of S	6.02×10^{23} atoms of S
1 mole of water (H_2O)	6.02×10^{23} molecules of H_2O
1 mole of vitamin C ($C_6H_8O_6$)	6.02×10^{23} molecules of vitamin C
1 mole of NaCl	6.02×10^{23} formula units of NaCl

LOOKING AHEAD

REVIEW

Writing Numbers in Scientific Notation (1.5)
Counting Significant Figures (2.2)
Writing Conversion Factors from Equalities (2.5)
Using Conversion Factors (2.6)

CORE CHEMISTRY SKILL

Converting Particles to Moles

One mole of sulfur contains 6.02×10^{23} sulfur atoms.

We use Avogadro's number as a conversion factor to convert between the moles of a substance and the number of particles it contains.

$$\frac{6.02 \times 10^{23} \text{ particles}}{1 \text{ mole}} \quad \text{and} \quad \frac{1 \text{ mole}}{6.02 \times 10^{23} \text{ particles}}$$

For example, we use Avogadro's number to convert 4.00 moles of sulfur to atoms of sulfur.

$$4.00 \cancel{\text{moles S}} \times \frac{6.02 \times 10^{23} \text{ atoms S}}{1 \cancel{\text{mole S}}} = 2.41 \times 10^{24} \text{ atoms of S}$$

Avogadro's number as a conversion factor

We also use Avogadro's number to convert 3.01×10^{24} molecules of CO_2 to moles of CO_2.

$$3.01 \times 10^{24} \cancel{\text{molecules CO}_2} \times \frac{1 \text{ mole CO}_2}{6.02 \times 10^{23} \cancel{\text{molecules CO}_2}} = 5.00 \text{ moles of CO}_2$$

Avogadro's number as a conversion factor

ENGAGE

Why is 0.20 mole of aluminum a small number, but the number 1.2×10^{23} atoms of aluminum in 0.20 mole is a large number?

In calculations that convert between moles and particles, the number of moles will be a small number compared to the number of atoms or molecules, which will be large as shown in Sample Problem 7.1.

SAMPLE PROBLEM 7.1 Calculating the Number of Molecules

TRY IT FIRST

How many molecules are present in 1.75 moles of carbon dioxide, CO_2?

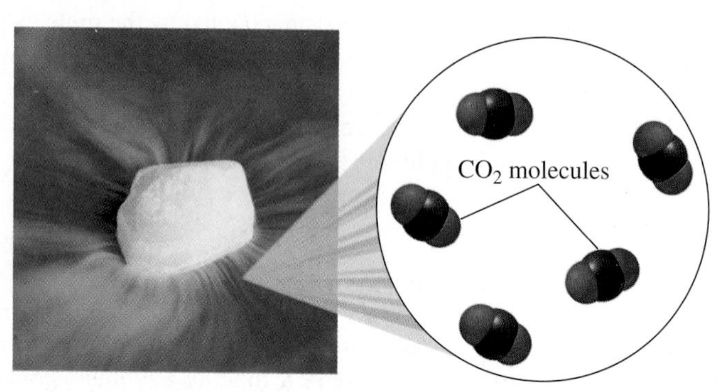

The solid form of carbon dioxide is known as "dry ice."

SOLUTION GUIDE

STEP 1 State the given and needed quantities.

ANALYZE THE PROBLEM	Given	Need	Connect
	1.75 moles of CO_2	molecules of CO_2	Avogadro's number

STEP 2 Write a plan to convert moles to atoms or molecules.

moles of CO_2 — Avogadro's number → molecules of CO_2

STEP 3 Use Avogadro's number to write conversion factors.

$$1 \text{ mole of CO}_2 = 6.02 \times 10^{23} \text{ molecules of CO}_2$$

$$\frac{6.02 \times 10^{23} \text{ molecules CO}_2}{1 \text{ mole CO}_2} \quad \text{and} \quad \frac{1 \text{ mole CO}_2}{6.02 \times 10^{23} \text{ molecules CO}_2}$$

STEP 4 Set up the problem to calculate the number of particles.

$$1.75 \text{ ~~moles CO}_2\text{~~} \times \frac{6.02 \times 10^{23} \text{ molecules } CO_2}{1 \text{ ~~mole CO}_2\text{~~}} = 1.05 \times 10^{24} \text{ molecules of } CO_2$$

STUDY CHECK 7.1

How many moles of water, H_2O, contain 2.60×10^{23} molecules of water?

ANSWER

0.432 mole of H_2O

TEST

Try Practice Problems 7.1 to 7.4

Moles of Elements in a Chemical Compound

We have seen that the subscripts in the chemical formula of a compound indicate the number of atoms of each type of element in the compound. For example, aspirin, $C_9H_8O_4$, is a drug used to reduce pain and inflammation in the body. Using the subscripts in the chemical formula of aspirin shows that there are 9 carbon atoms, 8 hydrogen atoms, and 4 oxygen atoms in each molecule of aspirin. The subscripts also tell us the number of moles of each element in 1 mole of aspirin: 9 moles of C atoms, 8 moles of H atoms, and 4 moles of O atoms.

ENGAGE

Why does 1 mole of $Zn(C_2H_3O_2)_2$, a dietary supplement, contain 1 mole of Zn, 4 moles of C, 6 moles of H, and 4 moles of O?

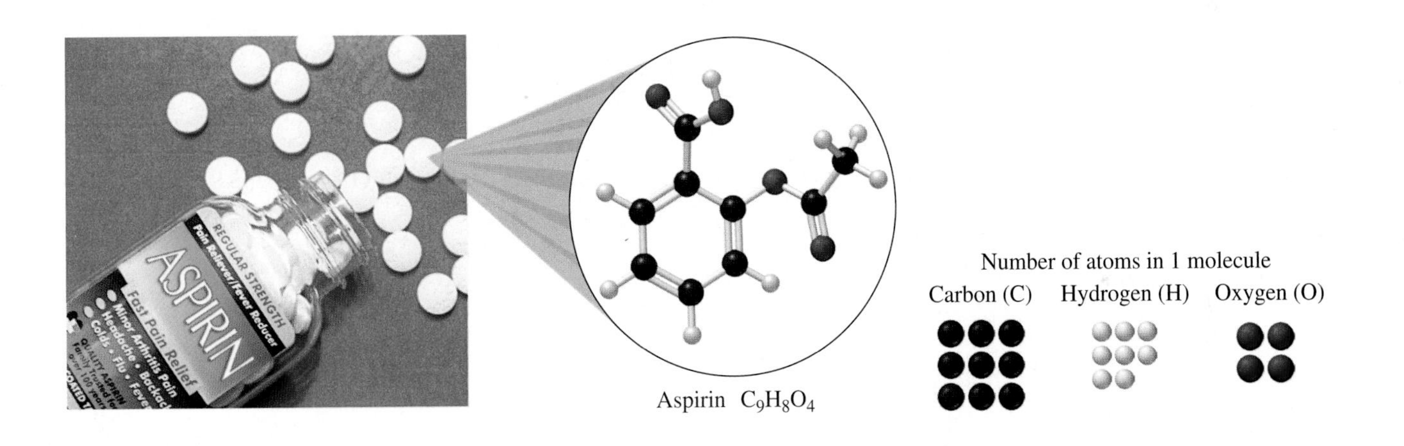

Aspirin $C_9H_8O_4$

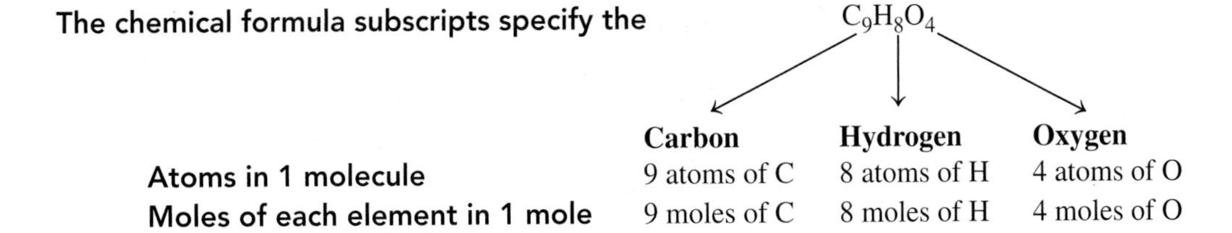

The chemical formula subscripts specify the
$C_9H_8O_4$
Carbon
Hydrogen
Oxygen
Atoms in 1 molecule
9 atoms of C
8 atoms of H
4 atoms of O
Moles of each element in 1 mole
9 moles of C
8 moles of H
4 moles of O

Using a Chemical Formula to Write Conversion Factors

Using the subscripts from the formula, $C_9H_8O_4$, we can write the conversion factors for each of the elements in 1 mole of aspirin:

$$\frac{9 \text{ moles C}}{1 \text{ mole } C_9H_8O_4} \quad \frac{8 \text{ moles H}}{1 \text{ mole } C_9H_8O_4} \quad \frac{4 \text{ moles O}}{1 \text{ mole } C_9H_8O_4}$$

$$\frac{1 \text{ mole } C_9H_8O_4}{9 \text{ moles C}} \quad \frac{1 \text{ mole } C_9H_8O_4}{8 \text{ moles H}} \quad \frac{1 \text{ mole } C_9H_8O_4}{4 \text{ moles O}}$$

SAMPLE PROBLEM 7.2 Calculating the Moles of an Element in a Compound

TRY IT FIRST

How many moles of carbon are present in 1.50 moles of aspirin, $C_9H_8O_4$?

SOLUTION GUIDE

STEP 1 State the given and needed quantities.

ANALYZE THE PROBLEM	Given	Need	Connect
	1.50 moles of aspirin, $C_9H_8O_4$	moles of C	subscripts in formula

STEP 2 Write a plan to convert moles of a compound to moles of an element.

moles of $C_9H_8O_4$ → Subscript → moles of C

STEP 3 Write the equalities and conversion factors using subscripts.

$$1 \text{ mole of } C_9H_8O_4 = 9 \text{ moles of C}$$

$$\frac{9 \text{ moles C}}{1 \text{ mole } C_9H_8O_4} \quad \text{and} \quad \frac{1 \text{ mole } C_9H_8O_4}{9 \text{ moles C}}$$

STEP 4 Set up the problem to calculate the moles of an element.

$$1.50 \cancel{\text{moles } C_9H_8O_4} \times \frac{9 \text{ moles C}}{1 \cancel{\text{mole } C_9H_8O_4}} = 13.5 \text{ moles of C}$$

STUDY CHECK 7.2

How many moles of aspirin, $C_9H_8O_4$, contain 0.480 mole of O?

ANSWER

0.120 mole of aspirin

TEST

Try Practice Problems 7.5 to 7.10

PRACTICE PROBLEMS

7.1 The Mole

LEARNING GOAL Use Avogadro's number to determine the number of particles in a given number of moles.

7.1 What is a mole?

7.2 What is Avogadro's number?

7.3 Calculate each of the following:
- **a.** number of C atoms in 0.500 mole of C
- **b.** number of SO_2 molecules in 1.28 moles of SO_2
- **c.** moles of Fe in 5.22×10^{22} atoms of Fe
- **d.** moles of C_2H_6O in 8.50×10^{24} molecules of C_2H_6O

7.4 Calculate each of the following:
- **a.** number of Li atoms in 4.5 moles of Li
- **b.** number of CO_2 molecules in 0.0180 mole of CO_2
- **c.** moles of Cu in 7.8×10^{21} atoms of Cu
- **d.** moles of C_2H_6 in 3.75×10^{23} molecules of C_2H_6

7.5 Calculate each of the following quantities in 2.00 moles of H_3PO_4:
- **a.** moles of H
- **b.** moles of O
- **c.** atoms of P
- **d.** atoms of O

7.6 Calculate each of the following quantities in 0.185 mole of $C_6H_{14}O$:
- **a.** moles of C
- **b.** moles of O
- **c.** atoms of H
- **d.** atoms of C

Clinical Applications

7.7 Quinine, $C_{20}H_{24}N_2O_2$, is a component of tonic water and bitter lemon.
- **a.** How many moles of H are in 1.5 moles of quinine?
- **b.** How many moles of C are in 5.0 moles of quinine?
- **c.** How many moles of N are in 0.020 mole of quinine?

7.8 Aluminum sulfate, $Al_2(SO_4)_3$, is used in some antiperspirants.
- **a.** How many moles of O are present in 3.0 moles of $Al_2(SO_4)_3$?
- **b.** How many moles of aluminum ions (Al^{3+}) are present in 0.40 mole of $Al_2(SO_4)_3$?
- **c.** How many moles of sulfate ions (SO_4^{2-}) are present in 1.5 moles of $Al_2(SO_4)_3$?

7.9 Naproxen, found in Aleve, is used to treat the pain and inflammation caused by arthritis. Naproxen has a formula of $C_{14}H_{14}O_3$.

a. How many moles of C are present in 2.30 moles of naproxen?

b. How many moles of H are present in 0.444 mole of naproxen?

c. How many moles of O are present in 0.0765 mole of naproxen?

7.10 Benadryl is an over-the-counter drug used to treat allergy symptoms. The formula of Benadryl is $C_{17}H_{21}NO$.

a. How many moles of C are present in 0.733 mole of Benadryl?

b. How many moles of H are present in 2.20 moles of Benadryl?

c. How many moles of N are present in 1.54 moles of Benadryl?

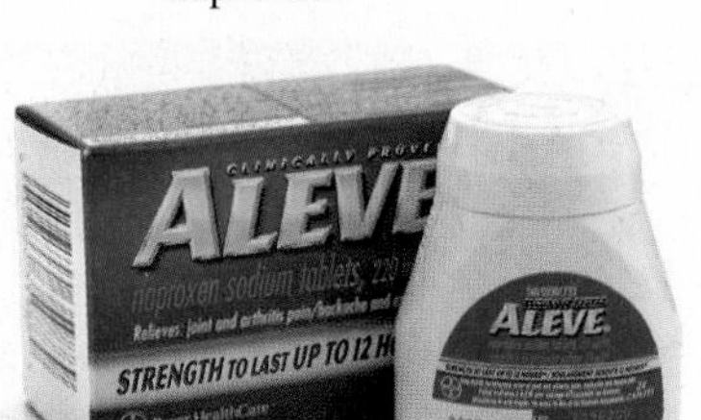

Naproxen is used to treat pain and inflammation caused by arthritis.

7.2 Molar Mass

LEARNING GOAL Given the chemical formula of a substance, calculate its molar mass.

A single atom or molecule is much too small to weigh, even on the most accurate balance. In fact, it takes a huge number of atoms or molecules to make enough of a substance for you to see. An amount of water that contains Avogadro's number of water molecules is only a few sips. However, in the laboratory, we can use a balance to weigh out Avogadro's number of particles for 1 mole of substance.

For any element, the quantity called **molar mass** is the quantity in grams that equals the atomic mass of that element. We are counting 6.02×10^{23} atoms of an element when we weigh out the number of grams equal to its molar mass. For example, carbon has an atomic mass of 12.01 on the periodic table. This means 1 mole of carbon atoms has a mass of 12.01 g. Then to obtain 1 mole of carbon atoms, we would need to weigh out 12.01 g of carbon. The molar mass of carbon is found by looking at its atomic mass on the periodic table.

12.01 g of C atoms

1 mole of C atoms

6.02×10^{23} atoms of C

47 Ag 107.9	6 C 12.01	16 S 32.07
1 mole of silver atoms has a mass of 107.9 g	1 mole of carbon atoms has a mass of 12.01 g	1 mole of sulfur atoms has a mass of 32.07 g

CORE CHEMISTRY SKILL
Calculating Molar Mass

Molar Mass of a Compound

To determine the molar mass of a compound, multiply the molar mass of each element by its subscript in the formula and add the results as shown in Sample Problem 7.3. **In this text, we round the molar mass of an element to the hundredths place (0.01) or use at least four significant figures for calculations.**

FIGURE 7.1 shows some 1-mole quantities of substances.

1-Mole Quantities

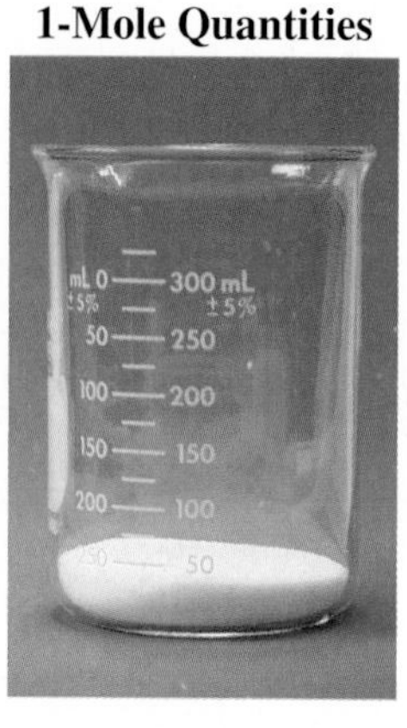

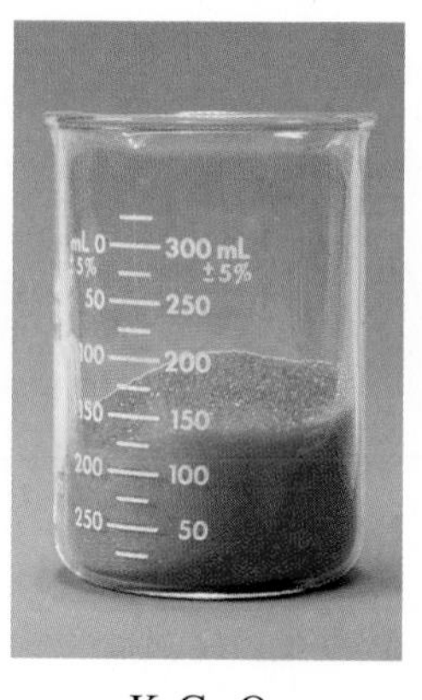

S Fe NaCl $K_2Cr_2O_7$ $C_{12}H_{22}O_{11}$

FIGURE 7.1 1-mole samples: sulfur, S (32.07 g); iron, Fe (55.85 g); salt, NaCl (58.44 g); potassium dichromate, $K_2Cr_2O_7$ (294.2 g); and sucrose, $C_{12}H_{22}O_{11}$ (342.3 g).

Q How is the molar mass for $K_2Cr_2O_7$ obtained?

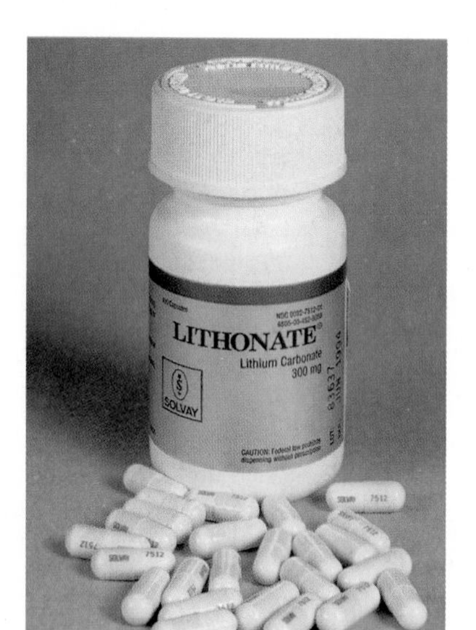

Lithium carbonate is used to treat bipolar disorder.

SAMPLE PROBLEM 7.3 Calculating Molar Mass

TRY IT FIRST

Calculate the molar mass for lithium carbonate, Li_2CO_3, used to treat bipolar disorder.

SOLUTION GUIDE

ANALYZE THE PROBLEM	Given	Need	Connect
	formula Li_2CO_3	molar mass of Li_2CO_3	periodic table

STEP 1 Obtain the molar mass of each element.

$$\frac{6.941 \text{ g Li}}{1 \text{ mole Li}} \qquad \frac{12.01 \text{ g C}}{1 \text{ mole C}} \qquad \frac{16.00 \text{ g O}}{1 \text{ mole O}}$$

STEP 2 Multiply each molar mass by the number of moles (subscript) in the formula.

Grams from 2 moles of Li:

$$2 \cancel{\text{moles Li}} \times \frac{6.941 \text{ g Li}}{1 \cancel{\text{mole Li}}} = 13.88 \text{ g of Li}$$

Grams from 1 mole of C:

$$1 \cancel{\text{mole C}} \times \frac{12.01 \text{ g C}}{1 \cancel{\text{mole C}}} = 12.01 \text{ g of C}$$

Grams from 3 moles of O:

$$3 \cancel{\text{moles O}} \times \frac{16.00 \text{ g O}}{1 \cancel{\text{mole O}}} = 48.00 \text{ g of O}$$

STEP 3 Calculate the molar mass by adding the masses of the elements.

2 moles of Li	= 13.88 g of Li
1 mole of C	= 12.01 g of C
3 moles of O	= 48.00 g of O
Molar mass of Li_2CO_3	= 73.89 g

STUDY CHECK 7.3

Calculate the molar mass for salicylic acid, $C_7H_6O_3$, which is used to treat skin conditions such as acne, psoriasis, and dandruff.

ANSWER

138.12 g

TEST

Try Practice Problems 7.11 to 7.20

PRACTICE PROBLEMS

7.2 Molar Mass

LEARNING GOAL Given the chemical formula of a substance, calculate its molar mass.

7.11 Calculate the molar mass for each of the following:
a. Cl_2 **b.** $C_3H_6O_3$ **c.** $Mg_3(PO_4)_2$

7.12 Calculate the molar mass for each of the following:
a. O_2 **b.** KH_2PO_4 **c.** $Fe(ClO_4)_3$

7.13 Calculate the molar mass for each of the following:
a. AlF_3 **b.** $C_2H_4Cl_2$ **c.** SnF_2

7.14 Calculate the molar mass for each of the following:
a. $C_4H_8O_4$ **b.** $Ga_2(CO_3)_3$ **c.** $KBrO_4$

Clinical Applications

7.15 Calculate the molar mass for each of the following:
a. KCl, salt substitute
b. $C_6H_5NO_2$, niacin (vitamin B_3) needed for cell metabolism
c. $C_{19}H_{20}FNO_3$, Paxil, antidepressant

7.16 Calculate the molar mass for each of the following:
a. $FeSO_4$, iron supplement
b. Al_2O_3, absorbent and abrasive
c. $C_7H_5NO_3S$, saccharin, artificial sweetener

7.17 Calculate the molar mass for each of the following:
a. $Al_2(SO_4)_3$, antiperspirant
b. $KC_4H_5O_6$, cream of tartar
c. $C_{16}H_{19}N_3O_5S$, amoxicillin, antibiotic

7.18 Calculate the molar mass for each of the following:
a. C_3H_8O, rubbing alcohol
b. $(NH_4)_2CO_3$, baking powder
c. $Zn(C_2H_3O_2)_2$, zinc supplement

7.19 Calculate the molar mass for each of the following:
a. $C_8H_9NO_2$, acetaminophen used in Tylenol
b. $Ca_3(C_6H_5O_7)_2$, calcium supplement
c. $C_{17}H_{18}FN_3O_3$, Cipro, used to treat a range of bacterial infections

7.20 Calculate the molar mass for each of the following:
a. $CaSO_4$, calcium sulfate, used to make casts to protect broken bones
b. $C_6H_{12}N_2O_4Pt$, Carboplatin, used in chemotherapy
c. $C_{12}H_{18}O$, Propofol, used to induce anesthesia during surgery

7.3 Calculations Using Molar Mass

LEARNING GOAL Use molar mass to convert between grams and moles.

The molar mass of an element is one of the most useful conversion factors in chemistry because it converts moles of a substance to grams, or grams to moles. For example, 1 mole of silver has a mass of 107.9 g. To express the molar mass of Ag as an equality, we write

$$1 \text{ mole of Ag} = 107.9 \text{ g of Ag}$$

From this equality for the molar mass, two conversion factors can be written as

$$\frac{107.9 \text{ g Ag}}{1 \text{ mole Ag}} \quad \text{and} \quad \frac{1 \text{ mole Ag}}{107.9 \text{ g Ag}}$$

CORE CHEMISTRY SKILL

Using Molar Mass as a Conversion Factor

Sample Problem 7.4 shows how the molar mass of silver is used as a conversion factor.

Silver metal is used to make jewelry.

SAMPLE PROBLEM 7.4 Converting Moles to Grams

TRY IT FIRST

Silver metal is used in the manufacture of tableware, mirrors, jewelry, and dental alloys. If the design for a piece of jewelry requires 0.750 mole of silver, how many grams of silver are needed?

SOLUTION GUIDE

STEP 1 State the given and needed quantities.

ANALYZE THE PROBLEM	Given	Need	Connect
	0.750 mole of Ag	grams of Ag	molar mass

STEP 2 Write a plan to convert moles to grams.

moles of Ag → Molar mass → grams of Ag

STEP 3 Determine the molar mass and write conversion factors.

$$1 \text{ mole of Ag} = 107.9 \text{ g of Ag}$$

$$\frac{107.9 \text{ g Ag}}{1 \text{ mole Ag}} \text{ and } \frac{1 \text{ mole Ag}}{107.9 \text{ g Ag}}$$

STEP 4 Set up the problem to convert moles to grams.

$$0.750 \cancel{\text{mole Ag}} \times \frac{107.9 \text{ g Ag}}{1 \cancel{\text{mole Ag}}} = 80.9 \text{ g of Ag}$$

STUDY CHECK 7.4

A dentist orders 24.4 g of gold (Au) to prepare dental crowns and fillings. Calculate the number of moles of gold in the order.

ANSWER

0.124 mole of Au

Writing Conversion Factors for the Molar Mass of a Compound

The conversion factors for a compound are also written from the molar mass. For example, the molar mass of the compound H_2O is written

$$1 \text{ mole of } H_2O = 18.02 \text{ g of } H_2O$$

From this equality, conversion factors for the molar mass of H_2O are written as

$$\frac{18.02 \text{ g } H_2O}{1 \text{ mole } H_2O} \text{ and } \frac{1 \text{ mole } H_2O}{18.02 \text{ g } H_2O}$$

We can now change from moles to grams, or grams to moles, using the conversion factors derived from the molar mass of a compound shown in Sample Problem 7.5. (Remember, you must determine the molar mass first.)

SAMPLE PROBLEM 7.5 Converting Mass of a Compound to Moles

TRY IT FIRST

A box of salt contains 737 g of NaCl. How many moles of NaCl are present in the box?

SOLUTION GUIDE

STEP 1 State the given and needed quantities.

ANALYZE THE PROBLEM	Given	Need	Connect
	737 g of NaCl	moles of NaCl	molar mass

STEP 2 Write a plan to convert grams to moles.

grams of NaCl → Molar mass → moles of NaCl

STEP 3 Determine the molar mass and write conversion factors.

$$(1 \times 22.99) + (1 \times 35.45) = 58.44 \text{ g/mole}$$

$$1 \text{ mole of NaCl} = 58.44 \text{ g of NaCl}$$

$$\frac{58.44 \text{ g NaCl}}{1 \text{ mole NaCl}} \text{ and } \frac{1 \text{ mole NaCl}}{58.44 \text{ g NaCl}}$$

STEP 4 Set up the problem to convert grams to moles.

$$737 \cancel{\text{g NaCl}} \times \frac{1 \text{ mole NaCl}}{58.44 \cancel{\text{g NaCl}}} = 12.6 \text{ moles of NaCl}$$

STUDY CHECK 7.5

One tablet of an antacid contains 680. mg of $CaCO_3$. How many moles of $CaCO_3$ are present?

ANSWER

0.006 79 or 6.79×10^{-3} mole of $CaCO_3$

Table salt is sodium chloride, NaCl.

TEST

Try Practice Problems 7.21 to 7.34

FIGURE 7.2 gives a summary of the calculations to show the connections between the moles of a compound, its mass in grams, the number of molecules (or formula units if ionic), and the moles and atoms of each element in that compound.

ENGAGE

Why are there more grams of chlorine than grams of fluorine in 1 mole of Freon-12, CCl_2F_2?

Mass		Moles		Particles
Grams of element	← Molar Mass (g/mole) →	**Moles of element**	← Avogadro's Number →	**Atoms (or ions)**
		↕ Formula Subscripts		↕ Formula Subscripts
Grams of compound	← Molar Mass (g/mole) →	**Moles of compound**	← Avogadro's Number →	**Molecules (or formula units)**

FIGURE 7.2 The moles of a compound are related to its mass in grams by molar mass, to the number of molecules (or formula units) by Avogadro's number, and to the moles of each element by the subscripts in the formula.

Q What steps are needed to calculate the number of atoms of H in 5.00 g of CH_4?

PRACTICE PROBLEMS

7.3 Calculations Using Molar Mass

LEARNING GOAL Use molar mass to convert between grams and moles.

7.21 Calculate the mass, in grams, for each of the following:
- **a.** 1.50 moles of Na
- **b.** 2.80 moles of Ca
- **c.** 0.125 mole of CO_2
- **d.** 0.0485 mole of Na_2CO_3
- **e.** 7.14×10^2 moles of PCl_3

7.22 Calculate the mass, in grams, for each of the following:
- **a.** 5.12 moles of Al
- **b.** 0.75 mole of Cu
- **c.** 3.52 moles of $MgBr_2$
- **d.** 0.145 mole of C_2H_6O
- **e.** 2.08 moles of $(NH_4)_2SO_4$

7.23 Calculate the mass, in grams, in 0.150 mole of each of the following:
- **a.** Ne
- **b.** I_2
- **c.** Na_2O
- **d.** $Ca(NO_3)_2$
- **e.** C_6H_{14}

7.24 Calculate the mass, in grams, in 2.28 moles of each of the following:
- **a.** N_2
- **b.** SO_3
- **c.** $C_3H_6O_3$
- **d.** $Mg(HCO_3)_2$
- **e.** SF_6

7.25 Calculate the number of moles in each of the following:
- **a.** 82.0 g of Ag
- **b.** 0.288 g of C
- **c.** 15.0 g of ammonia, NH_3
- **d.** 7.25 g of CH_4
- **e.** 245 g of Fe_2O_3

7.26 Calculate the number of moles in each of the following:
- **a.** 85.2 g of Ni
- **b.** 144 g of K
- **c.** 6.4 g of H_2O
- **d.** 308 g of $BaSO_4$
- **e.** 252.8 g of fructose, $C_6H_{12}O_6$

7.27 Calculate the number of moles in 25.0 g of each of the following:
- **a.** He
- **b.** O_2
- **c.** $Al(OH)_3$
- **d.** Ga_2S_3
- **e.** C_4H_{10}, butane

7.28 Calculate the number of moles in 4.00 g of each of the following:
- **a.** Au
- **b.** SnO_2
- **c.** CS_2
- **d.** Ca_3N_2
- **e.** $C_6H_8O_6$, vitamin C

Clinical Applications

7.29 Chloroethane, C_2H_5Cl, is used to diagnose dead tooth nerves.
- **a.** How many moles are in 34.0 g of chloroethane?
- **b.** How many grams are in 1.50 moles of chloroethane?

7.30 Allyl sulfide, $(C_3H_5)_2S$, gives garlic, onions, and leeks their characteristic odor.

The characteristic odor of garlic is due to a sulfur-containing compound.

- **a.** How many moles are in 23.2 g of allyl sulfide?
- **b.** How many grams are in 0.75 mole of allyl sulfide?

7.31
- **a.** The compound $MgSO_4$, Epsom salts, is used to soothe sore feet and muscles. How many grams will you need to prepare a bath containing 5.00 moles of Epsom salts?
- **b.** Potassium iodide, KI, is used as an expectorant. How many grams are in 0.450 mole of potassium iodide?

7.32
- **a.** Cyclopropane, C_3H_6, is an anesthetic given by inhalation. How many grams are in 0.25 mole of cyclopropane?
- **b.** The sedative Demerol hydrochloride has the formula $C_{15}H_{22}ClNO_2$. How many grams are in 0.025 mole of Demerol hydrochloride?

7.33 Dinitrogen oxide (or nitrous oxide), N_2O, also known as laughing gas, is widely used as an anesthetic in dentistry.
- **a.** How many grams are in 1.50 moles of dinitrogen oxide?
- **b.** How many moles are in 34.0 g of dinitrogen oxide?

7.34 Chloroform, $CHCl_3$, was formerly used as an anesthetic but its use was discontinued due to respiratory and cardiac failure.
- **a.** How many grams are in 0.122 mole of chloroform?
- **b.** How many moles are in 26.7 g of chloroform?

REVIEW

Writing Ionic Formulas (6.2)
Naming Ionic Compounds (6.3)
Writing the Names and Formulas for Molecular Compounds (6.5)

7.4 Equations for Chemical Reactions

LEARNING GOAL Write a balanced chemical equation from the formulas of the reactants and products for a reaction; determine the number of atoms in the reactants and products.

Chemical reactions occur everywhere. The fuel in a car burns with oxygen to make a car move and run the air conditioner. When we cook food or bleach our hair, a chemical reaction takes place. In our bodies, chemical reactions convert food into molecules that build muscles and move them. In the leaves of trees and plants, carbon dioxide and water are converted into carbohydrates. Chemical equations are used by chemists to describe chemical reactions. In every chemical equation, the atoms in the reacting substances, called *reactants*, are rearranged to give new substances called *products*.

A *chemical change* occurs when a substance is converted into one or more new substances that have different formulas and different properties. For example, when silver tarnishes, the shiny silver metal (Ag) reacts with sulfur (S) to become the dull, black substance we call tarnish (Ag_2S) (see **FIGURE 7.3**).

A chemical change:
the tarnishing of silver

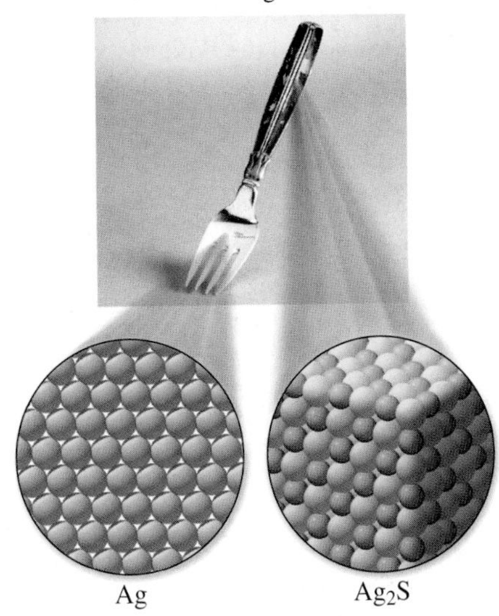

FIGURE 7.3 A chemical change produces new substances with new properties.

Q Why is the formation of tarnish a chemical change?

A *chemical reaction* always involves chemical change because atoms of the reacting substances form new combinations with new properties. For example, a chemical reaction takes place when a piece of iron (Fe) combines with oxygen (O_2) in the air to produce a new substance, rust (Fe_2O_3), which has a reddish-brown color. During a chemical change, new properties become visible, which are an indication that a chemical reaction has taken place (see **TABLE 7.2**).

TABLE 7.2 Types of Evidence of a Chemical Reaction

1. Change in color	**2.** Formation of a gas (bubbles)
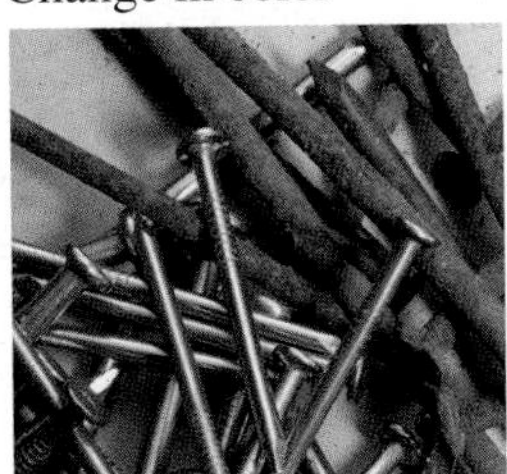	
Fe Fe_2O_3 Iron nails change color when they react with oxygen to form rust.	Bubbles (gas) form when $CaCO_3$ reacts with acid.
3. Formation of a solid (precipitate)	**4.** Heat (or a flame) produced or heat absorbed
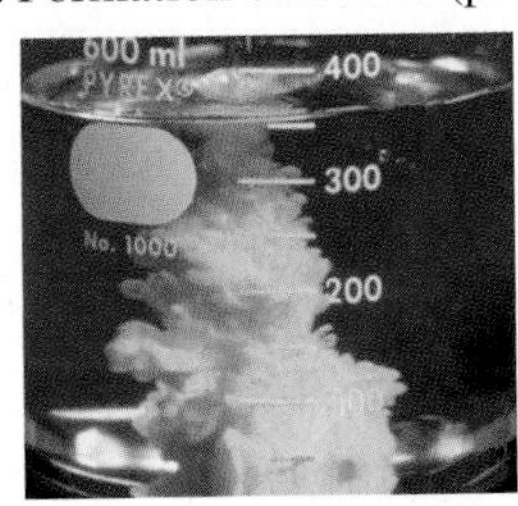	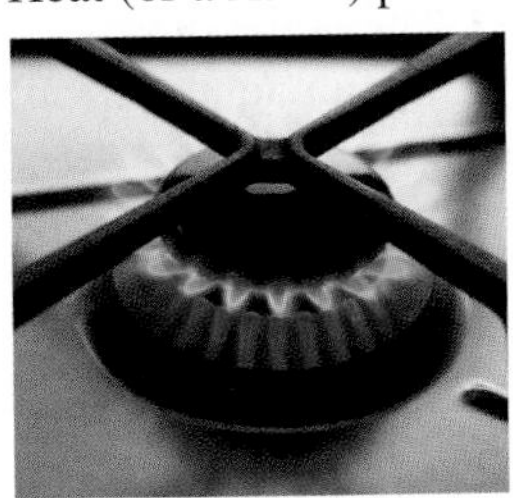
A yellow solid forms when potassium iodide is added to lead nitrate.	Methane gas burns in the air with a hot flame.

Writing a Chemical Equation

When you build a model airplane, prepare a new recipe, or mix a medication, you follow a set of directions. These directions tell you what materials to use and the products you will obtain. In chemistry, a *chemical equation* tells us the materials we need and the products that will form.

Suppose you work in a bicycle shop, assembling wheels and frames into bicycles. You could represent this process by a simple equation:

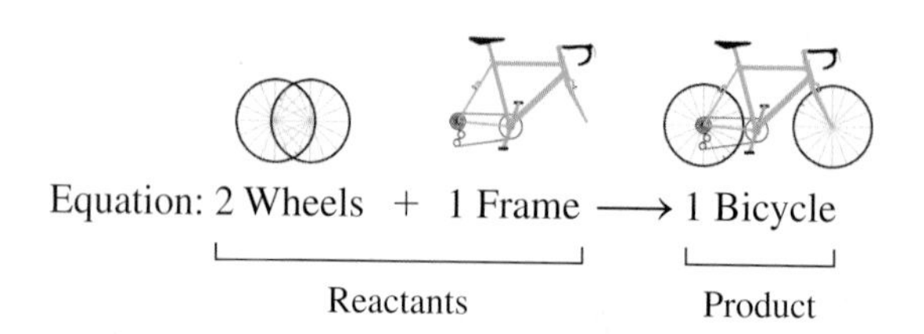

When you burn charcoal in a grill, the carbon in the charcoal combines with oxygen to form carbon dioxide. We can represent this reaction by a chemical equation.

ENGAGE

What is the evidence for a chemical change in the reaction of carbon and oxygen to form carbon dioxide?

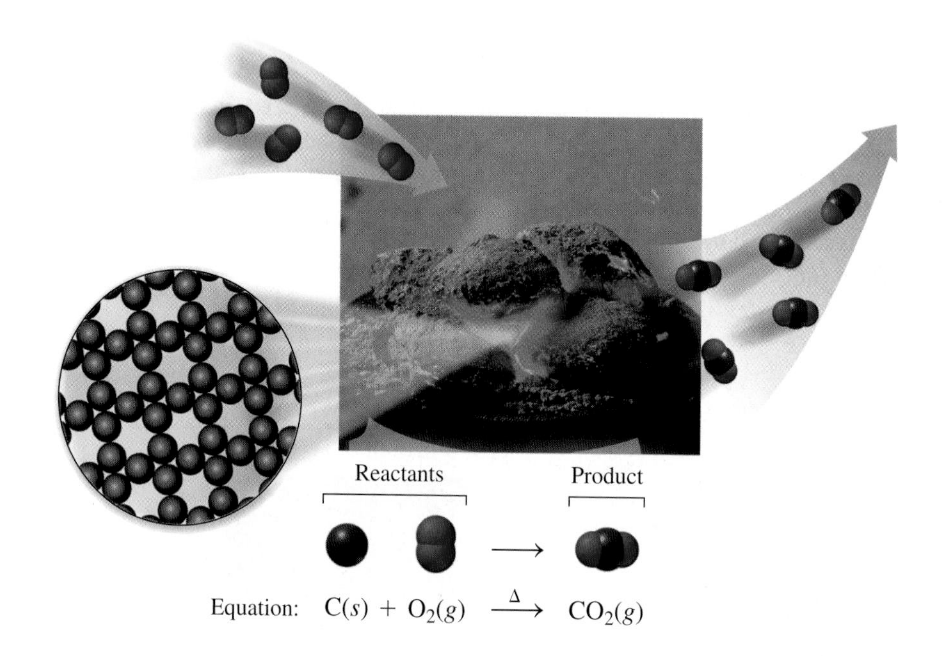

TABLE 7.3 Some Symbols Used in Writing Equations

Symbol	Meaning
+	Separates two or more formulas
⟶	Reacts to form products
(s)	Solid
(l)	Liquid
(g)	Gas
(aq)	Aqueous
$\xrightarrow{\Delta}$	Reactants are heated

In a **chemical equation**, the formulas of the **reactants** are written on the left of the arrow and the formulas of the **products** on the right. When there are two or more formulas on the same side, they are separated by plus (+) signs. The chemical equation for burning carbon is *balanced* because there is one carbon atom and two oxygen atoms in both the reactants and the products.

Generally, each formula in an equation is followed by an abbreviation, in parentheses, that gives the physical state of the substance: solid (s), liquid (l), or gas (g). If a substance is dissolved in water, it is an aqueous (aq) solution. The delta sign (Δ) indicates that heat was used to start the reaction. **TABLE 7.3** summarizes some of the symbols used in equations.

Identifying a Balanced Chemical Equation

When a chemical reaction takes place, the bonds between the atoms of the reactants are broken and new bonds are formed to give the products. All atoms are conserved, which means that atoms cannot be gained, lost, or changed into other types of atoms during a chemical reaction. Every chemical reaction must be written as a **balanced equation**, which shows the same number of atoms for each element in the reactants as well as in the products.

Now consider the reaction in which hydrogen reacts with oxygen to form water written as follows:

$H_2(g) + O_2(g) \longrightarrow H_2O(g)$ Not balanced

In the *balanced* equation, there are whole numbers called **coefficients** in front of the formulas. On the reactant side, the coefficient of 2 in front of the H_2 formula represents two molecules of hydrogen, which is 4 atoms of H. A coefficient of 1 is understood for O_2, which gives 2 atoms of O. On the product side, the coefficient of 2 in front of the H_2O formula represents 2 molecules of water. Because the coefficient of 2 multiplies all the atoms in H_2O, there are 4 hydrogen atoms and 2 oxygen atoms in the products. Because there are the same number of hydrogen atoms and oxygen atoms in the reactants as in the products, we know that the equation is *balanced*. This illustrates the *Law of Conservation of Matter*, which states that matter cannot be created or destroyed during a chemical reaction.

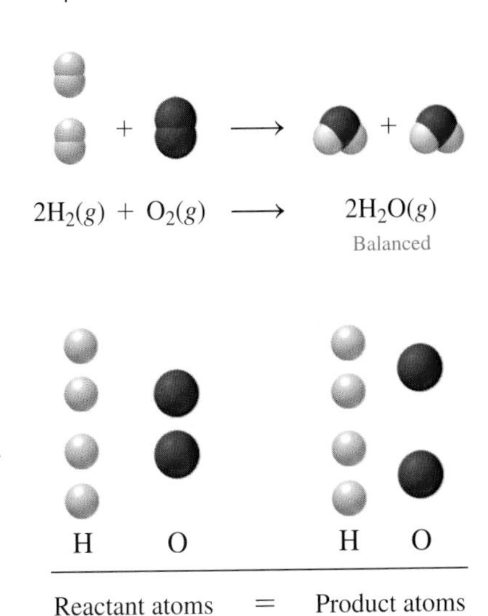

SAMPLE PROBLEM 7.6 Number of Atoms in Balanced Chemical Equations

TRY IT FIRST

Indicate the number of each type of atom in the following balanced chemical equation:

$Fe_2S_3(s) + 6HCl(aq) \longrightarrow 2FeCl_3(aq) + 3H_2S(g)$

	Reactants	Products
Fe		
S		
H		
Cl		

SOLUTION

The total number of atoms in each formula is obtained by multiplying the coefficient by each subscript in a chemical formula.

	Reactants	Products
Fe	2 (1 × 2)	2 (2 × 1)
S	3 (1 × 3)	3 (3 × 1)
H	6 (6 × 1)	6 (3 × 2)
Cl	6 (6 × 1)	6 (2 × 3)

STUDY CHECK 7.6

When ethane, C_2H_6, burns in oxygen, the products are carbon dioxide and water. The balanced chemical equation is written as

$2C_2H_6(g) + 7O_2(g) \xrightarrow{\Delta} 4CO_2(g) + 6H_2O(g)$

Calculate the number of each type of atom in the reactants and in the products.

ANSWER

In both the reactants and products, there are 4 C atoms, 12 H atoms, and 14 O atoms.

TEST

Try Practice Problems 7.35 and 7.36

Balancing a Chemical Equation

CORE CHEMISTRY SKILL

Balancing a Chemical Equation

The chemical reaction that occurs in the flame of a gas burner you use in the laboratory or a gas cooktop is the reaction of methane gas, CH_4, and oxygen to produce carbon dioxide and water. We now show the process of balancing a chemical equation in Sample Problem 7.7.

SAMPLE PROBLEM 7.7 Writing and Balancing a Chemical Equation

TRY IT FIRST

The chemical reaction of methane gas (CH_4) and oxygen gas (O_2) produces the gases carbon dioxide (CO_2) and water (H_2O). Write a balanced chemical equation for this reaction.

SOLUTION GUIDE

	Given	Need	Connect
ANALYZE THE PROBLEM	reactants, products	balanced equation	equal numbers of atoms in reactants and products

STEP 1 **Write an equation using the correct formulas for the reactants and products.**

$$CH_4(g) + O_2(g) \xrightarrow{\Delta} CO_2(g) + H_2O(g)$$

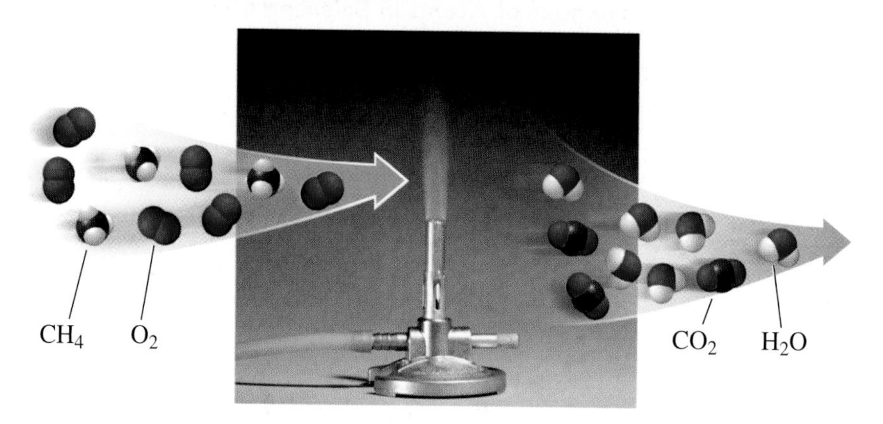

STEP 2 **Count the atoms of each element in the reactants and products.** When we count the atoms on the reactant side and the atoms on the product side, we see that there are more H atoms in the reactants and more O atoms in the products.

$$CH_4(g) + O_2(g) \xrightarrow{\Delta} CO_2(g) + H_2O(g)$$

Reactants	Products	
1 C atom	1 C atom	Balanced
4 H atoms	2 H atoms	Not balanced
2 O atoms	3 O atoms	Not balanced

STEP 3 **Use coefficients to balance each element.** We will start by balancing the H atoms in CH_4 because it has the most atoms. By placing a coefficient of 2 in front of the formula for H_2O, a total of 4 H atoms in the products is obtained. *Only use coefficients to balance an equation. Do not change any of the subscripts: This would alter the chemical formula of a reactant or product.*

$$CH_4(g) + O_2(g) \xrightarrow{\Delta} CO_2(g) + 2H_2O(g)$$

Reactants	Products	
1 C atom	1 C atom	Balanced
4 H atoms	4 H atoms	Balanced
2 O atoms	4 O atoms	Not balanced

We can balance the O atoms on the reactant side by placing a coefficient of 2 in front of the formula O_2. There are now 4 O atoms in both the reactants and products.

$$CH_4(g) + 2O_2(g) \xrightarrow{\Delta} CO_2(g) + 2H_2O(g)$$ Balanced

STEP 4 Check the final equation to confirm it is balanced.

$$CH_4(g) + 2O_2(g) \xrightarrow{\Delta} CO_2(g) + 2H_2O(g)$$ The equation is balanced.

Reactants	Products	
1 C atom	1 C atom	Balanced
4 H atoms	4 H atoms	Balanced
4 O atoms	4 O atoms	Balanced

In a balanced chemical equation, the coefficients must be the *lowest possible whole numbers*. Suppose you had obtained the following for the balanced equation:

$$2CH_4(g) + 4O_2(g) \xrightarrow{\Delta} 2CO_2(g) + 4H_2O(g)$$ Incorrect

Although there are equal numbers of atoms on both sides of the equation, this is not written correctly. To obtain coefficients that are the lowest whole numbers, we divide all the coefficients by 2.

ENGAGE

How do you check that a chemical equation is balanced?

STUDY CHECK 7.7

Balance the following chemical equation:

$$Al(s) + Cl_2(g) \longrightarrow AlCl_3(s)$$

ANSWER

$$2Al(s) + 3Cl_2(g) \longrightarrow 2AlCl_3(s)$$

Equations with Polyatomic Ions

Sometimes an equation contains the same polyatomic ion in both the reactants and the products. Then we can balance the polyatomic ions as a group on both sides of the equation, as shown in Sample Problem 7.8.

SAMPLE PROBLEM 7.8 Balancing Chemical Equations with Polyatomic Ions

TRY IT FIRST

Balance the following chemical equation:

$$Na_3PO_4(aq) + MgCl_2(aq) \longrightarrow Mg_3(PO_4)_2(s) + NaCl(aq)$$

SOLUTION GUIDE

	Given	Need	Connect
ANALYZE THE PROBLEM	reactants, products	balanced equation	equal numbers of atoms in reactants and products

STEP 1 Write an equation using the correct formulas for the reactants and products.

$$Na_3PO_4(aq) + MgCl_2(aq) \longrightarrow Mg_3(PO_4)_2(s) + NaCl(aq)$$ Not balanced

STEP 2 **Count the atoms of each element in the reactants and products.** When we compare the number of ions in the reactants and products, we find that the equation is not balanced. In this equation, we can balance the phosphate ion as a group of atoms because it appears on both sides of the equation.

ENGAGE

What is the evidence for a chemical reaction when $Na_3PO_4(aq)$ and $MgCl_2(aq)$ are mixed?

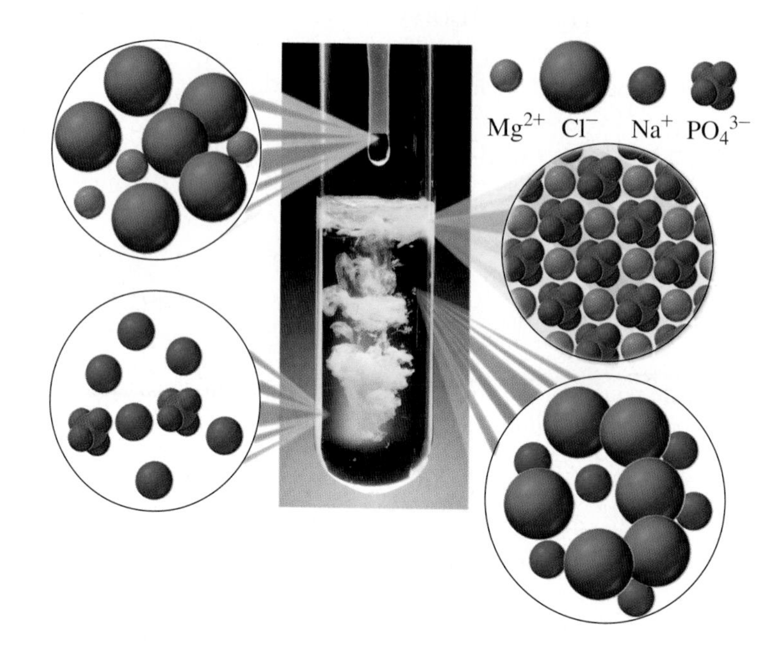

$$Na_3PO_4(aq) + MgCl_2(aq) \longrightarrow Mg_3(PO_4)_2(s) + NaCl(aq)$$

Reactants	Products	
3 Na^+	1 Na^+	Not balanced
1 $\mathbf{PO_4^{3-}}$	2 $\mathbf{PO_4^{3-}}$	Not balanced
1 Mg^{2+}	3 Mg^{2+}	Not balanced
2 Cl^-	1 Cl^-	Not balanced

STEP 3 **Use coefficients to balance each element.** We begin with the formula that has the highest subscript values, which in this equation is $Mg_3(PO_4)_2$. The subscript 3 in $Mg_3(PO_4)_2$ is used as a coefficient for $MgCl_2$ to balance magnesium. The subscript 2 in $Mg_3(PO_4)_2$ is used as a coefficient for Na_3PO_4 to balance the phosphate ion.

$$2Na_3PO_4(aq) + 3MgCl_2(aq) \longrightarrow Mg_3(PO_4)_2(s) + NaCl(aq)$$

Reactants	Products	
6 Na^+	1 Na^+	Not balanced
2 $\mathbf{PO_4^{3-}}$	2 $\mathbf{PO_4^{3-}}$	Balanced
3 Mg^{2+}	3 Mg^{2+}	Balanced
6 Cl^-	1 Cl^-	Not balanced

In the reactants and products, we see that the sodium and chloride ions are not yet balanced. A coefficient of 6 is placed in front of the NaCl to balance the equation.

$$2Na_3PO_4(aq) + 3MgCl_2(aq) \longrightarrow Mg_3(PO_4)_2(s) + 6NaCl(aq)$$

STEP 4 **Check the final equation to confirm it is balanced.**

$$2Na_3PO_4(aq) + 3MgCl_2(aq) \longrightarrow Mg_3(PO_4)_2(s) + 6NaCl(aq) \quad \text{Balanced}$$

Reactants	Products	
6 Na^+	6 Na^+	Balanced
2 PO_4^{3-}	2 PO_4^{3-}	Balanced
3 Mg^{2+}	3 Mg^{2+}	Balanced
6 Cl^-	6 Cl^-	Balanced

STUDY CHECK 7.8

Balance the following chemical equation:

$$Pb(NO_3)_2(aq) + AlBr_3(aq) \longrightarrow PbBr_2(s) + Al(NO_3)_3(aq)$$

ANSWER

$$3Pb(NO_3)_2(aq) + 2AlBr_3(aq) \longrightarrow 3PbBr_2(s) + 2Al(NO_3)_3(aq)$$

TEST

Try Practice Problems 7.37 to 7.40

PRACTICE PROBLEMS

7.4 Equations for Chemical Reactions

LEARNING GOAL Write a balanced chemical equation from the formulas of the reactants and products for a reaction; determine the number of atoms in the reactants and products.

7.35 Determine whether each of the following chemical equations is balanced or not balanced:
- **a.** $S(s) + O_2(g) \longrightarrow SO_3(g)$
- **b.** $2Ga(s) + 3Cl_2(g) \longrightarrow 2GaCl_3(s)$
- **c.** $H_2(g) + O_2(g) \longrightarrow H_2O(g)$
- **d.** $C_3H_8(g) + 5O_2(g) \xrightarrow{\Delta} 3CO_2(g) + 4H_2O(g)$

7.36 Determine whether each of the following chemical equations is balanced or not balanced:
- **a.** $PCl_3(s) + Cl_2(g) \longrightarrow PCl_5(s)$
- **b.** $CO(g) + 2H_2(g) \longrightarrow CH_4O(g)$
- **c.** $2KClO_3(s) \longrightarrow 2KCl(s) + O_2(g)$
- **d.** $Mg(s) + N_2(g) \longrightarrow Mg_3N_2(s)$

7.37 Balance each of the following chemical equations:
- **a.** $N_2(g) + O_2(g) \longrightarrow NO(g)$
- **b.** $HgO(s) \xrightarrow{\Delta} Hg(l) + O_2(g)$
- **c.** $Fe(s) + O_2(g) \longrightarrow Fe_2O_3(s)$
- **d.** $Na(s) + Cl_2(g) \longrightarrow NaCl(s)$

7.38 Balance each of the following chemical equations:
- **a.** $Ca(s) + Br_2(l) \longrightarrow CaBr_2(s)$
- **b.** $P_4(s) + O_2(g) \longrightarrow P_4O_{10}(s)$
- **c.** $Sb_2S_3(s) + HCl(aq) \longrightarrow SbCl_3(aq) + H_2S(g)$
- **d.** $Fe_2O_3(s) + C(s) \longrightarrow Fe(s) + CO(g)$

7.39 Balance each of the following chemical equations:
- **a.** $Mg(s) + AgNO_3(aq) \longrightarrow Mg(NO_3)_2(aq) + Ag(s)$
- **b.** $Al(s) + CuSO_4(aq) \longrightarrow Al_2(SO_4)_3(aq) + Cu(s)$
- **c.** $Pb(NO_3)_2(aq) + NaCl(aq) \longrightarrow PbCl_2(s) + NaNO_3(aq)$
- **d.** $Al(s) + HCl(aq) \longrightarrow H_2(g) + AlCl_3(aq)$

7.40 Balance each of the following chemical equations:
- **a.** $Zn(s) + HNO_3(aq) \longrightarrow H_2(g) + Zn(NO_3)_2(aq)$
- **b.** $Al(s) + H_2SO_4(aq) \longrightarrow H_2(g) + Al_2(SO_4)_3(aq)$
- **c.** $K_2SO_4(aq) + BaCl_2(aq) \longrightarrow BaSO_4(s) + KCl(aq)$
- **d.** $CaCO_3(s) \longrightarrow CaO(s) + CO_2(g)$

7.5 Types of Chemical Reactions

LEARNING GOAL Identify a chemical reaction as a combination, decomposition, single replacement, double replacement, or combustion.

CORE CHEMISTRY SKILL

Classifying Types of Chemical Reactions

A great number of chemical reactions occur in nature, in biological systems, and in the laboratory. However, there are some general patterns that help us classify most reactions into five general types.

Combination Reactions

In a **combination reaction**, two or more elements or compounds bond to form one product. For example, sulfur and oxygen combine to form the product sulfur dioxide.

Combination

Two or more reactants | combine to yield | a single product

+

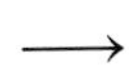

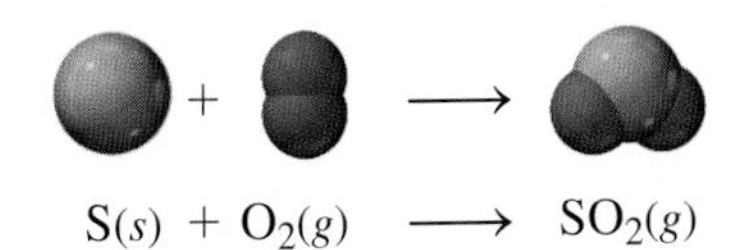

$$S(s) + O_2(g) \longrightarrow SO_2(g)$$

In **FIGURE 7.4**, the elements magnesium and oxygen combine to form a single product, which is the ionic compound magnesium oxide formed from Mg^{2+} and O^{2-} ions.

$$2Mg(s) + O_2(g) \longrightarrow 2MgO(s)$$

In other examples of combination reactions, elements or compounds combine to form a single product.

$$N_2(g) + 3H_2(g) \longrightarrow 2NH_3(g)$$
$$Cu(s) + S(s) \longrightarrow CuS(s)$$
$$MgO(s) + CO_2(g) \longrightarrow MgCO_3(s)$$

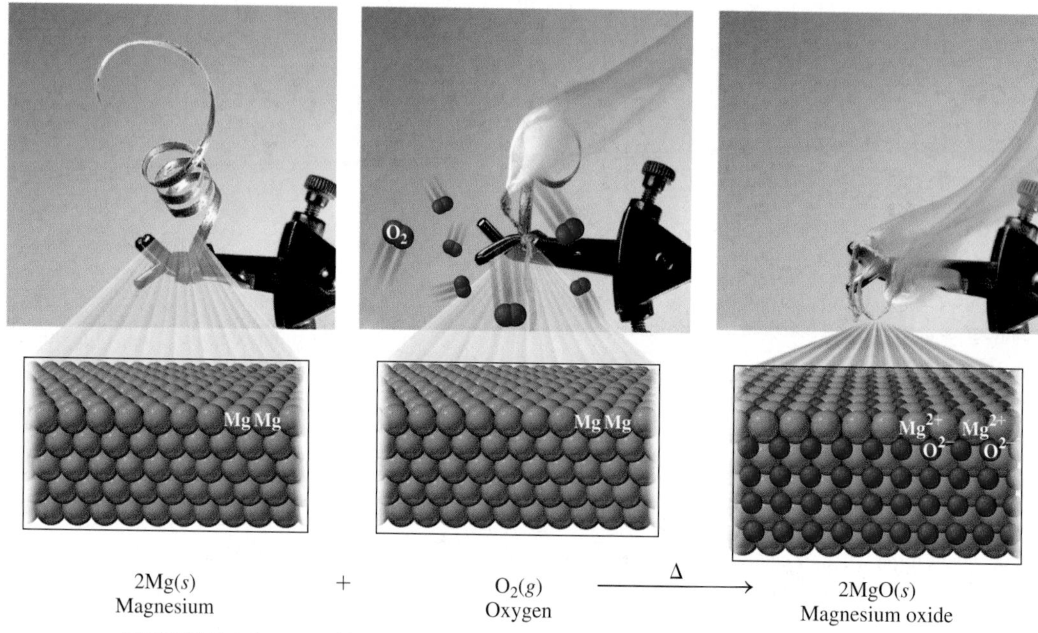

FIGURE 7.4 In a combination reaction, two or more substances combine to form one substance as product.

Q What happens to the atoms in the reactants in a combination reaction?

Decomposition Reactions

Decomposition

A reactant splits into two or more products

AB ⟶ A + B

In a **decomposition reaction**, a reactant splits into two or more simpler products. For example, when mercury(II) oxide is heated, the compound breaks apart into mercury atoms and oxygen (see **FIGURE 7.5**).

$$2HgO(s) \xrightarrow{\Delta} 2Hg(l) + O_2(g)$$

In another example of a decomposition reaction, when calcium carbonate is heated, it breaks apart into simpler compounds of calcium oxide and carbon dioxide.

$$CaCO_3(s) \xrightarrow{\Delta} CaO(s) + CO_2(g)$$

Replacement Reactions

In a replacement reaction, elements in a compound are replaced by other elements. In a **single replacement reaction**, a reacting element switches places with an element in the other reacting compound.

In the single replacement reaction shown in **FIGURE 7.6**, zinc replaces hydrogen in hydrochloric acid, $HCl(aq)$.

Single replacement

One element replaces another element

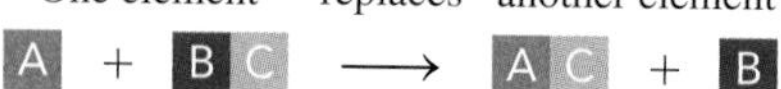

$$Zn(s) + 2HCl(aq) \longrightarrow H_2(g) + ZnCl_2(aq)$$

In another single replacement reaction, chlorine replaces bromine in the compound potassium bromide.

$$Cl_2(g) + 2KBr(s) \longrightarrow 2KCl(s) + Br_2(l)$$

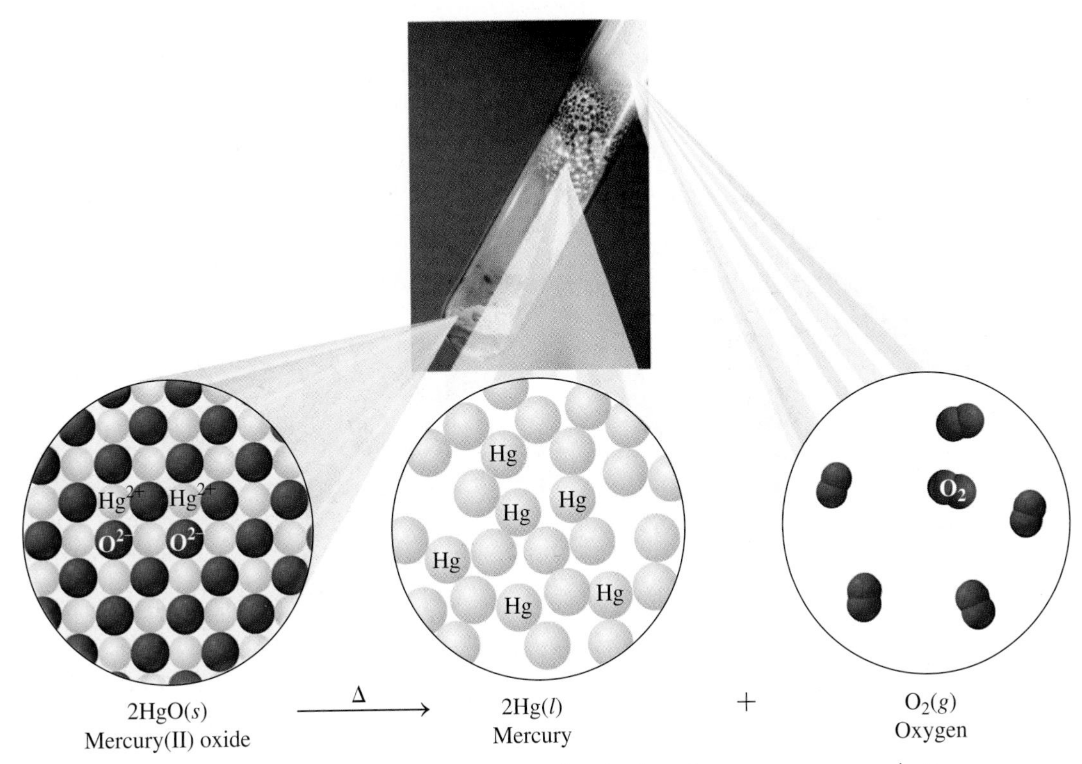

FIGURE 7.5 In a decomposition reaction, one reactant breaks down into two or more products.

Q How do the differences in the reactant and products classify this as a decomposition reaction?

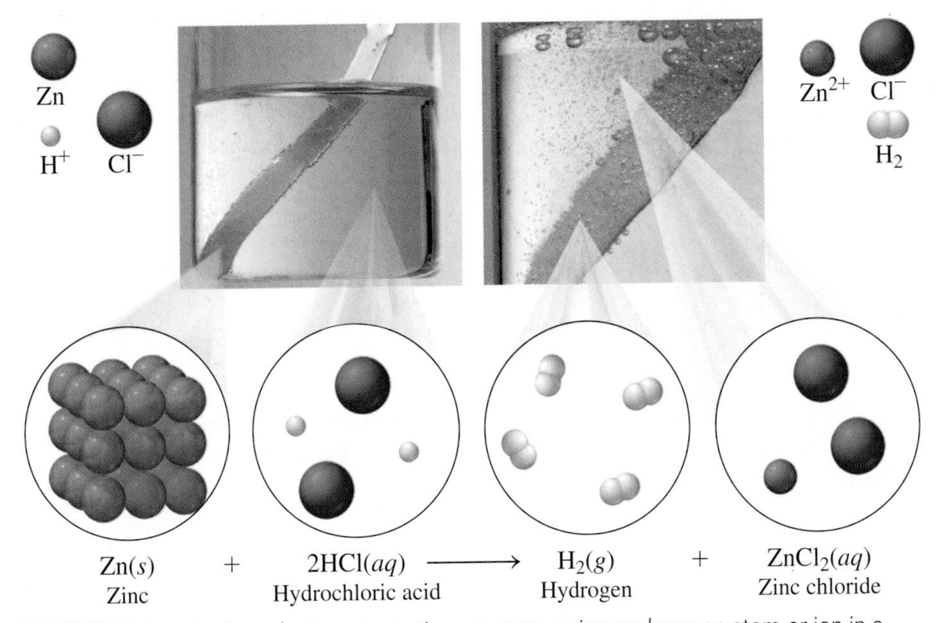

FIGURE 7.6 In a single replacement reaction, an atom or ion replaces an atom or ion in a compound.

Q What changes in the formulas of the reactants identify this equation as a single replacement?

In a **double replacement reaction**, the positive ions in the reacting compounds switch places.

In the reaction shown in **FIGURE 7.7**, barium ions change places with sodium ions in the reactants to form sodium chloride and a white solid precipitate of barium sulfate. The formulas of the products depend on the charges of the ions.

$$BaCl_2(aq) + Na_2SO_4(aq) \longrightarrow BaSO_4(s) + 2NaCl(aq)$$

When sodium hydroxide and hydrochloric acid (HCl) react, sodium and hydrogen ions switch places, forming water and sodium chloride.

$$NaOH(aq) + HCl(aq) \longrightarrow H_2O(l) + NaCl(aq)$$

Double replacement

Two elements replace each other

$$AB + CD \longrightarrow AD + CB$$

ENGAGE

How can you distinguish a single replacement reaction from a double replacement reaction?

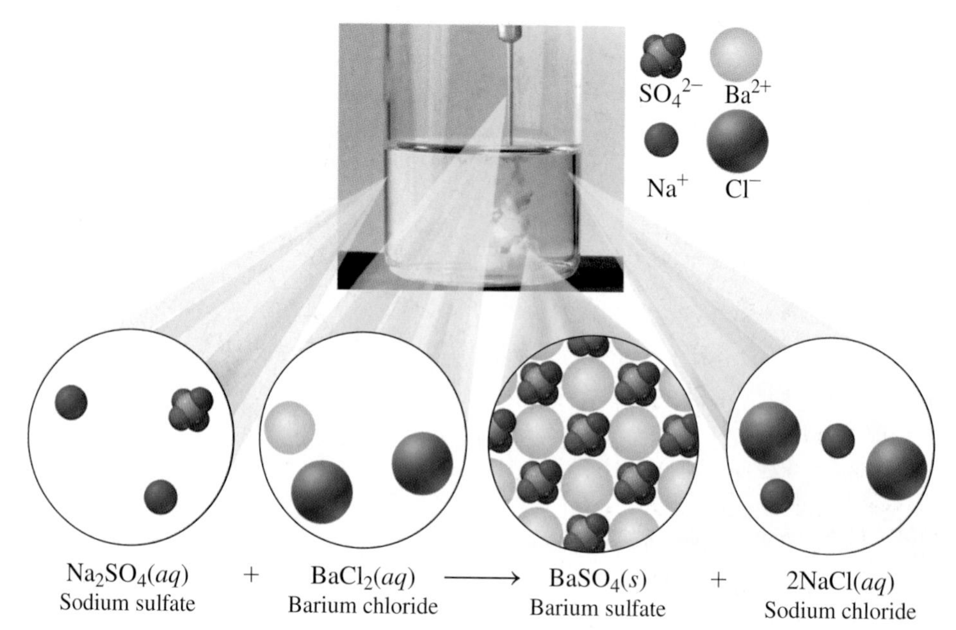

FIGURE 7.7 In a double replacement reaction, the positive ions in the reactants replace each other.

Q How do the changes in the formulas of the reactants identify this equation as a double replacement reaction?

Combustion Reactions

In a combustion reaction, a candle burns using the oxygen in the air.

The burning of a candle and the burning of fuel in the engine of a car are examples of combustion reactions. In a **combustion reaction**, a carbon-containing compound, usually a fuel, burns in oxygen gas to produce the gases carbon dioxide (CO_2), water (H_2O), and energy in the form of heat or a flame. For example, methane gas (CH_4) undergoes combustion when used to cook our food on a gas cooktop and to heat our homes. In the equation for the combustion of methane, each element in the fuel (CH_4) forms a compound with oxygen.

$$\underset{\text{Methane}}{CH_4(g)} + 2O_2(g) \xrightarrow{\Delta} CO_2(g) + 2H_2O(g) + \text{energy}$$

The balanced equation for the combustion of propane (C_3H_8) is

$$C_3H_8(g) + 5O_2(g) \xrightarrow{\Delta} 3CO_2(g) + 4H_2O(g) + \text{energy}$$

Propane is the fuel used in portable heaters and gas barbecues. Gasoline, a mixture of liquid hydrocarbons, is the fuel that powers our cars, lawn mowers, and snow blowers.

TABLE 7.4 summarizes the reaction types and gives examples.

TABLE 7.4 Summary of Reaction Types

Reaction Type	Example
Combination	
$A + B \longrightarrow AB$	$Ca(s) + Cl_2(g) \longrightarrow CaCl_2(s)$
Decomposition	
$AB \longrightarrow A + B$	$Fe_2S_3(s) \longrightarrow 2Fe(s) + 3S(s)$
Single Replacement	
$A + BC \longrightarrow AC + B$	$Cu(s) + 2AgNO_3(aq) \longrightarrow 2Ag(s) + Cu(NO_3)_2(aq)$
Double Replacement	
$AB + CD \longrightarrow AD + CB$	$BaCl_2(aq) + K_2SO_4(aq) \longrightarrow BaSO_4(s) + 2KCl(aq)$
Combustion	
$C_XH_Y + ZO_2(g) \xrightarrow{\Delta} XCO_2(g) + \frac{Y}{2}H_2O(g) + \text{energy}$	$CH_4(g) + 2O_2(g) \xrightarrow{\Delta} CO_2(g) + 2H_2O(g) + \text{energy}$

SAMPLE PROBLEM 7.9 Identifying Reactions

TRY IT FIRST

Classify each of the following as a combination, decomposition, single replacement, double replacement, or combustion reaction:

a. $2Fe_2O_3(s) + 3C(s) \longrightarrow 3CO_2(g) + 4Fe(s)$
b. $2KClO_3(s) \xrightarrow{\Delta} 2KCl(s) + 3O_2(g)$
c. $C_2H_4(g) + 3O_2(g) \xrightarrow{\Delta} 2CO_2(g) + 2H_2O(g) + \text{energy}$

SOLUTION

a. In this single replacement reaction, a C atom replaces Fe in Fe_2O_3 to form the compound CO_2 and Fe atoms.
b. When one reactant breaks down to produce two products, the reaction is decomposition.
c. The reaction of a carbon compound with oxygen to produce carbon dioxide, water, and energy makes this a combustion reaction.

STUDY CHECK 7.9

Nitrogen gas (N_2) and oxygen gas (O_2) react to form nitrogen dioxide gas. Write the balanced chemical equation using the correct chemical formulas of the reactants and product, and identify the reaction type.

ANSWER

$N_2(g) + 2O_2(g) \longrightarrow 2NO_2(g)$ Combination

TEST
Try Practice Problems 7.41 to 7.44

PRACTICE PROBLEMS

7.5 Types of Chemical Reactions

LEARNING GOAL Identify a reaction as a combination, decomposition, single replacement, double replacement, or combustion.

7.41 Classify each of the following as a combination, decomposition, single replacement, double replacement, or combustion reaction:
a. $2Al_2O_3(s) \xrightarrow{\Delta} 4Al(s) + 3O_2(g)$
b. $Br_2(l) + BaI_2(s) \longrightarrow BaBr_2(s) + I_2(s)$
c. $2C_2H_2(g) + 5O_2(g) \xrightarrow{\Delta} 4CO_2(g) + 2H_2O(g)$
d. $BaCl_2(aq) + K_2CO_3(aq) \longrightarrow BaCO_3(s) + 2KCl(aq)$
e. $Pb(s) + O_2(g) \longrightarrow PbO_2(s)$

7.42 Classify each of the following as a combination, decomposition, single replacement, double replacement, or combustion reaction:
a. $H_2(g) + Br_2(l) \longrightarrow 2HBr(g)$
b. $AgNO_3(aq) + NaCl(aq) \longrightarrow AgCl(s) + NaNO_3(aq)$
c. $2H_2O_2(aq) \longrightarrow 2H_2O(l) + O_2(g)$
d. $Zn(s) + CuCl_2(aq) \longrightarrow Cu(s) + ZnCl_2(aq)$
e. $C_5H_8(g) + 7O_2(g) \xrightarrow{\Delta} 5CO_2(g) + 4H_2O(g)$

7.43 Classify each of the following as a combination, decomposition, single replacement, double replacement, or combustion reaction:
a. $4Fe(s) + 3O_2(g) \longrightarrow 2Fe_2O_3(s)$
b. $Mg(s) + 2AgNO_3(aq) \longrightarrow 2Ag(s) + Mg(NO_3)_2(aq)$
c. $CuCO_3(s) \xrightarrow{\Delta} CuO(s) + CO_2(g)$
d. $Al_2(SO_4)_3(aq) + 6KOH(aq) \longrightarrow 2Al(OH)_3(s) + 3K_2SO_4(aq)$
e. $C_4H_8(g) + 6O_2(g) \xrightarrow{\Delta} 4CO_2(g) + 4H_2O(g)$

7.44 Classify each of the following as a combination, decomposition, single replacement, double replacement, or combustion reaction:
a. $CuO(s) + 2HCl(aq) \longrightarrow CuCl_2(aq) + H_2O(l)$
b. $2Al(s) + 3Br_2(l) \longrightarrow 2AlBr_3(s)$
c. $2C_2H_2(g) + 5O_2(g) \xrightarrow{\Delta} 4CO_2(g) + 2H_2O(g)$
d. $Fe_2O_3(s) + 3C(s) \longrightarrow 2Fe(s) + 3CO(g)$
e. $C_6H_{12}O_6(aq) \longrightarrow 2C_2H_6O(aq) + 2CO_2(g)$

CHEMISTRY LINK TO HEALTH

Incomplete Combustion: Toxicity of Carbon Monoxide

When a propane heater, fireplace, or woodstove is used in a closed room, there must be adequate ventilation. If the supply of oxygen is limited, incomplete combustion from burning gas, oil, or wood produces carbon monoxide. The incomplete combustion of methane in natural gas is written

$$2CH_4(g) + \underset{\text{Limited oxygen supply}}{3O_2(g)} \xrightarrow{\Delta} \underset{\text{Carbon monoxide}}{2CO(g)} + 4H_2O(g) + \text{energy}$$

Carbon monoxide (CO) is a colorless, odorless, poisonous gas. When inhaled, CO passes into the bloodstream, where it attaches to hemoglobin, which reduces the amount of oxygen (O_2) reaching the cells. As a result, a person can experience a reduction in exercise capability, visual perception, and manual dexterity.

Hemoglobin is the protein that transports O_2 in the blood. When the amount of hemoglobin bound to CO (COHb) is about 10%, a person may experience shortness of breath, mild headache, and drowsiness. Heavy smokers can have levels of COHb in their blood as high as 9%. When as much as 30% of the hemoglobin is bound to CO, a person may experience more severe symptoms, including dizziness, mental confusion, severe headache, and nausea. If 50% or more of the hemoglobin is bound to CO, a person could become unconscious and die if not treated immediately with oxygen.

7.6 Oxidation–Reduction Reactions

LEARNING GOAL Define the terms oxidation and reduction; identify the reactants oxidized and reduced.

Rust forms when the oxygen in the air reacts with iron.

Perhaps you have never heard of an oxidation and reduction reaction. However, this type of reaction has many important applications in your everyday life. When you see a rusty nail, tarnish on a silver spoon, or corrosion on metal, you are observing oxidation.

$$4Fe(s) + 3O_2(g) \longrightarrow 2Fe_2O_3(s) \quad \text{Fe is oxidized}$$
Rust

When we turn the lights on in our automobiles, an oxidation–reduction reaction within the car battery provides the electricity. On a cold, wintry day, we might build a fire. As the wood burns, oxygen combines with carbon and hydrogen to produce carbon dioxide, water, and heat. In the previous Section, we called this a combustion reaction, but it is also an *oxidation–reduction reaction*. When we eat foods with starches in them, the starches break down to give glucose, which is oxidized in our cells to give us energy along with carbon dioxide and water. Every breath we take provides oxygen to carry out oxidation in our cells.

$$C_6H_{12}O_6(aq) + 6O_2(g) \longrightarrow 6CO_2(g) + 6H_2O(l) + \text{energy}$$
Glucose

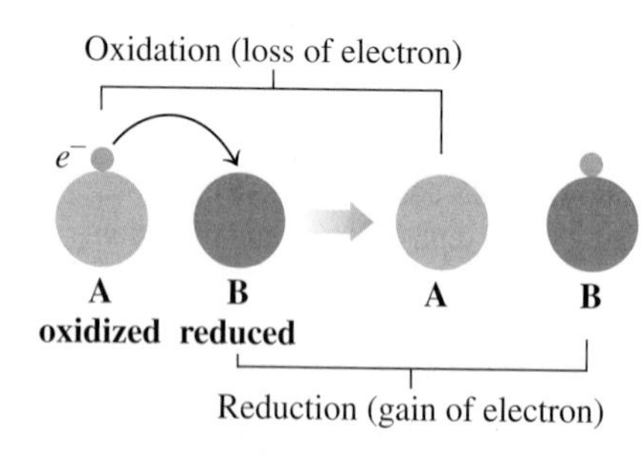

Oxidation–Reduction Reactions

In an **oxidation–reduction reaction** (*redox*), electrons are transferred from one substance to another. If one substance loses electrons, another substance must gain electrons. **Oxidation** is defined as the *loss* of electrons; **reduction** is the *gain* of electrons.

One way to remember these definitions is to use the following acronym:

OIL RIG

Oxidation **I**s **L**oss of electrons

Reduction **I**s **G**ain of electrons

In general, atoms of metals lose electrons to form positive ions, whereas nonmetals gain electrons to form negative ions. Now we can say that metals are oxidized and nonmetals are reduced.

The green color that appears on copper surfaces from weathering, known as *patina*, is a mixture of $CuCO_3$ and CuO. We can now look at the oxidation and reduction reactions that take place when copper metal reacts with oxygen in the air to produce copper(II) oxide.

$$2Cu(s) + O_2(g) \longrightarrow 2CuO(s)$$

The element Cu in the reactants has a charge of 0, but in the CuO product, it is present as Cu^{2+}, which has a 2+ charge. Because the Cu atom lost two electrons, the charge is more positive. This means that Cu was oxidized in the reaction.

$$Cu^0(s) \longrightarrow Cu^{2+}(s) + 2\,e^- \quad \text{Oxidation: loss of electrons by Cu}$$

The green patina on copper is due to oxidation.

At the same time, the element O in the reactants has a charge of 0, but in the CuO product, it is present as O^{2-}, which has a 2− charge. Because the O atom has gained two electrons, the charge is more negative. This means that O was reduced in the reaction.

$$O_2^0(g) + 4\,e^- \longrightarrow 2O^{2-}(s) \quad \text{Reduction: gain of electrons by O}$$

Thus, the overall equation for the formation of CuO involves an oxidation and a reduction that occur simultaneously. In every oxidation and reduction, the number of electrons lost must be equal to the number of electrons gained. Therefore, we multiply the oxidation reaction of Cu by 2. Canceling the 4 e^- on each side, we obtain the overall oxidation–reduction equation for the formation of CuO.

CORE CHEMISTRY SKILL

Identifying Oxidized and Reduced Substances

$$2Cu(s) \longrightarrow 2Cu^{2+}(s) + \cancel{4e^-} \quad \text{Oxidation}$$

$$O_2(g) + \cancel{4e^-} \longrightarrow 2O^{2-}(s) \quad \text{Reduction}$$

$$2Cu(s) + O_2(g) \longrightarrow 2CuO(s) \quad \text{Oxidation–reduction equation}$$

As we see in the next reaction between zinc and copper(II) sulfate, there is always an oxidation with every reduction (see **FIGURE 7.8**). We write the equation to show the atoms and ions.

$$\mathbf{Zn}(s) + \mathbf{Cu^{2+}}(aq) + SO_4^{2-}(aq) \longrightarrow \mathbf{Zn^{2+}}(aq) + SO_4^{2-}(aq) + \mathbf{Cu}(s)$$

In this reaction, Zn atoms lose two electrons to form Zn^{2+}. The increase in positive charge indicates that Zn is oxidized. At the same time, Cu^{2+} gains two electrons. The decrease in charge indicates that Cu is reduced. The SO_4^{2-} ions are *spectator ions*, which are present in both the reactants and products and do not change.

$$Zn(s) \longrightarrow Zn^{2+}(aq) + 2\,e^- \quad \text{Oxidation of Zn}$$

$$Cu^{2+}(aq) + 2\,e^- \longrightarrow Cu(s) \quad \text{Reduction of } Cu^{2+}$$

In this single replacement reaction, zinc was oxidized and copper(II) ion was reduced.

Reduced		Oxidized
Na	Oxidation: Lose e^- →	$Na^+ + e^-$
Ca		$Ca^{2+} + 2\,e^-$
$2Br^-$	← Reduction: Gain e^-	$Br_2 + 2\,e^-$
Fe^{2+}		$Fe^{3+} + e^-$

Oxidation is a loss of electrons; reduction is a gain of electrons.

FIGURE 7.8 In this single replacement reaction, Zn(s) is oxidized to $Zn^{2+}(aq)$ when it provides two electrons to reduce $Cu^{2+}(aq)$ to Cu(s):

$$Zn(s) + CuSO_4(aq) \longrightarrow ZnSO_4(aq) + Cu(s)$$

Q In the oxidation, does Zn(s) lose or gain electrons?

ENGAGE

How can you determine that Cu^{2+} is reduced in this reaction?

Oxidation and Reduction in Biological Systems

Oxidation may also involve the addition of oxygen or the loss of hydrogen, and reduction may involve the loss of oxygen or the gain of hydrogen. In the cells of the body, oxidation of organic (carbon) compounds involves the transfer of hydrogen atoms (H), which are composed of electrons and protons. For example, the oxidation of a typical biochemical molecule can involve the transfer of two hydrogen atoms (or $2H^+$ and $2\ e^-$) to a hydrogen ion acceptor such as the coenzyme FAD (flavin adenine dinucleotide). The coenzyme is reduced to $FADH_2$.

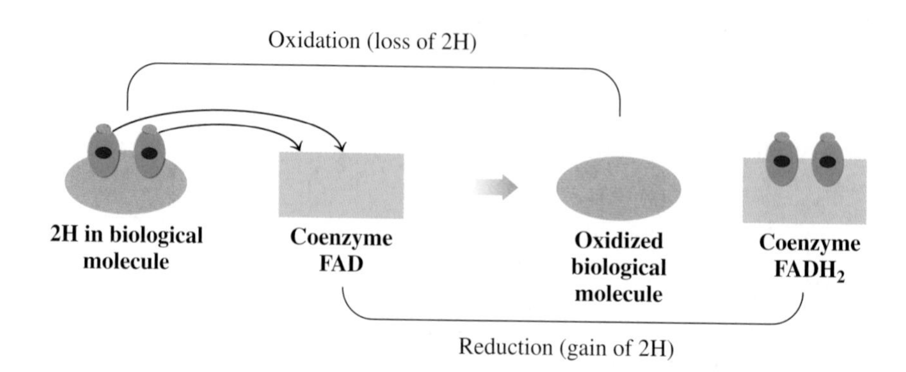

In many biochemical oxidation–reduction reactions, the transfer of hydrogen atoms is necessary for the production of energy in the cells. For example, methyl alcohol (CH_4O), a poisonous substance, is metabolized in the body by the following reactions:

$$\underset{\text{Methyl alcohol}}{CH_4O} \longrightarrow \underset{\text{Formaldehyde}}{CH_2O} + 2H \quad \text{Oxidation: loss of H atoms}$$

The formaldehyde can be oxidized further, this time by the addition of oxygen, to produce formic acid.

$$\underset{\text{Formaldehyde}}{2CH_2O} + O_2 \longrightarrow \underset{\text{Formic acid}}{2CH_2O_2} \quad \text{Oxidation: addition of O atoms}$$

Finally, formic acid is oxidized to carbon dioxide and water.

$$\underset{\text{Formic acid}}{2CH_2O_2} + O_2 \longrightarrow 2CO_2 + 2H_2O \quad \text{Oxidation: addition of O atoms}$$

The intermediate products of the oxidation of methyl alcohol are quite toxic, causing blindness and possibly death as they interfere with key reactions in the cells of the body.

In summary, we find that the particular definition of oxidation and reduction we use depends on the process that occurs in the reaction. All these definitions are summarized in **TABLE 7.5**. Oxidation always involves a loss of electrons, but it may also be seen as an addition of oxygen or the loss of hydrogen atoms. A reduction always involves a gain of electrons and may also be seen as the loss of oxygen or the gain of hydrogen.

TABLE 7.5 Characteristics of Oxidation and Reduction

Always Involves	May Involve
Oxidation	
Loss of electrons	Addition of oxygen Loss of hydrogen
Reduction	
Gain of electrons	Loss of oxygen Gain of hydrogen

TEST

Try Practice Problems 7.45 to 7.52

PRACTICE PROBLEMS

7.6 Oxidation–Reduction Reactions

LEARNING GOAL Define the terms oxidation and reduction; identify the reactants oxidized and reduced.

7.45 Identify each of the following as an oxidation or a reduction:
- **a.** $Na^+(aq) + e^- \longrightarrow Na(s)$
- **b.** $Ni(s) \longrightarrow Ni^{2+}(aq) + 2\ e^-$
- **c.** $Cr^{3+}(aq) + 3\ e^- \longrightarrow Cr(s)$
- **d.** $2H^+(aq) + 2\ e^- \longrightarrow H_2(g)$

7.46 Identify each of the following as an oxidation or a reduction:
- **a.** $O_2(g) + 4\ e^- \longrightarrow 2O^{2-}(aq)$
- **b.** $Ag(s) \longrightarrow Ag^+(aq) + e^-$
- **c.** $Fe^{3+}(aq) + e^- \longrightarrow Fe^{2+}(aq)$
- **d.** $2Br^-(aq) \longrightarrow Br_2(l) + 2\ e^-$

7.47 In each of the following, identify the reactant that is oxidized and the reactant that is reduced:
- **a.** $Zn(s) + Cl_2(g) \longrightarrow ZnCl_2(s)$
- **b.** $Cl_2(g) + 2NaBr(aq) \longrightarrow 2NaCl(aq) + Br_2(l)$
- **c.** $2PbO(s) \longrightarrow 2Pb(s) + O_2(g)$
- **d.** $2Fe^{3+}(aq) + Sn^{2+}(aq) \longrightarrow 2Fe^{2+}(aq) + Sn^{4+}(aq)$

7.48 In each of the following, identify the reactant that is oxidized and the reactant that is reduced:

a. $2Li(s) + F_2(g) \longrightarrow 2LiF(s)$

b. $Cl_2(g) + 2KI(aq) \longrightarrow I_2(s) + 2KCl(aq)$

c. $2Al(s) + 3Sn^{2+}(aq) \longrightarrow 3Sn(s) + 2Al^{3+}(aq)$

d. $Fe(s) + CuSO_4(aq) \longrightarrow Cu(s) + FeSO_4(aq)$

Clinical Applications

7.49 In the mitochondria of human cells, energy is provided by the oxidation and reduction reactions of the iron ions in the cytochromes in electron transport. Identify each of the following as an oxidation or a reduction:

a. $Fe^{3+} + e^- \longrightarrow Fe^{2+}$

b. $Fe^{2+} \longrightarrow Fe^{3+} + e^-$

7.50 Chlorine (Cl_2) is a strong germicide used to disinfect drinking water and to kill microbes in swimming pools. If the product is Cl^-, was the elemental chlorine oxidized or reduced?

7.51 When linoleic acid, an unsaturated fatty acid, reacts with hydrogen, it forms a saturated fatty acid. Is linoleic acid oxidized or reduced in the hydrogenation reaction?

$$\underset{\text{Linoleic acid}}{C_{18}H_{32}O_2} + 2H_2 \longrightarrow C_{18}H_{36}O_2$$

7.52 In one of the reactions in the citric acid cycle, which provides energy, succinic acid is converted to fumaric acid.

$$\underset{\text{Succinic acid}}{C_4H_6O_4} \longrightarrow \underset{\text{Fumaric acid}}{C_4H_4O_4} + 2H$$

The reaction is accompanied by a coenzyme, flavin adenine dinucleotide (FAD).

$$FAD + 2H \longrightarrow FADH_2$$

a. Is succinic acid oxidized or reduced?

b. Is FAD oxidized or reduced?

c. Why would the two reactions occur together?

7.7 Mole Relationships in Chemical Equations

LEARNING GOAL Use a mole–mole factor from a balanced chemical equation to calculate the number of moles of another substance in the reaction.

In any chemical reaction, the total amount of matter in the reactants is equal to the total amount of matter in the products. Thus, the total mass of all the reactants must be equal to the total mass of all the products. This is known as the *Law of Conservation of Mass*, which states that there is no change in the total mass of the substances reacting in a chemical reaction. Thus, no material is lost or gained as original substances are changed to new substances.

For example, tarnish (Ag_2S) forms when silver reacts with sulfur to form silver sulfide.

$$2Ag(s) + S(s) \longrightarrow Ag_2S(s)$$

$2Ag(s)$ + $S(s)$ $\longrightarrow$ $Ag_2S(s)$

Mass of reactants = Mass of product

In the chemical reaction of Ag and S, the mass of the reactants is the same as the mass of the product, Ag_2S.

In this reaction, the number of silver atoms that reacts is twice the number of sulfur atoms. When 200 silver atoms react, 100 sulfur atoms are required. However, in the actual chemical reaction, many more atoms of both silver and sulfur would react. If we are dealing with moles of silver and sulfur, then the coefficients in the equation can be interpreted in terms of moles. Thus, 2 moles of silver react with 1 mole of sulfur to form 1 mole of Ag_2S. Because the molar mass of each can be determined, the moles of Ag, S, and Ag_2S can also be stated in terms of mass in grams of each. Thus, 215.8 g of Ag and 32.1 g of S react to form 247.9 g of Ag_2S. The total mass of the reactants (247.9 g) is equal to the mass of product (247.9 g). The various ways in which a chemical equation can be interpreted are seen in **TABLE 7.6**.

TABLE 7.6 Information Available from a Balanced Equation

	Reactants		Products
Equation	$2\,Ag(s)$	$+\ S(s)$	$\longrightarrow Ag_2S(s)$
Atoms	2 Ag atoms	+ 1 S atom	$\longrightarrow$ 1 Ag_2S formula unit
	200 Ag atoms	+ 100 S atoms	$\longrightarrow$ 100 Ag_2S formula units
Avogadro's Number of Atoms	$2(6.02 \times 10^{23})$ Ag atoms	$+\ 1(6.02 \times 10^{23})$ S atoms	$\longrightarrow 1(6.02 \times 10^{23})$ Ag_2S formula units
Moles	2 moles of Ag	+ 1 mole of S	$\longrightarrow$ 1 mole of Ag_2S
Mass (g)	2(107.9 g) of Ag	+ 1(32.07 g) of S	$\longrightarrow$ 1(247.9 g) of Ag_2S
Total Mass (g)	247.9 g		$\longrightarrow$ 247.9 g

Mole–Mole Factors from a Balanced Equation

When iron reacts with sulfur, the product is iron(III) sulfide.

$$2Fe(s) + 3S(s) \longrightarrow Fe_2S_3(s)$$

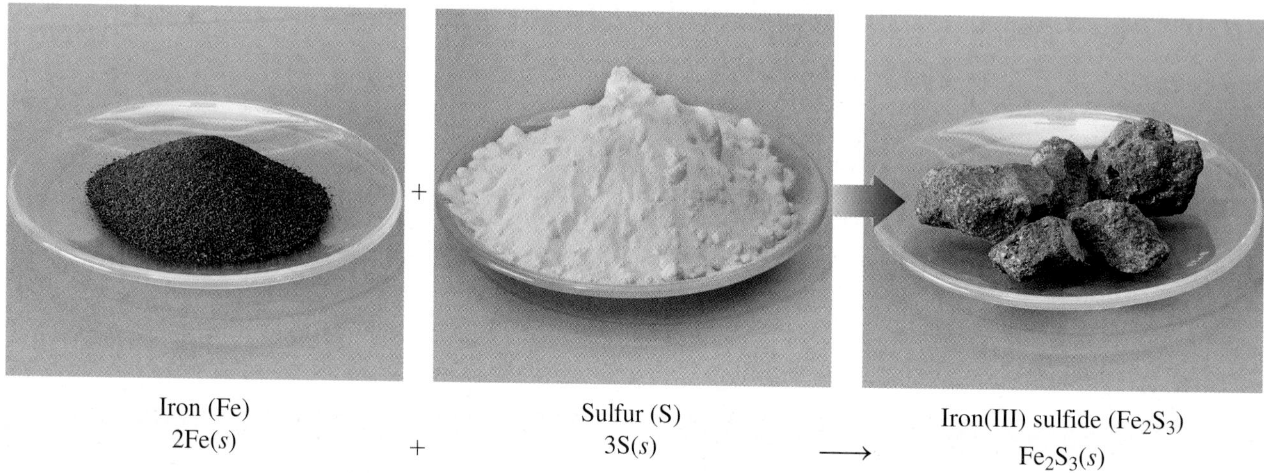

Iron (Fe) $2Fe(s)$ + Sulfur (S) $3S(s)$ $\longrightarrow$ Iron(III) sulfide (Fe_2S_3) $Fe_2S_3(s)$

In the chemical reaction of Fe and S, the mass of the reactants is the same as the mass of the product, Fe_2S_3.

From the balanced chemical equation, we see that 2 moles of iron reacts with 3 moles of sulfur to form 1 mole of iron(III) sulfide. Actually, any amount of iron or sulfur may be used, but the *ratio* of iron reacting with sulfur will always be the same. From the coefficients, we can write **mole–mole factors** between reactants and between reactants and products. The coefficients used in the mole–mole factors are *exact* numbers; they do not limit the number of significant figures.

Fe and S: $\dfrac{2\text{ moles Fe}}{3\text{ moles S}}$ and $\dfrac{3\text{ moles S}}{2\text{ moles Fe}}$

Fe and Fe_2S_3: $\dfrac{2\text{ moles Fe}}{1\text{ mole }Fe_2S_3}$ and $\dfrac{1\text{ mole }Fe_2S_3}{2\text{ moles Fe}}$

S and Fe_2S_3: $\dfrac{3\text{ moles S}}{1\text{ mole }Fe_2S_3}$ and $\dfrac{1\text{ mole }Fe_2S_3}{3\text{ moles S}}$

TEST

Try Practice Problems 7.53 and 7.54

CORE CHEMISTRY SKILL

Using Mole–Mole Factors

Using Mole–Mole Factors in Calculations

Whenever you prepare a recipe, adjust an engine for the proper mixture of fuel and air, or prepare medicines in a pharmaceutical laboratory, you need to know the proper amounts of reactants to use and how much of the product will form. Now that we have written all the possible conversion factors for the balanced equation $2Fe(s) + 3S(s) \longrightarrow Fe_2S_3(s)$, we will use those mole–mole factors in a chemical calculation in Sample Problem 7.10.

SAMPLE PROBLEM 7.10 Calculating Moles of a Reactant

TRY IT FIRST

In the chemical reaction of iron and sulfur, how many moles of sulfur are needed to react with 1.42 moles of iron?

$$2Fe(s) + 3S(s) \longrightarrow Fe_2S_3(s)$$

SOLUTION GUIDE

STEP 1 State the given and needed quantities (moles).

	Given	Need	Connect
ANALYZE THE PROBLEM	1.42 moles of Fe	moles of S	mole–mole factor
	Equation		
	$2Fe(s) + 3S(s) \longrightarrow Fe_2S_3(s)$		

STEP 2 Write a plan to convert the given to the needed quantity (moles).

moles of Fe → Mole–mole factor → moles of S

STEP 3 Use coefficients to write mole–mole factors.

2 moles of Fe = 3 moles of S

$$\frac{2 \text{ moles Fe}}{3 \text{ moles S}} \quad \text{and} \quad \frac{3 \text{ moles S}}{2 \text{ moles Fe}}$$

STEP 4 Set up the problem to give the needed quantity (moles).

$$1.42 \text{ moles Fe} \times \frac{3 \text{ moles S}}{2 \text{ moles Fe}} = 2.13 \text{ moles of S}$$

Three SFs — Exact (both numerator and denominator) — Three SFs

STUDY CHECK 7.10

Using the equation in Sample Problem 7.10, calculate the number of moles of iron needed to react with 2.75 moles of sulfur.

ANSWER

1.83 moles of iron

TEST

Try Practice Problems 7.55 to 7.58

PRACTICE PROBLEMS

7.7 Mole Relationships in Chemical Equations

LEARNING GOAL Use a mole–mole factor from a balanced chemical equation to calculate the number of moles of another substance in the reaction.

7.53 Write all of the mole–mole factors for each of the following chemical equations:

a. $2SO_2(g) + O_2(g) \longrightarrow 2SO_3(g)$

b. $4P(s) + 5O_2(g) \longrightarrow 2P_2O_5(s)$

7.54 Write all of the mole–mole factors for each of the following chemical equations:

a. $2Al(s) + 3Cl_2(g) \longrightarrow 2AlCl_3(s)$

b. $4HCl(g) + O_2(g) \longrightarrow 2Cl_2(g) + 2H_2O(g)$

7.55 The chemical reaction of hydrogen with oxygen produces water.

$$2H_2(g) + O_2(g) \longrightarrow 2H_2O(g)$$

a. How many moles of O_2 are required to react with 2.6 moles of H_2?

b. How many moles of H_2 are needed to react with 5.0 moles of O_2?

c. How many moles of H_2O form when 2.5 moles of O_2 reacts?

7.56 Ammonia is produced by the chemical reaction of hydrogen and nitrogen.

$$N_2(g) + 3H_2(g) \longrightarrow 2NH_3(g)$$
Ammonia

a. How many moles of H_2 are needed to react with 1.8 moles of N_2?
b. How many moles of N_2 reacted if 0.60 mole of NH_3 is produced?
c. How many moles of NH_3 are produced when 1.4 moles of H_2 reacts?

7.57 Carbon disulfide and carbon monoxide are produced when carbon is heated with sulfur dioxide.

$$5C(s) + 2SO_2(g) \xrightarrow{\Delta} CS_2(l) + 4CO(g)$$

a. How many moles of C are needed to react with 0.500 mole of SO_2?
b. How many moles of CO are produced when 1.2 moles of C reacts?
c. How many moles of SO_2 are needed to produce 0.50 mole of CS_2?
d. How many moles of CS_2 are produced when 2.5 moles of C reacts?

7.58 In the acetylene torch, acetylene gas (C_2H_2) burns in oxygen to produce carbon dioxide, water, and energy.

$$2C_2H_2(g) + 5O_2(g) \xrightarrow{\Delta} 4CO_2(g) + 2H_2O(g)$$

a. How many moles of O_2 are needed to react with 2.40 moles of C_2H_2?
b. How many moles of CO_2 are produced when 3.5 moles of C_2H_2 reacts?
c. How many moles of C_2H_2 are needed to produce 0.50 mole of H_2O?
d. How many moles of CO_2 are produced from 0.100 mole of O_2?

7.8 Mass Calculations for Chemical Reactions

REVIEW
Using Significant Figures in Calculations (2.3)

CORE CHEMISTRY SKILL
Converting Grams to Grams

LEARNING GOAL Given the mass in grams of a substance in a reaction, calculate the mass in grams of another substance in the reaction.

When we have the balanced chemical equation for a reaction, we can use the mass of one of the substances (A) in the reaction to calculate the mass of another substance (B) in the reaction. However, the calculations require us to convert the mass of A to moles of A using the molar mass of A. Then we use the mole–mole factor that links substance A to substance B, which we obtain from the coefficients in the balanced equation. This mole–mole factor (B/A) will convert the moles of A to moles of B. Then the molar mass of B is used to calculate the grams of substance B.

Substance A | **Substance B**

grams of A → Molar mass A → moles of A → Mole–mole factor B/A → moles of B → Molar mass B → grams of B

A mixture of acetylene and oxygen undergoes combustion during the welding of metals.

SAMPLE PROBLEM 7.11 Calculating Mass of Product

TRY IT FIRST

When acetylene, C_2H_2, burns in oxygen, high temperatures are produced that are used for welding metals.

$$2C_2H_2(g) + 5O_2(g) \xrightarrow{\Delta} 4CO_2(g) + 2H_2O(g)$$

How many grams of CO_2 are produced when 54.6 g of C_2H_2 is burned?

SOLUTION GUIDE

STEP 1 State the given and needed quantities (grams).

ANALYZE THE PROBLEM	Given	Need	Connect
	54.6 g of C_2H_2	grams of CO_2	molar masses, mole–mole factor
	Equation		
	$2C_2H_2(g) + 5O_2(g) \xrightarrow{\Delta} 4CO_2(g) + 2H_2O(g)$		

STEP 2 Write a plan to convert the given to the needed quantity (grams).

grams of C_2H_2 → Molar mass → moles of C_2H_2 → Mole–mole factor → moles of CO_2 → Molar mass → grams of CO_2

STEP 3 Use coefficients to write mole–mole factors; write molar masses.

1 mole of C_2H_2 = 26.04 g of C_2H_2

$$\frac{26.04 \text{ g } C_2H_2}{1 \text{ mole } C_2H_2} \quad \text{and} \quad \frac{1 \text{ mole } C_2H_2}{26.04 \text{ g } C_2H_2}$$

2 moles of C_2H_2 = 4 moles of CO_2

$$\frac{2 \text{ moles } C_2H_2}{4 \text{ moles } CO_2} \quad \text{and} \quad \frac{4 \text{ moles } CO_2}{2 \text{ moles } C_2H_2}$$

1 mole of CO_2 = 44.01 g of CO_2

$$\frac{44.01 \text{ g } CO_2}{1 \text{ mole } CO_2} \quad \text{and} \quad \frac{1 \text{ mole } CO_2}{44.01 \text{ g } CO_2}$$

STEP 4 Set up the problem to give the needed quantity (grams).

$$\underset{\text{Three SFs}}{54.6 \text{ g } \cancel{C_2H_2}} \times \overset{\text{Exact}}{\underset{\text{Four SFs}}{\frac{1 \text{ } \cancel{\text{mole } C_2H_2}}{26.04 \text{ } \cancel{\text{g } C_2H_2}}}} \times \overset{\text{Exact}}{\underset{\text{Exact}}{\frac{4 \text{ } \cancel{\text{moles } CO_2}}{2 \text{ } \cancel{\text{moles } C_2H_2}}}} \times \overset{\text{Four SFs}}{\underset{\text{Exact}}{\frac{44.01 \text{ g } CO_2}{1 \text{ } \cancel{\text{mole } CO_2}}}} = \underset{\text{Three SFs}}{185 \text{ g of } CO_2}$$

STUDY CHECK 7.11

Using the equation in Sample Problem 7.11, calculate the grams of CO_2 that can be produced when 25.0 g of O_2 reacts.

ANSWER

27.5 g of CO_2

TEST

Try Practice Problems 7.59 to 7.66

INTERACTIVE VIDEO

Problem 7.65

PRACTICE PROBLEMS

7.8 Mass Calculations for Chemical Reactions

LEARNING GOAL Given the mass in grams of a substance in a reaction, calculate the mass in grams of another substance in the reaction.

7.59 Sodium reacts with oxygen to produce sodium oxide.

$$4Na(s) + O_2(g) \longrightarrow 2Na_2O(s)$$

a. How many grams of Na_2O are produced when 57.5 g of Na reacts?
b. If you have 18.0 g of Na, how many grams of O_2 are needed for the reaction?
c. How many grams of O_2 are needed in a reaction that produces 75.0 g of Na_2O?

7.60 Nitrogen gas reacts with hydrogen gas to produce ammonia.

$$N_2(g) + 3H_2(g) \longrightarrow 2NH_3(g)$$

a. If you have 3.64 g of H_2, how many grams of NH_3 can be produced?
b. How many grams of H_2 are needed to react with 2.80 g of N_2?
c. How many grams of NH_3 can be produced from 12.0 g of H_2?

7.61 Ammonia and oxygen react to form nitrogen and water.

$$4NH_3(g) + 3O_2(g) \longrightarrow 2N_2(g) + 6H_2O(g)$$

a. How many grams of O_2 are needed to react with 13.6 g of NH_3?
b. How many grams of N_2 can be produced when 6.50 g of O_2 reacts?
c. How many grams of H_2O are formed from the reaction of 34.0 g of NH_3?

7.62 Iron(III) oxide reacts with carbon to give iron and carbon monoxide.

$$Fe_2O_3(s) + 3C(s) \longrightarrow 2Fe(s) + 3CO(g)$$

a. How many grams of C are required to react with 16.5 g of Fe_2O_3?
b. How many grams of CO are produced when 36.0 g of C reacts?
c. How many grams of Fe can be produced when 6.00 g of Fe_2O_3 reacts?

7.63 Nitrogen dioxide and water react to produce nitric acid, HNO_3, and nitrogen oxide.

$$3NO_2(g) + H_2O(l) \longrightarrow 2HNO_3(aq) + NO(g)$$

a. How many grams of H_2O are needed to react with 28.0 g of NO_2?
b. How many grams of NO are produced from 15.8 g of H_2O?
c. How many grams of HNO_3 are produced from 8.25 g of NO_2?

7.64 Calcium cyanamide, $CaCN_2$, reacts with water to form calcium carbonate and ammonia.

$$CaCN_2(s) + 3H_2O(l) \longrightarrow CaCO_3(s) + 2NH_3(g)$$

a. How many grams of H_2O are needed to react with 75.0 g of $CaCN_2$?
b. How many grams of NH_3 are produced from 5.24 g of $CaCN_2$?
c. How many grams of $CaCO_3$ form if 155 g of H_2O reacts?

7.65 When solid lead(II) sulfide reacts with oxygen gas, the products are solid lead(II) oxide and sulfur dioxide gas.
a. Write the balanced chemical equation for the reaction.
b. How many grams of oxygen are required to react with 29.9 g of lead(II) sulfide?
c. How many grams of sulfur dioxide can be produced when 65.0 g of lead(II) sulfide reacts?
d. How many grams of lead(II) sulfide are used to produce 128 g of lead(II) oxide?

7.66 When the gases dihydrogen sulfide and oxygen react, they form the gases sulfur dioxide and water vapor.
a. Write the balanced chemical equation for the reaction.
b. How many grams of oxygen are needed to react with 2.50 g of dihydrogen sulfide?
c. How many grams of sulfur dioxide can be produced when 38.5 g of oxygen reacts?
d. How many grams of oxygen are needed to produce 55.8 g of water vapor?

REVIEW
Using Energy Units (3.4)

7.9 Energy in Chemical Reactions

LEARNING GOAL Describe exothermic and endothermic reactions and factors that affect the rate of a reaction.

For a chemical reaction to take place, the molecules of the reactants must collide with each other and have the proper orientation and energy. Even when a collision has the proper orientation, there still must be sufficient energy to break the bonds of the reactants. The **activation energy** is the amount of energy required to break the bonds between atoms of the reactants. If the energy of a collision is less than the activation energy, the molecules bounce apart without reacting. Many collisions occur, but only a few actually lead to the formation of product.

The concept of activation energy is analogous to climbing over a hill. To reach a destination on the other side, we must expend energy to climb to the top of the hill. Once we are at the top, we can easily run down the other side. The energy needed to get us from our starting point to the top of the hill would be the activation energy.

Three Conditions Required for a Reaction to Occur

1. **Collision** The reactants must collide.
2. **Orientation** The reactants must align properly to break and form bonds.
3. **Energy** The collision must provide the energy of activation.

Exothermic Reactions

The high temperature of the thermite reaction has been used to cut or weld railroad tracks.

In every chemical reaction, heat is absorbed or released as reactants are converted to products. The *heat of reaction* is the difference between the energy of the reactants and the energy of the products. In an **exothermic reaction** (*exo* means "out"), the energy of the products is lower than the energy of the reactants. Thus, heat is released in exothermic reactions. For example, in the thermite reaction, the reaction of aluminum and iron(III) oxide produces so much heat that temperatures of 2500 °C can be reached. The thermite reaction has been used to cut or weld railroad tracks. In the equation for an exothermic reaction, the heat of reaction is written on the same side as the products.

Exothermic, Heat Released

$$2Al(s) + Fe_2O_3(s) \longrightarrow 2Fe(s) + Al_2O_3(s) + 850\ kJ$$ Heat is a product

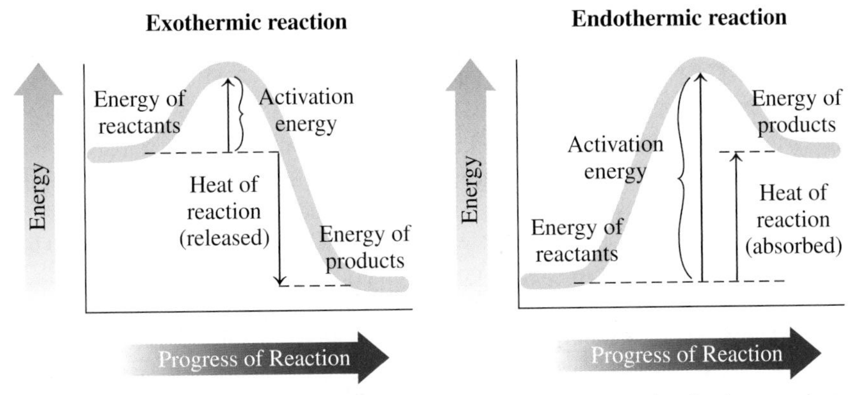

The activation energy is the energy needed to convert reacting molecules into products.

Endothermic Reactions

In **endothermic reaction** (*endo* means "within"), the energy of the products is higher than that of the reactants. Thus, heat is absorbed in endothermic reactions. For example, when hydrogen and iodine react to form hydrogen iodide, heat must be absorbed. In the equation for an endothermic reaction, the heat of reaction is written on the same side as the reactants.

Endothermic, Heat Absorbed

$H_2(g) + I_2(g) + 55kJ \longrightarrow 2HI(g)$ Heat is a reactant

Reaction	Energy Change	Heat in the Equation
Exothermic	Heat released	Product side
Endothermic	Heat absorbed	Reactant side

TEST

Try Practice Problems 7.67 to 7.72

CHEMISTRY LINK TO HEALTH

Cold Packs and Hot Packs

In a hospital, at a first-aid station, or at an athletic event, an instant *cold pack* may be used to reduce swelling from an injury, remove heat from inflammation, or decrease capillary size to lessen the effect of hemorrhaging. Inside the plastic container of a cold pack, there is a compartment containing solid ammonium nitrate (NH_4NO_3) that is separated from a compartment containing water. The pack is activated when it is hit or squeezed hard enough to break the walls between the compartments and cause the ammonium nitrate to mix with the water (shown as H_2O over the reaction arrow). In an endothermic process, 1 mole of NH_4NO_3 that dissolves absorbs 26 kJ of heat. The temperature drops to about 4 to 5 °C to give a cold pack that is ready to use.

Endothermic Reaction in a Cold Pack

$$NH_4NO_3(s) + 26\text{ kJ} \xrightarrow{H_2O} NH_4NO_3(aq)$$

Hot packs are used to relax muscles, lessen aches and cramps, and increase circulation by expanding capillary size. Constructed in the same way as cold packs, a hot pack contains a salt such as $CaCl_2$. When 1 mole of $CaCl_2$ dissolves in water, 82 kJ are released as heat. The temperature increases as much as 66 °C to give a hot pack that is ready to use.

Exothermic Reaction in a Hot Pack

$$CaCl_2 \xrightarrow{H_2O} CaCl_2(aq) + 82\text{ kJ}$$

Cold packs use an endothermic reaction.

Rate of Reaction

The *rate* (or speed) *of reaction* is determined by measuring the amount of a reactant used up, or the amount of a product formed, in a certain period of time. Reactions with low activation energies go faster than reactions with high activation energies. Some reactions go very fast, while others are very slow. For any reaction, the rate is affected by changes in temperature, changes in the concentration of the reactants, and the addition of catalysts.

Temperature

At higher temperatures, the increase in kinetic energy of the reactants makes them move faster and collide more often, and it provides more collisions with the required energy of activation. Reactions almost always go faster at higher temperatures. For every 10 °C increase in temperature, most reaction rates approximately double. If we want food to cook faster, we raise the temperature. When body temperature rises, there is an increase in the pulse rate, rate of breathing, and metabolic rate. On the other hand, we slow down a reaction by lowering the temperature. For example, we refrigerate perishable foods to make them last longer. In some cardiac surgeries, body temperature is lowered to 28 °C so the heart can be stopped and less oxygen is required by the brain. This is also the reason why some people have survived submersion in icy lakes for long periods of time.

TABLE 7.7 Factors That Increase Reaction Rate

Factor	Reason
Increase temperature	More collisions, more collisions with energy of activation
Increase reactant concentration	More collisions
Add a catalyst	Lowers energy of activation

Concentrations of Reactants

The rate of a reaction also increases when reactants are added. Then there are more collisions between the reactants and the reaction goes faster (see **TABLE 7.7**). For example, a patient having difficulty breathing may be given a breathing mixture with a higher oxygen content than the atmosphere. The increase in the number of oxygen molecules in the lungs increases the rate at which oxygen combines with hemoglobin. The increased rate of oxygenation of the blood means that the patient can breathe more easily.

$$\underset{\text{Hemoglobin}}{Hb(aq)} + \underset{\text{Oxygen}}{O_2(g)} \longrightarrow \underset{\text{Oxyhemoglobin}}{HbO_2(aq)}$$

Catalysts

Another way to speed up a reaction is to lower the energy of activation. This can be done by adding a **catalyst**. Earlier, we discussed the energy required to climb a hill. If instead we find a tunnel through the hill, we do not need as much energy to get to the other side. A catalyst acts by providing an alternate pathway with a lower energy requirement. As a result, more collisions form product successfully. Catalysts have found many uses in industry. In the production of margarine, the reaction of hydrogen with vegetable oils is normally very slow. However, when finely divided platinum is present as a catalyst, the reaction occurs rapidly. In the body, biocatalysts called enzymes make most metabolic reactions proceed at the rates necessary for proper cellular activity.

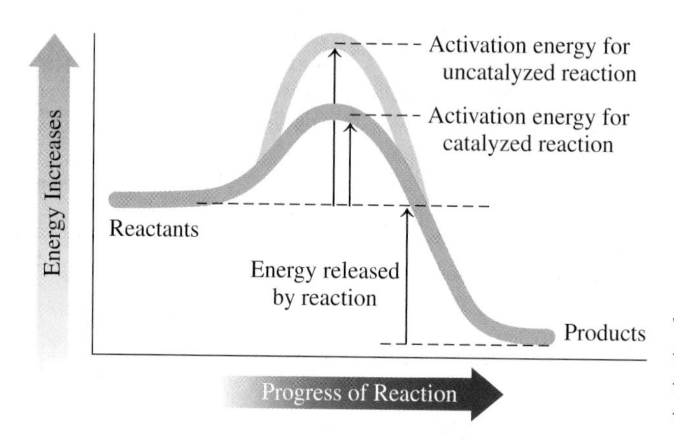

When a catalyst lowers the activation energy, the reaction occurs at a faster rate.

SAMPLE PROBLEM 7.12 Reactions and Rates

TRY IT FIRST

In the reaction of 1 mole of solid carbon with oxygen gas, the energy of the carbon dioxide gas produced is 393 kJ lower than the energy of the reactants.

a. Is the reaction exothermic or endothermic?
b. Write the balanced equation for the reaction, including the heat of reaction.
c. Indicate how an increase in the number of O_2 molecules will change the rate of the reaction.

SOLUTION

a. When the products have a lower energy than the reactants, the reaction is exothermic.
b. $C(s) + O_2(g) \longrightarrow CO_2(g) + 393\ \text{kJ}$
c. An increase in the number of O_2 molecules will increase the rate of the reaction.

STUDY CHECK 7.12

How does lowering the temperature affect the rate of reaction?

ANSWER

The rate of reaction will decrease because the number of collisions between reacting particles will be fewer, and a smaller number of the collisions that do occur will have sufficient activation energy.

TEST

Try Practice Problems 7.73 to 7.76

PRACTICE PROBLEMS

7.9 Energy in Chemical Reactions

LEARNING GOAL Describe exothermic and endothermic reactions and factors that affect the rate of a reaction.

7.67 **a.** Why do chemical reactions require energy of activation?
b. In an exothermic reaction, is the energy of the products higher or lower than that of the reactants?
c. Draw an energy diagram for an exothermic reaction.

7.68 **a.** What is measured by the heat of reaction?
b. In an endothermic reaction, is the energy of the products higher or lower than that of the reactants?
c. Draw an energy diagram for an endothermic reaction.

7.69 Classify each of the following as exothermic or endothermic:
a. A reaction releases 550 kJ.
b. The energy level of the products is higher than that of the reactants.
c. The metabolism of glucose in the body provides energy.

7.70 Classify each of the following as exothermic or endothermic:
a. The energy level of the products is lower than that of the reactants.
b. In the body, the synthesis of proteins requires energy.
c. A reaction absorbs 125 kJ.

7.71 Classify each of the following as exothermic or endothermic:
a. $CH_4(g) + 2O_2(g) \xrightarrow{\Delta} CO_2(g) + 2H_2O(g) + 802\ \text{kJ}$
b. $Ca(OH)_2(s) + 65.3\ \text{kJ} \longrightarrow CaO(s) + H_2O(l)$
c. $2Al(s) + Fe_2O_3(s) \longrightarrow Al_2O_3(s) + 2Fe(s) + 850\ \text{kJ}$

7.72 Classify each of the following as exothermic or endothermic:
a. $C_3H_8(g) + 5O_2(g) \xrightarrow{\Delta} 3CO_2(g) + 4H_2O(g) + 2220\ \text{kJ}$
b. $2Na(s) + Cl_2(g) \longrightarrow 2NaCl(s) + 819\ \text{kJ}$
c. $PCl_5(g) + 67\ \text{kJ} \longrightarrow PCl_3(g) + Cl_2(g)$

7.73 **a.** What is meant by the rate of a reaction?
b. Why does bread grow mold more quickly at room temperature than in the refrigerator?

7.74 **a.** How does a catalyst affect the activation energy?
b. Why is pure oxygen used in respiratory distress?

7.75 How would each of the following change the rate of the reaction shown here?

$$2SO_2(g) + O_2(g) \longrightarrow 2SO_3(g)$$

a. adding some $SO_2(g)$
b. increasing the temperature
c. adding a catalyst
d. removing some $O_2(g)$

7.76 How would each of the following change the rate of the reaction shown here?

$$2NO(g) + 2H_2(g) \longrightarrow N_2(g) + 2H_2O(g)$$

a. adding some $NO(g)$
b. decreasing the temperature
c. removing some $H_2(g)$
d. adding a catalyst

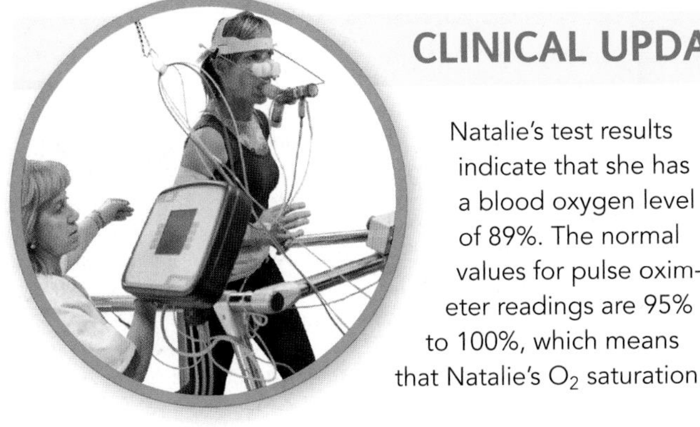

CLINICAL UPDATE Improving Natalie's Overall Fitness

Natalie's test results indicate that she has a blood oxygen level of 89%. The normal values for pulse oximeter readings are 95% to 100%, which means that Natalie's O_2 saturation is low. Thus, Natalie does not have an adequate amount of O_2 in her blood, and may be *hypoxic*. This may be the reason she has noticed a shortness of breath and a dry cough. Her doctor diagnosed her with *interstitial lung disease*, which is scarring of the tissue of the lungs.

Angela teaches Natalie to inhale and exhale slower and deeper to fill the lungs with more air and thus more oxygen. Angela also develops a workout program with the goal of increasing Natalie's overall fitness level. During the exercises, Angela continues to monitor Natalie's heart rate, blood O_2 level, and blood pressure to ensure that Natalie is exercising at a level that will enable her to become stronger without breaking down muscle due to a lack of oxygen.

Low-intensity exercises are used at the beginning of Natalie's exercise program.

Clinical Applications

7.77 **a.** During cellular respiration, aqueous $C_6H_{12}O_6$ (glucose) in the cells undergoes a reaction with oxygen gas to form gaseous carbon dioxide and liquid water. Write and balance the chemical equation for reaction of glucose in the human body.

b. In plants, carbon dioxide gas and liquid water are converted to aqueous glucose ($C_6H_{12}O_6$) and oxygen gas. Write and balance the chemical equation for production of glucose in plants.

7.78 Fatty acids undergo reaction with oxygen gas and form gaseous carbon dioxide and liquid water when utilized for energy in the body.

a. Write and balance the equation for the combustion of the fatty acid aqueous capric acid, $C_{10}H_{20}O_2$.

b. Write and balance the equation for the combustion of the fatty acid aqueous myristic acid, $C_{14}H_{28}O_2$.

CONCEPT MAP

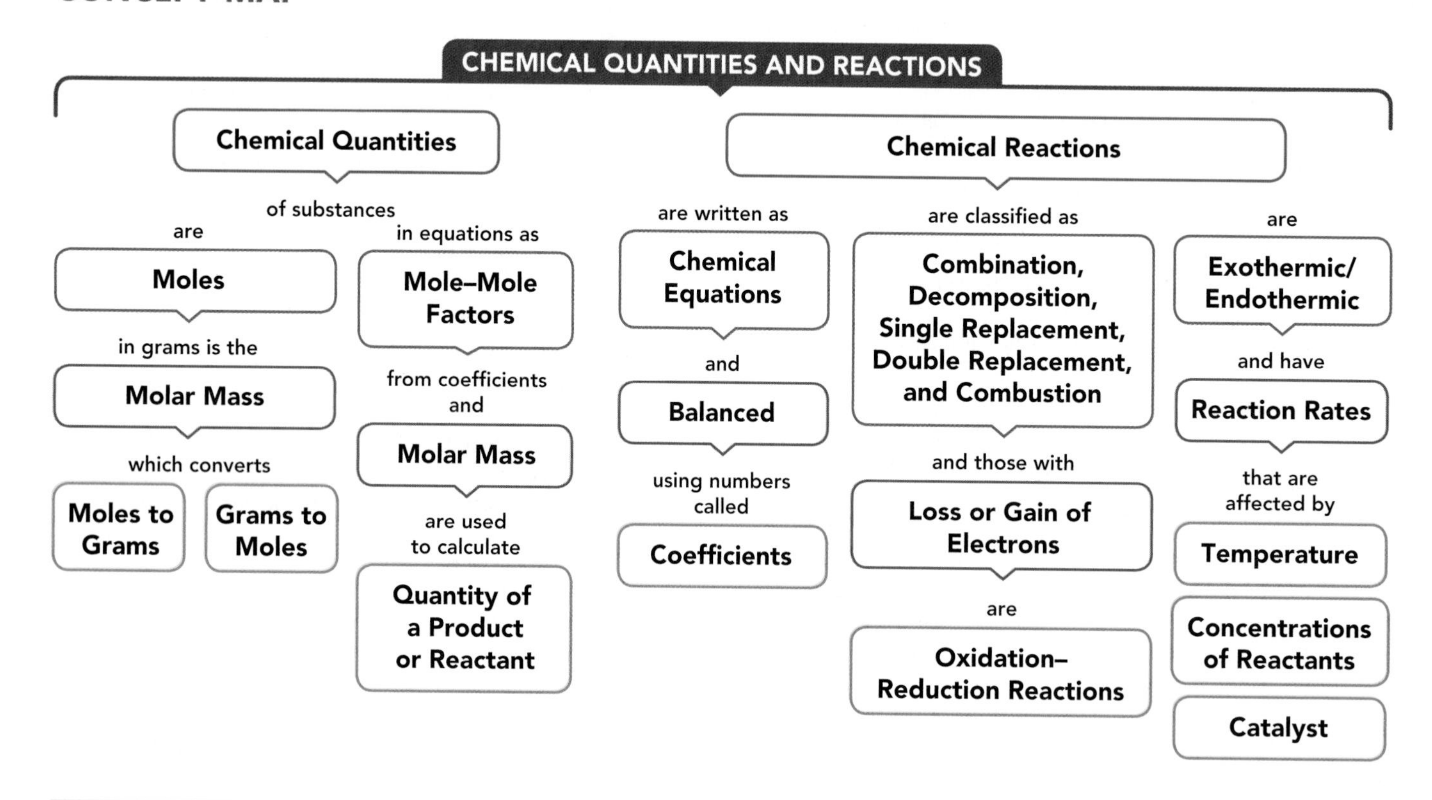

CHAPTER REVIEW

7.1 The Mole

LEARNING GOAL Use Avogadro's number to determine the number of particles in a given number of moles.

- One mole of an element contains 6.02×10^{23} atoms; 1 mole of a compound contains 6.02×10^{23} molecules or formula units.

7.2 Molar Mass

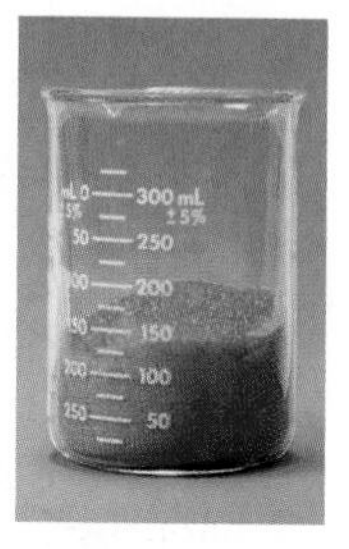

LEARNING GOAL Given the chemical formula of a substance, calculate its molar mass.

- The molar mass (g/mole) of any substance is the mass in grams equal numerically to its atomic mass, or the sum of the atomic masses, which have been multiplied by their subscripts in a formula.

7.3 Calculations Using Molar Mass

LEARNING GOAL Use molar mass to convert between grams and moles.

- The molar mass is used as a conversion factor to change a quantity in grams to moles or to change a given number of moles to grams.

7.4 Equations for Chemical Reactions

LEARNING GOAL Write a balanced chemical equation from the formulas of the reactants and products for a reaction; determine the number of atoms in the reactants and products.

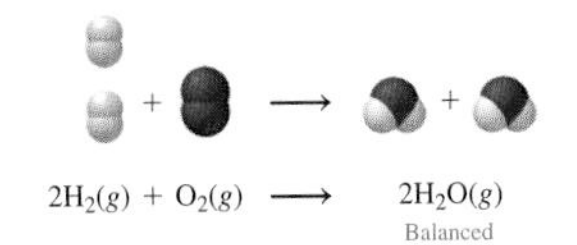

- A chemical reaction occurs when the atoms of the initial substances rearrange to form new substances.
- A chemical equation shows the formulas of the substances that react on the left side of a reaction arrow and the products that form on the right side of the reaction arrow.
- A chemical equation is balanced by writing coefficients, small whole numbers, in front of formulas to equalize the atoms of each of the elements in the reactants and the products.

7.5 Types of Chemical Reactions

LEARNING GOAL Identify a chemical reaction as a combination, decomposition, single replacement, double replacement, or combustion.

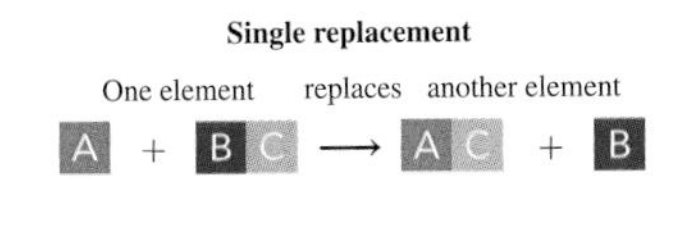

- Many chemical reactions can be organized by reaction type: combination, decomposition, single replacement, double replacement, or combustion.

7.6 Oxidation–Reduction Reactions

LEARNING GOAL Define the terms oxidation and reduction; identify the reactants oxidized and reduced.

Oxidation (loss of electron)

e^-

A oxidized B reduced A B

Reduction (gain of electron)

- When electrons are transferred in a reaction, it is an oxidation–reduction reaction.
- One reactant loses electrons, and another reactant gains electrons.
- Overall, the number of electrons lost and gained is equal.

7.7 Mole Relationships in Chemical Equations

LEARNING GOAL Use a mole–mole factor from a balanced chemical equation to calculate the number of moles of another substance in the reaction.

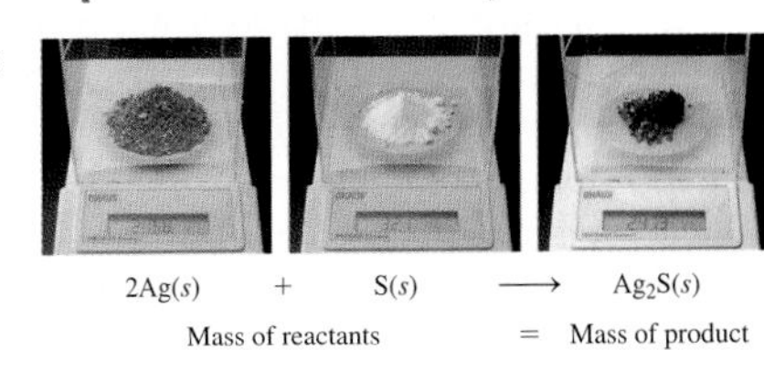

- In a balanced equation, the total mass of the reactants is equal to the total mass of the products.
- The coefficients in an equation describing the relationship between the moles of any two components are used to write mole–mole factors.
- When the number of moles for one substance is known, a mole–mole factor is used to find the moles of a different substance in the reaction.

7.8 Mass Calculations for Chemical Reactions

LEARNING GOAL Given the mass in grams of a substance in a reaction, calculate the mass in grams of another substance in the reaction.

- In calculations using equations, the molar masses of the substances and their mole–mole factors are used to change the number of grams of one substance to the corresponding grams of a different substance.

7.9 Energy in Chemical Reactions

LEARNING GOAL Describe exothermic and endothermic reactions and factors that affect the rate of a reaction.

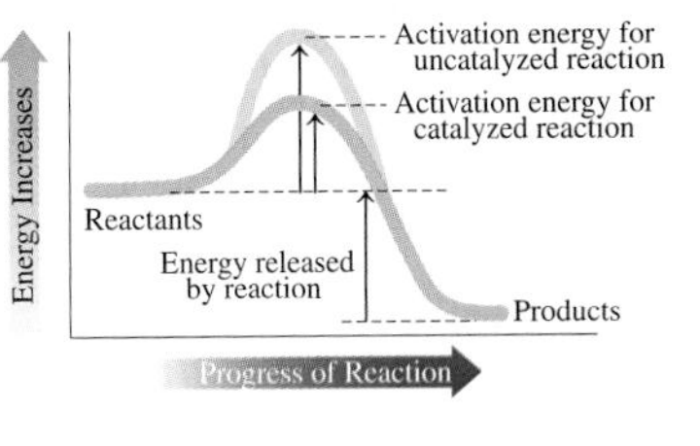

- In chemical reactions, the heat of reaction is the energy difference between the products and the reactants.
- In an exothermic reaction, heat is released because the energy of the products is lower than that of the reactants.
- In an endothermic reaction, heat is absorbed because the energy of the products is higher than that of the reactants.
- The rate of a reaction is the speed at which the reactants are converted to products.
- Increasing the concentrations of reactants, raising the temperature, or adding a catalyst can increase the rate of a reaction.

KEY TERMS

activation energy The energy needed upon collision to break apart the bonds of the reacting molecules.

Avogadro's number The number of items in a mole, equal to 6.02×10^{23}.

balanced equation The final form of a chemical equation that shows the same number of atoms of each element in the reactants and products.

catalyst A substance that increases the rate of reaction by lowering the activation energy.

chemical equation A shorthand way to represent a chemical reaction using chemical formulas to indicate the reactants and products and coefficients to show reacting ratios.

coefficients Whole numbers placed in front of the formulas to balance the number of atoms or moles of atoms of each element on both sides of an equation.

combination reaction A chemical reaction in which reactants combine to form a single product.

combustion reaction A chemical reaction in which a fuel containing carbon and hydrogen reacts with oxygen to produce CO_2, H_2O, and energy.

decomposition reaction A reaction in which a single reactant splits into two or more simpler substances.

double replacement reaction A reaction in which the positive ions in the reacting compounds exchange places.

endothermic reaction A reaction that requires heat; the energy of the products is higher than the energy of the reactants.

exothermic reaction A reaction that releases heat; the energy of the products is lower than the energy of the reactants.

formula unit The group of ions represented by the formula of an ionic compound.

molar mass The mass in grams of 1 mole of an element equal numerically to its atomic mass. The molar mass of a compound is equal to the sum of the masses of the elements in the formula.

mole A group of atoms, molecules, or formula units that contains 6.02×10^{23} of these items.

mole–mole factor A conversion factor that relates the number of moles of two compounds in an equation derived from their coefficients.

oxidation The loss of electrons by a substance. Biological oxidation may involve the addition of oxygen or the loss of hydrogen.

oxidation–reduction reaction A reaction in which the oxidation of one reactant is always accompanied by the reduction of another reactant.

products The substances formed as a result of a chemical reaction.

reactants The initial substances that undergo change in a chemical reaction.

reduction The gain of electrons by a substance. Biological reduction may involve the loss of oxygen or the gain of hydrogen.

single replacement reaction A reaction in which an element replaces a different element in a compound.

CORE CHEMISTRY SKILLS

The chapter Section containing each Core Chemistry Skill is shown in parentheses at the end of each heading.

Converting Particles to Moles (7.1)

- In chemistry, atoms, molecules, and ions are counted by the mole, a unit that contains 6.02×10^{23} items, which is Avogadro's number.
- For example, 1 mole of carbon contains 6.02×10^{23} atoms of carbon and 1 mole of H_2O contains 6.02×10^{23} molecules of H_2O.
- Avogadro's number is used to convert between particles and moles.

Example: How many moles of nickel contain 2.45×10^{24} Ni atoms?

Answer: $2.45 \times 10^{24}\ \cancel{\text{Ni atoms}} \times \dfrac{1\ \text{mole Ni}}{6.02 \times 10^{23}\ \cancel{\text{Ni atoms}}}$ (Three SFs; Exact / Three SFs)

$= 4.07$ moles of Ni (Three SFs)

Calculating Molar Mass (7.2)

- The molar mass of an element is its mass in grams equal numerically to its atomic mass.
- The molar mass of a compound is the sum of the molar mass of each element in its chemical formula multiplied by its subscript in the formula.

Pinene is a component of pine sap.

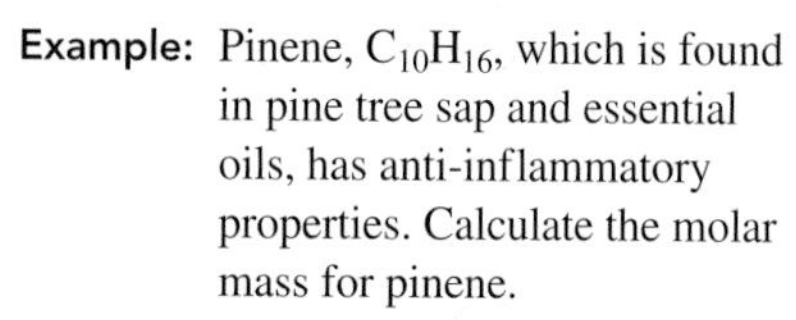

Example: Pinene, $C_{10}H_{16}$, which is found in pine tree sap and essential oils, has anti-inflammatory properties. Calculate the molar mass for pinene.

Answer: $10\ \cancel{\text{moles C}} \times \dfrac{12.01\ \text{g C}}{1\ \cancel{\text{mole C}}} = 120.1$ g of C

$16\ \cancel{\text{moles H}} \times \dfrac{1.008\ \text{g H}}{1\ \cancel{\text{mole H}}} = 16.13$ g of H

Molar mass of $C_{10}H_{16}$ = 136.2 g

Using Molar Mass as a Conversion Factor (7.3)

A bicycle with an aluminum frame.

- Molar mass is used as a conversion factor to convert between the moles and grams of a substance.

Example: The frame of a bicycle contains 6500 g of aluminum. How many moles of aluminum are in the bicycle frame?

Equality: 1 mole of Al = 26.98 g of Al

Conversion factors: $\dfrac{26.98\ \text{g Al}}{1\ \text{mole Al}}$ and $\dfrac{1\ \text{mole Al}}{26.98\ \text{g Al}}$

Answer: $6500\ \cancel{\text{g Al}} \times \dfrac{1\ \text{mole Al}}{26.98\ \cancel{\text{g Al}}} = 240$ moles of Al (Two SFs; Exact / Four SFs; Two SFs)

Balancing a Chemical Equation (7.4)

- In a *balanced* chemical equation, whole numbers called coefficients multiply each of the atoms in the chemical formulas so that the number of each type of atom in the reactants is equal to the number of the same type of atom in the products.

Example: Balance the following chemical equation:

$SnCl_4(s) + H_2O(l) \longrightarrow Sn(OH)_4(s) + HCl(aq)$ Not balanced

Answer: When we compare the atoms on the reactant side and the product side, we see that there are more Cl atoms in the reactants and more O and H atoms in the products.

To balance the equation, we need to use coefficients in front of the formulas containing the Cl atoms, H atoms, and O atoms.

- Place a 4 in front of the formula HCl to give 8 H atoms and 4 Cl atoms in the products.

$SnCl_4(s) + H_2O(l) \longrightarrow Sn(OH)_4(s) + 4HCl(aq)$

- Place a 4 in front of the formula H_2O to give 8 H atoms and 4 O atoms in the reactants.

$SnCl_4(s) + 4H_2O(l) \longrightarrow Sn(OH)_4(s) + 4HCl(aq)$

- The total number of Sn (1), Cl (4), H (8), and O (4) atoms is now equal on both sides of the equation. Thus, this equation is balanced.

Classifying Types of Chemical Reactions (7.5)

- Chemical reactions are classified by identifying general patterns in their equations.
- In a combination reaction, two or more elements or compounds combine to form one product.
- In a decomposition reaction, a single reactant splits into two or more products.
- In a single replacement reaction, an uncombined element takes the place of an element in a compound.
- In a double replacement reaction, the positive ions in the reacting compounds switch places.
- In a combustion reaction, a carbon-containing compound that is the fuel burns in oxygen to produce carbon dioxide (CO_2), water (H_2O), and energy.

Example: Classify the type of the following reaction:

$$2Al(s) + Fe_2O_3(s) \xrightarrow{\Delta} Al_2O_3(s) + 2Fe(l)$$

Answer: The iron in iron(III) oxide is replaced by aluminum, which makes this a single replacement reaction.

Identifying Oxidized and Reduced Substances (7.6)

- In an oxidation–reduction reaction (abbreviated *redox*), one reactant is oxidized when it loses electrons, and another reactant is reduced when it gains electrons.
- Oxidation is the *loss* of electrons; reduction is the *gain* of electrons.

Example: For the following redox reaction, identify the reactant that is oxidized and the reactant that is reduced:

$$Fe(s) + Cu^{2+}(aq) \longrightarrow Fe^{2+}(aq) + Cu(s)$$

Answer: $Fe^0(s) \longrightarrow Fe^{2+}(aq) + 2\,e^-$

Fe loses electrons; it is oxidized.

$Cu^{2+}(aq) + 2\,e^- \longrightarrow Cu^0(s)$

Cu^{2+} gains electrons; it is reduced.

Using Mole–Mole Factors (7.7)

Consider the balanced chemical equation

$$4Na(s) + O_2(g) \longrightarrow 2Na_2O(s)$$

- The coefficients in a balanced chemical equation represent the moles of reactants and the moles of products. Thus, 4 moles of Na react with 1 mole of O_2 to form 2 moles of Na_2O.
- From the coefficients, mole–mole factors can be written for any two substances as follows:

Na and O_2	$\frac{4\text{ moles Na}}{1\text{ mole }O_2}$	and	$\frac{1\text{ mole }O_2}{4\text{ moles Na}}$
Na and Na_2O	$\frac{4\text{ moles Na}}{2\text{ moles }Na_2O}$	and	$\frac{2\text{ moles }Na_2O}{4\text{ moles Na}}$
O_2 and Na_2O	$\frac{2\text{ moles }Na_2O}{1\text{ mole }O_2}$	and	$\frac{1\text{ mole }O_2}{2\text{ moles }Na_2O}$

- A mole–mole factor is used to convert the number of moles of one substance in the reaction to the number of moles of another substance in the reaction.

Example: How many moles of sodium are needed to produce 3.5 moles of sodium oxide?

Answer:

Given:	Need:	Connect:
3.5 moles of Na_2O	moles of Na	mole–mole factor

$$\underset{\text{Two SFs}}{3.5\text{ }\cancel{\text{moles }Na_2O}} \times \underset{\text{Exact}}{\overset{\text{Exact}}{\frac{4\text{ moles Na}}{2\text{ }\cancel{\text{moles }Na_2O}}}} = \underset{\text{Two SFs}}{7.0\text{ moles of Na}}$$

Converting Grams to Grams (7.8)

- When we have the balanced chemical equation for a reaction, we can use the mass of substance A and then calculate the mass of substance B. The process is as follows:
  - Use the molar mass of A to convert the mass, in grams, of A to moles of A.
  - Use the mole–mole factor that converts moles of A to moles of B.
  - Use the molar mass of B to calculate the mass, in grams, of B.

$$\text{grams of A} \xrightarrow{\text{Molar mass A}} \text{moles of A} \xrightarrow{\text{Mole–mole factor}} \text{moles of B} \xrightarrow{\text{Molar mass B}} \text{grams of B}$$

Example: How many grams of O_2 are needed to completely react with 14.6 g of Na?

$$4Na(s) + O_2(g) \longrightarrow 2Na_2O(s)$$

Answer:

$$\underset{\text{Three SFs}}{14.6\text{ }\cancel{\text{g Na}}} \times \underset{\text{Four SFs}}{\overset{\text{Exact}}{\frac{1\text{ }\cancel{\text{mole Na}}}{22.99\text{ }\cancel{\text{g Na}}}}} \times \underset{\text{Exact}}{\overset{\text{Exact}}{\frac{1\text{ }\cancel{\text{mole }O_2}}{4\text{ }\cancel{\text{moles Na}}}}} \times \underset{\text{Exact}}{\overset{\text{Four SFs}}{\frac{32.00\text{ g }O_2}{1\text{ }\cancel{\text{mole }O_2}}}} = \underset{\text{Three SFs}}{5.08\text{ g of }O_2}$$

UNDERSTANDING THE CONCEPTS

The chapter Sections to review are shown in parentheses at the end of each problem.

7.79 Using the models of the molecules (black = C, white = H, yellow = S, green = Cl), determine each of the following for models of compounds **1** and **2**: (7.1, 7.2, 7.3)

1.

2.

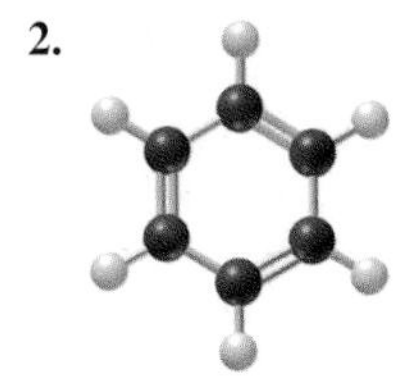

a. molecular formula **b.** molar mass
c. number of moles in 10.0 g

7.80 Using the models of the molecules (black = C, white = H, yellow = S, red = O), determine each of the following for models of compounds **1** and **2**: (7.1, 7.2, 7.3)

1.

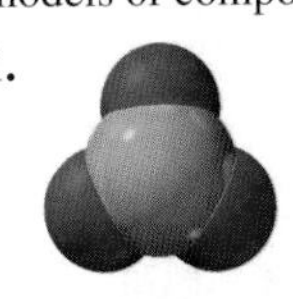

2.

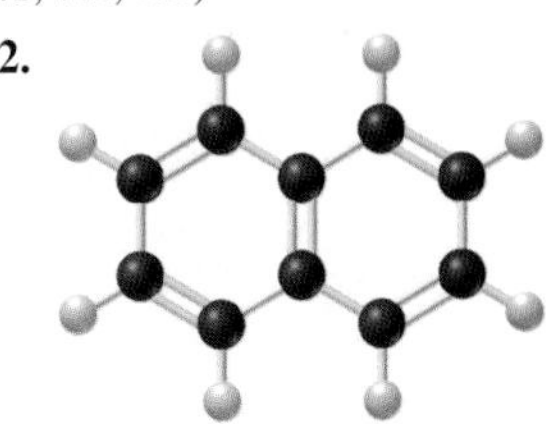

a. molecular formula **b.** molar mass
c. number of moles in 10.0 g

7.81 A dandruff shampoo contains dipyrithione, $C_{10}H_8N_2O_2S_2$, which acts as an antibacterial and antifungal agent. (7.1, 7.2, 7.3)

a. What is the molar mass of dipyrithione?
b. How many moles of dipyrithione are in 25.0 g?
c. How many moles of C are in 25.0 g of dipyrithione?
d. How many moles of dipyrithione contain 8.2×10^{24} atoms of N?

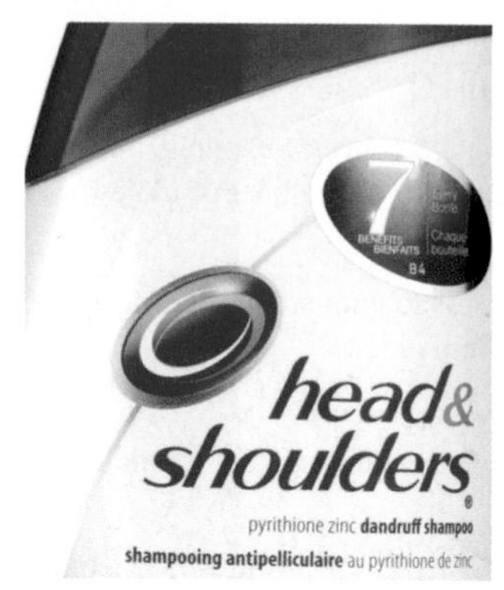

Dandruff shampoo contains dipyrithione.

7.82 Ibuprofen, an anti-inflammatory drug in Advil, has the formula $C_{13}H_{18}O_2$. (7.1, 7.2, 7.3)

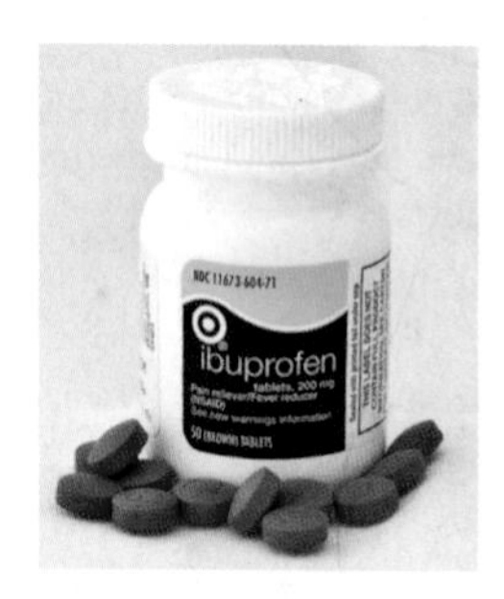

Ibuprofen is an anti-inflammatory.

a. What is the molar mass of ibuprofen?
b. How many grams of ibuprofen are in 0.525 mole?
c. How many moles of C are in 12.0 g of ibuprofen?
d. How many moles of ibuprofen contain 1.22×10^{23} atoms of C?

7.83 Balance each of the following by adding coefficients, and identify the type of reaction for each: (7.4, 7.5)

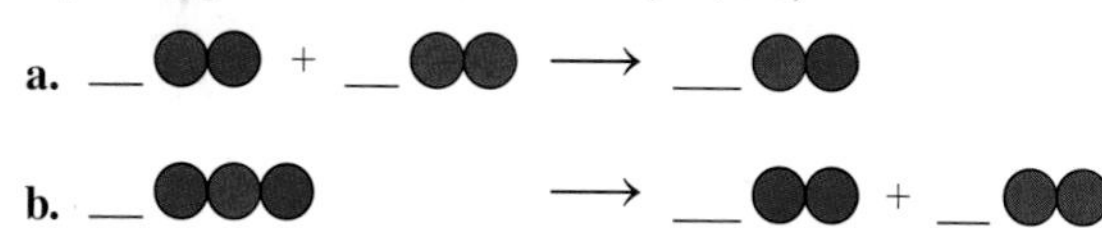

7.84 Balance each of the following by adding coefficients, and identify the type of reaction for each: (7.4, 7.5)

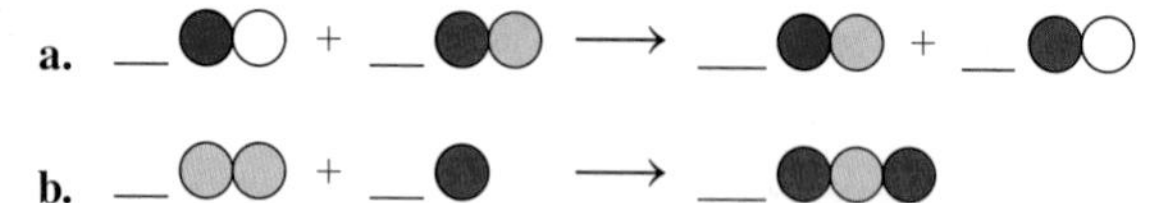

7.85 If red spheres represent oxygen atoms, blue spheres represent nitrogen atoms, and all the molecules are gases, (7.4, 7.5)

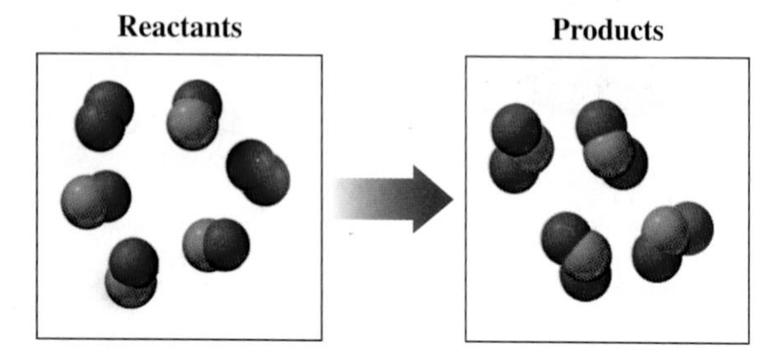

a. write the formula for each of the reactants and products.
b. write a balanced equation for the reaction.
c. indicate the type of reaction as combination, decomposition, single replacement, double replacement, or combustion.

7.86 If purple spheres represent iodine atoms, white spheres represent hydrogen atoms, and all the molecules are gases, (7.4, 7.5)

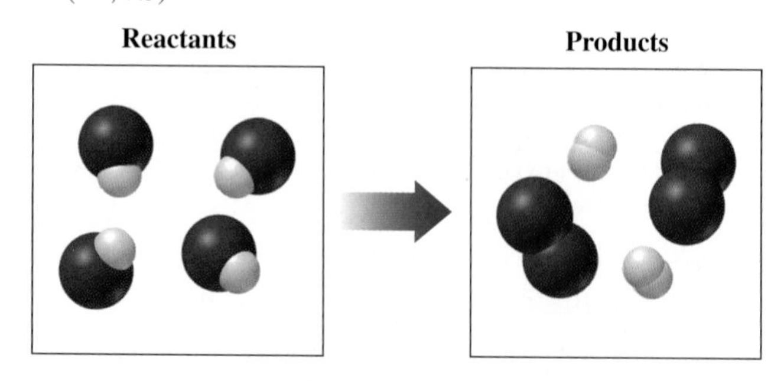

a. write the formula for each of the reactants and products.
b. write a balanced equation for the reaction.
c. indicate the type of reaction as combination, decomposition, single replacement, double replacement, or combustion.

7.87 If blue spheres represent nitrogen atoms, purple spheres represent iodine atoms and the reacting molecules are solid, while the products are gases, (7.4, 7.5)

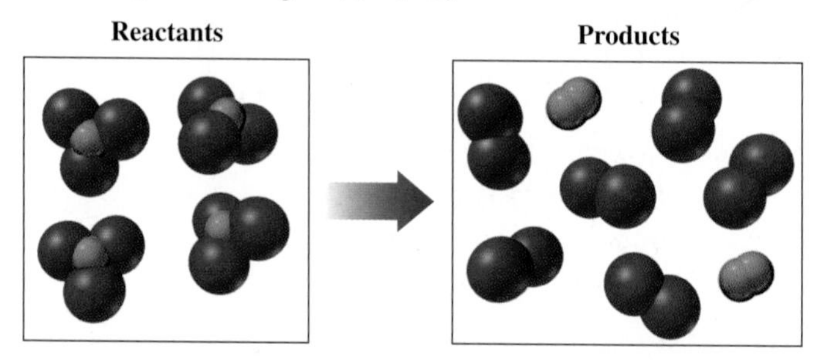

a. write the formula for each of the reactants and products.
b. write a balanced equation for the reaction.
c. indicate the type of reaction as combination, decomposition, single replacement, double replacement, or combustion.

7.88 If green spheres represent chlorine atoms, yellow-green spheres represent fluorine atoms, white spheres represent hydrogen atoms, and all the molecules are gases, (7.4, 7.5)

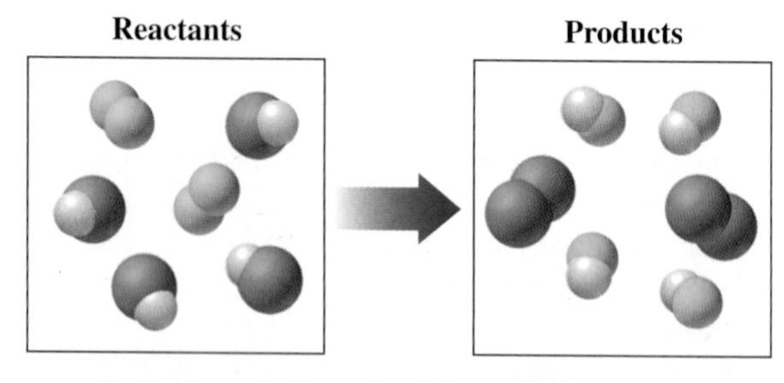

a. write the formula for each of the reactants and products.
b. write a balanced equation for the reaction.
c. indicate the type of reaction as combination, decomposition, single replacement, double replacement, or combustion.

7.89 If green spheres represent chlorine atoms, red spheres represent oxygen atoms, and all the molecules are gases, (7.4, 7.5)

Reactants **Products**

a. write the formula for each of the reactants and products.
b. write a balanced equation for the reaction.
c. indicate the type of reaction as combination, decomposition, single replacement, double replacement, or combustion.

7.90 If blue spheres represent nitrogen atoms, purple spheres represent iodine atoms, the reacting molecules are gases, and the products are solid, (7.4, 7.5)

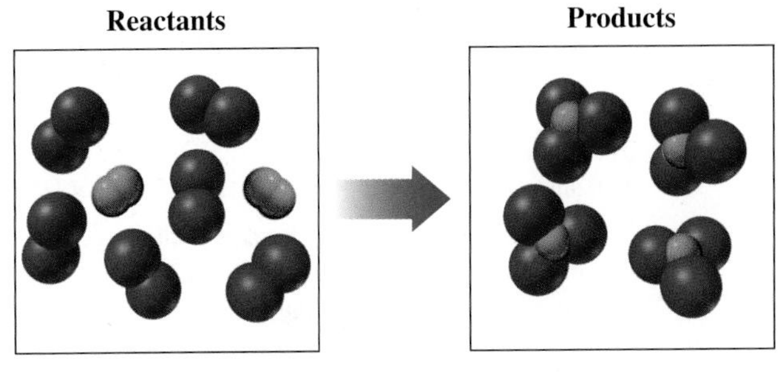

a. write the formula for each of the reactants and products.
b. write a balanced equation for the reaction.
c. indicate the type of reaction as combination, decomposition, single replacement, double replacement, or combustion.

ADDITIONAL PRACTICE PROBLEMS

7.91 Calculate the molar mass for each of the following: (7.2)
a. $ZnSO_4$, zinc sulfate, zinc supplement
b. $Ca(IO_3)_2$, calcium iodate, iodine source in table salt
c. $C_5H_8NNaO_4$, monosodium glutamate, flavor enhancer
d. $C_6H_{12}O_2$, isoamyl formate, used to make artificial fruit syrups

7.92 Calculate the molar mass for each of the following: (7.2)
a. $MgCO_3$, magnesium carbonate, used in antacids
b. $Au(OH)_3$, gold(III) hydroxide, used in gold plating
c. $C_{18}H_{34}O_2$, oleic acid, from olive oil
d. $C_{21}H_{26}O_5$, prednisone, anti-inflammatory

7.93 How many grams are in 0.150 mole of each of the following? (7.3)
a. K b. Cl_2 c. Na_2CO_3

7.94 How many grams are in 2.25 moles of each of the following? (7.3)
a. N_2 b. NaBr c. C_6H_{14}

7.95 How many moles are in 25.0 g of each of the following compounds? (7.3)
a. CO_2 b. Al_2O_3 c. $MgCl_2$

7.96 How many moles are in 4.00 g of each of the following compounds? (7.3)
a. NH_3 b. $Ca(NO_3)_2$ c. SO_3

7.97 Identify the type of reaction for each of the following as combination, decomposition, single replacement, double replacement, or combustion: (7.5)
a. A metal and a nonmetal form an ionic compound.
b. A compound of hydrogen and carbon reacts with oxygen to produce carbon dioxide and water.
c. Heating calcium carbonate produces calcium oxide and carbon dioxide.
d. Zinc replaces copper in $Cu(NO_3)_2$.

7.98 Identify the type of reaction for each of the following as combination, decomposition, single replacement, double replacement, or combustion: (7.5)
a. A compound breaks apart into its elements.
b. Copper and bromine form copper(II) bromide.
c. Iron(II) sulfite breaks down to iron(II) oxide and sulfur dioxide.
d. Silver ion from $AgNO_3(aq)$ forms a solid with bromide ion from $KBr(aq)$.

7.99 Balance each of the following chemical equations, and identify the type of reaction: (7.4, 7.5)
a. $NH_3(g) + HCl(g) \longrightarrow NH_4Cl(s)$
b. $C_4H_8(g) + O_2(g) \xrightarrow{\Delta} CO_2(g) + H_2O(g)$
c. $Sb(s) + Cl_2(g) \longrightarrow SbCl_3(s)$
d. $NI_3(s) \longrightarrow N_2(g) + I_2(g)$
e. $KBr(aq) + Cl_2(aq) \longrightarrow KCl(aq) + Br_2(l)$
f. $Fe(s) + H_2SO_4(aq) \longrightarrow H_2(g) + Fe_2(SO_4)_3(aq)$
g. $Al_2(SO_4)_3(aq) + NaOH(aq) \longrightarrow Al(OH)_3(s) + Na_2SO_4(aq)$

7.100 Balance each of the following chemical equations, and identify the type of reaction: (7.4, 7.5)
a. $Si_3N_4(s) \longrightarrow Si(s) + N_2(g)$
b. $Mg(s) + N_2(g) \longrightarrow Mg_3N_2(s)$
c. $Al(s) + H_3PO_4(aq) \longrightarrow H_2(g) + AlPO_4(aq)$
d. $C_3H_4(g) + O_2(g) \xrightarrow{\Delta} CO_2(g) + H_2O(g)$
e. $Cr_2O_3(s) + H_2(g) \longrightarrow Cr(s) + H_2O(g)$
f. $Al(s) + Cl_2(g) \longrightarrow AlCl_3(s)$
g. $MgCl_2(aq) + AgNO_3(aq) \longrightarrow AgCl(s) + Mg(NO_3)_2(aq)$

7.101 Identify each of the following as an oxidation or a reduction: (7.6)
a. $Zn^{2+}(aq) + 2\,e^- \longrightarrow Zn(s)$
b. $Al(s) \longrightarrow Al^{3+}(aq) + 3\,e^-$
c. $Pb(s) \longrightarrow Pb^{2+}(aq) + 2\,e^-$
d. $Cl_2(g) + 2\,e^- \longrightarrow 2Cl^-(aq)$

7.102 Identify each of the following as an oxidation or a reduction: (7.6)
a. $Mg^{2+}(aq) + 2\,e^- \longrightarrow Mg(s)$
b. $2I^-(aq) \longrightarrow I_2(s) + 2\,e^-$
c. $Li(s) \longrightarrow Li^+(aq) + e^-$
d. $Co^{2+}(aq) + 2\,e^- \longrightarrow Co(s)$

7.103 When ammonia (NH_3) gas reacts with fluorine gas, the gaseous products are dinitrogen tetrafluoride (N_2F_4) and hydrogen fluoride (HF). (7.4, 7.7, 7.8)
a. Write the balanced chemical equation.
b. How many moles of each reactant are needed to produce 4.00 moles of HF?
c. How many grams of F_2 are needed to react with 25.5 g of NH_3?
d. How many grams of N_2F_4 can be produced when 3.40 g of NH_3 reacts?

7.104 When nitrogen dioxide (NO_2) gas from car exhaust combines with water vapor in the air, it forms aqueous nitric acid (HNO_3), which causes acid rain, and nitrogen oxide gas. (7.4, 7.7, 7.8)

a. Write the balanced chemical equation.

b. How many moles of each product are produced from 0.250 mole of H_2O?

c. How many grams of HNO_3 are produced when 60.0 g of NO_2 completely reacts?

d. How many grams of NO_2 are needed to form 75.0 g of HNO_3?

7.105 Pentane gas, C_5H_{12}, undergoes combustion with oxygen gas to produce carbon dioxide and water gases. (7.4, 7.8)

a. Write the balanced chemical equation.

b. How many grams of C_5H_{12} are needed to produce 72 g of water?

c. How many grams of CO_2 are produced from 32.0 g of O_2?

7.106 Propane gas, C_3H_8, undergoes combustion with oxygen gas to produce carbon dioxide and water gases. Propane has a density of 2.02 g/L at room temperature. (7.4, 7.8)

a. Write the balanced chemical equation.

b. How many grams of H_2O form when 5.00 L of C_3H_8 reacts?

c. How many grams of H_2O can be produced from the reaction of 100. g of C_3H_8?

7.107 The equation for the formation of silicon tetrachloride from silicon and chlorine is (7.9)

$$Si(s) + 2Cl_2(g) \longrightarrow SiCl_4(g) + 157 \text{ kcal}$$

a. Is the formation of $SiCl_4$ an endothermic or exothermic reaction?

b. Is the energy of the product higher or lower than the energy of the reactants?

7.108 The equation for the formation of nitrogen oxide is (7.9)

$$N_2(g) + O_2(g) + 90.2 \text{ kJ} \longrightarrow 2NO(g)$$

a. Is the formation of NO an endothermic or exothermic reaction?

b. Is the energy of the product higher or lower than the energy of the reactants?

CHALLENGE PROBLEMS

The following problems are related to the topics in this chapter. However, they do not all follow the chapter order, and they require you to combine concepts and skills from several Sections. These problems will help you increase your critical thinking skills and prepare for your next exam.

7.109 At a winery, glucose ($C_6H_{12}O_6$) in grapes undergoes fermentation to produce ethanol (C_2H_6O) and carbon dioxide. (7.8)

$$\underset{\text{Glucose}}{C_6H_{12}O_6(aq)} \longrightarrow \underset{\text{Ethanol}}{2C_2H_6O(aq)} + 2CO_2(g)$$

Glucose in grapes ferments to produce ethanol.

a. How many grams of glucose are required to form 124 g of ethanol?

b. How many grams of ethanol would be formed from the reaction of 0.240 kg of glucose?

7.110 Gasohol is a fuel containing liquid ethanol (C_2H_6O) that burns in oxygen gas to give carbon dioxide and water gases. (7.4, 7.7, 7.8)

a. Write the balanced chemical equation.

b. How many moles of O_2 are needed to completely react with 4.0 moles of C_2H_6O?

c. If a car produces 88 g of CO_2, how many grams of O_2 are used up in the reaction?

d. If you burn 125 g of C_2H_6O, how many grams of CO_2 and H_2O can be produced?

7.111 Consider the following *unbalanced* equation: (7.4, 7.5, 7.7, 7.8)

$$Al(s) + O_2(g) \longrightarrow Al_2O_3(s)$$

a. Write the balanced chemical equation.

a. Identify the type of reaction.

c. How many moles of O_2 are needed to react with 4.50 moles of Al?

d. How many grams of Al_2O_3 are produced when 50.2 g of Al reacts?

e. When Al is reacted with 8.00 g of O_2, how many grams of Al_2O_3 can form?

7.112 A toothpaste contains 0.240% by mass sodium fluoride used to prevent dental caries and 5.0% by mass KNO_3, which decreases pain sensitivity. One tube contains 119 g of toothpaste. (7.1, 7.2)

NaF and KNO_3 are components found in toothpastes.

a. How many moles of NaF are in the tube of toothpaste?

b. How many fluoride ions, F^-, are in the tube of toothpaste?

c. How many grams of sodium ion, Na^+, are in 1.50 g of toothpaste?

d. How many KNO_3 formula units are in the tube of toothpaste?

7.113 During heavy exercise and workouts, lactic acid, $C_3H_6O_3$, accumulates in the muscles where it can cause pain and soreness. (7.1, 7.2)

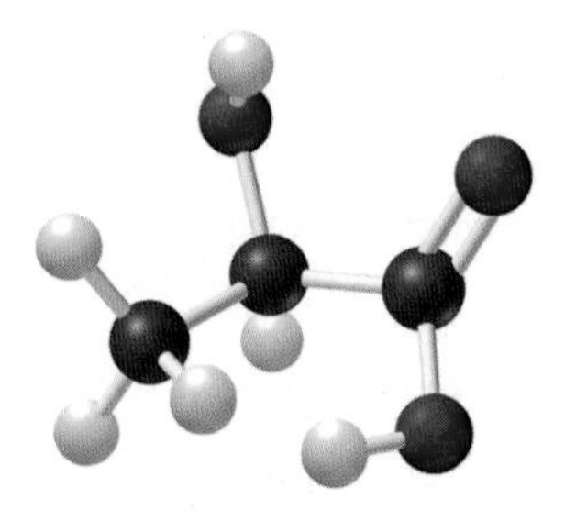

In the ball-and-stick model of lactic acid, black spheres = C, white spheres = H, and red spheres = O.

a. How many molecules are in 0.500 mole of lactic acid?
b. How many atoms of C are in 1.50 moles of lactic acid?
c. How many moles of lactic acid contain 4.5×10^{24} atoms of O?
d. What is the molar mass of lactic acid?

7.114 Ammonium sulfate, $(NH_4)_2SO_4$, is used in fertilizers to provide nitrogen for the soil. (7.1, 7.2)
a. How many formula units are in 0.200 mole of ammonium sulfate?
b. How many H atoms are in 0.100 mole of ammonium sulfate?
c. How many moles of ammonium sulfate contain 7.4×10^{25} atoms of N?
d. What is the molar mass of ammonium sulfate?

7.115 Acetylene gas, C_2H_2, used in welders' torches, releases 1300 kJ of heat when 1 mole of C_2H_2 undergoes combustion. (7.4, 7.7, 7.8, 7.9)
a. Write a balanced equation for the reaction, including the heat of reaction.
b. Is the reaction endothermic or exothermic?
c. How many moles of H_2O are produced when 2.00 moles of O_2 reacts?
d. How many grams of O_2 are needed to react with 9.80 g of C_2H_2?

7.116 Liquid methanol (CH_4O), which is used as a cooking fuel, burns with oxygen gas to produce the gases carbon dioxide and water. The reaction produces 363 kJ of heat per mole of methanol. (7.4, 7.7, 7.8, 7.9)
a. Write a balanced equation for the reaction, including the heat of reaction.
b. Is the reaction endothermic or exothermic?
c. How many moles of O_2 must react with 0.450 mole of CH_4O?
d. How many grams of CO_2 are produced when 78.0 g of CH_4O reacts?

ANSWERS

Answers to Selected Practice Problems

7.1 One mole contains 6.02×10^{23} atoms of an element, molecules of a molecular substance, or formula units of an ionic substance.

7.3 a. 3.01×10^{23} atoms of C
b. 7.71×10^{23} molecules of SO_2
c. 0.0867 mole of Fe
d. 14.1 moles of C_2H_6O

7.5 a. 6.00 moles of H b. 8.00 moles of O
c. 1.20×10^{24} atoms of P d. 4.82×10^{24} atoms of O

7.7 a. 36 moles of H b. 1.0×10^2 moles of C
c. 0.040 mole of N

7.9 a. 32.2 moles of C b. 6.22 moles of H
c. 0.230 mole of O

7.11 a. 70.90 g b. 90.08 g c. 262.9 g

7.13 a. 83.98 g b. 98.95 g c. 156.7 g

7.15 a. 74.55 g b. 123.11 g c. 329.4 g

7.17 a. 342.2 g b. 188.18 g c. 365.5 g

7.19 a. 151.16 g b. 489.4 g c. 331.4 g

7.21 a. 34.5 g b. 112 g c. 5.50 g
d. 5.14 g e. 9.80×10^4 g

7.23 a. 3.03 g b. 38.1 g c. 9.30 g
d. 24.6 g e. 12.9 g

7.25 a. 0.760 mole of Ag b. 0.0240 mole of C
c. 0.881 mole of NH_3 d. 0.452 mole of CH_4
e. 1.53 moles of Fe_2O_3

7.27 a. 6.25 moles of He b. 0.781 mole of O_2
c. 0.321 mole of $Al(OH)_3$ d. 0.106 mole of Ga_2S_3
e. 0.430 mole of C_4H_{10}

7.29 a. 0.527 mole b. 96.8 g

7.31 a. 602 g b. 74.7 g

7.33 a. 66.0 g b. 0.772 mole

7.35 a. not balanced b. balanced
c. not balanced d. balanced

7.37 a. $N_2(g) + O_2(g) \longrightarrow 2NO(g)$
b. $2HgO(s) \longrightarrow 2Hg(l) + O_2(g)$
c. $4Fe(s) + 3O_2(g) \longrightarrow 2Fe_2O_3(s)$
d. $2Na(s) + Cl_2(g) \longrightarrow 2NaCl(s)$

7.39 a. $Mg(s) + 2AgNO_3(aq) \longrightarrow Mg(NO_3)_2(aq) + 2Ag(s)$
b. $2Al(s) + 3CuSO_4(aq) \longrightarrow 3Cu(s) + Al_2(SO_4)_3(aq)$
c. $Pb(NO_3)_2(aq) + 2NaCl(aq) \longrightarrow PbCl_2(s) + 2NaNO_3(aq)$
d. $2Al(s) + 6HCl(aq) \longrightarrow 3H_2(g) + 2AlCl_3(aq)$

7.41 a. decomposition b. single replacement
c. combustion d. double replacement
e. combination

7.43 a. combination b. single replacement
c. decomposition d. double replacement
e. combustion

7.45 a. reduction b. oxidation
c. reduction d. reduction

7.47 a. Zn is oxidized, Cl_2 is reduced.
b. The Br^- in NaBr is oxidized, Cl_2 is reduced.
c. The O^{2-} in PbO is oxidized, the Pb^{2+} in PbO is reduced.
d. Sn^{2+} is oxidized, Fe^{3+} is reduced.

7.49 a. reduction b. oxidation

7.51 Linoleic acid gains hydrogen atoms and is reduced.

7.53 a. $\frac{2 \text{ moles } SO_2}{1 \text{ mole } O_2}$ and $\frac{1 \text{ mole } O_2}{2 \text{ moles } SO_2}$

$\frac{2 \text{ moles } SO_2}{2 \text{ moles } SO_3}$ and $\frac{2 \text{ moles } SO_3}{2 \text{ moles } SO_2}$

$\frac{2 \text{ moles } SO_3}{1 \text{ mole } O_2}$ and $\frac{1 \text{ mole } O_2}{2 \text{ moles } SO_3}$

b. $\frac{4 \text{ moles P}}{5 \text{ moles } O_2}$ and $\frac{5 \text{ moles } O_2}{4 \text{ moles P}}$

$\frac{4 \text{ moles P}}{2 \text{ moles } P_2O_5}$ and $\frac{2 \text{ moles } P_2O_5}{4 \text{ moles P}}$

$\frac{5 \text{ moles } O_2}{2 \text{ moles } P_2O_5}$ and $\frac{2 \text{ moles } P_2O_5}{5 \text{ moles } O_2}$

7.55 **a.** 1.3 moles of O_2 **b.** 10. moles of H_2
c. 5.0 moles of H_2O

7.57 **a.** 1.25 moles of C **b.** 0.96 mole of CO
c. 1.0 mole of SO_2 **d.** 0.50 mole of CS_2

7.59 **a.** 77.5 g of Na_2O **b.** 6.26 g of O_2 **c.** 19.4 g of O_2

7.61 **a.** 19.2 g of O_2 **b.** 3.79 g of N_2 **c.** 54.0 g of H_2O

7.63 **a.** 3.66 g of H_2O **b.** 26.3 g of NO **c.** 7.53 g of HNO_3

7.65 **a.** $2PbS(s) + 3O_2(g) \longrightarrow 2PbO(s) + 2SO_2(g)$
b. 6.00 g of O_2 **c.** 17.4 g of SO_2 **d.** 137 g of PbS

7.67 **a.** The energy of activation is the energy required to break the bonds of the reacting molecules.
b. In exothermic reactions, the energy of the products is lower than the energy of the reactants.
c.

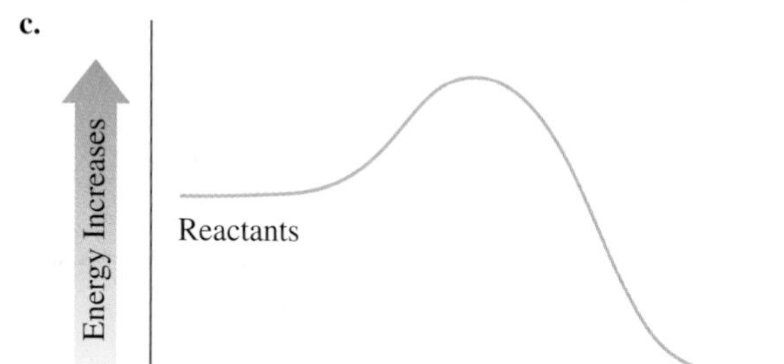

7.69 **a.** exothermic **b.** endothermic **c.** exothermic

7.71 **a.** exothermic **b.** endothermic **c.** exothermic

7.73 **a.** The rate of a reaction tells how fast the products are formed or how fast the reactants are consumed.
b. Reactions go faster at higher temperatures.

7.75 **a.** increase **b.** increase
c. increase **d.** decrease

7.77 **a.** $C_6H_{12}O_6(aq) + 6O_2(g) \longrightarrow 6CO_2(g) + 6H_2O(l)$
b. $6CO_2(g) + 6H_2O(l) \longrightarrow C_6H_{12}O_6(aq) + 6O_2(g)$

7.79 **a.** S_2Cl_2 **b.** 135.04 g/mole **c.** 0.0741 mole
a. C_6H_6 **b.** 78.11 g/mole **c.** 0.128 mole

7.81 **a.** 252.3 g/mole
b. 0.0991 mole of dipyrithione
c. 0.991 mole of C
d. 6.8 moles of dipyrithione

7.83 **a.** 1,1,2 combination **b.** 2,2,1 decomposition

7.85 **a.** reactants NO and O_2; product NO_2
b. $2NO(g) + O_2(g) \longrightarrow 2NO_2(g)$
c. combination

7.87 **a.** reactant NI_3; products N_2 and I_2
b. $2NI_3(s) \longrightarrow N_2(g) + 3I_2(g)$
c. decomposition

7.89 **a.** reactants Cl_2 and O_2; product OCl_2
b. $2Cl_2(g) + O_2(g) \longrightarrow 2OCl_2(g)$
c. combination

7.91 **a.** 161.48 g/mole **b.** 389.9 g/mole
c. 169.11 g/mole **d.** 116.16 g/mole

7.93 **a.** 5.87 g **b.** 10.6 g **c.** 15.9 g

7.95 **a.** 0.568 mole **b.** 0.245 mole **c.** 0.263 mole

7.97 **a.** combination **b.** combustion
c. decomposition **d.** single replacement

7.99 **a.** $NH_3(g) + HCl(g) \longrightarrow NH_4Cl(s)$ combination
b. $C_4H_8(g) + 6O_2(g) \xrightarrow{\Delta} 4CO_2(g) + 4H_2O(g)$ combustion
c. $2Sb(s) + 3Cl_2(g) \longrightarrow 2SbCl_3(s)$ combination
d. $2NI_3(s) \longrightarrow N_2(g) + 3I_2(g)$ decomposition
e. $2KBr(aq) + Cl_2(aq) \longrightarrow 2KCl(aq) + Br_2(l)$ single replacement
f. $2Fe(s) + 3H_2SO_4(aq) \longrightarrow 3H_2(g) + Fe_2(SO_4)_3(aq)$ single replacement
g. $Al_2(SO_4)_3(aq) + 6NaOH(aq) \longrightarrow 2Al(OH)_3(s) + 3Na_2SO_4(aq)$ double replacement

7.101 **a.** reduction **b.** oxidation **c.** oxidation **d.** reduction

7.103 **a.** $2NH_3(g) + 5F_2(g) \longrightarrow N_2F_4(g) + 6HF(g)$
b. 1.33 moles of NH_3 and 3.33 moles of F_2
c. 142 g of F_2
d. 10.4 g of N_2F_4

7.105 **a.** $C_5H_{12}(g) + 8O_2(g) \xrightarrow{\Delta} 5CO_2(g) + 6H_2O(g) + \text{energy}$
b. 48 g of pentane
c. 27.5 g of CO_2

7.107 **a.** exothermic **b.** lower

7.109 **a.** 242 g of glucose **b.** 123 g of ethanol

7.111 **a.** $4Al(s) + 3O_2(g) \longrightarrow 2Al_2O_3(s)$
b. combination **c.** 3.38 moles of oxygen
d. 94.9 g of Al_2O_3 **e.** 17.0 g of Al_2O_3

7.113 **a.** 3.01×10^{23} molecules **b.** 2.71×10^{24} atoms of C
c. 2.5 moles of lactic acid **d.** 90.08 g

7.115 **a.** $2C_2H_2(g) + 5O_2(g) \xrightarrow{\Delta} 4CO_2(g) + 2H_2O(g) + 2600\text{ kJ}$
b. exothermic
c. 0.800 mole of H_2O
d. 30.1 g of O_2

8 Gases

AFTER SOCCER PRACTICE, WHITNEY COMPLAINED that she was having difficulty breathing. Her father took her to the emergency room where she was seen by Sam, a respiratory therapist, who listened to Whitney's chest and then tested her breathing capacity using a spirometer. Based on her limited breathing capacity and the wheezing noise in her chest, Whitney was diagnosed with asthma.

Sam gave Whitney a nebulizer containing a bronchodilator that opens the airways and allows more air to go into the lungs. During the breathing treatment, he measured the amount of oxygen (O_2) in her blood and explained to Whitney and her father that air is a mixture of gases containing 78% nitrogen (N_2) gas and 21% O_2 gas. Because Whitney had difficulty obtaining sufficient oxygen, Sam gave her supplemental oxygen through an oxygen mask. Within a short period of time, Whitney's breathing returned to normal. The therapist then explained that the lungs work according to Boyle's law: The volume of the lungs increases upon inhalation, and the pressure decreases to allow air to flow in. However, during an asthma attack, the airways become restricted, and it becomes more difficult to expand the volume of the lungs.

CAREER Respiratory Therapist

Respiratory therapists assess and treat a range of patients, including premature infants whose lungs have not developed and asthmatics or patients with emphysema or cystic fibrosis. In assessing patients, they perform a variety of diagnostic tests including breathing capacity and concentrations of oxygen and carbon dioxide in a patient's blood, as well as blood pH. In order to treat patients, therapists provide oxygen or aerosol medications to the patient, as well as chest physiotherapy to remove mucus from their lungs. Respiratory therapists also educate patients on how to correctly use their inhalers.

CLINICAL UPDATE Exercise-Induced Asthma

Whitney's doctor prescribed an inhaled medication that opens up her airways before she starts exercise. In the **CLINICAL UPDATE Exercise-Induced Asthma**, page 279, you can view the impact of Whitney's medication and other treatments that help prevent exercise-induced asthma.

LOOKING AHEAD

REVIEW

Using Significant Figures in Calculations (2.3)

Writing Conversion Factors from Equalities (2.5)

Using Conversion Factors (2.6)

INTERACTIVE VIDEO

Kinetic Molecular Theory

ENGAGE

Use the kinetic molecular theory to explain why a gas completely fills a container of any size and shape.

TEST

Try Practice Problems 8.1 and 8.2

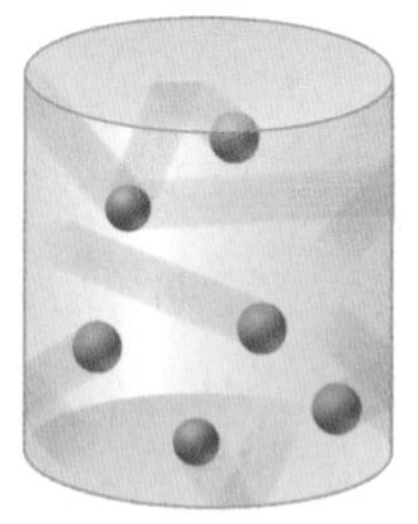

FIGURE 8.1 Gas particles moving in straight lines within a container exert pressure when they collide with the walls of the container.

Q Why does heating the container increase the pressure of the gas within it?

8.1 Properties of Gases

LEARNING GOAL Describe the kinetic molecular theory of gases and the units of measurement used for gases.

We all live at the bottom of a sea of gases called the atmosphere. The most important of these gases is oxygen, which constitutes about 21% of the atmosphere. Without oxygen, life on this planet would be impossible: Oxygen is vital to all life processes of plants and animals. Ozone (O_3), formed in the upper atmosphere by the interaction of oxygen with ultraviolet light, absorbs some of the harmful radiation before it can strike Earth's surface. The other gases in the atmosphere include nitrogen (78%), argon, carbon dioxide (CO_2), and water vapor. Carbon dioxide gas, a product of combustion and metabolism, is used by plants in photosynthesis, which produces the oxygen that is essential for humans and animals.

The behavior of gases is quite different from that of liquids and solids. Gas particles are far apart, whereas particles of both liquids and solids are held close together. A gas has no definite shape or volume and will completely fill any container. Because there are great distances between gas particles, a gas is less dense than a solid or liquid, and easy to compress. A model for the behavior of a gas, called the **kinetic molecular theory of gases**, helps us understand gas behavior.

Kinetic Molecular Theory of Gases

1. **A gas consists of small particles (atoms or molecules) that move randomly with high velocities.** Gas molecules moving in random directions at high speeds cause a gas to fill the entire volume of a container.
2. **The attractive forces between the particles of a gas are usually very small.** Gas particles are far apart and fill a container of any size and shape.
3. **The actual volume occupied by gas molecules is extremely small compared to the volume that the gas occupies.** The volume of the gas is considered equal to the volume of the container. Most of the volume of a gas is empty space, which allows gases to be easily compressed.
4. **Gas particles are in constant motion, moving rapidly in straight paths.** When gas particles collide, they rebound and travel in new directions. Every time they hit the walls of the container, they exert pressure. An increase in the number or force of collisions against the walls of the container causes an increase in the pressure of the gas.
5. **The average kinetic energy of gas molecules is proportional to the Kelvin temperature.** Gas particles move faster as the temperature increases. At higher temperatures, gas particles hit the walls of the container more often and with more force, producing higher pressures.

The kinetic molecular theory helps explain some of the characteristics of gases. For example, you can smell perfume when a bottle is opened on the other side of a room because its particles move rapidly in all directions. At room temperature, the molecules in the air are moving at about 450 m/s, which is 1000 mi/h. They move faster at higher temperatures and more slowly at lower temperatures. Sometimes tires and gas-filled containers explode when temperatures are too high. From the kinetic molecular theory, you know that gas particles move faster when heated, hit the walls of a container with more force, and cause a buildup of pressure inside a container.

When we talk about a gas, we describe it in terms of four properties: pressure, volume, temperature, and the amount of gas.

Pressure (*P*)

Gas particles are extremely small and move rapidly. When they hit the walls of a container, they exert a **pressure** (see **FIGURE 8.1**). If we heat the container, the molecules move faster and smash into the walls of the container more often and with increased force, thus increasing the pressure. The gas particles in the air, mostly oxygen and nitrogen, exert a pressure on us called **atmospheric pressure** (see **FIGURE 8.2**). As you go to higher altitudes, the

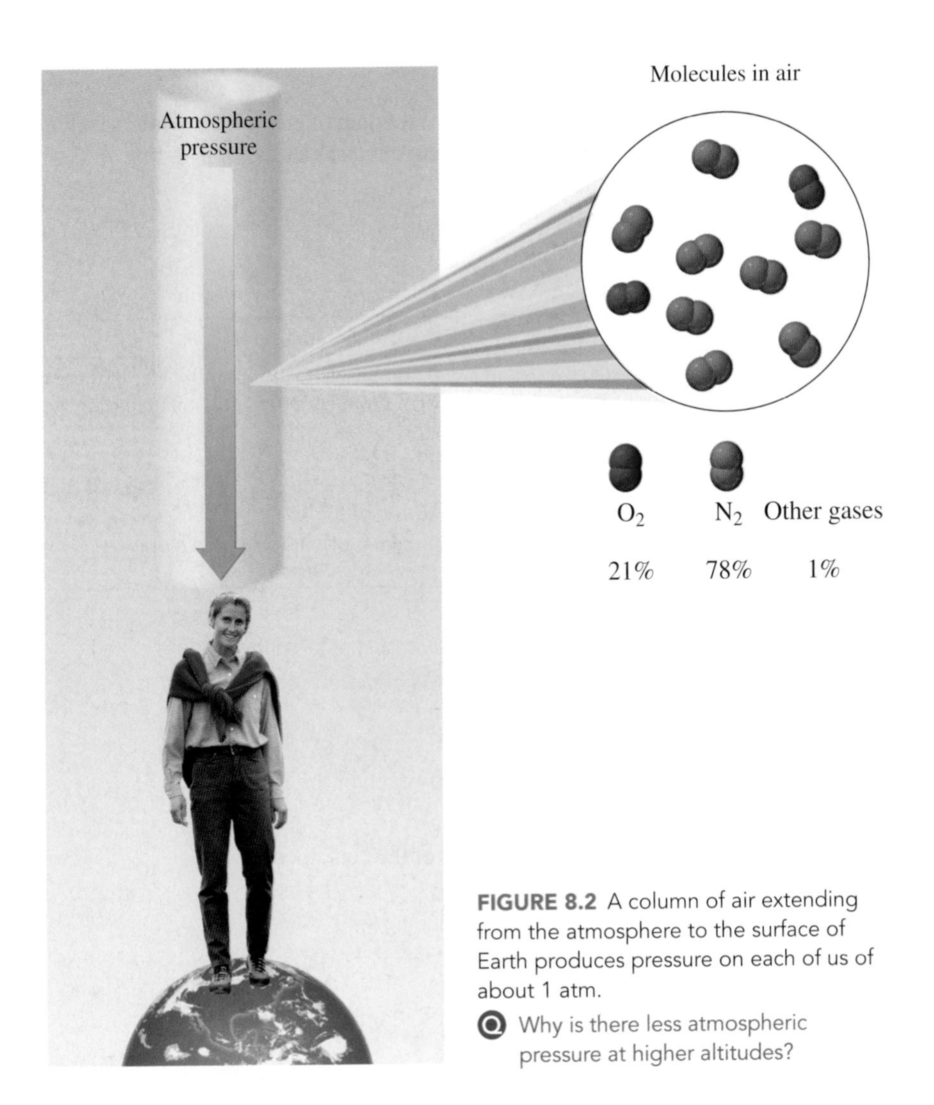

FIGURE 8.2 A column of air extending from the atmosphere to the surface of Earth produces pressure on each of us of about 1 atm.

Q Why is there less atmospheric pressure at higher altitudes?

atmospheric pressure is less because there are fewer particles in the air. The most common units used to measure gas pressure are the *atmosphere* (atm) and *millimeters of mercury* (mmHg). On a TV weather report, you may hear or see the atmospheric pressure given in inches of mercury, or in kilopascals in countries other than the United States. In a hospital, the unit torr or pounds per square inch (psi) may be used.

Volume (*V*)

The volume of gas equals the size of the container in which the gas is placed. When you inflate a tire or a basketball, you are adding more gas particles. The increase in the number of particles hitting the walls of the tire or basketball increases the volume. Sometimes, on a cold morning, a tire looks flat. The volume of the tire has decreased because a lower temperature decreases the speed of the molecules, which in turn reduces the force of their impacts on the walls of the tire. The most common units for volume measurement are liters (L) and milliliters (mL).

Temperature (*T*)

The temperature of a gas is related to the kinetic energy of its particles. For example, if we have a gas at 200 K and heat it to a temperature of 400 K, the gas particles will have twice the kinetic energy that they did at 200 K. This also means that the gas at 400 K exerts twice the pressure of the gas at 200 K, if the volume and amount of gas do not change. Although we measure gas temperature using a Celsius thermometer, all comparisons of gas behavior and all calculations related to temperature must use the Kelvin temperature scale. No one has quite created the conditions for absolute zero (0 K), but scientists predict that the particles will have zero kinetic energy and exert zero pressure at absolute zero.

Amount of Gas (n)

When you add air to a bicycle tire, you increase the amount of gas, which results in a higher pressure in the tire. Usually, we measure the amount of gas by its mass, in grams. In gas law calculations, we need to change the grams of gas to moles.

A summary of the four properties of a gas is given in **TABLE 8.1**.

TABLE 8.1 Properties That Describe a Gas

Property	Description	Units of Measurement
Pressure (P)	The force exerted by a gas against the walls of the container	atmosphere (atm); millimeter of mercury (mmHg); torr (Torr); pascal (Pa)
Volume (V)	The space occupied by a gas	liter (L); milliliter (mL)
Temperature (T)	The determining factor of the kinetic energy of gas particles	degree Celsius (°C); kelvin (K) *is required in calculations*
Amount (n)	The quantity of gas present in a container	gram (g); mole (n) *is required in calculations*

SAMPLE PROBLEM 8.1 Properties of Gases

TRY IT FIRST

Identify the property of a gas that is described by each of the following:

a. increases the kinetic energy of gas particles
b. the force of the gas particles hitting the walls of the container
c. the space that is occupied by a gas

SOLUTION

a. temperature
b. pressure
c. volume

STUDY CHECK 8.1

When helium is added to a balloon, the number of grams of helium increases. What property of a gas is described?

ANSWER

The mass, in grams, gives the amount of gas.

TEST

Try Practice Problems 8.3 and 8.4

CHEMISTRY LINK TO HEALTH

Measuring Blood Pressure

Your blood pressure is one of the vital signs a doctor or nurse checks during a physical examination. It actually consists of two separate measurements. Acting like a pump, the heart contracts to create the pressure that pushes blood through the circulatory system. During contraction, the blood pressure is at its highest; this is your *systolic* pressure. When the heart muscles relax, the blood pressure falls; this is your *diastolic* pressure. The normal range for systolic pressure is 100 to 200 mmHg. For diastolic pressure, it is 60 to 80 mmHg. These two measurements are usually expressed as a ratio such as 100/80. These values are somewhat higher in older people. When blood pressures are elevated, such as 140/90, there is a greater risk of stroke, heart attack, or kidney damage. Low blood pressure prevents the brain from receiving adequate oxygen, causing dizziness and fainting.

The blood pressures are measured by a *sphygmomanometer*, an instrument consisting of a stethoscope and an inflatable cuff connected to a tube of mercury called a manometer. After the cuff is wrapped around the upper arm, it is pumped up with air until it cuts off the flow of blood through the arm. With the stethoscope over the artery, the air is slowly released from the cuff, decreasing the pressure on the artery. When the blood flow first starts again in the artery, a noise can be heard through the stethoscope signifying the systolic blood pressure as the pressure shown on the

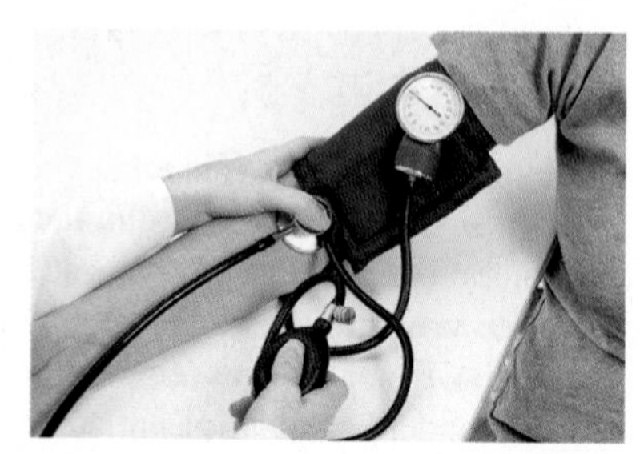

The measurement of blood pressure is part of a routine physical exam.

manometer. As air continues to be released, the cuff deflates until no sound is heard in the artery. A second pressure reading is taken at the moment of silence and denotes the diastolic pressure, the pressure when the heart is not contracting.

The use of digital blood pressure monitors is becoming more common. However, they have not been validated for use in all situations and can sometimes give inaccurate readings.

Measurement of Gas Pressure

When billions and billions of gas particles hit against the walls of a container, they exert pressure, which is a force acting on a certain area.

$$\text{Pressure } (P) = \frac{\text{force}}{\text{area}}$$

The atmospheric pressure can be measured using a barometer (see **FIGURE 8.3**). At a pressure of exactly 1 atmosphere (atm), a mercury column in an inverted tube would be *exactly* 760 mm high. One **atmosphere (atm)** is defined as *exactly* 760 mmHg (millimeters of mercury). One atmosphere is also 760 Torr, a pressure unit named to honor Evangelista Torricelli, the inventor of the barometer. Because the units of Torr and mmHg are equal, they are used interchangeably. One atmosphere is also equivalent to 29.9 in. of mercury (inHg).

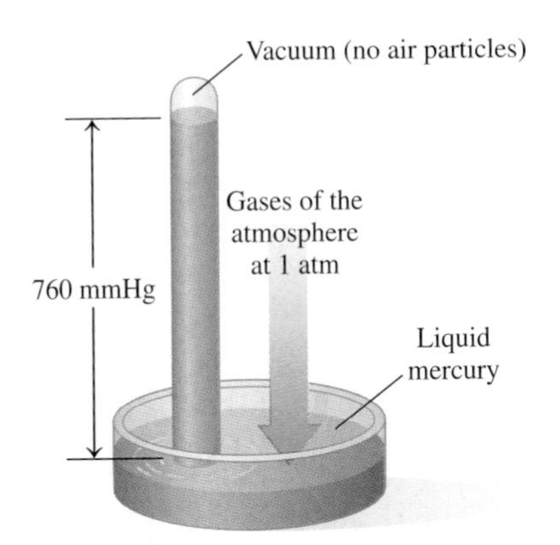

FIGURE 8.3 A barometer: The pressure exerted by the gases in the atmosphere is equal to the downward pressure of a mercury column in a closed glass tube. The height of the mercury column measured in mmHg is called atmospheric pressure.

Why does the height of the mercury column change from day to day?

1 atm = 760 mmHg = 760 Torr (exact)

1 atm = 29.9 inHg

1 mmHg = 1 Torr (exact)

In SI units, pressure is measured in pascals (Pa); 1 atm is equal to 101 325 Pa. Because a pascal is a very small unit, pressures are usually reported in kilopascals.

1 atm = 101 325 Pa = 101.325 kPa

The U.S. equivalent of 1 atm is 14.7 lb/in.^{2} (psi). When you use a pressure gauge to check the air pressure in the tires of a car, it may read 30 to 35 psi. This measurement is actually 30 to 35 psi above the pressure that the atmosphere exerts on the outside of the tire.

1 atm = 14.7 lb/in.^{2}

TABLE 8.2 summarizes the various units used in the measurement of pressure.

TABLE 8.2 Units for Measuring Pressure

Unit	Abbreviation	Unit Equivalent to 1 atm
atmosphere	atm	1 atm (exact)
millimeters of Hg	mmHg	760 mmHg (exact)
torr	Torr	760 Torr (exact)
inches of Hg	inHg	29.9 inHg
pounds per square inch	lb/in.^{2} (psi)	14.7 lb/in.^{2}
pascal	Pa	101 325 Pa
kilopascal	kPa	101.325 kPa

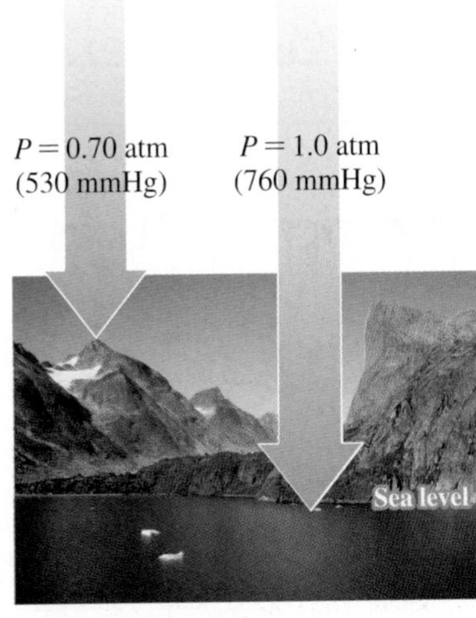

The atmospheric pressure decreases as the altitude increases.

Atmospheric pressure changes with variations in weather and altitude. On a hot, sunny day, the mercury column rises, indicating a higher atmospheric pressure. On a rainy day, the atmosphere exerts less pressure, which causes the mercury column to fall. In a weather report, this type of weather is called a *low-pressure system*. Above sea level, the density of the gases in the air decreases, which causes lower atmospheric pressures; the atmospheric pressure is greater than 760 mmHg at the Dead Sea because it is below sea level (see **TABLE 8.3**).

TABLE 8.3 Altitude and Atmospheric Pressure

Location	Altitude (km)	Atmospheric Pressure (mmHg)
Dead Sea	−0.40	800
Sea level	0.00	760
Los Angeles	0.09	752
Las Vegas	0.70	700
Denver	1.60	630
Mount Whitney	4.50	440
Mount Everest	8.90	253

Divers must be concerned about increasing pressures on their ears and lungs when they dive below the surface of the ocean. Because water is more dense than air, the pressure on a diver increases rapidly as the diver descends. At a depth of 33 ft below the surface of the ocean, an additional 1 atm of pressure is exerted by the water on a diver, which gives a total pressure of 2 atm. At 100 ft, there is a total pressure of 4 atm on a diver. The regulator that a diver uses continuously adjusts the pressure of the breathing mixture to match the increase in pressure.

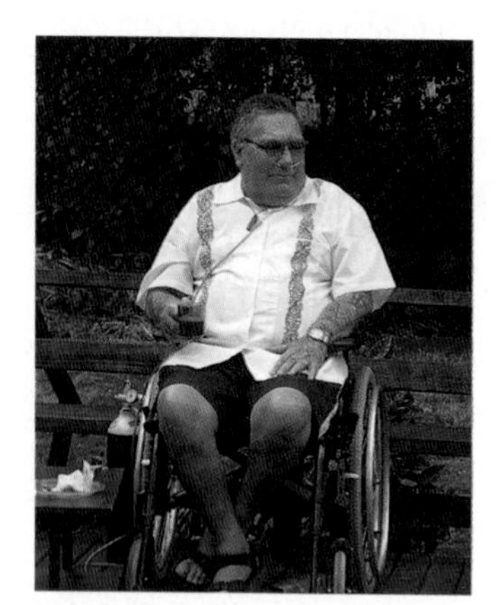

A patient with severe COPD obtains oxygen from an oxygen tank.

SAMPLE PROBLEM 8.2 Units of Pressure

TRY IT FIRST

The oxygen in a tank in the hospital respiratory unit has a pressure of 4820 mmHg. Calculate the pressure, in atmospheres, of the oxygen gas.

SOLUTION

The equality 1 atm = 760 mmHg can be written as two conversion factors:

$$\frac{760 \text{ mmHg}}{1 \text{ atm}} \quad \text{and} \quad \frac{1 \text{ atm}}{760 \text{ mmHg}}$$

Using the conversion factor that cancels mmHg and gives atm, we can set up the problem as

$$4820 \cancel{\text{mmHg}} \times \frac{1 \text{ atm}}{760 \cancel{\text{mmHg}}} = 6.34 \text{ atm}$$

STUDY CHECK 8.2

A tank of nitrous oxide (N_2O) used as an anesthetic has a pressure of 48 psi. What is that pressure in atmospheres?

ANSWER

3.3 atm

TEST

Try Practice Problems 8.5 to 8.8

PRACTICE PROBLEMS

8.1 Properties of Gases

LEARNING GOAL Describe the kinetic molecular theory of gases and the units of measurement used for gases.

8.1 Use the kinetic molecular theory of gases to explain each of the following:
- **a.** Gases move faster at higher temperatures.
- **b.** Gases can be compressed much more easily than liquids or solids.
- **c.** Gases have low densities.

8.2 Use the kinetic molecular theory of gases to explain each of the following:
- **a.** A container of nonstick cooking spray explodes when thrown into a fire.
- **b.** The air in a hot-air balloon is heated to make the balloon rise.
- **c.** You can smell the odor of cooking onions from far away.

8.3 Identify the property of a gas that is measured in each of the following:
- **a.** 350 K
- **b.** 125 mL
- **c.** 2.00 g of O_2
- **d.** 755 mmHg

8.4 Identify the property of a gas that is measured in each of the following:
a. 425 K **b.** 1.0 atm
c. 10.0 L **d.** 0.50 mole of He

8.5 Which of the following statements describe the pressure of a gas?
a. the force of the gas particles on the walls of the container
b. the number of gas particles in a container
c. 4.5 L of helium gas
d. 750 Torr
e. 28.8 lb/in.^{2}

8.6 Which of the following statements describe the pressure of a gas?
a. 350 K
b. the volume of the container
c. 3.00 atm
d. 0.25 mole of O_2
e. 101 kPa

Clinical Applications

8.7 A tank contains oxygen (O_2) at a pressure of 2.00 atm. What is the pressure in the tank in terms of the following units?
a. torr **b.** lb/in.^{2}
c. mmHg **d.** kPa

8.8 On a climb up Mount Whitney, the atmospheric pressure drops to 467 mmHg. What is the pressure in terms of the following units?
a. atm **b.** torr
c. inHg **d.** Pa

8.2 Pressure and Volume (Boyle's Law)

REVIEW
Solving Equations (1.4)

LEARNING GOAL Use the pressure–volume relationship (Boyle's law) to calculate the unknown pressure or volume when the temperature and amount of gas do not change.

Imagine that you can see air particles hitting the walls inside a bicycle tire pump. What happens to the pressure inside the pump as you push down on the handle? As the volume decreases, there is a decrease in the surface area of the container. The air particles are crowded together, more collisions occur, and the pressure increases within the container.

When a change in one property (in this case, volume) causes a change in another property (in this case, pressure), the properties are related. If the change occurs in opposite directions, the properties have an **inverse relationship**. The inverse relationship between the pressure and volume of a gas is known as **Boyle's law**. The law states that the volume (V) of a sample of gas changes inversely with the pressure (P) of the gas as long as there is no change in the temperature (T) or amount of gas (n), as illustrated in FIGURE 8.4.

If the volume or pressure of a gas changes without any change occurring in the temperature or in the amount of the gas, then the final pressure and volume will give the same PV product as the initial pressure and volume. Then we can set the initial and final PV products equal to each other. In the equation for Boyle's law, the initial pressure and volume are written as P_1 and V_1 and the final pressure and volume are written as P_2 and V_2.

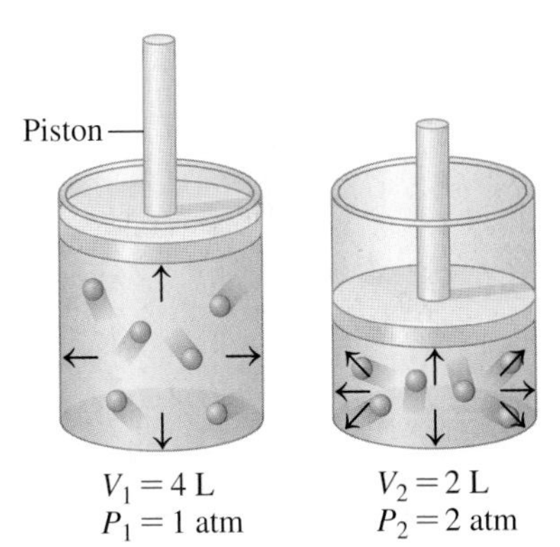

FIGURE 8.4 Boyle's law: As volume decreases, gas molecules become more crowded, which causes the pressure to increase. Pressure and volume are inversely related.

Q If the volume of a gas increases, what will happen to its pressure?

Boyle's Law

$P_1V_1 = P_2V_2$ No change in temperature and amount of gas

SAMPLE PROBLEM 8.3 Calculating Volume When Pressure Changes

TRY IT FIRST

When Whitney had her asthma attack, she was given oxygen through a face mask. The gauge on a 12-L tank of compressed oxygen reads 3800 mmHg. How many liters would this same gas occupy at a final pressure of 570 mmHg when temperature and amount of gas do not change?

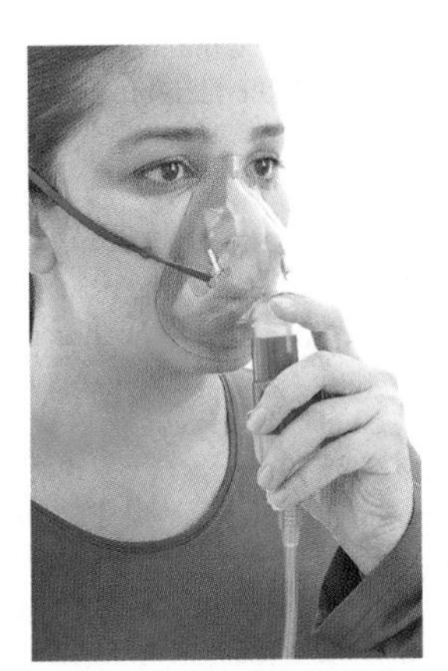

Oxygen therapy increases the oxygen available to the tissues of the body.

SOLUTION GUIDE

STEP 1 **State the given and needed quantities.** We place the gas data in a table by writing the initial pressure and volume as P_1 and V_1 and the final pressure and volume as P_2 and V_2. We see that the pressure decreases from 3800 mmHg to 570 mmHg. Using Boyle's law, we predict that the volume increases.

ENGAGE

Why does the pressure of a gas in a closed container decrease when the volume of the container increases?

	Given	Need	Connect
ANALYZE THE PROBLEM	$P_1 = 3800$ mmHg $P_2 = 570$ mmHg $V_1 = 12$ L **Factors that do not change:** T and n	V_2	Boyle's law, $P_1V_1 = P_2V_2$ **Predict:** P decreases, V increases

CORE CHEMISTRY SKILL

Using the Gas Laws

STEP 2 **Rearrange the gas law equation to solve for the unknown quantity.** For a PV relationship, we use Boyle's law and solve for V_2 by dividing both sides by P_2. According to Boyle's law, a decrease in the pressure will cause an increase in the volume when T and n do not change.

$$P_1V_1 = P_2V_2$$

$$\frac{P_1V_1}{P_2} = \frac{\cancel{P_2}V_2}{\cancel{P_2}}$$

$$V_2 = V_1 \times \frac{P_1}{P_2}$$

STEP 3 **Substitute values into the gas law equation and calculate.** When we substitute in the values with pressures in units of mmHg, the ratio of pressures (pressure factor) is greater than 1, which increases the volume as predicted.

$$V_2 = 12\text{ L} \times \frac{3800\ \cancel{\text{mmHg}}}{570\ \cancel{\text{mmHg}}} = 80.\text{ L}$$

Pressure factor increases volume

STUDY CHECK 8.3

In an underground gas reserve, a bubble of methane gas (CH_4) has a volume of 45.0 mL at 1.60 atm pressure. What volume, in milliliters, will the gas bubble occupy when it reaches the surface where the atmospheric pressure is 744 mmHg, if there is no change in the temperature and amount of gas?

ANSWER

73.5 mL

TEST

Try Practice Problems 8.9 to 8.24

CHEMISTRY LINK TO HEALTH

Pressure–Volume Relationship in Breathing

The importance of Boyle's law becomes apparent when you consider the mechanics of breathing. Our lungs are elastic, balloon-like structures contained within an airtight chamber called the thoracic cavity. The diaphragm, a muscle, forms the flexible floor of the cavity.

Inspiration

The process of taking a breath of air begins when the diaphragm contracts and the rib cage expands, causing an increase in the volume of the thoracic cavity. The elasticity of the lungs allows them to expand when the thoracic cavity expands. According to Boyle's law, the pressure inside the lungs decreases when their volume increases, causing the pressure inside the lungs to fall below the pressure of the atmosphere. This difference in pressures produces a *pressure gradient* between the lungs and the atmosphere. In a pressure gradient, molecules flow from an area of higher pressure to an area of lower pressure. During the inhalation phase of breathing, air flows into the lungs (*inspiration*), until the pressure within the lungs becomes equal to the pressure of the atmosphere.

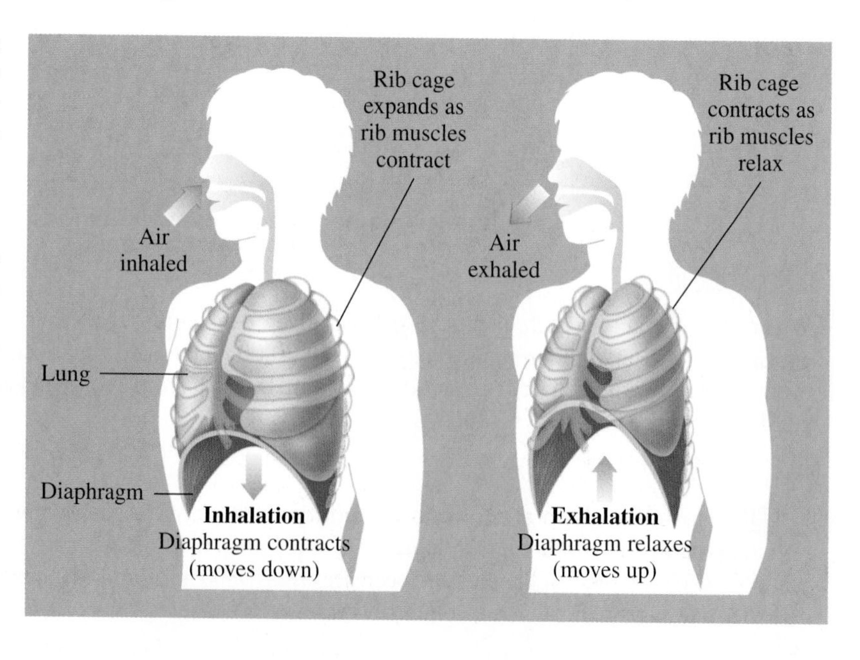

Expiration

Expiration, or the exhalation phase of breathing, occurs when the diaphragm relaxes and moves back up into the thoracic cavity to its resting position. The volume of the thoracic cavity decreases, which squeezes the lungs and decreases their volume. Now the pressure in the lungs is higher than the pressure of the atmosphere, so air flows out of the lungs. Thus, breathing is a process in which pressure gradients are continuously created between the lungs and the environment because of the changes in the volume and pressure.

PRACTICE PROBLEMS

8.2 Pressure and Volume (Boyle's Law)

LEARNING GOAL Use the pressure–volume relationship (Boyle's law) to calculate the unknown pressure or volume when the temperature and amount of gas do not change.

8.9 Why do scuba divers need to exhale air when they ascend to the surface of the water?

8.10 Why does a sealed bag of chips expand when you take it to a higher altitude?

8.11 The air in a cylinder with a piston has a volume of 220 mL and a pressure of 650 mmHg.

a. To obtain a higher pressure inside the cylinder when the temperature and amount of gas do not change, would the cylinder change as shown in **A** or **B**? Explain your choice.

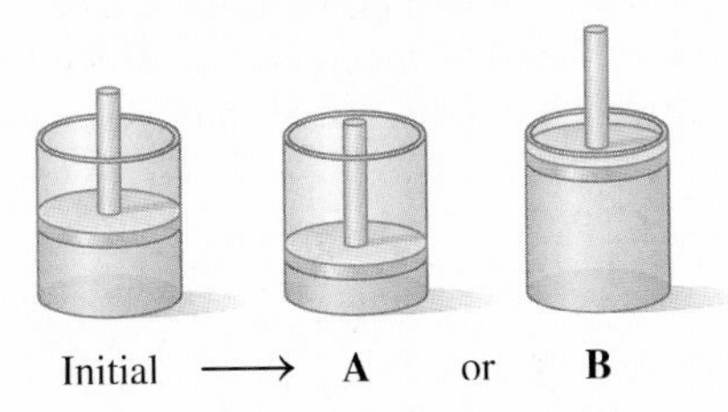

b. If the pressure inside the cylinder increases to 1.2 atm, what is the final volume, in milliliters, of the cylinder?

8.12 A balloon is filled with helium gas. When each of the following changes are made with no change in temperature, which of these diagrams (**A**, **B**, or **C**) shows the final volume of the balloon?

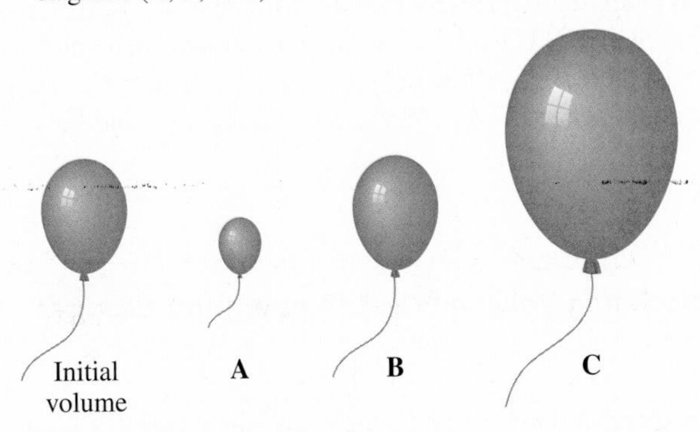

a. The balloon floats to a higher altitude where the outside pressure is lower.

b. The balloon is taken inside the house, but the atmospheric pressure does not change.

c. The balloon is put in a hyperbaric chamber in which the pressure is increased.

8.13 A gas with a volume of 4.0 L is in a closed container. Indicate the changes (*increases*, *decreases*, *does not change*) in its pressure when the volume undergoes the following changes at the same temperature and amount of gas:

a. The volume is compressed to 2.0 L.

b. The volume expands to 12 L.

c. The volume is compressed to 0.40 L.

8.14 A gas at a pressure of 2.0 atm is in a closed container. Indicate the changes (*increases*, *decreases*, *does not change*) in its volume when the pressure undergoes the following changes at the same temperature and amount of gas:

a. The pressure increases to 6.0 atm.

b. The pressure remains at 2.0 atm.

c. The pressure drops to 0.40 atm.

8.15 A 10.0-L balloon contains helium gas at a pressure of 655 mmHg. What is the final pressure, in millimeters of mercury, when the helium is placed in tanks that have the following volumes, if there is no change in temperature and amount of gas?

a. 20.0 L **b.** 2.50 L

c. 13 800 mL **d.** 1250 mL

8.16 The air in a 5.00-L tank has a pressure of 1.20 atm. What is the final pressure, in atmospheres, when the air is placed in tanks that have the following volumes, if there is no change in temperature and amount of gas?

a. 1.00 L **b.** 2500. mL

c. 750. mL **d.** 8.00 L

8.17 A sample of nitrogen (N_2) has a volume of 50.0 L at a pressure of 760. mmHg. What is the final volume, in liters, of the gas at each of the following pressures, if there is no change in temperature and amount of gas?

a. 725 mmHg **b.** 2.0 atm

c. 0.500 atm **d.** 850 Torr

8.18 A sample of methane (CH_4) has a volume of 25 mL at a pressure of 0.80 atm. What is the final volume, in milliliters, of the gas at each of the following pressures, if there is no change in temperature and amount of gas?

a. 0.40 atm **b.** 2.00 atm

c. 2500 mmHg **d.** 80.0 Torr

8.19 A sample of Ar gas has a volume of 5.40 L with an unknown pressure. The gas has a volume of 9.73 L when the pressure is 3.62 atm, with no change in temperature or amount of gas. What was the initial pressure, in atmospheres, of the gas?

8.20 A sample of Ne gas has a pressure of 654 mmHg with an unknown volume. The gas has a pressure of 345 mmHg when the volume is 495 mL, with no change in temperature or amount of gas. What was the initial volume, in milliliters, of the gas?

Clinical Applications

8.21 Cyclopropane, C_3H_6, is a general anesthetic. A 5.0-L sample has a pressure of 5.0 atm. What is the final volume, in liters, of this gas given to a patient at a pressure of 1.0 atm with no change in temperature and amount of gas?

8.22 A patient's oxygen tank holds 20.0 L of oxygen (O_2) at a pressure of 15.0 atm. What is the final volume, in liters, of this gas when it is released at a pressure of 1.00 atm with no change in temperature and amount of gas?

8.23 Use the words *inspiration* and *expiration* to describe the part of the breathing cycle that occurs as a result of each of the following:
a. The diaphragm contracts.
b. The volume of the lungs decreases.
c. The pressure within the lungs is less than that of the atmosphere.

8.24 Use the words *inspiration* and *expiration* to describe the part of the breathing cycle that occurs as a result of each of the following:
a. The diaphragm relaxes, moving up into the thoracic cavity.
b. The volume of the lungs expands.
c. The pressure within the lungs is higher than that of the atmosphere.

8.3 Temperature and Volume (Charles's Law)

LEARNING GOAL Use the temperature–volume relationship (Charles's law) to calculate the unknown temperature or volume when the pressure and amount of gas do not change.

As the gas in a hot-air balloon is heated, it expands.

Suppose that you are going to take a ride in a hot-air balloon. The captain turns on a propane burner to heat the air inside the balloon. As the air is heated, it expands and becomes less dense than the air outside, causing the balloon and its passengers to lift off. In 1787, Jacques Charles, a balloonist as well as a physicist, proposed that the volume of a gas is related to the temperature. This proposal became **Charles's law**, which states that the volume (V) of a gas is directly related to the temperature (T) when there is no change in the pressure (P) or amount (n) of gas. A **direct relationship** is one in which the related properties increase or decrease together. For two conditions, initial and final, we can write Charles's law as follows:

ENGAGE
Why does the volume of a gas increase when the temperature increases if pressure and amount of gas do not change?

Charles's Law

$$\frac{V_1}{T_1} = \frac{V_2}{T_2} \quad \text{No change in pressure and amount of gas}$$

All temperatures used in gas law calculations must be converted to their corresponding Kelvin (K) temperatures.

To determine the effect of changing temperature on the volume of a gas, the pressure and the amount of gas are not changed. If we increase the temperature of a gas sample, the volume of the container must increase (see **FIGURE 8.5**). If the temperature of the gas is decreased, the volume of the container must also decrease when pressure and the amount of gas do not change.

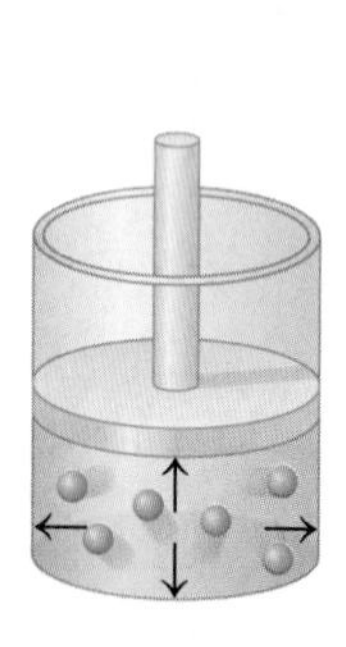

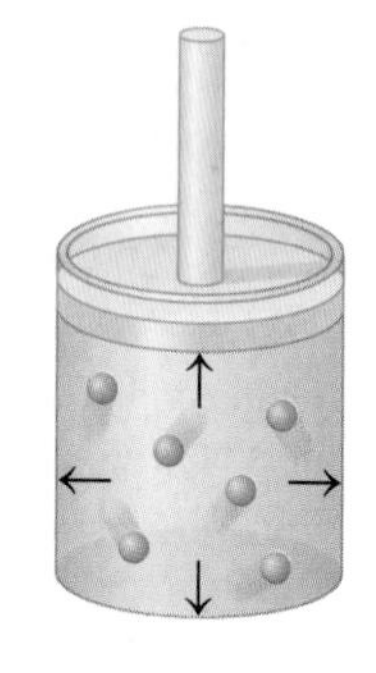

$T_1 = 200$ K $\quad T_2 = 400$ K
$V_1 = 1$ L $\quad V_2 = 2$ L

FIGURE 8.5 Charles's law: The Kelvin temperature of a gas is directly related to the volume of the gas when there is no change in the pressure and amount of gas.

If the temperature of a gas decreases, how will the volume change if the pressure and amount of gas do not change?

SAMPLE PROBLEM 8.4 Calculating Volume When Temperature Changes

TRY IT FIRST

Helium gas is used to inflate the abdomen during laparoscopic surgery. A sample of helium gas has a volume of 5.40 L and a temperature of 15 °C. What is the final volume, in liters, of the gas after the temperature has been increased to 42 °C when the pressure and amount of gas do not change?

SOLUTION GUIDE

STEP 1 **State the given and needed quantities.** We place the gas data in a table by writing the initial temperature and volume as T_1 and V_1 and the final temperature and volume as T_2 and V_2. We see that the temperature increases from 15 °C to 42 °C. Using Charles's law, we predict that the volume increases.

$$T_1 = 15\ °\text{C} + 273 = 288\ \text{K}$$

$$T_2 = 42\ °\text{C} + 273 = 315\ \text{K}$$

	Given	Need	Connect
ANALYZE THE PROBLEM	$T_1 = 288\text{ K}$ $T_2 = 315\text{ K}$ $V_1 = 5.40\text{ L}$	V_2	Charles's law, $\frac{V_1}{T_1} = \frac{V_2}{T_2}$
	Factors that do not change: P and n		**Predict:** T increases, V increases

STEP 2 Rearrange the gas law equation to solve for the unknown quantity. In this problem, we want to know the final volume (V_2) when the temperature increases. Using Charles's law, we solve for V_2 by multiplying both sides by T_2.

$$\frac{V_1}{T_1} = \frac{V_2}{T_2}$$

$$\frac{V_1}{T_1} \times T_2 = \frac{V_2}{\cancel{T_2}} \times \cancel{T_2}$$

$$V_2 = V_1 \times \frac{T_2}{T_1}$$

STEP 3 Substitute values into the gas law equation and calculate. From the table, we see that the temperature has increased. Because temperature is directly related to volume, the volume must increase. When we substitute in the values, we see that the ratio of the temperatures (temperature factor) is greater than 1, which increases the volume as predicted.

$$V_2 = 5.40\text{ L} \times \frac{315\text{ }\cancel{\text{K}}}{288\text{ }\cancel{\text{K}}} = 5.91\text{ L}$$

Temperature factor increases volume

STUDY CHECK 8.4

A mountain climber inhales air that has a temperature of −8 °C. If the final volume of air in the lungs is 569 mL at a body temperature of 37 °C, what was the initial volume of air, in milliliters, inhaled by the climber if the pressure and amount of gas do not change?

ANSWER

486 mL

TEST

Try Practice Problems 8.25 to 8.32

PRACTICE PROBLEMS

8.3 Temperature and Volume (Charles's Law)

LEARNING GOAL Use the temperature–volume relationship (Charles's law) to calculate the unknown temperature or volume when the pressure and amount of gas do not change.

8.25 Select the diagram that shows the final volume of a balloon when each of the following changes are made when the pressure and amount of gas do not change:

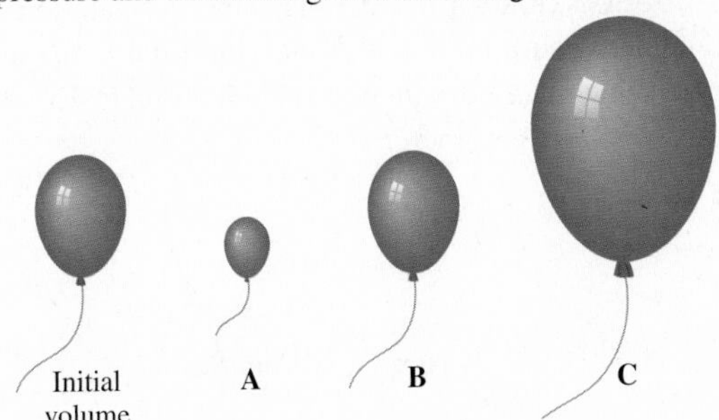

a. The temperature is changed from 100 K to 300 K. C

b. The balloon is placed in a freezer. A

c. The balloon is first warmed and then returned to its starting temperature. B

8.26 Indicate whether the final volume of gas in each of the following is the *same*, *larger*, or *smaller* than the initial volume, if pressure and amount of gas do not change:

a. A volume of 505 mL of air on a cold winter day at −15 °C is breathed into the lungs, where body temperature is 37 °C.

b. The heater used to heat the air in a hot-air balloon is turned off.

c. A balloon filled with helium at the amusement park is left in a car on a hot day.

8.27 A sample of neon initially has a volume of 2.50 L at 15 °C. What final temperature, in degrees Celsius, is needed to change the volume of the gas to each of the following, if P and n do not change?

a. 5.00 L b. 1250 mL c. 7.50 L d. 3550 mL

8.28 A gas has a volume of 4.00 L at 0 °C. What final temperature, in degrees Celsius, is needed to change the volume of the gas to each of the following, if P and n do not change?

a. 1.50 L b. 1200 mL c. 10.0 L d. 50.0 mL

8.29 A balloon contains 2500 mL of helium gas at 75 °C. What is the final volume, in milliliters, of the gas when the temperature changes to each of the following, if P and n do not change?

a. 55 °C b. 680. K c. −25 °C d. 240. K

8.30 An air bubble has a volume of 0.500 L at 18 °C. What is the final volume, in liters, of the gas when the temperature changes to each of the following, if P and n do not change?
a. 0 °C **b.** 425 K **c.** −12 °C **d.** 575 K

8.31 A gas sample has a volume of 0.256 L with an unknown temperature. The same gas has a volume of 0.198 L when the temperature is 32 °C, with no change in the pressure or amount of gas. What was the initial temperature, in degrees Celsius, of the gas?

8.32 A gas sample has a temperature of 22 °C with an unknown volume. The same gas has a volume of 456 mL when the temperature is 86 °C, with no change in the pressure or amount of gas. What was the initial volume, in milliliters, of the gas?

8.4 Temperature and Pressure (Gay-Lussac's Law)

LEARNING GOAL Use the temperature–pressure relationship (Gay-Lussac's law) to calculate the unknown temperature or pressure when the volume and amount of gas do not change.

FIGURE 8.6 Gay-Lussac's law: When the Kelvin temperature of a gas is doubled and the volume and amount of gas do not change, the pressure also doubles.

Q How does a decrease in the temperature of a gas affect its pressure when the volume and amount of gas do not change?

If we could observe the molecules of a gas as the temperature rises, we would notice that they move faster and hit the sides of the container more often and with greater force. If volume and amount of gas do not change, the pressure would increase. In the temperature–pressure relationship known as **Gay-Lussac's law**, the pressure of a gas is directly related to its Kelvin temperature. This means that an increase in temperature increases the pressure of a gas, and a decrease in temperature decreases the pressure of the gas as long as the volume and amount of gas do not change (see **FIGURE 8.6**).

Gay-Lussac's Law

$$\frac{P_1}{T_1} = \frac{P_2}{T_2} \quad \text{No change in volume and amount of gas}$$

All temperatures used in gas law calculations must be converted to their corresponding Kelvin (K) temperatures.

ENGAGE

Why does the pressure of a gas decrease when the temperature decreases if the volume and amount of gas do not change?

SAMPLE PROBLEM 8.5 Calculating Pressure When Temperature Changes

TRY IT FIRST

Home oxygen tanks can be dangerous if they are heated because they can explode. Suppose an oxygen tank has a pressure of 120 atm at a room temperature of 25 °C. If a fire in the room causes the temperature of the gas inside the oxygen tank to reach 402 °C, what is its pressure in atmospheres if the volume and amount of gas do not change? The oxygen tank may rupture if the pressure inside exceeds 180 atm. Would you expect it to rupture?

SOLUTION GUIDE

STEP 1 **State the given and needed quantities.** We place the gas data in a table by writing the initial temperature and pressure as T_1 and P_1 and the final temperature and pressure as T_2 and P_2. We see that the temperature increases from 25 °C to 402 °C. Using Gay-Lussac's law, we predict that the pressure increases.

$T_1 = 25\ °C + 273 = 298$ K
$T_2 = 402\ °C + 273 = 675$ K

	Given	Need	Connect
ANALYZE THE PROBLEM	$P_1 = 120$ atm $T_1 = 298$ K $T_2 = 675$ K **Factors that do not change:** V and n	P_2	Gay-Lussac's law, $\frac{P_1}{T_1} = \frac{P_2}{T_2}$ **Predict:** T increases, P increases

STEP 2 **Rearrange the gas law equation to solve for the unknown quantity.** Using Gay-Lussac's law, we solve for P_2 by multiplying both sides by T_2.

$$\frac{P_1}{T_1} = \frac{P_2}{T_2}$$

$$\frac{P_1}{T_1} \times T_2 = \frac{P_2}{\cancel{T_2}} \times \cancel{T_2}$$

$$P_2 = P_1 \times \frac{T_2}{T_1}$$

STEP 3 **Substitute values into the gas law equation and calculate.** When we substitute in the values, we see that the ratio of the temperatures (temperature factor) is greater than 1, which increases the pressure as predicted.

$$P_2 = 120 \text{ atm} \times \frac{675 \cancel{\text{K}}}{298 \cancel{\text{K}}} = 270 \text{ atm}$$

Temperature factor increases pressure

Because the calculated pressure of 270 atm exceeds the limit of 180 atm, we would expect the oxygen tank to rupture.

Cylinders of oxygen gas are placed in a hospital storage room.

STUDY CHECK 8.5

In a storage area of a hospital where the temperature has reached 55 °C, the pressure of oxygen gas in a 15.0-L steel cylinder is 965 Torr. To what temperature, in degrees Celsius, would the gas have to be cooled to reduce the pressure to 850 Torr, when the volume and the amount of the gas do not change?

ANSWER

16 °C

TEST

Try Practice Problems 8.33 to 8.40

PRACTICE PROBLEMS

8.4 Temperature and Pressure (Gay-Lussac's Law)

LEARNING GOAL Use the temperature–pressure relationship (Gay-Lussac's law) to calculate the unknown temperature or pressure when the volume and amount of gas do not change.

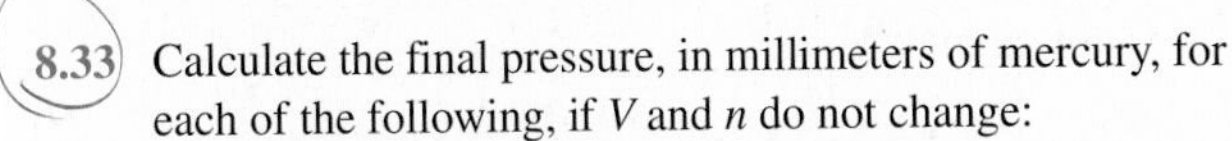

8.33 Calculate the final pressure, in millimeters of mercury, for each of the following, if V and n do not change:

a. A gas with an initial pressure of 1200 Torr at 155 °C is cooled to 0 °C.

b. A gas in an aerosol can at an initial pressure of 1.40 atm at 12 °C is heated to 35 °C.

8.34 Calculate the final pressure, in atmospheres, for each of the following, if V and n do not change:

a. A gas with an initial pressure of 1.20 atm at 75 °C is cooled to −32 °C.

b. A sample of N_2 with an initial pressure of 780. mmHg at −75 °C is heated to 28 °C.

8.35 Calculate the final temperature, in degrees Celsius, for each of the following, if V and n do not change:

a. A sample of xenon gas at 25 °C and 740. mmHg is cooled to give a pressure of 620. mmHg.

b. A tank of argon gas with a pressure of 0.950 atm at −18 °C is heated to give a pressure of 1250 Torr.

8.36 Calculate the final temperature, in degrees Celsius, for each of the following, if V and n do not change:

a. A sample of helium gas with a pressure of 250 Torr at 0 °C is heated to give a pressure of 1500 Torr.

b. A sample of air at 40 °C and 740. mmHg is cooled to give a pressure of 680. mmHg.

8.37 A gas sample has a pressure of 744 mmHg when the temperature is 22 °C. What is the final temperature, in degrees Celsius, when the pressure is 766 mmHg, with no change in the volume or amount of gas?

8.38 A gas sample has a pressure of 2.35 atm when the temperature is −15 °C. What is the final pressure, in atmospheres, when the temperature is 46 °C, with no change in the volume or amount of gas?

Clinical Applications

8.39 A tank contains isoflurane, an inhaled anesthetic, at a pressure of 1.8 atm and 5 °C. What is the pressure, in atmospheres, if the gas is warmed to a temperature of 22 °C, if V and n do not change?

8.40 Bacteria and viruses are inactivated by temperatures above 135 °C. An autoclave contains steam at 1.00 atm and 100 °C. What is the pressure, in atmospheres, when the temperature of the steam in the autoclave reaches 135 °C, if V and n do not change?

8.5 The Combined Gas Law

LEARNING GOAL Use the combined gas law to calculate the unknown pressure, volume, or temperature of a gas when changes in two of these properties are given and the amount of gas does not change.

All of the pressure–volume–temperature relationships for gases that we have studied may be combined into a single relationship called the **combined gas law**. This expression is useful for studying the effect of changes in two of these variables on the third as long as the amount of gas (number of moles) does not change.

Combined Gas Law

$$\frac{P_1V_1}{T_1} = \frac{P_2V_2}{T_2} \quad \text{No change in number of moles of gas}$$

By using the combined gas law, we can derive any of the gas laws by omitting those properties that do not change, as seen in **TABLE 8.4**.

ENGAGE

Why does the pressure of a gas decrease to one-fourth of its initial pressure when the volume of the gas doubles and the Kelvin temperature decreases by half, if the amount of gas does not change?

TABLE 8.4 Summary of Gas Laws

Combined Gas Law	Properties That Do Not Change	Relationship	Name of Gas Law
$\frac{P_1V_1}{\cancel{T_1}} = \frac{P_2V_2}{\cancel{T_2}}$	T, n	$P_1V_1 = P_2V_2$	Boyle's law
$\frac{\cancel{P_1}V_1}{T_1} = \frac{\cancel{P_2}V_2}{T_2}$	P, n	$\frac{V_1}{T_1} = \frac{V_2}{T_2}$	Charles's law
$\frac{P_1\cancel{V_1}}{T_1} = \frac{P_2\cancel{V_2}}{T_2}$	V, n	$\frac{P_1}{T_1} = \frac{P_2}{T_2}$	Gay-Lussac's law

SAMPLE PROBLEM 8.6 Using the Combined Gas Law

Under water, the pressure on a diver is greater than the atmospheric pressure.

TRY IT FIRST

A 25.0-mL bubble is released from a diver's air tank at a pressure of 4.00 atm and a temperature of 11 °C. What is the volume, in milliliters, of the bubble when it reaches the ocean surface where the pressure is 1.00 atm and the temperature is 18 °C? (Assume the amount of gas in the bubble does not change.)

SOLUTION GUIDE

STEP 1 **State the given and needed quantities.** We list the properties that change, which are the pressure, volume, and temperature. The property that does not change, which is the amount of gas, is shown below the table. The temperatures in degrees Celsius must be changed to kelvins.

$$T_1 = 11\ °\text{C} + 273 = 284\ \text{K}$$

$$T_2 = 18\ °\text{C} + 273 = 291\ \text{K}$$

ANALYZE THE PROBLEM	Given	Need	Connect
	$P_1 = 4.00$ atm $\quad P_2 = 1.00$ atm $V_1 = 25.0$ mL $T_1 = 284$ K $\quad T_2 = 291$ K **Factor that does not change:** n	V_2	combined gas law, $\frac{P_1V_1}{T_1} = \frac{P_2V_2}{T_2}$

STEP 2 **Rearrange the gas law equation to solve for the unknown quantity.** Using the combined gas law, we solve for V_2 by multiplying both sides by T_2 and dividing both sides by P_2.

$$\frac{P_1V_1}{T_1} = \frac{P_2V_2}{T_2}$$

$$\frac{P_1V_1}{T_1} \times \frac{T_2}{P_2} = \frac{\cancel{P_2}V_2}{\cancel{T_2}} \times \frac{\cancel{T_2}}{\cancel{P_2}}$$

$$V_2 = V_1 \times \frac{P_1}{P_2} \times \frac{T_2}{T_1}$$

STEP 3 **Substitute values into the gas law equation and calculate.** From the data table, we determine that both the pressure decrease and the temperature increase will increase the volume.

$$V_2 = 25.0 \text{ mL} \times \frac{4.00 \cancel{\text{atm}}}{1.00 \cancel{\text{atm}}} \times \frac{291 \cancel{\text{K}}}{284 \cancel{\text{K}}} = 102 \text{ mL}$$

Pressure factor increases volume | Temperature factor increases volume

However, in situations where the unknown value is decreased by one change but increased by the second change, it is difficult to predict the overall change for the unknown.

STUDY CHECK 8.6

A weather balloon is filled with 15.0 L of helium at a temperature of 25 °C and a pressure of 685 mmHg. What is the pressure, in millimeters of mercury, of the helium in the balloon in the upper atmosphere when the final temperature is −35 °C and the final volume becomes 34.0 L, if the amount of He does not change?

ANSWER

241 mmHg

TEST

Try Practice Problems 8.41 to 8.46

PRACTICE PROBLEMS

8.5 The Combined Gas Law

LEARNING GOAL Use the combined gas law to calculate the unknown pressure, volume, or temperature of a gas when changes in two of these properties are given and the amount of gas does not change.

8.41 Rearrange the variables in the combined gas law to solve for T_2.

8.42 Rearrange the variables in the combined gas law to solve for P_2.

8.43 A sample of helium gas has a volume of 6.50 L at a pressure of 845 mmHg and a temperature of 25 °C. What is the final pressure of the gas, in atmospheres, when the volume and temperature of the gas sample are changed to the following, if the amount of gas does not change?
a. 1850 mL and 325 K **b.** 2.25 L and 12 °C
c. 12.8 L and 47 °C

8.44 A sample of argon gas has a volume of 735 mL at a pressure of 1.20 atm and a temperature of 112 °C. What is the final volume of the gas, in milliliters, when the pressure and temperature of the gas sample are changed to the following, if the amount of gas does not change?
a. 658 mmHg and 281 K **b.** 0.55 atm and 75 °C
c. 15.4 atm and −15 °C

Clinical Applications

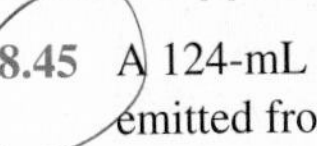

8.45 A 124-mL bubble of hot gas initially at 212 °C and 1.80 atm is emitted from an active volcano. What is the final temperature, in degrees Celsius, of the gas in the bubble outside the volcano if the final volume of the bubble is 138 mL and the pressure is 0.800 atm, if the amount of gas does not change?

8.46 A scuba diver 60 ft below the ocean surface inhales 50.0 mL of compressed air from a scuba tank at a pressure of 3.00 atm and a temperature of 8 °C. What is the final pressure of air, in atmospheres, in the lungs when the gas expands to 150.0 mL at a body temperature of 37 °C, if the amount of gas does not change?

8.6 Volume and Moles (Avogadro's Law)

LEARNING GOAL Use Avogadro's law to calculate the unknown amount or volume of a gas when the pressure and temperature do not change.

REVIEW

Using Molar Mass as a Conversion Factor (7.3)

In our study of the gas laws, we have looked at changes in properties for a specified amount (n) of gas. Now we consider how the properties of a gas change when there is a change in the number of moles or grams of the gas.

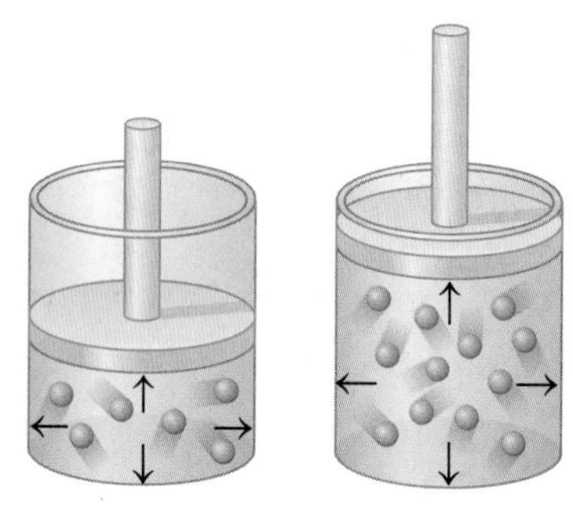

FIGURE 8.7 Avogadro's law: The volume of a gas is directly related to the number of moles of the gas. If the number of moles is doubled, the volume must double when temperature and pressure do not change.

Q If a balloon has a leak, what happens to its volume?

When you blow up a balloon, its volume increases because you add more air molecules. If the balloon has a small hole in it, air leaks out, causing its volume to decrease. In 1811, Amedeo Avogadro formulated **Avogadro's law**, which states that the volume of a gas is directly related to the number of moles of a gas when temperature and pressure do not change. For example, if the number of moles of a gas is doubled, then the volume will also double as long as we do not change the pressure or the temperature (see **FIGURE 8.7**). When pressure and temperature do not change, we can write Avogadro's law as follows:

Avogadro's Law

$$\frac{V_1}{n_1} = \frac{V_2}{n_2} \quad \text{No change in pressure and temperature}$$

SAMPLE PROBLEM 8.7 Calculating Volume for a Change in Moles

TRY IT FIRST

A weather balloon with a volume of 44 L is filled with 2.0 moles of helium. What is the final volume, in liters, if 3.0 moles of helium are added, to give a total of 5.0 moles of helium, if the pressure and temperature do not change?

SOLUTION GUIDE

STEP 1 **State the given and needed quantities.** We list those properties that change, which are volume and amount (moles). The properties that do not change, which are pressure and temperature, are shown below the table. Because there is an increase in the number of moles of gas, we can predict that the volume increases.

	Given	Need	Connect
ANALYZE THE PROBLEM	$V_1 = 44$ L $n_1 = 2.0$ moles $n_2 = 5.0$ moles	V_2	Avogadro's law, $\frac{V_1}{n_1} = \frac{V_2}{n_2}$
	Factors that do not change: P and T		**Predict:** n increases, V increases

STEP 2 **Rearrange the gas law equation to solve for the unknown quantity.** Using Avogadro's law, we can solve for V_2 by multiplying both sides of the equation by n_2.

$$\frac{V_1}{n_1} = \frac{V_2}{n_2}$$

$$\frac{V_1}{n_1} \times n_2 = \frac{V_2}{\cancel{n_2}} \times \cancel{n_2}$$

$$V_2 = V_1 \times \frac{n_2}{n_1}$$

STEP 3 **Substitute values into the gas law equation and calculate.** When we substitute in the values, we see that the ratio of the moles (mole factor) is greater than 1, which increases the volume as predicted.

$$V_2 = 44\text{ L} \times \frac{5.0 \cancel{\text{moles}}}{2.0 \cancel{\text{moles}}} = 110\text{ L}$$

Mole factor increases volume

STUDY CHECK 8.7

A sample containing 8.00 g of oxygen gas has a volume of 5.00 L. What is the volume, in liters, after 4.00 g of oxygen gas is added to the 8.00 g of oxygen in the balloon, if the temperature and pressure do not change?

ANSWER

7.50 L

ENGAGE

Why does the volume of 2.0 moles of neon double when 2.0 moles of helium gas are added, if the pressure and temperature of the gas mixture do not change?

TEST

Try Practice Problems 8.47 to 8.50

STP and Molar Volume

Using Avogadro's law, we can say that any two gases will have equal volumes if they contain the same number of moles of gas at the same temperature and pressure. To help us make comparisons between different gases, arbitrary conditions called *standard temperature* (273 K) and *standard pressure* (1 atm), together abbreviated **STP**, were selected by scientists:

The molar volume of a gas at STP is about the same as the volume of three basketballs.

STP Conditions

Standard temperature is *exactly* 0 °C (273 K).

Standard pressure is *exactly* 1 atm (760 mmHg).

At STP, one mole of any gas occupies a volume of 22.4 L, which is about the same as the volume of three basketballs. This volume, 22.4 L, of any gas is called the **molar volume** (see **FIGURE 8.8**).

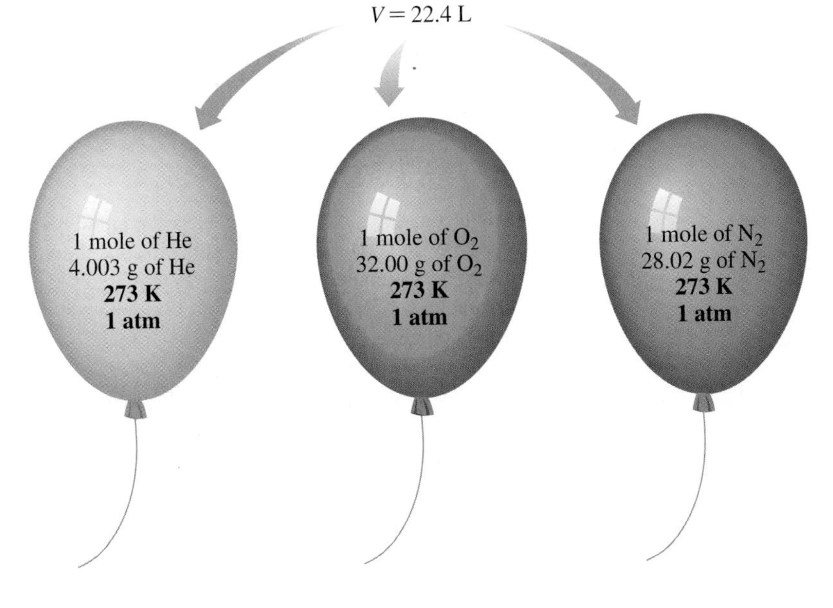

FIGURE 8.8 Avogadro's law indicates that 1 mole of any gas at STP has a volume of 22.4 L.

Q What volume of gas is occupied by 16.0 g of methane gas, CH_4, at STP?

ENGAGE

Why would the molar volume of a gas be greater than 22.4 L/mole if the pressure is 1 atm and the temperature is 100 °C?

When a gas is at STP conditions (0 °C and 1 atm), its molar volume can be used to write conversion factors between the number of moles of gas and its volume, in liters.

Molar Volume Conversion Factors

1 mole of gas = 22.4 L (STP)

$$\frac{22.4\text{ L (STP)}}{1\text{ mole gas}} \quad \text{and} \quad \frac{1\text{ mole gas}}{22.4\text{ L (STP)}}$$

SAMPLE PROBLEM 8.8 Using Molar Volume

TRY IT FIRST

What is the volume, in liters, of 64.0 g of O_2 gas at STP?

SOLUTION GUIDE

STEP 1 State the given and needed quantities.

	Given	Need	Connect
ANALYZE THE PROBLEM	64.0 g of $O_2(g)$ at STP	liters of O_2 gas at STP	molar mass, molar volume (STP)

STEP 2 Write a plan to calculate the needed quantity.

grams of O_2 → Molar mass → moles of O_2 → Molar volume → liters of O_2

STEP 3 Write the equalities and conversion factors including 22.4 L/mole at STP.

$$1 \text{ mole of } O_2 = 32.00 \text{ g of } O_2 \qquad \frac{32.00 \text{ g } O_2}{1 \text{ mole } O_2} \text{ and } \frac{1 \text{ mole } O_2}{32.00 \text{ g } O_2}$$

$$1 \text{ mole of } O_2 = 22.4 \text{ L of } O_2 \text{ (STP)} \qquad \frac{22.4 \text{ L } O_2 \text{ (STP)}}{1 \text{ mole } O_2} \text{ and } \frac{1 \text{ mole } O_2}{22.4 \text{ L } O_2 \text{ (STP)}}$$

STEP 4 Set up the problem with factors to cancel units.

$$64.0 \text{ g } O_2 \times \frac{1 \text{ mole } O_2}{32.00 \text{ g } O_2} \times \frac{22.4 \text{ L } O_2 \text{ (STP)}}{1 \text{ mole } O_2} = 44.8 \text{ L of } O_2 \text{ (STP)}$$

STUDY CHECK 8.8

How many grams of $Cl_2(g)$ are in 5.00 L of $Cl_2(g)$ at STP?

ANSWER

15.8 g of $Cl_2(g)$

TEST

Try Practice Problems 8.51 and 8.52

PRACTICE PROBLEMS

8.6 Volume and Moles (Avogadro's Law)

LEARNING GOAL Use Avogadro's law to calculate the unknown amount or volume of a gas when the pressure and temperature do not change.

8.47 What happens to the volume of a bicycle tire or a basketball when you use an air pump to add air?

8.48 Sometimes when you blow up a balloon and release it, it flies around the room. What is happening to the air in the balloon and its volume?

8.49 A sample containing 1.50 moles of Ne gas has an initial volume of 8.00 L. What is the final volume, in liters, when each of the following occurs and pressure and temperature do not change?
- **a.** A leak allows one-half of Ne atoms to escape.
- **b.** A sample of 3.50 moles of Ne is added to the 1.50 moles of Ne gas in the container.
- **c.** A sample of 25.0 g of Ne is added to the 1.50 moles of Ne gas in the container.

8.50 A sample containing 4.80 g of O_2 gas has an initial volume of 15.0 L. What is the final volume, in liters, when each of the following occurs and pressure and temperature do not change?
- **a.** A sample of 0.500 mole of O_2 is added to the 4.80 g of O_2 in the container.
- **b.** A sample of 2.00 g of O_2 is removed.
- **c.** A sample of 4.00 g of O_2 is added to the 4.80 g of O_2 gas in the container.

8.51 Use the molar volume to calculate each of the following at STP:
- **a.** the number of moles of O_2 in 44.8 L of O_2 gas
- **b.** the volume, in liters, occupied by 2.50 moles of N_2 gas
- **c.** the volume, in liters, occupied by 50.0 g of Ar gas
- **d.** the number of grams of H_2 in 1620 mL of H_2 gas

8.52 Use the molar volume to calculate each of the following at STP:
- **a.** the number of moles of CO_2 in 4.00 L of CO_2 gas
- **b.** the volume, in liters, occupied by 0.420 mole of He gas
- **c.** the volume, in liters, occupied by 6.40 g of O_2 gas
- **d.** the number of grams of Ne contained in 11.2 L of Ne gas

8.7 Partial Pressures (Dalton's Law)

LEARNING GOAL Use Dalton's law of partial pressures to calculate the total pressure of a mixture of gases.

Many gas samples are a mixture of gases. For example, the air you breathe is a mixture of mostly oxygen and nitrogen gases. In ideal gas mixtures, scientists observed that all gas particles behave in the same way. Therefore, the total pressure of the gases in a mixture is a result of the collisions of the gas particles regardless of what type of gas they are.

In a gas mixture, each gas exerts its **partial pressure**, which is the pressure it would exert if it were the only gas in the container. **Dalton's law** states that the total pressure of a gas mixture is the sum of the partial pressures of the gases in the mixture.

CORE CHEMISTRY SKILL

Calculating Partial Pressure

Dalton's Law

$$P_{\text{total}} = P_1 + P_2 + P_3 + \cdots$$

Total pressure of a gas mixture = Sum of the partial pressures of the gases in the mixture

Suppose we have two separate tanks, one filled with helium at a pressure of 2.0 atm and the other filled with argon at a pressure of 4.0 atm. When the gases are combined in a single tank with the same volume and temperature, the number of gas molecules, not the type of gas, determines the pressure in the container. There the pressure of the gas mixture would be 6.0 atm, which is the sum of their individual or partial pressures.

ENGAGE

Why does the combination of helium with a pressure of 2.0 atm and argon with a pressure of 4.0 atm produce a gas mixture with a total pressure of 6.0 atm?

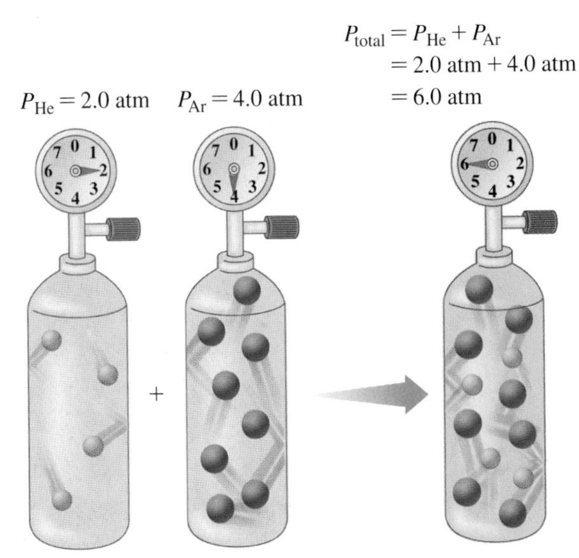

The total pressure of two gases is the sum of their partial pressures.

Air Is a Gas Mixture

The air you breathe is a mixture of gases. What we call the *atmospheric pressure* is actually the sum of the partial pressures of the gases in the air. **TABLE 8.5** lists partial pressures for the gases in air on a typical day.

TABLE 8.5 Typical Composition of Air

Gas	Partial Pressure (mmHg)	Percentage (%)
Nitrogen, N_2	594	78.2
Oxygen, O_2	160.	21.0
Carbon dioxide, CO_2; Argon, Ar; Water vapor, H_2O	6	0.8
Total air	760.	100

SAMPLE PROBLEM 8.9 Calculating the Partial Pressure of a Gas in a Mixture

TRY IT FIRST

A heliox breathing mixture of oxygen and helium is prepared for a patient with *chronic obstructive pulmonary disease* (COPD). The gas mixture has a total pressure of 7.00 atm. If the partial pressure of the oxygen in the tank is 1140 mmHg, what is the partial pressure, in atmospheres, of the helium in the breathing mixture?

SOLUTION GUIDE

	Given	Need	Connect
ANALYZE THE PROBLEM	$P_{total} = 7.00$ atm $P_{O_2} = 1140$ mmHg	partial pressure of He	Dalton's law

STEP 1 Write the equation for the sum of the partial pressures.

$$P_{total} = P_{O_2} + P_{He} \quad \text{Dalton's law}$$

STEP 2 Rearrange the equation to solve for the unknown pressure. To solve for the partial pressure of helium (P_{He}), we rearrange the equation to give the following:

$$P_{He} = P_{total} - P_{O_2}$$

Convert units to match.

$$P_{O_2} = 1140 \text{ mmHg} \times \frac{1 \text{ atm}}{760 \text{ mmHg}} = 1.50 \text{ atm}$$

STEP 3 Substitute known pressures into the equation and calculate the unknown pressure.

$P_{He} = P_{total} - P_{O_2}$

$P_{He} = 7.00\text{ atm} - 1.50\text{ atm} = 5.50\text{ atm}$

STUDY CHECK 8.9

An anesthetic consists of a mixture of cyclopropane gas, C_3H_6, and oxygen gas, O_2. If the mixture has a total pressure of 1.09 atm, and the partial pressure of the cyclopropane is 73 mmHg, what is the partial pressure, in millimeters of mercury, of the oxygen in the anesthetic?

ANSWER

755 mmHg

TEST

Try Practice Problems 8.53 to 8.60

CHEMISTRY LINK TO HEALTH

Hyperbaric Chambers

A burn patient may undergo treatment for burns and infections in a *hyperbaric chamber*, a device in which pressures can be obtained that are two to three times greater than atmospheric pressure. A greater oxygen pressure increases the level of dissolved oxygen in the blood and tissues, where it fights bacterial infections. High levels of oxygen are toxic to many strains of bacteria. The hyperbaric chamber may also be used during surgery, to help counteract carbon monoxide (CO) poisoning, and to treat some cancers.

The blood is normally capable of dissolving up to 95% of the oxygen. Thus, if the partial pressure of the oxygen in the hyperbaric chamber is 2280 mmHg (3 atm), about 2170 mmHg of oxygen can dissolve in the blood, saturating the tissues. In the treatment for carbon monoxide poisoning, oxygen at high pressure is used to displace the CO from the hemoglobin faster than breathing pure oxygen at 1 atm.

A patient undergoing treatment in a hyperbaric chamber must also undergo decompression (reduction of pressure) at a rate that slowly reduces the concentration of dissolved oxygen in the blood. If decompression is too rapid, the oxygen dissolved in the blood may form gas bubbles in the circulatory system.

Similarly, if a scuba diver does not decompress slowly, a condition called the "bends" may occur. While below the surface of the ocean, a diver uses a breathing mixture with higher pressures. If there is nitrogen in the mixture, higher quantities of nitrogen gas will dissolve in the blood. If the diver ascends to the surface too quickly, the dissolved nitrogen forms gas bubbles that can block a blood vessel and cut off the flow of blood in the joints and tissues of the body and be quite painful. A diver suffering from the bends is placed immediately into a hyperbaric chamber where pressure is first increased and then slowly decreased. The dissolved nitrogen can then diffuse through the lungs until atmospheric pressure is reached.

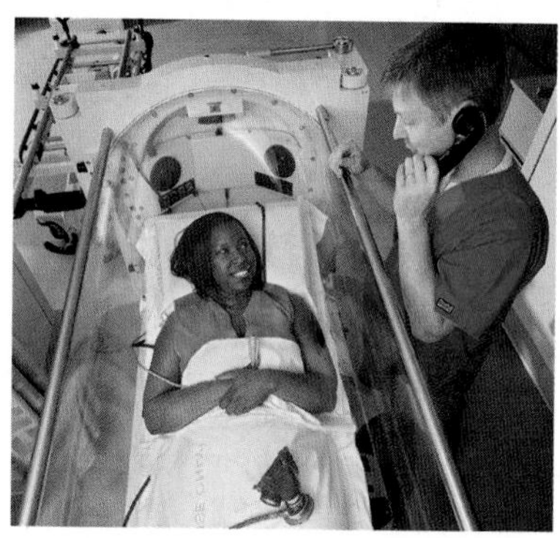

A hyperbaric chamber is used in the treatment of certain diseases.

PRACTICE PROBLEMS

8.7 Partial Pressures (Dalton's Law)

LEARNING GOAL Use Dalton's law of partial pressures to calculate the total pressure of a mixture of gases.

8.53 A typical air sample in the lungs contains oxygen at 100 mmHg, nitrogen at 573 mmHg, carbon dioxide at 40 mmHg, and water vapor at 47 mmHg. Why are these pressures called partial pressures?

8.54 Suppose a mixture contains helium and oxygen gases. If the partial pressure of helium is the same as the partial pressure of oxygen, what do you know about the number of helium atoms compared to the number of oxygen molecules? Explain.

8.55 In a gas mixture, the partial pressures are nitrogen 425 Torr, oxygen 115 Torr, and helium 225 Torr. What is the total pressure, in torr, exerted by the gas mixture?

8.56 In a gas mixture, the partial pressures are argon 415 mmHg, neon 75 mmHg, and nitrogen 125 mmHg. What is the total pressure, in mmHg, exerted by the gas mixture?

8.57 A gas mixture containing oxygen, nitrogen, and helium exerts a total pressure of 925 Torr. If the partial pressures are oxygen 425 Torr and helium 75 Torr, what is the partial pressure, in torr, of the nitrogen in the mixture?

8.58 A gas mixture containing oxygen, nitrogen, and neon exerts a total pressure of 1.20 atm. If helium added to the mixture increases the pressure to 1.50 atm, what is the partial pressure, in atmospheres, of the helium?

Clinical Applications

8.59 In certain lung ailments such as emphysema, there is a decrease in the ability of oxygen to diffuse into the blood.

a. How would the partial pressure of oxygen in the blood change?

b. Why does a person with severe emphysema sometimes use a portable oxygen tank?

8.60 A head injury can affect the ability of a person to ventilate (breathe in and out).

a. What would happen to the partial pressures of oxygen and carbon dioxide in the blood if a person cannot properly ventilate?

b. When a person who cannot breathe properly is placed on a ventilator, an air mixture is delivered at pressures that are alternately above the air pressure in the person's lung, and then below. How will this move oxygen into the lungs, and carbon dioxide out?

CLINICAL UPDATE Exercise-Induced Asthma

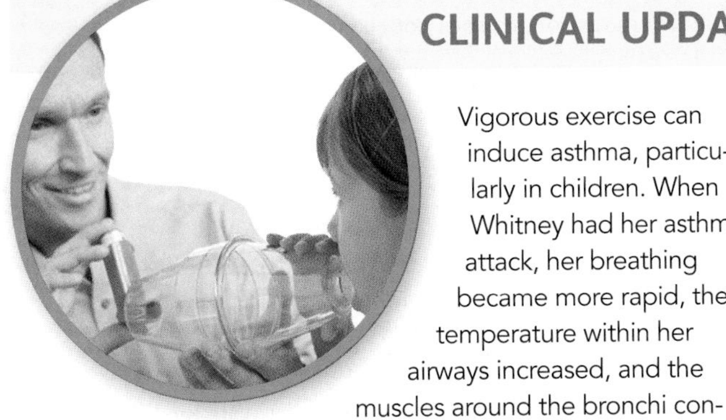

Vigorous exercise can induce asthma, particularly in children. When Whitney had her asthma attack, her breathing became more rapid, the temperature within her airways increased, and the muscles around the bronchi contracted, causing a narrowing of the airways. Whitney's symptoms, which may occur within 5 to 20 min after the start of vigorous exercise, include shortness of breath, wheezing, and coughing.

Whitney now does several things to prevent exercise-induced asthma. She uses a pre-exercise inhaled medication before she starts her activity. The medication relaxes the muscles that surround the airways and opens up the airways. Then she does a warm-up set of exercises. If pollen counts are high, she avoids exercising outdoors.

Clinical Applications

8.61 Whitney's lung capacity was measured as 3.2 L at a body temperature of 37 °C and a pressure of 745 mmHg. What is her lung capacity, in liters, at STP?

8.62 Using the answer from problem 8.61, how many grams of nitrogen are in Whitney's lungs at STP if air contains 78% nitrogen?

CONCEPT MAP

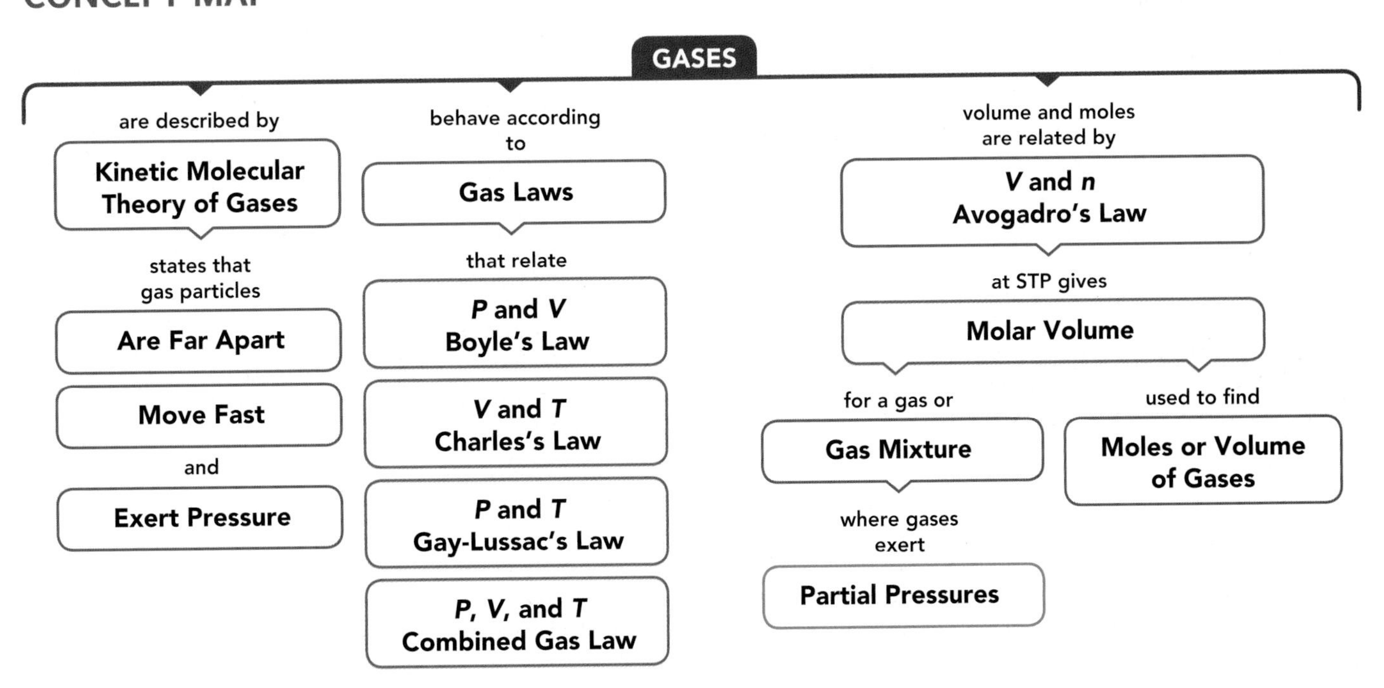

CHAPTER REVIEW

8.1 Properties of Gases

LEARNING GOAL Describe the kinetic molecular theory of gases and the units of measurement used for gases.

- In a gas, particles are so far apart and moving so fast that their attractions are negligible.
- A gas is described by the physical properties of pressure (P), volume (V), temperature (T), and amount in moles (n).
- A gas exerts pressure, the force of the gas particles striking the surface of a container.
- Gas pressure is measured in units such as torr, mmHg, atm, and Pa.

8.2 Pressure and Volume (Boyle's Law)

LEARNING GOAL Use the pressure–volume relationship (Boyle's law) to calculate the unknown pressure or volume when the temperature and amount of gas do not change.

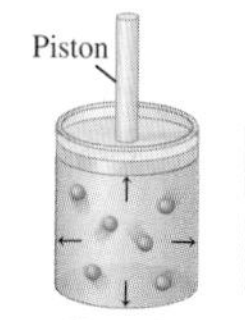

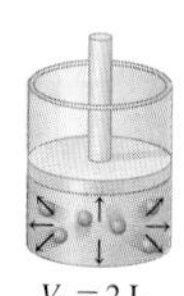

- The volume (V) of a gas changes inversely with the pressure (P) of the gas if there is no change in the temperature and the amount of gas.

$$P_1V_1 = P_2V_2$$

- The pressure increases if volume decreases; its pressure decreases if the volume increases.

8.3 Temperature and Volume (Charles's Law)

LEARNING GOAL Use the temperature–volume relationship (Charles's law) to calculate the unknown temperature or volume when the pressure and amount of gas do not change.

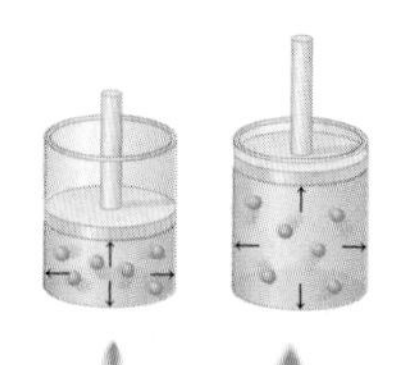

- The volume (V) of a gas is directly related to its Kelvin temperature (T) when there is no change in the pressure and the amount of gas.

$$\frac{V_1}{T_1} = \frac{V_2}{T_2}$$

- If the temperature of a gas increases, its volume increases; if its temperature decreases, the volume decreases.

8.4 Temperature and Pressure (Gay-Lussac's Law)

LEARNING GOAL Use the temperature–pressure relationship (Gay-Lussac's law) to calculate the unknown temperature or pressure when the volume and amount of gas do not change.

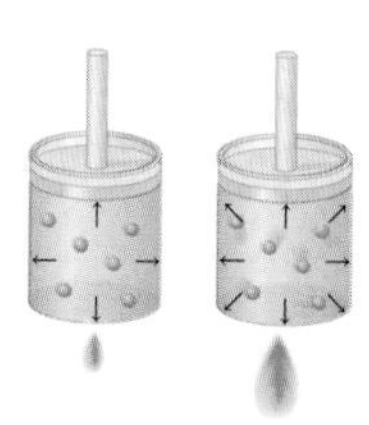

- The pressure (P) of a gas is directly related to its Kelvin temperature (T) when there is no change in the volume and the amount of the gas.

$$\frac{P_1}{T_1} = \frac{P_2}{T_2}$$

- As temperature of a gas increases, its pressure increases; if its temperature decreases, its pressure decreases.

8.5 The Combined Gas Law

LEARNING GOAL Use the combined gas law to calculate the unknown pressure, volume, or temperature of a gas when changes in two of these properties are given and the amount of gas does not change.

- The combined gas law is the relationship of pressure (P), volume (V), and temperature (T) when the amount of gas does not change.

$$\frac{P_1V_1}{T_1} = \frac{P_2V_2}{T_2}$$

- The combined gas law is used to determine the effect of changes in two of the variables on the third.

8.6 Volume and Moles (Avogadro's Law)

V = 22.4 L

1 mole of O_2
32.00 g of O_2
273 K
1 atm

LEARNING GOAL Use Avogadro's law to calculate the unknown amount or volume of a gas when the pressure and temperature do not change.

- The volume (V) of a gas is directly related to the number of moles (n) of the gas when the pressure and temperature of the gas do not change.

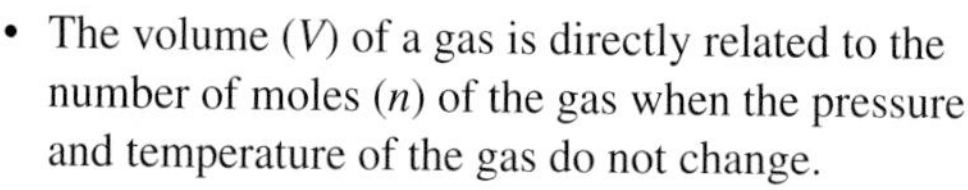

$$\frac{V_1}{n_1} = \frac{V_2}{n_2}$$

- If the moles of gas increase, the volume must increase; if the moles of gas decrease, the volume must decrease.
- At standard temperature (273 K) and standard pressure (1 atm), abbreviated STP, 1 mole of any gas has a volume of 22.4 L.

8.7 Partial Pressures (Dalton's Law)

LEARNING GOAL Use Dalton's law of partial pressures to calculate the total pressure of a mixture of gases.

- In a mixture of two or more gases, the total pressure is the sum of the partial pressures of the individual gases.

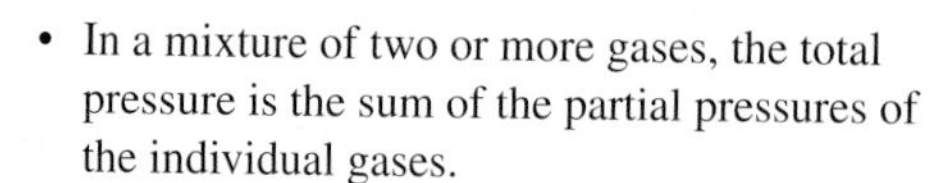

$$P_{total} = P_1 + P_2 + P_3 + \cdots$$

- The partial pressure of a gas in a mixture is the pressure it would exert if it were the only gas in the container.

KEY TERMS

atmosphere (atm) A unit equal to the pressure exerted by a column of mercury 760 mm high.

atmospheric pressure The pressure exerted by the atmosphere.

Avogadro's law A gas law stating that the volume of a gas is directly related to the number of moles of gas when pressure and temperature do not change.

Boyle's law A gas law stating that the pressure of a gas is inversely related to the volume when temperature and moles of the gas do not change.

Charles's law A gas law stating that the volume of a gas is directly related to the Kelvin temperature when pressure and moles of the gas do not change.

combined gas law A relationship that combines several gas laws relating pressure, volume, and temperature when the amount of gas does not change.

$$\frac{P_1V_1}{T_1} = \frac{P_2V_2}{T_2}$$

Dalton's law A gas law stating that the total pressure exerted by a mixture of gases in a container is the sum of the partial pressures that each gas would exert alone.

direct relationship A relationship in which two properties increase or decrease together.

Gay-Lussac's law A gas law stating that the pressure of a gas is directly related to the Kelvin temperature when the number of moles of a gas and its volume do not change.

inverse relationship A relationship in which two properties change in opposite directions.

kinetic molecular theory of gases A model used to explain the behavior of gases.

molar volume A volume of 22.4 L occupied by 1 mole of a gas at STP conditions of 0 °C (273 K) and 1 atm.

partial pressure The pressure exerted by a single gas in a gas mixture.

pressure The force exerted by gas particles that hit the walls of a container.

STP Standard conditions of exactly 0 °C (273 K) temperature and 1 atm pressure used for the comparison of gases.

CORE CHEMISTRY SKILLS

The chapter Section containing each Core Chemistry Skill is shown in parentheses at the end of each heading.

Using the Gas Laws (8.2)

- Boyle's law is one of the gas laws that shows the relationships between two properties of a gas.

$$P_1V_1 = P_2V_2 \quad \text{Boyle's law}$$

- When two of the four properties of a gas (P, V, T, or n) vary and the other two do not change, we list the initial and final conditions of each property in a table.

Example: A sample of helium gas (He) has a volume of 6.8 L and a pressure of 2.5 atm. What is the final volume, in liters, if it has a final pressure of 1.2 atm with no change in temperature and amount of gas?

Answer: Using Boyle's law, we can write the relationship for V_2, which we predict will increase.

$P_1 = 2.5$ atm $P_2 = 1.2$ atm V_2 Need
$V_1 = 6.8$ L

$$V_2 = V_1 \times \frac{P_1}{P_2}$$

$$V_2 = 6.8\text{ L} \times \frac{2.5 \text{ atm}}{1.2 \text{ atm}} = 14\text{ L}$$

Calculating Partial Pressure (8.7)

- In a gas mixture, each gas exerts its partial pressure, which is the pressure it would exert if it were the only gas in the container.
- Dalton's law states that the total pressure of a gas mixture is the sum of the partial pressures of the gases in the mixture.

$$P_{total} = P_1 + P_2 + P_3 + \cdots$$

Example: A gas mixture with a total pressure of 1.18 atm contains helium gas at a partial pressure of 465 mmHg and nitrogen gas. What is the partial pressure, in atmospheres, of the nitrogen gas?

Answer: Initially, we convert the partial pressure of helium gas from mmHg to atm.

$$465 \text{ mmHg} \times \frac{1 \text{ atm}}{760 \text{ mmHg}} = 0.612 \text{ atm of He gas}$$

Using Dalton's law, we solve for the needed quantity, P_{N_2} in atm.

$$P_{total} = P_{N_2} + P_{He}$$
$$P_{N_2} = P_{total} - P_{He}$$
$$P_{N_2} = 1.18 \text{ atm} - 0.612 \text{ atm} = 0.57 \text{ atm}$$

UNDERSTANDING THE CONCEPTS

The chapter Sections to review are shown in parentheses at the end of each problem.

8.63 Two flasks of equal volume and at the same temperature contain different gases. One flask contains 10.0 g of Ne, and the other flask contains 10.0 g of He. Is each of the following statements *true* or *false*? Explain. (8.1)

a. The flask that contains He has a higher pressure than the flask that contains Ne.

b. The densities of the gases are the same.

8.64 Two flasks of equal volume and at the same temperature contain different gases. One flask contains 5.0 g of O_2, and the other flask contains 5.0 g of H_2. Is each of the following statements *true* or *false*? Explain. (8.1)

a. Both flasks contain the same number of molecules.

b. The pressures in the flasks are the same.

8.65 At 100 °C, which of the following diagrams (**1**, **2**, or **3**) represents a gas sample that exerts the: (8.1)

a. lowest pressure? **b.** highest pressure?

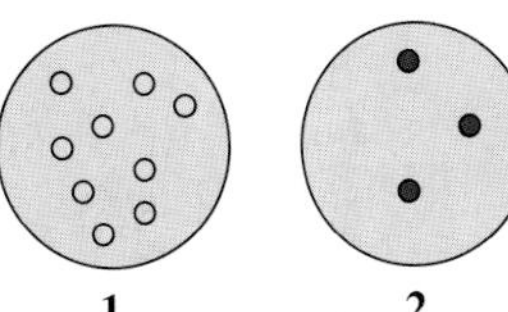

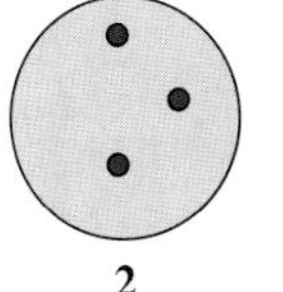

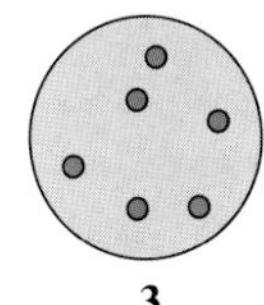

8.66 Indicate which diagram (**1**, **2**, or **3**) represents the volume of the gas sample in a flexible container when each of the following changes (**a** to **d**) takes place: (8.2, 8.3)

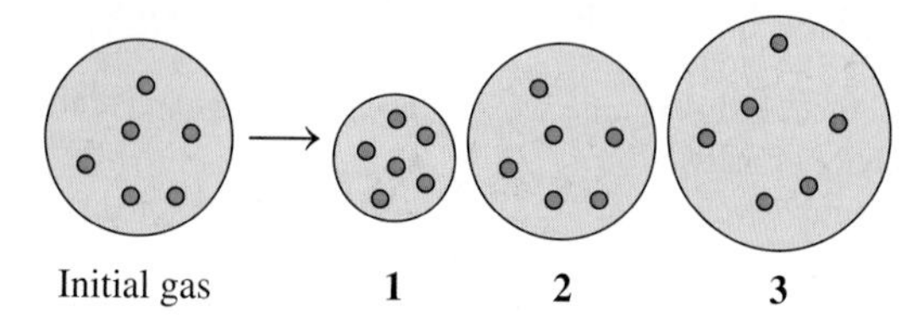

a. Temperature increases if pressure does not change.
b. Temperature decreases if pressure does not change.
c. Atmospheric pressure decreases if temperature does not change.
d. Doubling the atmospheric pressure and doubling the Kelvin temperature.

8.67 A balloon is filled with helium gas with a partial pressure of 1.00 atm and neon gas with a partial pressure of 0.50 atm. For each of the following changes (**a** to **e**) of the initial balloon, select the diagram (**A**, **B**, or **C**) that shows the final volume of the balloon: (8.2, 8.3, 8.6)

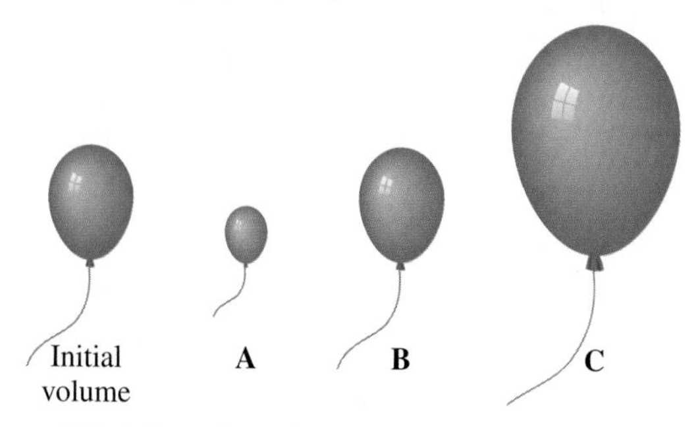

a. The balloon is put in a cold storage unit (P and n do not change).
b. The balloon floats to a higher altitude where the pressure is less (n and T do not change).
c. All of the neon gas is removed (T and P do not change).
d. The Kelvin temperature doubles and half of the gas atoms leak out (P does not change).
e. 2.0 moles of O_2 gas is added (T and P do not change).

8.68 Indicate if pressure *increases*, *decreases*, or *stays the same* in each of the following: (8.2, 8.4, 8.6)

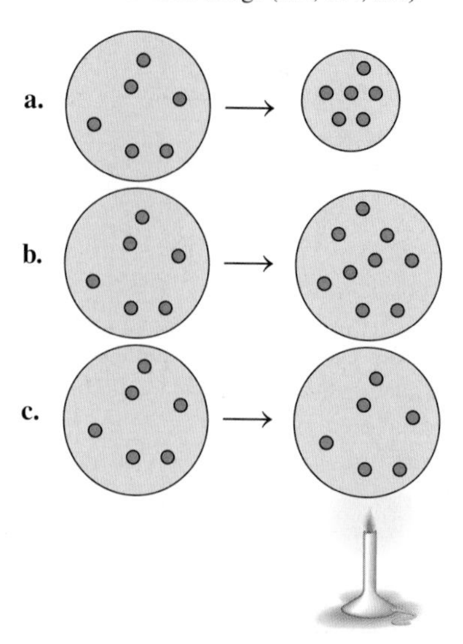

8.69 At a restaurant, a customer chokes on a piece of food. You put your arms around the person's waist and use your fists to push up on the person's abdomen, an action called the Heimlich maneuver. (8.2)
a. How would this action change the volume of the chest and lungs?
b. Why does it cause the person to expel the food item from the airway?

8.70 An airplane is pressurized with air to 650. mmHg. (8.7)
a. If air is 21% oxygen, what is the partial pressure of oxygen on the plane?
b. If the partial pressure of oxygen drops below 100. mmHg, passengers become drowsy. If this happens, oxygen masks are released. What is the total cabin pressure at which oxygen masks are dropped?

ADDITIONAL PRACTICE PROBLEMS

8.71 In 1783, Jacques Charles launched his first balloon filled with hydrogen gas, which he chose because it was lighter than air. If the balloon had a volume of 31 000 L, how many kilograms of hydrogen were needed to fill the balloon at STP? (8.6)

Jacques Charles used hydrogen to launch his balloon in 1783.

8.72 In problem 8.71, the balloon reached an altitude of 1000 m, where the pressure was 658 mmHg and the temperature was −8 °C. What was the volume, in liters, of the balloon at these conditions? (8.5)

8.73 A sample of hydrogen (H_2) gas at 127 °C has a pressure of 2.00 atm. At final temperature, in degrees Celsius, will the pressure of the H_2 decrease to 0.25 atm, if V and n do not change? (8.4)

8.74 A fire extinguisher has a pressure of 10. atm at 25 °C. What is the final pressure, in atmospheres, when the fire extinguisher is used at a temperature of 75 °C, if V and n do not change? (8.4)

8.75 A weather balloon has a volume of 750 L when filled with helium at 8 °C at a pressure of 380 Torr. What is the final volume, in liters, of the balloon, where the pressure is 0.20 atm and the temperature is −45 °C, when n does not change? (8.5)

8.76 During laparoscopic surgery, carbon dioxide gas is used to expand the abdomen to help create a larger working space. If 4.80 L of CO_2 gas at 18 °C at 785 mmHg is used, what is the final volume, in liters, of the gas at 37 °C and a pressure of 745 mmHg, if the amount of CO_2 does not change? (8.5)

8.77 A weather balloon is partially filled with helium to allow for expansion at high altitudes. At STP, a weather balloon is filled with enough helium to give a volume of 25.0 L. At an altitude of 30.0 km and −35 °C, it has expanded to 2460 L. The increase in volume causes it to burst and a small parachute returns the instruments to Earth. (8.5, 8.6)

a. How many grams of helium are added to the balloon?

b. What is the final pressure, in millimeters of mercury, of the helium inside the balloon when it bursts?

8.78 A mixture of nitrogen (N_2) and helium has a volume of 250 mL at 30 °C and a total pressure of 745 mmHg. (8.5, 8.6, 8.7)

a. If the partial pressure of helium is 32 mmHg, what is the partial pressure of the nitrogen?

b. What is the final volume, in liters, of the nitrogen at STP?

8.79 A gas mixture contains oxygen and argon at partial pressures of 0.60 atm and 425 mmHg. If nitrogen gas added to the sample increases the total pressure to 1250 Torr, what is the partial pressure, in torr, of the nitrogen added? (8.7)

8.80 What is the total pressure, in millimeters of mercury, of a gas mixture containing argon gas at 0.25 atm, helium gas at 350 mmHg, and nitrogen gas at 360 Torr? (8.7)

CHALLENGE PROBLEMS

The following problems are related to the topics in this chapter. However, they do not all follow the chapter order, and they require you to combine concepts and skills from several Sections. These problems will help you increase your critical thinking skills and prepare for your next exam.

8.81 Your spaceship has docked at a space station above Mars. The temperature inside the space station is a carefully controlled 24 °C at a pressure of 745 mmHg. A balloon with a volume of 425 mL drifts into the airlock where the temperature is −95 °C and the pressure is 0.115 atm. What is the final volume, in milliliters, of the balloon if n does not change and the balloon is very elastic? (8.5)

8.82 You are doing research on planet X. The temperature inside the space station is a carefully controlled 24 °C and the pressure is 755 mmHg. Suppose that a balloon, which has a volume of 850. mL inside the space station, is placed into the airlock, and floats out to planet X. If planet X has an atmospheric pressure of 0.150 atm and the volume of the balloon changes to 3.22 L, what is the temperature, in degrees Celsius, on planet X (n does not change)? (8.5)

8.83 A gas sample has a volume of 4250 mL at 15 °C and 745 mmHg. What is the final temperature, in degrees Celsius, after the sample is transferred to a different container with a volume of 2.50 L and a pressure of 1.20 atm if the amount of gas does not change? (8.5)

8.84 In the fermentation of glucose (wine making), 780 mL of CO_2 gas was produced at 37 °C and 1.00 atm. What is the final volume, in liters, of the gas when measured at 22 °C and 675 mmHg, if the amount of gas does not change? (8.5)

ANSWERS

Answers to Selected Practice Problems

8.1 a. At a higher temperature, gas particles have greater kinetic energy, which makes them move faster.
b. Because there are great distances between the particles of a gas, they can be pushed closer together and still remain a gas.
c. Gas particles are very far apart, which means that the mass of a gas in a certain volume is very small, resulting in a low density.

8.3 a. temperature b. volume c. amount d. pressure

8.5 Statements **a**, **d**, and **e** describe the pressure of a gas.

8.7 a. 1520 Torr b. 29.4 lb/in.^{2} c. 1520 mmHg d. 203 kPa

8.9 As a diver ascends to the surface, external pressure decreases. If the air in the lungs were not exhaled, its volume would expand and severely damage the lungs. The pressure in the lungs must adjust to changes in the external pressure.

8.11 a. The pressure is greater in cylinder **A**. According to Boyle's law, a decrease in volume pushes the gas particles closer together, which will cause an increase in the pressure.
b. 160 mL

8.13 a. increases b. decreases c. increases

8.15 a. 328 mmHg b. 2620 mmHg c. 475 mmHg d. 5240 mmHg

8.17 a. 52.4 L b. 25 L c. 100. L d. 45 L

8.19 6.52 atm

8.21 25 L of cyclopropane

8.23 a. inspiration b. expiration c. inspiration

8.25 a. C b. A c. B

8.27 a. 303 °C b. −129 °C c. 591 °C d. 136 °C

8.29 a. 2400 mL b. 4900 mL c. 1800 mL d. 1700 mL

8.31 121 °C

8.33 a. 770 mmHg b. 1150 mmHg

8.35 a. −23 °C b. 168 °C

8.37 31 °C

8.39 1.9 atm

8.41 $T_2 = T_1 \times \frac{P_2}{P_1} \times \frac{V_2}{V_1}$

8.43 a. 4.26 atm b. 3.07 atm c. 0.606 atm

8.45 −33 °C

8.47 The volume increases because the number of gas particles is increased.

8.49 a. 4.00 L b. 26.7 L c. 14.6 L

8.51 a. 2.00 moles of O_2 b. 56.0 L c. 28.0 L d. 0.146 g of H_2

8.53 In a gas mixture, the pressure that each gas exerts as part of the total pressure is called the partial pressure of that gas. Because the air sample is a mixture of gases, the total pressure is the sum of the partial pressures of each gas in the sample.

8.55 765 Torr

8.57 425 Torr

8.59 **a.** The partial pressure of oxygen will be lower than normal.
b. Breathing a higher concentration of oxygen will help to increase the supply of oxygen in the lungs and blood and raise the partial pressure of oxygen in the blood.

8.61 2.8 L

8.63 **a.** True. The flask containing helium has more moles of helium and thus more helium atoms.
b. True. The mass and volume of each are the same, which means the mass/volume ratio or density is the same in both flasks.

8.65 **a.** 2 **b.** 1

8.67 **a.** A **b.** C **c.** A **d.** B **e.** C

8.69 **a.** The volume of the chest and lungs is decreased.
b. The decrease in volume increases the pressure, which can dislodge the food in the trachea.

8.71 2.8 kg of H_2

8.73 −223 °C

8.75 1500 L of He

8.77 **a.** 4.47 g of helium **b.** 6.73 mmHg

8.79 370 Torr

8.81 2170 mL

8.83 −66 °C

9 Solutions

OUR KIDNEYS PRODUCE URINE, WHICH carries waste products and excess fluid from the body. They also reabsorb electrolytes such as potassium and produce hormones that regulate blood pressure and the levels of calcium in the blood. Diseases such as diabetes and high blood pressure can cause a decrease in kidney function. Symptoms of kidney malfunction include protein in the urine, an abnormal level of urea in the blood, frequent urination, and swollen feet. If kidney failure occurs, it may be treated with dialysis or transplantation.

Michelle suffers from kidney disease because of severe strep throat she contracted as a child. When her kidneys stopped functioning, Michelle was placed on dialysis three times a week. As she enters the dialysis unit, her dialysis nurse, Amanda, asks Michelle how she is feeling. Michelle indicates that she feels tired today and has considerable swelling around her ankles. Amanda informs her that these side effects occur because of her body's inability to regulate the amount of water in her cells. Amanda explains that the amount of water is regulated by the concentration of electrolytes in her body fluids and the rate at which waste products are removed from her body. Amanda explains that although water is essential for the many chemical reactions that occur in the body, the amount of water can become too high or too low because of various diseases and conditions. Because Michelle's kidneys no longer perform dialysis, she cannot regulate the amount of electrolytes or waste in her body fluids. As a result, she has an electrolyte imbalance and a buildup of waste products, so her body retains water. Amanda then explains that the dialysis machine does the work of her kidneys to reduce the high levels of electrolytes and waste products.

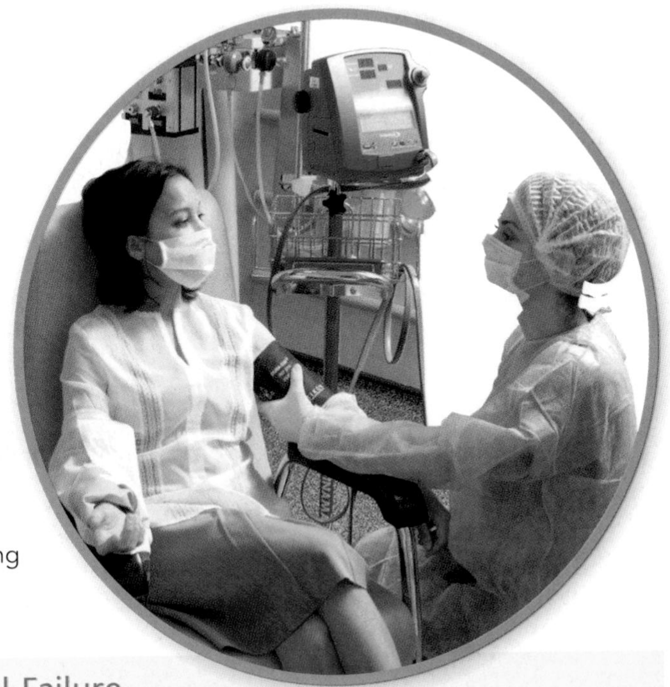

CAREER Dialysis Nurse

A dialysis nurse specializes in assisting patients with kidney disease undergoing dialysis. This requires monitoring the patient before, during, and after dialysis for any complications such as a drop in blood pressure or cramping. The dialysis nurse connects the patient to the dialysis unit via a dialysis catheter that is inserted into the neck or chest, which must be kept clean to prevent infection. A dialysis nurse must have considerable knowledge about how the dialysis machine functions to ensure that it is operating correctly at all times.

CLINICAL UPDATE Using Dialysis for Renal Failure

Michelle continues to have dialysis treatment three times a week. You can see more details of Michelle's dialysis in the **CLINICAL UPDATE Using Dialysis for Renal Failure**, page 314, and discover that dialysis requires 120 L of fluid containing electrolytes to adjust Michelle's blood level to that of normal serum.

LOOKING AHEAD

REVIEW

Identifying Polarity of Molecules and Intermolecular Forces (6.9)

A solution has at least one solute dispersed in a solvent.

9.1 Solutions

LEARNING GOAL Identify the solute and solvent in a solution; describe the formation of a solution.

Solutions are everywhere around us. Most of the gases, liquids, and solids we see are mixtures of at least one substance dissolved in another. There are different types of solutions. The air we breathe is a solution that is primarily oxygen and nitrogen gases. Carbon dioxide gas dissolved in water makes carbonated drinks. When we make solutions of coffee or tea, we use hot water to dissolve substances from coffee beans or tea leaves. The ocean is also a solution, consisting of many ionic compounds such as sodium chloride dissolved in water. In your medicine cabinet, the antiseptic tincture of iodine is a solution of iodine dissolved in ethanol.

A **solution** is a homogeneous mixture in which one substance, called the **solute**, is uniformly dispersed in another substance called the **solvent**. Because the solute and the solvent do not react with each other, they can be mixed in varying proportions. A solution of a little salt dissolved in water tastes slightly salty. When a large amount of salt is dissolved in water, the solution tastes very salty. Usually, the solute (in this case, salt) is the substance present in the lesser amount, whereas the solvent (in this case, water) is present in the greater amount. For example, in a solution composed of 5.0 g of salt and 50. g of water, salt is the solute and water is the solvent. In a solution, the particles of the solute are evenly dispersed among the molecules within the solvent (see **FIGURE 9.1**).

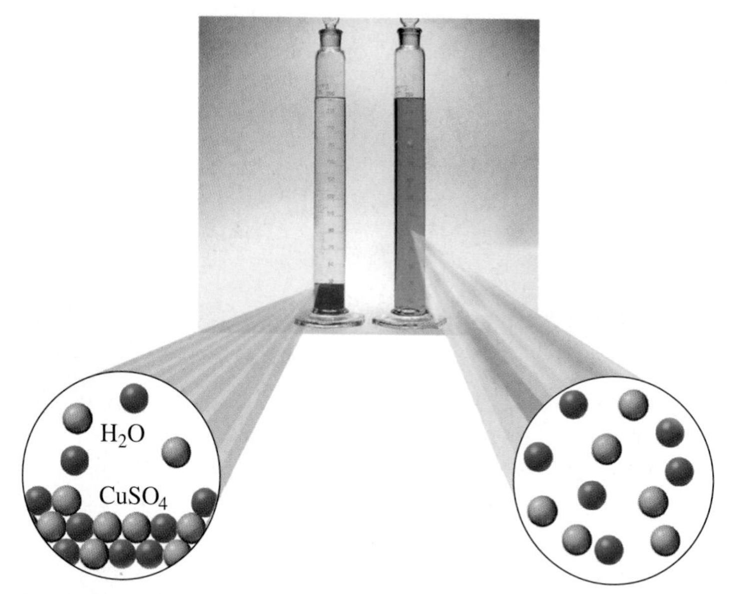

FIGURE 9.1 A solution of copper(II) sulfate ($CuSO_4$) forms as particles of solute dissolve, move away from the crystal, and become evenly dispersed among the solvent (water) molecules.

Q What does the uniform blue color in the graduated cylinder on the right indicate?

Types of Solutes and Solvents

Solutes and solvents may be solids, liquids, or gases. The solution that forms has the same physical state as the solvent. When sugar crystals are dissolved in water, the resulting sugar solution is liquid. Sugar is the solute, and water is the solvent. Soda water and soft drinks are prepared by dissolving carbon dioxide gas in water. The carbon dioxide gas is the solute, and water is the solvent. **TABLE 9.1** lists some solutes and solvents and their solutions.

TABLE 9.1 Some Examples of Solutions

Type	Example	Primary Solute	Solvent
Gas Solutions			
Gas in a gas	Air	$O_2(g)$	$N_2(g)$
Liquid Solutions			
Gas in a liquid	Soda water	$CO_2(g)$	$H_2O(l)$
	Household ammonia	$NH_3(g)$	$H_2O(l)$
Liquid in a liquid	Vinegar	$HC_2H_3O_2(l)$	$H_2O(l)$
Solid in a liquid	Seawater	$NaCl(s)$	$H_2O(l)$
	Tincture of iodine	$I_2(s)$	$C_2H_6O(l)$
Solid Solutions			
Solid in a solid	Brass	$Zn(s)$	$Cu(s)$
	Steel	$C(s)$	$Fe(s)$

TEST

Try Practice Problems 9.1 and 9.2

Water as a Solvent

Water is one of the most common solvents in nature. In the H_2O molecule, an oxygen atom shares electrons with two hydrogen atoms. Because oxygen is much more electronegative than hydrogen, the O—H bonds are polar. In each polar bond, the oxygen atom has a partial negative (δ^-) charge and the hydrogen atom has a partial positive (δ^+) charge. Because the shape of a water molecule is bent, not linear, its dipoles do not cancel out. Thus, water is polar, and is a *polar solvent.*

Attractive forces known as *hydrogen bonds* occur between molecules where partially positive hydrogen atoms are attracted to the partially negative atoms N, O, or F. As seen in the diagram, the hydrogen bonds are shown as a series of dots. Although hydrogen bonds are much weaker than covalent or ionic bonds, there are many of them linking water molecules together. Hydrogen bonds are important in the properties of biological compounds such as proteins, carbohydrates, and DNA.

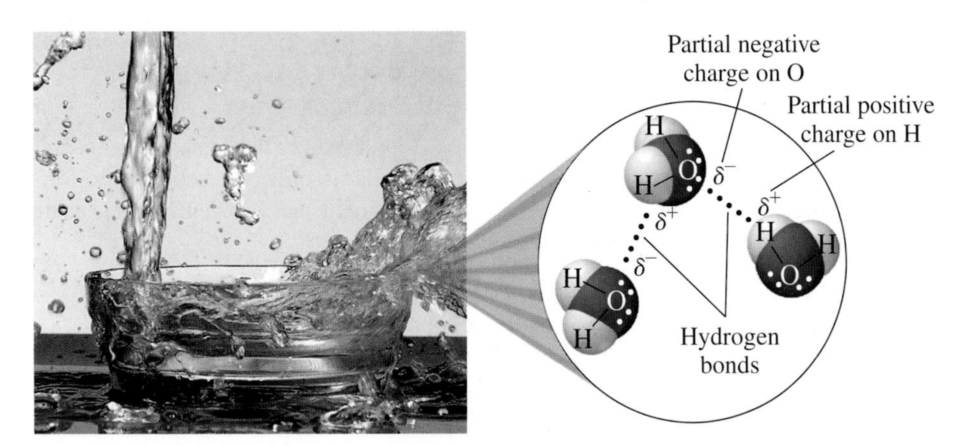

In water, hydrogen bonds form between an oxygen atom in one water molecule and the hydrogen atom in another.

Formation of Solutions

The interactions between solute and solvent will determine whether a solution will form. Initially, energy is needed to separate the particles in the solute and the solvent particles. Then energy is released as solute particles move between the solvent particles to form a solution. However, there must be attractions between the solute and the solvent particles to provide the energy for the initial separation. These attractions occur when the solute and

CHEMISTRY LINK TO HEALTH

Water in the Body

The average adult is about 60% water by mass, and the average infant about 75%. About 60% of the body's water is contained within the cells as intracellular fluids; the other 40% makes up extracellular fluids, which include the interstitial fluid in tissue and the plasma in the blood. These external fluids carry nutrients and waste materials between the cells and the circulatory system.

Typical water gain and loss during 24 hours

Water Gain	
Liquid	1000 mL
Food	1200 mL
Metabolism	300 mL
Total	2500 mL

Water Loss	
Urine	1500 mL
Perspiration	300 mL
Breath	600 mL
Feces	100 mL
Total	2500 mL

The water lost from the body is replaced by the intake of fluids.

Every day you lose between 1500 and 3000 mL of water from the kidneys as urine, from the skin as perspiration, from the lungs as you exhale, and from the gastrointestinal tract. Serious dehydration can occur in an adult if there is a 10% net loss in total body fluid; a 20% loss of fluid can be fatal. An infant suffers severe dehydration with only a 5 to 10% loss in body fluid.

Water loss is continually replaced by the liquids and foods in the diet and from metabolic processes that produce water in the cells of the body. **TABLE 9.2** lists the percentage by mass of water contained in some foods.

TABLE 9.2 Percentage of Water in Some Foods

Food	Water (% by mass)	Food	Water (% by mass)
Vegetables		**Meats/Fish**	
Carrot	88	Chicken, cooked	71
Celery	94	Hamburger, broiled	60
Cucumber	96	Salmon	71
Tomato	94		
Fruits		**Milk Products**	
Apple	85	Cottage cheese	78
Cantaloupe	91	Milk, whole	87
Orange	86	Yogurt	88
Strawberry	90		
Watermelon	93		

(a) (b) (c)

FIGURE 9.2 Like dissolves like. In each test tube, the lower layer is CH_2Cl_2 (more dense), and the upper layer is water (less dense). **(a)** CH_2Cl_2 is nonpolar and water is polar; the two layers do not mix. **(b)** The nonpolar solute I_2 (purple) is soluble in the nonpolar solvent CH_2Cl_2. **(c)** The ionic solute $Ni(NO_3)_2$ (green) is soluble in the polar solvent water.

In which layer would polar molecules of sucrose ($C_{12}H_{22}O_{11}$) be soluble?

the solvent have similar polarities. The expression "like dissolves like" is a way of saying that the polarities of a solute and a solvent must be similar in order for a solution to form (see **FIGURE 9.2**). In the absence of attractions between a solute and a solvent, there is insufficient energy to form a solution (see **TABLE 9.3**).

TABLE 9.3 Possible Combinations of Solutes and Solvents

Solutions Will Form		Solutions Will Not Form	
Solute	**Solvent**	**Solute**	**Solvent**
Polar	Polar	Polar	Nonpolar
Nonpolar	Nonpolar	Nonpolar	Polar

Solutions with Ionic and Polar Solutes

In ionic solutes such as sodium chloride, NaCl, there are strong ionic bonds between positively charged Na^+ ions and negatively charged Cl^- ions. In water, a polar solvent, the hydrogen bonds provide strong solvent–solvent attractions. When NaCl crystals are placed in water, partially negative oxygen atoms in water molecules attract positive Na^+ ions, and the partially positive hydrogen atoms in other water molecules attract negative Cl^- ions (see **FIGURE 9.3**). As soon as the Na^+ ions and the Cl^- ions form a solution, they undergo

hydration as water molecules surround each ion. Hydration of the ions diminishes their attraction to other ions and keeps them in solution.

In the equation for the formation of the NaCl solution, the solid and aqueous NaCl are shown with the formula H_2O over the arrow, which indicates that water is needed for the dissociation process but is not a reactant.

$$NaCl(s) \xrightarrow[\text{Dissociation}]{H_2O} Na^+(aq) + Cl^-(aq)$$

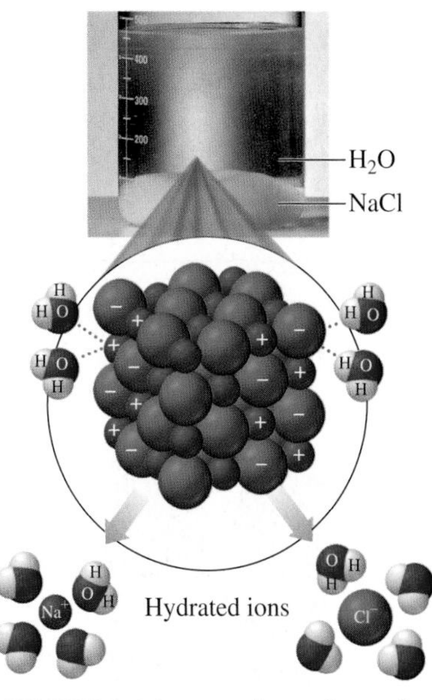

FIGURE 9.3 Ions on the surface of a crystal of NaCl dissolve in water as they are attracted to the polar water molecules that pull the ions into solution and surround them.

Q What helps keep Na^+ and Cl^- ions in solution?

In another example, we find that a polar molecular compound such as methanol, CH_3OH, is soluble in water because methanol has a polar —OH group that forms hydrogen bonds with water (see FIGURE 9.4). Polar solutes require polar solvents for a solution to form.

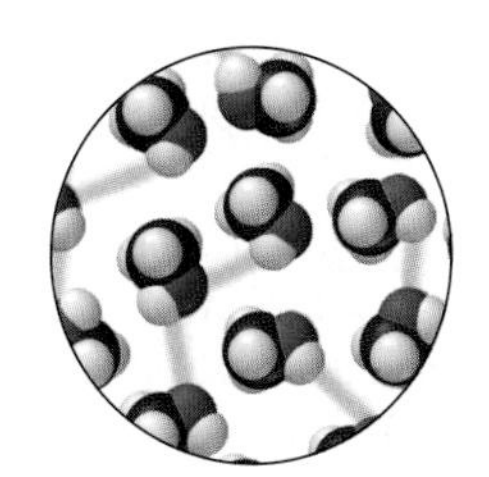

Methanol (CH_3OH) solute

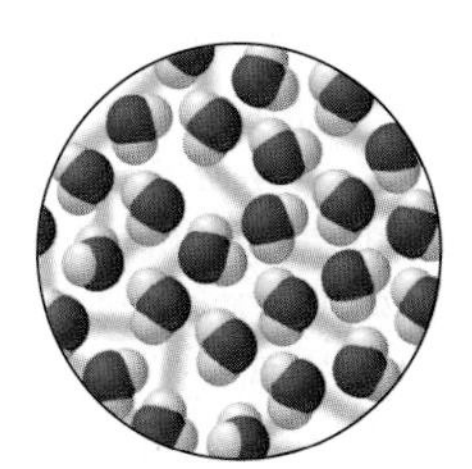

Water solvent

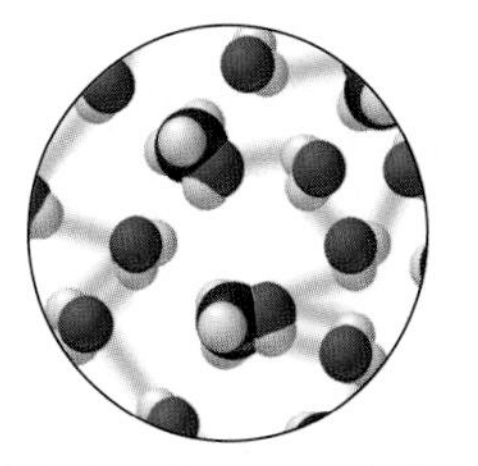

Methanol–water solution with hydrogen bonding

FIGURE 9.4 Polar molecules of methanol, CH_3OH, form hydrogen bonds with polar water molecules to form a methanol–water solution.

Q Why are there attractions between the solute methanol and the solvent water?

ENGAGE

Why does KCl form a solution with water, but nonpolar hexane (C_6H_{14}) does not form a solution with water?

Solutions with Nonpolar Solutes

Compounds containing nonpolar molecules, such as iodine (I_2), oil, or grease, do not dissolve in water because there are no attractions between the particles of a nonpolar solute and the polar solvent. Nonpolar solutes require nonpolar solvents for a solution to form.

TEST

Try Practice Problems 9.3 to 9.6

PRACTICE PROBLEMS

9.1 Solutions

LEARNING GOAL Identify the solute and solvent in a solution; describe the formation of a solution.

9.1 Identify the solute and the solvent in each solution composed of the following:
- **a.** 10.0 g of NaCl and 100.0 g of H_2O
- **b.** 50.0 mL of ethanol, C_2H_6O, and 10.0 mL of H_2O
- **c.** 0.20 L of O_2 and 0.80 L of N_2

9.2 Identify the solute and the solvent in each solution composed of the following:
- **a.** 10.0 mL of acetic acid and 200. mL of water
- **b.** 100.0 mL of water and 5.0 g of sugar
- **c.** 1.0 g of Br_2 and 50.0 mL of methylene chloride(l)

9.3 Describe the formation of an aqueous KI solution, when solid KI dissolves in water.

9.4 Describe the formation of an aqueous LiBr solution, when solid LiBr dissolves in water.

Clinical Applications

9.5 Water is a polar solvent and carbon tetrachloride (CCl_4) is a nonpolar solvent. In which solvent is each of the following, which is found or used in the body, more likely to be soluble?
- **a.** $CaCO_3$ (calcium supplement), ionic
- **b.** retinol (vitamin A), nonpolar
- **c.** sucrose (table sugar), polar
- **d.** cholesterol (lipid), nonpolar

9.6 Water is a polar solvent and hexane (C_6H_{14}) is a nonpolar solvent. In which solvent is each of the following, which is found or used in the body, more likely to be soluble?
- **a.** vegetable oil, nonpolar
- **b.** oleic acid (lipid), nonpolar
- **c.** niacin (vitamin B_3), polar
- **d.** $FeSO_4$ (iron supplement), ionic

REVIEW

Writing Conversion Factors from Equalities (2.5)
Using Conversion Factors (2.6)
Writing Positive and Negative Ions (6.1)

9.2 Electrolytes and Nonelectrolytes

LEARNING GOAL Identify solutes as electrolytes or nonelectrolytes.

Solutes can be classified by their ability to conduct an electrical current. When **electrolytes** dissolve in water, the process of *dissociation* separates them into ions forming solutions that conduct electricity. When **nonelectrolytes** dissolve in water, they do not separate into ions and their solutions do not conduct electricity.

To test solutions for the presence of ions, we can use an apparatus that consists of a battery and a pair of electrodes connected by wires to a light bulb. The light bulb glows when electricity can flow, which can only happen when electrolytes provide ions that move between the electrodes to complete the circuit.

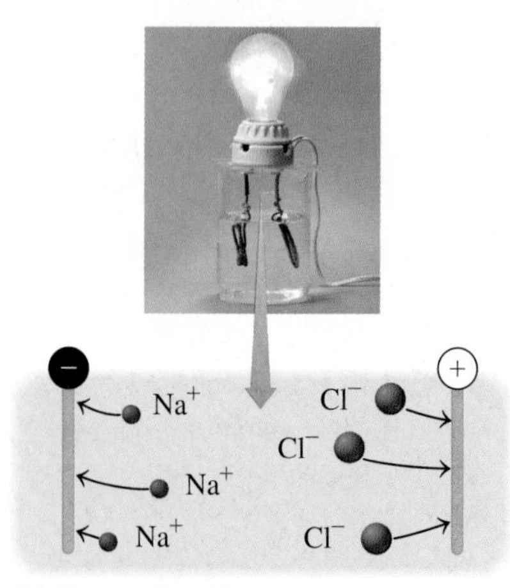

(a) Strong electrolyte

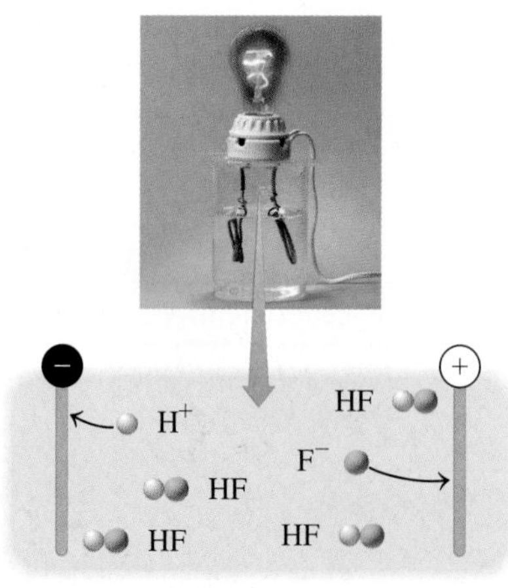

(b) Weak electrolyte

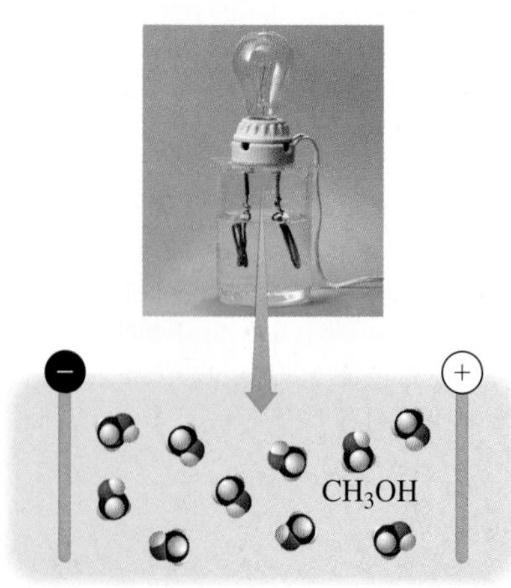

(c) Nonelectrolyte

Types of Electrolytes

Electrolytes can be further classified as *strong electrolytes* or *weak electrolytes*. For a **strong electrolyte**, such as sodium chloride (NaCl), there is 100% dissociation of the solute into ions. When the electrodes from the light bulb apparatus are placed in the NaCl solution, the light bulb glows very brightly.

In an equation for dissociation of a compound in water, the charges must balance. For example, magnesium nitrate dissociates to give one magnesium ion for every two nitrate ions. However, only the ionic bonds between Mg^{2+} and NO_3^- are broken, not the covalent bonds within the polyatomic ion. The equation for the dissociation of $Mg(NO_3)_2$ is written as follows:

$$Mg(NO_3)_2(s) \xrightarrow[\text{Dissociation}]{H_2O} Mg^{2+}(aq) + 2NO_3^-(aq)$$

A **weak electrolyte** is a compound that dissolves in water mostly as molecules. Only a few of the dissolved solute molecules undergo *dissociation*, producing a small number of ions in solution. Thus, solutions of weak electrolytes do not conduct electrical current as well as solutions of strong electrolytes. When the electrodes are placed in a solution of a weak electrolyte, the glow of the light bulb is very dim. In an aqueous solution of the weak electrolyte HF, a few HF molecules dissociate to produce H^+ and F^- ions. As more H^+ and F^- ions form, some recombine to give HF molecules. These forward and reverse reactions of molecules to ions and back again are indicated by two arrows between reactant and products that point in opposite directions:

$$HF(aq) \underset{\text{Recombination}}{\overset{\text{Dissociation}}{\rightleftharpoons}} H^+(aq) + F^-(aq)$$

A nonelectrolyte such as methanol (CH_4O) dissolves in water only as molecules, which do not dissociate. When electrodes of the light bulb apparatus are placed in a solution of a nonelectrolyte, the light bulb does not glow, because the solution does not contain ions and cannot conduct electricity.

$$CH_4O(l) \xrightarrow{H_2O} CH_4O(aq)$$

SAMPLE PROBLEM 9.1 Solutions of Electrolytes and Nonelectrolytes

TRY IT FIRST

Indicate whether solutions of each of the following contain only ions, only molecules, or mostly molecules and a few ions. Write the equation for the formation of a solution for each of the following:

a. $Na_2SO_4(s)$, a strong electrolyte
b. sucrose, $C_{12}H_{22}O_{11}(s)$, a nonelectrolyte
c. acetic acid, $HC_2H_3O_2(l)$, a weak electrolyte

TABLE 9.4 summarizes the classification of solutes in aqueous solutions.

TABLE 9.4 Classification of Solutes in Aqueous Solutions

Type of Solute	In Solution	Type(s) of Particles in Solution	Conducts Electricity?	Examples
Strong electrolyte	Dissociates completely	Only ions	Yes	Ionic compounds such as NaCl, KBr, $MgCl_2$, $NaNO_3$; bases such as NaOH, KOH; acids such as HCl, HBr, HI, HNO_3, $HClO_4$, H_2SO_4
Weak electrolyte	Dissociates partially	Mostly molecules and a few ions	Weakly	HF, H_2O, NH_3, $HC_2H_3O_2$ (acetic acid)
Nonelectrolyte	No dissociation	Only molecules	No	Carbon compounds such as CH_4O (methanol), C_2H_6O (ethanol), $C_{12}H_{22}O_{11}$ (sucrose), CH_4N_2O (urea)

SOLUTION

a. An aqueous solution of $Na_2SO_4(s)$ contains only the ions Na^+ and SO_4^{2-}.

$$Na_2SO_4(s) \xrightarrow{H_2O} 2Na^+(aq) + SO_4^{2-}(aq)$$

b. A nonelectrolyte such as sucrose, $C_{12}H_{22}O_{11}(s)$, produces only molecules when it dissolves in water.

$$C_{12}H_{22}O_{11}(s) \xrightarrow{H_2O} C_{12}H_{22}O_{11}(aq)$$

c. A weak electrolyte such as $HC_2H_3O_2(l)$ produces mostly molecules and a few ions when it dissolves in water.

$$HC_2H_3O_2(l) \overset{H_2O}{\rightleftharpoons} H^+(aq) + C_2H_3O_2^-(aq)$$

ENGAGE

Why does a solution of $LiNO_3$, a strong electrolyte, contain only ions whereas a solution of urea, CH_4N_2O, a nonelectrolyte, contains only molecules?

STUDY CHECK 9.1

Boric acid, $H_3BO_3(s)$, is a weak electrolyte. Would you expect a boric acid solution to contain only ions, only molecules, or mostly molecules and a few ions?

ANSWER

A solution of a weak electrolyte would contain mostly molecules and a few ions.

TEST

Try Practice Problems 9.7 to 9.14

Equivalents

Body fluids contain a mixture of electrolytes, such as Na^+, Cl^-, K^+, and Ca^{2+}. We measure each individual ion in terms of an **equivalent (Eq)**, which is the amount of that ion equal to 1 mole of positive or negative electrical charge. For example, 1 mole of Na^+ ions and 1 mole of Cl^- ions are each 1 equivalent because they each contain 1 mole of charge. For an ion with a charge of 2+ or 2−, there are 2 equivalents for each mole. Some examples of ions and equivalents are shown in **TABLE 9.5**.

TABLE 9.5 Equivalents of Electrolytes in Clinical Intravenous (IV) Solutions

Ion	Ionic Charge	Number of Equivalents in 1 Mole
Na^+, K^+, Li^+, NH_4^+	1+	1 Eq
Ca^{2+}, Mg^{2+}	2+	2 Eq
Fe^{3+}	3+	3 Eq
Cl^-, $C_2H_3O_2^-$ (acetate), $H_2PO_4^-$, $C_3H_5O_3^-$ (lactate)	1−	1 Eq
CO_3^{2-}, HPO_4^{2-}	2−	2 Eq
PO_4^{3-}, $C_6H_5O_7^{3-}$ (citrate)	3−	3 Eq

In any solution, the charge of the positive ions is always balanced by the charge of the negative ions. The concentrations of electrolytes in intravenous fluids are expressed in milliequivalents per liter (mEq/L); 1 Eq = 1000 mEq. For example, a solution containing 25 mEq/L of Na^+ and 4 mEq/L of K^+ has a total positive charge of 29 mEq/L. If Cl^- is the only anion, its concentration must be 29 mEq/L.

SAMPLE PROBLEM 9.2 Electrolyte Concentration

TRY IT FIRST

The laboratory tests for Michelle indicate that she has hypercalcemia with a blood calcium level of 8.8 mEq/L. How many moles of calcium ion are in 0.50 L of her blood?

SOLUTION GUIDE

STEP 1 State the given and needed quantities.

	Given	Need	Connect
ANALYZE THE PROBLEM	0.50 L, 8.8 mEq of Ca^{2+}/L	moles of Ca^{2+}	1 Eq = 1000 mEq, 1 mole of Ca^{2+} = 2 Eq of Ca^{2+}

STEP 2 Write a plan to calculate the moles.

liters of solution → Electrolyte concentration → milliequivalents of Ca^{2+} → Metric factor → equivalents of Ca^{2+} → Eq/mole → moles of Ca^{2+}

STEP 3 State the equalities and conversion factors.

1 L of solution = 8.8 mEq of Ca^{2+}

$$\frac{8.8 \text{ mEq } Ca^{2+}}{1 \text{ L solution}} \quad \text{and} \quad \frac{1 \text{ L solution}}{8.8 \text{ mEq } Ca^{2+}}$$

1 Eq = 1000 mEq

$$\frac{1000 \text{ mEq } Ca^{2+}}{1 \text{ Eq } Ca^{2+}} \quad \text{and} \quad \frac{1 \text{ Eq } Ca^{2+}}{1000 \text{ mEq } Ca^{2+}}$$

1 mole of Ca^{2+} = 2 Eq of Ca^{2+}

$$\frac{2 \text{ Eq } Ca^{2+}}{1 \text{ mole } Ca^{2+}} \quad \text{and} \quad \frac{1 \text{ mole } Ca^{2+}}{2 \text{ Eq } Ca^{2+}}$$

STEP 4 Set up the problem to calculate the number of moles.

$$\underset{\text{Two SFs}}{0.50 \cancel{\text{L}}} \times \underset{\text{Exact}}{\overset{\text{Two SFs}}{\frac{8.8 \cancel{\text{mEq } Ca^{2+}}}{1 \cancel{\text{L}}}}} \times \underset{\text{Exact}}{\overset{\text{Exact}}{\frac{1 \cancel{\text{Eq } Ca^{2+}}}{1000 \cancel{\text{mEq } Ca^{2+}}}}} \times \underset{\text{Exact}}{\overset{\text{Exact}}{\frac{1 \text{ mole } Ca^{2+}}{2 \cancel{\text{Eq } Ca^{2+}}}}} = \underset{\text{Two SFs}}{0.0022 \text{ mole of } Ca^{2+}}$$

STUDY CHECK 9.2

A lactated Ringer's solution for intravenous fluid replacement contains 109 mEq of Cl^- per liter of solution. If a patient received 1250 mL of Ringer's solution, how many moles of chloride ion were given?

ANSWER

0.136 mole of Cl^-

TEST

Try Practice Problems 9.15 to 9.22

CHEMISTRY LINK TO HEALTH

Electrolytes in Body Fluids

Electrolytes in the body play an important role in maintaining the proper function of the cells and organs in the body. Typically, the electrolytes sodium, potassium, chloride, and bicarbonate are measured in a blood test. Sodium ions regulate the water content in the body and are important in carrying electrical impulses through the nervous system. Potassium ions are also involved in the transmission of electrical impulses and play a role in the maintenance of a regular heartbeat. Chloride ions balance the charges of the positive ions and also control the balance of fluids in the body. Bicarbonate is important in maintaining the proper pH of the blood. Sometimes when vomiting, diarrhea, or sweating is excessive, the concentrations of certain electrolytes may decrease. Then fluids such as Pedialyte may be given to return electrolyte levels to normal.

The concentrations of electrolytes present in body fluids and in intravenous fluids given to a patient are often expressed in milliequivalents per liter (mEq/L) of solution. For example, one liter of Pedialyte contains the following electrolytes: Na^+ 45 mEq, Cl^- 35 mEq, K^+ 20 mEq, and citrate^{3-} 30 mEq.

TABLE 9.6 gives the concentrations of some typical electrolytes in blood plasma and various types of solutions. There is a charge balance because the total number of positive charges is equal to the total number of negative charges. The use of a specific intravenous solution depends on the nutritional, electrolyte, and fluid needs of the individual patient.

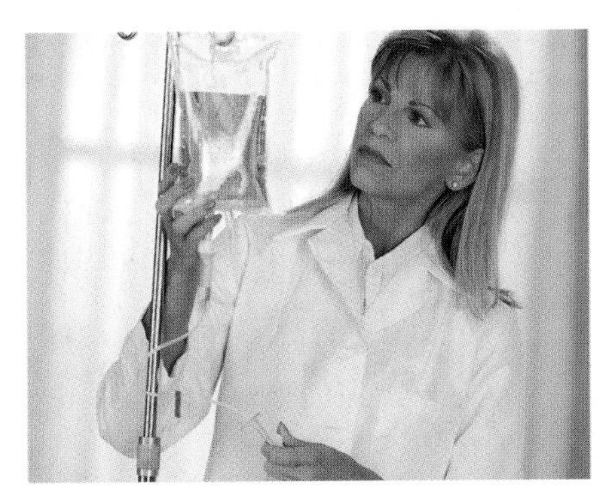

An intravenous solution is used to replace electrolytes in the body.

TABLE 9.6 Electrolytes in Blood Plasma and Selected Intravenous Solutions

	Normal Concentrations of Ions (mEq/L)				
	Blood Plasma	Normal Saline 0.9% NaCl	Lactated Ringer's Solution	Maintenance Solution	Replacement Solution (extracellular)
Purpose		Replaces fluid loss	Hydration	Maintains electrolytes and fluids	Replaces electrolytes
Cations					
Na^+	135–145	154	130	40	140
K^+	3.5–5.5		4	35	10
Ca^{2+}	4.5–5.5		3		5
Mg^{2+}	1.5–3.0				3
Total		154	137	75	158
Anions			7		
Acetate$^-$					47
Cl^-	95–105	154	109	40	103
HCO_3^-	22–28				
Lactate$^-$			28	20	
HPO_4^{2-}	1.8–2.3			15	
Citrate^{3-}					8
Total		154	137	75	158

PRACTICE PROBLEMS

9.2 Electrolytes and Nonelectrolytes

LEARNING GOAL Identify solutes as electrolytes or nonelectrolytes.

9.7 KF is a strong electrolyte, and HF is a weak electrolyte. How is the solution of KF different from that of HF?

9.8 NaOH is a strong electrolyte, and CH_3OH is a nonelectrolyte. How is the solution of NaOH different from that of CH_3OH?

9.9 Write a balanced equation for the dissociation of each of the following strong electrolytes in water:
a. KCl **b.** $CaCl_2$ **c.** K_3PO_4 **d.** $Fe(NO_3)_3$

9.10 Write a balanced equation for the dissociation of each of the following strong electrolytes in water:
a. LiBr **b.** $NaNO_3$ **c.** $CuCl_2$ **d.** K_2CO_3

9.11 Indicate whether aqueous solutions of each of the following solutes contain only ions, only molecules, or mostly molecules and a few ions:
a. acetic acid, $HC_2H_3O_2$, a weak electrolyte
b. NaBr, a strong electrolyte
c. fructose, $C_6H_{12}O_6$, a nonelectrolyte

9.12 Indicate whether aqueous solutions of each of the following solutes contain only ions, only molecules, or mostly molecules and a few ions:
a. NH_4Cl, a strong electrolyte
b. ethanol, C_2H_6O, a nonelectrolyte
c. HCN, hydrocyanic acid, a weak electrolyte

9.13 Classify the solute represented in each of the following equations as a strong, weak, or nonelectrolyte:

a. $K_2SO_4(s) \xrightarrow{H_2O} 2K^+(aq) + SO_4^{2-}(aq)$

b. $NH_3(g) + H_2O(l) \rightleftarrows NH_4^+(aq) + OH^-(aq)$

c. $C_6H_{12}O_6(s) \xrightarrow{H_2O} C_6H_{12}O_6(aq)$

9.14 Classify the solute represented in each of the following equations as a strong, weak, or nonelectrolyte:

a. $CH_4O(l) \xrightarrow{H_2O} CH_4O(aq)$

b. $MgCl_2(s) \xrightarrow{H_2O} Mg^{2+}(aq) + 2Cl^-(aq)$

c. $HClO(aq) \overset{H_2O}{\rightleftarrows} H^+(aq) + ClO^-(aq)$

9.15 Calculate the number of equivalents in each of the following:

a. 1 mole of K^+ **b.** 2 moles of OH^-
c. 1 mole of Ca^{2+} **d.** 3 moles of CO_3^{2-}

9.16 Calculate the number of equivalents in each of the following:

a. 1 mole of Mg^{2+} **b.** 0.5 mole of H^+
c. 4 moles of Cl^- **d.** 2 moles of Fe^{3+}

Clinical Applications

9.17 An intravenous saline solution contains 154 mEq/L each of Na^+ and Cl^-. How many moles each of Na^+ and Cl^- are in 1.00 L of the saline solution?

9.18 An intravenous solution to replace potassium loss contains 40. mEq/L each of K^+ and Cl^-. How many moles each of K^+ and Cl^- are in 1.5 L of the solution?

9.19 An intravenous solution contains 40. mEq/L of Cl^- and 15 mEq/L of HPO_4^{2-}. If Na^+ is the only cation in the solution, what is the Na^+ concentration, in milliequivalents per liter?

9.20 A Ringer's solution contains the following concentrations (mEq/L) of cations: Na^+ 147, K^+ 4, and Ca^{2+} 4. If Cl^- is the only anion in the solution, what is the Cl^- concentration, in milliequivalents per liter?

9.21 When Michelle's blood was tested, the chloride level was 0.45 g/dL.

a. What is this value in milliequivalents per liter?
b. According to Table 9.6, is this value above, below, or within the normal range?

9.22 After dialysis, the level of magnesium in Michelle's blood was 0.0026 g/dL.

a. What is this value in milliequivalents per liter?
b. According to Table 9.6, is this value above, below, or within the normal range?

9.3 Solubility

LEARNING GOAL Define solubility; distinguish between an unsaturated and a saturated solution. Identify an ionic compound as soluble or insoluble.

REVIEW
Interpreting Graphs (1.4)

The term *solubility* is used to describe the amount of a solute that can dissolve in a given amount of solvent. Many factors, such as the type of solute, the type of solvent, and the temperature, affect the solubility of a solute. **Solubility**, usually expressed in grams of solute in 100. g of solvent, is the maximum amount of solute that can be dissolved at a certain temperature. If a solute readily dissolves when added to the solvent, the solution does not contain the maximum amount of solute. We call this solution an **unsaturated solution**.

A solution that contains all the solute that can dissolve is a **saturated solution**. When a solution is saturated, the rate at which the solute dissolves becomes equal to the rate at which solid forms, a process known as *crystallization*. Then there is no further change in the amount of dissolved solute in solution.

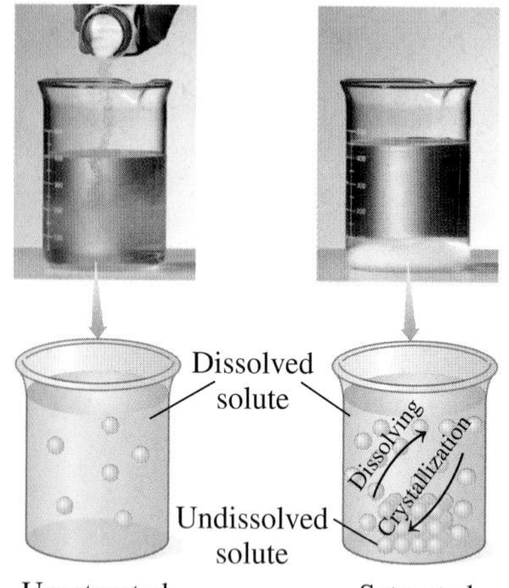

Additional solute can dissolve in an unsaturated solution, but not in a saturated solution.

$$\text{Solute} + \text{solvent} \underset{\text{Solute recrystallizes}}{\overset{\text{Solute dissolves}}{\rightleftarrows}} \text{saturated solution}$$

We can prepare a saturated solution by adding an amount of solute greater than that needed to reach solubility. Stirring the solution will dissolve the maximum amount of solute and leave the excess on the bottom of the container. Once we have a saturated solution, the addition of more solute will only increase the amount of undissolved solute.

SAMPLE PROBLEM 9.3 Saturated Solutions

TRY IT FIRST

At 20 °C, the solubility of KCl is 34 g/100. g of H_2O. In the laboratory, a student mixes 75 g of KCl with 200. g of H_2O at a temperature of 20 °C.

a. How much of the KCl will dissolve?
b. Is the solution saturated or unsaturated?
c. What is the mass, in grams, of any solid KCl left undissolved on the bottom of the container?

SOLUTION

a. At 20 °C, KCl has a solubility of 34 g of KCl in 100. g of water. Using the solubility as a conversion factor, we can calculate the maximum amount of KCl that can dissolve in 200. g of water as follows:

$$200.\ \text{g}\ \cancel{H_2O} \times \frac{34\ \text{g KCl}}{100.\ \text{g}\ \cancel{H_2O}} = 68\ \text{g of KCl}$$

b. Because 75 g of KCl exceeds the maximum amount (68 g) that can dissolve in 200. g of water, the KCl solution is saturated.

c. If we add 75 g of KCl to 200. g of water and only 68 g of KCl can dissolve, there is 7 g (75 g − 68 g) of solid (undissolved) KCl on the bottom of the container.

STUDY CHECK 9.3

At 40 °C, the solubility of KNO_3 is 65 g/100. g of H_2O. How many grams of KNO_3 will dissolve in 120 g of H_2O at 40 °C?

ANSWER

78 g of KNO_3

TEST

Try Practice Problems 9.23 to 9.28

CHEMISTRY LINK TO HEALTH

Gout and Kidney Stones: A Problem of Saturation in Body Fluids

The conditions of gout and kidney stones involve compounds in the body that exceed their solubility levels and form solid products. Gout affects adults, primarily men, over the age of 40. Attacks of gout may occur when the concentration of uric acid in blood plasma exceeds its solubility, which is 7 mg/100 mL of plasma at 37 °C. Insoluble deposits of needle-like crystals of uric acid can form in the cartilage, tendons, and soft tissues, where they cause painful gout attacks. They may also form in the tissues of the kidneys, where they can cause renal damage. High levels of uric acid in the body can be caused by an increase in uric acid production, failure of the kidneys to remove uric acid, or a diet with an overabundance of foods containing purines, which are metabolized to uric acid in the body. Foods in the diet that contribute to high levels of uric acid include certain meats, sardines, mushrooms, asparagus, and beans. Drinking alcoholic beverages may also significantly increase uric acid levels and bring about gout attacks.

Treatment for gout involves diet changes and drugs. Medications, such as probenecid, which helps the kidneys eliminate uric acid, or allopurinol, which blocks the production of uric acid by the body, may be useful.

Kidney stones are solid materials that form in the urinary tract. Most kidney stones are composed of calcium phosphate and calcium oxalate, although they can be solid uric acid. Insufficient water intake and high levels of calcium, oxalate, and phosphate in the urine can lead to the formation of kidney stones. When a kidney stone passes through the urinary tract, it causes considerable pain and discomfort, necessitating the use of painkillers and surgery. Sometimes ultrasound is used to break up kidney stones. Persons prone to kidney stones are advised to drink six to eight glasses of water every day to prevent saturation levels of minerals in the urine.

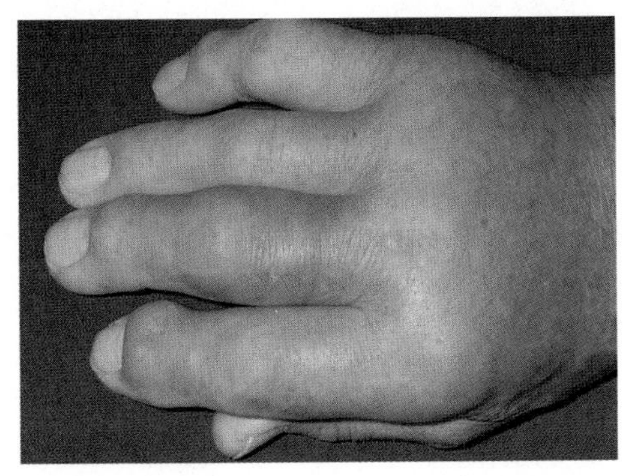

Gout occurs when uric acid exceeds its solubility in blood plasma.

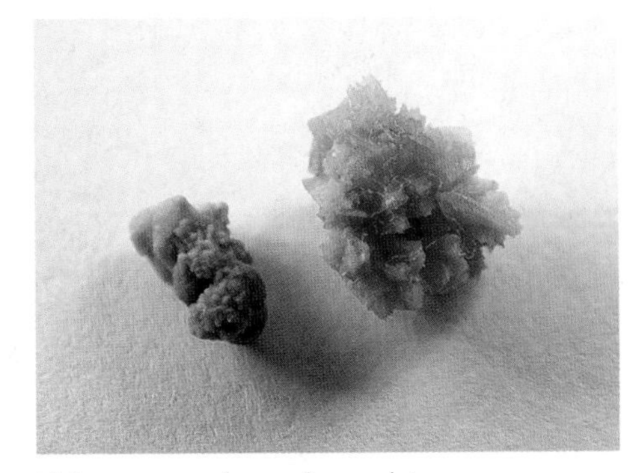

Kidney stones form when calcium phosphate exceeds its solubility.

Effect of Temperature on Solubility

The solubility of most solids is greater as temperature increases, which means that solutions usually contain more dissolved solute at higher temperatures. A few substances show little change in solubility at higher temperatures, and a few are less soluble (see **FIGURE 9.5**). For

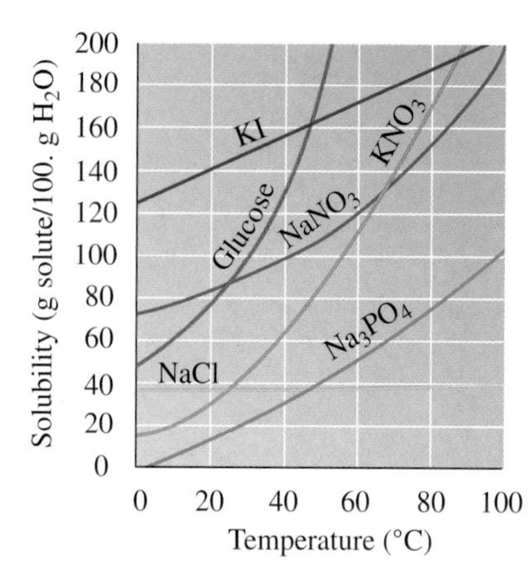

FIGURE 9.5 In water, most common solids are more soluble as the temperature increases.

Q Compare the solubility of $NaNO_3$ at 20 °C and 60 °C.

example, when you add sugar to iced tea, some undissolved sugar may form on the bottom of the glass. But if you add sugar to hot tea, many teaspoons of sugar are needed before solid sugar appears. Hot tea dissolves more sugar than does cold tea because the solubility of sugar is much greater at a higher temperature.

When a saturated solution is carefully cooled, it becomes a *supersaturated solution* because it contains more solute than the solubility allows. Such a solution is unstable, and if the solution is agitated or if a solute crystal is added, the excess solute will crystallize to give a saturated solution again.

Conversely, the solubility of a gas in water decreases as the temperature increases. At higher temperatures, more gas molecules have the energy to escape from the solution. Perhaps you have observed bubbles escaping from a cold carbonated soft drink as it warms. At high temperatures, bottles containing carbonated solutions may burst as more gas molecules leave the solution and increase the gas pressure inside the bottle. Biologists have found that increased temperatures in rivers and lakes cause the amount of dissolved oxygen to decrease until the warm water can no longer support a biological community. Electricity-generating plants are required to have their own ponds to use with their cooling towers to lessen the threat of thermal pollution in surrounding waterways.

Henry's Law

Henry's law states that the solubility of gas in a liquid is directly related to the pressure of that gas above the liquid. At higher pressures, there are more gas molecules available to enter and dissolve in the liquid. A can of soda is carbonated by using CO_2 gas at high pressure to increase the solubility of the CO_2 in the beverage. When you open the can at atmospheric pressure, the pressure on the CO_2 drops, which decreases the solubility of CO_2. As a result, bubbles of CO_2 rapidly escape from the solution. The burst of bubbles is even more noticeable when you open a warm can of soda.

TEST

Try Practice Problems 9.29 and 9.30

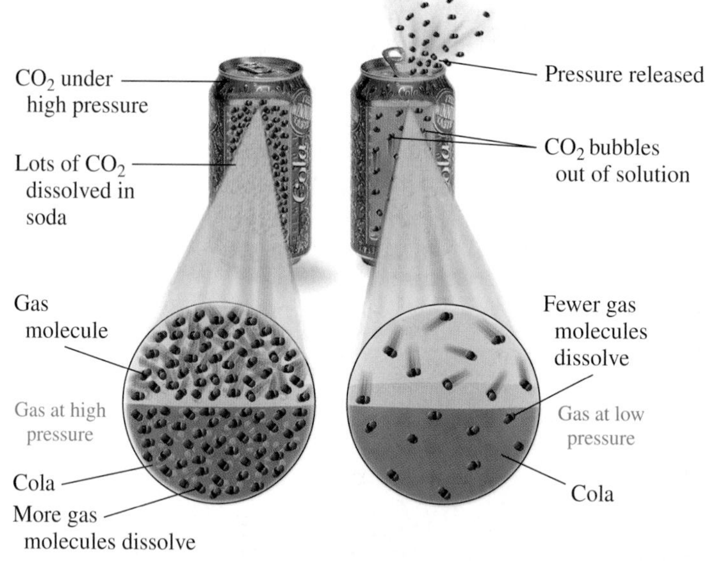

When the pressure of a gas above a solution decreases, the solubility of that gas in the solution also decreases.

Soluble and Insoluble Ionic Compounds

In our discussion up to now, we have considered ionic compounds that dissolve in water. However, some ionic compounds do not dissociate into ions and remain as solids even in contact with water. The **solubility rules** give some guidelines about the solubility of ionic compounds in water.

Ionic compounds that are soluble in water typically contain at least one of the ions in **TABLE 9.7**. *Only an ionic compound containing a soluble cation or anion will dissolve in water.* Most ionic compounds containing Cl^- are soluble, but AgCl, $PbCl_2$,

TABLE 9.7 Solubility Rules for Ionic Compounds in Water

An ionic compound is soluble in water if it contains one of the following:	
Positive Ions:	Li^+, Na^+, K^+, Rb^+, Cs^+, NH_4^+
Negative Ions:	NO_3^-, $C_2H_3O_2^-$
	Cl^-, Br^-, I^- except when combined with Ag^+, Pb^{2+}, or Hg_2^{2+}
	SO_4^{2-} except when combined with Ba^{2+}, Pb^{2+}, Ca^{2+}, Sr^{2+}, or Hg_2^{2+}
Ionic compounds that do not contain at least one of these ions are usually insoluble.	

and Hg_2Cl_2 are insoluble. Similarly, most ionic compounds containing SO_4^{2-} are soluble, but a few are insoluble. Most other ionic compounds are insoluble (see **FIGURE 9.6**). In an insoluble ionic compound, the ionic bonds between its positive and negative ions are too strong for the polar water molecules to break. We can use the solubility rules to predict whether a solid ionic compound would be soluble or not. **TABLE 9.8** illustrates the use of these rules.

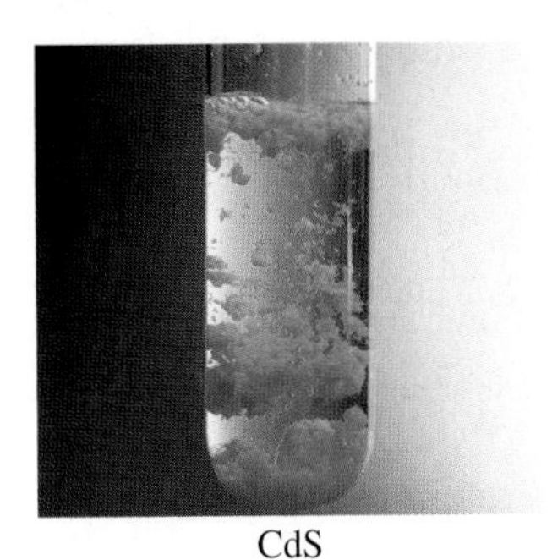
CdS

FeS

PbI_2

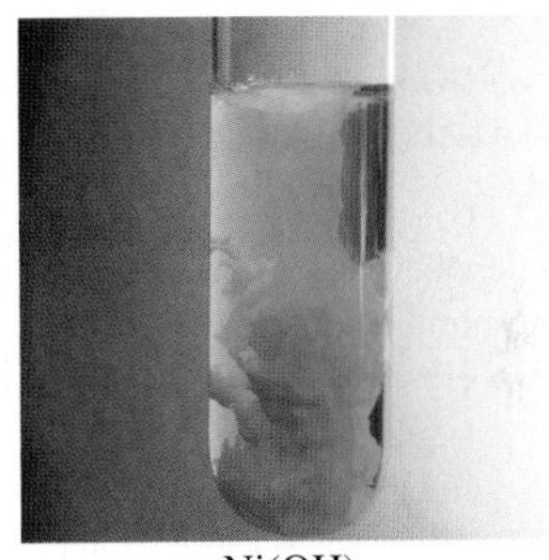
$Ni(OH)_2$

FIGURE 9.6 If an ionic compound contains a combination of a cation and an anion that are not soluble, that ionic compound is insoluble. For example, combinations of cadmium and sulfide, iron and sulfide, lead and iodide, and nickel and hydroxide do not contain any soluble ions. Thus, they form insoluble ionic compounds.

Q Why are each of these ionic compounds insoluble in water?

CORE CHEMISTRY SKILL
Using Solubility Rules

TABLE 9.8 Using Solubility Rules

Ionic Compound	Solubility in Water	Reasoning
K_2S	Soluble	Contains K^+
$Ca(NO_3)_2$	Soluble	Contains NO_3^-
$PbCl_2$	Insoluble	Is an insoluble chloride
NaOH	Soluble	Contains Na^+
$AlPO_4$	Insoluble	Contains no soluble ions

ENGAGE
Why is K_2S soluble in water whereas $PbCl_2$ is not soluble?

In medicine, insoluble $BaSO_4$ is used as an opaque substance to enhance X-rays of the gastrointestinal tract (see **FIGURE 9.7**). $BaSO_4$ is so insoluble that it does not dissolve in gastric fluids. Other ionic barium compounds cannot be used because they would dissolve in water, releasing Ba^{2+}, which is poisonous.

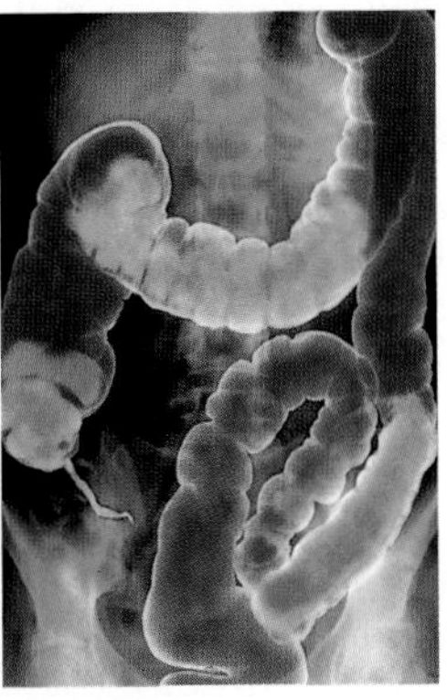
FIGURE 9.7 A barium sulfate-enhanced X-ray of the abdomen shows the lower gastrointestinal (GI) tract.

Q Is $BaSO_4$ a soluble or an insoluble substance?

SAMPLE PROBLEM 9.4 Soluble and Insoluble Ionic Compounds

TRY IT FIRST

Predict whether each of the following ionic compounds is soluble in water and explain why:

a. Na_3PO_4 **b.** $CaCO_3$

SOLUTION

a. The ionic compound Na_3PO_4 is soluble in water because any compound that contains Na^+ is soluble.

b. The ionic compound $CaCO_3$ is not soluble. The compound does not contain a soluble positive ion, which means that ionic compound containing Ca^{2+} and CO_3^{2-} is not soluble.

STUDY CHECK 9.4

In some electrolyte drinks, $MgCl_2$ is added to provide magnesium. Why would you expect $MgCl_2$ to be soluble in water?

ANSWER

$MgCl_2$ is soluble in water because ionic compounds that contain chloride are soluble unless they contain Ag^+, Pb^{2+}, or Hg_2^{2+}.

TEST

Try Practice Problems 9.31 and 9.32

PRACTICE PROBLEMS

9.3 Solubility

LEARNING GOAL Define solubility; distinguish between an unsaturated and a saturated solution. Identify an ionic compound as soluble or insoluble.

9.23 State whether each of the following refers to a saturated or an unsaturated solution:
a. A crystal added to a solution does not change in size.
b. A sugar cube completely dissolves when added to a cup of coffee.
c. A uric acid concentration of 4.6 mg/100 mL in the kidney does not cause gout.

9.24 State whether each of the following refers to a saturated or an unsaturated solution:
a. A spoonful of salt added to boiling water dissolves.
b. A layer of sugar forms on the bottom of a glass of tea as ice is added.
c. A kidney stone of calcium phosphate forms in the kidneys when urine becomes concentrated.

Use the following table for problems 9.25 to 9.28:

	Solubility (g/100. g H_2O)	
Substance	20 °C	50 °C
KCl	34	43
$NaNO_3$	88	110
$C_{12}H_{22}O_{11}$ (sugar)	204	260

9.25 Determine whether each of the following solutions will be saturated or unsaturated at 20 °C:
a. adding 25 g of KCl to 100. g of H_2O
b. adding 11 g of $NaNO_3$ to 25 g of H_2O
c. adding 400. g of sugar to 125 g of H_2O

9.26 Determine whether each of the following solutions will be saturated or unsaturated at 50 °C:
a. adding 25 g of KCl to 50. g of H_2O
b. adding 150. g of $NaNO_3$ to 75 g of H_2O
c. adding 80. g of sugar to 25 g of H_2O

9.27 A solution containing 80. g of KCl in 200. g of H_2O at 50 °C is cooled to 20 °C.
a. How many grams of KCl remain in solution at 20 °C?
b. How many grams of solid KCl crystallized after cooling?

9.28 A solution containing 80. g of $NaNO_3$ in 75 g of H_2O at 50 °C is cooled to 20 °C.
a. How many grams of $NaNO_3$ remain in solution at 20 °C?
b. How many grams of solid $NaNO_3$ crystallized after cooling?

9.29 Explain the following observations:
a. More sugar dissolves in hot tea than in iced tea.
b. Champagne in a warm room goes flat.
c. A warm can of soda has more spray when opened than a cold one.

9.30 Explain the following observations:
a. An open can of soda loses its "fizz" faster at room temperature than in the refrigerator.
b. Chlorine gas in tap water escapes as the sample warms to room temperature.
c. Less sugar dissolves in iced coffee than in hot coffee.

9.31 Predict whether each of the following ionic compounds is soluble in water:
a. LiCl **b.** PbS **c.** $BaCO_3$ **d.** K_2O **e.** $Fe(NO_3)_3$

9.32 Predict whether each of the following ionic compounds is soluble in water:
a. AgCl **b.** KI **c.** Na_2S **d.** Ag_2O **e.** $CaSO_4$

REVIEW

Writing Conversion Factors from Equalities (2.5)
Using Conversion Factors (2.6)
Using Molar Mass as a Conversion Factor (7.3)

9.4 Solution Concentrations

LEARNING GOAL Calculate the concentration of a solute in a solution; use concentration as a conversion factor to calculate the amount of solute or solution.

Our body fluids contain water and dissolved substances including glucose, urea, and electrolytes such as K^+, Na^+, Cl^-, Mg^{2+}, HCO_3^-, and HPO_4^{2-}. Proper amounts of each of these dissolved substances and water must be maintained in the body fluids. Small changes in electrolyte levels can seriously disrupt cellular processes and endanger our health. Solutions can be described by their **concentration**, which is the amount of solute in a specific amount of that solution as shown in TABLE 9.9. The amount of a solute may be expressed in units of grams, milliliters, or moles. The amount of a solution may be expressed in units of grams, milliliters, or liters.

CORE CHEMISTRY SKILL

Calculating Concentration

$$\text{Concentration of a solution} = \frac{\text{amount of solute}}{\text{amount of solution}}$$

TABLE 9.9 Summary of Types of Concentration Expressions and Their Units

Concentration Units	Mass Percent (m/m)	Volume Percent (v/v)	Mass/Volume Percent (m/v)	Molarity (M)
Solute	g	mL	g	mole
Solution	g	mL	mL	L

Mass Percent (m/m) Concentration

Mass percent (m/m) describes the mass of the solute in grams for 100. g of solution. The mass percent is calculated by dividing the mass of a solute by the mass of the solution multiplied by 100% to give the percentage. In the calculation of mass percent (m/m), the units of mass of the solute and solution must be the same. If the mass of the solute is given as grams, then the mass of the solution must also be grams. The mass of the solution is the sum of the mass of the solute and the mass of the solvent.

$$\text{Mass percent (m/m)} = \frac{\text{mass of solute (g)}}{\text{mass of solute (g)} + \text{mass of solvent (g)}} \times 100\%$$

$$= \frac{\text{mass of solute (g)}}{\text{mass of solution (g)}} \times 100\%$$

Add 8.00 g of KCl

Suppose we prepared a solution by mixing 8.00 g of KCl (solute) with 42.00 g of water (solvent). Together, the mass of the solute and mass of solvent give the mass of the solution (8.00 g + 42.00 g = 50.00 g). Mass percent is calculated by substituting the mass of the solute and the mass of the solution into the mass percent expression.

$$\frac{8.00 \text{ g KCl}}{50.00 \text{ g solution}} \times 100\% = 16.0\% \text{ (m/m) KCl solution}$$

$8.00 \text{ g KCl} + 42.00 \text{ g } H_2O$

Solute + Solvent

Add water until the solution has a mass of 50.00 g

When water is added to 8.00 g of KCl to form 50.00 g of a KCl solution, the mass percent concentration is 16.0% (m/m).

SAMPLE PROBLEM 9.5 Calculating Mass Percent (m/m) Concentration

TRY IT FIRST

What is the mass percent of NaOH in a solution prepared by dissolving 30.0 g of NaOH in 120.0 g of H_2O?

SOLUTION GUIDE

STEP 1 State the given and needed quantities.

	Given	Need	Connect
ANALYZE THE PROBLEM	30.0 g of NaOH, 120.0 g of H_2O	mass percent (m/m)	$\frac{\text{mass of solute}}{\text{mass of solution}} \times 100\%$

STEP 2 Write the concentration expression.

$$\text{Mass percent (m/m)} = \frac{\text{grams of solute}}{\text{grams of solution}} \times 100\%$$

STEP 3 **Substitute solute and solution quantities into the expression and calculate.** The mass of the solution is obtained by adding the mass of the solute and the mass of the solution.

$$\text{mass of solution} = 30.0 \text{ g NaOH} + 120.0 \text{ g } H_2O = 150.0 \text{ g of NaOH solution}$$

$$\text{Mass percent (m/m)} = \frac{\overset{\text{Three SFs}}{30.0 \text{ g NaOH}}}{\underset{\text{Four SFs}}{150.0 \text{ g solution}}} \times 100\%$$

$$= \underset{\text{Three SFs}}{20.0\%} \text{ (m/m) NaOH solution}$$

STUDY CHECK 9.5

What is the mass percent (m/m) of NaCl in a solution made by dissolving 2.0 g of NaCl in 56.0 g of H_2O?

ANSWER

3.4% (m/m) NaCl solution

TEST

Try Practice Problems 9.33 and 9.34

CORE CHEMISTRY SKILL

Using Concentration as a Conversion Factor

Using Mass Percent Concentration as a Conversion Factor

In the preparation of solutions, we often need to calculate the amount of solute or solution. Then the concentration of a solution is useful as a conversion factor as shown in Sample Problem 9.6.

SAMPLE PROBLEM 9.6 Using Mass Percent to Calculate Mass of Solute

TRY IT FIRST

The topical antibiotic ointment Neosporin is 3.5% (m/m) neomycin solution. How many grams of neomycin are in a tube containing 64 g of ointment?

SOLUTION GUIDE

STEP 1 **State the given and needed quantities.**

ANALYZE THE PROBLEM	Given	Need	Connect
	64 g of 3.5% (m/m) neomycin solution	grams of neomycin	mass percent factor $\frac{\text{g of solute}}{\text{100. g of solution}}$

STEP 2 **Write a plan to calculate the mass.**

grams of ointment → % (m/m) factor → grams of neomycin

STEP 3 **Write equalities and conversion factors.** The mass percent (m/m) indicates the grams of a solute in every 100. g of a solution. The mass percent (3.5% m/m) can be written as two conversion factors.

$$3.5 \text{ g of neomycin} = 100. \text{ g of ointment}$$

$$\frac{3.5 \text{ g neomycin}}{100. \text{ g ointment}} \quad \text{and} \quad \frac{100. \text{ g ointment}}{3.5 \text{ g neomycin}}$$

STEP 4 **Set up the problem to calculate the mass.**

$$\underset{\text{Two SFs}}{64 \text{ g ointment}} \times \underset{\text{Exact}}{\frac{\overset{\text{Two SFs}}{3.5 \text{ g neomycin}}}{100. \text{ g ointment}}} = \underset{\text{Two SFs}}{2.2 \text{ g of neomycin}}$$

STUDY CHECK 9.6

Calculate the grams of KCl in 225 g of an 8.00% (m/m) KCl solution.

ANSWER

18.0 g of KCl

ENGAGE

How is the mass percent (m/m) of a solution used to convert the mass of the solution to the grams of solute?

Volume Percent (v/v) Concentration

Because the volumes of liquids or gases are easily measured, the concentrations of their solutions are often expressed as **volume percent (v/v)**. The units of volume used in the ratio must be the same, for example, both in milliliters or both in liters.

$$\text{Volume percent (v/v)} = \frac{\text{volume of solute}}{\text{volume of solution}} \times 100\%$$

We interpret a volume percent as the volume of solute in 100. mL of solution. On a bottle of extract of vanilla, a label that reads alcohol 35% (v/v) means 35 mL of ethanol solute in 100. mL of vanilla solution.

The label indicates that vanilla extract contains 35% (v/v) alcohol.

SAMPLE PROBLEM 9.7 Calculating Volume Percent (v/v) Concentration

TRY IT FIRST

A bottle contains 59 mL of lemon extract solution. If the extract contains 49 mL of alcohol, what is the volume percent (v/v) of the alcohol in the solution?

SOLUTION GUIDE

STEP 1 State the given and needed quantities.

	Given	Need	Connect
ANALYZE THE PROBLEM	49 mL of alcohol, 59 mL of solution	volume percent (v/v)	$\frac{\text{volume of solute}}{\text{volume of solution}} \times 100\%$

STEP 2 Write the concentration expression.

$$\text{Volume percent (v/v)} = \frac{\text{volume of solute}}{\text{volume of solution}} \times 100\%$$

STEP 3 Substitute solute and solution quantities into the expression and calculate.

$$\text{Volume percent (v/v)} = \frac{\overset{\text{Two SFs}}{49 \text{ mL alcohol}}}{\underset{\text{Two SFs}}{59 \text{ mL solution}}} \times 100\% = \underset{\text{Two SFs}}{83\%} \text{ (v/v) alcohol solution}$$

Lemon extract is a solution of lemon flavor and alcohol.

STUDY CHECK 9.7

What is the volume percent (v/v) of Br_2 in a solution prepared by dissolving 12 mL of liquid bromine (Br_2) in the solvent carbon tetrachloride (CCl_4) to make 250 mL of solution?

ANSWER

4.8% (v/v) Br_2 in CCl_4

TEST

Try Practice Problems 9.35 and 9.36

Mass/Volume Percent (m/v) Concentration

Mass/volume percent (m/v) describes the mass of the solute in grams for exactly 100. mL of solution. In the calculation of mass/volume percent, the unit of mass of the solute is grams and the unit of the solution volume is milliliters.

$$\text{Mass/volume percent (m/v)} = \frac{\text{grams of solute}}{\text{milliliters of solution}} \times 100\%$$

The mass/volume percent is widely used in hospitals and pharmacies for the preparation of intravenous solutions and medicines. For example, a 5% (m/v) glucose solution contains 5 g of glucose in 100. mL of solution. The volume of solution represents the combined volumes of the glucose and H_2O.

SAMPLE PROBLEM 9.8 Calculating Mass/Volume Percent (m/v) Concentration

TRY IT FIRST

A potassium iodide solution may be used in a diet that is low in iodine. A KI solution is prepared by dissolving 5.0 g of KI in enough water to give a final volume of 250 mL. What is the mass/volume percent (m/v) of the KI solution?

SOLUTION GUIDE

STEP 1 State the given and needed quantities.

	Given	Need	Connect
ANALYZE THE PROBLEM	5.0 g of KI solute, 250 mL of KI solution	mass/volume percent (m/v)	$\frac{\text{mass of solute}}{\text{volume of solution}} \times 100\%$

STEP 2 Write the concentration expression.

$$\text{Mass/volume percent (m/v)} = \frac{\text{mass of solute}}{\text{volume of solution}} \times 100\%$$

STEP 3 Substitute solute and solution quantities into the expression and calculate.

$$\text{Mass/volume percent (m/v)} = \frac{\overset{\text{Two SFs}}{5.0\text{ g KI}}}{\underset{\text{Two SFs}}{250\text{ mL solution}}} \times 100\% = \underset{\text{Two SFs}}{2.0\%} \text{ (m/v) KI solution}$$

STUDY CHECK 9.8

What is the mass/volume percent (m/v) of NaOH in a solution prepared by dissolving 12 g of NaOH in enough water to make 220 mL of solution?

ANSWER

5.5% (m/v) NaOH solution

TEST

Try Practice Problems 9.37 and 9.38

SAMPLE PROBLEM 9.9 Using Mass/Volume Percent to Calculate Mass of Solute

TRY IT FIRST

A topical antibiotic is 1.0% (m/v) clindamycin. How many grams of clindamycin are in 60. mL of the 1.0% (m/v) solution?

SOLUTION GUIDE

STEP 1 State the given and needed quantities.

	Given	Need	Connect
ANALYZE THE PROBLEM	60. mL of 1.0% (m/v) clindamycin solution	grams of clindamycin	% (m/v) factor

STEP 2 Write a plan to calculate the mass.

milliliters of solution → % (m/v) factor → grams of clindamycin

STEP 3 Write equalities and conversion factors. The percent (m/v) indicates the grams of a solute in every 100. mL of a solution. The 1.0% (m/v) can be written as two conversion factors.

1.0 g of clindamycin = 100. mL of solution

$$\frac{1.0 \text{ g clindamycin}}{100. \text{ mL solution}} \quad \text{and} \quad \frac{100. \text{ mL solution}}{1.0 \text{ g clindamycin}}$$

STEP 4 Set up the problem to calculate the mass. The volume of the solution is converted to mass of solute using the conversion factor that cancels mL.

$$\underset{\text{Two SFs}}{60. \cancel{\text{mL solution}}} \times \overset{\text{Two SFs}}{\underset{\text{Exact}}{\frac{1.0 \text{ g clindamycin}}{100. \cancel{\text{mL solution}}}}} = \underset{\text{Two SFs}}{0.60 \text{ g of clindamycin}}$$

STUDY CHECK 9.9

In 2010, the FDA approved a 2.0% (m/v) morphine oral solution to treat severe or chronic pain. How many grams of morphine does a patient receive if 0.60 mL of 2.0% (m/v) morphine solution was ordered?

ANSWER

0.012 g of morphine

TEST

Try Practice Problems 9.39 to 9.44, 9.51 to 9.54

Molarity (M) Concentration

When chemists work with solutions, they often use **molarity (M)**, a concentration that states the number of moles of solute in exactly 1 L of solution.

$$\text{Molarity (M)} = \frac{\text{moles of solute}}{\text{liters of solution}}$$

The molarity of a solution can be calculated when we know the moles of solute and the volume of solution in liters. For example, if 1.0 mole of NaCl were dissolved in enough water to prepare 1.0 L of solution, the resulting NaCl solution has a molarity of 1.0 M. The abbreviation M indicates the units of mole per liter (mole/L).

$$\text{M} = \frac{\text{moles of solute}}{\text{liters of solution}} = \frac{1.0 \text{ mole NaCl}}{1 \text{ L solution}} = 1.0 \text{ M NaCl solution}$$

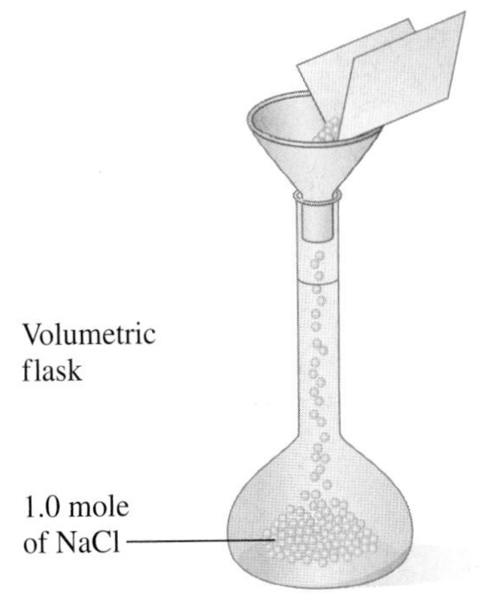

A 1.0 molar (M) NaCl solution

SAMPLE PROBLEM 9.10 Calculating Molarity

TRY IT FIRST

What is the molarity (M) of 60.0 g of NaOH in 0.250 L of NaOH solution?

SOLUTION GUIDE

STEP 1 State the given and needed quantities.

	Given	Need	Connect
ANALYZE THE PROBLEM	60.0 g of NaOH, 0.250 L of NaOH solution	molarity (mole/L)	molar mass of NaOH, $\frac{\text{moles of solute}}{\text{liters of solution}}$

To calculate the moles of NaOH, we need to write the equality and conversion factors for the molar mass of NaOH. Then the moles in 60.0 g of NaOH can be determined.

$$1 \text{ mole of NaOH} = 40.00 \text{ g of NaOH}$$
$$\frac{40.00 \text{ g NaOH}}{1 \text{ mole NaOH}} \text{ and } \frac{1 \text{ mole NaOH}}{40.00 \text{ g NaOH}}$$

$$\text{moles of NaOH} = 60.0 \cancel{\text{g NaOH}} \times \frac{1 \text{ mole NaOH}}{40.00 \cancel{\text{g NaOH}}}$$
$$= 1.50 \text{ moles of NaOH}$$
$$\text{volume of solution} = 0.250 \text{ L of NaOH solution}$$

STEP 2 Write the concentration expression.

$$\text{Molarity (M)} = \frac{\text{moles of solute}}{\text{liters of solution}}$$

STEP 3 Substitute solute and solution quantities into the expression and calculate.

$$\text{M} = \frac{\overset{\text{Three SFs}}{1.50 \text{ moles NaOH}}}{\underset{\text{Three SFs}}{0.250 \text{ L solution}}} = \frac{\overset{\text{Three SFs}}{6.00 \text{ moles NaOH}}}{\underset{\text{Exact}}{1 \text{ L solution}}} = \underset{\text{Three SFs}}{6.00 \text{ M NaOH solution}}$$

STUDY CHECK 9.10

What is the molarity of a solution that contains 75.0 g of KNO_3 dissolved in 0.350 L of solution?

ANSWER

2.12 M KNO_3 solution

TEST

Try Practice Problems 9.45 and 9.46

SAMPLE PROBLEM 9.11 Using Molarity to Calculate Volume of Solution

TRY IT FIRST

How many liters of a 2.00 M NaCl solution are needed to provide 67.3 g of NaCl?

SOLUTION GUIDE

STEP 1 State the given and needed quantities.

	Given	Need	Connect
ANALYZE THE PROBLEM	67.3 g of NaCl, 2.00 M NaCl solution	liters of NaCl solution	molar mass of NaCl, molarity

STEP 2 Write a plan to calculate the volume.

grams of NaCl → Molar mass → moles of NaCl → Molarity → liters of NaCl solution

STEP 3 Write equalities and conversion factors.

$$1 \text{ mole of NaCl} = 58.44 \text{ g of NaCl}$$
$$\frac{58.44 \text{ g NaCl}}{1 \text{ mole NaCl}} \text{ and } \frac{1 \text{ mole NaCl}}{58.44 \text{ g NaCl}}$$

$$1 \text{ L of NaCl solution} = 2.00 \text{ moles of NaCl}$$
$$\frac{2.00 \text{ moles NaCl}}{1 \text{ L NaCl solution}} \text{ and } \frac{1 \text{ L NaCl solution}}{2.00 \text{ moles NaCl}}$$

STEP 4 Set up the problem to calculate the volume.

$$\underset{\text{Three SFs}}{67.3 \cancel{\text{g NaCl}}} \times \underset{\text{Four SFs}}{\frac{\overset{\text{Exact}}{1 \cancel{\text{mole NaCl}}}}{58.44 \cancel{\text{g NaCl}}}} \times \underset{\text{Three SFs}}{\frac{\overset{\text{Exact}}{1 \text{ L NaCl solution}}}{2.00 \cancel{\text{moles NaCl}}}} = \underset{\text{Three SFs}}{0.576 \text{ L of NaCl solution}}$$

STUDY CHECK 9.11

How many milliliters of a 6.0 M HCl solution will provide 164 g of HCl?

ANSWER

750 mL of HCl solution

TEST

Try Practice Problems 9.47 to 9.50

A summary of percent concentrations and molarity, their meanings, and conversion factors are given in **TABLE 9.10**.

TABLE 9.10 Conversion Factors from Concentrations

Percent Concentration	Meaning	Conversion Factors
10% (m/m) KCl solution	10 g of KCl in 100. g of KCl solution	$\frac{10\text{ g KCl}}{100.\text{ g solution}}$ and $\frac{100.\text{ g solution}}{10\text{ g KCl}}$
12% (v/v) ethanol solution	12 mL of ethanol in 100. mL of ethanol solution	$\frac{12\text{ mL ethanol}}{100.\text{ mL solution}}$ and $\frac{100.\text{ mL solution}}{12\text{ mL ethanol}}$
5% (m/v) glucose solution	5 g of glucose in 100. mL of glucose solution	$\frac{5\text{ g glucose}}{100.\text{ mL solution}}$ and $\frac{100.\text{ mL solution}}{5\text{ g glucose}}$
Molarity	**Meaning**	**Conversion Factors**
6.0 M HCl solution	6.0 moles of HCl in 1 L of HCl solution	$\frac{6.0\text{ moles HCl}}{1\text{ L solution}}$ and $\frac{1\text{ L solution}}{6.0\text{ moles HCl}}$

INTERACTIVE VIDEO

Solutions

PRACTICE PROBLEMS

9.4 Solution Concentrations

LEARNING GOAL Calculate the concentration of a solute in a solution; use concentration as a conversion factor to calculate the amount of solute or solution.

9.33 Calculate the mass percent (m/m) for the solute in each of the following:
a. 25 g of KCl and 125 g of H_2O
b. 12 g of sucrose in 225 g of tea solution
c. 8.0 g of $CaCl_2$ in 80.0 g of $CaCl_2$ solution

9.34 Calculate the mass percent (m/m) for the solute in each of the following:
a. 75 g of NaOH in 325 g of NaOH solution
b. 2.0 g of KOH and 20.0 g of H_2O
c. 48.5 g of Na_2CO_3 in 250.0 g of Na_2CO_3 solution

9.35 A mouthwash contains 22.5% (v/v) alcohol. If the bottle of mouthwash contains 355 mL, what is the volume, in milliliters, of alcohol?

9.36 A bottle of champagne is 11% (v/v) alcohol. If there are 750 mL of champagne in the bottle, what is the volume, in milliliters, of alcohol?

9.37 What is the difference between a 5.00% (m/m) glucose solution and a 5.00% (m/v) glucose solution?

9.38 What is the difference between a 10.0% (v/v) methanol (CH_4O) solution and a 10.0% (m/m) methanol solution?

9.39 Calculate the mass/volume percent (m/v) for the solute in each of the following:
a. 75 g of Na_2SO_4 in 250 mL of Na_2SO_4 solution
b. 39 g of sucrose in 355 mL of a carbonated drink

9.40 Calculate the mass/volume percent (m/v) for the solute in each of the following:
a. 2.50 g of LiCl in 40.0 mL of LiCl solution
b. 7.5 g of casein in 120 mL of low-fat milk

9.41 Calculate the grams or milliliters of solute needed to prepare the following:
a. 50. g of a 5.0% (m/m) KCl solution
b. 1250 mL of a 4.0% (m/v) NH_4Cl solution
c. 250. mL of a 10.0% (v/v) acetic acid solution

9.42 Calculate the grams or milliliters of solute needed to prepare the following:
a. 150. g of a 40.0% (m/m) $LiNO_3$ solution
b. 450 mL of a 2.0% (m/v) KOH solution
c. 225 mL of a 15% (v/v) isopropyl alcohol solution

9.43 For each of the following solutions, calculate the:
a. grams of 25% (m/m) $LiNO_3$ solution that contains 5.0 g of $LiNO_3$
b. milliliters of 10.0% (m/v) KOH solution that contains 40.0 g of KOH
c. milliliters of 10.0% (v/v) formic acid solution that contains 2.0 mL of formic acid

9.44 For each of the following solutions, calculate the:
a. grams of 2.0% (m/m) NaCl solution that contains 7.50 g of NaCl
b. milliliters of 25% (m/v) NaF solution that contains 4.0 g of NaF
c. milliliters of 8.0% (v/v) ethanol solution that contains 20.0 mL of ethanol

9.45 Calculate the molarity of each of the following:
a. 2.00 moles of glucose in 4.00 L of a glucose solution
b. 4.00 g of KOH in 2.00 L of a KOH solution
c. 5.85 g of NaCl in 400. mL of a NaCl solution

9.46 Calculate the molarity of each of the following:
a. 0.500 mole of glucose in 0.200 L of a glucose solution
b. 73.0 g of HCl in 2.00 L of a HCl solution
c. 30.0 g of NaOH in 350. mL of a NaOH solution

9.47 Calculate the grams of solute needed to prepare each of the following:
  a. 2.00 L of a 1.50 M NaOH solution
  b. 4.00 L of a 0.200 M KCl solution
  c. 25.0 mL of a 6.00 M HCl solution

9.48 Calculate the grams of solute needed to prepare each of the following:
  a. 2.00 L of a 6.00 M NaOH solution
  b. 5.00 L of a 0.100 M $CaCl_2$ solution
  c. 175 mL of a 3.00 M $NaNO_3$ solution

9.49 For each of the following solutions, calculate the:
  a. liters of a 2.00 M KBr solution to obtain 3.00 moles of KBr
  b. liters of a 1.50 M NaCl solution to obtain 15.0 moles of NaCl
  c. milliliters of a 0.800 M $Ca(NO_3)_2$ solution to obtain 0.0500 mole of $Ca(NO_3)_2$

9.50 For each of the following solutions, calculate the:
  a. liters of a 4.00 M KCl solution to obtain 0.100 mole of KCl
  b. liters of a 6.00 M HCl solution to obtain 5.00 moles of HCl
  c. milliliters of a 2.50 M K_2SO_4 solution to obtain 1.20 moles of K_2SO_4

Clinical Applications

9.51 A patient receives 100. mL of 20.% (m/v) mannitol solution every hour.
  a. How many grams of mannitol are given in 1 h?
  b. How many grams of mannitol does the patient receive in 12 h?

9.52 A patient receives 250 mL of a 4.0% (m/v) amino acid solution twice a day.
  a. How many grams of amino acids are in 250 mL of solution?
  b. How many grams of amino acids does the patient receive in 1 day?

9.53 A patient needs 100. g of glucose in the next 12 h. How many liters of a 5% (m/v) glucose solution must be given?

9.54 A patient received 2.0 g of NaCl in 8 h. How many milliliters of a 0.90% (m/v) NaCl (saline) solution were delivered?

REVIEW
Solving Equations (1.4)

9.5 Dilution of Solutions

LEARNING GOAL Describe the dilution of a solution; calculate the unknown concentration or volume when a solution is diluted.

In chemistry and biology, we often prepare diluted solutions from more concentrated solutions. In a process called **dilution**, a solvent, usually water, is added to a solution, which increases the volume. As a result, the concentration of the solution decreases. In an everyday example, you are making a dilution when you add three cans of water to a can of concentrated orange juice.

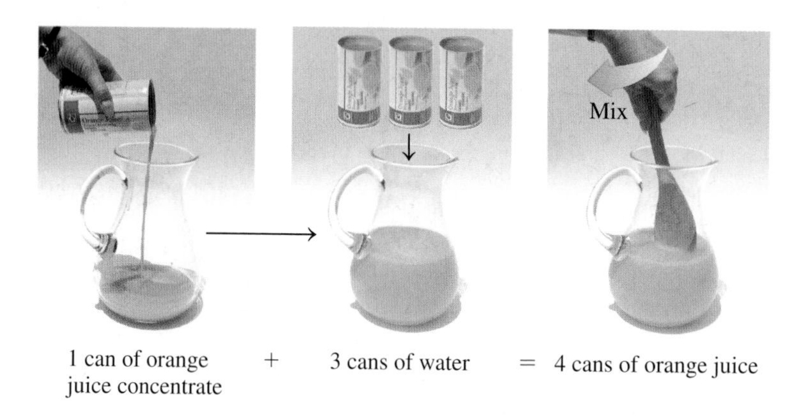

Although the addition of solvent increases the volume, the amount of solute does not change; it is the same in the concentrated solution and the diluted solution (see **FIGURE 9.8**).

$$\underset{\text{Concentrated solution}}{\text{Grams or moles of solute}} = \underset{\text{Diluted solution}}{\text{grams or moles of solute}}$$

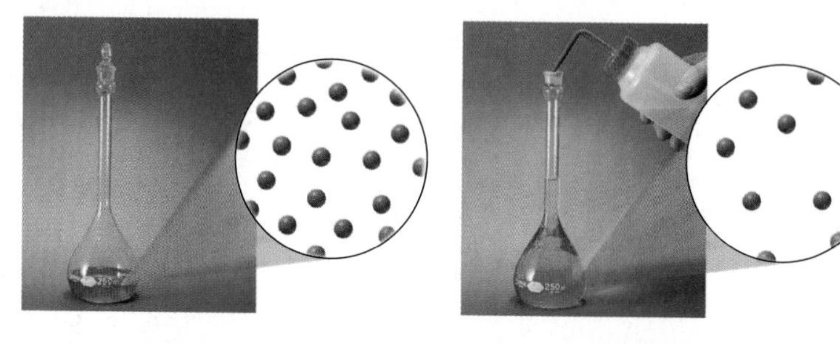

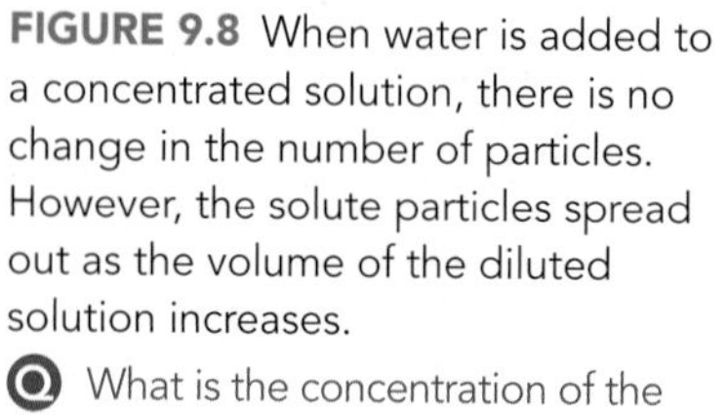

FIGURE 9.8 When water is added to a concentrated solution, there is no change in the number of particles. However, the solute particles spread out as the volume of the diluted solution increases.

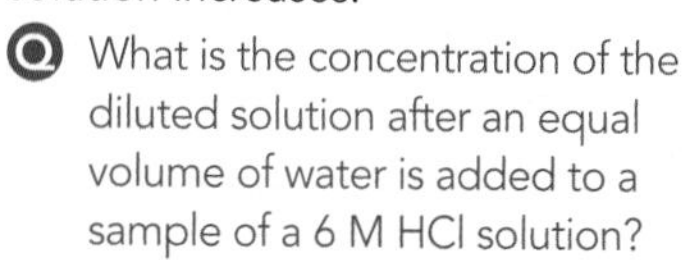

Q What is the concentration of the diluted solution after an equal volume of water is added to a sample of a 6 M HCl solution?

We can write this equality in terms of the concentration, C, and the volume, V. The concentration, C, may be percent concentration or molarity.

C_1V_1	=	C_2V_2
Concentrated solution		Diluted solution

If we are given any three of the four variables (C_1, C_2, V_1, or V_2), we can rearrange the dilution expression to solve for the unknown quantity as seen in Sample Problems 9.12 and 9.13.

SAMPLE PROBLEM 9.12 Molarity of a Diluted Solution

TRY IT FIRST

What is the molarity of a solution when 75.0 mL of a 4.00 M KCl solution is diluted to a volume of 500. mL?

SOLUTION GUIDE

STEP 1 Prepare a table of the concentrations and volumes of the solutions.

	Given		Need	Connect
ANALYZE THE PROBLEM	$C_1 = 4.00$ M $V_1 = 75.0$ mL	$V_2 = 500.$ mL	C_2	$C_1V_1 = C_2V_2$ V_2 increases, C_2 decreases

STEP 2 Rearrange the dilution expression to solve for the unknown quantity.

$$C_1V_1 = C_2V_2$$

$$\frac{C_1V_1}{V_2} = \frac{C_2\cancel{V_2}}{\cancel{V_2}} \quad \text{Divide both sides by } V_2$$

$$C_2 = C_1 \times \frac{V_1}{V_2}$$

STEP 3 Substitute the known quantities into the dilution expression and calculate.

$$C_2 = 4.00 \text{ M} \times \frac{75.0 \cancel{\text{mL}}}{500. \cancel{\text{mL}}} = 0.600 \text{ M (diluted KCl solution)}$$

Three SFs (4.00 M); Three SFs (75.0 mL); Three SFs (500. mL); Three SFs (0.600 M)
Volume factor decreases concentration

When the initial molarity (C_1) is multiplied by a ratio of the volumes (volume factor) that is less than 1, the molarity of the diluted solution (C_2) decreases as predicted in Step 1.

STUDY CHECK 9.12

What is the molarity of a solution when 50.0 mL of a 4.00 M KOH solution is diluted to 200. mL?

ANSWER

1.00 M KOH solution

TEST

Try Practice Problems 9.55 and 9.56

SAMPLE PROBLEM 9.13 Dilution of a Solution

TRY IT FIRST

A doctor orders 1000. mL of a 35.0% (m/v) dextrose solution. If you have a 50.0% (m/v) dextrose solution, how many milliliters would you use to prepare 1000. mL of 35.0% (m/v) dextrose solution?

SOLUTION GUIDE

STEP 1 Prepare a table of the concentrations and volumes of the solutions. For our problem analysis, we organize the solution data in a table, making sure that the units of concentration and volume are the same.

	Given		Need	Connect
ANALYZE THE PROBLEM	$C_1 = 50.0\%$ (m/v)	$C_2 = 35.0\%$ (m/v) $V_2 = 1000.$ mL	V_1	$C_1V_1 = C_2V_2$ C_1 increases, V_1 decreases

STEP 2 Rearrange the dilution expression to solve for the unknown quantity.

$$C_1 V_1 = C_2 V_2$$

$$\frac{\cancel{C_1} V_1}{\cancel{C_1}} = \frac{C_2 V_2}{C_1} \quad \text{Divide both sides by } C_1$$

$$V_1 = V_2 \times \frac{C_2}{C_1}$$

STEP 3 Substitute the known quantities into the dilution expression and calculate.

$$V_1 = \underset{\text{Four SFs}}{1000.\text{ mL}} \times \underset{\substack{\text{Three SFs} \\ \text{Concentration factor} \\ \text{decreases volume}}}{\overset{\text{Three SFs}}{\frac{35.0\%}{50.0\%}}} = \underset{\text{Three SFs}}{700.\text{ mL of dextrose solution}}$$

When the final volume (V_2) is multiplied by a ratio of the percent concentrations (concentration factor) that is less than 1, the initial volume (V_1) is less than the final volume (V_2) as predicted in Step 1.

STUDY CHECK 9.13

What initial volume of a 15% (m/v) mannose solution is needed to prepare 125 mL of a 3.0% (m/v) mannose solution?

ANSWER

25 mL of a 15% (m/v) mannose solution

TEST

Try Practice Problems 9.57 to 9.62

PRACTICE PROBLEMS

9.5 Dilution of Solutions

LEARNING GOAL Describe the dilution of a solution; calculate the unknown concentration or volume when a solution is diluted.

9.55 Calculate the final concentration of each of the following:

a. 2.0 L of a 6.0 M HCl solution is added to water so that the final volume is 6.0 L.

b. Water is added to 0.50 L of a 12 M NaOH solution to make 3.0 L of a diluted NaOH solution.

c. A 10.0-mL sample of a 25% (m/v) KOH solution is diluted with water so that the final volume is 100.0 mL.

d. A 50.0-mL sample of a 15% (m/v) H_2SO_4 solution is added to water to give a final volume of 250 mL.

9.56 Calculate the final concentration of each of the following:

a. 1.0 L of a 4.0 M HNO_3 solution is added to water so that the final volume is 8.0 L.

b. Water is added to 0.25 L of a 6.0 M NaF solution to make 2.0 L of a diluted NaF solution.

c. A 50.0-mL sample of an 8.0% (m/v) KBr solution is diluted with water so that the final volume is 200.0 mL.

d. A 5.0-mL sample of a 50.0% (m/v) acetic acid ($HC_2H_3O_2$) solution is added to water to give a final volume of 25 mL.

9.57 Determine the final volume, in milliliters, of each of the following:

a. a 1.5 M HCl solution prepared from 20.0 mL of a 6.0 M HCl solution

b. a 2.0% (m/v) LiCl solution prepared from 50.0 mL of a 10.0% (m/v) LiCl solution

c. a 0.500 M H_3PO_4 solution prepared from 50.0 mL of a 6.00 M H_3PO_4 solution

d. a 5.0% (m/v) glucose solution prepared from 75 mL of a 12% (m/v) glucose solution

9.58 Determine the final volume, in milliliters, of each of the following:

a. a 1.00% (m/v) H_2SO_4 solution prepared from 10.0 mL of a 20.0% H_2SO_4 solution

b. a 0.10 M HCl solution prepared from 25 mL of a 6.0 M HCl solution

c. a 1.0 M NaOH solution prepared from 50.0 mL of a 12 M NaOH solution

d. a 1.0% (m/v) $CaCl_2$ solution prepared from 18 mL of a 4.0% (m/v) $CaCl_2$ solution

9.59 Determine the initial volume, in milliliters, required to prepare each of the following:

a. 255 mL of a 0.200 M HNO_3 solution using a 4.00 M HNO_3 solution

b. 715 mL of a 0.100 M $MgCl_2$ solution using a 6.00 M $MgCl_2$ solution

c. 0.100 L of a 0.150 M KCl solution using an 8.00 M KCl solution

9.60 Determine the initial volume, in milliliters, required to prepare each of the following:

a. 20.0 mL of a 0.250 M KNO_3 solution using a 6.00 M KNO_3 solution

b. 25.0 mL of a 2.50 M H_2SO_4 solution using a 12.0 M H_2SO_4 solution

c. 0.500 L of a 1.50 M NH_4Cl solution using a 10.0 M NH_4Cl solution

Clinical Applications

9.61 You need 500. mL of a 5.0% (m/v) glucose solution. If you have a 25% (m/v) glucose solution on hand, how many milliliters do you need?

9.62 A doctor orders 100. mL of 2.0% (m/v) ibuprofen. If you have 8.0% (m/v) ibuprofen on hand, how many milliliters do you need?

9.6 Properties of Solutions

LEARNING GOAL Identify a mixture as a solution, a colloid, or a suspension. Describe how the number of particles in a solution affects the osmotic pressure.

The size and number of solute particles in different types of mixtures play an important role in determining the properties of that mixture.

Solutions

In the solutions discussed up to now, the solute was dissolved as small particles that are uniformly dispersed throughout the solvent to give a homogeneous solution. When you observe a solution, such as salt water, you cannot visually distinguish the solute from the solvent. The solution appears transparent, although it may have a color. The particles are so small that they go through filters and through *semipermeable membranes*. A semipermeable membrane allows solvent molecules such as water and very small solute particles to pass through, but does allow the passage of large solute molecules.

Colloids

The particles in a **colloid** are much larger than solute particles in a solution. Colloidal particles are large molecules, such as proteins, or groups of molecules or ions. Colloids are uniformly dispersed, like solutions, and cannot be separated by filters but can be separated by semipermeable membranes. **TABLE 9.11** lists several examples of colloids.

TABLE 9.11 Examples of Colloids

Colloid	Substance Dispersed	Dispersing Medium
Fog, clouds, hairsprays	Liquid	Gas
Dust, smoke	Solid	Gas
Shaving cream, whipped cream, soap suds	Gas	Liquid
Styrofoam, marshmallows	Gas	Solid
Mayonnaise, homogenized milk	Liquid	Liquid
Cheese, butter	Liquid	Solid
Blood plasma, paints (latex), gelatin	Solid	Liquid

Suspensions

Suspensions are heterogeneous, nonuniform mixtures that are very different from solutions or colloids. The particles of a suspension are so large that they can often be seen with the naked eye. They are trapped by filters and semipermeable membranes.

The weight of the suspended solute particles causes them to settle out soon after mixing. If you stir muddy water, it mixes but then quickly separates as the suspended particles settle to the bottom and leave clear liquid at the top. You can find suspensions among the medications in a hospital or in your medicine cabinet. These include Kaopectate, calamine lotion, antacid mixtures, and liquid penicillin. It is important to follow the instructions on the label that states "shake well before using" so that the particles form a suspension.

Water-treatment plants make use of the properties of suspensions to purify water. When chemicals such as aluminum sulfate or iron(III) sulfate are added to untreated water, they react with impurities to form large suspended particles called *floc*. In the water-treatment plant, a system of filters traps the suspended particles, but clean water passes through.

TABLE 9.12 compares the different types of mixtures and **FIGURE 9.9** illustrates some properties of solutions, colloids, and suspensions.

TEST

Try Practice Problems 9.63 and 9.64

TABLE 9.12 Comparison of Solutions, Colloids, and Suspensions

Type of Mixture	Type of Particle	Settling	Separation
Solution	Small particles such as atoms, ions, or small molecules	Particles do not settle	Particles cannot be separated by filters or semipermeable membranes
Colloid	Larger molecules or groups of molecules or ions	Particles do not settle	Particles can be separated by semipermeable membranes but not by filters
Suspension	Very large particles that may be visible	Particles settle rapidly	Particles can be separated by filters

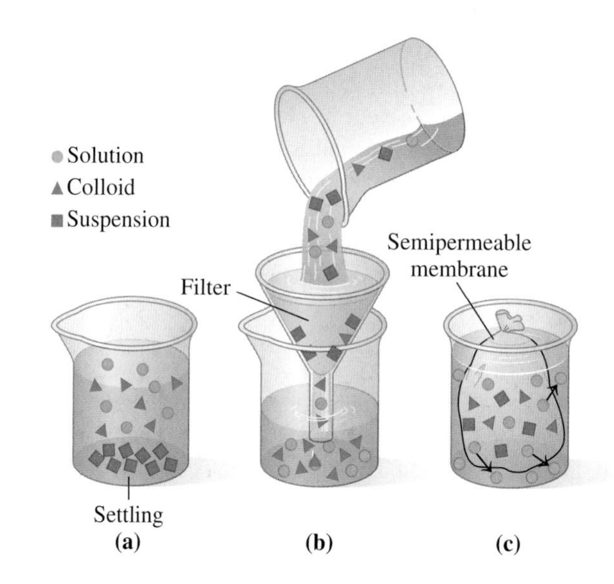

FIGURE 9.9 Properties of different types of mixtures: **(a)** suspensions settle out; **(b)** suspensions are separated by a filter; **(c)** solution particles go through a semipermeable membrane, but colloids and suspensions do not.

Q A filter can be used to separate suspension particles from a solution, but a semipermeable membrane is needed to separate colloids from a solution. Explain.

Osmotic Pressure

The movement of water into and out of the cells of plants as well as the cells of our bodies is an important biological process that depends on the solute concentration. In a process called **osmosis**, water molecules move through a semipermeable membrane from the solution with the lower concentration of solute into a solution with the higher solute concentration. In an osmosis apparatus, water is placed on one side of a semipermeable membrane and a sucrose (sugar) solution on the other side. The semipermeable membrane allows water molecules to flow back and forth but blocks the sucrose molecules because they cannot pass through the membrane. Because the sucrose solution has a higher solute concentration, more water molecules flow into the sucrose solution than out of the sucrose solution. The volume level of the sucrose solution rises as the volume level on the water side falls. The increase of water dilutes the sucrose solution to equalize (or attempt to equalize) the concentrations on both sides of the membrane.

Eventually the height of the sucrose solution creates sufficient pressure to equalize the flow of water between the two compartments. This pressure, called **osmotic pressure**, prevents the flow of additional water into the more concentrated solution. Then there is no further change in the volumes of the two solutions. The osmotic pressure depends on the concentration of solute particles in the solution. The greater the number of particles dissolved, the higher its osmotic pressure. In this example, the sucrose solution has a higher osmotic pressure than pure water, which has an osmotic pressure of zero.

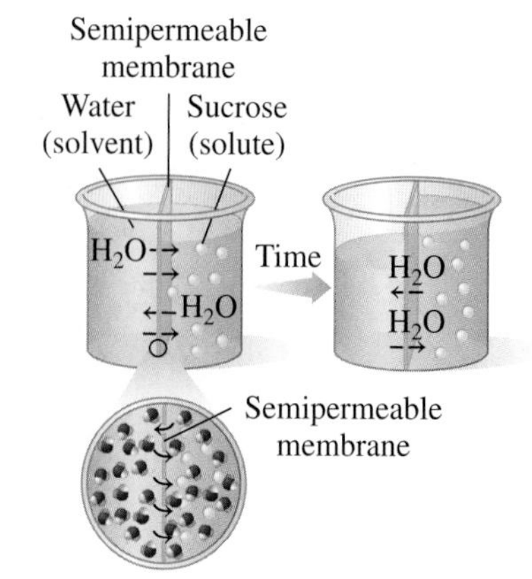

Water flows into the solution with a higher solute concentration until the flow of water becomes equal in both directions.

In a process called *reverse osmosis*, a pressure greater than the osmotic pressure is applied to a solution so that it is forced through a purification membrane. The flow of water is reversed because water flows from an area of lower water concentration to an area of higher water concentration. The molecules and ions in solution stay behind, trapped by the membrane, while water passes through the membrane. This process of reverse osmosis is used in desalination plants to obtain pure water from sea (salt) water. However, the pressure that must be applied requires so much energy that reverse osmosis is not yet an economical method for obtaining pure water in most parts of the world.

Isotonic Solutions

Because the cell membranes in biological systems are semipermeable, osmosis is an ongoing process. The solutes in body solutions such as blood, tissue fluids, lymph, and plasma all exert osmotic pressure. Most intravenous (IV) solutions used in a hospital are *isotonic solutions*, which exert the same osmotic pressure as body fluids such as blood. The percent concentration typically used in IV solutions is mass/volume percent (m/v), which is a type of percent concentration we have already discussed. The most typical isotonic solutions are 0.9% (m/v) NaCl solution, or 0.9 g of NaCl/100. mL of solution, and 5% (m/v) glucose, or 5 g of glucose/100. mL of solution. Although they do not contain the same kinds of particles, a 0.9% (m/v) NaCl solution as well as a 5% (m/v) glucose solution both have the same osmotic pressure. A red blood cell placed in an isotonic solution retains its volume because there is an equal flow of water into and out of the cell (see **FIGURE 9.10A**).

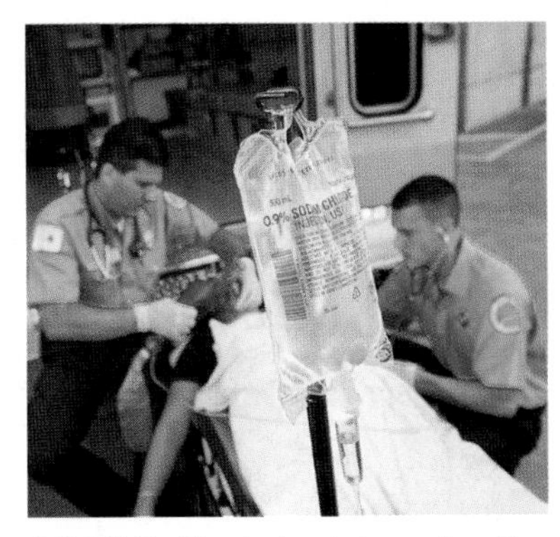

A 0.9% NaCl solution is isotonic with the solute concentration of the blood cells of the body.

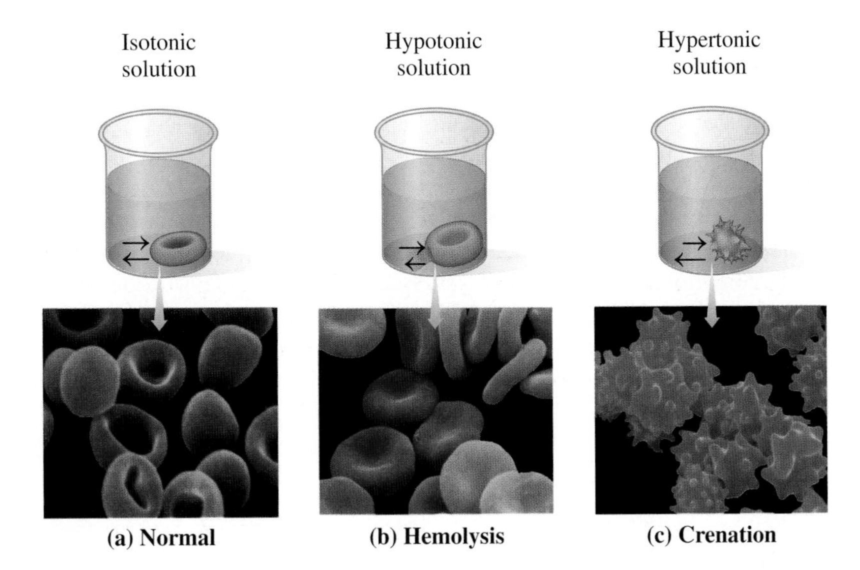

FIGURE 9.10 **(a)** In an isotonic solution, a red blood cell retains its normal volume. **(b)** Hemolysis: In a hypotonic solution, water flows into a red blood cell, causing it to swell and burst. **(c)** Crenation: In a hypertonic solution, water leaves the red blood cell, causing it to shrink.

Q What happens to a red blood cell placed in a 4% NaCl solution?

Hypotonic and Hypertonic Solutions

If a red blood cell is placed in a solution that is not isotonic, the differences in osmotic pressure inside and outside the cell can alter the volume of the cell. When a red blood cell is placed in a *hypotonic solution*, which has a lower solute concentration (*hypo* means "lower than"), water flows into the cell by osmosis. The increase in fluid causes the cell to swell, and possibly burst—a process called *hemolysis* (see Figure 9.10b). A similar process occurs when you place dehydrated food, such as raisins or dried fruit, in water. The water enters the cells, and the food becomes plump and smooth.

If a red blood cell is placed in a *hypertonic solution*, which has a higher solute concentration (*hyper* means "greater than"), water flows out of the cell into the hypertonic solution by osmosis. Suppose a red blood cell is placed in a 10% (m/v) NaCl solution. Because the osmotic pressure in the red blood cell is the same as a 0.9% (m/v) NaCl solution, the 10% (m/v) NaCl solution has a much greater osmotic pressure. As water leaves the cell, it shrinks, a process called *crenation* (see Figure 9.10c). A similar process occurs when making pickles, which uses a hypertonic salt solution that causes the cucumbers to shrivel as they lose water.

SAMPLE PROBLEM 9.14 Isotonic, Hypotonic, and Hypertonic Solutions

TRY IT FIRST

Describe each of the following solutions as isotonic, hypotonic, or hypertonic. Indicate whether a red blood cell placed in each solution will undergo hemolysis, crenation, or no change.

a. a 5% (m/v) glucose solution
b. a 0.2% (m/v) NaCl solution

SOLUTION

a. A 5% (m/v) glucose solution is isotonic. A red blood cell will not undergo any change.
b. A 0.2% (m/v) NaCl solution is hypotonic. A red blood cell will undergo hemolysis.

TEST

Try Practice Problems 9.65 to 9.72

STUDY CHECK 9.14

What will happen to a red blood cell placed in a 10% (m/v) glucose solution?

ANSWER

The red blood cell will shrink (crenate).

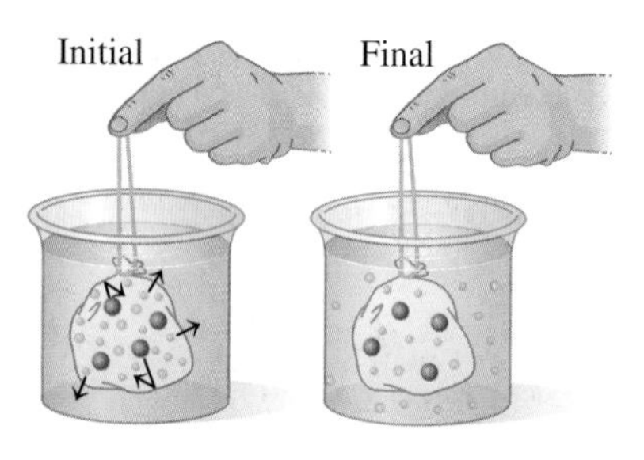

Solution particles such as Na^+, Cl^-, glucose

Colloidal particles such as protein, starch

Solution particles pass through a dialyzing membrane, but colloidal particles are retained.

Dialysis

Dialysis is a process that is similar to osmosis. In dialysis, a semipermeable membrane, called a dialyzing membrane, permits small solute molecules and ions as well as solvent water molecules to pass through, but it retains large particles, such as colloids. Dialysis is a way to separate solution particles from colloids.

Suppose we fill a cellophane bag with a solution containing NaCl, glucose, starch, and protein and place it in pure water. Cellophane is a dialyzing membrane, and the sodium ions, chloride ions, and glucose molecules will pass through it into the surrounding water. However, large colloidal particles, like starch and protein, remain inside. Water molecules will flow into the cellophane bag. Eventually the concentrations of sodium ions, chloride ions, and glucose molecules inside and outside the dialysis bag become equal. To remove more NaCl or glucose, the cellophane bag must be placed in a fresh sample of pure water.

CHEMISTRY LINK TO HEALTH

Dialysis by the Kidneys and the Artificial Kidney

The fluids of the body undergo dialysis by the membranes of the kidneys, which remove waste materials, excess salts, and water. In an adult, each kidney contains about 2 million nephrons. At the top of each nephron, there is a network of arterial capillaries called the *glomerulus*.

As blood flows into the glomerulus, small particles, such as amino acids, glucose, urea, water, and certain ions, will move through the capillary membranes into the nephron. As this solution moves through the nephron, substances still of value to the body (such as amino acids, glucose, certain ions, and 99% of the water) are reabsorbed. The major waste product, urea, is excreted in the urine.

Hemodialysis

If the kidneys fail to dialyze waste products, increased levels of urea can become life-threatening in a relatively short time. A person with kidney failure must use an artificial kidney, which cleanses the blood by *hemodialysis*.

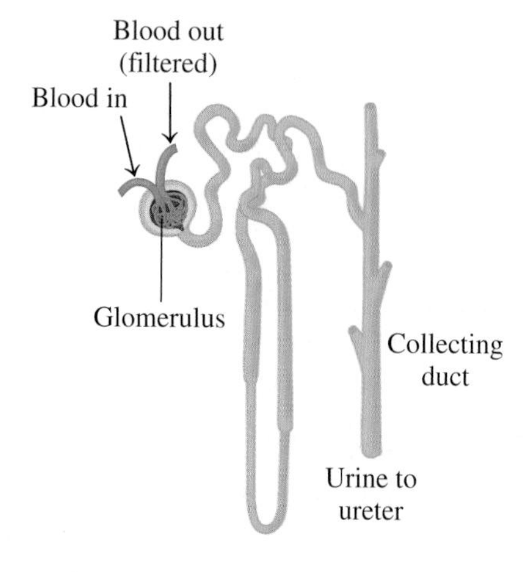

In the kidneys, each nephron contains a glomerulus where urea and waste products are removed to form urine.

A typical artificial kidney machine contains a large tank filled with water containing selected electrolytes. In the center of this dialyzing bath (dialysate), there is a dialyzing coil or membrane made of cellulose tubing. As the patient's blood flows through the dialyzing coil, the highly concentrated waste products dialyze out of the blood. No blood is lost because the membrane is not permeable to large particles such as red blood cells.

Dialysis patients do not produce much urine. As a result, they retain large amounts of water between dialysis treatments, which produces a strain on the heart. The intake of fluids for a dialysis patient may be restricted to as little as a few teaspoons of water a day. In the dialysis procedure, the pressure of the blood is increased as it circulates through the dialyzing coil so water can be squeezed out of the blood. For some dialysis patients, 2 to 10 L of water may be removed during one treatment. Dialysis patients have from two to three treatments a week, each treatment requiring about 5 to 7 h. Some of the newer treatments require less time. For many patients, dialysis is done at home with a home dialysis unit.

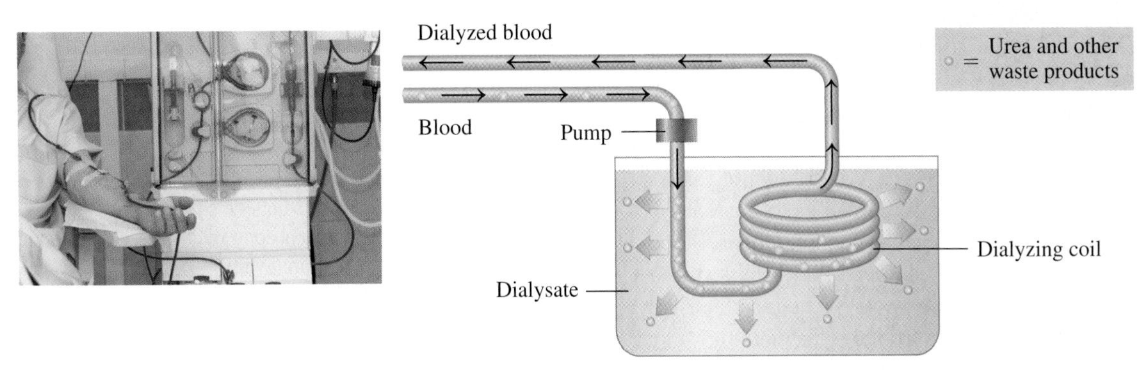

During dialysis, waste products and excess water are removed from the blood.

PRACTICE PROBLEMS

9.6 Properties of Solutions

LEARNING GOAL Identify a mixture as a solution, a colloid, or a suspension. Describe how the number of particles in a solution affects the osmotic pressure.

9.63 Identify each of the following as characteristic of a solution, colloid, or suspension:
a. a mixture that cannot be separated by a semipermeable membrane
b. a mixture that settles out upon standing

9.64 Identify each of the following as characteristic of a solution, colloid, or suspension:
a. particles of this mixture remain inside a semipermeable membrane but pass through filters
b. particles of solute in this mixture are very large and visible

9.65 A 10% (m/v) starch solution is separated from a 1% (m/v) starch solution by a semipermeable membrane. (Starch is a colloid.)
a. Which compartment has the higher osmotic pressure?
b. In which direction will water flow initially?
c. In which compartment will the volume level rise?

9.66 A 0.1% (m/v) albumin solution is separated from a 2% (m/v) albumin solution by a semipermeable membrane. (Albumin is a colloid.)
a. Which compartment has the higher osmotic pressure?
b. In which direction will water flow initially?
c. In which compartment will the volume level rise?

9.67 Indicate the compartment (**A** or **B**) that will increase in volume for each of the following pairs of solutions separated by a semipermeable membrane:

A	B
a. 5% (m/v) sucrose	10% (m/v) sucrose
b. 8% (m/v) albumin	4% (m/v) albumin
c. 0.1% (m/v) starch	10% (m/v) starch

9.68 Indicate the compartment (**A** or **B**) that will increase in volume for each of the following pairs of solutions separated by a semipermeable membrane:

A	B
a. 20% (m/v) starch	10% (m/v) starch
b. 10% (m/v) albumin	2% (m/v) albumin
c. 0.5% (m/v) sucrose	5% (m/v) sucrose

Clinical Applications

9.69 Are the following solutions isotonic, hypotonic, or hypertonic compared with a red blood cell?
a. distilled H_2O
b. 1% (m/v) glucose
c. 0.9% (m/v) NaCl
d. 15% (m/v) glucose

9.70 Will a red blood cell undergo crenation, hemolysis, or no change in each of the following solutions?
a. 1% (m/v) glucose
b. 2% (m/v) NaCl
c. 5% (m/v) glucose
d. 0.1% (m/v) NaCl

9.71 Each of the following mixtures is placed in a dialyzing bag and immersed in distilled water. Which substances will be found outside the bag in the distilled water?
a. NaCl solution
b. starch solution (colloid) and alanine (an amino acid) solution
c. NaCl solution and starch solution (colloid)
d. urea solution

9.72 Each of the following mixtures is placed in a dialyzing bag and immersed in distilled water. Which substances will be found outside the bag in the distilled water?
a. KCl solution and glucose solution
b. albumin solution (colloid)
c. albumin solution (colloid), KCl solution, and glucose solution
d. urea solution and NaCl solution

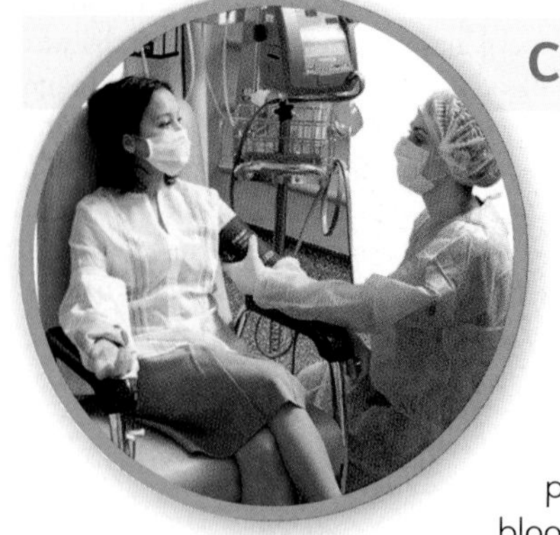

CLINICAL UPDATE Using Dialysis for Renal Failure

As a dialysis patient, Michelle has a 4-h dialysis treatment three times a week. When she arrives at the dialysis clinic, her weight, temperature, and blood pressure are taken and blood tests are done to determine the level of electrolytes and urea in her blood. In the dialysis center, tubes to the dialyzer are connected to the catheter she has had implanted. Blood is then pumped out of her body, through the dialyzer where it is filtered, and returned to her body. As Michelle's blood flows through the dialyzer, electrolytes from the dialysate move into her blood, and waste products in her blood move into the dialysate, which is continually renewed. To achieve normal serum electrolyte levels, dialysate fluid contains sodium, chloride, and magnesium levels that are equal to serum concentrations. These electrolytes are removed from the blood only if their concentrations are higher than normal. Typically, in dialysis patients, the potassium ion level is higher than normal. Therefore, initial dialysis may start with a low concentration of potassium ion in the dialysate. During dialysis excess fluid is removed by osmosis. A 4-h dialysis session requires at least 120 L of dialysis fluid. During dialysis, the electrolytes in the dialysate are adjusted until the electrolytes have the same levels as normal serum. Initially the dialysate solution prepared for Michelle contains the following: HCO_3^-, K^+, Na^+, Ca^{2+}, Mg^{2+}, Cl^-, glucose.

Clinical Applications

9.73 After her latest dialysis treatment, Michelle experienced vertigo and nausea. Michelle's doctor orders 0.075 g of chlorpromazine, which is used to treat nausea. If the stock solution is 2.5% (m/v), how many milliliters are administered?

9.74 Michelle's doctor orders 5.0 mg of compazine, which is used to treat vertigo. If the stock solution is 2.5% (m/v), how many milliliters are administered?

9.75 A $CaCl_2$ solution is given to increase blood levels of calcium. If a patient receives 5.0 mL of a 10.% (m/v) $CaCl_2$ solution, how many grams of $CaCl_2$ were given?

9.76 An intravenous solution of mannitol is used as a diuretic to increase the loss of sodium and chloride by a patient. If a patient receives 30.0 mL of a 25% (m/v) mannitol solution, how many grams of mannitol were given?

CONCEPT MAP

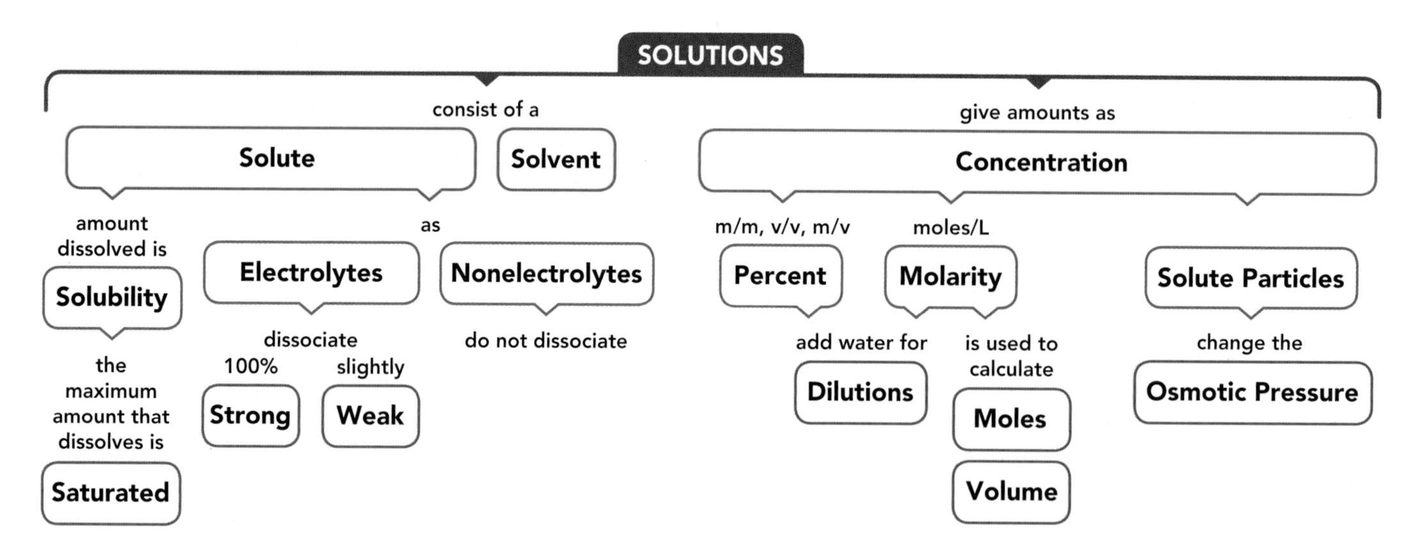

CHAPTER REVIEW

9.1 Solutions

LEARNING GOAL Identify the solute and solvent in a solution; describe the formation of a solution.

- A solution forms when a solute dissolves in a solvent.
- In a solution, the particles of solute are evenly dispersed in the solvent.

- The solute and solvent may be solid, liquid, or gas.
- The polar O—H bond leads to hydrogen bonding between water molecules.
- An ionic solute dissolves in water, a polar solvent, because the polar water molecules attract and pull the ions into solution, where they become hydrated.
- The expression *like dissolves like* means that a polar or an ionic solute dissolves in a polar solvent while a nonpolar solute dissolves in a nonpolar solvent.

9.2 Electrolytes and Nonelectrolytes

LEARNING GOAL Identify solutes as electrolytes or nonelectrolytes.

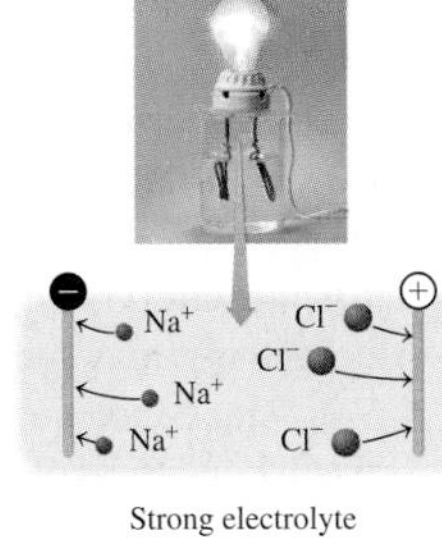

Strong electrolyte

- Substances that produce ions in water are called electrolytes because their solutions will conduct an electrical current.
- Strong electrolytes are completely dissociated, whereas weak electrolytes are only partially dissociated.
- Nonelectrolytes are substances that dissolve in water to produce only molecules and cannot conduct electrical currents.

9.3 Solubility

LEARNING GOAL Define solubility; distinguish between an unsaturated and a saturated solution. Identify an ionic compound as soluble or insoluble.

Dissolved solute
Undissolved solute
Dissolving
Crystallization
Saturated solution

- The solubility of a solute is the maximum amount of a solute that can dissolve in 100. g of solvent.
- A solution that contains the maximum amount of dissolved solute is a saturated solution.
- A solution containing less than the maximum amount of dissolved solute is unsaturated.
- An increase in temperature increases the solubility of most solids in water, but decreases the solubility of gases in water.
- Ionic compounds that are soluble in water usually contain Li^+, Na^+, K^+, NH_4^+, NO_3^-, or acetate, $C_2H_3O_2^-$.

9.4 Solution Concentrations

LEARNING GOAL Calculate the concentration of a solute in a solution; use concentration as a conversion factor to calculate the amount of solute or solution.

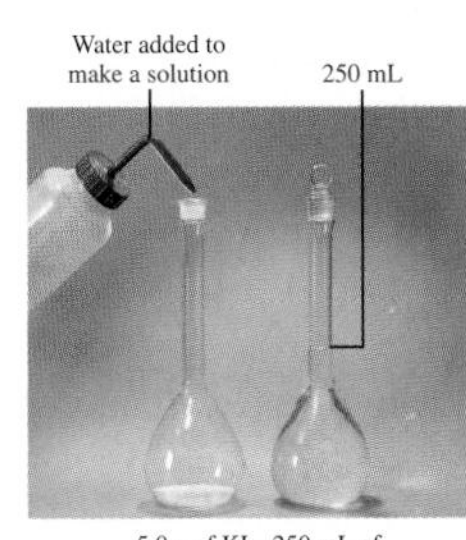

- Mass percent expresses the mass/mass (m/m) ratio of the mass of solute to the mass of solution multiplied by 100%.
- Percent concentration can also be expressed as volume/volume (v/v) and mass/volume (m/v) ratios.
- Molarity is the moles of solute per liter of solution.
- In calculations of grams or milliliters of solute or solution, the concentration is used as a conversion factor.
- Molarity (or moles/L) is written as a conversion factor to solve for moles of solute or volume of solution.

9.5 Dilution of Solutions

LEARNING GOAL Describe the dilution of a solution; calculate the unknown concentration or volume when a solution is diluted.

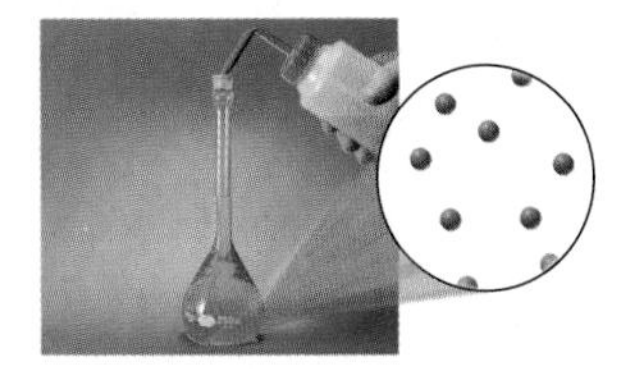

- In dilution, a solvent such as water is added to a solution, which increases its volume and decreases its concentration.

9.6 Properties of Solutions

LEARNING GOAL Identify a mixture as a solution, a colloid, or a suspension. Describe how the number of particles in a solution affects the osmotic pressure.

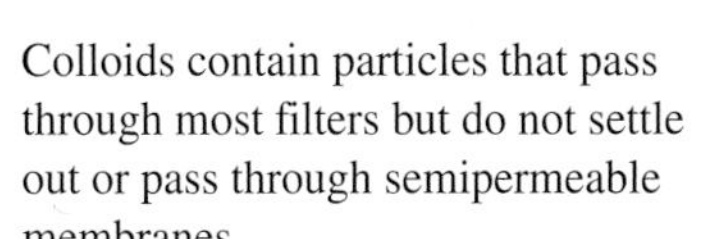

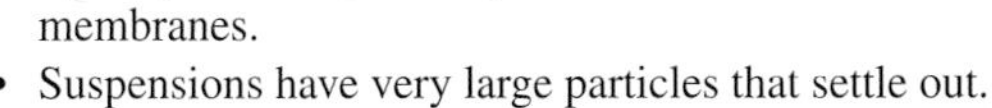

- Colloids contain particles that pass through most filters but do not settle out or pass through semipermeable membranes.
- Suspensions have very large particles that settle out.
- The particles in a solution increase the osmotic pressure.
- In osmosis, solvent (water) passes through a semipermeable membrane from a solution with a lower osmotic pressure (lower solute concentration) to a solution with a higher osmotic pressure (higher solute concentration).
- Isotonic solutions have osmotic pressures equal to that of body fluids.
- A red blood cell maintains its volume in an isotonic solution but swells in a hypotonic solution, and shrinks in a hypertonic solution.
- In dialysis, water and small solute particles pass through a dialyzing membrane, while larger particles are retained.

KEY TERMS

colloid A mixture having particles that are moderately large. Colloids pass through filters but cannot pass through semipermeable membranes.

concentration A measure of the amount of solute that is dissolved in a specified amount of solution.

dialysis A process in which water and small solute particles pass through a semipermeable membrane.

dilution A process by which water (solvent) is added to a solution to increase the volume and decrease (dilute) the solute concentration.

electrolyte A substance that produces ions when dissolved in water; its solution conducts electricity.

equivalent (Eq) The amount of positive or negative ion that supplies 1 mole of electrical charge.

Henry's law The solubility of a gas in a liquid is directly related to the pressure of that gas above the liquid.

hydration The process of surrounding dissolved ions by water molecules.

mass percent (m/m) The grams of solute in 100. g of solution.

mass/volume percent (m/v) The grams of solute in 100. mL of solution.

molarity (M) The number of moles of solute in exactly 1 L of solution.

nonelectrolyte A substance that dissolves in water as molecules; its solution does not conduct an electrical current.

osmosis The flow of a solvent, usually water, through a semipermeable membrane into a solution of higher solute concentration.

osmotic pressure The pressure that prevents the flow of water into the more concentrated solution.

saturated solution A solution containing the maximum amount of solute that can dissolve at a given temperature. Any additional solute will remain undissolved in the container.

solubility The maximum amount of solute that can dissolve in 100. g of solvent, usually water, at a given temperature.

solubility rules A set of guidelines that states whether an ionic compound is soluble or insoluble in water.
solute The component in a solution that is present in the lesser amount.
solution A homogeneous mixture in which the solute is made up of small particles (ions or molecules) that can pass through filters and semipermeable membranes.
solvent The substance in which the solute dissolves; usually the component present in greater amount.
strong electrolyte A compound that dissociates completely when it dissolves in water; its solution is a good conductor of electricity.
suspension A mixture in which the solute particles are large enough and heavy enough to settle out and be retained by both filters and semipermeable membranes.
unsaturated solution A solution that contains less solute than can be dissolved.
volume percent (v/v) The milliliters of solute in 100. mL of solution.
weak electrolyte A substance that produces only a few ions along with many molecules when it dissolves in water; its solution is a weak conductor of electricity.

CORE CHEMISTRY SKILLS

The chapter Section containing each Core Chemistry Skill is shown in parentheses at the end of each heading.

Using Solubility Rules (9.3)

- Ionic compounds that are soluble in water contain Li^+, Na^+, K^+, NH_4^+, NO_3^-, or $C_2H_3O_2^-$ (acetate).
- Most ionic compounds containing Cl^-, Br^-, or I^- are soluble, but if they contain Ag^+, Pb^{2+}, or Hg_2^{2+}, they are insoluble.
- Most ionic compounds containing SO_4^{2-} are soluble, but if they contain Ba^{2+}, Pb^{2+}, Ca^{2+}, Sr^{2+}, or Hg_2^{2+} they are insoluble.
- Most other ionic compounds including those containing the anions CO_3^{2-}, S^{2-}, PO_4^{3-}, or OH^- are insoluble.

Example: Determine if the ionic compounds Ag_3PO_4 and K_2CO_3 are soluble in water.

Answer: Ag_3PO_4 is not soluble in water because it does not contain an ion that makes it soluble. K_2CO_3 is soluble in water because it contains K^+, which makes it soluble.

Calculating Concentration (9.4)

The amount of solute dissolved in a certain amount of solution is called the concentration of the solution.

- Mass percent (m/m) $= \dfrac{\text{mass of solute}}{\text{mass of solution}} \times 100\%$
- Volume percent (v/v) $= \dfrac{\text{volume of solute}}{\text{volume of solution}} \times 100\%$
- Mass/volume percent (m/v) $= \dfrac{\text{grams of solute}}{\text{milliliters of solution}} \times 100\%$
- Molarity (M) $= \dfrac{\text{moles of solute}}{\text{liters of solution}}$

Example: What is the mass/volume percent (m/v) and the molarity (M) of 225 mL of a LiCl solution that contains 17.1 g of LiCl?

Answer:

$$\text{Mass/volume \% (m/v)} = \frac{\text{grams of solute}}{\text{milliliters of solution}} \times 100\% = \frac{17.1 \text{ g LiCl}}{225 \text{ mL solution}} \times 100\% = 7.60\% \text{ (m/v) LiCl solution}$$

$$\text{moles of LiCl} = 17.1 \text{ g LiCl} \times \frac{1 \text{ mole LiCl}}{42.39 \text{ g LiCl}} = 0.403 \text{ mole of LiCl}$$

$$\text{Molarity (M)} = \frac{\text{moles of solute}}{\text{liters of solution}} = \frac{0.403 \text{ mole LiCl}}{0.225 \text{ L solution}} = 1.79 \text{ M LiCl solution}$$

Using Concentration as a Conversion Factor (9.4)

- When we need to calculate the amount of solute or solution, we use the concentration as a conversion factor.
- For example, the concentration of a 4.50 M HCl solution means there are 4.50 moles of HCl in 1 L of HCl solution, which gives two conversion factors written as

$$\frac{4.50 \text{ moles HCl}}{1 \text{ L solution}} \quad \text{and} \quad \frac{1 \text{ L solution}}{4.50 \text{ moles HCl}}$$

Example: How many milliliters of a 4.50 M HCl solution will provide 1.13 moles of HCl?

Answer:

$$1.13 \text{ moles HCl} \times \frac{1 \text{ L solution}}{4.50 \text{ moles HCl}} \times \frac{1000 \text{ mL solution}}{1 \text{ L solution}} = 251 \text{ mL of HCl solution}$$

UNDERSTANDING THE CONCEPTS

The chapter Sections to review are shown in parentheses at the end of each problem.

9.77 Match the diagrams with the following: (9.1)
a. a polar solute and a polar solvent
b. a nonpolar solute and a polar solvent
c. a nonpolar solute and a nonpolar solvent

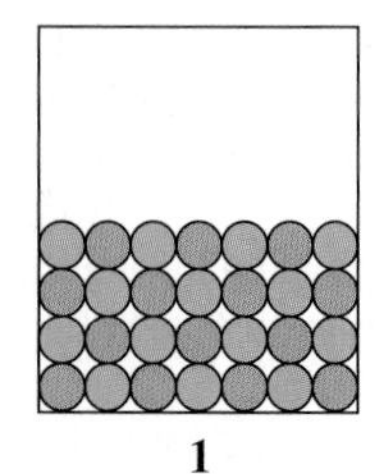

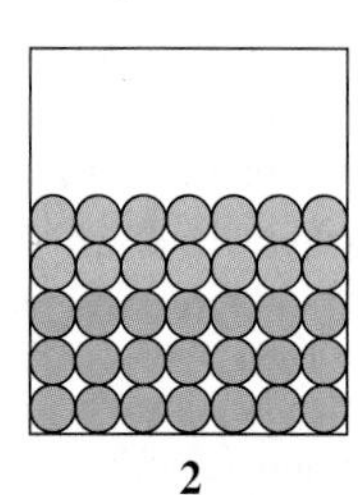

1 2

9.78 If all the solute is dissolved in diagram **1**, how would heating or cooling the solution cause each of the following changes? (9.3)
a. **2** to **3** **b.** **2** to **1**

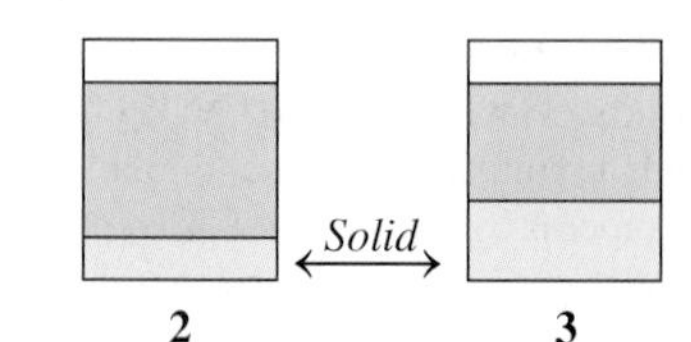

9.79 Select the diagram that represents the solution formed by a solute 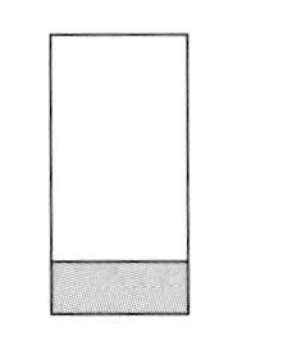 that is a (9.2)

a. nonelectrolyte
b. weak electrolyte
c. strong electrolyte

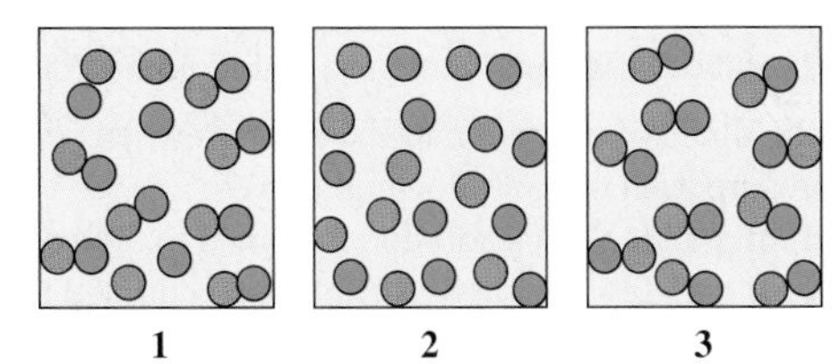

9.80 Select the container that represents the dilution of a 4% (m/v) KCl solution to give each of the following: (9.5)

a. a 2% (m/v) KCl solution
b. a 1% (m/v) KCl solution

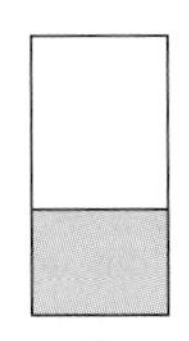
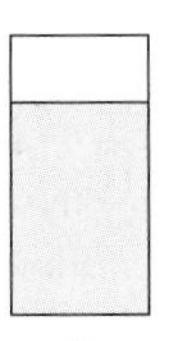

9.81 A pickle is made by soaking a cucumber in brine, a salt-water solution. What makes the smooth cucumber become wrinkled like a prune? (9.6)

9.82 Why do lettuce leaves in a salad wilt after a vinaigrette dressing containing salt is added? (9.6)

9.83 A semipermeable membrane separates two compartments, **A** and **B**. If the levels in **A** and **B** are equal initially, select the diagram that illustrates the final levels in **a** to **d**: (9.6)

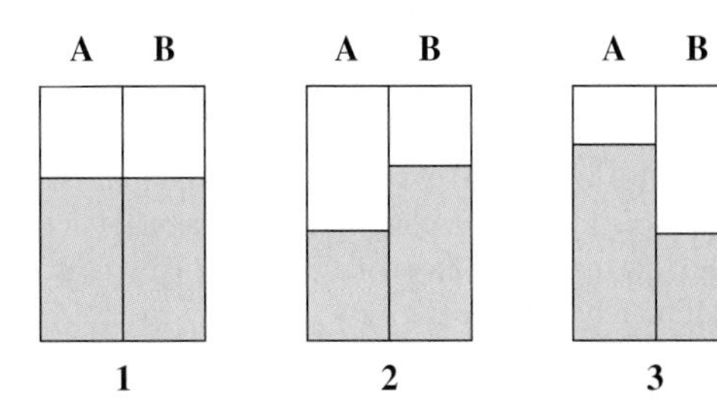

Solution in A	Solution in B
a. 2% (m/v) starch	8% (m/v) starch
b. 1% (m/v) starch	1% (m/v) starch
c. 5% (m/v) sucrose	1% (m/v) sucrose
d. 0.1% (m/v) sucrose	1% (m/v) sucrose

9.84 Select the diagram that represents the shape of a red blood cell when placed in each of the following **a** to **e**: (9.6)

Normal red blood cell

a. 0.9% (m/v) NaCl solution
b. 10% (m/v) glucose solution
c. 0.01% (m/v) NaCl solution
d. 5% (m/v) glucose solution
e. 1% (m/v) glucose solution

ADDITIONAL PRACTICE PROBLEMS

9.85 Why does iodine dissolve in hexane, but not in water? (9.1)

9.86 How do temperature and pressure affect the solubility of solids and gases in water? (9.3)

9.87 Potassium nitrate has a solubility of 32 g of KNO_3 in 100. g of H_2O at 20 °C. Determine if each of the following forms an unsaturated or saturated solution at 20 °C: (9.3)

a. adding 32 g of KNO_3 to 200. g of H_2O
b. adding 19 g of KNO_3 to 50. g of H_2O
c. adding 68 g of KNO_3 to 150. g of H_2O

9.88 Potassium chloride has a solubility of 43 g of KCl in 100. g of H_2O at 50 °C. Determine if each of the following forms an unsaturated or saturated solution at 50 °C: (9.3)

a. adding 25 g of KCl to 100. g of H_2O
b. adding 15 g of KCl to 25 g of H_2O
c. adding 86 g of KCl to 150. g of H_2O

9.89 Indicate whether each of the following ionic compounds is soluble or insoluble in water: (9.3)

a. KCl
b. $MgSO_4$
c. CuS
d. $AgNO_3$
e. $Ca(OH)_2$

9.90 Indicate whether each of the following ionic compounds is soluble or insoluble in water: (9.3)

a. $CuCO_3$
b. FeO
c. $Mg_3(PO_4)_2$
d. $(NH_4)_2SO_4$
e. $NaHCO_3$

9.91 If NaCl has a solubility of 36.0 g in 100. g of H_2O at 20 °C, how many grams of water are needed to prepare a saturated solution containing 80.0 g of NaCl? (9.3)

9.92 If the solid NaCl in a saturated solution of NaCl continues to dissolve, why is there no change in the concentration of the NaCl solution? (9.3)

9.93 Calculate the mass percent (m/m) of a solution containing 15.5 g of Na_2SO_4 and 75.5 g of H_2O. (9.4)

9.94 Calculate the mass percent (m/m) of a solution containing 26 g of K_2CO_3 and 724 g of H_2O. (9.4)

9.95 How many milliliters of a 12% (v/v) propyl alcohol solution would you need to obtain 4.5 mL of propyl alcohol? (9.4)

9.96 An 80-proof brandy is a 40.% (v/v) ethanol solution. The "proof" is twice the percent concentration of alcohol in the beverage. How many milliliters of alcohol are present in 750 mL of brandy? (9.4)

9.97 How many liters of a 12% (m/v) KOH solution would you need to obtain 86.0 g of KOH? (9.4)

9.98 How many liters of a 5.0% (m/v) glucose solution would you need to obtain 75 g of glucose? (9.4)

9.99 What is the molarity of a solution containing 8.0 g of NaOH in 400. mL of NaOH solution? (9.4)

9.100 What is the molarity of a solution containing 15.6 g of KCl in 274 mL of KCl solution? (9.4)

9.101 How many milliliters of a 1.75 M LiCl solution contain 15.2 g of LiCl? (9.4)

9.102 How many milliliters of a 1.50 M NaBr solution contain 75.0 g of NaBr? (9.4)

9.103 How many liters of a 2.50 M KNO_3 solution can be prepared from 60.0 g of KNO_3? (9.4)

9.104 How many liters of a 4.00 M NaCl solution will provide 25.0 g of NaCl? (9.4)

9.105 How many grams of solute are in each of the following solutions? (9.4)
a. 2.5 L of a 3.0 M $Al(NO_3)_3$ solution
b. 75 mL of a 0.50 M $C_6H_{12}O_6$ solution
c. 235 mL of a 1.80 M LiCl solution

9.106 How many grams of solute are in each of the following solutions? (9.4)
a. 0.428 L of a 0.450 M K_2CO_3 solution
b. 10.5 mL of a 2.50 M $AgNO_3$ solution
c. 28.4 mL of a 6.00 M H_3PO_4 solution

9.107 Calculate the final concentration of the solution when water is added to prepare each of the following: (9.5)
a. 25.0 mL of a 0.200 M NaBr solution is diluted to 50.0 mL
b. 15.0 mL of a 12.0% (m/v) K_2SO_4 solution is diluted to 40.0 mL
c. 75.0 mL of a 6.00 M NaOH solution is diluted to 255 mL

9.108 Calculate the final concentration of the solution when water is added to prepare each of the following: (9.5)
a. 25.0 mL of a 18.0 M HCl solution is diluted to 500. mL
b. 50.0 mL of a 15.0% (m/v) NH_4Cl solution is diluted to 125 mL
c. 4.50 mL of a 8.50 M KOH solution is diluted to 75.0 mL

9.109 What is the initial volume, in milliliters, needed to prepare each of the following diluted solutions? (9.5)
a. 250 mL of 3.0% (m/v) HCl from 10.0% (m/v) HCl
b. 500. mL of 0.90% (m/v) NaCl from 5.0% (m/v) NaCl
c. 350. mL of 2.00 M NaOH from 6.00 M NaOH

9.110 What is the initial volume, in milliliters, needed to prepare each of the following diluted solutions? (9.5)
a. 250 mL of 5.0% (m/v) glucose from 20.% (m/v) glucose
b. 45.0 mL of 1.0% (m/v) $CaCl_2$ from 5.0% (m/v) $CaCl_2$
c. 100. mL of 6.00 M H_2SO_4 from 18.0 M H_2SO_4

9.111 What is the final volume, in milliliters, when 25.0 mL of each of the following solutions is diluted to provide the given concentration? (9.5)
a. 10.0% (m/v) HCl solution to give a 2.50% (m/v) HCl solution
b. 5.00 M HCl solution to give a 1.00 M HCl solution
c. 6.00 M HCl solution to give a 0.500 M HCl solution

9.112 What is the final volume, in milliliters, when 5.00 mL of each of the following solutions is diluted to provide the given concentration? (9.5)
a. 20.0% (m/v) NaOH solution to give a 4.00% (m/v) NaOH solution
b. 0.600 M NaOH solution to give a 0.100 M NaOH solution
c. 16.0% (m/v) NaOH solution to give a 2.00% (m/v) NaOH solution

CHALLENGE PROBLEMS

The following problems are related to the topics in this chapter. However, they do not all follow the chapter order, and they require you to combine concepts and skills from several Sections. These problems will help you increase your critical thinking skills and prepare for your next exam.

9.113 In a laboratory experiment, a 10.0-mL sample of NaCl solution is poured into an evaporating dish with a mass of 24.10 g. The combined mass of the evaporating dish and NaCl solution is 36.15 g. After heating, the evaporating dish and dry NaCl have a combined mass of 25.50 g. (9.4)
a. What is the mass percent (m/m) of the NaCl solution?
b. What is the molarity (M) of the NaCl solution?
c. If water is added to 10.0 mL of the initial NaCl solution to give a final volume of 60.0 mL, what is the molarity of the diluted NaCl solution?

9.114 In a laboratory experiment, a 15.0-mL sample of KCl solution is poured into an evaporating dish with a mass of 24.10 g. The combined mass of the evaporating dish and KCl solution is 41.50 g. After heating, the evaporating dish and dry KCl have a combined mass of 28.28 g. (9.4)
a. What is the mass percent (m/m) of the KCl solution?
b. What is the molarity (M) of the KCl solution?
c. If water is added to 10.0 mL of the initial KCl solution to give a final volume of 60.0 mL, what is the molarity of the diluted KCl solution?

9.115 Potassium fluoride has a solubility of 92 g of KF in 100. g of H_2O at 18 °C. Determine if each of the following mixtures forms an unsaturated or saturated solution at 18 °C: (9.3)
a. adding 35 g of KF to 25 g of H_2O
b. adding 42 g of KF to 50. g of H_2O
c. adding 145 g of KF to 150. g of H_2O

9.116 Lithium chloride has a solubility of 55 g of LiCl in 100. g of H_2O at 25 °C. Determine if each of the following mixtures forms an unsaturated or saturated solution at 25 °C: (9.3)
a. adding 10 g of LiCl to 15 g of H_2O
b. adding 25 g of LiCl to 50. g of H_2O
c. adding 75 g of LiCl to 150. g of H_2O

9.117 A solution is prepared with 70.0 g of HNO_3 and 130.0 g of H_2O. The HNO_3 solution has a density of 1.21 g/mL. (9.4)
a. What is the mass percent (m/m) of the HNO_3 solution?
b. What is the total volume, in milliliters, of the solution?
c. What is the mass/volume percent (m/v) of the solution?
d. What is the molarity (M) of the solution?

9.118 A solution is prepared by dissolving 22.0 g of NaOH in 118.0 g of water. The NaOH solution has a density of 1.15 g/mL. (9.4)
a. What is the mass percent (m/m) of the NaOH solution?
b. What is the total volume, in milliliters, of the solution?
c. What is the mass/volume percent (m/v) of the solution?
d. What is the molarity (M) of the solution?

ANSWERS

Answers to Selected Practice Problems

9.1 **a.** NaCl, solute; water, solvent
b. water, solute; ethanol, solvent
c. oxygen, solute; nitrogen, solvent

9.3 The polar water molecules pull the K^+ and I^- ions away from the solid and into solution, where they are hydrated.

9.5 **a.** water **b.** CCl_4 **c.** water **d.** CCl_4

9.7 In a solution of KF, only the ions of K^+ and F^- are present in the solvent. In a HF solution, there are a few ions of H^+ and F^- present but mostly dissolved HF molecules.

9.9 **a.** $KCl(s) \xrightarrow{H_2O} K^+(aq) + Cl^-(aq)$

b. $CaCl_2(s) \xrightarrow{H_2O} Ca^{2+}(aq) + 2Cl^-(aq)$

c. $K_3PO_4(s) \xrightarrow{H_2O} 3K^+(aq) + PO_4^{3-}(aq)$

d. $Fe(NO_3)_3(s) \xrightarrow{H_2O} Fe^{3+}(aq) + 3NO_3^-(aq)$

9.11 **a.** mostly molecules and a few ions
b. only ions
c. only molecules

9.13 **a.** strong electrolyte **b.** weak electrolyte
c. nonelectrolyte

9.15 **a.** 1 Eq **b.** 2 Eq **c.** 2 Eq **d.** 6 Eq

9.17 0.154 mole of Na^+, 0.154 mole of Cl^-

9.19 55 mEq/L

9.21 **a.** 130 mEq/L **b.** above the normal range

9.23 **a.** saturated **b.** unsaturated
c. unsaturated

9.25 **a.** unsaturated **b.** unsaturated
c. saturated

9.27 **a.** 68 g of KCl **b.** 12 g of KCl

9.29 **a.** The solubility of solid solutes typically increases as temperature increases.
b. The solubility of a gas is less at a higher temperature.
c. Gas solubility is less at a higher temperature and the CO_2 pressure in the can is increased.

9.31 **a.** soluble **b.** insoluble **c.** insoluble
d. soluble **e.** soluble

9.33 **a.** 17% (m/m) KCl solution
b. 5.3% (m/m) sucrose solution
c. 10.% (m/m) $CaCl_2$ solution

9.35 79.9 mL of alcohol

9.37 A 5.00% (m/m) glucose solution can be made by adding 5.00 g of glucose to 95.00 g of water, whereas a 5.00% (m/v) glucose solution can be made by adding 5.00 g of glucose to enough water to make 100.0 mL of solution.

9.39 **a.** 30.% (m/v) Na_2SO_4 solution
b. 11% (m/v) sucrose solution

9.41 **a.** 2.5 g of KCl **b.** 50. g of NH_4Cl
c. 25.0 mL of acetic acid

9.43 **a.** 20. g of $LiNO_3$ solution
b. 400. mL of KOH solution
c. 20. mL of formic acid solution

9.45 **a.** 0.500 M glucose solution
b. 0.0356 M KOH solution
c. 0.250 M NaCl solution

9.47 **a.** 120. g of NaOH **b.** 59.6 g of KCl
c. 5.47 g of HCl

9.49 **a.** 1.50 L of KBr solution **b.** 10.0 L of NaCl solution
c. 62.5 mL of $Ca(NO_3)_2$ solution

9.51 **a.** 20. g of mannitol **b.** 240 g of mannitol

9.53 2 L of glucose solution

9.55 **a.** 2.0 M HCl solution
b. 2.0 M NaOH solution
c. 2.5% (m/v) KOH solution
d. 3.0% (m/v) H_2SO_4 solution

9.57 **a.** 80. mL of HCl solution
b. 250 mL of LiCl solution
c. 600. mL of H_3PO_4 solution
d. 180 mL of glucose solution

9.59 **a.** 12.8 mL of 4.00 M HNO_3 solution
b. 11.9 mL of 6.00 M $MgCl_2$ solution
c. 1.88 mL of 8.00 M KCl solution

9.61 1.0×10^2 mL

9.63 **a.** solution **b.** suspension

9.65 **a.** 10% (m/v) starch solution
b. from the 1% (m/v) starch solution into the 10% (m/v) starch solution
c. 10% (m/v) starch solution

9.67 **a.** B 10% (m/v) sucrose solution
b. A 8% (m/v) albumin solution
c. B 10% (m/v) starch solution

9.69 **a.** hypotonic **b.** hypotonic **c.** isotonic **d.** hypertonic

9.71 **a.** Na^+, Cl^- **b.** alanine **c.** Na^+, Cl^- **d.** urea

9.73 3.0 mL of chlorpromazine solution

9.75 0.50 g mL of $CaCl_2$

9.77 **a.** 1 **b.** 2 **c.** 1

9.79 **a.** 3 **b.** 1 **c.** 2

9.81 The skin of the cucumber acts like a semipermeable membrane, and the more dilute solution inside flows into the brine solution.

9.83 **a.** 2 **b.** 1 **c.** 3 **d.** 2

9.85 Because iodine is a nonpolar molecule, it will dissolve in hexane, a nonpolar solvent. Iodine does not dissolve in water because water is a polar solvent.

9.87 **a.** unsaturated solution **b.** saturated solution **c.** saturated solution

9.89 **a.** soluble **b.** soluble **c.** insoluble **d.** soluble **e.** insoluble

9.91 222 g of water

9.93 17.0% (m/m) Na_2SO_4 solution

9.95 38 mL of propyl alcohol solution

9.97 0.72 L of KOH solution

9.99 0.50 M NaOH solution

9.101 205 mL of LiCl solution

9.103 0.237 L of KNO_3 solution

9.105 **a.** 1600 g of $Al(NO_3)_3$ **b.** 6.8 g of $C_6H_{12}O_6$ **c.** 17.9 g of LiCl

9.107 **a.** 0.100 M NaBr solution **b.** 4.50% (m/v) K_2SO_4 solution **c.** 1.76 M NaOH solution

9.109 **a.** 75 mL of 10.0% (m/v) HCl solution **b.** 90. mL of 5.0% (m/v) NaCl solution **c.** 117 mL of 6.00 M NaOH solution

9.111 **a.** 100. mL **b.** 125 mL **c.** 300. mL

9.113 **a.** 11.6% (m/m) NaCl solution **b.** 2.40 M NaCl solution **c.** 0.400 M NaCl solution

9.115 **a.** saturated **b.** unsaturated **c.** saturated

9.117 **a.** 35.0% (m/m) HNO_3 solution **b.** 165 mL **c.** 42.4% (m/v) HNO_3 solution **d.** 6.73 M HNO_3 solution

COMBINING IDEAS from Chapters 7 to 9

CI.13 In the following diagram, blue spheres represent the element **A** and yellow spheres represent the element **B**: (6.5, 7.4, 7.5)

Reactants **Products**

a. Write the formulas for each of the reactants and products.
b. Write a balanced chemical equation for the reaction.
c. Indicate the type of reaction as combination, decomposition, single replacement, double replacement, or combustion.

CI.14 Automobile exhaust is a major cause of air pollution. One pollutant is the gas nitrogen oxide, which forms from nitrogen and oxygen gases in the air at the high temperatures in an automobile engine. Once emitted into the air, nitrogen oxide reacts with oxygen to produce nitrogen dioxide, a reddish brown gas with a sharp, pungent odor that makes up smog. One component of gasoline is heptane, C_7H_{16}, a liquid with a density of 0.684 g/mL. In one year, a typical automobile uses 550 gal of gasoline and produces 41 lb of nitrogen oxide. (7.4, 7.8, 8.6)

Two gases found in automobile exhaust are carbon dioxide and nitrogen oxide.

a. Write balanced chemical equations for the production of nitrogen oxide and nitrogen dioxide.
b. If all the nitrogen oxide emitted by one automobile is converted to nitrogen dioxide in the atmosphere, how many kilograms of nitrogen dioxide are produced in one year by a single automobile?
c. Write a balanced chemical equation for the combustion of heptane.
d. How many moles of gaseous CO_2 are produced from the gasoline used by the typical automobile in one year, assuming the gasoline is all heptane?
e. How many liters of carbon dioxide at STP are produced in one year from the gasoline used by the typical automobile?

CI.15 Bleach is often added to a wash to remove stains from clothes. The active ingredient in bleach is sodium hypochlorite (NaClO). A bleach solution can be prepared by bubbling chlorine gas into a solution of sodium hydroxide to produce liquid water and an aqueous solution of sodium hypochlorite and sodium chloride. A typical bottle of bleach has a volume of 1.42 gal, with a density of 1.08 g/mL and contains 282 g of NaClO. (6.2, 6.3, 7.4, 7.8, 8.6, 9.4)

The active component of bleach is sodium hypochlorite.

a. Is sodium hypochlorite an ionic or a molecular compound?
b. What is the mass/volume percent (m/v) of sodium hypochlorite in the bleach solution?
c. Write a balanced chemical equation for the preparation of a bleach solution.
d. How many liters of chlorine gas at STP are required to produce one bottle of bleach?

CI.16 The compound butyric acid gives rancid butter its characteristic odor. (7.1, 7.2)

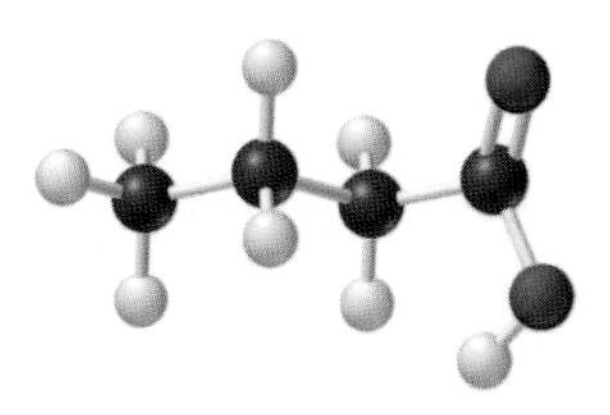

Butyric acid produces the characteristic odor of rancid butter.

a. If black spheres are carbon atoms, white spheres are hydrogen atoms, and red spheres are oxygen atoms, what is the molecular formula of butyric acid?
b. What is the molar mass of butyric acid?
c. How many grams of butyric acid contain 3.28×10^{23} atoms of oxygen?
d. How many grams of carbon are in 5.28 g of butyric acid?
e. Butyric acid has a density of 0.959 g/mL at 20 °C. How many moles of butyric acid are contained in 1.56 mL of butyric acid?
f. Identify the bonds C—C, C—H, and C—O in a molecule of butyric acid as polar covalent or nonpolar covalent.

CI.17 Methane is a major component of purified natural gas used for heating and cooking. When 1.0 mole of methane gas burns with oxygen to produce carbon dioxide and water, 883 kJ is produced. Methane gas has a density of 0.715 g/L at STP. For transport, the natural gas is cooled to −163 °C to form liquefied natural gas (LNG) with a density of 0.45 g/mL. A tank on a ship can hold 7.0 million gallons of LNG. (2.1, 2.7, 3.4, 6.7, 6.9, 7.7, 7.8, 7.9, 8.6)

An LNG carrier transports liquefied natural gas.

a. Draw the Lewis structure for methane, which has the formula CH_4.
b. What is the mass, in kilograms, of LNG (assume that LNG is all methane) transported in one tank on a ship?
c. What is the volume, in liters, of LNG (methane) from one tank when the LNG is converted to gas at STP?
d. Write a balanced chemical equation for the combustion of methane in a gas burner, including the heat of reaction.

Methane is the fuel burned in a gas cooktop.

e. How many kilograms of oxygen are needed to react with all of the methane provided by one tank of LNG?
f. How much heat, in kilojoules, is released from burning all of the methane in one tank of LNG?

Clinical Applications

CI.18 The active ingredient in Tums is calcium carbonate. One Tums tablet contains 500. mg of calcium carbonate. (6.3, 6.4, 7.2, 7.3)

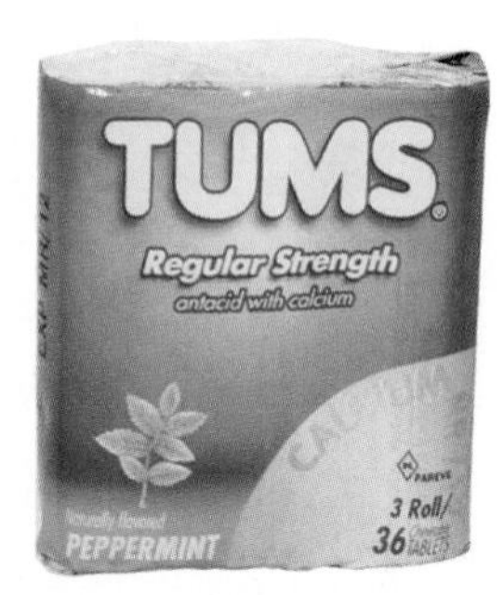

The active ingredient in Tums neutralizes excess stomach acid.

a. What is the formula of calcium carbonate?
b. What is the molar mass of calcium carbonate?
c. How many moles of calcium carbonate are in one roll of Tums that contains 12 tablets?
d. If a person takes two Tums tablets, how many grams of calcium are obtained?
e. If the daily recommended quantity of Ca^{2+} to maintain bone strength in older women is 1500 mg, how many Tums tablets are needed each day to supply the needed calcium?

CI.19 Tamiflu (oseltamivir), $C_{16}H_{28}N_2O_4$, is an antiviral drug used to treat influenza. The preparation of Tamiflu begins with the extraction of shikimic acid from the seedpods of star anise. From 2.6 g of star anise, 0.13 g of shikimic acid can be obtained and used to produce one capsule containing 75 mg of Tamiflu. The usual adult dosage for treatment of influenza is two capsules of Tamiflu daily for 5 days. (7.1, 7.2)

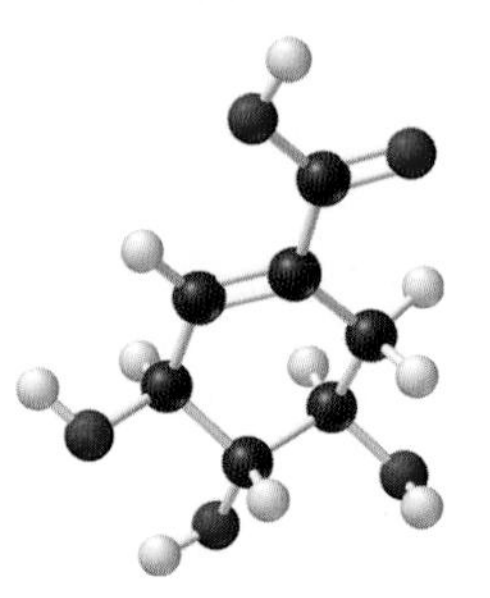

Shikimic acid is the basis for the antiviral drug in Tamiflu.

The spice called star anise is a plant source of shikimic acid.

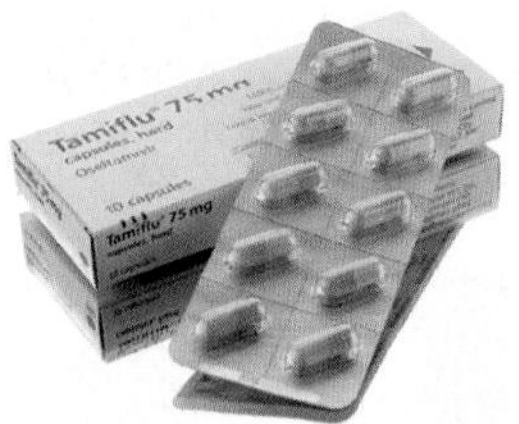

Each capsule contains 75 mg of Tamiflu.

a. If black spheres are carbon atoms, white spheres are hydrogen atoms, and red spheres are oxygen atoms, what is the formula of shikimic acid?
b. What is the molar mass of shikimic acid?
c. How many moles of shikimic acid are contained in 130 g of shikimic acid?
d. How many capsules containing 75 mg of Tamiflu could be produced from 155 g of star anise?
e. What is the molar mass of Tamiflu?
f. How many kilograms of Tamiflu would be needed to treat all the people in a city with a population of 500 000 if each person consumes two Tamiflu capsules a day for 5 days?

CI.20 In wine making, aqueous glucose ($C_6H_{12}O_6$) from grapes undergoes fermentation to produce liquid ethanol and carbon dioxide gas. A bottle of vintage port wine has a volume of 750 mL and contains 135 mL of ethanol (C_2H_6O). Ethanol has a density of 0.789 g/mL. In 1.5 lb of grapes, there is 26 g of glucose. (7.4, 7.8, 9.4)

When the sugar in grapes is fermented, ethanol is produced.

Port is a type of fortified wine that is produced in Portugal.

a. Calculate the volume percent (v/v) of ethanol in the port wine.
b. What is the molarity (M) of ethanol in the port wine?
c. Write a balanced chemical equation for the fermentation reaction of glucose.
d. How many grams of glucose are required to produce one bottle of port wine?
e. How many bottles of port wine can be produced from 1.0 ton of grapes (1 ton = 2000 lb)?

ANSWERS

CI.13 **a.** Reactants: A and B_2; Products: AB_3
b. $2A + 3B_2 \longrightarrow 2AB_3$
c. combination

CI.15 **a.** ionic
b. 5.25% (m/v)
c. $2NaOH(aq) + Cl_2(g) \longrightarrow NaClO(aq) + NaCl(aq) + H_2O(l)$
d. 84.9 L of chlorine gas

CI.17 **a.** H:C:H (electron-dot structure with H above and below C) or

$$\begin{array}{c} H \\ | \\ H-C-H \\ | \\ H \end{array}$$

b. 1.2×10^7 kg of LNG (methane)
c. 1.7×10^{10} L of LNG (methane) at STP
d. $CH_4(g) + 2O_2(g) \xrightarrow{\Delta} CO_2(g) + 2H_2O(g) + 883\text{ kJ}$
e. 4.8×10^7 kg of O_2
f. 6.6×10^{11} kJ

CI.19 **a.** $C_7H_{10}O_5$
b. 174.15 g/mole
c. 0.75 mole
d. 59 capsules
e. 312.4 g/mole
f. 400 kg (4×10^2 kg)

10 Acids and Bases and Equilibrium

A 30-YEAR-OLD MAN IS BROUGHT TO THE emergency room after an automobile accident. The emergency room nurses are tending to the patient, Larry, who is unresponsive. One of the nurses takes a blood sample, which is then sent to Brianna, a clinical laboratory technician, who begins the process of analyzing the pH, the partial pressures of O_2 and CO_2, and the concentrations of glucose and electrolytes.

Within minutes, Brianna determines that Larry's blood pH is 7.30 and the partial pressure of CO_2 gas is above the desired level. Blood pH is typically in the range of 7.35 to 7.45, and a value less than 7.35 indicates a state of acidosis. Respiratory acidosis occurs because an increase in the partial pressure of CO_2 gas in the bloodstream prevents the biochemical buffers in blood from making a change in the pH.

Brianna recognizes these signs and immediately contacts the emergency room to inform them that Larry's airway may be blocked. In the emergency room, they provide Larry with an IV containing bicarbonate to increase the blood pH and begin the process of unblocking his airway. Shortly afterward, Larry's airway is cleared, and his blood pH and partial pressure of CO_2 gas return to normal.

CAREER Clinical Laboratory Technician

Clinical laboratory technicians, also known as medical laboratory technicians, perform a wide variety of tests on body fluids and cells that help in the diagnosis and treatment of patients. These tests range from determining blood concentrations of glucose and cholesterol to determining drug levels in the blood for transplant patients or a patient undergoing treatment. Clinical laboratory technicians also prepare specimens in the detection of cancerous tumors and type blood samples for transfusions. Clinical laboratory technicians must also interpret and analyze the test results, which are then passed on to the physician.

CLINICAL UPDATE Acid Reflux Disease

After Larry was discharged from the hospital, he complained of a sore throat and dry cough, which his doctor diagnosed as acid reflux. You can view the symptoms of acid reflux disease (GERD) in the **CLINICAL UPDATE Acid Reflux Disease**, page 354, and learn about the pH changes in the stomach and how the condition is treated.

LOOKING AHEAD

REVIEW

Writing Ionic Formulas (6.2)

10.1 Acids and Bases

LEARNING GOAL Describe and name acids and bases.

Acids and bases are important substances in health, industry, and the environment. One of the most common characteristics of acids is their sour taste. Lemons and grapefruits taste sour because they contain acids such as citric and ascorbic acid (vitamin C). Vinegar tastes sour because it contains acetic acid. We produce lactic acid in our muscles when we exercise. Acid from bacteria turns milk sour in the production of yogurt and cottage cheese. We have hydrochloric acid in our stomachs that helps us digest food. Sometimes we take antacids, which are bases such as sodium bicarbonate or milk of magnesia, to neutralize the effects of too much stomach acid.

The term *acid* comes from the Latin word *acidus*, which means "sour." You are probably already familiar with the sour tastes of vinegar and lemons.

In 1887, Swedish chemist Svante Arrhenius was the first to describe **acids** as substances that produce hydrogen ions (H^+) when they dissolve in water. Because acids produce ions in water, they are electrolytes. For example, hydrogen chloride dissociates in water to give hydrogen ions, H^+, and chloride ions, Cl^-. The hydrogen ions give acids a sour taste, change the blue litmus indicator to red, and corrode some metals.

$$\underset{\text{Polar molecular compound}}{HCl(g)} \xrightarrow[\text{Dissociation}]{H_2O} \underset{\text{Hydrogen ion}}{H^+(aq)} + Cl^-(aq)$$

Citrus fruits are sour because of the presence of acids.

Naming Acids

Acids dissolve in water to produce hydrogen ions, along with a negative ion that may be a simple nonmetal anion or a polyatomic ion. When an acid dissolves in water to produce a hydrogen ion and a simple nonmetal anion, the prefix *hydro* is used before the name of the nonmetal, and its *ide* ending is changed to *ic acid*. For example, hydrogen chloride (HCl) dissolves in water to form HCl(*aq*), which is named hydrochloric acid. An exception is hydrogen cyanide (HCN), which as an acid is named hydrocyanic acid.

When an acid contains oxygen, it dissolves in water to produce a hydrogen ion and an oxygen-containing polyatomic anion. The most common form of an oxygen-containing acid has a name that ends with *ic acid*. The name of its polyatomic anion ends in *ate*. If the acid contains a polyatomic ion with an *ite* ending, its name ends in *ous acid*. The names of some common acids and their anions are listed in **TABLE 10.1**.

TABLE 10.1 Names of Common Acids and Their Anions

Acid	Name of Acid	Anion	Name of Anion
HCl	**Hydro**chlor**ic acid**	Cl^-	Chlor**ide**
HBr	**Hydro**brom**ic acid**	Br^-	Brom**ide**
HI	**Hydro**iod**ic acid**	I^-	Iod**ide**
HCN	**Hydro**cyan**ic acid**	CN^-	Cyan**ide**
HNO_3	Nitr**ic acid**	NO_3^-	Nitr**ate**
HNO_2	Nitr**ous acid**	NO_2^-	Nitr**ite**
H_2SO_4	Sulfur**ic acid**	SO_4^{2-}	Sulf**ate**
H_2SO_3	Sulfur**ous acid**	SO_3^{2-}	Sulf**ite**
H_2CO_3	Carbon**ic acid**	CO_3^{2-}	Carbon**ate**
$HC_2H_3O_2$	Acet**ic acid**	$C_2H_3O_2^-$	Acet**ate**
H_3PO_4	Phosphor**ic acid**	PO_4^{3-}	Phosph**ate**
H_3PO_3	Phosphor**ous acid**	PO_3^{3-}	Phosph**ite**
$HClO_3$	Chlor**ic acid**	ClO_3^-	Chlor**ate**
$HClO_2$	Chlor**ous acid**	ClO_2^-	Chlor**ite**

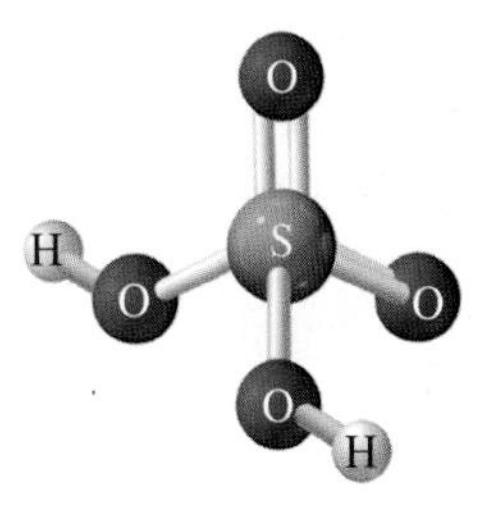

Sulfuric acid dissolves in water to produce one or two H^+ and an anion.

ENGAGE

Why is HBr named hydrobromic acid but $HBrO_3$ is named bromic acid?

The halogens in Group 7A (17) can form more than two oxygen-containing acids. For chlorine, the common form is chloric acid ($HClO_3$), which contains the chlorate polyatomic ion (ClO_3^-). For the acid that contains one more oxygen atom than the common form, the prefix *per* is used; $HClO_4$ is named *perchloric acid*. When the polyatomic ion in the acid has one oxygen atom less than the common form, the suffix *ous* is used. Thus, $HClO_2$ is named *chlorous acid*; it contains the chlorite ion (ClO_2^-). The prefix *hypo* is used for the acid that has two oxygen atoms less than the common form; HClO is named *hypochlorous acid*.

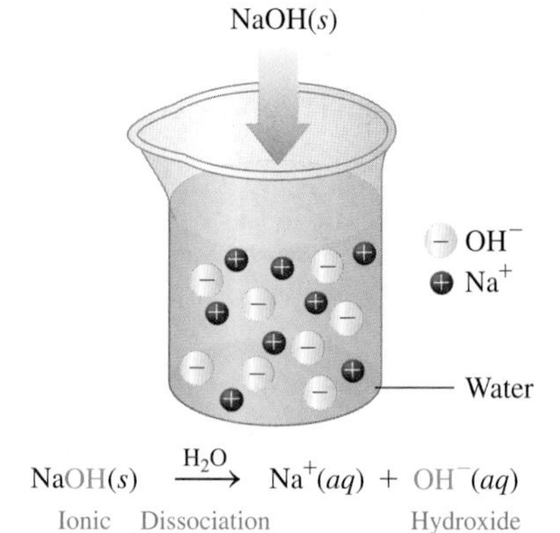

$$NaOH(s) \xrightarrow{H_2O} Na^+(aq) + OH^-(aq)$$

Ionic compound — Dissociation — Hydroxide ion

An Arrhenius base produces cations and OH^- anions in an aqueous solution.

Bases

You may be familiar with some household bases such as antacids, drain openers, and oven cleaners. According to the Arrhenius theory, **bases** are ionic compounds that dissociate into cations and hydroxide ions (OH^-) when they dissolve in water. They are another example of strong electrolytes. For example, sodium hydroxide is an Arrhenius base that dissociates completely in water to give sodium ions (Na^+) and hydroxide ions (OH^-).

Most Arrhenius bases are formed from Groups 1A (1) and 2A (2) metals, such as NaOH, KOH, LiOH, and $Ca(OH)_2$. The hydroxide ions (OH^-) give Arrhenius bases common characteristics, such as a bitter taste and a slippery feel. A base turns litmus indicator blue and phenolphthalein indicator pink. **TABLE 10.2** compares some characteristics of acids and bases.

TEST

Try Practice Problems 10.1 and 10.2

TABLE 10.2 Some Characteristics of Acids and Bases

Characteristic	Acids	Bases
Arrhenius	Produce H^+	Produce OH^-
Electrolyte	Yes	Yes
Taste	Sour	Bitter, chalky
Feel	May sting	Soapy, slippery
Litmus	Red	Blue
Phenolphthalein	Colorless	Pink
Neutralization	Neutralize bases	Neutralize acids

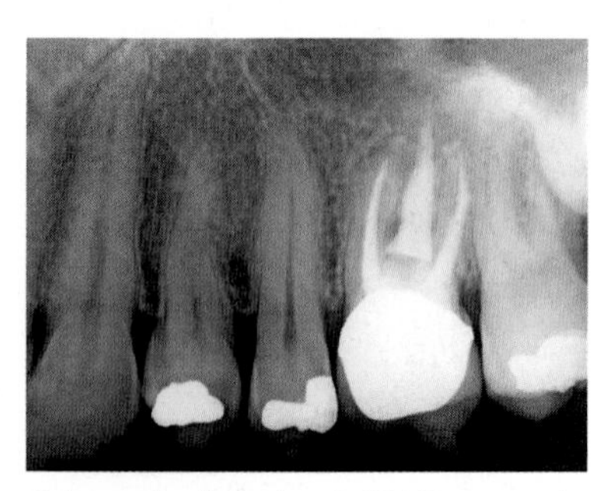

Calcium hydroxide, $Ca(OH)_2$, is used in dentistry as a filler for root canals.

Naming Bases

Typical Arrhenius bases are named as *hydroxides*.

Base	Name
LiOH	Lithium **hydroxide**
NaOH	Sodium **hydroxide**
KOH	Potassium **hydroxide**
$Ca(OH)_2$	Calcium **hydroxide**
$Al(OH)_3$	Aluminum **hydroxide**

A soft drink contains H_3PO_4 and H_2CO_3.

SAMPLE PROBLEM 10.1 Names and Formulas of Acids and Bases

TRY IT FIRST

a. Identify each of the following as an acid or a base and give its name:
  1. H_3PO_4, ingredient in soft drinks
  2. NaOH, ingredient in oven cleaner

b. Write the formula for each of the following:
  1. magnesium hydroxide, ingredient in antacids
  2. hydrobromic acid, used industrially to prepare bromide compounds

SOLUTION

a. **1.** acid, phosphoric acid
2. base, sodium hydroxide

b. **1.** $Mg(OH)_2$
2. HBr

STUDY CHECK 10.1

a. Identify as an acid or a base and give the name for H_2CO_3.
b. Write the formula for iron(III) hydroxide.

ANSWER

a. acid, carbonic acid
b. $Fe(OH)_3$

TEST

Try Practice Problems 10.3 to 10.6

PRACTICE PROBLEMS

10.1 Acids and Bases

LEARNING GOAL Describe and name acids and bases.

10.1 Indicate whether each of the following statements is characteristic of an acid, a base, or both:
a. has a sour taste
b. neutralizes bases
c. produces H^+ ions in water
d. is named barium hydroxide
e. is an electrolyte

10.2 Indicate whether each of the following statements is characteristic of an acid, a base, or both:
a. neutralizes acids
b. produces OH^- ions in water
c. has a slippery feel
d. conducts an electrical current in solution
e. turns litmus red

10.3 Name each of the following acids or bases:
a. HCl **b.** $Ca(OH)_2$ **c.** $HClO_4$
d. HNO_3 **e.** H_2SO_3 **f.** $HBrO_2$

10.4 Name each of the following acids or bases:
a. $Al(OH)_3$ **b.** HBr **c.** H_2SO_4
d. KOH **e.** HNO_2 **f.** $HClO_2$

10.5 Write formulas for each of the following acids and bases:
a. rubidium hydroxide **b.** hydrofluoric acid
c. phosphoric acid **d.** lithium hydroxide
e. ammonium hydroxide **f.** periodic acid

10.6 Write formulas for each of the following acids and bases:
a. barium hydroxide **b.** hydroiodic acid
c. bromic acid **d.** strontium hydroxide
e. acetic acid **f.** hypochlorous acid

10.2 Brønsted–Lowry Acids and Bases

LEARNING GOAL Identify conjugate acid–base pairs for Brønsted–Lowry acids and bases.

In 1923, J. N. Brønsted in Denmark and T. M. Lowry in Great Britain expanded the definition of acids and bases to include bases that do not contain OH^- ions. A **Brønsted–Lowry acid** can donate a hydrogen ion, H^+, and a **Brønsted–Lowry base** can accept a hydrogen ion.

A Brønsted–Lowry acid is a substance that donates H^+.

A Brønsted–Lowry base is a substance that accepts H^+.

A free hydrogen ion does not actually exist in water. Its attraction to polar water molecules is so strong that the H^+ bonds to a water molecule and forms a **hydronium ion, H_3O^+**.

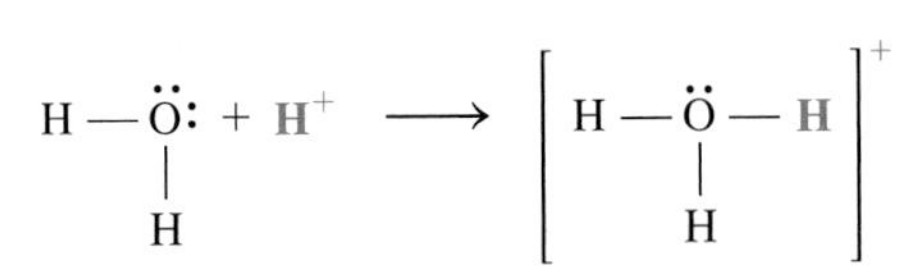

We can write the formation of a hydrochloric acid solution as a transfer of H^+ from hydrogen chloride to water. By accepting an H^+ in the reaction, water is acting as a base according to the Brønsted–Lowry concept.

$$HCl + H_2O \longrightarrow H_3O^+ + Cl^-$$

HCl	H₂O		H_3O^+	Cl^-
Hydrogen chloride	Water		Hydronium ion	Chloride ion
Acid (H^+ donor)	Base (H^+ acceptor)		Acidic solution	

In another reaction, ammonia (NH_3) acts as a base by accepting H^+ when it reacts with water. Because the nitrogen atom of NH_3 has a stronger attraction for H^+ than oxygen, water acts as an acid by donating H^+.

$$NH_3 + H_2O \rightleftarrows NH_4^+ + OH^-$$

NH_3	H_2O		NH_4^+	OH^-
Ammonia	Water		Ammonium ion	Hydroxide ion
Base (H^+ acceptor)	Acid (H^+ donor)		Basic solution	

SAMPLE PROBLEM 10.2 Acids and Bases

TRY IT FIRST

In each of the following equations, identify the reactant that is a Brønsted–Lowry acid and the reactant that is a Brønsted–Lowry base:

a. $HBr(aq) + H_2O(l) \longrightarrow H_3O^+(aq) + Br^-(aq)$
b. $CN^-(aq) + H_2O(l) \rightleftarrows HCN(aq) + OH^-(aq)$

SOLUTION

a. HBr, Brønsted–Lowry acid; H_2O, Brønsted–Lowry base
b. H_2O, Brønsted–Lowry acid; CN^-, Brønsted–Lowry base

STUDY CHECK 10.2

When HNO_3 reacts with water, water acts as a Brønsted–Lowry base. Write the equation for the reaction.

ANSWER

$HNO_3(aq) + H_2O(l) \longrightarrow H_3O^+(aq) + NO_3^-(aq)$

TEST

Try Practice Problems 10.7 to 10.12

CORE CHEMISTRY SKILL

Identifying Conjugate Acid–Base Pairs

Conjugate Acid–Base Pairs

According to the Brønsted–Lowry theory, a **conjugate acid–base pair** consists of molecules or ions related by the loss of one H^+ by an acid, and the gain of one H^+ by a base. Every acid–base reaction contains two conjugate acid–base pairs because an H^+ is transferred in both the forward and reverse directions. When an acid such as HF loses one H^+, the conjugate base F^- is formed. When the base H_2O gains an H^+, its conjugate acid, H_3O^+, is formed.

Because the overall reaction of HF is *reversible*, the conjugate acid H_3O^+ can donate H^+ to the conjugate base F^- and re-form the acid HF and the base H_2O. Using the relationship of loss and gain of one H^+, we can now identify the conjugate acid–base pairs as HF/F^- along with H_3O^+/H_2O.

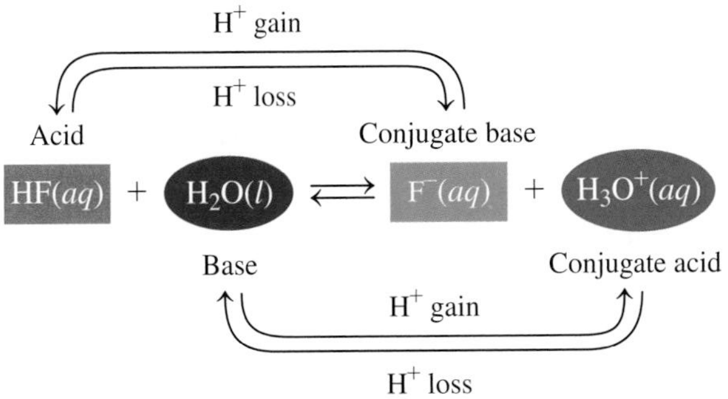

H^+ gain
H^+ loss
Acid   Conjugate base

$$HF(aq) + H_2O(l) \rightleftarrows F^-(aq) + H_3O^+(aq)$$

Base   Conjugate acid
H^+ gain
H^+ loss

HF, an acid, loses one H^+ to form its conjugate base F^-. Water acts as a base by gaining one H^+ to form its conjugate acid H_3O^+.

Conjugate acid–base pair

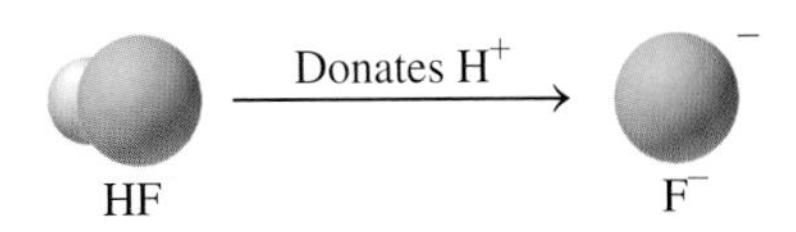

Conjugate acid–base pair

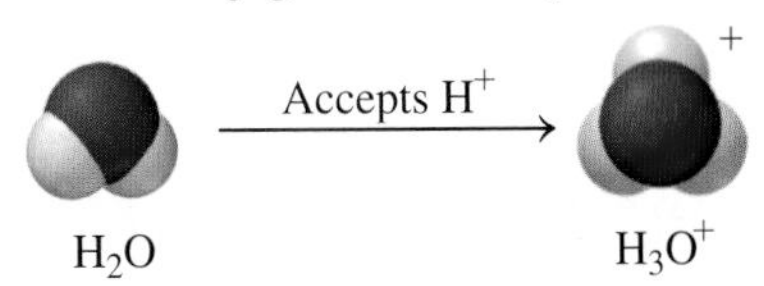

In another reaction, ammonia (NH_3) accepts H^+ from H_2O to form the conjugate acid NH_4^+ and conjugate base OH^-. Each of these conjugate acid–base pairs, NH_4^+/NH_3 and H_2O/OH^-, is related by the loss and gain of one H^+.

ENGAGE

Why is $HBrO_2$ the conjugate acid of BrO_2^-?

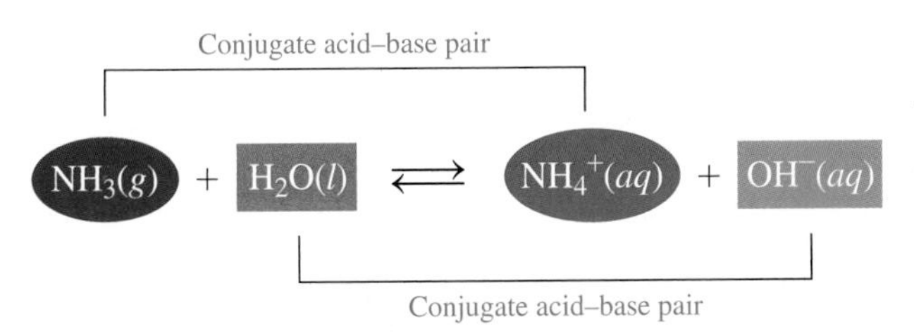

Ammonia, NH_3, acts as a base when it gains one H^+ to form its conjugate acid NH_4^+. Water acts as an acid by losing one H^+ to form its conjugate base OH^-.

In these two examples, we see that water can act as an acid when it donates H^+ or as a base when it accepts H^+. Substances that can act as both acids and bases are **amphoteric** or *amphiprotic*. For water, the most common amphoteric substance, the acidic or basic behavior depends on the other reactant. Water donates H^+ when it reacts with a stronger base, and it accepts H^+ when it reacts with a stronger acid. Another example of an amphoteric substance is bicarbonate (HCO_3^-). With a base, HCO_3^- acts as an acid and donates one H^+ to give CO_3^{2-}. However, when HCO_3^- reacts with an acid, it acts as a base and accepts one H^+ to form H_2CO_3.

H_3O^+, H_2CO_3 ← Acts as a base — H_2O, HCO_3^- — Acts as an acid → OH^-, CO_3^{2-}

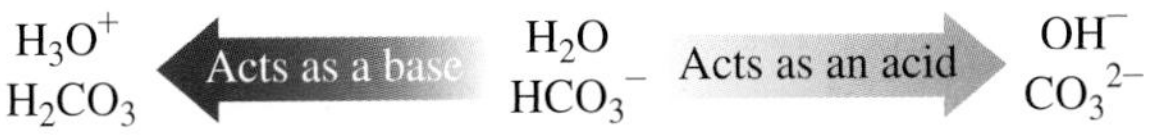

Amphoteric substances act as both acids and bases.

ENGAGE

Why can H_2O be both the conjugate base of H_3O^+ and the conjugate acid of OH^-?

SAMPLE PROBLEM 10.3 Identifying Conjugate Acid–Base Pairs

TRY IT FIRST

Identify the conjugate acid–base pairs in the following reaction:

$$HBr(aq) + NH_3(aq) \longrightarrow Br^-(aq) + NH_4^+(aq)$$

SOLUTION GUIDE

	Given	Need	Connect
ANALYZE THE PROBLEM	HBr Br^- NH_3 NH_4^+	conjugate acid–base pairs	lose/gain one H^+

STEP 1 **Identify the reactant that loses H^+ as the acid.** In the reaction, HBr donates H^+ to form the product Br^-. Thus HBr is the acid and Br^- is its conjugate base.

STEP 2 **Identify the reactant that gains H^+ as the base.** In the reaction, NH_3 gains H^+ to form the product NH_4^+. Thus, NH_3 is the base and NH_4^+ is its conjugate acid.

STEP 3 **Write the conjugate acid–base pairs.**

HBr/Br^- and NH_4^+/NH_3

STUDY CHECK 10.3

Identify the conjugate acid–base pairs in the following reaction:

$HCN(aq) + SO_4^{2-}(aq) \rightleftharpoons CN^-(aq) + HSO_4^-(aq)$

ANSWER

The conjugate acid–base pairs are HCN/CN^- and HSO_4^-/SO_4^{2-}.

TEST

Try Practice Problems 10.13 and 10.14

PRACTICE PROBLEMS

10.2 Brønsted–Lowry Acids and Bases

LEARNING GOAL Identify conjugate acid–base pairs for Brønsted–Lowry acids and bases.

10.7 Identify the reactant that is a Brønsted–Lowry acid and the reactant that is a Brønsted–Lowry base in each of the following:
- **a.** $HI(aq) + H_2O(l) \longrightarrow I^-(aq) + H_3O^+(aq)$
- **b.** $F^-(aq) + H_2O(l) \rightleftharpoons HF(aq) + OH^-(aq)$

10.8 Identify the reactant that is a Brønsted–Lowry acid and the reactant that is a Brønsted–Lowry base in each of the following:
- **a.** $CO_3^{2-}(aq) + H_2O(l) \rightleftharpoons HCO_3^-(aq) + OH^-(aq)$
- **b.** $H_2SO_4(aq) + H_2O(l) \longrightarrow HSO_4^-(aq) + H_3O^+(aq)$

10.9 Write the formula for the conjugate base for each of the following acids:
- **a.** HF
- **b.** H_2O
- **c.** $H_2PO_3^-$
- **d.** HSO_4^-

10.10 Write the formula for the conjugate base for each of the following acids:
- **a.** HCO_3^-
- **b.** $CH_3NH_3^+$
- **c.** HPO_4^{2-}
- **d.** HNO_2

10.11 Write the formula for the conjugate acid for each of the following bases:
- **a.** CO_3^{2-}
- **b.** H_2O
- **c.** $H_2PO_4^-$
- **d.** Br^-

10.12 Write the formula for the conjugate acid for each of the following bases:
- **a.** SO_4^{2-}
- **b.** CN^-
- **c.** NH_3
- **d.** ClO_2^-

10.13 Identify the Brønsted–Lowry acid–base pairs in each of the following equations:
- **a.** $H_2CO_3(aq) + H_2O(l) \rightleftharpoons HCO_3^-(aq) + H_3O^+(aq)$
- **b.** $NH_4^+(aq) + H_2O(l) \rightleftharpoons NH_3(aq) + H_3O^+(aq)$
- **c.** $HCN(aq) + NO_2^-(aq) \rightleftharpoons CN^-(aq) + HNO_2(aq)$

10.14 Identify the Brønsted–Lowry acid–base pairs in each of the following equations:
- **a.** $H_3PO_4(aq) + H_2O(l) \rightleftharpoons H_2PO_4^-(aq) + H_3O^+(aq)$
- **b.** $CO_3^{2-}(aq) + H_2O(l) \rightleftharpoons HCO_3^-(aq) + OH^-(aq)$
- **c.** $H_3PO_4(aq) + NH_3(aq) \rightleftharpoons H_2PO_4^-(aq) + NH_4^+(aq)$

10.3 Strengths of Acids and Bases

LEARNING GOAL Write equations for the dissociation of strong and weak acids and bases.

In the process called **dissociation**, an acid or a base separates into ions in water. The *strength* of an acid is determined by the moles of H_3O^+ that are produced for each mole of acid that dissolves. The *strength* of a base is determined by the moles of OH^- that are produced for each mole of base that dissolves. Strong acids and strong bases dissociate completely in water, whereas weak acids and weak bases dissociate only slightly, leaving most of the initial acid or base undissociated.

Strong and Weak Acids

Strong acids are examples of strong electrolytes because they donate H^+ so easily that their dissociation in water is essentially complete. For example, when HCl, a strong acid, dissociates in water, H^+ is transferred to H_2O; the resulting solution contains essentially only the ions H_3O^+ and Cl^-. We consider the reaction of HCl in H_2O as going 100% to products. Thus, one mole of a strong acid dissociates in water to yield one mole of H_3O^+ and one mole of its conjugate base. We write the equation for a strong acid such as HCl with a single arrow to the products.

$$HCl(g) + H_2O(l) \longrightarrow H_3O^+(aq) + Cl^-(aq)$$

There are only six common strong acids, which are stronger acids than H_3O^+. All other acids are weak. **TABLE 10.3** lists the relative strengths of acids and bases. **Weak acids** are weak electrolytes because they dissociate slightly in water, forming only a small amount

TABLE 10.3 Relative Strengths of Acids and Bases

Acid		Conjugate Base	
Strong Acids			
Hydroiodic acid	HI	I^-	Iodide ion
Hydrobromic acid	HBr	Br^-	Bromide ion
Perchloric acid	$HClO_4$	ClO_4^-	Perchlorate ion
Hydrochloric acid	HCl	Cl^-	Chloride ion
Sulfuric acid	H_2SO_4	HSO_4^-	Hydrogen sulfate ion
Nitric acid	HNO_3	NO_3^-	Nitrate ion
Hydronium ion	H_3O^+	H_2O	Water
Weak Acids			
Hydrogen sulfate ion	HSO_4^-	SO_4^{2-}	Sulfate ion
Phosphoric acid	H_3PO_4	$H_2PO_4^-$	Dihydrogen phosphate ion
Nitrous acid	HNO_2	NO_2^-	Nitrite ion
Hydrofluoric acid	HF	F^-	Fluoride ion
Acetic acid	$HC_2H_3O_2$	$C_2H_3O_2^-$	Acetate ion
Carbonic acid	H_2CO_3	HCO_3^-	Bicarbonate ion
Hydrosulfuric acid	H_2S	HS^-	Hydrogen sulfide ion
Dihydrogen phosphate ion	$H_2PO_4^-$	HPO_4^{2-}	Hydrogen phosphate ion
Ammonium ion	NH_4^+	NH_3	Ammonia
Hydrocyanic acid	HCN	CN^-	Cyanide ion
Bicarbonate ion	HCO_3^-	CO_3^{2-}	Carbonate ion
Methylammonium ion	$CH_3NH_3^+$	CH_3NH_2	Methylamine
Hydrogen phosphate ion	HPO_4^{2-}	PO_4^{3-}	Phosphate ion
Water	H_2O	OH^-	Hydroxide ion

Acid Strength Increases (upward) — Base Strength Increases (downward)

ENGAGE

Which is the weaker acid: H_2SO_4 or H_2S?

of H_3O^+ ions. A weak acid has a strong conjugate base, which is why the reverse reaction is more prevalent. Even at high concentrations, weak acids produce low concentrations of H_3O^+ ions (see **FIGURE 10.1**).

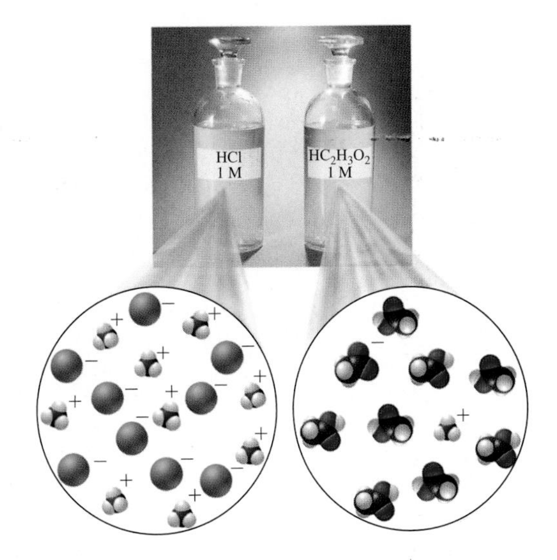

FIGURE 10.1 A strong acid such as HCl is completely dissociated, whereas a weak acid such as $HC_2H_3O_2$ contains mostly molecules and a few ions.

Q What is the difference between a strong acid and a weak acid?

Many of the products you use at home contain weak acids. Citric acid is a weak acid found in fruits and fruit juices such as lemons, oranges, and grapefruit. The vinegar used in salad dressings is typically a 5% (m/v) acetic acid ($HC_2H_3O_2$) solution. In water, a few $HC_2H_3O_2$ molecules donate H^+ to H_2O to form H_3O^+ ions and acetate ions ($C_2H_3O_2^-$). The reverse reaction also takes place, which converts the H_3O^+ ions and acetate ions ($C_2H_3O_2^-$) back to reactants. The formation of hydronium ions from vinegar is the reason we notice

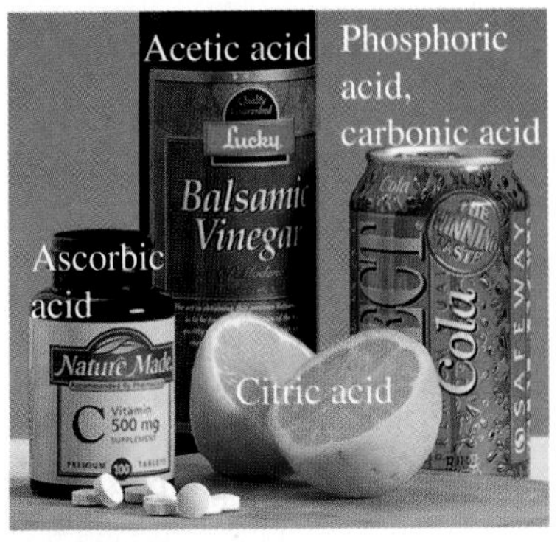

Weak acids are found in foods and household products.

the sour taste of vinegar. We write the equation for a weak acid in an aqueous solution with a double arrow to indicate that the forward and reverse reactions are at equilibrium.

$$HC_2H_3O_2(aq) + H_2O(l) \rightleftharpoons C_2H_3O_2^-(aq) + H_3O^+(aq)$$

Acetic acid — Acetate ion

Diprotic Acids

Some weak acids, such as carbonic acid, are *diprotic acids* that have two H^+, which dissociate one at a time. For example, carbonated soft drinks are prepared by dissolving CO_2 in water to form carbonic acid (H_2CO_3). A weak acid such as H_2CO_3 reaches equilibrium between the mostly undissociated H_2CO_3 molecules and the ions H_3O^+ and HCO_3^-.

$$H_2CO_3(aq) + H_2O(l) \rightleftharpoons H_3O^+(aq) + HCO_3^-(aq)$$

Carbonic acid — Bicarbonate ion (hydrogen carbonate)

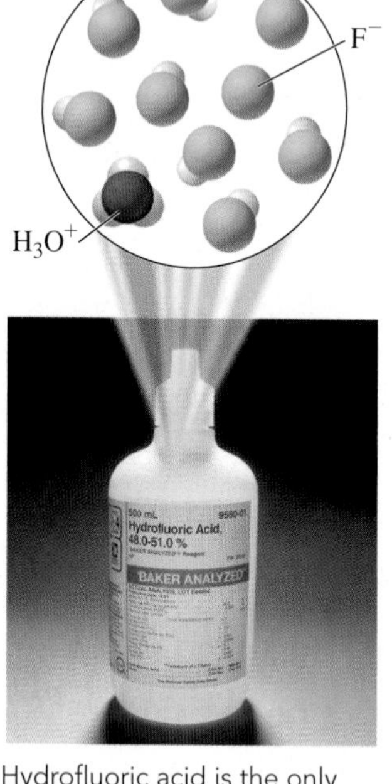

Hydrofluoric acid is the only halogen acid that is a weak acid.

Because HCO_3^- is also a weak acid, a second dissociation can take place to produce another hydronium ion and the carbonate ion (CO_3^{2-}).

$$HCO_3^-(aq) + H_2O(l) \rightleftharpoons H_3O^+(aq) + CO_3^{2-}(aq)$$

Bicarbonate ion (hydrogen carbonate) — Carbonate ion

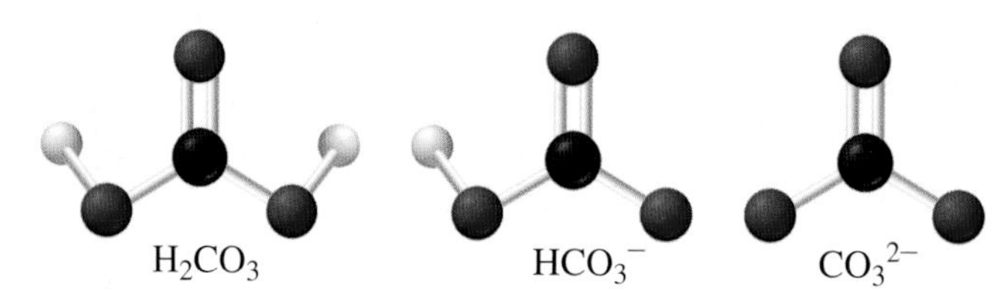

Carbonic acid, a weak acid, loses one H^+ to form hydrogen carbonate ion, which loses a second H^+ to form carbonate ion.

In summary, a strong acid such as HI in water dissociates completely to form an aqueous solution of the ions H_3O^+ and I^-. A weak acid such as HF dissociates only slightly in water to form an aqueous solution that consists mostly of HF molecules and a few H_3O^+ and F^- ions (see **FIGURE 10.2**).

Strong acid: $HI(aq) + H_2O(l) \longrightarrow H_3O^+(aq) + I^-(aq)$ Completely dissociated

Weak acid: $HF(aq) + H_2O(l) \rightleftharpoons H_3O^+(aq) + F^-(aq)$ Slightly dissociated

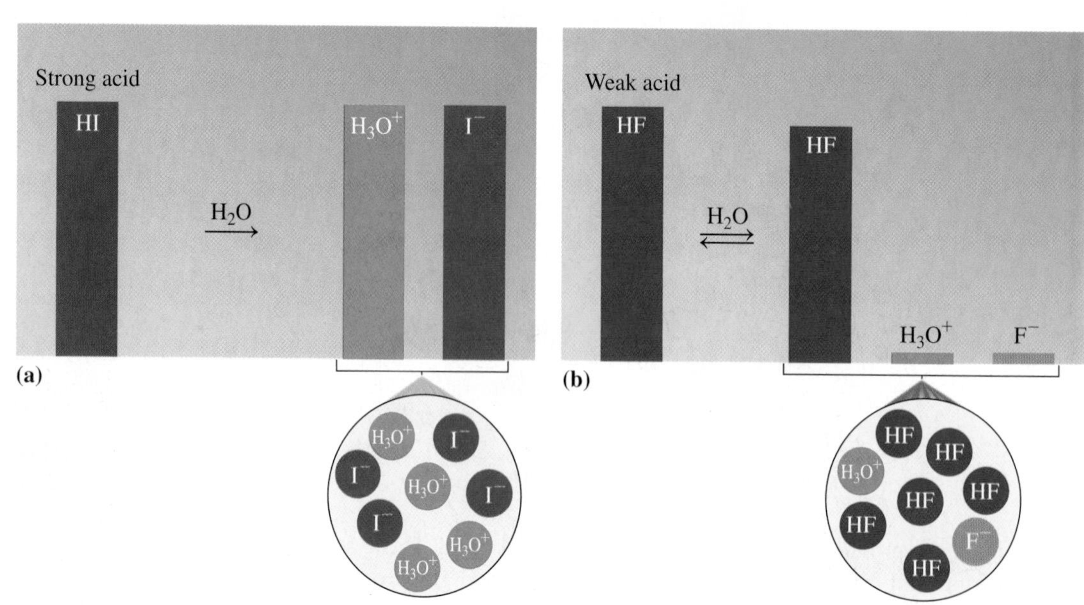

FIGURE 10.2 After dissociation in water, **(a)** the strong acid HI has high concentrations of H_3O^+ and I^-, and **(b)** the weak acid HF has a high concentration of HF and low concentrations of H_3O^+ and F^-.

Q How do the heights of H_3O^+ and F^- compare to the height of the weak acid HF in the bar diagram for HF?

Strong and Weak Bases

As strong electrolytes, **strong bases** dissociate completely in water. Because these strong bases are ionic compounds, they dissociate in water to give an aqueous solution of metal ions and hydroxide ions. The Group 1A (1) hydroxides are very soluble in water, which can give high concentrations of OH^- ions. For example, when KOH forms a KOH solution, it contains only the ions K^+ and OH^-. A few strong bases are less soluble in water, but what does dissolve dissociates completely as ions.

$$KOH(s) \xrightarrow{H_2O} K^+(aq) + OH^-(aq)$$

Strong Bases

Lithium hydroxide LiOH

Sodium hydroxide NaOH

Potassium hydroxide KOH

Rubidium hydroxide RbOH

Cesium hydroxide CsOH

Calcium hydroxide $Ca(OH)_2$*

Strontium hydroxide $Sr(OH)_2$*

Barium hydroxide $Ba(OH)_2$*

*Low solubility, but they dissociate completely.

Bases in household products are used to remove grease and to open drains.

TEST

Try Practice Problems 10.15 to 10.18

Sodium hydroxide, NaOH (also known as lye), is used in household products to remove grease in ovens and to clean drains. Because high concentrations of hydroxide ions cause severe damage to the skin and eyes, directions must be followed carefully when such products are used in the home, and use in the chemistry laboratory should be carefully supervised. If you spill an acid or a base on your skin or get some in your eyes, be sure to flood the area immediately with water for at least 10 minutes and seek medical attention.

Weak bases are weak electrolytes that are poor acceptors of hydrogen ions and produce very few ions in solution. A typical weak base, ammonia, NH_3, is found in window cleaners. In an aqueous solution, only a few ammonia molecules accept H^+ to form NH_4^+ and OH^-.

$$\underset{\text{Ammonia}}{NH_3(g)} + H_2O(l) \rightleftharpoons \underset{\text{Ammonium hydroxide}}{NH_4^+(aq) + OH^-(aq)}$$

Bases in Household Products

Weak Bases

Window cleaner, ammonia, NH_3

Bleach, NaClO

Laundry detergent, Na_2CO_3, Na_3PO_4

Toothpaste and baking soda, $NaHCO_3$

Baking powder, scouring powder, Na_2CO_3

Lime for lawns and agriculture, $CaCO_3$

Laxatives, antacids, $Mg(OH)_2$, $Al(OH)_3$

Strong Bases

Drain cleaner, oven cleaner, NaOH

PRACTICE PROBLEMS

10.3 Strengths of Acids and Bases

LEARNING GOAL Write equations for the dissociation of strong and weak acids and bases.

10.15 Using Table 10.3, identify the stronger acid in each of the following pairs:
a. HBr or HNO_2 **b.** H_3PO_4 or HSO_4^- **c.** HCN or H_2CO_3

10.16 Using Table 10.3, identify the stronger acid in each of the following pairs:
a. NH_4^+ or H_3O^+ **b.** H_2SO_4 or HCN **c.** H_2O or H_2CO_3

10.17 Using Table 10.3, identify the weaker acid in each of the following pairs:
a. HCl or HSO_4^-
b. HNO_2 or HF
c. HCO_3^- or NH_4^+

10.18 Using Table 10.3, identify the weaker acid in each of the following pairs:
a. HNO_3 or HCO_3^-
b. HSO_4^- or H_2O
c. H_2SO_4 or H_2CO_3

10.4 Acid–Base Equilibrium

LEARNING GOAL Use the concept of reversible reactions to explain acid–base equilibrium. Use Le Châtelier's principle to determine the effect on equilibrium concentrations when reaction conditions change.

As we have seen, reactants in acid–base reactions are not always completely converted to products because a reverse reaction takes place in which products form reactants. A **reversible reaction** proceeds in both the forward and reverse directions. That means there are two reactions taking place: One is the reaction in the forward direction, while the other is the reaction in the reverse direction. When molecules begin to react, the forward reaction occurs at a faster rate than the reverse reaction. As reactants are consumed and products accumulate, the rate of the forward reaction decreases and the rate of the reverse reaction increases.

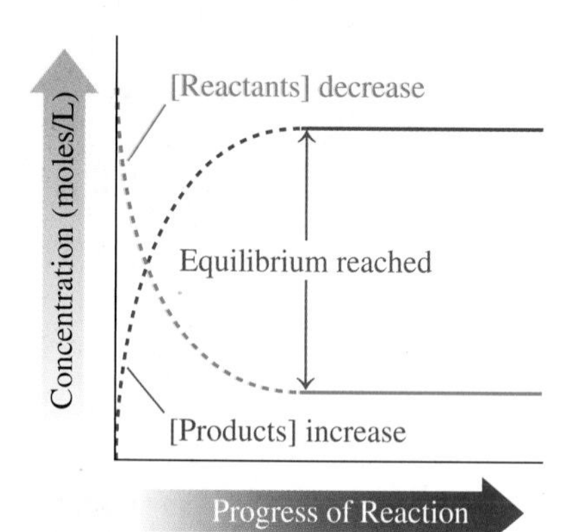

Equilibrium is reached when there are no further changes in the concentrations of reactants and products.

Equilibrium

Eventually, the rates of the forward and reverse reactions become equal. This means that the reactants form products at the same rate that the products form reactants. **Equilibrium** has been reached when no further change takes place in the concentrations of the reactants and products, even though the two reactions continue at equal but opposite rates.

Let us look at the reaction of the weak acid HF and H_2O as it proceeds to equilibrium. Initially, only the reactants HF and H_2O are present.

$$\text{Forward reaction:}\quad HF(aq) + H_2O(l) \longrightarrow F^-(aq) + H_3O^+(aq)$$

As F^- and H_3O^+ products build up, the rate of the reverse reaction increases, while the rate of the forward reaction decreases.

$$\text{Reverse reaction:}\quad F^-(aq) + H_3O^+(aq) \longrightarrow HF(aq) + H_2O(l)$$

Eventually, the rates of the forward and reverse reactions become equal, which means that equilibrium has been reached. Then the concentrations of the reactants and products remain constant even though the forward and reverse reactions continue. The forward and reverse reactions are usually shown together in a single equation, using a double arrow.

$$HF(aq) + H_2O(l) \underset{\text{Reverse reaction}}{\overset{\text{Forward reaction}}{\rightleftharpoons}} F^-(aq) + H_3O^+(aq)$$

ENGAGE

Why does the concentration of a reactant decrease before equilibrium is reached?

SAMPLE PROBLEM 10.4 Reversible Reactions and Equilibrium

TRY IT FIRST

Complete each of the following with *change* or *do not change*, *faster* or *slower*, *equal* or *not equal*:

a. Before equilibrium is reached, the concentrations of the reactants and products ______________.

b. Initially, reactants have a ______________ rate of reaction than the rate of reaction of the products.

c. At equilibrium, the rate of the forward reaction is ______________ to the rate of the reverse reaction.

SOLUTION

a. Before equilibrium is reached, the concentrations of the reactants and products *change*.

b. Initially, reactants have a *faster* rate of reaction than the rate of reaction of the products.

c. At equilibrium, the rate of the forward reaction is *equal* to the rate of the reverse reaction.

STUDY CHECK 10.4

Complete the following statement with *change* or *do not change*:

At equilibrium, the concentrations of the reactants and products ______________.

ANSWER

At equilibrium, the concentrations of the reactants and products *do not change*.

TEST

Try Practice Problems 10.19 to 10.22

Le Châtelier's Principle

CORE CHEMISTRY SKILL
Using Le Châtelier's Principle

When we alter the concentration of a reactant or product of a system at equilibrium, the rates of the forward and reverse reactions will no longer be equal. We say that a *stress* is placed on the equilibrium. **Le Châtelier's principle** states that when equilibrium is disturbed, the rates of the forward and reverse reactions change to relieve that stress and reestablish equilibrium.

Suppose we have two water tanks connected by a pipe. When the water levels in the tanks are equal, water flows in the forward direction from Tank A to Tank B at the same rate as it flows in the reverse direction from Tank B to Tank A. If we add more water to Tank A, there is an increase in the rate at which water flows from Tank A into Tank B, which is shown with a longer arrow. Equilibrium is reestablished when the water levels in both tanks become equal. The water levels are higher than before, but the water flows equally between Tank A and Tank B.

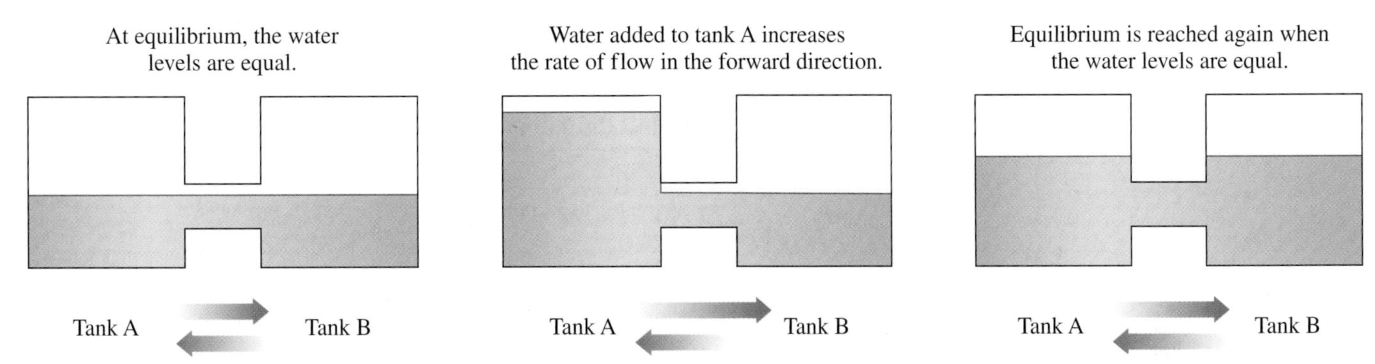

The stress of adding water to Tank A increases the rate of flow in the forward direction to reestablish equal water levels and equilibrium.

Effect of Concentration Changes on Equilibrium

We will use the reaction of HF and H_2O to illustrate how a change in concentration disturbs a reaction at equilibrium and how the system responds to that stress.

$$HF(aq) + H_2O(l) \rightleftharpoons F^-(aq) + H_3O^+(aq)$$

Suppose that more HF is added to the equilibrium mixture, which increases the concentration of HF. The system relieves this stress by increasing the rate of the forward reaction. According to Le Châtelier's principle, adding more reactant causes the system to *shift* in the direction of the products until equilibrium is reestablished.

Add HF

$$HF(aq) + H_2O(l) \rightleftharpoons F^-(aq) + H_3O^+(aq)$$

In another example, suppose that some HF is removed from the reaction mixture at equilibrium. The decrease in HF concentration slows the rate of the forward reaction. According to Le Châtelier's principle, the stress of removing some of the reactant causes the system to *shift* in the direction of the reactants until equilibrium is reestablished.

Remove HF

$$HF(aq) + H_2O(l) \rightleftharpoons F^-(aq) + H_3O^+(aq)$$

Stress is also placed on a system at equilibrium by changing the concentration of a product. For example, if more F^- is added to the equilibrium mixture, the rate of the reverse reaction increases as products are converted to reactants until equilibrium is reestablished. According to Le Châtelier's principle, the addition of more product causes the system to *shift* in the direction of the reactants.

Add F^-

$$HF(aq) + H_2O(l) \rightleftharpoons F^-(aq) + H_3O^+(aq)$$

In another example, suppose that some F^- is removed from the reaction mixture at equilibrium. The decrease in F^- concentration slows the rate of the reverse reaction. According to Le Châtelier's principle, the stress of removing some of the product causes the system to *shift* in the direction of the products until equilibrium is reestablished.

Remove F^-

$$HF(aq) + H_2O(l) \rightleftarrows F^-(aq) + H_3O^+(aq)$$

In summary, Le Châtelier's principle indicates that a stress caused by adding a substance at equilibrium is relieved when the system shifts the reaction away from that substance. When some of a substance is removed, the equilibrium system shifts in the direction of that substance. These features of Le Châtelier's principle are summarized in **TABLE 10.4**.

TABLE 10.4 Effect of Concentration Changes on Equilibrium

Stress	Change	Shift in the Direction of
Add reactant	Increases forward reaction rate	Products
Remove reactant	Decreases forward reaction rate	Reactants
Add product	Increases reverse reaction rate	Reactants
Remove product	Decreases reverse reaction rate	Products

CHEMISTRY LINK TO HEALTH

Oxygen–Hemoglobin Equilibrium and Hypoxia

The transport of oxygen involves an equilibrium between hemoglobin (Hb), oxygen, and oxyhemoglobin (HbO_2).

$$Hb(aq) + O_2(g) \rightleftarrows HbO_2(aq)$$

When the O_2 level is high in the alveoli of the lung, the reaction shifts in the direction of the product HbO_2. In the tissues where O_2 concentration is low, the reverse reaction releases the oxygen from the hemoglobin.

At normal atmospheric pressure, oxygen diffuses into the blood because the partial pressure of oxygen in the alveoli is higher than that in the blood. At an altitude above 8000 ft, a decrease in the atmospheric pressure results in a significant reduction in the partial pressure of oxygen, which means that less oxygen is available for the blood and body tissues. The fall in atmospheric pressure at higher altitudes decreases the partial pressure of inhaled oxygen. At an altitude of 18 000 ft, a person will obtain 29% less oxygen. When oxygen levels are lowered, a person may experience *hypoxia*, characterized by increased respiratory rate, headache, decreased mental acuteness, fatigue, decreased physical coordination, nausea, vomiting, and cyanosis. A similar problem occurs in persons with a history of lung disease that impairs gas diffusion in the alveoli or in persons with a reduced number of red blood cells, such as smokers.

According to Le Châtelier's principle, we see that a decrease in oxygen will shift the system in the direction of the reactants to reestablish equilibrium. Such a shift depletes the concentration of HbO_2 and causes the hypoxia.

$$Hb(aq) + O_2(g) \longleftarrow HbO_2(aq)$$

Immediate treatment of altitude sickness includes hydration, rest, and, if necessary, descending to a lower altitude. The adaptation to lowered oxygen levels requires about 10 days. During this time, the bone marrow increases red blood cell production, providing more red blood cells and more hemoglobin. A person living at a high altitude can have 50% more red blood cells than someone at sea level. This increase in hemoglobin causes a shift in the equilibrium back in the direction of HbO_2 product. Eventually, the higher concentration of HbO_2 will provide more oxygen to the tissues and the symptoms of hypoxia will lessen.

$$Hb(aq) + O_2(g) \longrightarrow HbO_2(aq)$$

For those who climb high mountains, it is important to stop and acclimatize for several days at increasing altitudes. At very high altitudes, it may be necessary to use an oxygen tank.

Hypoxia may occur at high altitudes where the oxygen concentration is lower.

SAMPLE PROBLEM 10.5 Effect of Changes in Concentration on Equilibrium

TRY IT FIRST

An important reaction in the body fluids is

$$H_2CO_3(aq) + H_2O(l) \rightleftharpoons HCO_3^-(aq) + H_3O^+(aq)$$

Use Le Châtelier's principle to predict whether the system shifts in the direction of products or reactants for each of the following:

a. adding more $H_2CO_3(aq)$
b. removing some $HCO_3^-(aq)$
c. adding more $H_3O^+(aq)$

SOLUTION

According to Le Châtelier's principle, when stress is applied to a reaction at equilibrium, the system shifts to relieve that stress.

a. When the concentration of the reactant H_2CO_3 increases, the equilibrium shifts in the direction of the products.
b. When the concentration of the product HCO_3^- decreases, the equilibrium shifts in the direction of the products.
c. When the concentration of the product H_3O^+ increases, the equilibrium shifts in the direction of the reactants.

STUDY CHECK 10.5

Using the reaction in Sample Problem 10.5, predict whether removing some $H_2CO_3(aq)$ causes the system to shift in the direction of products or reactants.

ANSWER

When some H_2CO_3 is removed, the equilibrium shifts in the direction of the reactants.

TEST

Try Practice Problems 10.23 and 10.24

PRACTICE PROBLEMS

10.4 Acid–Base Equilibrium

LEARNING GOAL Use the concept of reversible reactions to explain acid–base equilibrium. Use Le Châtelier's principle to determine the effect on equilibrium concentrations when reaction conditions change.

10.19 What is meant by the term reversible reaction?

10.20 When does a reversible reaction reach equilibrium?

10.21 Which of the following are at equilibrium?
a. The rate of the forward reaction is twice as fast as the rate of the reverse reaction.
b. The concentrations of the reactants and the products do not change.
c. The rate of the reverse reaction does not change.

10.22 Which of the following are not at equilibrium?
a. The rates of the forward and reverse reactions are equal.
b. The rate of the forward reaction does not change.
c. The concentrations of reactants and the products are not constant.

10.23 Use Le Châtelier's principle to predict whether each of the following changes causes the system to shift in the direction of products or reactants:

$$HCHO_2(aq) + H_2O(l) \rightleftharpoons CHO_2^-(aq) + H_3O^+(aq)$$

a. adding more $CHO_2^-(aq)$
b. removing some $HCHO_2(aq)$
c. removing some $H_3O^+(aq)$
d. adding more $HCHO_2(aq)$

10.24 Use Le Châtelier's principle to predict whether each of the following changes causes the system to shift in the direction of products or reactants:

$$HNO_2(aq) + H_2O(l) \rightleftharpoons NO_2^-(aq) + H_3O^+(aq)$$

a. adding more $HNO_2(aq)$
b. removing some $NO_2^-(aq)$
c. adding more $H_3O^+(aq)$
d. removing some $HNO_2(aq)$

REVIEW

Solving Equations (1.4)

10.5 Dissociation of Water

LEARNING GOAL Use the water dissociation expression to calculate the $[H_3O^+]$ and $[OH^-]$ in an aqueous solution.

In many acid–base reactions, water is *amphoteric*, which means that it can act either as an acid or as a base. In pure water, there is a forward reaction between two water molecules that transfers H^+ from one water molecule to the other. One molecule acts as an acid by losing H^+, and the water molecule that gains H^+ acts as a base. Every time H^+ is transferred between two water molecules, the products are one H_3O^+ and one OH^-, which react in the reverse direction to re-form two water molecules. Thus, equilibrium is reached between the conjugate acid–base pairs of water.

ENGAGE

Why is the $[H_3O^+]$ equal to the $[OH^-]$ in pure water?

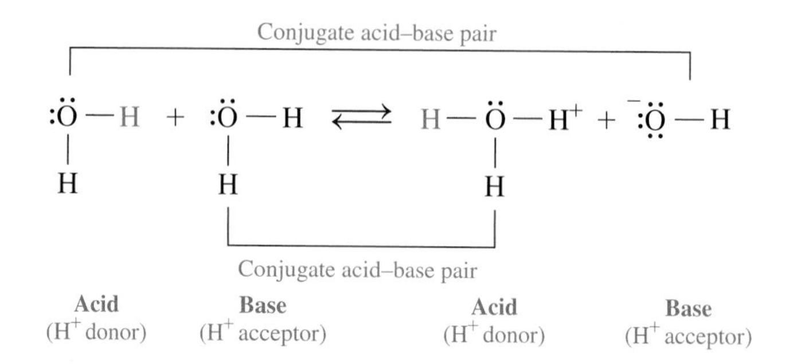

Writing the Water Dissociation Expression, K_w

In the equation for the dissociation of water, there is both a forward and a reverse reaction:

$$H_2O(l) + H_2O(l) \rightleftarrows H_3O^+(aq) + OH^-(aq)$$

Base, Acid, Conjugate acid, Conjugate base

Experiments have determined that, in pure water, the concentrations of H_3O^+ and OH^- at 25 °C are each 1.0×10^{-7} M. Square brackets are used to indicate the concentrations in moles per liter (M).

$$\text{Pure water} \quad [H_3O^+] = [OH^-] = 1.0 \times 10^{-7}\ M$$

When these concentrations are multiplied, we obtain the expression and value called the **water dissociation expression, K_w**. The concentration units are omitted in the K_w value.

$$K_w = [H_3O^+][OH^-]$$
$$= (1.0 \times 10^{-7})(1.0 \times 10^{-7}) = 1.0 \times 10^{-14}$$

TEST

Try Practice Problems 10.25 and 10.26

Neutral, Acidic, and Basic Solutions

The K_w value (1.0×10^{-14}) applies to any aqueous solution at 25 °C because all aqueous solutions contain both H_3O^+ and OH^- (see **FIGURE 10.3**). When the $[H_3O^+]$ and $[OH^-]$

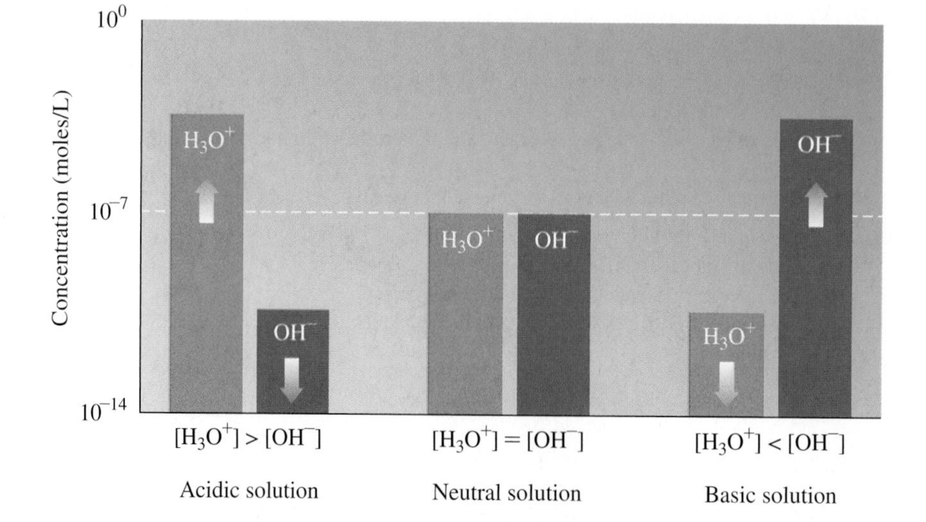

FIGURE 10.3 In a neutral solution, $[H_3O^+]$ and $[OH^-]$ are equal. In acidic solutions, the $[H_3O^+]$ is greater than the $[OH^-]$. In basic solutions, the $[OH^-]$ is greater than the $[H_3O^+]$.

Q Is a solution that has a $[H_3O^+]$ of 1.0×10^{-3} M acidic, basic, or neutral?

in a solution are equal, the solution is **neutral**. However, most solutions are not neutral; they have different concentrations of H_3O^+ and OH^-. If acid is added to water, there is an increase in $[H_3O^+]$ and a decrease in $[OH^-]$, which makes an acidic solution. If base is added, $[OH^-]$ increases and $[H_3O^+]$ decreases, which gives a basic solution. However, for any aqueous solution, whether it is neutral, acidic, or basic, the product $[H_3O^+][OH^-]$ is equal to K_w (1.0×10^{-14} at 25 °C) (see **TABLE 10.5**).

TABLE 10.5 Examples of $[H_3O^+]$ and $[OH^-]$ in Neutral, Acidic, and Basic Solutions

Type of Solution	$[H_3O^+]$	$[OH^-]$	K_w (25 °C)
Neutral	1.0×10^{-7} M	1.0×10^{-7} M	1.0×10^{-14}
Acidic	1.0×10^{-2} M	1.0×10^{-12} M	1.0×10^{-14}
Acidic	2.5×10^{-5} M	4.0×10^{-10} M	1.0×10^{-14}
Basic	1.0×10^{-8} M	1.0×10^{-6} M	1.0×10^{-14}
Basic	5.0×10^{-11} M	2.0×10^{-4} M	1.0×10^{-14}

TEST

Try Practice Problems 10.27 to 10.30

Using the K_w to Calculate $[H_3O^+]$ and $[OH^-]$ in a Solution

If we know the $[H_3O^+]$ of a solution, we can use the K_w to calculate $[OH^-]$. If we know the $[OH^-]$ of a solution, we can calculate $[H_3O^+]$ from their relationship in the K_w, as shown in Sample Problem 10.6.

$$K_w = [H_3O^+][OH^-]$$

$$[OH^-] = \frac{K_w}{[H_3O^+]} \qquad [H_3O^+] = \frac{K_w}{[OH^-]}$$

ENGAGE

If you know the $[H_3O^+]$ of a solution, how do you use the K_w to calculate the $[OH^-]$?

CORE CHEMISTRY SKILL

Calculating $[H_3O^+]$ and $[OH^-]$ in Solutions

SAMPLE PROBLEM 10.6 Calculating the $[H_3O^+]$ of a Solution

TRY IT FIRST

A vinegar solution has a $[OH^-] = 5.0 \times 10^{-12}$ M at 25 °C. What is the $[H_3O^+]$ of the vinegar solution? Is the solution acidic, basic, or neutral?

SOLUTION GUIDE

STEP 1 State the given and needed quantities.

ANALYZE THE PROBLEM	Given	Need	Connect
	$[OH^-] = 5.0 \times 10^{-12}$ M	$[H_3O^+]$	$K_w = [H_3O^+][OH^-]$

STEP 2 Write the K_w for water and solve for the unknown $[H_3O^+]$.

$$K_w = [H_3O^+][OH^-] = 1.0 \times 10^{-14}$$

Solve for $[H_3O^+]$ by dividing both sides by $[OH^-]$.

$$\frac{K_w}{[OH^-]} = \frac{[H_3O^+][\cancel{OH^-}]}{[\cancel{OH^-}]}$$

$$[H_3O^+] = \frac{1.0 \times 10^{-14}}{[OH^-]}$$

STEP 3 Substitute the known $[OH^-]$ into the equation and calculate.

$$[H_3O^+] = \frac{1.0 \times 10^{-14}}{[5.0 \times 10^{-12}]} = 2.0 \times 10^{-3}\text{ M}$$

Because the $[H_3O^+]$ of 2.0×10^{-3} M is larger than the $[OH^-]$ of 5.0×10^{-12} M, the solution is acidic.

ENGAGE

Why does the $[H_3O^+]$ of an aqueous solution increase if the $[OH^-]$ decreases?

STUDY CHECK 10.6

What is the $[H_3O^+]$ of an ammonia cleaning solution with $[OH^-] = 4.0 \times 10^{-4}$ M? Is the solution acidic, basic, or neutral?

ANSWER

$[H_3O^+] = 2.5 \times 10^{-11}$ M, basic

TEST

Try Practice Problems 10.31 to 10.34

PRACTICE PROBLEMS

10.5 Dissociation of Water

LEARNING GOAL Use the water dissociation expression to calculate the $[H_3O^+]$ and $[OH^-]$ in an aqueous solution.

10.25 Why are the concentrations of H_3O^+ and OH^- equal in pure water?

10.26 What is the meaning and value of K_w?

10.27 In an acidic solution, how does the concentration of H_3O^+ compare to the concentration of OH^-?

10.28 If a base is added to pure water, why does the $[H_3O^+]$ decrease?

10.29 Indicate whether each of the following solutions is acidic, basic, or neutral:

a. $[H_3O^+] = 2.0 \times 10^{-5}$ M
b. $[H_3O^+] = 1.4 \times 10^{-9}$ M
c. $[OH^-] = 8.0 \times 10^{-3}$ M
d. $[OH^-] = 3.5 \times 10^{-10}$ M

10.30 Indicate whether each of the following solutions is acidic, basic, or neutral:

a. $[H_3O^+] = 6.0 \times 10^{-12}$ M
b. $[H_3O^+] = 1.4 \times 10^{-4}$ M
c. $[OH^-] = 5.0 \times 10^{-12}$ M
d. $[OH^-] = 4.5 \times 10^{-2}$ M

Clinical Applications

10.31 Calculate the $[OH^-]$ of each aqueous solution with the following $[H_3O^+]$:

a. coffee, 1.0×10^{-5} M
b. soap, 1.0×10^{-8} M
c. cleanser, 5.0×10^{-10} M
d. lemon juice, 2.5×10^{-2} M

10.32 Calculate the $[OH^-]$ of each aqueous solution with the following $[H_3O^+]$:

a. oven cleaner, 1.0×10^{-12} M
b. milk of magnesia, 1.0×10^{-9} M
c. aspirin, 6.0×10^{-4} M
d. pancreatic juice, 4.0×10^{-9} M

10.33 Calculate the $[H_3O^+]$ of each aqueous solution with the following $[OH^-]$:

a. stomach acid, 2.5×10^{-13} M
b. urine, 2.0×10^{-9} M
c. orange juice, 5.0×10^{-11} M
d. bile, 2.5×10^{-6} M

10.34 Calculate the $[H_3O^+]$ of each aqueous solution with the following $[OH^-]$:

a. baking soda, 1.0×10^{-6} M
b. blood, 2.5×10^{-7} M
c. milk, 4.0×10^{-7} M
d. bleach, 2.1×10^{-3} M

10.6 The pH Scale

LEARNING GOAL Calculate the pH of a solution from $[H_3O^+]$; given the pH, calculate $[H_3O^+]$.

The proper level of acidity is necessary to evaluate the functioning of the lungs and kidneys, to control bacterial growth in foods, and to prevent the growth of pests in food crops. In the environment, the acidity, or pH, of rain, water, and soil can have significant effects. When rain becomes too acidic, it can dissolve marble statues and accelerate the corrosion of metals. In lakes and ponds, the acidity of water can affect the ability of plants and fish to survive. The acidity of soil around plants affects their growth. If the soil pH is too acidic or too basic, the roots of the plant cannot take up some nutrients. Most plants thrive in soil with a nearly neutral pH, although certain plants, such as orchids, camellias, and blueberries, require a more acidic soil.

Although we have expressed H_3O^+ and OH^- as molar concentrations, it is more convenient to describe the acidity of solutions using the *pH scale*. On this scale, a number between 0 and 14 represents the H_3O^+ concentration for common solutions. A neutral solution has a pH of 7.0 at 25 °C. An acidic solution has a pH less than 7.0; a basic solution has a pH greater than 7.0 (see **FIGURE 10.4**).

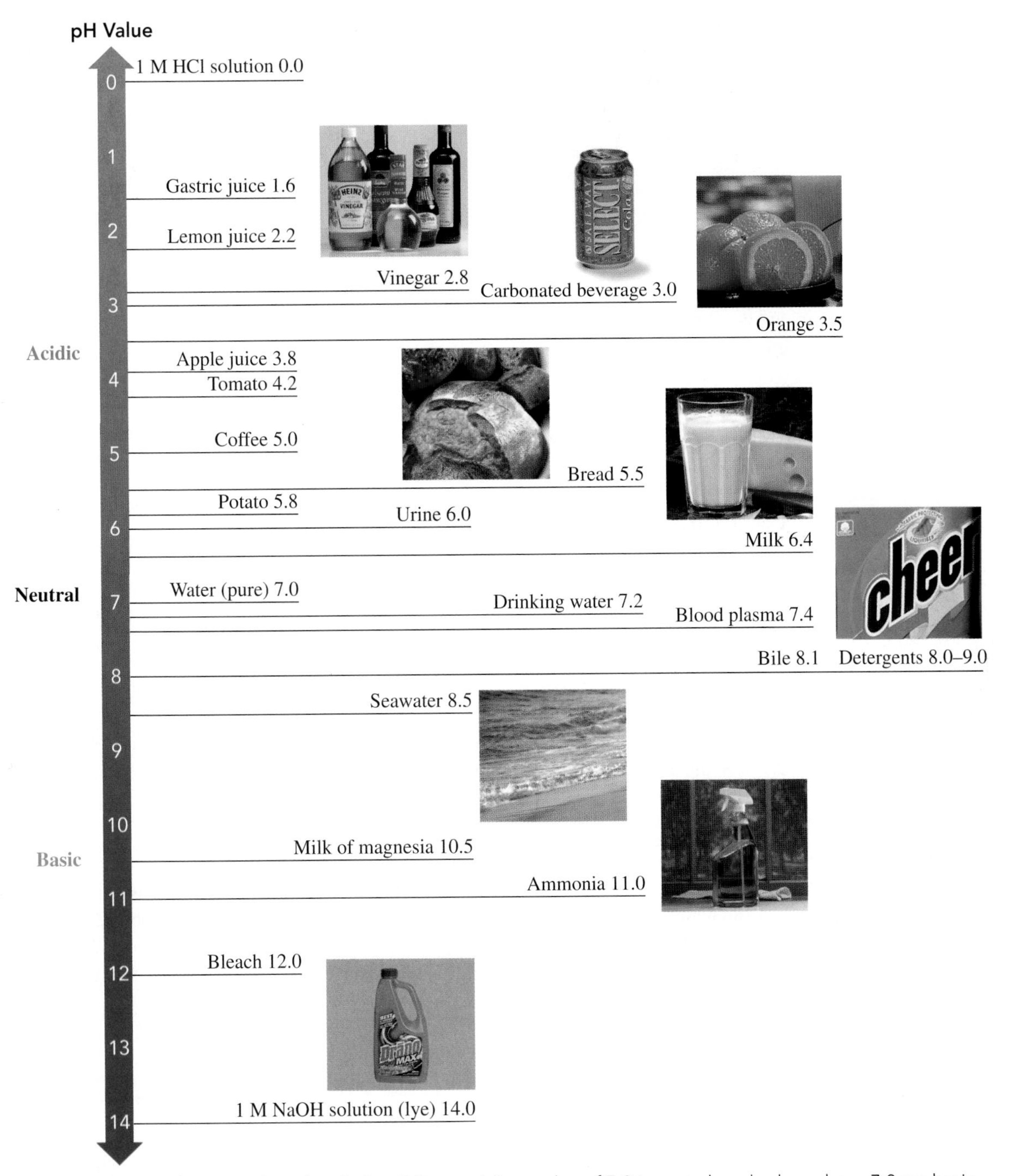

FIGURE 10.4 On the pH scale, values below 7.0 are acidic, a value of 7.0 is neutral, and values above 7.0 are basic.

Q Is apple juice an acidic, a basic, or a neutral solution?

Acidic solution	$pH < 7.0$	$[H_3O^+] > 1 \times 10^{-7}$ M
Neutral solution	$pH = 7.0$	$[H_3O^+] = 1 \times 10^{-7}$ M
Basic solution	$pH > 7.0$	$[H_3O^+] < 1 \times 10^{-7}$ M

When we relate acidity and pH, we are using an *inverse relationship*, which is when one component increases while the other component decreases. When an acid is added to pure water, the $[H_3O^+]$ (acidity) of the solution increases but its pH decreases. When a base is added to pure water, it becomes more basic, which means its acidity decreases but the pH increases.

In the laboratory, a pH meter is commonly used to determine the pH of a solution. There are also indicators and pH papers that turn specific colors when placed in solutions of different pH values. The pH is found by comparing the colors to a chart (see **FIGURE 10.5**).

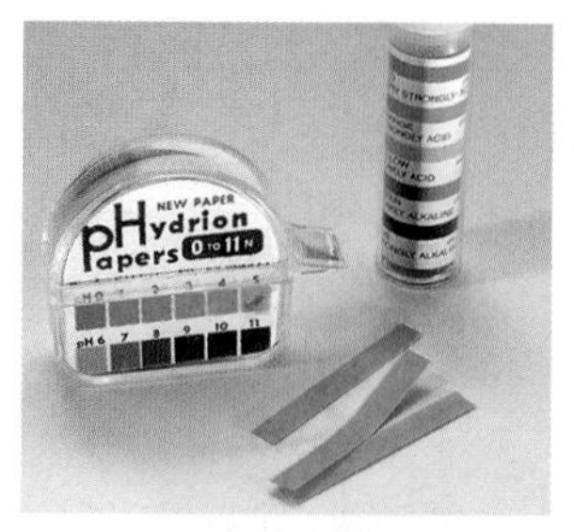

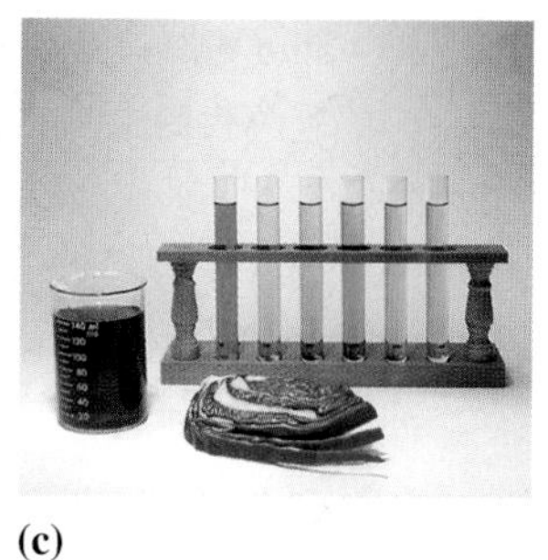

(a) (b) (c)

FIGURE 10.5 The pH of a solution can be determined using **(a)** a pH meter, **(b)** pH paper, and **(c)** indicators that turn different colors corresponding to different pH values.

Q If a pH meter reads 4.00, is the solution acidic, basic, or neutral?

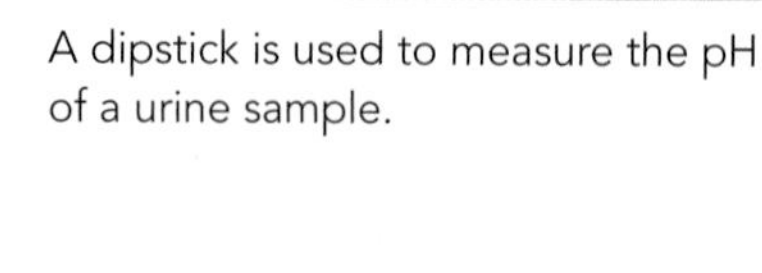

A dipstick is used to measure the pH of a urine sample.

SAMPLE PROBLEM 10.7 pH of Solutions

TRY IT FIRST

Consider the pH of the following body fluids:

Body Fluid	pH
Stomach acid	1.4
Pancreatic juice	8.4
Sweat	4.8
Urine	5.3
Cerebrospinal fluid	7.3

a. Place the pH values of the body fluids on the list in order of most acidic to most basic.

b. Which body fluid has the highest $[H_3O^+]$?

SOLUTION

a. The most acidic body fluid is the one with the lowest pH, and the most basic is the body fluid with the highest pH: stomach acid (1.4), sweat (4.8), urine (5.3), cerebrospinal fluid (7.3), pancreatic juice (8.4).

b. The body fluid with the highest $[H_3O^+]$ would have the lowest pH value, which is stomach acid.

STUDY CHECK 10.7

Which body fluid in Sample Problem 10.7 has the highest $[OH^-]$?

ANSWER

The body fluid with the highest $[OH^-]$ would have the highest pH value, which is pancreatic juice.

TEST

Try Practice Problems 10.35 and 10.36

KEY MATH SKILL

Calculating pH from $[H_3O^+]$

Calculating the pH of Solutions

The pH scale is a logarithmic scale that corresponds to the $[H_3O^+]$ of aqueous solutions. Mathematically, **pH** is the negative logarithm (base 10) of the $[H_3O^+]$.

$$pH = -\log[H_3O^+]$$

Essentially, the negative powers of 10 in the molar concentrations are converted to positive numbers. For example, a lemon juice solution with $[H_3O^+] = 1.0 \times 10^{-2}$ M has a pH of 2.00. This can be calculated using the pH equation:

$$pH = -\log[1.0 \times 10^{-2}]$$
$$pH = -(-2.00)$$
$$= 2.00$$

The number of *decimal places* in the pH value is the same as the number of significant figures in the $[H_3O^+]$. The number to the left of the decimal point in the pH value is the power of 10.

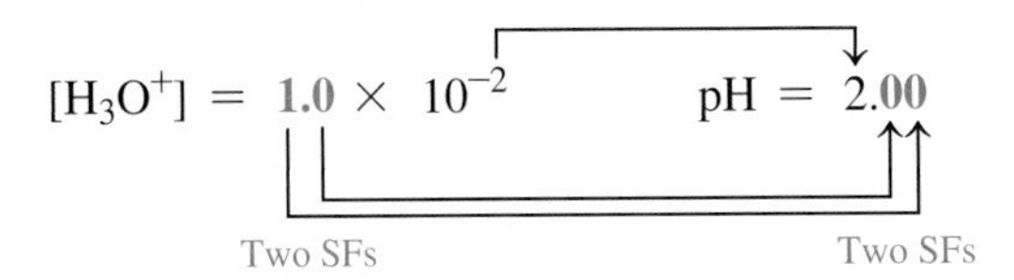

Because pH is a log scale, a change of one pH unit corresponds to a tenfold change in $[H_3O^+]$. It is important to note that the pH decreases as the $[H_3O^+]$ increases. For example, a solution with a pH of 2.00 has a $[H_3O^+]$ that is ten times greater than a solution with a pH of 3.00 and 100 times greater than a solution with a pH of 4.00. The pH of a solution is calculated from the $[H_3O^+]$ by using the *log* key and changing the sign as shown in Sample Problem 10.8.

ENGAGE

Explain why 6.00 but not 6.0 is the correct pH for $[H_3O^+] = 1.0 \times 10^{-6}$ M.

SAMPLE PROBLEM 10.8 Calculating pH from $[H_3O^+]$

TRY IT FIRST

Aspirin, which is acetylsalicylic acid, was the first nonsteroidal anti-inflammatory drug (NSAID) used to alleviate pain and fever. If a solution of aspirin has a $[H_3O^+] = 1.7 \times 10^{-3}$ M, what is the pH of the solution?

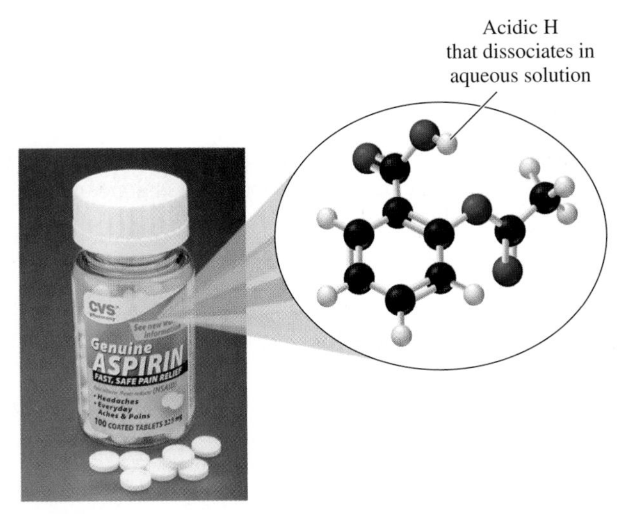

Aspirin, acetylsalicylic acid, is a weak acid.

SOLUTION GUIDE

STEP 1 State the given and needed quantities.

ANALYZE THE PROBLEM	Given	Need	Connect
	$[H_3O^+] = 1.7 \times 10^{-3}$ M	pH	pH equation

STEP 2 Enter the $[H_3O^+]$ into the pH equation and calculate.

$$pH = -\log[H_3O^+] = -\log[1.7 \times 10^{-3}]$$

Calculator Procedure	Calculator Display
1.7 (EE or EXP) (+/−) 3 (log) (+/−) (=) or (+/−) (log) 1.7 (EE or EXP) (+/−) 3 (=)	2.769551079

Be sure to check the instructions for your calculator. Different calculators can have different methods for pH calculation.

STEP 3 **Adjust the number of SFs on the *right* of the decimal point.** In a pH value, the number to the *left* of the decimal point is an *exact* number derived from the power of 10. Thus, the two SFs in the coefficient determine that there are two SFs after the decimal point in the pH value.

Coefficient		**Power of ten**	
1.7	×	10^{-3} M	$pH = -\log[1.7 \times 10^{-3}] = 2.77$
Two SFs		Exact	Exact, Two SFs

STUDY CHECK 10.8

What is the pH of bleach with $[H_3O^+] = 4.2 \times 10^{-12}$ M?

ANSWER

pH = 11.38

KEY MATH SKILL

Calculating $[H_3O^+]$ from pH

Calculating $[H_3O^+]$ from pH

If we are given the pH of the solution and asked to determine the $[H_3O^+]$, we need to reverse the calculation of pH.

$$[H_3O^+] = 10^{-pH}$$

For example, if the pH of a solution is 3.0, we can substitute it into this equation. The number of significant figures in $[H_3O^+]$ is equal to the number of decimal places in the pH value.

$$[H_3O^+] = 10^{-pH} = 10^{-3.0} = 1 \times 10^{-3}\text{ M}$$

For pH values that are not whole numbers, the calculation requires the use of the 10^x key, which is usually a *2nd function* key. On some calculators, this operation is done using the inverse log equation as shown in Sample Problem 10.9.

SAMPLE PROBLEM 10.9 Calculating $[H_3O^+]$ from pH

TRY IT FIRST

Calculate $[H_3O^+]$ for a urine sample, which has a pH of 7.5.

SOLUTION GUIDE

STEP 1 State the given and needed quantities.

ANALYZE THE PROBLEM	Given	Need	Connect
	pH = 7.5	$[H_3O^+]$	$[H_3O^+] = 10^{-pH}$

STEP 2 Enter the pH value into the inverse log equation and calculate.

$$[H_3O^+] = 10^{-pH} = 10^{-7.5}$$

Calculator Procedure

[2nd] [log] [+/−] 7.5 [=] or 7.5 [+/−] [2nd] [log] [=]

Calculator Display

3.16227766E−08

Be sure to check the instructions for your calculator. Different calculators can have different methods for this calculation.

STEP 3 Adjust the SFs for the coefficient. Because the pH value 7.5 has one digit to the *right* of the decimal point, the coefficient for $[H_3O^+]$ is written with one SF.

$$[H_3O^+] = 3 \times 10^{-8}\text{ M}$$

One SF

STUDY CHECK 10.9

What are the $[H_3O^+]$ and $[OH^-]$ of Diet Coke that has a pH of 3.17?

ANSWER

$[H_3O^+] = 6.8 \times 10^{-4}$ M, $[OH^-] = 1.5 \times 10^{-11}$ M

The pH of Diet Coke is 3.17.

A comparison of $[H_3O^+]$, $[OH^-]$, and their corresponding pH values is given in **TABLE 10.6**.

TABLE 10.6 A Comparison of $[H_3O^+]$, $[OH^-]$, and Corresponding pH Values

$[H_3O^+]$	pH	$[OH^-]$	
10^0	0	10^{-14}	
10^{-1}	1	10^{-13}	
10^{-2}	2	10^{-12}	
10^{-3}	3	10^{-11}	
10^{-4}	4	10^{-10}	Acidic
10^{-5}	5	10^{-9}	
10^{-6}	6	10^{-8}	
10^{-7}	7	10^{-7}	Neutral
10^{-8}	8	10^{-6}	
10^{-9}	9	10^{-5}	
10^{-10}	10	10^{-4}	Basic
10^{-11}	11	10^{-3}	
10^{-12}	12	10^{-2}	
10^{-13}	13	10^{-1}	
10^{-14}	14	10^0	

TEST

Try Practice Problems 10.37 to 10.44

CHEMISTRY LINK TO HEALTH

Stomach Acid, HCl

Gastric acid, which contains HCl, is produced by parietal cells that line the stomach. When the stomach expands with the intake of food, the gastric glands begin to secrete a strongly acidic solution of HCl. In a single day, a person may secrete 2000 mL of gastric juice, which contains hydrochloric acid, mucins, and the enzymes pepsin and lipase.

The HCl in the gastric juice activates a digestive enzyme from the chief cells called *pepsinogen* to form *pepsin*, which breaks down proteins in food entering the stomach. The secretion of HCl continues until the stomach has a pH of about 1.5, which is the optimum for activating the digestive enzymes without ulcerating the stomach lining. In addition, the low pH destroys bacteria that reach the stomach. Normally, large quantities of viscous mucus are secreted within the stomach to protect its lining from acid and enzyme damage. Gastric acid may also form under conditions of stress when the nervous system activates the production of HCl. As the contents of the stomach move into the small intestine, cells produce bicarbonate that neutralizes the gastric acid until the pH is about 5.

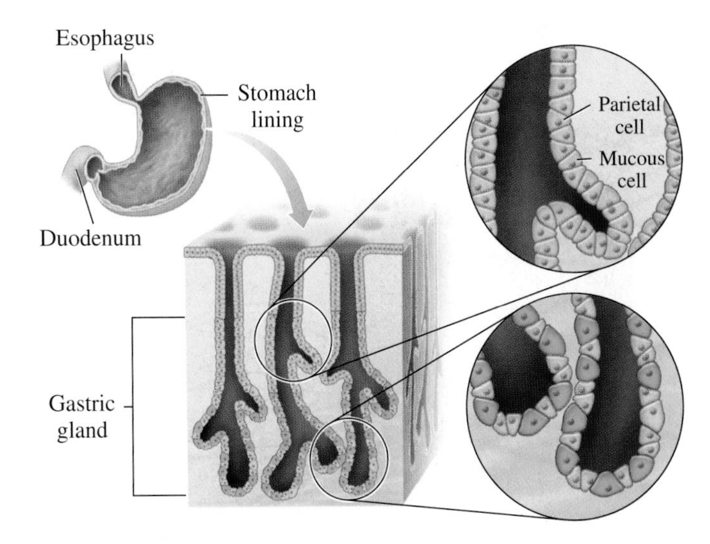

Parietal cells in the lining of the stomach secrete gastric acid HCl.

PRACTICE PROBLEMS

10.6 The pH Scale

LEARNING GOAL Calculate the pH of a solution from $[H_3O^+]$; given the pH, calculate $[H_3O^+]$.

10.35 State whether each of the following is acidic, basic, or neutral:
a. blood plasma, pH 7.38 **b.** vinegar, pH 2.8
c. drain cleaner, pH 11.2 **d.** coffee, pH 5.52
e. tomatoes, pH 4.2 **f.** chocolate cake, pH 7.6

10.36 State whether each of the following is acidic, basic, or neutral:
a. soda, pH 3.22 **b.** shampoo, pH 5.7
c. laundry detergent, pH 9.44 **d.** rain, pH 5.8
e. honey, pH 3.9 **f.** cheese, pH 4.9

10.37 Why does a neutral solution have a pH of 7.0?

10.38 If you know the $[OH^-]$, how can you determine the pH of a solution?

10.39 Calculate the pH of each solution given the following:
a. $[H_3O^+] = 1 \times 10^{-4}$ M **b.** $[H_3O^+] = 3 \times 10^{-9}$ M
c. $[OH^-] = 1 \times 10^{-5}$ M **d.** $[OH^-] = 2.5 \times 10^{-11}$ M
e. $[H_3O^+] = 6.7 \times 10^{-8}$ M **f.** $[OH^-] = 8.2 \times 10^{-4}$ M

10.40 Calculate the pH of each solution given the following:
a. $[H_3O^+] = 1 \times 10^{-8}$ M **b.** $[H_3O^+] = 5 \times 10^{-6}$ M
c. $[OH^-] = 1 \times 10^{-2}$ M **d.** $[OH^-] = 8.0 \times 10^{-3}$ M
e. $[H_3O^+] = 4.7 \times 10^{-2}$ M **f.** $[OH^-] = 3.9 \times 10^{-6}$ M

10.41 Complete the following table:

$[H_3O^+]$	$[OH^-]$	pH	Acidic, Basic, or Neutral?
	1.0×10^{-6} M		
		3.49	
2.8×10^{-5} M			

10.42 Complete the following table:

$[H_3O^+]$	$[OH^-]$	pH	Acidic, Basic, or Neutral?
		10.00	
			Neutral
6.4×10^{-12} M			

Clinical Applications

10.43 A patient with severe metabolic acidosis has a blood plasma pH of 6.92. What is the $[H_3O^+]$ of the blood plasma?

10.44 A patient with respiratory alkalosis has a blood plasma pH of 7.58. What is the $[H_3O^+]$ of the blood plasma?

REVIEW
Balancing a Chemical Equation (7.4)
Using Concentration as a Conversion Factor (9.4)

CORE CHEMISTRY SKILL
Writing Equations for Reactions of Acids and Bases

10.7 Reactions of Acids and Bases

LEARNING GOAL Write balanced equations for reactions of acids with metals, carbonates or bicarbonates, and bases; calculate the molarity or volume of an acid from titration information.

Typical reactions of acids and bases include the reactions of acids with metals, carbonate or bicarbonate ions, and bases. For example, when you drop an antacid tablet in water, the bicarbonate ion and citric acid in the tablet react to produce carbon dioxide bubbles, water, and a salt. A *salt* is an ionic compound that does not have H^+ as the cation or OH^- as the anion.

Acids and Metals

Acids react with certain metals to produce hydrogen gas (H_2) and a salt. Metals that react with acids include potassium, sodium, calcium, magnesium, aluminum, zinc, iron, and tin. In these replacement reactions, the metal ion replaces the hydrogen in the acid.

$$\underset{\text{Metal}}{Mg(s)} + \underset{\text{Acid}}{2HCl(aq)} \longrightarrow \underset{\text{Hydrogen}}{H_2(g)} + \underset{\text{Salt}}{MgCl_2(aq)}$$

$$\underset{\text{Metal}}{Zn(s)} + \underset{\text{Acid}}{2HNO_3(aq)} \longrightarrow \underset{\text{Hydrogen}}{H_2(g)} + \underset{\text{Salt}}{Zn(NO_3)_2(aq)}$$

Magnesium reacts rapidly with acid and forms H_2 gas and a salt of magnesium.

SAMPLE PROBLEM 10.10 Equations for Metals and Acids

TRY IT FIRST

Write a balanced equation for the reaction of Al(*s*) with HCl(*aq*).

ANALYZE THE PROBLEM	Given	Need	Connect
	Al, HCl	balanced equation	products: H_2 and a salt

SOLUTION GUIDE

STEP 1 **Write the reactants and products.** When a metal reacts with an acid, the products are hydrogen gas and a salt. The unbalanced equation is written as

$$Al(s) + HCl(aq) \longrightarrow H_2(g) + salt(aq)$$

STEP 2 **Write the formula of the salt.** When $Al(s)$ reacts, it forms Al^{3+}, which is balanced by $3Cl^-$ from HCl.

$$Al(s) + HCl(aq) \longrightarrow H_2(g) + AlCl_3(aq)$$

STEP 3 **Balance the equation.**

$$2Al(s) + 6HCl(aq) \longrightarrow 3H_2(g) + 2AlCl_3(aq)$$

STUDY CHECK 10.10

Write the balanced equation when $Ca(s)$ reacts with $HBr(aq)$.

ANSWER

$$Ca(s) + 2HBr(aq) \longrightarrow H_2(g) + CaBr_2(aq)$$

Acids React with Carbonates and Bicarbonates

When an acid is added to a carbonate or bicarbonate, the products are carbon dioxide gas, water, and a salt. The acid reacts with CO_3^{2-} or HCO_3^- to produce carbonic acid, H_2CO_3, which breaks down rapidly to CO_2 and H_2O.

$$\underset{\text{Acid}}{2HCl(aq)} + \underset{\text{Carbonate}}{Na_2CO_3(aq)} \longrightarrow \underset{\text{Carbon dioxide}}{CO_2(g)} + \underset{\text{Water}}{H_2O(l)} + \underset{\text{Salt}}{2NaCl(aq)}$$

$$\underset{\text{Acid}}{HBr(aq)} + \underset{\text{Bicarbonate}}{NaHCO_3(aq)} \longrightarrow \underset{\text{Carbon dioxide}}{CO_2(g)} + \underset{\text{Water}}{H_2O(l)} + \underset{\text{Salt}}{NaBr(aq)}$$

When sodium bicarbonate (baking soda) reacts with an acid (vinegar), the products are carbon dioxide gas, water, and a salt.

Acids and Hydroxides: Neutralization

Neutralization is a reaction between a strong or weak acid and a strong base to produce water and a salt. The H^+ of the acid and the OH^- of the base combine to form water. The salt is the combination of the cation from the base and the anion from the acid. We can write the following equation for the neutralization reaction between HCl and NaOH:

$$\underset{\text{Acid}}{HCl(aq)} + \underset{\text{Base}}{NaOH(aq)} \longrightarrow \underset{\text{Water}}{H_2O(l)} + \underset{\text{Salt}}{NaCl(aq)}$$

If we write the strong acid HCl and the strong base NaOH as ions, we see that H^+ reacts with OH^- to form water, leaving the ions Na^+ and Cl^- in solution.

$$H^+(aq) + Cl^-(aq) + Na^+(aq) + OH^-(aq) \longrightarrow H_2O(l) + Na^+(aq) + Cl^-(aq)$$

When we omit the *spectator ions* that do not change during the reaction (Na^+ and Cl^-), we obtain the net ionic equation.

$$H^+(aq) + \cancel{Cl^-(aq)} + \cancel{Na^+(aq)} + OH^-(aq) \longrightarrow H_2O(l) + \cancel{Na^+(aq)} + \cancel{Cl^-(aq)}$$

The net ionic equation for the neutralization of H^+ and OH^- to form H_2O is

$$H^+(aq) + OH^-(aq) \longrightarrow H_2O(l) \quad \text{Net ionic equation}$$

Balancing Neutralization Equations

In a neutralization reaction, one H^+ always reacts with one OH^-. Therefore, a neutralization equation may need coefficients to balance H^+ from the acid with the OH^- from the base as shown in Sample Problem 10.11.

SAMPLE PROBLEM 10.11 Balancing Equations for Reactions of Acids and Bases

TRY IT FIRST

Write the balanced equation for the neutralization of $HCl(aq)$ and $Ba(OH)_2(s)$.

ANALYZE THE PROBLEM	Given	Need	Connect
	HCl, $Ba(OH)_2$	balanced equation	neutralization products

SOLUTION GUIDE

STEP 1 Write the reactants and products.

$$HCl(aq) + Ba(OH)_2(s) \longrightarrow H_2O(l) + \text{salt}$$

STEP 2 Balance the H^+ in the acid with the OH^- in the base. Placing a coefficient of 2 in front of the HCl provides $2H^+$ for the $2OH^-$ from $Ba(OH)_2$.

$$2HCl(aq) + Ba(OH)_2(s) \longrightarrow H_2O(l) + \text{salt}$$

STEP 3 Balance the H_2O with the H^+ and the OH^-. Use a coefficient of 2 in front of H_2O to balance $2H^+$ and $2OH^-$.

$$2HCl(aq) + Ba(OH)_2(s) \longrightarrow 2H_2O(l) + \text{salt}$$

STEP 4 Write the formula of the salt from the remaining ions. Use the ions Ba^{2+} and $2Cl^-$ to write the formula for the salt as $BaCl_2$.

$$2HCl(aq) + Ba(OH)_2(s) \longrightarrow 2H_2O(l) + BaCl_2(aq)$$

STUDY CHECK 10.11

Write the balanced equation for the reaction between $H_2SO_4(aq)$ and $NaHCO_3(aq)$.

ANSWER

$$H_2SO_4(aq) + 2NaHCO_3(s) \longrightarrow 2CO_2(g) + 2H_2O(l) + Na_2SO_4(aq)$$

TEST

Try Practice Problems 10.45 to 10.50

CORE CHEMISTRY SKILL

Calculating Molarity or Volume of an Acid or Base in a Titration

Acid–Base Titration

Suppose we need to find the molarity of a solution of HCl, which has an unknown concentration. We can do this by a laboratory procedure called **titration** in which we neutralize an acid sample with a known amount of base. In a titration, we place a measured volume of the acid in a flask and add a few drops of an *indicator*, such as phenolphthalein. An indicator is a compound that dramatically changes color when the pH of the solution changes. In an acidic solution, phenolphthalein is colorless. Then we fill a buret with a NaOH solution of known molarity and carefully add NaOH solution to neutralize the acid in the flask (see **FIGURE 10.6**). We know that neutralization has taken place when the phenolphthalein in the solution changes from colorless to pink. This is called the neutralization *endpoint*. From the measured volume of the NaOH solution and its molarity, we calculate the number of moles of NaOH, the moles of acid, and use the measured volume of acid to calculate its concentration.

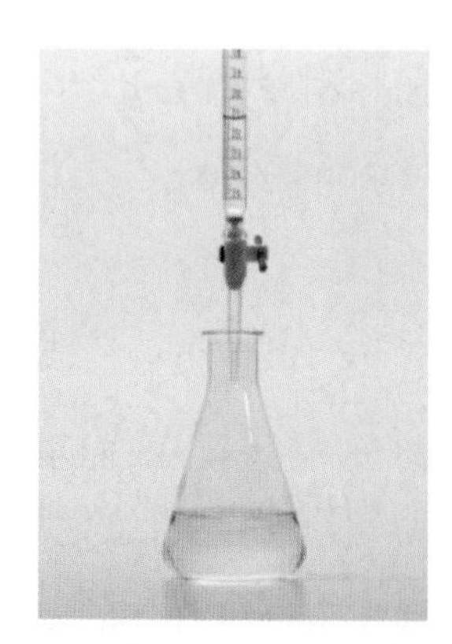

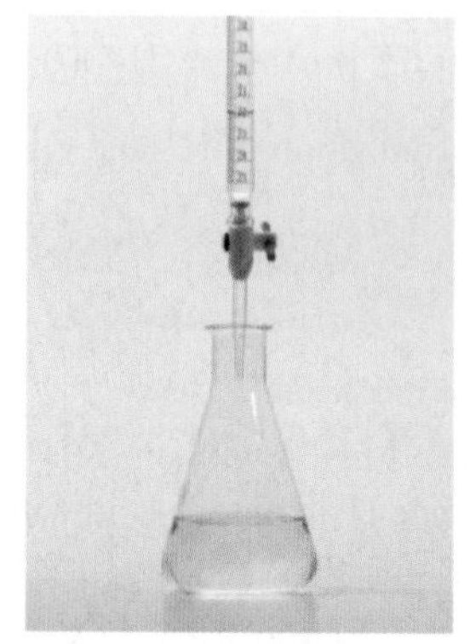

FIGURE 10.6 The titration of an acid. A known volume of an acid is placed in a flask with an indicator and titrated with a measured volume of a base solution, such as NaOH, to the neutralization endpoint.

Q What data is needed to determine the molarity of the acid in the flask?

SAMPLE PROBLEM 10.12 Titration of an Acid

TRY IT FIRST

A 25.0-mL (0.0250 L) sample of an HCl solution is placed in a flask with a few drops of phenolphthalein (indicator). If 32.6 mL of a 0.185 M NaOH solution is needed to reach the endpoint, what is the molarity of the HCl solution?

$$NaOH(aq) + HCl(aq) \longrightarrow H_2O(l) + NaCl(aq)$$

SOLUTION GUIDE

STEP 1 State the given and needed quantities and concentrations.

	Given	Need	Connect
ANALYZE THE PROBLEM	0.0250 L of HCl solution, 32.6 mL of 0.185 M NaOH solution	molarity of the HCl solution	molarity, mole–mole factor
	Neutralization Equation		
	$NaOH(aq) + HCl(aq) \longrightarrow H_2O(l) + NaCl(aq)$		

STEP 2 Write a plan to calculate the molarity.

L of NaOH solution → (Molarity) → moles of NaOH → (Mole–mole factor) → moles of HCl → (Divide by liters) → molarity of HCl solution

STEP 3 State equalities and conversion factors, including concentrations.

1 L of NaOH solution = 0.185 mole of NaOH

$$\frac{0.185 \text{ mole NaOH}}{1 \text{ L NaOH solution}} \text{ and } \frac{1 \text{ L NaOH solution}}{0.185 \text{ mole NaOH}}$$

1 mole of HCl = 1 mole of NaOH

$$\frac{1 \text{ mole HCl}}{1 \text{ mole NaOH}} \text{ and } \frac{1 \text{ mole NaOH}}{1 \text{ mole HCl}}$$

STEP 4 Set up the problem to calculate the needed quantity.

$$0.0326 \cancel{\text{L NaOH solution}} \times \frac{0.185 \cancel{\text{mole NaOH}}}{1 \cancel{\text{L NaOH solution}}} \times \frac{1 \text{ mole HCl}}{1 \cancel{\text{mole NaOH}}} = 0.006\,03 \text{ mole of HCl}$$

$$\text{molarity of HCl} = \frac{0.006\,03 \text{ mole HCl}}{0.0250 \text{ L solution}} = 0.241 \text{ M HCl solution}$$

STUDY CHECK 10.12

What is the molarity of an HCl solution, if 28.6 mL of a 0.175 M NaOH solution is needed to neutralize a 25.0-mL sample of the HCl solution?

ANSWER

0.200 M HCl solution

TEST

Try Practice Problems 10.51 to 10.54

INTERACTIVE VIDEO

Titration of an Acid

CHEMISTRY LINK TO HEALTH

Antacids

Antacids are substances used to neutralize excess stomach acid (HCl). Some antacids are mixtures of aluminum hydroxide and magnesium hydroxide. These hydroxides are not very soluble in water, so the levels of available OH^- are not damaging to the intestinal tract. However, aluminum hydroxide has the side effects of producing constipation and binding phosphate in the intestinal tract, which may cause weakness and loss of appetite. Magnesium hydroxide has a laxative effect. These side effects are less likely when a combination of the antacids is used.

$$Al(OH)_3(s) + 3HCl(aq) \longrightarrow 3H_2O(l) + AlCl_3(aq)$$

$$Mg(OH)_2(s) + 2HCl(aq) \longrightarrow 2H_2O(l) + MgCl_2(aq)$$

Some antacids use calcium carbonate to neutralize excess stomach acid. About 10% of the calcium is absorbed into the bloodstream, where it elevates the levels of serum calcium. Calcium carbonate is not recommended for patients who have peptic ulcers or a tendency to form kidney stones, which typically consist of an insoluble calcium salt.

$$CaCO_3(s) + 2HCl(aq) \longrightarrow CO_2(g) + H_2O(l) + CaCl_2(aq)$$

Still other antacids contain sodium bicarbonate. This type of antacid neutralizes excess gastric acid, increases blood pH, but also elevates sodium levels in the body fluids. It also is not recommended in the treatment of peptic ulcers.

$$NaHCO_3(s) + HCl(aq) \longrightarrow CO_2(g) + H_2O(l) + NaCl(aq)$$

The neutralizing substances in some antacid preparations are given in **TABLE 10.7**.

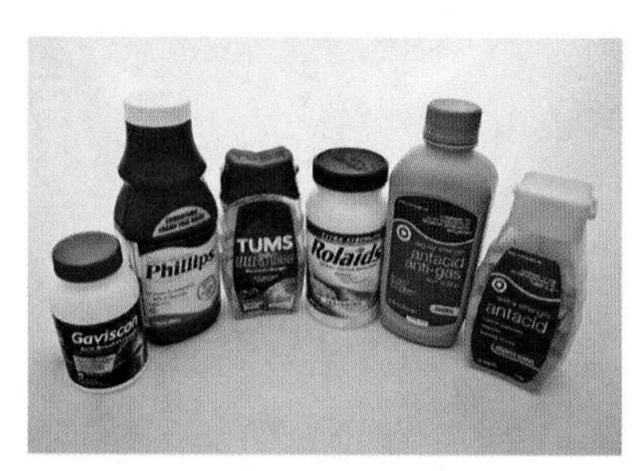

Antacids neutralize excess stomach acid.

TABLE 10.7 Basic Compounds in Some Antacids

Antacid	Base(s)
Amphojel	$Al(OH)_3$
Milk of magnesia	$Mg(OH)_2$
Mylanta, Maalox, Di-Gel, Gelusil, Riopan	$Mg(OH)_2$, $Al(OH)_3$
Bisodol, Rolaids	$CaCO_3$, $Mg(OH)_2$
Titralac, Tums, Pepto-Bismol	$CaCO_3$
Alka-Seltzer	$NaHCO_3$, $KHCO_3$

PRACTICE PROBLEMS

10.7 Reactions of Acids and Bases

LEARNING GOAL Write balanced equations for reactions of acids with metals, carbonates or bicarbonates, and bases; calculate the molarity or volume of an acid from titration information.

10.45 Complete and balance the equation for each of the following reactions:
- **a.** $ZnCO_3(s) + HBr(aq) \longrightarrow$
- **b.** $Zn(s) + HCl(aq) \longrightarrow$
- **c.** $HCl(aq) + NaHCO_3(s) \longrightarrow$
- **d.** $H_2SO_4(aq) + Mg(OH)_2(s) \longrightarrow$

10.46 Complete and balance the equation for each of the following reactions:
- **a.** $KHCO_3(s) + HCl(aq) \longrightarrow$
- **b.** $Ca(s) + H_2SO_4(aq) \longrightarrow$
- **c.** $H_2SO_4(aq) + Ca(OH)_2(s) \longrightarrow$
- **d.** $Na_2CO_3(s) + H_2SO_4(aq) \longrightarrow$

10.47 Balance each of the following neutralization equations:
- **a.** $HCl(aq) + Mg(OH)_2(s) \longrightarrow H_2O(l) + MgCl_2(aq)$
- **b.** $H_3PO_4(aq) + LiOH(aq) \longrightarrow H_2O(l) + Li_3PO_4(aq)$

10.48 Balance each of the following neutralization equations:
- **a.** $HNO_3(aq) + Ba(OH)_2(s) \longrightarrow H_2O(l) + Ba(NO_3)_2(aq)$
- **b.** $H_2SO_4(aq) + Al(OH)_3(s) \longrightarrow H_2O(l) + Al_2(SO_4)_3(aq)$

10.49 Write a balanced equation for the neutralization of each of the following:
- **a.** $H_2SO_4(aq)$ and $NaOH(aq)$ **b.** $HCl(aq)$ and $Fe(OH)_3(s)$
- **c.** $H_2CO_3(aq)$ and $Mg(OH)_2(s)$

10.50 Write a balanced equation for the neutralization of each of the following:
- **a.** $H_3PO_4(aq)$ and $NaOH(aq)$ **b.** $HI(aq)$ and $LiOH(aq)$
- **c.** $HNO_3(aq)$ and $Ca(OH)_2(s)$

10.51 What is the molarity of a solution of HCl if 5.00 mL of the HCl solution is titrated with 28.6 mL of a 0.145 M NaOH solution?

$$HCl(aq) + NaOH(aq) \longrightarrow H_2O(l) + NaCl(aq)$$

10.52 What is the molarity of an acetic acid solution if 25.0 mL of the $HC_2H_3O_2$ solution is titrated with 29.7 mL of a 0.205 M KOH solution?

$$HC_2H_3O_2(aq) + KOH(aq) \longrightarrow H_2O(l) + KC_2H_3O_2(aq)$$

10.53 If 32.8 mL of a 0.162 M NaOH solution is required to titrate 25.0 mL of a solution of H_2SO_4, what is the molarity of the H_2SO_4 solution?

$$H_2SO_4(aq) + 2KOH(aq) \longrightarrow 2H_2O(l) + K_2SO_4(aq)$$

10.54 If 38.2 mL of a 0.163 M KOH solution is required to titrate 25.0 mL of a solution of H_2SO_4, what is the molarity of the H_2SO_4 solution?

$$H_2SO_4(aq) + 2NaOH(aq) \longrightarrow 2H_2O(l) + Na_2SO_4(aq)$$

10.8 Buffers

LEARNING GOAL Describe the role of buffers in maintaining the pH of a solution.

The lungs and the kidneys are the primary organs that regulate the pH of body fluids, including blood and urine. Major changes in the pH of the body fluids can severely affect biological activities within the cells. *Buffers* are present to prevent large fluctuations in pH.

The pH of water and most solutions changes drastically when a small amount of acid or base is added. However, when an acid or base is added to a *buffer solution*, there is little change in pH. A **buffer solution** maintains the pH of a solution by neutralizing small amounts of added acid or base. In the human body, whole blood contains plasma, white blood cells and platelets, and red blood cells. Blood plasma contains buffers that maintain a consistent pH of about 7.4. If the pH of the blood plasma goes slightly above or below 7.4, changes in our oxygen levels and our metabolic processes can be drastic enough to cause death. Even though we obtain acids and bases from foods and cellular reactions, the buffers in the body absorb those compounds so effectively that the pH of our blood plasma remains essentially unchanged (see **FIGURE 10.7**).

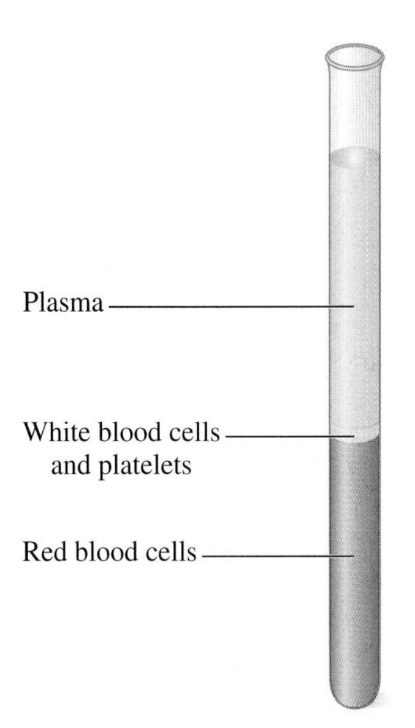

Whole blood consists of plasma, white blood cells and platelets, and red blood cells.

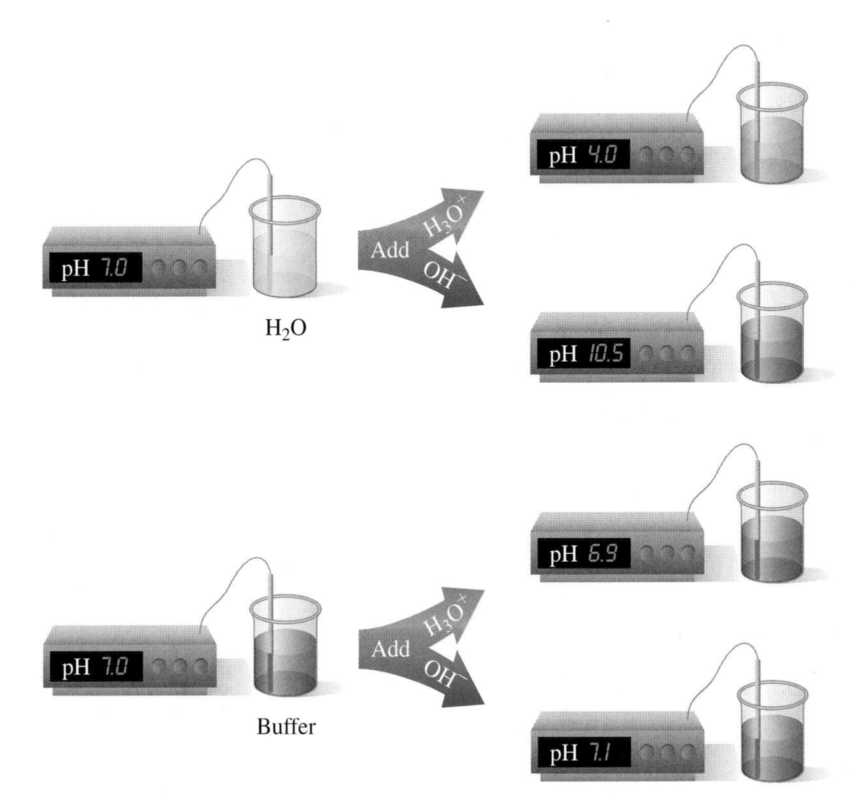

FIGURE 10.7 Adding an acid or a base to water changes the pH drastically, but a buffer resists pH change when small amounts of acid or base are added.

Q Why does the pH change several pH units when acid is added to water, but not when acid is added to a buffer?

In a buffer, an acid must be present to react with any OH^- that is added, and a base must be available to react with any added H_3O^+. However, that acid and base must not neutralize each other. Therefore, a combination of an acid–base conjugate pair is used in buffers. Most buffer solutions consist of nearly equal concentrations of a weak acid and a salt containing its conjugate base. Buffers may also contain a weak base and a salt of the weak base, which contains its conjugate acid.

ENGAGE

Why does a buffer require the presence of a weak acid or weak base and the salt of that weak acid or weak base?

For example, a typical buffer can be made from the weak acid acetic acid ($HC_2H_3O_2$) and its salt, sodium acetate ($NaC_2H_3O_2$). As a weak acid, acetic acid dissociates slightly in water to form H_3O^+ and a very small amount of $C_2H_3O_2^-$. The addition of its salt, sodium acetate, provides a much larger concentration of the acetate ion ($C_2H_3O_2^-$), which is necessary for its buffering capability.

$$\underset{\text{Large amount}}{HC_2H_3O_2(aq)} + H_2O(l) \rightleftharpoons H_3O^+(aq) + \underset{\text{Large amount}}{C_2H_3O_2^-(aq)}$$

We can now describe how this buffer solution maintains the $[H_3O^+]$. When a small amount of acid is added, the additional H_3O^+ combines with the acetate ion, $C_2H_3O_2^-$, causing the equilibrium to shift in the direction of the reactants, acetic acid and water. There will be a slight decrease in the $[C_2H_3O_2^-]$ and a slight increase in the $[HC_2H_3O_2]$, but both the $[H_3O^+]$ and pH are maintained (see **FIGURE 10.8**).

ENGAGE

Which part of a buffer neutralizes any H_3O^+ that is added?

$$HC_2H_3O_2(aq) + H_2O(l) \longleftarrow H_3O^+(aq) + C_2H_3O_2^-(aq)$$

Equilibrium shifts in the direction of the reactants

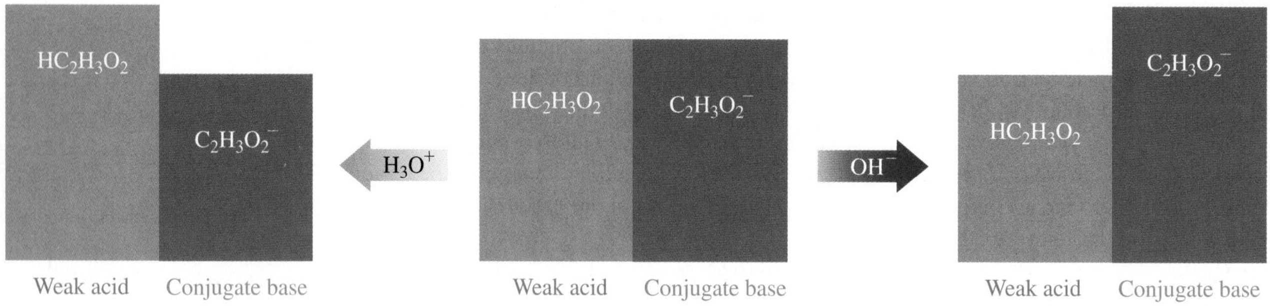

FIGURE 10.8 The buffer described here consists of about equal concentrations of acetic acid ($HC_2H_3O_2$) and its conjugate base, acetate ion ($C_2H_3O_2^-$). A small amount of H_3O^+ added to the buffer reacts with $C_2H_3O_2^-$, whereas a small amount of OH^- added to the buffer neutralizes $HC_2H_3O_2$. The pH of the solution is maintained as long as the added amounts of acid or base are small compared to the concentrations of the buffer components.

Q How does this acetic acid–acetate ion buffer maintain pH?

If a small amount of base is added to this same buffer solution, it is neutralized by the acetic acid, $HC_2H_3O_2$, which shifts the equilibrium in the direction of the products, water and acetate ion. The $[HC_2H_3O_2]$ decreases slightly and the $[C_2H_3O_2^-]$ increases slightly, but again the $[H_3O^+]$ and thus the pH of the solution are maintained.

$$HC_2H_3O_2(aq) + OH^-(aq) \longrightarrow H_2O(l) + C_2H_3O_2^-(aq)$$

Equilibrium shifts in the direction of the products

SAMPLE PROBLEM 10.13 Identifying Buffer Solutions

TRY IT FIRST

Indicate whether each of the following would make a buffer solution:

a. HCl, a strong acid, and NaCl
b. H_3PO_4, a weak acid
c. HF, a weak acid, and NaF

SOLUTION

a. No. A buffer requires a weak acid and a salt containing its conjugate base.
b. No. A weak acid is part of a buffer, but the salt containing the conjugate base of the weak acid is also needed.
c. Yes. This mixture would be a buffer since it contains a weak acid and a salt containing its conjugate base.

STUDY CHECK 10.13

A buffer is made from the weak acid $HCHO_2$ and its salt, $KCHO_2$. When H_3O^+ is added, is it neutralized by **(1)** the salt, **(2)** H_2O, **(3)** OH^-, or **(4)** the acid?

ANSWER

(1) the salt

TEST

Try Practice Problems 10.55 to 10.62

CHEMISTRY LINK TO HEALTH

Buffers in the Blood Plasma

The arterial blood plasma has a normal pH of 7.35 to 7.45. If changes in H_3O^+ lower the pH below 6.8 or raise it above 8.0, cells cannot function properly and death may result. In our cells, CO_2 is continually produced as an end product of cellular metabolism. Some CO_2 is carried to the lungs for elimination, and the rest dissolves in body fluids such as plasma and saliva, forming carbonic acid. As a weak acid, carbonic acid dissociates to give bicarbonate, HCO_3^-, and H_3O^+. More of the anion HCO_3^- is supplied by the kidneys to give an important buffer system in the body fluid—the H_2CO_3/HCO_3^- buffer.

$$CO_2(g) + H_2O(l) \rightleftarrows H_2CO_3(aq) \rightleftarrows H_3O^+(aq) + HCO_3^-(aq)$$

Excess H_3O^+ entering the body fluids reacts with the HCO_3^-, and excess OH^- reacts with the carbonic acid.

$$H_2CO_3(aq) + H_2O(l) \longleftarrow H_3O^+(aq) + HCO_3^-(aq)$$

Equilibrium shifts in the direction of the reactants

$$H_2CO_3(aq) + OH^-(aq) \longrightarrow H_2O(l) + HCO_3^-(aq)$$

Equilibrium shifts in the direction of the products

In the body, the concentration of carbonic acid is closely associated with the partial pressure of CO_2, P_{CO_2}. **TABLE 10.8** lists the normal values for arterial blood. If the CO_2 level rises, increasing $[H_2CO_3]$, the equilibrium shifts to produce more H_3O^+, which lowers the pH. This condition is called **acidosis**. Difficulty with ventilation or gas diffusion can lead to respiratory acidosis, which can happen in emphysema or when an accident or depressive drugs affect the medulla of the brain.

A lowering of the CO_2 level leads to a high blood pH, a condition called **alkalosis**. Excitement, trauma, or a high temperature may cause a person to hyperventilate, which expels large amounts of CO_2. As the partial pressure of CO_2 in the blood falls below normal, the equilibrium shifts from H_2CO_3 to CO_2 and H_2O. This shift decreases the $[H_3O^+]$ and raises the pH. The kidneys also regulate H_3O^+ and HCO_3^-, but they do so more slowly than the adjustment made by the lungs through ventilation.

TABLE 10.9 lists some of the conditions that lead to changes in the blood pH and some possible treatments.

TABLE 10.8 Normal Values for Blood Buffer in Arterial Blood

P_{CO_2}	40 mmHg
H_2CO_3	2.4 mmoles/L of plasma
HCO_3^-	24 mmoles/L of plasma
pH	7.35 to 7.45

TABLE 10.9 Acidosis and Alkalosis: Symptoms, Causes, and Treatments

Respiratory Acidosis: $CO_2\uparrow$ pH$\downarrow$	
Symptoms:	Failure to ventilate, suppression of breathing, disorientation, weakness, coma
Causes:	Lung disease blocking gas diffusion (e.g., emphysema, pneumonia, bronchitis, asthma); depression of respiratory center by drugs, cardiopulmonary arrest, stroke, poliomyelitis, or nervous system disorders
Treatment:	Correction of disorder, infusion of bicarbonate
Metabolic Acidosis: $H^+\uparrow$ pH$\downarrow$	
Symptoms:	Increased ventilation, fatigue, confusion
Causes:	Renal disease, including hepatitis and cirrhosis; increased acid production in diabetes mellitus, hyperthyroidism, alcoholism, and starvation; loss of alkali in diarrhea; acid retention in renal failure
Treatment:	Sodium bicarbonate given orally, dialysis for renal failure, insulin treatment for diabetic ketosis
Respiratory Alkalosis: $CO_2\downarrow$ pH$\uparrow$	
Symptoms:	Increased rate and depth of breathing, numbness, light-headedness, tetany
Causes:	Hyperventilation because of anxiety, hysteria, fever, exercise; reaction to drugs such as salicylate, quinine, and antihistamines; conditions causing hypoxia (e.g., pneumonia, pulmonary edema, heart disease)
Treatment:	Elimination of anxiety-producing state, rebreathing into a paper bag
Metabolic Alkalosis: $H^+\downarrow$ pH$\uparrow$	
Symptoms:	Depressed breathing, apathy, confusion
Causes:	Vomiting, diseases of the adrenal glands, ingestion of excess alkali
Treatment:	Infusion of saline solution, treatment of underlying diseases

PRACTICE PROBLEMS

10.8 Buffers

LEARNING GOAL Describe the role of buffers in maintaining the pH of a solution.

10.55 Which of the following represents a buffer system? Explain.

a. NaOH and NaCl
b. H_2CO_3 and $NaHCO_3$
c. HF and KF
d. KCl and NaCl

10.56 Which of the following represents a buffer system? Explain.

a. H_3PO_3
b. $NaNO_3$
c. $HC_2H_3O_2$ and $NaC_2H_3O_2$
d. HCl and NaOH

10.57 Consider the buffer system of hydrofluoric acid, HF, and its salt, NaF.

$$HF(aq) + H_2O(l) \rightleftharpoons H_3O^+(aq) + F^-(aq)$$

a. The purpose of this buffer system is to:
1. maintain [HF] **2.** maintain [F^-]
3. maintain pH

b. The salt of the weak acid is needed to:
1. provide the conjugate base
2. neutralize added H_3O^+
3. provide the conjugate acid

c. If OH^- is added, it is neutralized by:
1. the salt **2.** H_2O **3.** H_3O^+

d. When H_3O^+ is added, the equilibrium shifts in the direction of the:
1. reactants **2.** products
3. does not change

10.58 Consider the buffer system of nitrous acid, HNO_2, and its salt, $NaNO_2$.

$$HNO_2(aq) + H_2O(l) \rightleftharpoons H_3O^+(aq) + NO_2^-(aq)$$

a. The purpose of this buffer system is to:
1. maintain [HNO_2] **2.** maintain [NO_2^-]
3. maintain pH

b. The weak acid is needed to:
1. provide the conjugate base
2. neutralize added OH^-
3. provide the conjugate acid

c. If H_3O^+ is added, it is neutralized by:
1. the salt **2.** H_2O **3.** OH^-

d. When OH^- is added, the equilibrium shifts in the direction of the:
1. reactants **2.** products
3. does not change

Clinical Applications

10.59 Why would the pH of your blood plasma increase if you breathe fast?

10.60 Why would the pH of your blood plasma decrease if you hold your breath?

10.61 Someone with kidney failure excretes urine with large amounts of HCO_3^-. How would this loss of HCO_3^- affect the pH of the blood plasma?

10.62 Someone with severe diabetes obtains energy by the breakdown of fats, which produce large amounts of acidic substances. How would this affect the pH of the blood plasma?

CLINICAL UPDATE Acid Reflux Disease

Larry has not been feeling well lately. He tells his doctor that he has discomfort and a burning feeling in his chest, and a sour taste in his throat and mouth. At times, Larry says he feels bloated after a big meal, has a dry cough, is hoarse, and sometimes has a sore throat. He has tried antacids, but they do not bring any relief.

The doctor tells Larry that he thinks he has acid reflux. At the top of the stomach there is a valve, the lower esophageal sphincter, that normally closes after food passes through it. However, if the valve does not close completely, acid produced in the stomach to digest food can move up into the esophagus, a condition called *acid reflux*. The acid, which is hydrochloric acid, HCl, is produced in the stomach to kill bacteria and microorganisms, and to activate the enzymes we need to break down food.

If acid reflux occurs, the strong acid HCl comes in contact with the lining of the esophagus, where it causes irritation and produces a burning feeling in the chest. Sometimes the pain in the chest is called *heartburn*. If the HCl reflux goes high enough to reach the throat, a sour taste may be noticed in the mouth. If Larry's symptoms occur three or more times a week, he may have a chronic condition known as *acid reflux disease* or *gastroesophageal reflux disease* (GERD).

Larry's doctor orders an *esophageal pH test* in which the amount of acid entering the esophagus from the stomach is measured over 24 h. A probe that measures the pH is inserted into the lower esophagus above the esophageal sphincter. The pH measurements indicate a reflux episode each time the pH drops to 4 or less.

In the 24-h period, Larry has several reflux episodes and his doctor determines that he has chronic GERD. He and Larry discuss treatment for GERD, which includes eating smaller meals, not lying down for 3 h after eating, making dietary changes, and losing weight. Antacids may be used to neutralize the acid coming up from the stomach. Other medications known as *proton pump inhibitors* (PPIs), such as Prilosec and Nexium, may be used to suppress the production of HCl in the stomach (gastric parietal cells), which raises the pH in the stomach to between 4 and 5 and gives the esophagus time to heal. In severe GERD cases, an artificial valve may be created at the top of the stomach to strengthen the lower esophageal sphincter.

Esophagus
Lower esophageal sphincter closed
Lower esophageal sphincter open allowing reflux
Pylorus
Liquid
Stomach

In acid reflux disease, the lower esophageal sphincter opens, allowing acidic fluid from the stomach to enter the esophagus.

Clinical Applications

10.63 At rest, the $[H_3O^+]$ of the stomach fluid is 2.0×10^{-4} M. What is the pH of the stomach fluid?

10.64 When food enters the stomach, HCl is released and the $[H_3O^+]$ of the stomach fluid rises to 4×10^{-2} M. What is the pH of the stomach fluid?

10.65 In Larry's esophageal pH test, a pH value of 3.60 was recorded in the esophagus. What is the $[H_3O^+]$ in his esophagus?

10.66 After Larry had taken Nexium for 4 weeks, the pH in his stomach was raised to 4.52. What is the $[H_3O^+]$ in his stomach?

10.67 Write the balanced chemical equation for the neutralization reaction of stomach acid HCl with $CaCO_3$, an ingredient in some antacids.

10.68 Write the balanced chemical equation for the neutralization reaction of stomach acid HCl with $Al(OH)_3$, an ingredient in some antacids.

10.69 How many grams of $CaCO_3$ are required to neutralize 100. mL of stomach acid, which is equivalent to 0.0400 M HCl?

10.70 How many grams of $Al(OH)_3$ are required to neutralize 150. mL of stomach acid with a pH of 1.5?

CONCEPT MAP

ACIDS AND BASES AND EQUILIBRIUM

- **Acid** is a **H^+ Donor**
  - 100% → **Strong Acid**
  - small % → **Weak Acid**
- **Base** is a **H^+ Acceptor**
  - small % → **Weak Base**
  - 100 % → **Strong Base**
- Weak Acid / Weak Base involves **Reversible Reactions** that reach **Equilibrium** adjusts for change in concentration **Le Châtelier's Principle**
- **Dissociation of H_2O** gives **H_3O^+** and **OH^-**
  - H_3O^+ gives **$-\log[H_3O^+]$** is **pH**
  - product **$K_w = [H_3O^+][OH^-]$**
- undergo **Neutralization** to form **Salt** and **Water**
  - Salt of a **Weak Acid or Base** with its conjugate forms a **Buffer** to maintain **pH**

CHAPTER REVIEW

10.1 Acids and Bases

LEARNING GOAL Describe and name acids and bases.

- An Arrhenius acid produces H^+ and an Arrhenius base produces OH^- in aqueous solutions.

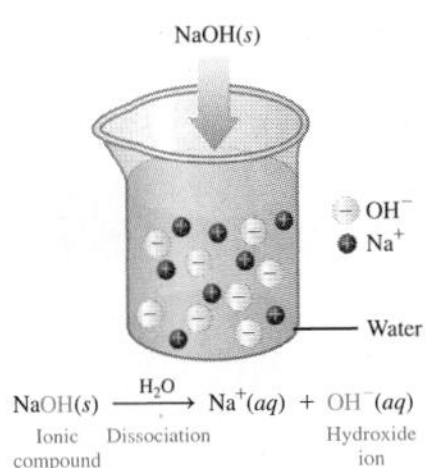

- Acids taste sour, may sting, and neutralize bases.
- Bases taste bitter, feel slippery, and neutralize acids.
- Acids containing a simple anion use a *hydro* prefix, whereas acids with oxygen-containing polyatomic anions are named as *ic* or *ous acids*.

10.2 Brønsted–Lowry Acids and Bases

LEARNING GOAL Identify conjugate acid–base pairs for Brønsted–Lowry acids and bases.

$$HCl + H_2O \longrightarrow H_3O^+ + Cl^-$$

Hydrogen chloride (Acid, H^+ donor) + Water (Base, H^+ acceptor) → Hydronium ion + Chloride ion (Acidic solution)

- According to the Brønsted–Lowry theory, acids are H^+ donors and bases are H^+ acceptors.
- A conjugate acid–base pair is related by the loss or gain of one H^+.
- For example, when the acid HF donates H^+, the F^- is its conjugate base. The other acid–base pair would be H_3O^+/H_2O.

$$HF(aq) + H_2O(l) \rightleftharpoons H_3O^+(aq) + F^-(aq)$$

10.3 Strengths of Acids and Bases

LEARNING GOAL Write equations for the dissociation of strong and weak acids and bases.

- Strong acids dissociate completely in water, and the H^+ is accepted by H_2O acting as a base.
- Weak acids dissociate slightly in water, producing only a small percentage of H_3O^+.
- Strong bases are hydroxides of Groups 1A (1) and 2A (2) that dissociate completely in water.
- An important weak base is ammonia, NH_3.

10.4 Acid–Base Equilibrium

LEARNING GOAL Use the concept of reversible reactions to explain acid–base equilibrium. Use Le Châtelier's principle to determine the effect on equilibrium concentrations when reaction conditions change.

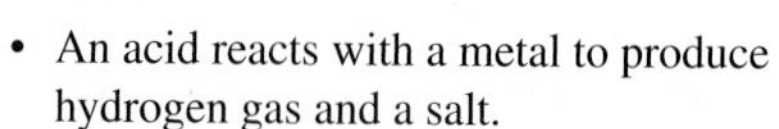

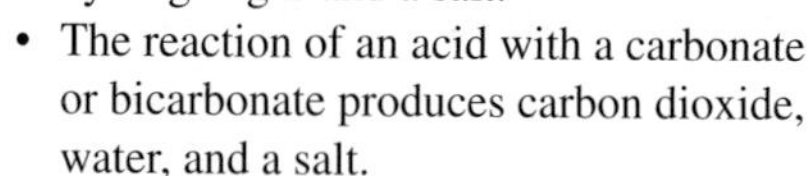

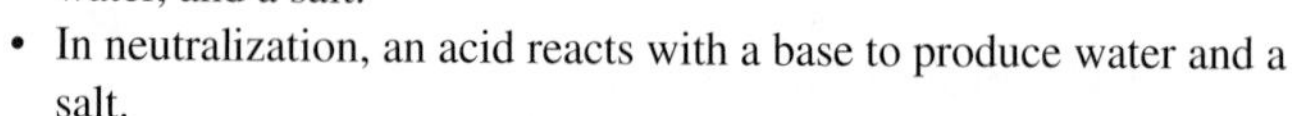

[Reactants] decrease
Equilibrium reached
[Products] increase
Concentration (moles/L)
Progress of Reaction

- Chemical equilibrium occurs in a reversible reaction when the rate of the forward reaction becomes equal to the rate of the reverse reaction.
- At equilibrium, no further change occurs in the concentrations of the reactants and products as the forward and reverse reactions continue.
- When reactants are removed or products are added to an equilibrium mixture, the system shifts in the direction of the reactants to reestablish equilibrium.
- When reactants are added or products are removed from an equilibrium mixture, the system shifts in the direction of the products to reestablish equilibrium.

10.5 Dissociation of Water

LEARNING GOAL Use the water dissociation expression to calculate the $[H_3O^+]$ and $[OH^-]$ in an aqueous solution.

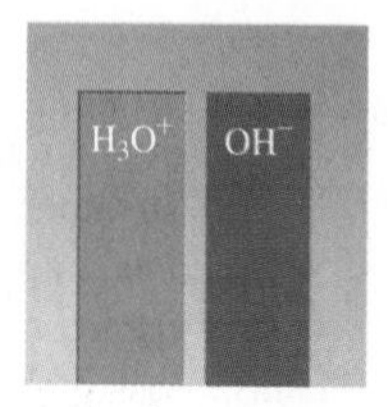

$[H_3O^+] = [OH^-]$

- In pure water, a few water molecules transfer H^+ to other water molecules, producing small, but equal, amounts of H_3O^+ and OH^-.
- In pure water, the molar concentrations of H_3O^+ and OH^- are each 1.0×10^{-7} M.
- The water dissociation expression, $K_w = [H_3O^+][OH^-] = 1.0 \times 10^{-14}$ at 25 °C.
- In acidic solutions, the $[H_3O^+]$ is greater than the $[OH^-]$.
- In basic solutions, the $[OH^-]$ is greater than the $[H_3O^+]$.

10.6 The pH Scale

LEARNING GOAL Calculate the pH of a solution from $[H_3O^+]$; given the pH, calculate $[H_3O^+]$.

- The pH scale is a range of numbers, typically from 0 to 14, which represents the $[H_3O^+]$ of the sofution.
- A neutral solution has a pH of 7.0. In acidic solutions, the pH is below 7.0; in basic solutions, the pH is above 7.0.
- Mathematically, pH is the negative logarithm of the hydronium ion concentration,

$$pH = -\log[H_3O^+].$$

10.7 Reactions of Acids and Bases

LEARNING GOAL Write balanced equations for reactions of acids with metals, carbonates or bicarbonates, and bases; calculate the molarity or volume of an acid from titration information.

- An acid reacts with a metal to produce hydrogen gas and a salt.
- The reaction of an acid with a carbonate or bicarbonate produces carbon dioxide, water, and a salt.
- In neutralization, an acid reacts with a base to produce water and a salt.
- In a titration, an acid sample is neutralized with a known amount of a base.
- From the volume and molarity of the base, the concentration of the acid is calculated.

10.8 Buffers

LEARNING GOAL Describe the role of buffers in maintaining the pH of a solution.

- A buffer solution resists changes in pH when small amounts of an acid or a base are added.
- A buffer contains either a weak acid and its salt or a weak base and its salt.
- In a buffer, the weak acid reacts with added OH^-, and the anion of the salt reacts with added H_3O^+.

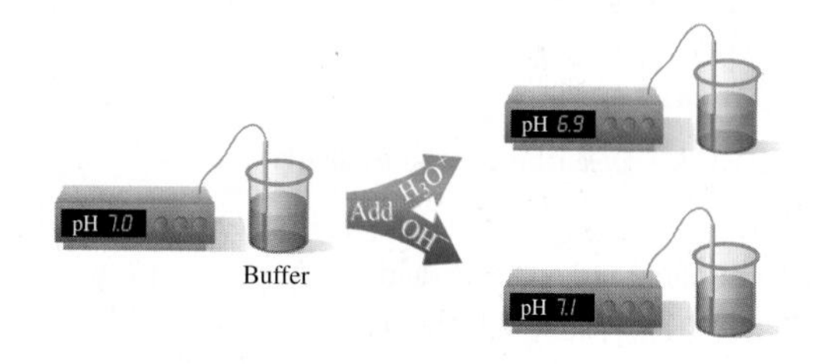

KEY TERMS

acid A substance that dissolves in water and produces hydrogen ions, according to the Arrhenius theory. All acids are hydrogen ion donors, according to the Brønsted–Lowry theory.

acidosis A physiological condition in which the blood pH is lower than 7.35.

alkalosis A physiological condition in which the blood pH is higher than 7.45.

amphoteric Substances that can act as either an acid or a base in water.

base A substance that dissolves in water and produces hydroxide ions (OH^-) according to the Arrhenius theory. All bases are hydrogen ion (H^+) acceptors, according to the Brønsted–Lowry theory.

Brønsted–Lowry acids and bases An acid is a hydrogen ion donor, and a base is a hydrogen ion acceptor.

buffer solution A solution of a weak acid and its conjugate base or a weak base and its conjugate acid that maintains the pH by neutralizing added acid or base.

conjugate acid–base pair An acid and a base that differ by one H^+. When an acid donates a hydrogen ion (H^+), the product is its conjugate base, which is capable of accepting a hydrogen ion in the reverse reaction.

dissociation The separation of an acid or a base into ions in water.

equilibrium The point at which the rate of forward and reverse reactions are equal so that no further change in concentrations of reactants and products takes place.

hydronium ion, H_3O^+ The ion formed by the attraction of H^+ to a water molecule.

Le Châtelier's principle When a stress is placed on a system at equilibrium, the system shifts to relieve that stress.

neutral The term that describes a solution with equal concentrations of H_3O^+ and OH^-.

neutralization A reaction between an acid and a base to form water and a salt.

pH A measure of the $[H_3O^+]$ in a solution; $pH = -\log[H_3O^+]$.

reversible reaction A reaction in which a forward reaction occurs from reactants to products, and a reverse reaction occurs from products back to reactants.

strong acid An acid that completely dissociates in water.

strong base A base that completely dissociates in water.

titration The addition of base to an acid sample to determine the concentration of the acid.

water dissociation expression, K_w The product of $[H_3O^+]$ and $[OH^-]$ in solution; $K_w = [H_3O^+][OH^-]$.

weak acid An acid that is a poor donor of H^+ and dissociates only slightly in water.

weak base A base that is a poor acceptor of H^+.

KEY MATH SKILLS

The chapter Section containing each Key Math Skill is shown in parentheses at the end of each heading.

Calculating pH from $[H_3O^+]$ (10.6)

- The pH of a solution is calculated from the negative log of the $[H_3O^+]$.

$$pH = -\log[H_3O^+]$$

Example: What is the pH of a solution that has $[H_3O^+] = 2.4 \times 10^{-11}$ M?

Answer: We substitute the given $[H_3O^+]$ into the pH equation and calculate the pH.

$$pH = -\log[H_3O^+] = -\log[2.4 \times 10^{-11}] = -(-10.62) = 10.62$$

Two decimal places in the pH equal the two SFs in the $[H_3O^+]$ coefficient.

Calculating $[H_3O^+]$ from pH (10.6)

- The calculation of $[H_3O^+]$ from the pH is done by reversing the pH calculation using the negative pH.

$$[H_3O^+] = 10^{-pH}$$

Example What is the $[H_3O^+]$ of a solution with a pH of 4.80?

Answer: $[H_3O^+] = 10^{-pH} = 10^{-4.80} = 1.6 \times 10^{-5}$ M

Two SFs in the $[H_3O^+]$ coefficient equal the two decimal places in the pH.

CORE CHEMISTRY SKILLS

The chapter Section containing each Core Chemistry Skill is shown in parentheses at the end of each heading.

Identifying Conjugate Acid–Base Pairs (10.2)

- According to the Brønsted–Lowry theory, a conjugate acid–base pair consists of molecules or ions related by the loss of one H^+ by an acid, and the gain of one H^+ by a base.
- Every acid–base reaction contains two conjugate acid–base pairs because an H^+ is transferred in both the forward and reverse directions.
- When an acid such as HF loses one H^+, the conjugate base F^- is formed. When H_2O acts as a base, it gains one H^+, which forms its conjugate acid, H_3O^+.

Example: Identify the conjugate acid–base pairs in the following reaction:

$$HSO_4^-(aq) + H_2O(l) \rightleftharpoons SO_4^{2-}(aq) + H_3O^+(aq)$$

Answer: $HSO_4^-(aq)$ (Acid) + $H_2O(l)$ (Base) $\rightleftharpoons$ $SO_4^{2-}(aq)$ (Conjugate base) + $H_3O^+(aq)$ (Conjugate acid)

Conjugate acid–base pairs: HSO_4^-/SO_4^{2-}, H_3O^+/H_2O

Using Le Châtelier's Principle (10.4)

Le Châtelier's principle states that when a system at equilibrium is disturbed by changes in concentration, the system will shift in the direction that will reduce that stress.

$$H_2S(aq) + H_2O(l) \rightleftharpoons H_3O^+(aq) + HS^-(aq)$$

Example: For each of the following changes at equilibrium, indicate whether the system shifts in the direction of products or reactants:

a. removing some $H_2S(aq)$
b. adding more $H_3O^+(aq)$

Answer: **a.** Removing a reactant shifts the system in the direction of the reactants.
b. Adding more product shifts the system in the direction of the reactants.

Calculating $[H_3O^+]$ and $[OH^-]$ in Solutions (10.5)

- For all aqueous solutions, the product of $[H_3O^+]$ and $[OH^-]$ is equal to the water dissociation expression, K_w.

$$K_w = [H_3O^+][OH^-]$$

- Because pure water contains equal numbers of OH^- ions and H_3O^+ ions each with molar concentrations of 1.0×10^{-7} M, the numerical value of K_w is 1.0×10^{-14} at 25 °C.

$$K_w = [H_3O^+][OH^-] = [1.0 \times 10^{-7}][1.0 \times 10^{-7}] = 1.0 \times 10^{-14}$$

- If we know the $[H_3O^+]$ of a solution, we can use the K_w expression to calculate the $[OH^-]$. If we know the $[OH^-]$ of a solution, we can calculate $[H_3O^+]$ using the K_w expression.

$$[OH^-] = \frac{K_w}{[H_3O^+]} \qquad [H_3O^+] = \frac{K_w}{[OH^-]}$$

Example: What is the $[OH^-]$ in a solution that has $[H_3O^+] = 2.4 \times 10^{-11}$ M? Is the solution acidic or basic?

Answer: We solve the K_w expression for $[OH^-]$ and substitute in the known values of K_w and $[H_3O^+]$.

$$[OH^-] = \frac{K_w}{[H_3O^+]} = \frac{1.0 \times 10^{-14}}{[2.4 \times 10^{-11}]} = 4.2 \times 10^{-4}\text{ M}$$

Because the $[OH^-]$ is greater than the $[H_3O^+]$, this is a basic solution.

Writing Equations for Reactions of Acids and Bases (10.7)

- Acids react with certain metals to produce hydrogen gas (H_2) and a salt.

$$\underset{\text{Metal}}{Mg(s)} + \underset{\text{Acid}}{2HCl(aq)} \longrightarrow \underset{\text{Hydrogen}}{H_2(g)} + \underset{\text{Salt}}{MgCl_2(aq)}$$

- When an acid is added to a carbonate or bicarbonate, the products are carbon dioxide gas, water, and a salt.

$$\underset{\text{Acid}}{2HCl(aq)} + \underset{\text{Carbonate}}{Na_2CO_3(aq)} \longrightarrow \underset{\text{Carbon dioxide}}{CO_2(g)} + \underset{\text{Water}}{H_2O(l)} + \underset{\text{Salt}}{2NaCl(aq)}$$

- Neutralization is a reaction between a strong or a weak acid and a strong base to produce water and a salt.

$$\underset{\text{Acid}}{HCl(aq)} + \underset{\text{Base}}{NaOH(aq)} \longrightarrow \underset{\text{Water}}{H_2O(l)} + \underset{\text{Salt}}{NaCl(aq)}$$

Example: Write the balanced chemical equation for the reaction of $ZnCO_3(s)$ and hydrobromic acid $HBr(aq)$.

Answer: $ZnCO_3(s) + 2HBr(aq) \longrightarrow CO_2(g) + H_2O(l) + ZnBr_2(aq)$

Calculating Molarity or Volume of an Acid or Base in a Titration (10.7)

- In a titration, a measured volume of acid is neutralized by a basic solution of known molarity.
- From the measured volume of the strong base solution required for titration and its molarity, the number of moles of base, the moles of acid, and the concentration of the acid are calculated.

Example: A 0.0150-L sample of a H_2SO_4 solution is titrated with 24.0 mL of a 0.245 M NaOH solution. What is the molarity of the H_2SO_4 solution?

$$H_2SO_4(aq) + 2NaOH(aq) \longrightarrow 2H_2O(l) + Na_2SO_4(aq)$$

Answer: $24.0\text{ mL NaOH solution} \times \frac{1\text{ L NaOH solution}}{1000\text{ mL NaOH solution}} \times \frac{0.245\text{ mole NaOH}}{1\text{ L NaOH solution}} \times \frac{1\text{ mole } H_2SO_4}{2\text{ moles NaOH}} = 0.002\,94\text{ mole of } H_2SO_4$

$$\text{Molarity (M)} = \frac{0.002\,94\text{ mole } H_2SO_4}{0.0150\text{ L } H_2SO_4\text{ solution}} = 0.196\text{ M } H_2SO_4\text{ solution}$$

UNDERSTANDING THE CONCEPTS

The chapter Sections to review are shown in parentheses at the end of each problem.

10.71 Identify each of the following as an acid or a base: (10.1)
a. H_2SO_4 **b.** RbOH **c.** $Ca(OH)_2$ **d.** HI

10.72 Identify each of the following as an acid or a base: (10.1)
a. $Sr(OH)_2$ **b.** H_2SO_3 **c.** $HC_2H_3O_2$ **d.** CsOH

10.73 Complete the following table: (10.2)

Acid	Conjugate Base
H_2O	
	CN^-
HNO_2	
	$H_2PO_4^-$

10.74 Complete the following table: (10.2)

Base	Conjugate Acid
	HS^-
	H_3O^+
NH_3	
HCO_3^-	

10.75 State whether each of the following solutions is acidic, basic, or neutral: (10.6)
a. rain, pH 5.2
b. tears, pH 7.5
c. tea, pH 3.8
d. cola, pH 2.5
e. photo developer, pH 12.0

10.76 State whether each of the following solutions is acidic, basic, or neutral: (10.6)
a. saliva, pH 6.8 **b.** urine, pH 5.9
c. pancreatic juice, pH 8.0 **d.** bile, pH 8.4
e. blood, pH 7.45

10.77 Determine if each of the following diagrams represents a strong acid or a weak acid. The acid has the formula HX. (10.3)

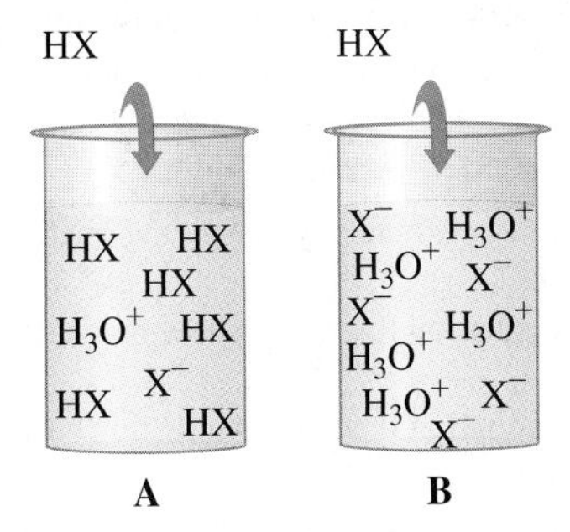

10.78 Adding a few drops of a strong acid to water will lower the pH appreciably. However, adding the same number of drops to a buffer does not appreciably alter the pH. Why? (10.8)

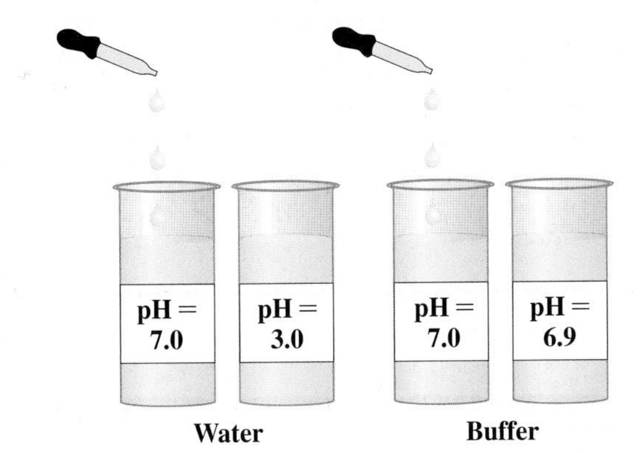

10.79 Sometimes, during stress or trauma, a person can start to hyperventilate. Then the person might breathe into a paper bag to avoid fainting. (10.8)
a. What changes occur in the blood pH during hyperventilation?
b. How does breathing into a paper bag help return blood pH to normal?

Breathing into a paper bag can help a person who is hyperventilating.

10.80 In the blood plasma, pH is maintained by the carbonic acid–bicarbonate buffer system. (10.8)
a. How is pH maintained when acid is added to the buffer system?
b. How is pH maintained when base is added to the buffer system?

ADDITIONAL PRACTICE PROBLEMS

10.81 Identify each of the following as an acid, base, or salt, and give its name: (10.1)
a. LiOH **b.** $Ca(NO_3)_2$ **c.** HBr
d. $Ba(OH)_2$ **e.** H_2CO_3

10.82 Identify each of the following as an acid, base, or salt, and give its name: (10.1)
a. H_3PO_4 **b.** $MgBr_2$ **c.** NH_3
d. $HClO_2$ **e.** NaCl

10.83 Using Table 10.3, identify the stronger acid in each of the following pairs: (10.3)
a. HF or HCN **b.** H_3O^+ or H_2S
c. HNO_2 or $HC_2H_3O_2$ **d.** H_2O or HCO_3^-

10.84 Using Table 10.3, identify the weaker acid in each of the following pairs: (10.3)
a. HNO_2 or H_2CO_3 **b.** HF or HCO_3^-
c. HBr or HSO_4^- **d.** NH_4^+ or H_3PO_4

10.85 Use Le Châtelier's principle to predict if each of the following changes causes the system to shift in the direction of products or reactants: (10.4)

$$H_2S(aq) + H_2O(l) \rightleftharpoons H_3O^+(aq) + HS^-(aq)$$

a. adding more $H_2S(aq)$ **b.** removing some $HS^-(aq)$
c. adding more $H_3O^+(aq)$ **d.** removing some $H_2S(aq)$

10.86 Use Le Châtelier's principle to predict if each of the following changes causes the system to shift in the direction of products or reactants: (10.4)

$$HClO(aq) + H_2O(l) \rightleftharpoons H_3O^+(aq) + ClO^-(aq)$$

a. removing some $HClO(aq)$ **b.** adding more $ClO^-(aq)$
c. removing some $H_3O^+(aq)$ **d.** adding more $HClO(aq)$

10.87 Determine the pH for the following solutions: (10.6)
a. $[H_3O^+] = 1.0 \times 10^{-11}$ M
b. $[H_3O^+] = 5.0 \times 10^{-2}$ M
c. $[OH^-] = 3.5 \times 10^{-4}$ M
d. $[OH^-] = 0.005$ M

10.88 Determine the pH for the following solutions: (10.6)
a. $[OH^-] = 1.0 \times 10^{-7}$ M
b. $[H_3O^+] = 4.2 \times 10^{-3}$ M
c. $[H_3O^+] = 0.0001$ M
d. $[OH^-] = 8.5 \times 10^{-9}$ M

10.89 Identify each of the solutions in problem 10.87 as acidic, basic, or neutral. (10.6)

10.90 Identify each of the solutions in problem 10.88 as acidic, basic, or neutral. (10.6)

10.91 Calculate the $[H_3O^+]$ and $[OH^-]$ for a solution with the following pH values: (10.6)
a. 3.40 **b.** 6.00 **c.** 8.0 **d.** 11.0 **e.** 9.20

10.92 Calculate the $[H_3O^+]$ and $[OH^-]$ for a solution with the following pH values: (10.6)
a. 10.0 **b.** 5.0 **c.** 7.00 **d.** 6.5 **e.** 1.82

10.93 Solution A has a pH of 4.0, and solution B has a pH of 6.0. (10.6)
a. Which solution is more acidic?
b. What is the $[H_3O^+]$ in each?
c. What is the $[OH^-]$ in each?

10.94 Solution X has a pH of 9.0, and solution Y has a pH of 7.0. (10.6)
a. Which solution is more acidic?
b. What is the $[H_3O^+]$ in each?
c. What is the $[OH^-]$ in each?

10.95 A 0.205 M NaOH solution is used to titrate 20.0 mL of a solution of H_2SO_4. If 45.6 mL of the NaOH solution is required to reach the endpoint, what is the molarity of the H_2SO_4 solution? (10.7)

$$H_2SO_4(aq) + 2NaOH(aq) \longrightarrow 2H_2O(l) + Na_2SO_4(aq)$$

10.96 A 10.0-mL sample of vinegar, which is an aqueous solution of acetic acid, $HC_2H_3O_2$, requires 16.5 mL of a 0.500 M NaOH solution to reach the endpoint in a titration. What is the molarity of the acetic acid solution? (10.7)

$$HC_2H_3O_2(aq) + NaOH(aq) \longrightarrow H_2O(l) + NaC_2H_3O_2(aq)$$

10.97 Calculate the volume, in milliliters, of a 0.150 M NaOH solution that will completely neutralize each of the following: (10.7)
a. 25.0 mL of a 0.288 M HCl solution
b. 10.0 mL of a 0.560 M H_2SO_4 solution
c. 5.00 mL of a 0.618 M HBr solution

10.98 Calculate the volume, in milliliters, of a 0.215 M KOH solution that will completely neutralize each of the following: (10.7)
a. 2.50 mL of a 0.825 M H_2SO_4 solution
b. 18.5 mL of a 0.560 M HNO_3 solution
c. 5.00 mL of a 3.18 M HCl solution

10.99 A buffer solution is made by dissolving H_3PO_4 and NaH_2PO_4 in water. (10.8)
a. Write an equation that shows how this buffer neutralizes added acid.
b. Write an equation that shows how this buffer neutralizes added base.

10.100 A buffer solution is made by dissolving $HC_2H_3O_2$ and $NaC_2H_3O_2$ in water. (10.8)
a. Write an equation that shows how this buffer neutralizes added acid.
b. Write an equation that shows how this buffer neutralizes added base.

CHALLENGE PROBLEMS

The following problems are related to the topics in this chapter. However, they do not all follow the chapter order, and they require you to combine concepts and skills from several Sections. These problems will help you increase your critical thinking skills and prepare for your next exam.

10.101 For each of the following: (10.2, 10.3)
1. H_2S **2.** H_3PO_4
a. Write the formula for the conjugate base.
b. Which is the weaker acid?

10.102 For each of the following: (10.2, 10.3)
1. HCN **2.** NH_4^+
a. Write the formula for the conjugate base.
b. Which is the stronger acid?

10.103 Identify the conjugate acid–base pairs in each of the following equations: (10.2)
a. $NH_3(aq) + HNO_3(aq) \longrightarrow NH_4^+(aq) + NO_3^-(aq)$
b. $H_2O(l) + HBr(aq) \longrightarrow H_3O^+(aq) + Br^-(aq)$

10.104 Identify the conjugate acid–base pairs in each of the following equations: (10.2)
a. $HNO_2(aq) + HS^-(aq) \rightleftarrows NO_2^-(aq) + H_2S(g)$
b. $HCl(aq) + OH^-(aq) \longrightarrow Cl^-(aq) + H_2O(l)$

10.105 Complete and balance each of the following: (10.7)
a. $ZnCO_3(s) + H_2SO_4(aq) \longrightarrow$
b. $Al(s) + HBr(aq) \longrightarrow$

10.106 Complete and balance each of the following: (10.7)
a. $H_3PO_4(aq) + Ca(OH)_2(s) \longrightarrow$
b. $KHCO_3(s) + HNO_3(aq) \longrightarrow$

10.107 Determine each of the following for a 0.050 M KOH solution: (10.5, 10.6, 10.7)
a. $[H_3O^+]$
b. pH
c. the balanced chemical equation for the reaction with H_2SO_4
d. milliliters of the KOH solution required to neutralize 40.0 mL of a 0.035 M H_2SO_4 solution

10.108 Determine each of the following for a 0.100 M HBr solution: (10.5, 10.6, 10.7)
a. $[H_3O^+]$
b. pH
c. the balanced chemical equation for the reaction with LiOH
d. milliliters of the HBr solution required to neutralize 36.0 mL of a 0.250 M LiOH solution

10.109 A 0.204 M NaOH solution is used to titrate 50.0 mL of an H_3PO_4 solution. (10.7)
a. Write the balanced chemical equation.
b. What is the molarity of the H_3PO_4 solution if 16.4 mL of the NaOH solution is required?

10.110 A 0.312 M KOH solution is used to titrate 15.0 mL of an H_2SO_4 solution. (10.7)
a. Write the balanced chemical equation.
b. What is the molarity of the H_2SO_4 solution if 28.2 mL of the KOH solution is required?

10.111 One of the most acidic lakes in the United States is Little Echo Pond in the Adirondacks in New York. Recently, this lake had a pH of 4.2, well below the recommended pH of 6.5. (10.6, 10.7)

A helicopter drops calcium carbonate on an acidic lake to increase its pH.

a. What are the $[H_3O^+]$ and $[OH^-]$ of Little Echo Pond?
b. What are the $[H_3O^+]$ and $[OH^-]$ of a lake that has a pH of 6.5?
c. One way to raise the pH (and restore aquatic life) is to add limestone ($CaCO_3$). How many grams of $CaCO_3$ are needed to neutralize 1.0 kL of the acidic water from Little Echo Pond if the acid is sulfuric acid?

$$H_2SO_4(aq) + CaCO_3(s) \longrightarrow CO_2(g) + H_2O(l) + CaSO_4(aq)$$

10.112 The daily output of stomach acid (gastric juice) is 1000 mL to 2000 mL. Prior to a meal, stomach acid (HCl) typically has a pH of 1.42. (10.6, 10.7)
a. What is the $[H_3O^+]$ of stomach acid?
b. One chewable tablet of the antacid Maalox contains 600. mg of $CaCO_3$. Write the neutralization equation, and calculate the milliliters of stomach acid neutralized by 2 tablets of Maalox.
c. The antacid milk of magnesia contains 400. mg of $Mg(OH)_2$ per teaspoon. Write the neutralization equation, and calculate the number of milliliters of stomach acid neutralized by 1 tablespoon of milk of magnesia (1 tablespoon = 3 teaspoons).

ANSWERS

Answers to Selected Practice Problems

10.1 a. acid b. acid c. acid
d. base e. both

10.3 a. hydrochloric acid b. calcium hydroxide
c. perchloric acid d. nitric acid
e. sulfurous acid f. bromous acid

10.5 a. RbOH b. HF
c. H_3PO_4 d. LiOH
e. NH_4OH f. HIO_4

10.7 a. HI is the acid (H^+ donor) and H_2O is the base (H^+ acceptor).
b. H_2O is the acid (H^+ donor) and F^- is the base (H^+ acceptor)

10.9 a. F^- b. OH^-
c. HPO_3^{2-} d. SO_4^{2-}

10.11 a. HCO_3^- b. H_3O^+
c. H_3PO_4 d. HBr

10.13 a. The conjugate acid–base pairs are H_2CO_3/HCO_3^- and H_3O^+/H_2O.
b. The conjugate acid–base pairs are NH_4^+/NH_3 and H_3O^+/H_2O.
c. The conjugate acid–base pairs are HCN/CN^- and HNO_2/NO_2^-.

10.15 a. HBr b. HSO_4^- c. H_2CO_3

10.17 a. HSO_4^- b. HF c. HCO_3^-

10.19 A reversible reaction is one in which a forward reaction converts reactants to products, whereas a reverse reaction converts products to reactants.

10.21 a. not at equilibrium b. at equilibrium
c. at equilibrium

10.23 a. The system shifts in the direction of the reactants.
b. The system shifts in the direction of the reactants.
c. The system shifts in the direction of the products.
d. The system shifts in the direction of the products.

10.25 In pure water, $[H_3O^+] = [OH^-]$ because one of each is produced every time an H^+ transfers from one water molecule to another.

10.27 In an acidic solution, the $[H_3O^+]$ is greater than the $[OH^-]$.

10.29 a. acidic b. basic
c. basic d. acidic

10.31 a. 1.0×10^{-9} M b. 1.0×10^{-6} M
c. 2.0×10^{-5} M d. 4.0×10^{-13} M

10.33 a. 4.0×10^{-2} M b. 5.0×10^{-6} M
c. 2.0×10^{-4} M d. 4.0×10^{-9} M

10.35 a. basic b. acidic c. basic
d. acidic e. acidic f. basic

10.37 In a neutral solution, the $[H_3O^+]$ is 1×10^{-7} M and the pH is 7.0, which is the negative value of the power of 10.

10.39 a. 4.0 b. 8.5 c. 9.0
d. 3.40 e. 7.17 f. 10.92

10.41

$[H_3O^+]$	$[OH^-]$	pH	Acidic, Basic, or Neutral?
1.0×10^{-8} M	1.0×10^{-6} M	8.00	Basic
3.2×10^{-4} M	3.1×10^{-11} M	3.49	Acidic
2.8×10^{-5} M	3.6×10^{-10} M	4.55	Acidic

10.43 1.2×10^{-7} M

10.45 a. $ZnCO_3(s) + 2HBr(aq) \longrightarrow CO_2(g) + H_2O(l) + ZnBr_2(aq)$
b. $Zn(s) + 2HCl(aq) \longrightarrow H_2(g) + ZnCl_2(aq)$
c. $HCl(aq) + NaHCO_3(s) \longrightarrow CO_2(g) + H_2O(l) + NaCl(aq)$
d. $H_2SO_4(aq) + Mg(OH)_2(s) \longrightarrow 2H_2O(l) + MgSO_4(aq)$

10.47 a. $2HCl(aq) + Mg(OH)_2(s) \longrightarrow 2H_2O(l) + MgCl_2(aq)$
b. $H_3PO_4(aq) + 3LiOH(aq) \longrightarrow 3H_2O(l) + Li_3PO_4(aq)$

10.49 a. $H_2SO_4(aq) + 2NaOH(aq) \longrightarrow 2H_2O(l) + Na_2SO_4(aq)$
b. $3HCl(aq) + Fe(OH)_3(s) \longrightarrow 3H_2O(l) + FeCl_3(aq)$
c. $H_2CO_3(aq) + Mg(OH)_2(s) \longrightarrow 2H_2O(l) + MgCO_3(s)$

10.51 0.830 M HCl solution

10.53 0.124 M H_2SO_4 solution

10.55 **b** and **c** are buffer systems; **b** contains the weak acid H_2CO_3 and its salt $NaHCO_3$; **c** contains HF, a weak acid, and its salt KF.

10.57 a. 3 **b.** 1 and 2
c. 3 **d.** 1

10.59 If you breathe fast, CO_2 is expelled and the equilibrium shifts to lower $[H_3O^+]$, which raises the pH.

10.61 If large amounts of HCO_3^- are lost, equilibrium shifts to higher $[H_3O^+]$, which lowers the pH.

10.63 pH = 3.70

10.65 2.5×10^{-4} M

10.67 $CaCO_3(s) + 2HCl(aq) \longrightarrow CO_2(g) + H_2O(l) + CaCl_2(aq)$

10.69 0.200 g of $CaCO_3$

10.71 a. acid **b.** base
c. base **d.** acid

10.73

Acid	Conjugate Base
H_2O	OH^-
HCN	CN^-
HNO_2	NO_2^-
H_3PO_4	$H_2PO_4^-$

10.75 a. acidic **b.** basic
c. acidic **d.** acidic
e. basic

10.77 a. weak acid **b.** strong acid

10.79 a. Hyperventilation will lower the CO_2 level in the blood, which lowers the $[H_2CO_3]$, which decreases the $[H_3O^+]$ and increases the blood pH.
b. Breathing into a bag increases the CO_2 level, increases the $[H_2CO_3]$, and increases the $[H_3O^+]$, which lowers the blood pH.

10.81 a. base, lithium hydroxide **b.** salt, calcium nitrate
c. acid, hydrobromic acid **d.** base, barium hydroxide
e. acid, carbonic acid

10.83 a. HF **b.** H_3O^+
c. HNO_2 **d.** HCO_3^-

10.85 a. System shifts in the direction of the products.
b. System shifts in the direction of the products.
c. System shifts in the direction of the reactants.
d. System shifts in the direction of the reactants.

10.87 a. 11.00 **b.** 1.30
c. 10.54 **d.** 11.7

10.89 a. basic **b.** acidic
c. basic **d.** basic

10.91 a. $[H_3O^+] = 4.0 \times 10^{-4}$ M, $[OH^-] = 2.5 \times 10^{-11}$ M
b. $[H_3O^+] = 1.0 \times 10^{-6}$ M, $[OH^-] = 1.0 \times 10^{-8}$ M
c. $[H_3O^+] = 1 \times 10^{-8}$ M, $[OH^-] = 1 \times 10^{-6}$ M
d. $[H_3O^+] = 1 \times 10^{-11}$ M, $[OH^-] = 1 \times 10^{-3}$ M
e. $[H_3O^+] = 6.3 \times 10^{-10}$ M, $[OH^-] = 1.6 \times 10^{-5}$ M

10.93 a. Solution A is more acidic than solution B.
b. Solution A: $[H_3O^+] = 1 \times 10^{-4}$ M,
Solution B: $[H_3O^+] = 1 \times 10^{-6}$ M
c. Solution A: $[OH^-] = 1 \times 10^{-10}$ M,
Solution B: $[OH^-] = 1 \times 10^{-8}$ M

10.95 0.234 M H_2SO_4 solution

10.97 a. 48.0 mL
b. 74.7 mL
c. 20.6 mL

10.99 a. Acid added:
$H_2PO_4^-(aq) + H_3O^+(aq) \longrightarrow H_2O(l) + H_3PO_4(aq)$
b. Base added:
$H_3PO_4(aq) + OH^-(aq) \longrightarrow H_2O(l) + H_2PO_4^-(aq)$

10.101 a. **1.** HCN **2.** $H_2PO_4^-$
b. H_2S

10.103 a. NH_4^+/NH_3 and HNO_3/NO_3^-
b. H_3O^+/H_2O and HBr/Br^-

10.105 a. $ZnCO_3(s) + H_2SO_4(aq) \longrightarrow CO_2(g) + H_2O(l) + ZnSO_4(aq)$
b. $2Al(s) + 6HBr(aq) \longrightarrow 3H_2(g) + 2AlBr_3(aq)$

10.107 a. 2.0×10^{-13} M
b. 12.70
c. $H_2SO_4(aq) + 2KOH(aq) \longrightarrow 2H_2O(l) + K_2SO_4(aq)$
d. 56 mL

10.109 a. $H_3PO_4(aq) + 3NaOH(aq) \longrightarrow 3H_2O(l) + Na_3PO_4(aq)$
b. 0.0224 M H_3PO_4 solution

10.111 a. $[H_3O^+] = 6 \times 10^{-5}$ M; $[OH^-] = 2 \times 10^{-10}$ M
b. $[H_3O^+] = 3 \times 10^{-7}$ M; $[OH^-] = 3 \times 10^{-8}$ M
c. 3 g of $CaCO_3$

11 Introduction to Organic Chemistry: Hydrocarbons

AT 4:35 A.M., A RESCUE CREW RESPONDED TO A call about a house fire. At the scene, Jack, a firefighter/emergency medical technician (EMT), finds Diane, a 62-year old woman, lying in the front yard of her house. In his assessment, Jack reports that Diane has second- and third-degree burns over 40% of her body as well as a broken leg. He places an oxygen re-breather mask on Diane to provide a high concentration of oxygen. Another firefighter/EMT, Nancy, begins dressing the burns with sterile water and cling film, a first aid material made of polyvinyl chloride, which does not stick to the skin and is protective. Jack and his crew transported Diane to the burn center for further treatment.

At the scene of the fire, arson investigators used trained dogs to find traces of accelerants and fuel. Gasoline, which is often found at arson scenes, is a mixture of organic molecules called alkanes. Alkanes or hydrocarbons are chains of carbon and hydrogen atoms. The alkanes present in gasoline consist of a mixture of compounds with five to eight carbon atoms in a chain. Alkanes are extremely combustible; they react with oxygen to form carbon dioxide, water, and large amounts of heat. Because alkanes undergo combustion reactions, they can be used to start arson fires.

CAREER Firefighter/Emergency Medical Technician

Firefighters/emergency medical technicians are first responders to fires, accidents, and other emergency situations. They are required to have an emergency medical technician certification in order to be able to treat seriously injured people. By combining the skills of a firefighter and an emergency medical technician, they increase the survival rates of the injured. The physical demands of firefighters are extremely high as they fight, extinguish, and prevent fires while wearing heavy protective clothing. They also train for and participate in firefighting drills, and maintain fire equipment so that it is always working and ready. Firefighters must also be knowledgeable about fire codes, arson, and the handling and disposal of hazardous materials. Since firefighters also provide emergency care for sick and injured people, they need to be aware of emergency medical and rescue procedures, as well as the proper methods for controlling the spread of infectious disease.

CLINICAL UPDATE Diane's Treatment in the Burn Unit

When Diane arrived at the hospital, she was diagnosed with second- and third-degree burns. You can see Diane's treatment in the **CLINICAL UPDATE Diane's Treatment in the Burn Unit**, page 389, and see the results of the arson investigation into the house fire.

LOOKING AHEAD

REVIEW

Drawing Lewis Structures (6.6)
Predicting Shape (6.8)

11.1 Organic Compounds

LEARNING GOAL Identify properties characteristic of organic or inorganic compounds.

At the beginning of the nineteenth century, scientists classified chemical compounds as inorganic or organic. An inorganic compound was a substance that was composed of minerals, and an organic compound was a substance that came from an organism, thus the use of the word *organic*. Early scientists thought that some type of "vital force," which could only be found in living cells, was required to synthesize an organic compound. This idea was shown to be incorrect in 1828, when German chemist Friedrich Wöhler synthesized urea, a product of protein metabolism, by heating an inorganic compound, ammonium cyanate.

$$NH_4CNO \xrightarrow{\text{Heat}} H_2N-\overset{\overset{\displaystyle O}{\|}}{C}-NH_2$$

Ammonium cyanate (inorganic) — Urea (organic)

Vegetable oil, a mixture of organic compounds, is not soluble in water.

Organic chemistry is the study of carbon compounds. The element carbon has a special role because many carbon atoms can bond together to give a vast array of molecular compounds. **Organic compounds** always contain carbon and hydrogen, and sometimes other nonmetals such as oxygen, sulfur, nitrogen, phosphorus, or a halogen. We find organic compounds in many common products we use every day, such as gasoline, medicines, shampoos, plastics, and perfumes. The food we eat is composed of organic compounds such as carbohydrates, fats, and proteins that supply us with fuel for energy and the carbon atoms needed to build and repair the cells of our bodies.

The formulas of organic compounds are written with carbon first, followed by hydrogen, and then any other elements. Organic compounds typically have low melting and boiling points, are not soluble in water, and are less dense than water. For example, vegetable oil, which is a mixture of organic compounds, does not dissolve in water but floats on top. Many organic compounds undergo combustion and burn vigorously in air. By contrast, many inorganic compounds have high melting and boiling points. Inorganic compounds that are ionic are usually soluble in water, and most do not burn in air. **TABLE 11.1** contrasts some of the properties associated with organic and inorganic compounds, such as propane, C_3H_8, and sodium chloride, NaCl (see **FIGURE 11.1**).

TABLE 11.1 Some Properties of Organic and Inorganic Compounds

Property	Organic	Example: C_3H_8	Inorganic	Example: NaCl
Elements Present	C and H, sometimes O, S, N, P, or Cl (F, Br, I)	C and H	Most metals and nonmetals	Na and Cl
Particles	Molecules	C_3H_8	Mostly ions	Na^+ and Cl^-
Bonding	Mostly covalent	Covalent	Many are ionic, some covalent	Ionic
Polarity of Bonds	Nonpolar, unless a strongly electronegative atom is present	Nonpolar	Most are ionic or polar covalent, a few are nonpolar covalent	Ionic
Melting Point	Usually low	−188 °C	Usually high	801 °C
Boiling Point	Usually low	−42 °C	Usually high	1413 °C
Flammability	High	Burns in air	Low	Does not burn
Solubility in Water	Not soluble unless a polar group is present	No	Most are soluble unless nonpolar	Yes

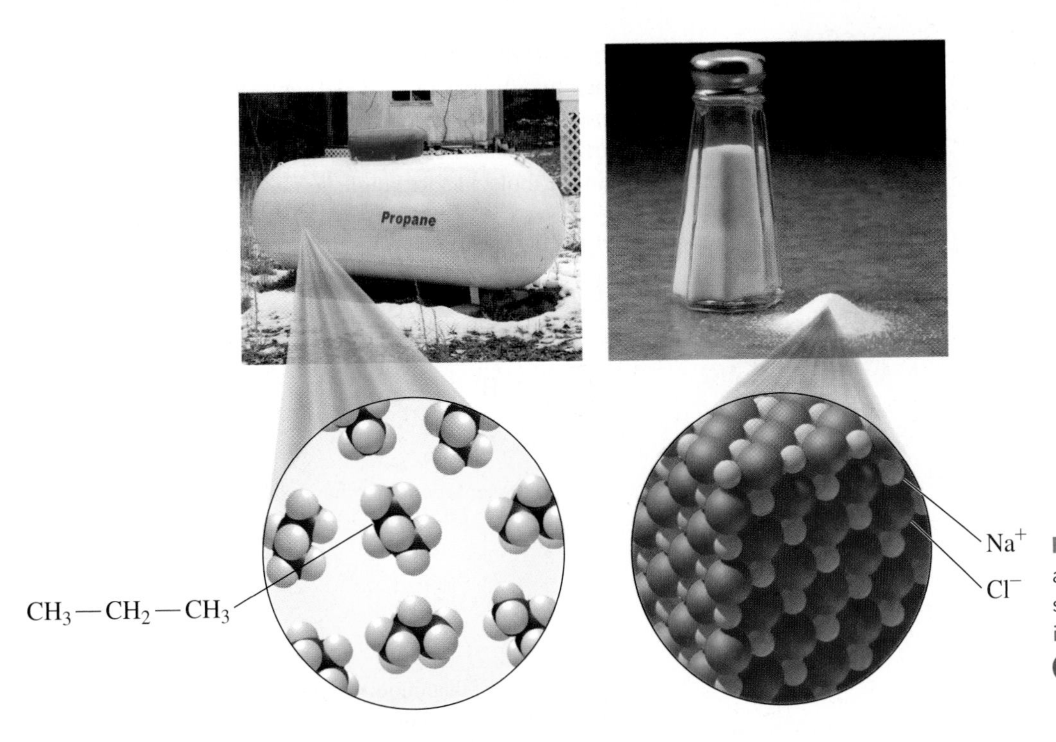

FIGURE 11.1 Propane, C_3H_8, is an organic compound, whereas sodium chloride, NaCl, is an inorganic compound.

Q Why is propane used as a fuel?

Representations of Carbon Compounds

Hydrocarbons are organic compounds that consist of only carbon and hydrogen. In organic molecules, every carbon atom has four bonds. In the simplest hydrocarbon, methane (CH_4), the carbon atom forms an octet by sharing its four valence electrons with four hydrogen atoms.

$$\cdot\dot{\underset{\cdot}{C}}\cdot + 4H\cdot \longrightarrow H:\overset{H}{\underset{H}{\ddot{\underset{\cdot\cdot}{C}}}}:H = H-\overset{H}{\underset{H}{\overset{|}{\underset{|}{C}}}}-H$$

Methane

The most accurate representation of methane is the three-dimensional *space-filling model* **(a)** in which spheres show the relative size and shape of all the atoms. Another type of three-dimensional representation is the *ball-and-stick model* **(b)**, where the atoms are shown as balls and the bonds between them are shown as sticks. In the ball-and-stick model of methane, CH_4, the covalent bonds from the carbon atom to each hydrogen atom are directed to the corners of a tetrahedron with bond angles of 109°. In the *wedge–dash model* **(c)**, the three-dimensional shape is represented by symbols of the atoms with lines for bonds in the plane of the page, wedges for bonds that project out from the page, and dashes for bonds that are behind the page.

However, the three-dimensional models are awkward to draw and view for more complex molecules. Therefore, it is more practical to use their corresponding two-dimensional formulas. The **expanded structural formula (d)** shows all of the atoms and the bonds connected to each atom. A **condensed structural formula (e)** shows the carbon atoms each grouped with the attached number of hydrogen atoms.

ENGAGE

Why does methane have a tetrahedral shape?

(a)

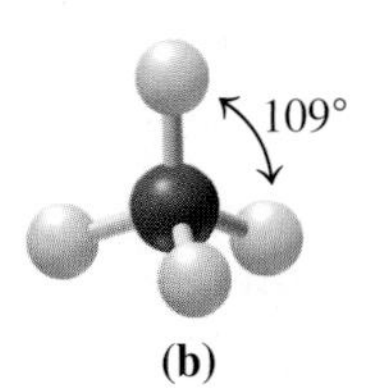

(b)

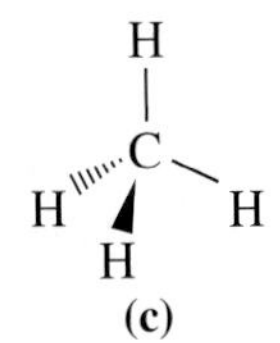

(c)

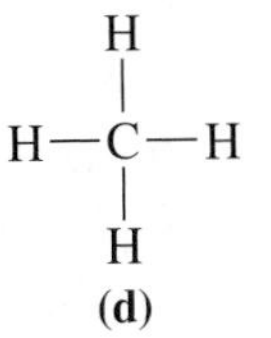

(d)

CH_4

(e)

Three-dimensional and two-dimensional representations of methane: **(a)** space-filling model, **(b)** ball-and-stick model, **(c)** wedge–dash model, **(d)** expanded structural formula, and **(e)** condensed structural formula.

TEST

Try Practice Problems 11.1 to 11.6

The hydrocarbon ethane with two carbon atoms and six hydrogen atoms can be represented by a similar set of three- and two-dimensional models and formulas in which each carbon atom is bonded to another carbon and three hydrogen atoms. As in methane, each carbon atom in ethane retains a tetrahedral shape. A hydrocarbon is referred to as a *saturated hydrocarbon* when all the bonds in the molecule are single bonds.

(a)

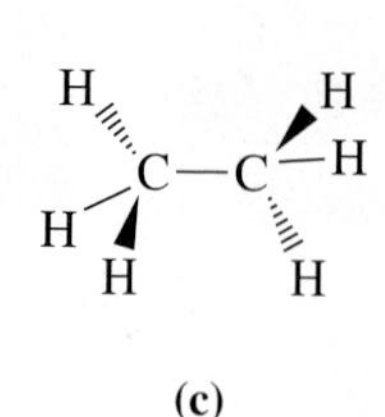

(b)

(c) wedge–dash model of ethane

(d) H—C—C—H with H above and below each C

(e) $CH_3—CH_3$

Three-dimensional and two-dimensional representations of ethane: **(a)** space-filling model, **(b)** ball-and-stick model, **(c)** wedge–dash model, **(d)** expanded structural formula, and **(e)** condensed structural formula.

PRACTICE PROBLEMS

11.1 Organic Compounds

LEARNING GOAL Identify properties characteristic of organic or inorganic compounds.

11.1 Identify each of the following as a formula of an organic or inorganic compound. For an organic compound, indicate if represented as molecular formula, expanded, or condensed structural formula:

a. KCl **b.** $CH_3—CH_2—CH_2—CH_3$

c. 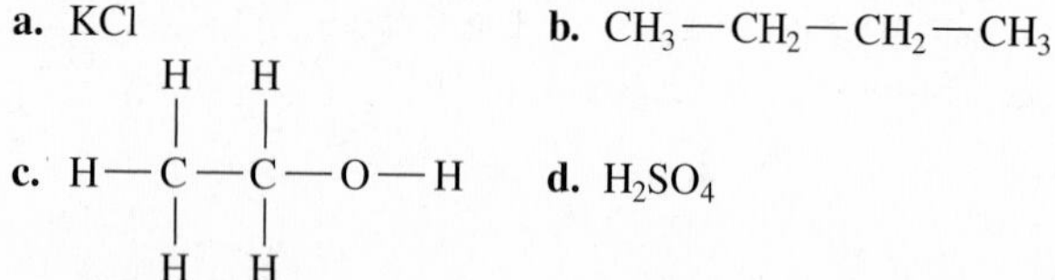 **d.** H_2SO_4

e. $CaCl_2$ **f.** C_3H_7Cl

11.2 Identify each of the following as a formula of an organic or inorganic compound. For an organic compound, indicate if represented as molecular formula, expanded, or condensed structural formula:

a. $C_6H_{12}O_6$ **b.** K_3PO_4

c. I_2 **d.** H—C—S—H (with H above and below C)

e. $CH_3—CH_2—CH_2—CH_2—CH_3$

f. C_4H_9Br

11.3 Identify each of the following properties as more typical of an organic or inorganic compound:

a. is soluble in water

b. has a low boiling point

c. contains carbon and hydrogen

d. contains ionic bonds

11.4 Identify each of the following properties as more typical of an organic or inorganic compound:

a. contains Li and F

b. is a gas at room temperature

c. contains covalent bonds

d. produces ions in water

11.5 Match each of the following physical and chemical properties with ethane, C_2H_6, or sodium bromide, NaBr:

a. boils at −89 °C

b. burns vigorously in air

c. is a solid at 250 °C

d. dissolves in water

11.6 Match each of the following physical and chemical properties with cyclohexane, C_6H_{12}, or calcium nitrate, $Ca(NO_3)_2$:

a. melts at 500 °C

b. is insoluble in water

c. does not burn in air

d. is a liquid at room temperature

11.2 Alkanes

LEARNING GOAL Write the IUPAC names and draw the condensed structural and line-angle formulas for alkanes and cycloalkanes.

More than 90% of the compounds in the world are organic compounds. The large number of carbon compounds is possible because the covalent bond between carbon atoms (C—C) is very strong, allowing carbon atoms to form long, stable chains.

The **alkanes** are a type of hydrocarbon in which the carbon atoms are connected only by single bonds. One of the most common uses of alkanes is as fuels. Methane, used in gas heaters and gas cooktops, is an alkane with one carbon atom. The alkanes ethane, propane, and butane contain two, three, and four carbon atoms, respectively, connected in a row or a *continuous chain*. As we can see, the names for alkanes end in *ane*. Such names are part of the **IUPAC system** (International Union of Pure and Applied Chemistry) used by chemists to name organic compounds. Alkanes with five or more carbon atoms in a chain are named using Greek prefixes: *pent* (5), *hex* (6), *hept* (7), *oct* (8), *non* (9), and *dec* (10) (see **TABLE 11.2**).

TABLE 11.2 IUPAC Names and Formulas of the First Ten Alkanes

Number of Carbon Atoms	IUPAC Name	Molecular Formula	Condensed Structural Formula	Line-Angle Formula
1	Methane	CH_4	CH_4	
2	Ethane	C_2H_6	CH_3-CH_3	
3	Propane	C_3H_8	$CH_3-CH_2-CH_3$	
4	Butane	C_4H_{10}	$CH_3-CH_2-CH_2-CH_3$	
5	Pentane	C_5H_{12}	$CH_3-CH_2-CH_2-CH_2-CH_3$	
6	Hexane	C_6H_{14}	$CH_3-CH_2-CH_2-CH_2-CH_2-CH_3$	
7	Heptane	C_7H_{16}	$CH_3-CH_2-CH_2-CH_2-CH_2-CH_2-CH_3$	
8	Octane	C_8H_{18}	$CH_3-CH_2-CH_2-CH_2-CH_2-CH_2-CH_2-CH_3$	
9	Nonane	C_9H_{20}	$CH_3-CH_2-CH_2-CH_2-CH_2-CH_2-CH_2-CH_2-CH_3$	
10	Decane	$C_{10}H_{22}$	$CH_3-CH_2-CH_2-CH_2-CH_2-CH_2-CH_2-CH_2-CH_2-CH_3$	

Condensed Structural and Line-Angle Formulas

CORE CHEMISTRY SKILL
Naming and Drawing Alkanes

In a condensed structural formula, each carbon atom and its attached hydrogen atoms are written as a group. A subscript indicates the number of hydrogen atoms bonded to each carbon atom.

Expanded		Condensed	Expanded		Condensed
H with C bonded to H above and below: $H-\overset{H}{\underset{H}{C}}-$	=	CH_3-	$-\overset{H}{\underset{H}{C}}-$	=	$-CH_2-$

When an organic molecule consists of a chain of three or more carbon atoms, the carbon atoms do not lie in a straight line. Rather, they are arranged in a zigzag pattern.

A simplified formula called the **line-angle formula** shows a zigzag line in which carbon atoms are represented as the ends of each line and as corners. For example, in the line-angle formula of pentane, each line in the zigzag drawing represents a single bond. The carbon atoms on the ends are bonded to three hydrogen atoms. However, the carbon atoms in the middle of the carbon chain are each bonded to two carbons and two hydrogen atoms as shown in Sample Problem 11.1.

ENGAGE
How does a line-angle formula represent an organic compound of carbon and hydrogen with single bonds?

SAMPLE PROBLEM 11.1 Drawing Formulas for an Alkane

TRY IT FIRST

Draw the expanded, condensed structural, and line-angle formulas for pentane.

SOLUTION GUIDE

	Given	Need	Connect
ANALYZE THE PROBLEM	pentane	expanded, condensed structural, and line-angle formulas	carbon chain, zigzag line

STEP 1 Draw the carbon chain. A molecule of pentane has five carbon atoms in a continuous chain.

$$C-C-C-C-C$$

STEP 2 Draw the expanded structural formula by adding the hydrogen atoms using single bonds to each of the carbon atoms.

$$\begin{array}{ccccccccccc} & & H & & H & & H & & H & & H & & \\ & & | & & | & & | & & | & & | & & \\ H & - & C & - & C & - & C & - & C & - & C & - & H \\ & & | & & | & & | & & | & & | & & \\ & & H & & H & & H & & H & & H & & \end{array}$$

STEP 3 Draw the condensed structural formula by combining the H atoms with each C atom.

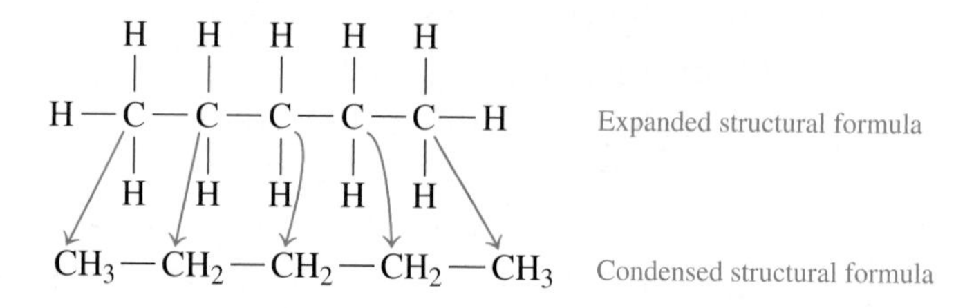

STEP 4 Draw the line-angle formula as a zigzag line in which the ends and corners represent C atoms.

$$CH_3-CH_2-CH_2-CH_2-CH_3$$ Condensed structural formula

Line-angle formula

STUDY CHECK 11.1

Draw the condensed structural formula and write the name for the following line-angle formula:

ANSWER

$CH_3-CH_2-CH_2-CH_2-CH_2-CH_2-CH_3$ heptane

Because an alkane has only single carbon–carbon bonds, the groups attached to each C are not in fixed positions. They can rotate freely around the bond connecting the carbon atoms. For example, butane can be drawn using a variety of structural formulas as shown in **TABLE 11.3**. All of these formulas represent the same compound with four carbon atoms.

Cycloalkanes

Hydrocarbons can also form cyclic or ring structures called **cycloalkanes**, which have two fewer hydrogen atoms than the corresponding alkanes. The simplest cycloalkane, cyclopropane (C_3H_6), has a ring of three carbon atoms bonded to six hydrogen atoms. Most often cycloalkanes are drawn using their line-angle formulas, which appear as simple geometric figures. As seen for alkanes, each corner of the line-angle formula for a cycloalkane represents a carbon atom. The ball-and-stick model and condensed structural and line-angle formulas for several cycloalkanes are shown in **TABLE 11.4**. A cycloalkane is named by adding the prefix *cyclo* to the name of the alkane with the same number of carbon atoms.

TABLE 11.3 Some Structural Formulas for Butane, C_4H_{10}

Expanded Structural Formula

$$\begin{array}{ccccccccc} & & H & & H & & H & & H & & \\ & & | & & | & & | & & | & & \\ H & - & C & - & C & - & C & - & C & - & H \\ & & | & & | & & | & & | & & \\ & & H & & H & & H & & H & & \end{array}$$

Condensed Structural Formulas

$CH_3-CH_2-CH_2-CH_3$

$$\begin{array}{ccc} CH_2 & - & CH_2 \\ | & & | \\ CH_3 & & CH_3 \end{array}$$

$$\begin{array}{ccc} CH_3 & & \\ | & & \\ CH_2 & - & CH_2 \\ & & | \\ & & CH_3 \end{array}$$

$$\begin{array}{c} CH_3 \\ | \\ CH_2 \\ | \\ CH_2 \\ | \\ CH_3 \end{array}$$

$$\begin{array}{ccc} CH_3 & - & CH_2 \\ & & | \\ & & CH_2 - CH_3 \end{array}$$

$$\begin{array}{c} CH_3 \\ | \\ CH_2-CH_2-CH_3 \end{array}$$

$$\begin{array}{ccccccc} CH_3 & & & & CH_2 & & \\ & \diagdown & & \diagup & & \diagdown & \\ & & CH_2 & & & & CH_3 \end{array}$$

Line-Angle Formulas

TABLE 11.4 Formulas of Some Common Cycloalkanes

Name	Cyclopropane	Cyclobutane	Cyclopentane	Cyclohexane
Ball-and-Stick Model		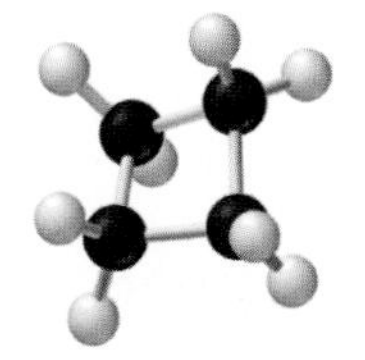	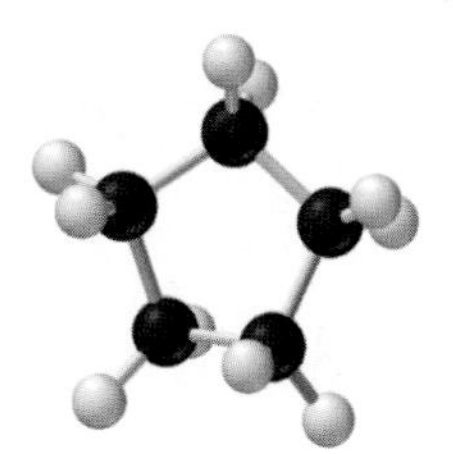	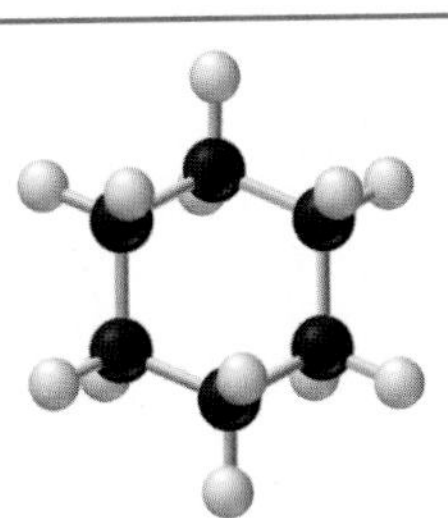
Condensed Structural Formula	CH_2 / H_2C-CH_2 (three-membered ring)	H_2C-CH_2 / H_2C-CH_2 (four-membered ring)	CH_2 / H_2C CH_2 / H_2C-CH_2 (five-membered ring)	CH_2 / H_2C CH_2 / H_2C CH_2 / CH_2 (six-membered ring)
Line-Angle Formula	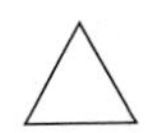			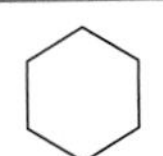

SAMPLE PROBLEM 11.2 Naming Alkanes and Cycloalkanes

TRY IT FIRST

Give the IUPAC name for each of the following:

a. (line-angle chain)

b. 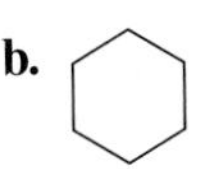

SOLUTION

a. A chain with eight carbon atoms is octane.

b. The ring of six carbon atoms is named cyclohexane.

STUDY CHECK 11.2

What is the IUPAC name of the following compound?

ANSWER

cyclobutane

TEST

Try Practice Problems 11.7 to 11.10

PRACTICE PROBLEMS

11.2 Alkanes

LEARNING GOAL Write the IUPAC names and draw the condensed structural and line-angle formulas for alkanes and cycloalkanes.

11.7 Give the IUPAC name for each of the following alkanes and cycloalkanes:

a. 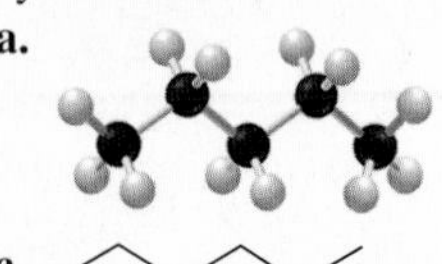

b. $CH_3—CH_3$

c.

d. 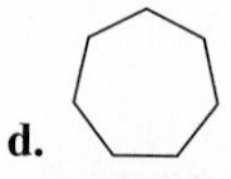

11.8 Give the IUPAC name for each of the following alkanes and cycloalkanes:

a. CH_4

b. 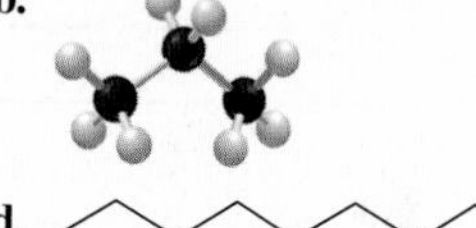

c.

d.

11.9 Draw the condensed structural formula for alkanes or the line-angle formula for cycloalkanes for each of the following:

a. methane

b. ethane

c. butane

d. cyclopropane

11.10 Draw the condensed structural formula for alkanes or the line-angle formula for cycloalkanes for each of the following:

a. propane

b. hexane

c. heptane

d. cyclohexane

11.3 Alkanes with Substituents

LEARNING GOAL Write the IUPAC names for alkanes with substituents and draw their condensed structural or line-angle formulas.

When an alkane has four or more carbon atoms, the atoms can be arranged so that a side group called a *branch* or **substituent** is attached to a carbon chain. For example, **FIGURE 11.2** shows two different ball-and-stick models for two compounds that have the same molecular formula, C_4H_{10}. One model is shown as a chain of four carbon atoms. In the other model, a carbon atom is attached as a branch or substituent to a carbon in a chain of three atoms. An alkane with at least one branch is called a *branched alkane*. When the two compounds have the same molecular formula but different arrangements of atoms, they are **structural isomers**.

ENGAGE

How can two or more structural isomers have the same molecular formula?

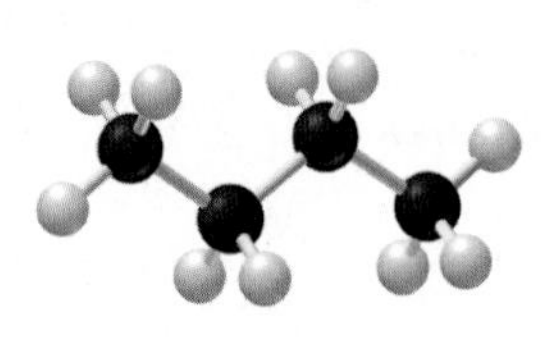

$CH_3—CH_2—CH_2—CH_3$

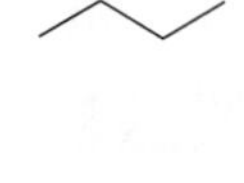

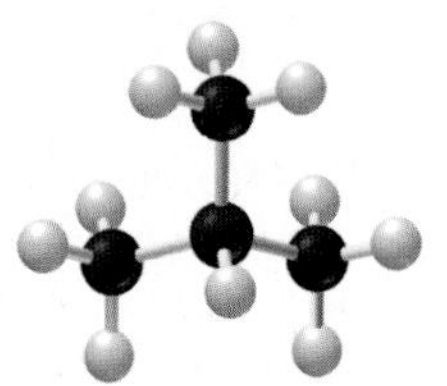

$$\begin{array}{c} CH_3 \\ | \\ CH_3—CH—CH_3 \end{array}$$

FIGURE 11.2 The structural isomers of C_4H_{10} have the same number and type of atoms but are bonded in a different order.

Q What makes these molecules structural isomers?

In another example, we can draw the condensed structural and line-angle formulas for three different isomers with the molecular formula C_5H_{12} as follows:

Isomers of C_5H_{12}

Condensed Structural Formula	$CH_3-CH_2-CH_2-CH_2-CH_3$	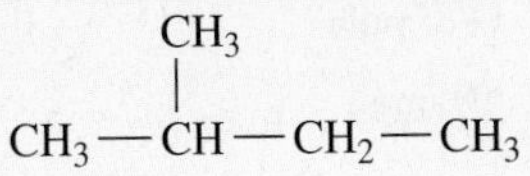 $CH_3-CH(CH_3)-CH_2-CH_3$	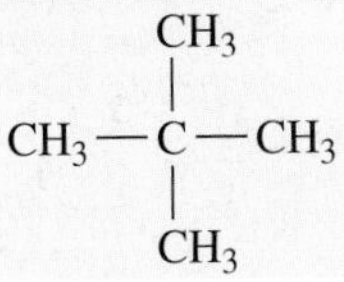 $CH_3-C(CH_3)_2-CH_3$
Line-Angle Formula		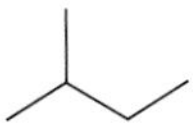	

SAMPLE PROBLEM 11.3 Structural Isomers

TRY IT FIRST

Identify each pair of formulas as structural isomers or the same molecule.

a. 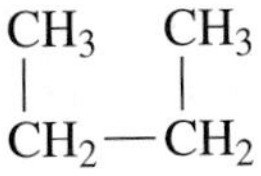 $CH_2(CH_3)-CH_2(CH_3)$ and $CH_2(CH_3)-CH_2-CH_3$

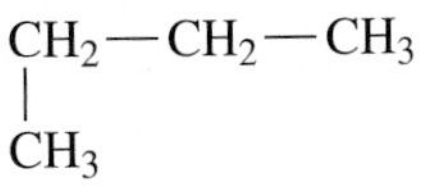

b. 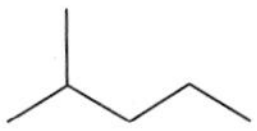 [line-angle formulas] and [line-angle formulas]

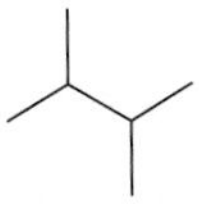

SOLUTION

a. When we count the number of C atoms and H atoms, they give the same molecular formula C_4H_{10}. Both of the structures are continuous four-carbon chains even though one or more of the $-CH_3$ ends are drawn above or below the horizontal part of the chain. Thus, both condensed structural formulas represent the same molecule.

b. When we count the number of C atoms and H atoms, they give the same molecular formula C_6H_{14}. The line-angle formula on the left has a five-carbon chain with a $-CH_3$ substituent on the second carbon of the chain. The line-angle formula on the right has a four-carbon chain with two $-CH_3$ substituents. Thus, there is a different order of bonding of atoms, which represents structural isomers.

STUDY CHECK 11.3

Why does the following line-angle formula represent a different structural isomer of the molecules in Sample Problem 11.3, part **b**?

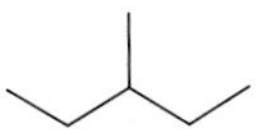

ANSWER

This line-angle formula represents a different structural isomer of C_6H_{14} because it has the same molecular formula but has a different arrangement of carbon atoms with a $-CH_3$ substituent on the third carbon of a five-carbon chain.

TEST

Try Practice Problems 11.11 and 11.12

Substituents in Alkanes

In the IUPAC names for alkanes, a carbon branch is named as an **alkyl group**, which is an alkane that is missing one hydrogen atom. The alkyl group is named by replacing the *ane* ending of the corresponding alkane name with *yl*. Alkyl groups cannot exist on their own: They must be attached to a carbon chain. When a halogen atom is attached to a carbon chain, it is named as a *halo* group: *fluoro*, *chloro*, *bromo*, or *iodo*. Some of the common groups attached to carbon chains are illustrated in **TABLE 11.5**.

TABLE 11.5 Formulas and Names of Some Common Substituents

Formula	CH_3-		CH_3-CH_2-	
Name	methyl		ethyl	
Formula	$CH_3-CH_2-CH_2-$		$CH_3-\underset{\vert}{CH}-CH_3$ (attached at central CH)	
Name	propyl		isopropyl	
Formula	$CH_3-CH_2-CH_2-CH_2-$		$CH_3-CH(CH_3)-CH_2-$	
Name	butyl		isobutyl	
Formula	F—	Cl—	Br—	I—
Name	fluoro	chloro	bromo	iodo

INTERACTIVE VIDEO
Naming Alkanes

Naming Alkanes with Substituents

In the IUPAC system of naming, a carbon chain is numbered to give the location of the substituents. Let's take a look at how we use the IUPAC system to name the alkane shown in Sample Problem 11.4.

ENGAGE
What part of the IUPAC name gives (a) the number of carbon atoms in the carbon chain and (b) the substituents on the carbon chain?

SAMPLE PROBLEM 11.4 Writing IUPAC Names for Alkanes with Substituents

TRY IT FIRST

Write the IUPAC name for the following alkane:

$$CH_3-\overset{\displaystyle CH_3}{\overset{\vert}{CH}}-CH_2-\underset{\displaystyle CH_3}{\underset{\vert}{\overset{\displaystyle Br}{\overset{\vert}{C}}}}-CH_2-CH_3$$

SOLUTION GUIDE

	Given	Need	Connect
ANALYZE THE PROBLEM	six-carbon chain, two methyl groups, one bromo group	IUPAC name	position of substituents on the carbon chain

STEP 1 Write the alkane name for the longest chain of carbon atoms.

$$CH_3-\overset{\displaystyle CH_3}{\overset{\vert}{CH}}-CH_2-\underset{\displaystyle CH_3}{\underset{\vert}{\overset{\displaystyle Br}{\overset{\vert}{C}}}}-CH_2-CH_3 \quad \text{hexane}$$

STEP 2 Number the carbon atoms from the end nearer a substituent.

$$CH_3-\overset{\displaystyle CH_3}{\overset{\vert}{CH}}-CH_2-\underset{\displaystyle CH_3}{\underset{\vert}{\overset{\displaystyle Br}{\overset{\vert}{C}}}}-CH_2-CH_3 \quad \text{hexane}$$

1 2 3 4 5 6

STEP 3 **Give the location and name for each substituent (alphabetical order) as a prefix to the name of the main chain.** The substituents are listed in alphabetical order (bromo first, then methyl). A hyphen is placed between the number and the substituent name. When there are two or more of the same substituent, a prefix (*di*, *tri*, *tetra*) is used and commas separate the numbers. However, prefixes are not used to determine the alphabetical order of the substituents.

$$\begin{array}{cccccc} & CH_3 & & Br & & \\ & | & & | & & \\ CH_3- & CH- & CH_2- & C- & CH_2- & CH_3 \\ & & & | & & \\ & & & CH_3 & & \\ 1 & 2 & 3 & 4 & 5 & 6 \end{array}$$

4-bromo-2,4-dimethylhexane

STUDY CHECK 11.4

Write the IUPAC name for the following compound:

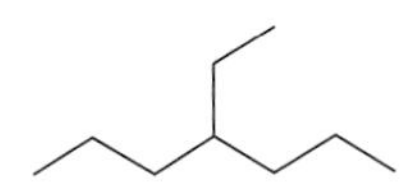

ANSWER

4-ethylheptane

TEST

Try Practice Problems 11.13 and 11.14

Drawing Structural Formulas for Alkanes with Substituents

The IUPAC name gives all the information needed to draw the condensed structural formula for an alkane. Suppose you are asked to draw the condensed structural formula for 2,3-dimethylbutane. The alkane name gives the number of carbon atoms in the longest chain. The other names indicate the substituents and where they are attached. We can break down the name in the following way:

2,3-Dimethylbutane				
2,3-	**Di**	**methyl**	**but**	**ane**
Substituents on carbons 2 and 3	two identical groups	$-CH_3$ alkyl groups	four C atoms in the main chain	single (C—C) bonds

SAMPLE PROBLEM 11.5 **Drawing Condensed Structural and Line-Angle Formulas from IUPAC Names**

TRY IT FIRST

Draw the condensed structural and line-angle formulas for 2,3-dimethylbutane.

SOLUTION GUIDE

	Given	Need	Connect
ANALYZE THE PROBLEM	2,3-dimethylbutane	condensed structural and line-angle formulas	four-carbon chain, two methyl groups

STEP 1 **Draw the main chain of carbon atoms.** For butane, we draw a chain and a zigzag line of four carbon atoms.

C—C—C—C

STEP 2 **Number the chain and place the substituents on the carbons indicated by the numbers.** The first part of the name indicates two methyl groups $-CH_3$: one on carbon 2 and one on carbon 3.

Methyl Methyl
CH_3 CH_3
C—C—C—C
1 2 3 4

1 2 3 4

STEP 3 **For the condensed structural formula, add the correct number of hydrogen atoms to give four bonds to each C atom.**

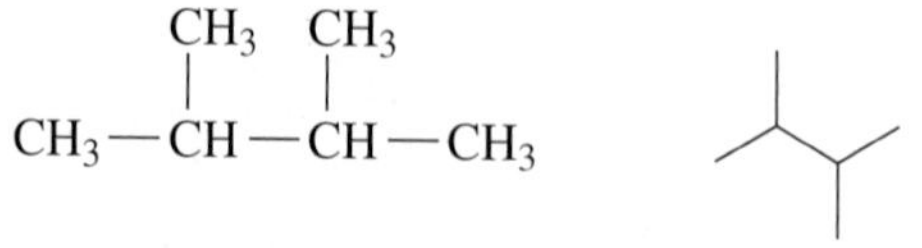

CH_3 CH_3
$CH_3-CH-CH-CH_3$

STUDY CHECK 11.5

Draw the condensed structural and line-angle formulas for 2-bromo-4-methylpentane.

ANSWER

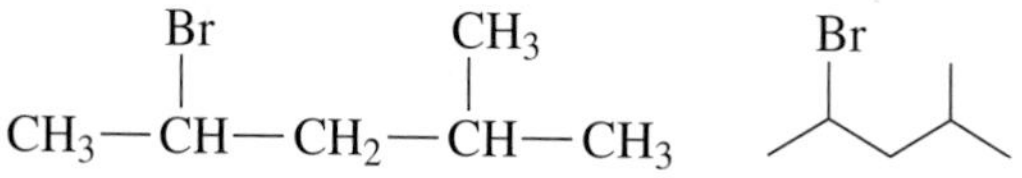

Br CH_3
$CH_3-CH-CH_2-CH-CH_3$

Br

TEST

Try Practice Problems 11.15 to 11.18

PRACTICE PROBLEMS

11.3 Alkanes with Substituents

LEARNING GOAL Write the IUPAC names for alkanes with substituents and draw their condensed structural or line-angle formulas.

11.11 Indicate whether each of the following pairs represent structural isomers or the same molecule:

a. 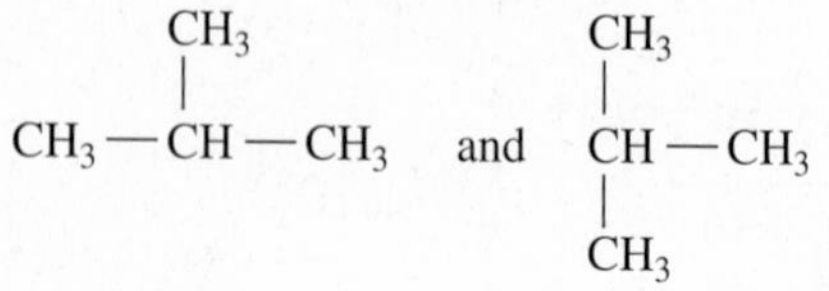 $CH_3-CH(CH_3)-CH_3$ and $CH(CH_3)-CH_3$ with CH_3 below

b. 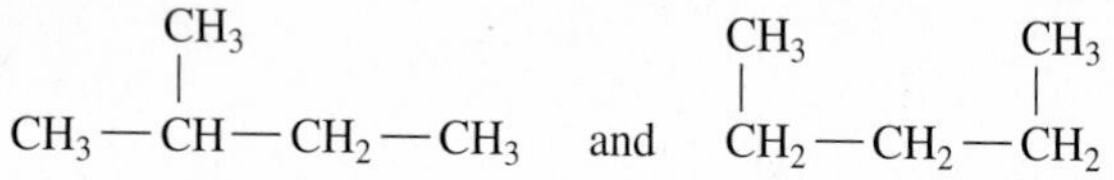 $CH_3-CH(CH_3)-CH_2-CH_3$ and $CH_2(CH_3)-CH_2-CH_2(CH_3)$

c. 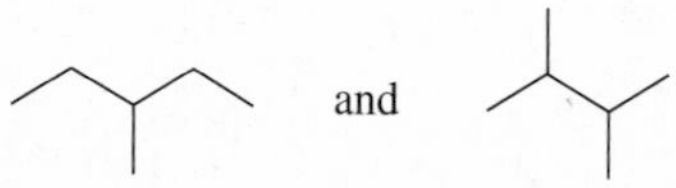 and

11.12 Indicate whether each of the following pairs represent structural isomers or the same molecule:

a. 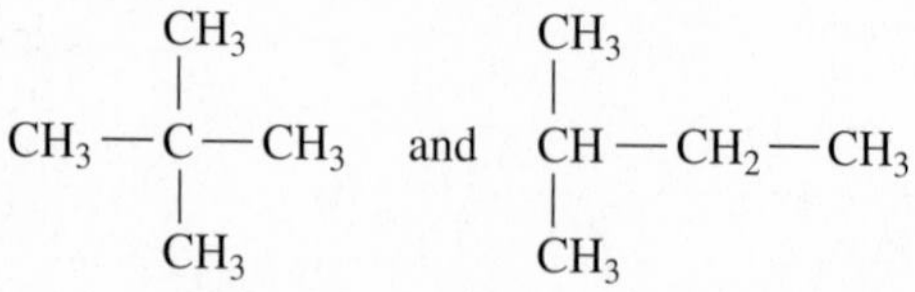 $CH_3-C(CH_3)_2-CH_3$ and $CH(CH_3)(CH_3)-CH_2-CH_3$

b. $CH_3-CH(CH_3)-CH(CH_3)-CH_2(CH_3)$ and $CH_3-CH(CH_3)-CH_2-CH(CH_3)-CH_3$

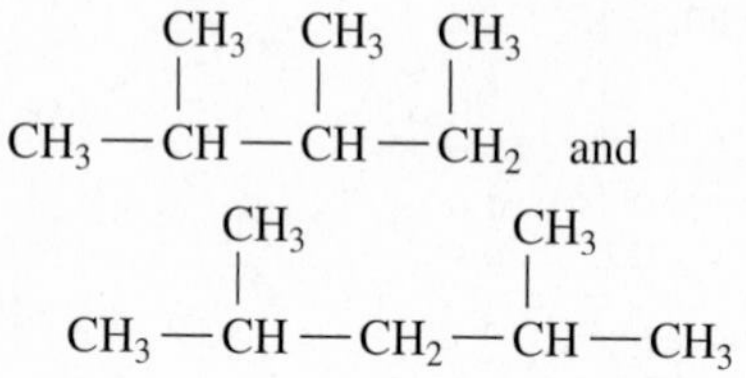

c. and

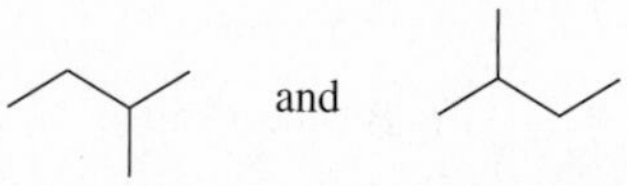

11.13 Give the IUPAC name for each of the following:

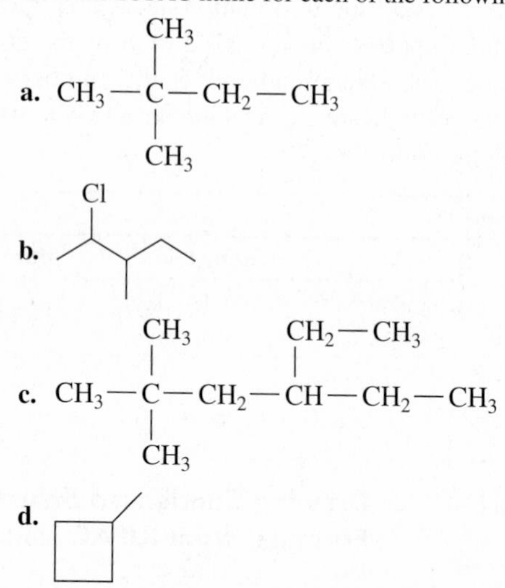

11.14 Give the IUPAC name for each of the following:

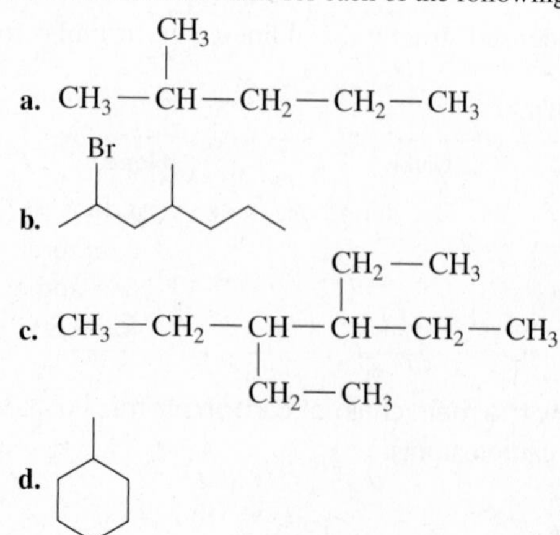

11.15 Draw the condensed structural formula for each of the following alkanes:
a. 3,3-dimethylpentane **b.** 2,3,5-trimethylhexane
c. 3-ethyl-2,5-dimethyloctane **d.** 1-bromo-2-chloroethane

11.16 Draw the condensed structural formula for each of the following alkanes:
a. 3-ethylpentane **b.** 3-ethyl-2-methylheptane
c. 4-ethyl-2,2-dimethyloctane **d.** 2-bromopropane

11.17 Draw the line-angle formula for each of the following:
a. 3-methylheptane **b.** ethylcyclopentane
c. bromocyclobutane **d.** 2,3-dichlorohexane

11.18 Draw the line-angle formula for each of the following:
a. 1-bromo-2-methylpentane **b.** methylcyclopropane
c. ethylcyclohexane **d.** 4-chlorooctane

11.4 Properties of Alkanes

LEARNING GOAL Identify the properties of alkanes and write a balanced chemical equation for combustion.

Many types of alkanes are the components of fuels that power our cars and oil that heats our homes. You may have used a mixture of hydrocarbons such as mineral oil as a laxative or petrolatum to soften your skin. The differences in uses of many of the alkanes result from their physical properties, including solubility and density.

The solid alkanes that make up waxy coatings on fruits and vegetables help retain moisture, inhibit mold, and enhance appearance.

Some Uses of Alkanes

The first four alkanes—methane, ethane, propane, and butane—are gases at room temperature and are widely used as heating fuels.

Alkanes having five to eight carbon atoms (pentane, hexane, heptane, and octane) are liquids at room temperature. They are highly volatile, which makes them useful in fuels such as gasoline.

Liquid alkanes with 9 to 17 carbon atoms have higher boiling points and are found in kerosene, diesel, and jet fuels. Motor oil is a mixture of high-molecular-weight liquid hydrocarbons and is used to lubricate the internal components of engines. Mineral oil is a mixture of liquid hydrocarbons and is used as a laxative and a lubricant. Alkanes with 18 or more carbon atoms are waxy solids at room temperature. Known as paraffins, they are used in waxy coatings added to fruits and vegetables to retain moisture, inhibit mold, and enhance appearance. Petrolatum, or Vaseline, is a semisolid mixture of hydrocarbons with more than 25 carbon atoms used in ointments and cosmetics and as a lubricant.

FIGURE 11.3 In oil spills, large quantities of oil spread out to form a thin layer on the ocean surface.

Q What physical properties cause oil to remain on the surface of water?

Solubility and Density

Alkanes are nonpolar, which makes them insoluble in water. However, they are soluble in nonpolar solvents such as other alkanes. Alkanes have densities from 0.62 g/mL to about 0.79 g/mL, which is less than the density of water (1.0 g/mL).

If there is an oil spill in the ocean, the alkanes in the oil, which do not mix with water, form a thin layer on the surface that spreads over a large area. In April 2010, an explosion on an oil-drilling rig in the Gulf of Mexico caused the largest oil spill in U.S. history. At its maximum, an estimated 10 million liters of oil was leaked every day (see **FIGURE 11.3**). If the crude oil reaches land, there can be considerable damage to beaches, shellfish, birds, and wildlife habitats. When animals such as birds are covered with oil, they must be cleaned quickly because ingestion of the hydrocarbons when they try to clean themselves is fatal.

Butane in a portable burner undergoes combustion.

Combustion of Alkanes

The carbon–carbon single bonds in alkanes are difficult to break, which makes them the least reactive family of organic compounds. However, alkanes burn readily in oxygen to produce carbon dioxide, water, and energy. For example, methane is the natural gas we use to cook our food on a gas cooktop and heat our homes.

$$CH_4 + 2O_2 \xrightarrow{\Delta} CO_2 + 2H_2O + \text{energy}$$

Butane is used in cooking, camping, and torches.

$$2C_4H_{10} + 13O_2 \xrightarrow{\Delta} 8CO_2 + 10H_2O + \text{energy}$$

TEST

Try Practice Problems 11.19 to 11.22

PRACTICE PROBLEMS

11.4 Properties of Alkanes

LEARNING GOAL Identify the properties of alkanes and write a balanced chemical equation for combustion.

11.19 Heptane, used as a solvent for rubber cement, has a density of 0.68 g/mL and boils at 98 °C.
- **a.** Draw the condensed structural and line-angle formulas for heptane.
- **b.** Is heptane a solid, liquid, or gas at room temperature?
- **c.** Is heptane soluble in water?
- **d.** Will heptane float on water or sink?
- **e.** Write the balanced chemical equation for the complete combustion of heptane.

11.20 Nonane has a density of 0.79 g/mL and boils at 151 °C.
- **a.** Draw the condensed structural and line-angle formulas for nonane.
- **b.** Is nonane a solid, liquid, or gas at room temperature?
- **c.** Is nonane soluble in water?
- **d.** Will nonane float on water or sink?
- **e.** Write the balanced chemical equation for the complete combustion of nonane.

11.21 Write the balanced chemical equation for the complete combustion of each of the following compounds:
a. ethane **b.** cyclopropane **c.** 2,3-dimethylhexane

11.22 Write the balanced chemical equation for the complete combustion of each of the following compounds:
a. hexane **b.** cyclopentane **c.** dimethylpropane

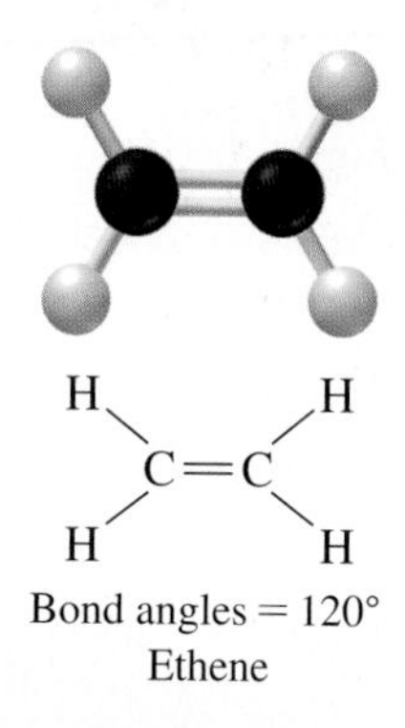

Bond angles = 120°
Ethene

H—C≡C—H
Bond angles = 180°
Ethyne

FIGURE 11.4 Ball-and-stick models of ethene and ethyne show the functional groups of double or triple bonds and the bond angles.

Q Why are these compounds called unsaturated hydrocarbons?

11.5 Alkenes and Alkynes

LEARNING GOAL Write the IUPAC names or draw the condensed structural or line-angle formulas for alkenes and alkynes.

We organize organic compounds into *classes* or *families* by their *functional groups*, which are groups of specific atoms. Compounds that contain the same functional group have similar physical and chemical properties. Identifying functional groups allows us to classify organic compounds according to their structure, to name compounds within each family, predict their chemical reactions, and draw the structures for their products.

Alkenes and *alkynes* are families of hydrocarbons that contain double and triple bonds as functional groups. They are called *unsaturated hydrocarbons* because they do not contain the maximum number of hydrogen atoms, as do alkanes. They react with hydrogen gas to increase the number of hydrogen atoms to become alkanes, which are *saturated hydrocarbons*.

Identifying Alkenes and Alkynes

Alkenes contain one or more carbon–carbon double bonds that form when adjacent carbon atoms share two pairs of valence electrons. Recall that *a carbon atom always forms four covalent bonds*. In the simplest alkene, ethene (C_2H_4), two carbon atoms are connected by a double bond and each is also attached to two H atoms. This gives each carbon atom in the double bond a trigonal planar arrangement with bond angles of 120°. As a result, the ethene molecule is flat because the carbon and hydrogen atoms all lie in the same plane (see **FIGURE 11.4**).

In an **alkyne**, a triple bond forms when two carbon atoms share three pairs of valence electrons. In the simplest alkyne, ethyne (C_2H_2), the two carbon atoms of the triple bond are each attached to one hydrogen atom, which gives a triple bond a linear geometry.

TEST

Try Practice Problems 11.23 and 11.24

Fruit is ripened with ethene, a plant hormone.

A mixture of acetylene and oxygen undergoes combustion during the welding of metals.

Ethene, more commonly called ethylene, is an important plant hormone involved in promoting the ripening of fruit. Commercially grown fruit, such as avocados, bananas, and tomatoes, are often picked before they are ripe. Before the fruit is brought to market, it is exposed to ethylene to accelerate the ripening process. Ethylene also accelerates the breakdown of cellulose in plants, which causes flowers to wilt and leaves to fall from trees. Ethyne, commonly called acetylene, is used in welding where it reacts with oxygen to produce flames with temperatures above 3300 °C.

Naming Alkenes and Alkynes

The IUPAC names for alkenes and alkynes are similar to those of alkanes. Using the alkane name with the same number of carbon atoms, the *ane* ending is replaced with *ene* for an alkene and *yne* for an alkyne (see **TABLE 11.6**). Cyclic alkenes are named as *cycloalkenes*.

TABLE 11.6 Comparison of Names for Alkanes, Alkenes, and Alkynes

Alkane	Alkene	Alkyne
CH_3-CH_3	$H_2C=CH_2$	$HC\equiv CH$
Ethane	Ethene (ethylene)	Ethyne (acetylene)
$CH_3-CH_2-CH_3$	$CH_3-CH=CH_2$	$CH_3-C\equiv CH$
Propane	Propene	Propyne

An example of naming an alkene is seen in Sample Problem 11.6.

SAMPLE PROBLEM 11.6 Naming Alkenes and Alkynes

TRY IT FIRST

Write the IUPAC name for the following:

$$\begin{array}{l} \quad\;\;\; CH_3 \\ \quad\;\;\;\; | \\ CH_3-CH-CH=CH-CH_3 \end{array}$$

SOLUTION GUIDE

	Given	Need	Connect
ANALYZE THE PROBLEM	five-carbon chain, double bond, methyl group	IUPAC name	replace the *ane* of the alkane name with *ene*

STEP 1 **Name the longest carbon chain that contains the double bond.** There are five carbon atoms in the longest carbon chain containing the double bond. Replace the *ane* in the corresponding alkane name with *ene* to give pentene.

$$\begin{array}{l} \quad\;\;\; CH_3 \\ \quad\;\;\;\; | \\ CH_3-CH-CH=CH-CH_3 \end{array} \qquad \text{pentene}$$

STEP 2 **Number the carbon chain starting from the end nearer the double bond.** Place the number of the first carbon in the double bond in front of the alkene name.

$$\begin{array}{l} \quad\;\;\; CH_3 \\ \quad\;\;\;\; | \\ \overset{5}{CH_3}-\overset{4}{CH}-\overset{3}{CH}=\overset{2}{CH}-\overset{1}{CH_3} \end{array} \qquad \text{2-pentene}$$

Alkenes with two or three carbons do not need numbers.

STEP 3 Give the location and name for each substituent (alphabetical order) as a prefix to the alkene name.

$$\underset{5}{CH_3}-\underset{4}{\overset{\overset{\displaystyle CH_3}{|}}{CH}}-\underset{3}{CH}=\underset{2}{CH}-\underset{1}{CH_3} \qquad \text{4-methyl-2-pentene}$$

ENGAGE

Why is the name 2-methyl-3-pentene incorrect for this compound?

STUDY CHECK 11.6

Draw the condensed structural formula for 1-chloro-3-hexyne.

ANSWER

$Cl-CH_2-CH_2-C\equiv C-CH_2-CH_3$

TEST

Try Practice Problems 11.25 to 11.28

Naming Cycloalkenes

Some alkenes called *cycloalkenes* have a double bond within a ring structure. If there is no substituent, the double bond does not need a number. If there is a substituent, the carbons in the double bond are numbered as 1 and 2, and the ring is numbered in the direction that will give the lower number to the substituent.

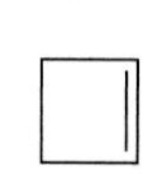

Cyclobutene

3-Methylcyclopentene

PRACTICE PROBLEMS

11.5 Alkenes and Alkynes

LEARNING GOAL Write the IUPAC names or draw the condensed structural or line-angle formulas for alkenes and alkynes.

11.23 Identify the following as alkanes, alkenes, cycloalkenes, or alkynes:

a. $H-\overset{\overset{\displaystyle H}{|}}{\underset{\underset{\displaystyle H}{|}}{C}}-\overset{\overset{\displaystyle H}{|}}{C}=\overset{\overset{\displaystyle H}{|}}{C}-H$ **b.** $CH_3-CH_2-C\equiv CH$

c. 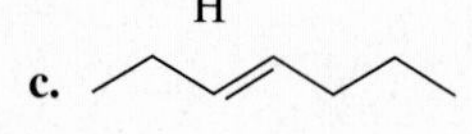 **d.** 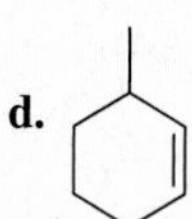

11.24 Identify the following as alkanes, alkenes, cycloalkenes, or alkynes:

a. 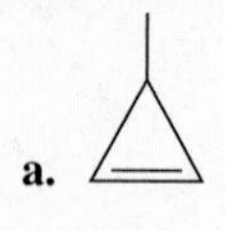 **b.**

c. $CH_3-\overset{\overset{\displaystyle CH_3}{|}}{C}=\underset{\underset{\displaystyle CH_3}{|}}{C}-CH_3$ **d.** $-C\equiv CH$

11.25 Give the IUPAC name for each of the following:

a. $H_2C=CH_2$

b. $CH_3-\overset{\overset{\displaystyle CH_3}{|}}{C}=CH_2$

c. $CH_3-CH_2-C\equiv C-CH_3$

d.

11.26 Give the IUPAC name for each of the following:

a. $H_2C=CH-CH_2-CH_3$

b. $CH_3-C\equiv C-CH_2-CH_2-\overset{\overset{\displaystyle CH_3}{|}}{CH}-CH_3$

c.

d. $CH_3-CH_2-CH=CH-CH_3$

11.27 Draw the condensed structural formula, or line-angle formula, if cyclic, for each of the following:

a. 1-pentene **b.** 2-methyl-1-butene

c. 3-methylcyclohexene **d.** 3-chloro-1-butyne

11.28 Draw the condensed structural formula, or line-angle formula, if cyclic, for each of the following:

a. 1-methylcyclopentene **b.** 1-bromo-3-hexyne

c. 3,4-dimethyl-1-pentene **d.** 2-methyl-2-hexene

11.6 Cis–Trans Isomers

LEARNING GOAL Draw the condensed structural formulas and give the names for the cis–trans isomers of alkenes.

In any alkene, the double bond is rigid, which means there is no rotation around the double bond. As a result, the atoms or groups are attached to the carbon atoms in the double bond on one side or the other, which gives two different structures called *geometric isomers* or *cis–trans isomers*.

For example, the formula for 2-butene can be drawn as two different molecules, which are cis–trans isomers. In the ball-and-stick models, the atoms bonded to the carbon atoms in the double bond have bond angles of 120° (see **FIGURE 11.5**). We add the prefix *cis* or *trans* to denote whether the atoms bonded to the double bond are on the same side or opposite sides. In the **cis isomer**, the CH_3— groups are on the same side of the double bond. In the **trans isomer**, the CH_3— groups are on opposite sides. *Trans* means "across," as in transcontinental; *cis* means "on this side."

As with any pair of cis–trans isomers, *cis*-2-butene and *trans*-2-butene are different compounds with different physical and chemical properties.

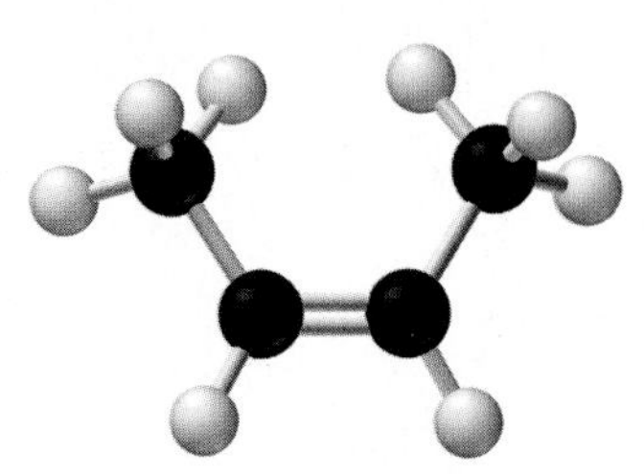

cis-2-Butene

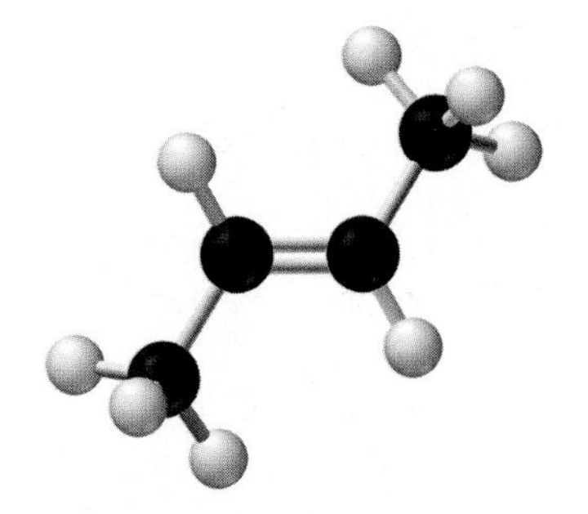

trans-2-Butene

FIGURE 11.5 Ball-and-stick models of the cis and trans isomers of 2-butene.

Q What feature in 2-butene accounts for the cis and trans isomers?

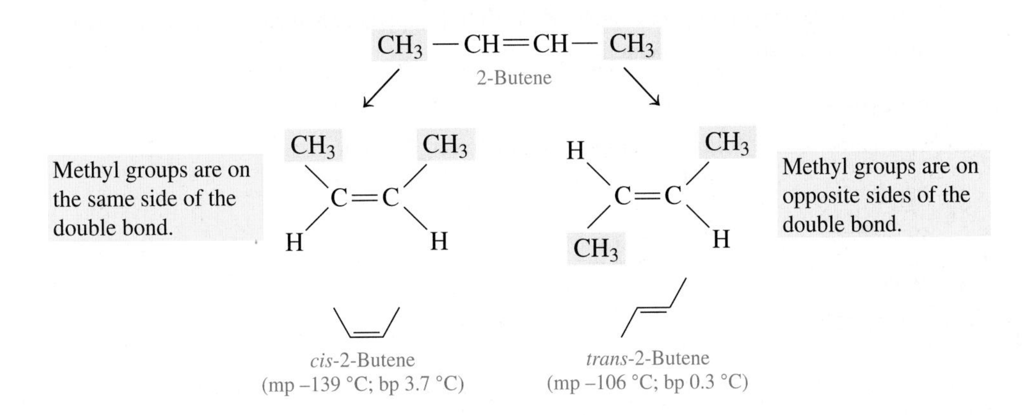

When the carbon atoms in the double bond are attached to two different atoms or groups of atoms, an alkene can have cis–trans isomers. For example, 1,2-dichloroethene can be drawn with cis and trans isomers because there is one H atom and one Cl atom attached to each carbon atom in the double bond. When you are asked to draw the formula for an alkene, it is important to consider the possibility of cis and trans isomers.

ENGAGE

Explain how an alkene can have two geometric isomers.

Cl—CH=CH—Cl

1,2-Dichloroethene

Chlorine atoms are on the same side of the double bond.

Cl Cl
C=C
H H

Cl H
C=C
H Cl

Chlorine atoms are on opposite sides of the double bond.

cis-1,2-Dichloroethene *trans*-1,2-Dichloroethene

When an alkene has identical groups on the same carbon atom of the double bond, cis and trans isomers cannot be drawn. For example, 1,1-dichloropropene has just one condensed structural formula without cis and trans isomers.

Cl CH3
C=C
Cl H

1,1-Dichloropropene

SAMPLE PROBLEM 11.7 Identifying Cis–Trans Isomers

TRY IT FIRST

Identify each of the following as the cis or trans isomer and give its name:

a.
```
Br          Cl
  \        /
   C == C
  /        \
H            H
```

b.
```
CH3         H
   \       /
    C == C
   /       \
H           CH2—CH3
```

SOLUTION

a. This is a cis isomer because the two halogen atoms attached to the carbon atoms of the double bond are on the same side. The name of the two-carbon alkene, starting with the bromo group on carbon 1, is *cis*-1-bromo-2-chloroethene.

b. This is a trans isomer because the two alkyl groups attached to the carbon atoms of the double bond are on opposite sides of the double bond. This isomer of the five-carbon alkene, is named *trans*-2-pentene.

STUDY CHECK 11.7

Give the name for the following compound, including cis or trans:

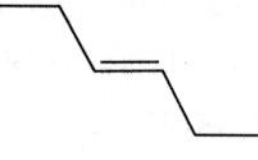

ANSWER

trans-3-hexene

TEST

Try Practice Problems 11.29 to 11.32

Modeling Cis–Trans Isomers

Because cis–trans isomerism is not easy to visualize, here are some things you can do to understand the difference in rotation around a single bond compared to a double bond and how it affects groups that are attached to the carbon atoms in the double bond.

Put the tips of your index fingers together. This is a model of a single bond. Consider the index fingers as a pair of carbon atoms, and think of your thumbs and other fingers as other parts of a carbon chain. While your index fingers are touching, twist your hands and change the position of your thumbs relative to each other. Notice how the relationship of your other fingers changes.

Cis-hands (cis-thumbs/fingers)

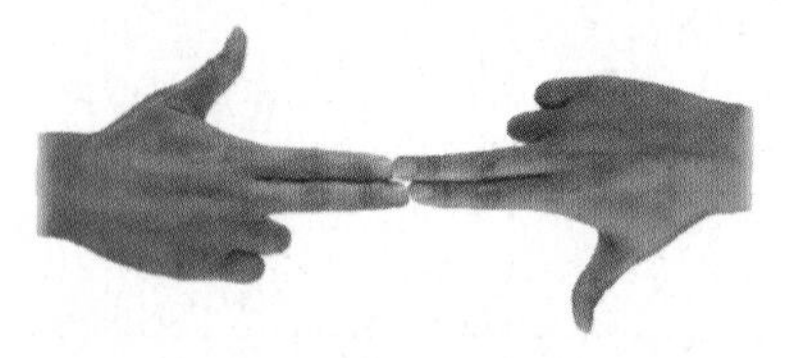

Trans-hands (trans-thumbs/fingers)

Now place the tips of your index fingers and middle fingers together in a model of a double bond. As you did before, twist your hands to move your thumbs away from each other. What happens? Can you change the location of your thumbs relative to each other without breaking the double bond? The difficulty of moving your hands with two fingers touching represents the lack of rotation about a double bond. You have made a model of a cis isomer when both thumbs point in the same direction. If you turn one hand over so one thumb points down and the other thumb points up, you have made a model of a trans isomer.

Using Gumdrops and Toothpicks to Model Cis–Trans Isomers

Obtain some toothpicks and yellow, green, and black gumdrops. The black gumdrops represent C atoms, the yellow gumdrops represent H atoms, and the green gumdrops represent Cl atoms. Place a toothpick between two black gumdrops. Use three more toothpicks to attach two yellow gumdrops and one green gumdrop to each black gumdrop carbon atom. Move one of the black gumdrops to show the rotation of the attached H and Cl atoms.

Remove a toothpick and yellow gumdrop from each black gumdrop. Place a second toothpick between the carbon atoms, which makes a double bond. Try to twist the double bond of toothpicks. Can you do it? When you observe the location of the green gumdrops, does the model you made represent a cis or trans isomer? Why? If your model is a cis isomer, how would you change it to a trans isomer? If your model is a trans isomer, how could you change it to a cis isomer?

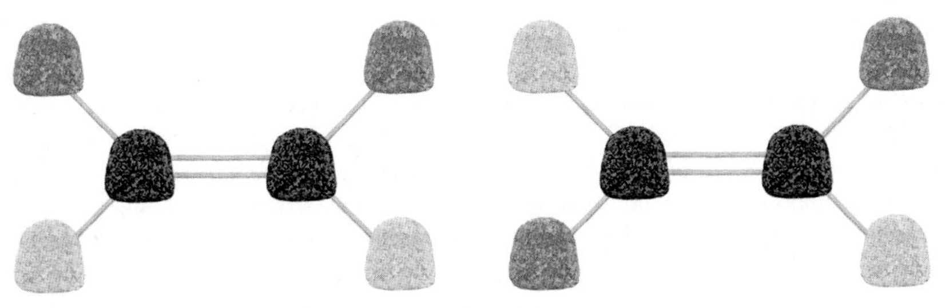

Models from gumdrops represent cis and trans isomers.

PRACTICE PROBLEMS

11.6 Cis–Trans Isomers

LEARNING GOAL Draw the condensed structural formulas and give the names for the cis–trans isomers of alkenes.

11.29 Give the IUPAC name for each of the following, using cis or trans prefixes, if needed:

a.

b.
CH_3-CH_2 H
C=C
H $CH_2-CH_2-CH_2-CH_3$

c.
H CH_3
C=C
H H

11.30 Give the IUPAC name for each of the following, using cis or trans prefixes, if needed:

a.
CH_3 CH_2-CH_3
C=C
H H

b.
CH_3 H
C=C
H $CH_2-CH_2-CH_2-CH_3$

c.

11.31 Draw the condensed structural formula for each of the following:
a. *trans*-1-bromo-2-chloroethene **b.** *cis*-2-hexene
c. *cis*-4-octene

11.32 Draw the condensed structural formula for each of the following:
a. *cis*-1,2-difluoroethene **b.** *trans*-2-pentene
c. *trans*-2-heptene

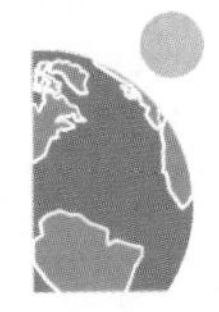

CHEMISTRY LINK TO THE ENVIRONMENT

Pheromones in Insect Communication

Many insects emit minute quantities of chemicals called *pheromones* to send messages to individuals of the same species. Some pheromones warn of danger, others call for defense, mark a trail, or attract the opposite sex. One of the most studied is bombykol, the sex pheromone produced by the female silkworm moth. The bombykol molecule contains one cis double bond and one trans double bond. Even a few nanograms of bombykol will attract male silkworm moths from distances of over 1 km. The effectiveness of many of these pheromones depends on the cis or trans configuration of the double bonds in the molecules. A certain species will respond to one isomer but not the other.

Pheromones allow insects to attract mates from a great distance.

Scientists are interested in synthesizing pheromones to use as nontoxic alternatives to pesticides. When placed in a trap, bombykol can be used to capture male silkworm moths. When a synthetic pheromone is released in a field, the males cannot locate the females, which disrupts the reproductive cycle. This technique has been successful with controlling the oriental fruit moth, the grapevine moth, and the pink bollworm.

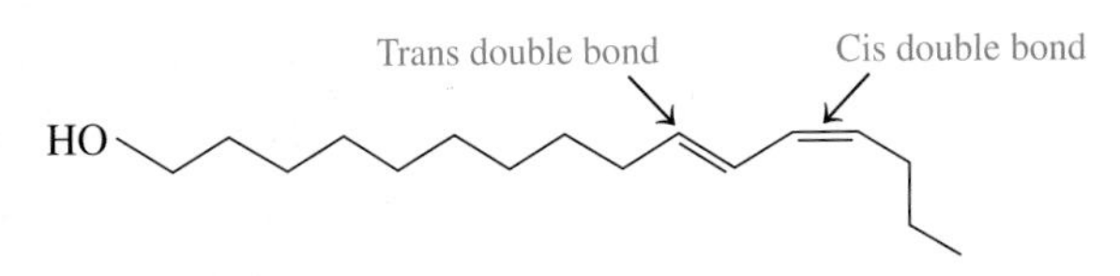

Bombykol, sex attractant for the silkworm moth

CHEMISTRY LINK TO HEALTH

Cis–Trans Isomers for Night Vision

The retinas of the eyes consist of two types of cells: rods and cones. The rods on the edge of the retina allow us to see in dim light, and the cones, in the center, produce our vision in bright light. In the rods, there is a substance called *rhodopsin* that absorbs light. Rhodopsin is composed of *cis*-11-retinal, an unsaturated compound, attached to a protein. When rhodopsin absorbs light, the *cis*-11-retinal isomer is converted to its trans isomer, which changes its shape. The trans form no longer fits the protein, and it separates from the protein. The change from the cis to trans isomer and its separation from the protein generate an electrical signal that the brain converts into an image.

An enzyme (isomerase) converts the trans isomer back to the *cis*-11-retinal isomer and the rhodopsin re-forms. If there is a deficiency of rhodopsin in the rods of the retina, *night blindness* may occur. One common cause is a lack of vitamin A in the diet. In our diet, we obtain vitamin A from plant pigments containing β-carotene, which is found in foods such as carrots, squash, and spinach. In the small intestine, the β-carotene is converted to vitamin A, which can be converted to *cis*-11-retinal or stored in the liver for future use. Without a sufficient quantity of retinal, not enough rhodopsin is produced to enable us to see adequately in dim light.

Cis–Trans Isomers of Retinal

Cis double bond — Trans double bond

cis-11-Retinal $\underset{\text{Isomerase}}{\overset{\text{Light}}{\rightleftarrows}}$ *trans*-11-Retinal

11.7 Addition Reactions for Alkenes

LEARNING GOAL Draw the condensed structural and line-angle formulas and give the names for the organic products of addition reactions of alkenes.

The most characteristic reaction of alkenes is the **addition** of atoms or groups of atoms to the carbon atoms in a double bond. Addition occurs because double bonds are easily broken, providing electrons to form new single bonds.

The addition reactions have different names that depend on the type of reactant we add to the alkene, as **TABLE 11.7** shows.

TABLE 11.7 Summary of Addition Reactions

Name of Addition Reaction	Reactants	Catalysts	Product
Hydrogenation	Alkene + H_2	Pt, Ni, or Pd	Alkane
Hydration	Alkene + H_2O	H^+ (strong acid)	Alcohol

Hydrogenation

CORE CHEMISTRY SKILL
Writing Equations for Hydrogenation and Hydration

In a reaction called **hydrogenation**, H atoms add to each of the carbon atoms in a double bond of an alkene. During hydrogenation, the double bonds are converted to single bonds in alkanes. A catalyst such as finely divided platinum (Pt), nickel (Ni), or palladium (Pd) is used to speed up the reaction. The general equation for hydrogenation can be written as follows:

$$\text{>C=C<} + \text{H—H} \xrightarrow{\text{Pt, Ni, or Pd}} \text{—}\overset{\text{H}}{\underset{|}{\overset{|}{\text{C}}}}\text{—}\overset{\text{H}}{\underset{|}{\overset{|}{\text{C}}}}\text{—}$$

Double bond (alkene) — Single bond (alkane)

Some examples of the hydrogenation of alkenes follow:

$$CH_3{-}CH{=}CH{-}CH_3 + H_2 \xrightarrow{Pt} CH_3{-}CH_2{-}CH_2{-}CH_3$$

2-Butene — Butane

Cyclohexene $+ H_2 \xrightarrow{Ni}$ Cyclohexane

SAMPLE PROBLEM 11.8 Writing Equations for Hydrogenation

TRY IT FIRST

Draw the line-angle formula for the product of the following hydrogenation reaction:

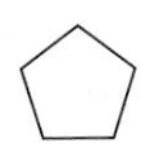 $+ H_2 \xrightarrow{Pt}$

SOLUTION

In an addition reaction, hydrogen adds to the double bond to give an alkane.

STUDY CHECK 11.8

Draw the condensed structural formula for the product of the hydrogenation of 2-methyl-1-butene, using a platinum catalyst.

ANSWER

$$\begin{array}{l} \quad\;\; CH_3 \\ \quad\;\;\; | \\ CH_3{-}CH{-}CH_2{-}CH_3 \end{array}$$

CHEMISTRY LINK TO HEALTH

Hydrogenation of Unsaturated Fats

Vegetable oils such as corn oil or safflower oil contain unsaturated fats composed of fatty acids that contain double bonds. The process of hydrogenation is used commercially to convert the double bonds in the unsaturated fats in vegetable oils to saturated fats such as margarine, which are more solid. Adjusting the amount of hydrogen added produces partially hydrogenated fats such as soft margarine, solid margarine in sticks, and shortenings, which are used in cooking. For example, oleic acid is a typical unsaturated fatty acid in olive oil and has a cis double bond at carbon 9. When oleic acid is hydrogenated, it is converted to stearic acid, a saturated fatty acid.

The unsaturated fats in vegetable oils are converted to saturated fats to make a more solid product.

Cis double bond — Oleic acid (found in olive oil and other unsaturated fats)

Single bond — Stearic acid (found in saturated fats)

Hydration

In **hydration**, an alkene reacts with water (H—OH). A hydrogen atom (H—) from water forms a bond with one carbon atom in the double bond, and the oxygen atom in —OH forms a bond with the other carbon. The reaction is catalyzed by a strong acid such as H_2SO_4, written as H^+. Hydration is used to prepare alcohols, which have the hydroxyl (—OH) functional group. When water (H—OH) adds to a symmetrical alkene, such as ethene, a single product is formed.

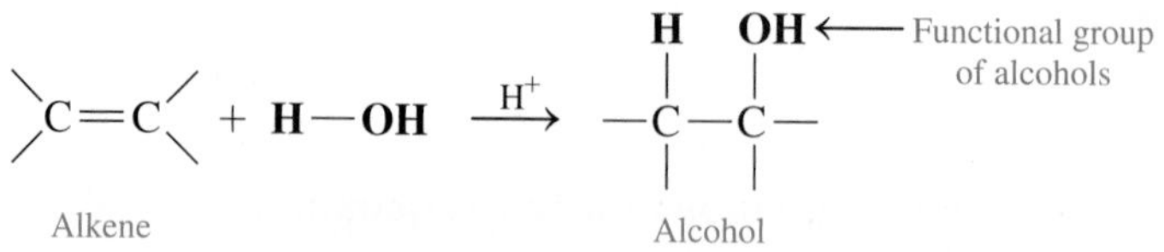

$$\mathrm{>C{=}C<} + \mathbf{H{-}OH} \xrightarrow{H^+} \mathrm{-\overset{\displaystyle H}{\underset{|}{\overset{|}{C}}}-\overset{\displaystyle OH}{\underset{|}{\overset{|}{C}}}-}$$

Alkene — Alcohol (OH ← Functional group of alcohols)

$$H_2C{=}CH_2 + \mathbf{H_2O} \xrightarrow{H^+} CH_3{-}CH_2{-}OH$$

Ethene — Ethanol (ethyl alcohol)

ENGAGE
Why do you need to determine if the double bond in an alkene is symmetrical or asymmetrical?

However, when H_2O adds to a double bond in an asymmetrical alkene, two products are possible. We can write the prevalent product by attaching the H— from H_2O to the carbon that has the *greater* number of H atoms and the —OH from H_2O to the other carbon atom from the double bond. In the following example, the H— from H_2O attaches to the end carbon of the double bond, which has more hydrogen atoms, and the —OH adds to the middle carbon atom:

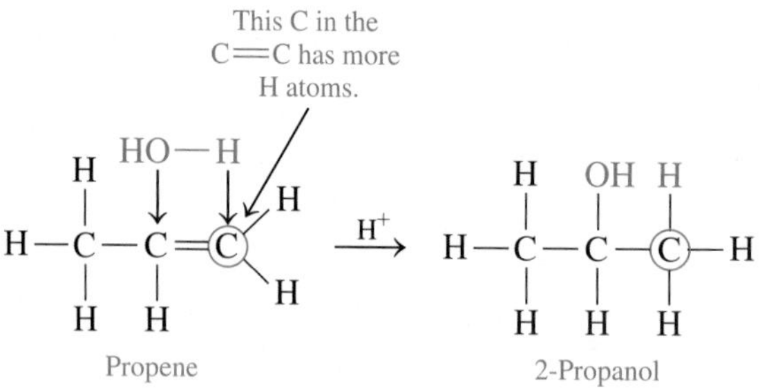

This C in the C=C has more H atoms.

$$CH_3{-}CH{=}CH_2 + HO{-}H \xrightarrow{H^+} CH_3{-}CH(OH){-}CH_3$$

Propene — 2-Propanol

SAMPLE PROBLEM 11.9 Hydration

TRY IT FIRST

Draw the condensed structural formula for the product that forms in the following hydration reaction:

$$CH_3{-}CH_2{-}CH_2{-}CH{=}CH_2 + H_2O \xrightarrow{H^+}$$

SOLUTION

The H— and —OH from water add to the carbon atoms in the double bond. The H— adds to the carbon with the *greater* number of hydrogen atoms, and the —OH bonds to the carbon with fewer H atoms.

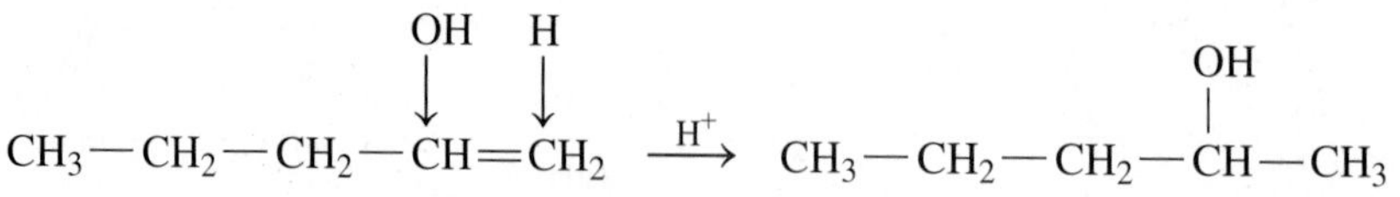

$$CH_3{-}CH_2{-}CH_2{-}CH{=}CH_2 \xrightarrow{H^+} CH_3{-}CH_2{-}CH_2{-}\underset{}{\overset{\displaystyle OH}{\overset{|}{C}H}}{-}CH_3$$

STUDY CHECK 11.9

Draw the condensed structural formula for the product obtained by the hydration of 2-methyl-2-pentene.

INTERACTIVE VIDEO
Addition to an Asymmetric Bond

ANSWER

$$CH_3-\underset{\displaystyle CH_3}{\overset{\displaystyle OH}{\overset{|}{\underset{|}{C}}}}-CH_2-CH_2-CH_3$$

TEST

Try Practice Problems 11.33 and 11.34

PRACTICE PROBLEMS

11.7 Addition Reactions for Alkenes

LEARNING GOAL Draw the condensed structural and line-angle formulas and give the names for the organic products of addition reactions of alkenes.

11.33 Draw the structural formula for the product in each of the following reactions:

a. $CH_3-CH_2-CH_2-CH=CH_2 + H_2 \xrightarrow{Pt}$

b. $H_2C=\overset{\overset{\displaystyle CH_3}{|}}{C}-CH_2-CH_3 + H_2O \xrightarrow{H^+}$

c. (line-angle alkene) $+ H_2 \xrightarrow{Pt}$

d. (line-angle cycloalkene) $+ H_2O \xrightarrow{H^+}$

11.34 Draw the structural formula for the product in each of the following reactions:

a. $CH_3-CH_2-CH=CH_2 + H_2O \xrightarrow{H^+}$

b. (line-angle alkene) $+ H_2 \xrightarrow{Pt}$

c. (line-angle cycloalkene) $+ H_2 \xrightarrow{Pt}$

d. (line-angle alkene) $+ H_2O \xrightarrow{H^+}$

11.8 Aromatic Compounds

LEARNING GOAL Describe the bonding in benzene; name aromatic compounds and draw their line-angle formulas.

In 1825, Michael Faraday isolated a hydrocarbon called **benzene**, which consists of a ring of six carbon atoms with one hydrogen atom attached to each carbon. Because many compounds containing benzene had fragrant odors, the family of benzene compounds became known as **aromatic compounds**. Some common examples of aromatic compounds that we use for flavor are anisole from anise, estragole from tarragon, and thymol from thyme.

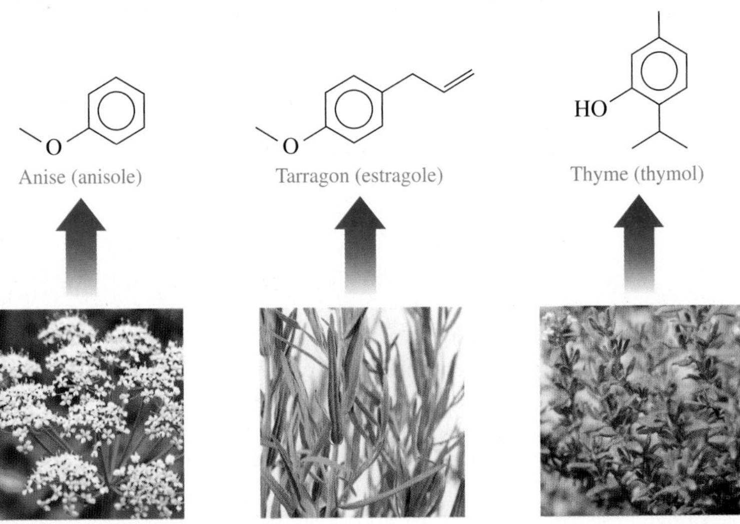

The aroma and flavor of the herbs anise, tarragon, and thyme are because of aromatic compounds.

In benzene, each carbon atom uses three valence electrons to bond to the hydrogen atom and two adjacent carbons. That leaves one valence electron, which scientists first thought was shared in a double bond with an adjacent carbon. In 1865, August Kekulé proposed that the carbon atoms in benzene were arranged in a flat ring with alternating single and double bonds between the adjacent carbon atoms. There are two possible structural representations of benzene in which the double bonds can form between two different carbon atoms. If there were double bonds as in alkenes, then benzene should be much more reactive than it is.

However, unlike the alkenes and alkynes, aromatic hydrocarbons do not easily undergo addition reactions. If their reaction behavior is quite different, they must also differ in how the atoms are bonded in their structures. Today, we know that the six electrons are shared equally among the six carbon atoms. This unique feature of benzene makes it especially stable. Benzene is most often represented as a line-angle formula, which shows a hexagon with a circle in the center. Some of the ways to represent benzene are shown as follows:

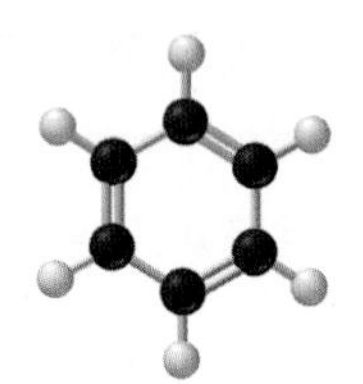

Equivalent structures for benzene

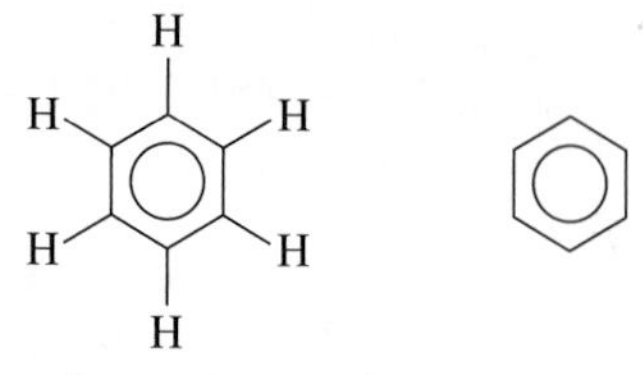

Structural formulas for benzene

Naming Aromatic Compounds

Many compounds containing benzene have been important in chemistry for many years and still use their common names. Toluene consists of benzene with a methyl group ($—CH_3$), aniline is benzene with an amino group ($—NH_2$), and phenol is benzene with a hydroxyl group ($—OH$). The names toluene, aniline, and phenol are allowed by IUPAC rules.

ENGAGE

How is the formula of toluene similar to and different from that of phenol?

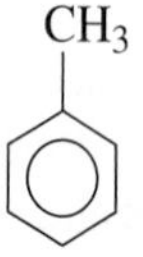

Toluene

Line-angle structure for toluene

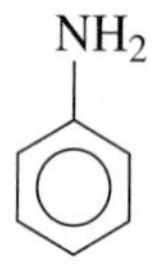

Aniline

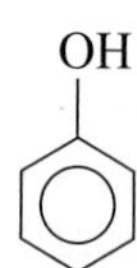

Phenol

When benzene has only one substituent, the ring is not numbered. When there are two or more substituents, the benzene ring is numbered to give the lowest numbers to the substituents.

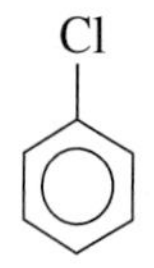

Chlorobenzene

1,2-Dichlorobenzene

1,3-Dichlorobenzene

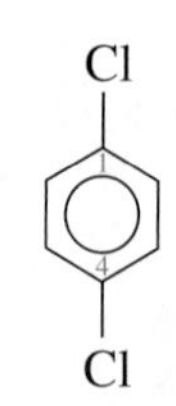

1,4-Dichlorobenzene

When aniline, phenol, or toluene has substituents, the carbon atom attached to the amine, hydroxyl, or methyl group is numbered as carbon 1 and the substituents are named alphabetically.

3-Bromoaniline

4-Bromo-2-chlorophenol

2,6-Dibromo-4-chlorotoluene

SAMPLE PROBLEM 11.10 Naming Aromatic Compounds

TRY IT FIRST

Give the IUPAC name for the following:

Cl

Br

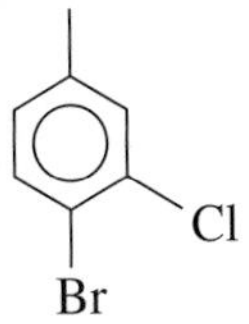

ANALYZE THE PROBLEM	Given	Need	Connect
	structural formula	IUPAC name	name of aromatic compound, number and list substituents

SOLUTION GUIDE

STEP 1 **Write the name for the aromatic compound.** A benzene ring with a methyl group is named toluene.

STEP 2 **If there are two or more substituents, number the aromatic ring starting from the substituent.** The methyl group of toluene is attached to carbon 1, and the ring is numbered to give the lower numbers.

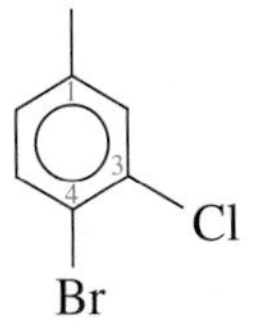

Cl

Br

STEP 3 **Name each substituent as a prefix.** Naming the substituents in alphabetical order, this aromatic compound is 4-bromo-3-chlorotoluene.

STUDY CHECK 11.10

Give the IUPAC name for the following:

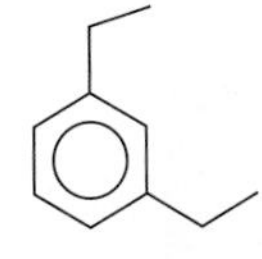

ANSWER

1,3-diethylbenzene

TEST

Try Practice Problems 11.35 to 11.38

CHEMISTRY LINK TO HEALTH

Some Common Aromatic Compounds

Aromatic compounds are common in nature and in medicine. Toluene is used as a reactant to make drugs, dyes, and explosives such as TNT (trinitrotoluene). The benzene ring is found in pain relievers such as aspirin and acetaminophen, and in flavorings such as vanillin.

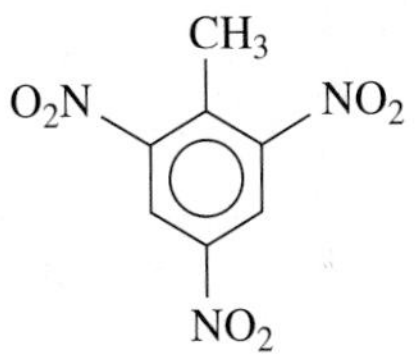

TNT (2,4,6-trinitrotoluene)

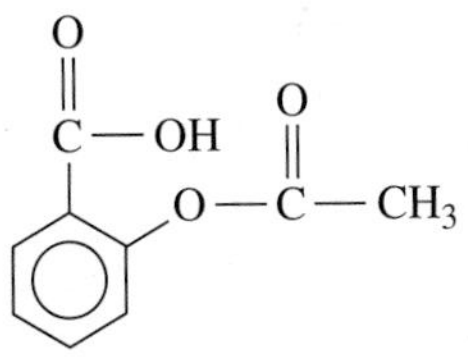

Aspirin

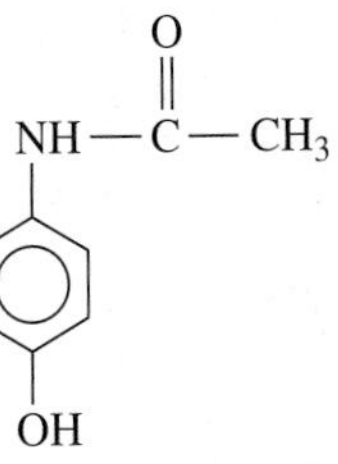

Acetaminophen

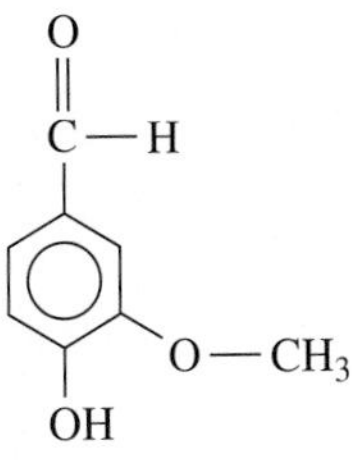

Vanillin

PRACTICE PROBLEMS

11.8 Aromatic Compounds

LEARNING GOAL Describe the bonding in benzene; name aromatic compounds and draw their line-angle formulas.

11.35 Give the IUPAC name for each of the following:

a. Cl

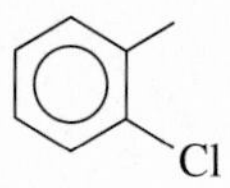

b.

c. Cl, Cl, Cl

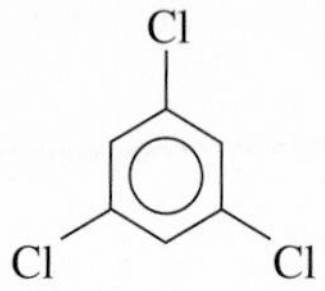

11.36 Give the IUPAC name for each of the following:

a. Br

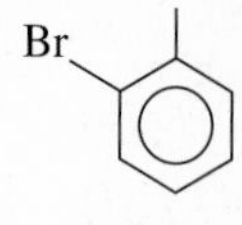

b. OH, Cl, Cl

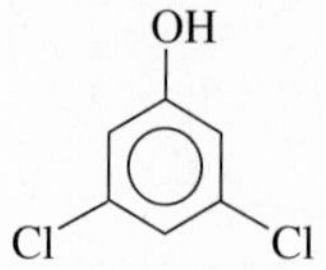

c. Br, Cl, Cl

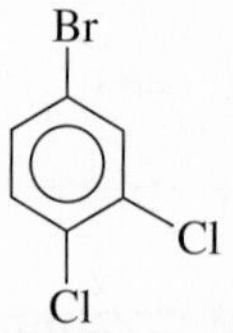

11.37 Draw the line-angle formula for each of the following compounds:

a. phenol **b.** 1,3-dichlorobenzene

c. 4-ethyltoluene

11.38 Draw the line-angle formula for each of the following compounds:

a. toluene **b.** 4-bromoaniline

c. 1,2,4-trichlorobenzene

CHEMISTRY LINK TO HEALTH

Polycyclic Aromatic Hydrocarbons (PAHs)

Large aromatic compounds known as *polycyclic aromatic hydrocarbons* (PAHs) are formed by fusing together two or more benzene rings edge to edge. In a fused-ring compound, neighboring benzene rings share two carbon atoms. Naphthalene, with two benzene rings, is well known for its use in mothballs; anthracene, with three rings, is used in the manufacture of dyes.

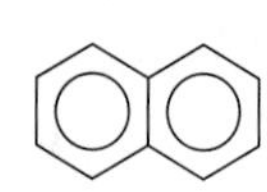

Naphthalene

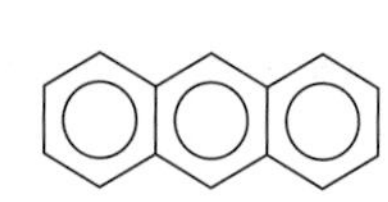

Anthracene

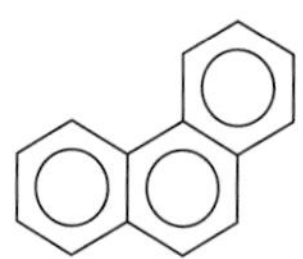

Phenanthrene

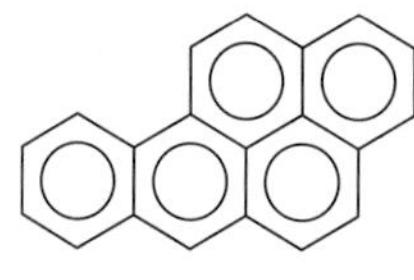

Benzo[*a*]pyrene

When a polycyclic compound contains phenanthrene, it may act as a carcinogen, a substance known to cause cancer.

Compounds containing five or more fused benzene rings such as benzo[*a*]pyrene are potent carcinogens. The molecules interact with the DNA in the cells, causing abnormal cell growth and cancer. Increased exposure to carcinogens increases the chance of DNA alterations in the cells. Benzo[*a*]pyrene, a product of combustion, has been identified in coal tar, tobacco smoke, barbecued meats, and automobile exhaust.

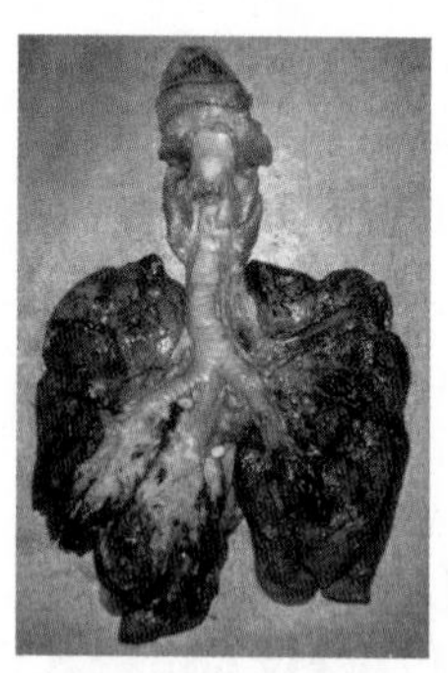

Aromatic compounds such as benzo[*a*]-pyrene are strongly associated with lung cancers.

CLINICAL UPDATE Diane's Treatment in the Burn Unit

When Diane arrives at the hospital, she is transferred to the ICU burn unit. She has second-degree burns that cause damage to the underlying layers of skin and third-degree burns that cause damage to all the layers of the skin. Because body fluids are lost when deep burns occur, a lactated Ringer's solution is administered. The most common complications of burns are related to infection. To prevent infection, her skin is covered with a topical antibiotic. The next day Diane is placed in a tank to remove dressings, lotions, and damaged tissue. Dressings and ointments are changed every eight hours. Over a period of 3 months, grafts of Diane's unburned skin will be used to cover burned areas.

The arson investigators determine that gasoline was the primary accelerant used to start the fire at Diane's house. Because there was a lot of paper and dry wood in the area, the fire spread quickly. Some of the hydrocarbons found in gasoline include hexane, heptane, octane, nonane, decane, and cyclohexane. Some other hydrocarbons in gasoline are 3-ethyltoluene, isopentane, and toluene.

Clinical Applications

11.39 Write the balanced chemical equation for the complete combustion of each of the following hydrocarbons found in gasoline:

a. octane
b. isopentane (2-methylbutane)
c. 3-ethyltoluene

11.40 Write the balanced chemical equation for the complete combustion of each of the following hydrocarbons found in gasoline:

a. decane
b. cyclohexane
c. toluene

CONCEPT MAP

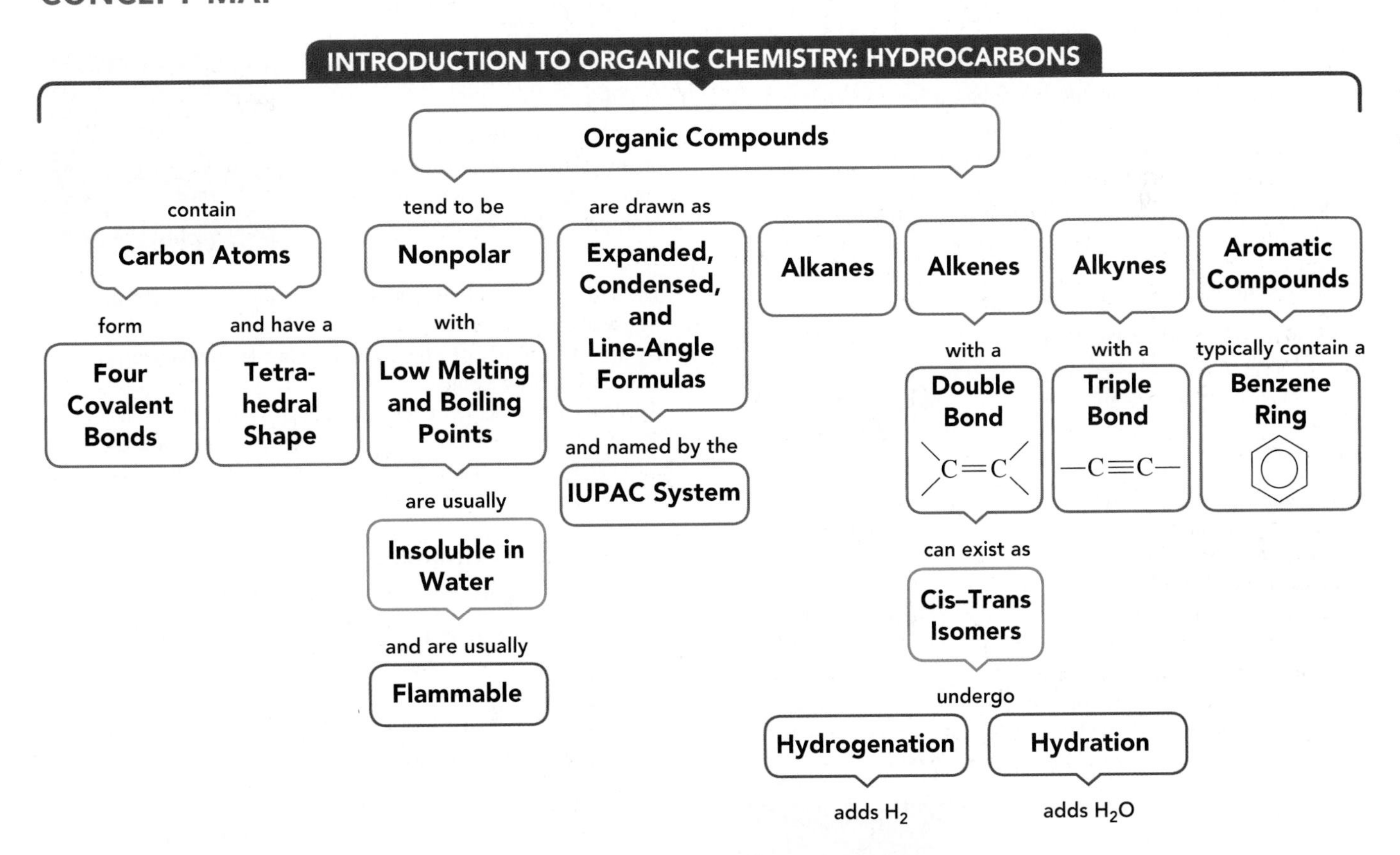

CHAPTER REVIEW

11.1 Organic Compounds

LEARNING GOAL Identify properties characteristic of organic or inorganic compounds.

- Organic compounds have covalent bonds, most form nonpolar molecules, have low melting points and low boiling points, are not very soluble in water, dissolve as molecules in solutions, and burn vigorously in air.
- Inorganic compounds are often ionic or contain polar covalent bonds and form polar molecules, have high melting and boiling points, are usually soluble in water, produce ions in water, and do not burn in air.
- In the simplest organic molecule, methane, CH_4, the C—H bonds that attach four hydrogen atoms to the carbon atom are directed to the corners of a tetrahedron with bond angles of 109°.
- In the expanded structural formula, a separate line is drawn for every bond.

11.2 Alkanes

LEARNING GOAL Write the IUPAC names and draw the condensed structural and line-angle formulas for alkanes and cycloalkanes.

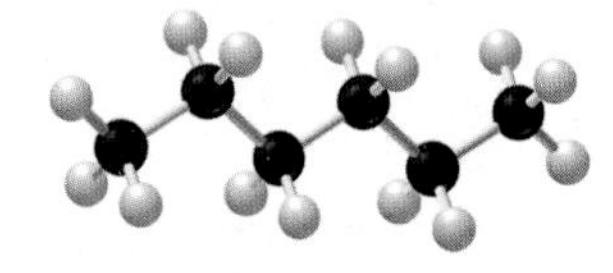

- Alkanes are hydrocarbons that have only C—C single bonds.
- A condensed structural formula depicts groups composed of each carbon atom and its attached hydrogen atoms.
- A line-angle formula represents the carbon skeleton as ends and corners of a zigzag line or geometric figure.
- The IUPAC system is used to name organic compounds by indicating the number of carbon atoms.
- The name of a cycloalkane is written by placing the prefix *cyclo* before the alkane name with the same number of carbon atoms.

11.3 Alkanes with Substituents

LEARNING GOAL Write the IUPAC names for alkanes with substituents and draw their condensed structural or line-angle formulas.

- Structural isomers are compounds with the same molecular formulas that differ in the order in which their atoms are bonded.
- Substituents, which are attached to an alkane chain, include alkyl groups and halogen atoms (F, Cl, Br, or I).
- In the IUPAC system, alkyl substituents have names such as methyl, propyl, and isopropyl; halogen atoms are named as fluoro, chloro, bromo, or iodo.

11.4 Properties of Alkanes

LEARNING GOAL Identify the properties of alkanes and write a balanced chemical equation for combustion.

- Alkanes, which are nonpolar molecules, are not soluble in water, and are usually less dense than water.
- Alkanes undergo combustion in which they react with oxygen to produce carbon dioxide, water, and energy.

11.5 Alkenes and Alkynes

LEARNING GOAL Write the IUPAC names or draw the condensed structural or line-angle formulas for alkenes and alkynes.

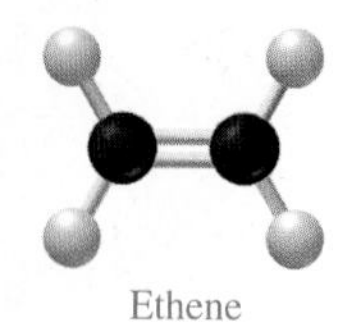

Ethene

- Alkenes are unsaturated hydrocarbons that contain carbon–carbon double bonds.
- Alkynes contain a carbon–carbon triple bond.
- The IUPAC names of alkenes end with *ene*, while alkyne names end with *yne*. The main chain is numbered from the end nearer the double or triple bond.

11.6 Cis–Trans Isomers

LEARNING GOAL Draw the condensed structural formulas and give the names for the cis–trans isomers of alkenes.

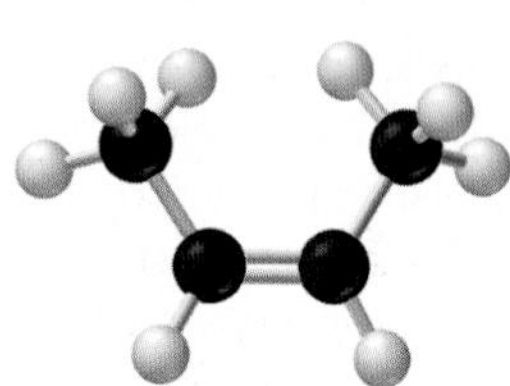

cis-2-Butene

- Geometric or cis–trans isomers of alkenes occur when the carbon atoms in the double bond are connected to different atoms or groups.
- In the cis isomer, the similar groups are on the same side of the double bond, whereas in the trans isomer they are connected on opposite sides of the double bond.

11.7 Addition Reactions for Alkenes

LEARNING GOAL Draw the condensed structural and line-angle formulas and give the names for the organic products of addition reactions of alkenes.

- The addition of small molecules to the double bond is a characteristic reaction of alkenes.
- Hydrogenation adds hydrogen atoms to the double bond of an alkene to yield an alkane.
- Hydration adds water to a double bond of an alkene to form an alcohol.
- When a different number of hydrogen atoms are attached to the carbons in a double bond, the H from water adds to the carbon with the greater number of H atoms.

11.8 Aromatic Compounds

$$CH_3-CH(CH_3)-CH_2-C_6H_4-CH(CH_3)-C(=O)-OH$$

Ibuprofen

LEARNING GOAL Describe the bonding in benzene; name aromatic compounds and draw their line-angle formulas.

- Most aromatic compounds contain benzene, C_6H_6, a cyclic structure containing six carbon atoms and six hydrogen atoms.
- The structure of benzene is represented as a hexagon with a circle in the center.
- Aromatic compounds are named using the IUPAC name benzene, although names such as toluene, aniline, and phenol are retained.

SUMMARY OF NAMING

Family	Structure	IUPAC Name	Common Name
Alkane	$CH_3—CH_2—CH_3$	Propane	
	$CH_3—CH(CH_3)—CH_2—CH_3$	2-Methylbutane	
Haloalkane	$CH_3—CH_2—CH_2—Cl$	1-Chloropropane	
Cycloalkane	(hexagon)	Cyclohexane	
Alkene	$CH_3—CH{=}CH_2$	Propene	Propylene
Cycloalkene	(triangle with double bond)	Cyclopropene	
Alkyne	$CH_3—C{\equiv}CH$	Propyne	
Aromatic	(benzene ring with methyl)	Toluene	

SUMMARY OF REACTIONS

The chapter Sections to review are shown after the name of each reaction.

Combustion (11.4)

$$\underset{\text{Propane}}{CH_3—CH_2—CH_3} + 5O_2 \xrightarrow{\Delta} 3CO_2 + 4H_2O + \text{energy}$$

Hydrogenation (11.7)

$$\underset{\text{Propene}}{H_2C{=}CH—CH_3} + H_2 \xrightarrow{Pt} \underset{\text{Propane}}{CH_3—CH_2—CH_3}$$

Hydration (11.7)

$$\underset{\text{Propene}}{H_2C{=}CH—CH_3} + H_2O \xrightarrow{H^+} \underset{\text{2-Propanol}}{CH_3—CH(OH)—CH_3}$$

KEY TERMS

addition A reaction in which atoms or groups of atoms bond to a carbon–carbon double bond. Addition reactions include the addition of hydrogen (hydrogenation) and water (hydration).

alkane A hydrocarbon that contains only single bonds between carbon atoms.

alkene A hydrocarbon that contains one or more carbon–carbon double bonds (C=C).

alkyl group An alkane minus one hydrogen atom. Alkyl groups are named like the alkanes except a *yl* ending replaces *ane*.

alkyne A hydrocarbon that contains one or more carbon–carbon triple bonds ($C{\equiv}C$).

aromatic compound A compound that contains the ring structure of benzene.

benzene A ring of six carbon atoms, each of which is attached to a hydrogen atom, C_6H_6.

cis isomer An isomer of an alkene in which similar groups in the double bond are on the same side.

condensed structural formula A structural formula that shows the arrangement of the carbon atoms in a molecule but groups each carbon atom with its bonded hydrogen atoms $—CH_3$, $—CH_2—$, or $—\overset{|}{C}H—$.

cycloalkane An alkane that is a ring or cyclic structure.

expanded structural formula A type of structural formula that shows the arrangement of the atoms by drawing each bond in the hydrocarbon.

hydration An addition reaction in which the components of water, H— and —OH, bond to the carbon–carbon double bond to form an alcohol.

hydrocarbon An organic compound that contains only carbon and hydrogen.

hydrogenation The addition of hydrogen (H_2) to the double bond of alkenes to yield alkanes.

IUPAC system A system for naming organic compounds devised by the International Union of Pure and Applied Chemistry.

line-angle formula A zigzag line in which carbon atoms are represented as the ends of each line and as corners.

organic compound A compound made of carbon that typically has covalent bonds, is nonpolar, has low melting and boiling points, is insoluble in water, and is flammable.

structural isomers Organic compounds in which identical molecular formulas have different arrangements of atoms.

substituent Groups of atoms such as an alkyl group or a halogen bonded to the main carbon chain or ring of carbon atoms.

trans isomer An isomer of an alkene in which similar groups in the double bond are on opposite sides.

CORE CHEMISTRY SKILLS

The chapter Section containing each Core Chemistry Skill is shown in parentheses at the end of each heading.

Naming and Drawing Alkanes (11.2)

- The alkanes ethane, propane, and butane contain two, three, and four carbon atoms, respectively, connected in a row or a *continuous* chain.
- Alkanes with five or more carbon atoms in a chain are named using the prefixes *pent* (5), *hex* (6), *hept* (7), *oct* (8), *non* (9), and *dec* (10).
- In the condensed structural formula, the carbon and hydrogen atoms on the ends are written as $-CH_3$ and the carbon and hydrogen atoms in the middle are written as $-CH_2-$.

Example: **a.** What is the name of $CH_3-CH_2-CH_2-CH_3$?
b. Draw the condensed structural formula for pentane.

Answer: **a.** An alkane with a four-carbon chain is named with the prefix *but* followed by *ane*, which is butane.
b. Pentane is an alkane with a five-carbon chain. The carbon atoms on the ends are attached to three H atoms each, and the carbon atoms in the middle are attached to two H each. $CH_3-CH_2-CH_2-CH_2-CH_3$

Writing Equations for Hydrogenation and Hydration (11.7)

- The addition of small molecules to the double bond is a characteristic reaction of alkenes.
- Hydrogenation adds hydrogen atoms to the double bond of an alkene to form an alkane.
- Hydration adds water to a double bond of an alkene to form an alcohol.
- When a different number of hydrogen atoms are attached to the carbons in a double bond, the H— from H_2O adds to the carbon atom with the greater number of H atoms, and the —OH adds to the other carbon atom from the double bond.

Example: **a.** Draw the condensed structural formula for 2-methyl-2-butene.
b. Draw the condensed structural formula for the product of the hydrogenation of 2-methyl-2-butene.
c. Draw the condensed structural formula for the product of the hydration of 2-methyl-2-butene.

Answer: **a.**

$$\begin{array}{c} \quad CH_3 \\ \quad | \\ CH_3-C=CH-CH_3 \end{array}$$

b. When H_2 adds to a double bond, the product is an alkane.

$$\begin{array}{c} \quad CH_3 \\ \quad | \\ CH_3-CH-CH_2-CH_3 \end{array}$$

c. When H_2O adds to a double bond, the product is an alcohol, which has an —OH group. The H— from H_2O adds to the carbon atom in the double bond that has the greater number of H atoms.

$$\begin{array}{c} \quad CH_3 \\ \quad | \\ CH_3-C-CH_2-CH_3 \\ \quad | \\ \quad OH \end{array}$$

UNDERSTANDING THE CONCEPTS

The chapter Sections to review are shown in parentheses at the end of each problem.

11.41 Match the following physical and chemical properties with potassium chloride, KCl, used in salt substitutes, or butane, C_4H_{10}, used in lighters: (11.1)

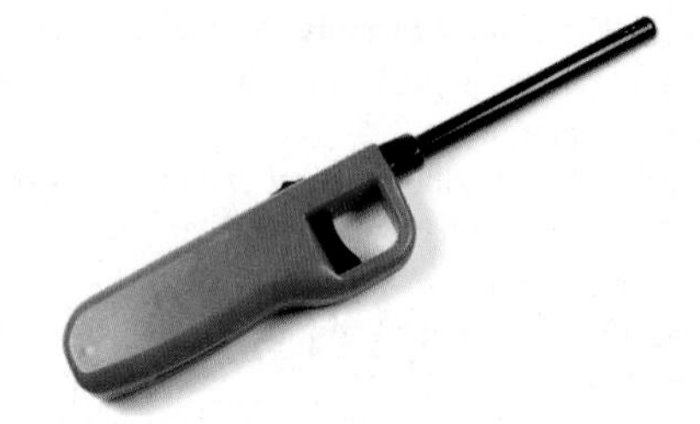

a. melts at -138 °C
b. burns vigorously in air
c. melts at 770 °C
d. contains ionic bonds
e. is a gas at room temperature

11.42 Match the following physical and chemical properties with octane, C_8H_{18}, found in gasoline, or magnesium sulfate, $MgSO_4$, also called Epsom salts: (11.1)
a. contains only covalent bonds
b. melts at 1124 °C
c. is insoluble in water
d. is a liquid at room temperature
e. is a strong electrolyte

11.43 Identify the compounds in each of the following pairs as structural isomers or not structural isomers: (11.3)

a. 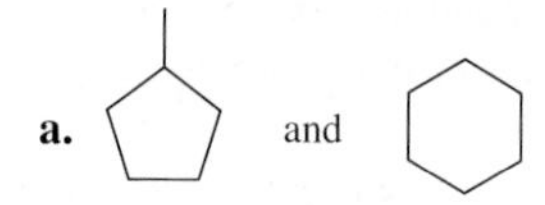

b. and

11.44 Identify the compounds in each of the following pairs as structural isomers or not structural isomers: (11.3)

a.

$$\begin{array}{lll} CH_2-CH_2-CH_2 & & \\ | \quad\quad\quad\quad\quad | & & \\ CH_3 \quad\quad\quad CH_2-CH_3 & \text{and} & \end{array} \quad \begin{array}{l} CH_3 \quad\quad\quad\quad\quad\quad\quad\quad CH_3 \\ | \quad\quad\quad\quad\quad\quad\quad\quad\quad\quad | \\ CH_2-CH_2-CH_2-CH_2 \end{array}$$

b. and

11.45 Convert each of the following line-angle formulas to a condensed structural formula and give its IUPAC name: (11.3)

a. b.

11.46 Convert each of the following line-angle formulas to a condensed structural formula and give its IUPAC name: (11.3)

a. b. Cl Br

ADDITIONAL PRACTICE PROBLEMS

11.47 Give the IUPAC name for each of the following: (11.3)

a. $$\begin{array}{c} CH_3 \\ | \\ CH_3-CH_2-C-CH_3 \\ | \\ CH_3 \end{array}$$

b. CH_3-CH_2-Cl

c. $$\begin{array}{l} \quad\quad\quad CH_3-CH_2 \quad\quad\quad Br \\ \quad\quad\quad\quad\quad | \quad\quad\quad\quad\quad\quad | \\ CH_3-CH_2-CH-CH_2-CH-CH_3 \end{array}$$

d.

11.48 Give the IUPAC name for each of the following: (11.3)

a.

b. $$\begin{array}{c} Br \\ | \\ Cl-CH_2-CH-CH_2-Br \end{array}$$

c. $$\begin{array}{l} \quad\quad\quad CH_3 \\ \quad\quad\quad | \\ CH_3-CH-CH-CH_3 \\ \quad\quad\quad\quad\quad | \\ \quad\quad\quad\quad\quad CH_2 \\ \quad\quad\quad\quad\quad | \\ \quad\quad\quad\quad\quad CH_2 \\ \quad\quad\quad\quad\quad | \\ \quad\quad\quad\quad\quad CH_3 \end{array}$$

d. $$\begin{array}{c} Cl \\ | \\ CH_3-CH_2-C-CH_2-CH_3 \\ | \\ CH_2 \\ | \\ CH_3 \end{array}$$

11.49 Give the IUPAC name (including cis or trans, if needed) for each of the following: (11.5, 11.6)

a. $$\begin{array}{ccc} CH_3 & & H \\ & C=C & \\ H & & CH_2-CH_3 \end{array}$$

b. Br

c.

11.50 Give the IUPAC name (including cis or trans, if needed) for each of the following: (11.5, 11.6)

a. $$\begin{array}{l} \quad\quad\quad\; CH_3 \\ \quad\quad\quad\; | \\ H_2C=C-CH_2-CH_2-CH_3 \end{array}$$

b.

c.

11.51 Identify the following pairs of structures as structural isomers, cis–trans isomers, or the same molecule: (11.5, 11.6)

a. Cl and Cl

b. $$\begin{array}{ccc} CH_3 & & H \\ & C=C & \\ H & & CH_3 \end{array} \quad \text{and} \quad \begin{array}{ccc} CH_3 & & CH_3 \\ & C=C & \\ H & & H \end{array}$$

11.52 Identify the following pairs of structures as structural isomers, cis–trans isomers, or the same molecule: (11.5, 11.6)

a. $$\begin{array}{l} H_2C=CH \\ \quad\quad\; | \\ \quad\quad\; CH_2-CH_2 \\ \quad\quad\quad\quad\quad | \\ \quad\quad\quad\quad\quad CH_3 \end{array} \quad \text{and} \quad CH_3-CH_2-CH_2-CH=CH_2$$

b. and

11.53 Name each of the following aromatic compounds: (11.8)

a. Cl

b. NH_2 Br

11.54 Name each of the following aromatic compounds: (11.8)

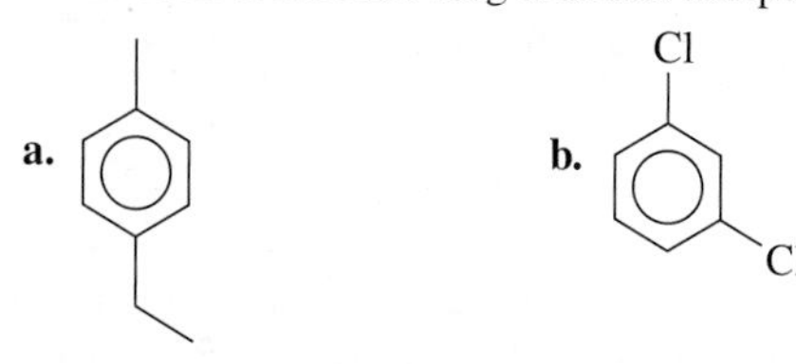

11.55 Draw the condensed structural or line-angle formula if cyclic, for each of the following: (11.3, 11.5, 11.6)
a. bromocyclopropane
b. 1,1-dibromo-2-pentyne
c. *cis*-2-heptene

11.56 Draw the condensed structural or line-angle formula if cyclic, for each of the following: (11.3, 11.5, 11.6)
a. ethylcyclopropane
b. 2-methylhexane
c. 2,3-dichloro-1-butene

11.57 Draw the cis and trans isomers for each of the following: (11.6)
a. 2-pentene
b. 3-hexene

11.58 Draw the cis and trans isomers for each of the following: (11.6)
a. 2-butene
b. 2-hexene

11.59 Draw the line-angle formula for each of the following: (11.8)
a. ethylbenzene
b. 2,5-dibromophenol
c. 3-chloroaniline

11.60 Draw the line-angle formula for each of the following: (11.8)
a. 3-chlorophenol
b. 1,4-dichlorobenzene
c. 2-bromotoluene

11.61 Write a balanced chemical equation for the complete combustion of each of the following: (11.4)
a. 2,2-dimethylpropane
b. cyclobutane
c. 2-hexene

11.62 Write a balanced chemical equation for the complete combustion of each of the following: (11.4)
a. heptane
b. 2-methyl-1-pentene
c. ethylcyclopentane

11.63 Give the name for the product from the hydrogenation of each of the following: (11.7)
a. 3-methyl-2-pentene
b. cyclohexene
c. propene

11.64 Give the name for the product from the hydrogenation of each of the following: (11.7)
a. 1-hexene
b. 2-methyl-2-pentene
c. cyclopropene

11.65 Draw the condensed structural or line-angle formula for the product of each of the following: (11.7)

a. (cyclopentene) $+ H_2 \xrightarrow{Ni}$

b. (1-hexene) $+ H_2O \xrightarrow{H^+}$

11.66 Draw the condensed structural or line-angle formula for the product of each of the following: (11.7)

a. (cyclopentene) $+ H_2O \xrightarrow{H^+}$

b. $CH_3—CH_2—CH_2—CH=CH_2 + H_2 \xrightarrow{Ni}$

CHALLENGE PROBLEMS

The following problems are related to the topics in this chapter. However, they do not all follow the chapter order, and they require you to combine concepts and skills from several Sections. These problems will help you increase your critical thinking skills and prepare for your next exam.

11.67 The density of pentane, a component of gasoline, is 0.63 g/mL. The heat of combustion for pentane is 845 kcal/mole. (7.2, 7.4, 7.7, 7.8, 8.6, 11.2, 11.4)
a. Write the balanced chemical equation for the complete combustion of pentane.
b. What is the molar mass of pentane?
c. How much heat is produced when 1 gal of pentane is burned (1 gal = 4 qt)?
d. How many liters of CO_2 at STP are produced from the complete combustion of 1 gal of pentane?

11.68 Acetylene (ethyne) gas reacts with oxygen and burns at 3300 °C in an acetylene torch. (7.4, 7.7, 8.6, 11.4, 11.5)
a. Write the balanced chemical equation for the complete combustion of acetylene.
b. What is the molar mass of acetylene?
c. How many grams of oxygen are needed to react with 8.5 L of acetylene gas at STP?
d. How many liters of CO_2 gas at STP are produced when 30.0 g of acetylene undergoes combustion?

Acetylene and oxygen burn at 3300 °C.

11.69 Draw the condensed structural formulas for all the possible alkane isomers that have a total of six carbon atoms and a four-carbon chain. (11.3)

11.70 Draw the condensed structural formulas for all the possible haloalkane isomers that have four carbon atoms and a bromine. (11.3)

11.71 Consider the compound propane. (7.4, 7.7, 11.2, 11.4)
- **a.** Draw the condensed structural formula for propane.
- **b.** Write the balanced chemical equation for the complete combustion of propane.
- **c.** How many grams of O_2 are needed to react with 12.0 L of propane gas at STP?
- **d.** How many grams of CO_2 would be produced from the reaction in part **c**?

11.72 Consider the compound ethylcyclopentane. (7.4, 7.7, 8.6, 11.3, 11.4)
- **a.** Draw the line-angle formula for ethylcyclopentane.
- **b.** Write the balanced chemical equation for the complete combustion of ethylcyclopentane.
- **c.** Calculate the grams of O_2 required for the combustion of 25.0 g of ethylcyclopentane.
- **d.** How many liters of CO_2 would be produced at STP from the reaction in part **c**?

11.73 Explosives used in mining contain TNT, or trinitrotoluene. (11.8)

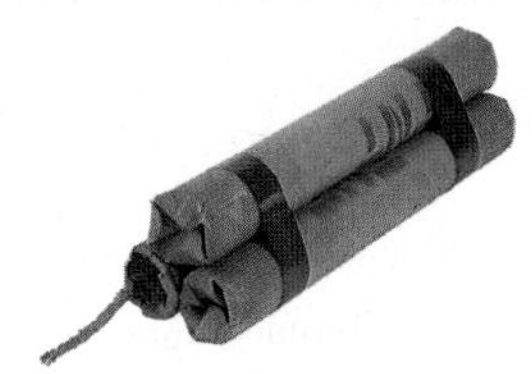

The TNT in explosives is used in mining.

- **a.** If the functional group *nitro* is $—NO_2$, draw the line-angle formula for 2,4,6-trinitrotoluene, one isomer of TNT.
- **b.** TNT is actually a mixture of isomers of trinitrotoluene. Draw the line-angle formulas for two other possible isomers.

11.74 Margarines are produced from the hydrogenation of vegetable oils, which contain unsaturated fatty acids. How many grams of hydrogen are required to completely saturate 75.0 g of oleic acid, $C_{18}H_{34}O_2$, which has one double bond? (7.7, 11.7)

Margarines are produced by hydrogenation of unsaturated fats.

ANSWERS

Answers to Selected Practice Problems

11.1
- **a.** inorganic
- **b.** organic, condensed structural formula
- **c.** organic, expanded structural formula
- **d.** inorganic
- **e.** inorganic
- **f.** organic, molecular formula

11.3 **a.** inorganic **b.** organic **c.** organic **d.** inorganic

11.5 **a.** ethane **b.** ethane **c.** NaBr **d.** NaBr

11.7 **a.** pentane **b.** ethane **c.** hexane **d.** cycloheptane

11.9
- **a.** CH_4
- **b.** $CH_3—CH_3$
- **c.** $CH_3—CH_2—CH_2—CH_3$
- **d.** (cyclopropane, triangle)

11.11
- **a.** same molecule
- **b.** structural isomers of C_5H_{12}
- **c.** structural isomers of C_6H_{14}

11.13 **a.** 2,2-dimethylbutane **b.** 2-chloro-3-methylpentane **c.** 4-ethyl-2,2-dimethylhexane **d.** methylcyclobutane

11.15
- **a.** $CH_3—CH_2—C(CH_3)_2—CH_2—CH_3$
- **b.** $CH_3—CH(CH_3)—CH(CH_3)—CH_2—CH(CH_3)—CH_3$
- **c.** $CH_3—CH(CH_3)—CH(CH_2—CH_3)—CH_2—CH(CH_3)—CH_2—CH_2—CH_3$
- **d.** $Br—CH_2—CH_2—Cl$

11.17 **a.** (line-angle formula) **b.** (line-angle formula) **c.** (line-angle formula, Br) **d.** (line-angle formula, Cl, Cl)

11.19
- **a.** $CH_3—CH_2—CH_2—CH_2—CH_2—CH_2—CH_3$ (line-angle formula)
- **b.** liquid
- **c.** no
- **d.** float
- **e.** $C_7H_{16} + 11O_2 \xrightarrow{\Delta} 7CO_2 + 8H_2O + \text{energy}$

11.21
- **a.** $2C_2H_6 + 7O_2 \xrightarrow{\Delta} 4CO_2 + 6H_2O + \text{energy}$
- **b.** $2C_3H_6 + 9O_2 \xrightarrow{\Delta} 6CO_2 + 6H_2O + \text{energy}$
- **c.** $2C_8H_{18} + 25O_2 \xrightarrow{\Delta} 16CO_2 + 18H_2O + \text{energy}$

11.23 **a.** alkene **b.** alkyne **c.** alkene **d.** cycloalkene

11.25 a. ethene **b.** methylpropene
c. 2-pentyne **d.** 3-methylcyclobutene

11.27 a. $H_2C{=}CH-CH_2-CH_2-CH_3$

CH_3
b. 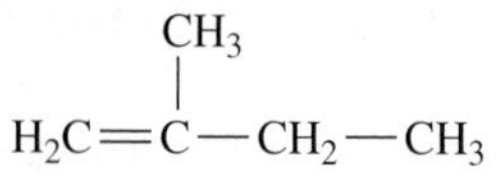 $H_2C{=}C-CH_2-CH_3$

c.

Cl
d. 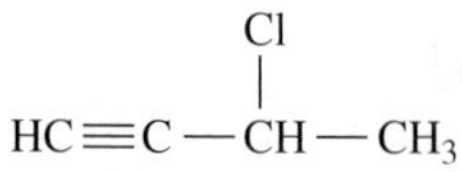 $HC{\equiv}C-CH-CH_3$

11.29 a. *cis*-3-heptene **b.** *trans*-3-octene
c. propene

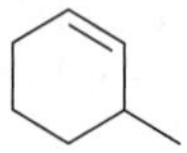

11.31 a. 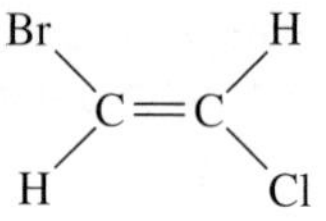
Br H
C=C
H Cl

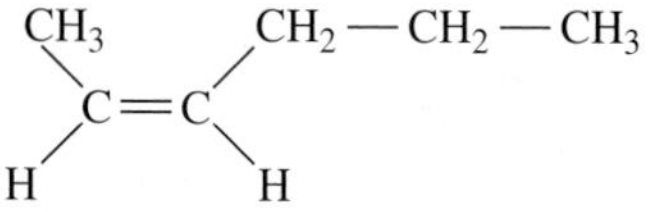

b.
CH_3 $CH_2-CH_2-CH_3$
C=C
H H

c.
$CH_3-CH_2-CH_2$ $CH_2-CH_2-CH_3$
C=C
H H

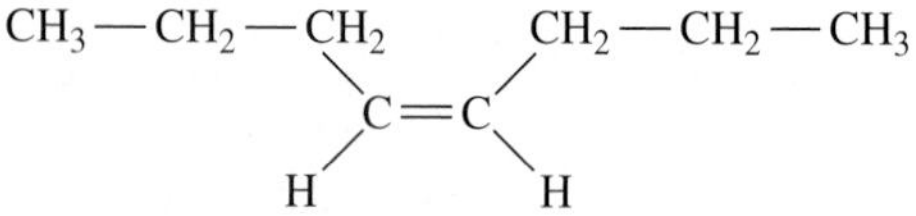

11.33 a. $CH_3-CH_2-CH_2-CH_2-CH_3$

CH_3
b. 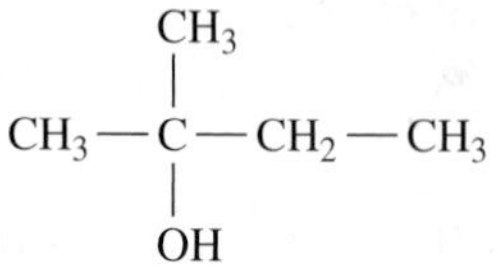 $CH_3-C-CH_2-CH_3$
OH

c.

d. 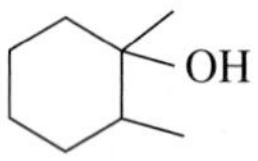 OH

11.35 a. 2-chlorotoluene **b.** ethylbenzene
c. 1,3,5-trichlorobenzene

11.37 a. 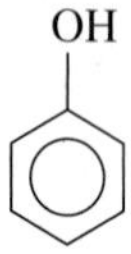 OH

b. 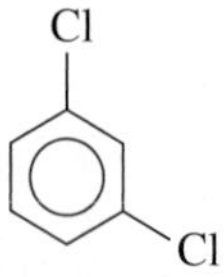 Cl, Cl

c.

11.39 a. $2C_8H_{18} + 25O_2 \xrightarrow{\Delta} 16CO_2 + 18H_2O + \text{energy}$
b. $C_5H_{12} + 8O_2 \xrightarrow{\Delta} 5CO_2 + 6H_2O + \text{energy}$
c. $C_9H_{12} + 12O_2 \xrightarrow{\Delta} 9CO_2 + 6H_2O + \text{energy}$

11.41 a. butane **b.** butane
c. potassium chloride **d.** potassium chloride
e. butane

11.43 a. structural isomers **b.** not structural isomers

CH_3
11.45 a. 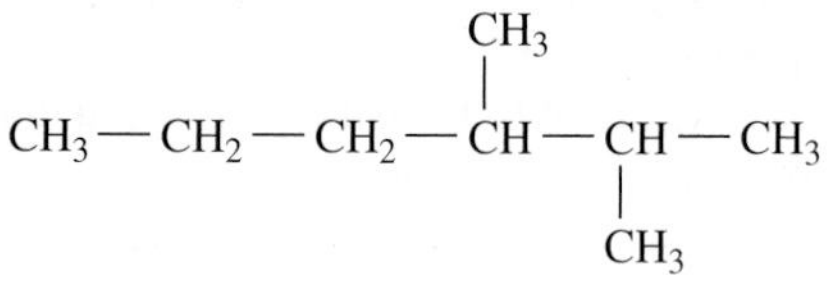 $CH_3-CH_2-CH_2-CH-CH-CH_3$
CH_3

2,3-Dimethylhexane

CH_3 CH_3
b. 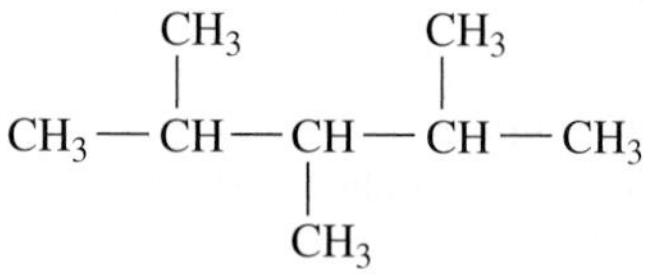 $CH_3-CH-CH-CH-CH_3$
CH_3

2,3,4-Trimethylpentane

11.47 a. 2,2-dimethylbutane **b.** chloroethane
c. 2-bromo-4-ethylhexane **d.** methylcyclopentane

11.49 a. *trans*-2-pentene
b. 4-bromo-5-methyl-1-hexene
c. 1-methylcyclopentene

11.51 a. structural isomers **b.** cis–trans isomers

11.53 a. 3-chlorotoluene **b.** 4-bromoaniline

11.55 a.  Br

Br
b. 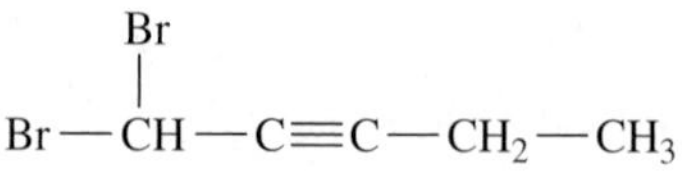 $Br-CH-C{\equiv}C-CH_2-CH_3$

c.
H H
C=C
CH_3 $CH_2-CH_2-CH_2-CH_3$

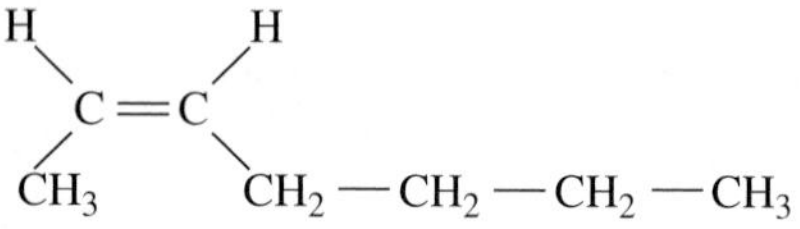

11.57 a. 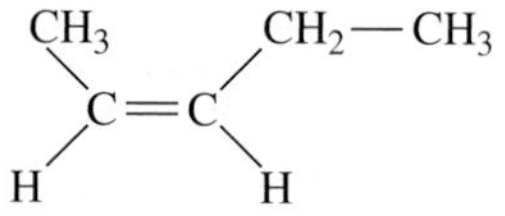
CH_3 CH_2-CH_3
C=C
H H
cis-2-Pentene

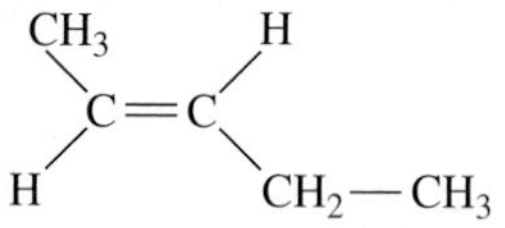

CH_3 H
C=C
H CH_2-CH_3
trans-2-Pentene

b. 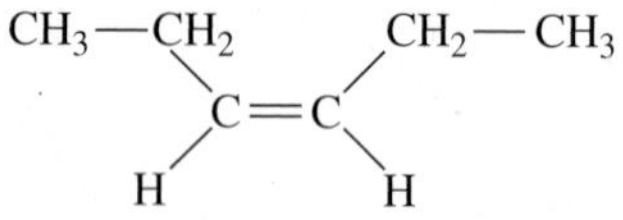
CH_3-CH_2 CH_2-CH_3
C=C
H H
cis-3-Hexene

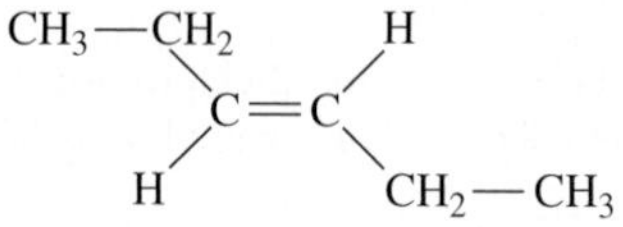

CH_3-CH_2 H
C=C
H CH_2-CH_3
trans-3-Hexene

11.59 a.

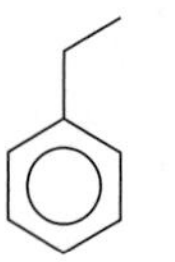

b. 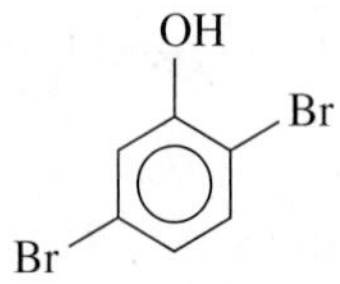 OH, Br, Br

c. 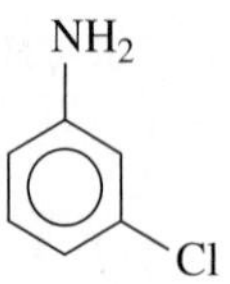 NH_2, Cl

11.61 a. $C_5H_{12} + 8O_2 \xrightarrow{\Delta} 5CO_2 + 6H_2O + \text{energy}$

b. $C_4H_8 + 6O_2 \xrightarrow{\Delta} 4CO_2 + 4H_2O + \text{energy}$

c. $C_6H_{12} + 9O_2 \xrightarrow{\Delta} 6CO_2 + 6H_2O + \text{energy}$

11.63 a. 3-methylpentane **b.** cyclohexane

c. propane

11.65 a.

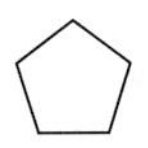

b. 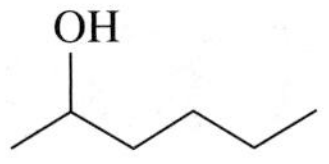 OH

11.67 a. $C_5H_{12} + 8O_2 \xrightarrow{\Delta} 5CO_2 + 6H_2O + \text{energy}$

b. 72.15 g/mole

c. 2.8×10^4 kcal

d. 3.7×10^3 L of CO_2 at STP

11.69

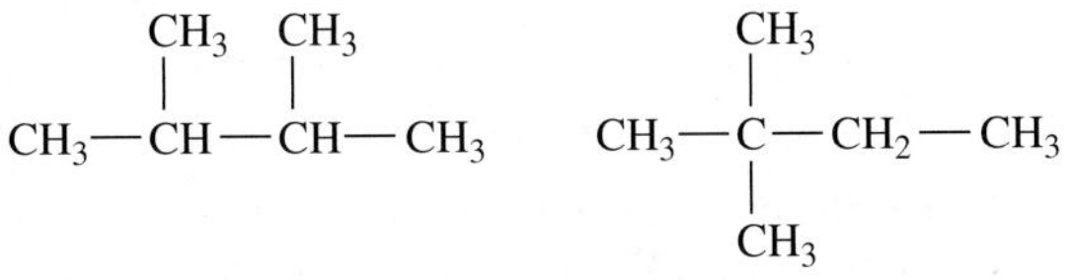

$$CH_3-\underset{}{\overset{CH_3}{\overset{|}{C}H}}-\overset{CH_3}{\overset{|}{C}H}-CH_3 \qquad CH_3-\underset{CH_3}{\underset{|}{\overset{CH_3}{\overset{|}{C}}}}-CH_2-CH_3$$

11.71 a. $CH_3-CH_2-CH_3$

b. $C_3H_8 + 5O_2 \xrightarrow{\Delta} 3CO_2 + 4H_2O + \text{energy}$

c. 85.7 g of O_2

d. 70.7 g of CO_2

11.73 a. 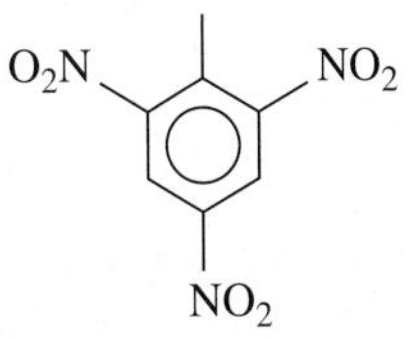 O_2N NO_2 NO_2

b.

O_2N NO_2 NO_2

O_2N NO_2 NO_2

NO_2 O_2N NO_2

NO_2 NO_2 NO_2

NO_2 O_2N NO_2

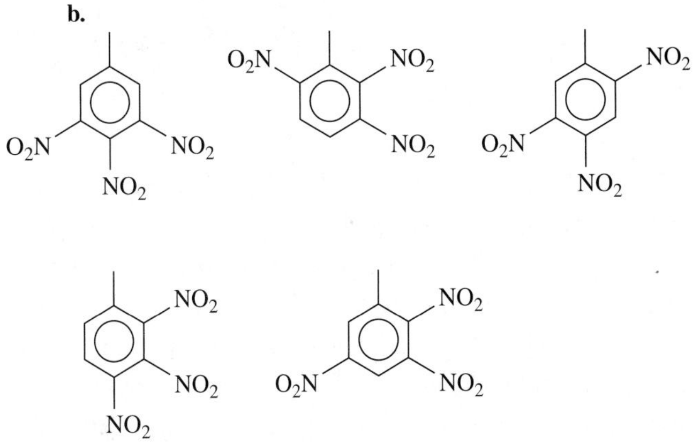

12 Alcohols, Thiols, Ethers, Aldehydes, and Ketones

RECENTLY, DIANA NOTICED THAT A MOLE ON her arm changed appearance. For many years, it was light brown in color, with a flat circular appearance. But over the last few weeks, the mole has become raised, with irregular borders, and has darkened. She called her dermatologist for an appointment. Diana told Margaret, a dermatology nurse, that she has been going to tanning salons and had about 20 tanning sessions the previous year. She loves being outside but does not always apply sunscreen while in the sun or at the beach. Diana said she has no family history of malignant melanoma.

The risk factors for melanoma include frequent exposure to sun, severe sunburns at an early age, skin type, and family history. During Diana's skin exam, Margaret looked for other suspicious moles with nonuniform borders, changes in color, and changes in size. To treat the mole on Diana's arm, Margaret numbed the area, then removed a sample of skin tissue, which she sent to a lab for evaluation. Because the results indicated the presence of malignant melanoma cells, a surgeon excised the entire mole including subcutaneous fat. Fortunately, the mole was not very large and no further treatment was needed. Margaret suggested that Diana return in six months for a follow-up skin check.

The number of cases of melanoma has been rising, unlike for many other types of cancer. Doctors think this change may be because of unprotected sun exposure, an increase in the use of tanning salons, and perhaps an increased awareness and detection of the disease.

CAREER Dermatology Nurse

A dermatology nurse performs many of the duties of dermatologists, including treating skin conditions, assisting in surgeries, performing biopsies and excisions, writing prescriptions, freezing skin lesions, and screening patients for skin cancer. To become a dermatology nurse, you first become a nurse or physician assistant and then specialize in dermatology. To be certified, the RN must have an additional two years of dermatology experience, with a minimum of 2000 h of work experience in dermatology, and pass an examination. Advanced training is available through colleges and universities, allied health schools, and medical schools.

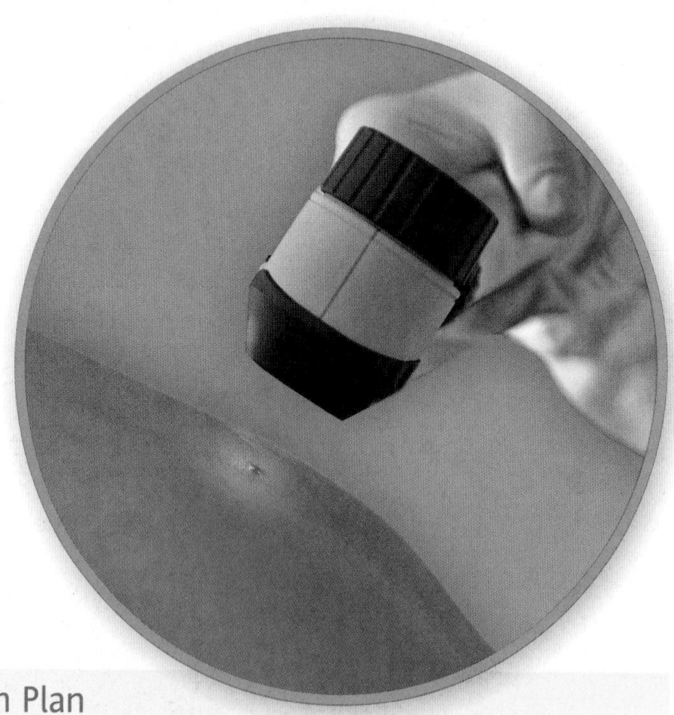

CLINICAL UPDATE Diana's Skin Protection Plan

Now, whenever Diana is in the sun, she wears a hat, a long-sleeved shirt, and uses a sunscreen. You can view the types of sunscreens in the **CLINICAL UPDATE Diana's Skin Protection Plan**, page 421, and identify the types of functional groups in the line-angle formula of each.

12.1 Alcohols, Phenols, Thiols, and Ethers

LEARNING GOAL Write the IUPAC and common names for alcohols, phenols, and thiols and the common names for ethers. Draw their condensed structural or line-angle formulas.

LOOKING AHEAD

Alcohols, which contain the *hydroxyl group* (—OH), are commonly found in nature and are used in industry and at home. For centuries, grains, vegetables, and fruits have been fermented to produce the ethanol present in alcoholic beverages. The hydroxyl group is important in biomolecules such as sugars and starches as well as in steroids such as cholesterol and estradiol. The phenols contain the hydroxyl group attached to a benzene ring. Thiols, which contain the —SH group, give the strong odors we associate with garlic and onions.

In an **alcohol**, the hydroxyl functional group (—OH) replaces a hydrogen atom in a hydrocarbon. Oxygen (O) atoms are shown in red in the ball-and-stick models (see **FIGURE 12.1**). In a **phenol**, the hydroxyl group replaces a hydrogen atom attached to a benzene ring. A **thiol** contains a sulfur atom, shown in yellow-green in the ball-and-stick model, which makes a thiol similar to an alcohol except that —OH is replaced by an —SH group. In an **ether**, the functional group consists of an oxygen atom, which is attached to two carbon atoms (—O—). Molecules of alcohols, phenols, thiols, and ethers have bent shapes around the oxygen or sulfur atom, similar to water.

REVIEW

Naming and Drawing Alkanes (11.2)

CORE CHEMISTRY SKILL

Identifying Functional Groups

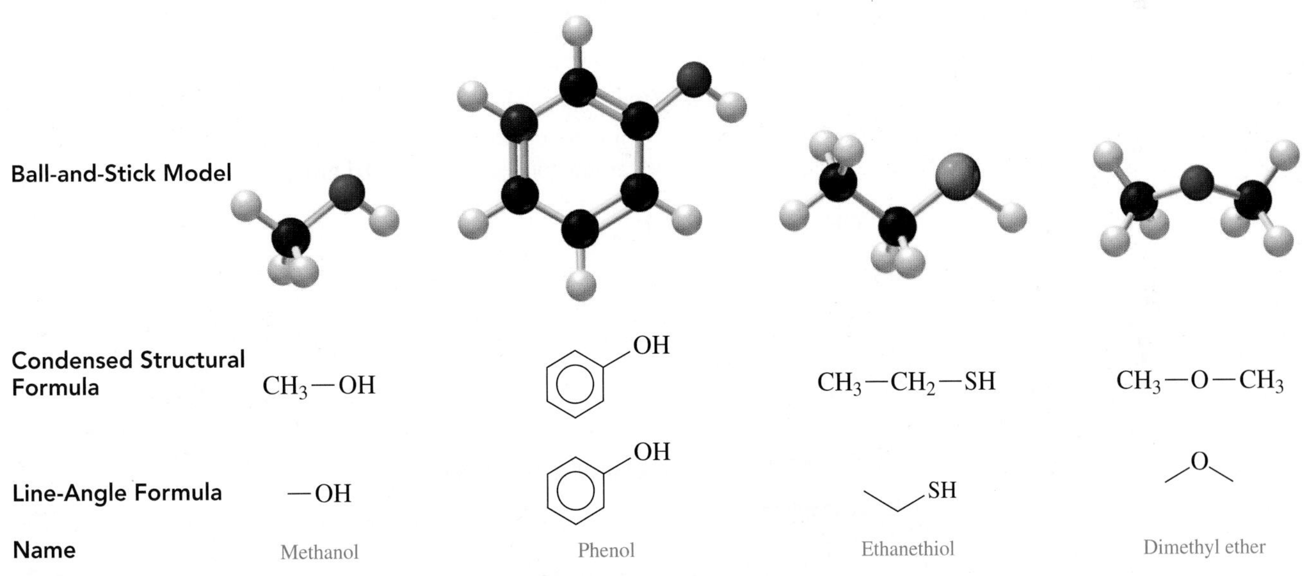

FIGURE 12.1 An alcohol or a phenol has a hydroxyl group (—OH) attached to carbon. A thiol has a thiol group (—SH) attached to carbon. An ether contains an oxygen atom (—O—) bonded to two carbon groups.

Q How is an alcohol different from a thiol?

Naming Alcohols

CORE CHEMISTRY SKILL

Naming Alcohols and Phenols

In the IUPAC system, an alcohol is named by replacing the *e* of the corresponding alkane name with *ol*. The common name of a simple alcohol uses the name of the alkyl group followed by *alcohol*.

$CH_3—H$	$CH_3—OH$	$CH_3—CH_2—H$	$CH_3—CH_2—OH$
Methane	Methanol (methyl alcohol)	Ethane	Ethanol (ethyl alcohol)

Alcohols with one or two carbon atoms do not require a number for the hydroxyl group. When an alcohol consists of a chain with three or more carbon atoms, the chain is numbered to give the position of the —OH group and any substituents on the chain.

$$\underset{3}{CH_3}-\underset{2}{CH_2}-\underset{1}{CH_2}-OH \qquad \underset{1}{CH_3}-\underset{2}{\overset{\overset{\displaystyle OH}{|}}{CH}}-\underset{3}{CH_3}$$

1-Propanol (propyl alcohol) — 2-Propanol (isopropyl alcohol)

We can also draw the line-angle formulas for alcohols as shown for 2-propanol and 2-butanol.

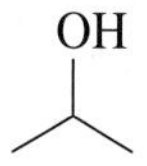

OH OH

2-Propanol 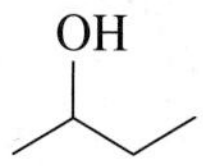 2-Butanol

A cyclic alcohol is named as a *cycloalkanol*. If there are substituents, the ring is numbered from carbon 1, which is the carbon attached to the —OH group. Compounds with no substituents on the ring do not require a number for the hydroxyl group.

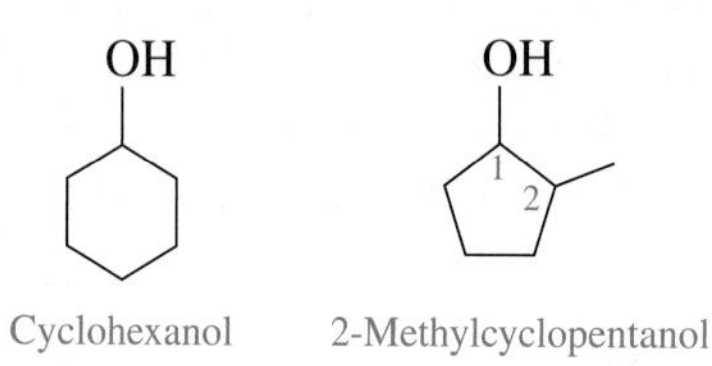

Cyclohexanol 2-Methylcyclopentanol

ENGAGE

If there are six carbon atoms in 2-hexanol, phenol, and cyclohexanol, how do the structures of the compounds differ?

Naming Phenols

The term *phenol* is the IUPAC name when a hydroxyl group (—OH) is bonded to a benzene ring. When there is a second substituent, the benzene ring is numbered starting from carbon 1, which is the carbon bonded to the —OH group.

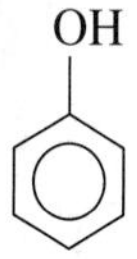

OH OH 1 3 Cl

Phenol 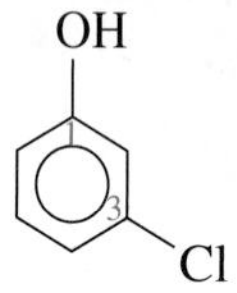 3-Chlorophenol

---

SAMPLE PROBLEM 12.1 Naming Alcohols

TRY IT FIRST

Give the IUPAC name for the following:

$$CH_3-\overset{\overset{\displaystyle CH_3}{|}}{CH}-CH_2-\overset{\overset{\displaystyle OH}{|}}{CH}-CH_3$$

SOLUTION GUIDE

	Given	Need	Connect
ANALYZE THE PROBLEM	five carbon chain, hydroxyl group, methyl group	IUPAC name	position of methyl and hydroxyl groups, replace *e* in alkane name with *ol*

STEP 1 Name the longest carbon chain attached to the —OH group by replacing the *e* in the corresponding alkane name with *ol*.

$$CH_3-\overset{\overset{\displaystyle CH_3}{|}}{CH}-CH_2-\overset{\overset{\displaystyle OH}{|}}{CH}-CH_3 \qquad \text{pentanol}$$

STEP 2 Number the chain starting at the end nearer to the —OH group. This carbon chain is numbered from right to left to give the position of the —OH group as carbon 2, which is shown as a prefix in the name 2-pentanol.

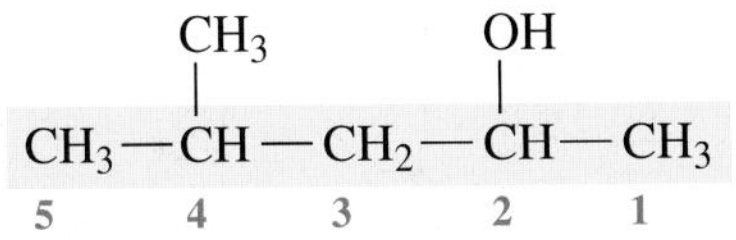

$$\underset{5}{CH_3}-\underset{4}{\overset{\overset{CH_3}{|}}{CH}}-\underset{3}{CH_2}-\underset{2}{\overset{\overset{OH}{|}}{CH}}-\underset{1}{CH_3}$$ 2-pentanol

STEP 3 Give the location and name for each substituent relative to the —OH group. With a methyl group on carbon 4, the compound is named 4-methyl-2-pentanol.

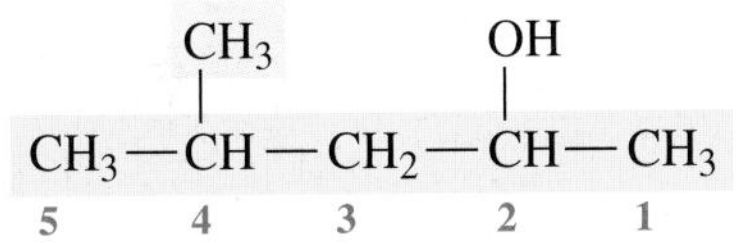

$$\underset{5}{CH_3}-\underset{4}{\overset{\overset{CH_3}{|}}{CH}}-\underset{3}{CH_2}-\underset{2}{\overset{\overset{OH}{|}}{CH}}-\underset{1}{CH_3}$$ 4-methyl-2-pentanol

STUDY CHECK 12.1

Give the IUPAC name for the following:

$$CH_3-\overset{\overset{Cl}{|}}{CH}-CH_2-CH_2-OH$$

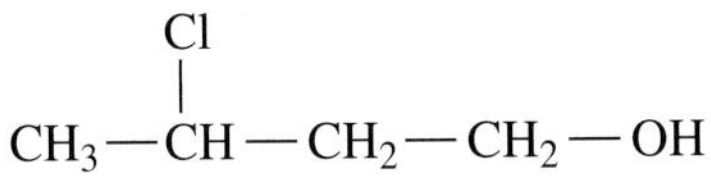

ANSWER

3-chloro-1-butanol

SAMPLE PROBLEM 12.2 Naming Phenols

TRY IT FIRST

Give the IUPAC name for the following:

OH
Br

(benzene ring with OH and Br on adjacent carbons)

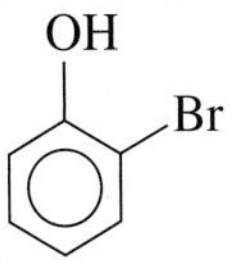

SOLUTION GUIDE

	Given	Need	Connect
ANALYZE THE PROBLEM	hydroxyl group bonded to benzene ring	IUPAC name	position of hydroxyl group on carbon 1, bromo group

STEP 1 Name an aromatic alcohol as a *phenol.*

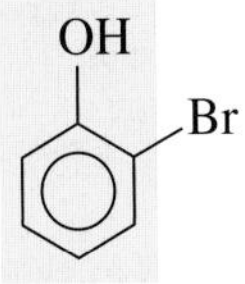

OH
Br

phenol

STEP 2 Number the chain starting at the end nearer to the —OH group. For a phenol, the carbon atom attached to the —OH group is carbon 1.

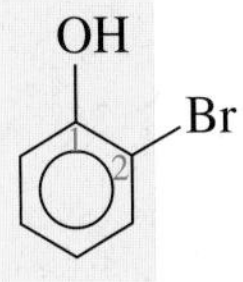

OH
Br
1 2

phenol

STEP 3 Give the location and name for each substituent relative to the —OH group.

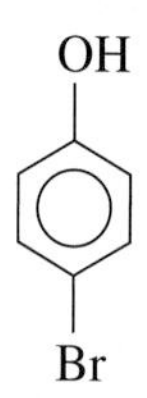

2-bromophenol

STUDY CHECK 12.2

Give the IUPAC name for the following:

OH (para) Br — benzene ring with OH at top and Br at bottom

ANSWER

4-bromophenol

TEST

Try Practice Problems 12.1 to 12.4

CHEMISTRY LINK TO HEALTH

Some Important Alcohols and Phenols

Methanol (*methyl alcohol*), the simplest alcohol, is found in many solvents and paint removers. If ingested, methanol is oxidized to formaldehyde, which can cause headaches, blindness, and death. Methanol is used to make plastics, medicines, and fuels. In car racing, it is used as a fuel because it is less flammable and has a higher octane rating than gasoline.

Ethanol (*ethyl alcohol*) has been known since prehistoric times as an intoxicating product formed by the fermentation of grains, sugars, and starches.

$$C_6H_{12}O_6 \xrightarrow{\text{Fermentation}} 2CH_3—CH_2—OH + 2CO_2$$

Glucose → Ethanol

Today, ethanol for commercial use is produced by reacting ethene and water at high temperatures and pressures. Ethanol is used as a solvent for perfumes, varnishes, and some medicines, such as tincture of iodine. Recent interest in alternative fuels has led to increased production of ethanol by the fermentation of sugars from grains such as corn, wheat, and rice. "Gasohol" is a mixture of ethanol and gasoline used as a fuel.

$$H_2C{=}CH_2 + H_2O \xrightarrow{300\ °C,\ 200\ \text{atm},\ H^+} CH_3—CH_2—OH$$

Ethene → Ethanol

1,2-Ethanediol (*ethylene glycol*) is used as an antifreeze in heating and cooling systems. It is also a solvent for paints, inks, and plastics, and it is used in the production of synthetic fibers such as Dacron. If ingested, it is extremely toxic. In the body, it is oxidized to oxalic acid, which forms insoluble salts in the kidneys that cause renal damage, convulsions, and death. Because its sweet taste is attractive to pets and children, ethylene glycol solutions must be carefully stored.

$$HO—CH_2—CH_2—OH \xrightarrow{[O]} HO—\overset{O}{\overset{\|}{C}}—\overset{O}{\overset{\|}{C}}—OH$$

1,2-Ethanediol (ethylene glycol) → Oxalic acid

1,2,3-Propanetriol (*glycerol* or *glycerin*), a trihydroxy alcohol, is a viscous liquid obtained from oils and fats during the production of soaps. The presence of several polar —OH groups makes it strongly attracted to water, a feature that makes glycerol useful as a skin softener in products such as skin lotions, cosmetics, shaving creams, and liquid soaps.

$$HO—CH_2—\underset{}{\overset{OH}{\overset{|}{CH}}}—CH_2—OH$$

1,2,3-Propanetriol (glycerol)

Bisphenol A (BPA) is used to make polycarbonate, a clear plastic that is used to manufacture beverage bottles, including baby bottles. Washing polycarbonate bottles with certain detergents or at high temperatures disrupts the polymer, causing small amounts of BPA to leach from the bottles. Because BPA is an estrogen mimic, there are concerns about the harmful effects from low levels of BPA. In 2008, Canada banned the use of polycarbonate baby bottles, which are now labeled "BPA free."

HO—(benzene ring)—C(CH_3)(CH_3)—(benzene ring)—OH

Bisphenol A (BPA)

Antifreeze (ethylene glycol) raises the boiling point and decreases the freezing point of water in a radiator.

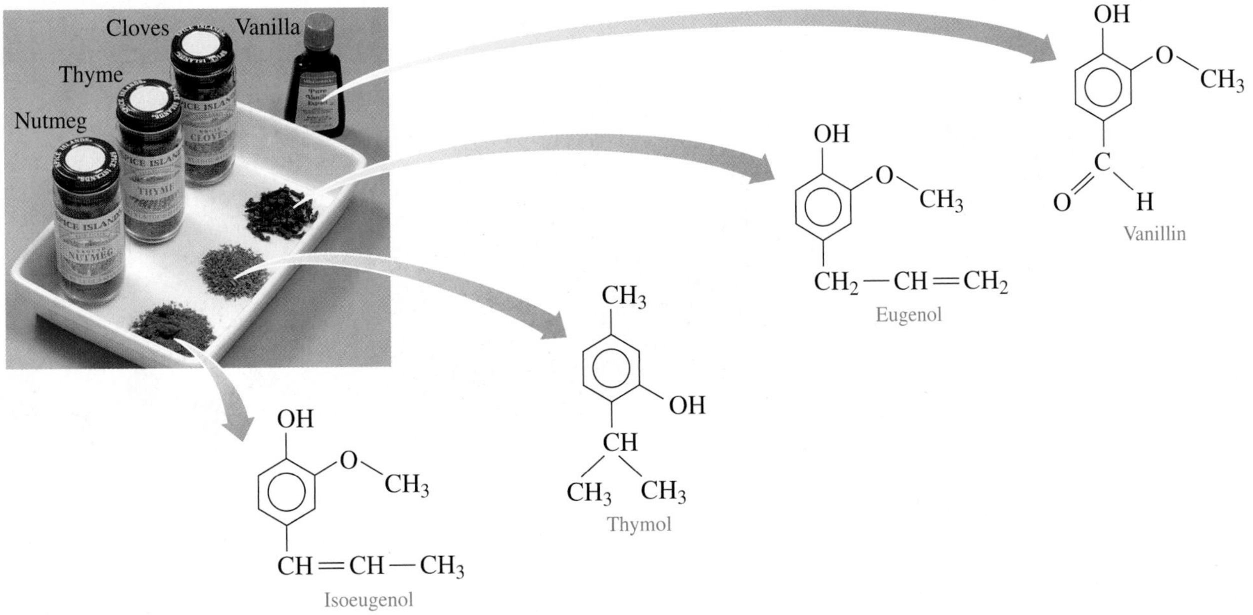

Cloves
Vanilla
Thyme
Nutmeg

OH, O—CH_3, $C(=O)H$
Vanillin

OH, O—CH_3, $CH_2—CH=CH_2$
Eugenol

CH_3, OH, $CH(CH_3)_2$
Thymol

OH, O—CH_3, $CH=CH—CH_3$
Isoeugenol

Phenols found in essential oils of plants produce the odor or flavor of the plant.

Thiols

Thiols are a family of sulfur-containing organic compounds that have a *thiol* group (—SH). In the IUPAC system, thiols are named by adding *thiol* to the alkane name of the longest carbon chain and numbering the carbon chain from the end nearer the —SH group.

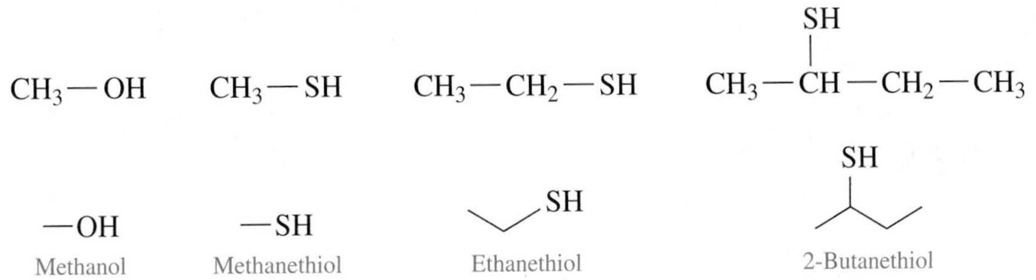

$CH_3—OH$	$CH_3—SH$	$CH_3—CH_2—SH$	$CH_3—CH(SH)—CH_2—CH_3$
—OH	—SH	⁄⁀SH	SH on a branched chain
Methanol	Methanethiol	Ethanethiol	2-Butanethiol

The spray of a skunk contains a mixture of thiols.

An important property of thiols is a strong, sometimes disagreeable, odor. To help us detect natural gas (methane) leaks, a small amount of a thiol is added to the gas supply, which is normally odorless. Thiols such as *trans*-2-butene-1-thiol give the pungent smell to the spray emitted when a skunk senses danger.

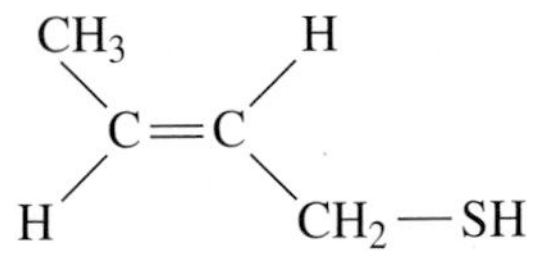

$CH_3(H)C=C(H)CH_2—SH$

trans-2-Butene-1-thiol
(in skunk spray)

Methanethiol is the characteristic odor of oysters, cheddar cheese, onions, and garlic. Garlic also contains 2-propene-1-thiol. The odor of onions is due to 1-propanethiol, which is also a lachrymator, a substance that makes eyes tear.

Thiols are sulfur-containing compounds that often have strong odors.

Dimethyl ether

Ethers

An ether contains an oxygen atom that is attached by single bonds to two carbon groups that are alkyls or aromatic groups. Ethers have a bent structure like water and alcohols.

ENGAGE

How does the functional group of an ether differ from that of an alcohol?

Naming Ethers

Most ethers have common names. The name of each alkyl or aromatic group attached to the oxygen atom is written in alphabetical order, followed by the word *ether*. In this text, we use only the common names of ethers.

TEST

Try Practice Problems 12.5 to 12.8

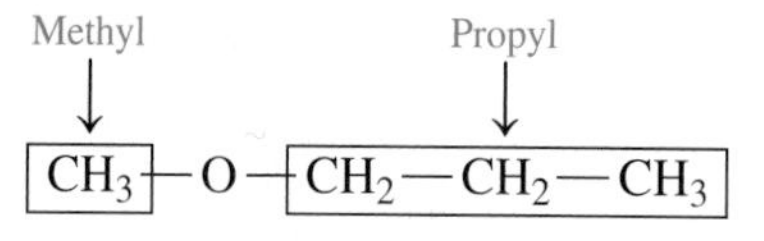

Common name: Methyl propyl ether

CHEMISTRY LINK TO HEALTH

Ethers as Anesthetics

Anesthesia is the loss of sensation and consciousness. A general anesthetic is a substance that blocks signals to the awareness centers in the brain so the person has a loss of memory, a loss of feeling pain, and an artificial sleep. The term *ether* has been associated with anesthesia because diethyl ether was the most widely used anesthetic for more than a hundred years. Although it is easy to administer, ether is very volatile and highly flammable. A small spark in the operating room could cause an explosion. Since the 1950s, anesthetics such as Forane (isoflurane), Ethrane (enflurane), Suprane (desflurane), and Sevoflurane have been developed that are not as flammable. Most of these anesthetics retain the ether group, but the addition of halogen atoms reduces the volatility and flammability of the ethers.

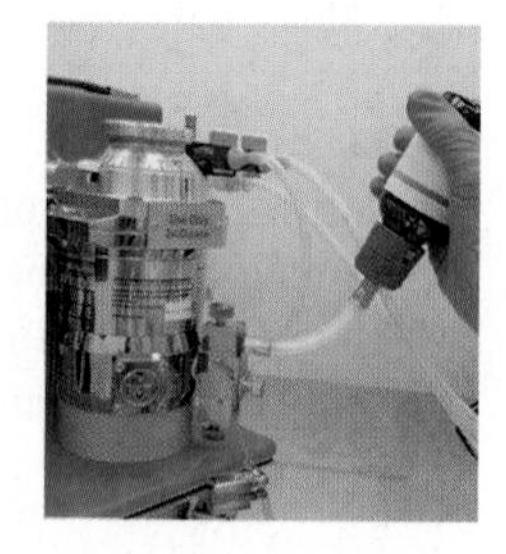

Isoflurane (Forane) is an inhaled anesthetic.

$CF_3-CHCl-O-CHF_2$

Forane (isoflurane)

$CHF_2-O-CF_2-CHFCl$

Ethrane (enflurane)

$CF_3-CHF-O-CHF_2$

Suprane (desflurane)

$(CF_3)_2CH-O-CH_2F$

Sevoflurane

PRACTICE PROBLEMS

12.1 Alcohols, Phenols, Thiols, and Ethers

LEARNING GOAL Write the IUPAC and common names for alcohols, phenols, and thiols and the common names for ethers. Draw their condensed structural or line-angle formulas.

12.1 Give the IUPAC name for each of the following:

a. CH_3-CH_2-OH

b. $CH_3-CH_2-\underset{|}{\overset{OH}{CH}}-CH_3$

c. 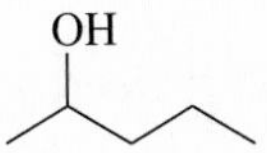 OH

d. 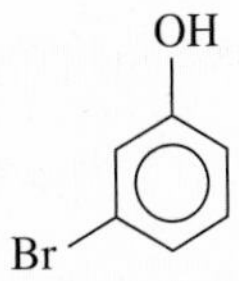 OH Br

12.2 Give the IUPAC name for each of the following:

a. 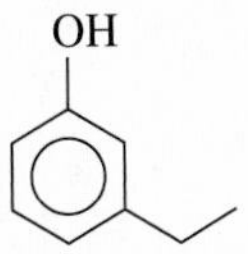 OH

b. $CH_3-CH_2-\overset{CH_3}{\overset{|}{CH}}-CH_2-OH$

c. 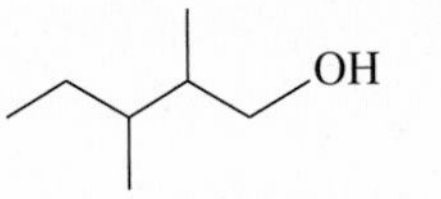 OH

d. 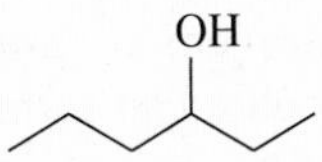 OH

12.3 Draw the condensed structural formula, or line-angle formula if cyclic, for each of the following:

a. propyl alcohol **b.** 3-pentanethiol

c. 2-methyl-2-butanol **d.** 4-bromophenol

12.4 Draw the condensed structural formula, or line-angle formula if cyclic, for each of the following:

a. ethyl alcohol **b.** 3-methyl-1-butanol

c. 1-propanethiol **d.** 2-chlorophenol

12.5 Give the common name for each of the following:

a. $CH_3-CH_2-CH_2-O-CH_2-CH_2-CH_3$

b. 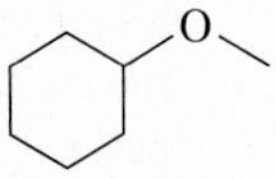 O

c. 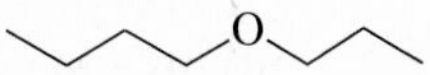 O

12.6 Give the common name for each of the following:

a. $CH_3-CH_2-O-CH_2-CH_2-CH_3$

b. 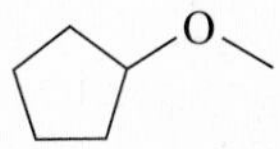 O

c. CH_3-O-CH_3

12.7 Draw the condensed structural formula, or line-angle formula if cyclic, for each of the following:

a. ethyl methyl ether **b.** cyclopropyl ethyl ether

12.8 Draw the condensed structural formula, or line-angle formula if cyclic, for each of the following:

a. diethyl ether **b.** cyclobutyl methyl ether

12.2 Properties of Alcohols

LEARNING GOAL Describe the classification of alcohols; describe the solubility of alcohols in water.

Alcohols are classified by the number of alkyl groups attached to the carbon atom bonded to the hydroxyl group (—OH). A **primary (1°) alcohol** has one alkyl group attached to the carbon atom bonded to the —OH group, a **secondary (2°) alcohol** has two alkyl groups, and a **tertiary (3°) alcohol** has three alkyl groups.

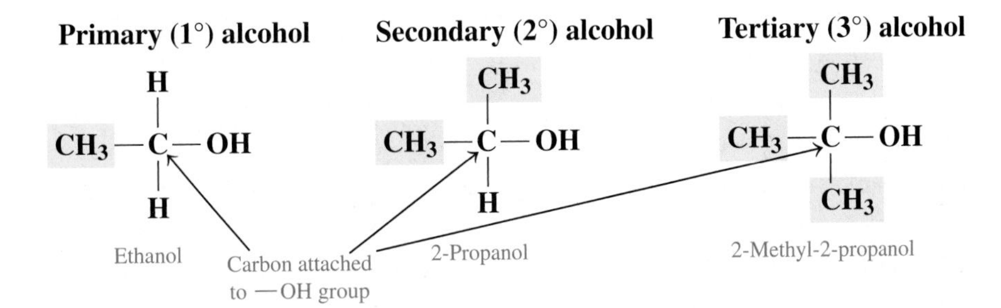

SAMPLE PROBLEM 12.3 Classifying Alcohols

TRY IT FIRST

Classify each of the following alcohols as primary (1°), secondary (2°), or tertiary (3°):

a. $CH_3-CH_2-CH_2-OH$

b. 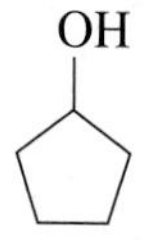

SOLUTION

a. The carbon atom bonded to the —OH group is attached to one alkyl group, which makes this a primary (1°) alcohol.

b. The carbon atom bonded to the —OH group is attached to three alkyl groups, which makes this a tertiary (3°) alcohol.

STUDY CHECK 12.3

Classify the following alcohol as primary (1°), secondary (2°), or tertiary (3°):

OH

(cyclopentane ring with OH)

ANSWER

secondary (2°)

TEST

Try Practice Problems 12.9 and 12.10

Solubility of Alcohols in Water

Hydrocarbons, which are composed of only carbon and hydrogen, are nonpolar and insoluble in water. However, the polar —OH group in an alcohol can form hydrogen bonds with the H and O atoms of water, which makes alcohols more soluble in water (see **FIGURE 12.2**).

Alcohols with one to three carbon atoms are *miscible* with water, which means any amount of the alcohol is completely soluble in water. However, the solubility provided by the polar —OH group decreases as the number of carbon atoms increases. Alcohols with four carbon atoms are slightly soluble in water, and alcohols with five or more carbon atoms are insoluble. **TABLE 12.1** compares the solubility of some alcohols.

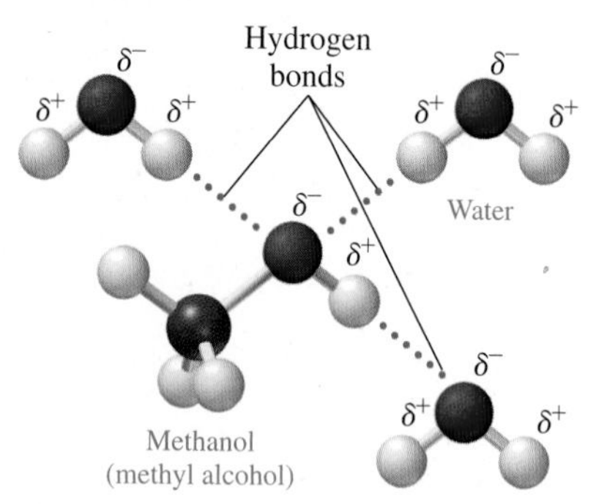

FIGURE 12.2 Methanol forms hydrogen bonds with water molecules.

Q Why is methanol more soluble in water than 1-pentanol?

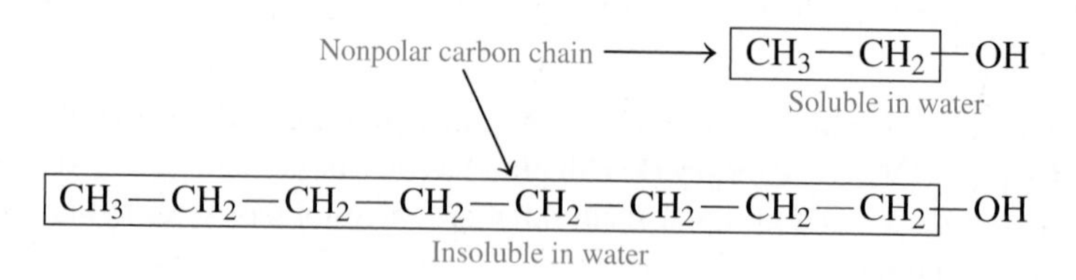

ENGAGE

What part of 1-hexanol makes it insoluble in water?

TEST

Try Practice Problems 12.11 to 12.14

TABLE 12.1 Solubility of Some Alcohols

Compound	Condensed Structural Formula	Number of Carbon Atoms	Solubility in Water
Methanol	CH_3-OH	1	Soluble
Ethanol	CH_3-CH_2-OH	2	Soluble
1-Propanol	$CH_3-CH_2-CH_2-OH$	3	Soluble
1-Butanol	$CH_3-CH_2-CH_2-CH_2-OH$	4	Slightly soluble
1-Pentanol	$CH_3-CH_2-CH_2-CH_2-CH_2-OH$	5	Insoluble

Solubility of Phenol

Phenol is slightly soluble in water because the —OH group can form hydrogen bonds with water molecules. In water, the —OH group of phenol dissociates slightly, which makes it a weak acid. In fact, an early name for phenol was *carbolic acid.*

$$\text{Phenol} + H_2O \rightleftarrows \text{Phenoxide ion} + H_3O^+$$

Phenol   Phenoxide ion

Phenol and Antiseptics

An *antiseptic* is a substance applied to the skin to kill microorganisms that cause infection. At one time, dilute solutions of phenol (carbolic acid) were used in hospitals as antiseptics. Joseph Lister (1827–1912) is considered a pioneer in antiseptic surgery and was the first to sterilize surgical instruments and dressings with phenol. Phenol was also used to disinfect wounds to prevent post-surgical infections such as gangrene. However, phenol is very corrosive and highly irritating to the skin; it can cause severe burns and ingestion can be fatal. Soon phenol solutions were replaced with other disinfectants. 4-Hexylresorcinol is a form of phenol used in topical antiseptics, throat lozenges, mouth washes, and throat sprays. Lysol, used to disinfect surfaces in a home or hospital, contains the antiseptics 2-phenylphenol and 2-benzyl-4-chlorophenol.

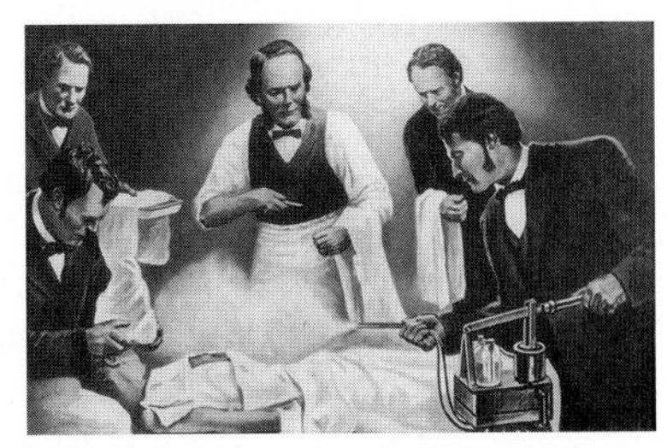

Joseph Lister was the first to use phenol to sterilize surgical instruments.

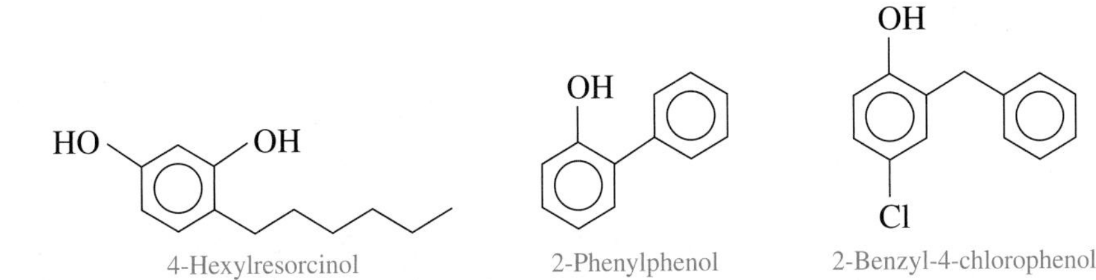

Lysol, used as a disinfectant, contains phenol compounds.

CHEMISTRY LINK TO HEALTH

Hand Sanitizers

When soap and water are not available, hand sanitizers may be used to kill most bacteria and viruses that spread colds and flu. As a gel or liquid solution, many hand sanitizers use ethanol or propanol as their active ingredient. Sanitizers also contain glycerin and propylene glycol to prevent the skin from drying. When children use hand sanitizers, they must be carefully supervised because the ingestion of a small amount can cause alcohol poisoning.

In an alcohol-containing sanitizer, the amount of ethanol is typically 60% (v/v) but can be as high as 85% (v/v). This amount of ethanol can make hand sanitizers a fire hazard in the home because ethanol is highly flammable. When ethanol undergoes combustion, it produces a transparent blue flame. When using an ethanol-containing sanitizer, it is important to rub hands together until they are completely dry. It is also recommended that sanitizers containing ethanol be placed in storage areas that are away from heat sources in the home.

Some sanitizers are alcohol-free, but often the active ingredient is triclosan, which contains aromatic, ether, and phenol functional groups. The Food and Drug Administration has banned triclosan in personal-care products because its use may promote the growth of antibiotic-resistant bacteria. Recent reports indicate that triclosan may disrupt the endocrine system and interfere with the function of estrogens, androgens, and thyroid hormones.

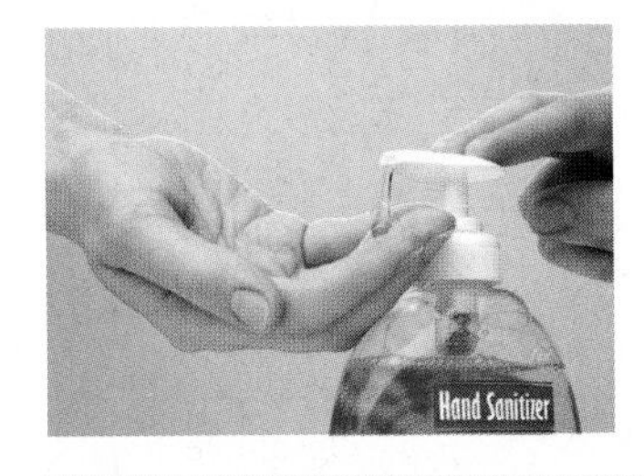

Hand sanitizers that contain ethanol or propanol are used to kill bacteria on the hands.

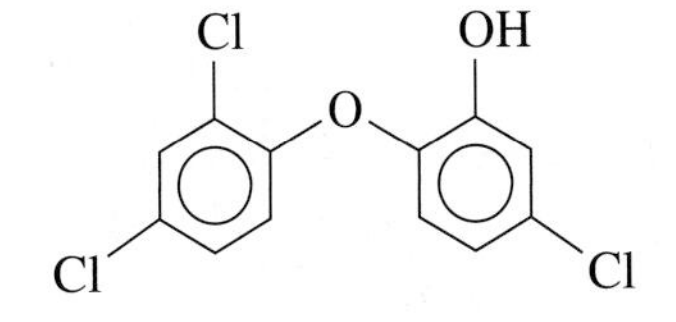

Triclosan, an antibacterial compound, is now banned in personal-care products.

PRACTICE PROBLEMS

12.2 Properties of Alcohols

LEARNING GOAL Describe the classification of alcohols; describe the solubility of alcohols in water.

12.9 Classify each of the following alcohols as primary (1°), secondary (2°), or tertiary (3°):

a. (line-angle formula) OH

b. $CH_3-CH_2-CH_2-CH_2-OH$

c. $CH_3-C(OH)(CH_3)-CH_2-CH_3$ **d.** (cyclobutane ring) OH

12.10 Classify each of the following alcohols as primary (1°), secondary (2°), or tertiary (3°):

a. (cyclopentane ring) OH **b.** (line-angle formula) OH

c. (benzene ring) OH **d.** $CH_3-CH_2-CH_2-C(CH_3)_2-OH$

12.11 Are each of the following soluble, slightly soluble, or insoluble in water? Explain.

a. CH_3-CH_2-OH
b. $CH_3-CH_2-CH_2-CH_2-CH_2-CH_2-OH$

12.12 Are each of the following soluble, slightly soluble, or insoluble in water? Explain.

a. $CH_3-CH_2-CH_2-OH$
b. $CH_3-CH_2-CH_2-CH_2-CH_3$

12.13 Give an explanation for each of the following observations:

a. Methanol is soluble in water, but ethane is not.
b. 1-Propanol is soluble in water, but ethyl methyl ether is only slightly soluble.

12.14 Give an explanation for each of the following observations:

a. Ethanol is soluble in water, but propane is not.
b. 1-Propanol is soluble in water, but 1-hexanol is not.

ENGAGE

How does the structure of an aldehyde differ from that of a ketone?

12.3 Aldehydes and Ketones

LEARNING GOAL Write the IUPAC and common names for aldehydes and ketones; draw their condensed structural or line-angle formulas. Describe the solubility of aldehydes and ketones in water.

Aldehydes and ketones contain a **carbonyl group** that consists of a carbon–oxygen double bond with two groups of atoms attached to the carbon at angles of 120°. The oxygen atom with two lone pairs of electrons is much more electronegative than the carbon atom. Therefore, the carbonyl group has a strong dipole with a partial negative charge (δ^-) on the oxygen and a partial positive charge (δ^+) on the carbon. The polarity of the carbonyl group strongly influences the physical and chemical properties of aldehydes and ketones.

In an **aldehyde**, the carbon of the carbonyl group is bonded to at least one hydrogen atom. That carbon may also be bonded to another hydrogen atom, a carbon of an alkyl group, or an aromatic ring (see FIGURE 12.3). The aldehyde group may be written as separate atoms or as —CHO, with the double bond understood. In a **ketone**, the carbonyl group is bonded to two alkyl groups or aromatic rings. The keto group (C=O) can be written as CO. A line-angle formula may also be used to represent an aldehyde or ketone.

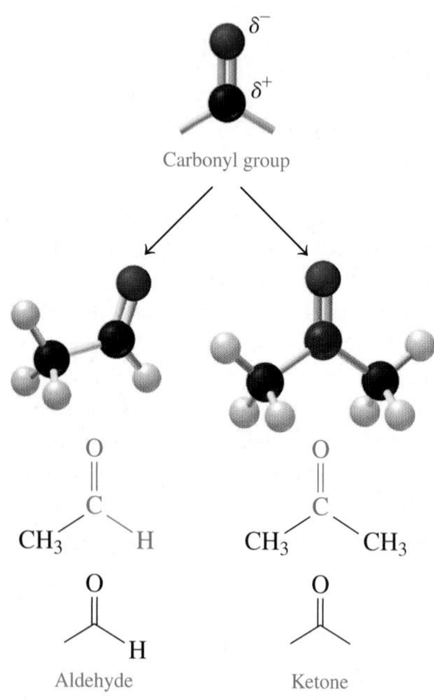

FIGURE 12.3 The carbonyl group is found in aldehydes and ketones.

Q If aldehydes and ketones both contain a carbonyl group, how can you differentiate between compounds from each family?

Formulas for C_3H_6O

Aldehyde

$$CH_3-CH_2-\overset{O}{\overset{\|}{C}}-H = CH_3-CH_2-CHO = \text{(line-angle formula)}$$

Ketone

$$CH_3-\overset{O}{\overset{\|}{C}}-CH_3 = CH_3-CO-CH_3 = \text{(line-angle formula)}$$

SAMPLE PROBLEM 12.4 Identifying Aldehydes and Ketones

TRY IT FIRST

Identify each of the following as an aldehyde or ketone:

a. O ‖ H **b.** O ‖ **c.** O ‖ H

SOLUTION

a. aldehyde **b.** ketone **c.** aldehyde

STUDY CHECK 12.4

Identify the following as an aldehyde or ketone:

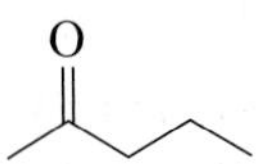

ANSWER

ketone

TEST

Try Practice Problems 12.15 and 12.16

Naming Aldehydes

CORE CHEMISTRY SKILL

Naming Aldehydes and Ketones

In the IUPAC system, an aldehyde is named by replacing the *e* of the corresponding alkane name with *al*. No number is needed for the aldehyde group because it always appears at the end of the chain. The aldehydes with carbon chains of one to four carbons are often referred to by their common names, which end in *aldehyde* (see **FIGURE 12.4**). The roots (*form*, *acet*, *propion*, and *butyr*) of these common names are derived from Latin or Greek words.

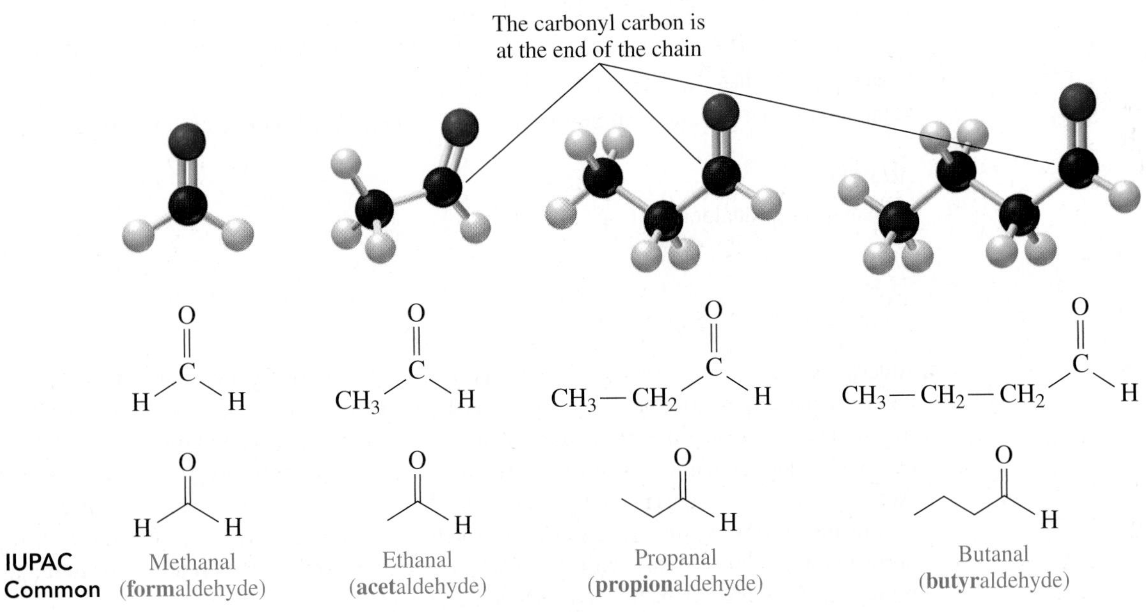

FIGURE 12.4 In the structures of aldehydes, the carbonyl group is always the end carbon.

Q Why is the carbon in the carbonyl group in aldehydes always at the end of the chain?

The IUPAC system names the aldehyde of benzene as benzaldehyde.

O ‖ H

Benzaldehyde

SAMPLE PROBLEM 12.5 Naming Aldehydes

TRY IT FIRST

Give the IUPAC name for the following:

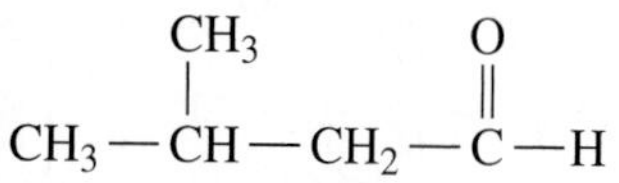

$$CH_3-\underset{}{\overset{\displaystyle CH_3 \atop |}{CH}}-CH_2-\overset{\displaystyle O \atop \|}{C}-H$$

SOLUTION GUIDE

	Given	Need	Connect
ANALYZE THE PROBLEM	four-carbon chain, carbonyl group, methyl substituent	IUPAC name	position of methyl group, replace e in alkane name with *al*

STEP 1 Name the longest carbon chain by replacing the e in the alkane name with *al*. The longest carbon chain containing the carbonyl group has four carbons, which is named butanal in the IUPAC system.

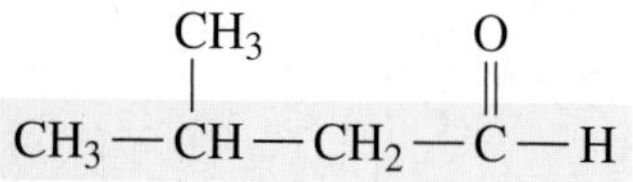

$$CH_3-\overset{\displaystyle CH_3 \atop |}{CH}-CH_2-\overset{\displaystyle O \atop \|}{C}-H \qquad \text{butanal}$$

STEP 2 Name and number any substituents by counting the carbonyl group as carbon 1. The substituent, which is the $-CH_3$ group on carbon 3, is methyl. The IUPAC name for this compound is 3-methylbutanal.

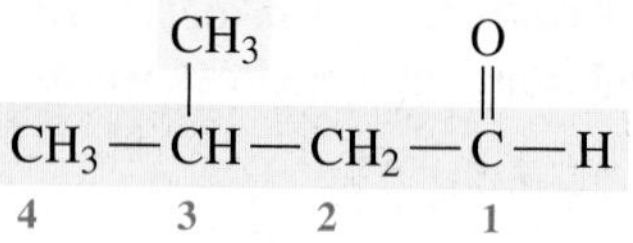

$$\underset{4}{CH_3}-\underset{3}{\overset{\displaystyle CH_3 \atop |}{CH}}-\underset{2}{CH_2}-\underset{1}{\overset{\displaystyle O \atop \|}{C}}-H \qquad \text{3-methylbutanal}$$

STUDY CHECK 12.5

What are the IUPAC and common names of the aldehyde with three carbon atoms?

ANSWER

propanal (propionaldehyde)

Naming Ketones

Aldehydes and ketones are some of the most important classes of organic compounds. Because they have played a major role in organic chemistry for more than a century, the common names for unbranched ketones are still in use. In the common names, the alkyl groups bonded to the carbonyl group are named as substituents and are listed alphabetically, followed by *ketone*. Acetone, which is another name for propanone, has been retained by the IUPAC system.

In the IUPAC system, the name of a ketone is obtained by replacing the *e* in the corresponding alkane name with *one*. Carbon chains with five carbon atoms or more are numbered from the end nearer the carbonyl group.

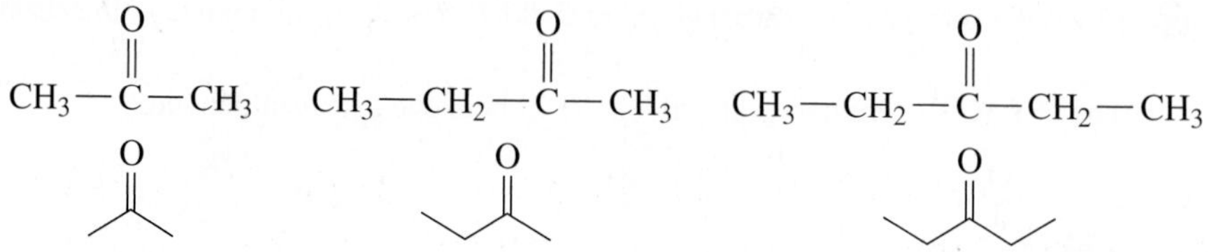

$$CH_3-\overset{\displaystyle O \atop \|}{C}-CH_3 \qquad CH_3-CH_2-\overset{\displaystyle O \atop \|}{C}-CH_3 \qquad CH_3-CH_2-\overset{\displaystyle O \atop \|}{C}-CH_2-CH_3$$

Propanone
(dimethyl ketone; acetone)

Butanone
(ethyl methyl ketone)

3-Pentanone
(diethyl ketone)

For cyclic ketones, the prefix *cyclo* is used in front of the ketone name. Any substituent is located by numbering the ring starting with the carbonyl carbon as carbon 1. The ring is numbered in the direction to give substituents the lowest possible numbers.

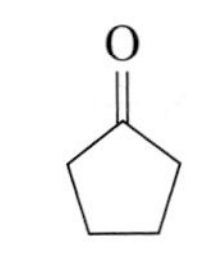

Cyclopentanone

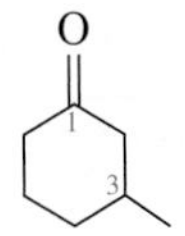

3-Methylcyclohexanone

SAMPLE PROBLEM 12.6 Naming Ketones

TRY IT FIRST

Give the IUPAC name for the following:

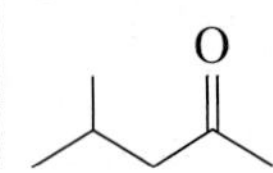

SOLUTION GUIDE

	Given	Need	Connect
ANALYZE THE PROBLEM	five-carbon chain, methyl substituent	IUPAC name	position of methyl and carbonyl groups, replace *e* in alkane name with *one*

STEP 1 **Name the longest carbon chain by replacing the *e* in the alkane name with *one*.** The longest chain has five carbon atoms, which is named pentanone.

STEP 2 **Number the carbon chain starting from the end nearer the carbonyl group and indicate its location.** Counting from the right, the carbonyl group is on carbon 2.

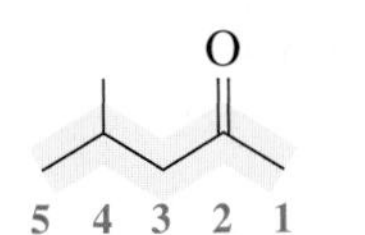

2-pentanone

STEP 3 **Name and number any substituents on the carbon chain.** Counting from the right, the methyl group is on carbon 4. The IUPAC name is 4-methyl-2-pentanone.

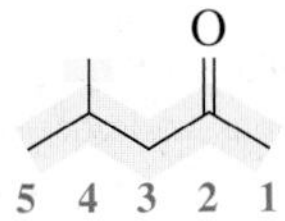

O

5 4 3 2 1

4-methyl-2-pentanone

STUDY CHECK 12.6

What is the common name of 3-hexanone?

ANSWER

ethyl propyl ketone

TEST

Try Practice Problems 12.17 to 12.22

Solubility of Aldehydes and Ketones in Water

Aldehydes and ketones contain a polar carbonyl group (carbon–oxygen double bond), which has a partially negative oxygen atom and a partially positive carbon atom. Because the electronegative oxygen atom forms hydrogen bonds with water molecules, aldehydes and ketones with one to four carbons are very soluble. However, aldehydes and ketones with

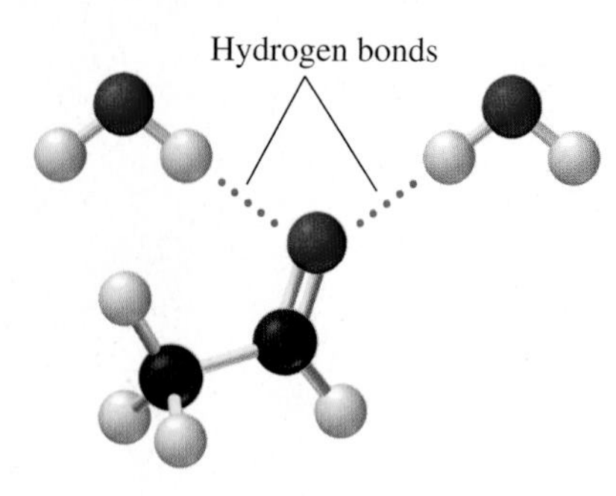

Ethanal (acetaldehyde) in water

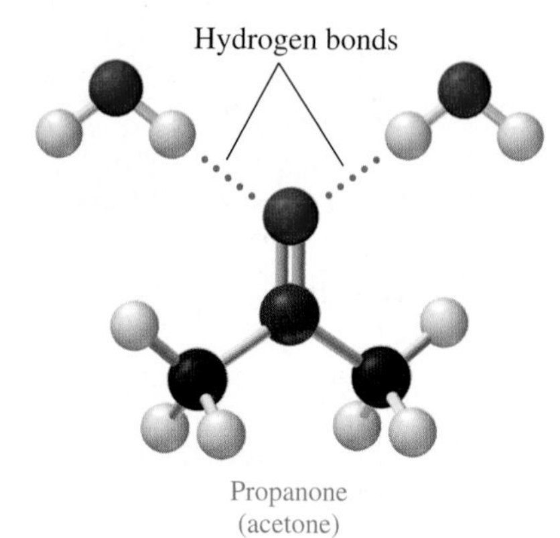

Propanone (acetone) in water

five or more carbon atoms are not soluble because longer hydrocarbon chains, which are nonpolar, diminish the solubility effect of the polar carbonyl group. **TABLE 12.2** compares the solubility of some aldehydes and ketones.

TABLE 12.2 Solubility of Selected Aldehydes and Ketones

Compound	Number of Carbon Atoms	Solubility in Water
Methanal (formaldehyde)	1	Soluble
Ethanal (acetaldehyde)	2	Soluble
Propanal (propionaldehyde)	3	Soluble
Propanone (acetone)	3	Soluble
Butanal (butyraldehyde)	4	Soluble
Butanone	4	Soluble
Pentanal	5	Slightly soluble
2-Pentanone	5	Slightly soluble
Hexanal	6	Insoluble
2-Hexanone	6	Insoluble

SAMPLE PROBLEM 12.7 Solubility of Ketones

TRY IT FIRST

Why is acetone soluble in water, but 2-hexanone is not?

SOLUTION

Acetone contains a carbonyl group with an electronegative oxygen atom that forms hydrogen bonds with water. 2-Hexanone also contains a carbonyl group, but its long, nonpolar hydrocarbon chain makes it insoluble in water.

STUDY CHECK 12.7

Would you expect hexanal to be more or less soluble in water than ethanal? Explain.

ANSWER

Hexanal, with its long hydrocarbon chain, would be less soluble than ethanal.

TEST

Try Practice Problems 12.23 and 12.24

CHEMISTRY LINK TO HEALTH

Some Important Aldehydes and Ketones

Formaldehyde, the simplest aldehyde, is a colorless gas with a pungent odor. An aqueous solution called *formalin*, which contains 40% formaldehyde, is used as a germicide and to preserve biological specimens. Industrially, it is a reactant in the synthesis of polymers used to make fabrics, insulation materials, carpeting, pressed wood products such as plywood, and plastics for kitchen counters. Exposure to formaldehyde fumes can irritate the eyes, nose, and upper respiratory tract.

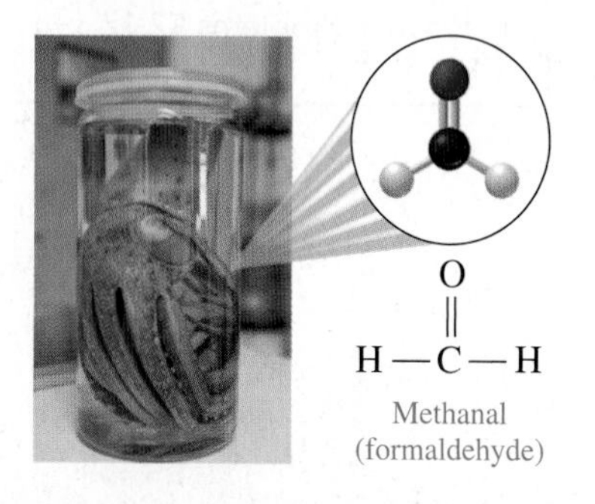

The simplest ketone, known as *acetone* or propanone (dimethyl ketone), is a colorless liquid with a mild odor that has wide use as a solvent in cleaning fluids, paint and nail polish removers, and rubber cement (see **FIGURE 12.5**). Acetone is extremely flammable, and care

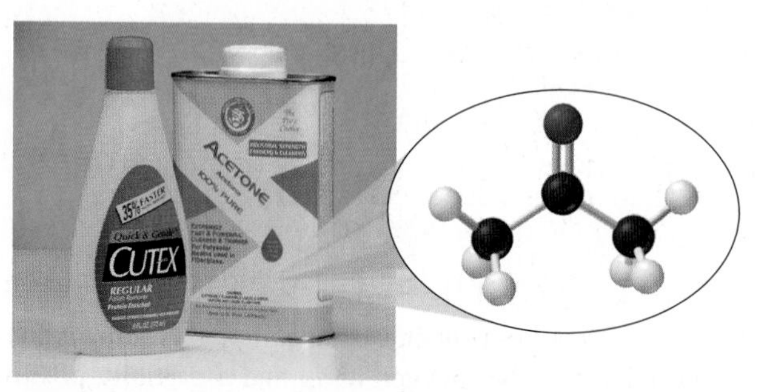

FIGURE 12.5 Acetone is used as a solvent in paint and nail polish removers.

Q What is the IUPAC name for acetone?

must be taken when using it. In the body, acetone may be produced in uncontrolled diabetes, fasting, and high-protein diets when large amounts of fats are metabolized for energy.

Muscone is a ketone used to make musk perfumes, and oil of spearmint contains carvone.

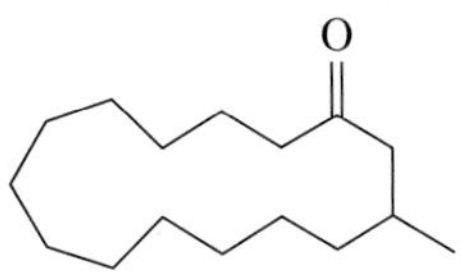

Muscone
(musk)

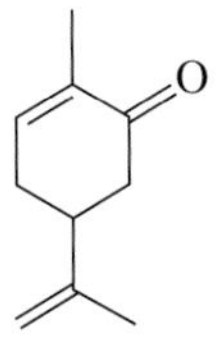

Carvone
(spearmint oil)

Several naturally occurring aromatic aldehydes are used to flavor food and as fragrances in perfumes. Benzaldehyde is found in almonds, vanillin in vanilla beans, and cinnamaldehyde in cinnamon.

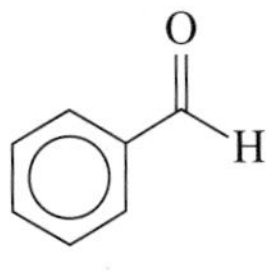

Benzaldehyde
(almond)

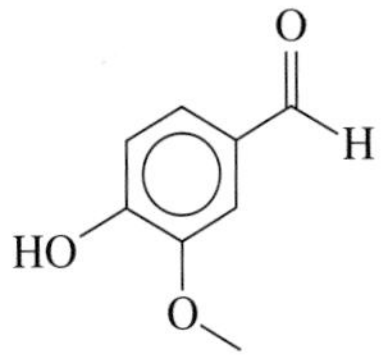

Vanillin
(vanilla)

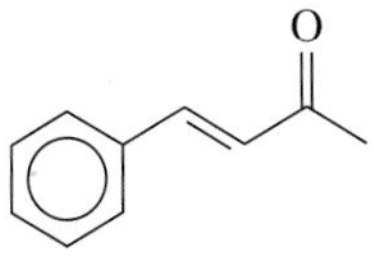

Cinnamaldehyde
(cinnamon)

PRACTICE PROBLEMS

12.3 Aldehydes and Ketones

LEARNING GOAL Write the IUPAC and common names for aldehydes and ketones; draw their condensed structural or line-angle formulas. Describe the solubility of aldehydes and ketones in water.

12.15 Identify each of the following compounds as an aldehyde or a ketone:

a. $CH_3-CH_2-\overset{\overset{\displaystyle O}{\|}}{C}-CH_3$

b.

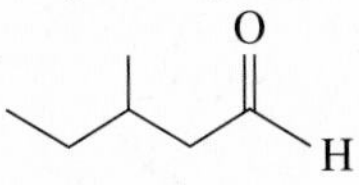

c.

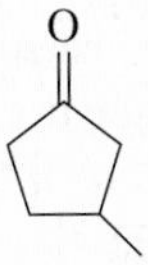

d.

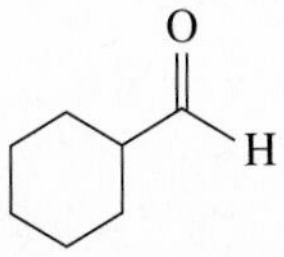

12.16 Identify each of the following compounds as an aldehyde or a ketone:

a.

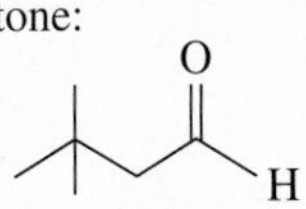

b. $CH_3-\overset{\overset{\displaystyle CH_3}{|}}{CH}-\overset{\overset{\displaystyle O}{\|}}{C}-H$

c.

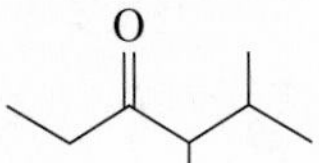

d. 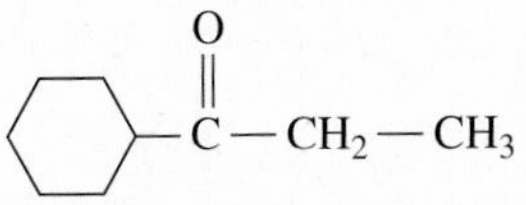 (cyclohexyl)$-\overset{\overset{\displaystyle O}{\|}}{C}-CH_2-CH_3$

12.17 Give the common name for each of the following:

a. $CH_3-\overset{\overset{\displaystyle O}{\|}}{C}-H$

b.

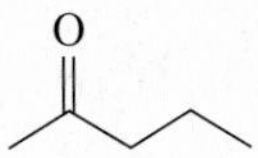

c. $H-\overset{\overset{\displaystyle O}{\|}}{C}-H$

12.18 Give the common name for each of the following:

a. $CH_3-\overset{\overset{\displaystyle O}{\|}}{C}-CH_2-CH_3$

b.

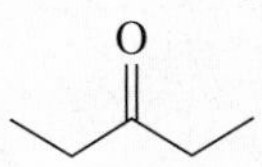

c. $CH_3-CH_2-\overset{\overset{\displaystyle O}{\|}}{C}-H$

12.19 Give the IUPAC name for each of the following:

a. $CH_3-CH_2-\overset{\overset{\displaystyle O}{\|}}{C}-H$

b. $CH_3-CH_2-\overset{\overset{\displaystyle O}{\|}}{C}-\overset{\overset{\displaystyle CH_3}{|}}{CH}-CH_3$

c.

d.

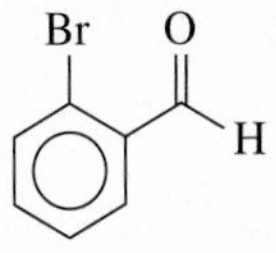

12.20 Give the IUPAC name for each of the following:

a. $CH_3-CH_2-\overset{\overset{\displaystyle CH_3}{|}}{CH}-CH_2-\overset{\overset{\displaystyle O}{\|}}{C}-H$

b. $CH_3-CH_2-CH_2-\overset{\overset{\displaystyle O}{\|}}{C}-CH_3$

c.

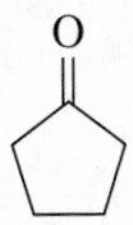

d.

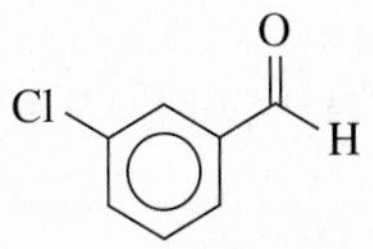

12.21 Draw the condensed structural formula for each of the following:

a. acetaldehyde
b. 4-chloro-2-pentanone
c. butyl methyl ketone
d. 3-methylpentanal

12.22 Draw the condensed structural formula for each of the following:

a. propionaldehyde
b. 2,3-dichlorobutanal
c. 2-methyl-3-hexanone
d. acetone

12.23 Which compound in each of the following pairs would be more soluble in water? Explain.

a. $CH_3-C(=O)-CH_2-CH_2-CH_3$ or $CH_3-C(=O)-C(=O)-CH_2-CH_3$

b. acetone or 2-pentanone

c. propanal or pentanal

12.24 Which compound in each of the following pairs would be more soluble in water? Explain.

a. $CH_3-CH_2-CH_3$ or CH_3-CH_2-CHO

b. propanone or 3-hexanone

c. butanal or hexanal

FIGURE 12.6 A flaming dessert is prepared using liquor that undergoes combustion.

Q What is the equation for the complete combustion of the ethanol in the liquor?

12.4 Reactions of Alcohols, Thiols, Aldehydes, and Ketones

LEARNING GOAL Write balanced chemical equations for the combustion, dehydration, and oxidation of alcohols. Write balanced chemical equations for the reduction of aldehydes and ketones.

Alcohols, similar to hydrocarbons, undergo combustion in the presence of oxygen. For example, in a restaurant, a flaming dessert may be prepared by pouring liquor on fruit or ice cream and lighting it (see **FIGURE 12.6**). The combustion of the ethanol in the liquor proceeds as follows:

$$CH_3-CH_2-OH + 3O_2 \xrightarrow{\Delta} 2CO_2 + 3H_2O + \text{energy}$$

Dehydration of Alcohols to Form Alkenes

In a **dehydration** reaction, alcohols lose a water molecule when they are heated at a high temperature (180 °C) with an acid catalyst such as H_2SO_4. During the dehydration of an alcohol, the components H— and —OH are removed from *adjacent carbon atoms of the same alcohol* to produce a water molecule. A double bond forms between the same two carbon atoms to produce an alkene product.

CORE CHEMISTRY SKILL

Writing Equations for the Dehydration of Alcohols

$$-\overset{H}{C}-\overset{OH}{C}- \xrightarrow{H^+,\ \text{heat}} C=C + \mathbf{H_2O}$$

Alcohol — Alkene — Water

ENGAGE

If 2-pentanol is dehydrated, what type of organic product forms?

Examples

$$H-\underset{H}{\overset{H}{C}}-\underset{H}{\overset{OH}{C}}-H \xrightarrow{H^+,\ \text{heat}} H_2C=CH_2 + \mathbf{H_2O}$$

Ethanol — Ethene

$$\text{Cyclopentanol (OH, H on adjacent carbons)} \xrightarrow{H^+,\ \text{heat}} \text{Cyclopentene} + \mathbf{H_2O}$$

Cyclopentanol — Cyclopentene

TEST

Try Practice Problems 12.25 and 12.26

SAMPLE PROBLEM 12.8 Dehydration of Alcohols

TRY IT FIRST

Draw the condensed structural formula for the alkene produced by the dehydration of the following:

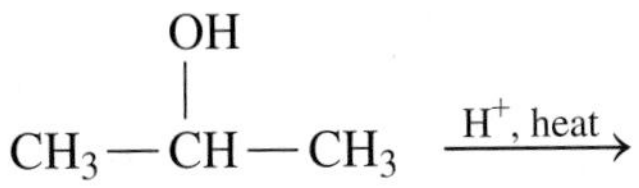

$$CH_3-\underset{\Large |}{\overset{OH}{CH}}-CH_3 \xrightarrow{H^+,\ heat}$$

SOLUTION

In a dehydration, the —OH is removed from carbon 2 and one H— is removed from an adjacent carbon, which forms a double bond.

$$CH_3-CH=CH_2$$

STUDY CHECK 12.8

What is the name of the alkene produced by the dehydration of cyclohexanol?

ANSWER

cyclohexene

TEST

Try Practice Problems 12.27 and 12.28

Oxidation of Primary Alcohols

CORE CHEMISTRY SKILL

Writing Equations for the Oxidation of Alcohols

In organic chemistry, an **oxidation** involves the addition of oxygen or a loss of hydrogen atoms. The oxidation of a primary alcohol produces an aldehyde, which contains a double bond between carbon and oxygen. For example, the oxidation of methanol and ethanol occurs by removing two hydrogen atoms, one from the —OH group and another from the carbon that is bonded to the —OH group. The reaction is written with the symbol [O] over the arrow to indicate that O is obtained from an oxidizing agent, such as $KMnO_4$ or $K_2Cr_2O_7$.

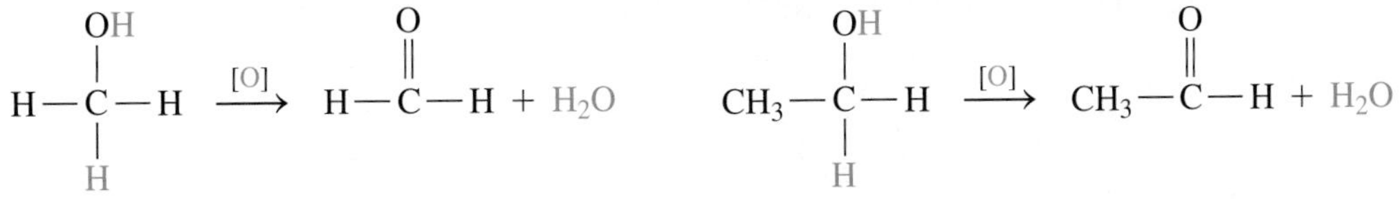

$$H-\underset{H}{\overset{OH}{C}}-H \xrightarrow{[O]} H-\overset{O}{\overset{\|}{C}}-H + H_2O \qquad CH_3-\underset{H}{\overset{OH}{C}}-H \xrightarrow{[O]} CH_3-\overset{O}{\overset{\|}{C}}-H + H_2O$$

Methanol (methyl alcohol) → Methanal (formaldehyde); Ethanol (ethyl alcohol) → Ethanal (acetaldehyde)

Aldehydes oxidize further by the addition of another O to form carboxylic acids, which have three carbon–oxygen bonds. This reaction occurs so readily that it is often difficult to isolate the aldehyde product during the oxidation reaction.

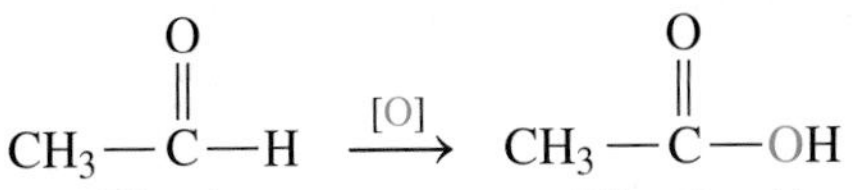

$$CH_3-\overset{O}{\overset{\|}{C}}-H \xrightarrow{[O]} CH_3-\overset{O}{\overset{\|}{C}}-OH$$

Ethanal (acetaldehyde) → Ethanoic acid (acetic acid)

Oxidation of Secondary Alcohols

In the oxidation of secondary alcohols, the products are ketones. Two hydrogen atoms are removed, one from the —OH group and another from the carbon bonded to the —OH group. The result is a ketone that has the carbon–oxygen double bond attached to alkyl or

ENGAGE

Why is it more difficult to oxidize a ketone than to oxidize an aldehyde?

aromatic groups on both sides. There is no further oxidation of a ketone because there are no hydrogen atoms attached to the carbon of the ketone group.

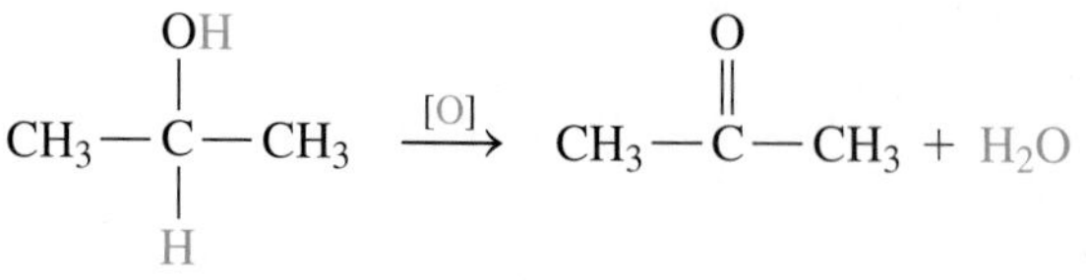

$$CH_3-\underset{\underset{H}{|}}{\overset{\overset{OH}{|}}{C}}-CH_3 \xrightarrow{[O]} CH_3-\overset{\overset{O}{\|}}{C}-CH_3 + H_2O$$

2-Propanol (isopropyl alcohol) — Propanone (dimethyl ketone; acetone)

Tertiary alcohols do not oxidize readily because there is no hydrogen atom on the carbon bonded to the —OH group. Because C—C bonds are usually too strong to oxidize, tertiary alcohols resist oxidation.

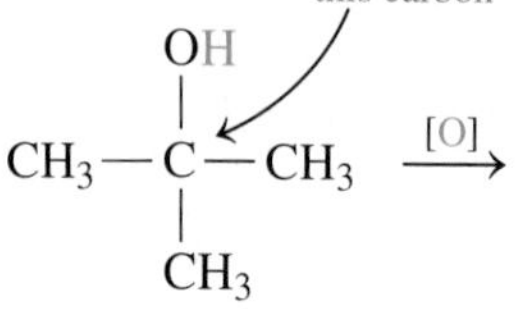

$$CH_3-\underset{\underset{CH_3}{|}}{\overset{\overset{OH}{|}}{C}}-CH_3 \xrightarrow{[O]} \text{No oxidation product readily forms}$$

No hydrogen on this carbon

No double bond forms

Alcohol (3°)

SAMPLE PROBLEM 12.9 Oxidation of Alcohols

TRY IT FIRST

Classify each of the following alcohols as primary (1°), secondary (2°), or tertiary (3°); draw the condensed structural or line-angle formula for the aldehyde or ketone formed by its oxidation:

a. 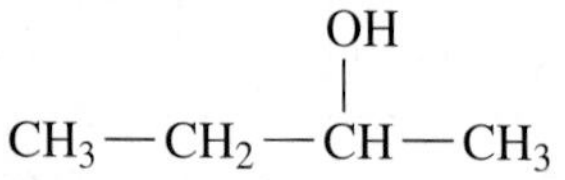 $CH_3-CH_2-\overset{\overset{OH}{|}}{CH}-CH_3$

b.  (line-angle formula: four-carbon chain ending in OH)

SOLUTION

a. This is a secondary (2°) alcohol, which can oxidize to a ketone.

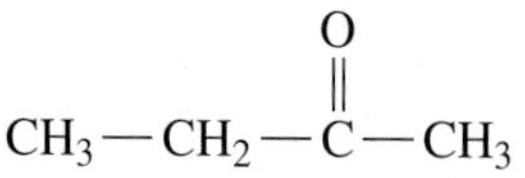

$$CH_3-CH_2-\overset{\overset{O}{\|}}{C}-CH_3$$

b. This is a primary (1°) alcohol, which can oxidize to an aldehyde.

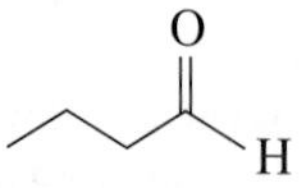

(line-angle formula: three-carbon chain ending in C(=O)H)

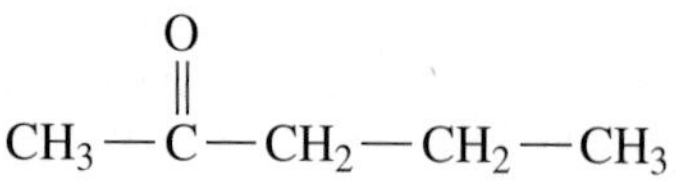

INTERACTIVE VIDEO

Oxidation of Alcohols

STUDY CHECK 12.9

Draw the condensed structural formula for the product of the oxidation of 2-pentanol.

ANSWER

$$CH_3-\overset{\overset{O}{\|}}{C}-CH_2-CH_2-CH_3$$

TEST

Try Practice Problems 12.29 to 12.32

Oxidation of Thiols

Thiols also undergo oxidation by the loss of hydrogen atoms from each of two —SH groups. The oxidized product contains a *disulfide bond* —S—S—. Much of the protein in hair is cross-linked by disulfide bonds, which occur mostly between the side chains of the amino acids of cysteine, which contain thiol groups.

$$\text{Protein Chain}—CH_2—SH + HS—CH_2—\text{Protein Chain} \xrightarrow{[O]}$$

Cysteine side groups

$$\text{Protein Chain}—CH_2—S—S—CH_2—\text{Protein Chain} + H_2O$$

Disulfide bond

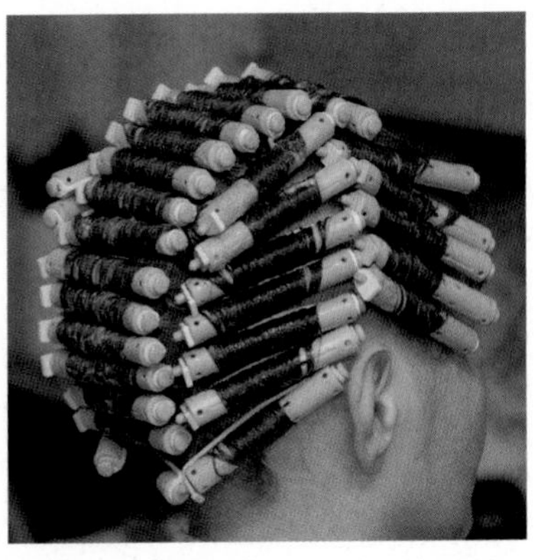

Proteins in the hair take new shapes when disulfide bonds are reduced and oxidized.

When a person has his or her hair permanently curled or straightened, a reducing substance is used to break the disulfide bonds. While the hair is wrapped around curlers or straightened, an oxidizing substance is applied that causes new disulfide bonds to form between different parts of the protein hair strands, which gives the hair a new shape.

$$CH_3—S—H + H—S—CH_3 \xrightarrow{[O]} CH_3—S—S—CH_3 + H_2O$$

Methanethiol — Dimethyl disulfide

CHEMISTRY LINK TO HEALTH

Oxidation of Alcohol in the Body

Ethanol is the most commonly abused drug in the United States. When ingested in small amounts, ethanol may produce a feeling of euphoria in the body despite the fact that it is a depressant. In the liver, enzymes such as alcohol dehydrogenase oxidize ethanol to acetaldehyde, a substance that impairs mental and physical coordination. If the blood alcohol concentration exceeds 0.4%, coma or death may occur. **TABLE 12.3** gives some of the typical behaviors exhibited at various blood alcohol levels.

$$CH_3—CH_2—OH \xrightarrow{[O]} CH_3—\overset{\overset{\large O}{\|}}{C}—H \xrightarrow{[O]} 2CO_2 + H_2O$$

Ethanol (ethyl alcohol) — Ethanal (acetaldehyde)

The acetaldehyde produced from ethanol in the liver is further oxidized to acetic acid, which is converted to carbon dioxide and water in the citric acid cycle. Thus, the enzymes in the liver can eventually break down ethanol, but the aldehyde and carboxylic acid intermediates can cause considerable damage while they are present within the cells of the liver.

A person weighing 150 lb requires about one hour to metabolize the alcohol in 12 ounces of beer. However, the rate of metabolism of ethanol varies between nondrinkers and drinkers. Typically, nondrinkers and social drinkers can metabolize 12 to 15 mg of ethanol/dL of blood in one hour, but an alcoholic can metabolize as much as 30 mg of ethanol/dL in one hour. Some effects of alcohol metabolism include an increase in liver lipids (fatty liver), gastritis, pancreatitis, ketoacidosis, alcoholic hepatitis, and psychological disturbances.

TABLE 12.3 Typical Behaviors Exhibited by a 150-lb Person Consuming Alcohol

Number of Beers (12 oz) or Glasses of Wine (5 oz) in 1 h	Blood Alcohol Level (% m/v)	Typical Behavior
1	0.025	Slightly dizzy, talkative
2	0.050	Euphoria, loud talking and laughing
4	0.10	Loss of inhibition, loss of coordination, drowsiness, legally intoxicated in most states
8	0.20	Intoxicated, quick to anger, exaggerated emotions
12	0.30	Unconscious
16 to 20	0.40 to 0.50	Coma and death

When alcohol is present in the blood, it evaporates through the lungs. Thus, the percentage of alcohol in the lungs can be used to calculate the blood alcohol concentration (BAC). Several devices are used to measure the BAC. When a Breathalyzer is used, a suspected drunk driver exhales through a mouthpiece into a solution containing the orange Cr^{6+} ion. Any alcohol present in the exhaled air is oxidized, which reduces the orange Cr^{6+} to a green Cr^{3+}.

$$CH_3—CH_2—OH + Cr^{6+} \xrightarrow{[O]} CH_3—\overset{\overset{\large O}{\|}}{C}—OH + Cr^{3+}$$

Ethanol — Orange — Ethanoic acid — Green

A Breathalyzer test is used to determine blood alcohol level.

The Alcosensor uses the oxidation of alcohol in a fuel cell to generate an electric current that is measured. The Intoxilyzer measures the amount of light absorbed by the alcohol molecules.

Sometimes alcoholics are treated with a drug called Antabuse (disulfiram), which prevents the oxidation of acetaldehyde to acetic acid. As a result, acetaldehyde accumulates in the blood, which causes nausea, profuse sweating, headache, dizziness, vomiting, and respiratory difficulties. Because of these unpleasant side effects, the person is less likely to use alcohol.

Methanol Poisoning

Methanol (methyl alcohol), CH_3OH, is a highly toxic alcohol present in products such as windshield washer fluid, Sterno, and paint strippers. Methanol is rapidly absorbed in the gastrointestinal tract. In the liver, it is oxidized to formaldehyde and then formic acid, a substance that causes nausea, severe abdominal pain, and blurred vision. Blindness can occur because the intermediate products destroy the retina of the eye. As little as 4 mL of methanol can produce blindness. The formic acid, which is not readily eliminated from the body, lowers blood pH so severely that just 30 mL of methanol can lead to coma and death.

$$\underset{\substack{\text{Methanol}\\ \text{(methyl alcohol)}}}{CH_3-OH} \xrightarrow{[O]} \underset{\substack{\text{Methanal}\\ \text{(formaldehyde)}}}{H-\overset{\overset{\displaystyle O}{\|}}{C}-H} \xrightarrow{[O]} \underset{\substack{\text{Methanoic acid}\\ \text{(formic acid)}}}{H-\overset{\overset{\displaystyle O}{\|}}{C}-OH}$$

The treatment for methanol poisoning involves giving sodium bicarbonate to neutralize the formic acid in the blood. In some cases, ethanol is given intravenously to the patient. The enzymes in the liver pick up ethanol molecules to oxidize instead of methanol molecules. This process gives time for the methanol to be eliminated via the lungs without the formation of its dangerous oxidation products.

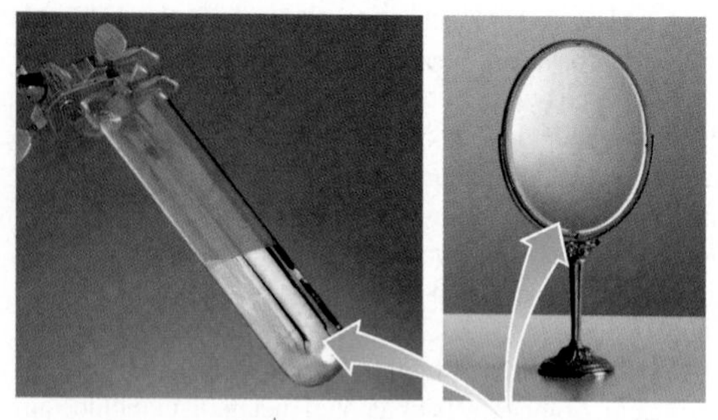

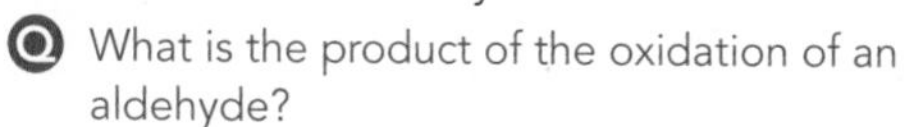

FIGURE 12.7 In Tollens' test, a silver mirror forms when the oxidation of an aldehyde reduces silver ion to metallic silver. The silvery surface of a mirror is formed in a similar way.

What is the product of the oxidation of an aldehyde?

Tollens' Test

The ease of oxidation of aldehydes allows certain mild oxidizing agents to oxidize the aldehyde functional group without oxidizing other functional groups. In the laboratory, the **Tollens' test** uses a solution of $Ag^+(AgNO_3)$ and ammonia, which oxidizes aldehydes, but not ketones. The silver ion is reduced and forms a "silver mirror" on the inside of the container. Commercially, a similar process is used to make mirrors by applying a mixture of $AgNO_3$, ammonia, and glucose on glass with a spray gun (see **FIGURE 12.7**).

$$\underset{\substack{\text{Ethanal}\\ \text{(acetaldehyde)}}}{CH_3-\overset{\overset{\displaystyle O}{\|}}{C}-H} + \underset{\substack{\text{Tollens'}\\ \text{reagent}}}{2Ag^+} \xrightarrow{[O]} \underset{\text{Silver mirror}}{2Ag(s)} + \underset{\substack{\text{Ethanoic acid}\\ \text{(acetic acid)}}}{CH_3-\overset{\overset{\displaystyle O}{\|}}{C}-OH}$$

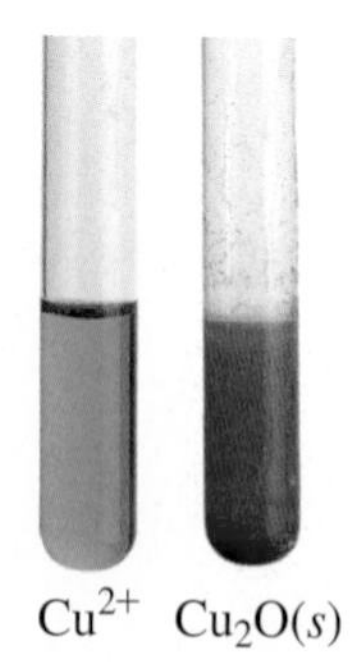

FIGURE 12.8 The blue Cu^{2+} in Benedict's solution forms a brick-red solid of Cu_2O in a positive test for many sugars and aldehydes.

Which test tube indicates that glucose is present?

Another test, called **Benedict's test**, gives a positive result with compounds that have an aldehyde functional group and an adjacent hydroxyl group. When Benedict's solution containing $Cu^{2+}(CuSO_4)$ is added to this type of aldehyde and heated, a brick-red solid of Cu_2O forms (see **FIGURE 12.8**). The test is negative with simple aldehydes and ketones.

$$\underset{\text{2-Hydroxypropanal}}{CH_3-\overset{\overset{\displaystyle OH}{|}}{CH}-\overset{\overset{\displaystyle O}{\|}}{C}-H} + \underset{\substack{\text{Benedict's}\\ \text{reagent}}}{2Cu^{2+}} \longrightarrow \underset{\substack{\text{Brick-red}\\ \text{solid}}}{Cu_2O(s)} + \underset{\text{2-Hydroxypropanoic acid}}{CH_3-\overset{\overset{\displaystyle OH}{|}}{CH}-\overset{\overset{\displaystyle O}{\|}}{C}-OH}$$

Because glucose contains an aldehyde with an adjacent hydroxyl group, Benedict's reagent can be used to determine the presence of glucose in blood or urine.

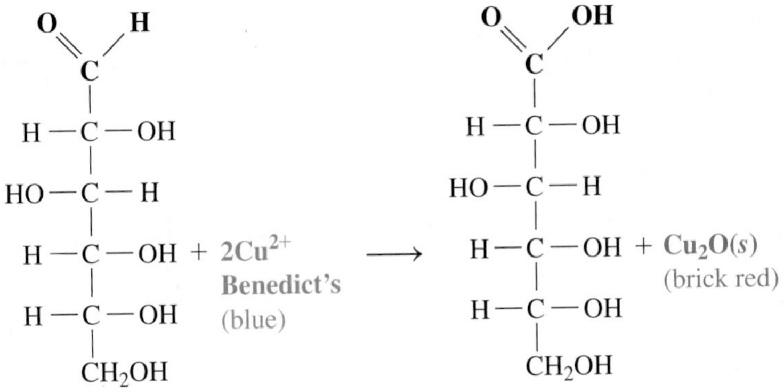

$$\text{D-Glucose (CHO–CHOH–CHOH–CHOH–CHOH–CH}_2\text{OH)} + 2Cu^{2+}\ \text{(Benedict's, blue)} \longrightarrow \text{D-Gluconic acid (COOH–CHOH–CHOH–CHOH–CHOH–CH}_2\text{OH)} + Cu_2O(s)\ \text{(brick red)}$$

Reduction of Aldehydes and Ketones

Aldehydes and ketones are reduced by sodium borohydride ($NaBH_4$) or hydrogen (H_2). In the **reduction** of organic compounds, there is a decrease in the number of carbon–oxygen bonds by the addition of hydrogen or the loss of oxygen. Aldehydes reduce to primary alcohols, and ketones reduce to secondary alcohols. A catalyst such as nickel, platinum, or palladium is needed for the addition of hydrogen to the carbonyl group.

Aldehydes Reduce to Primary Alcohols

$$CH_3-CH_2-\overset{\overset{\large O}{\|}}{C}-H + \mathbf{H_2} \xrightarrow{Pt} CH_3-CH_2-\underset{\underset{\large \mathbf{H}}{|}}{\overset{\overset{\large \mathbf{OH}}{|}}{C}}-H$$

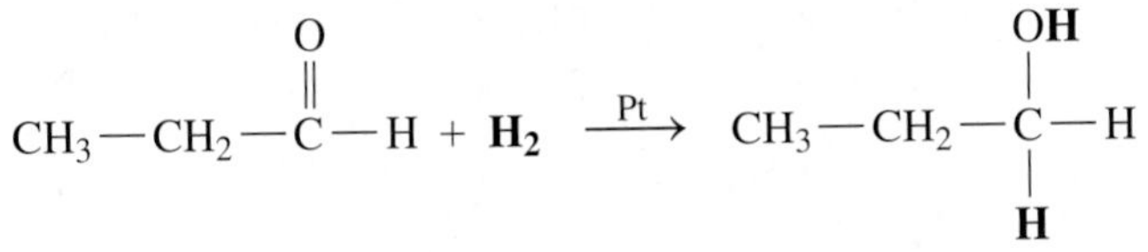

Propanal (propionaldehyde) → 1-Propanol (1° alcohol) (propyl alcohol)

Ketones Reduce to Secondary Alcohols

$$CH_3-\overset{\overset{\large O}{\|}}{C}-CH_3 + \mathbf{H_2} \xrightarrow{Ni} CH_3-\underset{\underset{\large \mathbf{H}}{|}}{\overset{\overset{\large \mathbf{OH}}{|}}{C}}-CH_3$$

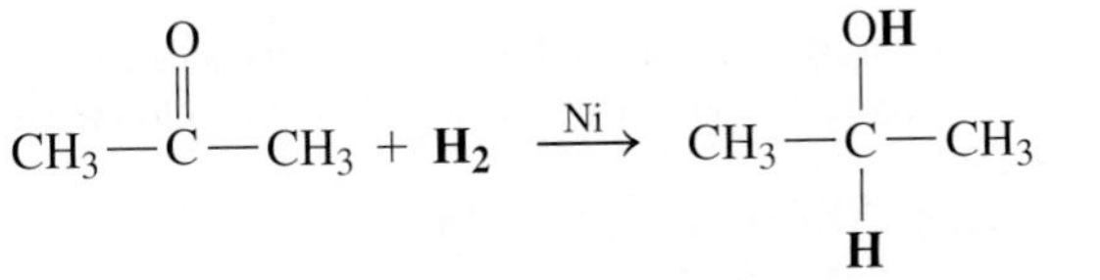

Propanone (dimethyl ketone) → 2-Propanol (2° alcohol) (isopropyl alcohol)

ENGAGE

What is the product of the reduction of a ketone?

SAMPLE PROBLEM 12.10 Reduction of Carbonyl Groups

TRY IT FIRST

Write a balanced chemical equation for the reduction of cyclopentanone using hydrogen in the presence of a nickel catalyst.

SOLUTION

The reacting molecule is a cyclic ketone that has five carbon atoms. During the reduction, hydrogen atoms add to the carbon and oxygen in the carbonyl group, which reduces the ketone to the corresponding secondary alcohol.

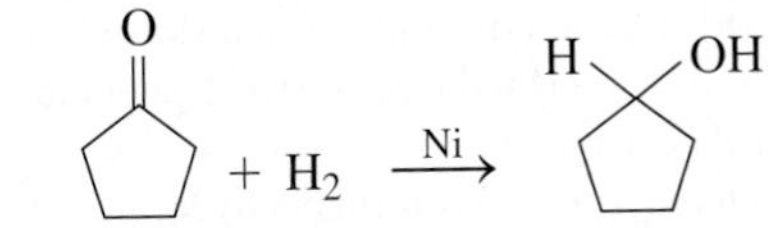

$$\text{Cyclopentanone} + H_2 \xrightarrow{Ni} \text{Cyclopentanol}$$

STUDY CHECK 12.10

Write a balanced chemical equation for the reduction of butanal using hydrogen in the presence of a platinum catalyst.

ANSWER

$$CH_3-CH_2-CH_2-\overset{\overset{\displaystyle O}{\|}}{C}-H + H_2 \xrightarrow{Pt} CH_3-CH_2-CH_2-CH_2-OH$$

TEST

Try Practice Problems 12.33 and 12.34

PRACTICE PROBLEMS

12.4 Reactions of Alcohols, Thiols, Aldehydes, and Ketones

LEARNING GOAL Write balanced chemical equations for the combustion, dehydration, and oxidation of alcohols. Write balanced chemical equations for the reduction of aldehydes and ketones.

12.25 Write the balanced chemical equation for the complete combustion of each of the following compounds:
a. methanol **b.** 2-butanol

12.26 Write the balanced chemical equation for the complete combustion of each of the following compounds:
a. 1-propanol **b.** 3-hexanol

12.27 Draw the condensed structural or line-angle formula for the alkene produced by each of the following dehydration reactions:

a. $CH_3-CH_2-CH_2-CH_2-OH \xrightarrow{H^+,\ heat}$

b. OH (cyclopentanol) $\xrightarrow{H^+,\ heat}$

c. OH (line-angle) $\xrightarrow{H^+,\ heat}$

12.28 Draw the condensed structural or line-angle formula for the alkene produced by each of the following dehydration reactions:

a. $CH_3-CH_2-OH \xrightarrow{H^+,\ heat}$

b. OH (line-angle) $\xrightarrow{H^+,\ heat}$

c. OH (cyclohexanol) $\xrightarrow{H^+,\ heat}$

12.29 Draw the condensed structural or line-angle formula for the aldehyde or ketone formed when each of the following alcohols is oxidized [O] (if no reaction, write *none*):

a. $CH_3-CH_2-CH_2-OH$

b. $CH_3-\underset{}{\overset{\overset{\displaystyle OH}{|}}{C}H}-CH_2-CH_2-CH_2-CH_3$

c. OH (cyclohexanol)

d. OH (line-angle)

12.30 Draw the condensed structural or line-angle formula for the aldehyde or ketone formed when each of the following alcohols is oxidized [O] (if no reaction, write *none*):

a. $CH_3-\overset{\overset{\displaystyle CH_3}{|}}{C}H-CH_2-OH$

b. $CH_3-CH_2-\underset{\underset{\displaystyle CH_3}{|}}{\overset{\overset{\displaystyle OH}{|}}{C}}-CH_3$

c. OH (cyclobutanol)

d. OH (line-angle)

12.31 Draw the condensed structural formulas for the aldehyde and carboxylic acid produced when each of the following is oxidized:

a. $CH_3-CH_2-CH_2-CH_2-CH_2-OH$

b. $CH_3-\overset{\overset{\displaystyle CH_3}{|}}{C}H-CH_2-CH_2-OH$

c. 1-butanol

12.32 Draw the condensed structural formulas for the aldehyde and carboxylic acid produced when each of the following is oxidized:

a. OH (cyclobutylmethanol)

b. CH_3-OH

c. 3-chloro-1-propanol

12.33 Draw the condensed structural formula for the alcohol formed when each of the following is reduced by hydrogen in the presence of a nickel catalyst:
a. butyraldehyde **b.** acetone
c. hexanal **d.** 2-methyl-3-pentanone

12.34 Draw the condensed structural formula for the alcohol formed when each of the following is reduced by hydrogen in the presence of a nickel catalyst:
a. ethyl propyl ketone **b.** formaldehyde
c. 3-chlorocyclopentanone **d.** 2-pentanone

CLINICAL UPDATE Diana's Skin Protection Plan

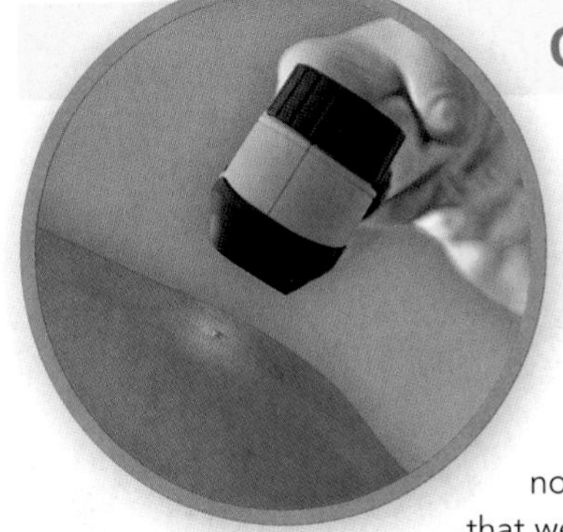

After six months, Diana returns to the dermatology office for a follow-up skin check. There has been no change in the skin where the mole was excised and no other moles were found that were suspicious. Margaret reminded Diana to limit her exposure to the sun, especially between 10 A.M. and 3 P.M., and to wear protective clothing, including a hat, a long-sleeved shirt, and pants that cover her legs. On all exposed skin, Margaret tells Diana to use broad-spectrum sunscreen with an SPF of at least 15 every day. Sunscreen absorbs UV light radiation when the skin is exposed to sunlight and thus helps protect against sunburn. The SPF (sun protection factor) number, which ranges from 2 to 100, gives the amount of time for protected skin to sunburn compared to the time for unprotected skin to sunburn. The principal ingredients in sunscreens are usually aromatic molecules with carbonyl groups such as oxybenzone and avobenzone.

Sunscreen absorbs UV light, which protects against sunburn.

Clinical Applications

12.35 Oxybenzone is an effective sunscreen whose structural formula is shown.

a. What functional groups are in oxybenzone?

b. What is the molecular formula and molar mass of oxybenzone?

c. If a bottle of sunscreen containing 178 mL has 6.0% (m/v) oxybenzone, how many grams of oxybenzone are present?

Oxybenzone

12.36 Avobenzone is a common ingredient in sunscreen. Its structural formula is shown.

a. What functional groups are in avobenzone?

b. What is the molecular formula and molar mass of avobenzone?

c. If a bottle of sunscreen containing 236 mL has 3.0% (m/v) avobenzone, how many grams of avobenzone are present?

Avobenzone

CONCEPT MAP

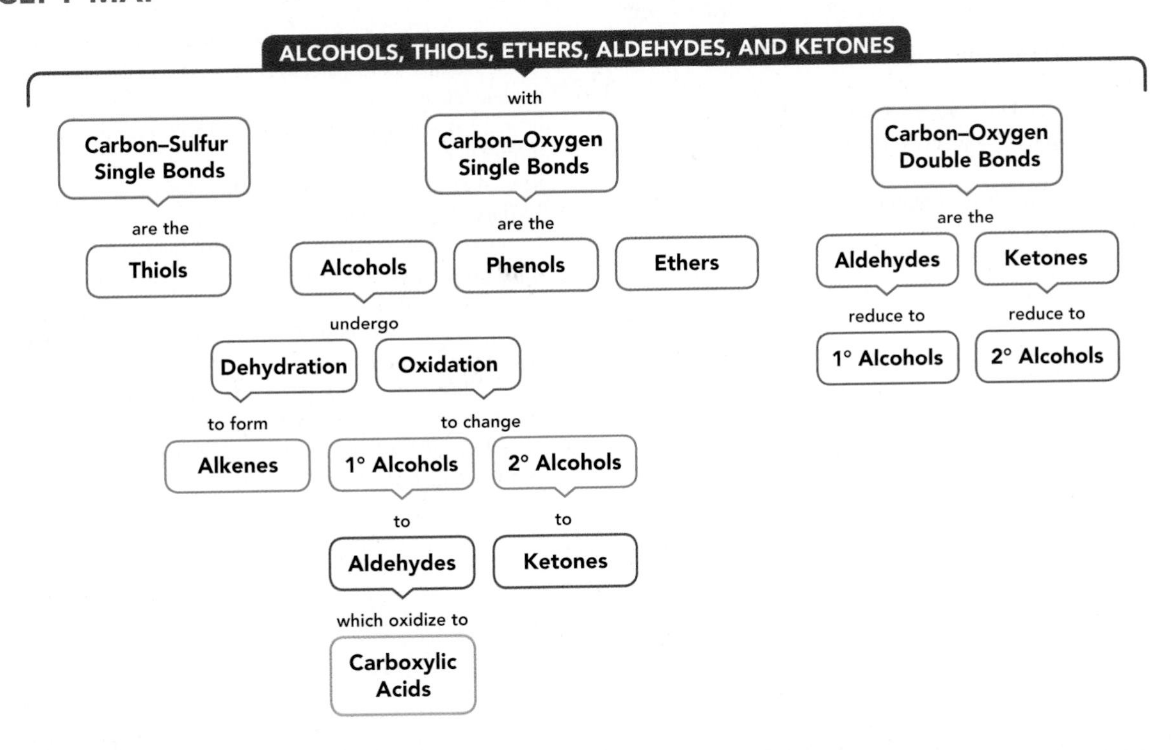

CHAPTER REVIEW

12.1 Alcohols, Phenols, Thiols, and Ethers

LEARNING GOAL Write the IUPAC and common names for alcohols, phenols, and thiols and the common names for ethers. Draw their condensed structural or line-angle formulas.

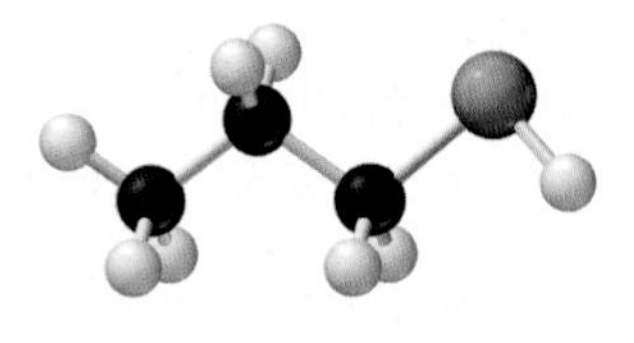

$CH_3—CH_2—CH_2—SH$
1-Propanethiol
(onions)

- The functional group of an alcohol is the hydroxyl group (—OH) bonded to a carbon chain.
- In a phenol, the hydroxyl group is bonded to an aromatic ring.
- In thiols, the functional group is —SH, which is analogous to the —OH group of alcohols.
- In the IUPAC system, the names of alcohols have *ol* endings, and the location of the —OH group is given by numbering the carbon chain.
- A cyclic alcohol is named as a cycloalkanol.
- Simple alcohols are generally named by their common names, with the alkyl name preceding the term alcohol.
- An aromatic alcohol is named as a phenol.
- In an ether, an oxygen atom is connected by single bonds to two alkyl or aromatic groups.
- In the common names of ethers, the alkyl or aromatic groups are listed alphabetically, followed by the name ether.

12.2 Properties of Alcohols

LEARNING GOAL Describe the classification of alcohols; describe the solubility of alcohols in water.

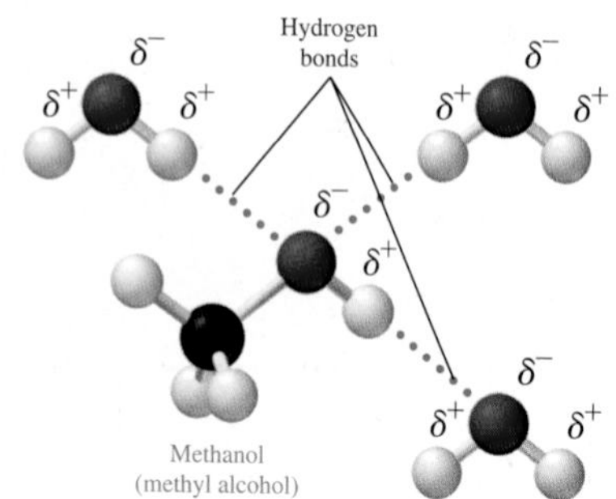

Methanol
(methyl alcohol)

- Alcohols are classified according to the number of alkyl groups bonded to the carbon that holds the —OH group.
- In a primary (1°) alcohol, one group is attached to the hydroxyl carbon.
- In a secondary (2°) alcohol, two groups are attached.
- In a tertiary (3°) alcohol, there are three groups bonded to the hydroxyl carbon.
- Short-chain alcohols can hydrogen bond with water, which makes them soluble.

12.3 Aldehydes and Ketones

LEARNING GOAL Write the IUPAC and common names for aldehydes and ketones; draw their condensed structural or line-angle formulas. Describe the solubility of aldehydes and ketones in water.

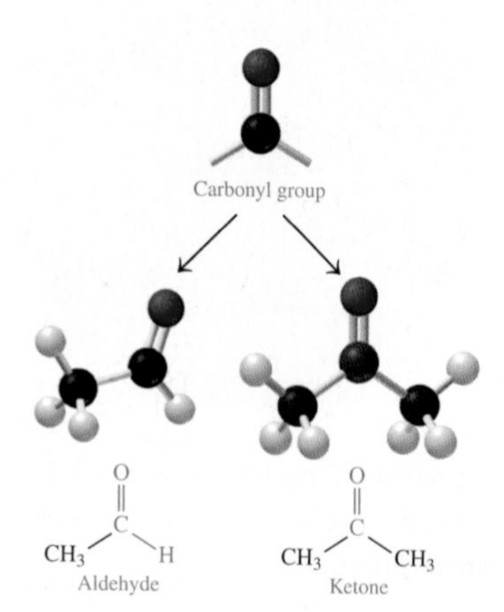

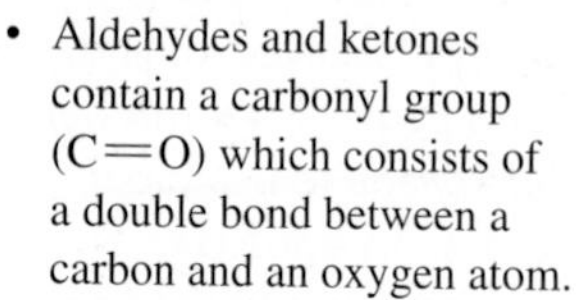

Aldehyde: $CH_3—C(=O)—H$
Ketone: $CH_3—C(=O)—CH_3$

- Aldehydes and ketones contain a carbonyl group (C=O) which consists of a double bond between a carbon and an oxygen atom.
- In aldehydes, the carbonyl group appears at the end of carbon chains attached to at least one hydrogen atom.
- In ketones, the carbonyl group occurs between two alkyl or aromatic groups.
- In the IUPAC system, the *e* in the corresponding alkane name is replaced with *al* for aldehydes and *one* for ketones.
- For ketones with more than four carbon atoms in the main chain, the carbonyl group is numbered to show its location.
- Many of the simple aldehydes and ketones use common names.
- Because they contain a polar carbonyl group, aldehydes and ketones can hydrogen bond with water molecules, which makes carbonyl compounds with one to four carbon atoms soluble in water.

12.4 Reactions of Alcohols, Thiols, Aldehydes, and Ketones

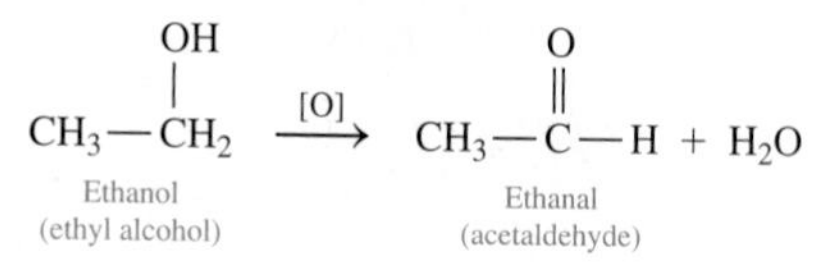

$$CH_3—CH_2—OH \xrightarrow{[O]} CH_3—C(=O)—H + H_2O$$
Ethanol (ethyl alcohol) → Ethanal (acetaldehyde)

LEARNING GOAL Write balanced chemical equations for the combustion, dehydration, and oxidation of alcohols. Write balanced chemical equations for the reduction of aldehydes and ketones.

- Alcohols undergo combustion with O_2 to form CO_2, H_2O, and energy.
- At high temperatures, alcohols dehydrate in the presence of an acid to yield alkenes.
- Primary alcohols are oxidized to aldehydes, which can oxidize further to carboxylic acids.
- Secondary alcohols are oxidized to ketones.
- Tertiary alcohols do not oxidize.
- Thiols undergo oxidation to form disulfides.
- Aldehydes are easily oxidized to carboxylic acids, but ketones do not oxidize further.
- Aldehydes, but not ketones, are oxidized by Tollens' reagent to give silver mirrors.
- In Benedict's test, aldehydes with adjacent hydroxyl groups reduce blue Cu^{2+} to give a brick-red Cu_2O solid.
- Aldehydes and ketones can be reduced with H_2 in the presence of a catalyst to give alcohols.

SUMMARY OF NAMING

Family	Structure	IUPAC Name	Common Name
Alcohol	CH_3-OH	Methanol	Methyl alcohol
Phenol	C_6H_5-OH (benzene ring—OH)	Phenol	Phenol
Thiol	CH_3-SH	Methanethiol	
Ether	CH_3-O-CH_3		Dimethyl ether
Aldehyde	$H-\overset{O}{\overset{\Vert}{C}}-H$	Methanal	Formaldehyde
Ketone	$CH_3-\overset{O}{\overset{\Vert}{C}}-CH_3$	Propanone	Acetone; dimethyl ketone

SUMMARY OF REACTIONS

The chapter Sections to review are shown after the name of each reaction.

Combustion of Alcohols (12.4)

$$\underset{\text{Ethanol}}{CH_3-CH_2-OH} + \underset{\text{Oxygen}}{3O_2} \xrightarrow{\Delta} \underset{\text{Carbon dioxide}}{2CO_2} + \underset{\text{Water}}{3H_2O} + \text{energy}$$

Dehydration of Alcohols to Form Alkenes (12.4)

$$\underset{\text{1-Propanol}}{CH_3-CH_2-CH_2-OH} \xrightarrow{H^+,\ \text{heat}} \underset{\text{Propene}}{CH_3-CH{=}CH_2} + H_2O$$

Oxidation of Primary Alcohols to Form Aldehydes (12.4)

$$\underset{\text{Ethanol}}{CH_3-\overset{OH}{\overset{|}{CH_2}}} \xrightarrow{[O]} \underset{\text{Ethanal}}{CH_3-\overset{O}{\overset{\Vert}{C}}-H} + H_2O$$

Oxidation of Secondary Alcohols to Form Ketones (12.4)

$$\underset{\text{2-Propanol}}{CH_3-\overset{OH}{\overset{|}{CH}}-CH_3} \xrightarrow{[O]} \underset{\text{Propanone}}{CH_3-\overset{O}{\overset{\Vert}{C}}-CH_3} + H_2O$$

Oxidation of Aldehydes to Form Carboxylic Acids (12.4)

$$\underset{\text{Ethanal}}{CH_3-\overset{O}{\overset{\Vert}{C}}-H} \xrightarrow{[O]} \underset{\text{Ethanoic acid}}{CH_3-\overset{O}{\overset{\Vert}{C}}-OH}$$

Reduction of Aldehydes to Form Primary Alcohols (12.4)

$$\underset{\text{Ethanal}}{CH_3-\overset{O}{\overset{\Vert}{C}}-H} + H_2 \xrightarrow{Pt} \underset{\text{Ethanol}}{CH_3-\overset{OH}{\overset{|}{CH_2}}}$$

Reduction of Ketones to Form Secondary Alcohols (12.4)

$$\underset{\text{Propanone}}{CH_3-\overset{O}{\overset{\Vert}{C}}-CH_3} + H_2 \xrightarrow{Ni} \underset{\text{2-Propanol}}{CH_3-\overset{OH}{\overset{|}{CH}}-CH_3}$$

KEY TERMS

alcohol An organic compound that contains the hydroxyl functional group (—OH) attached to a carbon chain.

aldehyde An organic compound with a carbonyl functional group bonded to at least one hydrogen.

$$-\overset{O}{\overset{\Vert}{C}}-H = -CHO$$

Benedict's test A test for aldehydes with adjacent hydroxyl groups in which Cu^{2+} ($CuSO_4$) ions in Benedict's reagent are reduced to a brick-red solid of Cu_2O.

carbonyl group A functional group that contains a carbon–oxygen double bond (C=O).

dehydration A reaction that removes water from an alcohol, in the presence of an acid, to form alkenes at high temperatures.

ether An organic compound in which an oxygen atom is bonded to two carbon groups that are alkyl or aromatic.

ketone An organic compound in which the carbonyl functional group is bonded to two alkyl or aromatic groups.

$$-\overset{O}{\overset{\Vert}{C}}- = -CO-$$

oxidation The loss of two hydrogen atoms from a reactant to give a more oxidized compound: primary alcohols oxidize to aldehydes, secondary alcohols oxidize to ketones. An oxidation can also be the addition of an oxygen atom or an increase in the number of carbon–oxygen bonds.

phenol An organic compound that has a hydroxyl group (—OH) attached to a benzene ring.

primary (1°) alcohol An alcohol that has one alkyl group bonded to the carbon atom with the —OH group.

reduction A gain of hydrogen when a ketone reduces to a 2° alcohol and an aldehyde reduces to a 1° alcohol.

secondary (2°) alcohol An alcohol that has two alkyl groups bonded to the carbon atom with the —OH group.

tertiary (3°) alcohol An alcohol that has three alkyl groups bonded to the carbon atom with the —OH group.

thiol An organic compound that contains a thiol group (—SH).

Tollens' test A test for aldehydes in which Ag^+ ions in Tollens' reagent is reduced to metallic silver forming a "silver mirror" on the walls of the container.

CORE CHEMISTRY SKILLS

The chapter Section containing each Core Chemistry Skill is shown in parentheses at the end of each heading.

Identifying Functional Groups (12.1)

- Functional groups are specific groups of atoms in organic compounds, which undergo characteristic chemical reactions.
- Organic compounds with the same functional group have similar properties and reactions.

Example: Identify the functional group in the following:

$$CH_3—CH_2—CH_2—CH_2—OH$$

Answer: The hydroxyl group (—OH) makes it an alcohol.

Naming Alcohols and Phenols (12.1)

- In the IUPAC system, an alcohol is named by replacing the *e* in the alkane name with *ol*.
- Simple alcohols are named with the alkyl name preceding the term *alcohol*.
- The carbon chain is numbered from the end nearer the —OH group and the location of the —OH is given in front of the name.
- A cyclic alcohol is named as a cycloalkanol.
- An aromatic alcohol is named as a phenol.

Example: Give the IUPAC name for

```
        CH3
        |
CH3 — CH — CH2 — OH
```

Answer: 2-methyl-1-propanol

Naming Aldehydes and Ketones (12.3)

- In the IUPAC system, an aldehyde is named by replacing the *e* in the alkane name with *al* and a ketone by replacing the *e* with *one*.
- The position of a substituent on an aldehyde is indicated by numbering the carbon chain from the carbonyl group and given in front of the name.
- For a ketone, the carbon chain is numbered from the end nearer the carbonyl group.

Example: Give the IUPAC name for

```
        CH3          O
        |            ||
CH3 — CH — CH2 — C — CH3
```

Answer: 4-methyl-2-pentanone

Writing Equations for the Dehydration of Alcohols (12.4)

- At high temperatures, alcohols dehydrate in the presence of an acid to yield alkenes.

Example: Draw the condensed structural formula for the organic product from the dehydration of 2-methyl-1-butanol.

```
          CH3
          |
H2C == C — CH2 — CH3
```

Answer: $H_2C{=}C(CH_3)—CH_2—CH_3$

Writing Equations for the Oxidation of Alcohols (12.4)

- Primary alcohols oxidize to aldehydes, which can oxidize further to carboxylic acids.
- Secondary alcohols oxidize to ketones.
- Tertiary alcohols do not oxidize.

Example: Draw the condensed structural formula for the product of the oxidation of 2-butanol.

```
        O
        ||
CH3 — C — CH2 — CH3
```

Answer: $CH_3—C(=O)—CH_2—CH_3$

UNDERSTANDING THE CONCEPTS

The chapter Sections to review are shown in parentheses at the end of each problem.

12.37 Gingerol is a pungent compound found in ginger. Identify the functional groups in gingerol. (12.1, 12.3)

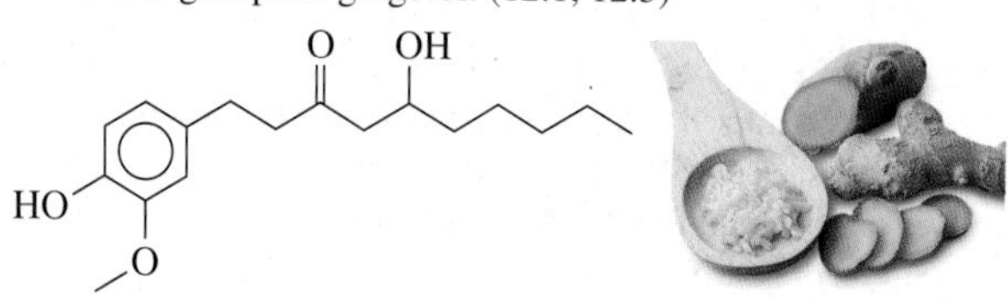

Gingerol

12.38 The compound frambinone has the taste of raspberries and has been used in weight loss. Identify the functional groups in frambinone. (12.1, 12.3)

```
                                     O
                                     ||
HO —⟨benzene⟩— CH2 — CH2 — C — CH3
```

Frambinone

12.39 A compound called resveratrol is an antioxidant, found in the skin of grapes. Identify the functional groups in resveratrol. (12.1, 12.3)

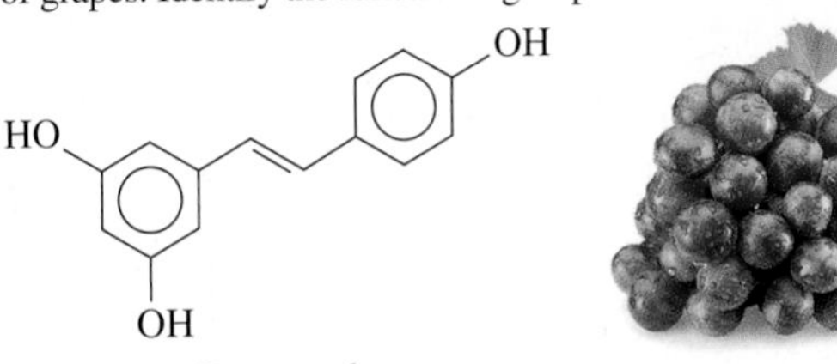

Resveratrol

12.40 A compound called cinnamaldehyde is found in cinnamon. Identify the functional groups in cinnamaldehyde. (12.1, 12.3)

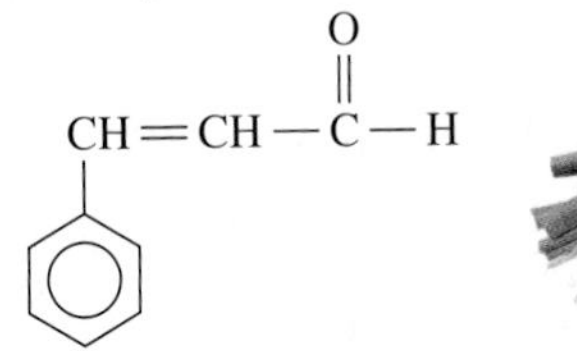

Cinnamaldehyde

12.41 Which of the following will give a positive Tollens' test? (12.4)

a. propanal
b. ethanol
c. ethyl methyl ether

12.42 Which of the following will give a positive Tollens' test? (12.4)

a. 1-propanol
b. 2-propanol
c. hexanal

ADDITIONAL PRACTICE PROBLEMS

12.43 Classify each of the following alcohols as primary (1°), secondary (2°), or tertiary (3°): (12.2)

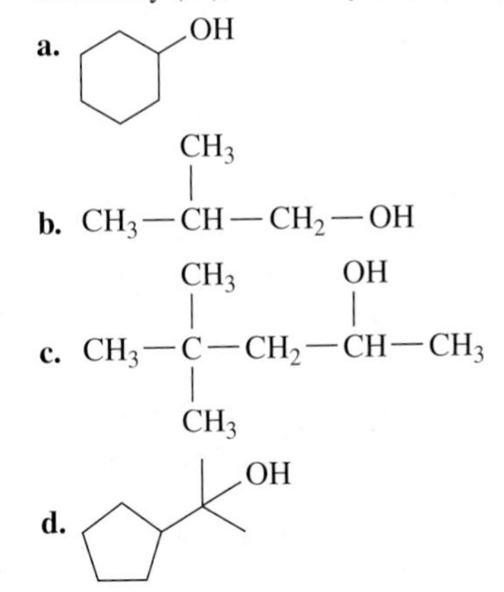

12.44 Classify each of the following alcohols as primary (1°), secondary (2°), or tertiary (3°): (12.2)

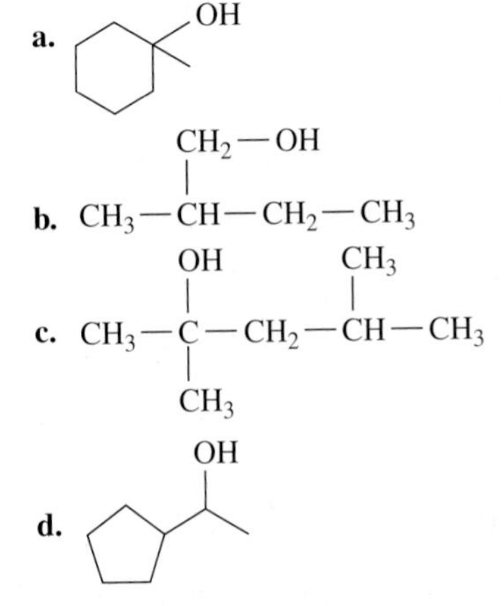

12.45 Give the IUPAC name for each of the following alcohols and phenols: (12.1)

a.

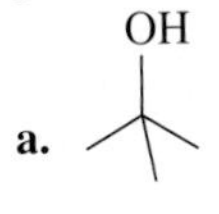

b. 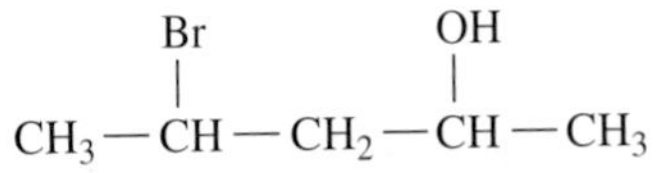 CH_3—CH(Br)—CH_2—CH(OH)—CH_3

c.

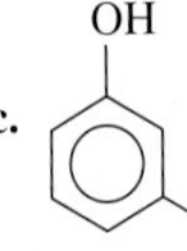

12.46 Give the IUPAC name for each of the following alcohols and phenols: (12.1)

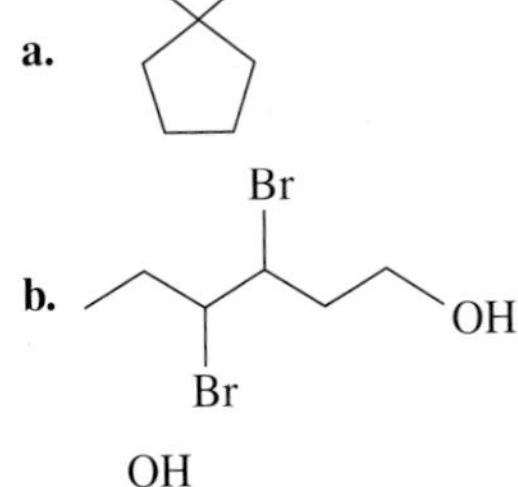

c.

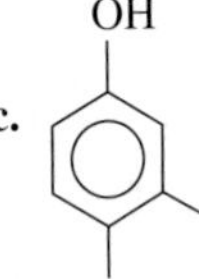

12.47 Draw the condensed structural or line-angle formula for each of the following: (12.1)

a. 4-chlorophenol
b. 2-methyl-3-pentanol
c. 2-methyl-1-propanol
d. 3-methyl-2-butanol

12.48 Draw the condensed structural or line-angle formula for each of the following: (12.1)

a. 3-methyl-1-hexanol
b. 2-pentanol
c. methyl propyl ether
d. 2,4-dibromophenol

12.49 Which compound in each pair would be more soluble in water? Explain. (12.2)

a. butane or 1-propanol
b. 2-propanol or diethyl ether
c. ethanol or 1-hexanol

12.50 Which compound in each pair would be more soluble in water? Explain. (12.2)

a. ethane or ethanol **b.** 2-propanol or 2-pentanol
c. methyl propyl ether or 1-butanol

12.51 Draw the condensed structural or line-angle formula for the alkene, aldehyde, or ketone product of each of the following reactions: (12.4)

a. $CH_3-CH_2-CH_2-OH \xrightarrow{H^+, \text{ heat}}$

b. 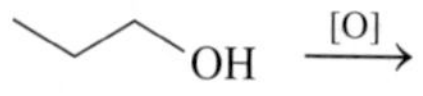 (line-angle structure) OH $\xrightarrow{[O]}$

c. $CH_3-CH_2-CH(OH)-CH_2-CH_3 \xrightarrow{H^+, \text{ heat}}$

d. 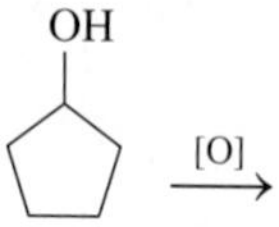 (cyclopentanol) OH $\xrightarrow{[O]}$

12.52 Draw the condensed structural or line-angle formula for the alkene, aldehyde, or ketone product of each of the following reactions: (12.4)

a. 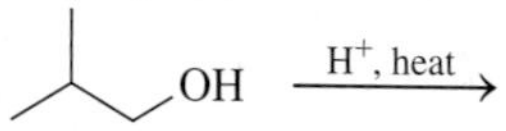 (line-angle structure) OH $\xrightarrow{H^+, \text{ heat}}$

b. $CH_3-CH(CH_3)-CH(OH)-CH_3 \xrightarrow{[O]}$

c. 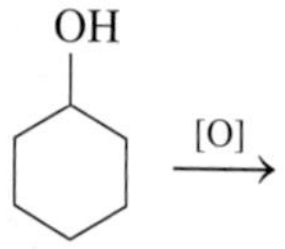 (cyclohexanol) OH $\xrightarrow{[O]}$

d. $CH_3-CH_2-CH_2-CH(OH)-CH_3 \xrightarrow{[O]}$

12.53 Draw the condensed structural or line-angle formula for the alcohol produced when hydrogen and a nickel catalyst reduce each of the following: (12.4)

a. $CH_3-CH_2-\overset{O}{\overset{\|}{C}}-CH_3$

b. 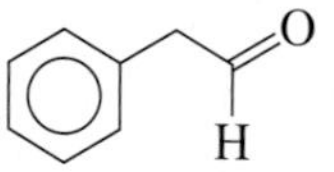 (benzene ring)–CH₂–C(=O)H

c. (cyclopropyl)–C(=O)–

12.54 Draw the condensed structural or line-angle formula for the alcohol produced when hydrogen and a nickel catalyst reduce each of the following: (12.4)

a. $CH_3-\overset{O}{\overset{\|}{C}}-H$

b. 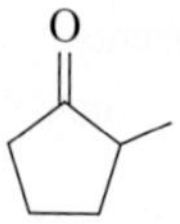 (2-methylcyclopentanone line-angle structure) O

c. $H-\overset{O}{\overset{\|}{C}}-H$

12.55 Give the IUPAC name for each of the following: (12.3)

a. 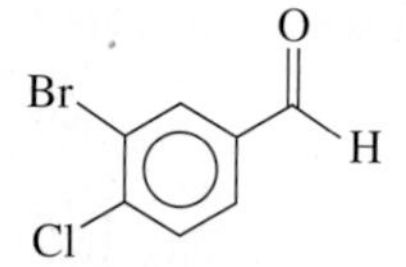 (benzene ring with Br, Cl, and C(=O)H)

b. $Cl-CH_2-CH_2-\overset{O}{\overset{\|}{C}}-H$

c. 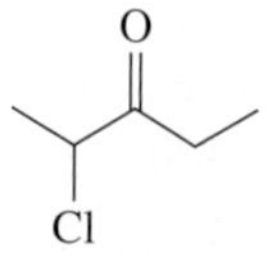 (line-angle structure with O and Cl)

12.56 Give the IUPAC name for each of the following: (12.3)

a. 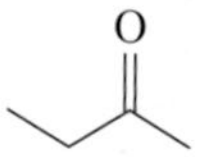 (line-angle structure with O)

b. 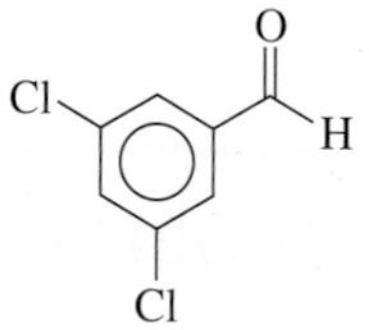 (benzene ring with two Cl and C(=O)H)

c. $CH_3-CH(CH_3)-CH(CH_3)-CH_2-\overset{O}{\overset{\|}{C}}-H$

12.57 Draw the condensed structural or line-angle formula, if cyclic, for each of the following: (12.3)

a. 4-chlorobenzaldehyde **b.** 3-chloropropionaldehyde
c. ethyl methyl ketone **d.** 3-methylhexanal

12.58 Draw the condensed structural or line-angle formula, if cyclic, for each of the following: (12.3)

a. formaldehyde **b.** 2-chlorobutanal
c. 3-methyl-2-hexanone **d.** 3,5-dimethylhexanal

12.59 Which of the following aldehydes or ketones are soluble in water? (12.3)

a. $CH_3-CH_2-\overset{O}{\overset{\|}{C}}-H$

b. $CH_3-\overset{O}{\overset{\|}{C}}-CH_3$

c. $CH_3-CH_2-\overset{O}{\overset{\|}{C}}-CH_2-CH_2-CH_3$

12.60 Which of the following aldehydes or ketones are soluble in water? (12.3)

a. $CH_3-CH_2-\overset{O}{\overset{\|}{C}}-CH_3$

b. $CH_3-\overset{O}{\overset{\|}{C}}-H$

c. $CH_3-CH_2-CH(CH_3)-CH_2-CH_2-\overset{O}{\overset{\|}{C}}-H$

12.61 Draw the condensed structural or line-angle formula for the ketone or carboxylic acid product when each of the following is oxidized: (12.4)

a. $CH_3-CH_2-CH_2-OH$

b. 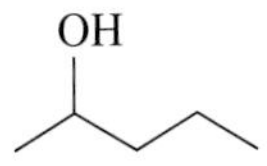 OH

c. $CH_3-CH_2-CH_2-\overset{\overset{\displaystyle O}{\|}}{C}-H$

d. 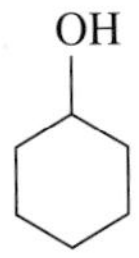 OH

12.62 Draw the condensed structural or line-angle formula for the ketone or carboxylic acid product when each of the following is oxidized: (12.4)

a. $CH_3-CH_2-CH_2-CH_2-OH$

b. $CH_3-CH_2-\overset{\overset{\displaystyle OH}{|}}{CH}-CH_3$

c. 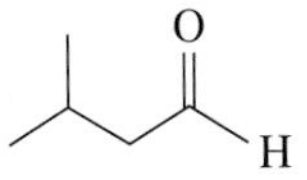 O H

d. 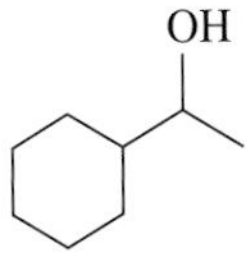 OH

CHALLENGE PROBLEMS

The following problems are related to the topics in this chapter. However, they do not all follow the chapter order, and they require you to combine concepts and skills from several Sections. These problems will help you increase your critical thinking skills and prepare for your next exam.

12.63 Draw the condensed structural formulas and give the IUPAC names for all the alcohols that have the formula $C_5H_{12}O$. (12.1)

12.64 Draw the condensed structural formulas and give the IUPAC names for all the aldehydes and ketones that have the formula $C_5H_{10}O$. (12.3)

12.65 A compound with the formula C_4H_8O is synthesized from 2-methyl-1-propanol and oxidizes easily to give a carboxylic acid. Draw the condensed structural formula and give the IUPAC name for the compound. (12.3, 12.4)

12.66 A compound with the formula $C_5H_{10}O$ oxidizes to give 3-pentanone. Draw the condensed structural formula and give the IUPAC name for the compound. (12.3, 12.4)

12.67 Compound **A** is a primary alcohol whose formula is C_3H_8O. When compound **A** is heated with strong acid, it dehydrates to form compound **B** (C_3H_6). When compound **A** is oxidized, compound **C** (C_3H_6O) forms. Draw the condensed structural formulas and give the IUPAC names for compounds **A**, **B**, and **C**. (12.3, 12.4)

12.68 Compound **X** is a secondary alcohol whose formula is C_3H_8O. When compound **X** is heated with strong acid, it dehydrates to form compound **Y** (C_3H_6). When compound **X** is oxidized, compound **Z** (C_3H_6O) forms, which cannot be oxidized further. Draw the condensed structural formulas and give the IUPAC names for compounds **X**, **Y**, and **Z**. (12.3, 12.4)

ANSWERS

Answers to Selected Practice Problems

12.1 **a.** ethanol **b.** 2-butanol
c. 2-pentanol **d.** 3-bromophenol

12.3 **a.** $CH_3-CH_2-CH_2-OH$

b. $CH_3-CH_2-\overset{\overset{\displaystyle SH}{|}}{CH}-CH_2-CH_3$

c. $CH_3-\underset{\underset{\displaystyle CH_3}{|}}{\overset{\overset{\displaystyle OH}{|}}{C}}-CH_2-CH_3$

d.

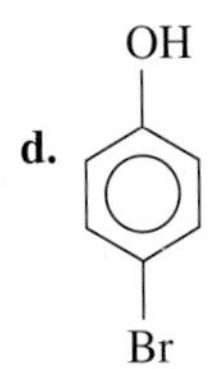

12.5 **a.** dipropyl ether **b.** cyclohexyl methyl ether
c. butyl propyl ether

12.7 **a.** $CH_3-CH_2-O-CH_3$

b. 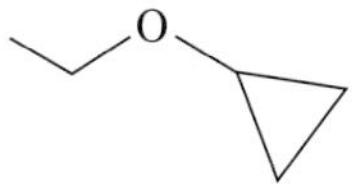 O

12.9 **a.** 1° **b.** 1° **c.** 3° **d.** 2°

12.11 **a.** soluble **b.** insoluble

12.13 **a.** Methanol can form hydrogen bonds with water, but ethane cannot.

b. 1-Propanol is more soluble because it can form more hydrogen bonds.

12.15 **a.** ketone **b.** aldehyde
c. ketone **d.** aldehyde

12.17 **a.** acetaldehyde **b.** methyl propyl ketone
c. formaldehyde

12.19 **a.** propanal
b. 2-methyl-3-pentanone
c. 3,4-dimethylcyclohexanone
d. 2-bromobenzaldehyde

12.21 a. $CH_3-\overset{O}{\overset{\|}{C}}-H$

b. $CH_3-\overset{O}{\overset{\|}{C}}-CH_2-\overset{Cl}{\overset{|}{CH}}-CH_3$

c. $CH_3-\overset{O}{\overset{\|}{C}}-CH_2-CH_2-CH_2-CH_3$

d. $CH_3-CH_2-\overset{CH_3}{\overset{|}{CH}}-CH_2-\overset{O}{\overset{\|}{C}}-H$

12.23 a. $CH_3-\overset{O}{\overset{\|}{C}}-\overset{O}{\overset{\|}{C}}-CH_2-CH_3$; more hydrogen bonding

b. acetone; lower number of carbon atoms

c. propanal; lower number of carbon atoms

12.25 a. $2CH_4O + 3O_2 \longrightarrow 2CO_2 + 4H_2O$ + energy

b. $C_4H_{10}O + 6O_2 \longrightarrow 4CO_2 + 5H_2O$ + energy

12.27 a. $CH_3-CH_2-CH{=}CH_2$

b.

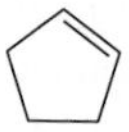

c.

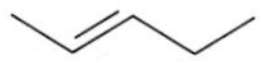

12.29 a. $CH_3-CH_2-\overset{O}{\overset{\|}{C}}-H$

b. $CH_3-\overset{O}{\overset{\|}{C}}-CH_2-CH_2-CH_2-CH_3$

c. 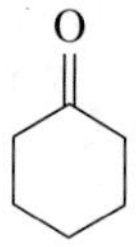 O

d. 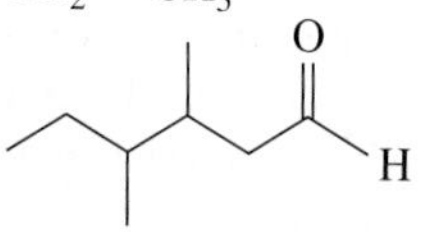 O H

12.31 a. $CH_3-CH_2-CH_2-CH_2-\overset{O}{\overset{\|}{C}}-H$;

$CH_3-CH_2-CH_2-CH_2-\overset{O}{\overset{\|}{C}}-OH$

b. $CH_3-\overset{CH_3}{\overset{|}{CH}}-CH_2-\overset{O}{\overset{\|}{C}}-H$;

$CH_3-\overset{CH_3}{\overset{|}{CH}}-CH_2-\overset{O}{\overset{\|}{C}}-OH$

c. $CH_3-CH_2-CH_2-\overset{O}{\overset{\|}{C}}-H$;

$CH_3-CH_2-CH_2-\overset{O}{\overset{\|}{C}}-OH$

12.33 a. $CH_3-CH_2-CH_2-CH_2-OH$

b. $CH_3-\overset{OH}{\overset{|}{CH}}-CH_3$

c. $CH_3-CH_2-CH_2-CH_2-CH_2-CH_2-OH$

d. $CH_3-\overset{CH_3}{\overset{|}{CH}}-\overset{OH}{\overset{|}{CH}}-CH_2-CH_3$

12.35 a. ether, phenol, ketone, aromatic

b. $C_{14}H_{12}O_3$, 228.2 g/mole

c. 11 g of oxybenzone

12.37 phenol, ether, ketone, alcohol

12.39 phenol, alkene

12.41 Compound **a** will give a positive Tollens' test.

12.43 a. 2° **b.** 1°

c. 2° **d.** 3°

12.45 a. 2-methyl-2-propanol **b.** 4-bromo-2-pentanol

c. 3-methylphenol

12.47 a.  OH, Cl

b. $CH_3-\overset{CH_3}{\overset{|}{CH}}-\overset{OH}{\overset{|}{CH}}-CH_2-CH_3$

c. $CH_3-\overset{CH_3}{\overset{|}{CH}}-CH_2-OH$

d. $CH_3-\overset{OH}{\overset{|}{CH}}-\overset{CH_3}{\overset{|}{CH}}-CH_3$

12.49 a. 1-Propanol because it can form hydrogen bonds with its polar hydroxyl group.

b. 2-Propanol because an alcohol forms more hydrogen bonds than an ether.

c. Ethanol because alcohols with one to four carbons are soluble in water.

12.51 a. $CH_3-CH{=}CH_2$

b. 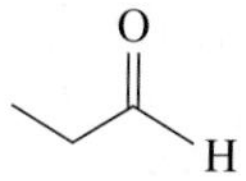 O H

c. $CH_3-CH{=}CH-CH_2-CH_3$

d.  O

12.53 a. $CH_3-CH_2-\overset{OH}{\overset{|}{CH}}-CH_3$

b. 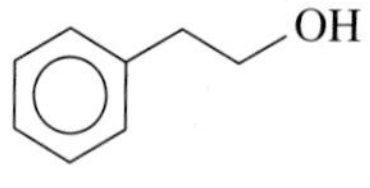 OH

c.  OH

12.55 a. 3-bromo-4-chlorobenzaldehyde

b. 3-chloropropanal

c. 2-chloro-3-pentanone

12.57 a. 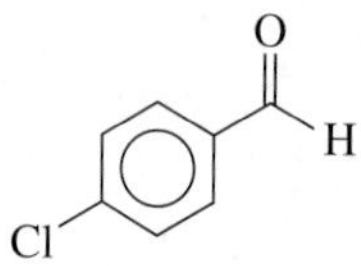 O H Cl

b. $Cl-CH_2-CH_2-\overset{O}{\overset{\|}{C}}-H$

c. $CH_3-CH_2-C(=O)-CH_3$

d. $CH_3-CH_2-CH_2-CH(CH_3)-CH_2-C(=O)-H$

12.59 **a** and **b**

12.61 a. $CH_3-CH_2-C(=O)-OH$

b. 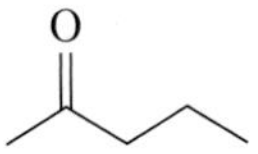 O

c. $CH_3-CH_2-CH_2-C(=O)-OH$

d. 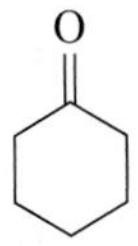 O

12.63 $CH_3-CH_2-CH_2-CH_2-CH_2-OH$ 1-Pentanol

$CH_3-CH(OH)-CH_2-CH_2-CH_3$ 2-Pentanol

$CH_3-CH_2-CH(OH)-CH_2-CH_3$ 3-Pentanol

$HO-CH_2-CH(CH_3)-CH_2-CH_3$ 2-Methyl-1-butanol

$HO-CH_2-CH_2-CH(CH_3)-CH_3$ 3-Methyl-1-butanol

$CH_3-C(CH_3)(OH)-CH_2-CH_3$ 2-Methyl-2-butanol

$CH_3-CH(OH)-CH(CH_3)-CH_3$ 3-Methyl-2-butanol

$CH_3-C(CH_3)_2-CH_2-OH$ 2,2-Dimethyl-1-propanol

12.65 $CH_3-CH(CH_3)-C(=O)-H$ 2-Methylpropanal

12.67 $CH_3-CH_2-CH_2-OH$ 1-Propanol
A

$CH_3-CH=CH_2$ Propene
B

$CH_3-CH_2-C(=O)-H$ Propanal
C

COMBINING IDEAS from Chapters 10 to 12

CI.21 A metal (M) completely reacts with 34.8 mL of a 0.520 M HCl solution to form $MCl_3(aq)$ and $H_2(g)$. (4.1, 7.2, 7.4, 7.7, 8.6, 10.7)

When a metal reacts with a strong acid, bubbles of hydrogen gas form.

a. Write the balanced chemical equation for the reaction of the metal M(*s*) and HCl(*aq*).
b. What volume, in milliliters, of H_2 is produced at STP?
c. How many moles of metal M reacted?
d. If the metal has a mass of 0.420 g, use your results from part **c** to determine the molar mass of the metal M.
e. What are the name and symbol of metal M in part **d**?
f. Write the balanced chemical equation using the symbol of the metal from part **e**.

CI.22 In a teaspoon (5.0 mL) of a common liquid antacid, there are 400. mg of $Mg(OH)_2$ and 400. mg of $Al(OH)_3$. A 0.080 M HCl solution, which is similar to stomach acid, is used to neutralize 5.0 mL of the liquid antacid. (9.5, 10.6, 10.7)

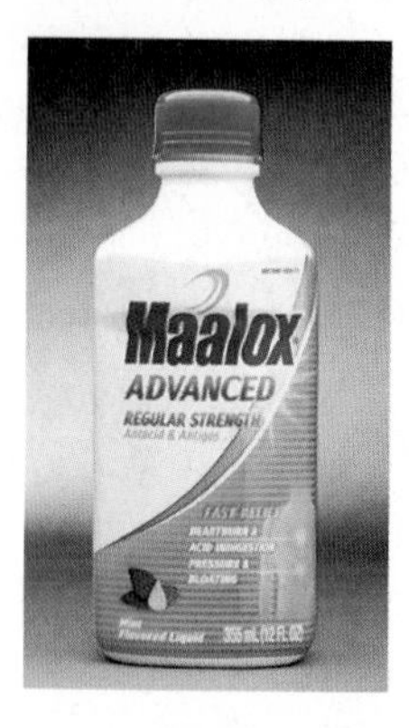

An antacid neutralizes stomach acid.

a. Write the balanced chemical equation for the neutralization of HCl and $Mg(OH)_2$.
b. Write the balanced chemical equation for the neutralization of HCl and $Al(OH)_3$.
c. What is the pH of the HCl solution?
d. How many milliliters of the HCl solution are needed to neutralize the $Mg(OH)_2$?
e. How many milliliters of the HCl solution are needed to neutralize the $Al(OH)_3$?

CI.23 Acetone (propanone), a clear liquid solvent with an acrid odor, is used to remove nail polish, paints, and resins. It has a low boiling point and is highly flammable. Acetone has a density of 0.786 g/mL. (6.7, 7.2, 7.4, 7.8, 12.3)

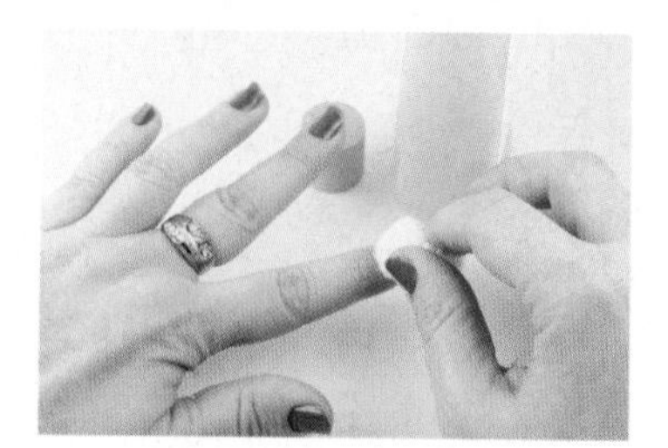
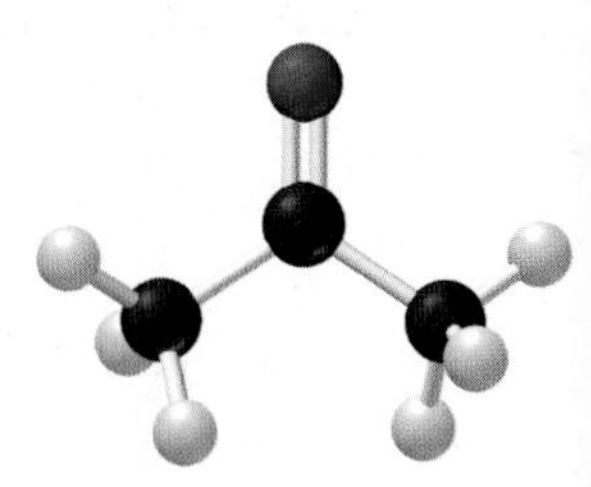

Acetone is a solvent used in nail polish remover.

a. What is the molecular formula of acetone?
b. What is the molar mass of acetone?
c. Identify the bonds C—C, C—H, and C—O in a molecule of acetone as polar covalent or nonpolar covalent.
d. Write the balanced chemical equation for the combustion of acetone.
e. How many grams of oxygen gas are needed to react with 15.0 mL of acetone?

CI.24 One of the components of gasoline is octane, C_8H_{18}, which has a density of 0.803 g/mL. The combustion of 1 mole of octane provides 5510 kJ. A hybrid car has a fuel tank with a capacity of 11.9 gal and a gas mileage of 45 mi/gal. (7.1, 7.3, 7.8, 11.2)

Octane is one of the components of gasoline.

a. Write the balanced chemical equation for the combustion of octane.
b. Is the combustion of octane endothermic or exothermic?
c. How many moles of octane are in one tank of fuel, assuming it is all octane?
d. If the total mileage of this hybrid car for one year is 24 500 mi, how many kilograms of carbon dioxide would be produced from the combustion of the fuel, assuming it is all octane?

CI.25 Butyraldehyde is a clear liquid solvent with an unpleasant odor. It has a low boiling point and is highly flammable. Butyraldehyde has a density of 0.802 g/mL. (7.4, 7.8, 8.6, 12.3, 12.4)

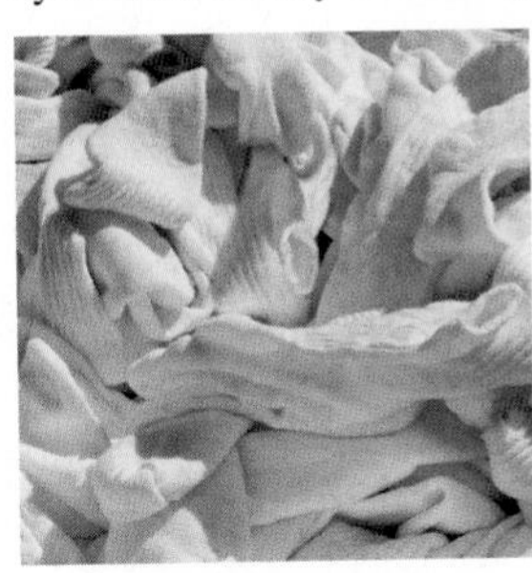

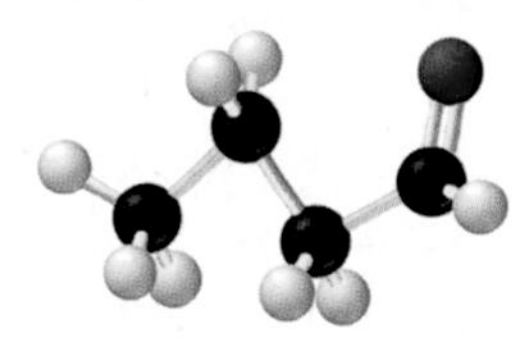

The unpleasant odor of old gym socks is due to butyraldehyde.

a. Draw the condensed structural formula for butyraldehyde.
b. Draw the line-angle formula for butyraldehyde.
c. What is the IUPAC name of butyraldehyde?
d. Draw the condensed structural formula for the alcohol that is produced when butyraldehyde is reduced.
e. Write the balanced chemical equation for the complete combustion of butyraldehyde.
f. How many grams of oxygen gas are needed to completely react with 15.0 mL of butyraldehyde?
g. How many liters of carbon dioxide gas are produced at STP from the reaction in part **f**?

CI.26 Ionone is a compound that gives violets their aroma. The small, edible, purple flowers of violets are used in salads and to make teas. Liquid ionone has a density of 0.935 g/mL. (7.2, 8.6, 11.5, 11.6, 12.3)

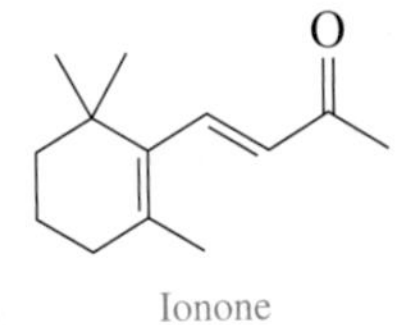

Ionone

The aroma of violets is due to ionone.

a. What functional groups are present in ionone?
b. Is the double bond on the side chain cis or trans?
c. What are the molecular formula and molar mass of ionone?
d. How many moles are in 2.00 mL of ionone?
e. When ionone reacts with hydrogen in the presence of a platinum catalyst, hydrogen adds to the double bonds and converts the ketone group to an alcohol. Draw the condensed structural formula and give the molecular formula for the product.
f. How many milliliters of hydrogen gas at STP are needed to completely react with 5.0 mL of ionone?

ANSWERS

CI.21 a. $2M(s) + 6HCl(aq) \longrightarrow 3H_2(g) + 2MCl_3(aq)$
b. 203 mL of H_2
c. 6.03×10^{-3} mole of M
d. 69.7 g/mole
e. gallium; Ga
f. $2Ga(s) + 6HCl(aq) \longrightarrow 3H_2(g) + 2GaCl_3(aq)$

CI.23 a. C_3H_6O
b. 58.08 g/mole
c. nonpolar covalent bonds: C—C, C—H; polar covalent bond: C—O
d. $C_3H_6O + 4O_2 \xrightarrow{\Delta} 3CO_2 + 3H_2O + \text{energy}$
e. 26.0 g of O_2

CI.25 a. $CH_3-CH_2-CH_2-\overset{\displaystyle O}{\overset{\|}{C}}-H$
b. (line-angle structure with O and H)
c. butanal
d. $CH_3-CH_2-CH_2-CH_2-OH$
e. $2C_4H_8O + 11O_2 \xrightarrow{\Delta} 8CO_2 + 8H_2O + \text{energy}$
f. 29.4 g of O_2
g. 14.9 L of CO_2

13 Carbohydrates

DURING HER ANNUAL PHYSICAL EXAMINATION, Kate, a 64-year old woman, reported that she was bothered by blurry vision, frequent need to urinate, and in the past year she had gained 22 lb. She had tried to lose weight and increase her exercise for the past 6 months without success. Her diet was high in carbohydrates. For dinner, Kate typically ate two cups of pasta and three to four slices of bread with butter or olive oil. She also ate eight to ten pieces of fresh fruit per day at meals and as snacks. A physical examination showed that her fasting blood glucose level was 178 mg/dL, indicating type 2 diabetes. She was referred to the diabetes specialty clinic, where she met Paula, a diabetes nurse. Paula explained to Kate that complex carbohydrates are long chains of glucose molecules that we obtain from ingesting breads and grains. They are broken down in the body into glucose, which is a simple carbohydrate or a monosaccharide.

CAREER Diabetes Nurse

Diabetes nurses teach patients about diabetes, so they can self-manage and control their condition. This includes education on proper diets and nutrition for both diabetic and pre-diabetic patients. Diabetes nurses help patients learn to monitor their medication and blood sugar levels and look for symptoms like diabetic nerve damage and vision loss. Diabetes nurses may also work with patients who have been hospitalized because of complications from their disease. This requires a thorough knowledge of the endocrine system, as this system is often involved with obesity and other diseases, resulting in some overlap between diabetes and endocrinology nursing.

CLINICAL UPDATE Kate's Program for Type 2 Diabetes

Now Kate uses a glucose meter to check her blood glucose level. You can view her change in diet in the **CLINICAL UPDATE Kate's Program for Type 2 Diabetes**, page 461, and see how the change in eating habits and increase in exercise helped Kate to decrease her blood sugar and lose weight.

13.1 Carbohydrates

LEARNING GOAL Classify a monosaccharide as an aldose or a ketose, and indicate the number of carbon atoms.

We have many carbohydrates in our food. There are polysaccharides called starches in bread and pasta. The table sugar used to sweeten cereal, tea, or coffee is sucrose, a disaccharide that consists of two simple sugars, glucose and fructose. **Carbohydrates** such as table sugar, lactose in milk, and cellulose are all made of carbon, hydrogen, and oxygen. Simple sugars, which have formulas of $C_n(H_2O)_n$, were once thought to be hydrates of carbon, thus the name *carbohydrate*. In a series of reactions called *photosynthesis*, energy from the Sun is used to combine the carbon atoms from carbon dioxide (CO_2) and the hydrogen and oxygen atoms of water (H_2O) into the carbohydrate glucose.

$$6CO_2 + 6H_2O + \text{energy} \underset{\text{Respiration}}{\overset{\text{Photosynthesis}}{\rightleftharpoons}} \underset{\text{Glucose}}{C_6H_{12}O_6} + 6O_2$$

In the body, glucose is oxidized in a series of metabolic reactions known as *respiration*, which releases chemical energy to do work in the cells. Carbon dioxide and water are produced and returned to the atmosphere. The combination of photosynthesis and respiration is called the *carbon cycle*, in which energy from the Sun is stored in plants by photosynthesis and made available to us when the carbohydrates in our diets are metabolized (see **FIGURE 13.1**).

FIGURE 13.1 During photosynthesis, energy from the Sun combines CO_2 and H_2O to form glucose ($C_6H_{12}O_6$) and O_2. During respiration in the body, carbohydrates are oxidized to CO_2 and H_2O, while energy is produced.

Q What are the reactants and products of respiration?

Types of Carbohydrates

The simplest carbohydrates are the **monosaccharides**. A monosaccharide cannot be split or hydrolyzed into smaller carbohydrates. One of the most common carbohydrates, glucose, $C_6H_{12}O_6$, is a monosaccharide. A **disaccharide** consists of two monosaccharide units joined together, which can be split into two monosaccharide units. For example, ordinary table sugar, sucrose, $C_{12}H_{22}O_{11}$, is a disaccharide that can be split by water (hydrolysis) in the presence of an acid or an enzyme to give one molecule of glucose and one molecule of another monosaccharide, fructose.

$$\underset{\text{Sucrose}}{C_{12}H_{22}O_{11}} + H_2O \xrightarrow{H^+ \text{ or enzyme}} \underset{\text{Glucose}}{C_6H_{12}O_6} + \underset{\text{Fructose}}{C_6H_{12}O_6}$$

A **polysaccharide** is a carbohydrate that contains many monosaccharide units, which is called a *polymer*. In the presence of an acid or an enzyme, a polysaccharide can be completely hydrolyzed to yield many monosaccharide molecules.

LOOKING AHEAD

REVIEW

Identifying Functional Groups (12.1)

Carbohydrates contained in foods such as pasta and bread provide energy for the body.

TEST

Try Practice Problems 13.1 and 13.2

TEST

Try Practice Problems 13.3 and 13.4

Example		Type of Carbohydrate		Products of Hydrolysis
	Honey contains the monosaccharides fructose and glucose.	Monosaccharide + H_2O	$\xrightarrow{H^+ \text{ or enzyme}}$	no hydrolysis
	Milk contains lactose, which is a disaccharide.	Disaccharide + H_2O	$\xrightarrow{H^+ \text{ or enzyme}}$	two monosaccharide molecules
	Cotton consists of the polysaccharide cellulose.	Polysaccharide + many H_2O	$\xrightarrow{H^+ \text{ or enzyme}}$	many monosaccharide molecules

TEST

Try Practice Problems 13.5 and 13.6

Monosaccharides

A monosaccharide contains several hydroxyl groups attached to a chain of three to seven carbon atoms that also contains an aldehyde or a ketone group. Thus, a monosaccharide is known as a polyhydroxy aldehyde or polyhydroxy ketone. In an **aldose**, the first carbon in the chain is an aldehyde, whereas a **ketose** has a ketone as the second carbon atom.

Erythrulose, used in sunless tanning lotions, reacts with amino acids or proteins in the skin to produce a bronze color.

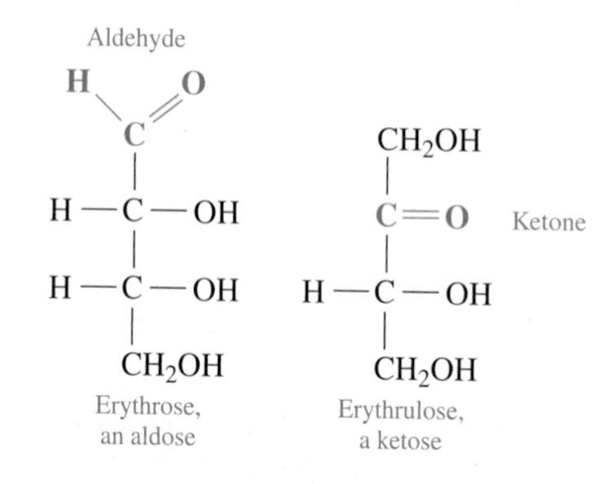

Monosaccharides are also classified by the number of carbon atoms. A monosaccharide with three carbon atoms is a *triose*, one with four carbon atoms is a *tetrose*; a *pentose* has five carbons, and a *hexose* contains six carbons. We can use both classification systems to indicate the aldehyde or ketone group and the number of carbon atoms. An aldopentose is a five-carbon monosaccharide that is an aldehyde; a ketohexose is a six-carbon monosaccharide that is a ketone.

TEST

Try Practice Problems 13.7 and 13.8

Glyceraldehyde (aldotriose)

Threose (aldotetrose)

Ribose (aldopentose)

Fructose (ketohexose)

ENGAGE

Why is fructose classified as a ketohexose?

SAMPLE PROBLEM 13.1 Monosaccharides

TRY IT FIRST

Classify each of the following monosaccharides as an aldopentose, ketopentose, aldohexose, or ketohexose:

a.

$$\begin{array}{c} CH_2OH \\ | \\ C{=}O \\ | \\ H{-}C{-}OH \\ | \\ H{-}C{-}OH \\ | \\ CH_2OH \end{array}$$

Ribulose

b.

$$\begin{array}{c} H \diagdown \; \diagup\!\!\diagup O \\ C \\ | \\ H{-}C{-}OH \\ | \\ HO{-}C{-}H \\ | \\ H{-}C{-}OH \\ | \\ H{-}C{-}OH \\ | \\ CH_2OH \end{array}$$

Glucose

SOLUTION

a. Ribulose has five carbon atoms (pentose) and is a ketone, which makes it a ketopentose.
b. Glucose has six carbon atoms (hexose) and is an aldehyde, which makes it an aldohexose.

STUDY CHECK 13.1

Classify the following monosaccharide, erythrose, as an aldotetrose, ketotetrose, aldopentose, or ketopentose:

$$\begin{array}{c} H \diagdown \; \diagup\!\!\diagup O \\ C \\ | \\ H{-}C{-}OH \\ | \\ H{-}C{-}OH \\ | \\ CH_2OH \end{array}$$

ANSWER

aldotetrose

TEST

Try Practice Problems 13.9 and 13.10

PRACTICE PROBLEMS

13.1 Carbohydrates

LEARNING GOAL Classify a monosaccharide as an aldose or a ketose, and indicate the number of carbon atoms.

13.1 What reactants are needed for photosynthesis and respiration?

13.2 What is the relationship between photosynthesis and respiration?

13.3 What is a monosaccharide? A disaccharide?

13.4 What is a polysaccharide?

13.5 What functional groups are found in all monosaccharides?

13.6 What is the difference between an aldose and a ketose?

13.7 What are the functional groups and number of carbons in a ketopentose?

13.8 What are the functional groups and number of carbons in an aldohexose?

Clinical Applications

13.9 Classify each of the following monosaccharides as an aldopentose, ketopentose, aldohexose, or ketohexose:

a. Psicose is present in low amounts in foods.

$$\begin{array}{c} CH_2OH \\ | \\ C{=}O \\ | \\ H{-}C{-}OH \\ | \\ H{-}C{-}OH \\ | \\ H{-}C{-}OH \\ | \\ CH_2OH \end{array}$$

Psicose

b. Lyxose is a component of bacterial glycolipids.

$$\begin{array}{c} H \diagdown \; \diagup\!\!\diagup O \\ C \\ | \\ HO{-}C{-}H \\ | \\ HO{-}C{-}H \\ | \\ H{-}C{-}OH \\ | \\ CH_2OH \end{array}$$

Lyxose

13.10 Classify each of the following monosaccharides as an aldopentose, ketopentose, aldohexose, or ketohexose:

a. A solution of xylose is given to test its absorption by the intestines.

H—C=O
H—C—OH
HO—C—H
H—C—OH
CH_2OH
Xylose

b. Tagatose, found in fruit, is similar in sweetness to sugar.

CH_2OH
C=O
HO—C—H
HO—C—H
H—C—OH
CH_2OH
Tagatose

13.2 Chiral Molecules

LEARNING GOAL Identify chiral and achiral carbon atoms in an organic molecule. Identify D and L enantiomers.

INTERACTIVE VIDEO
Chirality

Everything has a mirror image. If you hold your right hand up to a mirror, you see its mirror image, which matches your left hand (see **FIGURE 13.2**). If you turn your palms toward each other, one hand is the mirror image of the other. If you look at the palms of your hands, your thumbs are on opposite sides. If you then place your right hand over your left hand, you cannot match up all the parts of the hands: palms, backs, thumbs, and little fingers.

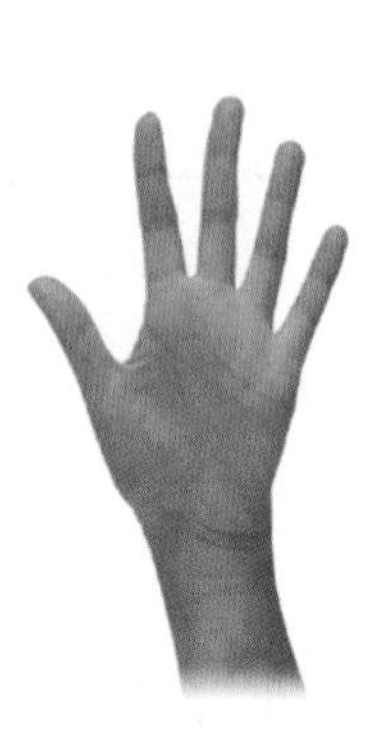
Left hand

Mirror image of right hand

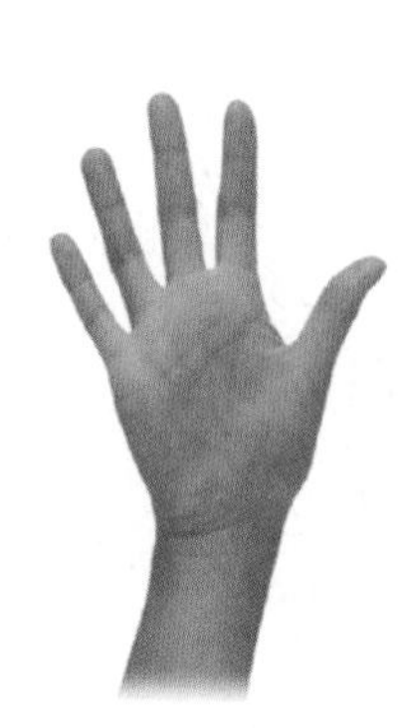
Right hand

FIGURE 13.2 The left and right hands are chiral because they have mirror images that cannot be superimposed on each other.

Q Why are your shoes chiral objects?

The thumbs and little fingers can be matched, but then the palms or backs of your hands are facing each other. Your hands are mirror images that cannot be superimposed on each other. When the mirror images cannot be completely matched, they are *nonsuperimposable*.

Objects such as hands that have nonsuperimposable mirror images are **chiral** (pronunciation *kai-ral*). Left and right shoes are chiral; left- and right-handed golf clubs are chiral. When we think of how difficult it is to put a left-hand glove on our right hand, put a right shoe on our left foot, or use left-handed scissors if we are right-handed, we begin to realize that certain properties of mirror images are very different.

When the mirror image of an object is identical and can be superimposed on the original, it is *achiral*. For example, the mirror image of a plain drinking glass is identical to the original glass, which means the mirror image can be superimposed on the glass (see **FIGURE 13.3**).

Many important compounds in medicine are chiral, including medications such as ibuprofen, penicillin, epinephrine, and morphine. Most of the compounds in biochemistry—carbohydrates, fats, amino acids, proteins, and DNA—are also chiral.

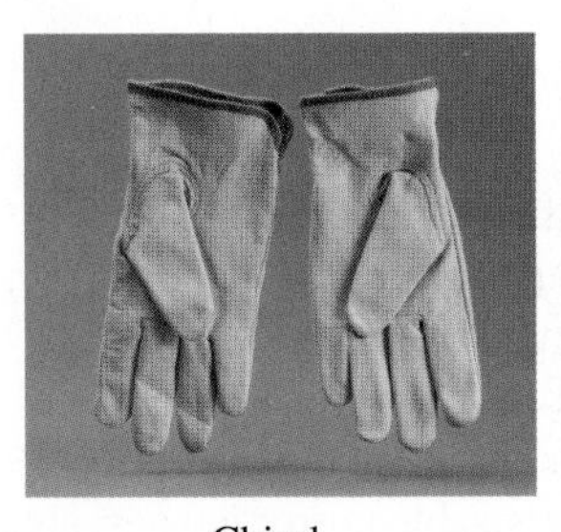

Chiral Achiral Chiral

FIGURE 13.3 Everyday objects such as gloves and shoes are chiral, but a plain glass is achiral.

Q Why are some of the objects chiral and others achiral?

Chiral Carbon Atoms

CORE CHEMISTRY SKILL

Identifying Chiral Molecules

A carbon compound is chiral if it has at least one carbon atom bonded to *four different atoms or groups*. This type of carbon atom is called a **chiral carbon** because there are two different ways that it can bond to four atoms or groups of atoms. The resulting structures are nonsuperimposable mirror images or *stereoisomers*. When two or more chiral structures have the same molecular formula, but differ in the three-dimensional arrangements of atoms in space, they are called **stereoisomers**. Let's look at the mirror images of a carbon bonded to four different atoms (see FIGURE 13.4). If we line up the hydrogen and iodine atoms in the mirror images, the bromine and chlorine atoms appear on opposite sides. No matter how we turn the models, we cannot align all four atoms at the same time. When stereoisomers cannot be superimposed, they are called **enantiomers**.

ENGAGE

Why does 3-methylhexane have a chiral carbon whereas 2-methylhexane does not?

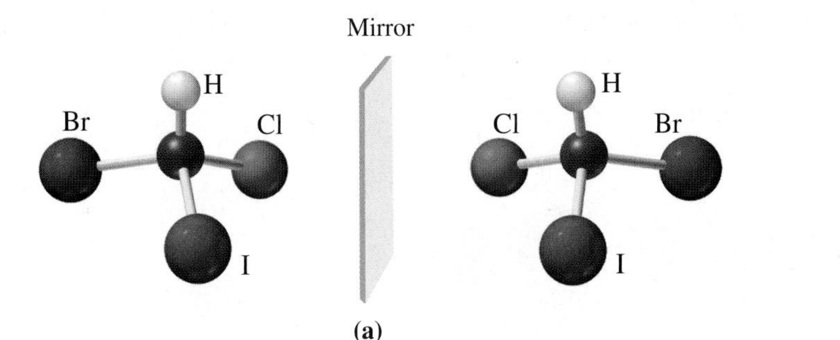

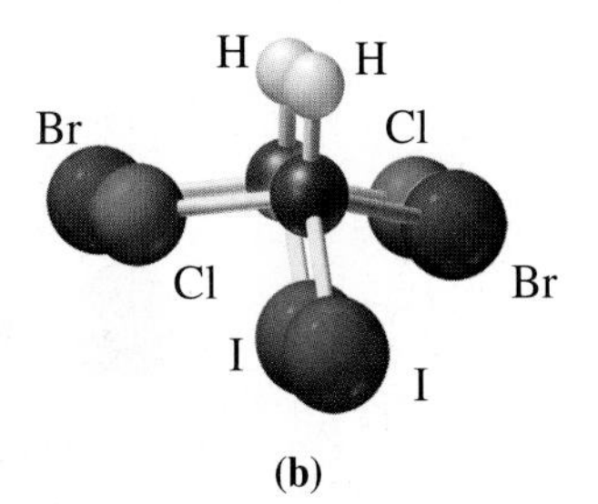

FIGURE 13.4 **(a)** The enantiomers of a chiral molecule are mirror images. **(b)** The enantiomers of a chiral molecule cannot be superimposed on each other.

Q Why is the carbon atom in this compound chiral?

If a molecule with two or more identical atoms bonded to the same atom is rotated, the atoms can be superimposed and the mirror images represent the same structure (see FIGURE 13.5).

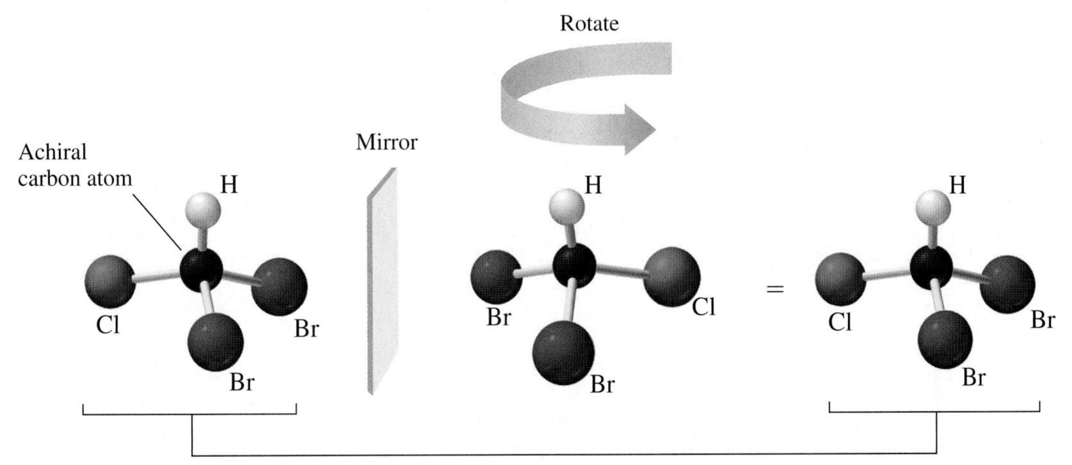

FIGURE 13.5 The mirror images of an achiral compound can be superimposed on each other.

Q Why can the mirror images of this compound be superimposed?

Modeling Chiral Objects

Achiral Objects

Obtain some toothpicks and orange, yellow, green, purple, and black gumdrops. Place four toothpicks into a black gumdrop making the ends of toothpicks the corners of a tetrahedron. Attach the gumdrops to the toothpicks: two orange, one green, and one yellow.

Using another black gumdrop, make a second model that is the mirror image of the original model. Now rotate one of the models and try to superimpose it on the other model. Are the models superimposable? If achiral objects have superimposable mirror images, are these models chiral or achiral?

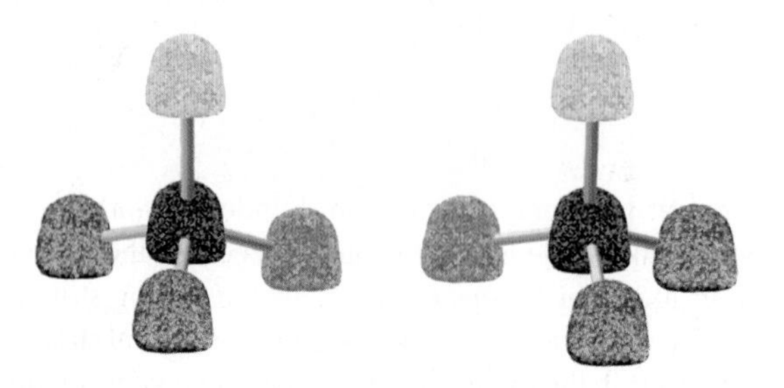

Chiral Objects

Using one of the original models, replace one orange gumdrop with a purple gumdrop. Now there are four different colors of gumdrops attached to the black gumdrop. Make its mirror image by replacing one orange gumdrop with a purple one on the second model. Now rotate one of the models, and try to superimpose it on the other model. Are the models superimposable? If chiral objects have nonsuperimposable mirror images, are these models chiral or achiral?

SAMPLE PROBLEM 13.2 Chiral Carbons

TRY IT FIRST

For each of the following, indicate whether the carbon in red is chiral or achiral:

a. Glycerol is used in soaps, creams, and hair care products.

```
      H   OH  H
      |   |   |
HO —  C — C — C — OH
      |   |   |
      H   H   H
```

Glycerol

b. Ibuprofen is used to relieve fever and pain.

```
                          O
                          ‖
      H   H               C — OH
      |   |               |
CH3 — C — C — (benzene) — C — CH3
      |   |               |
      CH3 H               H
```

Ibuprofen

SOLUTION

a. Achiral. Boxes are drawn around the groups of atoms attached to the red C to determine chirality. Two of the substituents on the carbon in red are the same.

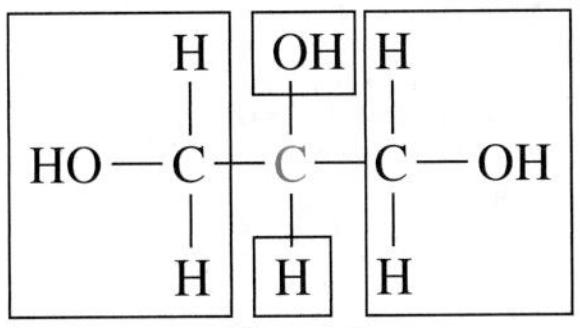

```
      H  OH  H
      |   |  |
  HO—C — C — C—OH
      |   |  |
      H   H  H
```

Glycerol

b. Chiral. The carbon in red is bonded to four different groups.

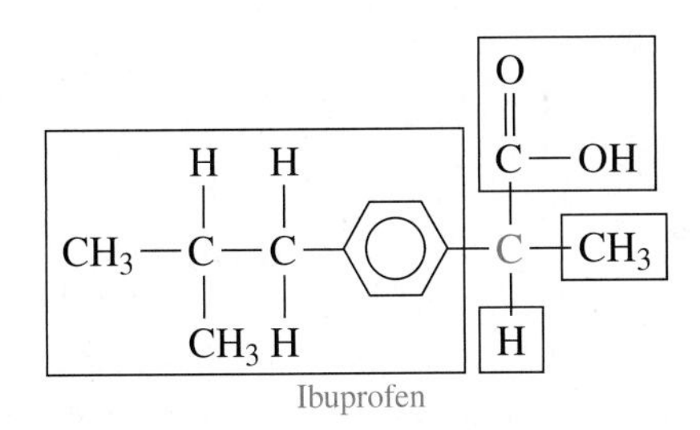

Ibuprofen

STUDY CHECK 13.2

Identify the chiral carbon of penicillamine, which is used in the treatment of rheumatoid arthritis.

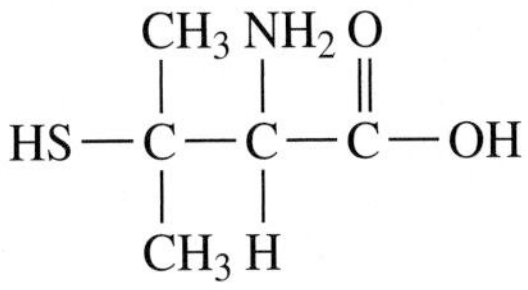

```
     CH3 NH2 O
      |   |  ||
  HS—C — C — C—OH
      |   |
     CH3  H
```

ANSWER

```
     CH3 NH2 O
      |   |  ||
  HS—C — C — C—OH
      |   |↖
     CH3  H   Chiral carbon
```

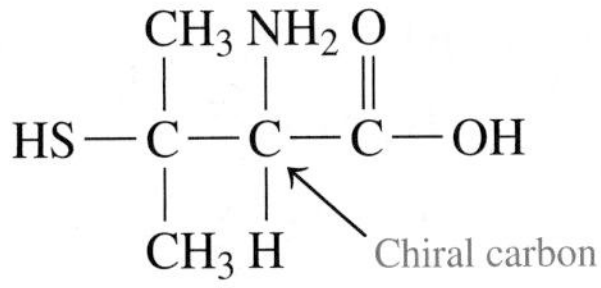

TEST

Try Practice Problems 13.11 and 13.12, 13.19 and 13.20

Drawing Fischer Projections

Emil Fischer devised a simplified system for drawing isomers that shows the arrangements of the atoms around the chiral carbons. Fischer received the Nobel Prize in Chemistry in 1902 for his contributions to carbohydrate and protein chemistry. Now we use his model, called a **Fischer projection**, to represent a three-dimensional structure of enantiomers. Vertical lines represent bonds that project backward from a carbon atom and horizontal lines represent bonds that project forward. In this model, the most highly oxidized carbon is placed at the top, and the intersections of vertical and horizontal lines represent a carbon atom that is usually chiral.

For the simplest aldose, glyceraldehyde, the only chiral carbon is the middle carbon. In the Fischer projection, the carbonyl group, which is the most highly oxidized group, is drawn at the top above the chiral carbon and the —CH_2OH group is drawn at the bottom. The —H and —OH groups can be drawn at each end of a horizontal line, but in two different ways. The isomer that has the —OH group drawn to the left of the chiral atom is designated as the L isomer. The isomer with the —OH group drawn to the right of the chiral carbon represents the D isomer (see **FIGURE 13.6**).

Fischer projections can also be drawn for compounds that have two or more chiral carbons. For example, in erythrose, both of the carbon atoms at the intersections are chiral. Then the designation as a D or L isomer is determined by the position of the —OH group attached to the chiral carbon *farthest from the carbonyl group.*

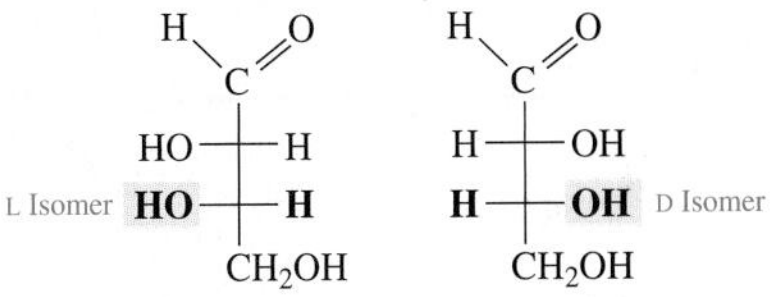

```
            H   O              H   O
             \ //               \ //
              C                  C
          HO—+—H             H—+—OH
L Isomer  HO—+—H             H—+—OH  D Isomer
            CH2OH              CH2OH
         L-Erythrose        D-Erythrose
```

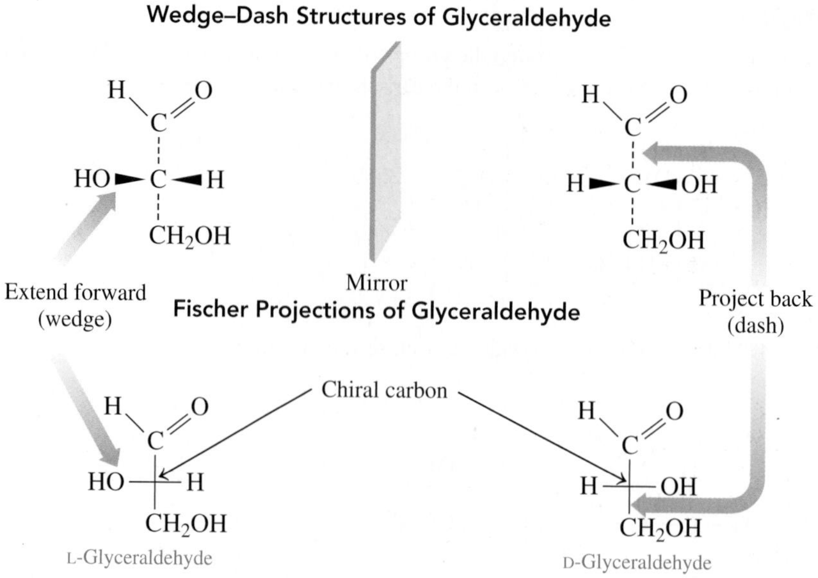

Wedge–Dash Structures of Glyceraldehyde

Extend forward (wedge)

Mirror

Project back (dash)

Fischer Projections of Glyceraldehyde

Chiral carbon

L-Glyceraldehyde

D-Glyceraldehyde

FIGURE 13.6 In the Fischer projection for glyceraldehyde, the chiral carbon atom is at the center, with horizontal lines for bonds that project forward and vertical lines for bonds that point away.

Q Why does glyceraldehyde have only one chiral carbon atom?

SAMPLE PROBLEM 13.3 Fischer Projections

TRY IT FIRST

Identify each of the following as the D or L isomer:

a. Fischer projection: CH_2OH at top; HO on left, H on right; CH_3 at bottom

b. Fischer projection: CHO (H–C=O) at top; H on left, OH on right; CH_3 at bottom

c. Fischer projection: CHO (H–C=O) at top; HO on left, H on right; H on left, OH on right; CH_2OH at bottom

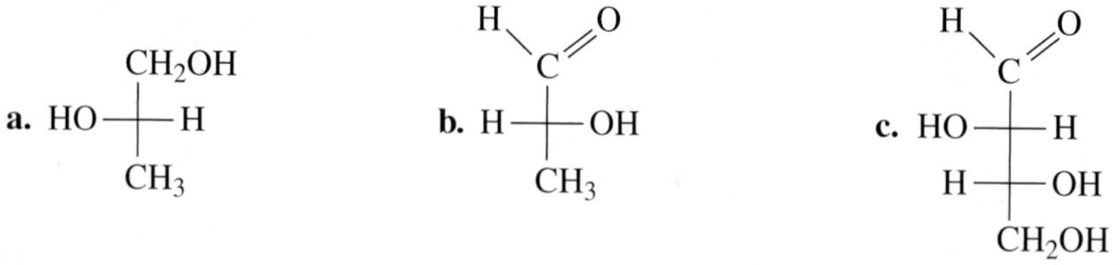

SOLUTION

a. When the —OH group is drawn to the left of the chiral carbon, it is the L isomer.

b. When the —OH group is drawn to the right of the chiral carbon, it is the D isomer.

c. When the —OH group on the chiral carbon farthest from the top of the Fischer projection is drawn to the right, it is the D isomer.

STUDY CHECK 13.3

Identify each of the following as the D or L isomer:

a. Fischer projection: CHO (H–C=O) at top; HO on left, H on right; CH_3 at bottom

b. Fischer projection: CHO (H–C=O) at top; HO on left, H on right; HO on left, H on right; H on left, OH on right; CH_2OH at bottom

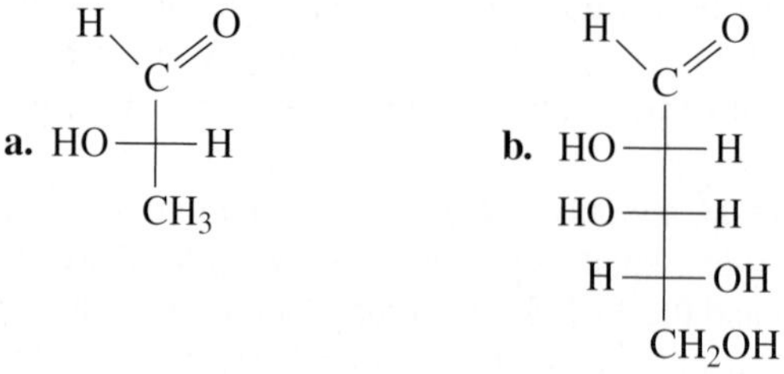

ANSWER

a. L isomer **b.** D isomer

TEST

Try Practice Problems 13.13 to 13.18

CHEMISTRY LINK TO HEALTH

Enantiomers in Biological Systems

Molecules in nature also have mirror images, and often one stereoisomer has a different biological effect than the other one. For example, the behavior of nicotine and adrenaline (epinephrine) depends on only one of their enantiomers. One enantiomer of nicotine is more toxic than the other. Only one enantiomer of epinephrine causes the constriction of blood vessels.

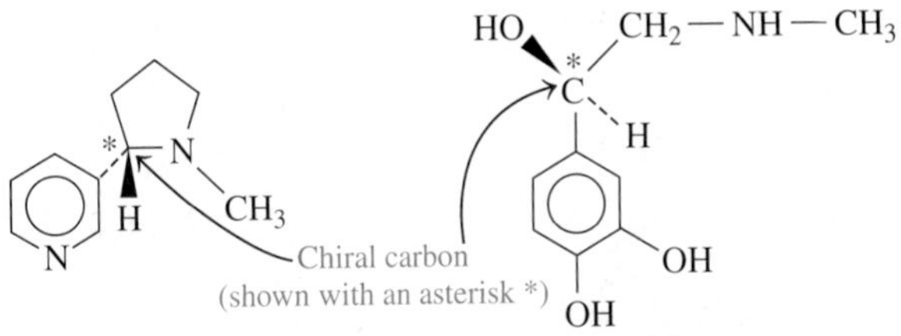

Nicotine

Adrenaline (epinephrine)

For some compounds, one enantiomer has a certain odor, and the other enantiomer has a completely different odor. For example, the oils that give the scent of spearmint and caraway seeds are both composed of carvone. However, carvone has one chiral carbon in the carbon ring indicated by an asterisk, which gives carvone two enantiomers.

Olfactory receptors in the nose detect these enantiomers as two different odors. One smells and tastes like spearmint, whereas its mirror image has the odor and taste of caraway in rye bread. Thus, our senses of smell and taste are responsive to the chirality of molecules.

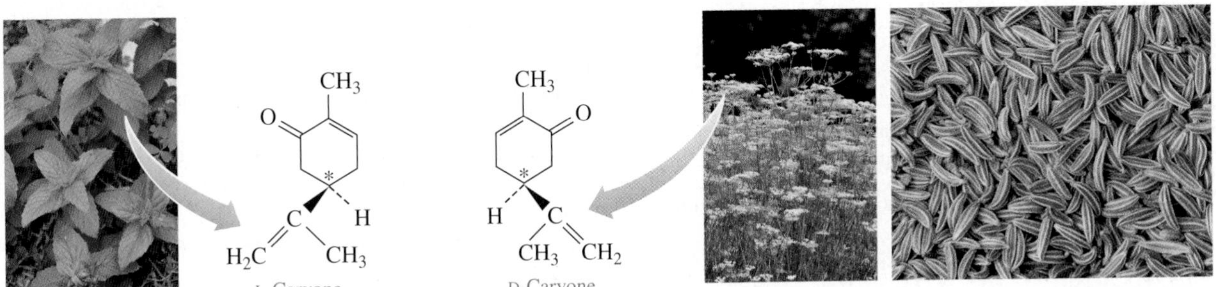

Spearmint

L-Carvone from spearmint

D-Carvone from caraway plant and seeds

Caraway plant and seeds

A substance used to treat Parkinson's disease is L-dopa, which is converted to dopamine in the brain, where it raises the serotonin level. However, the D-dopa enantiomer is not effective for the treatment of Parkinson's disease.

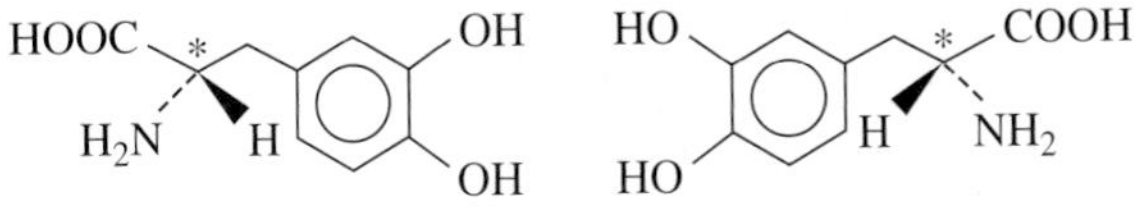

L-Dopa

D-Dopa

Many compounds in biological systems have only one active enantiomer because the enzymes and cell surface receptors on which metabolic reactions take place are themselves chiral. The chiral receptor fits the arrangement of the substituents in only one enantiomer; its mirror image does not fit properly (see **FIGURE 13.7**).

Today, drug researchers are using *chiral technology* to produce the active enantiomers of chiral drugs. Chiral catalysts are being designed that direct the formation of just one enantiomer rather than both. The active forms of several enantiomers are now being produced, such as

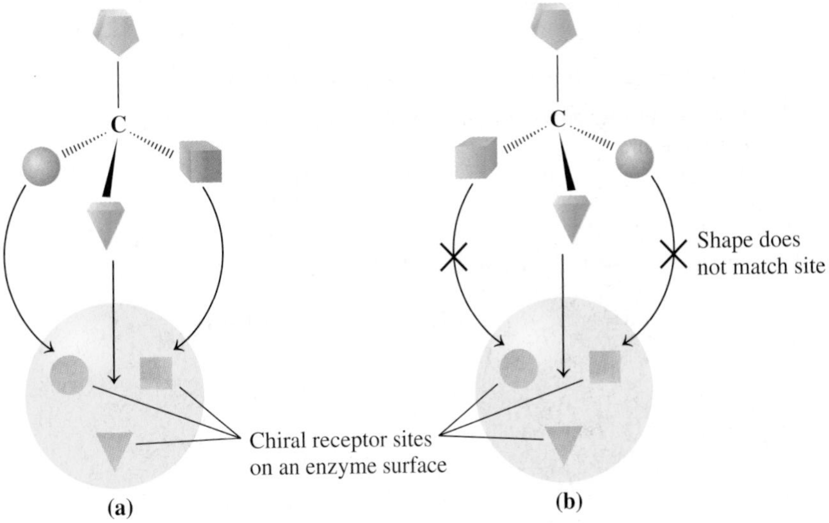

FIGURE 13.7 **(a)** The substituents on the biologically active enantiomer bind to all the sites on a chiral receptor; **(b)** its enantiomer does not bind properly and is not active biologically.

Q Why don't all the substituents of the mirror image of the active enantiomer fit into a chiral receptor site?

naproxen and ibuprofen, which are nonsteroidal anti-inflammatory drugs used to relieve pain, fever, and inflammation caused by osteoarthritis and tendinitis. The benefits of producing only the active enantiomer include using a lower dose, enhancing activity, reducing interactions with other drugs, and eliminating possible harmful side effects from the enantiomer.

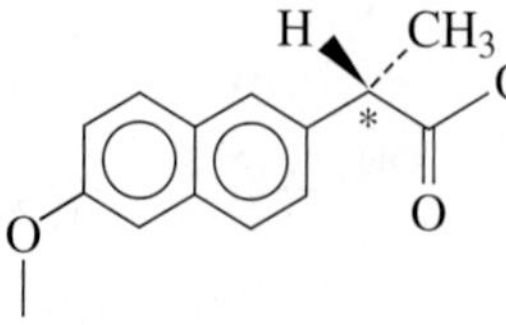

Active form of naproxen

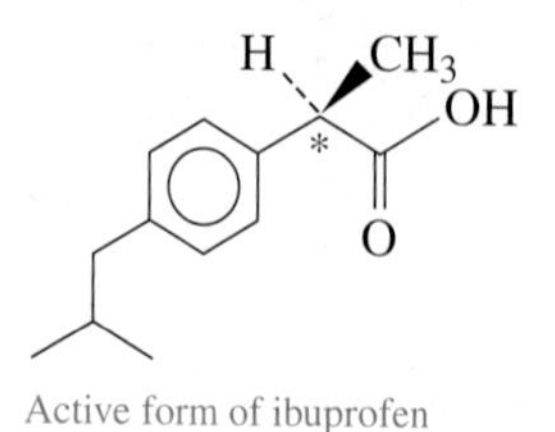

Active form of ibuprofen

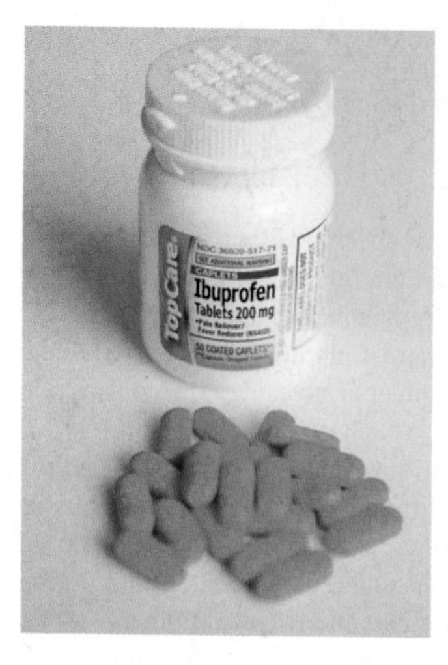

One enantiomer of ibuprofen is the active form, but its mirror image is inactive.

PRACTICE PROBLEMS

13.2 Chiral Molecules

LEARNING GOAL Identify chiral and achiral carbon atoms in an organic molecule. Identify D and L enantiomers.

13.11 Identify each of the following structures as chiral or achiral. If chiral, indicate the chiral carbon.

a. $CH_3-CH(OH)-CH_3$

b. (skeletal structure with Br)

c. $CH_3-CH(Br)-C(=O)-H$

d. $CH_3-CH_2-C(OH)(CH_3)-CH_3$

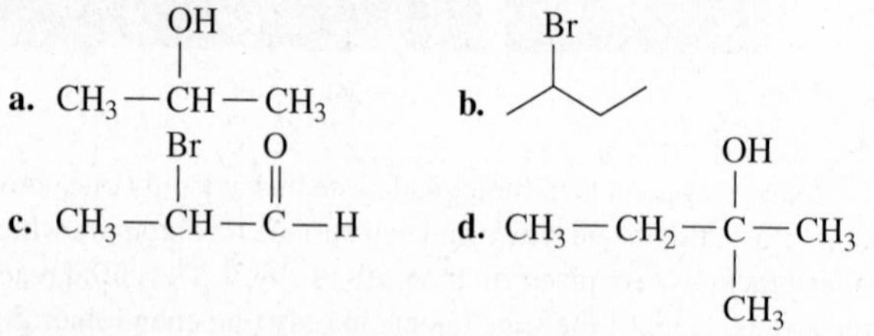

13.12 Identify each of the following structures as chiral or achiral. If chiral, indicate the chiral carbon.

a. $CH_3-C(Cl)(CH_3)-CH_2-CH(Cl)-CH_3$

b. (skeletal structure with two OH groups)

c. $CH_3-C(Br)=CH-CH_3$

d. $Br-CH_2-CH(Cl)-CH_3$

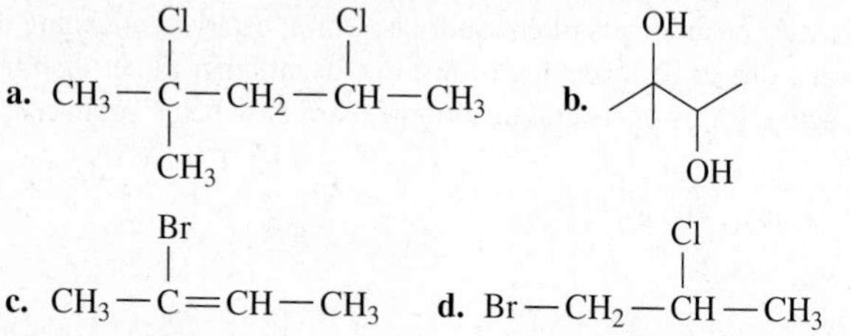

13.13 Draw the Fischer projection for each of the following wedge–dash structures:

a. H–C=O; HO ▸ C ◂ Br; CH_3

b. H–C=O; Cl ▸ C ◂ Br; OH

c. H–C=O; HO ▸ C ◂ H; CH_2CH_3

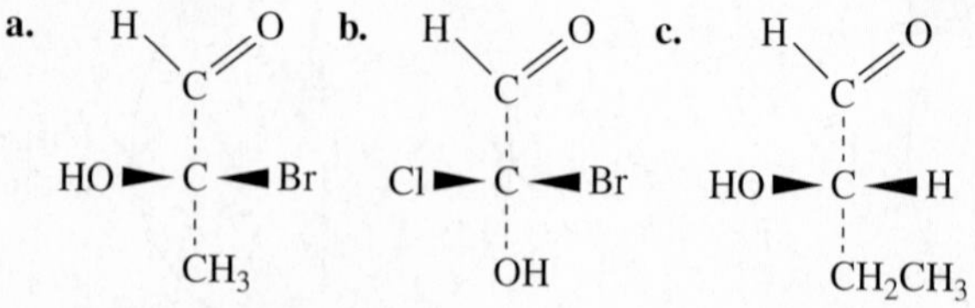

13.14 Draw the Fischer projection for each of the following wedge–dash structures:

a. H–C=O; HO ▸ C ◂ Br; CH_2OH

b. H–C=O; HO ▸ C ◂ H; CH_2OH

c. H–C=O; H ▸ C ◂ OH; CH_2OH

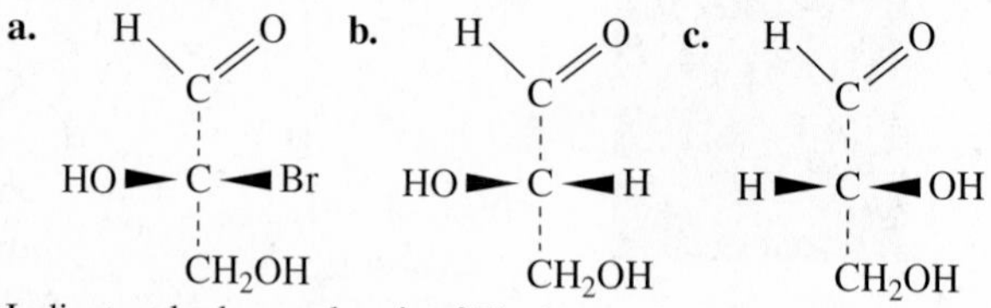

13.15 Indicate whether each pair of Fischer projections represents enantiomers or identical structures.

a. (CH_3 top; Br left, Cl right; CH_3 bottom) and (CH_3 top; Cl left, Br right; CH_3 bottom)

b. (H–C=O top; HO left, H right; CH_3 bottom) and (H–C=O top; H left, OH right; CH_3 bottom)

c. (CH_3 top; Cl left, Br right; H bottom) and (CH_3 top; Br left, Cl right; H bottom)

d. (HO–C=O top; H left, OH right; CH_3 bottom) and (HO–C=O top; HO left, H right; CH_3 bottom)

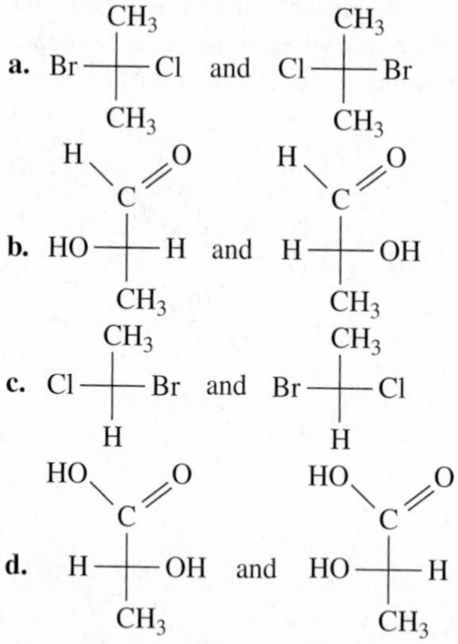

13.16 Indicate whether each pair of Fischer projections represents enantiomers or identical structures.

a. (CH_2OH top; Br left, Cl right; CH_3 bottom) and (CH_2OH top; Cl left, Br right; CH_3 bottom)

b. (H–C=O top; H left, H right; CH_3 bottom) and (H–C=O top; H left, H right; CH_3 bottom)

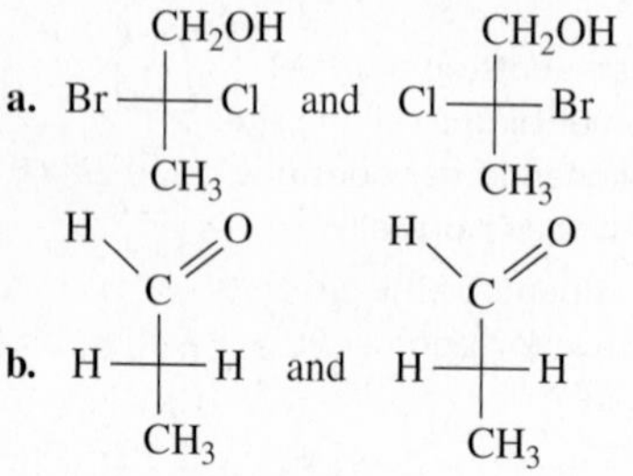

c. CH_3 / H—C—OH / CH_2CH_3 and CH_3 / HO—C—H / CH_2CH_3

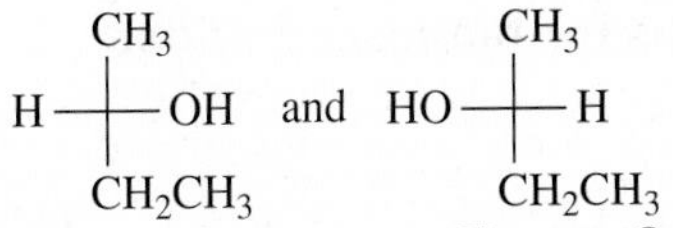

d. H—C(=O) / H—C—NH_2 / CH_3 and H—C(=O) / H_2N—C—H / CH_3

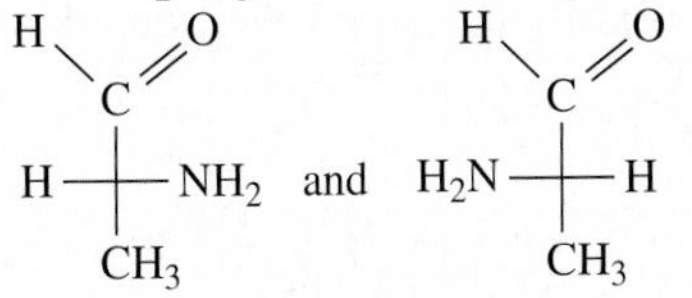

13.17 Identify each of the following as D or L:

a. CH_2OH / H—C—OH / CH_3

b. H—C(=O) / H—C—OH / CH_2OH

c. H—C(=O) / H—C—OH / HO—C—H / CH_2OH

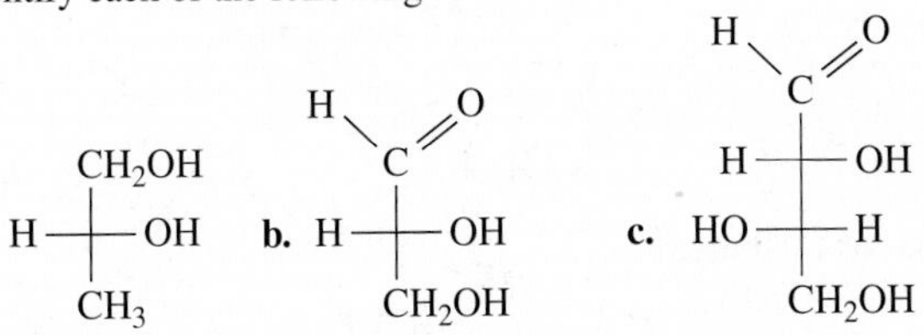

13.18 Identify each of the following as D or L:

a. H—C(=O) / HO—C—H / CH_2OH

b. H—C(=O) / HO—C—H / HO—C—H / CH_2OH

c. HO—C(=O) / H—C—OH / CH_2OH

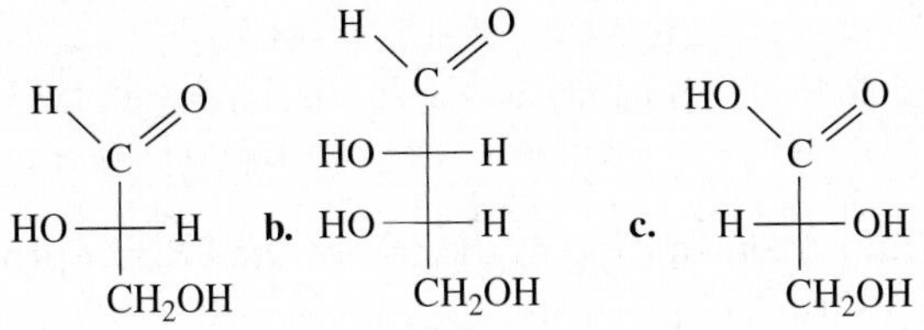

Clinical Applications

13.19 Identify the chiral carbon in each of the following compounds:

a. citronellol; one enantiomer has the odor of geranium

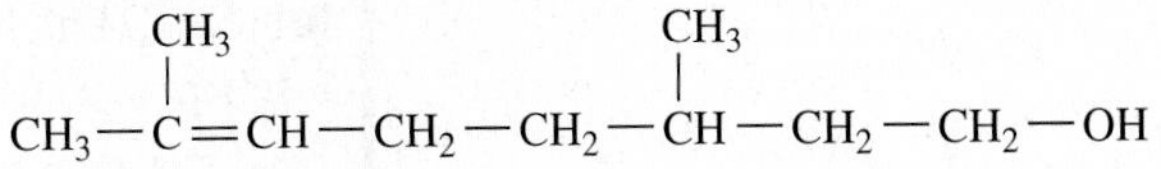

$CH_3—C(CH_3){=}CH—CH_2—CH_2—CH(CH_3)—CH_2—CH_2—OH$

b. alanine, an amino acid

$H_2N—CH(CH_3)—C(=O)—OH$

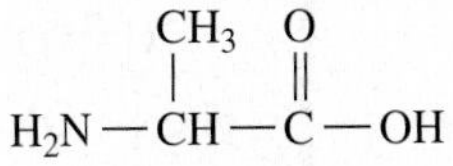

13.20 Identify the chiral carbon in each of the following compounds:

a. amphetamine (Benzedrine), stimulant, used in the treatment of hyperactivity

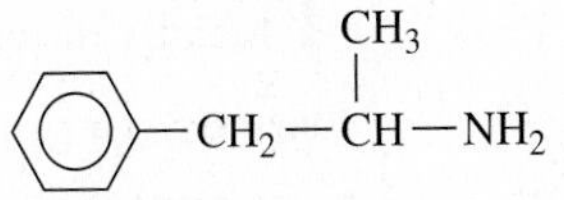

$C_6H_5—CH_2—CH(CH_3)—NH_2$

b. norepinephrine, increases blood pressure and nerve transmission

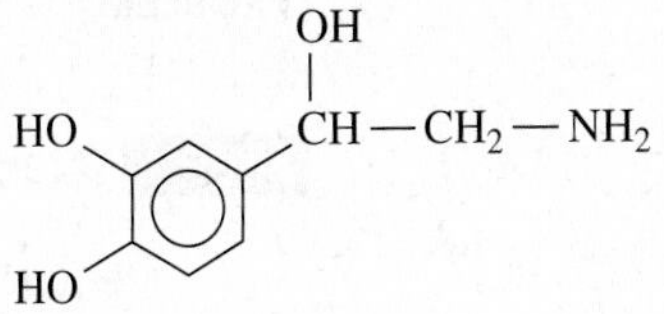

$(HO)_2C_6H_3—CH(OH)—CH_2—NH_2$

13.3 Fischer Projections of Monosaccharides

LEARNING GOAL Identify or draw the D and L configurations of the Fischer projections for common monosaccharides.

The most common monosaccharides contain five or six carbon atoms with several chiral carbons. Each Fischer projection of a monosaccharide can be drawn as a mirror image, which gives a pair of enantiomers. The following are the Fischer projections for the D and L enantiomers of ribose, a five-carbon monosaccharide, and the D and L enantiomers of glucose, a six-carbon monosaccharide. The vertical carbon chain is numbered starting from the top carbon.

In each pair of mirror images, all of the —OH groups on chiral carbon atoms are reversed so that they appear on the opposite sides of the molecule. For example, in L-ribose, all of the —OH groups drawn on the left side of the vertical line are drawn on the right side in the mirror image D-ribose.

CORE CHEMISTRY SKILL

Identifying D and L Fischer Projections for Carbohydrates

L-Ribose: H—C(=O) (1) / HO—C2—H / HO—C3—H / HO—C4—H / CH_2OH (5)

D-Ribose: H—C(=O) (1) / H—C2—OH / H—C3—OH / H—C4—OH / CH_2OH (5)

Chiral atom farthest from the carbonyl group

L-Glucose: H—C(=O) (1) / HO—C2—H / H—C3—OH / HO—C4—H / HO—C5—H / CH_2OH (6)

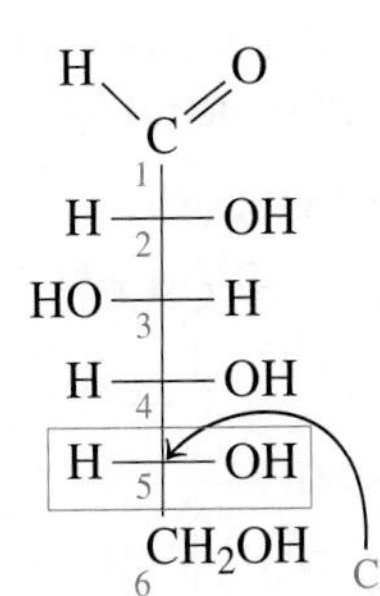

D-Glucose: H—C(=O) (1) / H—C2—OH / HO—C3—H / H—C4—OH / H—C5—OH / CH_2OH (6)

Chiral atom farthest from the carbonyl group

ENGAGE

Which of the chiral carbon atoms in a carbohydrate determines the D or L isomer?

SAMPLE PROBLEM 13.4 Fischer Projections for Monosaccharides

TRY IT FIRST

Ribulose, which is used in various brands of artificial sweeteners, has the following Fischer projection:

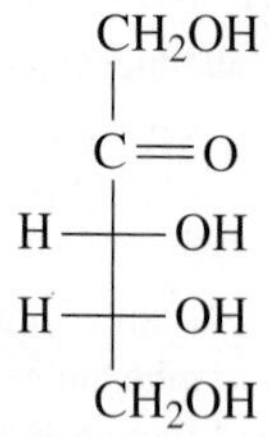

```
    CH2OH
     |
     C=O
  H -+- OH
  H -+- OH
    CH2OH
```

Identify the compound as D- or L-ribulose.

SOLUTION GUIDE

	Given	Need	Connect
ANALYZE THE PROBLEM	Fischer projection of ribulose	identify as D or L ribulose	chiral carbon farthest from carbonyl group

STEP 1 Number the carbon chain starting at the top of the Fischer projection.

```
    CH2OH
   1 |
     C=O
   2 |
  H -+- OH
     3
  H -+- OH
     4
    CH2OH
     5
```

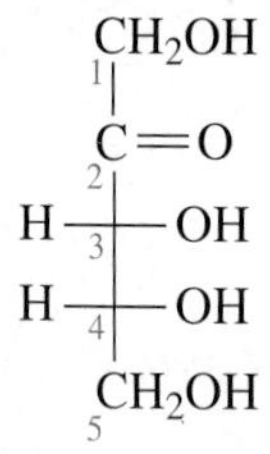

STEP 2 Locate the chiral carbon farthest from the top of the Fischer projection. The chiral carbon farthest from the top is carbon 4.

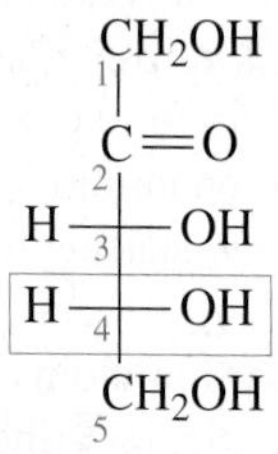

```
    CH2OH
   1 |
     C=O
   2 |
  H -+- OH
     3
 [H -+- OH]
     4
    CH2OH
     5
```

STEP 3 Identify the position of the —OH group as D or L. In this Fischer projection, the —OH group is on the right of carbon 4, which makes it D-ribulose.

STUDY CHECK 13.4

Draw and name the Fischer projection for the mirror image of the ribulose in Sample Problem 13.4.

ANSWER

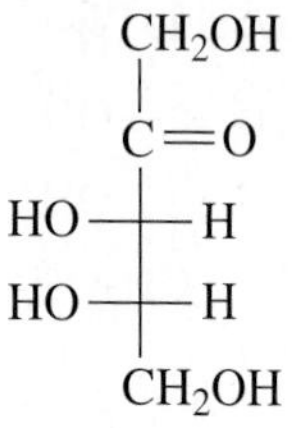

```
     CH2OH
      |
      C=O
  HO -+- H
  HO -+- H
     CH2OH
```

L-Ribulose

TEST

Try Practice Problems 13.21 to 13.28

Some Important Monosaccharides

D-Glucose, D-galactose, and D-fructose are the most important monosaccharides. They are all hexoses with the molecular formula $C_6H_{12}O_6$ and are isomers of each other. Although we can draw Fischer projections for their D and L enantiomers, the D enantiomers are more commonly found in nature and used in the cells of the body. The Fischer projections for the D enantiomers are drawn as follows:

Carbon	D-Glucose	D-Galactose	D-Fructose
1	H–C=O	H–C=O	CH_2OH
2	H—OH	H—OH	C=O
3	HO—H	HO—H	HO—H
4	H—OH	HO—H	H—OH
5	H—OH	H—OH	H—OH
6	CH_2OH	CH_2OH	CH_2OH

The most common hexose, **D-glucose**, $C_6H_{12}O_6$, also known as dextrose and blood sugar, is found in fruits, vegetables, corn syrup, and honey. D-Glucose is a building block of the disaccharides sucrose, lactose, and maltose, and polysaccharides such as amylose, cellulose, and glycogen.

D-Galactose, $C_6H_{12}O_6$, is obtained from the disaccharide lactose, which is found in milk and milk products. D-Galactose is important in the cellular membranes of the brain and nervous system. The only difference in the Fischer projections of D-glucose and D-galactose is the arrangement of the —OH group on carbon 4.

In contrast to D-glucose and D-galactose, **D-fructose**, $C_6H_{12}O_6$, is a ketohexose. The structure of D-fructose differs from glucose at carbons 1 and 2 by the location of the carbonyl group. D-Fructose is the sweetest of the carbohydrates, almost twice as sweet as sucrose (table sugar). This characteristic makes D-fructose popular with dieters because less fructose, and therefore fewer calories, is needed to provide a pleasant taste. D-Fructose, also called levulose and fruit sugar, is found in fruit juices and honey.

D-Fructose is also obtained as one of the hydrolysis products of sucrose, the disaccharide known as table sugar. High-fructose corn syrup (HFCS) is a sweetener produced when enzymes convert the glucose in corn syrup to fructose. When the fructose is mixed with corn syrup containing only glucose, the sweetener HFCS is produced. HFCS with 42% fructose is used in bakery goods, and 55% fructose is used in soft drinks.

The sweet taste of honey is because of the monosaccharides D-glucose and D-fructose.

ENGAGE

What are some differences in the Fischer projections of D-glucose, D-galactose, and D-fructose?

TEST

Try Practice Problems 13.29 and 13.30

A food label shows that high-fructose corn syrup is the sweetener in this product.

CHEMISTRY LINK TO HEALTH

Hyperglycemia and Hypoglycemia

Kate's doctor ordered an oral glucose tolerance test (OGTT) to evaluate her body's ability to return to normal glucose concentrations (70 to 90 mg/dL) in response to the ingestion of a specified amount of glucose (dextrose). After Kate fasted for 12 h, she drinks a solution containing 75 g of glucose. A blood sample is taken immediately, followed by more blood samples each half-hour for 2 h, and then every hour for a total of 5 h. After her test, Kate was told that her blood glucose was 178 mg/dL, which indicates hyperglycemia. The prefix *hyper* means above or over; *hypo* means below or under. The term *glyc* or *gluco* refers to "sugar." Thus, the blood sugar level in *hyperglycemia* is above normal and in *hypoglycemia* it is below normal.

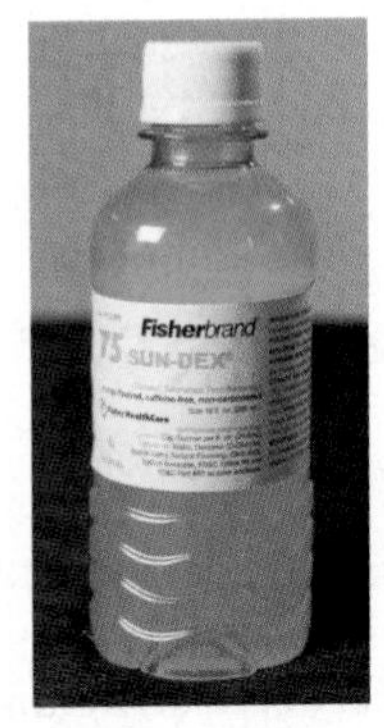

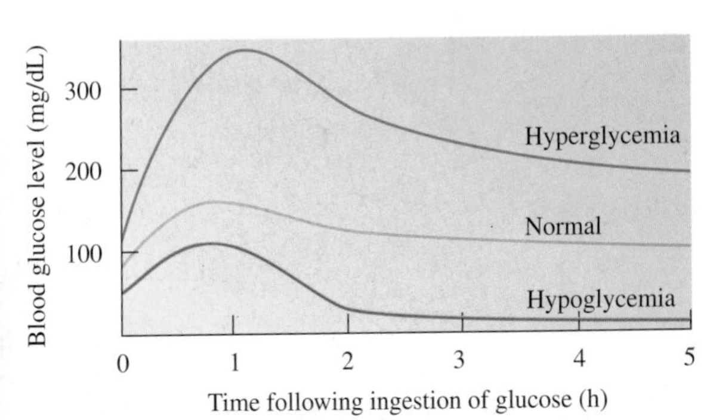

A glucose solution is given to determine blood glucose levels.

An example of a disease that can cause hyperglycemia is *type 2 diabetes*, which occurs when the pancreas is unable to produce sufficient quantities of insulin. As a result, glucose levels in the body fluids can rise as high as 350 mg/dL. Kate's symptoms of type 2 diabetes include thirst, excessive urination, and increased appetite. In older adults, type 2 diabetes is sometimes a consequence of excessive weight gain.

When a person is hypoglycemic, the blood glucose level rises and then decreases rapidly to levels as low as 40 mg/dL. In some cases, hypoglycemia is caused by overproduction of insulin by the pancreas. Low blood glucose can cause dizziness, general weakness, and muscle tremors. A diet may be prescribed that consists of several small meals high in protein and low in carbohydrate. Some hypoglycemic patients are finding success with diets that include more complex carbohydrates rather than simple sugars.

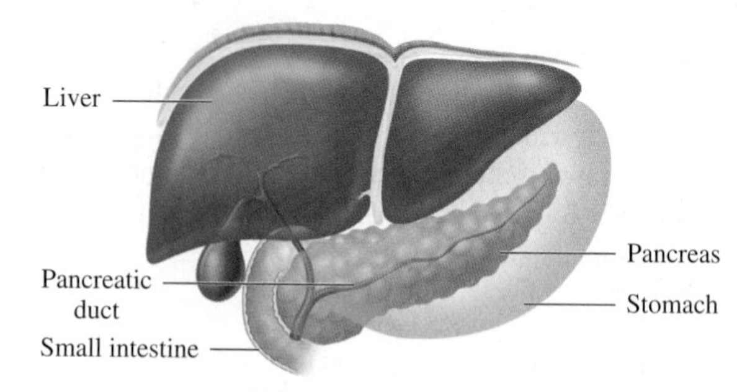

The insulin produced in the pancreas is needed in the digestive system for the metabolism of glucose.

PRACTICE PROBLEMS

13.3 Fischer Projections of Monosaccharides

LEARNING GOAL Identify or draw the D and L configurations of the Fischer projections for common monosaccharides.

13.21 Identify each of the following as the D or L enantiomer:

a.
H–C=O
HO—H
H—OH
CH_2OH
Threose

b.
CH_2OH
C=O
HO—H
H—OH
CH_2OH
Xylulose

c.
H–C=O
H—OH
H—OH
HO—H
HO—H
CH_2OH
Mannose

d.
H–C=O
H—OH
H—OH
H—OH
H—OH
CH_2OH
Allose

13.22 Identify each of the following as the D or L enantiomer:

a.
H–C=O
HO—H
H—OH
H—OH
CH_2OH
Arabinose

b.
CH_2OH
C=O
HO—H
H—OH
HO—H
CH_2OH
Sorbose

c.
H–C=O
HO—H
HO—H
H—OH
CH_2OH
Lyxose

d.
H–C=O
HO—H
HO—H
HO—H
CH_2OH
Ribose

13.23 Draw the Fischer projection for the other enantiomer of **a** to **d** in problem 13.21.

13.24 Draw the Fischer projection for the other enantiomer of **a** to **d** in problem 13.22.

13.25 Draw the Fischer projections for D-glucose and L-glucose.

13.26 Draw the Fischer projections for D-fructose and L-fructose.

Clinical Applications

13.27 An infant with galactosemia can utilize D-glucose in milk but not D-galactose. How does the Fischer projection of D-galactose differ from that of D-glucose?

13.28 D-Fructose is the sweetest monosaccharide. How does the Fischer projection of D-fructose differ from that of D-glucose?

13.29 Identify the monosaccharide that fits each of the following descriptions:
a. is also called blood sugar
b. is not metabolized in a condition known as galactosemia

13.30 Identify a monosaccharide that fits each of the following descriptions:
a. found in high blood levels in diabetes
b. is also called fruit sugar

13.4 Haworth Structures of Monosaccharides

LEARNING GOAL Draw and identify the Haworth structures for monosaccharides.

CORE CHEMISTRY SKILL
Drawing Haworth Structures

Up to now, we have drawn the Fischer projections for monosaccharides as open chains. However, the most stable form of pentoses and hexoses are five- or six-atom rings. These rings, known as **Haworth structures**, are produced from the reaction of a carbonyl group and a hydroxyl group in the *same* molecule. We will now show how to draw the Haworth structure for D-glucose from its Fischer projection.

Drawing Haworth Structures

To convert a Fischer projection to a Haworth structure, turn the Fischer projection clockwise by 90°. The —H and —OH groups on the right of the vertical carbon chain are now below the horizontal carbon chain. Those on the left of the open chain are now above the horizontal carbon chain.

Turn clockwise by 90°

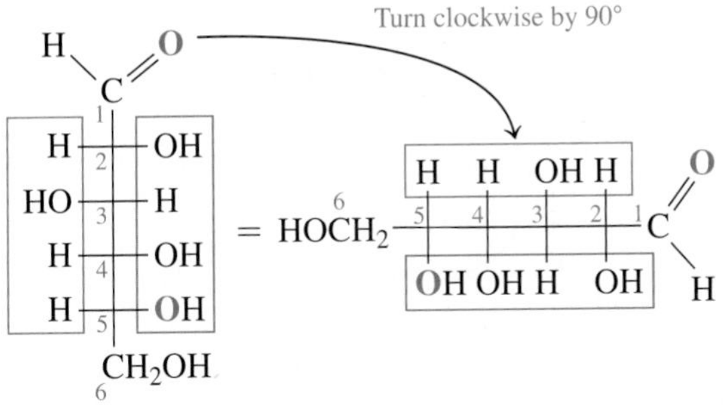

D-Glucose (open chain)

With carbons 2 and 3 as the base of a hexagon, move the remaining carbons upward. Rotate the groups on carbon 5 so that the —OH group is close to carbon 1. To complete the Haworth structure, draw a bond from the oxygen of the —OH group on carbon 5 to carbon 1. In the line-angle structures, the carbon atoms in the ring are drawn as corners.

ENGAGE

Why is an —OH group drawn on carbon 1 in the Haworth structure of an aldohexose?

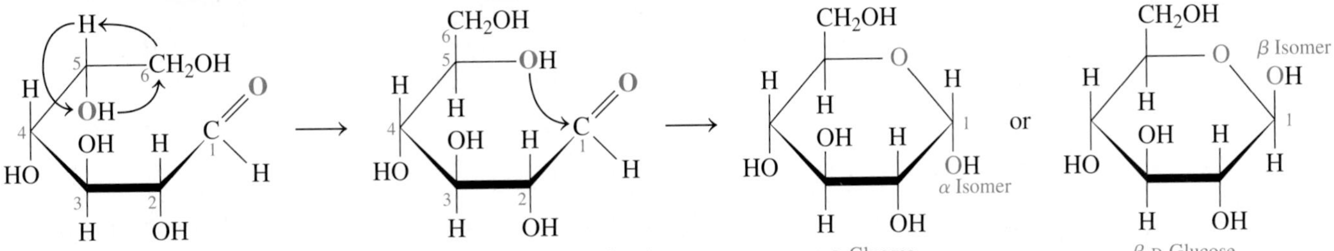

Rotation of groups on carbon 5 → Carbon-5 oxygen bonds to carbon 1 → α-D-Glucose (α Isomer) or β-D-Glucose (β Isomer)

Because the —OH group on the new chiral carbon (1) can be drawn above or below the plane of the Haworth structure, there are two isomers of D-glucose. In the α (alpha) isomer, the —OH group is drawn below the plane of the ring. In the β (beta) isomer, the —OH group is drawn above the plane of the ring. The carbon atoms in the ring are drawn as corners.

Mutarotation of α- and β-D-Glucose

In an aqueous solution, the Haworth structure of α-D-glucose opens to give the open chain of D-glucose, which has an aldehyde group. At any given time, there is only a trace amount of the open chain because it closes quickly to form a stable cyclic structure. However, when the open chain closes again it can form β-D-glucose. In this process called *mutarotation*, each isomer converts to the open chain and back again. As the ring opens and closes, the —OH group on carbon 1 can form either the α or β isomer. An aqueous glucose solution contains a mixture of 36% α-D-glucose and 64% β-D-glucose.

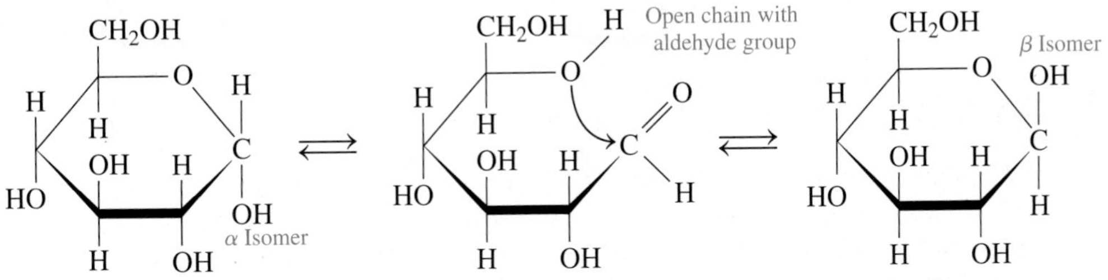

α-D-Glucose (36%) ⇌ D-Glucose open chain (trace) (Open chain with aldehyde group) ⇌ β-D-Glucose (64%)

Haworth Structures of Galactose

Galactose is an aldohexose that differs from glucose only in the arrangement of the —OH group on carbon 4. Thus, its Haworth structure is similar to glucose, except that the —OH group on carbon 4 is drawn above the ring. Galactose also exists as α and β isomers.

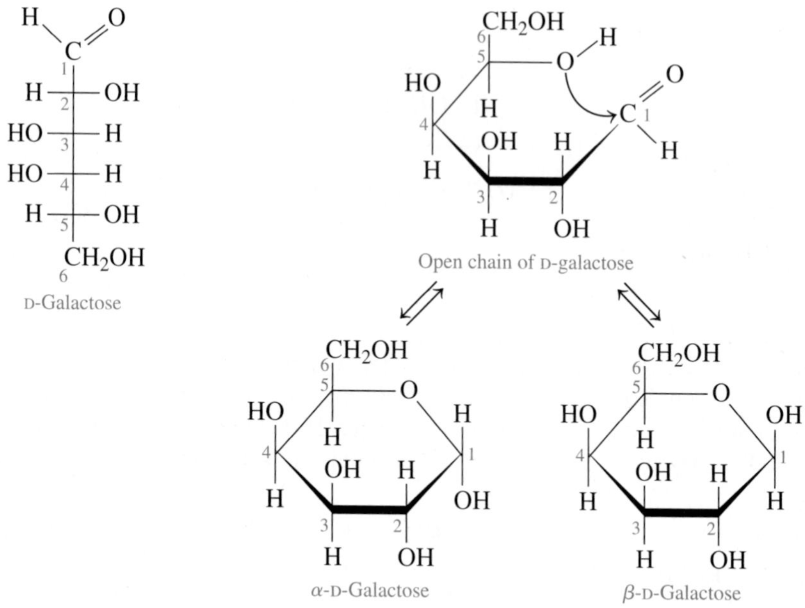

D-Galactose

Open chain of D-galactose

α-D-Galactose

β-D-Galactose

Haworth Structures of Fructose

In contrast to glucose and galactose, fructose is a ketohexose. The Haworth structure for fructose is a five-atom ring with carbon 2 at the right corner of the ring. The cyclic structure forms when the —OH group on carbon 5 bonds to carbon 2 in the carbonyl group. The new —OH group on carbon 2 gives the α and β isomers of fructose.

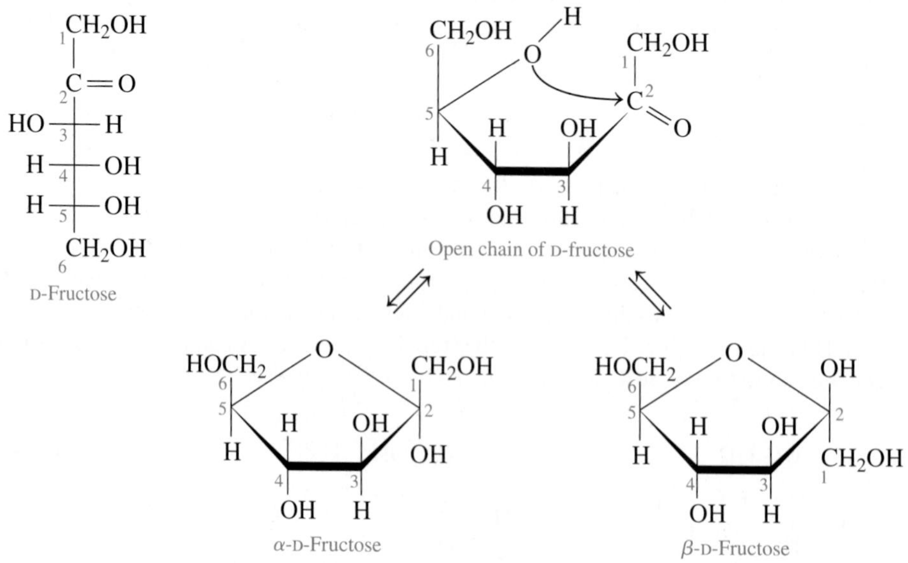

D-Fructose

Open chain of D-fructose

α-D-Fructose

β-D-Fructose

SAMPLE PROBLEM 13.5 Drawing Haworth Structures for Sugars

TRY IT FIRST

D-Mannose, a carbohydrate found in immunoglobulins, has the following Fischer projection. Draw the Haworth structure for β-D-mannose.

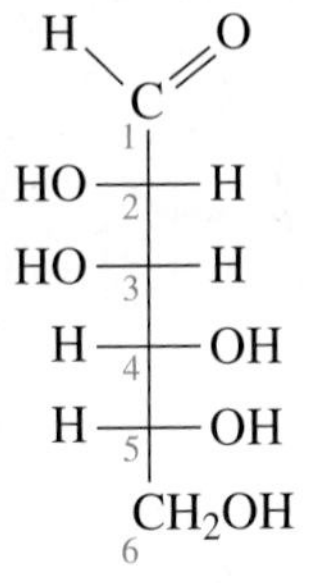

D-Mannose

SOLUTION GUIDE

STEP 1 Turn the Fischer projection clockwise by 90°.

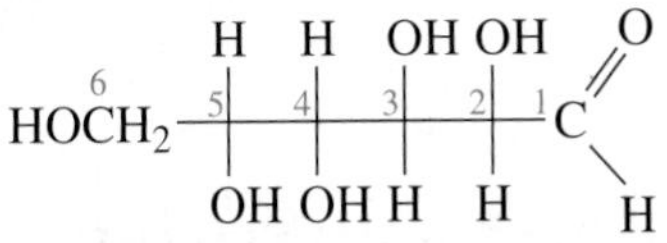

STEP 2 Fold the horizontal carbon chain into a hexagon, rotate the groups on carbon 5, and bond the O on carbon 5 to carbon 1.

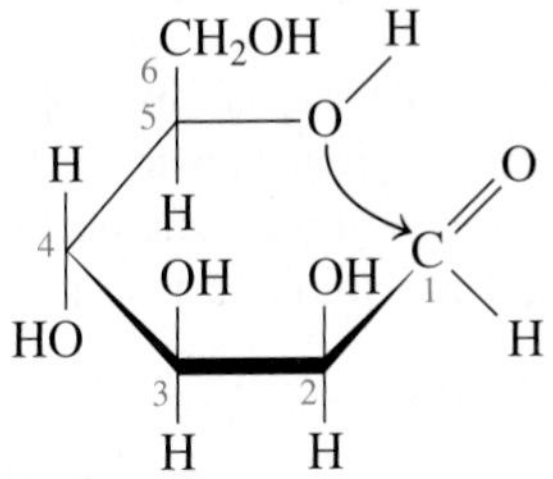

STEP 3 Draw the new —OH group on carbon 1 above the ring to give the β isomer.

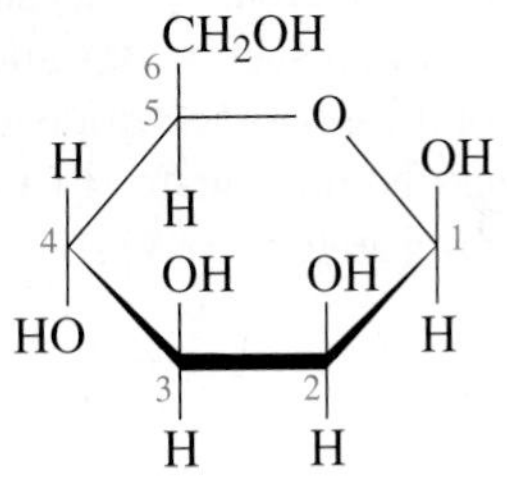

STUDY CHECK 13.5

Draw the Haworth structure for α-D-mannose.

ANSWER

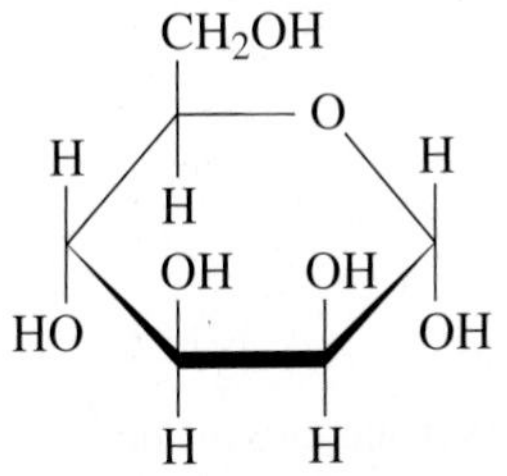

TEST

Try Practice Problems 13.31 to 13.36

PRACTICE PROBLEMS

13.4 Haworth Structures of Monosaccharides

LEARNING GOAL Draw and identify the Haworth structures for monosaccharides.

13.31 What are the kind and number of atoms in the ring portion of the Haworth structure of glucose?

13.32 What are the kind and number of atoms in the ring portion of the Haworth structure of fructose?

13.33 Draw the Haworth structures for α- and β-D-glucose.

13.34 Draw the Haworth structures for α- and β-D-fructose.

13.35 Identify each of the following as the α or β isomer:

a.

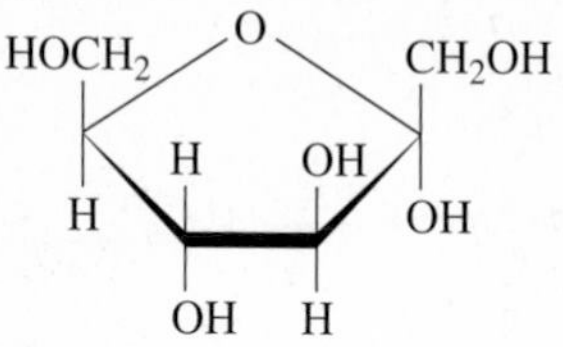

b.

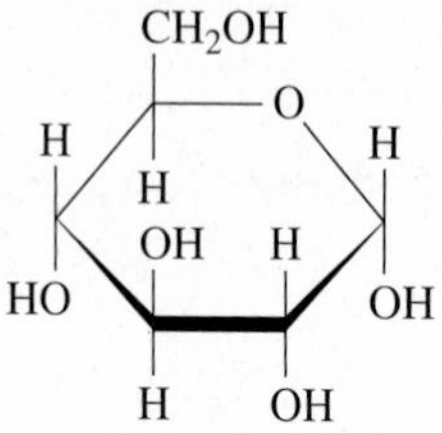

13.36 Identify each of the following as the α or β isomer:

a.

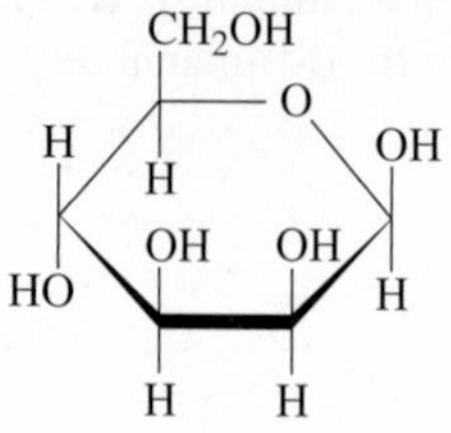

b.

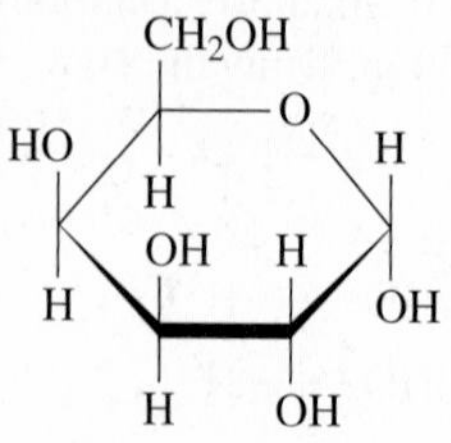

REVIEW

Writing Equations for the Oxidation of Alcohols (12.4)

ENGAGE

When is a monosaccharide called a reducing sugar?

13.5 Chemical Properties of Monosaccharides

LEARNING GOAL Identify the products of oxidation or reduction of monosaccharides; determine whether a carbohydrate is a reducing sugar.

Monosaccharides contain functional groups that can undergo chemical reactions. In an aldose, the aldehyde group can be oxidized to a carboxylic acid. The carbonyl group in both an aldose and a ketose can be reduced to give a hydroxyl group. The hydroxyl groups can react with other compounds to form a variety of derivatives that are important in biological structures.

Oxidation of Monosaccharides

Although monosaccharides exist mostly in cyclic forms, a small amount of the open-chain form is always present, which provides an aldehyde group. An aldehyde group with an adjacent hydroxyl can be oxidized to a carboxylic acid by an oxidizing agent such as Benedict's reagent. The sugar acids are named by replacing the *ose* ending of the monosaccharide with *onic acid.* Then the Cu^{2+} is reduced to Cu^{+}, which forms a brick-red precipitate of Cu_2O. A carbohydrate that reduces another substance is called a **reducing sugar**.

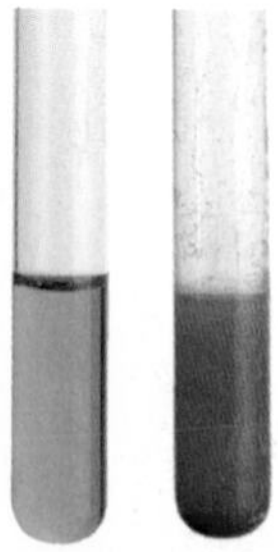

Benedict's solution gives a brick-red precipitate with reducing sugars.

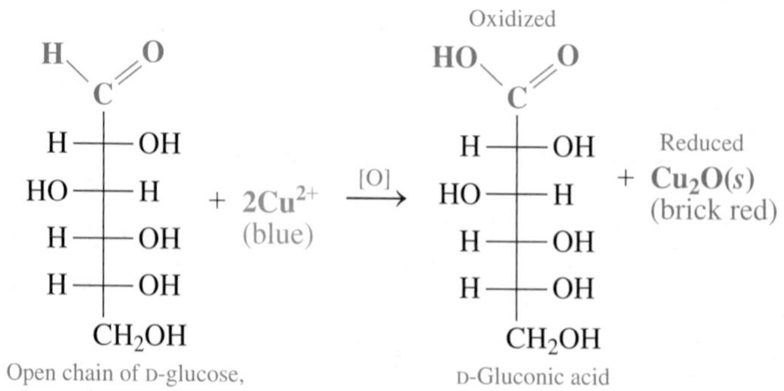

Open chain of D-glucose, a reducing sugar

D-Gluconic acid

Fructose, a ketohexose, is also a reducing sugar. Usually a ketone cannot be oxidized. However, in a basic Benedict's solution, a rearrangement occurs between the ketone group on carbon 2 and the hydroxyl group on carbon 1. As a result, fructose is converted to glucose, which produces an aldehyde group with an adjacent hydroxyl that can be oxidized.

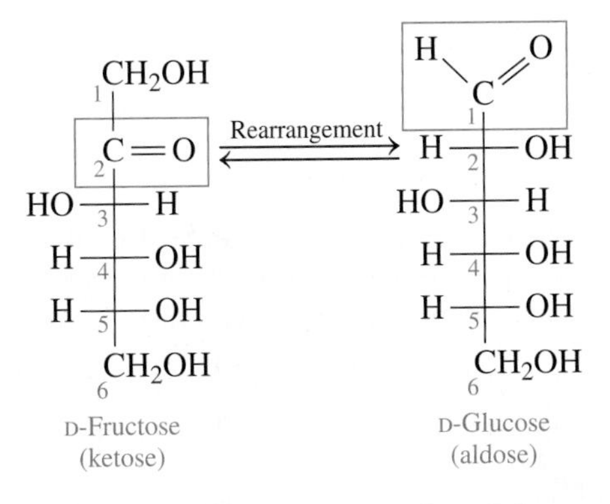

D-Fructose (ketose)

D-Glucose (aldose)

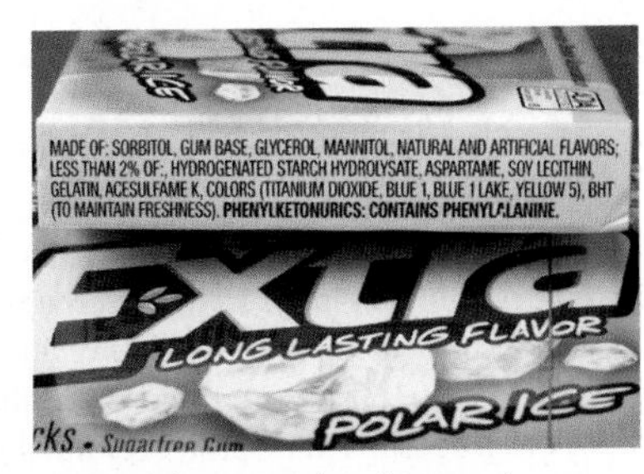

D-glucitol or D-sorbitol is used as a low-calorie sweetener in diet drinks and sugarless gum.

Reduction of Monosaccharides

The reduction of the carbonyl group in monosaccharides produces sugar alcohols, which are also called *alditols*. D-Glucose is reduced to D-glucitol, better known as D-sorbitol. The sugar alcohols are named by replacing the *ose* ending of the monosaccharide with *itol*. Sugar alcohols such as D-sorbitol, D-xylitol from D-xylose, and D-mannitol from D-mannose are used as sweeteners in many sugar-free products such as diet drinks and sugarless gum as well as products for people with diabetes. Some people experience some discomfort such as gas and diarrhea from the ingestion of sugar alcohols. The development of cataracts in diabetics is attributed to the accumulation of D-sorbitol in the lens of the eye.

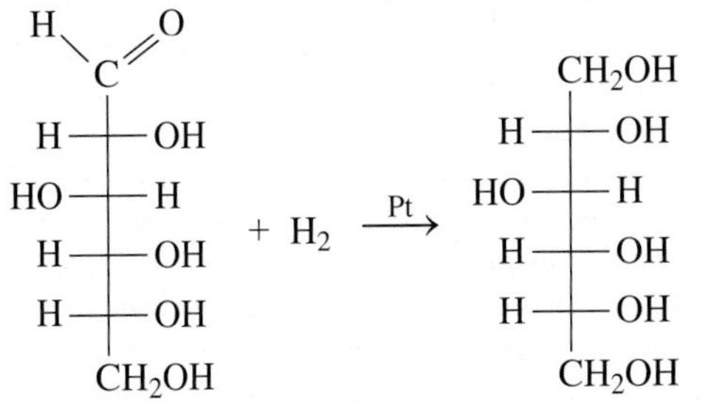

H C O
H—OH
HO—H
H—OH
H—OH
CH_2OH

D-Glucose

$+ H_2 \xrightarrow{Pt}$

CH_2OH
H—OH
HO—H
H—OH
H—OH
CH_2OH

D-Glucitol or D-Sorbitol

SAMPLE PROBLEM 13.6 Reducing Sugars

TRY IT FIRST

Draw and name the oxidation product of D-altrose. Why is D-altrose a reducing sugar?

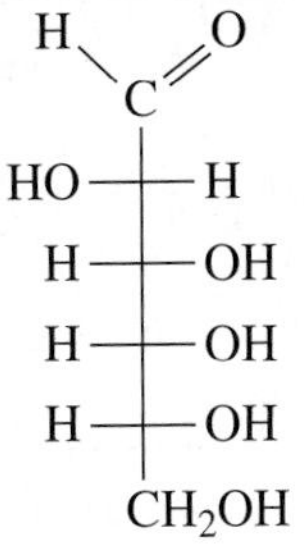

H C O
HO—H
H—OH
H—OH
H—OH
CH_2OH

D-Altrose

SOLUTION

HO C O
HO—H
H—OH
H—OH
H—OH
CH_2OH

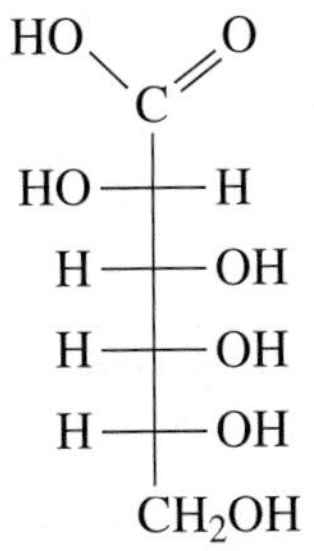

D-Altronic acid

To name the sugar acid produced by oxidation, we replace the *ose* ending of D-altrose with *onic acid* to give D-altronic acid. A carbohydrate that reduces Cu^{2+} to Cu^{+} is called a reducing sugar.

STUDY CHECK 13.6

A solution containing Benedict's reagent turns brick red with a urine sample. What might this result indicate?

ANSWER

The brick-red color of the Benedict's reagent shows a high level of reducing sugar (probably glucose) in the urine, which may indicate type 2 diabetes.

TEST

Try Practice Problems 13.37 to 13.40

CHEMISTRY LINK TO HEALTH

Testing for Glucose

Normally, blood glucose flows through the kidneys and is reabsorbed into the bloodstream. However, if the blood plasma level exceeds about 160 mg of glucose/dL, the kidneys cannot reabsorb all of the glucose, and it spills over into the urine, a condition known as *glucosuria*. A symptom of diabetes is a high level of glucose in the urine.

Benedict's test can be used to determine the presence of glucose in urine. The amount of copper(I) oxide (Cu_2O) formed is proportional to the amount of reducing sugar present in the urine. Low to moderate levels of reducing sugar turn the solution green; solutions with high glucose levels turn Benedict's reagent yellow or brick red.

In another clinical test, the presence of glucose in the urine can be detected using a paper strip containing 2-methylaniline and two enzymes, glucose oxidase and peroxidase. When the paper strip comes in contact with any glucose in the urine, glucose and oxygen are converted by glucose oxidase to gluconic acid and hydrogen peroxide.

Then peroxidase converts the hydrogen peroxide into a compound that reacts with 2-methylaniline in the test strip to form a color ranging from green to a dark brown. The intensity of the color depends on the amount of glucose present in the urine. **TABLE 13.1** lists some colors associated with the concentration of glucose in the urine.

$$H_2O_2 + \text{dye (2-methylaniline)} \xrightarrow{\text{Peroxidase}} \text{colored products}$$

Oxidized

H–C=O; H–C–OH; HO–C–H; H–C–OH; H–C–OH; CH_2OH (D-Glucose) $+ O_2$ $\xrightarrow{\text{Glucose oxidase}}$ HO–C=O; H–C–OH; HO–C–H; H–C–OH; H–C–OH; CH_2OH (D-Gluconic acid) $+ H_2O_2$ (Hydrogen peroxide)

D-Glucose

D-Gluconic acid

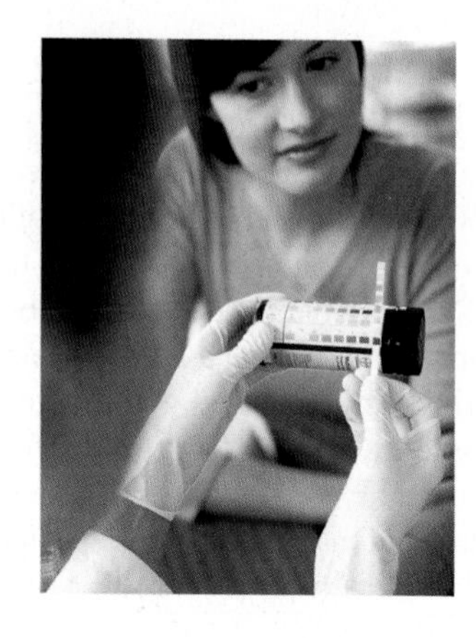

The color of the dye 2-methylaniline on a test strip determines the glucose level in urine.

NH_2

2-Methylaniline

TABLE 13.1 Glucose Oxidase Test Results

	Glucose Present in Urine	
Color	**% (m/v)**	**mg/dL**
Blue	0	0
Blue-green	0.10	100
Green	0.25	250
Green-brown	0.50	500
Brown	1.00	1000
Dark brown	2.00	2000

PRACTICE PROBLEMS

13.5 Chemical Properties of Monosaccharides

LEARNING GOAL Identify the products of oxidation or reduction of monosaccharides; determine whether a carbohydrate is a reducing sugar.

13.37 Draw the Fischer projection for the oxidation and the reduction products of D-xylose. What are the names of the sugar acid and the sugar alcohol produced?

H–C=O; H–C–OH; HO–C–H; H–C–OH; CH_2OH

D-Xylose

13.38 Draw the Fischer projection for the oxidation and the reduction products of D-mannose. What are the names of the sugar acid and the sugar alcohol produced?

H–C=O; HO–C–H; HO–C–H; H–C–OH; H–C–OH; CH_2OH

D-Mannose

13.39 Draw the Fischer projection for the oxidation and the reduction products of D-arabinose. What are the names of the sugar acid and the sugar alcohol produced?

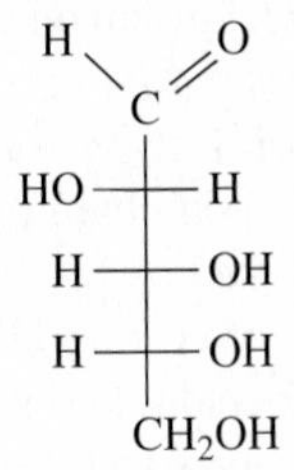

H O
C
HO — H
H — OH
H — OH
CH_2OH
D-Arabinose

13.40 Draw the Fischer projection for the oxidation and the reduction products of D-ribose. What are the names of the sugar acid and the sugar alcohol produced?

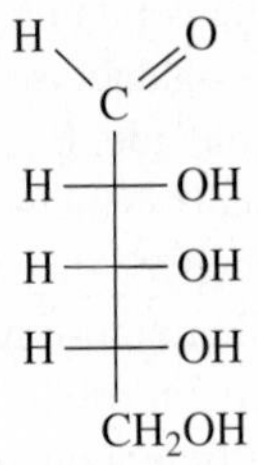

H O
C
H — OH
H — OH
H — OH
CH_2OH
D-Ribose

13.6 Disaccharides

LEARNING GOAL Describe the monosaccharide units and linkages in disaccharides.

A disaccharide is composed of two monosaccharides linked together. The most common disaccharides are maltose, lactose, and sucrose. When two monosaccharides combine in a dehydration reaction, the product is a disaccharide. The reaction occurs between the hydroxyl group on carbon 1 and one of the hydroxyl groups on a second monosaccharide.

$$\text{Glucose} + \text{glucose} \xrightarrow{\text{Maltose synthase}} \text{maltose} + H_2O$$
$$\text{Glucose} + \text{galactose} \xrightarrow{\text{Lactose synthase}} \text{lactose} + H_2O$$
$$\text{Glucose} + \text{fructose} \xrightarrow{\text{Sucrose synthase}} \text{sucrose} + H_2O$$

ENGAGE
Which disaccharide contains only glucose?

Maltose, or malt sugar, is obtained from starch and is found in germinating grains. When maltose in barley and other grains is hydrolyzed by yeast enzymes, glucose is obtained, which can undergo fermentation to give ethanol. Maltose is used in cereals, candies, and the brewing of beverages.

In the Haworth structure of a disaccharide, a **glycosidic bond** connects two monosaccharides. In maltose, a glycosidic bond forms between the —OH groups of carbons 1 and 4 of two α-D-glucose molecules with a loss of a water molecule. The glycosidic bond in maltose is designated as an α(1→4) linkage to show that an alpha —OH group on carbon 1 is joined to carbon 4 of the second glucose molecule. Because the second glucose molecule

CH_2OH … O … H, H, OH, H, HO, OH (1) + HO (4) … CH_2OH … O … H, H, OH, H, OH

α-D-Glucose   α-D-Glucose

↓

CH_2OH … O … (1) O (4) … CH_2OH … O … OH + H_2O

α Isomer

α(1→4)-Glycosidic bond

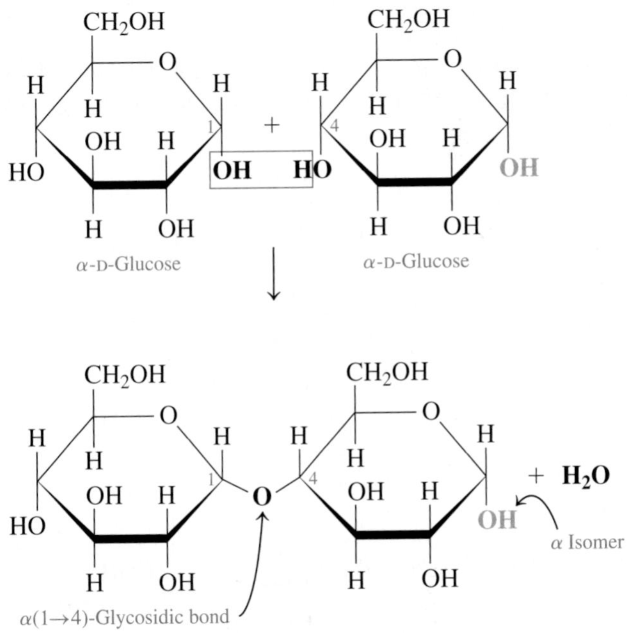

α-Maltose, a disaccharide

ENGAGE
How is α-maltose different from β-maltose?

still has a free —OH group on carbon 1, it can form an open chain, which allows maltose to form both α and β isomers. The open chain provides an aldehyde group that can be oxidized, making maltose a reducing sugar.

Lactose, milk sugar, is a disaccharide found in milk and milk products (see **FIGURE 13.8**). The bond in lactose is a $\beta(1\rightarrow4)$-glycosidic bond because the —OH group on carbon 1 of β-D-galactose forms a glycosidic bond with the —OH group on carbon 4 of a D-glucose molecule. Because D-glucose still has a free —OH group on carbon 1, it can form an open chain, which allows lactose to form both α and β isomers. The open chain provides an aldehyde group that can be oxidized, making lactose a reducing sugar.

Lactose makes up 6 to 8% of human milk and about 4 to 5% of cow's milk, and it is used in products that attempt to duplicate mother's milk. When a person does not produce sufficient quantities of the enzyme lactase, which is needed to hydrolyze lactose, it remains undigested when it enters the colon. Then bacteria in the colon digest the lactose in a fermentation process that creates large amounts of gas including carbon dioxide and methane, which cause bloating and abdominal cramps. In some commercial milk products, lactase has already been added to break down lactose.

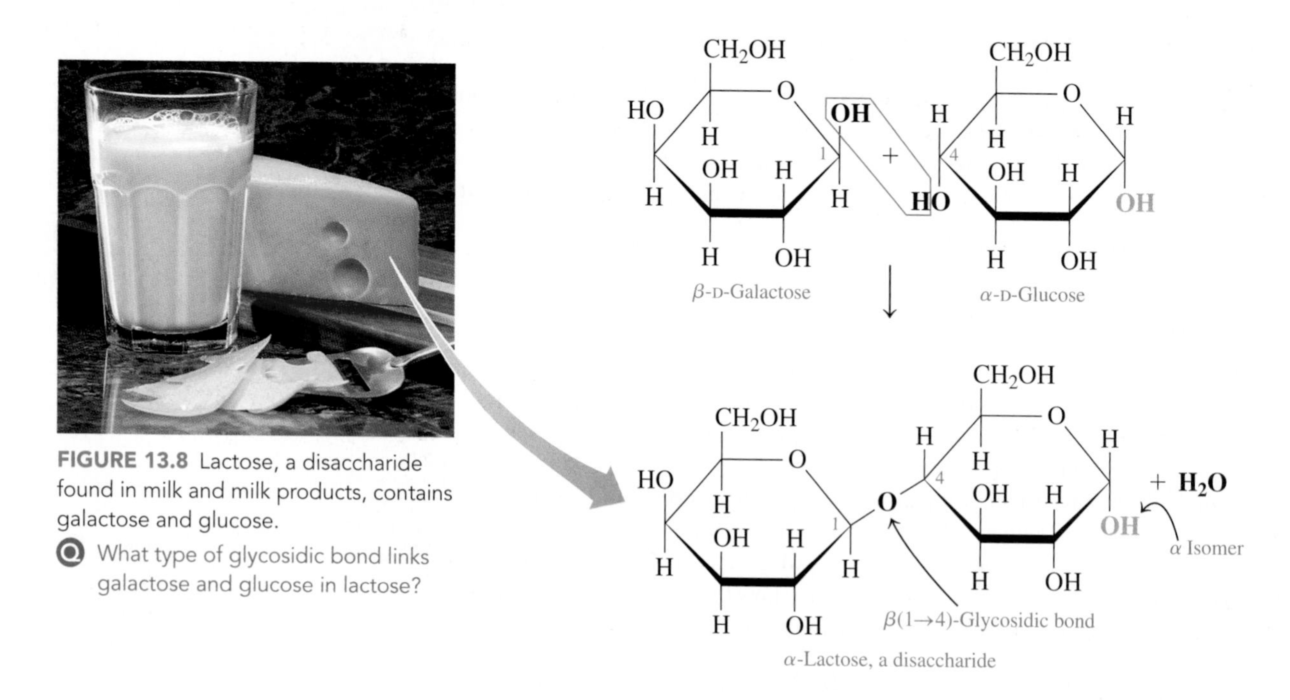

FIGURE 13.8 Lactose, a disaccharide found in milk and milk products, contains galactose and glucose.

Q What type of glycosidic bond links galactose and glucose in lactose?

Sucrose consists of an α-D-glucose and a β-D-fructose molecule joined by an $\alpha,\beta(1\rightarrow2)$-glycosidic bond (see **FIGURE 13.9**). Unlike maltose and lactose, the glycosidic

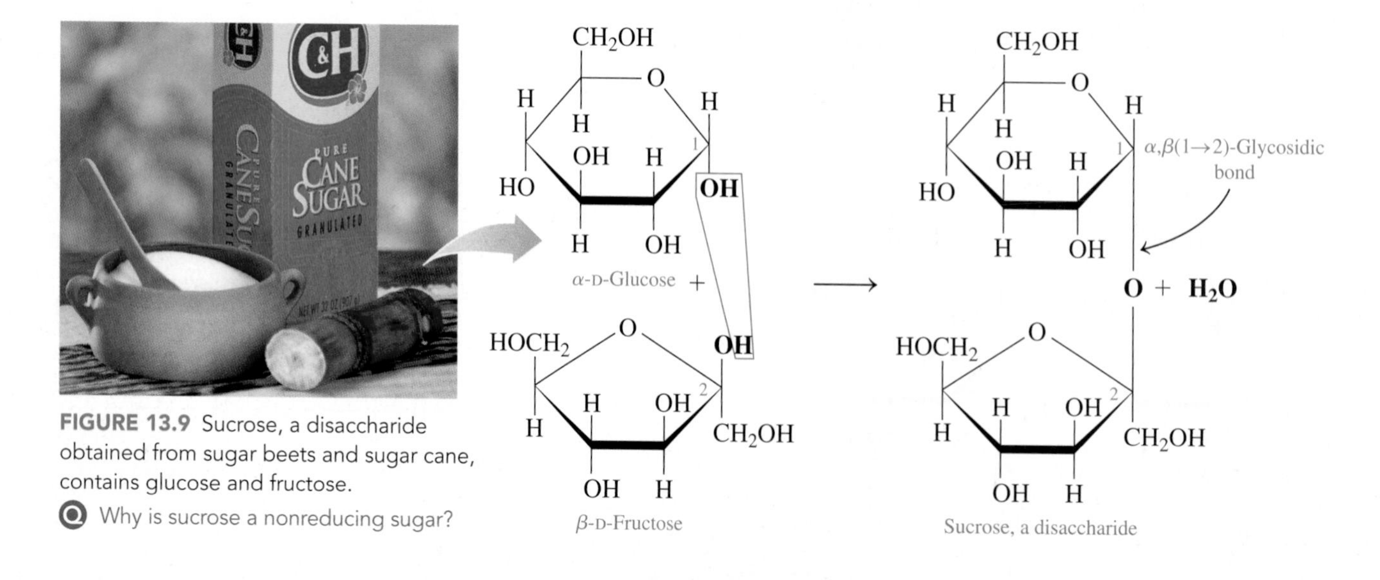

FIGURE 13.9 Sucrose, a disaccharide obtained from sugar beets and sugar cane, contains glucose and fructose.

Q Why is sucrose a nonreducing sugar?

bond in sucrose is between carbon 1 of glucose and carbon 2 of fructose. Thus, sucrose cannot form an open chain and cannot be oxidized. Sucrose cannot react with Benedict's reagent and is not a reducing sugar.

The sugar we use to sweeten our cereal, coffee, or tea is sucrose. Most of the sucrose for table sugar comes from sugar cane (20% by mass) or sugar beets (15% by mass). Both the raw and refined forms of sugar are sucrose. Some estimates indicate that each person in the United States consumes an average of 68 kg (150 lb) of sucrose every year, either by itself or in a variety of food products.

CHEMISTRY LINK TO HEALTH

How Sweet Is My Sweetener?

Although many of the monosaccharides and disaccharides taste sweet, they differ considerably in their degree of sweetness. Dietetic foods contain sweeteners that are noncarbohydrate or carbohydrates that are sweeter than sucrose. Some examples of sweeteners compared with sucrose are shown in **TABLE 13.2**.

Sucralose, which is known as Splenda, is made from sucrose by replacing some of the hydroxyl groups with chlorine atoms.

Sucralose (Splenda)

Aspartame, which is marketed as NutraSweet and Equal, is used in a large number of sugar-free products. It is a noncarbohydrate sweetener made of aspartate and a methyl ester of phenylalanine. It does have some caloric value, but it is so sweet that only a very small quantity is needed. However, phenylalanine, one of the breakdown products, poses a danger to anyone who cannot metabolize it properly, a condition called *phenylketonuria* (PKU).

From aspartate From phenylalanine

Aspartame (NutraSweet)

Another artificial sweetener, Neotame, is a modification of the aspartame structure. The addition of a large alkyl group to the amine group prevents enzymes from breaking the amide bond between aspartate and phenylalanine. Thus, phenylalanine is not produced when Neotame is used as a sweetener. Very small amounts of Neotame are needed because it is about 10 000 times sweeter than sucrose.

Large alkyl group to modify Aspartame From aspartate From phenylalanine

Neotame

TABLE 13.2 Relative Sweetness of Sugars and Artificial Sweeteners

Type	Sweetness Relative to Sucrose (= 100)
Monosaccharides	
Galactose	30
Glucose	75
Fructose	175
Disaccharides	
Lactose	16
Maltose	33
Sucrose	100
Sugar Alcohols	
Sorbitol	60
Maltitol	80
Xylitol	100
Noncarbohydrate Sweeteners	
Stevia	15 000
Aspartame	18 000
Saccharin	45 000
Sucralose	60 000
Neotame	1 000 000
Advantame	2 000 000

Advantame is a recently developed sweetener used in desserts, beverages, jams, toppings, and chewing gum. It is synthesized from aspartame and isovanillin and 20 000 times sweeter than sucrose.

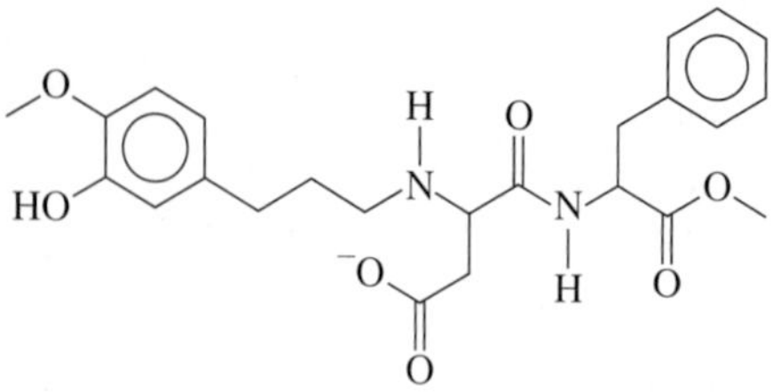

From isovanillin From aspartate From phenylalanine

Advantame

Saccharin, which is marketed as Sweet'N Low, has been used as a noncarbohydrate artificial sweetener for the past 35 years.

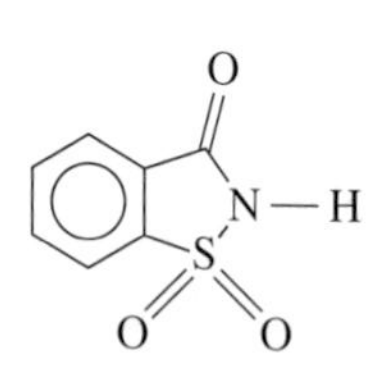

Saccharin (Sweet'N Low)

Artificial sweeteners are used as sugar substitutes.

Stevia is a sugar substitute obtained from the leaves of a plant *Stevia rebaudiana*. It is composed of steviol glycosides that are about 150 times sweeter than sucrose. Stevia has been used to sweeten tea and medicines in South America for 1500 years.

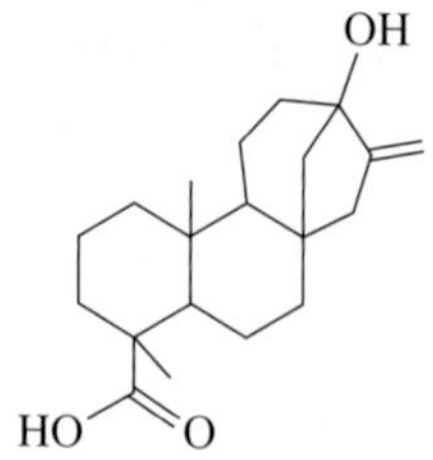

Steviol

CHEMISTRY LINK TO HEALTH

Blood Types and Carbohydrates

Every individual's blood can be typed as one of four blood groups: A, B, AB, and O. Although there is some variation among ethnic groups in the United States, the incidence of blood types in the general population is about 43% O, 40% A, 12% B, and 5% AB.

The blood types A, B, and O are determined by terminal saccharides attached to the surface of red blood cells. Blood type O has three terminal monosaccharides: *N*-acetylglucosamine, galactose, and fucose.

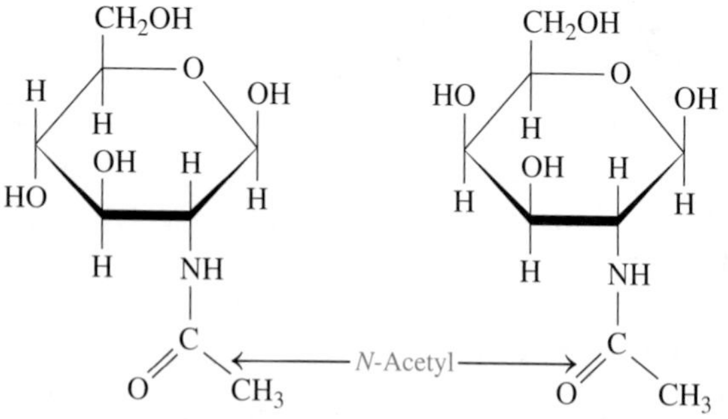

N-Acetyl

N-Acetylglucosamine (*N*-AcGlu) *N*-Acetylgalactosamine (*N*-AcGal) L-Fucose (Fuc) 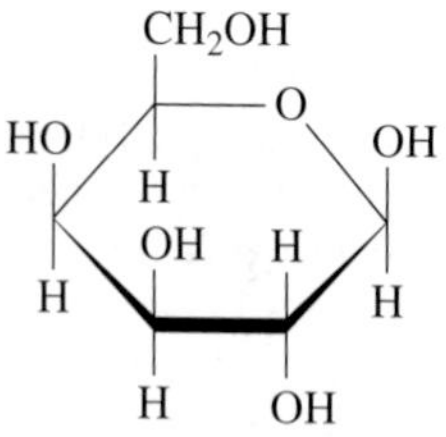 D-Galactose (Gal)

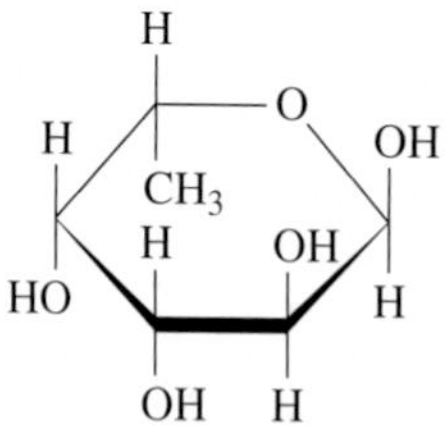

Terminal Saccharides for Each Blood Type

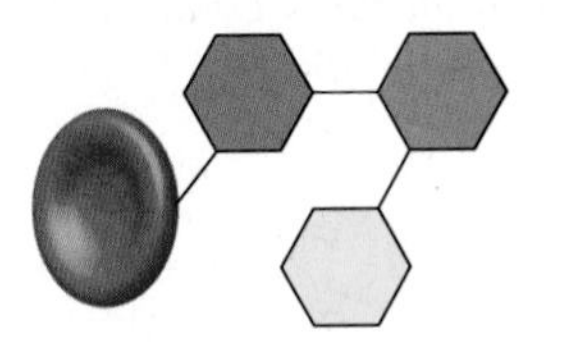

Type O

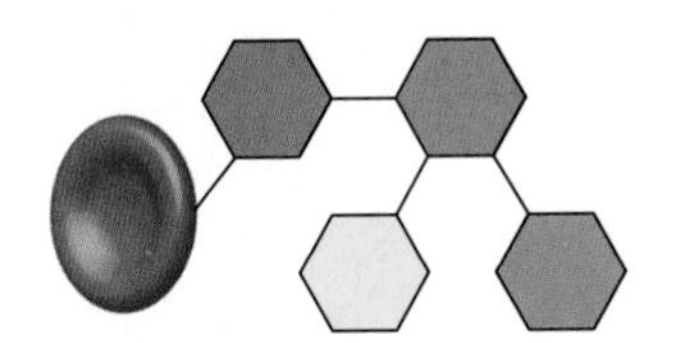

Type A

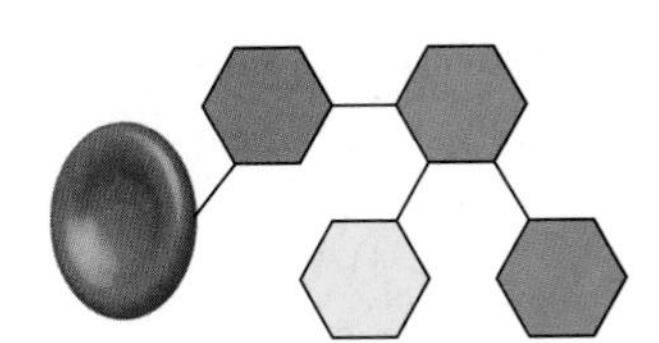

Type B

Red blood cell surface

N-Acetylglucosamine

Fucose

N-Acetylgalactosamine

Galactose

Blood type A contains the same three monosaccharides, but in addition, a molecule of *N*-acetylgalactosamine is attached to galactose in the saccharide chain. Blood type B also contains the same three monosaccharides, but in addition, a second molecule of galactose is attached to the saccharide chain. Blood type AB consists of the same monosaccharides found in blood types A and B. The structures of these monosaccharides are as follows:

Because type O blood has only the three common terminal monosaccharides, a person with type O produces antibodies against blood types A, B, and AB. However, persons with blood types A, B, and AB can receive type O blood. Thus, persons with type O blood are *universal donors*. Because type AB blood contains all the terminal monosaccharides, a person with type AB blood produces no antibodies to type A, B, or O blood. Persons with type AB blood are *universal recipients*.

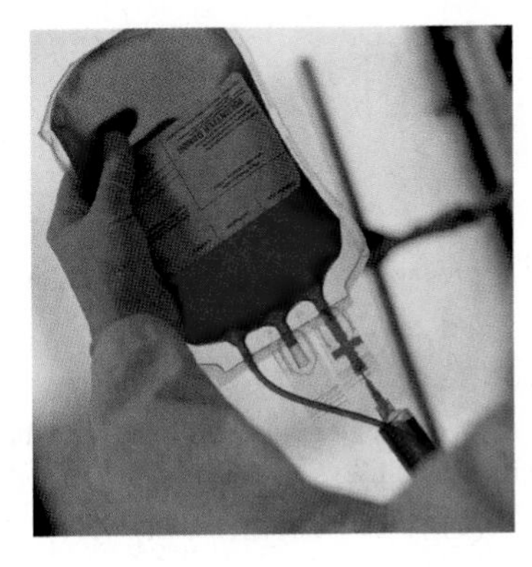

Blood from a donor is screened to make sure that there is an exact match with the blood type of the recipient.

SAMPLE PROBLEM 13.7 Glycosidic Bonds in Disaccharides

TRY IT FIRST

Melibiose is a disaccharide that is 30 times sweeter than sucrose.

a. What are the monosaccharide units in melibiose?
b. What is the glycosidic link in melibiose?
c. Is this the α or β isomer of melibiose?
d. Is melibiose a reducing sugar?

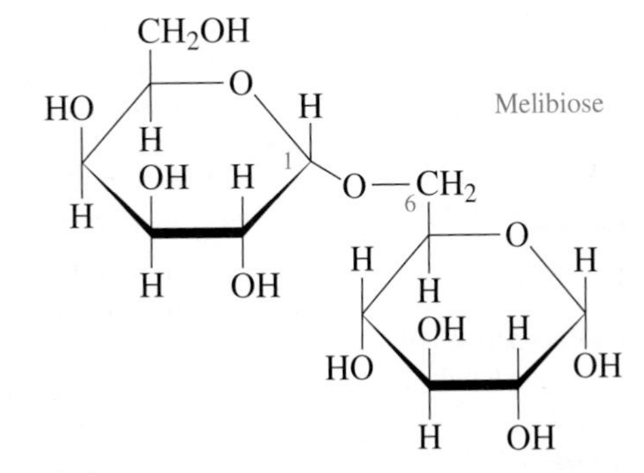

SOLUTION

a. First monosaccharide (left)	When the —OH group on carbon 4 is above the plane, it is D-galactose. When the —OH group on carbon 1 is below the plane, it is α-D-galactose.
Second monosaccharide (right)	When the —OH group on carbon 4 is below the plane, it is α-D-glucose.
b. Type of glycosidic bond	The —OH group at carbon 1 of α-D-galactose bonds with the —OH group on carbon 6 of glucose, which makes it an $\alpha(1\rightarrow6)$-glycosidic bond.
c. Name of disaccharide	The —OH group on carbon 1 of glucose is below the plane, which is α-melibiose.

d. Glucose on the right can open to form an aldehyde that can be oxidized, which makes melibiose a reducing sugar.

STUDY CHECK 13.7

Cellobiose is a disaccharide composed of two D-glucose molecules connected by a $\beta(1\rightarrow4)$-glycosidic linkage. Draw the Haworth structure for β-cellobiose.

ANSWER

TEST

Try Practice Problems 13.41 to 13.46

PRACTICE PROBLEMS

13.6 Disaccharides

LEARNING GOAL Describe the monosaccharide units and linkages in disaccharides.

13.41 For each of the following, give the monosaccharide units produced by hydrolysis, the type of glycosidic bond, and the name of the disaccharide including α or β:

a.

b.

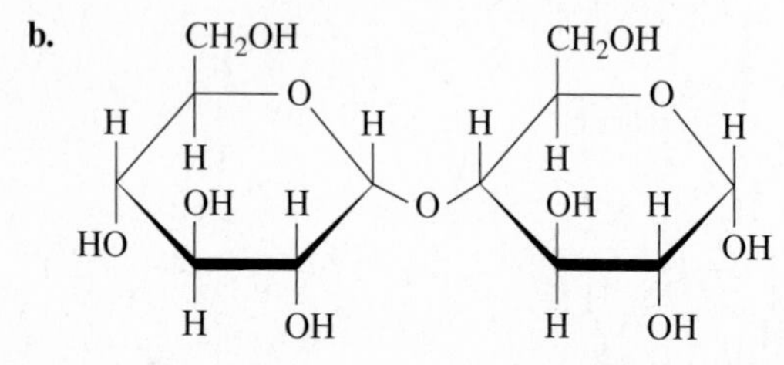

13.42 For each of the following, give the monosaccharide units produced by hydrolysis, the type of glycosidic bond, and the name of the disaccharide including α or β:

a.

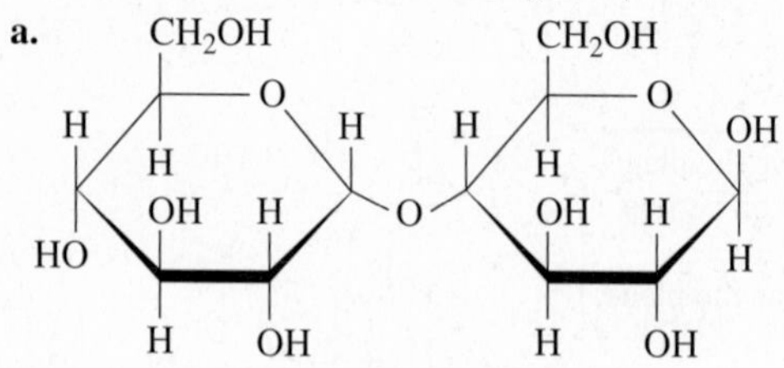

b.

13.43 Indicate whether each disaccharide in Problem 13.41 is a reducing sugar or not.

13.44 Indicate whether each disaccharide in Problem 13.42 is a reducing sugar or not.

Clinical Applications

13.45 Identify the disaccharide that fits each of the following descriptions:

a. ordinary table sugar
b. found in milk and milk products
c. also called malt sugar
d. hydrolysis gives galactose and glucose

13.46 Identify the disaccharide that fits each of the following descriptions:

a. not a reducing sugar
b. composed of two glucose units
c. also called milk sugar
d. hydrolysis gives glucose and fructose

13.7 Polysaccharides

LEARNING GOAL Describe the structural features of amylose, amylopectin, glycogen, and cellulose.

A polysaccharide is a polymer of many monosaccharides joined together. Four important polysaccharides—*amylose, amylopectin, cellulose*, and *glycogen*—are all polymers of D-glucose that differ only in the type of glycosidic bonds and the amount of branching in the molecule.

Starch

ENGAGE

What types of glycosidic bonds are present in amylose and amylopectin?

Starch, a storage form of glucose in plants, is found as insoluble granules in rice, wheat, potatoes, beans, and cereals. Starch is composed of two kinds of polysaccharides, amylose and amylopectin. **Amylose**, which makes up about 20% of starch, consists of 250 to 4000 α-D-glucose molecules connected by $\alpha(1\rightarrow4)$-glycosidic bonds in a continuous chain. Sometimes called a straight-chain polymer, polymers of amylose are actually coiled in helical fashion (see **FIGURE 13.10a**).

$\alpha(1\rightarrow4)$-Glycosidic bond

(a) Unbranched chain of amylose

Amylose

Glucose monomers

Amylopectin, glycogen

$\alpha(1\rightarrow6)$-Glycosidic bond to branch

$\alpha(1\rightarrow4)$-Glycosidic bond

(b) Branched chain of amylopectin

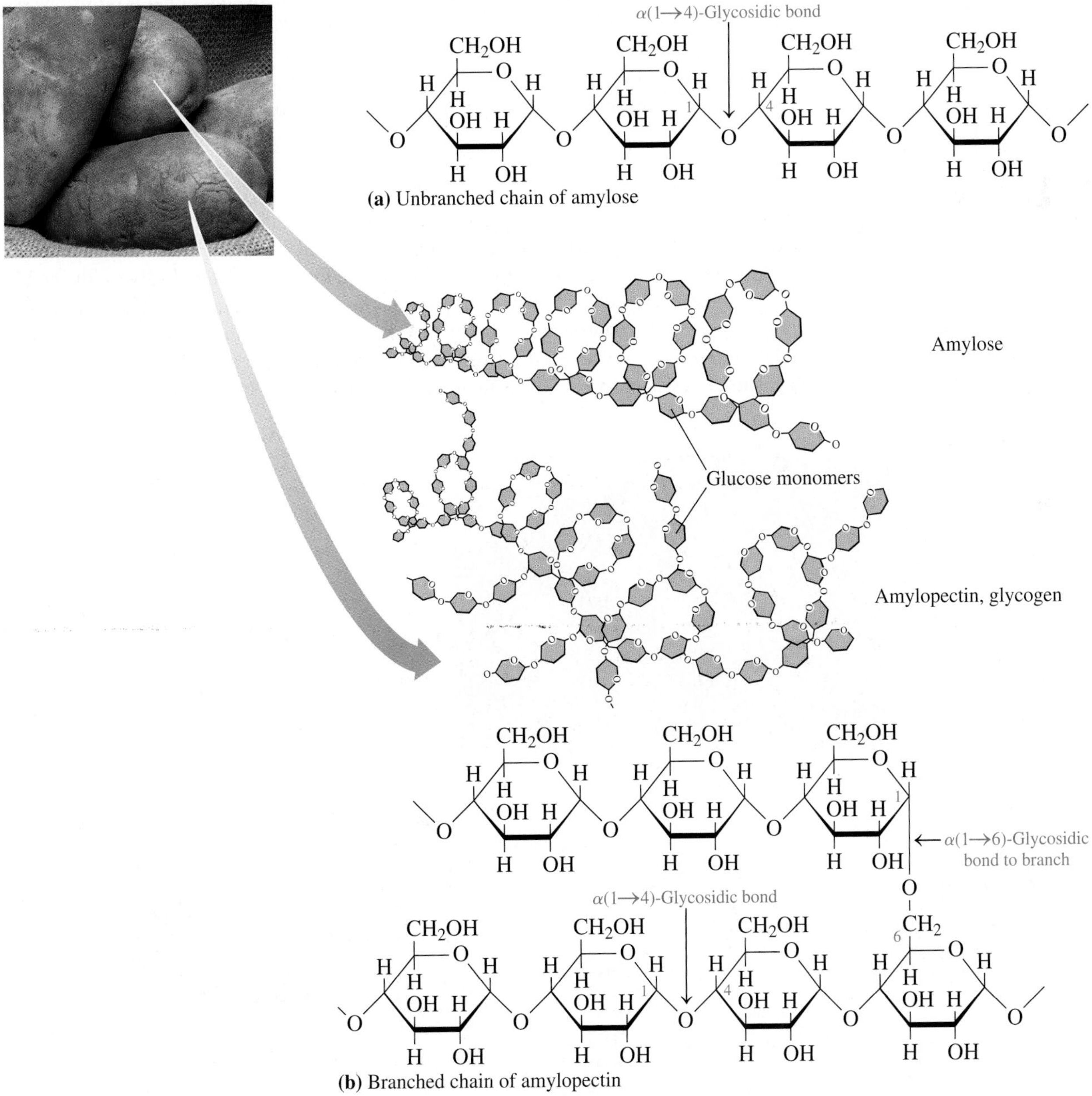

FIGURE 13.10 The structure of amylose **(a)** is a straight-chain polysaccharide of glucose units, and the structure of amylopectin **(b)** is a branched chain of glucose.

Q What are the two types of glycosidic bonds that link glucose molecules in amylopectin?

Amylopectin, which makes up as much as 80% of starch, is a branched-chain polysaccharide. Like amylose, the glucose molecules are connected by $\alpha(1\rightarrow4)$-glycosidic bonds. However, at about every 25 glucose units, there is a branch of glucose molecules attached by an $\alpha(1\rightarrow6)$-glycosidic bond between carbon 1 of the branch and carbon 6 in the main chain (see Figure 13.10b).

Starches hydrolyze in water and acid to give smaller saccharides, called *dextrins*, which then hydrolyze to maltose and finally glucose. In our bodies, these complex carbohydrates are digested by the enzymes amylase (in saliva) and maltase (in the intestine). The glucose obtained provides about 50% of our nutritional calories.

$$\text{Amylose, amylopectin} \xrightarrow{\text{H}^+ \text{ or amylase}} \text{dextrins} \xrightarrow{\text{H}^+ \text{ or amylase}} \text{maltose} \xrightarrow{\text{H}^+ \text{ or maltase}} \text{many D-glucose units}$$

Glycogen

Glycogen, or animal starch, is a polymer of glucose that is stored in the liver and muscle of animals. It is hydrolyzed in our cells at a rate that maintains the blood level of glucose and provides energy between meals. The structure of glycogen is very similar to that of amylopectin found in plants, except that glycogen is more highly branched. In glycogen, the glucose units are joined by $\alpha(1\rightarrow4)$-glycosidic bonds, and branches occurring about every 10 to 15 glucose units are attached by $\alpha(1\rightarrow6)$-glycosidic bonds.

Cellulose

Cellulose is the major structural material of wood and plants. Cotton is almost pure cellulose. In cellulose, glucose molecules form a long unbranched chain similar to that of amylose. However, the glucose units in cellulose are linked by $\beta(1\rightarrow4)$-glycosidic bonds. The cellulose chains do not form coils like amylose but are aligned in parallel rows that are held in place by hydrogen bonds between hydroxyl groups in adjacent chains, making cellulose insoluble in water. This gives a rigid structure to the cell walls in wood and fiber that is more resistant to hydrolysis than are the starches (see **FIGURE 13.11**).

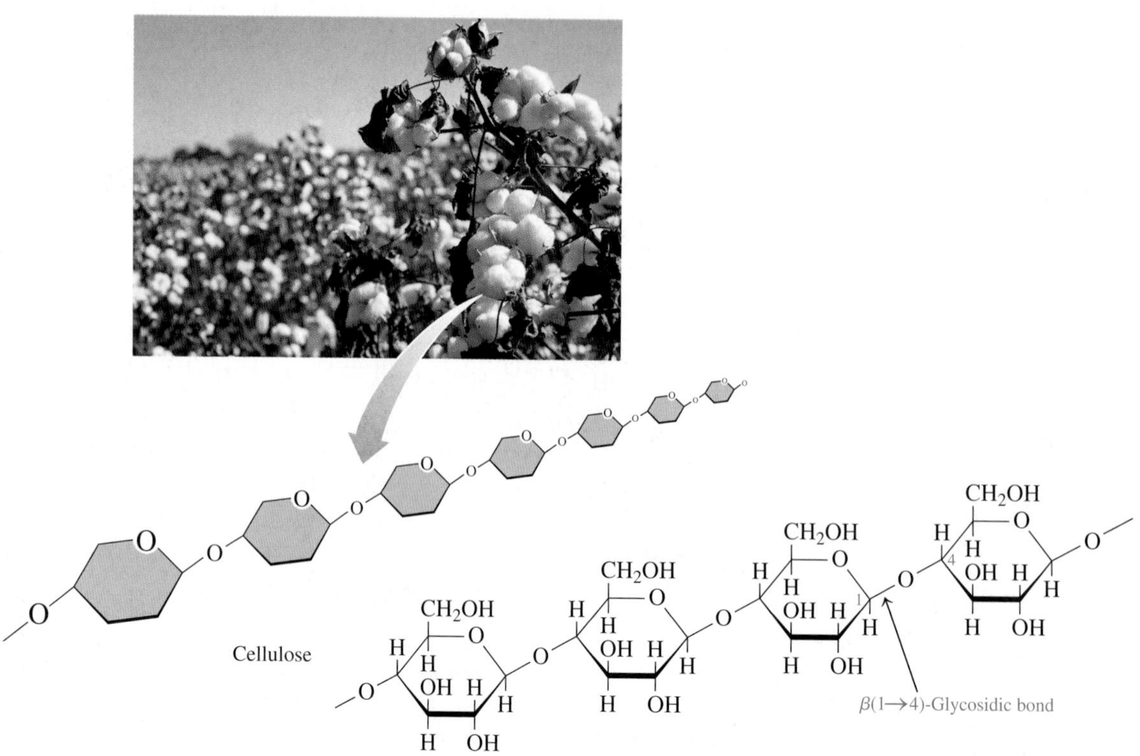

FIGURE 13.11 The polysaccharide cellulose is composed of glucose units connected by $\beta(1\rightarrow4)$-glycosidic bonds.

Q Why are humans unable to digest cellulose?

Humans have an enzyme called α-amylase in saliva and pancreatic juices that hydrolyzes the $\alpha(1\rightarrow4)$-glycosidic bonds of starches, but not the $\beta(1\rightarrow4)$-glycosidic bonds of cellulose. Thus, humans cannot digest cellulose. Animals such as horses, cows, and goats can obtain glucose from cellulose because their digestive systems contain bacteria that provide enzymes such as cellulase to hydrolyze $\beta(1\rightarrow4)$-glycosidic bonds.

SAMPLE PROBLEM 13.8 Structures of Polysaccharides

TRY IT FIRST

Give the name of one or more polysaccharides described by each of the following:

a. a polysaccharide that is stored in the liver and muscle tissues
b. an unbranched polysaccharide containing $\beta(1\rightarrow4)$-glycosidic bonds
c. a branched polysaccharide containing $\alpha(1\rightarrow4)$- and $\alpha(1\rightarrow6)$-glycosidic bonds

SOLUTION

a. glycogen **b.** cellulose **c.** amylopectin, glycogen

STUDY CHECK 13.8

Cellulose and amylose are both unbranched glucose polymers. How do they differ?

ANSWER

Cellulose contains glucose units connected by $\beta(1\rightarrow4)$-glycosidic bonds, whereas the glucose units in amylose are connected by $\alpha(1\rightarrow4)$-glycosidic bonds.

TEST

Try Practice Problems 13.47 to 13.50

PRACTICE PROBLEMS

13.7 Polysaccharides

LEARNING GOAL Describe the structural features of amylose, amylopectin, glycogen, and cellulose.

13.47 Describe the similarities and differences in the following:
a. amylose and amylopectin
b. amylopectin and glycogen

13.48 Describe the similarities and differences in the following:
a. amylose and cellulose
b. cellulose and glycogen

Clinical Applications

13.49 Give the name of one or more polysaccharides that matches each of the following descriptions:
a. not digestible by humans
b. the storage form of carbohydrates in plants
c. contains only $\alpha(1\rightarrow4)$-glycosidic bonds
d. the most highly branched polysaccharide

13.50 Give the name of one or more polysaccharides that matches each of the following descriptions:
a. the storage form of carbohydrates in animals
b. contains only $\beta(1\rightarrow4)$-glycosidic bonds
c. contains both $\alpha(1\rightarrow4)$- and $\alpha(1\rightarrow6)$-glycosidic bonds
d. produces maltose during digestion

CLINICAL UPDATE Kate's Program for Type 2 Diabetes

At Kate's next appointment, Paula showed Kate how to use a glucose meter. She instructed Kate to measure her blood glucose level twice a day before and after breakfast and dinner. Paula explains to Kate that her pre-meal blood glucose level should be 110 mg/dL or less, and if it increases by more than 50 mg/dL, she needs to lower the amount of carbohydrates she consumes.

Kate and Paula proceed to plan several meals. Because a meal should contain about 45 to 60 g of carbohydrates, they combined fruits and vegetables that have high and low levels of carbohydrates in the same meal to stay within the recommended range. Kate and Paula also discuss the fact that complex carbohydrates in the body take longer to break down into glucose, and therefore raise the blood sugar level more gradually.

Kate increased her exercise to walking 30 minutes twice a day. She began to change her diet by eating six small meals a day consisting of more fruits and vegetables without starch such as green beans and broccoli with small servings of whole grains and of chicken or fish. Kate also decreased the amounts of breads and pasta because she knew carbohydrates would raise her blood sugar.

After three months, Kate reported that she had lost 10 lb and that her blood glucose had dropped to 146 mg/dL. Her blurry vision had improved, and her need to urinate had decreased.

Clinical Applications

13.51 Kate's blood volume is 3.9 L. Before treatment, if her blood glucose was 178 mg/dL, how many grams of glucose were in her blood?

13.52 Kate's blood volume is 3.9 L. After three months of diet and exercise, if her blood glucose is 146 mg/dL, how many grams of glucose are in her blood?

13.53 For breakfast, Kate had 1 cup of orange juice (23 g carbohydrate), 2 slices of wheat toast (24 g carbohydrate), 2 tablespoons of grape jam (26 g carbohydrate), and coffee with sugar substitute (0 g carbohydrate).

a. Has Kate remained within the limit of 45 to 60 g of carbohydrate?

b. Using the energy value of 4 kcal/g for carbohydrate, calculate the total kilocalories from carbohydrates in Kate's breakfast, rounded to the tens place.

13.54 The next day, Kate had 1 cup of cereal (15 g carbohydrate) with skim milk (7 g carbohydrate), 1 banana (17 g carbohydrate), and 1/2 cup of orange juice (12 g carbohydrate) for breakfast.

a. Has Kate remained within the limit of 45 to 60 g of carbohydrate?

b. Using the energy value of 4 kcal/g for carbohydrate, calculate the total kilocalories from carbohydrates in Kate's breakfast, rounded to the tens place.

CONCEPT MAP

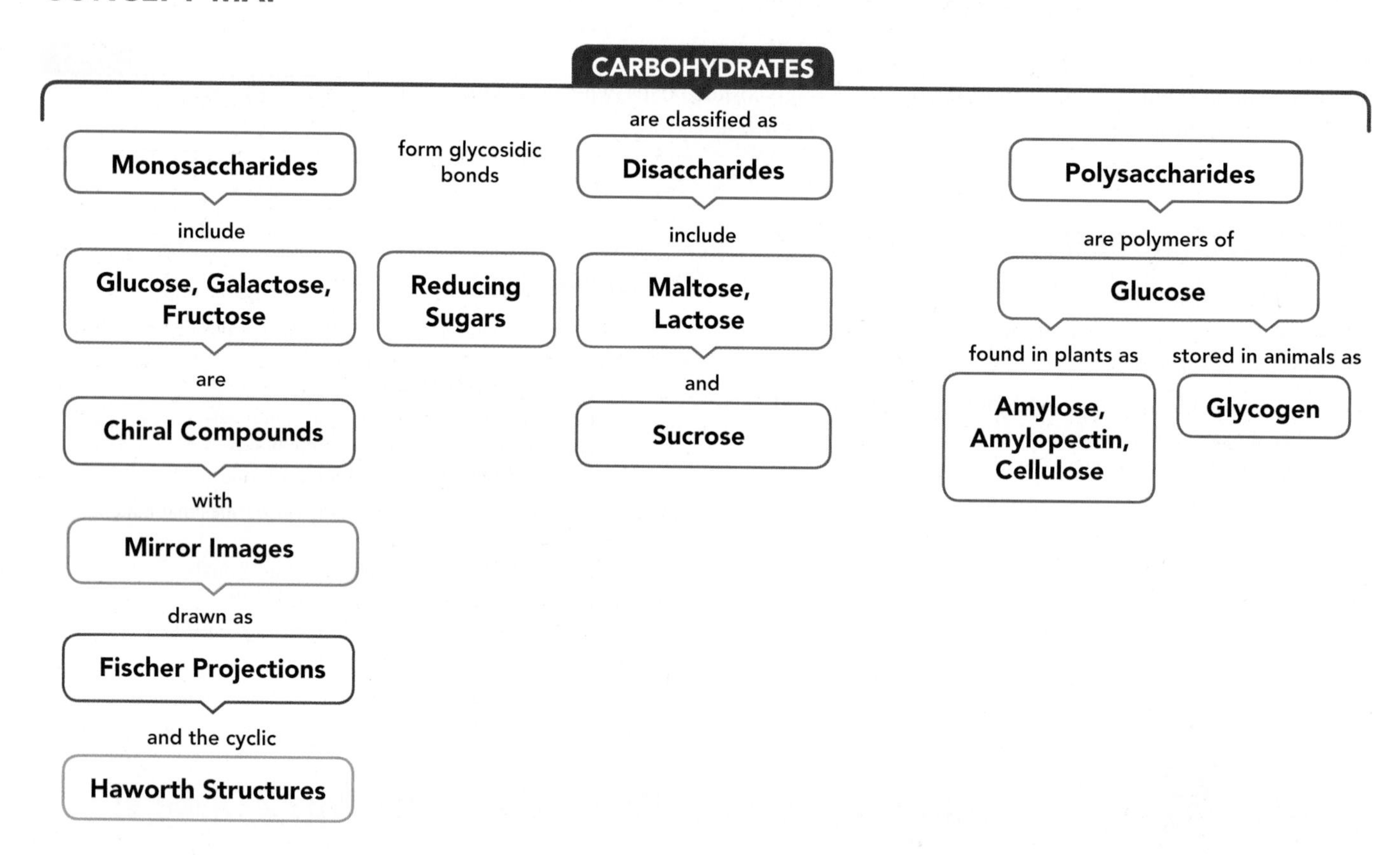

CHAPTER REVIEW

13.1 Carbohydrates

LEARNING GOAL Classify a monosaccharide as an aldose or a ketose, and indicate the number of carbon atoms.

- Carbohydrates are classified as monosaccharides (simple sugars), disaccharides (two monosaccharide units), or polysaccharides (many monosaccharide units).
- Monosaccharides are polyhydroxy aldehydes (aldoses) or ketones (ketoses).

Aldehyde

H—C=O
H—C—OH
H—C—OH
CH_2OH

Erythrose, an aldose

- Monosaccharides are also classified by their number of carbon atoms: triose, tetrose, pentose, or hexose.

13.2 Chiral Molecules

LEARNING GOAL Identify chiral and achiral carbon atoms in an organic molecule. Identify D and L enantiomers.

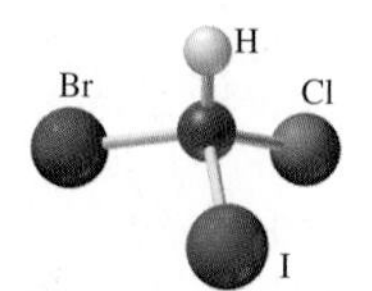

- Chiral molecules are molecules with mirror images that cannot be superimposed on each other. These types of stereoisomers are called enantiomers.
- A chiral molecule must have at least one chiral carbon, which is a carbon bonded to four different atoms or groups of atoms.
- The Fischer projection is a simplified way to draw the arrangements of atoms by placing the carbon atoms at the intersection of vertical and horizontal lines.
- The mirror images are labeled D or L to differentiate between enantiomers.

13.3 Fischer Projections of Monosaccharides

LEARNING GOAL Identify or draw the D and L configurations of the Fischer projections for common monosaccharides.

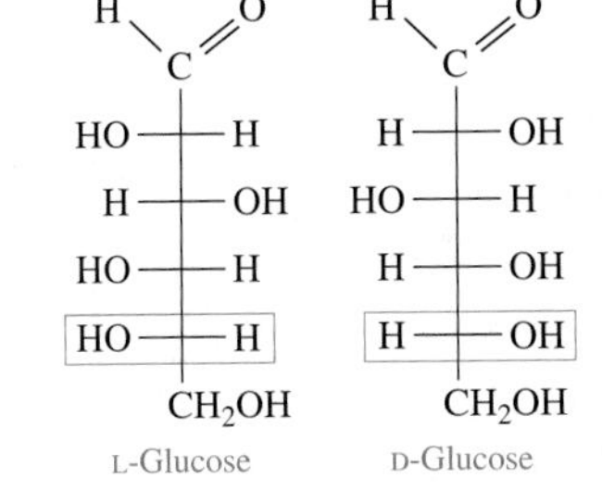

L-Glucose D-Glucose

- In a D enantiomer, the —OH group is on the right of the chiral carbon farthest from the carbonyl group; the —OH group is on the left in the L enantiomer.
- Important monosaccharides are the aldohexoses, glucose and galactose, and the ketohexose, fructose.

13.4 Haworth Structures of Monosaccharides

LEARNING GOAL Draw and identify the Haworth structures for monosaccharides.

α-D-Glucose

- The predominant form of a monosaccharide is a ring of five or six atoms.
- The cyclic structure forms when an —OH group (usually the one on carbon 5 in hexoses) reacts with the carbonyl group of the same molecule.
- The formation of a new hydroxyl group on carbon 1 gives α and β isomers of the cyclic monosaccharide.

13.5 Chemical Properties of Monosaccharides

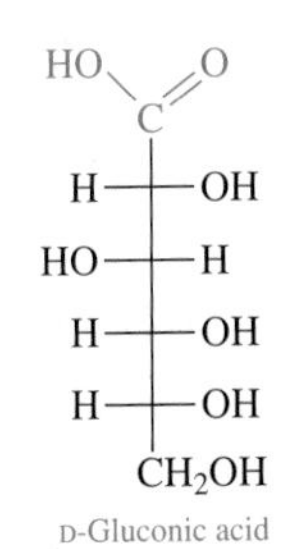

D-Gluconic acid

LEARNING GOAL Identify the products of oxidation or reduction of monosaccharides; determine whether a carbohydrate is a reducing sugar.

- The aldehyde group in an aldose can be oxidized to a carboxylic acid, while the carbonyl group in an aldose or a ketose can be reduced to give a hydroxyl group.
- Monosaccharides that are reducing sugars have an aldehyde group in the open chain that can be oxidized.

13.6 Disaccharides

LEARNING GOAL Describe the monosaccharide units and linkages in disaccharides.

- Disaccharides are two monosaccharide units joined together by a glycosidic bond.
- In the common disaccharides maltose, lactose, and sucrose, there is at least one glucose unit.
- Maltose and lactose form α and β isomers, which makes them reducing sugars.
- Sucrose does not have α and β isomers and is not a reducing sugar.

13.7 Polysaccharides

LEARNING GOAL Describe the structural features of amylose, amylopectin, glycogen, and cellulose.

- Polysaccharides are polymers of monosaccharide units.
- Amylose is an unbranched chain of glucose with α(1→4)-glycosidic bonds, and amylopectin is a branched polymer of glucose with α(1→4)- and α(1→6)-glycosidic bonds.
- Glycogen is similar to amylopectin with more branching.
- Cellulose is also a polymer of glucose, but in cellulose, the glycosidic bonds are β(1→4)-bonds.

SUMMARY OF CARBOHYDRATES

Carbohydrate	Found In	
Monosaccharides		
Glucose	Fruit juices, honey, corn syrup	
Galactose	Lactose hydrolysis	
Fructose	Fruit juices, honey, sucrose hydrolysis	
Disaccharides		**Monosaccharide Components**
Maltose	Germinating grains, starch hydrolysis	Glucose + glucose
Lactose	Milk, yogurt, ice cream	Glucose + galactose
Sucrose	Sugar cane, sugar beets	Glucose + fructose
Polysaccharides		
Amylose	Rice, wheat, grains, cereals	Unbranched polymer of glucose joined by α(1→4)-glycosidic bonds
Amylopectin	Rice, wheat, grains, cereals	Branched polymer of glucose joined by α(1→4)- and α(1→6)-glycosidic bonds
Glycogen	Liver, muscles	Highly branched polymer of glucose joined by α(1→4)- and α(1→6)-glycosidic bonds
Cellulose	Plant fiber, bran, beans, celery	Unbranched polymer of glucose joined by β(1→4)-glycosidic bonds

SUMMARY OF REACTIONS

The chapter Sections to review are shown after the name of each reaction.

Oxidation and Reduction of Monosaccharides (13.5)

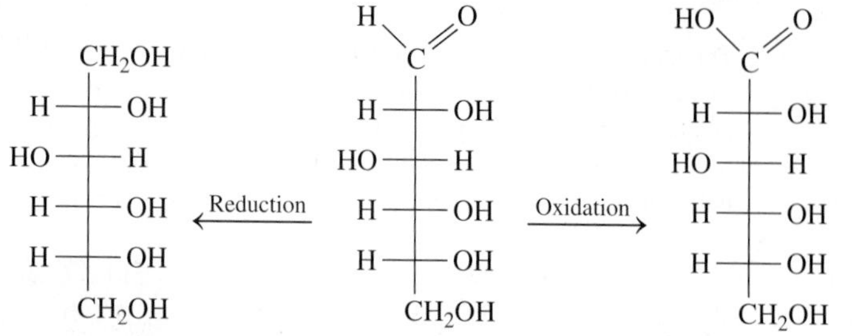

D-Glucitol $\xleftarrow{\text{Reduction}}$ D-Glucose $\xrightarrow{\text{Oxidation}}$ D-Gluconic acid

Formation of Disaccharides (13.6)

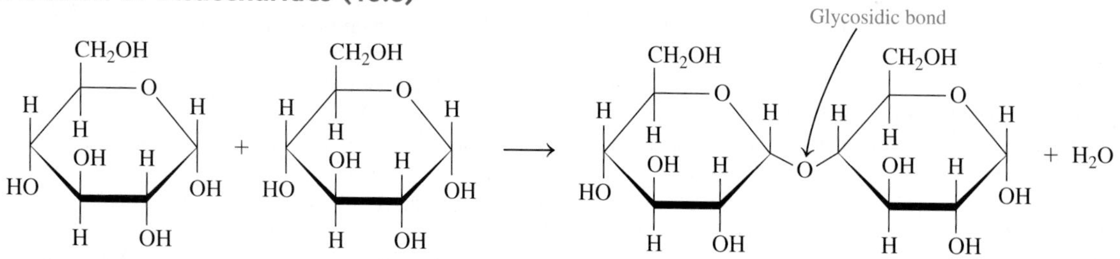

Monosaccharide + Monosaccharide $\longrightarrow$ Disaccharide + H_2O

Glycosidic bond

Hydrolysis of Polysaccharides (13.7)

Amylose, amylopectin $\xrightarrow{H^+ \text{ or enzymes}}$ many D-glucose units

KEY TERMS

aldose A monosaccharide that contains an aldehyde group.

amylopectin A branched-chain polymer of starch composed of glucose units joined by $\alpha(1\rightarrow4)$- and $\alpha(1\rightarrow6)$-glycosidic bonds.

amylose An unbranched polymer of starch composed of glucose units joined by $\alpha(1\rightarrow4)$-glycosidic bonds.

carbohydrate A simple or complex sugar composed of carbon, hydrogen, and oxygen.

cellulose An unbranched polysaccharide composed of glucose units linked by $\beta(1\rightarrow4)$-glycosidic bonds that cannot be hydrolyzed by the human digestive system.

chiral An object or molecule that has a nonsuperimposable mirror image.

chiral carbon A carbon atom that is bonded to four different atoms or groups.

disaccharide A carbohydrate composed of two monosaccharides joined by a glycosidic bond.

enantiomers Chiral compounds that are mirror images that cannot be superimposed.

Fischer projection A system for drawing stereoisomers; an intersection of a vertical and horizontal line represents a carbon atom. A vertical line represents bonds that project backward from a carbon atom and a horizontal line represents bonds that project forward. The most highly oxidized carbon is at the top.

fructose A monosaccharide that is also called levulose and fruit sugar and is found in honey and fruit juices; it is combined with glucose in sucrose.

galactose A monosaccharide that occurs combined with glucose in lactose.

glucose An aldohexose found in fruits, vegetables, corn syrup, and honey that is also known as blood sugar and dextrose. The most prevalent monosaccharide in the diet. Most polysaccharides are polymers of glucose.

glycogen A polysaccharide formed in the liver and muscles for the storage of glucose as an energy reserve. It is composed of glucose in a highly branched polymer joined by $\alpha(1\rightarrow4)$- and $\alpha(1\rightarrow6)$-glycosidic bonds.

glycosidic bond The bond that forms when the hydroxyl group of one monosaccharide reacts with the hydroxyl group of another monosaccharide. It is the type of bond that links monosaccharides in di- or polysaccharides.

Haworth structure The ring structure of a monosaccharide.

ketose A monosaccharide that contains a ketone group.

lactose A disaccharide consisting of glucose and galactose found in milk and milk products.

maltose A disaccharide consisting of two glucose units; it is obtained from the hydrolysis of starch and is found in germinating grains.

monosaccharide A polyhydroxy compound that contains an aldehyde or ketone group.

polysaccharide A polymer of many monosaccharide units, usually glucose. Polysaccharides differ in the types of glycosidic bonds and the amount of branching in the polymer.

reducing sugar A carbohydrate with an aldehyde group capable of reducing the Cu^{2+} in Benedict's reagent.

stereoisomers Isomers that have atoms bonded in the same order, but with different arrangements in space.

sucrose A disaccharide composed of glucose and fructose; a nonreducing sugar, commonly called table sugar or "sugar."

CORE CHEMISTRY SKILLS

The chapter Section containing each Core Chemistry Skill is shown in parentheses at the end of each heading.

Identifying Chiral Molecules (13.2)

- Chiral molecules are molecules with mirror images that cannot be superimposed. These types of stereoisomers are called enantiomers.
- A chiral molecule must have at least one chiral carbon, which is a carbon bonded to four different atoms or groups of atoms.

Example: Identify each of the following as chiral or achiral:

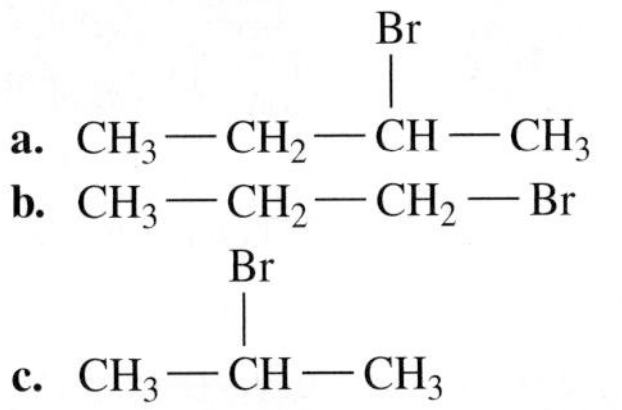

a. $CH_3-CH_2-CH(Br)-CH_3$

b. $CH_3-CH_2-CH_2-Br$

c. $CH_3-CH(Br)-CH_3$

Answer: a is chiral; **b** and **c** are achiral.

Identifying D and L Fischer Projections for Carbohydrates (13.3)

- The Fischer projection is a simplified way to draw the arrangements of atoms by placing the carbon atoms at the intersection of vertical and horizontal lines.
- The names of the mirror images are labeled D or L to differentiate between enantiomers of carbohydrates.

Example: Identify each of the following as the D or L enantiomer:

a.
H–C=O
H—|—OH
HO—|—H
H—|—OH
CH_2OH

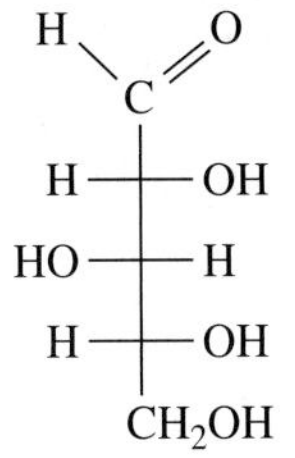

b.
CH_2OH
C=O
H—|—OH
HO—|—H
HO—|—H
CH_2OH

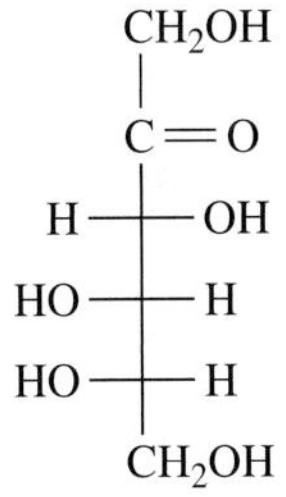

Answer: In **a**, the —OH group on the chiral carbon farthest from the carbonyl group is on the right; it is the D enantiomer.

In **b**, the —OH group on the chiral carbon farthest from the carbonyl group is on the left; it is the L enantiomer.

Drawing Haworth Structures (13.4)

- The Haworth structure shows the ring structure of a monosaccharide.
- Groups on the right side of the Fischer projection of the monosaccharide are below the plane of the ring, those on the left are above the plane.
- The new —OH group that forms is the α isomer if it is below the plane of the ring, or the β isomer if it is above the plane.

Example: Draw the Haworth structure for β-D-idose.

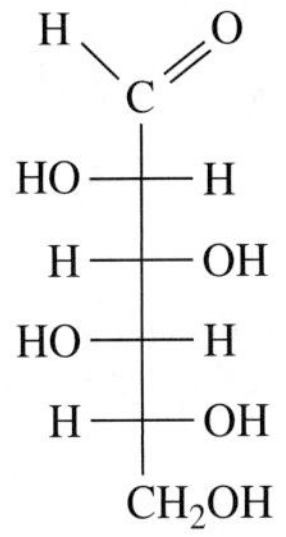

H–C=O
HO—|—H
H—|—OH
HO—|—H
H—|—OH
CH_2OH

D-Idose

Answer: In the β-isomer, the new —OH group is above the plane of the ring.

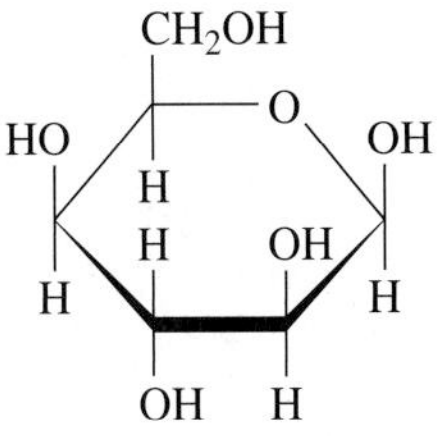

CH_2OH
O
HO H OH
H OH
H H
OH H

UNDERSTANDING THE CONCEPTS

The chapter Sections to review are shown in parentheses at the end of each problem.

13.55 Isomaltose, obtained from the breakdown of starch, has the following Haworth structure: (13.4, 13.5, 13.6)

CH_2OH
O
H H
H
OH H
HO O
H OH CH_2
O
H H
H
OH H
HO OH
H OH

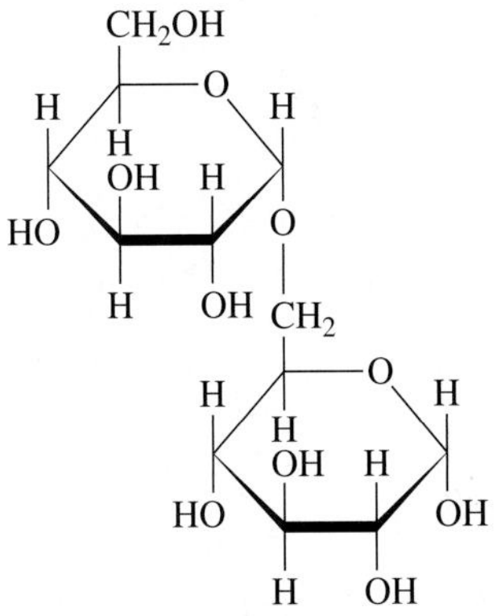

Isomaltose

a. Is isomaltose a mono-, di-, or polysaccharide?

b. What are the monosaccharides in isomaltose?

c. What is the glycosidic link in isomaltose?

d. Is this the α or β isomer of isomaltose?

e. Is isomaltose a reducing sugar?

13.56 Sophorose, a carbohydrate found in certain types of beans, has the following Haworth structure: (13.4, 13.5, 13.6)

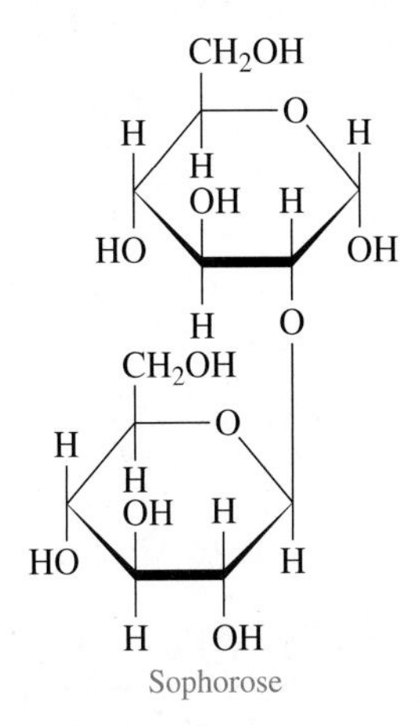

Sophorose

a. Is sophorose a mono-, di-, or polysaccharide?
b. What are the monosaccharides in sophorose?
c. What is the glycosidic link in sophorose?
d. Is this the α or β isomer of sophorose?
e. Is sophorose a reducing sugar?

13.57 Melezitose, a carbohydrate secreted by insects, has the following Haworth structure: (13.4, 13.5, 13.6)

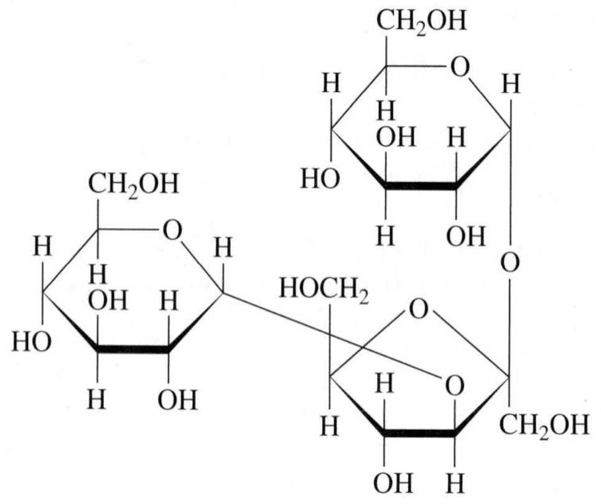

Melezitose

a. Is melezitose a mono-, di-, or trisaccharide?
b. What monosaccharides are present in melezitose?
c. Is melezitose a reducing sugar?

13.58 Raffinose, found in Australian manna and in cottonseed meal, has the following Haworth structure: (13.4, 13.5, 13.6)

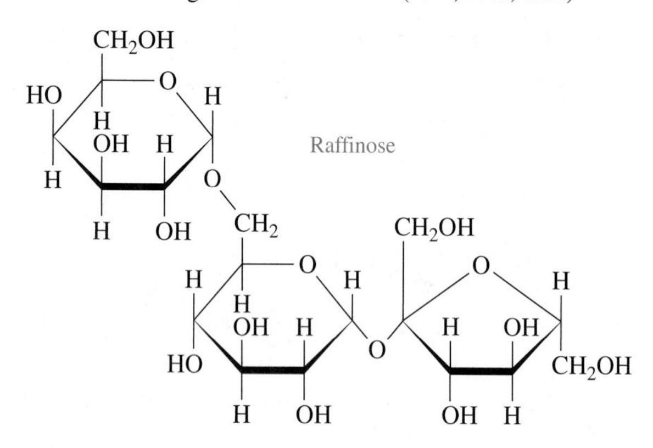
Raffinose

a. Is raffinose a mono-, di-, or trisaccharide?
b. What monosaccharides are present in raffinose?
c. Is raffinose a reducing sugar?

13.59 What are the disaccharides and polysaccharides present in each of the following? (13.6, 13.7)

(a)

(b)

13.60 What are the disaccharides and polysaccharides present in each of the following? (13.6, 13.7)

(a)

(b)

ADDITIONAL PRACTICE PROBLEMS

13.61 Identify the chiral carbons, if any, in each of the following compounds: (13.2)

a. 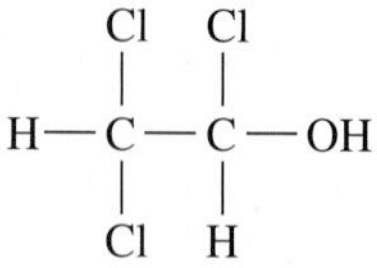 $H-C(Cl)(Cl)-C(Cl)(H)-OH$

b. $CH_3-C(H)=C(CH_3)-CH_3$

c. $HO-CH_2-CH(OH)-CH_2-OH$

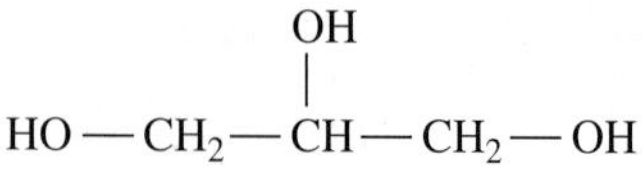

d. 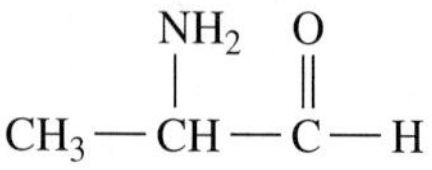 $CH_3-CH(NH_2)-C(=O)-H$

e. 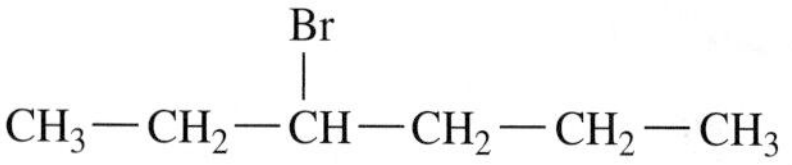 $CH_3-CH_2-CH(Br)-CH_2-CH_2-CH_3$

13.62 Identify the chiral carbons, if any, in each of the following compounds: (13.2)

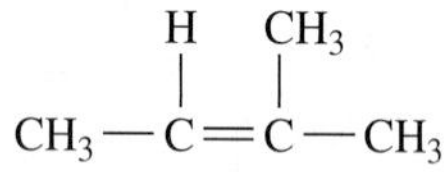

a. 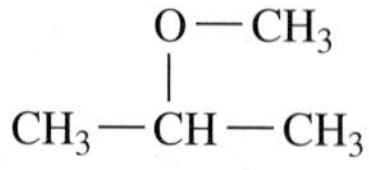 $CH_3-CH(O-CH_3)-CH_3$

b. 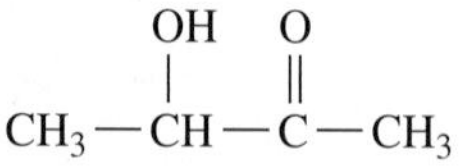 $CH_3-CH(OH)-C(=O)-CH_3$

c. $CH_3-C(OH)(OH)-CH_3$

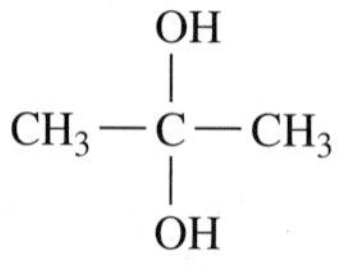

d. 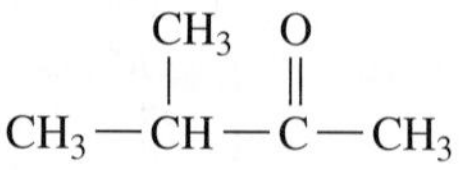 $CH_3-CH(CH_3)-C(=O)-CH_3$

e. 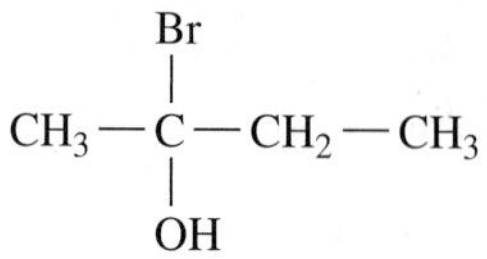 $CH_3-C(Br)(OH)-CH_2-CH_3$

13.63 Identify each of the following pairs of Fischer projections as enantiomers or identical compounds: (13.2, 13.3)

a. CH_2OH / H—|—OH / CH_2OH and CH_2OH / HO—|—H / CH_2OH

b. H, C=O / H—|—OH / CH_2OH and H, C=O / HO—|—H / CH_2OH

c. CH_2OH / Cl—|—H / CH_3 and CH_2OH / H—|—Cl / CH_3

d. OH / H—|—OH / CH_3 and OH / HO—|—H / CH_3

13.64 Identify each of the following pairs of Fischer projections as enantiomers or identical compounds: (13.2, 13.3)

a. CH_2OH / H—|—Cl / CH_2CH_3 and CH_2OH / Cl—|—H / CH_2CH_3

b. CH_2OH / H—|—OH / CH_3 and CH_2OH / HO—|—H / CH_3

c. CH_2OH / H—|—Cl / CH_3 and CH_2OH / H—|—Cl / CH_3

d. H, C=O / H—|—OH / CH_3 and H, C=O / HO—|—H / CH_3

13.65 What are the differences in the Fischer projections of D-fructose and D-galactose? (13.3)

13.66 What are the differences in the Fischer projections of D-glucose and D-fructose? (13.3)

13.67 What are the differences in the Fischer projections of D-galactose and L-galactose? (13.3)

13.68 What are the differences in the Haworth structures of α-D-glucose and β-D-glucose? (13.4)

13.69 The sugar D-gulose is a sweet-tasting syrup. (13.3, 13.4)

H, C=O
H—|—OH
H—|—OH
HO—|—H
H—|—OH
CH_2OH

D-Gulose

a. Draw the Fischer projection for L-gulose.

b. Draw the Haworth structures for α- and β-D-gulose.

13.70 Use the Fischer projection for D-gulose in problem 13.69 to answer each of the following: (13.3, 13.5)

a. Draw the Fischer projection and name the product formed by the reduction of D-gulose.

b. Draw the Fischer projection and name the product formed by the oxidation of D-gulose.

13.71 D-Sorbitol, a sweetener found in seaweed and berries, contains only hydroxyl functional groups. When D-sorbitol is oxidized, it forms D-glucose. Draw the Fischer projection for D-sorbitol. (13.3, 13.5)

13.72 D-Erythritol is 70% as sweet as sucrose and contains only hydroxyl functional groups. When D-erythritol is oxidized it forms D-erythrose. Draw the Fischer projection for D-erythritol. (13.3, 13.5)

13.73 If α-galactose is dissolved in water, β-galactose is eventually present. Explain how this occurs. (13.5)

13.74 Why are lactose and maltose reducing sugars, but sucrose is not? (13.6)

CHALLENGE PROBLEMS

The following problems are related to the topics in this chapter. However, they do not all follow the chapter order, and they require you to combine concepts and skills from several Sections. These problems will help you increase your critical thinking skills and prepare for your next exam.

13.75 α-Cellobiose is a disaccharide obtained from the hydrolysis of cellulose. It is quite similar to maltose except it has a $\beta(1\rightarrow4)$-glycosidic bond. Draw the Haworth structure for α-cellobiose. (13.4, 13.6)

13.76 The disaccharide trehalose found in mushrooms is composed of two α-D-glucose molecules joined by an $\alpha(1\rightarrow1)$-glycosidic bond. Draw the Haworth structure for trehalose. (13.4, 13.6)

13.77 Gentiobiose is found in saffron. (13.4, 13.5, 13.6)

a. Gentiobiose contains two glucose molecules linked by a $\beta(1\rightarrow6)$-glycosidic bond. Draw the Haworth structure for β-gentiobiose.

b. Is gentiobiose a reducing sugar? Explain.

13.78 Identify the Fischer projection **A** to **D** that matches each of the following: (13.1, 13.3)

a. the L enantiomer of mannose

b. a ketopentose

c. an aldopentose

d. a ketohexose

A: H, C=O / H—|—OH / HO—|—H / H—|—OH / CH_2OH

B: H, C=O / H—|—OH / H—|—OH / HO—|—H / HO—|—H / CH_2OH

C: CH_2OH / C=O / HO—|—H / H—|—OH / CH_2OH

D: CH_2OH / C=O / H—|—OH / HO—|—H / HO—|—H / CH_2OH

ANSWERS

Answers to Selected Practice Problems

13.1 Photosynthesis requires CO_2, H_2O, and the energy from the Sun. Respiration requires O_2 from the air and glucose from foods.

13.3 Monosaccharides can be a chain of three to eight carbon atoms, one in a carbonyl group as an aldehyde or ketone, and the rest attached to hydroxyl groups. A monosaccharide cannot be split or hydrolyzed into smaller carbohydrates. A disaccharide consists of two monosaccharide units joined together that can be split.

13.5 Hydroxyl groups are found in all monosaccharides along with a carbonyl on the first or second carbon that gives an aldehyde or ketone functional group.

13.7 A ketopentose contains hydroxyl and ketone functional groups and has five carbon atoms.

13.9 **a.** ketohexose **b.** aldopentose

13.11 **a.** achiral **b.** chiral

Br (Chiral carbon)

c. chiral

$CH_3-CH(Br)-C(=O)-H$ (Chiral carbon)

d. achiral

13.13 **a.** H–C=O; HO–C–Br; CH_3 **b.** H–C=O; Cl–C–Br; OH **c.** H–C=O; HO–C–H; CH_2CH_3

13.15 **a.** identical **b.** enantiomers **c.** enantiomers **d.** enantiomers

13.17 **a.** D **b.** D **c.** L

13.19 **a.** $CH_3-C(CH_3)=CH-CH_2-CH_2-CH(CH_3)-CH_2-CH_2-OH$ (Chiral carbon)

b. $H_2N-CH(CH_3)-C(=O)-OH$ (Chiral carbon)

13.21 **a.** D **b.** D **c.** L **d.** D

13.23 **a.** H–C=O; H–C–OH; HO–C–H; CH_2OH

b. CH_2OH; C=O; H–C–OH; HO–C–H; CH_2OH

c. H–C=O; HO–C–H; HO–C–H; H–C–OH; H–C–OH; CH_2OH

d. H–C=O; HO–C–H; HO–C–H; HO–C–H; HO–C–H; CH_2OH

13.25 D-Glucose: H–C=O; H–C–OH; HO–C–H; H–C–OH; H–C–OH; CH_2OH

L-Glucose: H–C=O; HO–C–H; H–C–OH; HO–C–H; HO–C–H; CH_2OH

13.27 In D-galactose, the —OH group on carbon 4 extends to the left. In D-glucose, this —OH group goes to the right.

13.29 **a.** glucose **b.** galactose

13.31 In the cyclic structure of glucose, there are five carbon atoms and an oxygen atom.

13.33

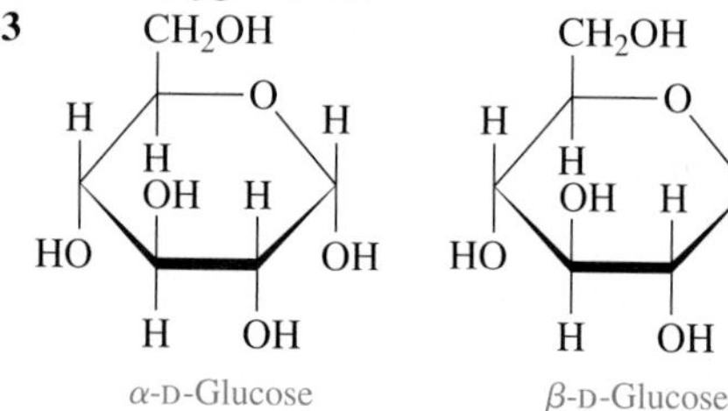

α-D-Glucose β-D-Glucose

13.35 **a.** α isomer **b.** α isomer

13.37 Oxidation product: HO–C=O; H–C–OH; HO–C–H; H–C–OH; CH_2OH — D-Xylonic acid

Reduction product: CH_2OH; H–C–OH; HO–C–H; H–C–OH; CH_2OH — D-Xylitol

13.39 Oxidation product: HO–C=O; HO–C–H; H–C–OH; H–C–OH; CH_2OH — D-Arabinonic acid

Reduction product: CH_2OH; HO–C–H; H–C–OH; H–C–OH; CH_2OH — D-Arabitol

13.41 **a.** galactose and glucose; β(1→4)-glycosidic bond; β-lactose
b. glucose and glucose; α(1→4)-glycosidic bond; α-maltose

13.43 **a.** is a reducing sugar **b.** is a reducing sugar

13.45 **a.** sucrose **b.** lactose **c.** maltose **d.** lactose

13.47 **a.** Amylose is an unbranched polymer of glucose units joined by α(1→4)-glycosidic bonds; amylopectin is a branched polymer of glucose joined by α(1→4)- and α(1→6)-glycosidic bonds.
b. Amylopectin, which is produced in plants, is a branched polymer of glucose, joined by α(1→4)- and α(1→6)-glycosidic bonds. The branches in amylopectin occur about every 25 glucose units. Glycogen, which is produced in animals, is a highly branched polymer of glucose, joined by α(1→4)- and α(1→6)-glycosidic bonds. The branches in glycogen occur about every 10 to 15 glucose units.

13.49 a. cellulose **b.** amylose, amylopectin
c. amylose **d.** glycogen

13.51 6.9 of glucose

13.53 a. Kate's breakfast had 73 g of carbohydrate. She still needs to cut down the amount of carbohydrate.
b. 290 kcal

13.55 a. disaccharide **b.** α-D-glucose
c. $\alpha(1\rightarrow6)$-glycosidic bond **d.** α
e. is a reducing sugar

13.57 a. trisaccharide
b. two glucose and one fructose
c. Melezitose has no free —OH groups on the glucose or fructose molecules; like sucrose, it is not a reducing sugar.

13.59 a. sucrose **b.** cellulose

13.61 a. $H-C(Cl)_2-C(Cl)(H)-OH$ (Cl, Cl above; Cl, H below) ← Chiral carbon **b.** none

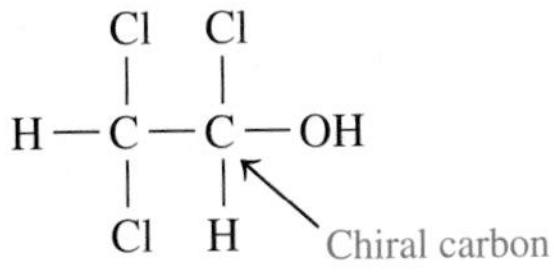

c. none **d.** $CH_3-CH(NH_2)-C(=O)-H$ ← Chiral carbon

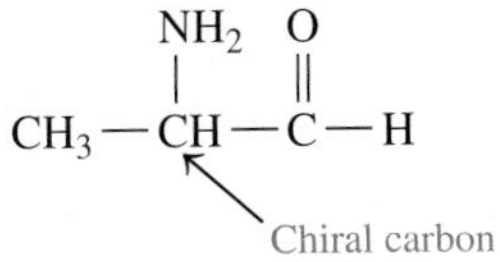

e. $CH_3-CH_2-CH(Br)-CH_2-CH_2-CH_3$ ← Chiral carbon

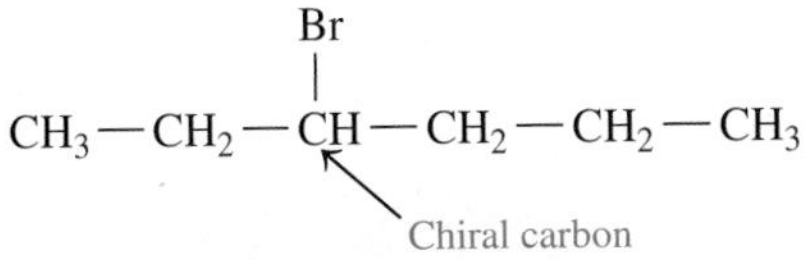

13.63 a. identical **b.** enantiomers
c. enantiomers **d.** identical

13.65 D-Fructose is a ketohexose, whereas D-galactose is an aldohexose. In the Fischer projection of D-galactose, the —OH group on carbon 4 is drawn on the left; in fructose, the —OH group is drawn on the right.

13.67 L-Galactose is the mirror image of D-galactose. In the Fischer projection of D-galactose, the —OH groups on carbon 2 and carbon 5 are drawn on the right side, but they are drawn on the left for carbon 3 and carbon 4. In L-galactose, the —OH groups are reversed; carbons 2 and 5 have —OH groups drawn on the left, and carbons 3 and 4 have —OH groups drawn on the right.

13.69 a.

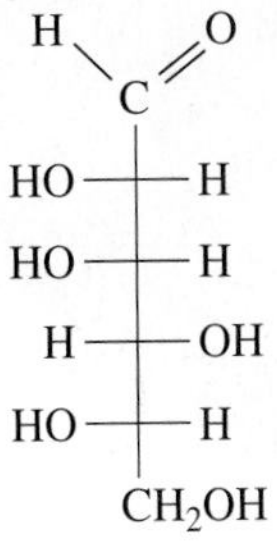

```
  H     O
   \\  //
     C
HO --+-- H
HO --+-- H
 H --+-- OH
HO --+-- H
   CH2OH
```

L-Gulose

b.

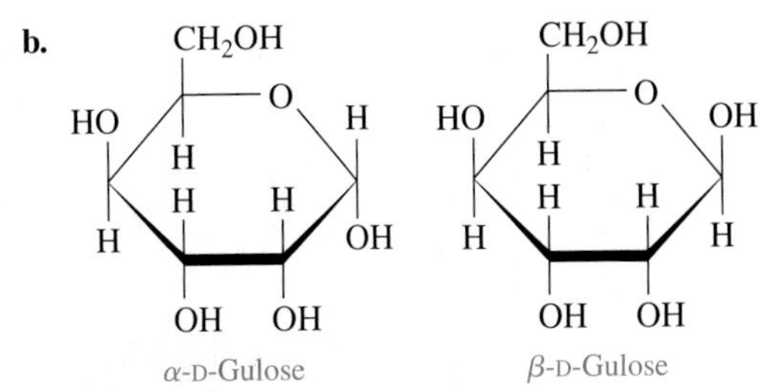

α-D-Gulose β-D-Gulose

13.71

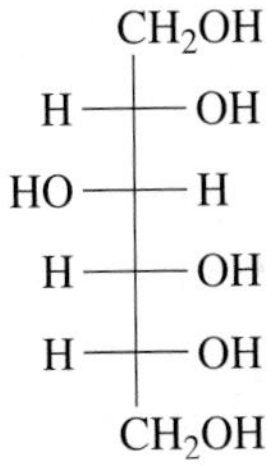

```
   CH2OH
 H --+-- OH
HO --+-- H
 H --+-- OH
 H --+-- OH
   CH2OH
```

13.73 When α-galactose forms an open-chain structure, it can close to form either α- or β-galactose.

13.75

13.77 a.

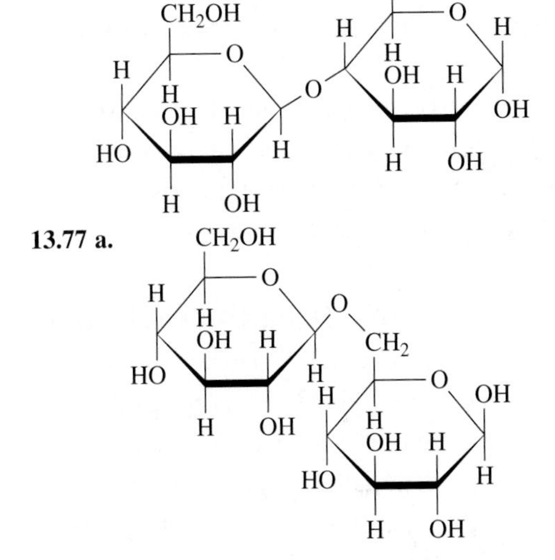

b. Yes. Gentiobiose is a reducing sugar. The ring on the right can open up to form an aldehyde that can be oxidized.

14 Carboxylic Acids, Esters, Amines, and Amides

LANCE, AN ENVIRONMENTAL HEALTH practitioner, collects soil and water samples at a ranch to test for the presence of any pesticides and pharmaceuticals. Ranchers use pesticides to increase food production and pharmaceuticals to treat and prevent animal-related diseases. Due to the common use of these chemicals, they may pass into the soil and water supply, potentially contaminating the environment and causing health problems.

Recently, the sheep on the ranch were treated with fenbendazole to destroy any gastrointestinal worms. Fenbendazole contains several functional groups: aromatic rings, an ester, an amide, and amines.

Lance detects small amounts of fenbendazole in the soil. He advises the rancher to decrease the dosage he administers to his sheep in order to reduce the amounts currently being detected in the soil. Lance then indicates he will be back in a month to retest the soil and water.

Fenbendazole

CAREER Environmental Health Practitioner

Environmental health practitioners (EHPs) monitor environmental pollution to protect the public's health. By using specialized equipment, EHPs measure pollution levels in soil, air, and water, as well as noise and radiation levels. EHPs can specialize in a specific area, such as air quality or hazardous and solid waste. For instance, air quality experts monitor indoor air for allergens, mold, and toxins; they measure outdoor air pollutants created by businesses, vehicles, and agriculture. Since EHPs obtain samples with potentially hazardous materials, they must be knowledgeable about safety protocols and wear personal protective equipment. EHPs also recommend methods to diminish various pollutants, and may assist in cleanup and remediation efforts.

CLINICAL UPDATE Testing Soil and Water Samples for Chemicals

When Lance returns to the ranch, the owners are using insecticide spray to treat an infestation of flies and maggots. You can see the chemicals used in the **CLINICAL UPDATE Testing Soil and Water Samples for Chemicals**, page 497, and see the structures of these insecticides and how Lance will analyze their levels in the environment.

14.1 Carboxylic Acids

LEARNING GOAL Write the IUPAC and common names for carboxylic acids; draw their condensed structural or line-angle formulas.

LOOKING AHEAD

REVIEW

Naming and Drawing Alkanes (11.2)

Carboxylic acids are weak acids. They have a sour or tart taste, produce hydronium ions in water, and neutralize bases. You encounter carboxylic acids when you use a salad dressing containing vinegar, which is a solution of acetic acid and water, or experience the sour taste of citric acid in a grapefruit or lemon.

In a **carboxylic acid**, the carbon atom of a carbonyl group is attached to a hydroxyl group that forms a **carboxyl group**. The carboxyl functional group may be attached to an alkyl group or an aromatic group. Some ways to represent the carboxyl group in carboxylic acids are shown for propanoic acid.

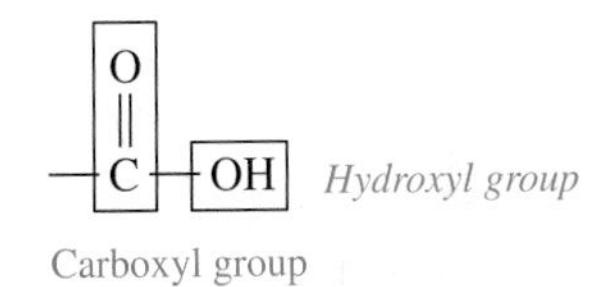

$CH_3-CH_2-\overset{\displaystyle O}{\overset{\|}{C}}-OH$ (line-angle formula) OH CH_3-CH_2-COOH

Propanoic acid
(propionic acid)

IUPAC Names of Carboxylic Acids

The IUPAC name of a carboxylic acid replaces the *e* of the corresponding alkane name with *oic acid*. If there are substituents, the carbon chain is numbered beginning with the carboxyl carbon.

CORE CHEMISTRY SKILL

Naming Carboxylic Acids

$H-\overset{\displaystyle O}{\overset{\|}{C}}-OH$ $CH_3-CH_2-\overset{\displaystyle O}{\overset{\|}{C}}-OH$ $CH_3-\overset{\displaystyle CH_3}{\overset{|}{CH}}-\overset{\displaystyle O}{\overset{\|}{C}}-OH$

Methanoic acid   Propanoic acid   2-Methylpropanoic acid

The simplest aromatic carboxylic acid is named benzoic acid. With the carboxyl carbon bonded to carbon 1, the ring is numbered in the direction that gives substituents the smallest possible numbers.

A red ant sting contains formic acid, which irritates the skin.

(Benzene ring with C(=O)OH)   (Benzene ring with C(=O)OH at carbon 1, Cl at carbons 3 and 4)

Benzoic acid   3,4-Dichlorobenzoic acid

Common Names of Carboxylic Acids

Many carboxylic acids are still named by their common names, which use prefixes: *form*, *acet*, *propion*, *butyr*. These prefixes are related to the natural sources of the simple carboxylic acids. For example, formic acid is injected under the skin from bee or red ant stings and other insect bites. Acetic acid is the oxidation product of the ethanol in wines and apple cider. A solution of acetic acid and water is known as vinegar. Butyric acid gives the foul odor to rancid butter (see **TABLE 14.1**).

The sour taste of vinegar is due to ethanoic acid (acetic acid).

TABLE 14.1 IUPAC and Common Names of Selected Carboxylic Acids

Condensed Structural Formula	Line-Angle Formula	IUPAC Name	Common Name	Ball-and-Stick Model
$H-C(=O)-OH$		Methanoic acid	Formic acid	
$CH_3-C(=O)-OH$		Ethanoic acid	Acetic acid	
$CH_3-CH_2-C(=O)-OH$		Propanoic acid	Propionic acid	
$CH_3-CH_2-CH_2-C(=O)-OH$		Butanoic acid	Butyric acid	

SAMPLE PROBLEM 14.1 Naming Carboxylic Acids

TRY IT FIRST

Write the IUPAC name for the following:

O

OH

SOLUTION GUIDE

ANALYZE THE PROBLEM	Given	Need	Connect
	four-carbon chain, methyl substituent	IUPAC name	position of methyl group, replace e in alkane name with *oic acid*

STEP 1 Identify the longest carbon chain and replace the e in the corresponding alkane name with *oic acid*. The longest chain contains four carbon atoms, which is butanoic acid.

O

OH

butanoic acid

STEP 2 Name and number any substituents by counting the carboxyl group as carbon 1. The substituent on carbon 2 is methyl. The IUPAC name for this compound is 2-methylbutanoic acid.

O

OH

4 3 2 1

2-methylbutanoic acid

STUDY CHECK 14.1

Write the IUPAC name for the following:

$$CH_3-CH(Cl)-CH_2-CH(Cl)-C(=O)-OH$$

ANSWER

2,4-dichloropentanoic acid

TEST

Try Practice Problems 14.1 to 14.6

PRACTICE PROBLEMS

14.1 Carboxylic Acids

LEARNING GOAL Write the IUPAC and common names for carboxylic acids; draw their condensed structural or line-angle formulas.

14.1 What carboxylic acid is responsible for the pain of an ant sting?

14.2 What carboxylic acid is found in vinegar?

14.3 Write the IUPAC and common name, if any, for each of the following carboxylic acids:

a. $CH_3-\overset{\overset{\large O}{\|}}{C}-OH$ **b.** O, OH **c.** O, OH **d.** O, OH, Br

14.4 Write the IUPAC and common name, if any, for each of the following carboxylic acids:

a. $H-\overset{\overset{\large O}{\|}}{C}-OH$ **b.** O, OH, Br

c. O, OH, Cl **d.** $CH_3-\overset{\overset{\large CH_3}{|}}{\underset{\underset{\large CH_3}{|}}{C}}-CH_2-\overset{\overset{\large O}{\|}}{C}-OH$

14.5 Draw the condensed structural formulas for **a** and **b** and line-angle formulas for **c** and **d**:
a. 3,4-dibromobutanoic acid
b. 2-chloropropanoic acid
c. benzoic acid
d. heptanoic acid

14.6 Draw the condensed structural formulas for **a** and **b** and line-angle formulas for **c** and **d**:
a. 2,3-dimethylpentanoic acid
b. 3,3-dichloropropanoic acid
c. 3-ethylbenzoic acid
d. hexanoic acid

14.2 Properties of Carboxylic Acids

Writing Equations for Reactions of Acids and Bases (10.7)

LEARNING GOAL Describe the solubility, dissociation, and neutralization of carboxylic acids.

Carboxylic acids are among the most polar organic compounds because their functional group consists of two polar groups: a hydroxyl group ($-OH$) and a carbonyl group ($C=O$). The $-OH$ group is similar to the functional group in alcohols, and the $C=O$ double bond is similar to that of aldehydes and ketones.

Two polar groups

$H-C(=O^{\delta-})^{\delta+}-O^{\delta-}-H^{\delta+}$ $\quad CH_3-CH_2-C(=O^{\delta-})^{\delta+}-O^{\delta-}-H^{\delta+}$

Solubility in Water

Carboxylic acids with one to five carbons are soluble in water because the carboxyl group forms hydrogen bonds with several water molecules (see **FIGURE 14.1**). However, as the length of the hydrocarbon chain increases, the nonpolar portion reduces the solubility of the carboxylic acid in water. Carboxylic acids having more than five carbons are not very soluble in water. **TABLE 14.2** lists the solubility for some selected carboxylic acids.

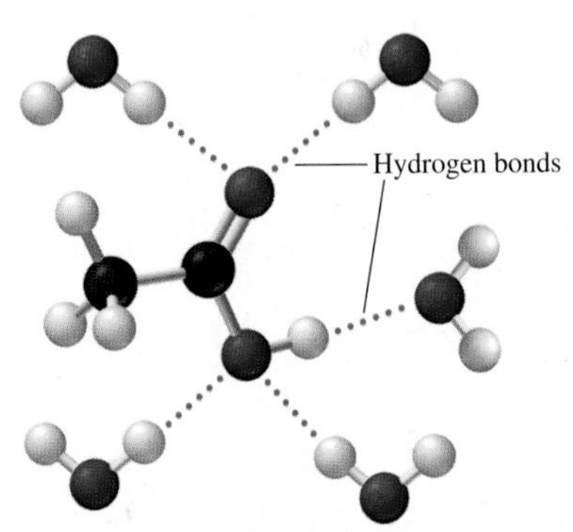

FIGURE 14.1 Acetic acid forms hydrogen bonds with water molecules.
Q Why do the atoms in the carboxyl group hydrogen bond with water molecules?

Acidity of Carboxylic Acids

An important property of carboxylic acids is their dissociation in water. When a carboxylic acid dissociates in water, a hydrogen ion is transferred to a water molecule to form a negatively charged **carboxylate ion** and a hydronium ion (H_3O^+). Carboxylic acids are more acidic than most other organic compounds including phenols. However, they are weak acids because only a small percentage ($<1\%$) of the carboxylic acid molecules dissociate in water. Carboxylic acids can lose hydrogen ions because the negative charge of the carboxylate

TABLE 14.2 Solubility of Selected Carboxylic Acids

IUPAC Name	Condensed Structural Formula	Solubility in Water
Methanoic acid	$H-\overset{\overset{\displaystyle O}{\|\|}}{C}-OH$	Soluble
Ethanoic acid	$CH_3-\overset{\overset{\displaystyle O}{\|\|}}{C}-OH$	Soluble
Propanoic acid	$CH_3-CH_2-\overset{\overset{\displaystyle O}{\|\|}}{C}-OH$	Soluble
Butanoic acid	$CH_3-CH_2-CH_2-\overset{\overset{\displaystyle O}{\|\|}}{C}-OH$	Soluble
Pentanoic acid	$CH_3-CH_2-CH_2-CH_2-\overset{\overset{\displaystyle O}{\|\|}}{C}-OH$	Soluble
Hexanoic acid	$CH_3-CH_2-CH_2-CH_2-CH_2-\overset{\overset{\displaystyle O}{\|\|}}{C}-OH$	Slightly soluble
Benzoic acid	$C_6H_5-\overset{\overset{\displaystyle O}{\|\|}}{C}-OH$	Slightly soluble

anion is stabilized by the two oxygen atoms. By comparison, alcohols do not lose the H in their —OH group because they cannot stabilize the negative charge that would result on a single oxygen atom.

TEST

Try Practice Problems 14.7 and 14.8

Carboxylic Acid **Carboxylate Ion**

$$CH_3-\overset{\overset{\displaystyle O}{\|}}{C}-OH + H_2O \rightleftarrows CH_3-\overset{\overset{\displaystyle O}{\|}}{C}-O^- + H_3O^+$$

Ethanoic acid (acetic acid) — Ethanoate ion (acetate ion) — Hydronium ion

SAMPLE PROBLEM 14.2 Dissociation of Carboxylic Acids in Water

TRY IT FIRST

Write the balanced chemical equation for the dissociation of propanoic acid in water.

SOLUTION

	Given	Need	Connect
ANALYZE THE PROBLEM	propanoic acid, H_2O	dissociation equation	products: propanoate ion, H_3O^+

The dissociation of propanoic acid produces a carboxylate ion and a hydronium ion.

$$CH_3-CH_2-\overset{\overset{\displaystyle O}{\|}}{C}-OH + H_2O \rightleftarrows CH_3-CH_2-\overset{\overset{\displaystyle O}{\|}}{C}-O^- + H_3O^+$$

Propanoic acid (propionic acid) — Propanoate ion (propionate ion)

STUDY CHECK 14.2

Write the balanced chemical equation for the dissociation of methanoic acid in water.

ANSWER

$$\mathrm{H-\overset{\overset{\displaystyle O}{\|}}{C}-OH + H_2O \rightleftarrows H-\overset{\overset{\displaystyle O}{\|}}{C}-O^- + H_3O^+}$$

TEST

Try Practice Problems 14.9 and 14.10

Neutralization of Carboxylic Acids

Because carboxylic acids are weak acids, they are completely neutralized by strong bases such as NaOH and KOH. The products are a **carboxylate salt** and water. The carboxylate ion is named by replacing the *ic acid* ending of the acid name with *ate*.

$$\mathrm{H-\overset{\overset{\displaystyle O}{\|}}{C}-\mathbf{OH} + Na\mathbf{OH} \longrightarrow H-\overset{\overset{\displaystyle O}{\|}}{C}-O^-\ Na^+ + \mathbf{H_2O}}$$

Methanoic acid (formic acid) → Sodium methanoate (sodium formate)

$$\mathrm{C_6H_5-\overset{\overset{\displaystyle O}{\|}}{C}-\mathbf{OH} + K\mathbf{OH} \longrightarrow C_6H_5-\overset{\overset{\displaystyle O}{\|}}{C}-O^-\ K^+ + \mathbf{H_2O}}$$

Benzoic acid → Potassium benzoate

Carboxylate salts are often used as preservatives and flavor enhancers in soups and seasonings.

Sodium propionate, a preservative, is added to cheeses, bread, and other bakery items to inhibit the spoilage of the food by microorganisms. Sodium benzoate, an inhibitor of mold and bacteria, is added to juices, margarine, relishes, salads, and jams. Monosodium glutamate (MSG) is added to meats, fish, vegetables, and bakery items to enhance flavor, although it causes headaches in some people.

$$\mathrm{CH_3-CH_2-\overset{\overset{\displaystyle O}{\|}}{C}-O^-\ Na^+}$$

Sodium propanoate (sodium propionate)

$$\mathrm{C_6H_5-\overset{\overset{\displaystyle O}{\|}}{C}-O^-\ Na^+}$$

Sodium benzoate

$$\mathrm{HO-\overset{\overset{\displaystyle O}{\|}}{C}-CH_2-CH_2-\overset{\overset{\displaystyle NH_2}{|}}{CH}-\overset{\overset{\displaystyle O}{\|}}{C}-O^-\ Na^+}$$

Monosodium glutamate

Carboxylate salts are ionic compounds with strong attractions between positively charged metal ions such as Li^+, Na^+, and K^+ and the negatively charged carboxylate ion. Like most salts, the carboxylate salts are solids at room temperature, have high melting points, and are usually soluble in water.

SAMPLE PROBLEM 14.3 Neutralization of a Carboxylic Acid

TRY IT FIRST

Write the balanced chemical equation for the neutralization of propanoic acid with sodium hydroxide.

SOLUTION

	Given	Need	Connect
ANALYZE THE PROBLEM	propanoic acid, NaOH	neutralization equation	products: sodium propanoate, H_2O

A chemical equation for the neutralization of a carboxylic acid includes the reactants, a carboxylic acid and a base, and the products, a carboxylate salt and water.

$$CH_3-CH_2-\overset{\overset{\displaystyle O}{\|}}{C}-OH + NaOH \longrightarrow CH_3-CH_2-\overset{\overset{\displaystyle O}{\|}}{C}-O^- Na^+ + H_2O$$

Propanoic acid (propionic acid) | Sodium hydroxide | Sodium propanoate (sodium propionate)

STUDY CHECK 14.3

What carboxylic acid will produce potassium butanoate when it is neutralized by KOH?

ANSWER

butanoic acid

TEST

Try Practice Problems 14.11 to 14.14

CHEMISTRY LINK TO HEALTH

Carboxylic Acids in Metabolism

Several carboxylic acids are part of the metabolic processes within our cells. For example, during glycolysis, a molecule of glucose is broken down into two molecules of pyruvic acid, or actually, its carboxylate ion, pyruvate. During strenuous exercise when oxygen levels are low (anaerobic), pyruvic acid is reduced to give lactic acid or the lactate ion.

$$CH_3-\overset{\overset{\displaystyle O}{\|}}{C}-\overset{\overset{\displaystyle O}{\|}}{C}-OH + 2H \xrightarrow{\text{Reduction}} CH_3-\overset{\overset{\displaystyle OH}{|}}{CH}-\overset{\overset{\displaystyle O}{\|}}{C}-OH$$

Pyruvic acid → Lactic acid

During exercise, pyruvic acid is converted to lactic acid in the muscles.

In the *citric acid cycle*, also called the *Krebs cycle*, di- and tricarboxylic acids are oxidized and decarboxylated (loss of CO_2) to produce energy for the cells of the body. These carboxylic acids are normally referred to by their common names. At the start of the citric acid cycle, citric acid with six carbons is converted to five-carbon α-ketoglutaric acid. Citric acid is the acid that gives the sour taste to citrus fruits such as lemons and grapefruits.

Citric acid: $HOOC-CH_2-C(OH)(COOH)-CH_2-COOH \xrightarrow{[O]}$ α-Ketoglutaric acid: $HOOC-CH_2-CH_2-C(=O)-COOH + CO_2$

The citric acid cycle continues as α-ketoglutaric acid loses CO_2 to give a four-carbon succinic acid. Then a series of reactions converts succinic acid to oxaloacetic acid. We see that some of the functional groups we have studied along with reactions such as hydration and oxidation are part of the metabolic processes that take place in our cells.

Succinic acid: $HOOC-CH_2-CH_2-COOH \xrightarrow{[O]}$ Fumaric acid: $HOOC-CH=CH-COOH \xrightarrow{H_2O}$ Malic acid: $HOOC-CH(OH)-CH_2-COOH \xrightarrow{[O]}$ Oxaloacetic acid: $HOOC-C(=O)-CH_2-COOH$

At the pH of the aqueous environment in the cells, the carboxylic acids are dissociated, which means it is actually the carboxylate ions that take part in the reactions of the citric acid cycle. For example, in water, succinic acid is in equilibrium with its carboxylate ion, succinate.

$$HOOC-CH_2-CH_2-COOH + 2H_2O \rightleftarrows {}^-OOC-CH_2-CH_2-COO^- + 2H_3O^+$$

Succinic acid → Succinate ion

Citric acid gives the sour taste to citrus fruits.

PRACTICE PROBLEMS

14.2 Properties of Carboxylic Acids

LEARNING GOAL Describe the solubility, dissociation, and neutralization of carboxylic acids.

14.7 Identify the compound in each group that is most soluble in water. Explain.
- **a.** propanoic acid, hexanoic acid, benzoic acid
- **b.** pentane, 1-hexanol, propanoic acid

14.8 Identify the compound in each group that is most soluble in water. Explain.
- **a.** butanone, butanoic acid, butane
- **b.** ethanoic acid (acetic acid), hexanoic acid, octanoic acid

14.9 Write the balanced chemical equation for the dissociation of each of the following carboxylic acids in water:
- **a.** pentanoic acid
- **b.** acetic acid

14.10 Write the balanced chemical equation for the dissociation of each of the following carboxylic acids in water:
- **a.** $CH_3-CH(CH_3)-C(=O)-OH$
- **b.** butanoic acid

14.11 Write the balanced chemical equation for the reaction of each of the following carboxylic acids with NaOH:
- **a.** formic acid
- **b.** 3-chloropropanoic acid
- **c.** benzoic acid

14.12 Write the balanced chemical equation for the reaction of each of the following carboxylic acids with KOH:
- **a.** acetic acid
- **b.** 2-methylbutanoic acid
- **c.** 4-chlorobenzoic acid

14.13 Write the IUPAC and common names, if any, of the carboxylate salts produced in problem 14.11.

14.14 Write the IUPAC and common names, if any, of the carboxylate salts produced in problem 14.12.

14.3 Esters

LEARNING GOAL Write the IUPAC and common names for an ester; write the balanced chemical equation for the formation of an ester.

When a carboxylic acid reacts with an alcohol, an **ester** and water are produced when the —H of the carboxylic acid is replaced by an alkyl group. Fats and oils are esters of glycerol and fatty acids, which are long-chain carboxylic acids. Esters produce the pleasant aromas and flavors of many fruits, such as bananas, strawberries, and oranges.

Carboxylic Acid	Ester
$CH_3-C(=O)-OH$	$CH_3-C(=O)-O-CH_3$
Ethanoic acid (acetic acid)	Methyl ethanoate (methyl acetate)

Esterification

In a reaction called **esterification**, an ester is produced when a carboxylic acid and an alcohol react in the presence of an acid catalyst (usually H_2SO_4) and heat. In esterification, the —OH group from the carboxylic acid and the —H from the alcohol are removed and combine to form water. An excess of the alcohol is used to shift the equilibrium in the direction of the formation of the ester product.

$$CH_3-C(=O)-OH + H-O-CH_3 \xrightleftharpoons{H^+,\ heat} CH_3-C(=O)-O-CH_3 + H-OH$$

Ethanoic acid (acetic acid) + Methanol (methyl alcohol) → Methyl ethanoate (methyl acetate)

Pentyl ethanoate provides the flavor and odor of bananas.

For example, pentyl ethanoate, which is the ester responsible for the flavor and odor of bananas, can be prepared using ethanoic acid and 1-pentanol. The equation for this esterification is written as

$$\underset{\text{Ethanoic acid (acetic acid)}}{CH_3-\overset{\overset{\displaystyle O}{\|}}{C}-\mathbf{OH}} + \underset{\text{1-Pentanol (pentyl alcohol)}}{\mathbf{H}-O-CH_2-CH_2-CH_2-CH_2-CH_3} \xrightleftharpoons{H^+,\ heat}$$

$$\underset{\text{Pentyl ethanoate (pentyl acetate)}}{CH_3-\overset{\overset{\displaystyle O}{\|}}{C}-O-CH_2-CH_2-CH_2-CH_2-CH_3} + \mathbf{H_2O}$$

SAMPLE PROBLEM 14.4 Writing Esterification Equations

TRY IT FIRST

An ester that has the smell of pineapple can be synthesized from butanoic acid and methanol. Write the balanced chemical equation for the formation of this ester.

SOLUTION

	Given	Need	Connect
ANALYZE THE PROBLEM	butanoic acid, methanol	esterification equation	products: methyl butanoate, H_2O

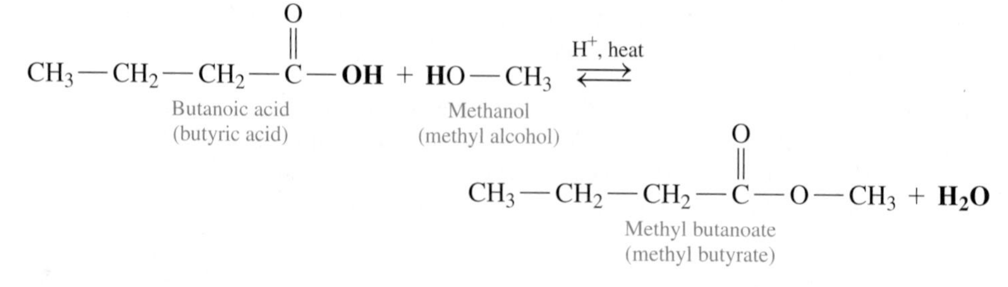

$$\underset{\text{Butanoic acid (butyric acid)}}{CH_3-CH_2-CH_2-\overset{\overset{\displaystyle O}{\|}}{C}-\mathbf{OH}} + \underset{\text{Methanol (methyl alcohol)}}{\mathbf{HO}-CH_3} \xrightleftharpoons{H^+,\ heat}$$

$$\underset{\text{Methyl butanoate (methyl butyrate)}}{CH_3-CH_2-CH_2-\overset{\overset{\displaystyle O}{\|}}{C}-O-CH_3} + \mathbf{H_2O}$$

The flavor and odor of plums is provided by an ester from methanoic acid and 1-butanol.

STUDY CHECK 14.4

The ester that smells like plums can be synthesized from methanoic acid and 1-butanol. Write the balanced chemical equation for the formation of this ester.

ANSWER

$$H-\overset{\overset{\displaystyle O}{\|}}{C}-OH + HO-CH_2-CH_2-CH_2-CH_3 \xrightleftharpoons{H^+,\ heat}$$

$$H-\overset{\overset{\displaystyle O}{\|}}{C}-O-CH_2-CH_2-CH_2-CH_3 + H_2O$$

TEST

Try Practice Problems 14.15 to 14.18

Naming Esters

The name of an ester consists of two words, which are derived from the names of the alcohol and the acid in that ester. The first word indicates the *alkyl* part from the alcohol. The second word is the name of the *carboxylate* from the carboxylic acid. The IUPAC names of esters use the IUPAC names of the acids, while the common names of esters use the common names of the acids. Let's take a look at the following ester, which has a pleasant, fruity odor. We start by separating the ester bond to identify the alkyl part from the alcohol and the carboxylate part from the acid. Then we name the ester as an alkyl carboxylate.

$$\underbrace{CH_3-\overset{\displaystyle O}{\overset{\|}{C}}}_{\text{From carboxylic acid (carboxylate)}}-\underbrace{O-CH_3}_{\text{From alcohol (alkyl)}}$$

$$CH_3-\overset{\displaystyle O}{\overset{\|}{C}}-OH + HO-CH_3$$

					Ester name
IUPAC	Ethanoic acid	+	Methanol	=	Methyl ethanoate
(common)	(acetic acid)	+	(methyl alcohol)	=	(methyl acetate)

Methyl ethanoate (methyl acetate)

The following examples of some typical esters show the IUPAC names, as well as the common names, of esters.

$$CH_3-\overset{\displaystyle O}{\overset{\|}{C}}-O-CH_2-CH_2$$

Ethyl ethanoate (ethyl acetate)

$$CH_3-CH_2-\overset{\displaystyle O}{\overset{\|}{C}}-O-CH_3$$

Methyl propanoate (methyl propionate)

Ethyl benzoate

CHEMISTRY LINK TO HEALTH

Salicylic Acid from a Willow Tree

For many centuries, relief from pain and fever was obtained by chewing on the leaves or a piece of bark from the willow tree. By the 1800s, chemists discovered that salicin was the agent in the bark responsible for the relief of pain. However, the body converts salicin to salicylic acid, which has a carboxyl group and a hydroxyl group that irritates the stomach lining. In 1899, the Bayer chemical company in Germany produced an ester of salicylic acid and acetic acid, called acetylsalicylic acid (aspirin), which is less irritating. In some aspirin preparations, a buffer is added to neutralize the carboxylic acid group. Today, aspirin is used as an analgesic (pain reliever), antipyretic (fever reducer), and anti-inflammatory agent. Many people take a daily low-dose aspirin, which has been found to lower the risk of heart attack and stroke.

The discovery of salicin in the leaves and bark of the willow tree led to the development of aspirin.

Salicylic acid + Acetic acid $\underset{}{\overset{H^+,\ heat}{\rightleftarrows}}$ Acetylsalicylic acid (aspirin) + H_2O

Oil of wintergreen, or methyl salicylate, has a pungent, minty odor and flavor. Because it can pass through the skin, methyl salicylate is used in skin ointments, where it acts as a counterirritant, producing heat to soothe sore muscles.

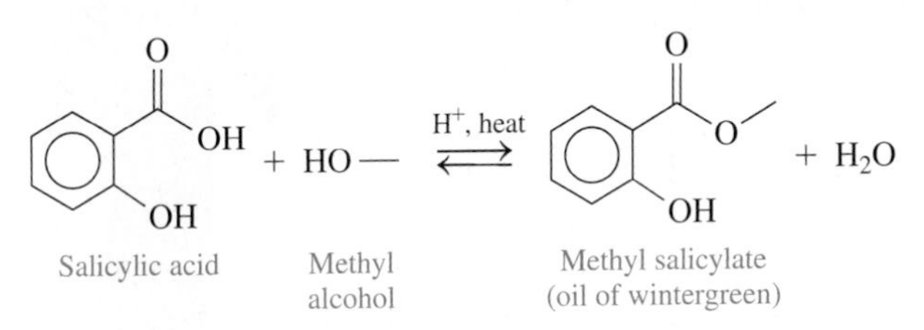

Ointments containing methyl salicylate are used to soothe sore muscles.

SAMPLE PROBLEM 14.5 Naming Esters

TRY IT FIRST

What are the IUPAC and common names of the following ester?

$$CH_3—CH_2—\overset{\overset{\large O}{\|}}{C}—O—CH_2—CH_2—CH_3$$

SOLUTION GUIDE

ANALYZE THE PROBLEM	Given	Need	Connect
	ester	IUPAC name, common name	write the alkyl name for the alcohol, change *ic acid* to *ate*

STEP 1 Write the name for the carbon chain from the alcohol as an *alkyl* group.

$$CH_3—CH_2—\overset{\overset{\large O}{\|}}{C}—O—CH_2—CH_2—CH_3 \quad \text{propyl}$$

STEP 2 Change the *ic acid* of the acid name to *ate*.

From carboxylic acid

From alcohol

$$CH_3—CH_2—\overset{\overset{\large O}{\|}}{C}—O—CH_2—CH_2—CH_3$$

propyl propanoate

(propyl propionate)

The odor of grapes is due to an ester.

STUDY CHECK 14.5

Write the IUPAC name for the following ester, which gives the odor and flavor to grapes.

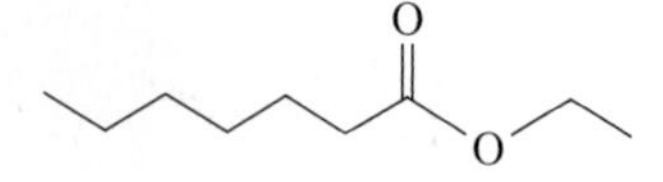

ANSWER

ethyl heptanoate

TEST

Try Practice Problems 14.19 and 14.20

CHEMISTRY LINK TO THE ENVIRONMENT

Plastics

Terephthalic acid (an aromatic acid with two carboxyl groups) is produced in large quantities for the manufacture of polyesters such as Dacron. When terephthalic acid reacts with ethylene glycol, ester bonds form on both ends of the molecules, allowing many molecules to combine into a long molecule.

$$\text{Terephthalic acid} + \text{Ethylene glycol} \underset{}{\overset{H^+,\ \text{heat}}{\rightleftarrows}}$$

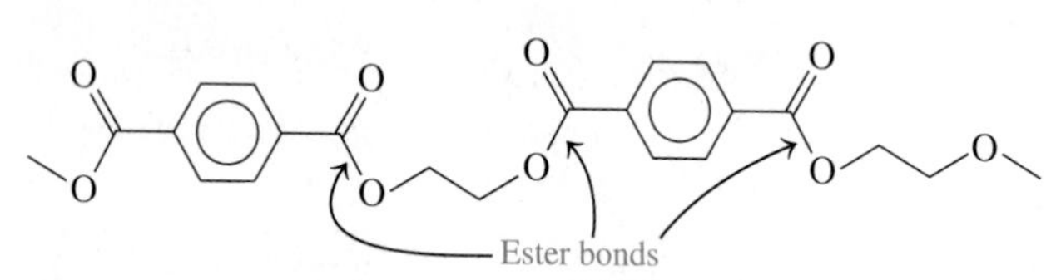

A section of the polyester Dacron

Dacron, a synthetic material first produced by DuPont in the 1960s, is a polyester used to make permanent press fabrics, carpets, and clothes. Permanent press is a chemical process in which fabrics are permanently shaped and treated for wrinkle resistance. In medicine, artificial blood vessels and valves are made of Dacron, which is biologically inert and does not clot the blood.

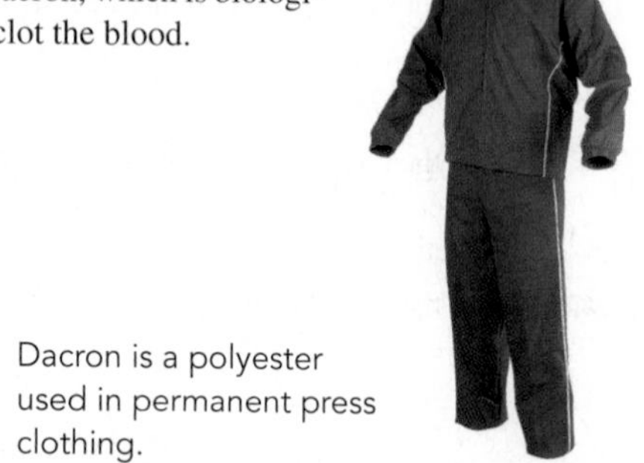

Dacron is a polyester used in permanent press clothing.

The polyester can also be made into a film called Mylar and a plastic known as PETE (polyethyleneterephthalate). PETE is used for plastic soft drink and water bottles as well as for peanut butter jars, containers of salad dressings, shampoos, and dishwashing liquids. Today, PETE (recycling symbol "1") is the most widely recycled of all the plastics. Every year, more than 1.5×10^9 lb (6.8×10^8 kg) of PETE is being recycled. After PETE is separated from other plastics, it is used to make useful items, including polyester fabric for T-shirts and coats, carpets, fill for sleeping bags, doormats, and containers for tennis balls.

Polyester, in the form of the plastic PETE, is used to make soft drink bottles.

Esters in Plants

Many of the fragrances of perfumes and flowers and the flavors of fruits are due to esters. Small esters are volatile, so we can smell them, and they are soluble in water, so we can taste them. Several of these, with their flavor and odor, are listed in **TABLE 14.3**.

TEST

Try Practice Problems 14.21 to 14.24

TABLE 14.3 Some Esters in Fruits and Flavorings

Condensed Structural Formula and Name	Flavor/Odor
$CH_3-\overset{O}{\overset{\|}{C}}-O-CH_2-CH_2-CH_3$ Propyl ethanoate (propyl acetate)	Pears
$CH_3-\overset{O}{\overset{\|}{C}}-O-CH_2-CH_2-CH_2-CH_2-CH_3$ Pentyl ethanoate (pentyl acetate)	Bananas
$CH_3-\overset{O}{\overset{\|}{C}}-O-CH_2-CH_2-CH_2-CH_2-CH_2-CH_2-CH_2-CH_3$ Octyl ethanoate (octyl acetate)	Oranges
$CH_3-CH_2-CH_2-\overset{O}{\overset{\|}{C}}-O-CH_2-CH_3$ Ethyl butanoate (ethyl butyrate)	Pineapples
$CH_3-CH_2-CH_2-\overset{O}{\overset{\|}{C}}-O-CH_2-CH_2-CH_2-CH_2-CH_3$ Pentyl butanoate (pentyl butyrate)	Apricots

Esters such as ethyl butanoate provide the odor and flavor of many fruits such as pineapples.

PRACTICE PROBLEMS

14.3 Esters

LEARNING GOAL Write the IUPAC and common names for an ester; write the balanced chemical equation for the formation of an ester.

14.15 Draw the condensed structural formula for the ester formed when each of the following reacts with methyl alcohol:

a. acetic acid **b.** pentanoic acid

14.16 Draw the condensed structural formula for the ester formed when each of the following reacts with ethyl alcohol:

a. formic acid **b.** propionic acid

14.17 Draw the condensed structural or line-angle formula for the ester formed in each of the following reactions:

a.
$$CH_3-CH_2-CH_2-CH_2-\overset{O}{\overset{\|}{C}}-OH + HO-\underset{}{\overset{CH_3}{\overset{|}{C}H}}-CH_3 \xrightleftharpoons{H^+,\ heat}$$

b. 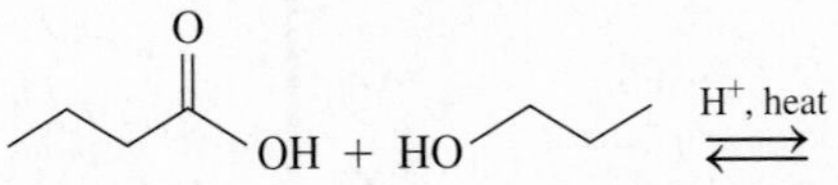

14.18 Draw the condensed structural or line-angle formula for the ester formed in each of the following reactions:

a. $CH_3-CH_2-\overset{O}{\overset{\|}{C}}-OH + HO-CH_3 \xrightleftharpoons{H^+, \text{ heat}}$

b. (benzene ring)–C(=O)–OH + HO–(butyl line-angle chain) $\xrightleftharpoons{H^+, \text{ heat}}$

14.19 Write the IUPAC and common names, if any, for each of the following:

a. $H-\overset{O}{\overset{\|}{C}}-O-CH_3$

b. $CH_3-CH_2-\overset{O}{\overset{\|}{C}}-O-CH_2-CH_3$

c. (line-angle formula: CH₃–C(=O)–O–propyl)

14.20 Write the IUPAC and common names, if any, for each of the following:

a. (line-angle formula: butanoyl–C(=O)–O–ethyl)

b. (line-angle formula: hexanoyl–C(=O)–O–methyl)

c. $CH_3-CH_2-\overset{O}{\overset{\|}{C}}-O-CH_2-CH_2-CH_2-CH_3$

14.21 Draw the condensed structural formulas for **a** and **b** and line-angle formulas for **c** and **d**:

a. propyl butyrate **b.** butyl formate
c. ethyl pentanoate **d.** methyl propanoate

14.22 Draw the condensed structural formulas for **a** and **b** and line-angle formulas for **c** and **d**:

a. hexyl acetate **b.** ethyl methanoate
c. propyl benzoate **d.** methyl octanoate

14.23 What is the ester responsible for the flavor and odor of the following fruit?

a. banana **b.** orange **c.** apricot

14.24 What flavor would you notice if you smelled or tasted the following?

a. ethyl butanoate **b.** propyl acetate **c.** pentyl acetate

14.4 Hydrolysis of Esters

LEARNING GOAL Draw the condensed structural or line-angle formulas for the products from acid and base hydrolysis of esters.

Esters undergo **hydrolysis** when they react with water in the presence of an acid or a base. Therefore, hydrolysis is the reverse of the esterification reaction.

CORE CHEMISTRY SKILL
Hydrolyzing Esters

Acid Hydrolysis of Esters

In *acid hydrolysis*, water reacts with an ester in the presence of a strong acid, usually H_2SO_4 or HCl, to form a carboxylic acid and an alcohol. A water molecule provides the $-OH$ group to convert the carbonyl group of the ester to a carboxyl group. When hydrolysis of biological esters occurs in the cells, an enzyme replaces the acid as the catalyst.

$$CH_3-\overset{O}{\overset{\|}{C}}-O-CH_3 + H-OH \xrightleftharpoons{H^+, \text{ heat}} CH_3-\overset{O}{\overset{\|}{C}}-OH + HO-CH_3$$

Methyl ethanoate (methyl acetate) | Water | Ethanoic acid (acetic acid) | Methanol (methyl alcohol)

SAMPLE PROBLEM 14.6 Acid Hydrolysis of Esters

TRY IT FIRST

Aspirin (acetylsalicylic acid) that has been stored for a long time may undergo acid hydrolysis in the presence of water and heat. Write a balanced chemical equation for the acid hydrolysis of aspirin.

(line-angle structure of aspirin)

Aspirin
(acetylsalicylic acid)

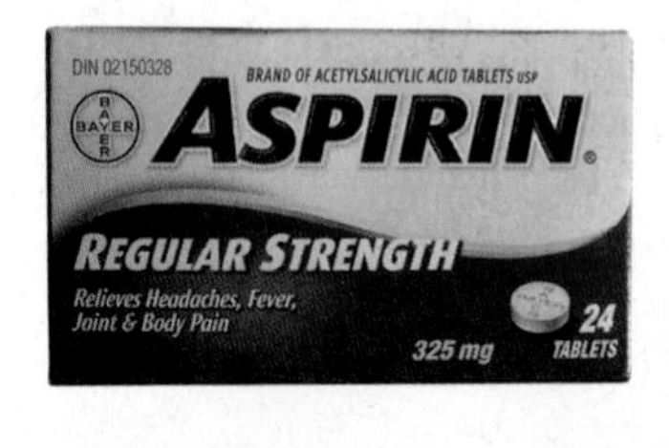

Aspirin stored in a warm, humid place may undergo hydrolysis.

SOLUTION

ANALYZE THE PROBLEM	Given	Need	Connect
	aspirin and water	acid hydrolysis equation	products: carboxylic acid, alcohol

For the hydrolysis products, write the formula of the carboxylic acid by adding —OH to the carbonyl group and —H to complete the alcohol group. The acetic acid in the products gives the odor of vinegar to the aspirin that has hydrolyzed.

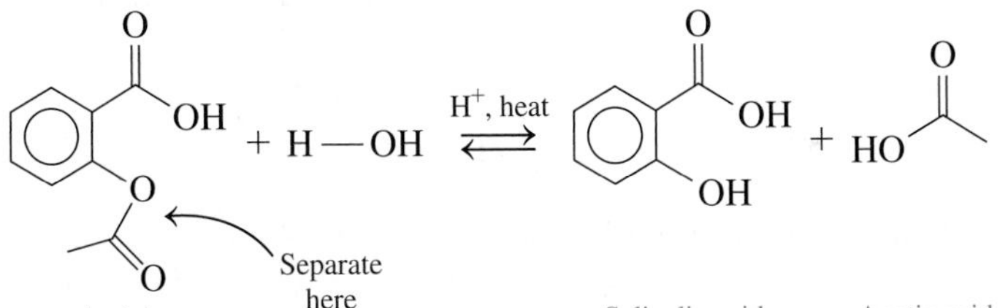

Aspirin + H—OH $\underset{}{\overset{H^+,\ heat}{\rightleftarrows}}$ Salicylic acid + Acetic acid

Separate here

STUDY CHECK 14.6

What are the IUPAC and common names of the products from the acid hydrolysis of ethyl propanoate (ethyl propionate)?

ANSWER

propanoic acid and ethanol (propionic acid and ethyl alcohol)

Base Hydrolysis of Esters (Saponification)

When an ester undergoes hydrolysis with a strong base such as NaOH or KOH, the products are the carboxylate salt and the corresponding alcohol. This base hydrolysis is also called *saponification*.

Ester + strong base $\longrightarrow$ carboxylate salt + alcohol

$$CH_3-\overset{O}{\overset{\|}{C}}-O-CH_3 + \mathbf{NaOH} \xrightarrow{Heat} CH_3-\overset{O}{\overset{\|}{C}}-O^-\ Na^+ + HO-CH_3$$

Methyl ethanoate (methyl acetate) | Sodium hydroxide | Sodium ethanoate (sodium acetate) | Methanol (methyl alcohol)

TEST

Try Practice Problems 14.25 and 14.26

SAMPLE PROBLEM 14.7 **Base Hydrolysis of Esters**

TRY IT FIRST

Ethyl acetate is a solvent used for fingernail polish, plastics, and lacquers. Write the balanced chemical equation for the hydrolysis of ethyl acetate with NaOH.

Ethyl acetate is the solvent in fingernail polish.

SOLUTION

ANALYZE THE PROBLEM	Given	Need	Connect
	ethyl acetate, NaOH	base hydrolysis equation	products: carboxylate ion, alcohol

The hydrolysis of ethyl acetate with NaOH gives the carboxylate salt, sodium acetate, and ethyl alcohol.

$$CH_3-\overset{O}{\overset{\|}{C}}-O-CH_2-CH_3 + NaOH \xrightarrow{Heat} CH_3-\overset{O}{\overset{\|}{C}}-O^-\ Na^+ + HO-CH_2-CH_3$$

Ethyl ethanoate (ethyl acetate) | Sodium ethanoate (sodium acetate) | Ethanol (ethyl alcohol)

INTERACTIVE VIDEO
Study Check 14.7

TEST
Try Practice Problems 14.27 and 14.28

STUDY CHECK 14.7

Draw the line-angle formulas for the products from the hydrolysis of methyl benzoate with KOH.

ANSWER

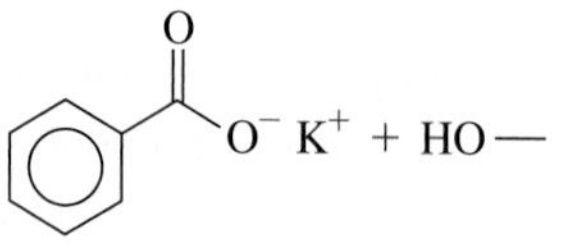

(benzoate line-angle structure) $O^- K^+$ + HO—

PRACTICE PROBLEMS

14.4 Hydrolysis of Esters

LEARNING GOAL Draw the condensed structural or line-angle formulas for the products from acid and base hydrolysis of esters.

14.25 What are the products of the acid hydrolysis of an ester?

14.26 What are the products of the base hydrolysis of an ester?

14.27 Draw the condensed structural or line-angle formulas for the products from the acid- or base-catalyzed hydrolysis of each of the following:

a. $CH_3—CH_2—\overset{O}{\overset{\|}{C}}—O—CH_3 + NaOH \xrightarrow{Heat}$

b. 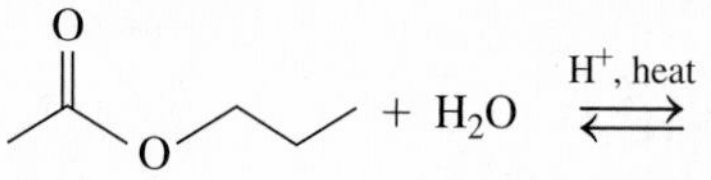 (line-angle ester) $+ H_2O \overset{H^+, heat}{\rightleftarrows}$

c. $CH_3—CH_2—CH_2—\overset{O}{\overset{\|}{C}}—O—CH_2—CH_3 + H_2O \overset{H^+, heat}{\rightleftarrows}$

d. 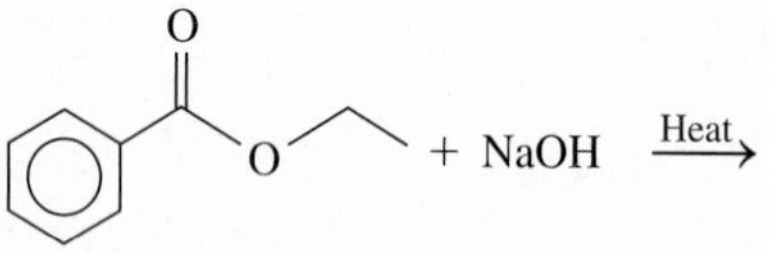 (line-angle ester with benzene ring) $+ NaOH \xrightarrow{Heat}$

14.28 Draw the condensed structural or line-angle formulas for the products from the acid- or base-catalyzed hydrolysis of each of the following:

a. $CH_3—CH_2—\overset{O}{\overset{\|}{C}}—O—CH_2—CH_2—CH_2—CH_3 + H_2O \overset{H^+, heat}{\rightleftarrows}$

b. 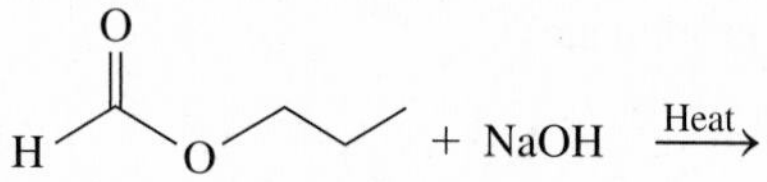 H (line-angle ester) $+ NaOH \xrightarrow{Heat}$

c. $CH_3—CH_2—\overset{O}{\overset{\|}{C}}—O—CH_3 + H_2O \overset{H^+, heat}{\rightleftarrows}$

d. 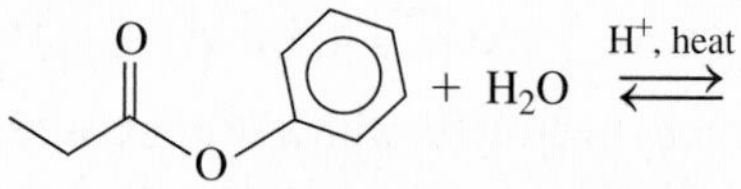 (line-angle ester with benzene ring) $+ H_2O \overset{H^+, heat}{\rightleftarrows}$

14.5 Amines

LEARNING GOAL Write the common names for amines; draw the condensed structural or line-angle formulas when given their names. Describe the solubility, dissociation, and neutralization of amines in water.

Amines and amides are organic compounds that contain nitrogen. Many nitrogen-containing compounds are important to life as components of amino acids, proteins, and nucleic acids (DNA and RNA). Many amines exhibit strong physiological activity and are used in medicine as decongestants, anesthetics, and sedatives. Examples include dopamine, histamine, epinephrine, and amphetamine. Alkaloids such as caffeine, nicotine, cocaine, and digitalis, which have powerful physiological activity, are naturally occurring amines obtained from plants.

Naming and Classifying Amines

Amines are derivatives of ammonia (NH_3), in which one or more hydrogen atoms are replaced with alkyl or aromatic groups. In methylamine, a methyl group replaces one hydrogen atom in ammonia. The bonding of two methyl groups gives dimethylamine, and the three methyl groups in trimethylamine replace all the hydrogen atoms in ammonia.

The common names of amines are often used when the alkyl groups bonded to the nitrogen atom are not branched. Then the alkyl groups are listed in alphabetical order. The prefixes *di* and *tri* are used to indicate two and three identical substituents.

$\boxed{CH_3}—NH_2$ — Methylamine

$\boxed{CH_3}—N(H)—\boxed{CH_3}$ — Dimethylamine

$\boxed{CH_3—CH_2—CH_2}—N(\boxed{CH_3})—\boxed{CH_2—CH_3}$ — Ethylmethylpropylamine

Amines are classified by counting the number of carbon atoms directly bonded to the nitrogen atom. In a primary (1°) amine, the nitrogen atom is bonded to one alkyl group. In a secondary (2°) amine, the nitrogen atom is bonded to two alkyl groups. In a tertiary (3°) amine, the nitrogen atom is bonded to three alkyl groups (see **FIGURE 14.2**).

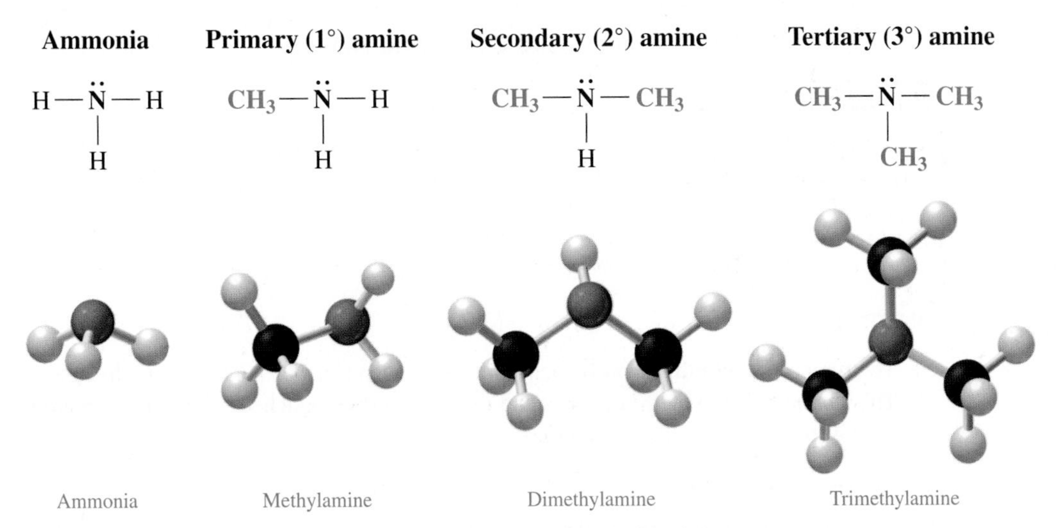

FIGURE 14.2 Amines have one or more carbon atoms bonded to the N atom.

Q How many carbon atoms are bonded to the nitrogen atom in dimethylamine?

Line-Angle Formulas for Amines

We can draw line-angle formulas for amines just as we have for other organic compounds. In the line-angle formula for an amine, we show the hydrogen atoms bonded to the N atom.

NH_2 — Propylamine, Primary (1°) amine

N, H — Ethylpropylamine, Secondary (2°) amine

N — Diethylmethylamine, Tertiary (3°) amine

Aromatic Amines

The aromatic amines use the name *aniline*, which is approved by IUPAC. Alkyl groups attached to the nitrogen of aniline are named with the prefix *N*- followed by the alkyl name.

NH_2 — Aniline

NH_2, 1, 4, Br — 4-Bromoaniline

H, N — *N*-Methylaniline

Aniline is used to make many dyes such as indigo, which give color to wool, cotton, and silk fibers, as well as blue jeans. It is also used to make polyurethane and in the synthesis of the pain reliever acetaminophen.

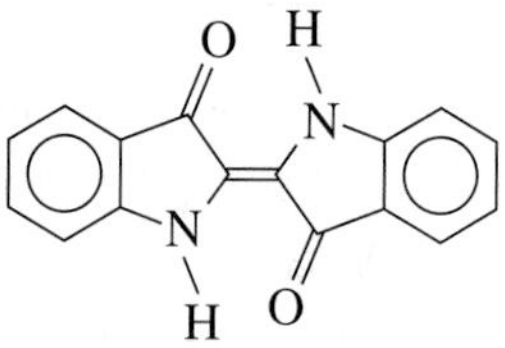

Indigo

Indigo used in blue dyes can be obtained from tropical plants such as *Indigofera tinctoria*.

SAMPLE PROBLEM 14.8 Naming and Classifying Amines

TRY IT FIRST

Classify each of the following amines as primary (1°), secondary (2°), or tertiary (3°), and write the common name for each:

a. $CH_3—CH_2—NH_2$

b. $CH_3—N(CH_3)—CH_2—CH_3$

SOLUTION

a. This amine has one ethyl group attached to the nitrogen atom (1°); its name is ethylamine.
b. This amine has two methyl groups and one ethyl group attached to the nitrogen atom (3°); its name is ethyldimethylamine.

STUDY CHECK 14.8

Draw the line-angle formula for methylpropylamine and classify as primary (1°), secondary (2°), or tertiary (3°).

ANSWER

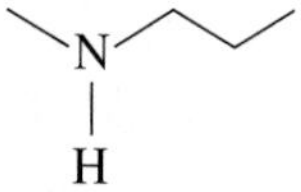

secondary (2°)

TEST

Try Practice Problems 14.29 to 14.34

CHEMISTRY LINK TO HEALTH

Amines in Health and Medicine

In response to allergic reactions or injury to cells, the body increases the production of histamine, which causes blood vessels to dilate and increases the permeability of the cells. Redness and swelling occur in the area. Administering an antihistamine such as diphenhydramine helps block the effects of histamine.

In the body, hormones called *biogenic amines* carry messages between the central nervous system and nerve cells. Epinephrine (adrenaline) and norepinephrine (noradrenaline) are released by the adrenal medulla in "fight-or-flight" situations to raise the blood glucose level and move the blood to the muscles. Used in remedies for colds, hay fever, and asthma, norepinephrine contracts the capillaries in the mucous membranes of the respiratory passages. The prefix *nor* in a drug name means there is one less $CH_3—$ group on the nitrogen atom. Parkinson's disease is a result of a deficiency in another biogenic amine called dopamine.

Produced synthetically, amphetamines (known as "uppers") are stimulants of the central nervous system much like epinephrine, but they also increase cardiovascular activity and depress the appetite.

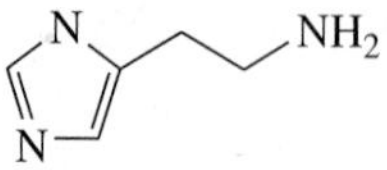

Histamine

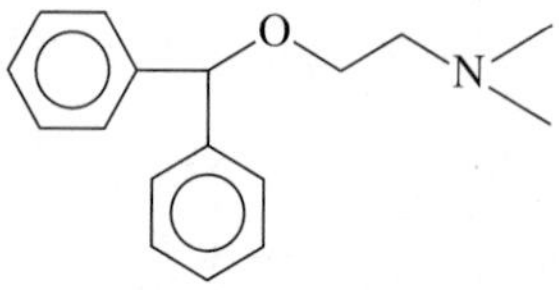

Diphenhydramine

They are sometimes used to bring about weight loss, but they can cause chemical dependency. Benzedrine and Neo-Synephrine (phenylephrine) are used in medications to reduce respiratory congestion from colds, hay fever, and asthma. Sometimes, benzedrine is taken to combat the desire to sleep, but it has side effects. Methedrine is used to treat depression and in the illegal form is known as "speed," "meth," or "crystal meth." The prefix *meth* means that there is one more $CH_3—$ group on the nitrogen atom.

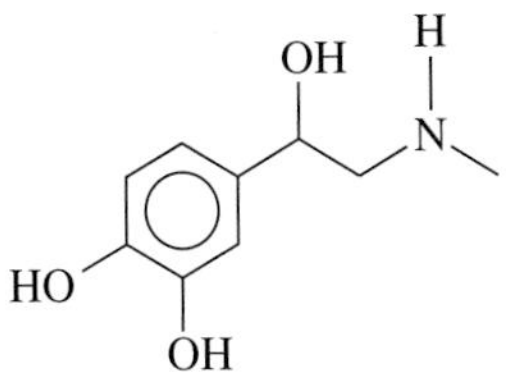

Epinephrine (adrenaline)

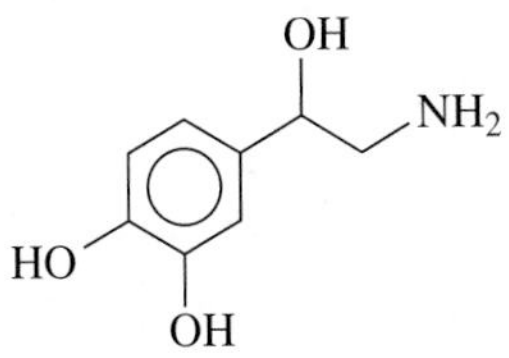

Norepinephrine (noradrenaline)

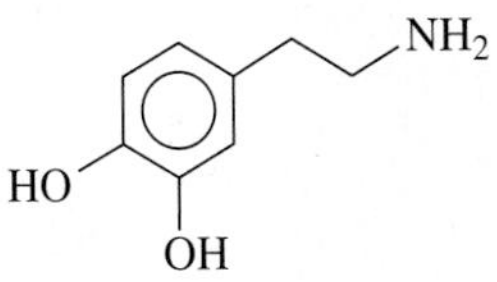

Dopamine

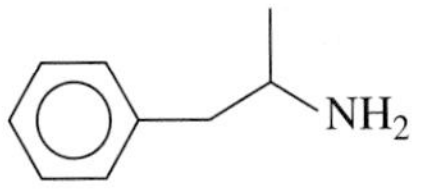

Benzedrine (amphetamine)

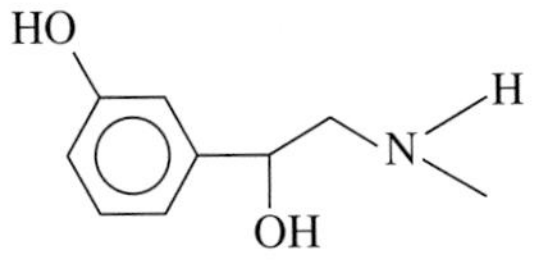

Neo-Synephrine (phenylephrine)

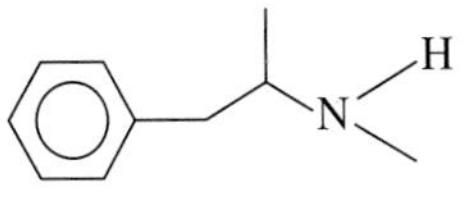

Methamphetamine (methedrine)

Solubility of Amines in Water

Because amines contain a polar N—H bond, they form hydrogen bonds with water. In primary (1°) amines, $—NH_2$ can form more hydrogen bonds than the secondary (2°) amines. A tertiary (3°) amine, which has no hydrogen on the nitrogen atom, can form only hydrogen bonds with water from the N atom in the amine to the H of a water molecule. Like alcohols, the smaller amines, including tertiary ones, are soluble because they form hydrogen bonds with water. However, in amines with more than six carbon atoms, the effect of hydrogen bonding is diminished. Then the nonpolar hydrocarbon chains of the amine decreases its solubility in water (see **FIGURE 14.3**).

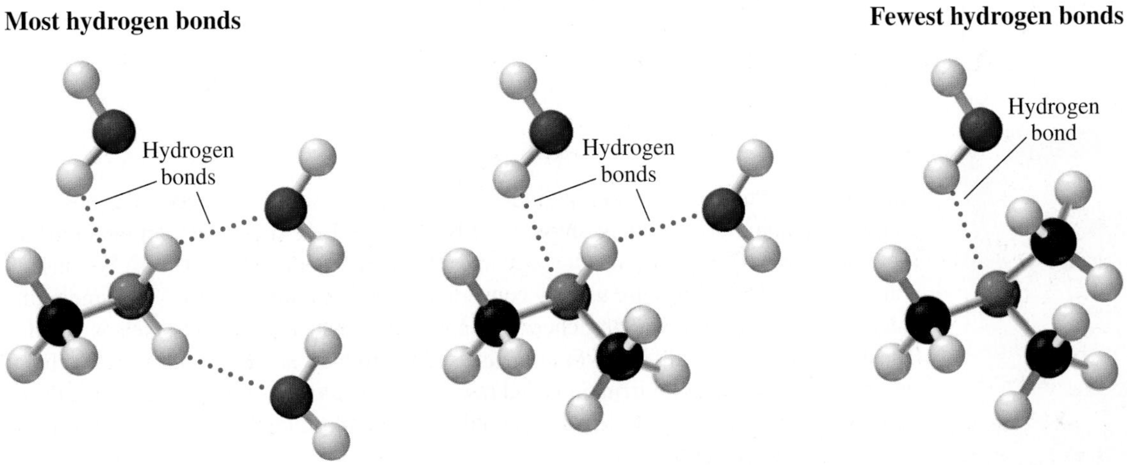

Most hydrogen bonds **Fewest hydrogen bonds**

Hydrogen bonds

Hydrogen bonds

Hydrogen bond

FIGURE 14.3 Primary, secondary, and tertiary amines form hydrogen bonds with water molecules, but primary amines form the most and tertiary amines form the least.

Q Why do tertiary (3°) amines form fewer hydrogen bonds with water than primary amines?

Amines React as Bases in Water

Ammonia (NH_3) acts as a Brønsted–Lowry base because it accepts H^+ from water to produce an ammonium ion (NH_4^+) and a hydroxide ion (OH^-).

$$\ddot{N}H_3 + H_2O \rightleftarrows NH_4^+ + OH^-$$

Ammonia — Ammonium ion — Hydroxide ion

The amines in fish react with the acid in lemon juice to neutralize the "fishy" odor.

In water, amines act as Brønsted–Lowry bases because the lone electron pair on the nitrogen atom accepts a hydrogen ion from water and produces alkylammonium and hydroxide ions.

Reaction of a Primary Amine with Water

$$\underset{\text{Methylamine}}{CH_3-\ddot{N}H_2} + H_2O \rightleftharpoons \underset{\text{Methylammonium ion}}{CH_3-\overset{+}{N}H_3} + \underset{\text{Hydroxide ion}}{OH^-}$$

Reaction of a Secondary Amine with Water

$$\underset{\text{Dimethylamine}}{CH_3-\ddot{N}H(CH_3)} + H_2O \rightleftharpoons \underset{\text{Dimethylammonium ion}}{CH_3-\overset{+}{N}H_2(CH_3)} + \underset{\text{Hydroxide ion}}{OH^-}$$

Ammonium Salts

When you squeeze lemon juice on fish, the "fishy" odor is removed by converting the amines to their ammonium salts. In a *neutralization reaction*, an amine acts as a base and reacts with an acid to form an **ammonium salt**. The lone pair of electrons on the nitrogen atom accepts H^+ from an acid to give an ammonium salt; no water is formed. An ammonium salt is named by using its alkylammonium ion name, followed by the name of the negative ion.

Neutralization of an Amine

$$\underset{\text{Methylamine}}{CH_3-\ddot{N}H_2} + HCl \longrightarrow \underset{\text{Methylammonium chloride}}{CH_3-\overset{+}{N}H_3\ Cl^-}$$

$$\underset{\text{Dimethylamine}}{CH_3-\ddot{N}H(CH_3)} + HCl \longrightarrow \underset{\text{Dimethylammonium chloride}}{CH_3-\overset{+}{N}H_2(CH_3)\ Cl^-}$$

TEST

Try Practice Problems 14.35 to 14.38

Properties of Ammonium Salts

Ammonium salts are ionic compounds with strong attractions between the positively charged ammonium ion and an anion, usually chloride. Like most salts, ammonium salts are solids at room temperature, odorless, and soluble in water and body fluids. For this reason, amines used as drugs are usually converted to their ammonium salts. The ammonium salt of ephedrine is used as a bronchodilator and in decongestant products such as Sudafed. The ammonium salt of diphenhydramine is used in products such as Benadryl for relief of itching and pain from skin irritations and rashes (see **FIGURE 14.4**). In pharmaceuticals, the naming of the ammonium salt follows an older method of giving the amine name followed by the name of the acid.

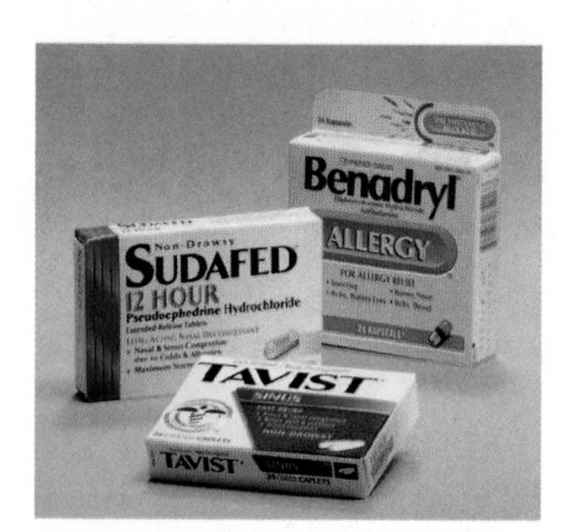

FIGURE 14.4 Decongestants and products that relieve itch and skin irritations often contain ammonium salts.

Why are ammonium salts used in drugs rather than the biologically active amines?

Ephedrine hydrochloride
Ephedrine HCl
Sudafed

Diphenhydramine hydrochloride
Diphenylhydramine HCl
Benadryl

When an ammonium salt reacts with a strong base such as NaOH, it is converted back to the amine, which is also called the free amine or free base.

$$CH_3{-}\overset{+}{N}H_3\ Cl^- + NaOH \longrightarrow CH_3{-}NH_2 + NaCl + H_2O$$

Cocaine is typically extracted from coca leaves using an acidic HCl solution to give a white, solid ammonium salt, which is cocaine hydrochloride. It is the salt of cocaine (cocaine hydrochloride) that is used illegally on the street. "Crack cocaine" is the free amine or free base of the amine obtained by treating the cocaine hydrochloride with NaOH and ether, a process known as "free-basing." The solid product is known as "crack cocaine" because it makes a crackling noise when heated. The free amine is rapidly absorbed when smoked and gives stronger highs than the cocaine hydrochloride, which makes crack cocaine more addictive.

Coca leaves are a source of cocaine.

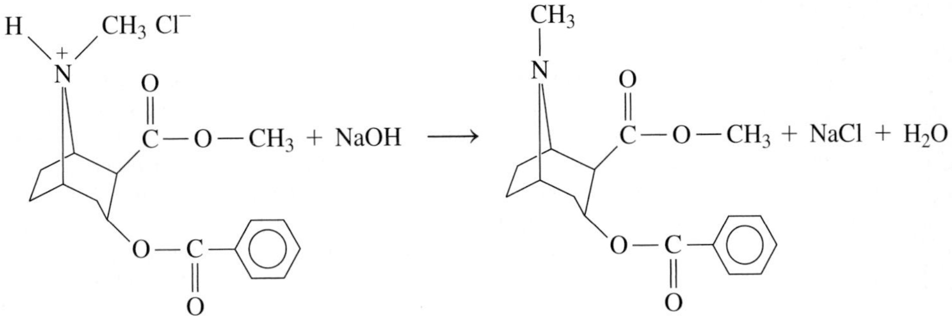

Cocaine hydrochloride + NaOH ⟶ Cocaine ("free base"; crack cocaine) + NaCl + H_2O

Cocaine hydrochloride

Cocaine ("free base"; crack cocaine)

SAMPLE PROBLEM 14.9 Reactions of Amines

TRY IT FIRST

Write the balanced chemical equation that shows ethylamine

a. acting as a weak base in water

b. neutralized by HCl

SOLUTION

a. In water, ethylamine acts as a weak base by accepting a hydrogen ion from water to produce ethylammonium and hydroxide ions.

$$CH_3{-}CH_2{-}NH_2 + H{-}OH \rightleftharpoons CH_3{-}CH_2{-}\overset{+}{N}H_3 + OH^-$$

b. In a reaction with an acid, ethylamine acts as a weak base by accepting a hydrogen ion from HCl to produce ethylammonium chloride.

$$CH_3{-}CH_2{-}NH_2 + HCl \longrightarrow CH_3{-}CH_2{-}\overset{+}{N}H_3\ Cl^-$$

STUDY CHECK 14.9

Draw the condensed structural formula for the salt formed by the reaction of trimethylamine and HCl.

ANSWER

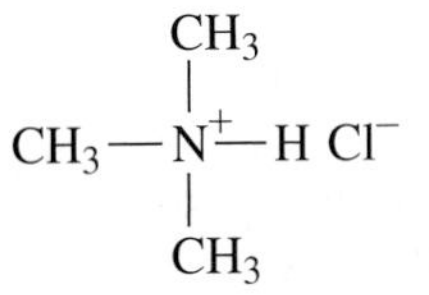

$$(CH_3)_3\overset{+}{N}{-}H\ Cl^-$$

CHEMISTRY LINK TO THE ENVIRONMENT

Alkaloids: Amines in Plants

Alkaloids are physiologically active nitrogen-containing compounds produced by plants. The term *alkaloid* refers to the "alkali-like" or basic characteristics of amines. Certain alkaloids are used in anesthetics, in antidepressants, and as stimulants, and many are habit forming.

Some of the pungent aroma and taste we associate with black pepper is due to piperidine. The fruit from the black pepper plant is dried and ground to give the black pepper we use to season our foods.

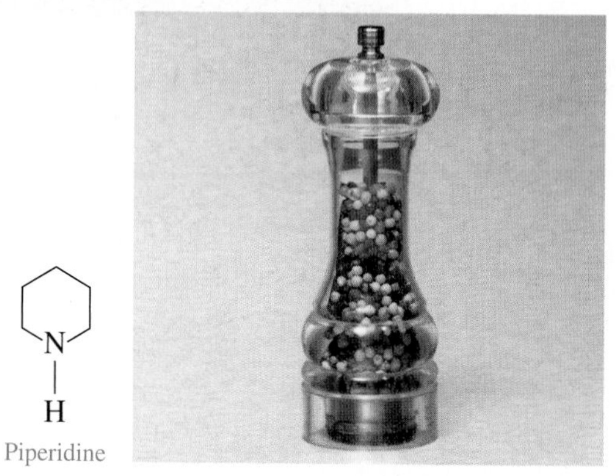

The aroma of black pepper is due to piperidine, an amine.

As a stimulant, nicotine increases the level of adrenaline in the blood, which increases the heart rate and blood pressure. Nicotine is addictive because it activates pleasure centers in the brain. Coniine, which is obtained from hemlock, is extremely toxic.

Nicotine

Coniine

Caffeine is a central nervous system stimulant. Present in coffee, tea, soft drinks, energy drinks, chocolate, and cocoa, caffeine increases alertness, but it may cause nervousness and insomnia. Caffeine is also used in certain pain relievers to counteract the drowsiness caused by an antihistamine.

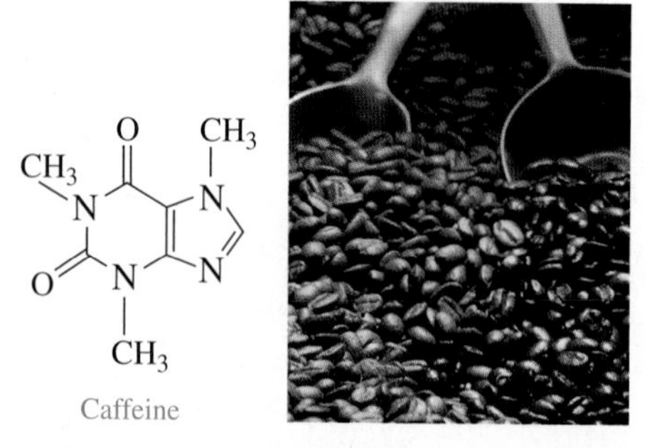

Caffeine is a stimulant found in coffee, tea, energy drinks, and chocolate.

Several alkaloids are used in medicine. Quinine, obtained from the bark of the cinchona tree, has been used in the treatment of malaria since the 1600s. Atropine from nightshade (belladonna) is used in low concentrations to accelerate slow heart rates and as an anesthetic for eye examinations.

Quinine

Atropine

For many centuries morphine and codeine, alkaloids found in the oriental poppy plant, have been used as effective painkillers. Codeine, which is structurally similar to morphine, is used in some prescription painkillers and cough syrups. Heroin, obtained by a chemical modification of morphine, is strongly addicting and is not used medically. The structure of the prescription drug OxyContin (oxycodone) used to relieve severe pain is similar to heroin. Today, there are an increasing number of deaths from OxyContin abuse because its physiological effects are also similar to those of heroin.

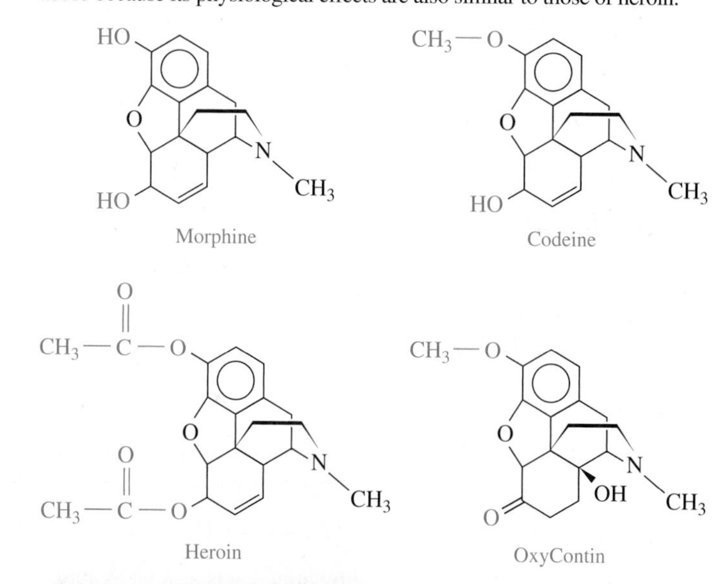

The green, unripe poppy seed capsule contains a milky sap (opium) that is the source of the alkaloids morphine and codeine.

PRACTICE PROBLEMS

14.5 Amines

LEARNING GOAL Write the common names for amines; draw the condensed structural or line-angle formulas when given their names. Describe the solubility, dissociation, and neutralization of amines in water.

14.29 Write the common name for each of the following:

a. $CH_3-CH_2-CH_2-NH_2$

b. 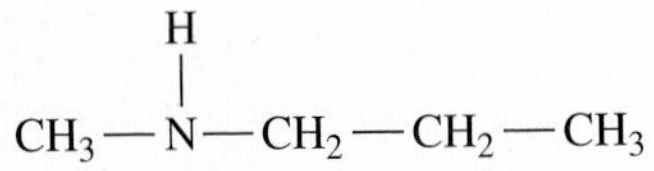 $CH_3-N(H)-CH_2-CH_2-CH_3$

c. 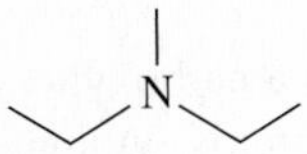 N (line-angle structure)

14.30 Write the common name for each of the following:

a. 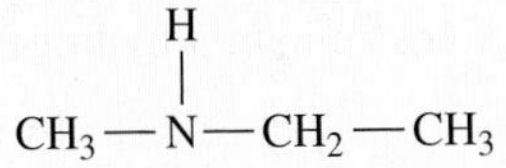 $CH_3-N(H)-CH_2-CH_3$

b. 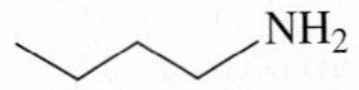 NH_2 (line-angle structure)

c. 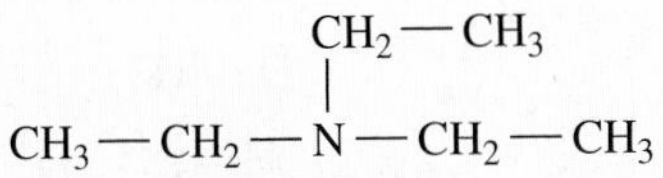 $CH_3-CH_2-N(CH_2-CH_3)-CH_2-CH_3$

14.31 Draw the condensed structural formula, or line-angle formula if cyclic, for each of the following amines:

a. ethylamine **b.** *N*-methylaniline

c. butylpropylamine

14.32 Draw the condensed structural formula, or line-angle formula if cyclic, for each of the following amines:

a. diethylamine **b.** 4-chloroaniline

c. *N,N*-diethylaniline

14.33 Classify each of the following amines as primary (1°), secondary (2°), or tertiary (3°):

a. $CH_3-CH_2-CH_2-NH_2$

b. 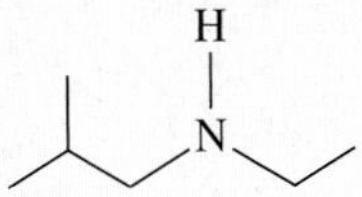 H, N (line-angle structure)

c. 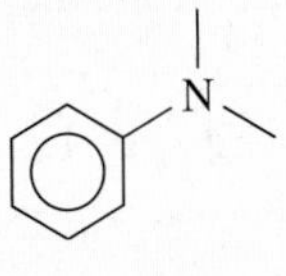 N (line-angle structure)

d. 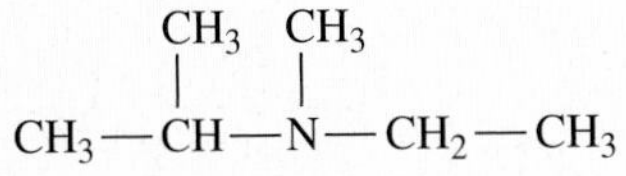 $CH_3-CH(CH_3)-N(CH_3)-CH_2-CH_3$

14.34 Classify each of the following amines as primary (1°), secondary (2°), or tertiary (3°):

a. 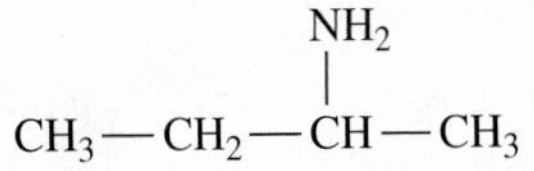 $CH_3-CH_2-CH(NH_2)-CH_3$

b. 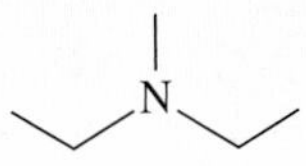 N (line-angle structure)

c. 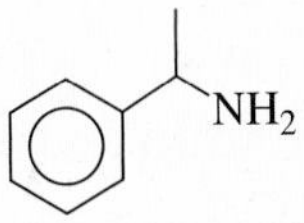 NH_2 (line-angle structure)

d. 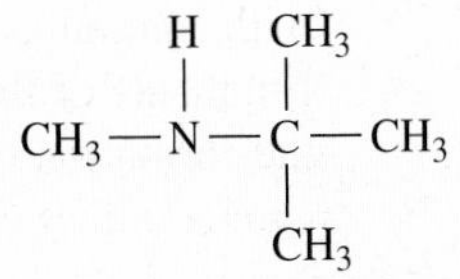 $CH_3-N(H)-C(CH_3)_2-CH_3$

14.35 Indicate if each of the following is soluble in water. Explain.

a. $CH_3-CH_2-NH_2$

b. 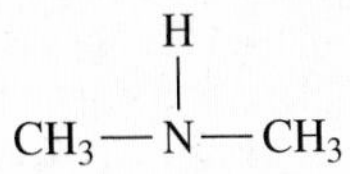 $CH_3-N(H)-CH_3$

c. 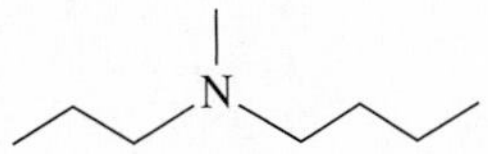 N (line-angle structure)

d. 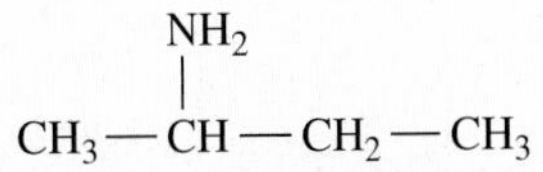 $CH_3-CH(NH_2)-CH_2-CH_3$

14.36 Indicate if each of the following is soluble in water. Explain.

a. $CH_3-CH_2-CH_2-NH_2$

b. 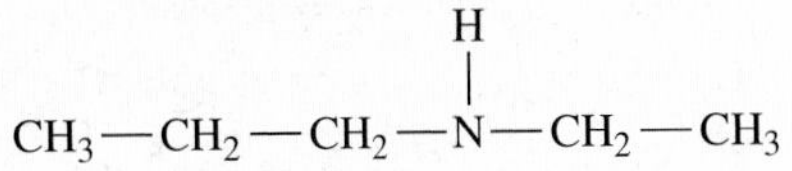 $CH_3-CH_2-CH_2-N(H)-CH_2-CH_3$

c. 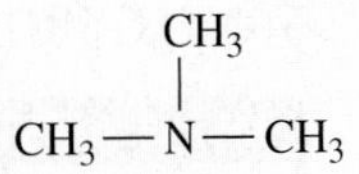 $CH_3-N(CH_3)-CH_3$

d. 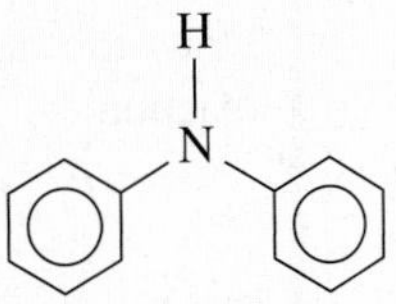 H, N (line-angle structure)

14.37 Write the balanced chemical equations for the (1) reaction of each of the following amines with water and (2) neutralization with HCl:

a. methylamine **b.** dimethylamine

c. aniline

14.38 Write the balanced chemical equations for the (1) reaction of each of the following amines with water and (2) neutralization with HBr:

a. ethylmethylamine **b.** propylamine

c. *N*-methylaniline

14.6 Amides

LEARNING GOAL Write the IUPAC and common names for amides and draw the condensed structural or line-angle formulas for the products of formation and hydrolysis.

The **amides** are derivatives of carboxylic acids in which a nitrogen group replaces the hydroxyl group.

Carboxylic Acid **Amide**

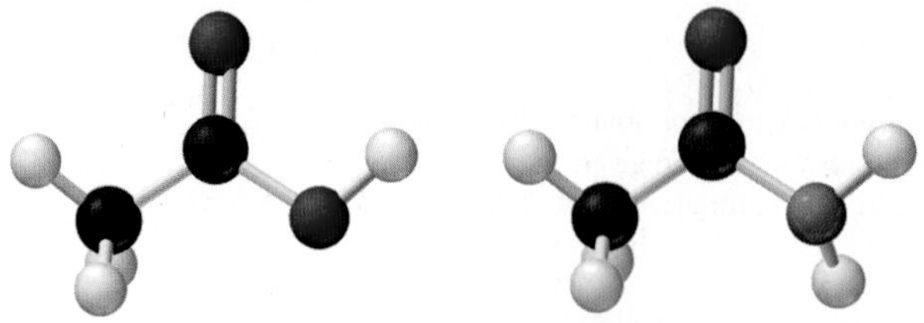

Ethanoic acid (acetic acid)

Ethanamide (acetamide)

CORE CHEMISTRY SKILL

Forming Amides

Preparation of Amides

An amide is produced in a reaction called *amidation*, in which a carboxylic acid reacts with ammonia or a primary or secondary amine. A molecule of water is eliminated, and the fragments of the carboxylic acid and amine molecules join to form the amide, much like the formation of an ester. Because a tertiary amine does not contain a hydrogen atom, it cannot undergo amidation.

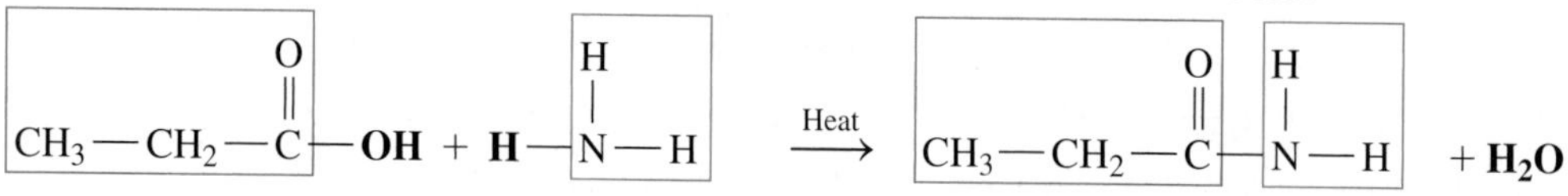

Amide bond

$$CH_3-CH_2-\overset{\overset{\large O}{\|}}{C}-\mathbf{OH} + \mathbf{H}-\overset{\overset{\large H}{|}}{N}-H \xrightarrow{\text{Heat}} CH_3-CH_2-\overset{\overset{\large O}{\|}}{C}-\overset{\overset{\large H}{|}}{N}-H + \mathbf{H_2O}$$

Propanoic acid (propionic acid) — Ammonia — Propanamide (propionamide)

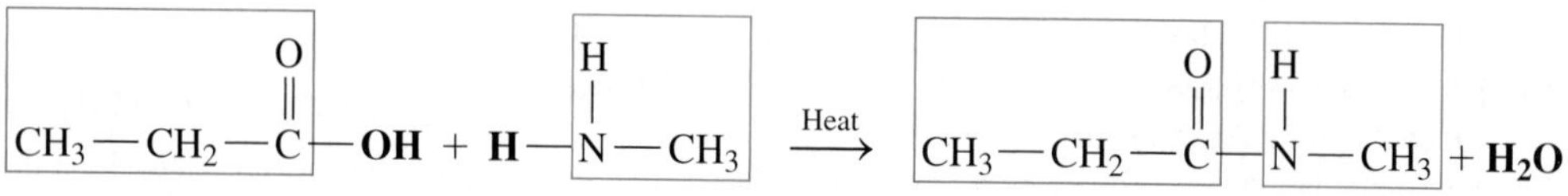

$$CH_3-CH_2-\overset{\overset{\large O}{\|}}{C}-\mathbf{OH} + \mathbf{H}-\overset{\overset{\large H}{|}}{N}-CH_3 \xrightarrow{\text{Heat}} CH_3-CH_2-\overset{\overset{\large O}{\|}}{C}-\overset{\overset{\large H}{|}}{N}-CH_3 + \mathbf{H_2O}$$

Propanoic acid (propionic acid) — Methylamine — *N*-Methylpropanamide (*N*-methylpropionamide)

SAMPLE PROBLEM 14.10 Formation of Amides

TRY IT FIRST

Draw the condensed structural formula for the amide produced from the following reaction:

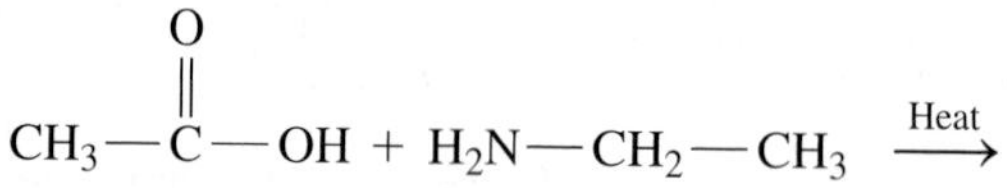

$$CH_3-\overset{\overset{\large O}{\|}}{C}-OH + H_2N-CH_2-CH_3 \xrightarrow{\text{Heat}}$$

SOLUTION

Remove the —OH group from the acid and the —H from the amine to form water. Draw the amide by attaching the carbonyl group from the acid to the nitrogen atom of the amine.

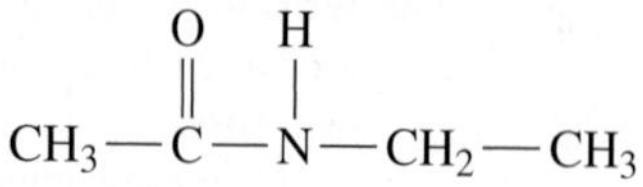

$$CH_3-\overset{\overset{\large O}{\|}}{C}-\overset{\overset{\large H}{|}}{N}-CH_2-CH_3$$

STUDY CHECK 14.10

Draw the condensed structural formulas for the carboxylic acid and amine needed to prepare the following amide. (*Hint*: Separate the N and C=O of the amide group, and add —H and —OH to give the original amine and carboxylic acid.)

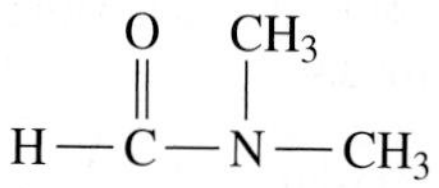

$$H-\overset{\overset{\large O}{\|}}{C}-\overset{\overset{\large CH_3}{|}}{N}-CH_3$$

ANSWER

$$\text{H}-\overset{\overset{\displaystyle \text{O}}{\|}}{\text{C}}-\text{OH} \quad \text{and} \quad \text{H}-\overset{\overset{\displaystyle \text{CH}_3}{|}}{\text{N}}-\text{CH}_3$$

TEST

Try Practice Problems 14.39 and 14.40

Naming Amides

In the IUPAC and common names for amides, the *oic acid* or *ic acid* from the corresponding carboxylic acid name is replaced with *amide*. When alkyl groups are attached to the nitrogen atom, the prefix *N*- or *N,N*- precedes the name of the amide, depending on whether there are one or two groups. We can diagram the name of an amide in the following way:

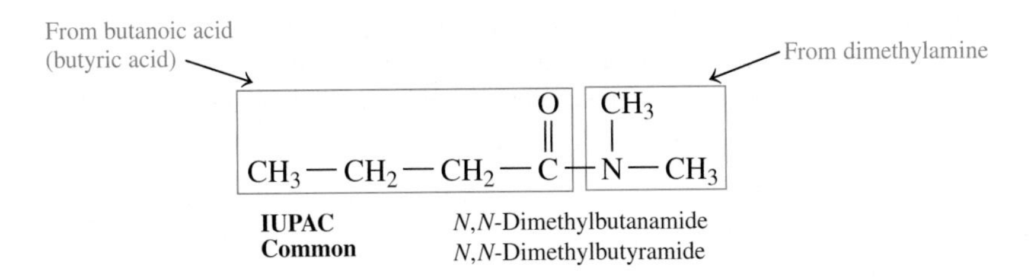

$$\text{H}-\overset{\overset{\displaystyle \text{O}}{\|}}{\text{C}}-\text{NH}_2 \qquad \text{CH}_3-\overset{\overset{\displaystyle \text{O}}{\|}}{\text{C}}-\text{NH}_2 \qquad \text{CH}_3-\text{CH}_2-\overset{\overset{\displaystyle \text{O}}{\|}}{\text{C}}-\overset{\overset{\displaystyle \text{H}}{|}}{\text{N}}-\text{CH}_3$$

Methanamide (formamide) | Ethanamide (acetamide) | *N*-Methylpropanamide (*N*-methylpropionamide) | Benzamide

SAMPLE PROBLEM 14.11 Naming Amides

TRY IT FIRST

Write the IUPAC name for the following amide:

$$\text{CH}_3-\text{CH}_2-\text{CH}_2-\overset{\overset{\displaystyle \text{O}}{\|}}{\text{C}}-\overset{\overset{\displaystyle \text{H}}{|}}{\text{N}}-\text{CH}_2-\text{CH}_3$$

SOLUTION GUIDE

ANALYZE THE PROBLEM	Given	Need	Connect
	amide	IUPAC name	replace *oic acid* in acid name with *amide*, alkyl substituent has prefix *N*-

STEP 1 Replace *oic acid* in the carboxylic acid name with *amide*.

$$\text{CH}_3-\text{CH}_2-\text{CH}_2-\overset{\overset{\displaystyle \text{O}}{\|}}{\text{C}}-\overset{\overset{\displaystyle \text{H}}{|}}{\text{N}}-\text{CH}_2-\text{CH}_3 \qquad \text{butanamide}$$

STEP 2 Name each substituent on the N atom using the prefix *N*- and the alkyl name.

$$\text{CH}_3-\text{CH}_2-\text{CH}_2-\overset{\overset{\displaystyle \text{O}}{\|}}{\text{C}}-\overset{\overset{\displaystyle \text{H}}{|}}{\text{N}}-\text{CH}_2-\text{CH}_3 \qquad N\text{-ethyl butanamide}$$

STUDY CHECK 14.11

Draw the line-angle formula for *N,N*-dimethylbenzamide.

ANSWER

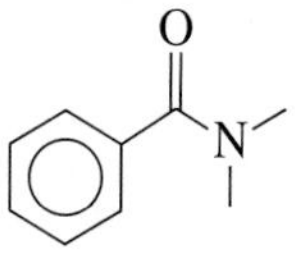

TEST

Try Practice Problems 14.41 to 14.44

Solubility of Amides in Water

The amides do not have the properties of bases that we saw for the amines. The amides with one to five carbon atoms are soluble in water because they can hydrogen bond with water molecules. However, in amides with more than five carbon atoms, the effect of hydrogen bonding is diminished as the longer carbon chain decreases the solubility of an amide in water.

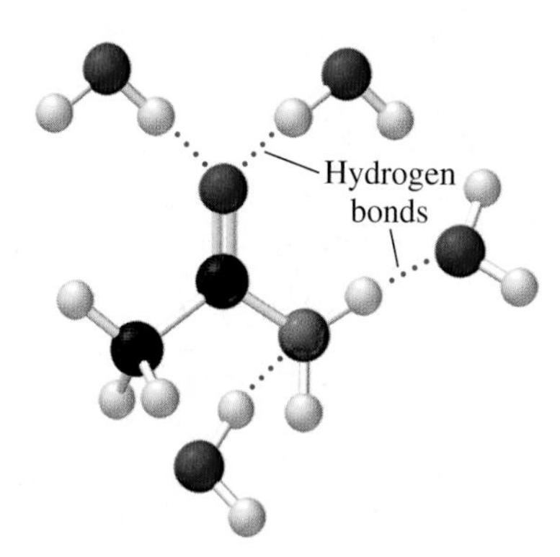

CHEMISTRY LINK TO HEALTH

Amides in Health and Medicine

The simplest natural amide is urea, an end product of protein metabolism in the body. The kidneys remove urea from the blood and provide for its excretion in urine. If the kidneys malfunction, urea is not removed and builds to a toxic level, a condition called *uremia*. Urea is also used as a component of fertilizer, to increase nitrogen in the soil.

$$H_2N-\overset{\overset{\large O}{\|}}{C}-NH_2$$

Urea

Tylenol with acetaminophen is an aspirin substitute.

Aspirin substitutes contain phenacetin or acetaminophen, which is used in Tylenol. Like aspirin, acetaminophen reduces fever and pain, but it has little anti-inflammatory effect.

Phenacetin

Many barbiturates are cyclic amides of barbituric acid that act as sedatives in small dosages or sleep inducers in larger dosages. They are often habit forming. Barbiturate drugs include phenobarbital (Luminal) and pentobarbital (Nembutal).

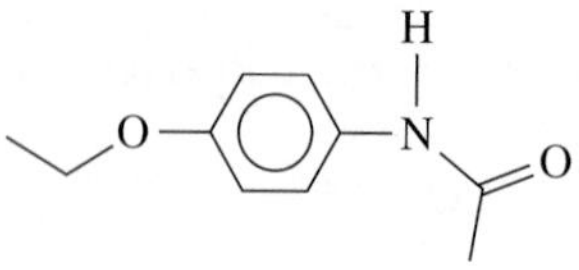

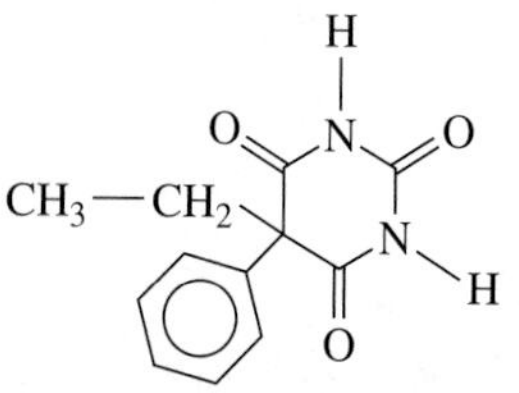

Phenobarbital (Luminal)

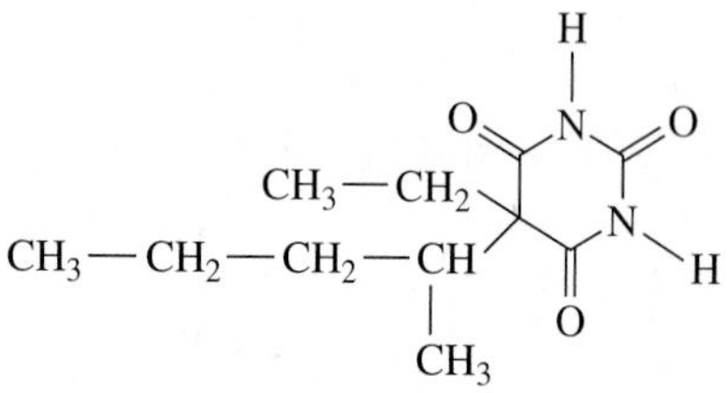

Pentobarbital (Nembutal)

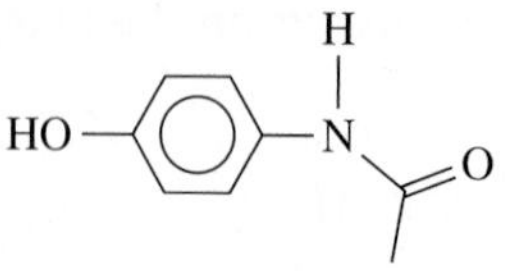

Acetaminophen

Hydrolysis of Amides

Amides undergo hydrolysis when water is added to the amide bond to split the molecule. When an acid is used, the hydrolysis products of an amide are the carboxylic acid and the ammonium salt. In base hydrolysis, the amide produces the carboxylate salt and ammonia or an amine.

Acid Hydrolysis of Amides

$$\underset{\text{Ethanamide (acetamide)}}{\underset{\textbf{Amide}}{CH_3-\overset{\overset{\displaystyle O}{\|}}{C}-\overset{\overset{\displaystyle H}{|}}{N}-H}} + H_2O + HCl \xrightarrow{\text{Heat}} \underset{\text{Ethanoic acid (acetic acid)}}{\underset{\textbf{Carboxylic Acid}}{CH_3-\overset{\overset{\displaystyle O}{\|}}{C}-OH}} + \underset{\text{Ammonium chloride}}{\underset{\textbf{Ammonium Salt}}{H-\overset{\overset{\displaystyle H}{|}}{\underset{\underset{\displaystyle H}{|}}{N^+}}-H\;Cl^-}}$$

Base Hydrolysis of Amides

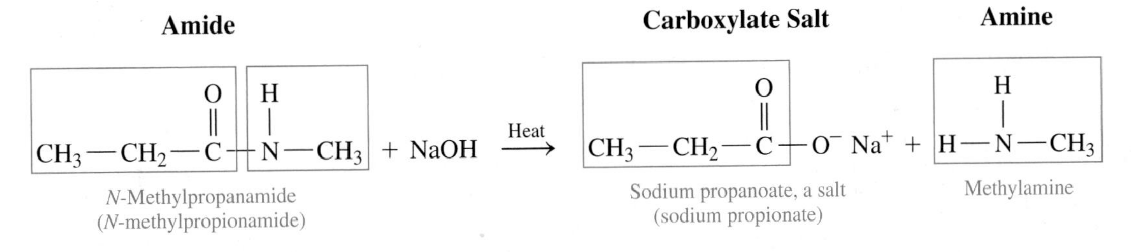

N-Methylpropanamide (*N*-methylpropionamide)

Sodium propanoate, a salt (sodium propionate)

Methylamine

SAMPLE PROBLEM 14.12 Base Hydrolysis of Amides

TRY IT FIRST

Draw the condensed structural formulas for the products from the hydrolysis of *N*-methylpentanamide with NaOH.

SOLUTION

ANALYZE THE PROBLEM	Given	Need	Connect
	N-methylpentanamide, NaOH	products: carboxylate salt, amine	base hydrolysis of an amide

In the hydrolysis of the amide, the amide bond is broken between the carboxyl carbon atom and the nitrogen atom. When a base such as NaOH is used, the products are the carboxylate salt and an amine.

$$\underset{N\text{-Methylpentanamide}}{CH_3-CH_2-CH_2-CH_2-\overset{\overset{\displaystyle O}{\|}}{C}-\overset{\overset{\displaystyle H}{|}}{N}-CH_3} \xrightarrow[\text{NaOH}]{\text{Heat}}$$

$$\underset{\text{Sodium pentanoate}}{CH_3-CH_2-CH_2-CH_2-\overset{\overset{\displaystyle O}{\|}}{C}-O^-\,Na^+} + \underset{\text{Methylamine}}{H-\overset{\overset{\displaystyle H}{|}}{N}-CH_3}$$

STUDY CHECK 14.12

Draw the condensed structural formulas for the products from the hydrolysis of *N*-methylbutyramide with HBr.

ANSWER

$$CH_3{-}CH_2{-}CH_2{-}\overset{O}{\overset{\|}{C}}{-}OH + CH_3{-}\overset{+}{N}H_3\ Br^-$$

TEST

Try Practice Problems 14.45 and 14.46

PRACTICE PROBLEMS

14.6 Amides

LEARNING GOAL Write the IUPAC and common names for amides and draw the condensed structural or line-angle formulas for the products of formation and hydrolysis.

14.39 Draw the condensed structural or line-angle formula for the amide formed in each of the following reactions:

a. $CH_3{-}\overset{O}{\overset{\|}{C}}{-}OH + NH_3 \xrightarrow{\text{Heat}}$

b. $CH_3{-}\overset{O}{\overset{\|}{C}}{-}OH + H_2N{-}CH_2{-}CH_3 \xrightarrow{\text{Heat}}$

c. 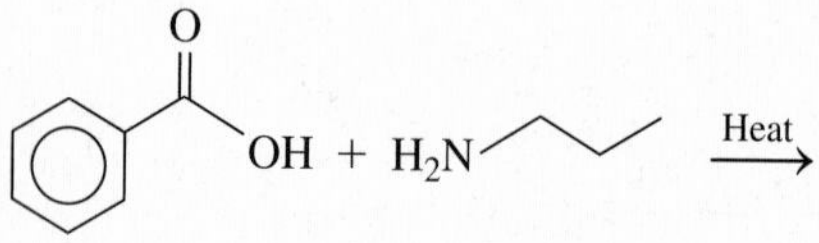 (benzoic acid, line-angle) OH + H_2N (propyl, line-angle) $\xrightarrow{\text{Heat}}$

14.40 Draw the condensed structural or line-angle formula for the amide formed in each of the following reactions:

a. $CH_3{-}CH_2{-}CH_2{-}CH_2{-}\overset{O}{\overset{\|}{C}}{-}OH + NH_3 \xrightarrow{\text{Heat}}$

b. 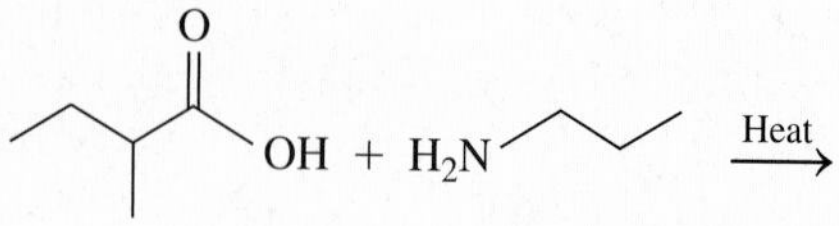 (2-methylbutanoic acid, line-angle) OH + H_2N (propyl, line-angle) $\xrightarrow{\text{Heat}}$

c. 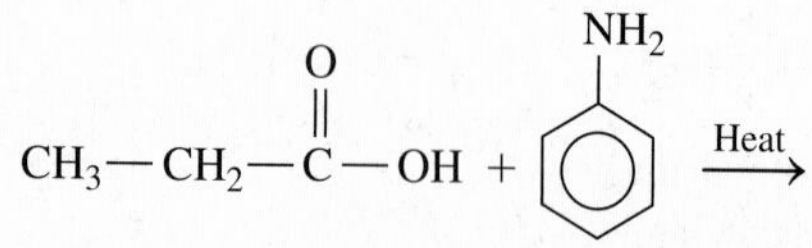 $CH_3{-}CH_2{-}\overset{O}{\overset{\|}{C}}{-}OH$ + (aniline, benzene ring with NH_2) $\xrightarrow{\text{Heat}}$

14.41 Write the IUPAC and common names, if any, for each of the following amides:

a. $CH_3{-}\overset{O}{\overset{\|}{C}}{-}NH_2$

b. 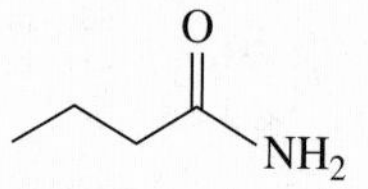 (line-angle butanamide) NH_2

c. $H{-}\overset{O}{\overset{\|}{C}}{-}\overset{H}{\overset{|}{N}}{-}CH_2{-}CH_3$

14.42 Write the IUPAC and common names, if any, for each of the following amides:

a. $CH_3{-}CH_2{-}CH_2{-}CH_2{-}\overset{O}{\overset{\|}{C}}{-}NH_2$

b. 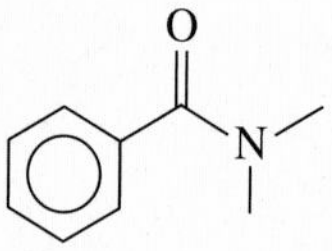 (benzene ring–C(=O)–N with two methyl groups, line-angle)

c. (line-angle hexanamide) NH_2

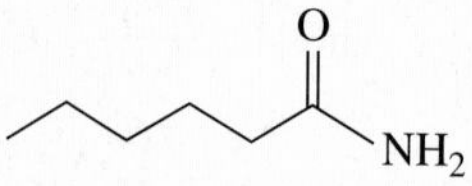

14.43 Draw the condensed structural formula for each of the following amides:

a. propionamide
b. 2-methylpentanamide
c. methanamide
d. *N,N*-dimethylethanamide

14.44 Draw the condensed structural formula for each of the following amides:

a. heptanamide
b. 3-chlorobenzamide
c. 3-methylbutyramide
d. *N*-ethylpropanamide

14.45 Draw the condensed structural or line-angle formulas for the products from the hydrolysis of each of the following amides with HCl:

a. 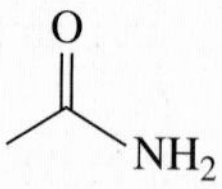 (line-angle ethanamide) NH_2

b. $CH_3{-}CH_2{-}\overset{O}{\overset{\|}{C}}{-}NH_2$

c. $CH_3{-}CH_2{-}CH_2{-}\overset{O}{\overset{\|}{C}}{-}\overset{H}{\overset{|}{N}}{-}CH_3$

d. 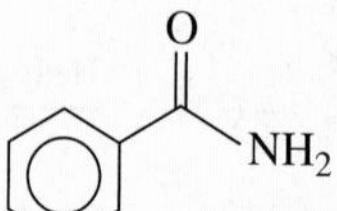 (benzene ring–C(=O)) NH_2

e. *N*-ethylpentanamide

14.46 Draw the condensed structural or line-angle formulas for the products from the hydrolysis of each of the following amides with NaOH:

a. O, NH_2

b. $CH_3-CH_2-CH_2-\overset{O}{\overset{\|}{C}}-\overset{CH_2-CH_3}{\overset{|}{N}}-CH_2-CH_3$

c. O, N

d. $CH_3-\overset{Cl}{\overset{|}{CH}}-\overset{O}{\overset{\|}{C}}-\overset{CH_3}{\overset{|}{N}}-CH_2-CH_3$

e. *N*-propylbenzamide

CLINICAL UPDATE Testing Soil and Water Samples for Chemicals

Lance, an environmental health practitioner, returns to the sheep ranch to obtain more soil and water samples. When he arrives at the ranch, he notices that the owners are spraying the sheep because they have an infestation of flies and maggots. Lance is told that the insecticide they are spraying on the sheep is called dicyclanil, a potent insect growth regulator. Lance is also informed that the sheep are being treated with enrofloxacin to counteract a respiratory infection.

Because high levels of these chemicals could be hazardous if they exceed acceptable environmental standards, Lance collectes samples of soil and water to test. He will send the test results to the rancher, who will use those results to adjust the levels of the drugs, order protective equipment, and do a cleanup, if necessary.

A soil bag is filled with soil from areas where sheep were sprayed with dicyclanil.

Clinical Applications

14.47 a. Identify the functional groups in dicyclanil.

b. The recommended application for dicyclanil for an adult sheep is 65 mg/kg of body mass. If dicyclanil is supplied in a spray with a concentration of 50. mg/mL, how many milliliters of the spray are required to treat a 70.-kg adult sheep?

14.48 a. Identify the functional groups in enrofloxacin.

b. The recommended dose for enrofloxacin for sheep is 30. mg/kg of body mass for 5 days. Enrofloxacin is supplied in 50.-mL vials with a concentration of 100. mg/mL. How many vials are needed to treat a 64-kg sheep for 5 days?

Dicyclanil

Enrofloxacin

CONCEPT MAP

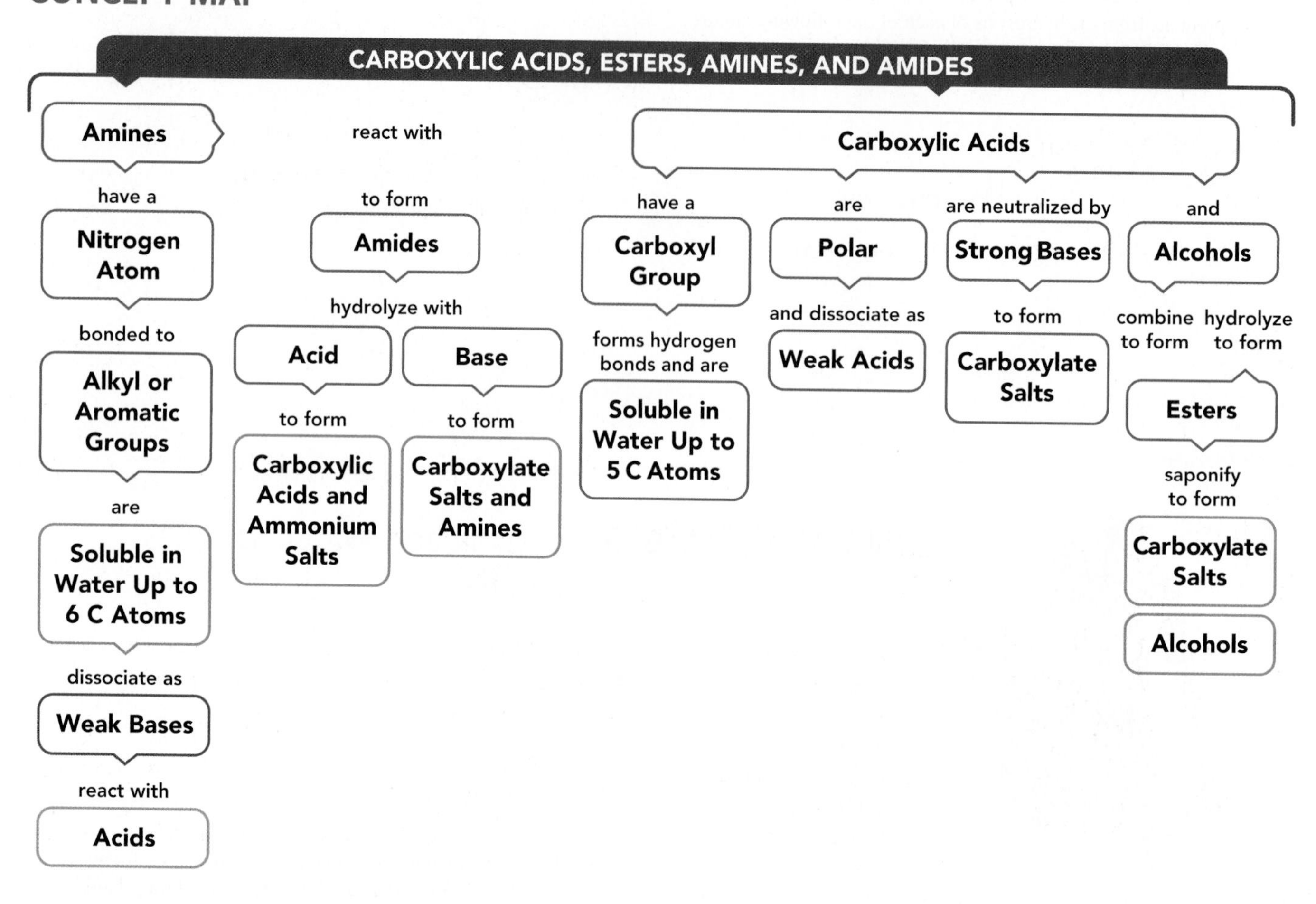

CHAPTER REVIEW

14.1 Carboxylic Acids

LEARNING GOAL Write the IUPAC and common names for carboxylic acids; draw their condensed structural or line-angle formulas.

- A carboxylic acid contains the carboxyl functional group, which is a hydroxyl group connected to a carbonyl group.
- The IUPAC name of a carboxylic acid is obtained by replacing the *e* in the alkane name with *oic acid.*
- The common names of carboxylic acids with one to four carbon atoms are formic acid, acetic acid, propionic acid, and butyric acid.

14.2 Properties of Carboxylic Acids

LEARNING GOAL Describe the solubility, dissociation, and neutralization of carboxylic acids.

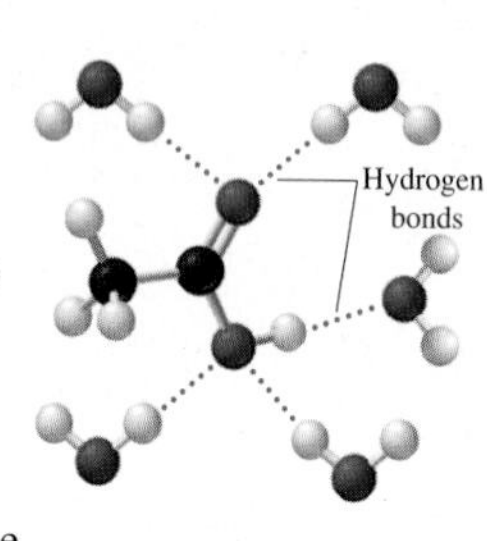

- The carboxyl group contains polar bonds of O—H and C=O, which makes a carboxylic acid with one to five carbon atoms soluble in water.
- As weak acids, carboxylic acids dissociate slightly by donating a hydrogen ion to water to form carboxylate and hydronium ions.
- Carboxylic acids are neutralized by base, producing a carboxylate salt and water.

14.3 Esters

LEARNING GOAL Write the IUPAC and common names for an ester; write the balanced chemical equation for the formation of an ester.

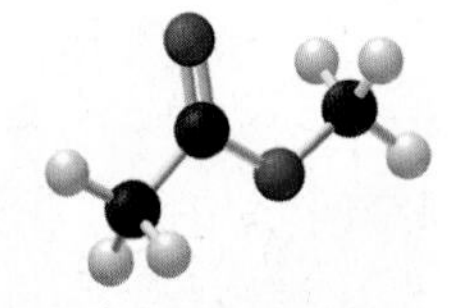
Methyl ethanoate (methyl acetate)

- In an ester, an alkyl or aromatic group replaces the H of the hydroxyl group of a carboxylic acid.
- In the presence of a strong acid, a carboxylic acid reacts with an alcohol to produce an ester. A molecule of water is removed: —OH from the carboxylic acid and —H from the alcohol molecule.
- The names of esters consist of two words: the alkyl group from the alcohol and the name of the carboxylate obtained by replacing *ic acid* with *ate.*

14.4 Hydrolysis of Esters

LEARNING GOAL Draw the condensed structural or line-angle formulas for the products from acid and base hydrolysis of esters.

- Esters undergo acid hydrolysis by adding water to produce the carboxylic acid and an alcohol.
- Esters undergo base hydrolysis, or saponification, to produce the carboxylate salt and an alcohol.

14.5 Amines

LEARNING GOAL Write the common names for amines; draw the condensed structural or line-angle formulas when given their names. Describe the solubility, dissociation, and neutralization of amines in water.

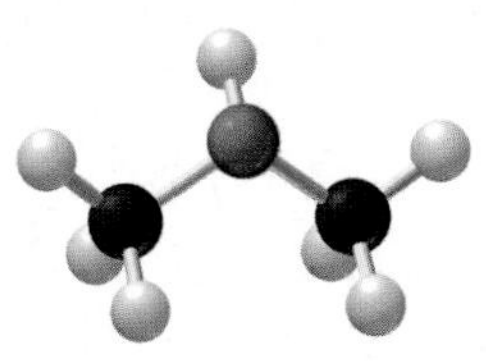
Dimethylamine

- In the common names of simple amines, the alkyl groups are listed alphabetically followed by *amine.*
- A nitrogen atom attached to one, two, or three alkyl or aromatic groups forms a primary (1°), secondary (2°), or tertiary (3°) amine.
- Amines with up to six carbon atoms are soluble in water.
- In water, amines act as weak bases to produce ammonium and hydroxide ions.
- When amines react with acids, they form ammonium salts. As ionic compounds, ammonium salts are solids, soluble in water, and odorless.

14.6 Amides

LEARNING GOAL Write the IUPAC and common names for amides and draw the condensed structural or line-angle formulas for the products of formation and hydrolysis.

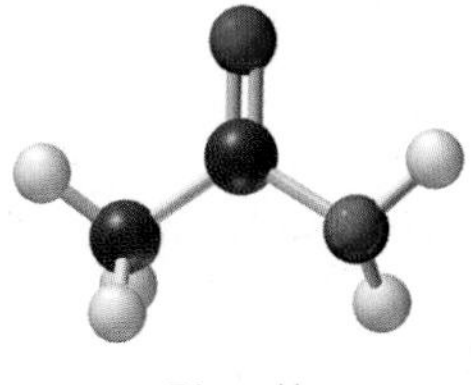
Ethanamide (acetamide)

- Amides are derivatives of carboxylic acids in which the hydroxyl group is replaced by $-NH_2$ or a primary or secondary amine group.
- Amides are formed when carboxylic acids react with ammonia or primary or secondary amines in the presence of heat.
- Amides are named by replacing the *oic acid* or *ic acid* from the carboxylic acid name with *amide.* Any carbon group attached to the nitrogen atom is named using the *N-* prefix.
- Hydrolysis of an amide by an acid produces the carboxylic acid and an ammonium salt.
- Hydrolysis of an amide by a base produces the carboxylate salt and an amine.

SUMMARY OF NAMING

Family	Structure	IUPAC Name	Common Name
Carboxylic acid	$CH_3-\overset{\overset{\large O}{\|\|}}{C}-OH$	Ethanoic acid	Acetic acid
Ester	$CH_3-\overset{\overset{\large O}{\|\|}}{C}-O-CH_3$	Methyl ethanoate	Methyl acetate
Amine	$CH_3-CH_2-NH_2$		Ethylamine
Amide	$CH_3-\overset{\overset{\large O}{\|\|}}{C}-NH_2$	Ethanamide	Acetamide

SUMMARY OF REACTIONS

The chapter Sections to review are shown after the name of each reaction.

Dissociation of a Carboxylic Acid in Water (14.2)

$$\underset{\text{(acetic acid)}}{\underset{\text{Ethanoic acid}}{CH_3-\overset{\overset{O}{\|}}{C}-OH}} + H_2O \rightleftarrows \underset{\text{(acetate ion)}}{\underset{\text{Ethanoate ion}}{CH_3-\overset{\overset{O}{\|}}{C}-O^-}} + \underset{\text{ion}}{\underset{\text{Hydronium}}{H_3O^+}}$$

Neutralization of a Carboxylic Acid (14.2)

$$\underset{\text{(propionic acid)}}{\underset{\text{Propanoic acid}}{CH_3-CH_2-\overset{\overset{O}{\|}}{C}-OH}} + \underset{\text{hydroxide}}{\underset{\text{Sodium}}{NaOH}} \longrightarrow \underset{\text{(sodium propionate)}}{\underset{\text{Sodium propanoate}}{CH_3-CH_2-\overset{\overset{O}{\|}}{C}-O^-\,Na^+}} + H_2O$$

Esterification: Carboxylic Acid and an Alcohol (14.3)

$$CH_3-\overset{\overset{\displaystyle O}{\|}}{C}-OH + HO-CH_3 \xrightleftharpoons{H^+,\ heat} CH_3-\overset{\overset{\displaystyle O}{\|}}{C}-O-CH_3 + H_2O$$

Ethanoic acid (acetic acid) + Methanol (methyl alcohol) → Methyl ethanoate (methyl acetate)

Acid Hydrolysis of an Ester (14.4)

$$CH_3-\overset{\overset{\displaystyle O}{\|}}{C}-O-CH_3 + H_2O \xrightleftharpoons{H^+,\ heat} CH_3-\overset{\overset{\displaystyle O}{\|}}{C}-OH + HO-CH_3$$

Methyl ethanoate (methyl acetate) → Ethanoic acid (acetic acid) + Methanol (methyl alcohol)

Base Hydrolysis of an Ester: Saponification (14.4)

$$CH_3-CH_2-\overset{\overset{\displaystyle O}{\|}}{C}-O-CH_3 + NaOH \xrightarrow{Heat} CH_3-CH_2-\overset{\overset{\displaystyle O}{\|}}{C}-O^-\ Na^+ + HO-CH_3$$

Methyl propanoate (methyl propionate) + Sodium hydroxide → Sodium propanoate (sodium propionate) + Methanol (methyl alcohol)

Dissociation of an Amine in Water (14.5)

$$CH_3-\ddot{N}H_2 + H_2O \rightleftharpoons CH_3-\overset{+}{N}H_3 + OH^-$$

Methylamine → Methylammonium ion + Hydroxide ion

Formation of an Ammonium Salt (14.5)

$$CH_3-\ddot{N}H_2 + HCl \longrightarrow CH_3-\overset{+}{N}H_3\ Cl^-$$

Methylamine → Methylammonium chloride

Amidation: Carboxylic Acid and an Amine (14.6)

$$CH_3-CH_2-\overset{\overset{\displaystyle O}{\|}}{C}-OH + H-\overset{\overset{\displaystyle H}{|}}{N}-H \xrightarrow{Heat} CH_3-CH_2-\overset{\overset{\displaystyle O}{\|}}{C}-\overset{\overset{\displaystyle H}{|}}{N}-H + H_2O$$

Propanoic acid (propionic acid) + Ammonia → Propanamide (propionamide)

Acid Hydrolysis of an Amide (14.6)

$$CH_3-\overset{\overset{\displaystyle O}{\|}}{C}-NH_2 + H_2O + HCl \xrightarrow{Heat} CH_3-\overset{\overset{\displaystyle O}{\|}}{C}-OH + NH_4^+\ Cl^-$$

Ethanamide (acetamide) → Ethanoic acid (acetic acid) + Ammonium chloride

Base Hydrolysis of an Amide (14.6)

$$CH_3-CH_2-\overset{\overset{\displaystyle O}{\|}}{C}-\overset{\overset{\displaystyle H}{|}}{N}-CH_3 + NaOH \xrightarrow{Heat} CH_3-CH_2-\overset{\overset{\displaystyle O}{\|}}{C}-O^-\ Na^+ + H_2N-CH_3$$

N-Methylpropanamide (*N*-methylpropionamide) → Sodium propanoate (sodium propionate) + Methylamine

KEY TERMS

amide An organic compound containing the carbonyl group attached to an amino group or a substituted nitrogen atom.

$$-\overset{\overset{\displaystyle O}{\|}}{C}-NH_2 \qquad -\overset{\overset{\displaystyle O}{\|}}{C}-\overset{|}{N}- \qquad \text{Amide}$$

amine An organic compound containing a nitrogen atom attached to one, two, or three hydrocarbon groups.

$$-\overset{|}{N}-$$

ammonium salt An ionic compound produced from an amine and an acid.

carboxyl group A functional group found in carboxylic acids composed of carbonyl and hydroxyl groups.

$$-\overset{\overset{\displaystyle O}{\|}}{C}-OH \qquad \text{Carboxyl group}$$

carboxylate ion The anion produced when a carboxylic acid donates a hydrogen ion to water.

carboxylate salt A carboxylate ion and the metal ion from the base that is the product of neutralization of a carboxylic acid.

carboxylic acid An organic compound containing the carboxyl group.

ester An organic compound in which an alkyl group replaces the hydrogen atom in a carboxylic acid.

$$-\overset{\overset{\displaystyle O}{\|}}{C}-O-\underset{|}{\overset{|}{C}}- \qquad \text{Ester}$$

esterification The formation of an ester from a carboxylic acid and an alcohol, with the elimination of a molecule of water in the presence of an acid catalyst.

hydrolysis The splitting of a molecule by the addition of water. In acid, esters hydrolyze to produce a carboxylic acid and an alcohol. Amides yield the corresponding carboxylic acid and amine or their salts.

CORE CHEMISTRY SKILLS

The chapter Section containing each Core Chemistry Skill is shown in parentheses at the end of each heading.

Naming Carboxylic Acids (14.1)

- The IUPAC name of a carboxylic acid is obtained by replacing the *e* in the corresponding alkane name with *oic acid.*
- The common names of carboxylic acids with one to four carbon atoms are formic acid, acetic acid, propionic acid, and butyric acid.

Example: Write the IUPAC name for the following:

$$CH_3-\overset{\overset{\displaystyle CH_3}{|}}{CH}-CH_2-\underset{\underset{\displaystyle CH_3}{|}}{\overset{\overset{\displaystyle Cl}{|}}{C}}-CH_2-\overset{\overset{\displaystyle O}{\|}}{C}-OH$$

Answer: 3-chloro-3,5-dimethylhexanoic acid

Hydrolyzing Esters (14.4)

- Esters undergo acid hydrolysis by adding water to produce the carboxylic acid and an alcohol.
- Esters undergo base hydrolysis, or saponification, to produce the carboxylate salt and an alcohol.

Example: Draw the condensed structural formulas for the products from the **(a)** acid and **(b)** base hydrolysis (NaOH) of ethyl butanoate.

Answer:

a. $CH_3-CH_2-CH_2-\overset{\overset{\displaystyle O}{\|}}{C}-OH + HO-CH_2-CH_3$

b. $CH_3-CH_2-CH_2-\overset{\overset{\displaystyle O}{\|}}{C}-O^- \; Na^+ + HO-CH_2-CH_3$

Forming Amides (14.6)

- Amides are formed when carboxylic acids react with ammonia or primary or secondary amines in the presence of heat.

Example: Draw the condensed structural formula for the amide product of the reaction of 3-methylbutanoic acid and ethylamine.

Answer:

$$CH_3-\overset{\overset{\displaystyle CH_3}{|}}{CH}-CH_2-\overset{\overset{\displaystyle O}{\|}}{C}-\overset{\overset{\displaystyle H}{|}}{N}-CH_2-CH_3$$

UNDERSTANDING THE CONCEPTS

The chapter Sections to review are shown in parentheses at the end of each problem.

14.49 Draw the condensed structural formulas and write the IUPAC names for two structural isomers of the carboxylic acids that have the molecular formula $C_4H_8O_2$. (14.1)

14.50 Draw the condensed structural formulas and write the IUPAC names for four structural isomers of the esters that have the molecular formula $C_5H_{10}O_2$. (14.3)

14.51 The ester methyl butanoate has the odor and flavor of strawberries. (14.3, 14.4)

a. Draw the condensed structural formula for methyl butanoate.

b. Write the IUPAC name for the carboxylic acid and the alcohol used to prepare methyl butanoate.

c. Write the balanced chemical equation for the acid hydrolysis of methyl butanoate.

14.52 Methyl benzoate, which smells like pineapple guava, is used to train detection dogs. (14.3, 14.4)

a. Draw the line-angle formula for methyl benzoate.

b. Write the IUPAC name for the carboxylic acid and alcohol used to prepare methyl benzoate.

c. Write the balanced chemical equation for the acid hydrolysis of methyl benzoate.

14.53 Phenylephrine is the active ingredient in some nose sprays used to reduce swelling of nasal membranes. Identify the functional groups in phenylephrine. (14.1, 14.3, 14.5, 14.6)

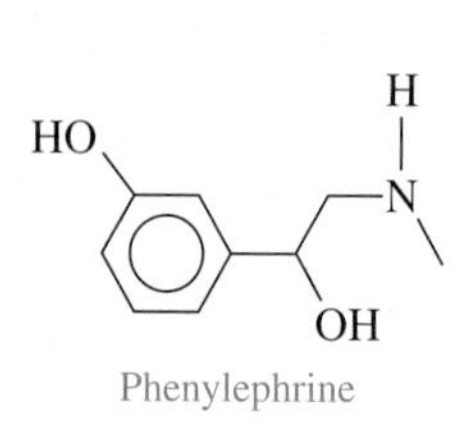

Phenylephrine

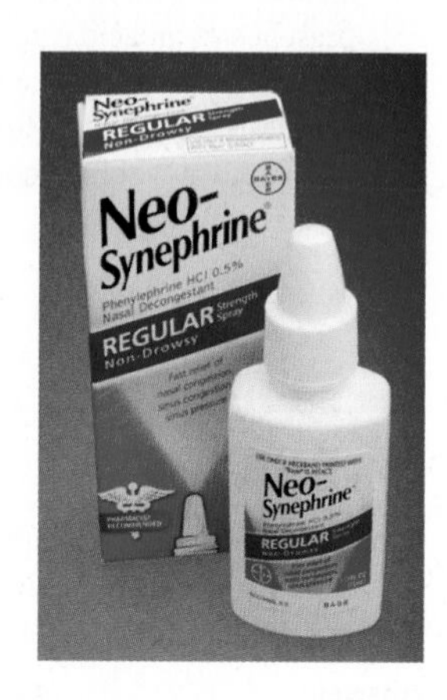

14.54 Melatonin is a naturally occurring compound in plants and animals, where it regulates the biological time clock. Melatonin is sometimes used to counteract jet lag. Identify the functional groups in melatonin. (14.1, 14.3, 14.5, 14.6)

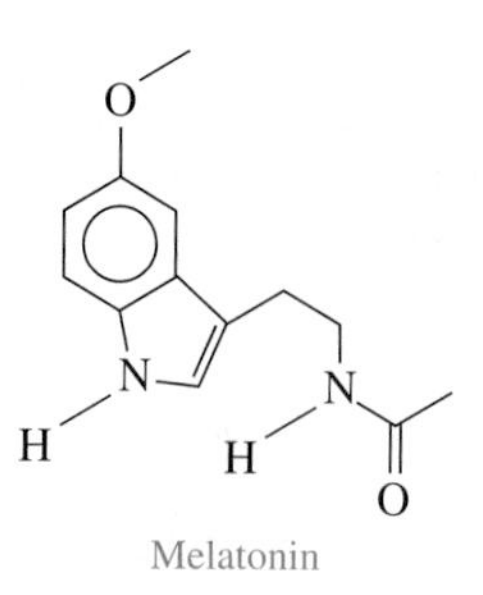

Melatonin

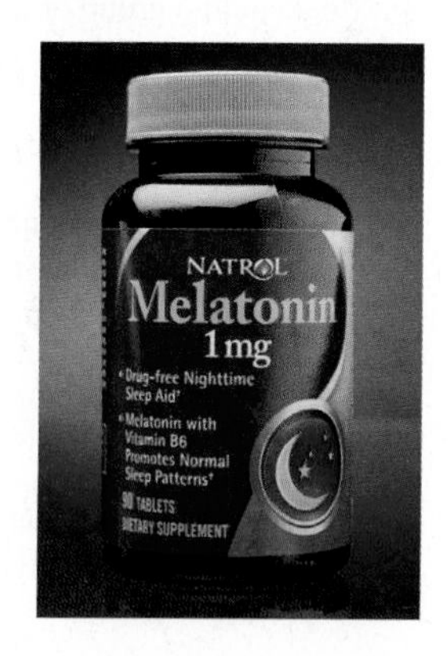

14.55 There are four amine isomers with the molecular formula C_3H_9N. Draw their condensed structural formulas, write the common name, and classify each as a primary (1°), secondary (2°), or tertiary (3°) amine. (14.5)

14.56 There are four amide isomers with the molecular formula C_3H_7NO. Draw their condensed structural formulas and write the IUPAC name for each. (14.6)

ADDITIONAL PRACTICE PROBLEMS

14.57 Write the IUPAC and common names, if any, for each of the following: (14.1, 14.3)

a. $CH_3-CH(CH_3)-CH_2-C(=O)-OH$

b. (line-angle formula: O, O)

c. $CH_3-CH_2-C(=O)-O-CH_2-CH_3$

d. (line-angle formula: O, OH, Cl)

e. (line-angle formula: O, OH)

14.58 Write the IUPAC and common names, if any, for each of the following: (14.1, 14.3)

a. $CH_3-CH(CH_3)-CH_2-CH_2-C(=O)-OH$

b. (line-angle formula: Cl, O, OH, Cl)

c. (line-angle formula: O, O)

d. $CH_3-CH_2-CH_2-C(=O)-O-CH_3$

e. (line-angle formula: O, O)

14.59 Draw the condensed structural formulas for **a** and **b** and line-angle formulas for **c** and **d**: (14.1, 14.3)

a. methyl acetate

b. 2,2-dichlorobutanoic acid

c. 3-bromopentanoic acid

d. butyl benzoate

14.60 Draw the condensed structural formulas for **a** and **b** and line-angle formulas for **c** and **d**:

a. pentyl formate
b. 5-bromopentanoic acid
c. 3,5-dimethylhexanoic acid
d. butyl propanoate

14.61 Draw the condensed structural or line-angle formulas for the products of the following: (14.2, 14.3, 14.6)

a. $CH_3-CH_2-\overset{O}{\overset{\|}{C}}-OH + KOH \longrightarrow$

b. $CH_3-CH_2-\overset{O}{\overset{\|}{C}}-OH + HO-CH_3 \xrightleftharpoons{H^+,\ heat}$

c. (benzoic acid, O, OH) $+ HO$ (ethanol) $\xrightleftharpoons{H^+,\ heat}$

d. $CH_3-CH_2-\overset{O}{\overset{\|}{C}}-OH + H_2N-CH_3 \xrightarrow{Heat}$

14.62 Draw the condensed structural or line-angle formulas for the products of the following: (14.2, 14.3, 14.6)

a. $CH_3-\overset{O}{\overset{\|}{C}}-OH + NaOH \longrightarrow$

b. (2-methylpropanoic acid, O, OH) $+ H_2O \longrightarrow$

c. $CH_3-\overset{CH_3}{\overset{|}{CH}}-\overset{O}{\overset{\|}{C}}-OH + HO-CH_3 \xrightleftharpoons{H^+,\ heat}$

d. $H-\overset{O}{\overset{\|}{C}}-OH +$ (aniline, NH_2) $\xrightarrow{Heat}$

14.63 Draw the condensed structural or line-angle formulas for the products of the following: (14.4, 14.5)

a. $CH_3-CH_2-\overset{O}{\overset{\|}{C}}-O-\overset{CH_3}{\overset{|}{CH}}-CH_3 + H_2O \xrightleftharpoons{H^+,\ heat}$

b. (ester, O, O) $+ H_2O \xrightleftharpoons{H^+,\ heat}$

c. (benzamide, O, N, H) $+ NaOH \xrightarrow{Heat}$

14.64 Draw the condensed structural or line-angle formulas for the products of the following: (14.4, 14.6)

a. $CH_3-CH_2-\overset{O}{\overset{\|}{C}}-O-\overset{CH_3}{\overset{|}{CH}}-CH_3 + NaOH \xrightarrow{Heat}$

b. (ester, O, O) $+ H_2O \xrightleftharpoons{H^+,\ heat}$

c. $CH_3-CH_2-\overset{O}{\overset{\|}{C}}-NH_2 + NaOH \xrightarrow{Heat}$

14.65 Write the common name and classify each of the following compounds as primary (1°), secondary (2°), or tertiary (3°): (14.5)

a. $H-\overset{CH_3}{\overset{|}{N}}-CH_2-CH_3$

b. (benzene ring, N, H, ethyl)

c. (chain) NH_2

14.66 Write the common name and classify each of the following compounds as primary (1°), secondary (2°), or tertiary (3°): (14.5)

a. (benzene ring, N, methyls)

b. (cyclopentane, NH_2)

c. $CH_3-CH_2-CH_2-\overset{H}{\overset{|}{N}}-CH_2-CH_2-CH_3$

14.67 Draw the condensed structural or line-angle formula if cyclic, for each of the following: (14.5)

a. dimethylamine
b. cyclohexylamine
c. dimethylammonium chloride
d. *N*-propylaniline

14.68 Draw the condensed structural or line-angle formula if cyclic, for each of the following: (14.5)

a. dibutylamine
b. triethylamine
c. *N,N*-dipropylaniline
d. ethylmethylammonium bromide

14.69 Draw the condensed structural or line-angle formulas for the products of the following: (14.5)

a. $CH_3-CH_2-NH_2 + H_2O \rightleftharpoons$
b. $CH_3-CH_2-NH_2 + HCl \longrightarrow$

14.70 Draw the condensed structural or line-angle formulas for the products of the following: (14.5)

a. $CH_3-CH_2-\overset{H}{\overset{|}{N}}-CH_3 + H_2O \rightleftharpoons$

b. $CH_3-CH_2-\overset{H}{\overset{|}{N}}-CH_3 + HCl \longrightarrow$

14.71 Write the IUPAC name for each of the following: (14.6)

a. $CH_3-\overset{O}{\overset{\|}{C}}-\overset{H}{\overset{|}{N}}-CH_2-CH_3$

b. $CH_3-CH_2-\overset{O}{\overset{\|}{C}}-NH_2$

c. [line-angle structure with C=O and NH_2]

14.72 Write the IUPAC name for each of the following: (14.6)

a. $CH_3-CH_2-CH_2-CH_2-\overset{O}{\overset{\|}{C}}-NH_2$

b. [line-angle structure: amide with N-methyl and N-propyl]

c. $CH_3-CH_2-\overset{O}{\overset{\|}{C}}-\overset{CH_3}{\overset{|}{N}}-CH_3$

14.73 Draw the condensed structural or line-angle formulas for the products from the hydrolysis of each of the following: (14.6)

a. $CH_3-\overset{O}{\overset{\|}{C}}-\overset{CH_3}{\overset{|}{N}}-H + H_2O + HCl \xrightarrow{\text{Heat}}$

b. $H-\overset{O}{\overset{\|}{C}}-NH_2 + NaOH \xrightarrow{\text{Heat}}$

c. [line-angle structure]$NH_2 + H_2O + HCl \xrightarrow{\text{Heat}}$

14.74 Draw the condensed structural or line-angle formulas for the products from the hydrolysis of each of the following: (14.6)

a. [line-angle structure with N] $+ NaOH \xrightarrow{\text{Heat}}$

b. $H-\overset{O}{\overset{\|}{C}}-NH_2 + H_2O + HCl \xrightarrow{\text{Heat}}$

c. $CH_3-CH_2-CH_2-CH_2-\overset{O}{\overset{\|}{C}}-NH_2 + NaOH \xrightarrow{\text{Heat}}$

14.75 Voltaren is indicated for acute and chronic treatment of the symptoms of rheumatoid arthritis. Identify the functional groups in Voltaren. (14.1, 14.3, 14.5, 14.6)

[structure: O^- Na^+, O, H, N, Cl, Cl]

Voltaren

14.76 Toradol is used in dentistry to relieve pain. Identify the functional groups in Toradol. (14.1, 14.3, 14.5, 14.6)

[structure: O, N, O, OH]

Toradol

CHALLENGE PROBLEMS

The following problems are related to the topics in this chapter. However, they do not all follow the chapter order, and they require you to combine concepts and skills from several Sections. These problems will help you increase your critical thinking skills and prepare for your next exam.

14.77 Draw the line-angle formula and write the IUPAC name for each of the following: (14.1)

a. A carboxylic acid that has the formula $C_6H_{12}O_2$, with no substituents

b. A carboxylic acid that has the formula $C_6H_{12}O_2$, with one ethyl substituent

14.78 Draw the line-angle formula and write the IUPAC name for each of the following: (14.1)

a. A carboxylic acid that has the formula $C_5H_{10}O_2$, with no substituents

b. A carboxylic acid that has the formula $C_5H_{10}O_2$, with two methyl substituents

14.79 Propyl acetate is the ester that gives the odor and smell of pears. (9.4, 10.6, 14.3, 14.4)

a. Draw the condensed structural formula for propyl acetate.

b. Write the balanced chemical equation for the formation of propyl acetate.

c. Write the balanced chemical equation for the acid hydrolysis of propyl acetate.

d. Write the balanced chemical equation for the base hydrolysis of propyl acetate with NaOH.

e. How many milliliters of a 0.208 M NaOH solution is needed to completely hydrolyze (saponify) 1.58 g of propyl acetate?

14.80 Ethyl octanoate is a flavor component of mangoes. (9.4, 10.6, 14.3, 14.4)

a. Draw the condensed structural formula for ethyl octanoate.

b. Write the balanced chemical equation for the formation of ethyl octanoate.

c. Write the balanced chemical equation for the acid hydrolysis of ethyl octanoate.

d. Write the balanced chemical equation for the base hydrolysis of ethyl octanoate with NaOH.

e. How many milliliters of a 0.315 M NaOH solution is needed to completely hydrolyze (saponify) 2.84 g of ethyl octanoate?

14.81 Novocain, a local anesthetic, is the ammonium salt of procaine. (14.5)

$$H_2N-C_6H_4-\overset{\displaystyle O}{\overset{\|}{C}}-O-CH_2-CH_2-N(CH_2-CH_3)_2$$

Procaine

a. Draw the condensed structural formula for the ammonium salt (procaine hydrochloride) formed when procaine reacts with HCl. (*Hint*: The tertiary amine reacts with HCl.)

b. Why is procaine hydrochloride used rather than procaine?

14.82 Lidocaine (xylocaine) is used as a local anesthetic and cardiac depressant. (14.5)

$$(CH_3)_2C_6H_3-\underset{\displaystyle H}{N}-\overset{\displaystyle O}{\overset{\|}{C}}-CH_2-N(CH_2-CH_3)_2$$

Lidocaine (xylocaine)

a. Draw the condensed structural formula for the ammonium salt formed when lidocaine reacts with HCl.

b. Why is the ammonium salt of lidocaine used rather than the amine?

ANSWERS

Answers to Selected Practice Problems

14.1 methanoic acid (formic acid)

14.3 **a.** ethanoic acid (acetic acid) **b.** butanoic acid (butyric acid) **c.** 3-methylhexanoic acid **d.** 3-bromobenzoic acid

14.5 **a.** $Br-CH_2-\overset{\displaystyle Br}{\overset{|}{C}H}-CH_2-\overset{\displaystyle O}{\overset{\|}{C}}-OH$

b. $CH_3-\overset{\displaystyle Cl}{\overset{|}{C}H}-\overset{\displaystyle O}{\overset{\|}{C}}-OH$

c. $C_6H_5-\overset{\displaystyle O}{\overset{\|}{C}}-OH$

d. $CH_3-(CH_2)_5-\overset{\displaystyle O}{\overset{\|}{C}}-OH$

14.7 **a.** Propanoic acid has the smallest carbon chain, which makes it most soluble.

b. Propanoic acid forms more hydrogen bonds with water, which makes it most soluble.

14.9 **a.** $CH_3-CH_2-CH_2-CH_2-\overset{\displaystyle O}{\overset{\|}{C}}-OH + H_2O \rightleftarrows CH_3-CH_2-CH_2-CH_2-\overset{\displaystyle O}{\overset{\|}{C}}-O^- + H_3O^+$

b. $CH_3-\overset{\displaystyle O}{\overset{\|}{C}}-OH + H_2O \rightleftarrows CH_3-\overset{\displaystyle O}{\overset{\|}{C}}-O^- + H_3O^+$

14.11 **a.** $H-\overset{\displaystyle O}{\overset{\|}{C}}-OH + NaOH \longrightarrow H-\overset{\displaystyle O}{\overset{\|}{C}}-O^-\ Na^+ + H_2O$

b. $Cl-CH_2-CH_2-\overset{\displaystyle O}{\overset{\|}{C}}-OH + NaOH \longrightarrow Cl-CH_2-CH_2-\overset{\displaystyle O}{\overset{\|}{C}}-O^-\ Na^+ + H_2O$

c. $C_6H_5-\overset{\displaystyle O}{\overset{\|}{C}}-OH + NaOH \longrightarrow C_6H_5-\overset{\displaystyle O}{\overset{\|}{C}}-O^-\ Na^+ + H_2O$

14.13 **a.** sodium methanoate (sodium formate)

b. sodium 3-chloropropanoate

c. sodium benzoate

14.15 **a.** $CH_3-\overset{\displaystyle O}{\overset{\|}{C}}-O-CH_3$

b. $CH_3-CH_2-CH_2-CH_2-\overset{\displaystyle O}{\overset{\|}{C}}-O-CH_3$

14.17 **a.** $CH_3-CH_2-CH_2-CH_2-\overset{\displaystyle O}{\overset{\|}{C}}-O-\overset{\displaystyle CH_3}{\overset{|}{C}H}-CH_3$

b. $CH_3-CH_2-CH_2-\overset{\displaystyle O}{\overset{\|}{C}}-O-CH_2-CH_2-CH_3$

14.19 a. methyl methanoate (methyl formate)
b. ethyl propanoate (ethyl propionate)
c. propyl ethanoate (propyl acetate)

14.21 a. $CH_3-CH_2-CH_2-\overset{O}{\overset{\|}{C}}-O-CH_2-CH_2-CH_3$

b. $H-\overset{O}{\overset{\|}{C}}-O-CH_2-CH_2-CH_2-CH_3$

c. 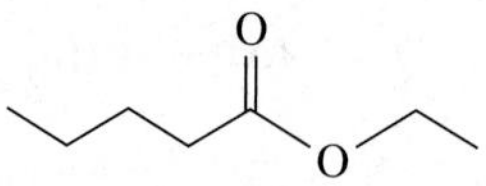 (skeletal structure: O, O)

d. 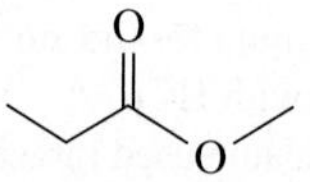 (skeletal structure: O, O)

14.23 a. pentyl ethanoate (pentyl acetate)
b. octyl ethanoate (octyl acetate)
c. pentyl butanoate (pentyl butyrate)

14.25 The products of the acid hydrolysis of an ester are a carboxylic acid and an alcohol.

14.27 a. $CH_3-CH_2-\overset{O}{\overset{\|}{C}}-O^- \; Na^+ + HO-CH_3$

b. 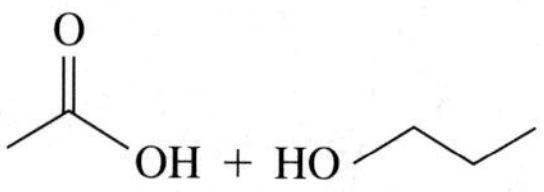 OH + HO

c. $CH_3-CH_2-CH_2-\overset{O}{\overset{\|}{C}}-OH + HO-CH_2-CH_3$

d. 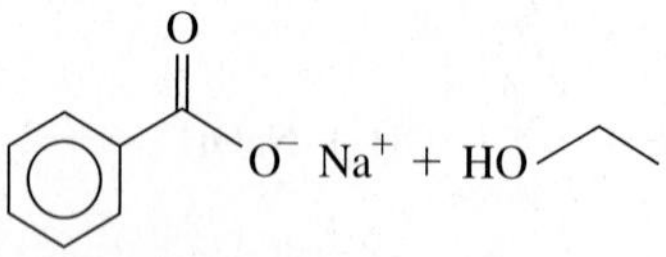 $O^- \; Na^+$ + HO

14.29 a. propylamine
b. methylpropylamine
c. diethylmethylamine

14.31 a. $CH_3-CH_2-NH_2$

b. 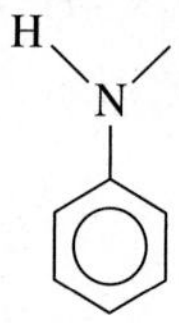 H, N (structure)

c. $CH_3-CH_2-CH_2-CH_2-\overset{H}{\overset{|}{N}}-CH_2-CH_2-CH_3$

14.33 a. 1° **b.** 2° **c.** 3° **d.** 3°

14.35 a. Yes, amines with fewer than seven carbon atoms hydrogen bond with water molecules and are soluble in water.
b. Yes, amines with fewer than seven carbon atoms hydrogen bond with water molecules and are soluble in water.
c. No, an amine with eight carbon atoms is insoluble in water.
d. Yes, amines with fewer than seven carbon atoms hydrogen bond with water molecules and are soluble in water.

14.37 a. $CH_3-NH_2 + HCl \longrightarrow CH_3-\overset{+}{N}H_3 + Cl^-$

b. $CH_3-\overset{H}{\overset{|}{N}}-CH_3 + HCl \longrightarrow CH_3-\overset{+}{N}H_2-CH_3 \; Cl^-$

c. 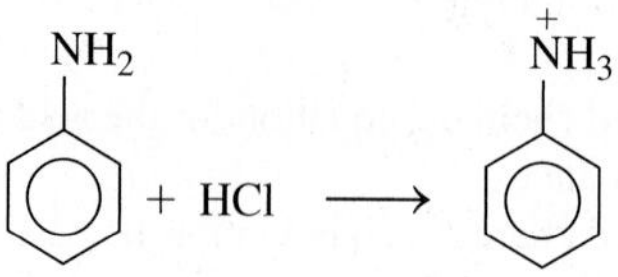 NH_2 + HCl $\longrightarrow$ $\overset{+}{N}H_3 \; Cl^-$

14.39 a. $CH_3-\overset{O}{\overset{\|}{C}}-NH_2$

b. $CH_3-\overset{O}{\overset{\|}{C}}-\overset{H}{\overset{|}{N}}-CH_2-CH_3$

c. 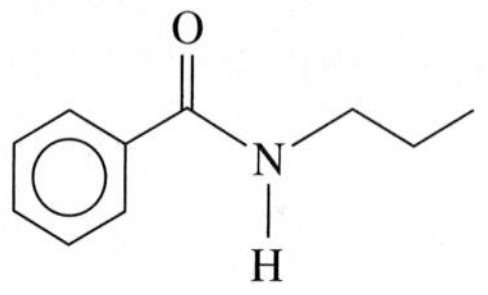 (structure: O, N, H)

14.41 a. ethanamide (acetamide)
b. butanamide (butyramide)
c. *N*-ethylmethanamide (*N*-ethylformamide)

14.43 a. $CH_3-CH_2-\overset{O}{\overset{\|}{C}}-NH_2$

b. $CH_3-CH_2-CH_2-\overset{CH_3}{\overset{|}{C}H}-\overset{O}{\overset{\|}{C}}-NH_2$

c. $H-\overset{O}{\overset{\|}{C}}-NH_2$

d. $CH_3-\overset{O}{\overset{\|}{C}}-\overset{CH_3}{\overset{|}{N}}-CH_3$

14.45 a. 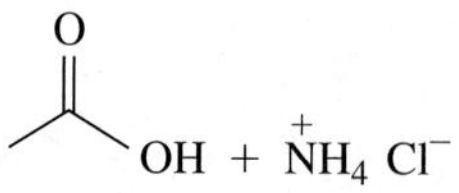 OH + $\overset{+}{N}H_4 \; Cl^-$

b. $CH_3-CH_2-\overset{O}{\overset{\|}{C}}-OH + \overset{+}{N}H_4 \; Cl^-$

c. $CH_3-CH_2-CH_2-\overset{O}{\overset{\|}{C}}-OH + CH_3-\overset{+}{N}H_3 \; Cl^-$

d. 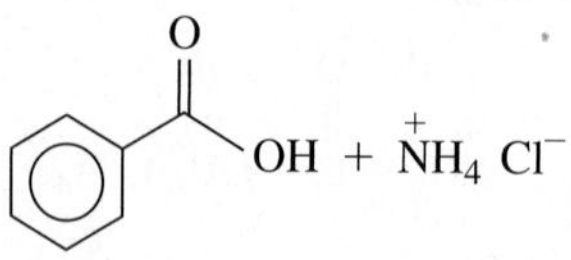 OH + $\overset{+}{N}H_4 \; Cl^-$

e. $CH_3-CH_2-CH_2-CH_2-\overset{O}{\overset{\|}{C}}-OH$
$+ CH_3-CH_2-\overset{+}{N}H_3 \; Cl^-$

14.47 a. amine **b.** 91 mL

14.49

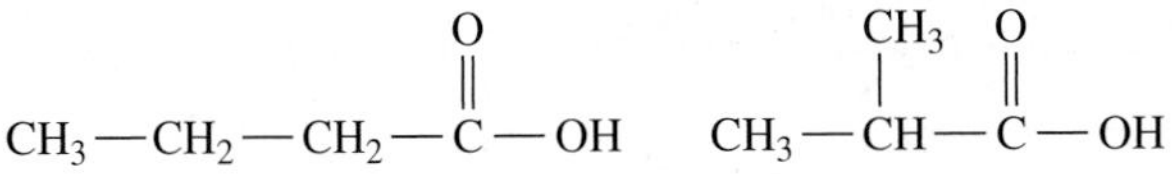

$$CH_3-CH_2-CH_2-\overset{O}{\overset{\|}{C}}-OH$$

Butanoic acid

$$CH_3-\underset{}{\overset{CH_3}{\overset{|}{CH}}}-\overset{O}{\overset{\|}{C}}-OH$$

2-Methylpropanoic acid

14.51 a. 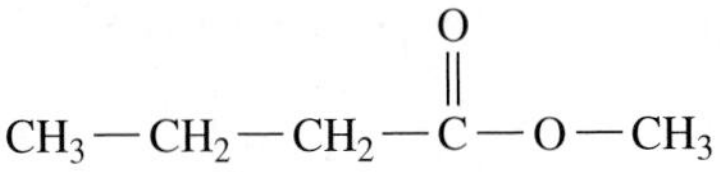 $CH_3-CH_2-CH_2-\overset{O}{\overset{\|}{C}}-O-CH_3$

b. butanoic acid and methanol

c. 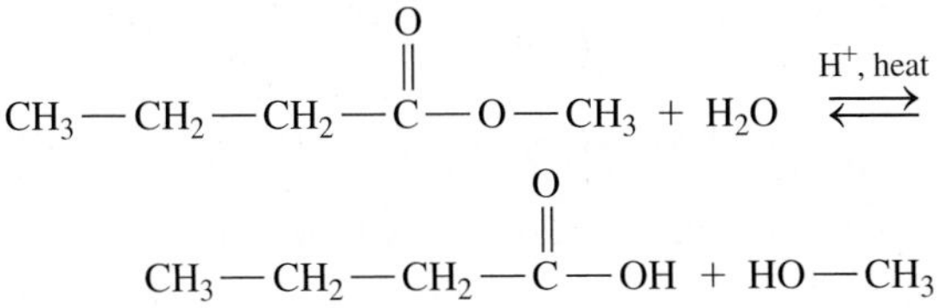 $CH_3-CH_2-CH_2-\overset{O}{\overset{\|}{C}}-O-CH_3 + H_2O \xrightleftharpoons{H^+,\ heat}$

$$CH_3-CH_2-CH_2-\overset{O}{\overset{\|}{C}}-OH + HO-CH_3$$

14.53 phenol, alcohol, amine

14.55 $CH_3-CH_2-CH_2-NH_2$

Propylamine (1°)

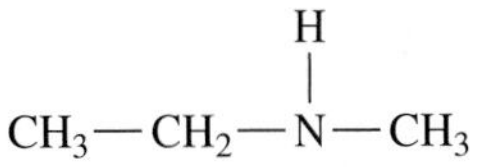

$$CH_3-CH_2-\overset{H}{\overset{|}{N}}-CH_3$$

Ethylmethylamine (2°)

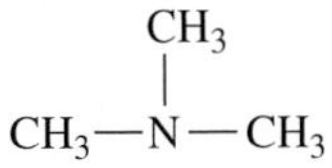

$$CH_3-\overset{CH_3}{\overset{|}{N}}-CH_3$$

Trimethylamine (3°)

$$CH_3-\overset{CH_3}{\overset{|}{CH}}-NH_2$$

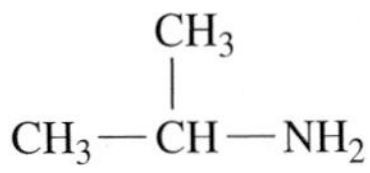

Isopropylamine (1°)

14.57 a. 3-methylbutanoic acid
b. ethyl benzoate
c. ethyl propanoate (ethyl propionate)
d. 2-chlorobenzoic acid
e. pentanoic acid

14.59 a. 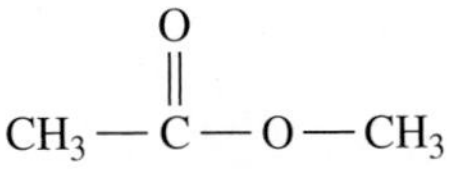 $CH_3-\overset{O}{\overset{\|}{C}}-O-CH_3$

b. $CH_3-CH_2-\underset{Cl}{\underset{|}{\overset{Cl}{\overset{|}{C}}}}-\overset{O}{\overset{\|}{C}}-OH$

c. Br, O, OH

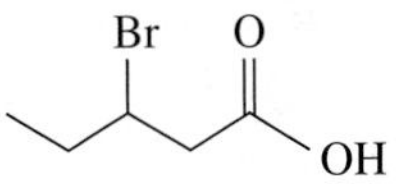

d. 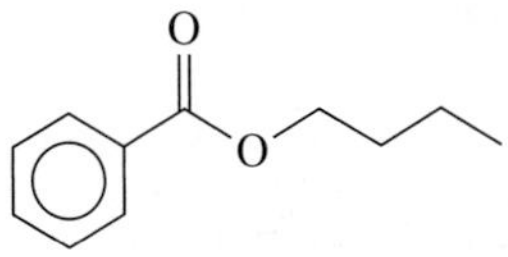 O, O

14.61 a. 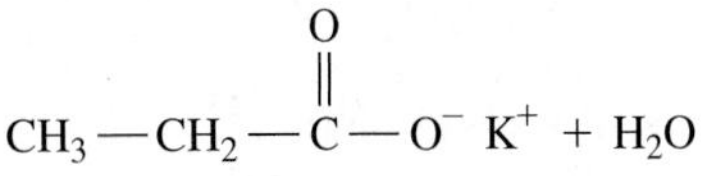 $CH_3-CH_2-\overset{O}{\overset{\|}{C}}-O^- \ K^+ + H_2O$

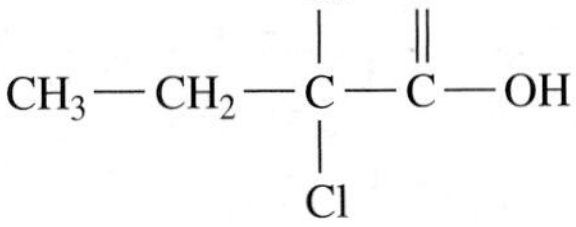

b. $CH_3-CH_2-\overset{O}{\overset{\|}{C}}-O-CH_3 + H_2O$

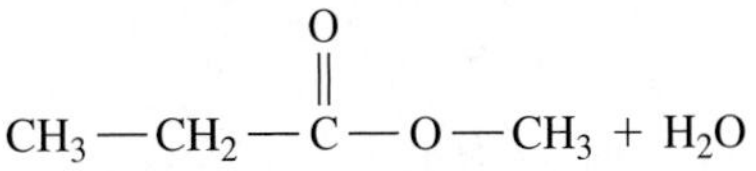

c. O, O $+ H_2O$

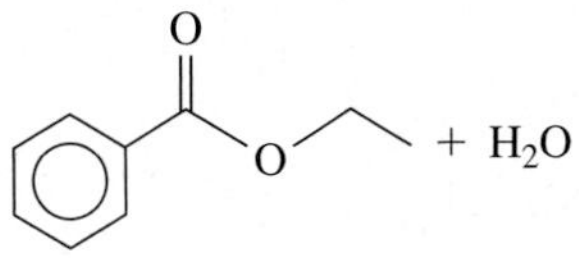

d. 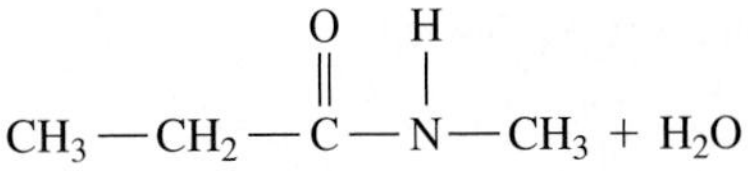 $CH_3-CH_2-\overset{O}{\overset{\|}{C}}-\overset{H}{\overset{|}{N}}-CH_3 + H_2O$

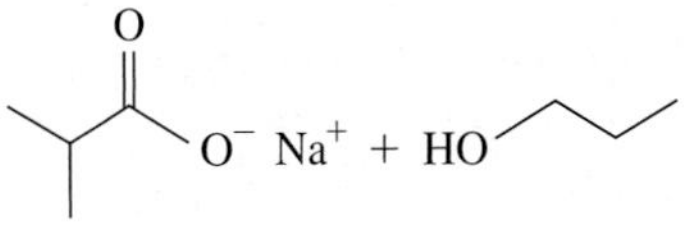

14.63 a. 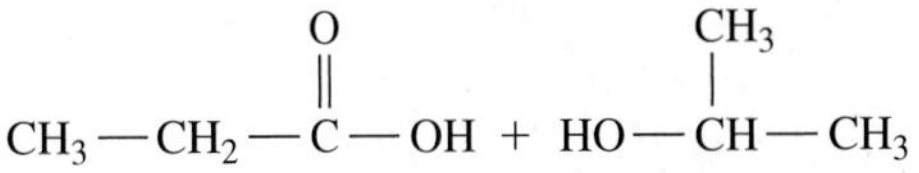 $CH_3-CH_2-\overset{O}{\overset{\|}{C}}-OH + HO-\overset{CH_3}{\overset{|}{CH}}-CH_3$

b. O, $O^- \ Na^+ + HO$

c. O, $O^- \ Na^+ + H_2N$

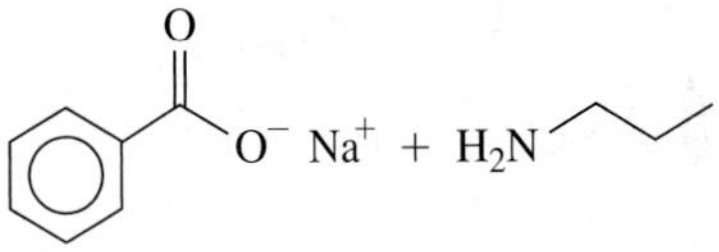

14.65 a. ethylmethylamine (2°) **b.** *N*-ethylaniline (2°)
c. butylamine (1°)

14.67 a. 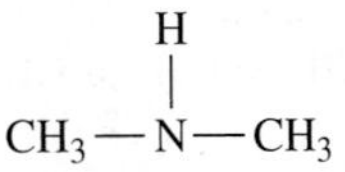 $CH_3-\overset{H}{\overset{|}{N}}-CH_3$

b. 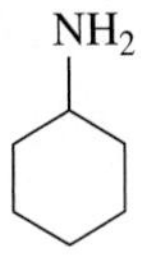 NH_2

c. 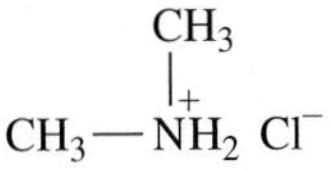 $CH_3-\overset{CH_3}{\overset{|}{\overset{+}{N}H_2}} \ Cl^-$

d. 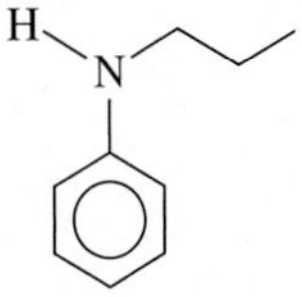 H, N

14.69 a. $CH_3-CH_2-\overset{+}{N}H_3 + OH^-$
b. $CH_3-CH_2-\overset{+}{N}H_3 \ Cl^-$

14.71 a. *N*-ethylethanamide
b. propanamide
c. 3-methylbutanamide

14.73 a. $CH_3{-}\overset{\overset{\displaystyle O}{\|}}{C}{-}OH + CH_3{-}\overset{+}{N}H_3\ Cl^-$

b. $H{-}\overset{\overset{\displaystyle O}{\|}}{C}{-}O^-\ Na^+ + NH_3$

c. 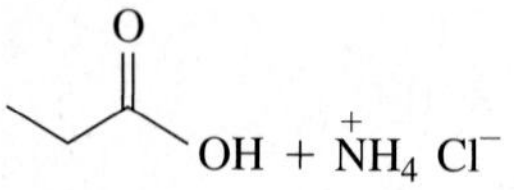 (skeletal structure of propanoic acid) $OH + \overset{+}{N}H_4\ Cl^-$

14.75 aromatic, amine, carboxylate salt

14.77 a. 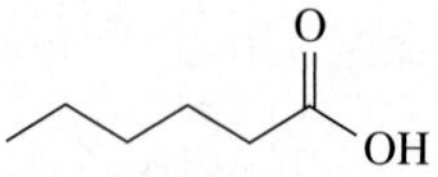 (skeletal structure) OH

Hexanoic acid

b. 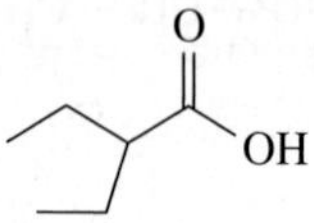 (skeletal structure) OH

2-Ethylbutanoic acid

14.79 a. $CH_3{-}\overset{\overset{\displaystyle O}{\|}}{C}{-}O{-}CH_2{-}CH_2{-}CH_3$

b. $CH_3{-}\overset{\overset{\displaystyle O}{\|}}{C}{-}OH + HO{-}CH_2{-}CH_2{-}CH_3 \xrightleftharpoons{H^+,\ heat}$

$CH_3{-}\overset{\overset{\displaystyle O}{\|}}{C}{-}O{-}CH_2{-}CH_2{-}CH_3 + H_2O$

c. $CH_3{-}\overset{\overset{\displaystyle O}{\|}}{C}{-}O{-}CH_2{-}CH_2{-}CH_3 + H_2O \xrightleftharpoons{H^+,\ heat}$

$CH_3{-}\overset{\overset{\displaystyle O}{\|}}{C}{-}OH + HO{-}CH_2{-}CH_2{-}CH_3$

d. $CH_3{-}\overset{\overset{\displaystyle O}{\|}}{C}{-}O{-}CH_2{-}CH_2{-}CH_3 + NaOH \xrightarrow{Heat}$

$CH_3{-}\overset{\overset{\displaystyle O}{\|}}{C}{-}O^-\ Na^+ + HO{-}CH_2{-}CH_2{-}CH_3$

e. 74.4 mL of a 0.208 M NaOH solution

14.81 a. 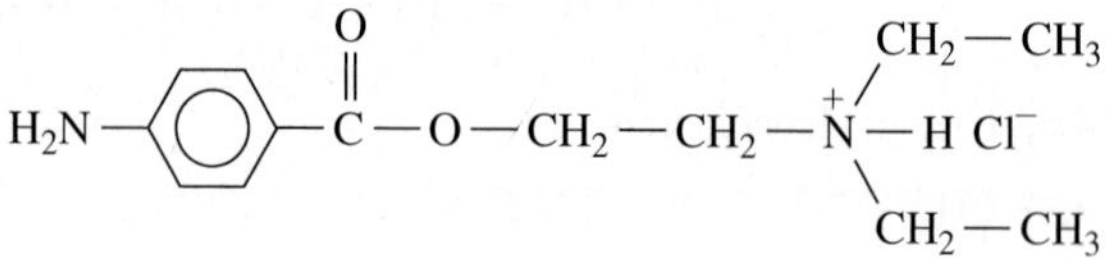 $H_2N{-}C_6H_4{-}\overset{\overset{\displaystyle O}{\|}}{C}{-}O{-}CH_2{-}CH_2{-}\overset{+}{N}(CH_2{-}CH_3)_2{-}H\ Cl^-$

b. The ammonium salt (Novocain) is more soluble in body fluids than procaine.

15 Lipids

REBECCA LEARNED OF HER FAMILY'S HYPER-cholesterolemia after her mother died of a heart attack at 44. Her mother had a total cholesterol level that was three times the normal level. Rebecca learned that the lumps under her mother's skin and around the cornea of her eyes, called *xanthomas*, were caused by cholesterol that was stored throughout her body. In adults with familial hypercholesterolemia (FH), cholesterol levels in the blood may be greater than 300 mg/dL. At the lipid clinic, Rebecca's blood tests showed that she had a total cholesterol level of 420 mg/dL. Rebecca was diagnosed with FH. She learned that from birth, cholesterol has been accumulating throughout her arteries. FH is caused by a genetic mutation that blocks the removal of cholesterol from the blood, resulting in the formation of deposits on the arterial walls.

After her diagnosis, Rebecca met with Susan, a clinical lipid specialist at the lipid clinic where she informed Rebecca about managing risk factors that can lead to heart attack and stroke. Susan told Rebecca about medications she would prescribe, and then followed the results.

CAREER Clinical Lipid Specialist

Clinical lipid specialists work with patients who have lipid disorders such as high cholesterol, high triglycerides, coronary heart disease, obesity, and FH. At a lipid clinic, a clinical lipid specialist reviews a patient's lipid profile, which includes total cholesterol, high-density lipoprotein (HDL), and low-density lipoprotein (LDL). If a lipid disorder is identified, the lipid specialist assesses the patient's current diet and exercise program. The lipid specialist diagnoses and determines treatment including dietary changes such as reducing salt intake, increasing the amount of fiber in the diet, and lowering the amount of fat. He or she also discusses drug therapy using lipid-lowering medications that remove LDL-cholesterol to help patients achieve and maintain good health.

Allied health professionals such as nurses, nurse practitioners, pharmacists, and dietitians are certified by a clinical lipid specialist program for specialized care of patients with lipid disorders.

CLINICAL UPDATE Rebecca's Program to Lower Cholesterol

When Rebecca returns to the lipid clinic, she is told that her diet and exercise are not lowering her cholesterol level sufficiently. You can view the medications that Susan prescribes in the **CLINICAL UPDATE Rebecca's Program to Lower Cholesterol**, page 538, identify the functional groups, and calculate Rebecca's cholesterol level several months later.

LOOKING AHEAD

ENGAGE

What is one feature that is characteristic of all lipids?

15.1 Lipids

LEARNING GOAL Describe the classes of lipids.

Lipids are a family of biomolecules that have the common property of being soluble in organic solvents but not in water. The word *lipid* comes from the Greek word *lipos*, meaning "fat" or "lard." Typically, the lipid content of a cell can be extracted using a nonpolar solvent such as ether or chloroform. Lipids are an important feature in cell membranes and steroid hormones.

Types of Lipids

Within the lipid family, there are specific structures that distinguish the different types of lipids. Lipids such as waxes, triacylglycerols, glycerophospholipids, and sphingolipids are esters that can be hydrolyzed to give fatty acids along with other molecules. Triacylglycerols and glycerophospholipids contain the alcohol *glycerol*, whereas sphingolipids contain the amino alcohol *sphingosine*. Steroids, which have a completely different structure, do not contain fatty acids and cannot be hydrolyzed. Steroids are characterized by the *steroid nucleus* of four fused carbon rings. **FIGURE 15.1** illustrates the types and general structure of lipids we discuss in this chapter.

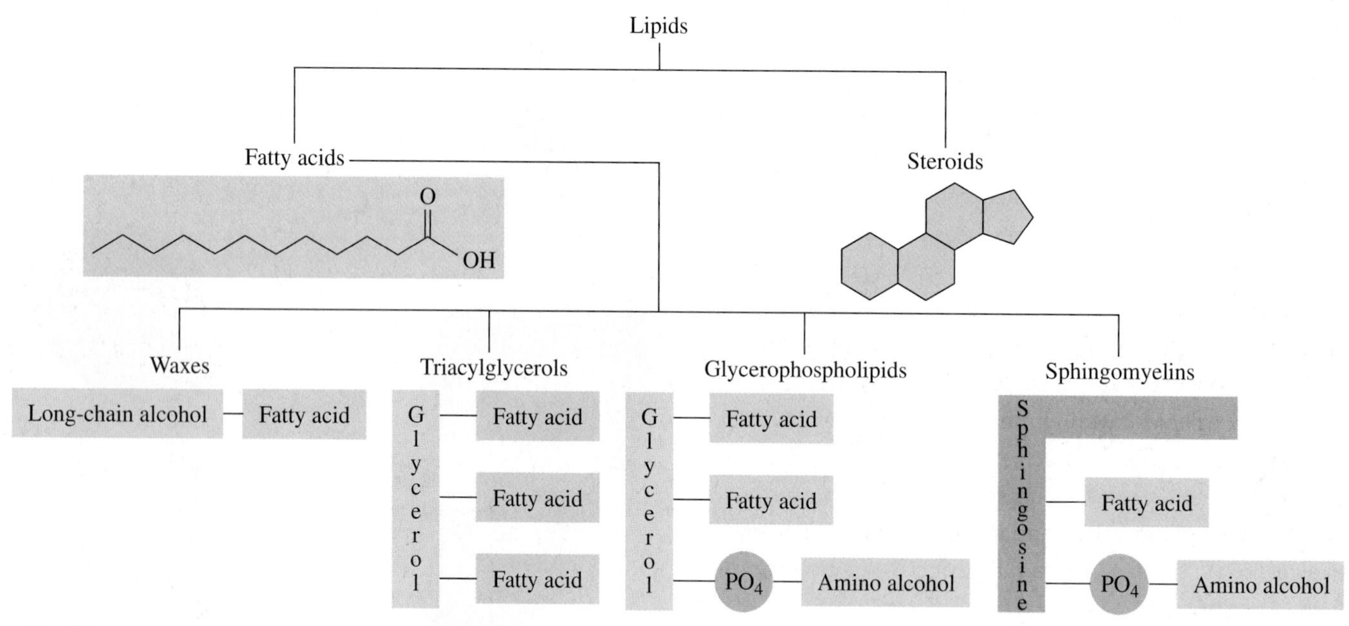

TEST

Try Practice Problems 15.1 to 15.4

FIGURE 15.1 Lipids are naturally occurring biomolecules, which are soluble in organic solvents but not in water.

Q What property do waxes, triacylglycerols, and steroids have in common?

PRACTICE PROBLEMS

15.1 Lipids

LEARNING GOAL Describe the classes of lipids.

15.1 Lipids are not soluble in water. Are lipids polar or nonpolar molecules?

15.2 Which of the following solvents might be used to dissolve an oil stain?

a. water **b.** CCl_4 **c.** diethyl ether
d. benzene **e.** NaCl solution

Clinical Applications

15.3 What are some functions of lipids in the body?

15.4 What are some of the different kinds of lipids?

15.2 Fatty Acids

LEARNING GOAL Draw the condensed structural or line-angle formula for a fatty acid and identify it as saturated or unsaturated.

A **fatty acid** contains a long, unbranched carbon chain with a carboxylic acid group at one end. Although the carboxylic acid part is hydrophilic, the long hydrophobic carbon chain makes fatty acids insoluble in water.

Most naturally occurring fatty acids have an even number of carbon atoms, usually between 12 and 20. An example of a fatty acid is lauric acid, a 12-carbon acid found in coconut oil. The shorthand notation description for lauric acid is 12:0; 12 for the number of carbon atoms and 0 for the number of double bonds.

CORE CHEMISTRY SKILL
Identifying Fatty Acids

Drawing Structural Formulas for Lauric Acid (12:0)

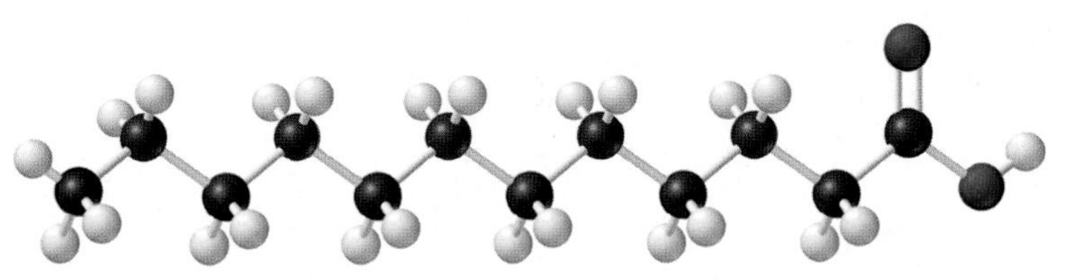

Ball-and-stick model of lauric acid

$$CH_3{-}(CH_2)_{10}{-}\overset{\overset{\displaystyle O}{\|}}{C}{-}OH \qquad CH_3{-}(CH_2)_{10}{-}COOH$$

$$CH_3{-}CH_2{-}CH_2{-}CH_2{-}CH_2{-}CH_2{-}CH_2{-}CH_2{-}CH_2{-}CH_2{-}CH_2{-}C({=}O){-}OH$$

Condensed structural formulas of lauric acid

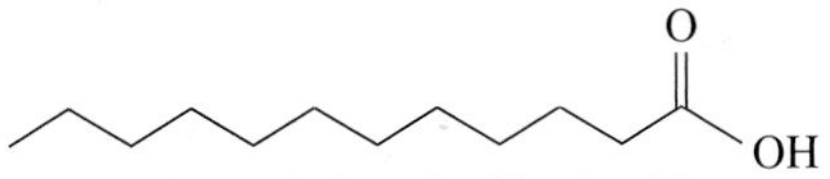

O
OH

Line-angle formula of lauric acid

A **saturated fatty acid (SFA)** contains only carbon–carbon single bonds, which make the properties of a long-chain fatty acid similar to those of an alkane. An **unsaturated fatty acid (UFA)** contains one or more carbon–carbon double bonds. In a **monounsaturated fatty acid (MUFA)**, the long carbon chain has one double bond, which makes its properties similar to those of an alkene. A **polyunsaturated fatty acid (PUFA)** has at least two carbon–carbon double bonds. **TABLE 15.1** lists some of the typical fatty acids in lipids. In the lipids of plants and animals, about half of the fatty acids are saturated and half are unsaturated.

TEST
Try Practice Problems 15.5 and 15.6

Cis and Trans Isomers of Unsaturated Fatty Acids

Unsaturated fatty acids can be drawn as *cis* and *trans* isomers. For example, oleic acid, an 18-carbon monounsaturated fatty acid found in olives, has one double bond starting at

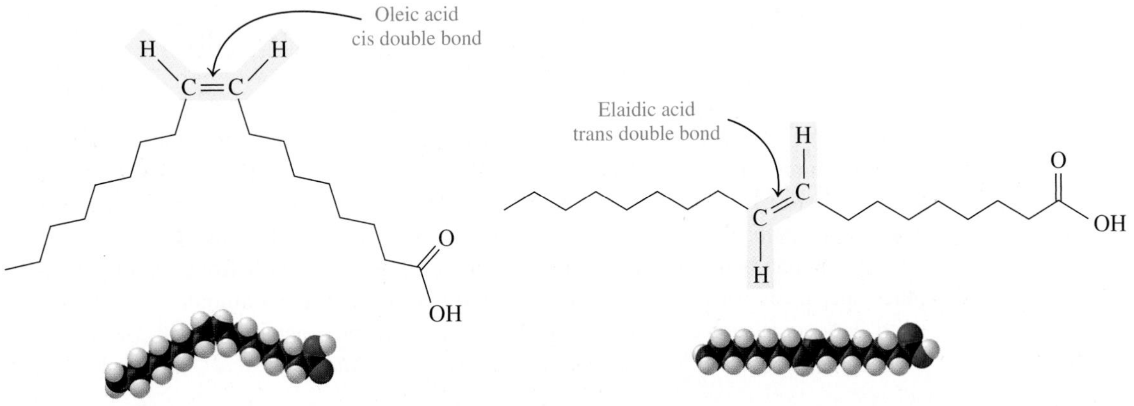

Oleic acid
cis double bond

H H
C=C
O
OH

Elaidic acid
trans double bond

H
C=C
H
O
OH

Almost all naturally occurring unsaturated fatty acids have one or more cis double bonds.

TABLE 15.1 Structures and Melting Points of Common Fatty Acids

Name	Carbon Atoms: Double Bonds	Present in	Melting Point (°C)	Structures
Saturated Fatty Acids				
Lauric acid	12:0	Coconut	44	$CH_3-(CH_2)_{10}-COOH$ 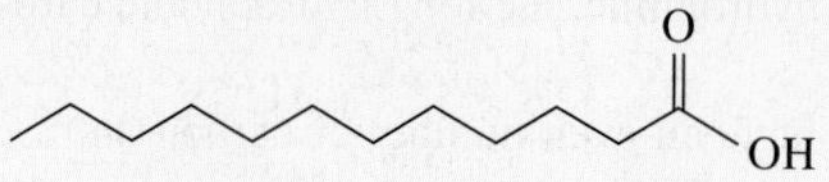
Myristic acid	14:0	Nutmeg	55	$CH_3-(CH_2)_{12}-COOH$ 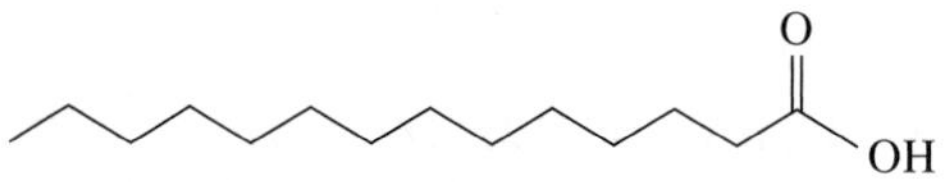
Palmitic acid	16:0	Palm	63	$CH_3-(CH_2)_{14}-COOH$ 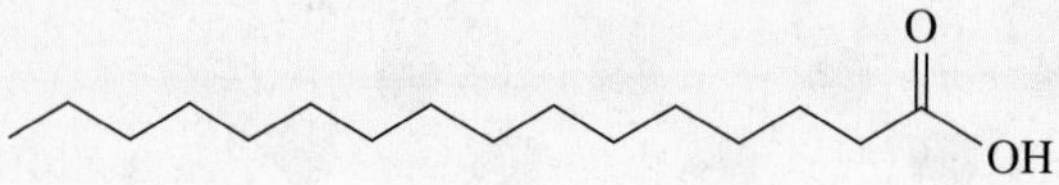
Stearic acid	18:0	Animal fat	69	$CH_3-(CH_2)_{16}-COOH$ 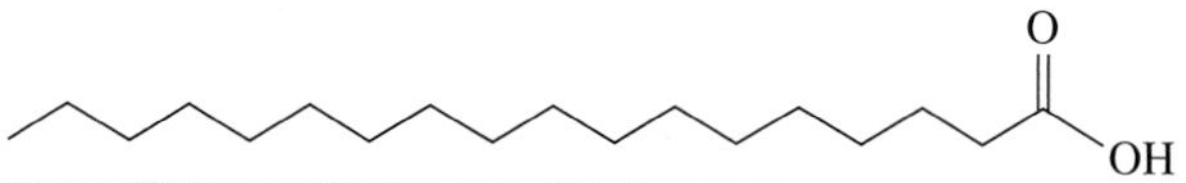
Monounsaturated Fatty Acids				
Palmitoleic acid	16:1	Butter	0	$CH_3-(CH_2)_5-CH=CH-(CH_2)_7-COOH$ 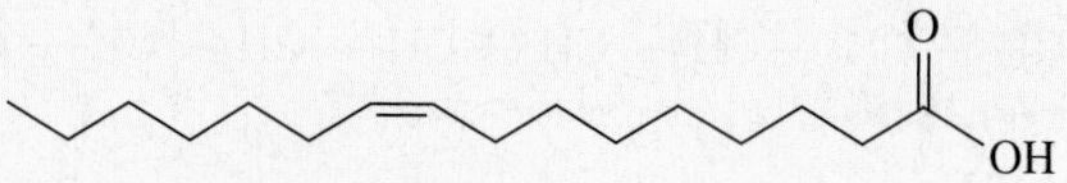
Oleic acid	18:1	Olives, pecan, grapeseed	14	$CH_3-(CH_2)_7-CH=CH-(CH_2)_7-COOH$ 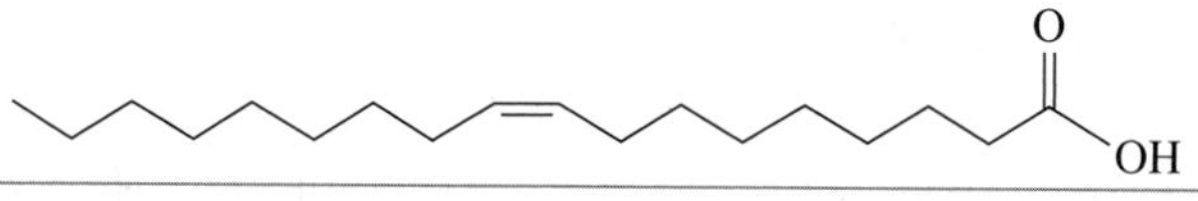
Polyunsaturated Fatty Acids				
Linoleic acid	18:2	Soybean, safflower, sunflower	−5	$CH_3-(CH_2)_4-CH=CH-CH_2-CH=CH-(CH_2)_7-COOH$ 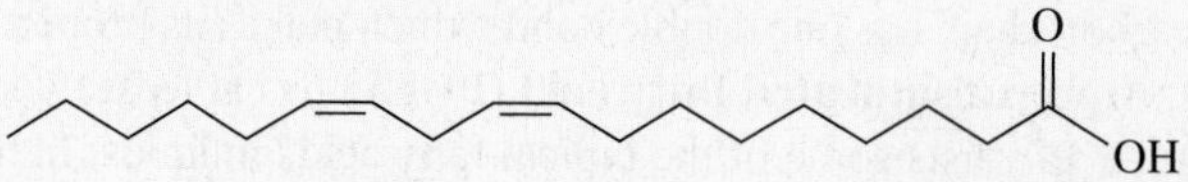
Linolenic acid	18:3	Corn	−11	$CH_3-CH_2-CH=CH-CH_2-CH=CH-CH_2-CH=CH-(CH_2)_7-COOH$ 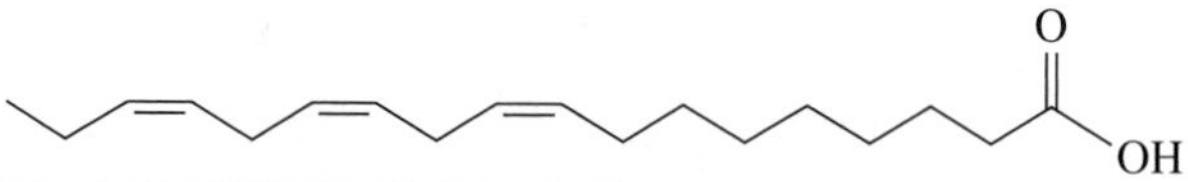
Arachidonic acid	20:4	Meat, eggs, fish	−50	$CH_3-(CH_2)_3-(CH_2-CH=CH)_4-(CH_2)_3-COOH$ 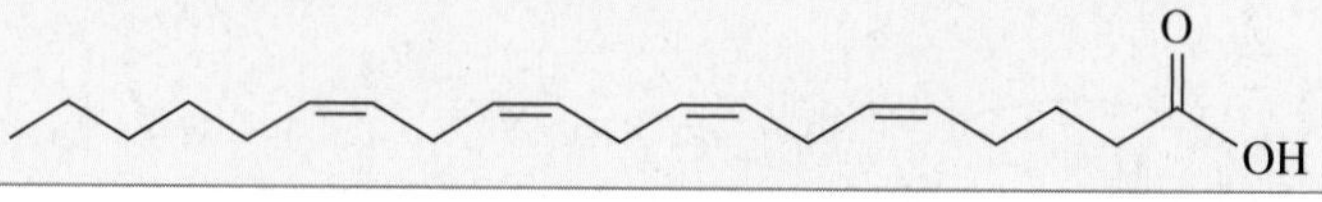

carbon 9. We can show its cis and trans structures using line-angle formulas. The cis structure is the more prevalent isomer found in naturally occurring unsaturated fatty acids. In the cis isomer, the carbon chain has a "kink" at the double bond site. The trans isomer of oleic acid, *elaidic acid*, is a straight chain without a kink at the double bond site.

The human body is capable of synthesizing some fatty acids from carbohydrates or other fatty acids. However, humans cannot synthesize sufficient amounts of polyunsaturated fatty acids such as linoleic acid, linolenic acid, and arachidonic acid. Because they must be obtained from the diet, they are known as *essential fatty acids* (EFAs). In infants, a deficiency of EFAs can cause skin dermatitis. However, the role of fatty acids in adult nutrition is not well understood. Adults do not usually have a deficiency of EFAs.

Physical Properties of Fatty Acids

TEST

Try Practice Problems 15.7 to 15.10

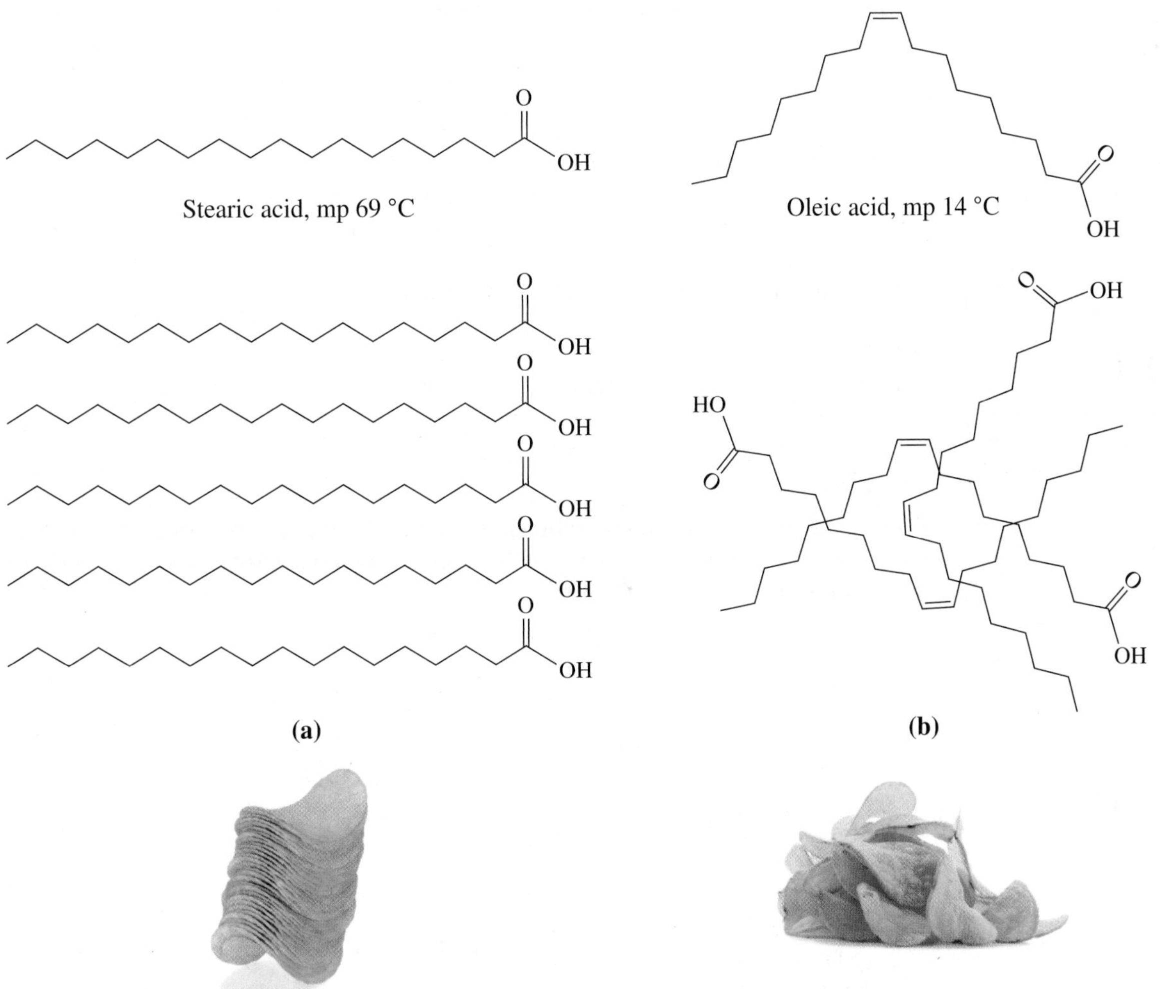

The saturated fatty acids fit closely together in a regular pattern, which allows many dispersion forces between the carbon chains. These normally weak forces of attraction become important when molecules of fatty acids are close together. As a result, a significant amount of energy and higher temperatures are required to separate the fatty acids before melting occurs. As the length of the carbon chain increases, more interactions occur between the fatty acids, requiring higher temperatures to melt. Saturated fatty acids are usually solids at room temperature (see **FIGURE 15.2**).

In unsaturated fatty acids, the cis double bonds cause the carbon chain to bend or kink, giving the molecules an irregular shape. As a result, unsaturated fatty acids cannot stack as closely as saturated fatty acids, and thus have fewer dispersion forces between carbon chains. We might think of saturated fatty acids as regularly shaped chips that stack closely together in a can. Similarly, unsaturated fatty acids would be like irregularly shaped chips that do not fit closely together. Consequently, less energy is required to separate these molecules, making the melting points of unsaturated fatty acids lower than those of saturated fatty acids. Most unsaturated fatty acids are liquids at room temperature.

Stearic acid, mp 69 °C

Oleic acid, mp 14 °C

(a)

(b)

FIGURE 15.2 **(a)** In saturated fatty acids, the molecules fit closely together to give high melting points. **(b)** In unsaturated fatty acids, molecules cannot fit closely together, resulting in lower melting points.

Q Why does the cis double bond affect the melting points of unsaturated fatty acids?

SAMPLE PROBLEM 15.1 Structures and Properties of Fatty Acids

TRY IT FIRST

Consider the line-angle formula for vaccenic acid, a fatty acid found in dairy products.

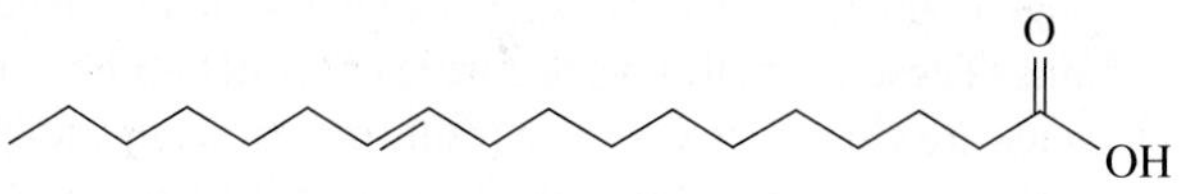

a. Why is this substance an acid?
b. How many carbon atoms are in vaccenic acid?
c. Is the fatty acid saturated, monounsaturated, or polyunsaturated?
d. Give the shorthand notation for the number of carbon atoms and double bonds in vaccenic acid.
e. Would it be soluble in water?

SOLUTION

a. Vaccenic acid contains a carboxylic acid group.
b. It contains 18 carbon atoms.
c. It is a monounsaturated fatty acid (MUFA).
d. 18:1
e. No. Its long hydrocarbon chain makes it insoluble in water.

STUDY CHECK 15.1

Palmitoleic acid is a fatty acid with the following condensed structural formula:

$$CH_3-(CH_2)_5-CH{=}CH-(CH_2)_7-\overset{\overset{\displaystyle O}{\|}}{C}-OH$$

a. How many carbon atoms are in palmitoleic acid?
b. Is the fatty acid saturated, monounsaturated, or polyunsaturated?
c. Give the shorthand notation for the number of carbon atoms and double bonds in palmitoleic acid.
d. Is it most likely to be solid or liquid at room temperature?

ANSWER

a. 16 **b.** monounsaturated **c.** 16:1 **d.** liquid

TEST

Try Practice Problems 15.11 and 15.12

Prostaglandins

Prostaglandins are hormone-like substances produced in small amounts in most cells of the body. The prostaglandins, also known as *eicosanoids*, are formed from arachidonic acid, the polyunsaturated fatty acid with 20 carbon atoms (*eicos* is the Greek word for 20). Swedish

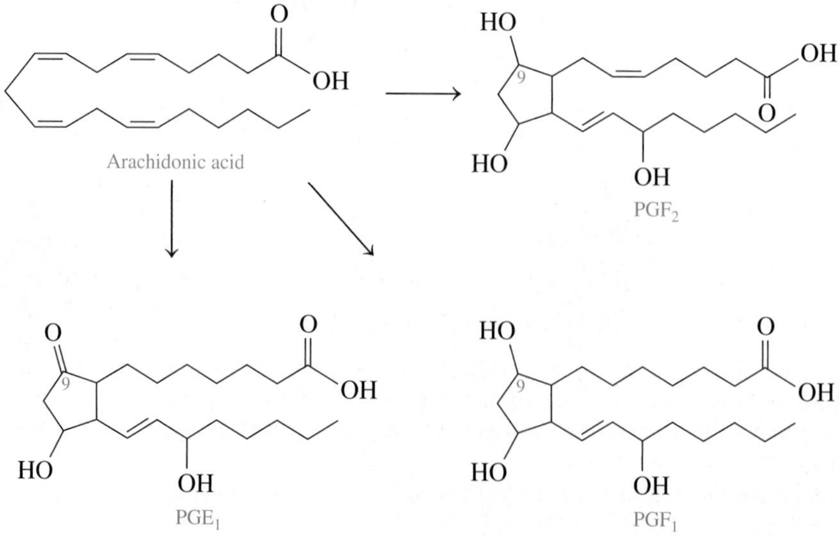

Arachidonic acid

PGF_2

PGE_1

PGF_1

chemists first discovered prostaglandins and named them "prostaglandin E" (soluble in ether) and "prostaglandin F" (soluble in phosphate buffer or *fosfat* in Swedish). The various kinds of prostaglandins differ by the substituents attached to the five-carbon ring. Prostaglandin E (PGE) has a ketone group on carbon 9, whereas prostaglandin F (PGF) has a hydroxyl group. The number of double bonds is shown as a subscript 1 or 2.

Although prostaglandins are broken down quickly, they have potent physiological effects. Some prostaglandins increase blood pressure, and others lower blood pressure. Other prostaglandins stimulate contraction and relaxation in the smooth muscle of the uterus. When tissues are injured, arachidonic acid is converted to prostaglandins that produce inflammation and pain in the area.

ENGAGE

What are some functions of prostaglandins in the body?

Arachidonic acid

Analgesics

PGF_2

Pain, fever, inflammation

When tissues are injured, prostaglandins are produced, which cause pain and inflammation.

The treatment of pain, fever, and inflammation is based on inhibiting the enzymes that convert arachidonic acid to prostaglandins. Several nonsteroidal anti-inflammatory drugs (NSAIDs), such as aspirin, block the production of prostaglandins and in doing so decrease pain and inflammation and reduce fever (antipyretics). Ibuprofen has similar anti-inflammatory and analgesic effects. Other NSAIDs include naproxen (Aleve and Naprosyn), ketoprofen (Actron), and nabumetone (Relafen). Although NSAIDs are helpful, their long-term use can result in liver, kidney, and gastrointestinal damage.

TEST

Try Practice Problems 15.13 to 15.16

Aspirin

Ibuprofen (Advil, Motrin)

Naproxen (Aleve, Naprosyn)

CHEMISTRY LINK TO HEALTH

Omega-3 Fatty Acids in Fish Oils

Because unsaturated fats are now recognized as being more beneficial to health than saturated fats, American diets have changed to include more unsaturated fats and less saturated fatty acids. This change is a response to research that indicates that atherosclerosis and heart disease are associated with high levels of fats in the diet. However, the Inuit people of Alaska have a diet with high levels of unsaturated fats as well as high levels of blood cholesterol, but a very low occurrence of atherosclerosis and heart attacks. The fats in the Inuit diet are primarily unsaturated fats from fish, rather than saturated fats from land animals.

Both fish and vegetable oils have high levels of unsaturated fats. The fatty acids in vegetable oils are *omega-6 fatty acids* (ω-6 fatty acids), in which the first double bond occurs at carbon 6 counting from the methyl end of the carbon chain. Omega is the last letter in the Greek alphabet and is used to denote the end. Two common omega-6 acids are linoleic acid (LA) and arachidonic acid (AA). However, the fatty acids in fish oils are mostly the omega-3 type, in which the first double bond occurs at the third carbon counting from the methyl group. Three common *omega-3 fatty acids* (or ω-3 fatty acids) in fish are linolenic acid (ALA), eicosapentaenoic acid (EPA), and docosahexaenoic acid (DHA).

In atherosclerosis and heart disease, cholesterol forms plaques that adhere to the walls of the blood vessels. Blood pressure rises as blood has to squeeze through a smaller opening in the blood vessel. As more plaque forms, there is also a possibility of blood clots blocking the blood vessels and causing a heart attack. Omega-3 fatty acids lower the tendency of blood platelets to stick together, thereby reducing the possibility of blood clots. However, high levels of omega-3 fatty acids can increase bleeding if the ability of the platelets to form blood clots is reduced too much. It does seem that a diet that includes fish such as salmon, tuna, and herring can provide higher amounts of the omega-3 fatty acids, which help lessen the possibility of developing heart disease.

Cold water fish are a good source of omega-3 fatty acids.

Omega-6 Fatty Acids

Omega-6 fatty acid

Linoleic acid (LA)

1 2 3 4 5 6

O

OH

Arachidonic acid (AA)

O

OH

Omega-3 Fatty Acids

Omega-3 fatty acid

Linolenic acid (ALA)

1 2 3

O

OH

Eicosapentaenoic acid (EPA)

O

OH

Docosahexaenoic acid (DHA)

O

OH

TEST

Try Practice Problems 15.17 and 15.18

PRACTICE PROBLEMS

15.2 Fatty Acids

LEARNING GOAL Draw the condensed structural or line-angle formula for a fatty acid and identify it as saturated or unsaturated.

15.5 Describe some similarities and differences in the structures of a saturated fatty acid and an unsaturated fatty acid.

15.6 Stearic acid and linoleic acid each have 18 carbon atoms. Why does stearic acid melt at 69 °C but linoleic acid melts at −5 °C?

15.7 Draw the line-angle formula for each of the following fatty acids:
a. palmitic acid **b.** oleic acid

15.8 Draw the line-angle formula for each of the following fatty acids:
a. stearic acid **b.** linoleic acid

15.9 For each of the following fatty acids, give the shorthand notation for the number of carbon atoms and double bonds, and classify as saturated, monounsaturated, or polyunsaturated:
a. lauric acid **b.** linolenic acid
c. palmitoleic acid **d.** stearic acid

15.10 For each of the following fatty acids, give the shorthand notation for the number of carbon atoms and double bonds, and classify as saturated, monounsaturated, or polyunsaturated:
a. linoleic acid **b.** palmitic acid
c. myristic acid **d.** oleic acid

15.11 How does the structure of a fatty acid with a cis double bond differ from the structure of a fatty acid with a trans double bond?

15.12 How does the double bond influence the dispersion forces that can form between the hydrocarbon chains of fatty acids?

Clinical Applications

15.13 Compare the structures and functional groups of arachidonic acid and prostaglandin PGE_1.

15.14 Compare the structures and functional groups of PGF_1 and PGF_2.

15.15 What are some effects of prostaglandins in the body?

15.16 How does an anti-inflammatory drug reduce inflammation?

15.17 What is the difference in the location of the first double bond in an omega-3 and an omega-6 fatty acid (see Chemistry Link to Health "Omega-3 Fatty Acids in Fish Oils")?

15.18 What are some sources of omega-3 and omega-6 fatty acids (see Chemistry Link to Health "Omega-3 Fatty Acids in Fish Oils")?

15.3 Waxes and Triacylglycerols

LEARNING GOAL Draw the condensed structural or line-angle formula for a wax or triacylglycerol produced by the reaction of a fatty acid and an alcohol or glycerol.

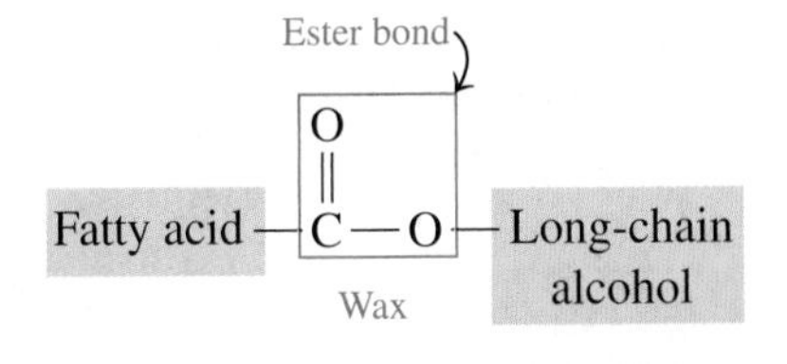

Waxes are found in many plants and animals. Natural waxes are found on the surface of fruits, and on the leaves and stems of plants where they help prevent loss of water and damage from pests. Waxes on the skin, fur, and feathers of animals provide a waterproof coating. A **wax** is an ester of a long-chain fatty acid and a long-chain alcohol, each containing from 14 to 30 carbon atoms.

The formulas of some common waxes are given in **TABLE 15.2**. Beeswax obtained from honeycombs and carnauba wax from palm trees are used to give a protective coating to furniture, cars, and floors. Jojoba wax is used in making candles and cosmetics such as lipstick. Lanolin, a mixture of waxes obtained from wool, is used in hand and facial lotions to aid retention of water, which softens the skin.

Waxes are esters of long-chain alcohols and fatty acids.

TABLE 15.2 Some Typical Waxes

Type	Condensed Structural Formula	Source	Uses
Beeswax	$CH_3—(CH_2)_{14}—\overset{O}{\overset{\parallel}{C}}—O—(CH_2)_{29}—CH_3$	Honeycomb	Candles, shoe polish, wax paper
Carnauba wax	$CH_3—(CH_2)_{24}—\overset{O}{\overset{\parallel}{C}}—O—(CH_2)_{29}—CH_3$	Brazilian palm tree	Waxes for furniture, cars, floors, shoes
Jojoba wax	$CH_3—(CH_2)_{18}—\overset{O}{\overset{\parallel}{C}}—O—(CH_2)_{19}—CH_3$	Jojoba bush	Candles, soaps, cosmetics

Jojoba wax is obtained from the seed of the jojoba bush.

Triacylglycerols

In the body, fatty acids are stored as **triacylglycerols**, also called *triglycerides*, which are triesters of glycerol (a trihydroxy alcohol) and fatty acids. The general formula of a triacylglycerol follows:

TEST

Try Practice Problems 15.19 and 15.20

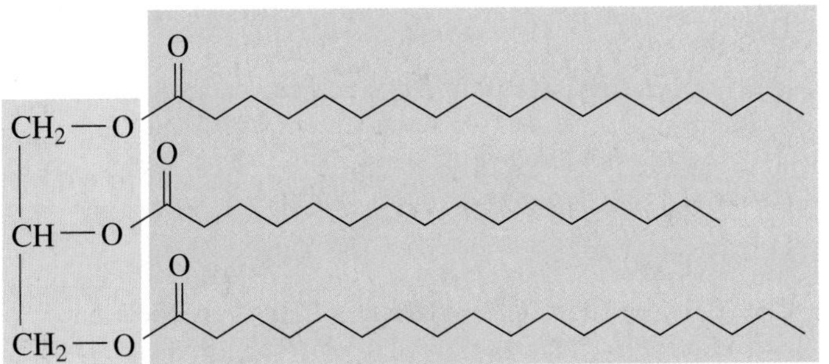

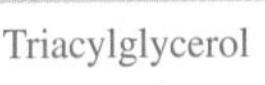

Triacylglycerol

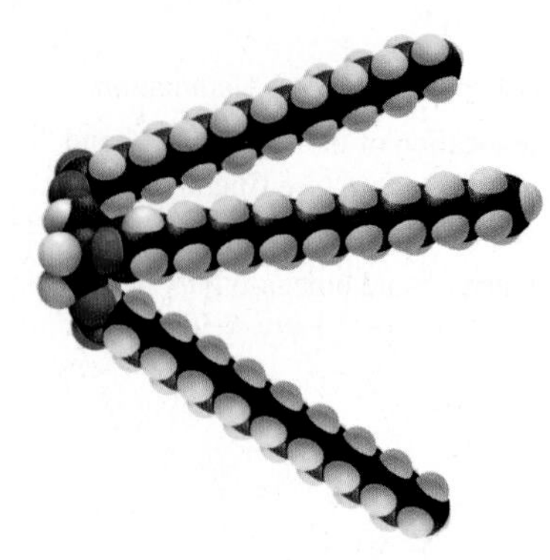
Tristearin consists of glycerol with three ester bonds to stearic acid molecules.

In a triacylglycerol, three hydroxyl groups of glycerol form ester bonds with the carboxyl groups of three fatty acids. For example, glycerol and three molecules of stearic acid form a triacylglycerol. In the name, glycerol is named *glyceryl* and the fatty acids are named as carboxylates. For example, stearic acid is named as stearate, which gives the name glyceryl tristearate. The common name of this compound is tristearin.

$$
\begin{array}{llcl}
CH_2{-}O{-}H + HO{-}\overset{\overset{\displaystyle O}{\|}}{C}{-}(CH_2)_{16}{-}CH_3 & & & CH_2{-}\mathbf{O}\overset{\text{Ester bond}}{\downarrow}\overset{\overset{\displaystyle \mathbf{O}}{\|}}{C}{-}(CH_2)_{16}{-}CH_3 \\
| & & & | \\
CH{-}O{-}H + HO{-}\overset{\overset{\displaystyle O}{\|}}{C}{-}(CH_2)_{16}{-}CH_3 & \longrightarrow & & CH{-}\mathbf{O}{-}\overset{\overset{\displaystyle \mathbf{O}}{\|}}{C}{-}(CH_2)_{16}{-}CH_3 + 3H_2O \\
| & & & | \\
CH_2{-}O{-}H + HO{-}\overset{\overset{\displaystyle O}{\|}}{C}{-}(CH_2)_{16}{-}CH_3 & & & CH_2{-}\mathbf{O}{-}\overset{\overset{\displaystyle \mathbf{O}}{\|}}{C}{-}(CH_2)_{16}{-}CH_3 \\
\text{Glycerol} \quad \text{3 Stearic acid molecules} & & & \text{Glyceryl tristearate (tristearin)}
\end{array}
$$

ENGAGE
What functional group is present in all fats (triacylglycerols)?

Most naturally occurring triacylglycerols contain glycerol bonded to two or three different fatty acids, typically palmitic acid, oleic acid, linoleic acid, and stearic acid. For example, a mixed triacylglycerol might be made from stearic acid, oleic acid, and palmitic acid. One possible structure for this mixed triacylglycerol follows:

CORE CHEMISTRY SKILL
Drawing Structures for Triacylglycerols

$$
\begin{array}{ll}
CH_2{-}O{-}\overset{\overset{\displaystyle O}{\|}}{C}{-}(CH_2)_{16}{-}CH_3 & \text{Stearic acid} \\
| & \\
CH{-}O{-}\overset{\overset{\displaystyle O}{\|}}{C}{-}(CH_2)_7{-}CH{=}CH{-}(CH_2)_7{-}CH_3 & \text{Oleic acid} \\
| & \\
CH_2{-}O{-}\overset{\overset{\displaystyle O}{\|}}{C}{-}(CH_2)_{14}{-}CH_3 & \text{Palmitic acid} \\
\text{A mixed triacylglycerol} &
\end{array}
$$

SAMPLE PROBLEM 15.2 Drawing the Structure for a Triacylglycerol

TRY IT FIRST

Draw the condensed structural formula for glyceryl tripalmitoleate (tripalmitolein), which is used in cosmetic creams and lotions.

Triacylglycerols are used to thicken creams and lotions.

SOLUTION GUIDE

	Given	Need	Connect
ANALYZE THE PROBLEM	glyceryl tripalmitoleate (tripalmitolein)	condensed structural formula	ester bonds of glycerol and three fatty acids

STEP 1 Draw the condensed structural formulas for glycerol and the fatty acids.

$$
\begin{array}{ll}
CH_2{-}OH + HO{-}\overset{\overset{\displaystyle O}{\|}}{C}{-}(CH_2)_7{-}CH{=}CH{-}(CH_2)_5{-}CH_3 & \text{Palmitoleic acid} \\
| & \\
CH{-}OH + HO{-}\overset{\overset{\displaystyle O}{\|}}{C}{-}(CH_2)_7{-}CH{=}CH{-}(CH_2)_5{-}CH_3 & \text{Palmitoleic acid} \\
| & \\
CH_2{-}OH + HO{-}\overset{\overset{\displaystyle O}{\|}}{C}{-}(CH_2)_7{-}CH{=}CH{-}(CH_2)_5{-}CH_3 & \text{Palmitoleic acid} \\
\text{Glycerol} &
\end{array}
$$

STEP 2 Form ester bonds between the hydroxyl groups on glycerol and the carboxyl groups on each fatty acid.

$$\begin{array}{l} CH_2-O-\overset{\overset{\displaystyle O}{\|}}{C}-(CH_2)_7-CH=CH-(CH_2)_5-CH_3 \\ | \\ CH-O-\overset{\overset{\displaystyle O}{\|}}{C}-(CH_2)_7-CH=CH-(CH_2)_5-CH_3 \\ | \\ CH_2-O-\overset{\overset{\displaystyle O}{\|}}{C}-(CH_2)_7-CH=CH-(CH_2)_5-CH_3 \end{array}$$

Glyceryl tripalmitoleate (tripalmitolein)

TEST

Try Practice Problems 15.21 to 15.24

STUDY CHECK 15.2

Draw the line-angle formula for the triacylglycerol containing three molecules of myristic acid (14:0).

ANSWER

$$\begin{array}{l} CH_2-O-\overset{\overset{\displaystyle O}{\|}}{C}\text{(line-angle chain)} \\ | \\ CH-O-\overset{\overset{\displaystyle O}{\|}}{C}\text{(line-angle chain)} \\ | \\ CH_2-O-\overset{\overset{\displaystyle O}{\|}}{C}\text{(line-angle chain)} \end{array}$$

Triacylglycerols are the major form of energy storage for animals. Animals that hibernate eat large quantities of plants, seeds, and nuts that are high in calories. Prior to hibernation, these animals, such as polar bears, gain as much as 14 kg per week. As the external temperature drops, the animal goes into hibernation. The body temperature drops to nearly freezing, and cellular activity, respiration, and heart rate are drastically reduced. Animals that live in extremely cold climates hibernate for 4 to 7 months. During this time, stored fat is the only source of energy.

Prior to hibernation, a polar bear eats food with a high caloric content.

Melting Points of Fats and Oils

A **fat** is a triacylglycerol that is solid at room temperature, and it usually comes from animal sources such as meat, whole milk, butter, and cheese. An **oil** is a triacylglycerol that is usually a liquid at room temperature and is obtained from a plant source. Olive oil and peanut oil are monounsaturated because they contain large amounts of oleic acid. Oils from corn, cottonseed, safflower seed, and sunflower seed are polyunsaturated because they contain large amounts of fatty acids with two or more double bonds (see **FIGURE 15.3**).

Saturated fatty acids have higher melting points than unsaturated fatty acids because they pack together more tightly. Animal fats usually contain more saturated fatty acids than do vegetable oils. Therefore, the melting points of animal fats are higher than those of vegetable oils.

Palm oil and coconut oil are solids at room temperature because they contain large amounts of saturated fatty acids. Coconut oil is 92% saturated fats, about half of which is lauric acid, which has 12 carbon atoms rather than 18 carbon atoms found in the stearic acid of animal sources. Thus coconut oil has a melting point that is higher than typical vegetable oils, but not as high as fats from animal sources that contain stearic acid. The percentages of saturated, monounsaturated, and polyunsaturated fatty acids in some typical fats and oils are listed in **FIGURE 15.4**.

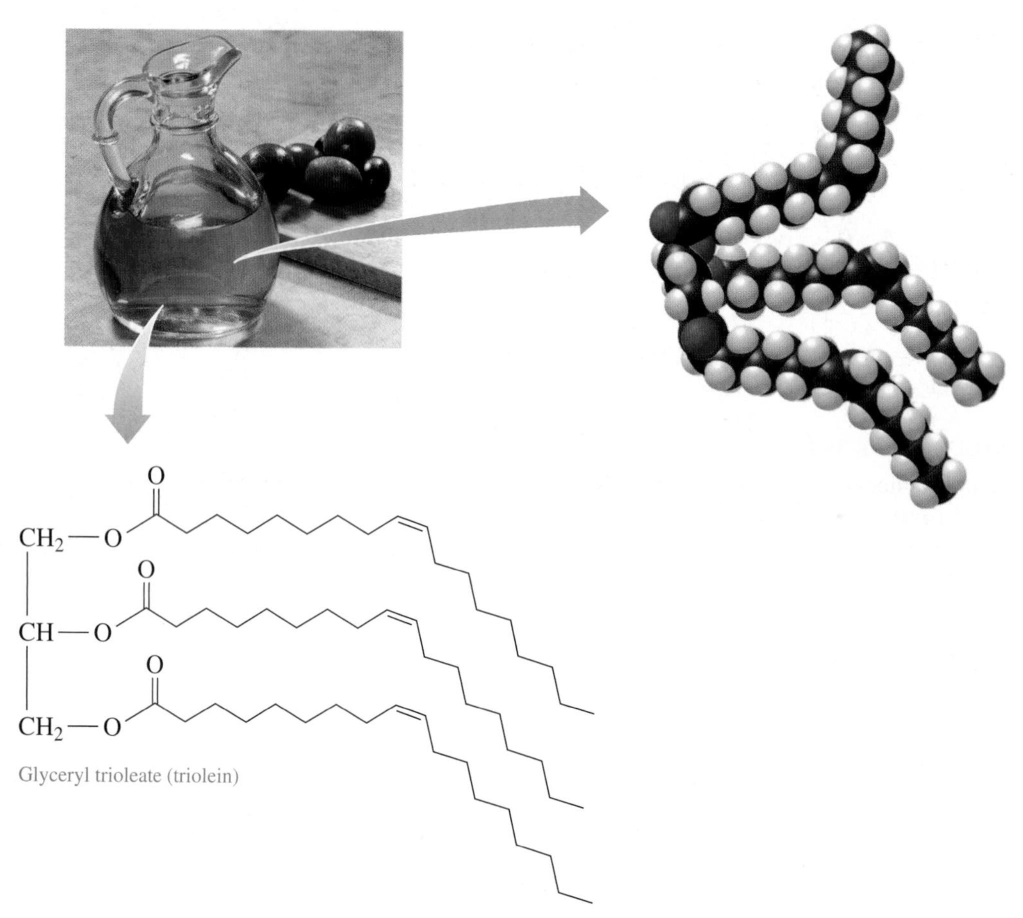

FIGURE 15.3 Vegetable oils such as olive oil, corn oil, and safflower oil contain unsaturated fats.
Q Why is olive oil a liquid at room temperature?

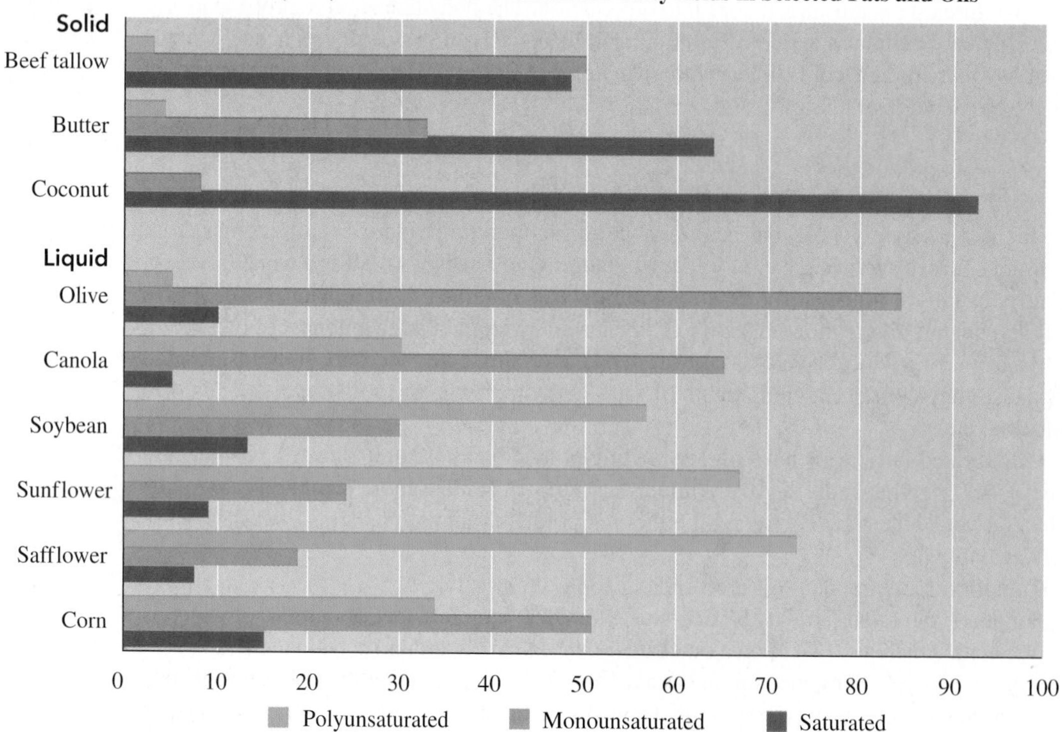

TEST

Try Practice Problems 15.25 to 15.28

FIGURE 15.4 Vegetable oils are liquids at room temperature because they have a higher percentage of unsaturated fatty acids than do animal fats.
Q Why is butter a solid at room temperature, whereas canola oil is a liquid?

PRACTICE PROBLEMS

15.3 Waxes and Triacylglycerols

LEARNING GOAL Draw the condensed structural or line-angle formula for a wax or triacylglycerol produced by the reaction of a fatty acid and an alcohol or glycerol.

15.19 Draw the condensed structural formula for the ester in beeswax that is formed from myricyl alcohol, $CH_3—(CH_2)_{29}—OH$, and palmitic acid.

15.20 Draw the condensed structural formula for the ester in jojoba wax that is formed from arachidic acid, a 20-carbon saturated fatty acid, and 1-docosanol, $CH_3—(CH_2)_{21}—OH$.

15.21 Draw the condensed structural formula for a triacylglycerol that contains stearic acid and glycerol.

15.22 Draw the condensed structural formula for a mixed triacylglycerol that contains two palmitic acid molecules and one oleic acid molecule on the center carbon of glycerol.

Clinical Applications

15.23 Caprylic acid is an 8-carbon saturated fatty acid that occurs in coconut oil (10%) and palm kernel oil (4%). Draw the line-angle formula for glyceryl tricaprylate (tricaprylin).

15.24 Linoleic acid is a 18-carbon unsaturated fatty acid with two double bonds that occurs in many vegetable oils including safflower oil (75%) and poppyseed oil (75%). Draw the line-angle formula for glyceryl trilinoleate (trilinolein).

15.25 Safflower oil is polyunsaturated, whereas olive oil is monounsaturated. Why would safflower oil have a lower melting point than olive oil?

15.26 Olive oil is monounsaturated, whereas butter fat is saturated. Why does olive oil have a lower melting point than butter fat?

15.27 How does the percentage of monounsaturated and polyunsaturated fatty acids in sunflower oil compare to that of safflower oil?

15.28 How does the percentage of monounsaturated and polyunsaturated fatty acids in olive oil compare to that of canola oil?

15.4 Chemical Properties of Triacylglycerols

LEARNING GOAL Draw the condensed structural or line-angle formula for the products of a triacylglycerol that undergoes hydrogenation, hydrolysis, or saponification.

REVIEW

Writing Equations for Hydrogenation and Hydration (11.7)

Hydrolyzing Esters (14.4)

The chemical reactions of triacylglycerols involve the hydrogenation of the double bonds in the fatty acids, and the hydrolysis and saponification of the ester bonds between glycerol and the fatty acids.

Hydrogenation

In the **hydrogenation** reaction, hydrogen gas is bubbled through the heated oil typically in the presence of a nickel catalyst. As a result, H atoms add to one or more carbon–carbon double bonds to form carbon–carbon single bonds.

$$—CH{=}CH— + \mathbf{H_2} \xrightarrow{Ni} —\overset{\mathbf{H}}{\underset{H}{C}}—\overset{\mathbf{H}}{\underset{H}{C}}—$$

For example, when hydrogen adds to all of the double bonds of glyceryl trioleate (triolein) using a nickel catalyst, the product is the saturated fat glyceryl tristearate (tristearin).

$$\begin{array}{l} CH_2—O—\overset{O}{\overset{\|}{C}}—(CH_2)_7—CH{=}CH—(CH_2)_7—CH_3 \\ | \\ CH—O—\overset{O}{\overset{\|}{C}}—(CH_2)_7—CH{=}CH—(CH_2)_7—CH_3 \\ | \\ CH_2—O—\overset{O}{\overset{\|}{C}}—(CH_2)_7—CH{=}CH—(CH_2)_7—CH_3 \end{array} + 3H_2 \xrightarrow{Ni} \begin{array}{l} CH_2—O—\overset{O}{\overset{\|}{C}}—(CH_2)_7—CH_2—CH_2—(CH_2)_7—CH_3 \\ | \\ CH—O—\overset{O}{\overset{\|}{C}}—(CH_2)_7—CH_2—CH_2—(CH_2)_7—CH_3 \\ | \\ CH_2—O—\overset{O}{\overset{\|}{C}}—(CH_2)_7—CH_2—CH_2—(CH_2)_7—CH_3 \end{array}$$

Double bonds (arrow to CH=CH) — Single bonds (arrow to $CH_2—CH_2$)

Glyceryl trioleate (triolein) — Glyceryl tristearate (tristearin)

CORE CHEMISTRY SKILL
Drawing the Products for the Hydrogenation, Hydrolysis, and Saponification of a Triacylglycerol

In commercial hydrogenation, the addition of hydrogen is stopped before all the double bonds in a liquid vegetable oil become saturated. Complete hydrogenation gives a brittle product, whereas the partial hydrogenation of a liquid vegetable oil changes it to a soft, semisolid fat. By controlling the amount of hydrogen, manufacturers can produce various types of products such as soft margarines, solid stick margarines, and solid shortenings (see **FIGURE 15.5**). Although these products now contain more saturated fatty acids than the original oils, they contain no cholesterol, unlike similar products from animal sources, such as butter and lard.

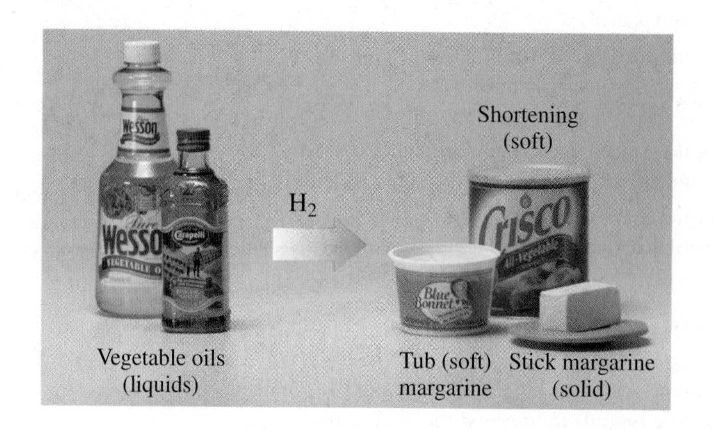

FIGURE 15.5 Many soft margarines, stick margarines, and solid shortenings are produced by the partial hydrogenation of vegetable oils.

Q How does hydrogenation change the structure of the fatty acids in the vegetable oils?

CHEMISTRY LINK TO HEALTH

Converting Unsaturated Fats to Saturated Fats: Hydrogenation

During the early 1900s, margarine became a popular replacement for the highly saturated fats such as butter and lard. Margarine is produced by partially hydrogenating the unsaturated fats in vegetable oils such as safflower oil, corn oil, canola oil, cottonseed oil, and sunflower oil.

Hydrogenation and Trans Fats

In vegetable oils, the unsaturated fats usually contain cis double bonds. As hydrogenation occurs, double bonds are converted to single bonds. However, a small amount of the cis double bonds are converted to trans double bonds because they are more stable, which causes a change in the overall structure of the fatty acids. If the label on a product states that the oils have been "partially hydrogenated," that product will also contain trans fatty acids. In the United States, it is estimated that 2 to 4% of our total calories comes from trans fatty acids.

The concern about trans fatty acids is that their altered structure may make them behave like saturated fatty acids in the body. In the 1980s, research indicated that trans fatty acids have an effect on blood cholesterol similar to that of saturated fats, although study results vary. Several studies reported that trans fatty acids raise the levels of LDL-cholesterol and lower the levels of HDL-cholesterol. (LDL and HDL are described in Section 15.6).

cis-Fatty acid

H_2/Ni

Ni catalyst

H_2 Isomerization

Addition of H_2

Undesired side product (*trans*-fatty acid)

Desired saturated product

Foods containing naturally occurring trans fats include milk, beef, and eggs. Foods that contain trans fatty acids from the hydrogenation process include fried foods, bread, baked goods, cookies, crackers, chips, stick and soft margarines, and vegetable shortening. The American Heart Association recommends that margarine should have no more than 2 g of saturated fat per tablespoon and a liquid vegetable oil should be the first ingredient. They also recommend the use of soft margarine, which is only slightly hydrogenated and therefore has fewer trans fatty acids. Currently, the amount of trans fats is included on the Nutrition Facts label on food products. In the United States, a food label for a product that has less than 0.5 g of trans fat in one serving can read "0 g of trans fat."

The best advice may be to reduce total fat in the diet by using fats and oils sparingly, cooking with little or no fat, substituting olive oil or canola oil for other oils, and limiting the use of coconut oil and palm oil, which are high in saturated fatty acids.

Nutrition Facts	
Serving Size 5 oz. (144g)	
Servings Per Container 4	
Amount Per Serving	
Calories 310	Calories from Fat 100
	% Daily Value*
Total Fat 15g	21%
Saturated Fat 2.6g	17%
Trans Fat 1g	
Cholesterol 118mg	39%
Sodium 560mg	28%
Total Carbohydrate 12g	4%
Dietary Fiber 1g	4%
Sugars 1g	
Protein 24g	

Nutrition Facts label includes the grams of trans fat in one serving.

Hydrolysis

Triacylglycerols are hydrolyzed (split by water) in the presence of strong acids such as HCl or H_2SO_4, or digestive enzymes called *lipases*. The products of hydrolysis of the ester bonds are glycerol and three fatty acids. The polar glycerol is soluble in water, but the fatty acids with their long hydrocarbon chains are not.

Water adds to ester bonds

$$\begin{array}{l} CH_2-O-\overset{O}{\overset{\|}{C}}-(CH_2)_{14}-CH_3 \\ | \\ CH-O-\overset{O}{\overset{\|}{C}}-(CH_2)_{14}-CH_3 + \mathbf{3H_2O} \\ | \\ CH_2-O-\overset{O}{\overset{\|}{C}}-(CH_2)_{14}-CH_3 \end{array} \xrightarrow[\text{lipase}]{H^+ \text{ or}} \begin{array}{l} CH_2-\mathbf{OH} \quad \mathbf{HO}-\overset{O}{\overset{\|}{C}}-(CH_2)_{14}-CH_3 \\ | \\ CH-\mathbf{OH} + \mathbf{HO}-\overset{O}{\overset{\|}{C}}-(CH_2)_{14}-CH_3 \\ | \\ CH_2-\mathbf{OH} \quad \mathbf{HO}-\overset{O}{\overset{\|}{C}}-(CH_2)_{14}-CH_3 \end{array}$$

Glyceryl tripalmitate (tripalmitin) — Glycerol — 3 Palmitic acid molecules

Saponification

Saponification occurs when a fat is heated with a strong base such as NaOH to form glycerol and the sodium salts of the fatty acids, which is soap. When NaOH is used, a solid soap is produced that can be molded into a desired shape; KOH produces a softer, liquid soap. An oil that is polyunsaturated produces a softer soap. Names like "coconut" or "avocado shampoo" tell you the sources of the oil used in the reaction.

$$\text{Fat or oil} + \text{strong base} \xrightarrow{\text{Heat}} \text{glycerol} + \text{salts of fatty acids (soap)}$$

$$\begin{array}{l} CH_2-O-\overset{O}{\overset{\|}{C}}-(CH_2)_{14}-CH_3 \\ | \\ CH-O-\overset{O}{\overset{\|}{C}}-(CH_2)_{14}-CH_3 + \mathbf{3NaOH} \\ | \\ CH_2-O-\overset{O}{\overset{\|}{C}}-(CH_2)_{14}-CH_3 \end{array} \xrightarrow{\text{Heat}} \begin{array}{l} CH_2-\mathbf{OH} \quad \mathbf{Na^+\ ^-O}-\overset{O}{\overset{\|}{C}}-(CH_2)_{14}-CH_3 \\ | \\ CH-\mathbf{OH} + \mathbf{Na^+\ ^-O}-\overset{O}{\overset{\|}{C}}-(CH_2)_{14}-CH_3 \\ | \\ CH_2-\mathbf{OH} \quad \mathbf{Na^+\ ^-O}-\overset{O}{\overset{\|}{C}}-(CH_2)_{14}-CH_3 \end{array}$$

Glyceryl tripalmitate (tripalmitin) — Glycerol — 3 Sodium palmitate (soap)

The reactions for fatty acids and triacylglycerols that are similar to the reactions of hydrogenation of alkenes, esterification, hydrolysis, and saponification are summarized in TABLE 15.3.

TABLE 15.3 Summary of Organic and Lipid Reactions

Reaction	Organic Reactants and Products	Lipid Reactants and Products
Esterification	Carboxylic acid + alcohol $\xrightarrow{H^+, \text{heat}}$ ester + water	3 Fatty acids + glycerol $\xrightarrow{\text{Enzyme}}$ triacylglycerol (fat) + 3 water
Hydrogenation	Alkene (double bond) + hydrogen $\xrightarrow{Pt}$ alkane (single bonds)	Unsaturated fat (double bonds) + hydrogen $\xrightarrow{Ni}$ saturated fat (single bonds)
Hydrolysis	Ester + water $\xrightarrow{H^+, \text{heat}}$ carboxylic acid + alcohol	Triacylglycerol (fat) + 3 water $\xrightarrow{\text{Enzyme}}$ 3 fatty acids + glycerol
Saponification	Ester + sodium hydroxide $\xrightarrow{\text{Heat}}$ sodium salt of carboxylic acid + alcohol	Triacylglycerol (fat) + 3 sodium hydroxide $\xrightarrow{\text{Heat}}$ 3 sodium salts of fatty acid (soap) + glycerol

SAMPLE PROBLEM 15.3 Reactions of Lipids During Digestion

TRY IT FIRST

Write the equation for the reaction catalyzed by lipase enzymes that hydrolyze glyceryl trilaurate (trilaurin) during the digestion process.

SOLUTION

ANALYZE THE PROBLEM	Given	Need	Connect
	glyceryl trilaurate, lipases	chemical equation for hydrolysis	products: glycerol and three fatty acids

$$\begin{array}{l} CH_2-O-\overset{\overset{O}{\|}}{C}-(CH_2)_{10}-CH_3 \\ | \\ CH-O-\overset{\overset{O}{\|}}{C}-(CH_2)_{10}-CH_3 \\ | \\ CH_2-O-\overset{\overset{O}{\|}}{C}-(CH_2)_{10}-CH_3 \end{array} + 3H_2O \xrightarrow{\text{Lipase}} \begin{array}{l} CH_2-OH \\ | \\ CH-OH \\ | \\ CH_2-OH \end{array} + \begin{array}{l} HO-\overset{\overset{O}{\|}}{C}-(CH_2)_{10}-CH_3 \\ HO-\overset{\overset{O}{\|}}{C}-(CH_2)_{10}-CH_3 \\ HO-\overset{\overset{O}{\|}}{C}-(CH_2)_{10}-CH_3 \end{array}$$

Glyceryl trilaurate (trilaurin) — Glycerol — 3 Lauric acid molecules

STUDY CHECK 15.3

What is the name of the product formed when a triacylglycerol containing oleic acid (18:1) and linoleic acid (18:2) is completely hydrogenated?

ANSWER

glyceryl tristearate (tristearin)

TEST

Try Practice Problems 15.29 to 15.38

PRACTICE PROBLEMS

15.4 Chemical Properties of Triacylglycerols

LEARNING GOAL Draw the condensed structural or line-angle formula for the products of a triacylglycerol that undergoes hydrogenation, hydrolysis, or saponification.

15.29 Identify each of the following processes as hydrogenation, hydrolysis, or saponification and give the products:

a. the reaction of palm oil with KOH

b. the reaction of glyceryl trilinoleate from safflower oil with water and HCl

15.30 Identify each of the following processes as hydrogenation, hydrolysis, or saponification and give the products:

a. the reaction of corn oil and hydrogen (H_2) with a nickel catalyst

b. the reaction of glyceryl tristearate with water in the presence of lipase enzyme

15.31 Use condensed structural formulas to write the balanced chemical equation for the hydrogenation of glyceryl tripalmitoleate, a fat containing glycerol and three palmitoleic acid molecules.

15.32 Use condensed structural formulas to write the balanced chemical equation for the hydrogenation of glyceryl trilinolenate, a fat containing glycerol and three linolenic acid molecules.

15.33 Use line-angle formulas to write the balanced chemical equation for the acid hydrolysis of glyceryl trimyristate (trimyristin).

15.34 Use line-angle formulas to write the balanced chemical equation for the acid hydrolysis of glyceryl trioleate (triolein).

15.35 Use condensed structural formulas to write the balanced chemical equation for the NaOH saponification of glyceryl trimyristate (trimyristin).

15.36 Use condensed structural formulas to write the balanced chemical equation for the NaOH saponification of glyceryl trioleate (triolein).

15.37 Draw the condensed structural formula for the product of the hydrogenation of the following triacylglycerol:

$$CH_2-O-\overset{O}{\overset{\|}{C}}-(CH_2)_{16}-CH_3$$
$$CH-O-\overset{O}{\overset{\|}{C}}-(CH_2)_7-CH{=}CH-(CH_2)_7-CH_3$$
$$CH_2-O-\overset{O}{\overset{\|}{C}}-(CH_2)_{16}-CH_3$$

15.38 Draw the condensed structural formulas for all the products that would be obtained when the triacylglycerol in problem 15.37 undergoes complete hydrolysis.

15.5 Phospholipids

LEARNING GOAL Draw the structure of a phospholipid containing glycerol or sphingosine.

The **phospholipids** are a family of lipids similar in structure to triacylglycerols; they include glycerophospholipids and sphingomyelins. In a **glycerophospholipid**, two fatty acids form ester bonds with the first and second hydroxyl groups of glycerol. The third hydroxyl group forms an ester with phosphoric acid, which forms another phosphoester bond with an amino alcohol. In a *sphingomyelin*, sphingosine replaces glycerol. We can compare the general structures of a triacylglycerol, a glycerophospholipid, and a sphingomyelin as follows:

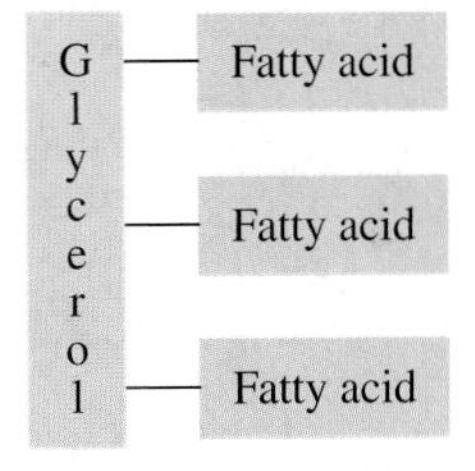

Triacylglycerol

Glycerol — Fatty acid, Fatty acid, PO_4 — Amino alcohol

Glycerophospholipid

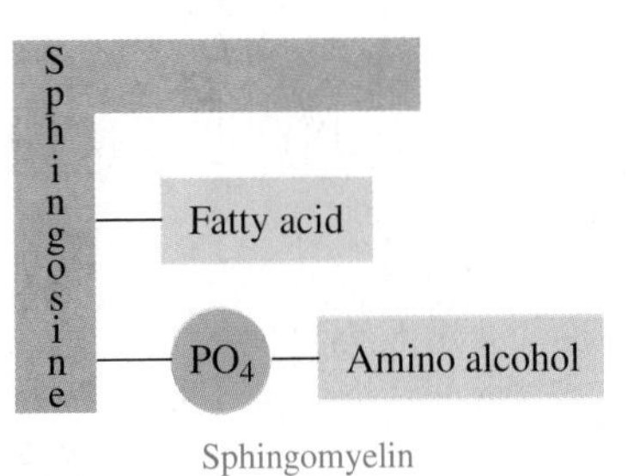

Sphingomyelin

Amino Alcohols

Three amino alcohols found in glycerophospholipids are choline, serine, and ethanolamine. In the body, at a physiological pH of 7.4, these amino alcohols are ionized.

$$HO-CH_2-CH_2-\overset{+}{N}(CH_3)_2-CH_3$$

Choline

$$HO-CH_2-\underset{}{CH}(\overset{+}{N}H_3)-\overset{O}{\overset{\|}{C}}-O^-$$

Serine

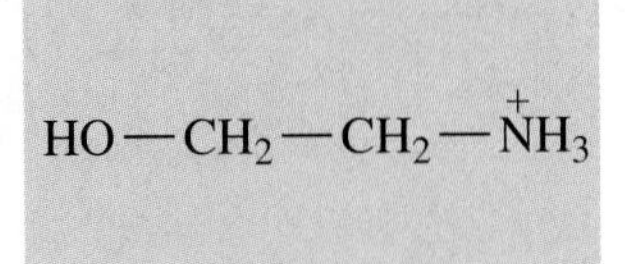

Ethanolamine

Lecithins and *cephalins* are two types of glycerophospholipids that are particularly abundant in brain and nerve tissue as well as in egg yolk, wheat germ, and yeast. Lecithins contain choline, and cephalins usually contain ethanolamine and sometimes serine. In the following structural formulas, the fatty acid that is used in each example is palmitic acid:

TEST

Try Practice Problems 15.39 and 15.40

$$\begin{array}{l} CH_2-O-C(=O)-(CH_2)_{14}-CH_3 \\ CH-O-C(=O)-(CH_2)_{14}-CH_3 \\ CH_2-O-P(=O)(O^-)-O-CH_2-CH_2-\overset{+}{N}(CH_3)_2-CH_3 \end{array}$$

Nonpolar fatty acids

Polar

Choline

A lecithin

Phosphatidylcholine with two palmitic fatty acids (16:0)

$$\begin{array}{l} CH_2-O-C(=O)-(CH_2)_{14}-CH_3 \\ CH-O-C(=O)-(CH_2)_{14}-CH_3 \\ CH_2-O-P(=O)(O^-)-O-CH_2-CH_2-\overset{+}{N}H_3 \end{array}$$

Ethanolamine

A cephalin

Phosphatidylethanolamine with two palmitic fatty acids (16:0)

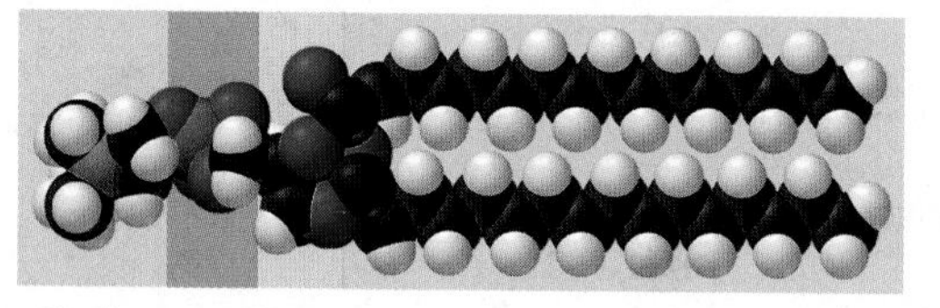

Glycerophospholipids contain both polar and nonpolar regions, which allow them to interact with both polar and nonpolar substances. The ionized amino alcohol and phosphate portion, called "the head," is polar and strongly attracted to water (see **FIGURE 15.6**). The hydrocarbon chains of the two fatty acids are the nonpolar "tails" of the glycerophospholipid, which are soluble in other nonpolar substances, mostly lipids.

$$CH_3-\overset{+}{N}(CH_3)_2-CH_2-CH_2-OH \quad HO-P(=O)(O^-)-OH \quad HO-CH_2-CH(OH)-CH_2-OH \quad HO-C(=O)-R$$

Choline

Phosphoric acid

Glycerol

Fatty acids

(a) Components of a typical glycerophospholipid

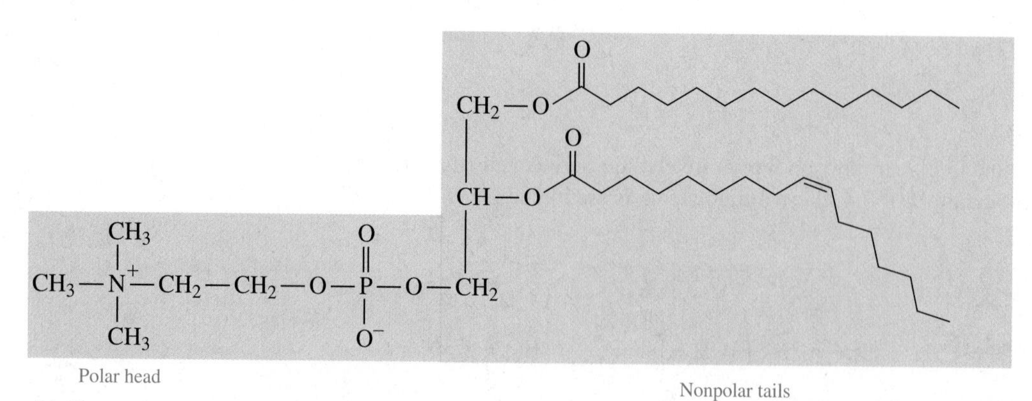

(b) Glycerophospholipid

(c) Simplified way to draw a glycerophospholipid

FIGURE 15.6 **(a)** The components of a typical glycerophospholipid: an amino alcohol, phosphoric acid, glycerol, and two fatty acids. **(b)** In a glycerophospholipid, a polar "head" contains the ionized amino alcohol and phosphate, while the hydrocarbon chains of two fatty acids make up the nonpolar "tails." **(c)** A simplified drawing indicates the polar and nonpolar regions.

Q Why are glycerophospholipids polar?

Snake venom is produced by the modified saliva glands of poisonous snakes. When a snake bites, venom is ejected through the fang of the snake. The venom of the eastern diamondback rattlesnake and the Indian cobra contains *phospholipases*, which are enzymes that catalyze the hydrolysis of the fatty acid on the center carbon of glycerophospholipids in the red blood cells. The resulting *lysophospholipids*, cause breakdown of the red blood cell membranes. This makes them permeable to water, which causes hemolysis of the red blood cells.

Poisonous snake venom contains phospholipases that hydrolyze phospholipids in red blood cells.

SAMPLE PROBLEM 15.4 Drawing Glycerophospholipid Structures

TRY IT FIRST

Draw the condensed structural formula for the cephalin that contains glycerol, two stearic acids (18:0), phosphate, and serine (ionized).

SOLUTION

	Given	Need	Connect
ANALYZE THE PROBLEM	cephalin, two stearic acids (18:0), phosphate, serine (ionized)	condensed structural formula	ester bonds with fatty acids, phosphoester bond with amino alcohol

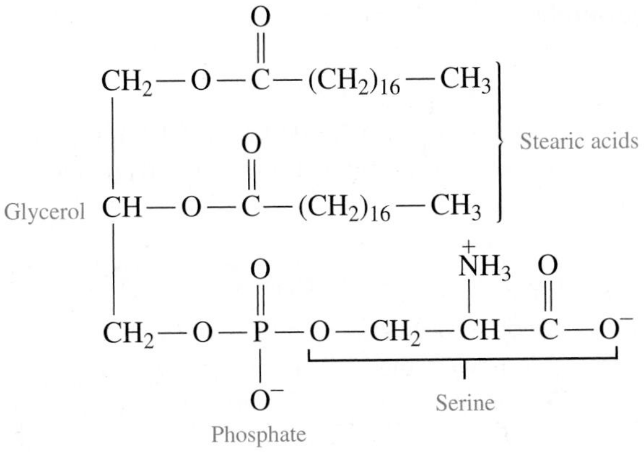

$$
\begin{array}{l}
\text{Glycerol} \\
CH_2-O-\overset{\overset{\displaystyle O}{\|}}{C}-(CH_2)_{16}-CH_3 \quad \text{(Stearic acids)} \\
\;| \\
CH-O-\overset{\overset{\displaystyle O}{\|}}{C}-(CH_2)_{16}-CH_3 \\
\;| \\
CH_2-O-\underset{\underset{\displaystyle O^-}{|}}{\overset{\overset{\displaystyle O}{\|}}{P}}-\underbrace{O-CH_2-\overset{\overset{\displaystyle \overset{+}{N}H_3}{|}}{CH}-\overset{\overset{\displaystyle O}{\|}}{C}-O^-}_{\text{Serine}}
\end{array}
$$

Phosphate

STUDY CHECK 15.4

Draw the condensed structural formula for the lecithin that contains glycerol, two myristic acids (14:0), phosphate, and choline (ionized).

ANSWER

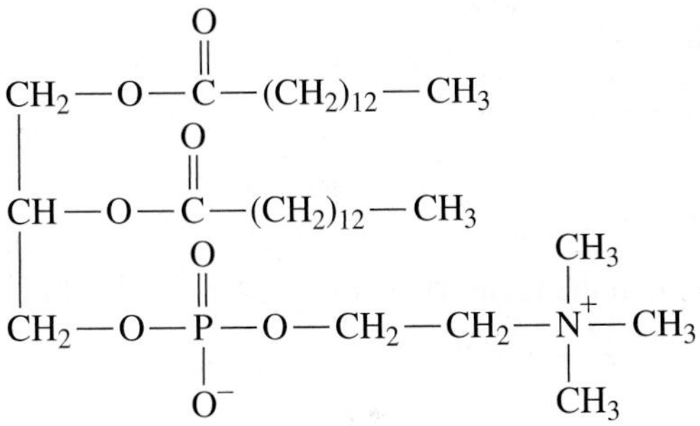

$$
\begin{array}{l}
CH_2-O-\overset{\overset{\displaystyle O}{\|}}{C}-(CH_2)_{12}-CH_3 \\
\;| \\
CH-O-\overset{\overset{\displaystyle O}{\|}}{C}-(CH_2)_{12}-CH_3 \\
\;| \\
CH_2-O-\underset{\underset{\displaystyle O^-}{|}}{\overset{\overset{\displaystyle O}{\|}}{P}}-O-CH_2-CH_2-\underset{\underset{\displaystyle CH_3}{|}}{\overset{\overset{\displaystyle CH_3}{|}}{\overset{+}{N}}}-CH_3
\end{array}
$$

Sphingosine, found in sphingomyelins and other sphingolipids, is a long-chain amino alcohol. In a **sphingomyelin**, the amine group of sphingosine forms an amide bond to a fatty acid and the hydroxyl group forms an ester bond with phosphate, which forms another phosphoester bond to choline or ethanolamine. The sphingomyelins are abundant in the white matter of the myelin sheath, a coating surrounding the nerve cells that increases the speed of nerve impulses and insulates and protects the nerve cells.

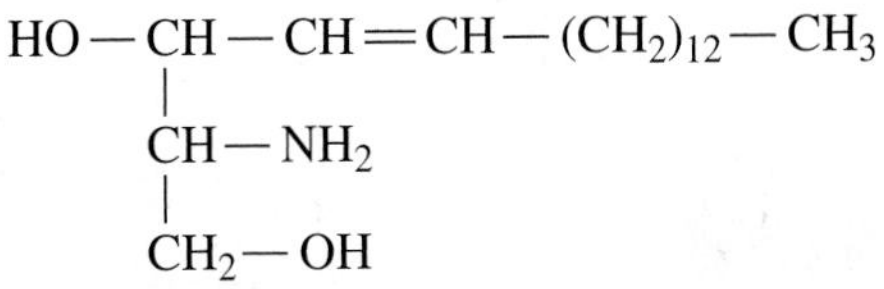

$$
\begin{array}{l}
HO-CH-CH=CH-(CH_2)_{12}-CH_3 \\
\quad\;\; | \\
\quad\; CH-NH_2 \\
\quad\;\; | \\
\quad\; CH_2-OH
\end{array}
$$

Sphingosine

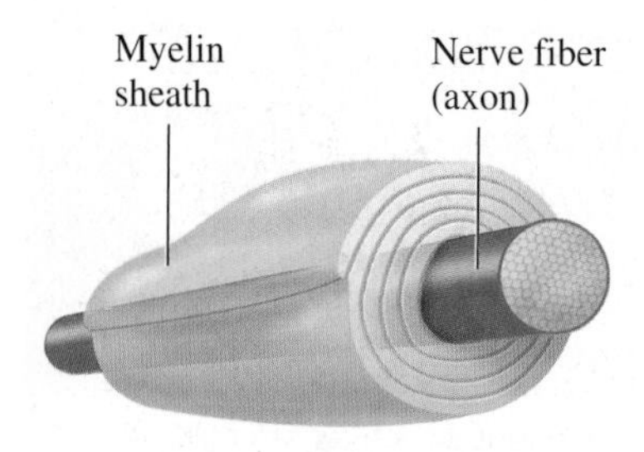

Normal myelin sheath

Damaged myelin sheath

When the myelin sheath loses sphingomyelin, it deteriorates and the transmission of nerve signals is impaired.

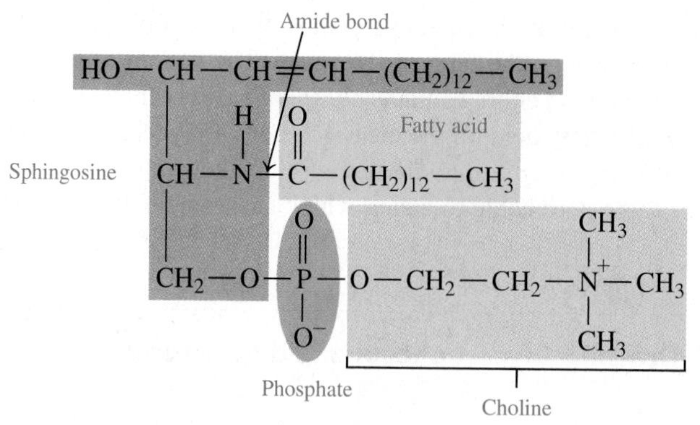

A sphingomyelin containing sphingosine, myristic acid, and choline

In multiple sclerosis, sphingomyelin is lost from the myelin sheath. As the disease progresses, the myelin sheath deteriorates. Scars form on the neurons and impair the transmission of nerve signals. The symptoms of multiple sclerosis include various levels of muscle weakness with loss of coordination and vision, depending on the amount of damage. The cause of multiple sclerosis is not known, although some researchers suggest that a virus is involved. Several studies also suggest that adequate levels of vitamin D may lessen the severity or lower the risk of developing multiple sclerosis.

ENGAGE

How is the polarity of a sphingomyelin different from that of a triacylglycerol?

SAMPLE PROBLEM 15.5 Sphingomyelin

TRY IT FIRST

A sphingomyelin found in eggs contains sphingosine, palmitic acid (16:0), phosphate, and choline (ionized). Draw the condensed structural formula for this sphingomyelin.

SOLUTION

	Given	Need	Connect
ANALYZE THE PROBLEM	sphingosine, palmitic acid (16:0), phosphate, choline (ionized)	condensed structural formula of the sphingomyelin	amide bond with fatty acid, phosphoester bond with amino alcohol

$HO-CH-CH=CH-(CH_2)_{12}-CH_3$ (Sphingosine)

$CH-NH-C(=O)-(CH_2)_{14}-CH_3$ (Palmitic acid)

$CH_2-O-P(=O)(O^-)-O-CH_2-CH_2-\overset{+}{N}(CH_3)_3$ (Phosphate, Choline)

Sphingosine — Palmitic acid — Phosphate — Choline

STUDY CHECK 15.5

Stearic acid (18:0) is found in sphingomyelin in the brain. Draw the condensed structural formula for this sphingomyelin using ethanolamine (ionized).

ANSWER

$HO-CH-CH=CH-(CH_2)_{12}-CH_3$

$CH-NH-C(=O)-(CH_2)_{16}-CH_3$

$CH_2-O-P(=O)(O^-)-O-CH_2-CH_2-\overset{+}{N}H_3$

TEST

Try Practice Problems 15.41 to 15.46

CHEMISTRY LINK TO HEALTH

Infant Respiratory Distress Syndrome (IRDS)

When an infant is born, an important key to its survival is proper lung function. In the lungs, there are many tiny air sacs called *alveoli*, where the exchange of O_2 and CO_2 takes place. Upon birth of a mature infant, surfactant is released into the lung tissues where it lowers the surface tension in the alveoli, which helps the air sacs inflate. The production of a *pulmonary surfactant*, which is a mixture of phospholipids including lecithin and sphingomyelin produced by specific lung cells, occurs in a fetus after 24 to 28 weeks of pregnancy. If an infant is born before 28 weeks of gestation, the low level of surfactant and immature lung development lead to a high risk of *infant respiratory distress syndrome* (IRDS). Without sufficient surfactant, the air sacs collapse and have to reopen with each breath. As a result, alveoli cells are damaged, less oxygen is taken in, and more carbon dioxide is retained, which can lead to hypoxia and acidosis.

One way to determine the maturity of the lungs of a fetus is to measure the *lecithin–sphingomyelin (L/S) ratio*. A ratio of 2.5 indicates mature fetal lung function, an L/S ratio of 2.4 to 1.6 indicates a low risk, and a ratio of less than 1.5 indicates a high risk of IRDS. Before the initiation of an early delivery, the L/S ratio of the amniotic fluid is measured. If the L/S ratio is low, steroids may be given to the mother to assist the lung development and production of surfactant in the fetus. Once a premature infant is born, treatment includes the use of steroids to help maturation of the lungs, the application of surfactants, and the administration of supplemental oxygen with ventilation to help minimize damage to the lungs.

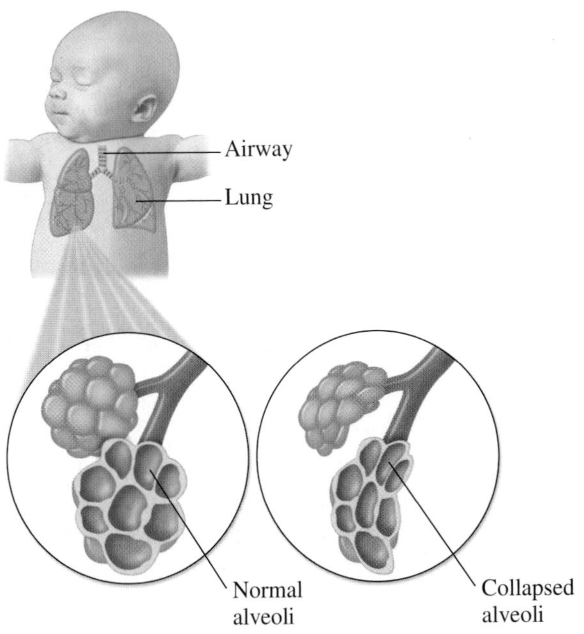

Without sufficient surfactant in the lungs of a premature infant, the alveoli collapse, which decreases pulmonary function.

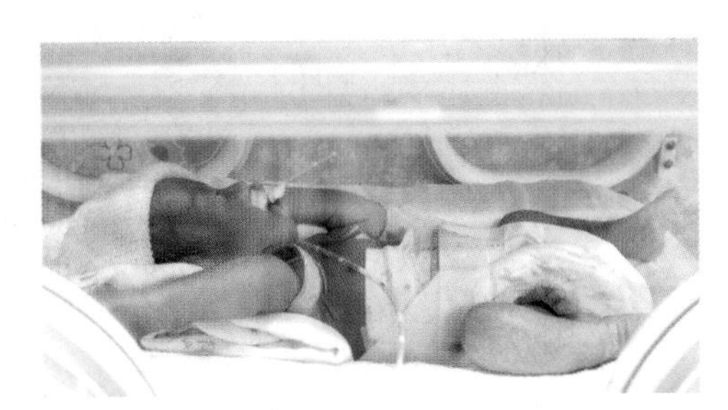

A premature infant with respiratory distress is treated with a surfactant and oxygen.

PRACTICE PROBLEMS

15.5 Phospholipids

LEARNING GOAL Draw the structure of a phospholipid containing glycerol or sphingosine.

15.39 Describe the similarities and differences between triacylglycerols and glycerophospholipids.

15.40 Describe the similarities and differences between lecithins and cephalins.

15.41 Draw the condensed structural formula for the cephalin that contains glycerol, two palmitic acids, phosphate, and ethanolamine (ionized).

15.42 Draw the condensed structural formula for the lecithin that contains glycerol, two palmitic acids, phosphate, and choline (ionized).

Clinical Applications

15.43 Identify the following glycerophospholipid, which is found in the nerves and spinal cord in the body, as a lecithin or cephalin, and list its components:

$$CH_2-O-\overset{\overset{O}{\|}}{C}-(CH_2)_7-CH{=}CH-(CH_2)_7-CH_3$$
$$CH-O-\overset{\overset{O}{\|}}{C}-(CH_2)_{16}-CH_3$$
$$CH_2-O-\overset{\overset{O}{\|}}{\underset{\underset{O^-}{|}}{P}}-O-CH_2-CH_2-\overset{+}{N}H_3$$

15.44 Identify the following glycerophospholipid, which helps conduct nerve impulses in the body, as a lecithin or cephalin, and list its components:

$$CH_2-O-\overset{\overset{O}{\|}}{C}-(CH_2)_{14}-CH_3$$
$$CH-O-\overset{\overset{O}{\|}}{C}-(CH_2)_{16}-CH_3$$
$$CH_2-O-\overset{\overset{O}{\|}}{\underset{\underset{O^-}{|}}{P}}-O-CH_2-CH_2-\overset{\overset{CH_3}{|}}{\underset{\underset{CH_3}{|}}{\overset{+}{N}}}-CH_3$$

15.45 Identify the following features of this phospholipid, which is abundant in the myelin sheath that surrounds nerve cells:

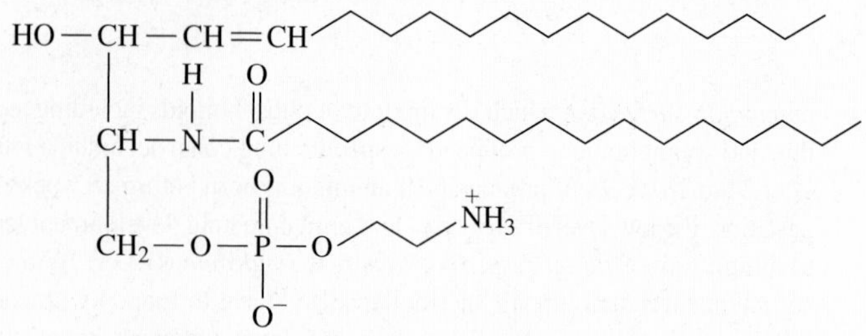

HO—CH—CH=CH

CH—N(H)—C(=O)

CH_2—O—P(=O)(O$^-$)—O—…$\overset{+}{N}H_3$

a. Is the phospholipid formed from glycerol or sphingosine?
b. What is the fatty acid?
c. What type of bond connects the fatty acid?
d. What is the amino alcohol?

15.46 Identify the following features of this phospholipid, which is needed for the brain and nerve tissues:

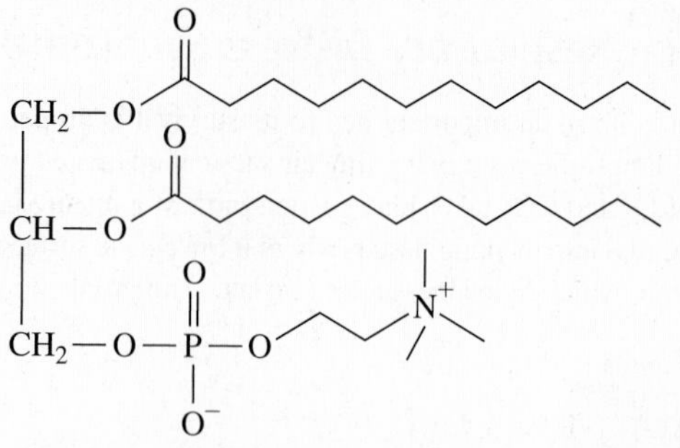

CH_2—O—C(=O)

CH—O—C(=O)

CH_2—O—P(=O)(O$^-$)—O—…N^+

a. Is the phospholipid formed from glycerol or sphingosine?
b. What is the fatty acid?
c. What type of bond connects the fatty acid?
d. What is the amino alcohol?

15.6 Steroids: Cholesterol, Bile Salts, and Steroid Hormones

LEARNING GOAL Draw the structures of steroids.

Steroids are compounds containing the *steroid nucleus*, which consists of three cyclohexane rings and one cyclopentane ring fused together. The four rings in the steroid nucleus are designated A, B, C, and D. The carbon atoms are numbered beginning with the carbons in ring A, and in steroids like cholesterol, ending with two methyl groups.

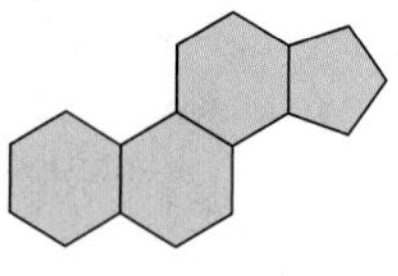

Steroid nucleus

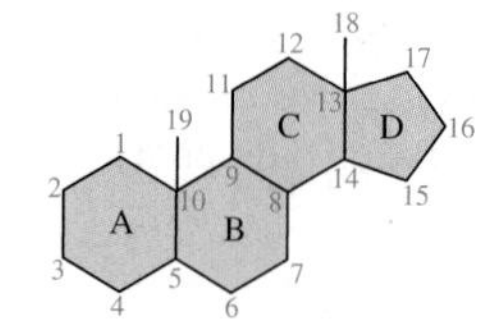

Steroid numbering system

Cholesterol

CORE CHEMISTRY SKILL
Identifying the Steroid Nucleus

Cholesterol, which is one of the most important and abundant steroids in the body, is a *sterol* because it contains an oxygen atom as a hydroxyl group (—OH) on carbon 3. Like many steroids, cholesterol has a double bond between carbon 5 and carbon 6, methyl groups at carbon 10 and carbon 13, and a carbon chain at carbon 17. In other steroids, the oxygen atom forms a carbonyl group (C=O) at carbon 3.

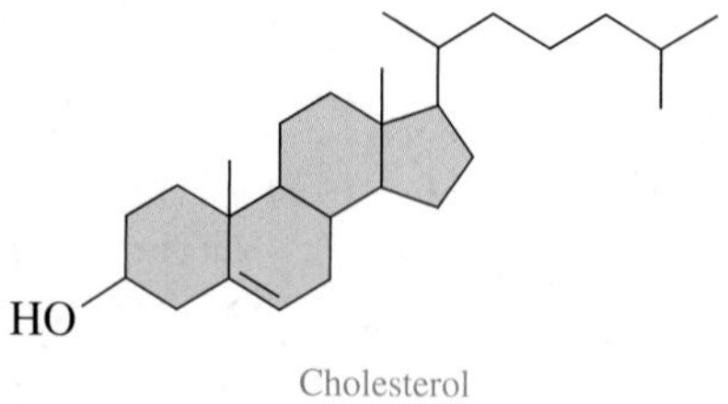

HO

Cholesterol

Cholesterol is a component of cellular membranes, myelin sheath, and brain and nerve tissue. It is also found in the liver and bile salts; large quantities of it are found in the skin, and some of it becomes vitamin D when the skin is exposed to direct sunlight. In the adrenal gland, cholesterol is used to synthesize steroid hormones. The liver synthesizes cholesterol for the body from fats, carbohydrates, and proteins. Additional cholesterol is obtained from meat, milk, and eggs in the diet. There is no cholesterol in vegetable and plant products. The cholesterol contents of some typical foods are listed in **TABLE 15.4**.

TABLE 15.4 Cholesterol Content of Some Foods

Food	Serving Size	Cholesterol (mg)
Liver (beef)	3 oz	370
Large egg	1	200
Lobster	3 oz	175
Fried chicken	$3\frac{1}{2}$ oz	130
Hamburger	3 oz	85
Chicken (no skin)	3 oz	75
Fish (salmon)	3 oz	40
Whole milk	1 cup	35
Butter	1 tablespoon	30
Skim milk	1 cup	5
Margarine	1 tablespoon	0

Cholesterol in the Body

If a diet is high in cholesterol, the liver produces less cholesterol. A typical daily American diet includes 400 to 500 mg of cholesterol, one of the highest in the world. The American Heart Association has recommended that we consume no more than 300 mg of cholesterol a day. Researchers suggest that saturated fats and cholesterol are associated with diseases such as diabetes; cancers of the breast, pancreas, and colon; and atherosclerosis. In atherosclerosis, deposits of a protein–lipid complex (plaque) accumulate in the coronary blood vessels, restricting the flow of blood to the tissue and causing necrosis (death) of the tissue (see **FIGURE 15.7**). In the heart, plaque accumulation could result in a *myocardial infarction* (heart attack). Other factors that may also increase the risk of heart disease are family history, lack of exercise, smoking, obesity, diabetes, gender, and age.

Clinically, cholesterol levels are considered elevated if the total plasma cholesterol level exceeds 200 mg/dL. Saturated fats in the diet may stimulate the production of cholesterol by the liver. A diet that is low in foods containing cholesterol and saturated fats appears to be helpful in reducing the plasma cholesterol level. The American Institute for Cancer Research (AICR) has recommended that our diet contain more fiber and starch by adding more vegetables, fruits, whole grains, and moderate amounts of foods with low levels of fat and cholesterol such as fish, poultry, lean meats, and low-fat dairy products. The AICR also suggests that we limit our intake of foods high in cholesterol such as eggs, nuts, French fries, fatty or organ meats, cheeses, butter, and coconut and palm oil.

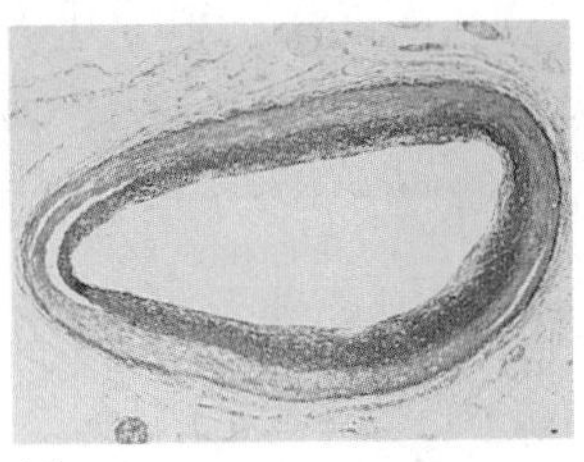

(a)

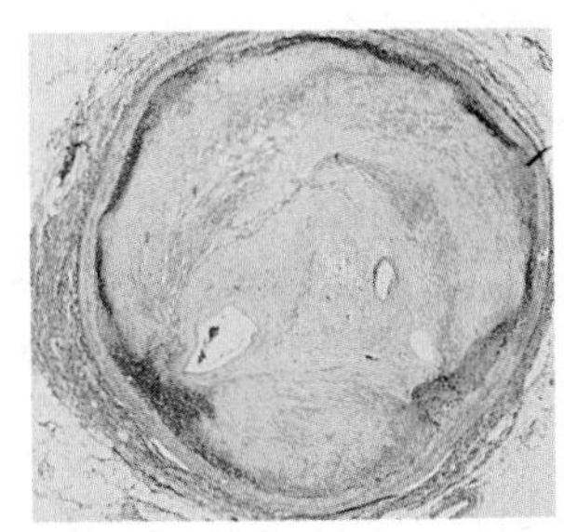
(b)

FIGURE 15.7 Excess cholesterol forms plaque that can block an artery, resulting in a heart attack. **(a)** A cross section of a normal, open artery shows no buildup of plaque. **(b)** A cross section of an artery that is almost completely clogged by atherosclerotic plaque.

Q What property of cholesterol would cause it to form deposits along the coronary arteries?

SAMPLE PROBLEM 15.6 Cholesterol

TRY IT FIRST

Refer to the structure of cholesterol for each of the following questions:

a. What part of cholesterol is the steroid nucleus?
b. What features have been added to the steroid nucleus in cholesterol?
c. What classifies cholesterol as a sterol?

SOLUTION

a. The four fused rings form the steroid nucleus.
b. The cholesterol molecule contains a hydroxyl group (—OH) on the first ring, one double bond in the second ring, methyl groups ($—CH_3$) at carbons 10 and 13, and a branched carbon chain at carbon 17.
c. The hydroxyl group determines the sterol classification.

TEST

Try Practice Problems 15.47 and 15.48

STUDY CHECK 15.6

Why is cholesterol in the lipid family?

ANSWER

Cholesterol is soluble in organic solvents but not in water; it is classified with the lipid family.

Bile Salts

The *bile salts* in the body are synthesized from cholesterol in the liver and stored in the gallbladder. When bile is secreted into the small intestine, the bile salts mix with the water-insoluble fats and oils in our diets. The bile salts with their nonpolar and polar regions act much like soaps, breaking down large globules of fat into smaller droplets. The smaller droplets containing fat have a larger surface area to react with lipases, which are the enzymes that digest fat. The bile salts also help with the absorption of cholesterol into the intestinal mucosa.

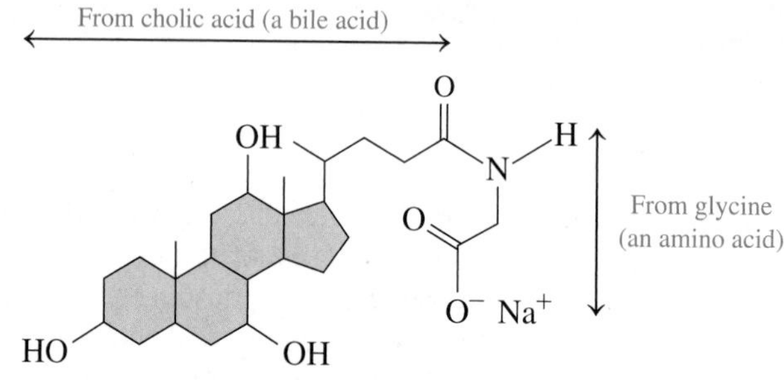

Sodium glycocholate (a bile salt)

FIGURE 15.8 Gallstones form in the gallbladder when cholesterol levels are high.

Q What type of steroid is stored in the gallbladder?

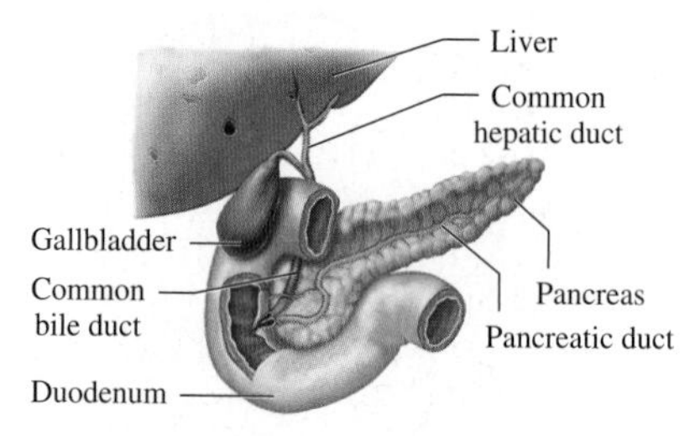

Gallstones formed in the gallbladder usually pass through the bile duct into the duodenum. If too large, they may obstruct the bile duct, causing pain and blocking bile.

When large amounts of cholesterol accumulate in the gallbladder, cholesterol can become solid, which forms gallstones (see **FIGURE 15.8**). Gallstones are composed of almost 100% cholesterol, with some calcium salts, fatty acids, and glycerophospholipids. Normally, small stones pass through the bile duct into the duodenum, the first part of the small intestine immediately beyond the stomach. If a large stone passes into the bile duct, it can get stuck, and the pain can be severe. If the gallstone obstructs the duct, bile cannot be excreted. Then bile pigments known as bilirubin will not be able to pass through the bile duct into the duodenum. They will back up into the liver and be excreted via the blood, causing jaundice (*hyperbilirubinemia*), which gives a yellow color to the skin and the whites of the eyes.

Lipoproteins: Transporting Lipids

In the body, lipids must be moved through the bloodstream to tissues where they are stored, used for energy, or used to make hormones. However, most lipids are nonpolar and insoluble in the aqueous environment of blood. They are made more soluble by combining them with phospholipids and proteins to form water-soluble complexes called **lipoproteins**. In general, lipoproteins are spherical particles with an outer surface of polar proteins and phospholipids that surround hundreds of nonpolar molecules of triacylglycerols and cholesteryl esters (see **FIGURE 15.9**). Cholesteryl esters are the prevalent form of cholesterol in the blood. They are formed by the esterification of the hydroxyl group in cholesterol with a fatty acid.

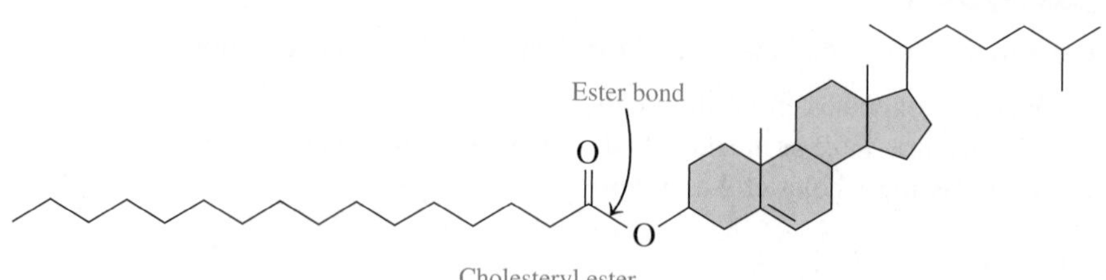

Cholesteryl ester

There are a variety of lipoproteins, which differ in density, lipid composition, and function. They include chylomicrons, very-low-density lipoproteins (VLDLs), low-density lipoproteins (LDLs), and high-density lipoproteins (HDLs). The density of the lipoproteins increases as the percentage of protein in each type increases (see **TABLE 15.5**).

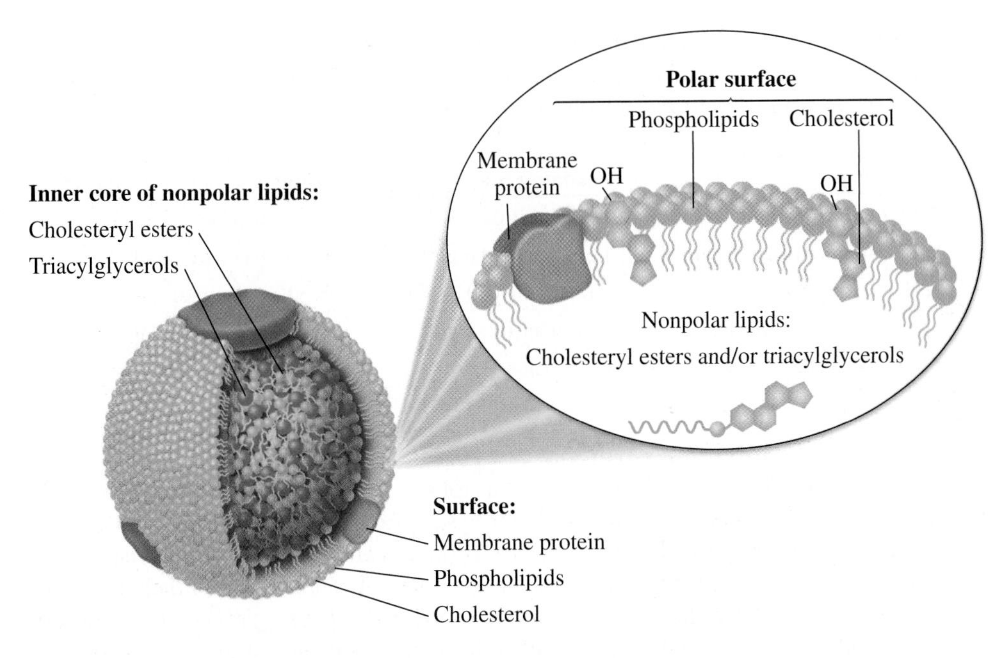

FIGURE 15.9 A spherical lipoprotein particle surrounds nonpolar lipids with polar lipids and protein for transport to body cells.

Q Why are the polar components on the surface of a lipoprotein particle and the nonpolar components at the center?

TABLE 15.5 Composition and Properties of Plasma Lipoproteins

	Chylomicrons	VLDL	LDL	HDL
Density (g/mL)	0.940	0.940–1.006	1.006–1.063	1.063–1.210
		Composition (% by mass)		
Type of Lipid				
Triacylglycerols	86	55	6	4
Phospholipids	7	18	22	24
Cholesterol	2	7	8	2
Cholesteryl esters	3	12	42	15
Protein	2	8	22	55

Two important lipoproteins are the LDL and HDL, which transport cholesterol. The LDL carries cholesterol to the tissues where it can be used for the synthesis of cell membranes and steroid hormones. When the LDL exceeds the amount of cholesterol needed by the tissues, the LDL deposits cholesterol in the arteries (plaque), which can restrict blood flow and increase the risk of developing heart disease and/or myocardial infarctions (heart attacks). This is why LDL is called "bad" cholesterol. The HDL picks up cholesterol from the tissues and carries it to the liver, where it can be converted to bile salts, which are eliminated from the body. This is why HDL is called "good" cholesterol. Other lipoproteins include chylomicrons that carry triacylglycerols from the intestines to the liver, muscle, and adipose tissues, and VLDL that carries the triacylglycerols synthesized in the liver to the adipose tissues for storage (see **FIGURE 15.10**).

Because high cholesterol levels are associated with the onset of atherosclerosis and heart disease, a doctor may order a *lipid panel* as part of a health examination. A lipid panel

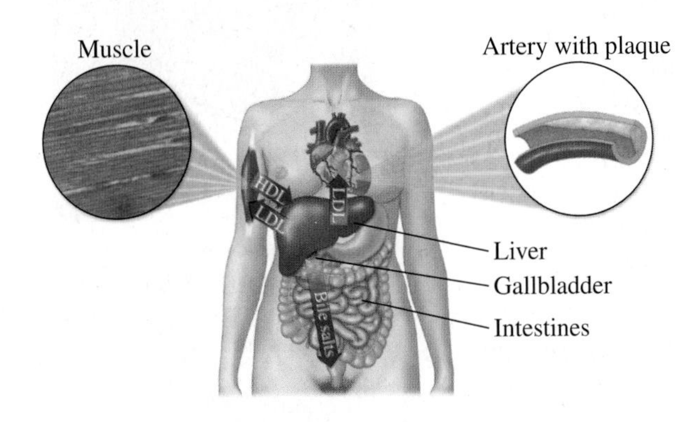

FIGURE 15.10 High- and low-density lipoproteins transport cholesterol between the tissues and the liver.

Q What type of lipoprotein transports cholesterol to the liver?

is a blood test that measures serum lipid levels including cholesterol, triglycerides, HDL, and LDL. The results of a lipid panel are used to evaluate a patient's risk of heart disease and to help a doctor determine the type of treatment needed.

TEST

Try Practice Problems 15.49 to 15.54

Lipid Panel	Recommended Level	Greater Risk of Heart Disease
Total Cholesterol	Less than 200 mg/dL	Greater than 240 mg/dL
Triglycerides (triacylglycerols)	Less than 150 mg/dL	Greater than 200 mg/dL
HDL ("good" cholesterol)	Greater than 60 mg/dL	Less than 40 mg/dL
LDL ("bad" cholesterol)	Less than 100 mg/dL	Greater than 160 mg/dL
Cholesterol/HDL Ratio	Less than 4	Greater than 7

Steroid Hormones

The word *hormone* comes from the Greek "to arouse" or "to excite." Hormones are chemical messengers that serve as a communication system from one part of the body to another. The *steroid* hormones, which include the sex hormones and the adrenocortical hormones, are closely related in structure to cholesterol and depend on cholesterol for their synthesis.

Two of the male sex hormones, *testosterone* and *androsterone*, promote the growth of muscle and facial hair, and the maturation of the male sex organs and of sperm.

The *estrogens*, a group of female sex hormones, direct the development of female sexual characteristics: the uterus increases in size, fat is deposited in the breasts, and the pelvis broadens. *Progesterone* prepares the uterus for the implantation of a fertilized egg. If an egg is not fertilized, the levels of progesterone and estrogen drop sharply, and menstruation follows. Synthetic forms of the female sex hormones are used in birth control pills. As with other kinds of steroids, side effects include weight gain and a greater risk of forming blood clots. The structures of some steroid hormones are shown below:

ENGAGE

How does testosterone differ from estradiol?

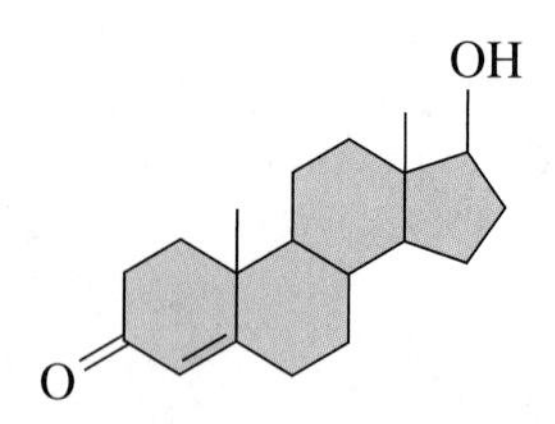

Testosterone (androgen)
(produced in testes)

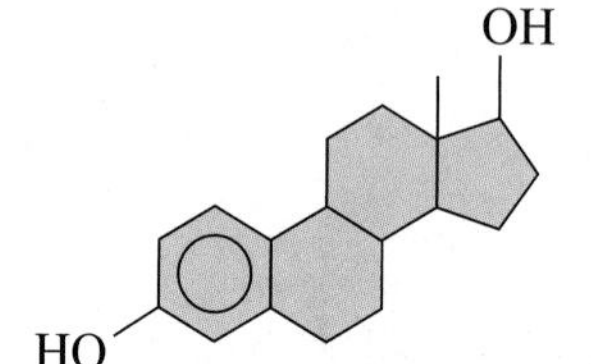

Estradiol (estrogen)
(produced in ovaries)

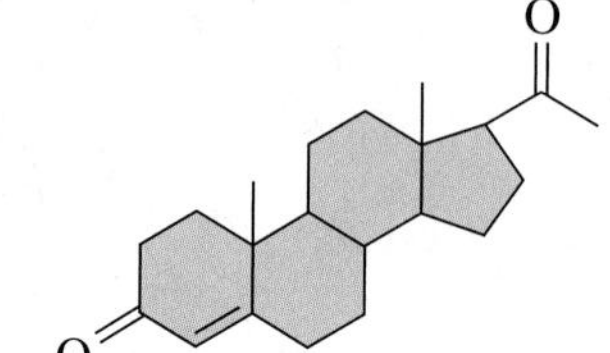

Progesterone
(produced in ovaries)

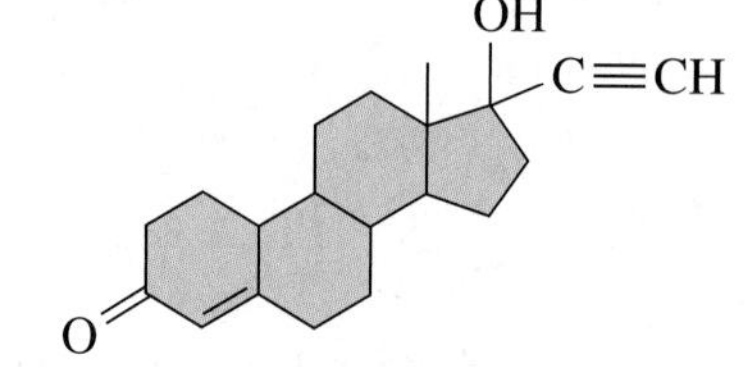

Norethindrone
(synthetic progestin)

CHEMISTRY LINK TO HEALTH

Anabolic Steroids

Some of the physiological effects of testosterone are to increase muscle mass and decrease body fat. Derivatives of testosterone called *anabolic steroids* that enhance these effects have been synthesized. Although they have some medical uses, anabolic steroids have been used in high dosages by some athletes in an effort to increase muscle mass. Such use is banned by most sports organizations.

Use of anabolic steroids in attempting to improve athletic strength can cause numerous side effects: in males—a reduction in testicle size, low sperm count and infertility, male pattern baldness, and breast development; in females—facial hair, deepening of the voice, male pattern baldness, breast atrophy, and menstrual dysfunction. Possible long-term consequences of anabolic steroid use in both men and women include liver disease and tumors, depression, and heart complications, with an increased risk of prostate cancer for men.

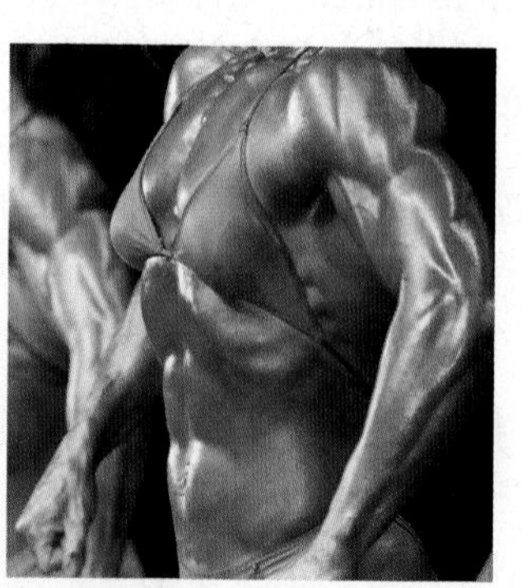

High doses of anabolic steroids, which are banned by sports organizations, are used to increase muscle mass.

The structures of some anabolic steroids are shown below:

Methandienone

Oxandrolone

Nandrolone

Stanozolol

Adrenal Corticosteroids

The adrenal glands, located on the top of each kidney, produce a large number of compounds known as the *corticosteroids*. *Cortisone* increases the blood glucose level and stimulates the synthesis of glycogen in the liver. *Aldosterone* is responsible for the regulation of electrolytes and water balance by the kidneys. *Cortisol* is released under stress to increase blood sugar and regulate carbohydrate, fat, and protein metabolism. Synthetic corticosteroid drugs such as *prednisone* are derived from cortisone and used medically for reducing inflammation and treating asthma and rheumatoid arthritis, although health problems can result from long-term use.

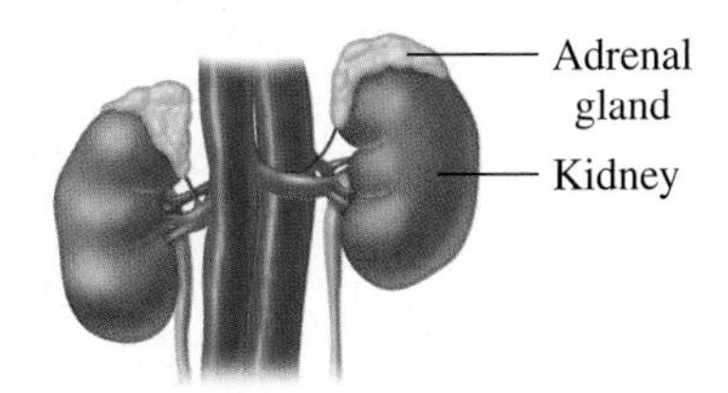

The adrenal glands on the kidneys produce corticosteroids.

Cortisone
(produced in adrenal gland)

Aldosterone (mineralocorticoid)
(produced in adrenal gland)

Cortisol
(produced in adrenal cortex)

Prednisone
(synthetic corticoid)

TEST

Try Practice Problems 15.55 to 15.58

PRACTICE PROBLEMS

15.6 Steroids: Cholesterol, Bile Salts, and Steroid Hormones

LEARNING GOAL Draw the structures of steroids.

15.47 Draw the structure for the steroid nucleus.

15.48 Draw the structure for cholesterol.

Clinical Applications

15.49 What is the function of bile salts in digestion?

15.50 Why are lipoproteins needed to transport lipids in the bloodstream?

15.51 How do chylomicrons differ from VLDL?

15.52 How does LDL differ from HDL?

15.53 Why is LDL called "bad" cholesterol?

15.54 Why is HDL called "good" cholesterol?

15.55 What are the similarities and differences between the steroid hormones estradiol and testosterone?

15.56 What are the similarities and differences between the adrenal hormone cortisone and the synthetic corticoid prednisone?

15.57 Which of the following are steroid hormones?

a. cholesterol **b.** cortisol
c. estradiol **d.** testosterone

15.58 Which of the following are adrenal corticosteroids?

a. prednisone **b.** aldosterone
c. cortisol **d.** testosterone

15.7 Cell Membranes

LEARNING GOAL Describe the composition and function of the lipid bilayer in cell membranes.

The membrane of a cell separates the contents of a cell from the external fluids. It is *semipermeable* so that nutrients can enter the cell and waste products can leave. The main components of a cell membrane are the glycerophospholipids and sphingolipids. Earlier in this chapter, we saw that the phospholipids consist of a nonpolar region, or hydrocarbon "tail," with two long-chain fatty acids and a polar region, or ionic "head" of phosphate and an ionized amino alcohol.

In a cell (plasma) membrane, two layers of phospholipids are arranged with their hydrophilic heads at the outer and inner surfaces of the membrane, and their hydrophobic tails in the center. This double layer arrangement of phospholipids is called a **lipid bilayer** (see FIGURE 15.11). The outer layer of phospholipids is in contact with the external fluids, and the inner layer is in contact with the internal contents of the cell.

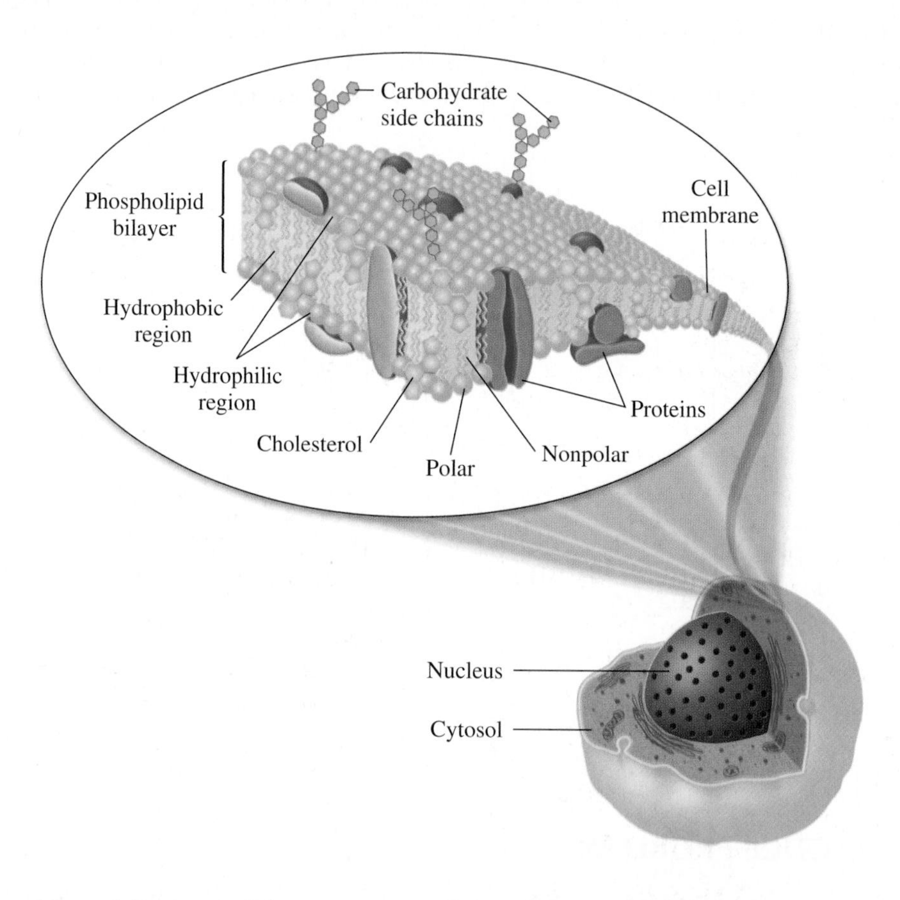

FIGURE 15.11 In the fluid mosaic model of a cell membrane, proteins and cholesterol are embedded in a lipid bilayer of phospholipids. The bilayer forms a barrier with polar heads at the membrane surfaces and the nonpolar tails in the center away from the water.

Q What types of fatty acids are found in the phospholipids of the lipid bilayer?

INTERACTIVE VIDEO
Membrane Structure

Most of the phospholipids in the lipid bilayer contain unsaturated fatty acids. Because of the kinks in the carbon chains at the cis double bonds, the phospholipids do not fit closely together. As a result, the lipid bilayer is not a rigid, fixed structure, but one that is dynamic and fluid-like. For this reason, the model of biological membranes is referred to as the **fluid mosaic model** of membranes. It is called a *mosaic* because the bilayer also contains proteins, carbohydrates, and cholesterol molecules.

In the fluid mosaic model, proteins known as *peripheral proteins* emerge on just one of the surfaces, outer or inner. The *integral proteins* extend through the entire lipid bilayer and appear on both surfaces of the membrane. Some proteins and lipids on the outer surface of the cell membrane are attached to carbohydrates. These carbohydrate chains project into the surrounding fluid environment where they are responsible for cell recognition and communication with chemical messengers such as hormones and neurotransmitters. In animals, cholesterol molecules embedded among the phospholipids make up 20 to 25% of the lipid bilayer. Because cholesterol molecules are large and rigid, they reduce the flexibility of the lipid bilayer and add strength to the cell membrane.

SAMPLE PROBLEM 15.7 Lipid Bilayer in Cell Membranes

TRY IT FIRST

Describe the role of phospholipids in the lipid bilayer.

SOLUTION

Phospholipids consist of polar and nonpolar parts. In a cell membrane, an alignment of the nonpolar sections toward the center with the polar sections on the outside produces a barrier that prevents the contents of a cell from mixing with the fluids on the outside of the cell.

STUDY CHECK 15.7

What is the function of cholesterol in the cell membrane?

ANSWER

Cholesterol adds strength and rigidity to the cell membrane.

TEST

Try Practice Problems 15.59 to 15.64

Transport through Cell Membranes

Although a nonpolar membrane separates aqueous solutions, it is necessary that certain substances can enter and leave the cell. The main function of a cell membrane is to allow the movement (transport) of ions and molecules on one side of the membrane to the other side. This transport of materials into and out of a cell is accomplished in several ways.

Diffusion (Passive) Transport

In the simplest transport mechanism called *diffusion* or *passive transport*, molecules can diffuse from a higher concentration to a lower concentration. For example, small molecules such as O_2, CO_2, urea, and water diffuse via passive transport through cell membranes. If their concentrations are greater outside the cell than inside, they diffuse into the cell. If their concentrations are higher within the cell, they diffuse out of the cell.

Facilitated Transport

In *facilitated transport*, proteins that extend from one side of the bilayer membrane to the other provide a channel through which certain substances can diffuse more rapidly than by passive diffusion to meet cellular needs. These protein channels allow transport of chloride ion (Cl^-), bicarbonate ion (HCO_3^-), and glucose molecules in and out of the cell.

Active Transport

Certain ions such as K^+, Na^+, and Ca^{2+} move across a cell membrane *against* their concentration gradients. For example, the K^+ concentration is greater inside a cell, and the Na^+ concentration is greater outside. However, in the conduction of nerve impulses and contraction of muscles, K^+ moves into the cell, and Na^+ moves out by a process known as *active transport*.

ENGAGE

Why are ions like Na^+ or K^+ unable pass through the lipid bilayer?

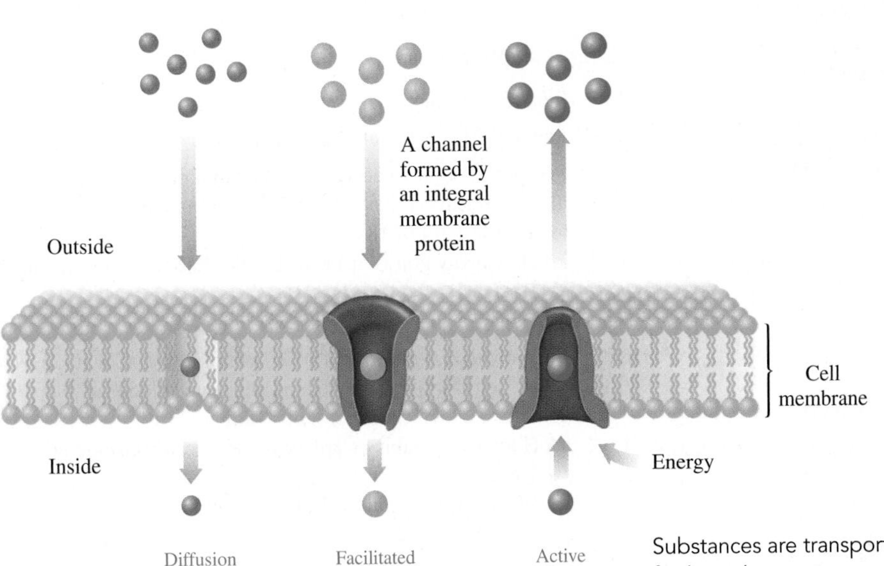

Substances are transported across a cell membrane by either diffusion, facilitated transport, or active transport.

TEST

Try Practice Problems 15.65 and 15.66

PRACTICE PROBLEMS

15.7 Cell Membranes

LEARNING GOAL Describe the composition and function of the lipid bilayer in cell membranes.

Clinical Applications

15.59 What is the function of the lipid bilayer in a cell membrane?

15.60 Describe the structure of a lipid bilayer.

15.61 How do molecules of cholesterol affect the structure of cell membranes?

15.62 How do the unsaturated fatty acids in the phospholipids affect the structure of cell membranes?

15.63 Where are proteins located in cell membranes?

15.64 What is the difference between peripheral and integral proteins?

15.65 Identify the type of transport described by each of the following:

a. A molecule moves through a protein channel.

b. O_2 moves into the cell from a higher concentration outside the cell.

15.66 Identify the type of transport described by each of the following:

a. An ion moves from low to high concentration in the cell.

b. Carbon dioxide moves through a cell membrane.

CLINICAL UPDATE Rebecca's Program to Lower Cholesterol

At the lipid clinic, Susan told Rebecca how to maintain a diet with less beef and chicken, more fish with omega-3 oils, low-fat dairy products, no egg yolks, and no coconut or palm oils. Rebecca maintained her new diet containing lower quantities of fats and increased quantities of fiber for the next year. She also increased her exercise using a treadmill and cycling and lost 35 lb. When Rebecca returned to the lipid clinic to check her progress with Susan, a new set of blood tests indicated that her total cholesterol had dropped from 420 to 390 mg/dL.

Because Rebecca's changes in diet and exercise did not sufficiently lower her cholesterol level, Susan prescribed a medication to help lower blood cholesterol levels. The most common medications used are the statins lovastatin (Mevacor), pravastatin (Pravachol), simvastatin (Zocor), atorvastatin (Lipitor), and rosuvastatin (Crestor). For some FH patients, statin therapy may be combined with fibrates such as gemfibrozil (Lopid) or fenofibrate (TriCor) that reduce the synthesis of enzymes that break down fats in the blood.

Susan prescribed pravastatin (Pravachol), 80 mg once a day, which was effective. Later, Susan added fenofibrate (TriCor), one 145-mg tablet, each day to the Pravachol. Rebecca understands that her medications, diet and exercise plan are a lifelong process. Because Rebecca was diagnosed with FH, both her children were tested for FH. Her older son was diagnosed with FH, and is also being treated with a statin. Her younger son does not have FH.

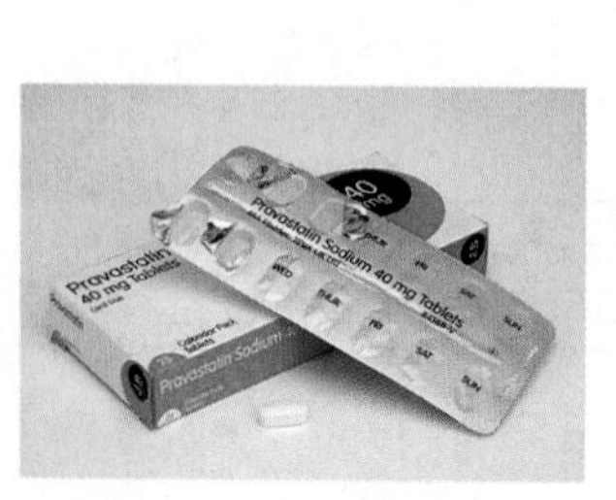

Pravastatin (Pravachol)

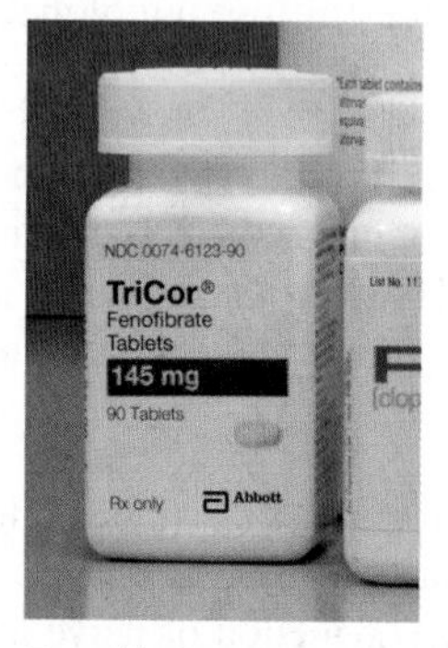

Fenofibrate (TriCor)

Clinical Applications

15.67 Identify the functional groups in Pravachol.

15.68 Identify the functional groups in TriCor.

15.69 Six months after Rebecca started using Pravachol, her blood test showed that a 5.0-mL blood sample contained a total cholesterol of 18 mg.

a. How many grams of Pravachol did Rebecca consume in 1 week?

b. What was her total cholesterol in mg/dL?

15.70 Five months after Rebecca added the fibrate TriCor to the statin, her blood test showed that a 5.0-mL blood sample contained a total cholesterol of 14 mg.

a. How many grams of TriCor did Rebecca consume in 1 week?

b. What was her total cholesterol in mg/dL?

CONCEPT MAP

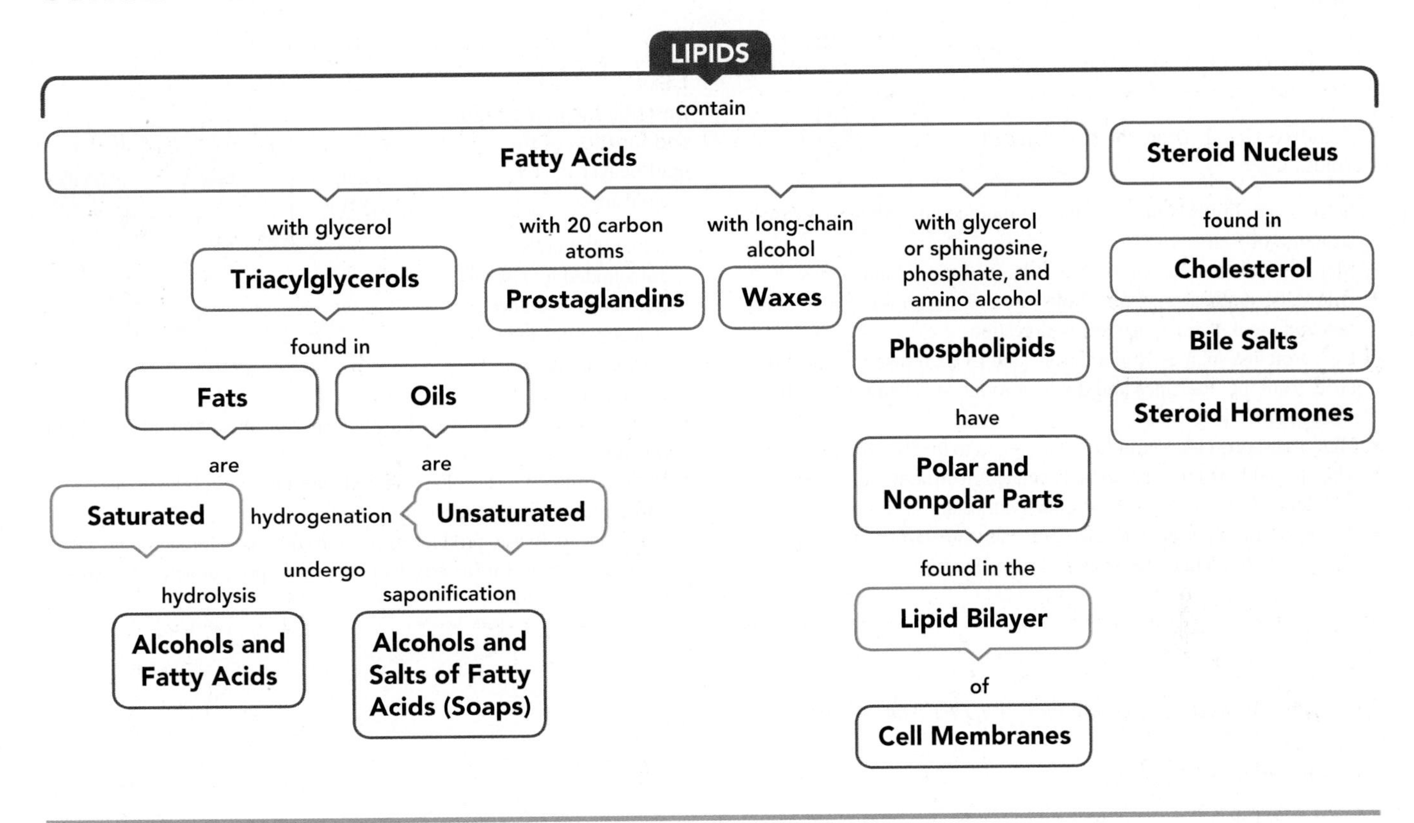

CHAPTER REVIEW

15.1 Lipids

LEARNING GOAL Describe the classes of lipids.

- Lipids are biomolecules that are not soluble in water.
- Classes of lipids include waxes, triacylglycerols, glycerophospholipids, sphingolipids, and steroids.

15.2 Fatty Acids

LEARNING GOAL Draw the condensed structural or line-angle formula for a fatty acid and identify it as saturated or unsaturated.

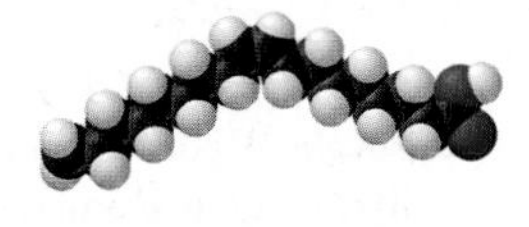

- Fatty acids are unbranched carboxylic acids that typically contain an even number (12 to 20) of carbon atoms.
- Fatty acids may be saturated, monounsaturated with one double bond, or polyunsaturated with two or more double bonds.
- The double bonds in unsaturated fatty acids are almost always cis.

15.3 Waxes and Triacylglycerols

LEARNING GOAL Draw the condensed structural or line-angle formula for a wax or triacylglycerol produced by the reaction of a fatty acid and an alcohol or glycerol.

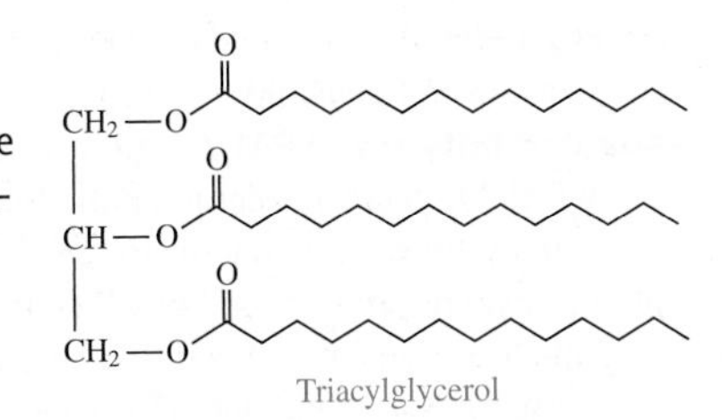

Triacylglycerol

- A wax is an ester of a long-chain fatty acid and a long-chain alcohol.
- The triacylglycerols are esters of glycerol with three long-chain fatty acids.
- Fats contain more saturated fatty acids and have higher melting points than most vegetable oils.

15.4 Chemical Properties of Triacylglycerols

LEARNING GOAL Draw the condensed structural or line-angle formula for the products of a triacylglycerol that undergoes hydrogenation, hydrolysis, or saponification.

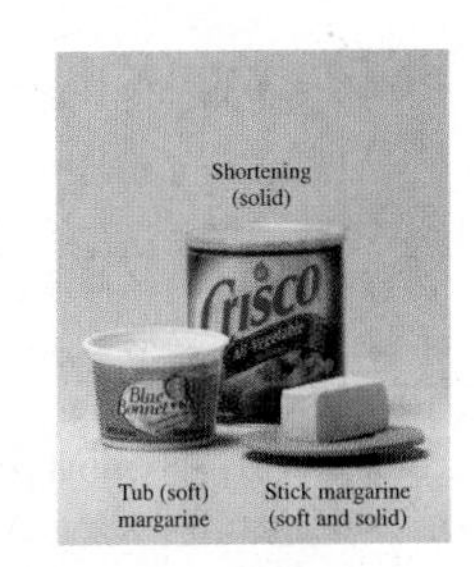

- The hydrogenation of unsaturated fatty acids of a triacylglycerol converts double bonds to single bonds.
- The hydrolysis of the ester bonds in triacylglycerols in the presence of a strong acid produces glycerol and fatty acids.
- In saponification, a triacylglycerol heated with a strong base produces glycerol and the salts of the fatty acids (soap).

15.5 Phospholipids

LEARNING GOAL Draw the structure of a phospholipid containing glycerol or sphingosine.

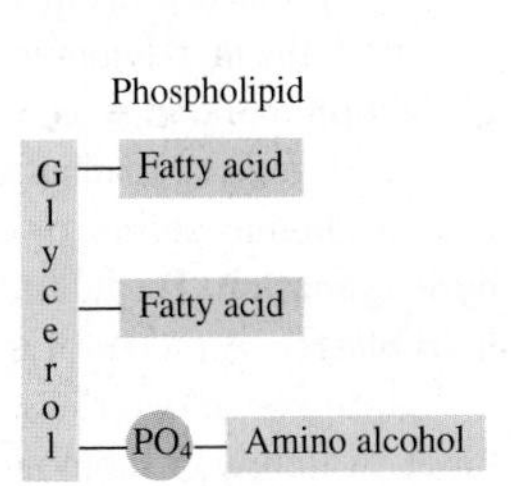

- Glycerophospholipids are esters of glycerol with two fatty acids and a phosphate group attached to an amino alcohol.

- In a sphingomyelin, the amino alcohol sphingosine forms an amide bond with a fatty acid, and phosphoester bonds to phosphate and an amino alcohol.

15.6 Steroids: Cholesterol, Bile Salts, and Steroid Hormones

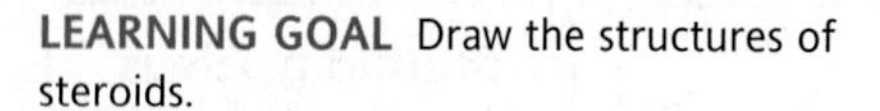

LEARNING GOAL Draw the structures of steroids.

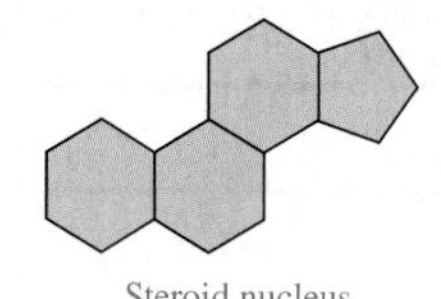

Steroid nucleus

- Steroids are lipids containing the steroid nucleus, which is a fused structure of four rings.
- Steroids include cholesterol, bile salts, and steroid hormones.
- Bile salts, synthesized from cholesterol, mix with water-insoluble fats and break them apart during digestion.
- Lipoproteins, such as chylomicrons and LDL, transport triacylglycerols from the intestines and the liver to fat cells and muscles for storage and energy.
- HDL transports cholesterol from the tissues to the liver for elimination.
- The steroid hormones are closely related in structure to cholesterol and depend on cholesterol for their synthesis.
- The sex hormones, such as estrogen and testosterone, are responsible for sexual characteristics and reproduction.
- The adrenal corticosteroids, such as aldosterone and cortisone, regulate water balance and glucose levels in the cells, respectively.

15.7 Cell Membranes

LEARNING GOAL Describe the composition and function of the lipid bilayer in cell membranes.

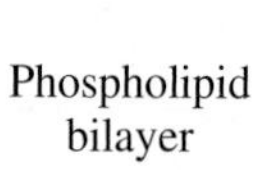

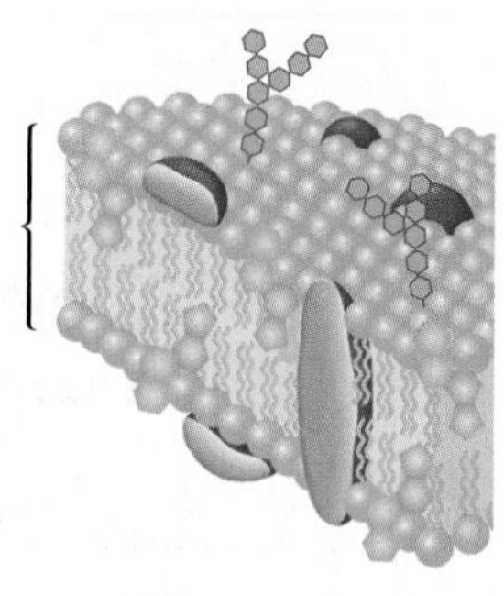

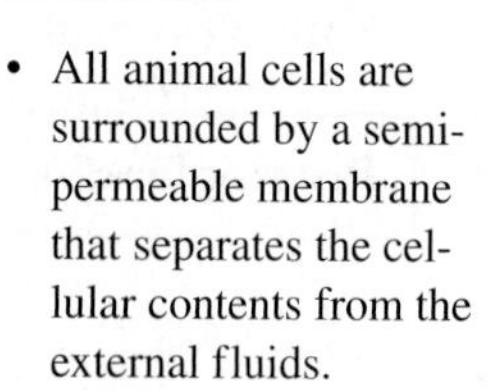

- All animal cells are surrounded by a semipermeable membrane that separates the cellular contents from the external fluids.
- The membrane is composed of two rows of phospholipids in a lipid bilayer.
- Proteins and cholesterol are embedded in the lipid bilayer, and carbohydrates are attached to its surface.
- Nutrients and waste products move through the cell membrane using passive transport (diffusion), facilitated transport, or active transport.

SUMMARY OF REACTIONS

The chapter Sections to review are shown after the name of each reaction.

Esterification (15.3)

$$\text{Glycerol} + 3 \text{ fatty acid molecules} \xrightarrow{\text{Enzyme}} \text{triacylglycerol} + 3H_2O$$

Hydrogenation of Triacylglycerols (15.4)

$$\text{Triacylglycerol (unsaturated)} + H_2 \xrightarrow{\text{Ni}} \text{triacylglycerol (saturated)}$$

Hydrolysis of Triacylglycerols (15.4)

$$\text{Triacylglycerol} + 3H_2O \xrightarrow{H^+ \text{ or lipase}} \text{glycerol} + 3 \text{ fatty acid molecules}$$

Saponification of Triacylglycerols (15.4)

$$\text{Triacylglycerol} + 3NaOH \xrightarrow{\text{Heat}} \text{glycerol} + 3 \text{ sodium salts of fatty acids (soaps)}$$

KEY TERMS

cholesterol The most prevalent of the steroid compounds; needed for cellular membranes and the synthesis of vitamin D, hormones, and bile salts.

fat A triacylglycerol that is solid at room temperature and usually comes from animal sources.

fatty acid A long-chain carboxylic acid found in many lipids.

fluid mosaic model The concept that cell membranes are lipid bilayer structures that contain an assortment of polar lipids and proteins in a dynamic, fluid arrangement.

glycerophospholipid A polar lipid of glycerol attached to two fatty acids and a phosphate group connected to an amino alcohol such as choline, serine, or ethanolamine.

hydrogenation The addition of hydrogen to unsaturated fats.

lipid bilayer A model of a cell membrane in which phospholipids are arranged in two rows.

lipids A family of biomolecules that is nonpolar in nature and not soluble in water; includes prostaglandins, waxes, triacylglycerols, phospholipids, and steroids.

lipoprotein A polar complex composed of a combination of nonpolar lipids with glycerophospholipids and proteins to form a polar complex that can be transported through body fluids.

monounsaturated fatty acid (MUFA) A fatty acid with one double bond.

oil A triacylglycerol that is a liquid at room temperature and is obtained from a plant source.

phospholipid A polar lipid of glycerol or sphingosine attached to fatty acids and a phosphate group connected to an amino alcohol.

polyunsaturated fatty acid (PUFA) A fatty acid that contains two or more double bonds.

prostaglandin A compound derived from arachidonic acid that regulates several physiological processes.

saturated fatty acid (SFA) A fatty acid that has no double bonds, which has a higher melting point than unsaturated lipids, and is usually solid at room temperature.

sphingomyelin A compound in which the amine group of sphingosine forms an amide bond to a fatty acid and the hydroxyl group forms an ester bond with phosphate, which forms another phosphoester bond to an amino alcohol.

steroid Type of lipid containing a multicyclic ring system.

triacylglycerol A lipid composed of three fatty acids bonded through ester bonds to glycerol, a trihydroxy alcohol.

unsaturated fatty acid (UFA) A fatty acid that contains one or more carbon–carbon double bonds, which has a lower melting point than saturated fatty acids, and is usually liquid at room temperature.

wax The ester of a long-chain alcohol and a long-chain fatty acid.

CORE CHEMISTRY SKILLS

The chapter Section containing each Core Chemistry Skill is shown in parentheses at the end of each heading.

Identifying Fatty Acids (15.2)

Fatty acids are unbranched carboxylic acids that typically contain an even number (12 to 20) of carbon atoms.

- Fatty acids may be saturated, monounsaturated with one double bond, or polyunsaturated with two or more double bonds.
- The double bonds in unsaturated fatty acids are almost always cis.

Example: State the number of carbon atoms, saturated or unsaturated, and name of the following:

Answer: 16 carbon atoms, saturated, palmitic acid

Drawing Structures for Triacylglycerols (15.3)

- Triacylglycerols are esters of glycerol with three long-chain fatty acids.

Example: Draw the line-angle formula and name the triacylglycerol formed from glycerol and palmitic acid.

Answer:

Glyceryl tripalmitate (tripalmitin)

Drawing the Products for the Hydrogenation, Hydrolysis, and Saponification of a Triacylglycerol (15.4)

- The hydrogenation of unsaturated fatty acids of a triacylglycerol in the presence of a Ni catalyst converts double bonds to single bonds.
- The hydrolysis of the ester bonds in triacylglycerols in the presence of a strong acid or digestive enzyme produces glycerol and fatty acids.
- In saponification, a triacylglycerol heated with a strong base produces glycerol and the salts of the fatty acids (soap).

Example: Identify each of the following as hydrogenation, hydrolysis, or saponification and state the products:

a. the reaction of palm oil with NaOH
b. the reaction of glyceryl trilinoleate from corn oil with water in the presence of an acid catalyst
c. the reaction of corn oil and H_2 using a nickel catalyst

Answer: **a.** The reaction of palm oil with NaOH is saponification, and the products are glycerol and the sodium salts of the fatty acids, which is soap.
b. In hydrolysis, glyceryl trilinoleate reacts with water in the presence of an acid catalyst, which splits the ester bonds to produce glycerol and three molecules of linoleic acid.
c. In hydrogenation, H_2 adds to double bonds in corn oil, which produces a more saturated, and thus more solid, fat.

Identifying the Steroid Nucleus (15.6)

- The steroid nucleus consists of three cyclohexane rings and one cyclopentane ring fused together.

Example: Why are cholesterol, sodium glycocholate (a bile salt), and cortisone (a corticosteroid) all considered steroids?

Answer: They all contain the steroid nucleus of three six-carbon rings and one five-carbon ring fused together.

UNDERSTANDING THE CONCEPTS

The chapter Sections to review are shown in parentheses at the end of each problem.

15.71 Palm oil has a high level of glyceryl tripalmitate (tripalmitin). Draw the condensed structural formula for glyceryl tripalmitate. (15.2, 15.3)

The fruit from palm trees are a source of palm oil.

15.72 Jojoba wax in candles consists of a 20-carbon saturated fatty acid and a 20-carbon saturated alcohol. Draw the condensed structural formula for jojoba wax. (15.2, 15.3)

Candles contain jojoba wax.

Clinical Applications

15.73 Identify each of the following as a saturated, monounsaturated, polyunsaturated, omega-3, or omega-6 fatty acid: (15.2)

a. $CH_3-(CH_2)_7-CH=CH-(CH_2)_7-\overset{\overset{\large O}{\|}}{C}-OH$

b. linoleic acid

c. $CH_3-CH_2-(CH=CH-CH_2)_5-CH_2-CH_2-\overset{\overset{\large O}{\|}}{C}-OH$

15.74 Identify each of the following as a saturated, monounsaturated, polyunsaturated, omega-3, or omega-6 fatty acid: (15.2)

a.

$CH_3-(CH_2)_4-CH=CH-CH_2-CH=CH-(CH_2)_7-\overset{\overset{\large O}{\|}}{C}-OH$

b. linolenic acid

c. $CH_3-(CH_2)_{14}-\overset{\overset{\large O}{\|}}{C}-OH$

Salmon is a good source of omega-3 unsaturated fatty acids.

ADDITIONAL PRACTICE PROBLEMS

15.75 Among the ingredients in lipstick are beeswax, carnauba wax, hydrogenated vegetable oils, and glyceryl tricaprate (tricaprin). (15.1, 15.2, 15.3)

a. What types of lipids are these?

b. Draw the condensed structural formula for glyceryl tricaprate (tricaprin). Capric acid is a saturated 10-carbon fatty acid.

15.76 Because peanut oil floats on the top of peanut butter, the peanut oil in many brands of peanut butter is hydrogenated and the solid is mixed into the peanut butter to give a product that does not separate. If a triacylglycerol in peanut oil that contains one oleic acid and two linoleic acids is completely hydrogenated, draw the condensed structural formula for the product. (15.3, 15.4)

Clinical Applications

15.77 The total kilocalories and grams of fat for some typical meals at fast-food restaurants are listed here. Calculate the number of kilocalories and the percentage of total kilocalories from fat (1 gram of fat = 9 kcal). Round answers to the tens place. (15.2, 15.3)

a. a chicken dinner, 830 kcal, 46 g of fat

b. a quarter-pound cheeseburger, 520 kcal, 29 g of fat

c. pepperoni pizza (three slices), 560 kcal, 18 g of fat

15.78 The total kilocalories and grams of fat for some typical meals at fast-food restaurants are listed here. Calculate the number of kilocalories and the percentage of total kilocalories from fat (1 gram of fat = 9 kcal). Round answers to the tens place. (15.2, 15.3)

a. a beef burrito, 470 kcal, 21 g of fat

b. deep-fried fish (three pieces), 480 kcal, 28 g of fat

c. a jumbo hot dog, 180 kcal, 18 g of fat

15.79 Identify each of the following as a fatty acid, soap, triacylglycerol, wax, glycerophospholipid, sphingolipid, or steroid: (15.1, 15.2, 15.3, 15.5, 15.6)

a. beeswax
b. cholesterol
c. lecithin
d. glyceryl tripalmitate (tripalmitin)
e. sodium stearate
f. safflower oil

15.80 Identify each of the following as a fatty acid, soap, triacylglycerol, wax, glycerophospholipid, sphingolipid, or steroid: (15.1, 15.2, 15.3, 15.5, 15.6)

a. sphingomyelin
b. whale blubber
c. adipose tissue
d. progesterone
e. cortisone
f. stearic acid

15.81 Identify the components (**1** to **6**) contained in each of the following lipids (**a** to **d**): (15.1, 15.2, 15.3, 15.5, 15.6)

1. glycerol
2. fatty acid
3. phosphate
4. amino alcohol
5. steroid nucleus
6. sphingosine

a. estrogen
b. cephalin
c. wax
d. triacylglycerol

15.82 Identify the components (**1** to **6**) contained in each of the following lipids (**a** to **d**): (15.1, 15.2, 15.3, 15.5, 15.6)

1. glycerol
2. fatty acid
3. phosphate
4. amino alcohol
5. steroid nucleus
6. sphingosine

a. glycerophospholipid
b. sphingomyelin
c. aldosterone
d. linoleic acid

15.83 Which of the following are found in cell membranes? (15.7)

a. cholesterol
b. triacylglycerols
c. carbohydrates

15.84 Which of the following are found in cell membranes? (15.7)

a. proteins
b. waxes
c. phospholipids

CHALLENGE PROBLEMS

The following problems are related to the topics in this chapter. However, they do not all follow the chapter order, and they require you to combine concepts and skills from several Sections. These problems will help you increase your critical thinking skills and prepare for your next exam.

15.85 Draw the condensed structural formula for a glycerophospholipid that contains glycerol, two stearic acids, phosphate, and ethanolamine (ionized). (15.2, 15.5)

15.86 Sunflower seed oil can be used to make margarine. A triacylglycerol in sunflower seed oil contains two linoleic acids and one oleic acid. (15.2, 15.3, 15.5)

a. Draw the condensed structural formulas for two isomers of the triacylglycerol in sunflower seed oil.

b. Using one of the isomers, write the reaction that takes place when sunflower seed oil is used to make solid margarine.

Sunflower oil is obtained from the seeds of the sunflower.

Clinical Applications

15.87 Match the lipoprotein (**1** to **4**) with its description (**a** to **d**). (15.6)

1. chylomicrons **2.** VLDL
3. LDL **4.** HDL

a. "good" cholesterol
b. transports most of the cholesterol to the cells
c. carries triacylglycerols from the intestine to the fat cells
d. transports cholesterol to the liver

15.88 Match the lipoprotein (**1** to **4**) with its description (**a** to **d**). (15.6)

1. chylomicrons **2.** VLDL
3. LDL **4.** HDL

a. has the greatest abundance of protein
b. "bad" cholesterol
c. carries triacylglycerols synthesized in the liver to the muscles
d. has the lowest density

15.89 A sink drain can become clogged with solid fat such as glyceryl tristearate (tristearin). (7.4, 7.7, 7.8, 9.4, 15.3, 15.4)

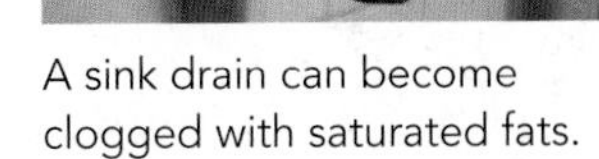

A sink drain can become clogged with saturated fats.

a. How would adding lye (NaOH) to the sink drain remove the blockage?

b. Write a balanced chemical equation for the reaction that occurs.

c. How many milliliters of a 0.500 M NaOH solution are needed to completely react with 10.0 g of glyceryl tristearate (tristearin)?

15.90 One of the triacylglycerols in olive oil is glyceryl tripalmitoleate (tripalmitolein). (7.4, 7.7, 7.8, 8.6, 9.4, 15.3, 15.4)

Olive oil contains glyceryl tripalmitoleate (tripalmitolein).

a. Draw the condensed structural formula for glyceryl tripalmitoleate (tripalmitolein).

b. How many liters of H_2 gas at STP are needed to completely saturate 100 g of glyceryl tripalmitoleate (tripalmitolein)?

c. How many milliliters of a 0.250 M NaOH solution are needed to completely react with 100 g of glyceryl tripalmitoleate (tripalmitolein)?

15.91 1.00 mole of glyceryl trioleate (triolein) is completely hydrogenated. (7.2, 7.3, 8.6, 15.3, 15.4)

a. Draw the condensed structural formula for the product.
b. How many moles of hydrogen are required?
c. How many grams of hydrogen are required?
d. How many liters of hydrogen gas are needed if the reaction is run at STP?

ANSWERS

Answers to Selected Practice Problems

15.1 Because lipids are not soluble in water, a polar solvent, they are nonpolar molecules.

15.3 Lipids are an important part of cell membranes and steroid hormones.

15.5 All fatty acids contain a long chain of carbon atoms with a carboxylic acid group. Saturated fatty acids contain only carbon–carbon single bonds; unsaturated fatty acids contain one or more double bonds.

15.7 **a.** palmitic acid

b. oleic acid

15.9 **a.** (12:0), saturated **b.** (18:3), polyunsaturated
c. (16:1), monounsaturated **d.** (18:0), saturated

15.11 In a cis fatty acid, the hydrogen atoms are on the same side of the double bond, which produces a kink in the carbon chain. In a trans fatty acid, the hydrogen atoms are on opposite sides of the double bond, which gives a carbon chain without any kink.

15.13 Arachidonic acid and PGE_1 are both carboxylic acids with 20 carbon atoms. The differences are that arachidonic acid has four cis double bonds and no other functional groups, whereas PGE_1 has one trans double bond, one ketone functional group, and two hydroxyl functional groups. In addition, a part of the PGE_1 chain forms a cyclopentane ring.

15.15 Prostaglandins raise or lower blood pressure, stimulate contraction and relaxation of smooth muscle, and may cause inflammation and pain.

15.17 In an omega-3 fatty acid, there is a double bond beginning at carbon 3, counting from the methyl group, whereas in an omega-6 fatty acid, there is a double bond beginning at carbon 6, counting from the methyl group.

15.19 $CH_3-(CH_2)_{14}-\overset{\overset{\large O}{\|}}{C}-O-(CH_2)_{29}-CH_3$

15.21

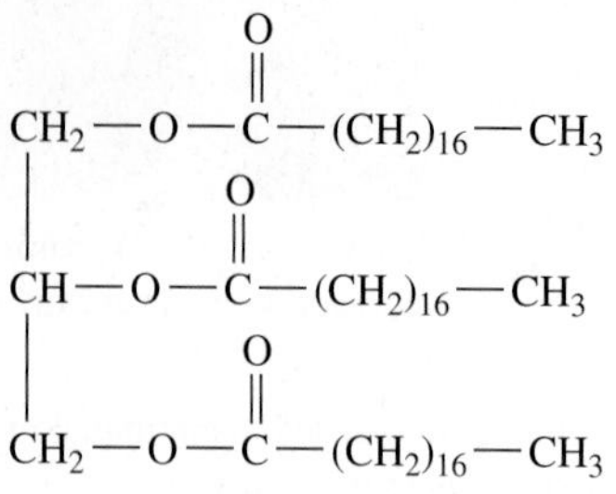

$$\begin{array}{l} CH_2-O-\overset{\overset{O}{\|}}{C}-(CH_2)_{16}-CH_3 \\ | \\ CH-O-\overset{\overset{O}{\|}}{C}-(CH_2)_{16}-CH_3 \\ | \\ CH_2-O-\overset{\overset{O}{\|}}{C}-(CH_2)_{16}-CH_3 \end{array}$$

15.23

$$\begin{array}{l} CH_2-O \quad (\text{ester with O}) \\ | \\ CH-O \quad (\text{ester with O}) \\ | \\ CH_2-O \quad (\text{ester with O}) \end{array}$$

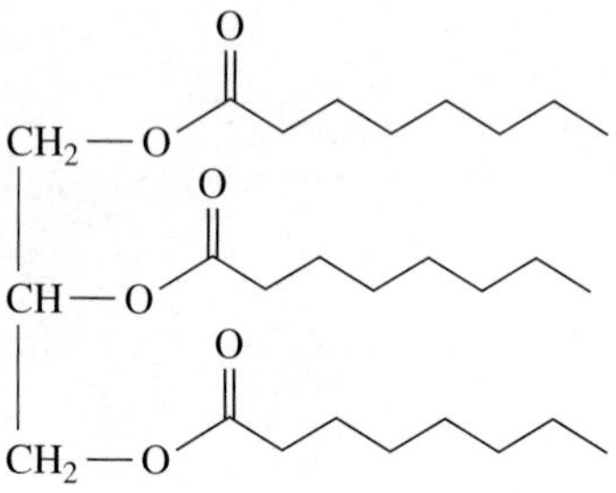

15.25 Safflower oil has a lower melting point since it contains fatty acids with two or three double bonds; olive oil contains a large amount of oleic acid, which has only one double bond (monounsaturated).

15.27 Sunflower oil has about 24% monounsaturated fats, whereas safflower oil has about 18%. Sunflower oil has about 66% polyunsaturated fats, whereas safflower oil has about 73%.

15.29 **a.** The reaction of palm oil with KOH is saponification; the products are glycerol and the potassium salts of the fatty acids, which are soaps.

b. The reaction of glyceryl trilinoleate from safflower oil with water and HCl is hydrolysis, which splits the ester bonds to produce glycerol and three molecules of linoleic acid.

15.31

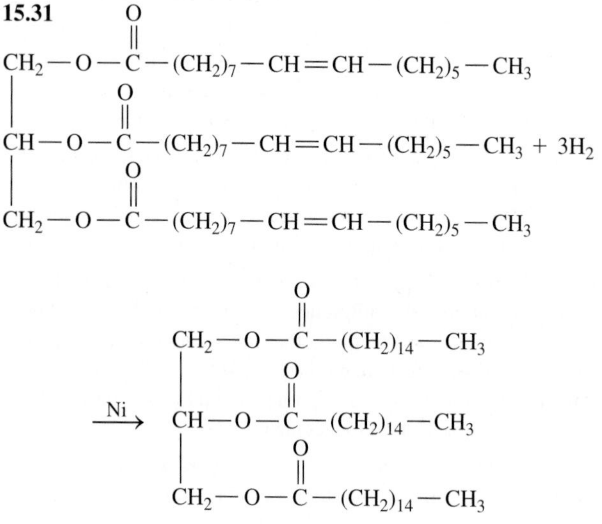

$$\begin{array}{l} CH_2-O-\overset{\overset{O}{\|}}{C}-(CH_2)_7-CH{=}CH-(CH_2)_5-CH_3 \\ | \\ CH-O-\overset{\overset{O}{\|}}{C}-(CH_2)_7-CH{=}CH-(CH_2)_5-CH_3 + 3H_2 \\ | \\ CH_2-O-\overset{\overset{O}{\|}}{C}-(CH_2)_7-CH{=}CH-(CH_2)_5-CH_3 \end{array}$$

$$\xrightarrow{Ni} \begin{array}{l} CH_2-O-\overset{\overset{O}{\|}}{C}-(CH_2)_{14}-CH_3 \\ | \\ CH-O-\overset{\overset{O}{\|}}{C}-(CH_2)_{14}-CH_3 \\ | \\ CH_2-O-\overset{\overset{O}{\|}}{C}-(CH_2)_{14}-CH_3 \end{array}$$

15.33

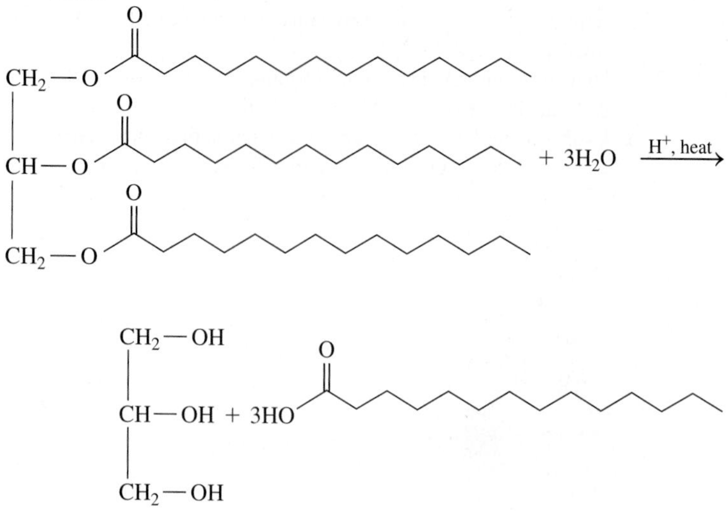

$$\begin{array}{l} CH_2-O \quad (\text{ester with O}) \\ | \\ CH-O \quad (\text{ester with O}) \\ | \\ CH_2-O \quad (\text{ester with O}) \end{array} + 3H_2O \xrightarrow{H^+,\ heat}$$

$$\begin{array}{l} CH_2-OH \\ | \\ CH-OH + 3HO \quad (\text{with O}) \\ | \\ CH_2-OH \end{array}$$

15.35

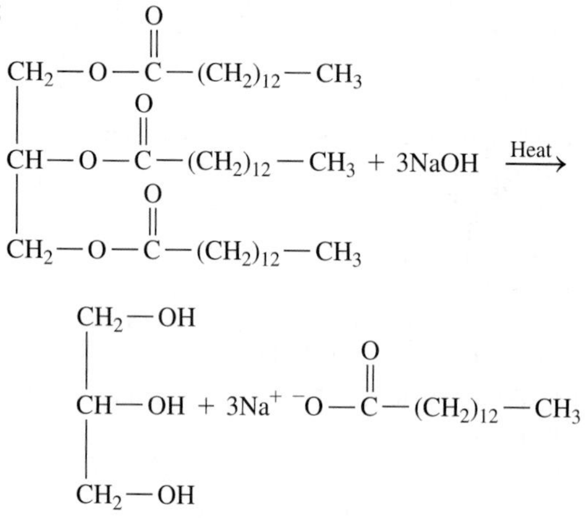

$$\begin{array}{l} CH_2-O-\overset{\overset{O}{\|}}{C}-(CH_2)_{12}-CH_3 \\ | \\ CH-O-\overset{\overset{O}{\|}}{C}-(CH_2)_{12}-CH_3 + 3NaOH \xrightarrow{Heat} \\ | \\ CH_2-O-\overset{\overset{O}{\|}}{C}-(CH_2)_{12}-CH_3 \end{array}$$

$$\begin{array}{l} CH_2-OH \\ | \\ CH-OH + 3Na^+\ {}^-O-\overset{\overset{O}{\|}}{C}-(CH_2)_{12}-CH_3 \\ | \\ CH_2-OH \end{array}$$

15.37

$$\begin{array}{l} CH_2-O-\overset{\overset{O}{\|}}{C}-(CH_2)_{16}-CH_3 \\ | \\ CH-O-\overset{\overset{O}{\|}}{C}-(CH_2)_{16}-CH_3 \\ | \\ CH_2-O-\overset{\overset{O}{\|}}{C}-(CH_2)_{16}-CH_3 \end{array}$$

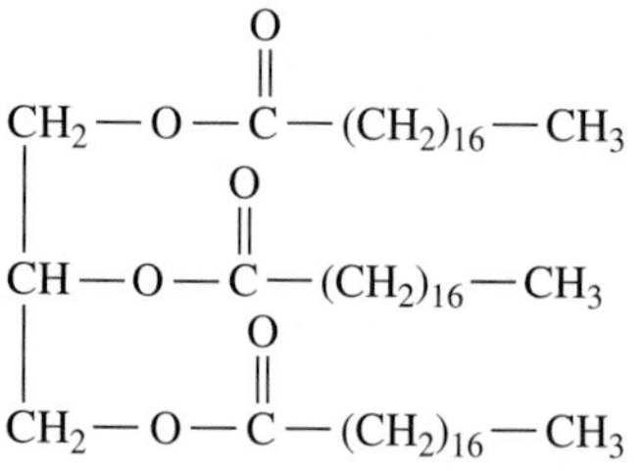

15.39 A triacylglycerol consists of glycerol and three fatty acids. A glycerophospholipid also contains glycerol, but has only two fatty acids. The hydroxyl group on the third carbon of glycerol is attached by a phosphoester bond to an amino alcohol.

15.41

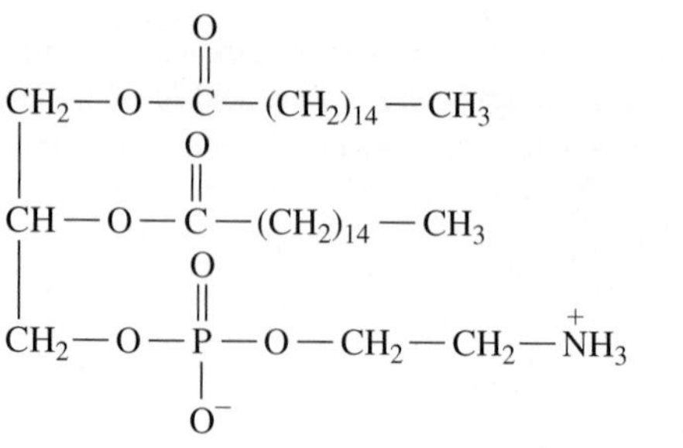

$$\begin{array}{l} CH_2-O-\overset{\overset{O}{\|}}{C}-(CH_2)_{14}-CH_3 \\ | \\ CH-O-\overset{\overset{O}{\|}}{C}-(CH_2)_{14}-CH_3 \\ | \\ CH_2-O-\underset{\underset{O^-}{|}}{\overset{\overset{O}{\|}}{P}}-O-CH_2-CH_2-\overset{+}{N}H_3 \end{array}$$

15.43 This glycerophospholipid is a cephalin. It contains glycerol, oleic acid, stearic acid, phosphate, and ethanolamine.

15.45 a. sphingosine **b.** palmitic acid
c. an amide bond **d.** ethanolamine

15.47

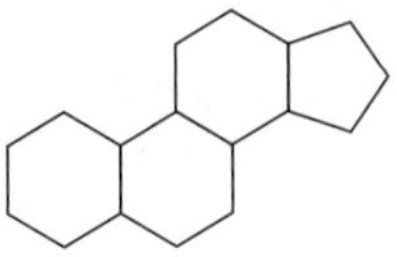

15.49 Bile salts act to emulsify fat globules, allowing the fat to be more easily digested.

15.51 Chylomicrons have a lower density than VLDLs. They pick up triacylglycerols from the intestine, whereas VLDLs transport triacylglycerols synthesized in the liver.

15.53 "Bad" cholesterol is the cholesterol carried by LDL that can form deposits in the arteries called plaque, which narrows the arteries.

15.55 Both estradiol and testosterone contain the steroid nucleus and a hydroxyl group. Testosterone has a ketone group, a double bond, and two methyl groups. Estradiol has an aromatic ring, a hydroxyl group in place of the ketone, and a methyl group.

15.57 b, **c**, and **d**

15.59 The lipid bilayer in a cell membrane surrounds the cell and separates the contents of the cell from the external fluids.

15.61 Because the molecules of cholesterol are large and rigid, they reduce the flexibility of the lipid bilayer and add strength to the cell membrane.

15.63 The peripheral proteins in the membrane emerge on the inner or outer surface only, whereas the integral proteins extend through the membrane to both surfaces.

15.65 Substances move through cell membranes by diffusion (passive transport), facilitated transport, and active transport.

a. facilitated transport **b.** diffusion (passive transport)

15.67 alkene, alcohol, ester, carboxylic acid

15.69 a. 0.56 g of Pravachol **b.** 360 mg/dL

15.71

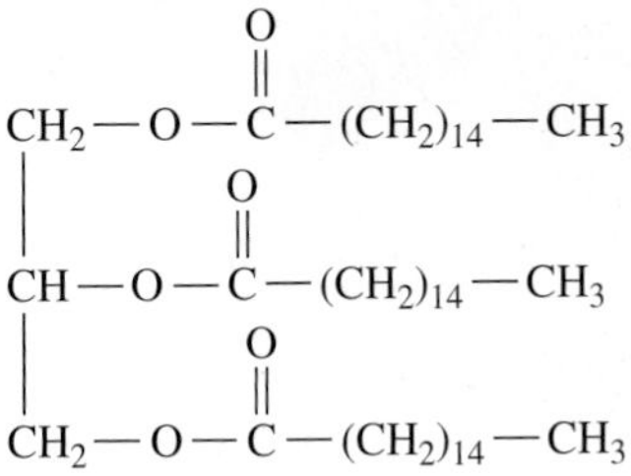

$CH_2-O-\overset{O}{\overset{\|}{C}}-(CH_2)_{14}-CH_3$
$|$
$CH-O-\overset{O}{\overset{\|}{C}}-(CH_2)_{14}-CH_3$
$|$
$CH_2-O-\overset{O}{\overset{\|}{C}}-(CH_2)_{14}-CH_3$

15.73 a. monounsaturated **b.** polyunsaturated, omega-6
c. polyunsaturated, omega-3

15.75 a. Beeswax and carnauba are waxes. Vegetable oil and glyceryl tricaprate (tricaprin) are triacylglycerols.

b.

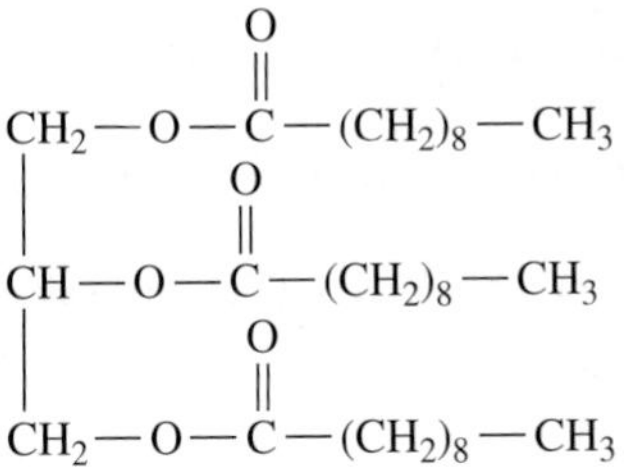

$CH_2-O-\overset{O}{\overset{\|}{C}}-(CH_2)_8-CH_3$
$|$
$CH-O-\overset{O}{\overset{\|}{C}}-(CH_2)_8-CH_3$
$|$
$CH_2-O-\overset{O}{\overset{\|}{C}}-(CH_2)_8-CH_3$

Glyceryl tricaprate (tricaprin)

15.77 a. 410 kcal from fat; 49% fat
b. 260 kcal from fat; 50.% fat
c. 160 kcal from fat; 29% fat

15.79 a. wax **b.** steroid
c. glycerophospholipid **d.** triacylglycerol
e. soap **f.** triacylglycerol

15.81 a. 5 **b.** 1, 2, 3, 4
c. 2 **d.** 1, 2

15.83 a and **c**

15.85

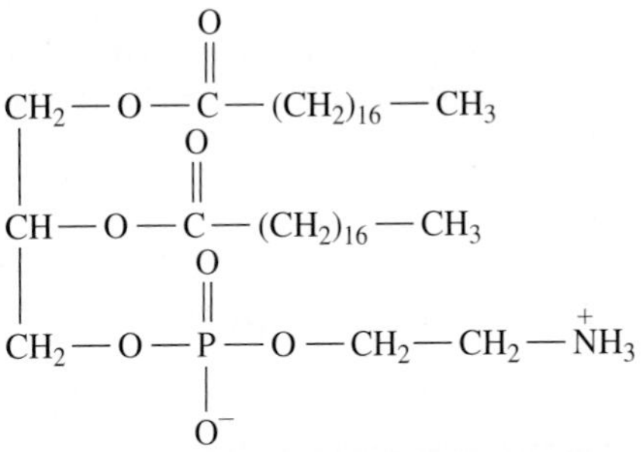

$CH_2-O-\overset{O}{\overset{\|}{C}}-(CH_2)_{16}-CH_3$
$|$
$CH-O-\overset{O}{\overset{\|}{C}}-(CH_2)_{16}-CH_3$
$|$
$CH_2-O-\overset{O}{\overset{\|}{P}}-O-CH_2-CH_2-\overset{+}{N}H_3$
$\quad\quad\quad\quad\ |$
$\quad\quad\quad\quad O^-$

15.87 a. (4) HDL **b.** (3) LDL
c. (1) chylomicrons **d.** (4) HDL

15.89 a. Adding NaOH would saponify lipids such as glyceryl tristearate (tristearin), forming glycerol and salts of the fatty acids that are soluble in water and would wash down the drain.

b.

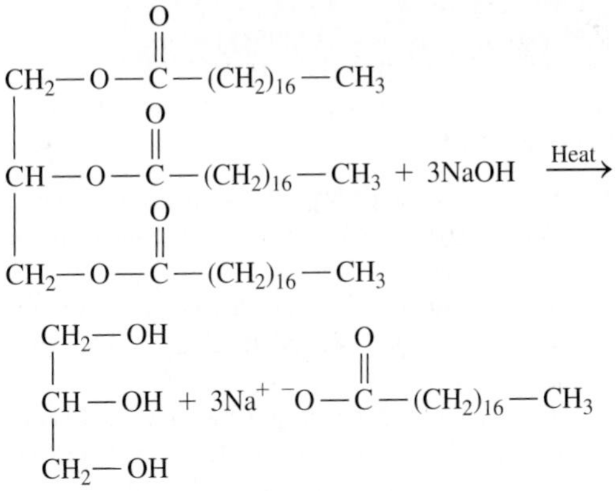

$CH_2-O-\overset{O}{\overset{\|}{C}}-(CH_2)_{16}-CH_3$
$|$
$CH-O-\overset{O}{\overset{\|}{C}}-(CH_2)_{16}-CH_3 + 3NaOH \xrightarrow{Heat}$
$|$
$CH_2-O-\overset{O}{\overset{\|}{C}}-(CH_2)_{16}-CH_3$

CH_2-OH
$|$
$CH-OH + 3Na^+ \ {}^-O-\overset{O}{\overset{\|}{C}}-(CH_2)_{16}-CH_3$
$|$
CH_2-OH

Glycerol Salts of stearic acid

c. 67.3 mL of a 0.500 M NaOH solution

15.91 a.

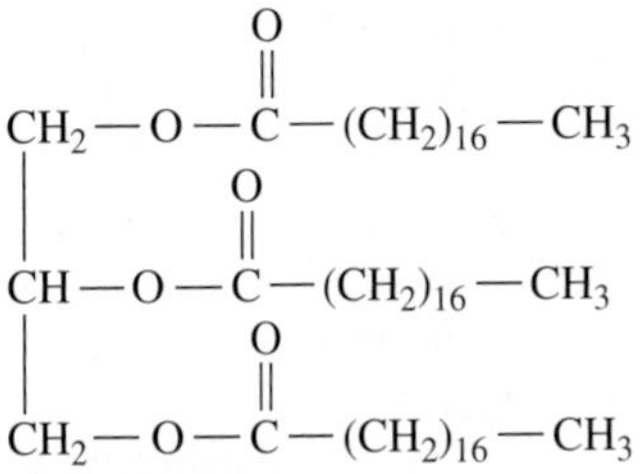

$CH_2-O-\overset{O}{\overset{\|}{C}}-(CH_2)_{16}-CH_3$
$|$
$CH-O-\overset{O}{\overset{\|}{C}}-(CH_2)_{16}-CH_3$
$|$
$CH_2-O-\overset{O}{\overset{\|}{C}}-(CH_2)_{16}-CH_3$

b. 3.00 moles of H_2
c. 6.05 g of H_2
d. 67.2 L of H_2

COMBINING IDEAS from Chapters 13 to 15

CI.27 The plastic known as PETE (polyethyleneterephthalate) is used to make plastic soft drink bottles and containers for salad dressing, shampoos, and dishwashing liquids. PETE is made from terephthalic acid and ethylene glycol. Today, PETE is the most widely recycled of all the plastics. In one year, 1.5×10^9 lb of PETE are recycled. After it is separated from other plastics, PETE can be used in polyester fabric, door mats, and tennis ball containers. The density of PETE is 1.38 g/mL. (2.6, 2.7, 14.3)

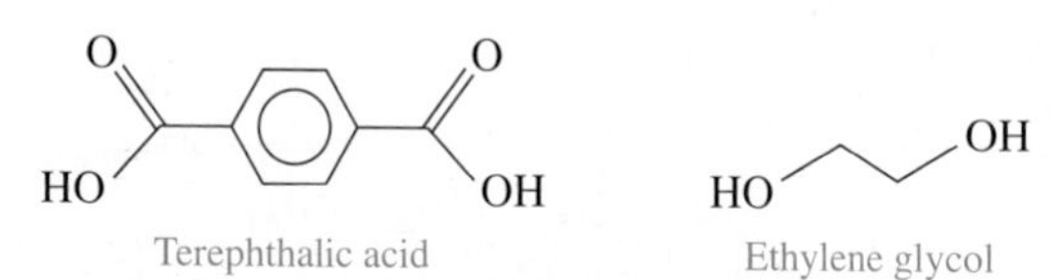

Plastic bottles made of PETE are ready to be recycled.

a. Draw the line-angle formula for the ester formed from one molecule of terephthalic acid and one molecule of ethylene glycol.
b. Draw the line-angle formula for the product formed when a second molecule of ethylene glycol reacts with the ester you drew in part **a**.
c. How many kilograms of PETE are recycled in one year?
d. What volume, in liters, of PETE is recycled in one year?
e. Suppose a landfill holds 2.7×10^7 L of recycled PETE. If all the PETE that is recycled in a year were placed instead in landfills, how many would it fill?

CI.28 Using the Internet, look up the condensed structural formulas for the following medicinal drugs and list the functional groups in the compounds: (12.1, 12.3, 14.1, 14.3, 14.5, 14.6)

a. baclofen, a muscle relaxant
b. anethole, a licorice flavoring agent in anise and fennel
c. alibendol, an antispasmodic drug
d. pargyline, an antihypertensive drug
e. naproxen, a nonsteroidal anti-inflammatory drug

CI.29 The insect repellent DEET is an amide that can be made from the amidation of 3-methylbenzoic acid with diethylamine. A 6.0-fl oz can of DEET repellent contains 25% DEET (m/v) (1 qt = 32 fl oz). (7.2, 7.3, 9.4, 14.6)

a. Draw the line-angle formula for DEET.
b. Give the molecular formula for DEET.
c. What is the molar mass of DEET?
d. How many grams of DEET are in one spray can?
e. How many molecules of DEET are in one spray can?

DEET is used in insect repellent.

CI.30 Glyceryl trimyristate (trimyristin) is found in the seeds of the nutmeg (*Myristica fragrans*). Trimyristin is used as a lubricant and fragrance in soaps and shaving creams. Isopropyl myristate is used to increase absorption of skin creams. Draw the condensed structural formula for each of the following: (15.2, 15.3, 15.4)

Nutmeg contains high levels of glyceryl trimyristate.

a. myristic acid (14:0)
b. glyceryl trimyristate (trimyristin)
c. isopropyl myristate
d. products from the hydrolysis of glyceryl trimyristate with an acid catalyst
e. products from the saponification of glyceryl trimyristate with KOH
f. reactant and product for oxidation of myristyl alcohol to myristic acid

CI.31 Panose is a trisaccharide that is being considered as a possible sweetener by the food industry. (13.4, 13.5, 13.6)

Panose

A B C

a. What are the monosaccharide units **A**, **B**, and **C** in panose?
b. What type of glycosidic bond connects monosaccharides **A** and **B**?
c. What type of glycosidic bond connects monosaccharides **B** and **C**?
d. Is the structure drawn as α- or β-panose?
e. Why would panose be a reducing sugar?

CI.32 Hyaluronic acid (HA) is a natural component of eye and joint fluid as well as skin and cartilage. Due to the ability of HA to absorb water, it is used in skin-care products and injections to smooth wrinkles and for treatment of arthritis. The repeating disaccharide units in HA consist of D-gluconic acid and *N*-acetyl-D-glucosamine. *N*-Acetyl-D-glucosamine is an amide derived from acetic acid and D-glucosamine, in which an amine group ($-NH_2$) replaces the hydroxyl on carbon 2 of D-glucose. Another natural polymer called chitin is found in the shells of crabs and lobsters. Chitin is made of repeating units of *N*-acetyl-D-glucosamine connected by $\beta(1\rightarrow4)$-glycosidic bonds. (13.4, 13.5, 13.6)

Hyaluronic acid

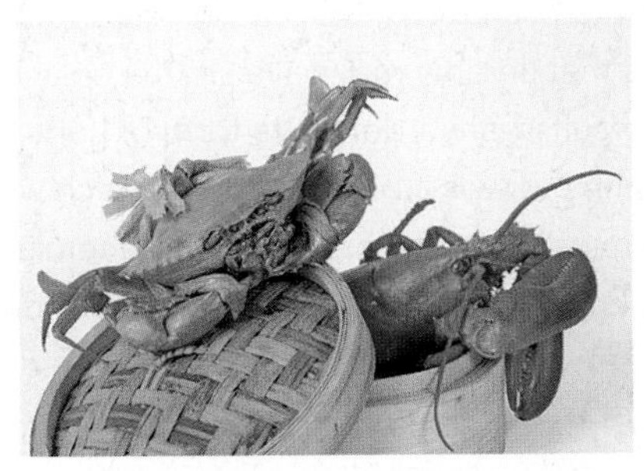
The shells of crabs and lobsters contain chitin.

a. Draw the Haworth structure for the product of the oxidation reaction of the hydroxyl group on carbon 6 in β-D-glucose to form β-D-gluconic acid.
b. Draw the Haworth structure for β-D-glucosamine.
c. Draw the Haworth structure for the amide of β-D-glucosamine and acetic acid.
d. What are the two types of glycosidic bonds that link the monosaccharides in hyaluronic acid?
e. Draw the structure for a section of chitin with two *N*-acetyl-D-glucosamine units linked by $\beta(1\rightarrow4)$-glycosidic bonds.

ANSWERS

CI.27

a.

b.

c. 6.8×10^8 kg of PETE
d. 4.9×10^8 L of PETE
e. 18 landfills

CI.29 a.

b. $C_{12}H_{17}NO$
c. 191.3 g/mole
d. 44 g of DEET
e. 1.4×10^{23} molecules

CI.31 a. **A**, **B**, and **C** are all glucose.
b. An $\alpha(1\rightarrow6)$-glycosidic bond links **A** and **B**.
c. An $\alpha(1\rightarrow4)$-glycosidic bond links **B** and **C**.
d. β-panose
e. Panose is a reducing sugar because it has a free hydroxyl group on carbon 1 of structure **C**, which allows glucose **C** to form an aldehyde.

16 Amino Acids, Proteins, and Enzymes

JEREMY, A 9-MONTH-OLD BOY RECENTLY adopted from Africa, is experiencing fever, along with painful swelling of his hands and feet. His mother makes an appointment to see Emma, his physician assistant. Emma suspects that Jeremy may have sickle-cell anemia, a disease that results from a defective form of hemoglobin in the blood. Emma draws a blood sample, which will be tested to determine if Jeremy has sickle-cell anemia. In the meantime, Emma prescribes over-the-counter pain relievers for Jeremy and suggests that his mother increase his fluid intake.

Hemoglobin is a protein that transports oxygen in the blood. Proteins are large biological molecules that carry out many different functions in living organisms. Proteins are made of smaller molecules, called amino acids, linked together in a chain. When the order of amino acids in a protein is altered, the result may be a defective protein that is no longer able to effectively carry out its original function. In the case of sickle-cell anemia, an alteration in the amino acids of the hemoglobin protein leads to patients experiencing anemia (a deficiency of red blood cells) and its accompanying symptoms.

CAREER Physician Assistant

A physician assistant, commonly referred to as a PA, helps a doctor by examining and treating patients, as well as prescribing medications. Many physician assistants take on the role of the primary caregiver. Their duties would also include obtaining patient medical records and histories, diagnosing illnesses, educating and counseling patients, and referring the patient, when needed, to a specialist. Because of this diversity, physician assistants must be knowledgeable about a variety of medical conditions. Physician assistants may also help the doctor during major surgery.

Physician assistants can work in clinics, hospitals, health maintenance organizations, private practices, or take on a more administrative role that involves hiring new PAs and acting as a representative for the hospital and patient.

CLINICAL UPDATE Jeremy's Diagnosis and Treatment for Sickle-Cell Anemia

Jeremy's blood test indicates that he has sickle-cell anemia. In the **CLINICAL UPDATE Jeremy's Diagnosis and Treatment for Sickle-Cell Anemia**, page 576, you can see how sickle-cell anemia is diagnosed, and how Jeremy is treated with medication.

16.1 Proteins and Amino Acids

LEARNING GOAL Classify proteins by their functions. Give the name and abbreviations for an amino acid and draw its structure at physiological pH.

Proteins are one of the most prevalent types of molecules in living organisms. In fact, researchers derived the name "protein" from the Greek word *proteios*, meaning "first," to indicate the central roles that proteins play in living organisms. For example, there are proteins that form structural components such as cartilage, muscles, hair, and nails. Wool, silk, feathers, and horns in animals are made of proteins (see **FIGURE 16.1**). Proteins that function as enzymes regulate biological reactions such as digestion and cellular metabolism. Other proteins, such as hemoglobin and myoglobin, transport oxygen in the blood and muscle. **TABLE 16.1** gives examples of proteins that are classified by their functions.

FIGURE 16.1 The horns of animals are made of proteins.

Q What class of protein would be in horns?

TEST

Try Practice Problems 16.1 and 16.2

TABLE 16.1 Classification of Some Proteins and Their Functions

Class of Protein	Function	Examples
Structural	Provide structural components	*Collagen* is in tendons and cartilage. *Keratin* is in hair, skin, wool, and nails.
Contractile	Make muscles move	*Myosin* and *actin* contract muscle fibers.
Transport	Carry essential substances throughout the body	*Hemoglobin* transports oxygen. *Lipoproteins* transport lipids.
Storage	Store nutrients	*Casein* stores protein in milk. *Ferritin* stores iron in the spleen and liver.
Hormone	Regulate body metabolism and the nervous system	*Insulin* regulates blood glucose level. *Growth hormone* regulates body growth.
Enzyme	Catalyze biochemical reactions in the cells	*Sucrase* catalyzes the hydrolysis of sucrose. *Trypsin* catalyzes the hydrolysis of proteins.
Protection	Recognize and destroy foreign substances	*Immunoglobulins* stimulate immune responses.

Although there are many types of proteins with many different functions, all proteins are made from the same building blocks. Proteins are formed when smaller molecules called amino acids link together in a chain. The specific order of the amino acids in the chain determines how the protein will fold into its three-dimensional shape, which determines the protein's function.

Just how important are the amino acids in determining the function of a protein? Sickle-cell anemia is a disease caused by an abnormality in one of the subunits of the hemoglobin protein. This abnormal protein causes red blood cells (RBCs) to change from a rounded shape to a crescent shape, like a sickle, which interferes with their ability to transport adequate quantities of oxygen. Patients who suffer from sickle-cell anemia experience fatigue, shortness of breath, dizziness, headaches, coldness in the hands and feet, and even jaundice. All of these problems are caused by the substitution of *one* amino acid out of more than 800 in the hemoglobin molecule. To understand why one amino acid can have such a significant effect on the structure and function of a large protein molecule, we need to look at the structures of amino acids.

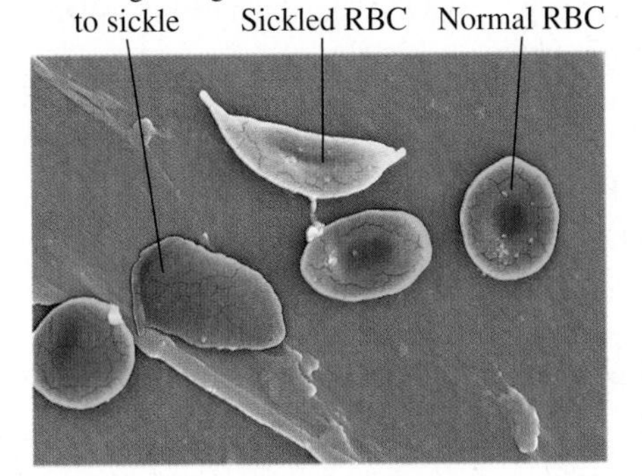

Red blood cells containing normal hemoglobin have a rounded shape, whereas red blood cells containing sickled hemoglobin have a crescent or sickle shape.

Amino Acids

Proteins are composed of molecular building blocks called *amino acids*. There are many amino acids in nature; however, there are only 20 amino acids commonly found in the proteins of living organisms. Every **amino acid** has a central carbon atom called the α carbon bonded to two α functional groups: an ammonium group ($-NH_3^+$) and a carboxylate group ($-COO^-$). The α carbon is also bonded to a hydrogen atom and a side chain called an R group. The differences in the 20 amino acids present in human proteins are due to the unique characteristics of the R groups. For example, alanine has a methyl, $-CH_3$, as its R group.

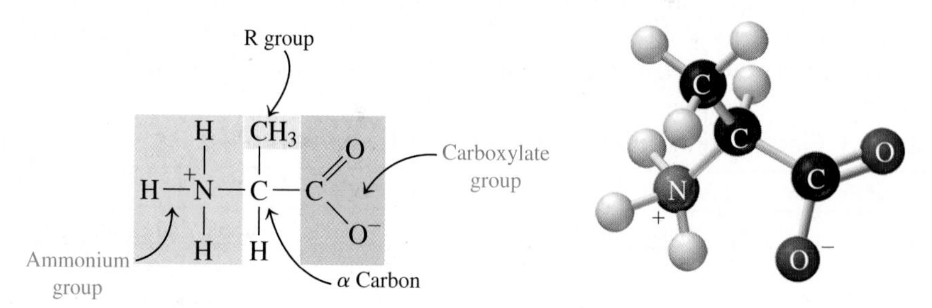

Ionized form of alanine

Ball-and-stick model of alanine

Classification of Amino Acids

We can now classify amino acids using their specific R groups, which determine their properties in aqueous solution. The **nonpolar amino acids** have hydrogen, alkyl, or aromatic R groups, which make them *hydrophobic* ("water fearing"). The **polar amino acids** have R groups that interact with water, which makes them *hydrophilic* ("water loving"). There are three groups of polar amino acids. The R groups of *polar neutral amino acids* contain hydroxyl ($-OH$), thiol ($-SH$), or amide ($-CONH_2$) groups. The R group of a polar **acidic amino acid** contains a carboxylate group ($-COO^-$). The R group of a polar **basic amino acid** contains an amino group, which ionizes to give an ammonium ion. The names and structures of the 20 alpha amino acids commonly found in proteins, along with their three-letter and one-letter abbreviations, are shown at physiological pH (7.4) in **TABLE 16.2**.

SAMPLE PROBLEM 16.1 Polarity of Amino Acids

TRY IT FIRST

Classify each of the following amino acids as nonpolar or polar. If polar, indicate if the R group is neutral, acidic, or basic. Indicate if each would be hydrophobic or hydrophilic.

a. valine **b.** asparagine

ENGAGE

Why is leucine classified as a nonpolar amino acid, whereas threonine is classified as a polar amino acid?

SOLUTION

a. The R group in valine is an alkyl group consisting of C atoms and H atoms, which makes valine a nonpolar amino acid that is hydrophobic.

b. The R group in asparagine contains an amide group ($-CONH_2$), which makes asparagine a polar amino acid that is neutral and hydrophilic.

STUDY CHECK 16.1

Classify the amino acid lysine as nonpolar or polar. If polar, indicate if the R group is neutral, acidic, or basic and if it would be hydrophobic or hydrophilic.

ANSWER

The R group in lysine contains an amino group, which makes lysine a polar amino acid that is basic and hydrophilic.

TEST

Try Practice Problems 16.3 and 16.4, 16.7 and 16.8

TABLE 16.2 Structures, Names, and Abbreviations of 20 Common Amino Acids at Physiological pH (7.4)

Nonpolar Amino Acids (Hydrophobic)

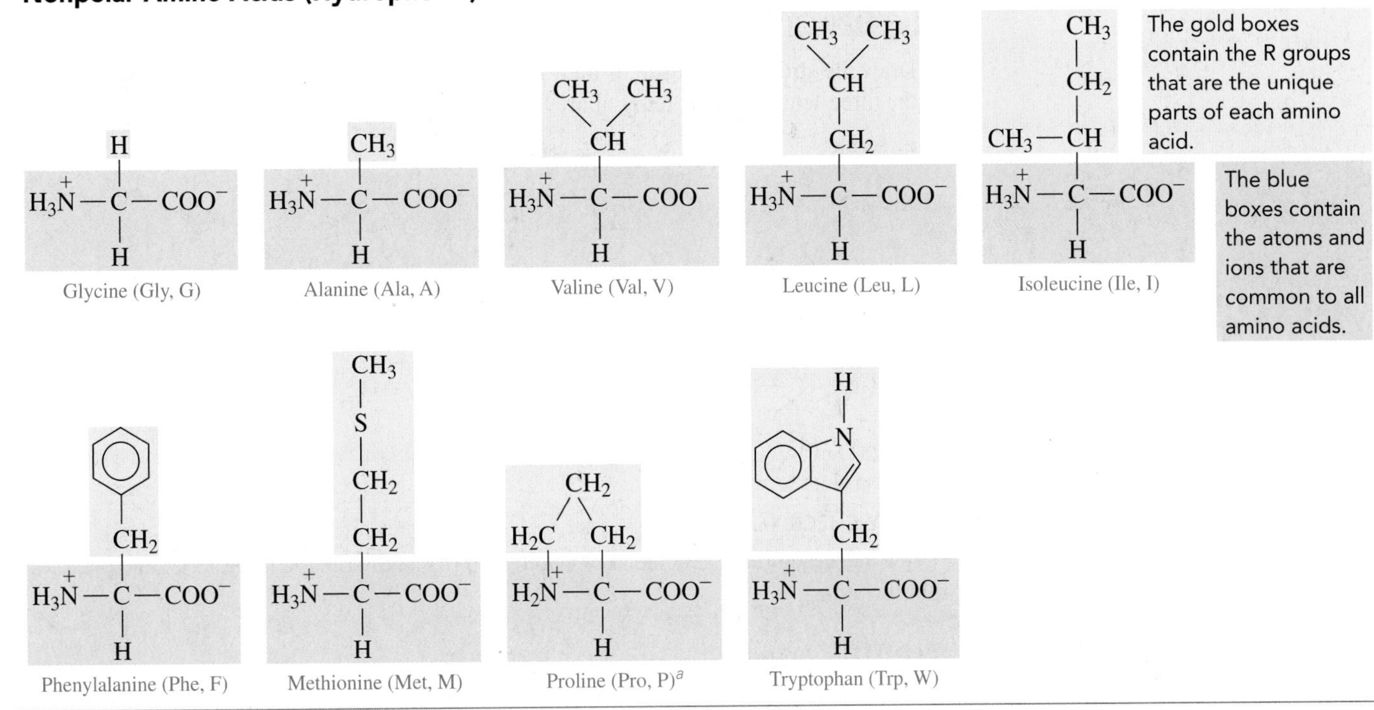

Polar Neutral Amino Acids (Hydrophilic)

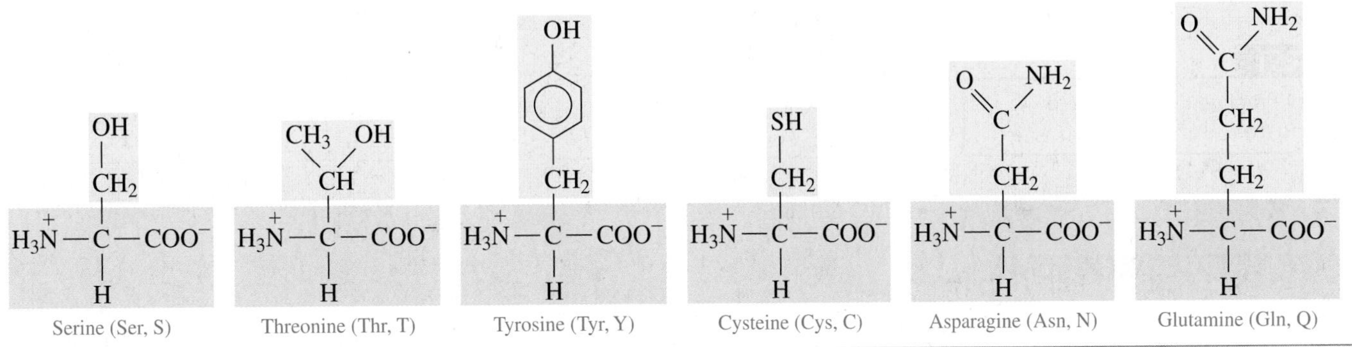

Polar Acidic Amino Acids (Hydrophilic)

Aspartate (Asp, D)

Glutamate (Glu, E)

Polar Basic Amino Acids (Hydrophilic)

Histidine (His, H)[b]

Lysine (Lys, K)

Arginine (Arg, R)

[a] Proline is a cyclic amino acid because its R group bonds to the nitrogen atom attached to the α carbon.

[b] At physiological pH, some histidine molecules have a positively charged R group, while other histidine molecules have a neutral R group. The molecule with the positively charged R group is shown here.

SAMPLE PROBLEM 16.2 Structural Formulas of Amino Acids

TRY IT FIRST

Draw the structure for each of the following amino acids at physiological pH, and write the three-letter and one-letter abbreviations:

a. serine **b.** aspartate

SOLUTION

CORE CHEMISTRY SKILL

Drawing the Structure for an Amino Acid at Physiological pH

$$\text{a. } \begin{array}{c} \text{OH} \\ | \\ \text{CH}_2\ \ \text{O} \\ |\quad\ \ \| \\ \text{H}_3\overset{+}{\text{N}}-\text{C}-\text{C}-\text{O}^- \\ | \\ \text{H} \end{array} \qquad \text{b. } \begin{array}{c} \text{O}\ \ \ \text{O}^- \\ \diagdown\ \diagup \\ \text{C} \\ | \\ \text{CH}_2\ \ \text{O} \\ |\quad\ \ \| \\ \text{H}_3\overset{+}{\text{N}}-\text{C}-\text{C}-\text{O}^- \\ | \\ \text{H} \end{array}$$

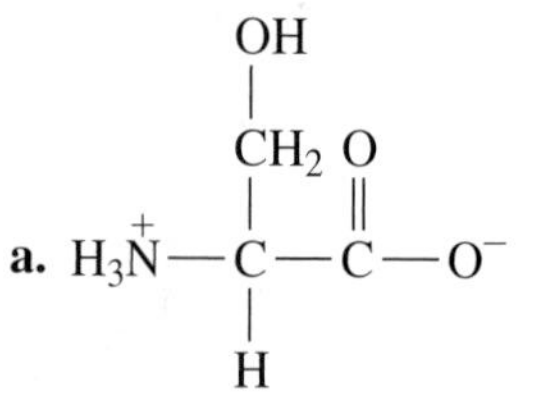

Serine (Ser, S) 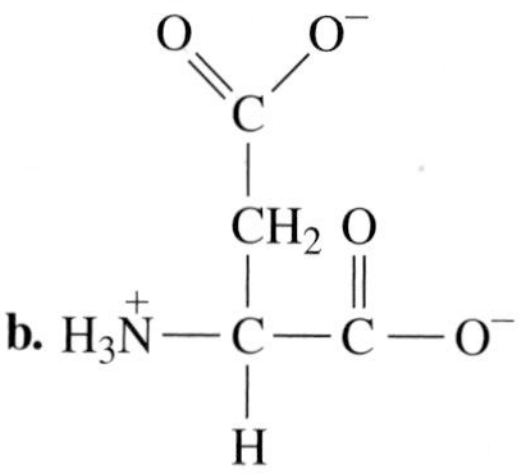 Aspartate (Asp, D)

STUDY CHECK 16.2

Draw the structure for leucine at physiological pH, and write the three-letter and one-letter abbreviations.

ANSWER

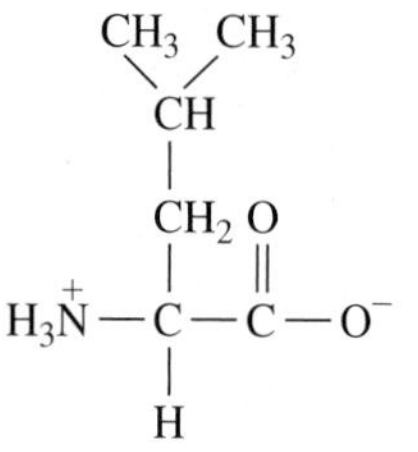

$$\begin{array}{c} \text{CH}_3\ \ \text{CH}_3 \\ \diagdown\ \diagup \\ \text{CH} \\ | \\ \text{CH}_2\ \ \text{O} \\ |\quad\ \ \| \\ \text{H}_3\overset{+}{\text{N}}-\text{C}-\text{C}-\text{O}^- \\ | \\ \text{H} \end{array}$$

Leucine (Leu, L)

TEST

Try Practice Problems 16.5 and 16.6, 16.9 and 16.10

PRACTICE PROBLEMS

16.1 Proteins and Amino Acids

LEARNING GOAL Classify proteins by their functions. Give the name and abbreviations for an amino acid and draw its structure at physiological pH.

16.1 Classify each of the following proteins according to its function:
a. hemoglobin, oxygen carrier in the blood
b. collagen, a major component of tendons and cartilage
c. keratin, a protein found in hair
d. amylases that catalyze the hydrolysis of starch

16.2 Classify each of the following proteins according to its function:
a. insulin, a protein needed for glucose utilization
b. antibodies that disable foreign proteins
c. casein, milk protein
d. lipases that catalyze the hydrolysis of lipids

16.3 What functional groups are found in all α-amino acids?

16.4 How does the polarity of the R group in leucine compare to the R group in serine?

16.5 Draw the structure for each of the following amino acids at physiological pH:
a. glycine **b.** T **c.** glutamate **d.** Phe

16.6 Draw the structure for each of the following amino acids at physiological pH:
a. lysine **b.** proline **c.** V **d.** Tyr

16.7 Classify each of the amino acids in problem 16.5 as polar or nonpolar. If polar, indicate if the R group is neutral, acidic, or basic. Indicate if each is hydrophobic or hydrophilic.

16.8 Classify each of the amino acids in problem 16.6 as polar or nonpolar. If polar, indicate if the R group is neutral, acidic, or basic. Indicate if each is hydrophobic or hydrophilic.

16.9 Give the name for the amino acid represented by each of the following abbreviations:
a. Ala **b.** Q **c.** K **d.** Cys

16.10 Give the name for the amino acid represented by each of the following abbreviations:
a. Trp **b.** M **c.** Pro **d.** G

16.2 Proteins: Primary Structure

REVIEW

Forming Amides (14.6)

LEARNING GOAL Draw the condensed structural formula for a peptide and give its name. Describe the primary structure for a protein.

Now that we know about the structure of individual amino acids, we can look at how amino acids link together to form a protein. A **peptide bond** is an amide bond that forms when the $-COO^-$ group of one amino acid reacts with the $-NH_3^+$ group of the next amino acid. The linking of two or more amino acids by peptide bonds forms a **peptide**. An O atom is removed from the carboxylate end of one amino acid, and two H atoms are removed from the ammonium end of the other amino acid, which produces water. Two amino acids form a *dipeptide*, three amino acids form a *tripeptide*, and four amino acids form a *tetrapeptide*. A chain of five amino acids is a *pentapeptide*, and longer chains of amino acids are *polypeptides*.

We can write the amidation reaction for the formation of a dipeptide formed between glycine and alanine, glycylalanine (Gly–Ala, GA) (see **FIGURE 16.2**). During the amidation,

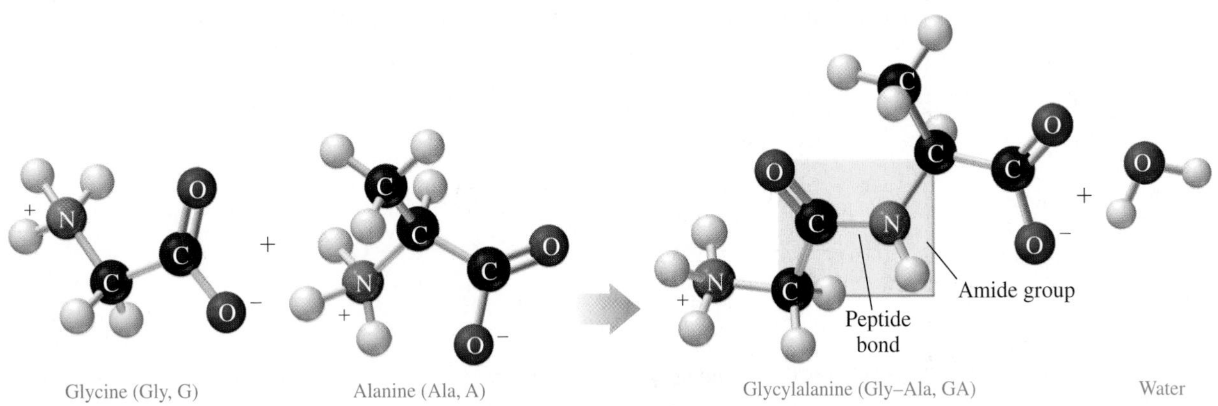

FIGURE 16.2 A peptide bond between glycine and alanine forms the dipeptide glycylalanine.

Q What functional groups in glycine and alanine form the peptide bond?

the O atom removed from the carboxylate group of glycine combines with two H atoms from the $-NH_3^+$ group of alanine to produce H_2O. The amino acid written on the left, glycine, has a free (unbonded) $-NH_3^+$ group, which makes it the **N-terminus**. In the peptide, the amino acid written on the right, alanine, has a free (unbonded) $-COO^-$ group, which makes it the **C-terminus**. The dipeptide forms when the carbonyl group in glycine bonds to the N atom in the $-NH_3^+$ group of alanine.

```
                                                              N-terminus   Peptide bond    C-terminus
      H   H   O              H   CH3 O                        H   H   O           CH3 O
      |+  |   ||             |+  |   ||                       |+  |   ||          |   ||
  H—N—C—C—O⁻  +  H—N—C—C—O⁻   ⟶    H—N—C—C—N—C—C—O⁻  + H2O
      |   |                  |   |                            |   |       |   |
      H   H                  H   H                            H   H       H   H
      Glycine                Alanine                      Glycylalanine (Gly–Ala, GA)
```

Naming Peptides

By convention, peptides are drawn and named from N-terminus to C-terminus. With the exception of the amino acid at the C-terminus, the names of all the other amino acids in a peptide end with *yl*. For example, a tripeptide consisting of alanine at the N-terminus, glycine, and serine at the C-terminus is named as one word: alan**yl**glyc**yl**serine. For convenience, the order of amino acids in the peptide is often written as the sequence of three-letter or one-letter abbreviations.

N-terminus | C-terminus

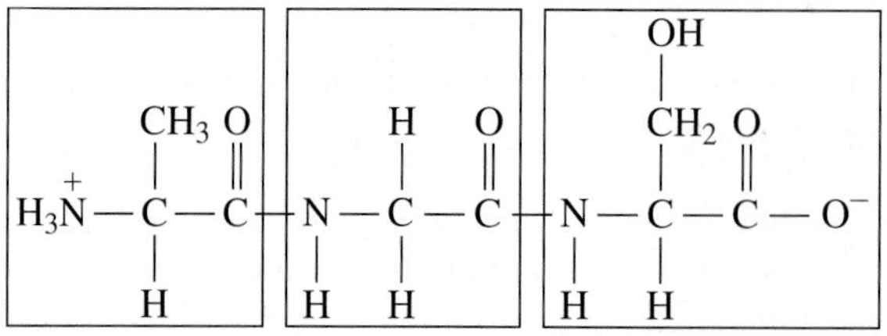

$H_3\overset{+}{N}-CH(CH_3)-C(=O)-NH-CH_2-C(=O)-NH-CH(CH_2OH)-C(=O)-O^-$

From alanine alanyl | From glycine glycyl | From serine serine

Alanylglycylserine
(Ala–Gly–Ser, AGS)

SAMPLE PROBLEM 16.3 Identifying a Tripeptide

TRY IT FIRST

Answer the questions for the tripeptide that is shown below:

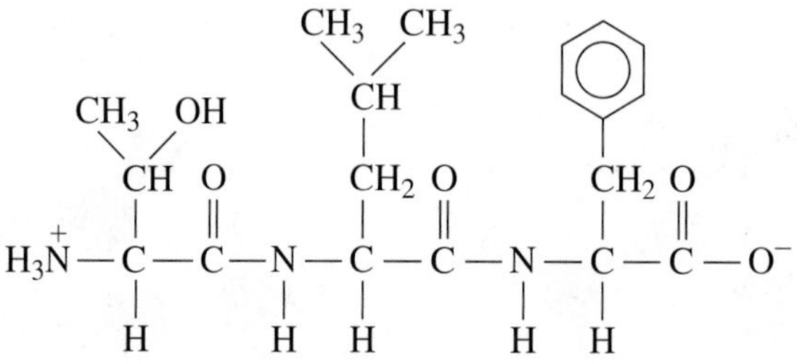

$H_3\overset{+}{N}-CH(CH(CH_3)OH)-C(=O)-NH-CH(CH_2CH(CH_3)_2)-C(=O)-NH-CH(CH_2C_6H_5)-C(=O)-O^-$

a. What is the amino acid at the N-terminus? What is the amino acid at the C-terminus?

b. Use the three-letter and one-letter abbreviations of amino acids to give the amino acid order in the tripeptide.

c. What is the name of the tripeptide?

SOLUTION

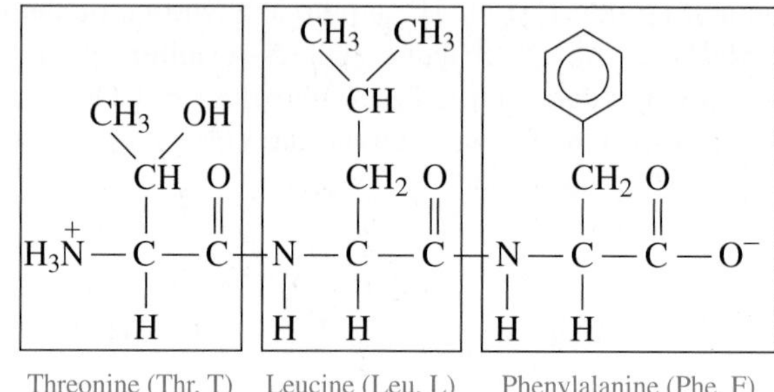

Threonine (Thr, T) N-terminus | Leucine (Leu, L) | Phenylalanine (Phe, F) C-terminus

a. Threonine is the amino acid at the N-terminus; phenylalanine is the amino acid at the C-terminus.

b. Thr–Leu–Phe; TLF

c. threonylleucylphenylalanine

STUDY CHECK 16.3

A tripeptide has the abbreviation: Pro–His–Met.

a. What is the amino acid at the N-terminus?

b. What is the amino acid at the C-terminus?

c. What is the name of the tripeptide?

ANSWER

a. proline **b.** methionine **c.** prolylhistidylmethionine

CHEMISTRY LINK TO HEALTH

Essential Amino Acids and Complete Proteins

Of the 20 common amino acids used to build the proteins in the body, only 11 can be synthesized in the body. The other 9 amino acids, listed in **TABLE 16.3**, are *essential amino acids* that must be obtained from the proteins in the diet.

Complete proteins, which contain all of the essential amino acids, are found in most animal products such as eggs, milk, meat, fish, and poultry. However, gelatin and plant proteins such as grains, beans, and nuts are *incomplete proteins* because they are deficient in one or more of the essential amino acids. Diets that rely on plant foods for protein must contain a variety of protein sources to obtain all the essential amino acids. For example, a diet of rice and beans contains all the essential amino acids because they are *complementary protein* sources.

Complete proteins such as eggs, milk, meat, and fish contain all of the essential amino acids. Incomplete proteins from plants such as grains, beans, and nuts are deficient in one or more essential amino acids.

Rice contains the methionine and tryptophan that are deficient in beans, while beans contain the lysine that is lacking in rice (see **TABLE 16.4**).

TABLE 16.3 Essential Amino Acids for Adults

Histidine (His, H)	Phenylalanine (Phe, F)
Isoleucine (Ile, I)	Threonine (Thr, T)
Leucine (Leu, L)	Tryptophan (Trp, W)
Lysine (Lys, K)	Valine (Val, V)
Methionine (Met, M)	

TABLE 16.4 Amino Acid Deficiencies in Selected Vegetables and Grains

Food Source	Amino Acid Deficiency
Eggs, milk, meat, fish, poultry	None
Wheat, rice, oats	Lysine
Corn	Lysine, tryptophan
Beans	Methionine, tryptophan
Peas, peanuts	Methionine
Almonds, walnuts	Lysine, tryptophan
Soy	Methionine

Primary Structure of a Protein

A **protein** is a polypeptide of 50 or more amino acids that has biological activity. The **primary structure** of a protein is the particular sequence of amino acids held together by peptide bonds from N- to C-terminus. For example, a hormone that stimulates the thyroid to release thyroxin is a tripeptide with the amino acid sequence Glu–His–Pro, EHP. Although five other amino acid sequences of the same three amino acids are possible, such as His–Pro–Glu or Pro–His–Glu, they do not produce hormonal activity. In the figure below, the R groups of the amino acids are colored red. The atoms colored black are known as the *backbone* of the peptide or protein, which is the repeating sequence of the **N** in the ammonium group, the **C** from α carbon, and the **C** from the carboxylate group (**—N—C—C—N—C—C—N—C—C—**). Thus the biological function of peptides and proteins depends on the specific sequence of the amino acids.

TEST

Try Practice Problems 16.15 and 16.16

ENGAGE

How can two peptides with exactly the same number and types of amino acids have different primary structures?

$$\overset{+}{H_3N}-\underset{\text{Glu, E}}{\underset{|}{CH}}(CH_2CH_2COO^-)-\overset{O}{\overset{\|}{C}}-\underset{H}{\underset{|}{N}}-\underset{\text{His, H}}{CH}(CH_2\text{-imidazolium})-\overset{O}{\overset{\|}{C}}-N-\underset{\text{Pro, P}}{CH}-\overset{O}{\overset{\|}{C}}-O^-$$

The first protein to have its primary structure determined was insulin, which was accomplished by Frederick Sanger in 1953. Since that time, scientists have determined the amino acid sequences of thousands of proteins. Insulin is a hormone that regulates the glucose level in the blood. In the primary structure of human insulin, there are two polypeptide

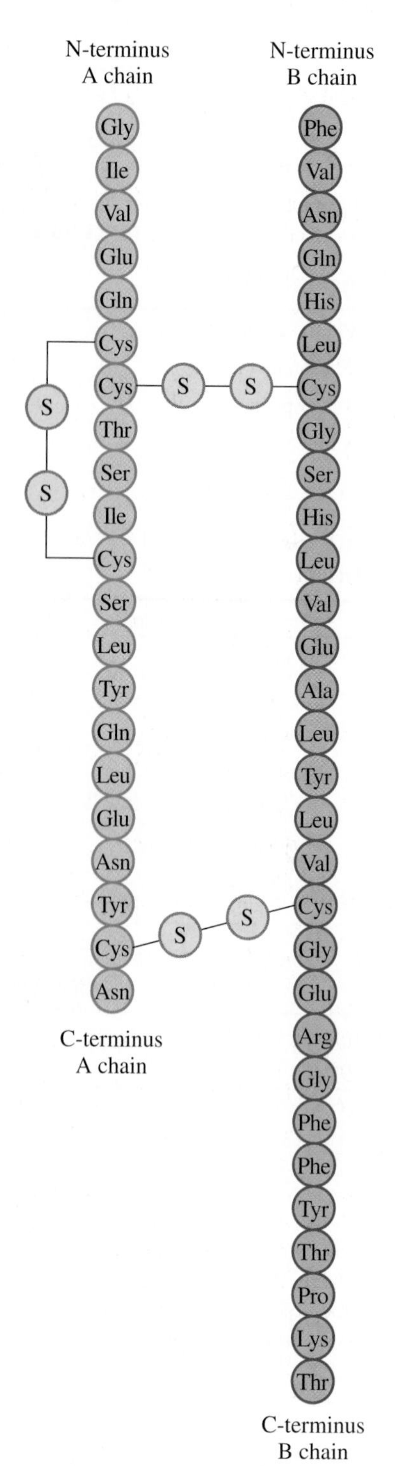

FIGURE 16.3 The sequence of amino acids in human insulin is its primary structure.

What kinds of bonds occur in the primary structure of a protein?

chains. In chain A, there are 21 amino acids, and chain B has 30 amino acids. The polypeptide chains are held together by *disulfide bonds* formed by the thiol groups of the cysteine amino acids in each of the chains (see **FIGURE 16.3**). Today, human insulin with this exact structure is produced through genetic engineering.

SAMPLE PROBLEM 16.4 Drawing a Peptide

TRY IT FIRST

Draw the structure and give the name for the tripeptide Gly–Ser–Met.

SOLUTION GUIDE

	Given	Need	Connect
ANALYZE THE PROBLEM	Gly–Ser–Met	tripeptide structure, name	peptide bonds, amino acid names

STEP 1 Draw the structure for each amino acid in the peptide, starting with the N-terminus.

```
                                                          CH3
                                                          |
                                                          S
                                                          |
                            OH                            CH2
                            |                             |
    H  H  O            H    CH2 O               H    CH2 O
   +|  |  ||          +|    |   ||             +|    |   ||
 H—N—C—C—O⁻        H—N—C—C—O⁻            H—N—C—C—O⁻
    |  |               |    |                   |    |
    H  H               H    H                   H    H
   Gly, G              Ser, S                  Met, M
  N-terminus                                 C-terminus
```

STEP 2 Remove the O atom from the carboxylate group of the N-terminus and two H atoms from the ammonium group in the adjacent amino acid. Repeat this process until the C-terminus is reached.

```
                                                  CH3
                                                  |
                                                  S
                                                  |
                          OH                      CH2
                          |                       |
    H  H  O               CH2 O                   CH2 O
   +|  |  ||              |   ||                  |   ||
 H—N—C—C            —N—C—C               —N—C—C—O⁻
    |  |               |  |                   |  |
    H  H               H  H                   H  H
   Gly, G              Ser, S                 Met, M
  N-terminus                                C-terminus
```

STEP 3 Use peptide bonds to connect the amino acids.

```
                                        CH3
                                        |
                                        S
                                        |
                       OH               CH2
                       |                |
    H  H  O            CH2 O            CH2 O
   +|  |  ||           |   ||           |   ||
 H—N—C—C—N—C—C—N—C—C—O⁻
    |  |       |  |        |  |
    H  H       H  H        H  H
 N-terminus                     C-terminus
               Gly–Ser–Met, GSM
```

The tripeptide is named by replacing the last syllable of each amino acid name with *yl*, starting with the N-terminus. The C-terminus retains its complete amino acid name.

N-terminus glycine is named glycyl
serine is named seryl
C-terminus methionine keeps its full name

The tripeptide is named glycylserylmethionine.

STUDY CHECK 16.4

Draw the structure and give the name for Phe–Thr, a section in glucagon, which is a peptide hormone that increases blood glucose levels.

ANSWER

$$H_3\overset{+}{N}-\underset{H}{\overset{CH_2C_6H_5}{C}}-\overset{O}{\overset{\|}{C}}-\underset{H}{N}-\underset{H}{\overset{CH(OH)CH_3}{C}}-\overset{O}{\overset{\|}{C}}-O^-$$

Phenylalanylthreonine

TEST
Try Practice Problems 16.11 to 16.14

CHEMISTRY LINK TO HEALTH

Polypeptides in the Body

Enkephalins and endorphins are natural painkillers produced in the body. They are polypeptides that bind to receptors in the brain to give relief from pain. This effect appears to be responsible for the "runner's high," for the temporary loss of pain when severe injury occurs, and for the analgesic effects of acupuncture.

The *enkephalins*, which are found in the thalamus and the spinal cord, are pentapeptides, the smallest molecules with opiate activity. The amino acid sequence of met-enkephalin is found in the longer amino acid sequence of the endorphins.

Four groups of *endorphins* have been identified: α-endorphin contains 16 amino acids, β-endorphin contains 31 amino acids, γ-endorphin has 17 amino acids, and δ-endorphin has 27 amino acids. Endorphins may produce their sedating effects by preventing the release of substance P, a polypeptide with 11 amino acids, which has been found to transmit pain impulses to the brain.

Two hormones produced by the pituitary gland are the nonapeptides (peptides with nine amino acids) oxytocin and vasopressin. Oxytocin stimulates uterine contractions in labor and vasopressin is an antidiuretic hormone that regulates blood pressure by adjusting the amount of water reabsorbed by the kidneys. The structures of these nonapeptides are very similar. Only the amino acids in positions 3 and 8 are different. However, the difference of two amino acids greatly affects how the two hormones function in the body.

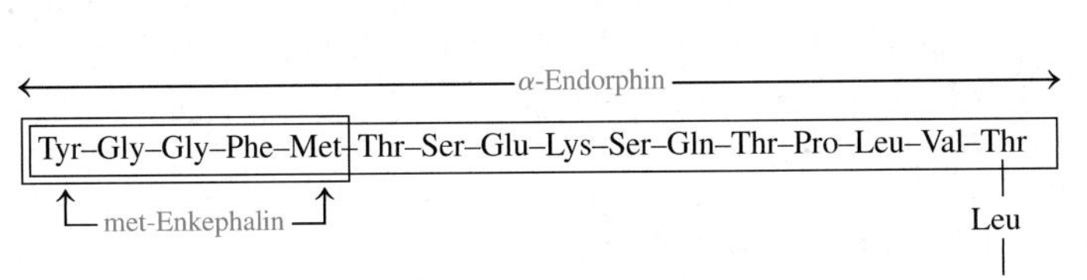

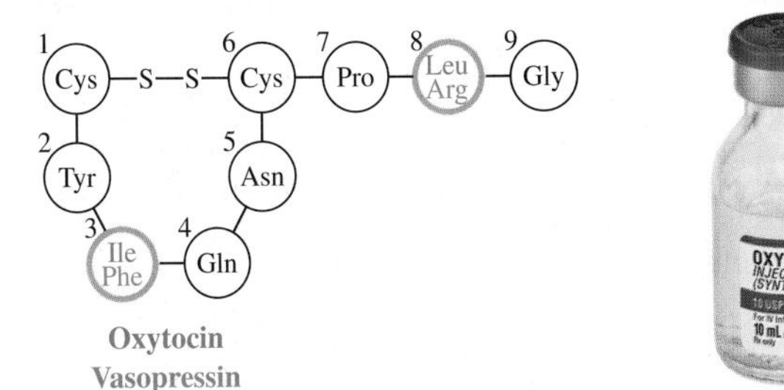

Oxytocin, a nonapeptide used to initiate labor, was the first hormone to be synthesized in the laboratory.

PRACTICE PROBLEMS

16.2 Proteins: Primary Structure

LEARNING GOAL Draw the condensed structural formula for a peptide and give its name. Describe the primary structure for a protein.

16.11 Draw the condensed structural formula for each of the following peptides, and give its three-letter and one-letter abbreviations:
a. alanylcysteine
b. serylphenylalanine
c. glycylalanylvaline
d. valylisoleucyltryptophan

16.12 Draw the condensed structural formula for each of the following peptides, and give its three-letter and one-letter abbreviations:
a. prolylaspartate
b. threonylleucine
c. methionylglutaminyllysine
d. histidylglycylglutamylisoleucine

Clinical Applications

16.13 Peptides isolated from rapeseed that may lower blood pressure have the following sequence of amino acids. Draw the structure for each peptide and write the one-letter abbreviations.
a. Arg–Ile–Tyr **b.** Val–Trp–Ile–Ser

16.14 Peptides from sweet potato with antioxidant properties have the following sequence of amino acids. Draw the structure for each peptide and write the one-letter abbreviations.
a. Asp–Cys–Gly–Tyr **b.** Asn–Tyr–Asp–Glu–Tyr

16.15 Explain why each of the following pairs are complementary proteins:
a. corn and peas **b.** rice and soy

16.16 Explain why each of the following pairs are complementary proteins:
a. beans and oats **b.** almonds and peanuts

16.3 Proteins: Secondary, Tertiary, and Quaternary Structures

LEARNING GOAL Describe the secondary, tertiary, and quaternary structures for a protein; describe the denaturation of a protein.

CORE CHEMISTRY SKILL
Identifying the Primary, Secondary, Tertiary, and Quaternary Structures of Proteins

The **secondary structure** of a protein describes the type of structure that forms when the atoms in the backbone of a protein or peptide form hydrogen bonds within a polypeptide or between polypeptide chains. The most common types of secondary structure are the *alpha helix* and the *beta-pleated sheet.*

Alpha Helix

In an **alpha helix (α helix)**, hydrogen bonds form between the oxygen of the C=O groups and the hydrogen of the N—H groups of the amide bonds in the next turn of the α helix (see FIGURE 16.4). The formation of many hydrogen bonds along the polypeptide chain gives the characteristic helical shape of a spiral staircase. All of the R groups of the different amino acids extend to the outside of the helix.

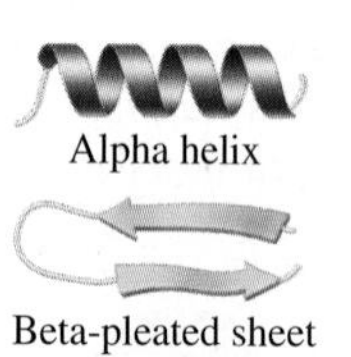

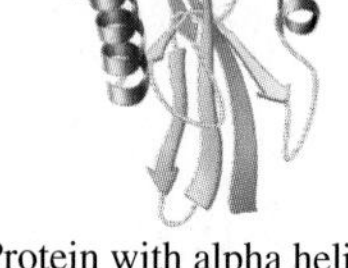

A ribbon model of a protein shows the regions of alpha helices and beta-pleated sheets.

Beta-Pleated Sheet

Another type of secondary structure found in proteins is the **beta-pleated sheet (β-pleated sheet)**. In a β-pleated sheet, hydrogen bonds form between the oxygen atoms in the carbonyl groups in one section of the polypeptide chain, and the hydrogen atoms in the N—H groups of the amide bonds in a nearby section of the polypeptide chain. A beta-pleated sheet can form between adjacent polypeptide chains or within the same polypeptide chain. The hydrogen bonds holding the β-pleated sheets tightly in place account for the strength and durability of proteins such as silk.

The tendency to form various kinds of secondary structures depends on the amino acids in a particular segment of the polypeptide chain. Typically, beta-pleated sheets contain mostly amino acids with small R groups such as glycine, valine, alanine, and serine, which extend above and below the beta-pleated sheet. The α-helical regions in a protein have higher amounts of amino acids with large R groups such as histidine, leucine, and methionine (see FIGURE 16.5).

TEST
Try Practice Problems 16.17 to 16.20

Collagen

Collagen, the most abundant protein in the body, makes up as much as one-third of all protein in vertebrates. It is found in connective tissue, blood vessels, skin, tendons, ligaments, the cornea of the eye, and cartilage. The strong structure of collagen is a result of three polypeptides woven together like a braid to form a **triple helix** (see FIGURE 16.6).

Collagen has a high content of glycine (33%), proline (22%), alanine (12%), and smaller amounts of hydroxyproline and hydroxylysine, which are modified forms of proline

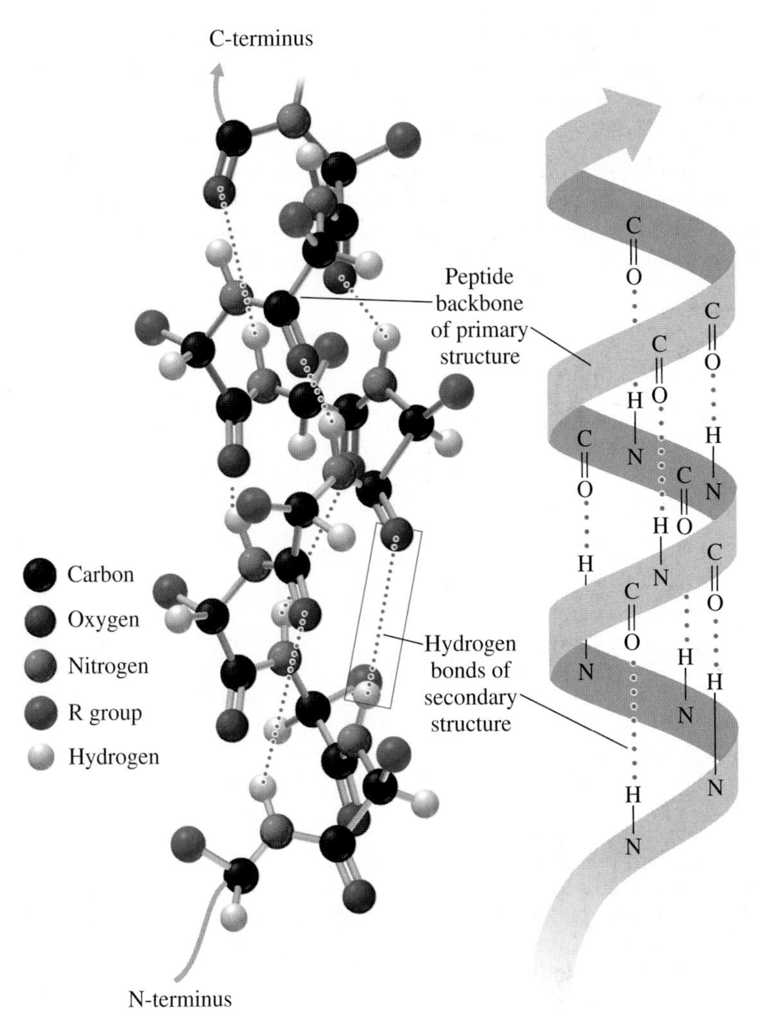

FIGURE 16.4 The α helix acquires a coiled shape from hydrogen bonds between the oxygen of the C=O group and the hydrogen of the N—H group in the next turn.

Q What are the partial charges of the H in N—H and the O in C=O that permit hydrogen bonds to form?

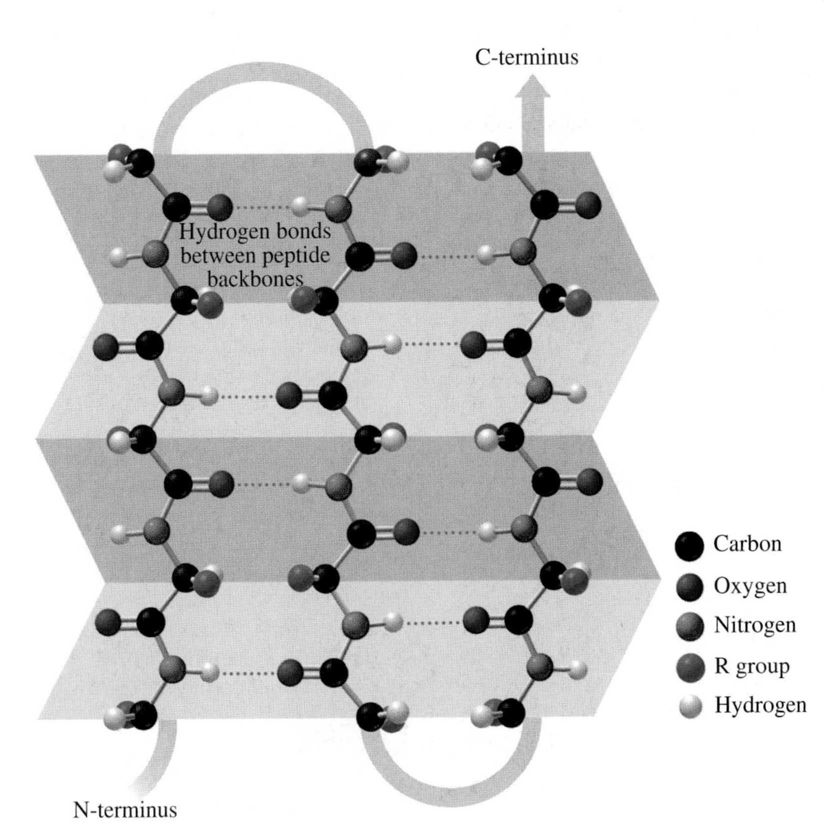

FIGURE 16.5 In the secondary structure of a β-pleated sheet, hydrogen bonds form between the peptide chains.

Q How do the hydrogen bonds in a β-pleated sheet differ from the hydrogen bonds in an α helix?

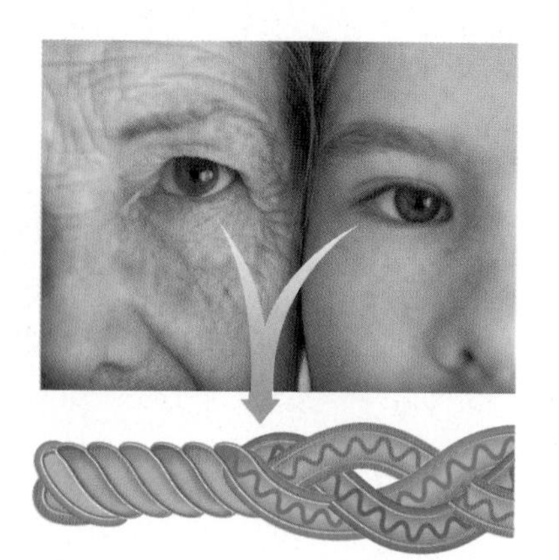

FIGURE 16.6 Collagen fibers are triple helices of polypeptide chains held together by hydrogen bonds.

Q What are some of the amino acids in collagen that form hydrogen bonds between the polypeptide chains?

and lysine. The —OH groups on these modified amino acids provide additional hydrogen bonds between the peptide chains to give strength to the collagen triple helix. When several triple helices wrap together as a braid, they form the fibrils that make up connective tissues and tendons. When a diet is deficient in vitamin C, collagen fibrils are weakened because the enzymes needed to form hydroxyproline and hydroxylysine require vitamin C. Because there are fewer —OH groups, there is less hydrogen bonding between collagen fibrils. Collagen becomes less elastic as a person ages because additional bonds form between the fibrils. Bones, cartilage, and tendons become more brittle, and wrinkles are seen as the skin loses elasticity.

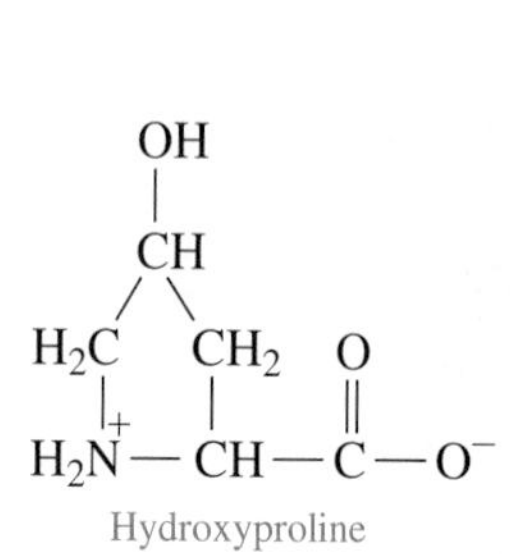

$$H_3\overset{+}{N}-CH(CH_2-CH_2-CH(OH)-CH_2-\overset{+}{N}H_3)-C(=O)-O^-$$

Hydroxylysine

Hydroxyproline and hydroxylysine provide additional hydrogen bonds in the triple helices of collagen.

CHEMISTRY LINK TO HEALTH

Protein Secondary Structures and Alzheimer's Disease

Alzheimer's disease is a form of dementia in which a person has increasing memory loss and inability to handle daily tasks. Alzheimer's usually occurs after age 60 and is irreversible. Although researchers are still investigating its causes, Alzheimer's patients have distinctly different brain tissue from people who do not have the disease. In the brain of a normal person, small beta-amyloid proteins, made up of 42 amino acids, exist in the alpha-helical form. In the brain of a person with Alzheimer's, the beta-amyloid proteins change shape from the normal alpha helices that are soluble, to sticky beta-pleated sheets, forming clusters of insoluble protein fragments called *plaques*. The diagnosis of Alzheimer's disease is based on the presence of plaques and neurofibrillary tangles in the neurons that affect the transmission of nerve signals.

There is no cure for Alzheimer's disease, but there are medications that can slow its progression or lessen symptoms for a limited time. Medications like donepezil (Aricept) and rivastigmine (Exelon) help keep the levels of nerve transmitters high in the brain, which improves learning and memory in patients.

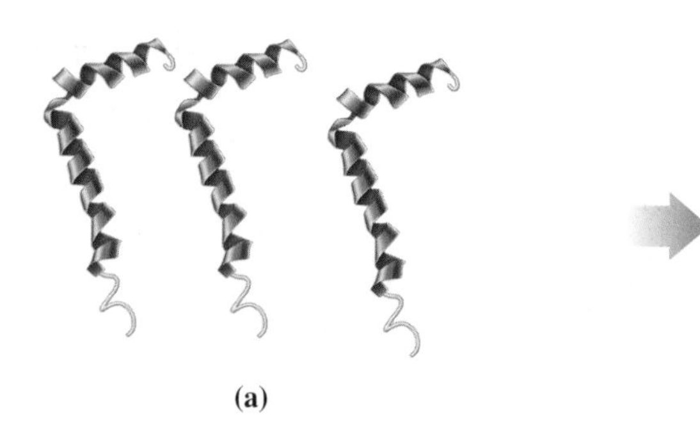

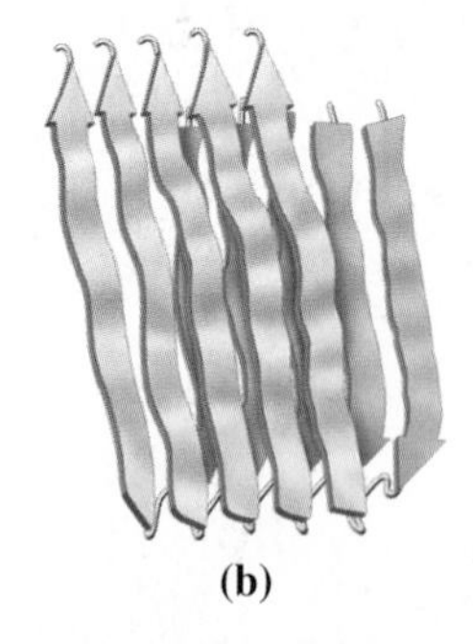

In patients with Alzheimer's disease, beta-amyloid proteins change from **(a)** a normal alpha-helical shape to **(b)** beta-pleated sheets that stick together and form plaques in the brain.

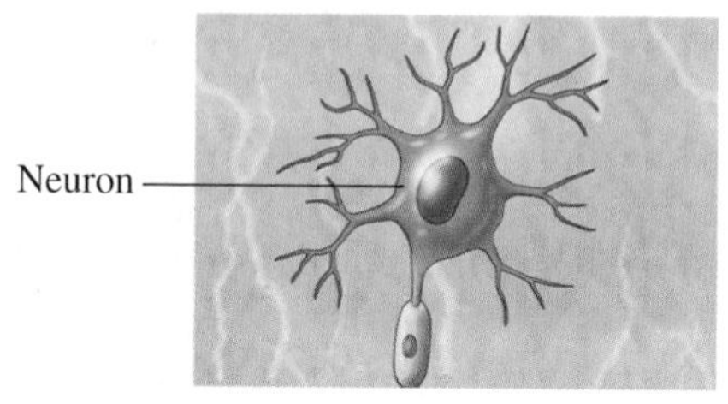

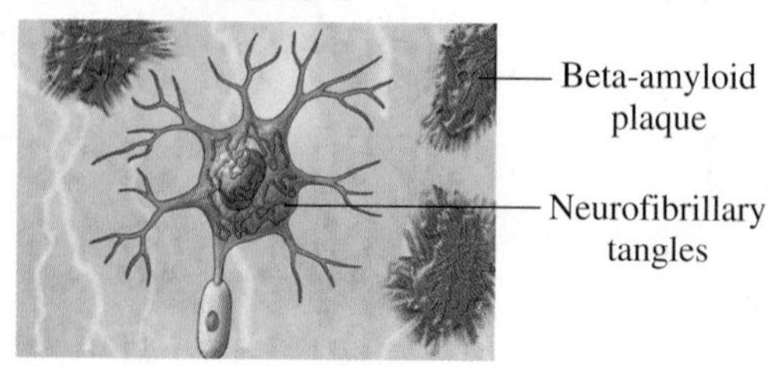

In an Alzheimer's brain, beta-amyloid plaques and neurofibrillary tangles damage the neurons and interfere with nerve signals.

Tertiary Structure

The **tertiary structure** of a protein involves attractions and repulsions between the R groups of the amino acids in the polypeptide chain. As interactions occur between different parts of the peptide chain, segments of the chain twist and bend until the protein acquires a specific three-dimensional shape. The tertiary structure of a protein is stabilized by interactions between the R groups of the amino acids in one region of the polypeptide chain and the R groups of amino acids in other regions of the protein (see **FIGURE 16.7**).

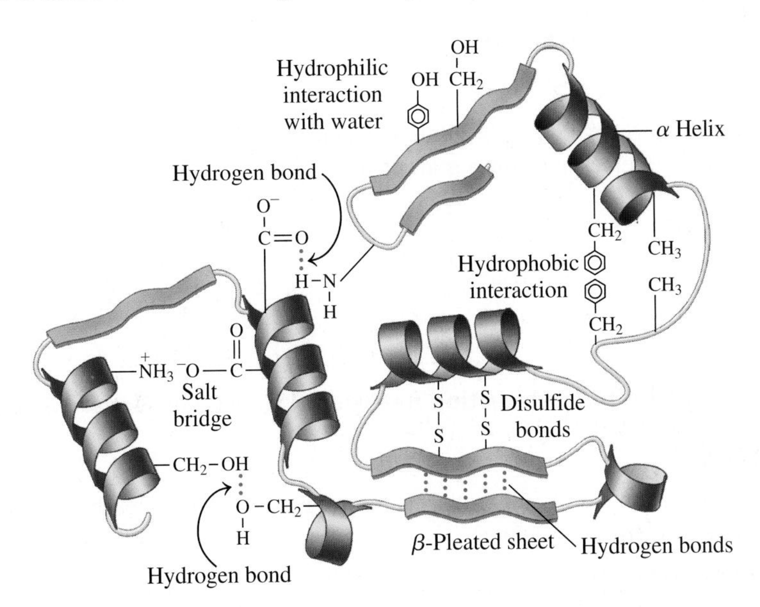

FIGURE 16.7 Interactions between amino acid R groups fold a protein into a specific three-dimensional shape called its tertiary structure.

Q Why would one section of a protein move to the center while another section remains on the surface of the tertiary structure?

1. **Hydrophobic interactions** are interactions between two nonpolar amino acids that have nonpolar R groups. Within a protein, the amino acids with nonpolar R groups are pushed away from the aqueous environment to form a hydrophobic center at the interior of the protein molecule.

$-CH_2-C_6H_5 \quad (CH_3)_2CH-CH_2-$

Phe Leu

2. **Hydrophilic interactions** are attractions between the external aqueous environment and the R groups of polar amino acids that are pulled to the outer surface of most proteins, where they form hydrogen bonds with water.

$-CH_2-OH$ with H_2O molecules

Ser

3. **Salt bridges** are ionic attractions between ionized R groups of polar basic and polar acidic amino acids. For example, the ionized R group of lysine, which has a positive charge, can form a salt bridge with the R group in aspartate, which has a negative charge.

$-CH_2-C(=O)O^- \quad H_3\overset{+}{N}-(CH_2)_4-$

Asp Lys

ENGAGE

Why is the interaction between arginine and aspartate in a tertiary structure called a salt bridge?

4. **Hydrogen bonds** form between the H of a polar R group and the O or N of another amino acid. For example, a hydrogen bond can form between the —OH groups of two serines or between the —OH of serine and the $—NH_2$ in the R group of asparagine.

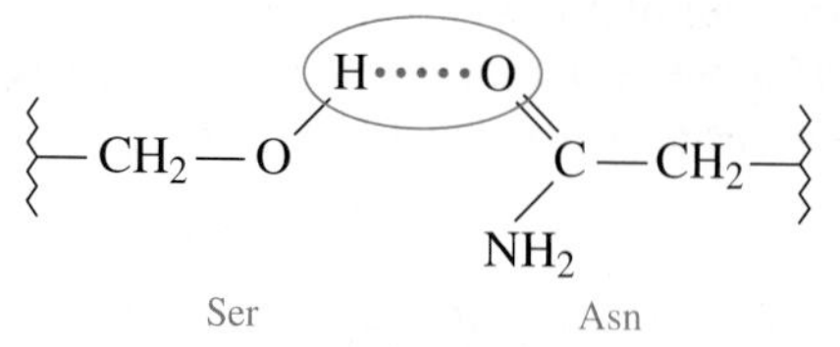

H·····O
$—CH_2—O$ $C—CH_2—$
NH_2
Ser Asn

5. **Disulfide bonds** (—S—S—) are covalent bonds that form between the —SH groups of cysteines in a polypeptide chain.

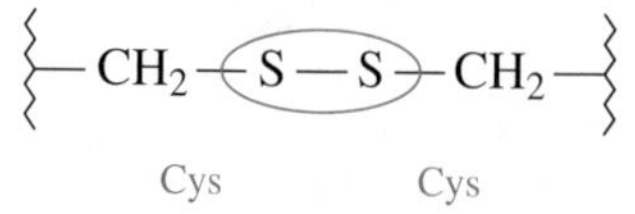

$—CH_2—S—S—CH_2—$
Cys Cys

SAMPLE PROBLEM 16.5 Interaction Between R Groups in Tertiary Structures

TRY IT FIRST

What type of interaction would you expect between the R groups of each of the following amino acids in a tertiary structure?

a. cysteine and cysteine
b. glutamate and lysine
c. tyrosine and water

SOLUTION

	Type of R Groups	Type of Interaction	Connect
ANALYZE THE PROBLEM	identify R groups on amino acids (see Table 16.2)	determine type of interaction between R groups (see Figure 16.7)	polarity of R groups

Type of R Groups	Type of Interaction
Nonpolar and nonpolar	Hydrophobic
Polar (neutral) and water	Hydrophilic
Polar (basic) $—NH_3^+$ and polar (acidic) $—COO^-$	Salt bridges
Polar (neutral) and polar (neutral) —OH and —NH— or $—NH_2$	Hydrogen bonds
—SH and —SH	Disulfide bonds

a. Two cysteines, each with an R group containing —SH, will form a disulfide bond.
b. The interaction of the $—COO^-$ in the R group of glutamate and the $—NH_3^+$ in the R group of lysine will form a salt bridge.
c. The R group in tyrosine has an —OH group that is attracted to water by hydrophilic interactions.

STUDY CHECK 16.5

Would you expect to find valine and leucine on the outside or the inside of the tertiary structure? Why?

ANSWER

Both are nonpolar and would be found on the inside of the tertiary structure.

TEST

Try Practice Problems 16.21 to 16.24

Quaternary Structure: Hemoglobin

While many proteins are biologically active as tertiary structures, some proteins require two or more tertiary structures, working together as a unit, to be biologically active. When a biologically active protein consists of two or more polypeptide chains or subunits, the structural level is referred to as a **quaternary structure**. Hemoglobin, a protein that transports oxygen in blood, consists of four polypeptide chains: two α-chains with 141 amino acids, and two β-chains with 146 amino acids. Although the α-chains and β-chains have different sequences of amino acids, they both form similar tertiary structures with similar shapes (see **FIGURE 16.8**).

In the quaternary structure, the subunits are held together by the same interactions that stabilize tertiary structures, such as hydrogen bonds, salt bridges, disulfide links, and hydrophobic interactions between R groups. Each subunit of the hemoglobin contains a heme group that binds oxygen. In the adult hemoglobin molecule, all four subunits ($\alpha_2\beta_2$) *must* be combined for hemoglobin to properly function as an oxygen carrier. Therefore, the complete quaternary structure of hemoglobin can bind and transport four molecules of oxygen.

TABLE 16.5 and **FIGURE 16.9** summarize the structural levels of proteins.

INTERACTIVE VIDEO

Different Levels of Protein Structure

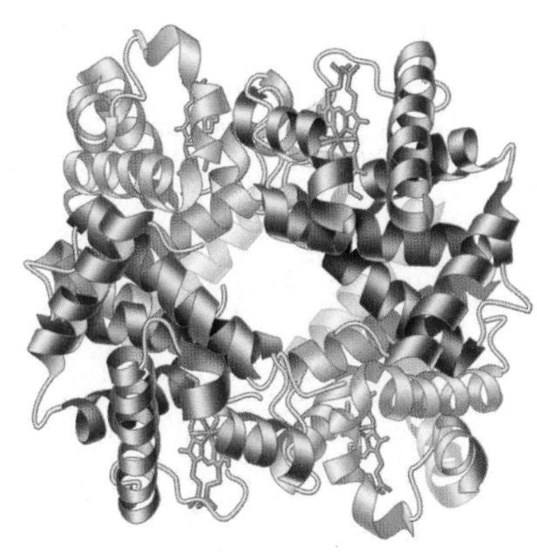

FIGURE 16.8 In the ribbon model of hemoglobin, the quaternary structure is made up of four polypeptide subunits, two (orange) are α-chains and two (red) are β-chains. The heme groups (green) in the four subunits bind oxygen.

Q What is the difference between a tertiary structure and a quaternary structure?

TABLE 16.5 Summary of Structural Levels in Proteins

Structural Level	Characteristics
Primary	Peptide bonds join amino acids in a specific sequence in a polypeptide.
Secondary	The α helix or β-pleated sheet forms by hydrogen bonding between the atoms in the peptide backbone.
Tertiary	A polypeptide folds into a compact, three-dimensional shape stabilized by interactions (hydrogen bonds, salt bridges, hydrophobic, hydrophilic, disulfide) between R groups of amino acids to form a biologically active protein.
Quaternary	Two or more protein subunits combine and are stabilized by interactions (hydrogen bonds, salt bridges, hydrophobic, hydrophilic, disulfide) to form a biologically active protein.

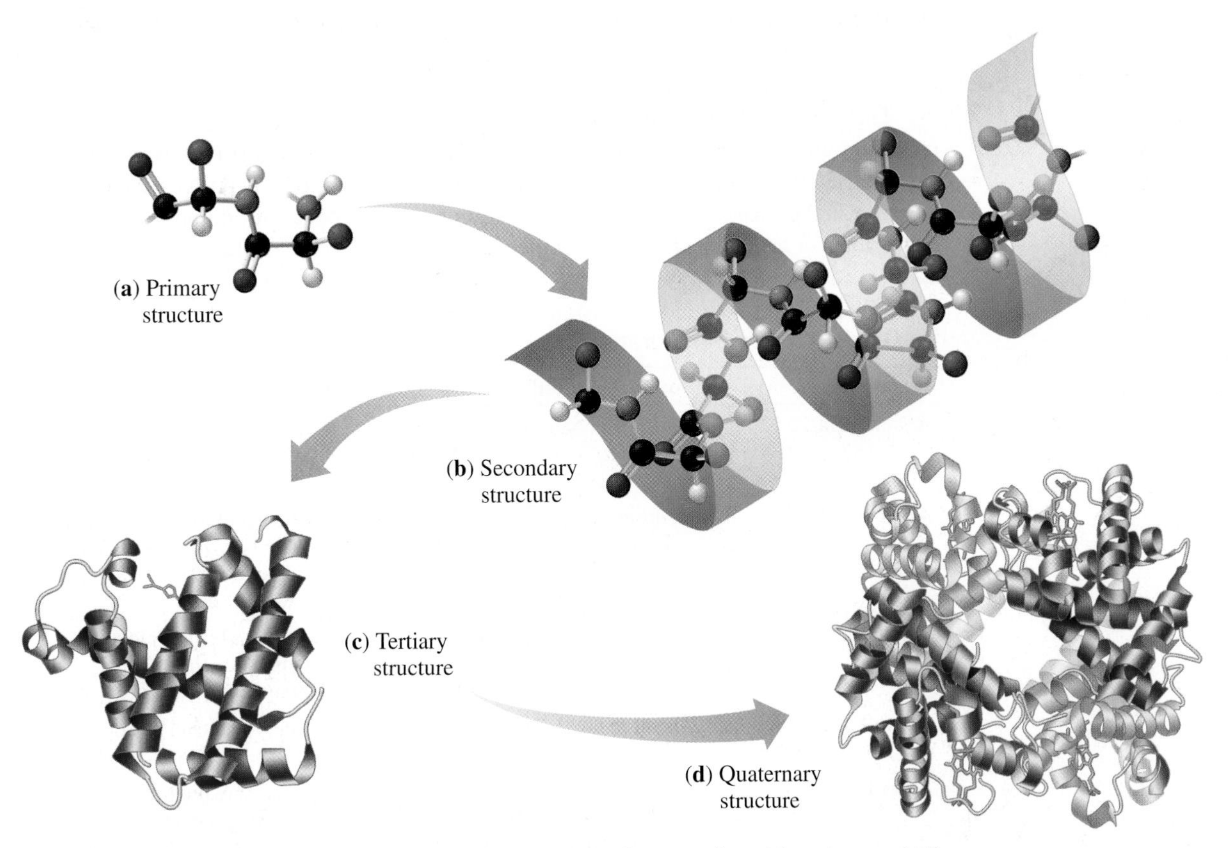

FIGURE 16.9 The structural levels of protein are **(a)** primary, **(b)** secondary, **(c)** tertiary, and **(d)** quaternary.

Q What is the difference between a primary structure and a tertiary structure?

SAMPLE PROBLEM 16.6 Identifying Protein Structure

TRY IT FIRST

Indicate whether the following conditions are responsible for primary, secondary, tertiary, or quaternary protein structures:

a. disulfide bonds that form between portions of a protein chain
b. peptide bonds that form a chain of amino acids
c. hydrogen bonds between the H of a peptide bond and the O of a peptide bond four amino acids away

SOLUTION

a. Disulfide bonds are a type of interaction between R groups found in the tertiary and quaternary levels of protein structure.
b. The peptide bonds in the sequence of amino acids in a polypeptide form the primary level of protein structure.
c. Hydrogen bonding between peptide bonds in the protein backbone forms the secondary level of protein structure.

STUDY CHECK 16.6

What structural level is represented by the interactions of the two subunits in insulin?

ANSWER

quaternary

TEST

Try Practice Problems 16.25 and 16.26

Denaturation of Proteins

Denaturation of a protein occurs when there is a change that disrupts the interactions that stabilize the secondary, tertiary, or quaternary structure. However, the covalent amide bonds of the primary structure are not affected.

The loss of secondary and tertiary structures occurs when conditions change, such as increasing the temperature or making the pH very acidic or basic. If the pH changes, the basic and acidic R groups lose their ionic charges and cannot form salt bridges, which causes a change in the shape of the protein. Denaturation can also occur by adding certain organic compounds or heavy metal ions or by mechanical agitation (see **TABLE 16.6**).

TABLE 16.6 Protein Denaturation

Denaturing Agent	Bonds Disrupted	Examples
Heat Above 50 °C	Hydrogen bonds; hydrophobic interactions between nonpolar R groups	Cooking food and autoclaving surgical items
Acids and Bases	Hydrogen bonds between polar R groups; salt bridges	Lactic acid from bacteria, which denatures milk protein in the preparation of yogurt and cheese
Organic Compounds	Hydrophobic interactions	Ethanol and isopropyl alcohol, which disinfect wounds and prepare the skin for injections
Heavy Metal Ions Ag^+, Pb^{2+}, and Hg^{2+}	Disulfide bonds in proteins by forming ionic bonds	Mercury and lead poisoning
Agitation	Hydrogen bonds and hydrophobic interactions by stretching polypeptide chains and disrupting stabilizing interactions	Whipped cream, meringue made from egg whites

When the interactions between the R groups are disrupted, a protein unfolds like a loose piece of spaghetti. With the loss of its overall shape (tertiary structure), the protein is no longer biologically active.

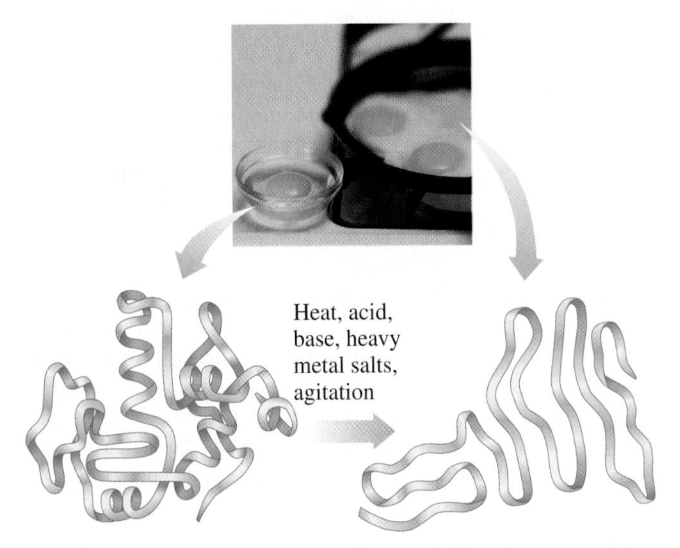

Denaturation of a protein occurs when the interactions of R groups that stabilize secondary, tertiary, or quarternary structures are disrupted.

TEST

Try Practice Problems 16.27 and 16.28

CHEMISTRY LINK TO HEALTH

Sickle-Cell Anemia

Sickle-cell anemia is a disease caused by an abnormality in the shape of one of the subunits of the hemoglobin protein. In the β-chain, the sixth amino acid, glutamate, which is polar acidic, is replaced by valine, a nonpolar amino acid.

Because valine has a nonpolar R group, it is attracted to the nonpolar regions within the beta hemoglobin chains of other hemoglobin molecules. Hydrophobic interactions cause sickle-cell hemoglobin molecules to stick together and form insoluble fibers of sickle-cell hemoglobin. The affected red blood cells change from a rounded shape to a crescent shape, like a sickle, which are rigid and cannot easily pass through capillaries. The sickled red blood cells clog capillaries, where they cause inflammation and pain as well as critically low oxygen levels in the affected tissues.

In sickle-cell anemia, both genes for the altered hemoglobin must be inherited. However, a few sickled cells are found in persons who carry one gene for sickle-cell hemoglobin, a condition that is also known to provide protection from malaria.

Polar acidic amino acid

Normal β-chain: Val–His–Leu–Thr–Pro–Glu–Glu–Lys–

Sickled β-chain: Val–His–Leu–Thr–Pro–Val–Glu–Lys–

Nonpolar amino acid

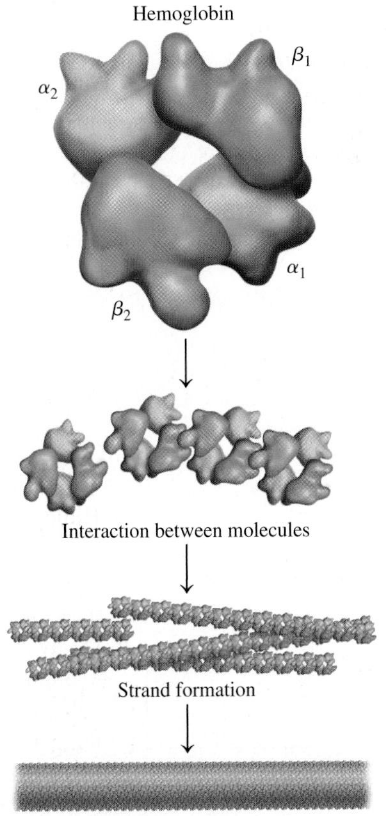

In sickle-cell hemoglobin, the substitution of a nonpolar amino acid causes hydrophobic interactions that produce insoluble fibers that change the shape of the red blood cell and block blood flow.

PRACTICE PROBLEMS

16.3 Proteins: Secondary, Tertiary, and Quaternary Structures

LEARNING GOAL Describe the secondary, tertiary, and quaternary structures for a protein; describe the denaturation of a protein.

16.17 What happens when a primary structure forms a secondary structure?

16.18 What are two types of secondary protein structure?

16.19 What is the difference in hydrogen bonding between an α helix and a β-pleated sheet?

16.20 In an α helix, how does hydrogen bonding occur between the amino acids in the polypeptide chain?

16.21 What type of interaction would you expect between the R groups of the following amino acids in a tertiary structure?
a. cysteine and cysteine
b. aspartate and lysine
c. serine and aspartate
d. leucine and leucine

16.22 What type of interaction would you expect between the R groups of the following amino acids in a quaternary structure?
a. phenylalanine and isoleucine
b. glutamate and arginine
c. asparagine and tyrosine
d. alanine and proline

16.23 A portion of a polypeptide chain contains the following sequence of amino acids:

—Leu—Val—Cys—Asp—

a. Which amino acids are likely to be found on the inside of the protein structure? Why?
b. Which amino acids would be found on the outside of the protein? Why?
c. How does the primary structure of a protein affect its tertiary structure?

16.24 In myoglobin, about one-half of the 153 amino acids have nonpolar R groups.
a. Where would you expect those amino acids to be located in the tertiary structure?
b. Where would you expect the polar R groups to be in the tertiary structure?
c. Why is myoglobin more soluble in water than silk or wool?

16.25 Indicate whether each of the following statements describes primary, secondary, tertiary, or quaternary protein structure:
a. R groups interact to form disulfide bonds or salt bridges.
b. Peptide bonds join amino acids in a polypeptide chain.
c. Several polypeptides in a beta-pleated sheet are held together by hydrogen bonds between adjacent chains.
d. Hydrogen bonding between amino acids in the same polypeptide gives a coiled shape to the protein.

16.26 Indicate whether each of the following statements describes primary, secondary, tertiary, or quaternary protein structure:
a. Hydrophobic R groups seeking a nonpolar environment move toward the inside of the folded protein.
b. Protein chains of collagen form a triple helix.
c. An active protein contains four tertiary subunits.
d. In sickle-cell anemia, valine replaces glutamate in the β-chain.

Clinical Applications

16.27 Indicate the changes in secondary and tertiary structural levels of proteins for each of the following:
a. An egg placed in water at 100 °C is soft boiled in 3 minutes.
b. Prior to giving an injection, the skin is wiped with an alcohol swab.
c. Surgical instruments are placed in a 120 °C autoclave.
d. During surgery, a wound is closed by cauterization (heat).

16.28 Indicate the changes in secondary and tertiary structural levels of proteins for each of the following:
a. Tannic acid is placed on a burn.
b. Milk is heated to 60 °C to make yogurt.
c. To avoid spoilage, seeds are treated with a solution of $HgCl_2$.
d. Hamburger is cooked at high temperatures to destroy *E. coli* bacteria that may cause intestinal illness.

16.4 Enzymes

LEARNING GOAL Describe enzymes and their role in enzyme-catalyzed reactions.

Biological catalysts known as **enzymes** are needed for most chemical reactions that take place in the body. A catalyst increases the reaction rate by changing the way a reaction takes place but is itself not changed at the end of the reaction. An uncatalyzed reaction in a cell would take place eventually, but not at a rate fast enough for survival. For example, the hydrolysis of proteins in our diet would eventually occur without a catalyst, but not fast enough to meet the body's requirements for amino acids. Most chemical reactions in our cells must occur at incredibly fast rates under the mild conditions of pH 7.4 and a body temperature of 37 °C.

TEST

Try Practice Problems 16.29 and 16.30

As catalysts, enzymes lower the activation energy for a chemical reaction (see FIGURE 16.10). Less energy is required to convert reactant molecules to products, which increases the rate of a biochemical reaction compared to the rate of the uncatalyzed reaction. For example, an enzyme in the blood called carbonic anhydrase catalyzes the rapid reaction of carbon dioxide and water to bicarbonate and H^+. In one second, one molecule of carbonic anhydrase catalyzes the reaction of about one million molecules of carbon dioxide.

$$CO_2 + H_2O \underset{}{\overset{\text{Carbonic anhydrase}}{\rightleftharpoons}} HCO_3^- + H^+$$

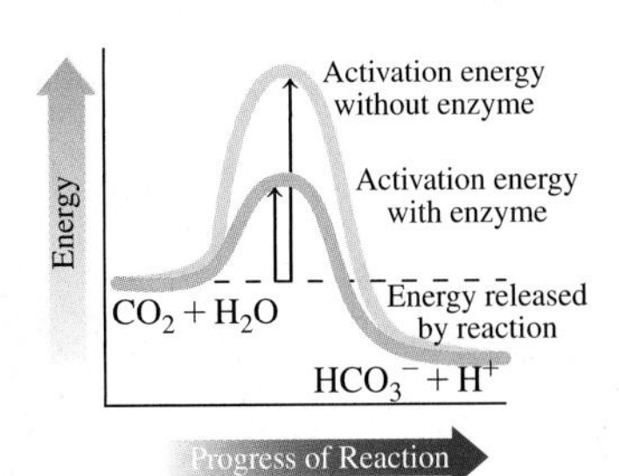

FIGURE 16.10 The enzyme carbonic anhydrase lowers the activation energy needed for the reversible reaction that converts CO_2 and H_2O to HCO_3^- and H^+.

Why are enzymes used in biological reactions?

Names and Classification of Enzymes

The names of enzymes describe the compound or the reaction that is catalyzed. The actual names of enzymes are derived by replacing the end of the name of the reaction or reacting compound with the suffix *ase*. For example, an *oxidase* catalyzes an oxidation reaction, and a *dehydrogenase* removes hydrogen atoms. The compound sucrose is hydrolyzed by the enzyme *sucrase*, and a lipid is hydrolyzed by a *lipase*. Some early known enzymes use names that end in the suffix *in*, such as *papain* found in papaya, *rennin* found in milk, and *pepsin* and *trypsin*, enzymes that catalyze the hydrolysis of proteins.

More recently, a systematic method of classifying and naming enzymes has been established. The name and class of each indicates the type of reaction it catalyzes. There are six main classes of enzymes, as described in **TABLE 16.7**.

TEST

Try Practice Problems 16.31 and 16.32

Nearly all enzymes are proteins. Each has a unique three-dimensional shape that recognizes and binds a small group of reacting molecules, which are called **substrates**. The tertiary structure of an enzyme plays an important role in how that enzyme catalyzes reactions. In a catalyzed reaction, an enzyme must first bind to a substrate in a way that favors catalysis. A typical enzyme is much larger than its substrate. However, within an enzyme's large tertiary structure is a region called the **active site**, in which the catalyzed reaction takes place. This active site is often a small pocket within the larger tertiary structure that closely fits the shape of the substrate (see **FIGURE 16.11**).

Within the active site, the R groups of specific amino acids interact with the functional groups of the substrate to form hydrogen bonds, salt bridges, or hydrophobic interactions. The active site of a particular enzyme fits the shape of only a few types of substrates, which makes an enzyme very specific about the type of substrate it binds.

TABLE 16.7 Classes of Enzymes

Class	Reaction Catalyzed	Examples
1. Oxidoreductases	Oxidation–reduction reactions	*Oxidases* oxidize a substance. *Reductases* reduce a substance. *Dehydrogenases* remove 2H atoms to form a double bond.
2. Transferases	Transfer a group between two compounds	*Transaminases* transfer amino groups. *Kinases* transfer phosphate groups.
3. Hydrolases	Hydrolysis reactions	*Proteases* hydrolyze peptide bonds in proteins. *Lipases* hydrolyze ester bonds in lipids. *Carbohydrases* hydrolyze glycosidic bonds in carbohydrates. *Phosphatases* hydrolyze phosphoester bonds. *Nucleases* hydrolyze nucleic acids.
4. Lyases	Add or remove groups involving a double bond without hydrolysis	*Carboxylases* add CO_2. *Deaminases* remove NH_3.
5. Isomerases	Rearrange atoms in a molecule to form an isomer	*Isomerases* convert cis to trans isomers or trans to cis isomers. *Epimerases* convert D to L stereoisomers or L to D.
6. Ligases	Form bonds between molecules using ATP energy	*Synthetases* combine two molecules.

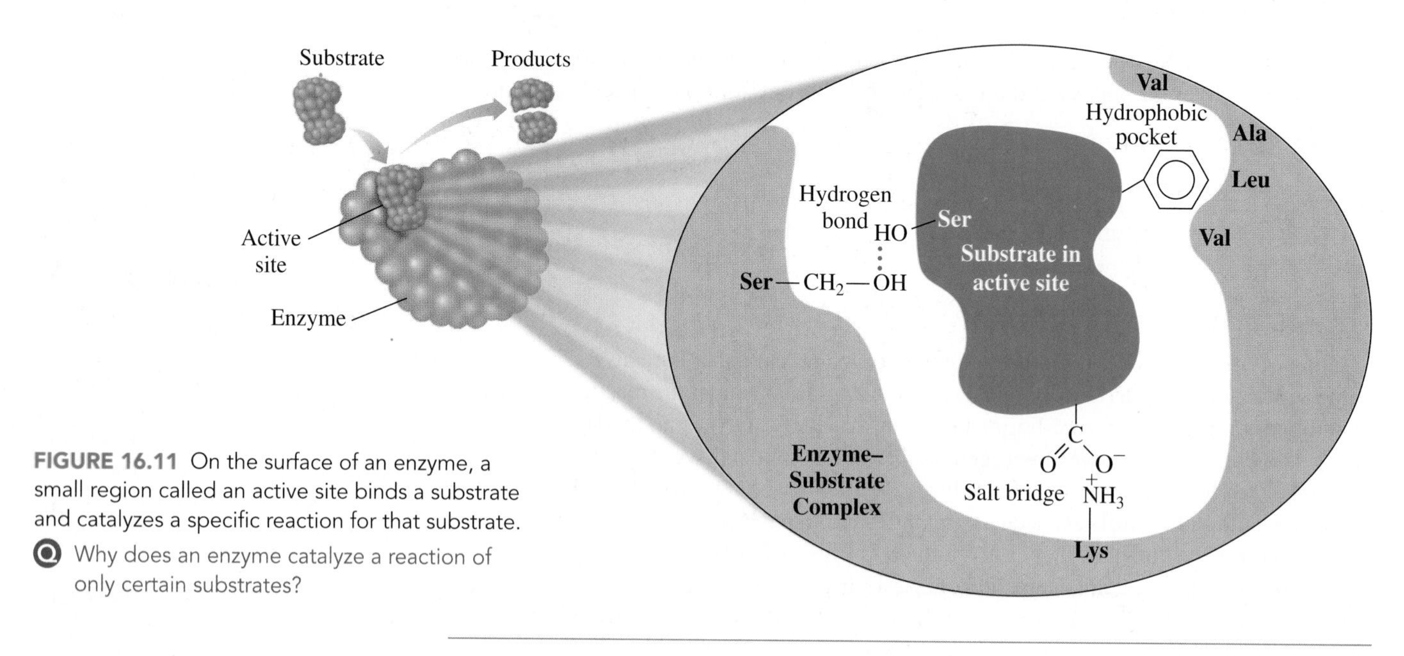

FIGURE 16.11 On the surface of an enzyme, a small region called an active site binds a substrate and catalyzes a specific reaction for that substrate.

Q Why does an enzyme catalyze a reaction of only certain substrates?

SAMPLE PROBLEM 16.7 Naming Enzymes

TRY IT FIRST

What chemical reaction would each of the following enzymes catalyze?

a. aminotransferase
b. lactate dehydrogenase (LDH)

SOLUTION

a. the transfer of an amino group
b. the removal of hydrogen from lactate

TEST

Try Practice Problems 16.33 and 16.34

STUDY CHECK 16.7

What is the name of the enzyme that catalyzes the hydrolysis of lipids?

ANSWER

lipase

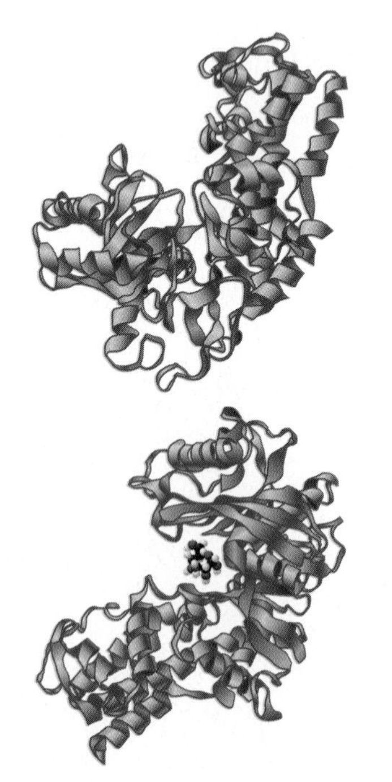

The ribbon model shows the change in the shape of the enzyme hexokinase as the glucose molecule binds to its active site.

Enzyme-Catalyzed Reactions

The combination of an enzyme and a substrate within the active site forms an **enzyme–substrate (ES) complex** that provides an alternative pathway with a lower activation energy. Within the active site, the amino acid R groups catalyze the reaction. Then the products are released from the enzyme so it can bind to another substrate molecule. We can write the catalyzed reaction of an enzyme (E) with a substrate (S) to form product (P) as follows:

$$\underset{\text{Enzyme and Substrate}}{E + S} \rightleftharpoons \underset{\text{Enzyme–Substrate Complex}}{ES} \longrightarrow \underset{\text{Enzyme and Product}}{E + P}$$

In the hydrolysis of the disaccharide lactose by the enzyme lactase, a molecule of lactose binds to the active site of lactase. In this ES complex, the glycosidic bond of lactose is in a position that is favorable for hydrolysis, which is the splitting by water of a large molecule into smaller parts. The R groups on the amino acids in the active site then catalyze the hydrolysis of sucrose, which produces the monosaccharides glucose and fructose. Because the structures of the products are no longer attracted to the active site, they are released, which allows sucrase to react with another lactose molecule.

$$\underset{\text{Lactase + Lactose}}{E + S} \rightleftharpoons \underset{\text{Lactase–Lactose Complex}}{\text{ES complex}} \longrightarrow \underset{\text{Lactase + Glucose + Galactose}}{E + P_1 + P_2}$$

Models of Enzyme Action

An early theory of enzyme action, called the *lock-and-key model*, described the active site as having a rigid, nonflexible shape. According to the lock-and-key model, the shape of the active site was analogous to a lock, and its substrate was the key that specifically fit that lock. However, this model was a static one that did not account for the fact that an enzyme is flexible and can change shape to adjust for the binding of a substrate to its active site.

In the dynamic model of enzyme action, called the **induced-fit model**, the flexibility of the active site allows it to adapt to the shape of the substrate. At the same time, the shape of the substrate may be modified to better fit the geometry of the active site. In the induced-fit model, substrate and enzyme work together to acquire a geometrical arrangement that lowers the activation energy (see **FIGURE 16.12**).

CORE CHEMISTRY SKILL
Describing Enzyme Action

ENGAGE
How can the induced-fit model be used to explain why some enzymes can interact with a range of similar substrates?

Galactose Glucose
Substrate (S) (lactose)
Active site
Enzyme (E) (lactase)
Enzyme–Substrate Complex (ES)
H_2O
Galactose Glucose
Products (P)

FIGURE 16.12 A flexible active site in lactase and the flexible substrate lactose adjust to provide the best fit for the hydrolysis reaction.
Q Why does the enzyme-catalyzed hydrolysis of lactose proceed faster than the hydrolysis of lactose in a test tube?

TEST
Try Practice Problems 16.35 to 16.38

CHEMISTRY LINK TO HEALTH

Isoenzymes as Diagnostic Tools

Isoenzymes are different forms of an enzyme that catalyze the same reaction in different cells or tissues of the body. They consist of quaternary structures with slight variations in the amino acids in the polypeptide subunits. For example, there are five isoenzymes of *lactate dehydrogenase* (LDH) that catalyze the conversion between lactate and pyruvate.

$$\underset{\text{Lactate}}{CH_3-CH(OH)-COO^-} + NAD^+ \underset{}{\overset{\text{Lactate dehydrogenase}}{\rightleftharpoons}} \underset{\text{Pyruvate}}{CH_3-C(=O)-COO^-} + NADH + H^+$$

Each LDH isoenzyme contains a mix of polypeptide subunits, M and H. In the liver and muscle, lactate is converted to pyruvate by the LDH_5 isoenzyme with four M subunits designated M_4. In the heart, the same reaction is catalyzed by the LDH_1 isoenzyme (H_4) containing four H subunits. Different combinations of the M and H subunits are found in the LDH isoenzymes of the brain, red blood cells, kidneys, and white blood cells.

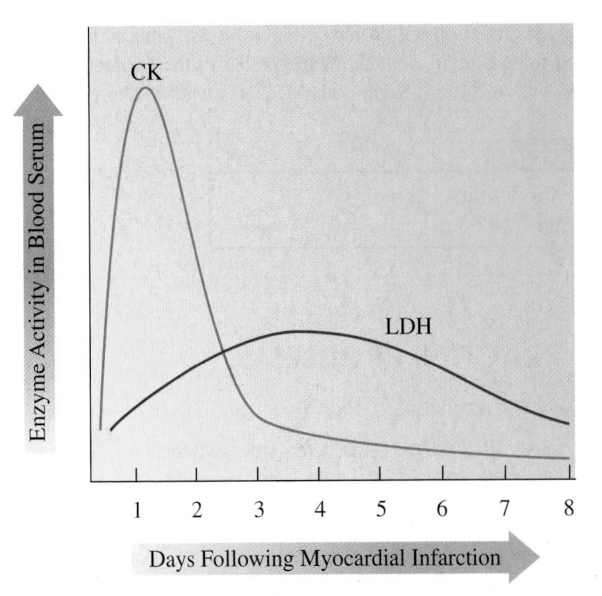

A myocardial infarction may be indicated by an increase in the levels of creatine kinase (CK) and lactate dehydrogenase (LDH).

Isoenzymes of Lactate Dehydrogenase	Highest Levels Found in the Following:
H M	
H_4 (LDH_1)	Heart, kidneys
H_3M (LDH_2)	Red blood cells, heart, kidney, brain
H_2M_2 (LDH_3)	Brain, lung, white blood cells
HM_3 (LDH_4)	Lung, skeletal muscle
M_4 (LDH_5)	Skeletal muscle, liver

The different isoenzymes of lactate dehydrogenase (LDH) indicate damage to different organs in the body.

The different forms of an enzyme allow a medical diagnosis of damage or disease to a particular organ or tissue. In healthy tissues, isoenzymes function within the cells. However, when a disease damages a particular organ, cells die, which releases their contents, including the isoenzymes, into the blood. Measurements of the elevated levels of specific isoenzymes in the blood serum help to identify the disease and its location in the body. For example, an elevation in serum LDH_5, M_4, indicates liver damage or disease. When a *myocardial infarction* (MI) or heart attack damages heart muscle, an increase in the level of LDH_1 (H_4) isoenzyme is detected in the blood serum (see **TABLE 16.8**).

TEST

Try Practice Problems 16.39 to 16.42

TABLE 16.8 Isoenzymes of Lactate Dehydrogenase and Creatine Kinase

Isoenzyme	Abundant in	Subunits
Lactate Dehydrogenase (LDH)		
LDH_1	Heart, kidneys	H_4
LDH_2	Red blood cells, heart, kidney, brain	H_3M
LDH_3	Brain, lung, white blood cells	H_2M_2
LDH_4	Lung, skeletal muscle	HM_3
LDH_5	Skeletal muscle, liver	M_4
Creatine Kinase (CK)		
CK_1	Brain, lung	BB
CK_2	Heart muscle	MB
CK_3	Skeletal muscle, red blood cells	MM

Another isoenzyme used diagnostically is creatine kinase (CK), which consists of two types of polypeptide subunits. Subunit B is prevalent in the brain, and subunit M predominates in muscle. Normally CK_3 (subunits MM) is present at low levels in the blood serum. However, in a patient who has suffered a myocardial infarction, the level of CK_2 (subunits MB) is elevated within 4 to 6 h and reaches a peak in about 24 h. **TABLE 16.9** lists some enzymes used to diagnose tissue damage and diseases of certain organs.

TABLE 16.9 Serum Enzymes Used in Diagnosis of Tissue Damage

Condition	Diagnostic Enzymes Elevated
Heart attack or liver disease (cirrhosis, hepatitis)	Lactate dehydrogenase (LDH) Aspartate transaminase (AST)
Heart attack	Creatine kinase (CK)
Hepatitis	Alanine transaminase (ALT)
Liver (carcinoma) or bone disease (rickets)	Alkaline phosphatase (ALP)
Pancreatic disease	Pancreatic amylase (PA) Cholinesterase (CE) Lipase (LPS)
Prostate carcinoma	Acid phosphatase (ACP) Prostate specific antigen (PSA)

PRACTICE PROBLEMS

16.4 Enzymes

LEARNING GOAL Describe enzymes and their role in enzyme-catalyzed reactions.

16.29 Why do chemical reactions in the body require enzymes?

16.30 How do enzymes make chemical reactions in the body proceed at faster rates?

16.31 What is the reactant for each of the following enzymes?
a. galactase **b.** lipase **c.** aspartase

16.32 What is the reactant for each of the following enzymes?
a. peptidase **b.** cellulase **c.** lactase

16.33 What is the name of the class of enzymes that would catalyze each of the following reactions?
a. hydrolysis of sucrose
b. addition of oxygen
c. converting glucose ($C_6H_{12}O_6$) to fructose ($C_6H_{12}O_6$)
d. moving an amino group from one molecule to another

16.34 What is the name of the class of enzymes that would catalyze each of the following reactions?
a. addition of water to a double bond
b. removing hydrogen atoms
c. splitting peptide bonds in proteins
d. converting a tertiary alcohol to a secondary alcohol

16.35 Match the terms (1) enzyme–substrate complex, (2) enzyme, and (3) substrate with each of the following:
a. has a tertiary structure that recognizes the substrate
b. the combination of an enzyme with the substrate
c. has a structure that fits the active site of an enzyme

16.36 Match the terms (1) active site, (2) lock-and-key model, and (3) induced-fit model with each of the following:
a. the portion of an enzyme where catalytic activity occurs
b. an active site that adapts to the shape of a substrate
c. an active site that has a rigid shape

16.37 **a.** Write an equation that represents an enzyme-catalyzed reaction.
b. How is the active site different from the whole enzyme structure?

16.38 a. How does an enzyme speed up the reaction of a substrate?
b. After the products have formed, what happens to the enzyme?

Clinical Applications

For problems 16.39 to 16.42, see Chemistry Link to Health "Isoenzymes as Diagnostic Tools."

16.39 What are isoenzymes?

16.40 How is the LDH isoenzyme in the heart different from the LDH isoenzyme in the liver?

16.41 A patient arrives in an emergency department complaining of chest pains. What enzymes would you test for in the blood serum?

16.42 A patient has elevated blood serum levels of LDH and AST. What condition might be indicated?

16.5 Factors Affecting Enzyme Activity

LEARNING GOAL Describe the effect of temperature, pH, and inhibitors on enzyme activity.

The **activity** of an enzyme describes how fast an enzyme catalyzes the reaction that converts a substrate to product. This activity is strongly affected by reaction conditions, which include temperature, pH, and the presence of inhibitors.

Temperature

Enzymes are very sensitive to temperature. At low temperatures, most enzymes show little activity because there is not a sufficient amount of energy for the catalyzed reaction to take place. At higher temperatures, enzyme activity increases as reacting molecules move faster to cause more collisions with enzymes. Enzymes are most active at **optimum temperature**, which is 37 °C, or body temperature, for most enzymes (see **FIGURE 16.13**).

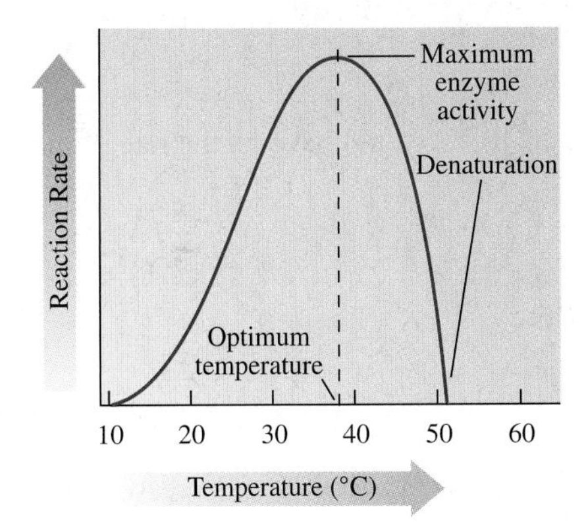

FIGURE 16.13 An enzyme attains maximum activity at its optimum temperature, usually 37 °C. Lower temperatures slow the rate of reaction, and temperatures above 50 °C denature most enzymes with a loss of catalytic activity.

Q Why is 37 °C the optimum temperature for many enzymes?

Thermophiles survive in the high temperatures (50 °C to 120 °C) of a hot spring.

At temperatures above 50 °C, there is enough energy to disrupt many of the interactions between the R groups in the protein. When too many of those interactions are disrupted, the tertiary structure, and thus the shape of most proteins, is destroyed, which causes a loss in enzyme activity. For this reason, equipment in hospitals and laboratories is sterilized in autoclaves where the high temperatures denature the enzymes in harmful bacteria.

Certain organisms, known as thermophiles, live in environments where temperatures range from 50 °C to 120 °C. In order to survive in these extreme conditions, thermophiles must have enzymes with tertiary structures that are not destroyed by such high temperatures. Some research shows that their enzymes are very similar to ordinary enzymes except they contain more hydrogen bonds and salt bridges that stabilize the tertiary structures at high temperatures and, thus, resist unfolding and the loss of enzymatic activity.

pH

Enzymes are most active at their **optimum pH**, the pH that maintains the proper tertiary structure of the protein (see **FIGURE 16.14**). If a pH value is above or below the optimum pH, the R group interactions are disrupted, which destroys the tertiary structure and the active site. As a result, the enzyme no longer binds substrate, and no catalysis occurs. Small changes in pH are reversible, which allows an enzyme to regain its structure and activity. However, large variations from optimum pH permanently destroy the structure of the enzyme.

Enzymes in most cells have optimum pH values at physiological pH around 7.4. However, enzymes in the stomach have a low optimum pH because they hydrolyze proteins at the acidic pH in the stomach. For example, pepsin, a digestive enzyme in the stomach, has an optimum pH of 1.5 to 2.0. Between meals, the pH in the stomach is 4 or 5 and pepsin shows little or no digestive activity. When food enters the stomach, the secretion of HCl lowers the pH to about 2, which activates pepsin. **TABLE 16.10** lists the optimum pH values for selected enzymes.

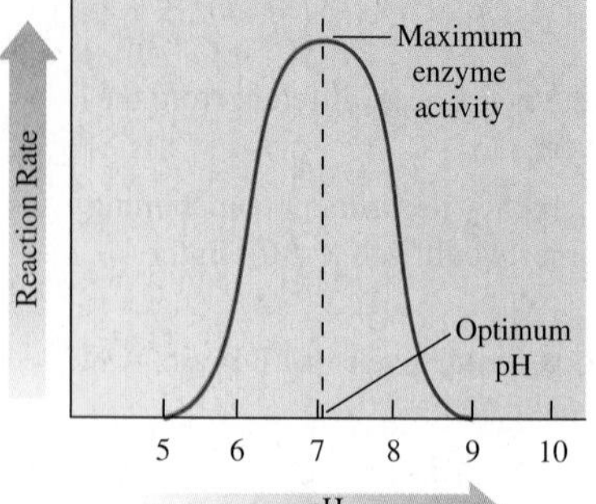

FIGURE 16.14 Enzymes are most active at their optimum pH. At a higher or lower pH, denaturation of the enzyme causes a loss of catalytic activity.

Q Why does the digestive enzyme pepsin have an optimum pH of 2?

TABLE 16.10 Optimum pH for Selected Enzymes

Enzyme	Location	Substrate	Optimum pH
Pepsin	Stomach	Peptide bonds	1.5–2.0
Lactase	GI tract	Lactose	6.0
Sucrase	Small intestine	Sucrose	6.2
Amylase	Pancreas	Amylose	6.7–7.0
Urease	Liver	Urea	7.0
Trypsin	Small intestine	Peptide bonds	7.7–8.0
Lipase	Pancreas	Lipid (ester bonds)	8.0
Arginase	Liver	Arginine	9.7

SAMPLE PROBLEM 16.8 Factors Affecting Enzymatic Activity

TRY IT FIRST

Describe the effect that lowering the temperature to 10 °C would have on the rate of the reaction catalyzed by urease.

$$\underset{\text{Urea}}{H_2N-\overset{\overset{\displaystyle O}{\|}}{C}-NH_2} + H_2O \xrightarrow{\text{Urease}} 2NH_3 + CO_2$$

SOLUTION

Because 10 °C is lower than the optimum temperature of 37 °C, there is a decrease in the rate of the reaction.

STUDY CHECK 16.8

If urease has an optimum pH of 7.0, what is the effect of lowering the pH to 3?

ANSWER

At a pH lower than the optimum pH, denaturation of the tertiary structure will decrease the activity of urease.

TEST

Try Practice Problems 16.43 to 16.46

Enzyme Inhibition

Many kinds of molecules called **inhibitors** cause enzymes to lose catalytic activity. Although inhibitors act differently, they all prevent the active site from binding with a substrate. An enzyme with a *reversible inhibitor* can regain enzymatic activity, but an enzyme attached to an *irreversible inhibitor* loses enzymatic activity permanently.

A **competitive inhibitor** has a chemical structure and polarity that is similar to that of the substrate. Thus, a competitive inhibitor competes with the substrate to bind with the active site on the enzyme. When the inhibitor occupies the active site, the substrate cannot bind to the enzyme and no reaction can occur (see **FIGURE 16.15**).

ENGAGE

What happens to the rate of an enzyme-catalyzed reaction when an inhibitor binds to the enzyme?

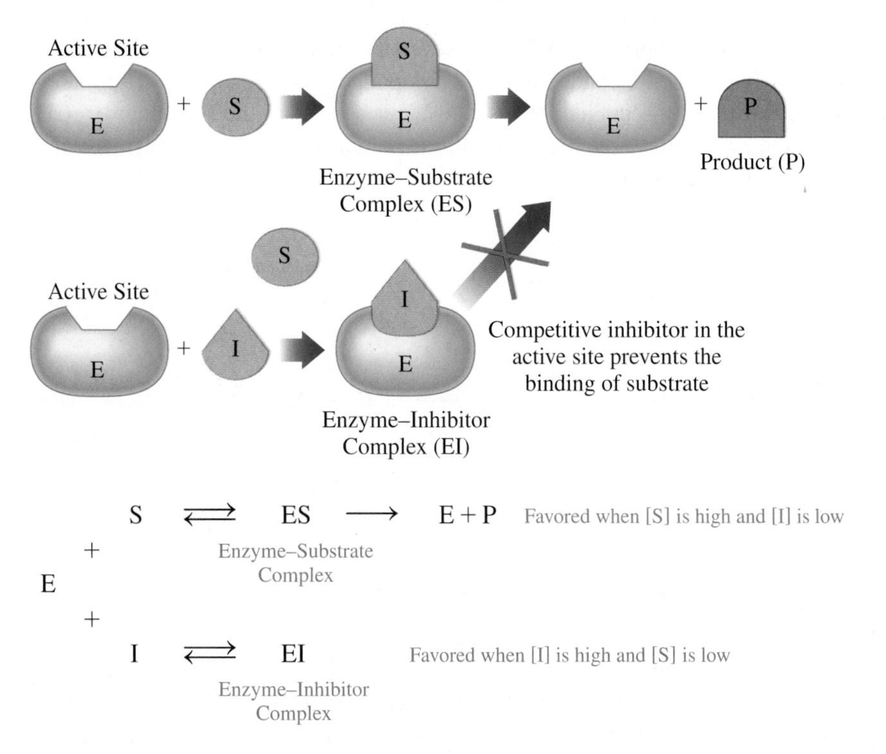

FIGURE 16.15 With a structure similar to the substrate for an enzyme, a competitive inhibitor also fits the active site and competes with the substrate when both are present.

Q Why does increasing the substrate concentration reverse the inhibition by a competitive inhibitor?

As long as the concentration of the inhibitor is substantial, there is a loss of enzyme activity. However, adding more substrate displaces the competitive inhibitor. As more enzyme molecules bind to substrate (ES), enzyme activity is regained.

The structure of a **noncompetitive inhibitor** does not resemble the substrate and does not compete for the active site. Instead, a noncompetitive inhibitor binds to a site on the enzyme that is not the active site. When the noncompetitive inhibitor is bonded to the enzyme, the shape of the enzyme is distorted. Inhibition occurs because the substrate cannot fit in the active site or it does not fit properly. Without the proper alignment of substrate with the amino acid side groups in the active site, no catalysis can take place (see **FIGURE 16.16**).

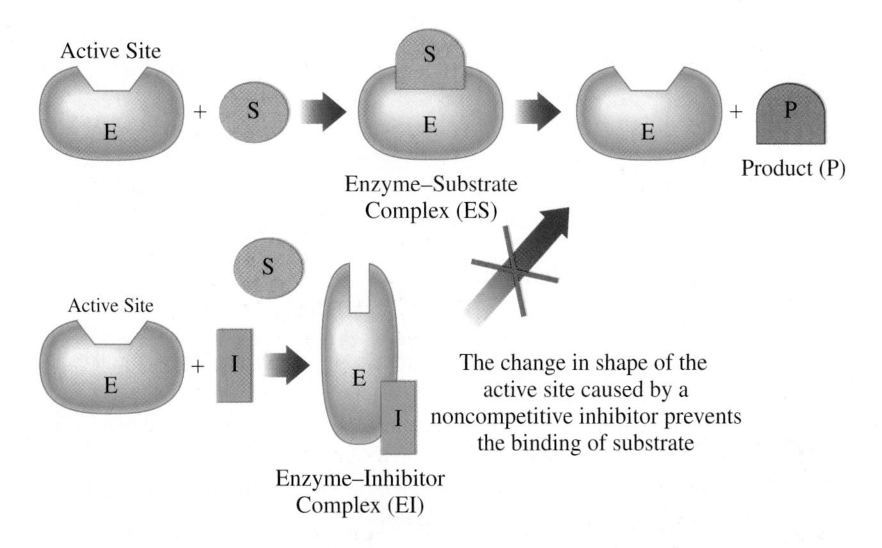

FIGURE 16.16 A noncompetitive inhibitor binds to an enzyme at a site other than the active site, which distorts the enzyme and prevents the proper binding and catalysis of the substrate at the active site.

Q Will an increase in substrate concentration reverse the inhibition by a noncompetitive inhibitor?

Because a noncompetitive inhibitor is not competing for the active site, the addition of more substrate does not reverse this type of inhibition. Examples of noncompetitive inhibitors are the heavy metal ions Pb^{2+}, Ag^{+}, and Hg^{2+} that bond with amino acid side groups such as $-COO^{-}$ or $-OH$. Catalytic activity is restored when chemical reagents remove the inhibitors.

Irreversible Inhibition

In irreversible inhibition, a molecule causes an enzyme to lose all enzymatic activity. Most irreversible inhibitors are toxic substances that destroy enzymes. Usually an irreversible inhibitor forms a covalent bond with an amino acid side group within the active site, which prevents the substrate from binding to the active site or prevents catalytic activity.

Insecticides and nerve gases act as irreversible inhibitors of acetylcholinesterase, an enzyme needed for nerve conduction. The compound DFP (diisopropyl fluorophosphate), an organophosphate insecticide, forms a covalent bond with the side chain $-CH_2OH$ of serine in the active site. When acetylcholinesterase is inhibited, the transmission of nerve impulses is blocked, and paralysis occurs.

TEST

Try Practice Problems 16.47 to 16.50

$$\text{(E)}-\mathbf{CH_2-OH} + F-P(=O)[O-CH(CH_3)_2]_2 \longrightarrow \text{(E)}-\mathbf{CH_2-O}-P(=O)[O-CH(CH_3)_2]_2 + \mathbf{HF}$$

Enzyme–Serine; Diisopropyl fluorophosphate (DFP); Enzyme–Serine complex covalently bonded to DFP

Antibiotics produced by bacteria, mold, or yeast are irreversible inhibitors used to stop bacterial growth. For example, penicillin inhibits an enzyme needed for the formation of cell walls in bacteria, but not human cell membranes. With an incomplete cell wall, bacteria cannot survive and the infection is stopped. However, some bacteria are resistant to penicillin because they produce penicillinase, an enzyme that breaks down penicillin. Over the years, derivatives of penicillin to which bacteria have not yet become resistant have been produced.

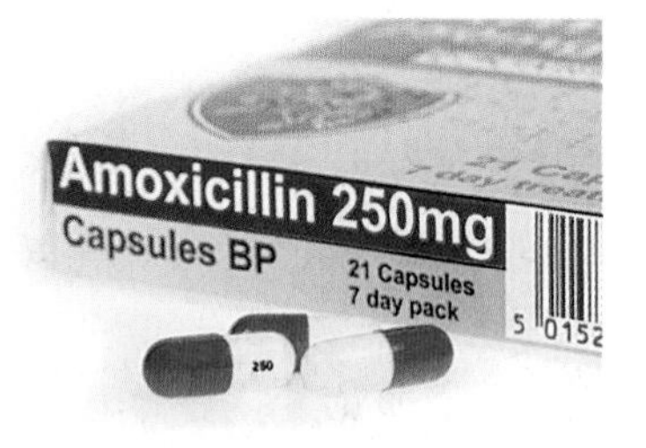

Amoxicillin is a derivative of the antibiotic penicillin.

R Groups for Penicillin Derivatives

Penicillin: R–C(=O)–NH– (β-lactam ring fused to thiazolidine ring with S, two CH_3, and COO^-)

Penicillin G: $C_6H_5-CH_2-$

Penicillin V: $C_6H_5-O-CH_2-$

Ampicillin: $C_6H_5-CH(NH_2)-$

Amoxicillin: $HO-C_6H_4-CH(NH_2)-$

A summary of competitive, noncompetitive, and irreversible inhibitors is shown in TABLE 16.11.

TABLE 16.11 Summary of Competitive, Noncompetitive, and Irreversible Inhibitors

Characteristics	Competitive	Noncompetitive	Irreversible
Shape of Inhibitor	Similar shape to the substrate	Does not have a similar shape to the substrate	Does not have a similar shape to the substrate
Binding to Enzyme	Competes for and binds at the active site	Binds away from the active site to change the shape of the enzyme and its activity	Forms a covalent bond with the enzyme
Reversibility	Adding more substrate reverses the inhibition	Not reversed by adding more substrate, but by a chemical change that removes the inhibitor	Permanent, not reversible

SAMPLE PROBLEM 16.9 Enzyme Inhibition

TRY IT FIRST

State the type of inhibition in the following:

a. The inhibitor has a structure that is similar to the substrate.
b. This inhibitor binds to the surface of the enzyme, changing its shape in such a way that it cannot bind to substrate.

SOLUTION

a. When an inhibitor has a structure similar to that of the substrate, it competes with the substrate for the active site. This type of inhibition is competitive inhibition, which is reversed by increasing the concentration of the substrate.
b. When an inhibitor binds to the surface of the enzyme, it changes the shape of the enzyme and the active site. This type of inhibition is noncompetitive inhibition because the inhibitor does not have a similar shape to the substrate and does not compete with the substrate for the active site.

STUDY CHECK 16.9

What type of inhibition occurs when Sarin, a nerve gas, forms a covalent bond with the R group of serine in the active site of acetylcholinesterase?

ANSWER

Because Sarin forms a covalent bond with an R group in the active site of the enzyme, the inhibition by Sarin is irreversible.

PRACTICE PROBLEMS

16.5 Factors Affecting Enzyme Activity

LEARNING GOAL Describe the effect of temperature, pH, and inhibitors on enzyme activity.

16.43 Trypsin, a peptidase that hydrolyzes polypeptides, functions in the small intestine at an optimum pH of 7.7 to 8.0. How is the rate of a trypsin-catalyzed reaction affected by each of the following conditions?
a. changing the pH to 3.0 **b.** running the reaction at 75 °C

16.44 Pepsin, a peptidase that hydrolyzes proteins, functions in the stomach at an optimum pH of 1.5 to 2.0. How is the rate of a pepsin-catalyzed reaction affected by each of the following conditions?
a. changing the pH to 5.0 **b.** running the reaction at 0 °C

16.45 The following graph shows the activity versus pH curves for pepsin, sucrase, and trypsin. Estimate the optimum pH for each.

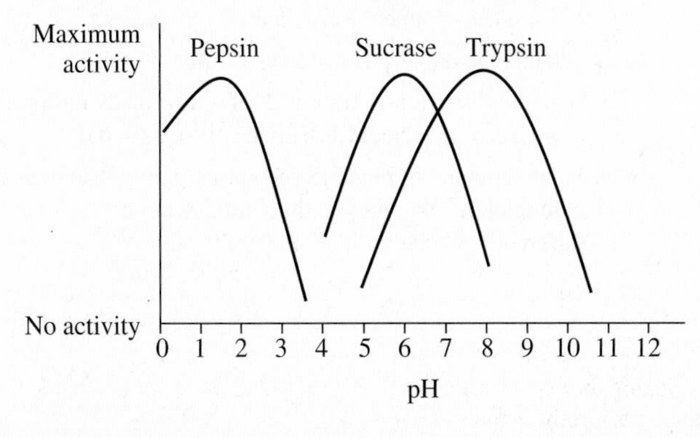

16.46 Refer to the graph in problem 16.45 to determine if the reaction rate in each condition will be at the optimum rate or not.
a. trypsin, pH 5.0 **b.** sucrase, pH 5.0 **c.** pepsin, pH 4.0
d. trypsin, pH 8.0 **e.** pepsin, pH 2.0

16.47 Indicate whether each of the following describes a competitive or a noncompetitive enzyme inhibitor:
a. The inhibitor has a structure similar to the substrate.
b. The effect of the inhibitor cannot be reversed by adding more substrate.
c. The inhibitor competes with the substrate for the active site.
d. The structure of the inhibitor is not similar to the substrate.
e. The addition of more substrate reverses the inhibition.

16.48 Oxaloacetate is an inhibitor of succinate dehydrogenase.

Succinate	Oxaloacetate
COO^-	COO^-
CH_2	CH_2
CH_2	$C{=}O$
COO^-	COO^-

a. Would you expect oxaloacetate to be a competitive or a noncompetitive inhibitor? Why?
b. Would oxaloacetate bind to the active site or elsewhere on the enzyme?
c. How would you reverse the effect of the inhibitor?

Clinical Applications

16.49 Methanol and ethanol are oxidized by alcohol dehydrogenase. In methanol poisoning, ethanol is given intravenously to prevent the formation of formaldehyde that has toxic effects.

a. Draw the condensed structural formulas for methanol and ethanol.

b. Would ethanol compete for the active site or bind to a different site?

c. Would ethanol be a competitive or noncompetitive inhibitor of alcohol dehydrogenase?

16.50 In humans, the antibiotic amoxicillin (a type of penicillin) is used to treat certain bacterial infections.

a. Does the antibiotic inhibit enzymes in humans?

b. Why does the antibiotic kill bacteria, but not humans?

c. Is amoxicillin a reversible or irreversible inhibitor?

CLINICAL UPDATE Jeremy's Diagnosis and Treatment for Sickle-Cell Anemia

Diagnosis of Sickle-Cell Anemia

When Jeremy and his mother return to the clinic, the physician assistant explains the result of the blood test. Proteins are used to identify and diagnose diseases such as sickle-cell anemia by *electrophoresis*. For Jeremy's diagnosis, normal hemoglobin, sickle-cell hemoglobin, and Jeremy's hemoglobin are placed on a gel and an electric current is applied. The proteins move along the gel based on their charge, size, and shape.

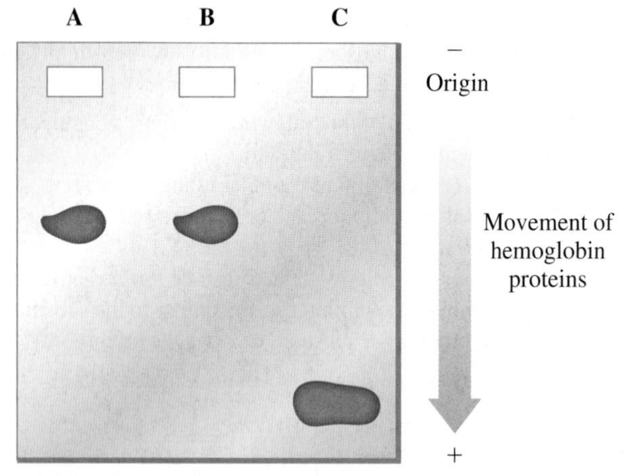

In electrophoresis, hemoglobin samples from Jeremy **(A)**, a person with sickle-cell anemia **(B)**, and a person with normal hemoglobin **(C)** are placed on a gel and an electrical field is applied. The differences in the movement of the abnormal and normal proteins allow the diagnosis of sickle-cell anemia.

Treatment of Sickle-Cell Anemia

By comparing the movement of Jeremy's hemoglobin with that of sickle-cell hemoglobin and normal hemoglobin, electrophoresis indicates that he does have sickle-cell anemia. Emma prescribes a daily dose of hydroxyurea for Jeremy. Hydroxyurea is a medication that stimulates the production of fetal hemoglobin, a variation of the hemoglobin molecule that binds more tightly to oxygen molecules than normal adult hemoglobin. Fetal hemoglobin is produced by a fetus *in utero*, but production of this molecule typically stops after birth.

Patients with sickle-cell anemia who take hydroxyurea daily experience less pain, fewer events in which blood vessels are obstructed by sickled red blood cells, and increased blood flow. The suggested daily dose for children is 20. mg of hydroxyurea per kg of body weight. Emma will monitor Jeremy regularly to determine if his hydroxyurea dose needs to be adjusted.

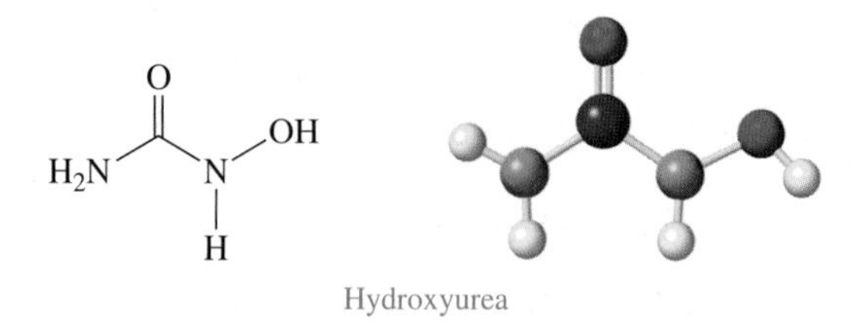

Hydroxyurea

Clinical Applications

16.51 What is the chemical formula for hydroxyurea?

16.52 What is the molar mass of hydroxyurea?

16.53 If Jeremy's current weight is 21 lbs, how many milligrams of hydroxyurea should Jeremy be given each day?

16.54 Fetal hemoglobin binds oxygen more tightly than does adult hemoglobin. Why do you think fetal hemoglobin functions differently than adult hemoglobin?

CONCEPT MAP

AMINO ACIDS, PROTEINS, AND ENZYMES

Proteins

- contain **Amino Acids** that have **Ammonium, Carboxylate, and R Groups**
- contain peptide bonds between **Amino Acids** in a specific order as **Primary Structure** and hydrogen bond as **Secondary Structure** with interactions that give **Tertiary and Quaternary Structures** are active as **Enzymes**
- are used for structure, transport, hormones, storage, and as **Enzymes** bind substrate at **Active Site** to form an **ES Complex** that is blocked by **Inhibitors**
- undergo **Denaturation** by **Heat, Acids, Bases, Organic Compounds** that change **Secondary or Tertiary Structures of Proteins**

CHAPTER REVIEW

16.1 Proteins and Amino Acids

LEARNING GOAL Classify proteins by their functions. Give the name and abbreviations for an amino acid and draw its structure at physiological pH.

- Some proteins are enzymes or hormones, whereas others are important in structure, transport, protection, storage, and muscle contraction.
- A group of 20 amino acids provides the molecular building blocks of proteins.
- Attached to the central α carbon of each amino acid are an ammonium group, a carboxylate group, and a unique R group.
- The R group gives an amino acid the property of being nonpolar, polar, acidic, or basic.

16.2 Proteins: Primary Structure

LEARNING GOAL Draw the condensed structural formula for a peptide and give its name. Describe the primary structure for a protein.

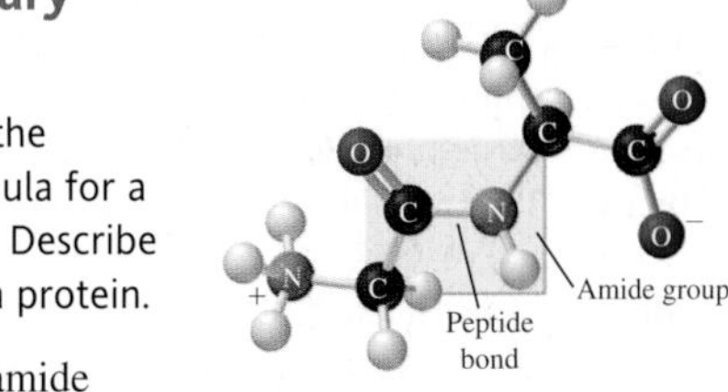

- Peptides form when an amide bond links the carboxylate group of one amino acid and the ammonium group of a second amino acid.
- Long chains of amino acids that are biologically active are called proteins.
- Essential amino acids must be obtained from the proteins in the diet; they cannot be made in the body.
- The primary structure of a protein is its sequence of amino acids joined by peptide bonds.
- Peptides are named from the N-terminus by replacing the *ine* or *ate* of each amino acid name with *yl* followed by the amino acid name of the C-terminus.

16.3 Proteins: Secondary, Tertiary, and Quaternary Structures

LEARNING GOAL Describe the secondary, tertiary, and quaternary structures for a protein; describe the denaturation of a protein.

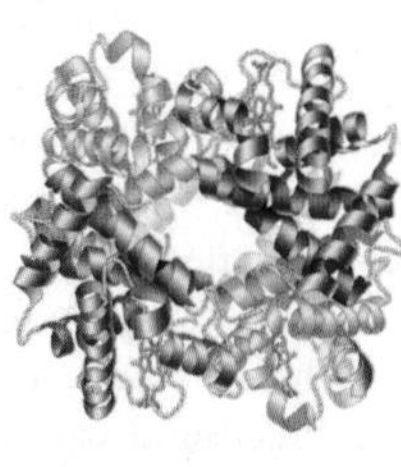

- In the secondary structure, hydrogen bonds between atoms in the peptide bonds produce a characteristic shape such as an α helix or a β-pleated sheet.
- A tertiary structure is stabilized by interactions that push amino acids with hydrophobic R groups to the center and pull amino acids with hydrophilic R groups to the surface, and by interactions between amino acids with R groups that form hydrogen bonds, disulfide bonds, and salt bridges.
- In a quaternary structure, two or more tertiary subunits are joined together for biological activity, held by the same interactions found in tertiary structures.
- Denaturation of a protein occurs when high temperatures, acids or bases, organic compounds, metal ions, or agitation destroy the secondary, tertiary, or quaternary structures of a protein with a loss of biological activity.

16.4 Enzymes

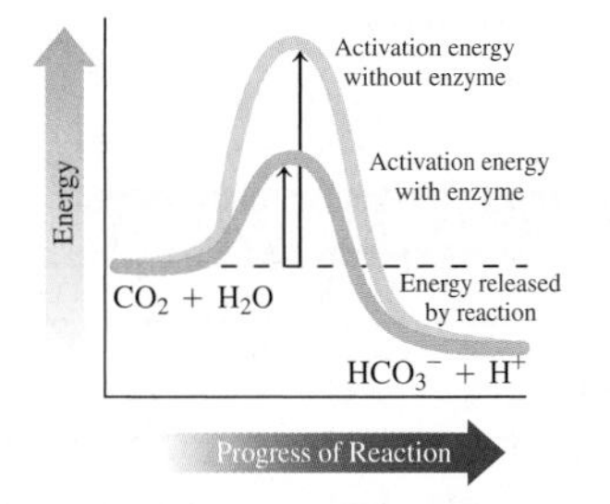

LEARNING GOAL Describe enzymes and their role in enzyme-catalyzed reactions.

- Most enzymes are proteins.
- Enzymes act as biological catalysts by lowering activation energy and accelerating the rate of cellular reactions.
- Within the tertiary structure of an enzyme, a small pocket called the active site binds the substrate.
- In the lock-and-key model, a substrate precisely fits the shape of the active site.
- In the induced-fit model, both the active site and the substrate undergo changes in their shapes to give the best fit for efficient catalysis.
- In the enzyme–substrate complex, catalysis takes place when amino acid R groups in the active site of an enzyme interact with a substrate.
- When the products of catalysis are released, the enzyme can bind to another substrate molecule.

16.5 Factors Affecting Enzyme Activity

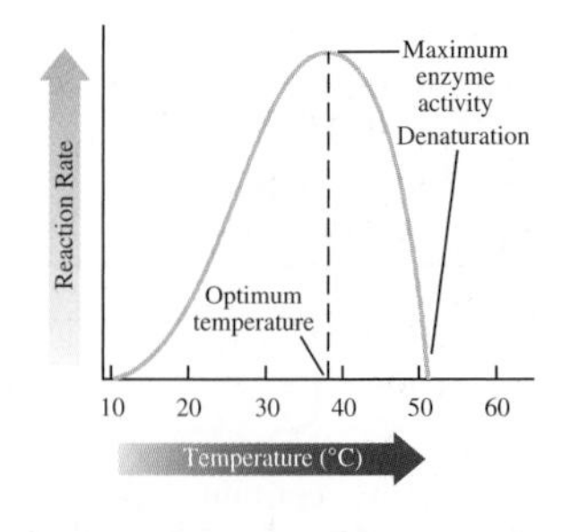

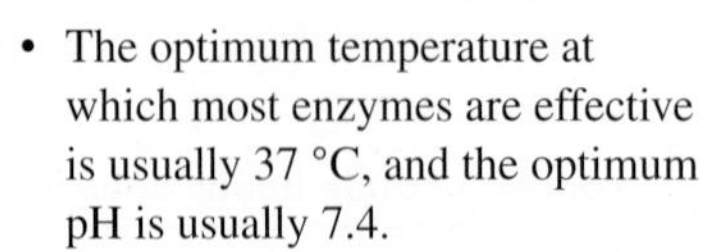

LEARNING GOAL Describe the effect of temperature, pH, and inhibitors on enzyme activity.

- The optimum temperature at which most enzymes are effective is usually 37 °C, and the optimum pH is usually 7.4.
- The rate of an enzyme-catalyzed reaction decreases as temperature and pH go above or below the optimum temperature and pH values.
- An inhibitor reduces the activity of an enzyme or makes it inactive.
- An inhibitor can be reversible or irreversible.
- A competitive inhibitor has a structure similar to the substrate and competes for the active site.
- A noncompetitive inhibitor attaches to the enzyme away from the active site, changing the shape of both the enzyme and its active site.
- An irreversible inhibitor forms a covalent bond within the active site that permanently prevents catalytic activity.

KEY TERMS

acidic amino acid An amino acid that has an R group with a carboxylate group ($-COO^-$).

active site A pocket in a part of the tertiary enzyme structure that binds substrate and catalyzes a reaction.

activity The rate at which an enzyme catalyzes the reaction that converts substrate to product.

α (alpha) helix A secondary level of protein structure, in which hydrogen bonds connect the N—H of one peptide bond with the C=O of a peptide bond farther along in the chain to form a coiled or corkscrew structure.

amino acid The building block of proteins, consisting of an ammonium group, a carboxylate group, and a unique R group attached to the α carbon.

basic amino acid An amino acid that contains an amine R group.

β (beta)-pleated sheet A secondary level of protein structure that consists of hydrogen bonds between peptide links in parallel polypeptide chains.

C-terminus The end amino acid in a peptide chain with a free $-COO^-$ group.

collagen The most abundant form of protein in the body, composed of fibrils of triple helices with hydrogen bonding between —OH groups of hydroxyproline and hydroxylysine.

competitive inhibitor A molecule with a structure similar to the substrate that inhibits enzyme action by competing for the active site.

denaturation The loss of secondary, tertiary, and quaternary protein structures, caused by heat, acids, bases, organic compounds, heavy metals, and/or agitation.

disulfide bond Covalent —S—S— bond that forms between the —SH groups of cysteines in a protein, which stabilizes the tertiary and quaternary structures.

enzyme A protein that catalyzes biological reactions.

enzyme–substrate (ES) complex An intermediate consisting of an enzyme that binds to a substrate in an enzyme-catalyzed reaction.

hydrogen bond Attraction between polar R groups such as —OH, $-NH_2$, and —COOH of amino acids in a polypeptide chain.

hydrophilic interaction The attraction between water and polar R groups on the outside of the protein.

hydrophobic interaction The attraction between nonpolar R groups on the inside of a protein.

induced-fit model A model of enzyme action in which the shapes of the substrate and its active site adjust to give an optimal fit.

inhibitor Substance that makes an enzyme inactive by interfering with its ability to react with a substrate.

isoenzymes Enzymes with different combinations of polypeptide subunits that catalyze the same reaction in different tissues of the body.

N-terminus The end amino acid in a peptide chain with a free $-NH_3^+$ group.

noncompetitive inhibitor An inhibitor that does not resemble the substrate, and attaches to the enzyme away from the active site to prevent binding of the substrate.

nonpolar amino acid Amino acid with nonpolar R groups that contains only C and H atoms.

optimum pH The pH at which an enzyme is most active.

optimum temperature The temperature at which an enzyme is most active.

peptide The combination of two or more amino acids joined by peptide bonds; dipeptide, tripeptide, and so on.

peptide bond The amide bond in peptides that joins the carboxylate group of one amino acid with the ammonium group in the next amino acid.

polar amino acid Amino acid with polar R groups that interact with water.

primary structure The specific sequence of the amino acids in a protein.

protein A polypeptide of 50 or more amino acids that has biological activity.

quaternary structure A protein structure in which two or more protein subunits form an active protein.

salt bridge The attraction between ionized R groups of basic and acidic amino acids in the tertiary structure of a protein.

secondary structure The formation of an α helix or a β-pleated sheet.

substrate The molecule that reacts in the active site in an enzyme-catalyzed reaction.

tertiary structure The folding of the secondary structure of a protein into a compact structure that is stabilized by the interactions of R groups.

triple helix The protein structure found in collagen, consisting of three polypeptide chains woven together like a braid.

CORE CHEMISTRY SKILLS

The chapter Section containing each Core Chemistry Skill is shown in parentheses at the end of each heading.

Drawing the Structure for an Amino Acid at Physiological pH (16.1)

- The central α carbon of each amino acid is bonded to an ammonium group ($-NH_3^+$), a carboxylate group ($-COO^-$), a hydrogen atom, and a unique R group.
- The R group gives an amino acid the property of being nonpolar, polar, acidic, or basic.

Example: Draw the structure for cysteine at physiological pH.

Answer:

$$H_3\overset{+}{N}-\underset{\displaystyle H}{\overset{\displaystyle \overset{\displaystyle SH}{|}\atop CH_2}{\overset{|}{\underset{|}{C}}}}-\overset{\displaystyle O}{\overset{\|}{C}}-O^-$$

Identifying the Primary, Secondary, Tertiary, and Quaternary Structures of Proteins (16.2, 16.3)

- The primary structure of a protein is the sequence of amino acids joined by peptide bonds.
- In the secondary structures of proteins, hydrogen bonds between atoms in the peptide bonds produce an alpha helix or a beta-pleated sheet.
- The tertiary structure of a protein is stabilized by R groups that form hydrogen bonds, disulfide bonds, and salt bridges, and hydrophobic R groups that move to the center and hydrophilic R groups that move to the surface.
- In a quaternary structure, two or more tertiary subunits are combined for biological activity, held by the same interactions found in tertiary structures.

Example: Identify the following as characteristic of the primary, secondary, tertiary, or quaternary structure of a protein:

a. The R groups of two amino acids interact to form a salt bridge.
b. Eight amino acids form peptide bonds.
c. A polypeptide forms an alpha helix.
d. Two amino acids with hydrophobic R groups move toward the inside of the folded protein.
e. A protein with biological activity contains four tertiary polypeptide subunits.

Answer: **a.** tertiary, quaternary **b.** primary **c.** secondary **d.** tertiary **e.** quaternary

Describing Enzyme Action (16.4)

- Enzymes are biological catalysts that lower the activation energy and accelerate the rate of cellular reactions.
- Within the tertiary structure of an enzyme, a small pocket called the active site binds the substrate.
- In the induced-fit model of enzyme action, the active site and the substrate change their shapes for efficient catalysis.
- When the products of catalysis are released, the enzyme can bind to another substrate molecule.

Example: Match the terms (1) enzyme, (2) substrate, (3) active site, and (4) induced-fit model with each of the following:

a. an active site that adapts to the shape of a substrate
b. has a structure that fits the active site of an enzyme
c. the portion of an enzyme where catalytic activity occurs
d. has a tertiary structure that recognizes the substrate

Answer: **a.** (4) induced-fit model **b.** (2) substrate **c.** (3) active site **d.** (1) enzyme

UNDERSTANDING THE CONCEPTS

The chapter Sections to review are shown in parentheses at the end of each problem.

16.55 Ethylene glycol ($HO-CH_2-CH_2-OH$) is a major component of antifreeze. If ingested, it is first converted to $HOOC-CHO$ (oxoethanoic acid) and then to $HOOC-COOH$ (oxalic acid), which is toxic. (16.4, 16.5)

Ethylene glycol is added to a radiator to prevent freezing and boiling.

a. What class of enzyme catalyzes the reactions described?
b. The treatment for the ingestion of ethylene glycol is an intravenous solution of ethanol. How might this help prevent toxic levels of oxalic acid in the body?

16.56 Adults who are lactose intolerant cannot break down the disaccharide in milk products. To help digest dairy food, a product known as Lactaid can be added to milk and the milk then refrigerated for 24 hours. (16.4, 16.5)

The disaccharide lactose is present in milk products.

a. What enzyme is present in Lactaid, and what is the major class of this enzyme?
b. What might happen to the enzyme if the Lactaid were stored at 55 °C?

16.57 Aspartame, which is used in artificial sweeteners, contains the following dipeptide: (16.1, 16.2)

$$H_3\overset{+}{N}-\underset{\underset{\displaystyle CH_2-C(=O)-O^-}{|}}{CH}-\overset{O}{\overset{\|}{C}}-\underset{H}{N}-\underset{\underset{\displaystyle CH_2-C_6H_5}{|}}{CH}-\overset{O}{\overset{\|}{C}}-O^-$$

Some artificial sweeteners contain aspartame, which is an ester of a dipeptide.

a. What are the amino acids in aspartame?
b. How would you name the dipeptide in aspartame?

16.58 Fresh pineapple contains the enzyme bromelain that hydrolyzes peptide bonds in proteins. (16.3, 16.4, 16.5)

Fresh pineapple contains the enzyme bromelain.

a. The directions for a gelatin (protein) dessert say not to add fresh pineapple. However, canned pineapple where pineapple is heated to high temperatures can be added. Why?
b. Fresh pineapple is used in a marinade to tenderize tough meat. Why?
c. What structural level of a protein does the bromelain enzyme destroy?

16.59 Identify the amino acids and type of interaction that occurs between the following R groups in tertiary protein structures: (16.1, 16.3)

a. $-CH_2-\overset{O}{\overset{\|}{C}}-NH_2$ and $HO-CH_2-$

b. $-CH_2-\overset{O}{\overset{\|}{C}}-O^-$ and $H_3\overset{+}{N}-(CH_2)_4-$

c. $-CH_2-SH$ and $HS-CH_2-$

d. $-CH_2-\overset{CH_3}{\overset{|}{CH}}-CH_3$ and CH_3-

16.60 What type of interaction would you expect between the following in a tertiary structure? (16.1, 16.3)

a. threonine and glutamine b. valine and alanine
c. arginine and glutamate

16.61 a. Draw the condensed structural formula for Ser–Lys–Asp. (16.1, 16.2, 16.3)
b. Would you expect to find this segment at the center or at the surface of a protein? Why?

16.62 a. Draw the condensed structural formula for Val–Ala–Leu. (16.1, 16.2, 16.3)
b. Would you expect to find this segment at the center or at the surface of a protein? Why?

16.63 Seeds and vegetables are often deficient in one or more essential amino acids. Using the following table, state whether each combination provides all of the essential amino acids: (16.2)

Source	Lysine	Tryptophan	Methionine
Oatmeal	No	Yes	Yes
Rice	No	Yes	Yes
Garbanzo beans	Yes	No	Yes
Lima beans	Yes	No	No
Cornmeal	No	No	Yes

a. rice and garbanzo beans b. lima beans and cornmeal
c. garbanzo beans and lima beans

16.64 Seeds and vegetables are often deficient in one or more essential amino acids. Using the table in problem 16.63, state whether each combination provides all of the essential amino acids. (16.2)

Oatmeal is deficient in the essential amino acid lysine.

a. rice and lima beans
b. rice and oatmeal
c. oatmeal and lima beans

ADDITIONAL PRACTICE PROBLEMS

16.65 What are some differences between each of the following pairs? (16.1, 16.2, 16.3)

a. secondary and tertiary protein structures
b. essential and nonessential amino acids
c. polar and nonpolar amino acids
d. dipeptides and tripeptides

16.66 What are some differences between each of the following pairs? (16.1, 16.2, 16.3)

a. a salt bridge and a disulfide bond
b. an α helix and collagen
c. α helix and β-pleated sheet
d. tertiary and quaternary structures of proteins

16.67 If glutamate were replaced by proline in a protein, how might the tertiary structure be affected? (16.1, 16.3)

16.68 If glycine were replaced by alanine in a protein, how might the tertiary structure be affected? (16.1, 16.3)

16.69 How do enzymes differ from catalysts used in chemical laboratories? (16.4)

16.70 Why do enzymes function only under mild conditions? (16.4, 16.5)

16.71 Lactase is an enzyme that hydrolyzes lactose to glucose and galactose. (16.4)
   a. What are the reactants and products of the reaction?
   b. Draw an energy diagram for the reaction with and without lactase.
   c. How does lactase make the reaction go faster?

16.72 Maltase is an enzyme that hydrolyzes maltose to two glucose molecules. (16.4)
   a. What are the reactants and products of the reaction?
   b. Draw an energy diagram for the reaction with and without maltase.
   c. How does maltase make the reaction go faster?

16.73 Indicate whether each of the following would be a substrate (S) or an enzyme (E): (16.4)
   a. lactose **b.** lipase **c.** amylase
   d. trypsin **e.** pyruvate **f.** transaminase

16.74 Indicate whether each of the following would be a substrate (S) or an enzyme (E): (16.4)
   a. glucose **b.** hydrolase **c.** maleate isomerase
   d. alanine **e.** amylose **f.** lactase

16.75 Give the substrate of each of the following enzymes: (16.4)
   a. urease **b.** succinate dehydrogenase
   c. aspartate transaminase **d.** tyrosinase

16.76 Give the substrate of each of the following enzymes: (16.4)
   a. maltase **b.** fructose oxidase
   c. phenolase **d.** sucrase

16.77 How would the lock-and-key model explain that sucrase hydrolyzes sucrose, but not lactose? (16.4)

16.78 How does the induced-fit model of enzyme action allow an enzyme to catalyze a reaction of a group of substrates? (16.4)

16.79 If a blood test indicates a high level of LDH and CK, what could be the cause? (16.4)

16.80 If a blood test indicates a high level of ALT, what could be the cause? (16.4)

CHALLENGE PROBLEMS

The following problems are related to the topics in this chapter. However, they do not all follow the chapter order, and they require you to combine concepts and skills from several Sections. These problems will help you increase your critical thinking skills and prepare for your next exam.

16.81 Consider the amino acids lysine, valine, and aspartate in an enzyme. State which of these amino acids have R groups that would: (16.1, 16.3)
   a. be found in hydrophobic regions
   b. be found in hydrophilic regions
   c. form hydrogen bonds
   d. form salt bridges

16.82 Consider the amino acids histidine, phenylalanine, and serine in an enzyme. State which of these amino acids have R groups that would: (16.1, 16.3)
   a. be found in hydrophobic regions
   b. be found in hydrophilic regions
   c. form hydrogen bonds
   d. form salt bridges

UNDERSTANDING PROTEIN STRUCTURES

In order to understand how proteins function, we need to know about their tertiary structures; but predicting the tertiary structures of proteins is a very challenging task. Proteins are large molecules consisting of hundreds to thousands of different amino acids. They can fold in many, many different ways.

Researchers at the University of Washington built an online game called Foldit that allows students to fold proteins into three-dimensional shapes. Game players learn about the rules that guide protein folding and then use these rules to fold provided primary structures into three-dimensional shapes. To try your hand at solving for the three-dimensional structure of a protein, go to https://fold.it/portal/.

ANSWERS

Answers to Selected Practice Problems

16.1 **a.** transport **b.** structural **c.** structural **d.** enzyme

16.3 All amino acids contain a carboxylate group and an ammonium group on the α carbon.

16.5 **a.** $H_3\overset{+}{N}-\underset{\displaystyle H}{\overset{\displaystyle H}{C}}-\overset{\displaystyle O}{\overset{\|}{C}}-O^-$

b. $H_3\overset{+}{N}-\underset{\displaystyle H}{\overset{\displaystyle CH(CH_3)OH}{C}}-\overset{\displaystyle O}{\overset{\|}{C}}-O^-$

c. $H_3\overset{+}{N}-\underset{\displaystyle H}{\overset{\displaystyle CH_2-CH_2-COO^-}{C}}-\overset{\displaystyle O}{\overset{\|}{C}}-O^-$

d. $H_3\overset{+}{N}-\underset{\displaystyle H}{\overset{\displaystyle CH_2-C_6H_5}{C}}-\overset{\displaystyle O}{\overset{\|}{C}}-O^-$

16.7 **a.** nonpolar, hydrophobic **b.** polar neutral, hydrophilic
c. polar acidic, hydrophilic **d.** nonpolar, hydrophobic

16.9 **a.** alanine **b.** glutamine **c.** lysine **d.** cysteine

16.11 **a.**

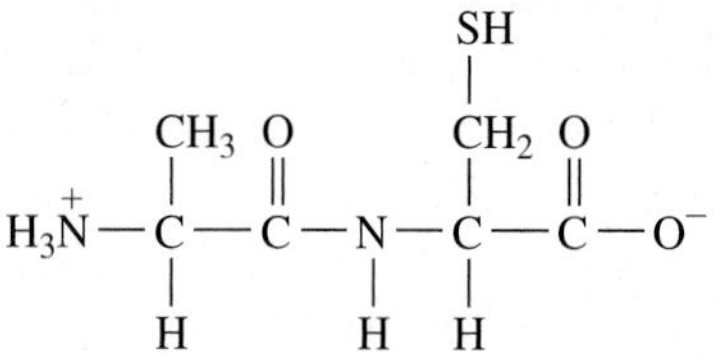

Ala–Cys, AC

b.

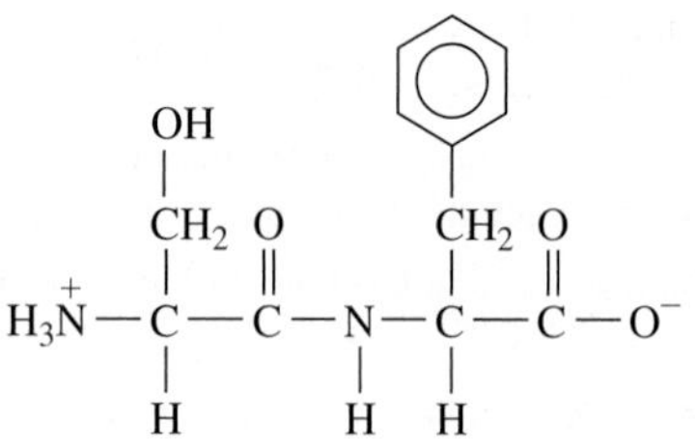

Ser–Phe, SF

c.

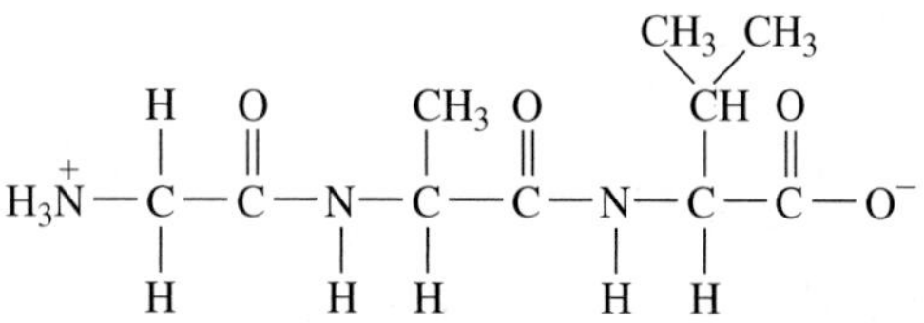

Gly–Ala–Val, GAV

d.

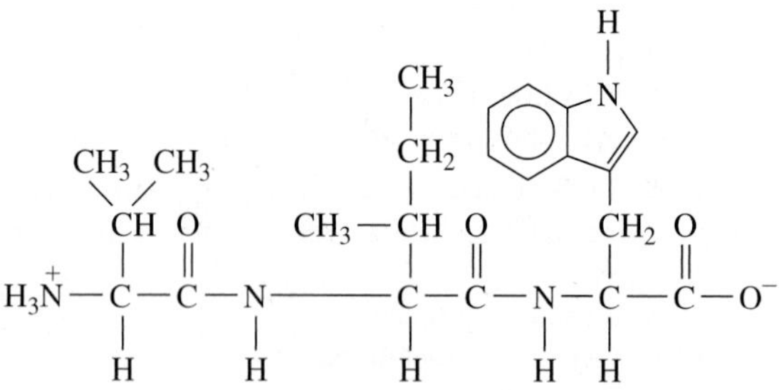

Val–Ile–Trp, VIW

16.13 **a.**

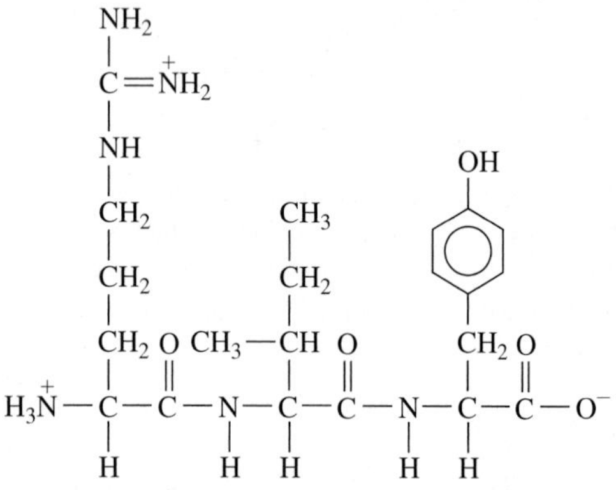

RIY

b.

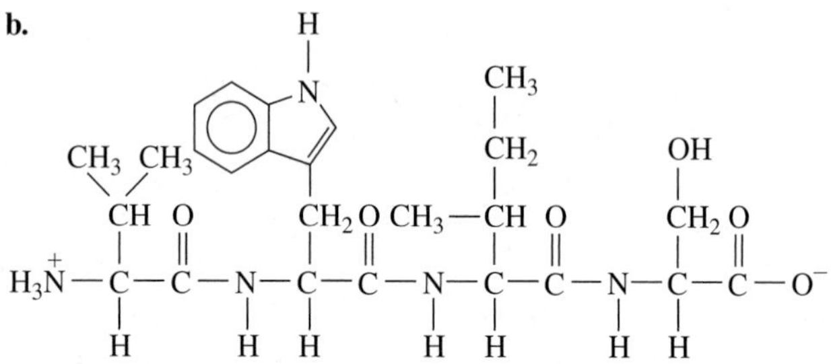

VWIS

16.15 **a.** Lysine and tryptophan are lacking in corn but supplied by peas; methionine is lacking in peas but supplied by corn.
b. Lysine is lacking in rice but supplied by soy; methionine is lacking in soy but supplied by rice.

16.17 The oxygen atoms of the carbonyl groups and the hydrogen atoms attached to the nitrogen atoms form alpha helices or beta-pleated sheets.

16.19 In the α helix, hydrogen bonds form between the carbonyl oxygen atom and the amino hydrogen atom in the next turn of the helical chain. In the β-pleated sheet, hydrogen bonds occur between parallel sections of a long polypeptide chain.

16.21 **a.** disulfide bond **b.** salt bridge
c. hydrogen bond **d.** hydrophobic interaction

16.23 **a.** Leucine and valine will be found on the inside of the protein because they have R groups that are hydrophobic.
b. The cysteine and aspartate would be on the outside of the protein because they have R groups that are polar.
c. The order of the amino acids (the primary structure) provides R groups that interact to determine the tertiary structure of the protein.

16.25 **a.** tertiary, quaternary **b.** primary
c. secondary **d.** secondary

16.27 **a.** Placing an egg in boiling water coagulates the proteins of the egg because the heat disrupts hydrogen bonds and hydrophobic interactions.
b. The alcohol on the swab coagulates the proteins of any bacteria present by forming hydrogen bonds as well as disrupting hydrophobic interactions.
c. The heat from an autoclave will coagulate the proteins of any bacteria on the surgical instruments by disrupting hydrogen bonds and hydrophobic interactions.
d. Heat will coagulate the surrounding proteins to close the wound by disrupting hydrogen bonds and hydrophobic interactions.

16.29 The chemical reactions can occur without enzymes, but the rates are too slow. Catalyzed reactions, which are many times faster, provide the amounts of products needed by the cell at a particular time.

16.31 **a.** galactose **b.** lipid **c.** aspartate

16.33 **a.** hydrolase **b.** oxidoreductase
c. isomerase **d.** transferase

16.35 **a.** (2) enzyme
b. (1) enzyme–substrate complex
c. (3) substrate

16.37 **a.** $E + S \rightleftharpoons ES \longrightarrow E + P$
b. The active site is a region or pocket within the tertiary structure of an enzyme that accepts the substrate, aligns the substrate for reaction, and catalyzes the reaction.

16.39 Isoenzymes are slightly different forms of an enzyme that catalyze the same reaction in different organs and tissues of the body.

16.41 A doctor might run tests for the enzymes CK, LDH, and AST to determine if the patient had a heart attack.

16.41 **a.** The rate would decrease.
b. The rate would decrease.

16.45 pepsin, pH 2; sucrase, pH 6; trypsin, pH 8

16.47 **a.** competitive **b.** noncompetitive
c. competitive **d.** noncompetitive
e. competitive

16.49 a. methanol, $CH_3—OH$; ethanol, $CH_3—CH_2—OH$
b. Ethanol has a structure similar to methanol and could compete for the active site.
c. Ethanol is a competitive inhibitor of alcohol dehydrogenase.

16.51 $CH_4N_2O_2$

16.53 76 g

16.55 a. oxidoreductase
b. At high concentration, ethanol, which acts as a competitive inhibitor of ethylene glycol, would saturate the alcohol dehydrogenase enzyme to allow ethylene glycol to be removed from the body without producing oxalic acid.

16.57 a. aspartate and phenylalanine
b. aspartylphenylalanine

16.59 a. asparagine and serine; hydrogen bond
b. aspartate and lysine; salt bridge
c. cysteine and cysteine; disulfide bond
d. leucine and alanine; hydrophobic interaction

16.61 a.

```
                        +
                       NH3
                        |
                       CH2
                        |
                       CH2
                        |              O   O-
         OH            CH2              \\ /
         |              |                C
         |             CH2               |
        CH2  O          |    O          CH2  O
  +      |   ||         |    ||          |   ||
H3N  —   C — C — N  —   C —  C — N   —   C — C — O-
         |       |      |        |       |
         H       H      H        H       H
```

b. This segment contains polar R groups, which would be found on the surface of a protein where they hydrogen bond with water.

16.63 a. yes **b.** no **c.** no

16.65 a. In the secondary structure of proteins, hydrogen bonds form a helix or a pleated sheet; the tertiary structure is determined by the interactions of R groups such as disulfide bonds, hydrogen bonds, and salt bridges.
b. Nonessential amino acids can be synthesized by the body; essential amino acids must be supplied by the diet.
c. Polar amino acids have hydrophilic R groups, whereas nonpolar amino acids have hydrophobic R groups.
d. Dipeptides contain two amino acids, whereas tripeptides contain three.

16.67 Glutamate is a polar acidic amino acid, whereas proline is nonpolar. The polar acidic R group of glutamate may form hydrogen bonds or salt bridges with another amino acid R group. It may also move to the outer surface of the protein where it would form hydrophilic interactions with water. However, the nonpolar proline would move to the hydrophobic center of the protein.

16.69 In chemical laboratories, reactions are often run at high temperatures using catalysts that are strong acids or bases. Enzymes are catalysts that are proteins and function only at mild temperature and pH. Catalysts used in chemistry laboratories are usually inorganic materials that can function at high temperatures and in strongly acidic or basic conditions.

16.71 a. The reactants are lactose and water and the products are glucose and galactose.
b.

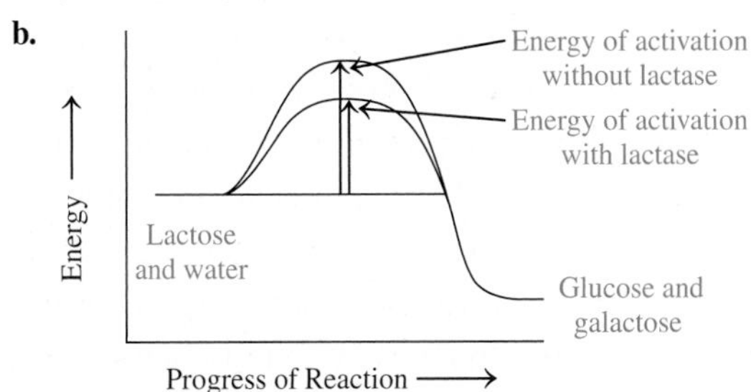

c. By lowering the energy of activation, the enzyme furnishes a lower energy pathway by which the reaction can take place.

16.73 a. S **b.** E **c.** E **d.** E **e.** S **f.** E

16.75 a. urea **b.** succinate **c.** aspartate **d.** tyrosine

16.77 Sucrose fits the shape of the active site in sucrase, but lactose does not.

16.79 A heart attack may be the cause.

16.81 a. valine **b.** lysine, aspartate
c. lysine, aspartate **d.** lysine, aspartate

17 Nucleic Acids and Protein Synthesis

ELLEN FOUND A PEA-SIZED LUMP IN HER breast. A needle biopsy confirms that Ellen has breast cancer. She undergoes a lumpectomy, during which the surgeon removes the tumor along with a small margin of surrounding normal tissue. The surgeon also makes an incision under her arm and removes the sentinel lymph node. Because cancer cells are found in the sentinel node, more lymph nodes are removed. He sends the excised tumor and lymph nodes to Lisa, a histology technician.

Lisa prepares the tissue sample to be viewed by a pathologist. Tissue preparation requires Lisa to cut the tissue into very thin sections, normally 0.001 mm, which are mounted onto microscope slides. She treats the tissue on the slides with a dye to stain the cells, which enables the pathologist to distinguish abnormal cells more easily.

When a person's DNA (deoxyribonucleic acid) is damaged, it may result in mutations that promote the abnormal cell growth found in cancer. Cancer as well as genetic diseases can be a result of mutations caused by environmental and hereditary factors.

CAREER Histology Technician

A histology technician studies the microscopic makeup of tissues, cells, and bodily fluids with the purpose of detecting and identifying the presence of a specific disease. They determine blood types and the concentrations of drugs and other substances in the blood. Sample preparation is a critical component of a histologist's job. The tissue samples are cut, using specialized equipment, into extremely thin sections, which are then mounted and stained using various dyes. The dyes provide contrast for the cells to be viewed and help highlight any abnormalities that may exist. Utilization of various dyes requires the histologist to be familiar with solution preparation and the handling of potentially hazardous chemicals.

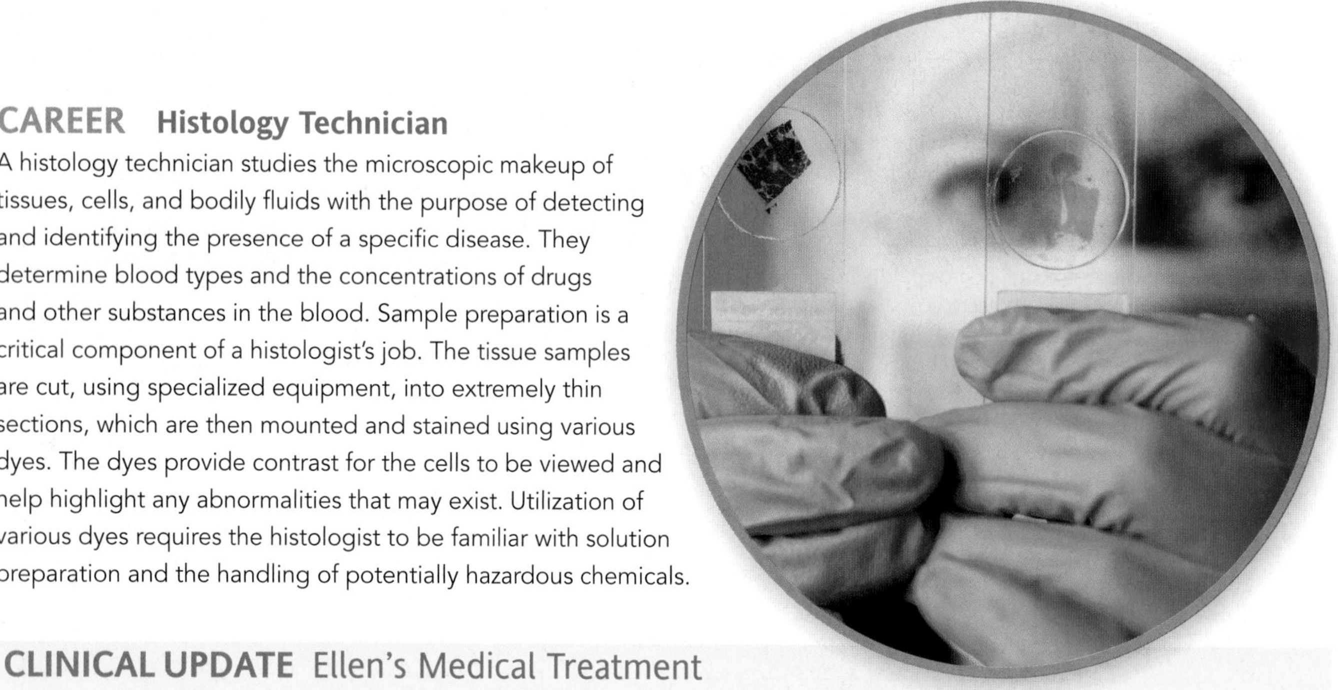

CLINICAL UPDATE Ellen's Medical Treatment Following Breast Cancer Surgery

You can learn more about the type of breast cancer that Ellen has, as well as some of the drugs that are available to treat this type of breast cancer, by reading the **CLINICAL UPDATE Ellen's Medical Treatment Following Breast Cancer Surgery**, page 611.

17.1 Components of Nucleic Acids

LEARNING GOAL Describe the bases and ribose sugars that make up the nucleic acids DNA and RNA.

Nucleic acids are large molecules found in the nuclei of cells that store information and direct activities for cellular growth and reproduction. There are two closely related types of nucleic acids: *deoxyribonucleic acid* (**DNA**) and *ribonucleic acid* (**RNA**). Deoxyribonucleic acid, the genetic material in the nucleus of a cell, contains all the information needed for the development of a complete living organism. Ribonucleic acid interprets the genetic information in DNA for the synthesis of protein.

Both DNA and RNA are composed of smaller units known as *nucleotides*, linked together in unbranched chains. Each nucleotide has three components: a base that contains nitrogen, a five-carbon sugar, and a phosphate group (see **FIGURE 17.1**). A DNA molecule may contain several million nucleotides; smaller RNA molecules may contain up to several thousand.

Bases

The nitrogen-containing **bases** in nucleic acids are derivatives of the heterocyclic amines *pyrimidine* or *purine*. A pyrimidine has a single ring with two nitrogen atoms, and a purine has two rings each with two nitrogen atoms. They are basic because the nitrogen atoms are H^+ acceptors. In DNA, the pyrimidine bases with single rings are cytosine (C) and thymine (T), and the purine bases with double rings are adenine (A) and guanine (G). RNA contains the same bases, except thymine (5-methyluracil) is replaced by uracil (U) (see **FIGURE 17.2**).

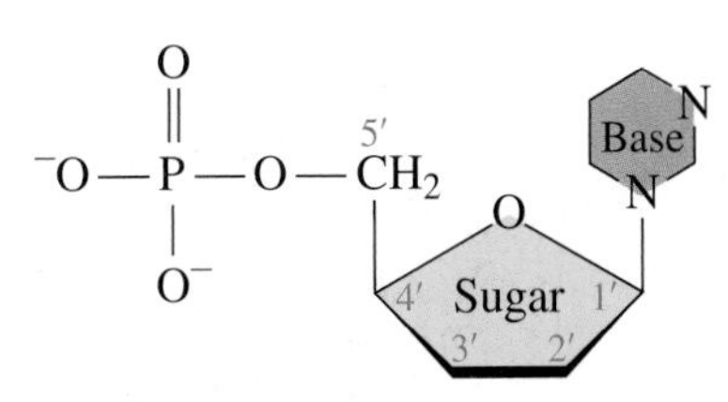

FIGURE 17.1 The general structure of a nucleotide includes a nitrogen-containing base, a sugar, and a phosphate group.

Q In a nucleotide, what types of groups are bonded to a five-carbon sugar?

Pyrimidine Bases in Nucleic Acids

Pyrimidine	Cytosine (C) (DNA and RNA)	Uracil (U) (RNA only)	Thymine (T) (DNA only)

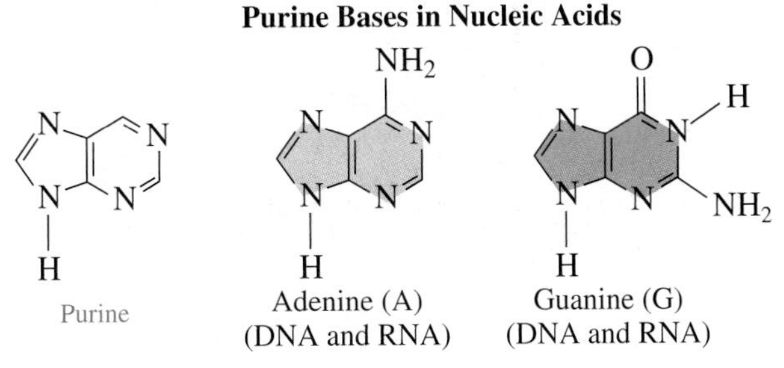

FIGURE 17.2 DNA contains the bases A, G, C, and T; RNA contains A, G, C, and U.

Q Which bases are found in DNA?

TEST

Try Practice Problems 17.1 to 17.4

Pentose Sugars

In RNA, the five-carbon sugar is *ribose*, which gives the letter R in the abbreviation RNA. The atoms in the pentose sugars are numbered with primes (1′, 2′, 3′, 4′, and 5′) to differentiate them from the atoms in the bases. In DNA, the five-carbon sugar is *deoxyribose*, which is similar to ribose except that there is no hydroxyl group (—OH) on C2′. The *deoxy* prefix means "without oxygen" and provides the letter D in DNA.

Pentose Sugars in Nucleic Acids

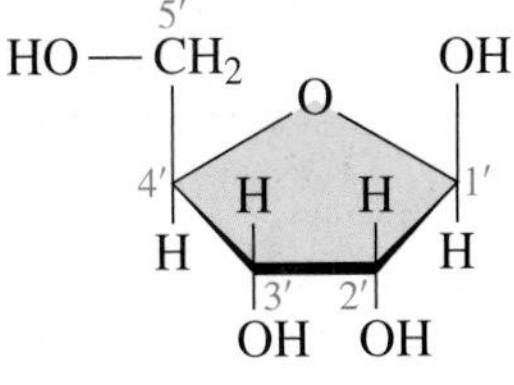

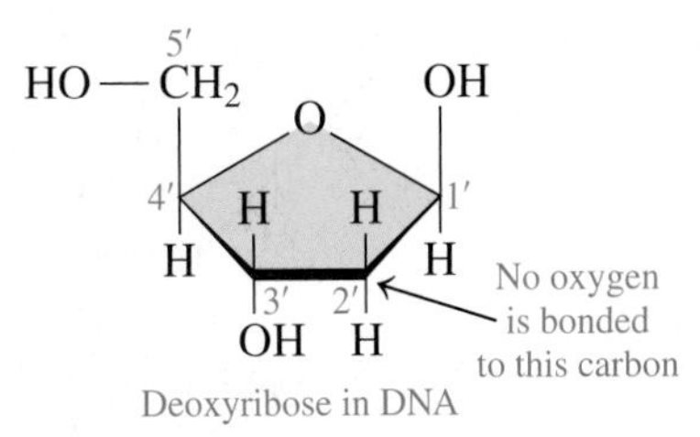

The five-carbon pentose sugar found in RNA is ribose and in DNA, deoxyribose.

Nucleosides and Nucleotides

A **nucleoside** is composed of one of the nitrogen-containing bases and one of the sugars, either ribose or deoxyribose. A nitrogen atom of the base is connected by a β-*N*-glycosidic bond to the C1′ of the sugar. For example, the combination of adenine, a purine, and ribose forms the nucleoside adenosine.

$$\text{Sugar} + \text{Base} \longrightarrow \text{Nucleoside} + H_2O$$

Adenine

Ribose

Adenosine

β-N-Glycosidic bond

$+ H_2O$

A base forms a β-*N*-glycosidic bond with a pentose sugar to form a nucleoside and water.

Nucleotides are produced when the C5′ hydroxyl group of ribose or deoxyribose in a nucleoside forms a phosphate ester. All the nucleotides in RNA and DNA are shown in **FIGURE 17.3**.

$$\text{Phosphate} + \text{Nucleoside} \longrightarrow \text{Nucleotide} + H_2O$$

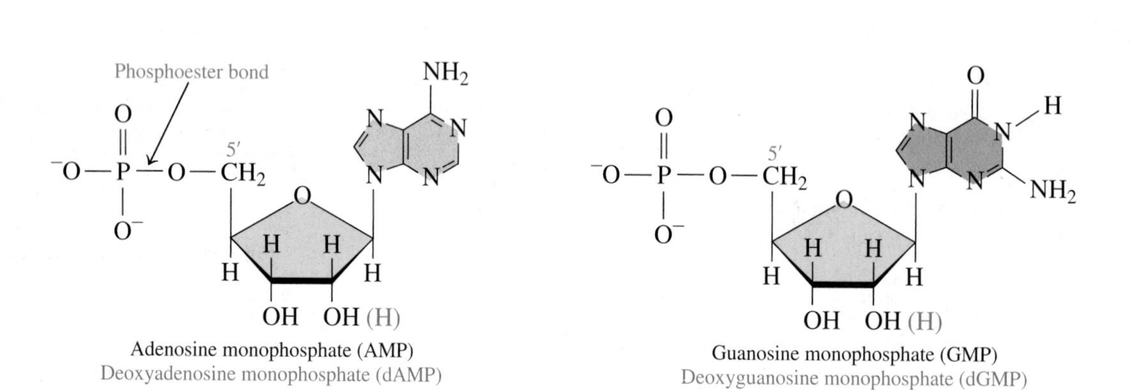

Cytidine monophosphate (CMP)
Deoxycytidine monophosphate (dCMP)

Uridine monophosphate (UMP)

Deoxythymidine monophosphate (dTMP)

FIGURE 17.3 The nucleotides of RNA (shown in black) are similar to those of DNA (shown in magenta), except in DNA the sugar is deoxyribose and deoxythymidine replaces uridine.

Q What are two differences in the nucleotides of RNA and DNA?

TABLE 17.1 summarizes the components in DNA and RNA.

TABLE 17.1 Components in DNA and RNA

Component	DNA	RNA
Bases	A, G, C, and T	A, G, C, and U
Sugar	Deoxyribose	Ribose
Nucleoside	Base + deoxyribose	Base + ribose
Nucleotide	Base + deoxyribose + phosphate	Base + ribose + phosphate
Nucleic Acid	Linear chain of deoxyribose nucleotides	Linear chain of ribose nucleotides

Naming Nucleosides and Nucleotides

The name of a nucleoside that contains a purine ends with *osine*, whereas a nucleoside that contains a pyrimidine ends with *idine*. The names of nucleosides of DNA add *deoxy* to the beginning of their names. The corresponding nucleotides in RNA and DNA are named by adding *monophosphate* to the end of the nucleoside name. Although the letters A, G, C, U, and T represent the bases, they are often used in the abbreviations of the respective nucleosides and nucleotides. The names of the bases, nucleosides, and nucleotides in DNA and RNA and their abbreviations are listed in **TABLE 17.2**.

TABLE 17.2 Nucleosides and Nucleotides in DNA and RNA

Base	Nucleosides	Nucleotides
DNA		
Adenine (A)	Deoxyadenosine (A)	Deoxyadenosine monophosphate (dAMP)
Guanine (G)	Deoxyguanosine (G)	Deoxyguanosine monophosphate (dGMP)
Cytosine (C)	Deoxycytidine (C)	Deoxycytidine monophosphate (dCMP)
Thymine (T)	Deoxythymidine (T)	Deoxythymidine monophosphate (dTMP)
RNA		
Adenine (A)	Adenosine (A)	Adenosine monophosphate (AMP)
Guanine (G)	Guanosine (G)	Guanosine monophosphate (GMP)
Cytosine (C)	Cytidine (C)	Cytidine monophosphate (CMP)
Uracil (U)	Uridine (U)	Uridine monophosphate (UMP)

ENGAGE

Is uridine monophosphate (UMP) found in DNA or RNA?

SAMPLE PROBLEM 17.1 Nucleotides

TRY IT FIRST

For each of the following nucleotides, identify the components and whether the nucleotide is found in DNA only, RNA only, or both DNA and RNA:

a. deoxyguanosine monophosphate (dGMP)
b. adenosine monophosphate (AMP)

SOLUTION

a. This nucleotide of deoxyribose, guanine, and a phosphate group is only found in DNA.
b. This nucleotide of ribose, adenine, and a phosphate group is only found in RNA.

STUDY CHECK 17.1

What is the name and abbreviation of the DNA nucleotide of cytosine?

ANSWER

deoxycytidine monophosphate (dCMP)

TEST

Try Practice Problems 17.5 to 17.14

PRACTICE PROBLEMS

17.1 Components of Nucleic Acids

LEARNING GOAL Describe the bases and ribose sugars that make up the nucleic acids DNA and RNA.

17.1 Identify each of the following bases as a purine or a pyrimidine:
a. thymine
b.

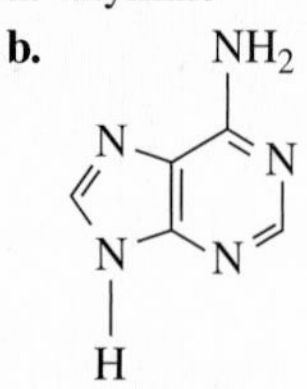

17.2 Identify each of the following bases as a purine or a pyrimidine:
a. guanine
b. NH_2
N
N
O
H

17.3 Identify each of the bases in problem 17.1 as a component of DNA only, RNA only, or both DNA and RNA.

17.4 Identify each of the bases in problem 17.2 as a component of DNA only, RNA only, or both DNA and RNA.

17.5 What are the names and abbreviations of the four nucleotides in DNA?

17.6 What are the names and abbreviations of the four nucleotides in RNA?

17.7 Identify each of the following as a nucleoside or a nucleotide:
a. adenosine
b. deoxycytidine
c. uridine
d. cytidine monophosphate

17.8 Identify each of the following as a nucleoside or a nucleotide:
a. deoxythymidine
b. guanosine
c. deoxyadenosine monophosphate
d. uridine monophosphate

17.9 State whether each of the following components is present in DNA only, RNA only, or both DNA and RNA:
a. phosphate
b. ribose
c. deoxycytidine monophosphate
d. adenine

17.10 State whether each of the following components is present in DNA only, RNA only, or both DNA and RNA:
a. deoxyribose
b. guanosine monophosphate
c. uracil
d. UMP

Clinical Applications

17.11 In the genetic disease *adenosine deaminase deficiency*, there is an accumulation of adenosine. Draw the condensed structural formula for deoxyadenosine monophosphate.

17.12 In the genetic disease *uridine monophosphate synthase deficiency*, symptoms include anemia, cardiac malformations, and infections. Draw the condensed structural formula for uridine monophosphate.

17.13 In Lesch–Nyhan syndrome, a deficiency of the enzyme guanine transferase causes an overproduction of uric acid. Draw the condensed structural formula for guanosine monophosphate.

17.14 A deficiency of the enzyme adenine transferase causes a lack of adenine for purine synthesis and a high level of adenine in the urine. Draw the condensed structural formula for adenosine monophosphate.

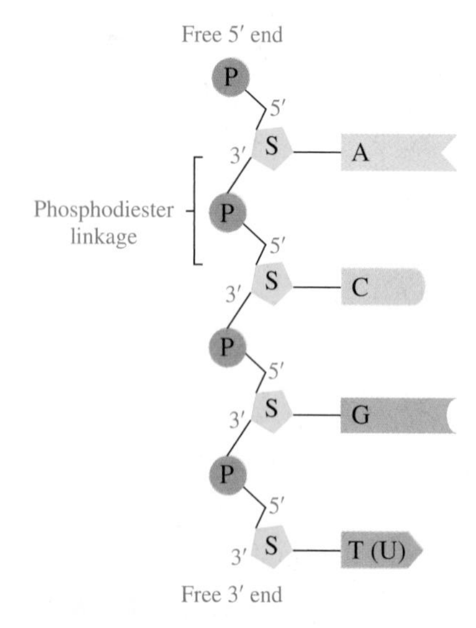

In the primary structure of nucleic acids, each sugar in a sugar–phosphate backbone is attached to a base.

17.2 Primary Structure of Nucleic Acids

LEARNING GOAL Describe the primary structures of RNA and DNA.

The **nucleic acids** are unbranched chains of many nucleotides in which the 3′ hydroxyl group of the sugar in one nucleotide bonds to the phosphate group on the 5′ carbon atom in the sugar of the next nucleotide. This connection between a phosphate and sugars in adjacent nucleotides is referred to as a **phosphodiester linkage**. As more nucleotides are added, a backbone forms that consists of alternating sugar and phosphate groups. The bases, which are attached to each sugar, extend out from the sugar–phosphate backbone.

Each nucleic acid has its own unique sequence of bases, which is known as its **primary structure**. It is this sequence of bases that carries the genetic information. In any nucleic acid, the sugar at one end has an unreacted or free 5′ phosphate terminal end, and the sugar at the other end has a free 3′ hydroxyl group.

A nucleic acid sequence is read from the sugar with the free 5′ phosphate to the sugar with the free 3′ hydroxyl group. The order of nucleotides in a nucleic acid is often written using the letters of the bases. For example, the nucleotide sequence starting with adenine (free 5′ phosphate end) in the section of RNA shown in **FIGURE 17.4** is A C G U.

Free 5′ phosphate

Phosphodiester linkage

Free 3′ hydroxyl

A phosphodiester linkage forms between the 3′ hydroxyl group in the sugar of one nucleotide and the phosphate group on the 5′ carbon atom in the sugar of the next nucleotide.

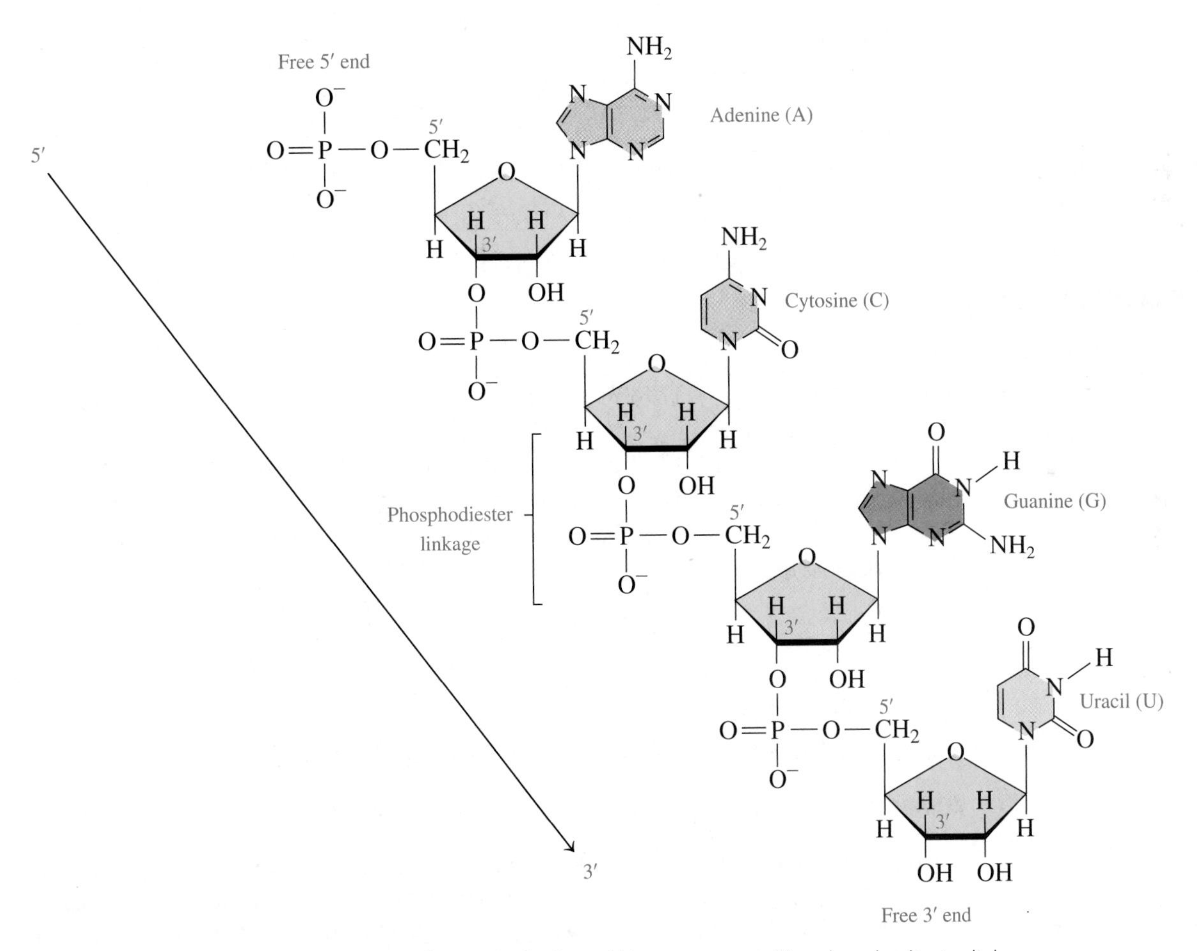

FIGURE 17.4 In the primary structure of RNA, A, C, G, and U are connected by phosphodiester linkages.
Q Where are the free 5′ phosphate and 3′ hydroxyl groups?

SAMPLE PROBLEM 17.2 Bonding of Nucleotides

TRY IT FIRST

Draw the condensed structural formula for an RNA dinucleotide formed by joining the 3′ OH group of adenosine monophosphate and the 5′ phosphate group of cytidine monophosphate.

SOLUTION

The dinucleotide is drawn by connecting the 3′ hydroxyl group on the adenosine monophosphate with the 5′ phosphate group on the cytidine monophosphate.

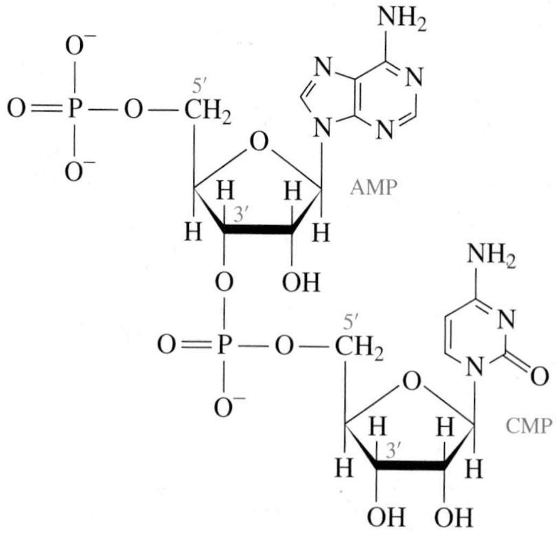

ENGAGE

How many phosphodiester linkages are in a dinucleotide?

STUDY CHECK 17.2

What type of linkage connects the two nucleotides in Sample Problem 17.2?

ANSWER

A phosphodiester linkage connects the 3′ OH group in the ribose of AMP and the 5′ carbon in the ribose of CMP.

TEST

Try Practice Problems 17.15 to 17.22

PRACTICE PROBLEMS

17.2 Primary Structure of Nucleic Acids

LEARNING GOAL Describe the primary structures of RNA and DNA.

17.15 What nucleic acid subunits are connected in a phosphodiester linkage in a polynucleotide?

17.16 What is the difference between the 3′ end and the 5′ end of a polynucleotide chain?

17.17 What components join together to form the backbone of a nucleic acid?

17.18 What component in the backbone of a nucleic acid is bonded to a nitrogen base?

17.19 What component in a nucleic acid determines the 5′ free end?

17.20 What component in a nucleic acid determines the 3′ free end?

17.21 Draw the condensed structural formula for the dinucleotide G C that would be in RNA.

17.22 Draw the condensed structural formula for the dinucleotide A T that would be in DNA.

17.3 DNA Double Helix and Replication

LEARNING GOAL Describe the double helix of DNA; describe the process of DNA replication.

During the 1940s, biologists determined that the bases in DNA from a variety of organisms had a specific relationship: the amount of adenine (A) was equal to the amount of thymine (T), and the amount of guanine (G) was equal to the amount of cytosine (C) (see **TABLE 17.3**). Eventually, scientists determined that adenine is paired (1:1) with thymine, and guanine is paired (1:1) with cytosine.

Number of purine molecules = Number of pyrimidine molecules
Adenine (A) = Thymine (T)
Guanine (G) = Cytosine (C)

In 1953, James Watson and Francis Crick proposed that DNA was a **double helix** that consisted of two polynucleotide strands winding about each other like a spiral staircase. The sugar–phosphate backbones are analogous to the outside railings of the stairs, with the bases arranged like steps along the inside. One strand goes from the 5′ to 3′ direction, and the other strand goes in the 3′ to 5′ direction.

TABLE 17.3 Percentages of Bases in the DNA of Selected Organisms

Organism	%A	%T	%G	%C
Human	30	30	20	20
Chicken	28	28	22	22
Salmon	28	28	22	22
Corn (maize)	27	27	23	23

Complementary Base Pairs

Each of the bases along one polynucleotide strand forms hydrogen bonds to only one specific base on the opposite DNA strand. Adenine forms hydrogen bonds to thymine only, and guanine bonds to cytosine only (see **FIGURE 17.5**). The pairs AT and GC are called **complementary base pairs**. Because of structural limitations, there are only two kinds of stable base pairs. The bases that bind utilizing two hydrogen bonds are adenine and thymine, and the bases that bind utilizing three hydrogen bonds are cytosine and guanine. *No other stable base pairs occur.* For example, adenine does not form hydrogen bonds with cytosine or guanine; cytosine does not form hydrogen bonds with adenine or thymine. This explains why DNA has equal amounts of A and T bases and equal amounts of G and C.

ENGAGE

Why does it require more energy to separate the strands of a DNA double helix containing GC pairs than it does to separate the strands containing AT pairs?

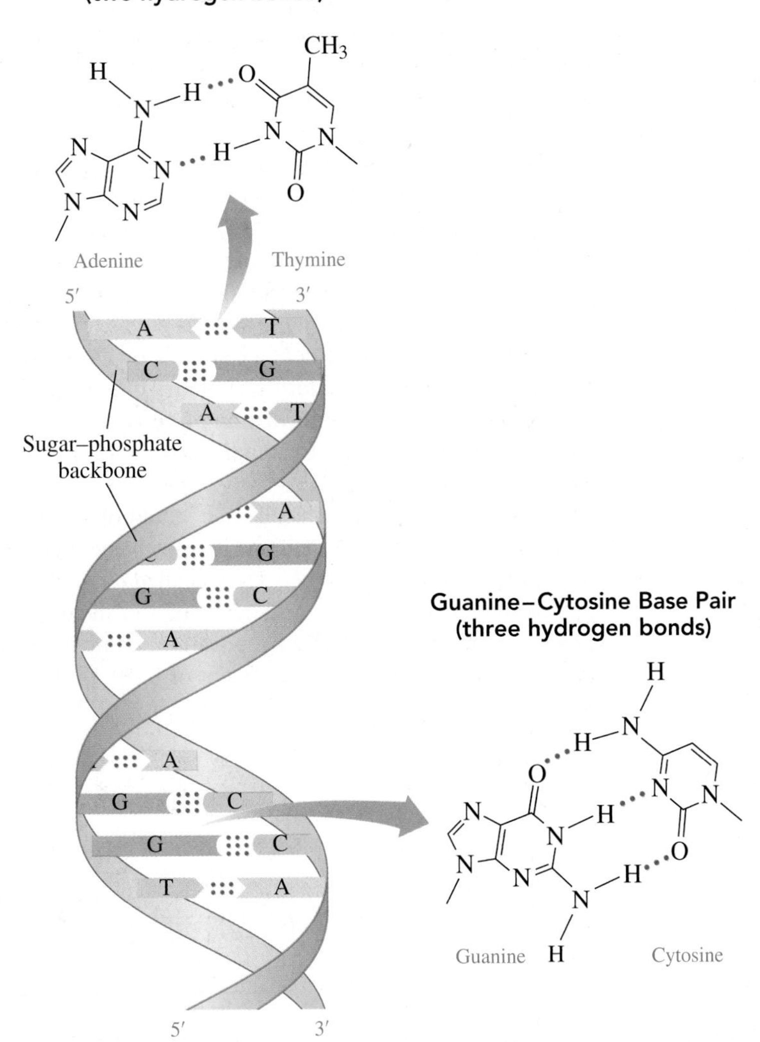

FIGURE 17.5 In the model shown, the sugar–phosphate backbone is represented by a ribbon with hydrogen bonds between complementary base pairs.

Q Why are GC base pairs more stable than AT base pairs?

SAMPLE PROBLEM 17.3 Complementary Base Pairs

TRY IT FIRST

Write the complementary base sequence for the following segment of a strand of DNA:
A C G A T C T

CORE CHEMISTRY SKILL

Writing the Complementary DNA Strand

SOLUTION

The complementary base pairs are AT and GC.

Original segment of DNA:	A	C	G	A	T	C	T
	⋮	⋮	⋮	⋮	⋮	⋮	⋮
Complementary segment:	T	G	C	T	A	G	A

STUDY CHECK 17.3

What sequence of bases is complementary to a DNA segment with a base sequence of G G T T A A C C?

TEST

Try Practice Problems 17.23 to 17.28

ANSWER

C C A A T T G G

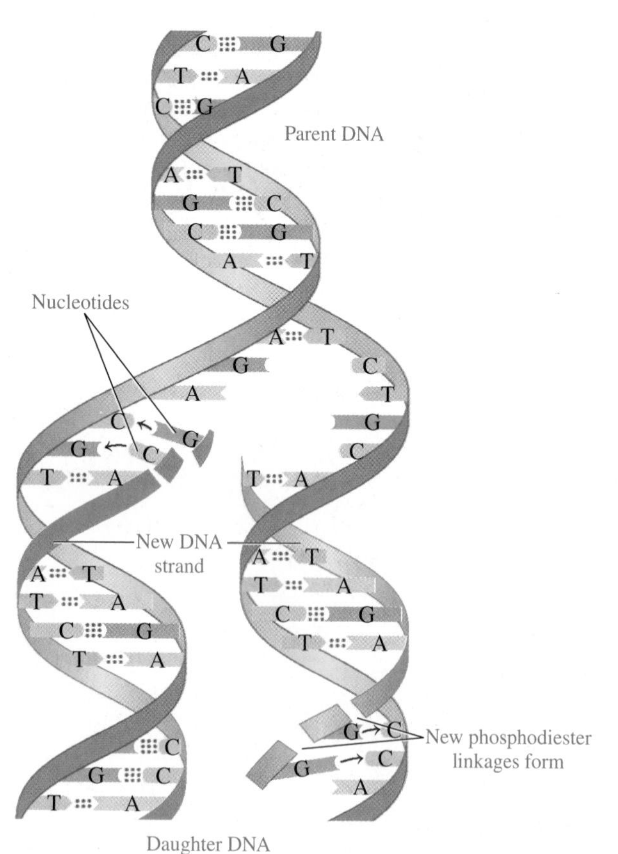

FIGURE 17.6 In DNA replication, the separate strands of the parent DNA are the templates for the synthesis of complementary strands, which produces two exact copies of DNA.

Q How many strands of the parent DNA are in each of the new copies of DNA?

DNA Replication

The function of DNA in cells of animals and plants as well as in bacteria is to preserve genetic information. As cells divide, copies of DNA are produced that transfer genetic information to the new cells.

In DNA **replication**, the strands in the original or *parent* DNA molecule separate to allow the synthesis of complementary DNA strands. The process begins when an enzyme called *helicase* catalyzes the unwinding of a portion of the double helix by breaking the hydrogen bonds between the complementary bases. The resulting single strands act as templates for the synthesis of new complementary strands of DNA (see **FIGURE 17.6**). As the complementary base pairs come together, *DNA polymerase* catalyzes the formation of phosphodiester linkages between the nucleotides. Eventually the entire double helix of the parent DNA is copied.

In each new DNA molecule, one strand of the double helix is from the original DNA, and one is a newly synthesized strand. This process produces two new DNAs called *daughter DNAs* that are identical to each other and exact copies of the original parent DNA. In the process of DNA replication, complementary base pairing ensures the correct placement of bases in the new DNA strands.

TEST

Try Practice Problems 17.29 to 17.32

PRACTICE PROBLEMS

17.3 DNA Double Helix and Replication

LEARNING GOAL Describe the double helix of DNA; describe the process of DNA replication.

17.23 List three structural characteristics of DNA.

17.24 What is meant by double helix?

17.25 How are the two strands of nucleic acid in DNA held together?

17.26 What is meant by complementary base pairing?

17.27 Write the base sequence in a complementary DNA segment if each original segment has the following base sequence:
a. A A A A A A
b. G G G G G G
c. A G T C C A G G T
d. C T G T A T A C G T T A

17.28 Write the base sequence in a complementary DNA segment if each original segment has the following base sequence:
a. T T T T T T
b. C C C C C C C C C
c. A T G G C A
d. A T A T G C G C T A A A

17.29 What is the function of the enzyme helicase in DNA replication?

17.30 What is the function of the enzyme DNA polymerase in DNA replication?

17.31 What process ensures that the replication of DNA produces identical copies?

17.32 How many daughter strands are formed during the replication of DNA?

17.4 RNA and Transcription

LEARNING GOAL Identify the different types of RNA; describe the synthesis of mRNA.

Ribonucleic acid, RNA, which makes up most of the nucleic acid found in the cell, is involved with transmitting the genetic information needed to operate the cell. Similar to DNA, RNA molecules are unbranched chains of nucleotides. However, RNA differs from DNA in several important ways:

1. The sugar in RNA is ribose rather than the deoxyribose found in DNA.
2. In RNA, the base uracil replaces thymine.
3. RNA molecules are single stranded, not double stranded.
4. RNA molecules are much smaller than DNA molecules.

Types of RNA

There are three major types of RNA in the cells: *messenger RNA*, *ribosomal RNA*, and *transfer RNA*. Ribosomal RNA (**rRNA**), the most abundant type of RNA, is combined with proteins to form ribosomes. Ribosomes, which are the sites for protein synthesis, consist of two subunits: a large subunit and a small subunit (see **FIGURE 17.7**). Cells that synthesize large numbers of proteins have thousands of ribosomes.

Messenger RNA (**mRNA**) carries genetic information from the DNA, located in the nucleus of the cell, to the ribosomes, located in the *cytosol*, the liquid outside the nucleus. A *gene* is a segment of DNA that produces a separate mRNA used to synthesize a protein needed in the cell.

Transfer RNA (**tRNA**), the smallest of the RNA molecules, interprets the genetic information in mRNA and brings specific amino acids to the ribosome for protein synthesis. Only tRNA can translate the genetic information in the mRNA into the amino acid sequence that makes a protein. There can be more than one tRNA for each of the 20 amino acids. The structures of all of the transfer RNAs are similar, consisting of 70 to 90 nucleotides. Hydrogen bonds between some complementary bases in the strand produce loops that give some double-stranded regions. The types of RNA molecules in humans are summarized in **TABLE 17.4**.

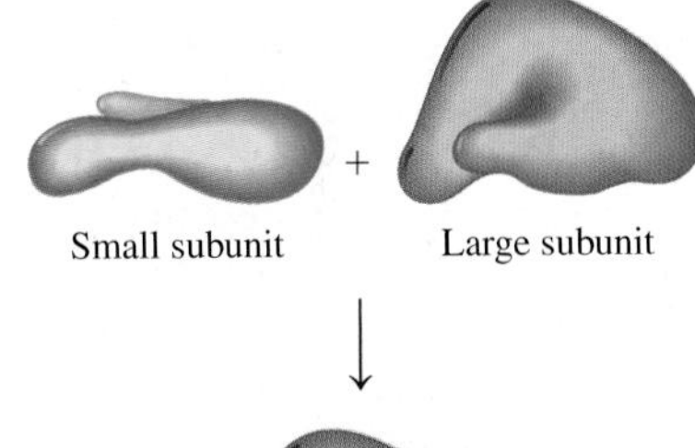

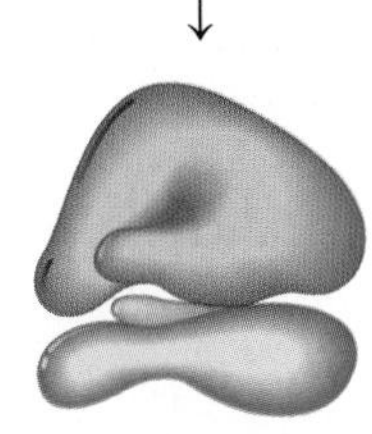

FIGURE 17.7 A typical ribosome consists of a small subunit and a large subunit. The subunit shapes shown contain both protein and rRNA.

Why would there be many thousands of ribosomes in a cell?

TABLE 17.4 Types of RNA Molecules in Humans

Type	Abbreviation	Percentage of Total RNA	Function in the Cell
Ribosomal RNA	rRNA	80	Major component of the ribosomes; site of protein synthesis
Messenger RNA	mRNA	5	Carries information for protein synthesis from the DNA to the ribosomes
Transfer RNA	tRNA	15	Brings specific amino acids to the site of protein synthesis

TEST

Try Practice Problems 17.33 to 17.36

Although the structure of tRNA in three dimensions is complex, we can draw tRNA as a two-dimensional cloverleaf (see **FIGURE 17.8a**). In the three-dimensional model, the RNA chain has more twists that shows the L-shape of tRNA (see **FIGURE 17.8b**). All tRNA molecules have a 3′ end with the nucleotide sequence ACC, which is known as the *acceptor stem*. An enzyme attaches an amino acid to the 3′ end of the acceptor stem by forming an ester bond with the free OH group of the acceptor stem. Each tRNA contains an **anticodon**, which is a series of three bases that complements three bases on mRNA.

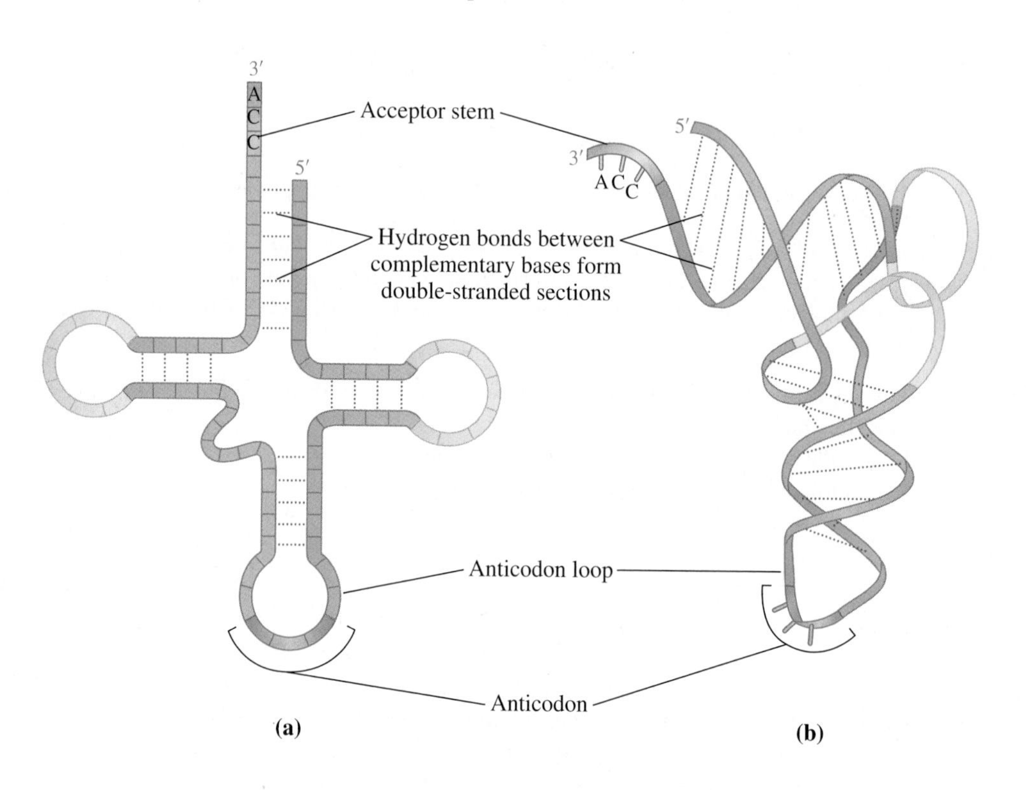

FIGURE 17.8 A typical tRNA molecule has an acceptor stem at the 3′ end that attaches to an amino acid and an anticodon loop that complements a codon on mRNA.

Q Why will different tRNAs have different bases in the anticodon loop?

RNA and Protein Synthesis

We now look at the overall processes involved in transferring genetic information encoded in the DNA to the production of proteins. In the nucleus, genetic information for the synthesis of a protein is copied from a gene in DNA to make mRNA, a process called **transcription**. The mRNA molecules move out of the nucleus into the cytosol, where they bind with the ribosomes. Then in a process called **translation**, tRNA molecules convert the information in the mRNA into amino acids, which are placed in the proper sequence to synthesize a protein (see **FIGURE 17.9**).

ENGAGE

How do the nucleotides found in the DNA template strand differ from the nucleotides found in a transcribed mRNA strand?

Transcription: Synthesis of mRNA

Transcription begins when the section of a DNA molecule that contains the gene to be copied unwinds. Within this unwound section of DNA, called a *transcription bubble*, the enzyme *RNA polymerase* uses one of the strands as a template to synthesize mRNA, using RNA bases that are complementary to the DNA template: C and G form pairs,

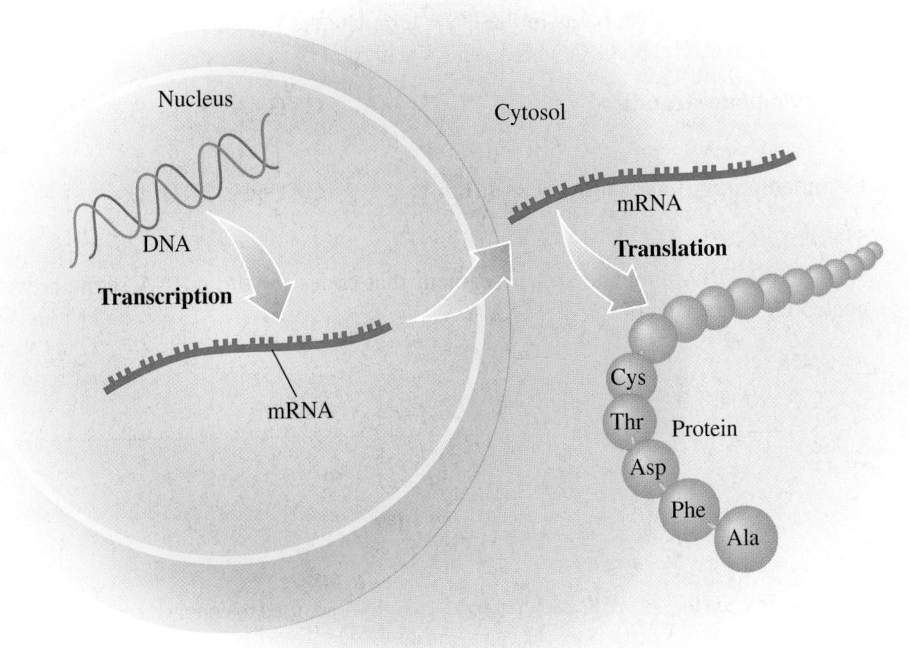

FIGURE 17.9 The genetic information in DNA is replicated in cell division and used to produce messenger RNA that codes for amino acids used in protein synthesis at the ribosomes.

Q What is the difference between transcription and translation?

T (in DNA) pairs with A (in mRNA), and A (in DNA) pairs with U (in mRNA). When the RNA polymerase reaches the termination site (a sequence of nucleotides that is a stop signal), transcription ends, and the new mRNA is released. The unwound portion of the DNA returns to its double-helix structure (see **FIGURE 17.10**).

TEST

Try Practice Problems 17.37 and 17.38

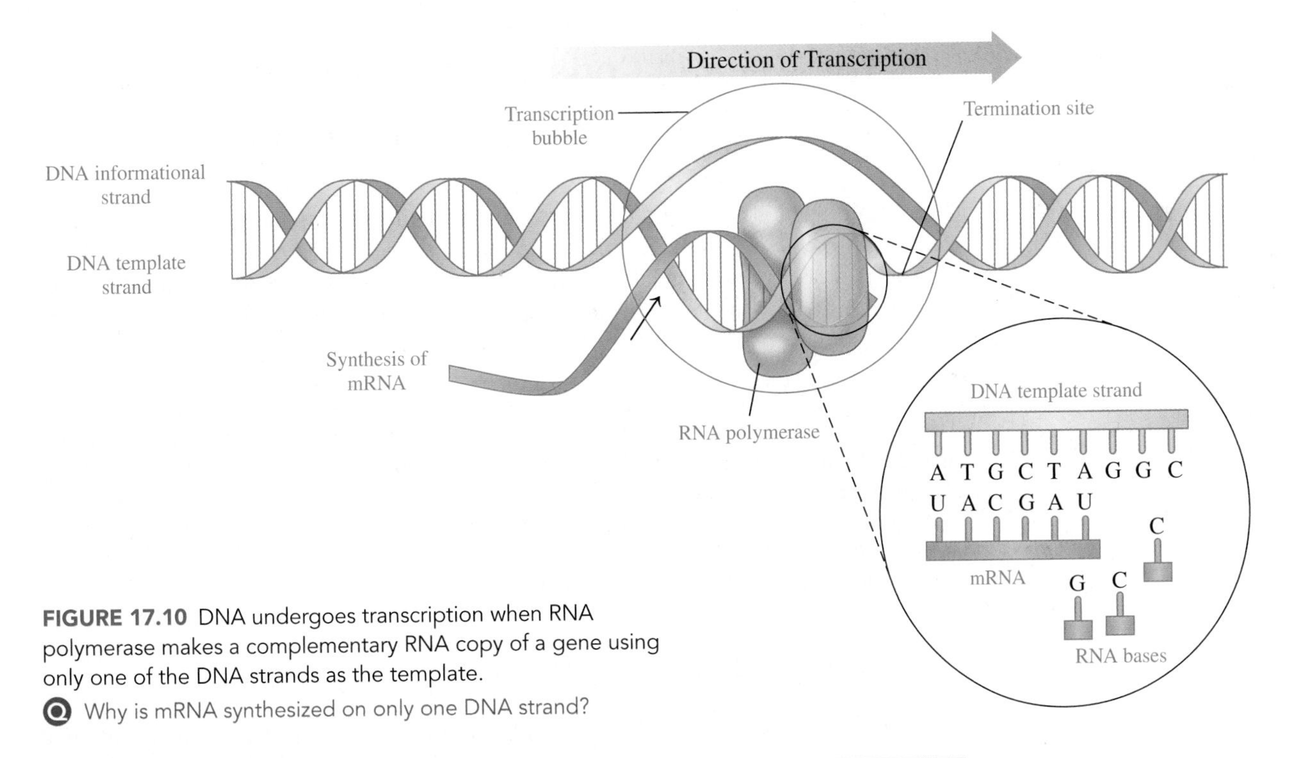

FIGURE 17.10 DNA undergoes transcription when RNA polymerase makes a complementary RNA copy of a gene using only one of the DNA strands as the template.

Q Why is mRNA synthesized on only one DNA strand?

SAMPLE PROBLEM 17.4 RNA Synthesis

TRY IT FIRST

The sequence of bases in a part of the DNA template strand is C G A T C A. What corresponding mRNA is produced?

CORE CHEMISTRY SKILL

Writing the mRNA Segment for a DNA Template

SOLUTION

To form the mRNA, the bases in the DNA template are paired with their complementary bases: G with C, C with G, T with A, and A with U.

DNA template strand: C G A T C A

Transcription ↓ ↓ ↓ ↓ ↓ ↓

Complementary base sequence in mRNA: G C U A G U

STUDY CHECK 17.4

What is the DNA template strand segment that codes for the mRNA segment with the nucleotide sequence G G G U U U A A A?

ANSWER

C C C A A A T T T

TEST

Try Practice Problems 17.39 and 17.40

PRACTICE PROBLEMS

17.4 RNA and Transcription

LEARNING GOAL Identify the different types of RNA; describe the synthesis of mRNA.

17.33 What are the three different types of RNA?

17.34 What are the functions of each type of RNA?

17.35 What is the composition of a ribosome?

17.36 What is the smallest RNA?

17.37 What is meant by the term "transcription"?

17.38 What bases in mRNA are used to complement the bases A, T, G, and C in DNA?

17.39 Write the corresponding section of mRNA produced from the following section of DNA template strand:

C C G A A G G T T C A C

17.40 Write the corresponding section of mRNA produced from the following section of DNA template strand:

T A C G G C A A G C T A

17.5 The Genetic Code and Protein Synthesis

LEARNING GOAL Use the genetic code to write the amino acid sequence for a segment of mRNA.

The overall function of the different types of RNA in the cell is to facilitate the task of synthesizing proteins. After the genetic information encoded in DNA is transcribed, mRNA molecules move out of the nucleus to the ribosomes in the cytosol. At the ribosomes, the genetic information in the mRNA is translated into a sequence of amino acids in protein.

Genetic Code

The **genetic code** consists of a series of three nucleotides (triplets) in mRNA called **codons** that specify the amino acids and their sequence in a protein. Early work on protein synthesis showed that repeating triplets of uracil (UUU) produced a polypeptide that contained only phenylalanine. Therefore, a sequence of UUU UUU UUU codes for three phenylalanines.

Codons in mRNA: UUU UUU UUU

↓ ↓ ↓

Amino acid sequence: Phe — Phe — Phe

Codons have been determined for all 20 amino acids. A total of 64 codons are possible from the triplet combinations of A, G, C, and U. Three of these, UGA, UAA, and UAG, are stop signals that code for the termination of protein synthesis. All the other three-base codons shown in **TABLE 17.5** specify amino acids. Thus, one amino acid can have several codons. For example, glycine has four codons: GGU, GGC, GGA, and GGG. The triplet AUG has two roles in protein synthesis. At the beginning of an mRNA, the codon AUG signals the start of protein synthesis. In the middle of a series of codons, the AUG codon specifies the amino acid methionine.

TEST

Try Practice Problems 17.41 to 17.46

TABLE 17.5 Codons in mRNA: The Genetic Code for Amino Acids

First Letter	Second Letter U		C		A		G		Third Letter
U	UUU	Phe (F)	UCU	Ser (S)	UAU	Tyr (Y)	UGU	Cys (C)	U
	UUC		UCC		UAC		UGC		C
	UUA	Leu (L)	UCA		UAA STOP[b]		UGA STOP[b]		A
	UUG		UCG		UAG STOP[b]		UGG Trp (W)		G
C	CUU	Leu (L)	CCU	Pro (P)	CAU	His (H)	CGU	Arg (R)	U
	CUC		CCC		CAC		CGC		C
	CUA		CCA		CAA	Gln (Q)	CGA		A
	CUG		CCG		CAG		CGG		G
A	AUU	Ile (I)	ACU	Thr (T)	AAU	Asn (N)	AGU	Ser (S)	U
	AUC		ACC		AAC		AGC		C
	AUA		ACA		AAA	Lys (K)	AGA	Arg (R)	A
	AUG START[a]/ Met (M)		ACG		AAG		AGG		G
G	GUU	Val (V)	GCU	Ala (A)	GAU	Asp (D)	GGU	Gly (G)	U
	GUC		GCC		GAC		GGC		C
	GUA		GCA		GAA	Glu (E)	GGA		A
	GUG		GCG		GAG		GGG		G

START[a] codon signals the initiation of a peptide chain.
STOP[b] codons signal the end of a peptide chain.

CORE CHEMISTRY SKILL
Writing the Amino Acid for an mRNA Codon

Translation

Once an mRNA is synthesized, it migrates out of the nucleus into the cytosol to the ribosomes. In the *translation* process, tRNA molecules, amino acids, and enzymes convert the mRNA codons into amino acids to build a protein.

Activation of tRNA

Each tRNA molecule contains a loop called the *anticodon*, which is a triplet of bases that complements a codon in mRNA. An amino acid is attached to the acceptor stem of each tRNA by an enzyme called *aminoacyl–tRNA synthetase*. Each amino acid has a different synthetase. Activation of tRNA occurs when aminoacyl–tRNA synthetase forms an ester bond between the carboxylate group of its amino acid and the hydroxyl group on the acceptor stem (see **FIGURE 17.11**). Each synthetase then checks the tRNA–amino acid combination and hydrolyzes any incorrect combinations.

TEST
Try Practice Problems 17.47 and 17.48

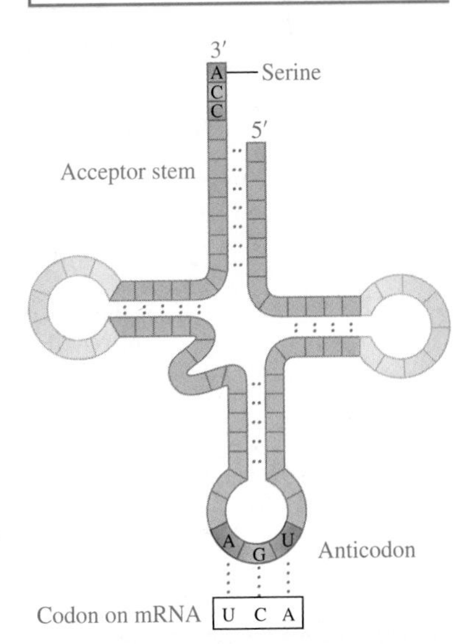

FIGURE 17.11 An activated tRNA with anticodon AGU bonds to serine at the acceptor stem.

What would be an anticodon on a methionine–tRNA?

Initiation and Chain Elongation

Protein synthesis begins when mRNA binds to a ribosome. The first codon in an mRNA is a *start codon*, AUG, which forms hydrogen bonds with methionine–tRNA. Another tRNA hydrogen bonds to the next codon, placing a second amino acid adjacent to methionine. A peptide bond forms between the C terminus of methionine and the N terminus of the second amino acid (see **FIGURE 17.12**). The initial tRNA detaches from the ribosome, which shifts to the next available codon, a process called *translocation*. During *chain elongation*, the ribosome moves along the mRNA from codon to codon so that the tRNAs can attach new amino acids to the growing polypeptide chain. Sometimes, a group of several ribosomes, called a polysome, translate the same strand of mRNA to produce several copies of the polypeptide at the same time.

Chain Termination

Eventually, a ribosome encounters a codon—UAA, UGA, or UAG—that has no corresponding tRNAs. These are *stop codons*, which signal the termination of polypeptide synthesis and the release of the polypeptide chain from the ribosome. The initial amino acid, methionine, is usually removed from the beginning of the polypeptide chain. The R groups on the amino acids in the new polypeptide chain form hydrogen bonds to give the secondary

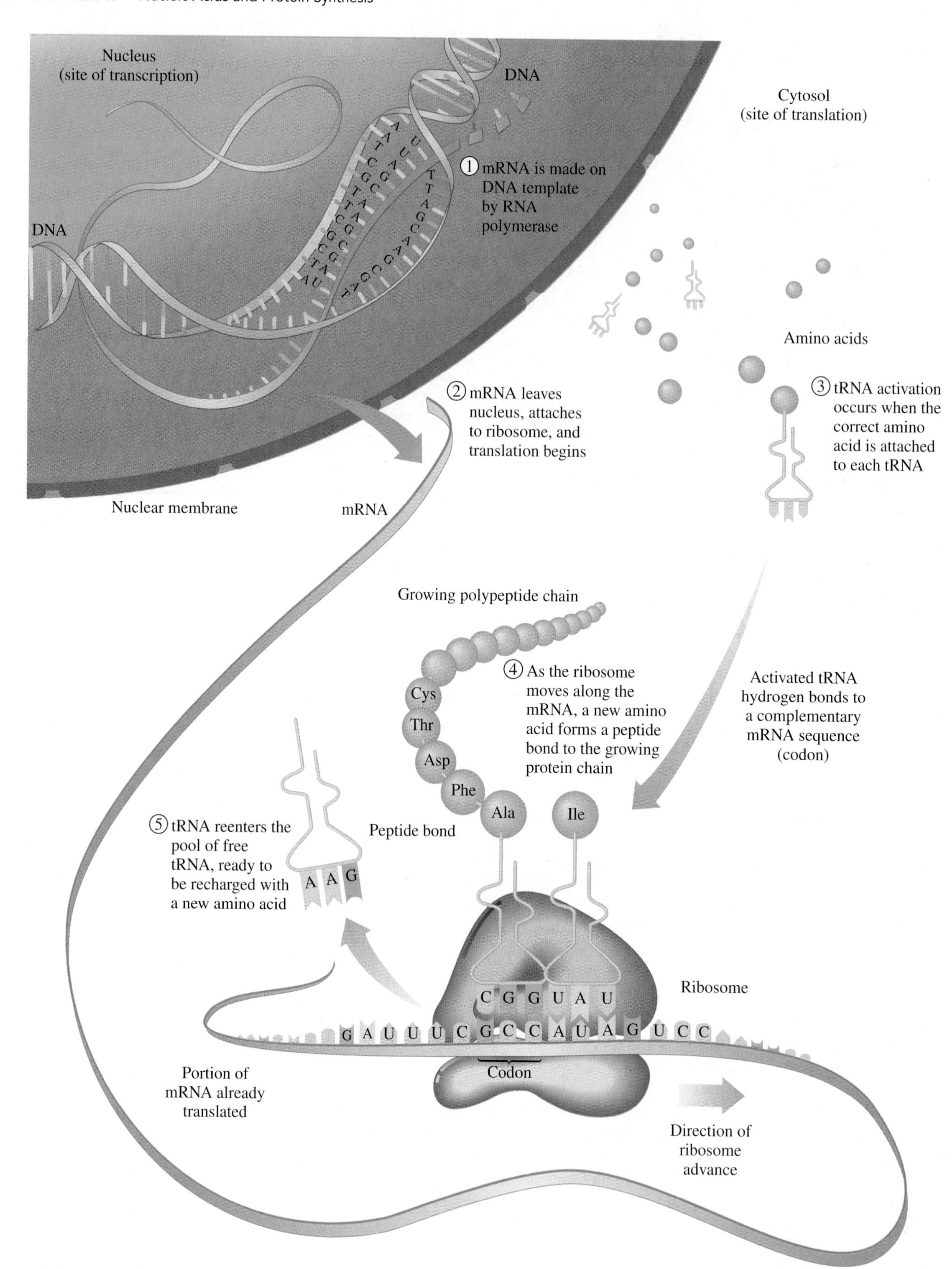

FIGURE 17.12 In the translation process, the mRNA synthesized by transcription attaches to a ribosome, and tRNAs pick up their amino acids, bind to the appropriate codon, and place their amino acids in a growing protein chain.

Q How is the correct amino acid placed in the protein chain?

INTERACTIVE VIDEO
Protein Synthesis

structures of α helices and β-pleated sheets and form interactions such as salt bridges and disulfide bonds to produce tertiary and quaternary structures, which make it a biologically active protein.

TABLE 17.6 summarizes the steps in protein synthesis.

TABLE 17.6 Steps in Protein Synthesis

Step	Site: Materials	Process
1. DNA Transcription	Nucleus: nucleotides, RNA polymerase	A DNA template is used to produce mRNA.
2. Activation of tRNA	Cytosol: amino acids, tRNAs, aminoacyl–tRNA synthetase	Molecules of tRNA pick up specific amino acids according to their anticodons.
3. Translation of mRNA	Ribosome: mRNA	mRNA binds to ribosomes where translation begins.
3a. and 3b. Initiation and Chain Elongation	Ribosome: Met–tRNA, mRNA, aminoacyl–tRNAs	A start codon binds the first tRNA carrying the amino acid methionine to the mRNA. Successive tRNAs bind to and detach from the ribosome as they add an amino acid to the polypeptide.
3c. Chain Termination	Ribosome: stop codon on mRNA	The protein is released from the ribosome.

TABLE 17.7 gives an example of corresponding nucleotide and amino acid sequences in protein synthesis.

TABLE 17.7 Complementary Sequences in DNA, mRNA, tRNA, and Peptides

Nucleus	
DNA informational strand	GCG AGT GGA TAC
DNA template strand	CGC TCA CCT ATG
Ribosome (cytosol)	
mRNA	GCG AGU GGA UAC
tRNA anticodons	CGC UCA CCU AUG
Polypeptide amino acids	Ala — Ser — Gly — Tyr

ENGAGE

What is one possible mRNA sequence that would code for the peptide with the amino acid sequence Cys–Ala–Arg?

CHEMISTRY LINK TO HEALTH

Many Antibiotics Inhibit Protein Synthesis

Several antibiotics stop bacterial infections by interfering with the synthesis of proteins needed by the bacteria. Some antibiotics act only on bacterial cells, binding to the ribosomes in bacteria but not those in human cells. A description of some of these antibiotics is given in **TABLE 17.8**.

TABLE 17.8 Antibiotics That Inhibit Protein Synthesis in Bacterial Cells

Antibiotic	Effect on Ribosomes to Inhibit Protein Synthesis
Chloramphenicol	Inhibits peptide bond formation and prevents the binding of tRNA
Erythromycin	Inhibits peptide chain growth by preventing the translocation of the ribosome along the mRNA
Puromycin	Causes release of an incomplete protein by ending the growth of the polypeptide early
Streptomycin	Prevents the proper attachment of the initial tRNA
Tetracycline	Prevents the binding of tRNA

SAMPLE PROBLEM 17.5 Protein Synthesis

TRY IT FIRST

Use three-letter and one-letter abbreviations to write the amino acid sequence for the peptide from the mRNA sequence of UCA AAA GCC CUU.

SOLUTION

Each of the codons specifies a particular amino acid. Using Table 17.5, we write a peptide with the following amino acid sequence:

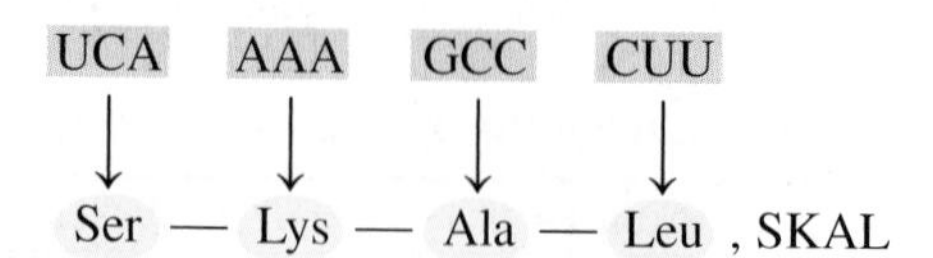

STUDY CHECK 17.5

Use three-letter and one-letter abbreviations to write the amino acid sequence for the peptide from the mRNA sequence of GGG AGC UGC GAG GUU.

ANSWER

Gly–Ser–Cys–Glu–Val, GSCEV

TEST

Try Practice Problems 17.49 to 17.58

PRACTICE PROBLEMS

17.5 The Genetic Code and Protein Synthesis

LEARNING GOAL Use the genetic code to write the amino acid sequence for a segment of mRNA.

17.41 What is a codon?

17.42 What is the genetic code?

17.43 What amino acid is coded for by each of the following mRNA codons?

a. CCA **b.** AAC **c.** GGU **d.** AGG

17.44 What amino acid is coded for by each of the following mRNA codons?

a. UCU **b.** UUC **c.** CGG **d.** GCA

17.45 When does the codon AUG signal the start of a protein? When does it code for the amino acid methionine?

17.46 The codons UGA, UAA, and UAG do not code for amino acids. What is their role as codons in mRNA?

17.47 What is the difference between a *codon* and an *anticodon*?

17.48 Why are there at least 20 different tRNAs?

17.49 What are the three steps of translation?

17.50 Where does protein synthesis take place?

17.51 Use three-letter and one-letter abbreviations to write the amino acid sequence for the peptide from each of the following mRNA sequences:

a. ACC ACA ACU **b.** UUU CCG UUC CCA
c. UAC GGG AGA UGU

17.52 Use three-letter and one-letter abbreviations to write the amino acid sequence for the peptide from each of the following mRNA sequences:

a. AAA CCC UUG GCC
b. CCU CGC AGC GGC UGA
c. AUG CAC AAG GAA GUA CUG

17.53 How is a peptide chain extended?

17.54 What is meant by "translocation"?

17.55 The following sequence is a portion of the DNA template strand:

GCT ATA CCA AAA

a. Write the corresponding mRNA segment.
b. What are the anticodons of the tRNAs?
c. Write the three-letter and one-letter abbreviations for this segment in the peptide chain.

17.56 The following sequence is a portion of the DNA template strand:

TGT GGG GTT ATT

a. Write the corresponding mRNA segment.
b. What are the anticodons of the tRNAs?
c. Write the three-letter and one-letter abbreviations for this segment in the peptide chain.

Clinical Applications

17.57 The following is a segment of the DNA template that codes for human insulin:

TTT GTG AAC CAA CAC CTG

a. Write the corresponding mRNA segment.
b. Write the three-letter and one-letter abbreviations for this corresponding peptide segment.

17.58 The following is a segment of the DNA template that codes for human insulin:

TGC GGC TCA CAC CTG GTG

a. Write the corresponding mRNA segment.
b. Write the three-letter and one-letter abbreviations for the corresponding peptide segment.

REVIEW

Identifying the Primary, Secondary, Tertiary, and Quaternary Structures of Proteins (16.2, 16.3)
Describing Enzyme Action (16.4)

17.6 Genetic Mutations

LEARNING GOAL Identify the type of change in DNA for a point mutation, a deletion mutation, and an insertion mutation.

A **mutation** is a change in the nucleotide sequence of DNA. Such a change may alter the sequence of amino acids, affecting the structure and function of a protein in a cell. Mutations may result from X-rays, overexposure to sun (ultraviolet [UV] light), chemicals called *mutagens*, and possibly some viruses. If a mutation occurs in a somatic cell (a cell other than a reproductive cell), the altered DNA is limited to that cell and its daughter cells. If the mutation affects DNA that controls the growth of the cell, cancer could result. If a mutation occurs in a germ cell (egg or sperm), then all DNA produced will contain the same genetic change. When a mutation severely alters proteins or enzymes, the new cells may not survive or the person may exhibit a disease or condition that is a result of a genetic defect.

Types of Mutations

Consider a triplet of bases CCG in the template strand of DNA, which produces the codon GGC in mRNA. At the ribosome, tRNA would place the amino acid glycine in the peptide chain (see **FIGURE 17.13a**). Now, suppose that T replaces the first C in the DNA triplet, which gives TCG as the triplet. Then the codon produced in the mRNA is AGC, which brings the tRNA with the amino acid serine to add to the peptide chain. The replacement of one base in the template strand of DNA with another is called a **point mutation**. When there is a change of a nucleotide in the codon, a different amino acid may be inserted into the polypeptide. However, if a point mutation does not change the amino acid, it is a *silent mutation*. A point mutation is the most common way in which mutations occur (see **FIGURE 17.13b**).

ENGAGE

Why does replacing the U in CGU with an A not change the primary structure of a protein?

In a **deletion mutation**, a base is deleted from the normal order of bases in the template strand of DNA. Suppose that an A is deleted from the triplet AAA, giving a new triplet of AAC (see **FIGURE 17.13c**). The next triplet becomes CGA rather than CCG, and so on. All the triplets shift by one base, which changes all the codons that follow and leads to a different sequence of amino acids from that point.

In an **insertion mutation**, a base is inserted into the normal order of bases in the template strand of DNA. Suppose a T is inserted into the triplet AAA, which gives a new triplet of AAT (see **FIGURE 17.13d**). The next triplet becomes ACC rather than CCG, and so on. All the triplets shift by one base, which changes all the codons that follow and leads to a different sequence of amino acids from that point.

SAMPLE PROBLEM 17.6 Mutations

TRY IT FIRST

An mRNA has the sequence of codons CCC AGA GCC. If a point mutation in the DNA changes the mRNA codon of AGA to GGA, how is the amino acid sequence affected in the resulting protein?

SOLUTION

The mRNA sequence CCC AGA GCC codes for the following amino acids: proline, arginine, and alanine. When the point mutation occurs, the new sequence of the mRNA codons is CCC GGA GCC, which codes for proline, glycine, and alanine. The polar basic amino acid arginine is replaced by the nonpolar amino acid glycine.

	Normal	After Point Mutation
mRNA codons:	CCC AGA GCC	CCC GGA GCC
Amino acid sequence:	Pro—Arg—Ala	Pro—Gly—Ala

STUDY CHECK 17.6

How might the protein made from the mutated mRNA in Sample Problem 17.6 be affected by this mutation?

ANSWER

Because the point mutation replaces a polar basic amino acid with a nonpolar neutral amino acid, the tertiary structure may be altered sufficiently to cause the resulting protein to be less effective or nonfunctional.

TEST

Try Practice Problems 17.59 to 17.66

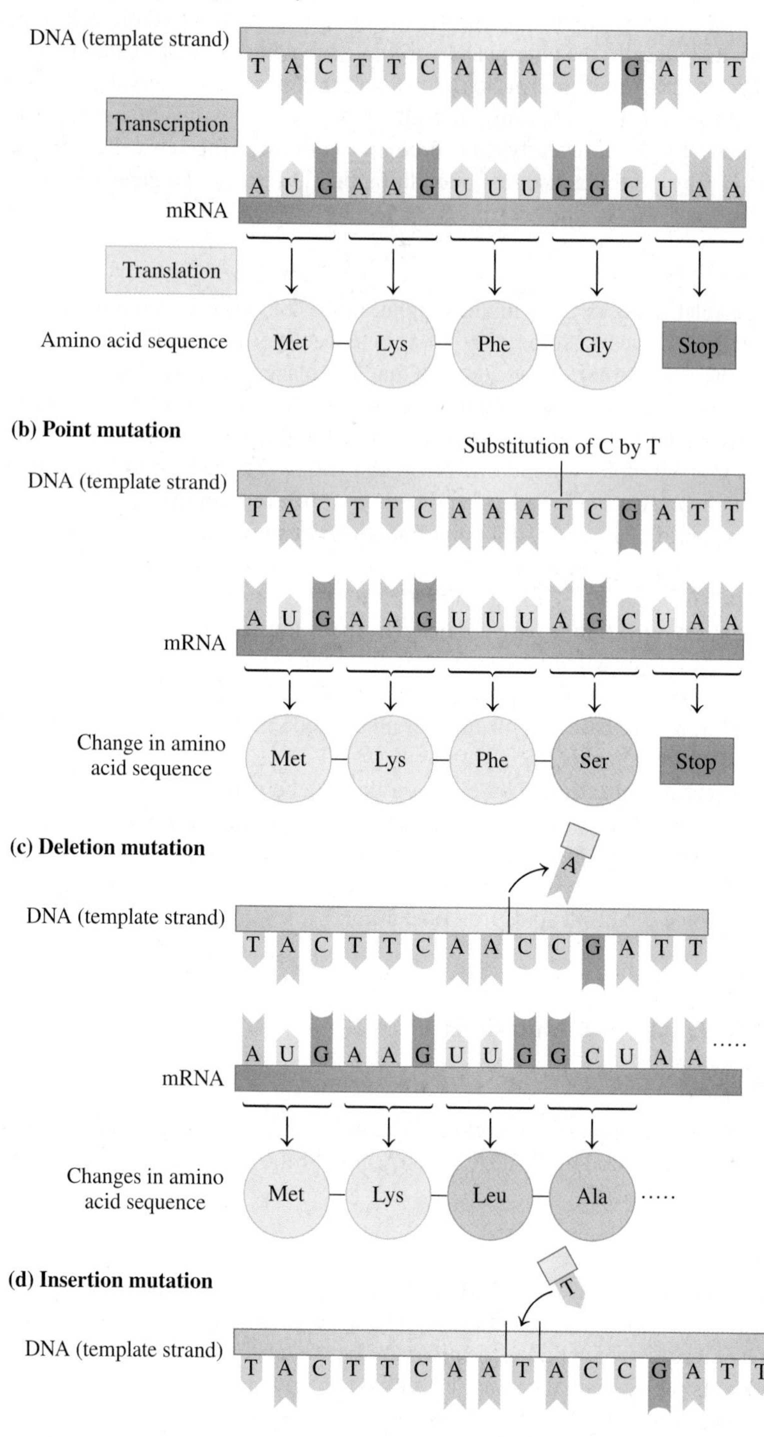

FIGURE 17.13 An alteration in the DNA template strand produces a change in the sequence of amino acids in the protein, which may result in a mutation. **(a)** A normal DNA leads to the correct amino acid order in a protein. **(b)** In a point mutation, the change of a base in DNA leads to a change in the mRNA codon and possibly a change in one amino acid. **(c)** The deletion of a base causes a deletion mutation, which changes the mRNA codons that follow the mutation and produces a different amino acid sequence. **(d)** The insertion of a base causes an insertion mutation, which changes the mRNA codons that follow the mutation and produces a different amino acid sequence.

Q When would a point mutation cause protein synthesis to stop?

Effect of Mutations

Some mutations do not cause a significant change in the primary structure of a protein and the protein is able to maintain biological activity. However, if the mutation causes a change to an amino acid critical to protein structure or function, the protein loses biological activity. If the protein is an enzyme, it may no longer bind to its substrate or react with the substrate at the active site. When an altered enzyme cannot catalyze a reaction, certain substances may accumulate until they act as poisons in the cell, or substances vital to survival may not be synthesized. If a defective enzyme occurs in a major metabolic pathway or is involved in the building of a cell membrane, the mutation can be lethal. When a protein deficiency is hereditary, the condition is called a **genetic disease**.

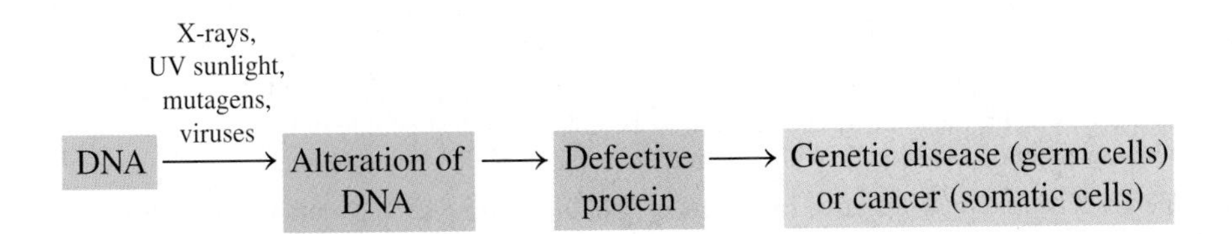

ENGAGE

Under what conditions can a DNA mutation result in a genetic disease?

Genetic Diseases

A genetic disease is the result of a defective enzyme caused by a mutation in its genetic code. For example, *phenylketonuria* (PKU) results when DNA cannot direct the synthesis of the enzyme phenylalanine hydroxylase, required for the conversion of phenylalanine to tyrosine. In an attempt to break down the phenylalanine, other enzymes in the cells convert it to phenylpyruvate. If phenylalanine and phenylpyruvate accumulate in the blood of an infant, it can lead to severe brain damage and mental retardation. If PKU is detected in a newborn baby, a diet is prescribed that eliminates all foods that contain phenylalanine. Preventing the buildup of phenylpyruvate ensures normal growth and development.

The amino acid tyrosine is needed in the formation of melanin, the pigment that gives the color to our skin and hair. If the enzyme that converts tyrosine to melanin is defective, no melanin is produced and a genetic disease known as *albinism* results. People and animals with no melanin have no skin, eye, or hair pigment (see **FIGURE 17.14**). **TABLE 17.9** lists some other common genetic diseases and the type of metabolism or area affected.

FIGURE 17.14 A peacock with albinism does not produce the melanin needed to make bright colors for its feathers.

Why are traits such as albinism related to the gene?

$C_6H_5-CH_2-CH(\overset{+}{N}H_3)-COO^-$ (Phenylalanine) $\longrightarrow$ $C_6H_5-CH_2-C(=O)-COO^-$ (Phenylpyruvate) → Phenylketonuria (PKU)

Phenylalanine $\downarrow$ (blocked: Phenylalanine hydroxylase)

$HO-C_6H_4-CH_2-CH(\overset{+}{N}H_3)-COO^-$ (Tyrosine) $\not\longrightarrow$ Melanin (pigments) → Albinism

TEST

Try Practice Problems 17.67 and 17.68

TABLE 17.9 Some Genetic Diseases

Genetic Disease	Result
Galactosemia	In galactosemia, the transferase enzyme required for the metabolism of galactose-1-phosphate is absent, resulting in the accumulation of galactose-1-phosphate, which leads to cataracts and mental retardation. Galactosemia occurs in about 1 in every 50 000 births.
Cystic fibrosis (CF)	Cystic fibrosis is caused by a mutation in the gene for the protein that regulates the production of stomach fluids and mucus. CF is one of the most common inherited diseases in children, in which thick mucus secretions make breathing difficult and block pancreatic function.
Down syndrome	Down syndrome is the leading cause of mental retardation, occurring in about 1 of every 800 live births; the mother's age strongly influences its occurrence. Mental and physical problems, including heart and eye defects, are the result of the formation of three chromosomes (trisomy), usually number 21, instead of a pair.
Familial hypercholesterolemia (FH)	Familial hypercholesterolemia occurs when there is a mutation of a gene on chromosome 19, which produces high cholesterol levels that lead to early coronary heart disease in people 30 to 40 years old.
Muscular dystrophy (MD) (Duchenne)	Muscular dystrophy, Duchenne form, is caused by a mutation in the X chromosome. This muscle-destroying disease appears at about age 5, with death by age 20, and occurs in about 1 of 10 000 males.
Huntington's disease (HD)	Huntington's disease affects the nervous system, leading to total physical impairment. It is the result of a mutation in a gene on chromosome 4, which can now be mapped to test people in families with a history of HD. There are about 30 000 people with Huntington's disease in the United States.
Sickle-cell anemia	Sickle-cell anemia is caused by a defective form of hemoglobin resulting from a mutation in a gene on chromosome 11. It decreases the oxygen-carrying ability of red blood cells, which take on a sickled shape, causing anemia and plugged capillaries from red blood cell aggregation. In the United States, about 72 000 people are affected by sickle-cell anemia.
Hemophilia	Hemophilia is the result of one or more defective blood clotting factors that lead to poor coagulation, excessive bleeding, and internal hemorrhages. There are about 20 000 hemophilia patients in the United States.
Tay–Sachs disease	Tay–Sachs disease is the result of a defective hexosaminidase A, which causes an accumulation of gangliosides and leads to mental retardation, loss of motor control, and early death.

PRACTICE PROBLEMS

17.6 Genetic Mutations

LEARNING GOAL Identify the type of change in DNA for a point mutation, a deletion mutation, and an insertion mutation.

17.59 What is a point mutation?

17.60 How does a point mutation for an enzyme affect the order of amino acids in that protein?

17.61 What is the effect of a deletion mutation on the amino acid sequence of a polypeptide?

17.62 How can a mutation decrease the activity of a protein?

17.63 How is protein synthesis affected if the normal base sequence TTT in the DNA template strand is changed to TTC?

17.64 How is protein synthesis affected if the normal base sequence CCC in the DNA template strand is changed to ACC?

17.65 Consider the following segment of mRNA produced by the normal order of DNA nucleotides:

ACA UCA CGG GUA

a. What is the amino acid order produced from this mRNA?
b. What is the amino acid order if a point mutation changes UCA to ACA?
c. What is the amino acid order if a point mutation changes CGG to GGG?
d. What happens to protein synthesis if a point mutation changes UCA to UAA?
e. What is the amino acid order if an insertion mutation adds a G to the beginning of the mRNA segment?
f. What is the amino acid order if a deletion mutation removes the A at the beginning of the mRNA segment?

17.66 Consider the following segment of mRNA produced by the normal order of DNA nucleotides:

CUU AAA CGA CAU

a. What is the amino acid order produced from this mRNA?
b. What is the amino acid order if a point mutation changes CUU to CCU?
c. What is the amino acid order if a point mutation changes CGA to AGA?
d. What happens to protein synthesis if a point mutation changes AAA to UAA?
e. What is the amino acid order if an insertion mutation adds a G to the beginning of the mRNA segment?
f. What is the amino acid order if a deletion mutation removes the C at the beginning of the mRNA segment?

Clinical Applications

17.67 **a.** A point mutation changes a codon in the mRNA for an enzyme from GCC to GCA. Why is there no change in the amino acid order in the protein?
b. In sickle-cell anemia, a point mutation in the mRNA for hemoglobin results in the replacement of glutamate with valine in the resulting hemoglobin molecule. Why does the replacement of one amino acid cause such a drastic change in biological function?

17.68 a. A point mutation in the mRNA for an enzyme results in the replacement of leucine with alanine in the resulting enzyme molecule. Why does this change in amino acids have little effect on the biological activity of the enzyme?

b. A point mutation in mRNA replaces cytosine in the codon UCA with adenine. How would this substitution affect the amino acid order in the protein?

17.7 Recombinant DNA

LEARNING GOAL Describe the preparation and uses of recombinant DNA.

Techniques in the field of genetic engineering permit scientists to cut and recombine DNA fragments to form **recombinant DNA**. The technology of recombinant DNA is used to produce important medicines like human insulin for diabetics, the antiviral substance interferon, blood clotting factor VIII, and human growth hormone.

Preparing Recombinant DNA

Much of the work with recombinant DNA is done with *Escherichia coli* (*E. coli*) bacteria, which contain small circular plasmids of DNA that can be easily isolated and replicated. Initially, *E. coli* cells are soaked in a detergent solution to disrupt the plasma membrane, releasing the plasmids. A *restriction enzyme* is used to cut the double strands of DNA between specific bases in the DNA sequence (see **FIGURE 17.15**). The same restriction enzymes are used to cut a piece of DNA called donor DNA from a gene of a different organism, such as the gene that produces insulin or growth hormone. When the donor DNA is mixed with the cut plasmids, the nucleotides in their "sticky ends" join to make *recombinant DNA*. The resulting altered plasmids containing the recombinant DNA are placed in a fresh culture of *E. coli* bacteria. The *E. coli* that take up the plasmids are then selected. As the recombined cells divide and replicate, they produce the protein from the inserted gene in the plasmids. **TABLE 17.10** lists some of the products developed through recombinant DNA technology that are now used therapeutically.

TEST

Try Practice Problems 17.69 to 17.74

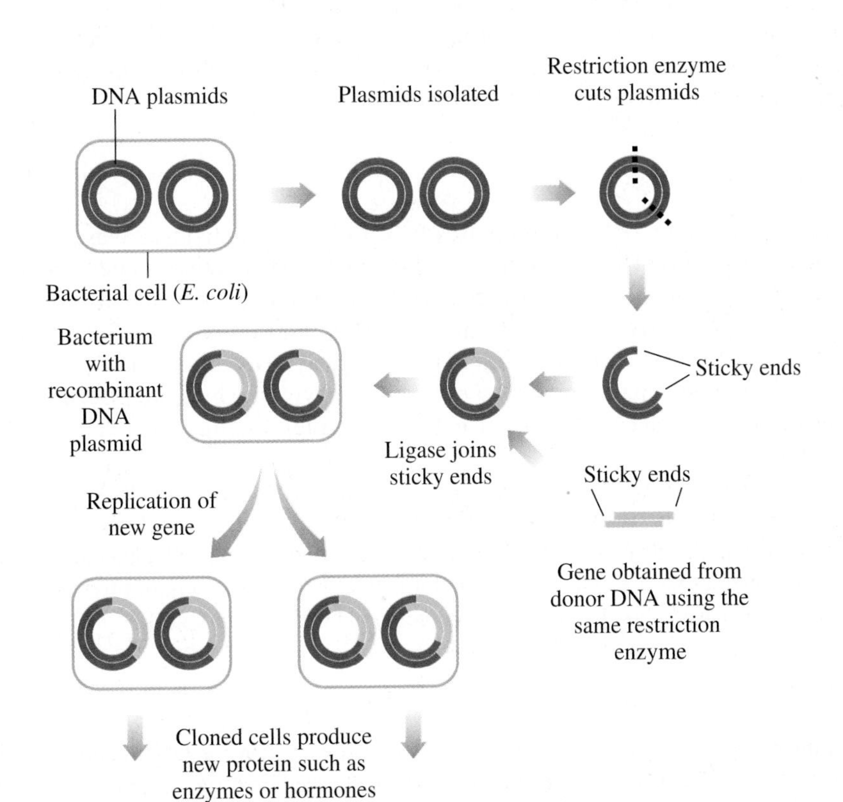

FIGURE 17.15 Recombinant DNA is formed by placing a gene from another organism in a plasmid DNA of the bacterium, which causes the bacterium to produce a nonbacterial protein such as insulin or growth hormone.

How can recombinant DNA help a person with a genetic disease?

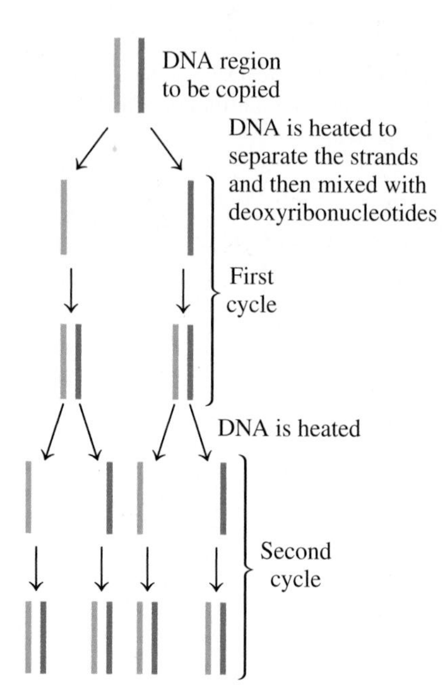

FIGURE 17.16 Each cycle of the polymerase chain reaction doubles the number of copies of the DNA section.

Q Why are the DNA strands heated at the start of each cycle?

TABLE 17.10 Therapeutic Products of Recombinant DNA

Product	Therapeutic Use
Human insulin	Treats diabetes
Erythropoietin (EPO)	Treats anemia; stimulates production of erythrocytes
Human growth hormone (HGH)	Stimulates growth
Interferon	Treats cancer and viral disease
Tumor necrosis factor (TNF)	Destroys tumor cells
Monoclonal antibody	Transports drugs needed to treat cancer and transplant rejection
Epidermal growth factor (EGF)	Stimulates healing of wounds and burns
Human blood clotting factor VIII	Treats hemophilia; allows blood to clot normally
Interleukin	Stimulates immune system; treats cancer
Prourokinase	Destroys blood clots; treats myocardial infarctions
Influenza vaccine	Prevents influenza
Hepatitis B virus (HBV) vaccine	Prevents viral hepatitis

Polymerase Chain Reaction

Before gene cloning can occur, a purified gene has to be isolated. In 1987, a process called the **polymerase chain reaction (PCR)** was developed. PCR made it possible to produce multiple copies of a gene (amplify) in a short time. In the PCR technique, a sequence of a DNA molecule is selected to copy, and the DNA is heated to separate the strands. The DNA strands are mixed with a heat-stable DNA polymerase and the four deoxyribonucleotides, and undergo repeated cycles of heating and cooling to produce complementary strands of the DNA section. After several cycles of the PCR process, millions of copies of the initial DNA section are produced (see FIGURE 17.16).

ENGAGE

How many cycles of PCR would be required to produce 16 copies of an original segment of DNA?

Genetic Testing

PCR allows screening for defective genes whose sequences are known. For example, there are several defects in two known breast cancer genes, called BRCA1 and BRCA2, that correlate to a higher risk of breast cancer. Patients are screened for the defects in these genes by using a DNA sample from their blood or saliva. PCR amplifies the defective genes while incorporating a fluorescent label that is visible if the test is positive.

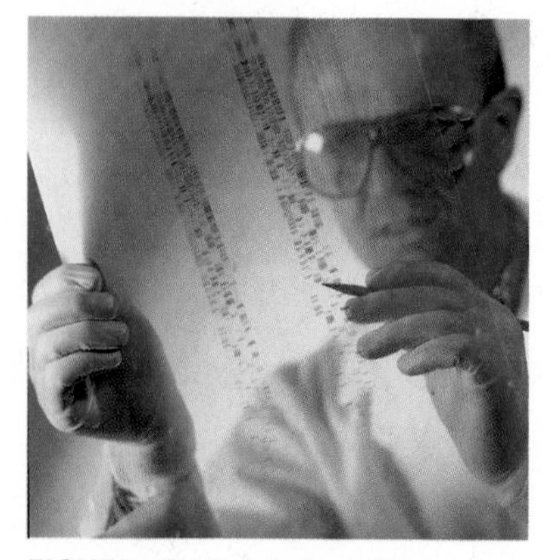

FIGURE 17.17 Bands on film represent DNA fingerprints that can be used to identify a person involved in a crime.

Q What causes DNA fragments to appear on X-ray film?

DNA Fingerprinting

DNA fingerprinting or *DNA profiling* uses PCR to identify individuals based on a set of 13 hereditary traits. The chance that two individuals of the same ethnic background have the same 13 genetic traits is 1 in 575 trillion unless they are identical twins. Only a very small sample size is needed from blood, skin, saliva, or semen. Fluorescent or radioactive isotopes are incorporated into the amplified DNA during the PCR process. The DNA is cut into smaller pieces by restriction enzymes, which are placed on a gel and separated using electrophoresis. The banding pattern on the gel is called a DNA fingerprint.

One application of DNA fingerprinting is in forensic science, where DNA samples are used to connect a suspect with a crime (see FIGURE 17.17). Recently, DNA fingerprinting has been used to gain the release of individuals who were wrongly convicted. Other applications of DNA fingerprinting are determining the biological parents of a child, establishing the identity of a deceased person, and matching recipients with organ donors.

TEST

Try Practice Problems 17.75 and 17.76

The Human Genome

The Human Genome Project, completed in 2003, showed that our DNA is composed of 3 billion bases and 21 000 genes coding for protein. These genes represent only 3% of the total DNA. Since then researchers have identified stretches of DNA that code for other RNA molecules. Much of our DNA regulates genes and serves as recognition sites for proteins. To date, almost 80% of our genome has been assigned a function, allowing scientists to better understand human diseases caused by errors in DNA replication, transcription, or regulation.

PRACTICE PROBLEMS

17.7 Recombinant DNA

LEARNING GOAL Describe the preparation and uses of recombinant DNA.

17.69 Why are *E. coli* bacteria used in recombinant DNA procedures?

17.70 What is a plasmid?

17.71 How are plasmids obtained from *E. coli*?

17.72 Why are restriction enzymes mixed with the plasmids?

17.73 How is a gene for a particular protein inserted into a plasmid?

17.74 What beneficial proteins are produced from recombinant DNA technology?

17.75 What is a DNA fingerprint?

17.76 Why is DNA polymerase useful in criminal investigations?

17.8 Viruses

LEARNING GOAL Describe the methods by which a virus infects a cell.

Viruses are small particles of 3 to 200 genes that cannot replicate without a host cell. A typical virus contains a nucleic acid, DNA or RNA, but not both, inside a protein coat. A virus does not have the necessary material such as nucleotides and enzymes to make proteins and grow. The only way a virus can replicate (make additional copies of itself) is to invade a host cell and take over the machinery and materials necessary for protein synthesis and growth. Some infections caused by viruses invading human cells are listed in **TABLE 17.11**. There are also viruses that attack bacteria, plants, and animals.

ENGAGE

What materials do viruses need from the host cell in order to replicate?

TEST

Try Practice Problems 17.77 and 17.78

TABLE 17.11 Some Diseases Caused by Viral Infection

Disease	Virus
Common cold	Coronavirus (over 100 types), rhinovirus (over 110 types)
Influenza	Orthomyxovirus
Warts	Papovavirus
Herpes	Herpesvirus
HPV	Human papilloma virus
Leukemia, cancers, AIDS	Retrovirus
Hepatitis	Hepatitis A virus (HAV), hepatitis B virus (HBV), hepatitis C virus (HCV)
Mumps	Paramyxovirus
Epstein–Barr	Epstein–Barr virus (EBV)
Chicken pox (shingles)	*Varicella zoster* virus (VZV)

A viral infection begins when an enzyme in the protein coat of the virus makes a hole in the outside of the host cell, allowing the viral nucleic acid to enter and mix with the materials in the host cell (see **FIGURE 17.18**). If the virus contains DNA, the host cell begins to replicate the viral DNA in the same way it would replicate normal DNA. Viral DNA produces viral RNA, and a protease processes proteins to produce a protein coat to form a viral particle that leaves the cell. The cell synthesizes so many virus particles that it eventually releases new viruses to infect more cells.

Vaccines are inactive forms of viruses that boost the immune response by causing the body to produce antibodies to the virus. Several childhood diseases, such as polio, mumps, chicken pox, and measles, can be prevented through the use of vaccines.

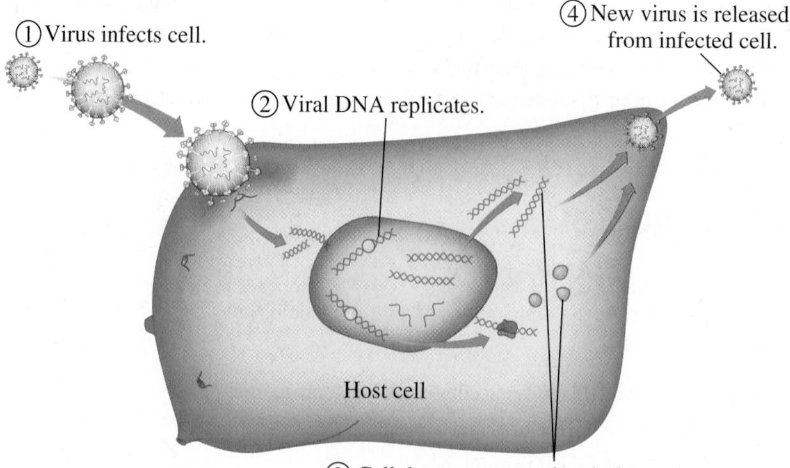

① Virus infects cell.
② Viral DNA replicates.
④ New virus is released from infected cell.
Host cell
③ Cellular enzymes make viral proteins and viral DNA, which assemble into new viruses.

FIGURE 17.18 After a virus attaches to the host cell, it injects its viral DNA and uses the host cell's machinery and materials to make viral mRNA, new viral DNA, and viral proteins. The newly assembled viruses are released to infect other cells.

Q Why does a virus need a host cell for replication?

TEST

Try Practice Problems 17.79 and 17.80

Reverse Transcription

A virus that contains RNA as its genetic material is a **retrovirus**. Once inside the host cell, it must first make viral DNA using a process known as *reverse transcription*. A retrovirus contains a polymerase enzyme called *reverse transcriptase* that uses the viral RNA template to synthesize complementary strands of DNA. Once produced, the single DNA strands form double-stranded DNA using the nucleotides present in the host cell. This newly formed viral DNA, called a *provirus*, integrates with the DNA of the host cell (see **FIGURE 17.19**).

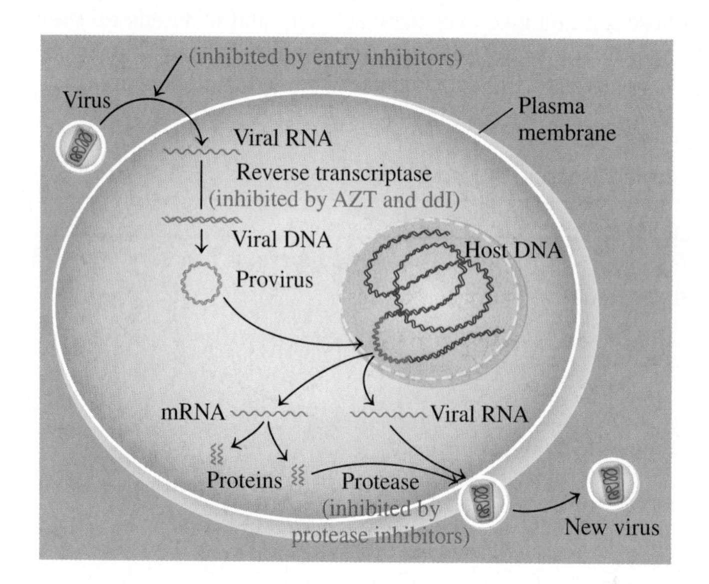

(inhibited by entry inhibitors)
Virus
Plasma membrane
Viral RNA
Reverse transcriptase
(inhibited by AZT and ddI)
Viral DNA
Provirus
Host DNA
mRNA
Viral RNA
Proteins
Protease
(inhibited by protease inhibitors)
New virus

FIGURE 17.19 After a retrovirus injects its viral RNA into a cell, it forms a DNA strand by reverse transcription. The single-stranded DNA forms a double-stranded DNA called a provirus, which is incorporated into the host cell DNA. When the cell replicates, the provirus produces the viral RNA needed to produce more virus particles.

Q What is reverse transcription?

AIDS

During the early 1980s, a disease called *acquired immune deficiency syndrome*, commonly known as AIDS, began to claim an alarming number of lives. We now know that the HIV virus (human immunodeficiency virus) causes the disease (see **FIGURE 17.20**). HIV is a retrovirus that infects and destroys T4 lymphocyte cells, which are involved in the immune response. After the HIV binds to receptors on the surface of a T4 cell, the virus injects viral RNA into the host cell. As a retrovirus, the genes of the viral RNA direct the formation of viral DNA, which is then incorporated into the host's genome so it can replicate as part of the host cell's DNA. The gradual depletion of T4 cells reduces the ability of the immune system to destroy harmful organisms. The AIDS syndrome is characterized by opportunistic infections such as *Pneumocystis carinii*, which causes pneumonia, and *Kaposi's sarcoma*, a skin cancer.

Treatment for AIDS is based on attacking the HIV at different points in its life cycle, including cell entry, reverse transcription, and protein synthesis. Nucleoside analogs mimic the structures of the nucleosides used for DNA synthesis and are able to successfully inhibit the reverse transcriptase enzyme. For example, the drug AZT (3′-azido-2′-deoxythymidine) is similar to thymidine, and ddI (2′,3′-dideoxyinosine) is similar to guanosine. Two other drugs are 2′,3′-dideoxycytidine (ddC) and 2′,3′-didehydro-2′,3′-dideoxythymidine (d4T). Such compounds are found in the "cocktails" that are providing extended remission of HIV infections. When a nucleoside analog is incorporated into viral DNA, the lack of a hydroxyl group on the 3′ carbon in the sugar prevents the formation of the sugar–phosphate bonds and stops the replication of the virus.

3′-Azido-2′-deoxythymidine (AZT)

2′,3′-Dideoxyinosine (ddI)

2′,3′-Dideoxycytidine (ddC)

2′,3′-Didehydro-2′,3′-dideoxythymidine (d4T)

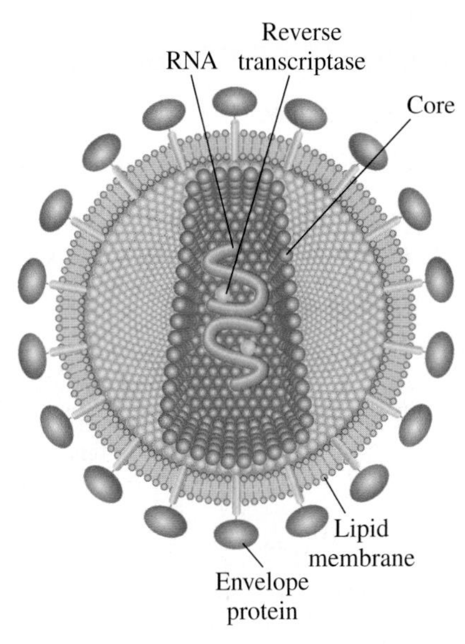

FIGURE 17.20 The HIV virus causes AIDS, which destroys the immune system in the body.

Q Is HIV a DNA virus or an RNA retrovirus?

Today, people with HIV and AIDS are treated with a combination of drugs that include entry inhibitors, reverse transcriptase inhibitors, and protease inhibitors. Entry inhibitors attach to the surface of either the lymphocyte or HIV virus, which blocks entry into the cell. Entry inhibitors include enfuvirtide (Fuzeon) and maraviroc (Selzentry). Protease inhibitors prevent the proper cutting and formation of proteins used by the virus to make more copies of its own proteins. Protease inhibitors include saquinavir (Invirase), ritonavir (Norvir), fosamprenavir (Lexiva), and several others. When patients become resistant to certain drugs, different combinations are used to prolong life expectancy.

TEST

Try Practice Problems 17.81 and 17.82

Lexiva metabolizes slowly to provide amprenavir, an HIV-protease inhibitor.

SAMPLE PROBLEM 17.7 Viruses

TRY IT FIRST

Why are viruses unable to replicate on their own?

SOLUTION

Viruses contain only packets of DNA or RNA, but not the necessary replication machinery that includes enzymes and nucleosides.

STUDY CHECK 17.7

What are the essential parts of a virus?

ANSWER

nucleic acid (DNA or RNA) and a protein coat

PRACTICE PROBLEMS

17.8 Viruses

LEARNING GOAL Describe the methods by which a virus infects a cell.

17.77 What type of genetic information is found in a virus?

17.78 Why do viruses need to invade a host cell?

Clinical Applications

17.79 A specific virus contains RNA as its genetic material.
- **a.** Why would reverse transcription be used in the life cycle of this type of virus?
- **b.** What is the name of this type of virus?

17.80 What is the purpose of a vaccine?

17.81 How do nucleoside analogs disrupt the life cycle of the HIV virus?

17.82 How do protease inhibitors disrupt the life cycle of the HIV virus?

CHEMISTRY LINK TO HEALTH

Cancer

Normally, cells in the body undergo an orderly and controlled cell division. When cells begin to grow and multiply without control, they invade neighboring cells and appear as a tumor. If these tumors are limited, they are benign. When they invade other tissues and interfere with normal functions of the body, the tumors are cancerous. Cancer can be caused by chemical and environmental substances, by ultraviolet or medical radiation, or by *oncogenic viruses*, which are associated with human cancers.

Some reports estimate that 70 to 80% of all human cancers are initiated by chemical and environmental substances. A *carcinogen* is any substance that increases the chance of inducing a tumor. Known carcinogens include aniline dyes, cigarette smoke, and asbestos. More than 90% of all persons with lung cancer are smokers. A carcinogen causes cancer by reacting with molecules in a cell, probably DNA, and altering the growth of that cell. Some known carcinogens are listed in **TABLE 17.12**.

Skin cancer has become one of the most prevalent forms of cancer. It appears that DNA damage in the areas of the skin exposed to ultraviolet radiation may eventually cause mutations. The cells lose their ability to control protein synthesis. This type of uncontrolled cell division becomes skin cancer. The incidence of *malignant melanoma*, one of the most serious skin cancers, has been rapidly increasing. Some possible factors for this increase may be the popularity of sun tanning as well as the reduction of the ozone layer, which absorbs much of the harmful radiation from sunlight.

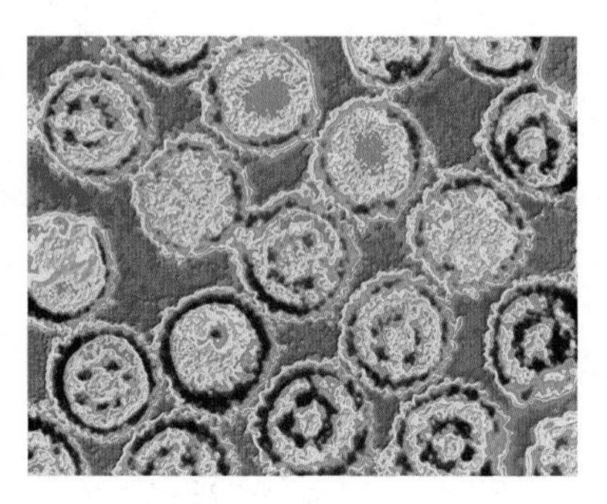

Epstein–Barr virus (EBV), herpesvirus 4, causes cancer in humans.

Oncogenic viruses cause cancer when cells are infected. Several viruses associated with human cancers are listed in **TABLE 17.13**. Some cancers such as retinoblastoma and breast cancer appear to occur more frequently within families. There is some indication that a missing or defective gene may be responsible.

TABLE 17.12 Some Chemical and Environmental Carcinogens

Carcinogen	Tumor Site
Aflatoxin	Liver
Aniline dyes	Bladder
Arsenic	Skin, lungs
Asbestos	Lungs, respiratory tract
Cadmium	Prostate, kidneys
Chromium	Lungs
Nickel	Lungs, sinuses
Nitrites	Stomach
Vinyl chloride	Liver

TABLE 17.13 Human Cancers Caused by Oncogenic Viruses

Virus	Disease
RNA Viruses	
Human T-cell lymphotropic virus–type 1 (HTLV-1)	Leukemia
DNA Viruses	
Epstein–Barr virus (EBV)	Burkitt's lymphoma (cancer of white blood B cells) Nasopharyngeal carcinoma Hodgkin's disease
Hepatitis B virus (HBV)	Liver cancer
Herpes simplex virus (HSV, type 2)	Cervical and uterine cancer
Papilloma virus (HPV)	Cervical and colon cancer, genital warts

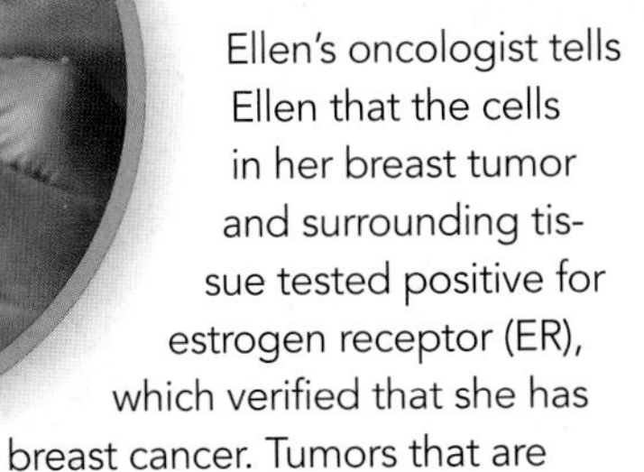

CLINICAL UPDATE Ellen's Medical Treatment Following Breast Cancer Surgery

Ellen's oncologist tells Ellen that the cells in her breast tumor and surrounding tissue tested positive for estrogen receptor (ER), which verified that she has breast cancer. Tumors that are estrogen positive require estrogen for their growth. Estrogen is a hormone that travels through the bloodstream and activates estrogen receptors in the cells of the breast and ovaries. The bonding of estrogen and the estrogen receptor to DNA increase the production of mammary cells and the number of potential mutations that can lead to cancer. High levels of the estrogen receptor appear in more than 60% of all breast cancer cases.

Because Ellen is 45 years old and has a family history of breast and ovarian cancer, she was also tested for altered genes BRCA1 and BRCA2. Normally, these genes suppress tumor growth by repairing DNA defects. However, if the mutated genes are inherited, a person's cells lose the ability to suppress tumor growth, and the risk for breast and ovarian cancer as well as other cancers becomes much greater. Both parents can carry a BRCA mutation and pass it on to their sons and daughters. Ellen is relieved that her test results for mutated BRCA1 and BRCA2 were negative.

After Ellen completed a radiation series, she discusses the choice of drugs available for the treatment of breast cancer. Drugs such as tamoxifen and raloxifene block the binding of estrogen to the estrogen receptor, which prevents the growth of cancers that are estrogen positive. Other drugs such as anastrozole (Arimidex) and letrozole (Femara) are aromatase inhibitors (AIs) that prevent the growth of new tumors by blocking the synthesis of aromatase (estrogen synthetase), an enzyme that produces estrogen. Aromatase inhibitors are used primarily for post-menopausal women whose ovaries no longer produce estrogen. Ellen and her oncologist agree upon the use of tamoxifen, which she will take for the next five years to prevent breast cancer from recurring after her surgery.

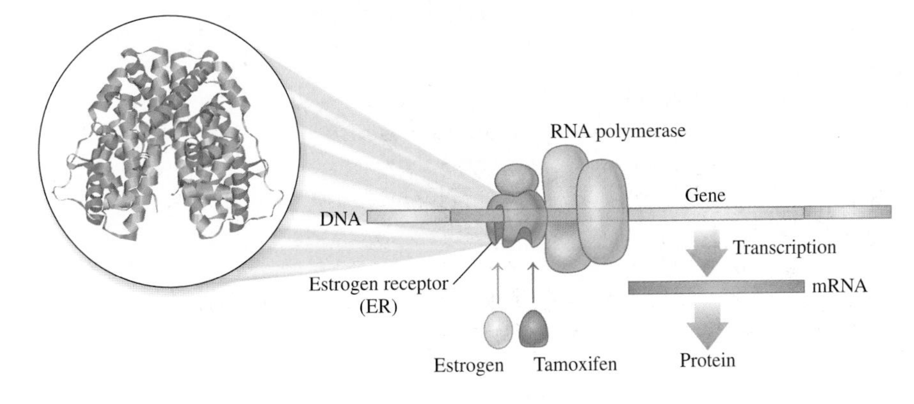

The estrogen receptor has a binding region for estrogen or tamoxifen.

Tamoxifen

Tamoxifen slows down transcription by blocking the binding of estrogen to estrogen receptors.

Clinical Applications

17.83 What are estrogen receptors?

17.84 How do estrogen and the estrogen receptor influence cancer growth of mammary cells?

17.85 How does tamoxifen reduce the potential of breast cancer?

17.86 How does letrozole reduce the potential of breast cancer?

17.87 The following is a segment of the template strand of human BRCA1 gene:

TGG AAT TAT CT**G** CTC TTC GCG

a. Write the corresponding mRNA segment.
b. Write the three-letter and one-letter abbreviations for the corresponding peptide segment.
c. If there is a point mutation in the fourth nucleotide triplet and A replaces **G**, what is the change, if any, in the amino acid sequence?

17.88 The following is a segment of the template strand of human BRCA1 gene:

ACA T**A**T TTT GCA AAT TTT GCA

a. Write the corresponding mRNA segment.
b. Write the three-letter and one-letter abbreviations for the corresponding peptide segment.
c. If there is a point mutation in the second nucleotide triplet and C replaces **A**, what is the change, if any, in the amino acid sequence?

CONCEPT MAP

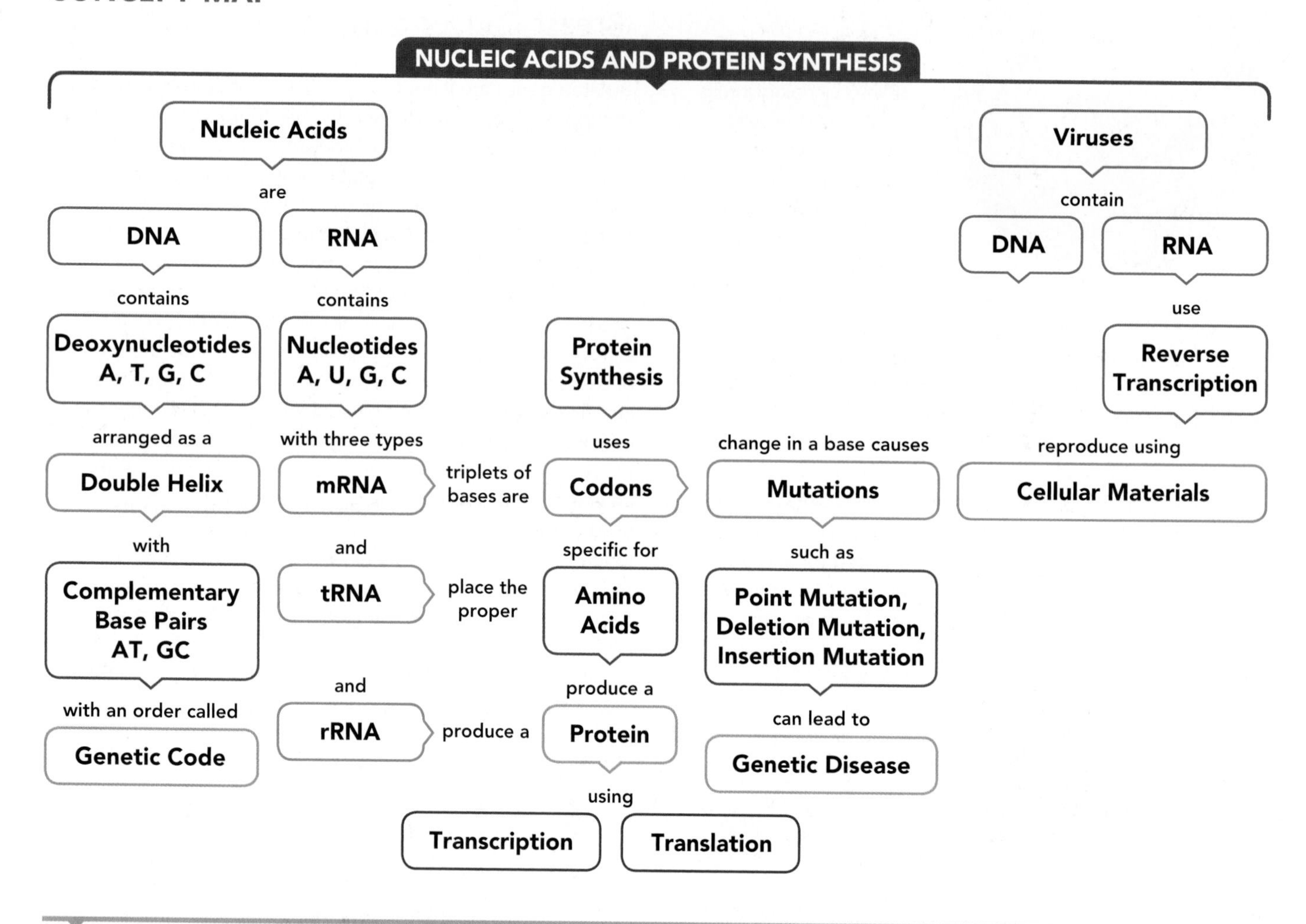

CHAPTER REVIEW

17.1 Components of Nucleic Acids

LEARNING GOAL Describe the bases and ribose sugars that make up the nucleic acids DNA and RNA.

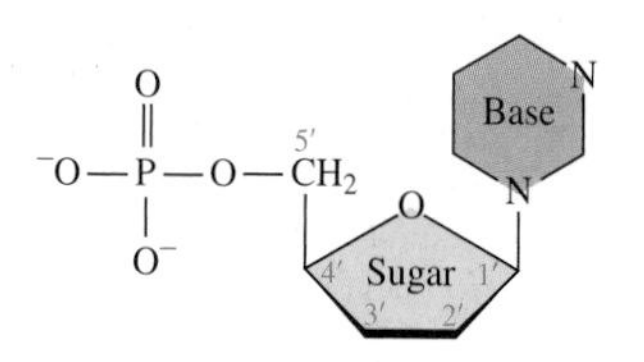

- Nucleic acids, such as deoxyribonucleic acid (DNA) and ribonucleic acid (RNA), are unbranched chains of nucleotides.
- A nucleoside is a combination of a pentose sugar and a base.
- A nucleotide is composed of three parts: a pentose sugar, a base, and a phosphate group.
- In DNA, the sugar is deoxyribose and the base can be adenine, thymine, guanine, or cytosine.
- In RNA, the sugar is ribose, and the base can be adenine, uracil, guanine, or cytosine.

17.2 Primary Structure of Nucleic Acids

LEARNING GOAL Describe the primary structures of RNA and DNA.

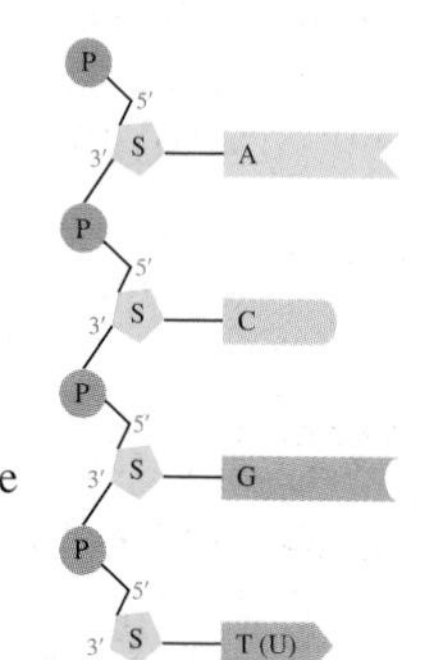

- Each nucleic acid has its own unique sequence of bases known as its primary structure.
- In a nucleic acid, the 3′ OH group of each ribose in RNA or deoxyribose in DNA forms a phosphodiester linkage to the phosphate group of the 5′ carbon atom of the sugar in the next nucleotide to give a backbone of alternating sugar and phosphate groups.
- There is a free 5′ phosphate at one end of the nucleic acid and a free 3′ OH group at the other end.

17.3 DNA Double Helix and Replication

LEARNING GOAL Describe the double helix of DNA; describe the process of DNA replication.

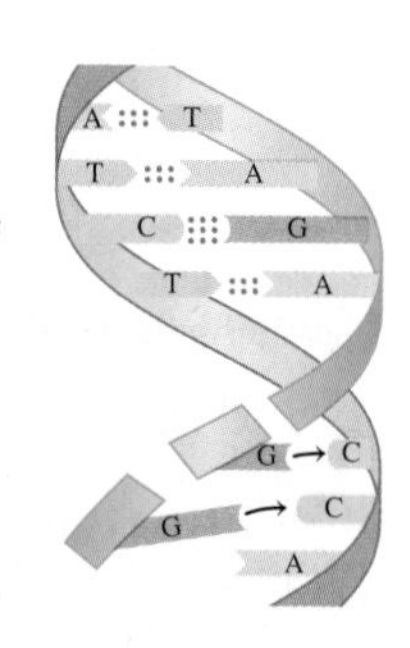

- A DNA molecule consists of two strands of nucleotides that are wound around each other like a spiral staircase.
- The two strands are held together by hydrogen bonds between complementary base pairs, A with T, and G with C.
- During DNA replication, DNA polymerase makes new DNA strands along each of the original DNA strands that serve as templates.
- Complementary base pairing ensures the correct pairing of bases to give identical copies of the original DNA.

17.4 RNA and Transcription

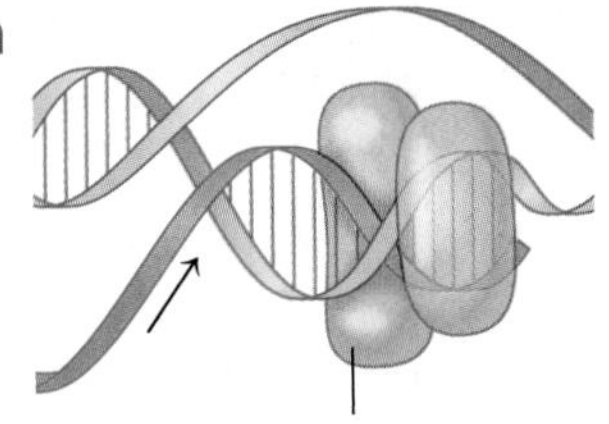

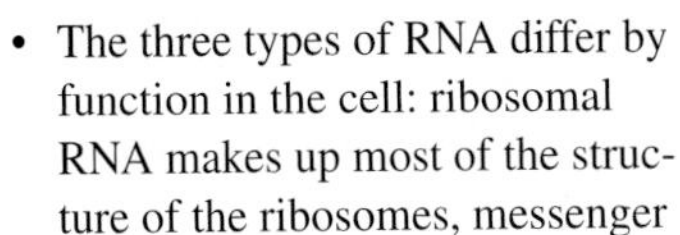

LEARNING GOAL Identify the different types of RNA; describe the synthesis of mRNA.

- The three types of RNA differ by function in the cell: ribosomal RNA makes up most of the structure of the ribosomes, messenger RNA carries genetic information from the DNA to the ribosomes, and transfer RNA places the correct amino acids in a growing peptide chain.
- Transcription is the process by which RNA polymerase produces mRNA from one strand of DNA.
- The bases in the mRNA are complementary to the DNA, except A in DNA is paired with U in RNA.
- The production of mRNA occurs when certain proteins are needed in the cell.

17.5 The Genetic Code and Protein Synthesis

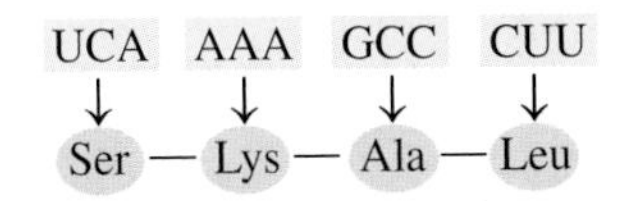

LEARNING GOAL Use the genetic code to write the amino acid sequence for a segment of mRNA.

- The genetic code consists of a series of codons, which are sequences of three bases that specify the order for the amino acids in a protein.
- There are 64 codons for the 20 amino acids, which means there are multiple codons for most amino acids.
- The codon AUG signals the start of transcription, and codons UAG, UGA, and UAA signal it to stop.
- Proteins are synthesized at the ribosomes in a translation process that includes three steps: initiation, chain elongation, and termination.
- During translation, tRNAs bring the appropriate amino acids to the ribosome, and peptide bonds form to join the amino acids in a peptide chain.
- When the polypeptide is released, it takes on its secondary and tertiary structures and becomes a functional protein in the cell.

17.6 Genetic Mutations

LEARNING GOAL Identify the type of change in DNA for a point mutation, a deletion mutation, and an insertion mutation.

- A genetic mutation is a change of one or more bases in the DNA sequence that may alter the structure and ability of the resulting protein to function properly.
- In a point mutation, one codon is altered, and in an insertion mutation or a deletion mutation, a base is added or removed, which changes all the codons after the base change.

17.7 Recombinant DNA

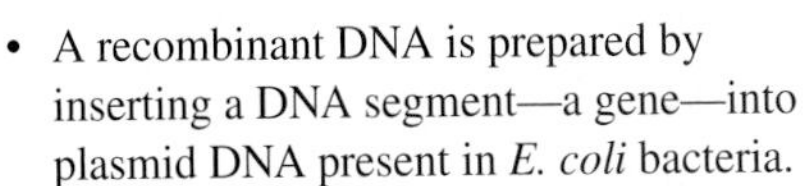

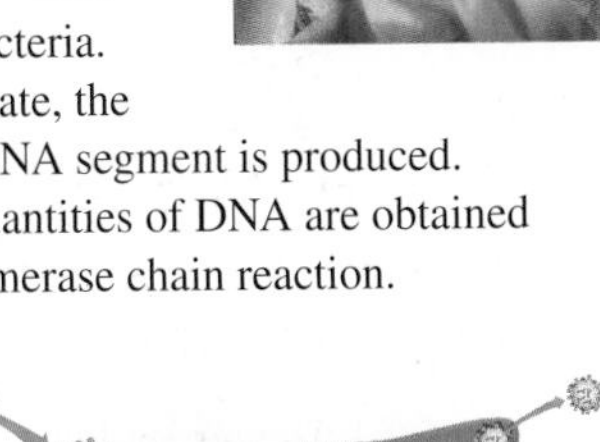

LEARNING GOAL Describe the preparation and uses of recombinant DNA.

- A recombinant DNA is prepared by inserting a DNA segment—a gene—into plasmid DNA present in *E. coli* bacteria.
- As the altered bacterial cells replicate, the protein expressed by the foreign DNA segment is produced.
- In criminal investigations, large quantities of DNA are obtained from smaller amounts by the polymerase chain reaction.

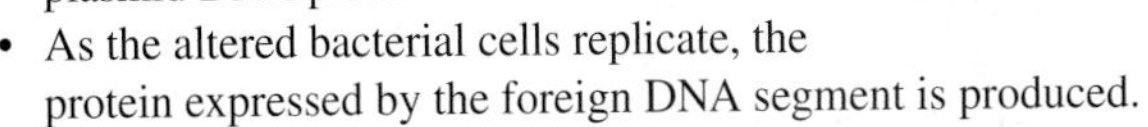

17.8 Viruses

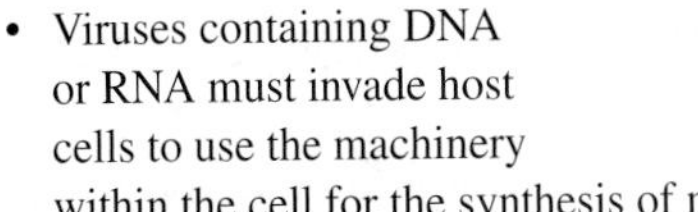

LEARNING GOAL Describe the methods by which a virus infects a cell.

- Viruses containing DNA or RNA must invade host cells to use the machinery within the cell for the synthesis of more viruses.
- For a retrovirus containing RNA, a viral DNA is synthesized by reverse transcription using the nucleotides and enzymes in the host cell.
- In the treatment of AIDS, entry inhibitors block the virus from entering the cell, nucleoside analogs inhibit the reverse transcriptase of the HIV virus, and protease inhibitors disrupt the catalytic activity of protease needed to produce proteins for the synthesis of more viruses.

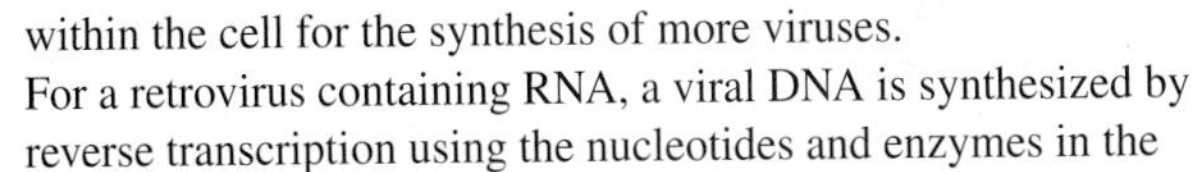

KEY TERMS

anticodon The triplet of bases in the center loop of tRNA that is complementary to a codon on mRNA.

base A nitrogen-containing compound found in DNA and RNA: adenine (A), thymine (T), cytosine (C), guanine (G), and uracil (U).

codon A sequence of three bases in mRNA that specifies a certain amino acid to be placed in a protein. A few codons signal the start or stop of protein synthesis.

complementary base pairs In DNA, adenine is always paired with thymine (A and T or T and A), and guanine is always paired with cytosine (G and C or C and G). In forming RNA, adenine is always paired with uracil (A and U or U and A).

deletion mutation A mutation that deletes a base from a DNA sequence.

DNA Deoxyribonucleic acid; the genetic material of all cells containing nucleotides with deoxyribose, phosphate, and the four bases: adenine, thymine, guanine, and cytosine.

double helix The helical shape of the double chain of DNA that is like a spiral staircase with a sugar–phosphate backbone on the outside and base pairs like stair steps on the inside.

genetic code The sequence of codons in mRNA that specifies the amino acid order for the synthesis of protein.

genetic disease A physical malformation or metabolic dysfunction caused by a mutation in the base sequence of DNA.

insertion mutation A mutation that inserts a base in a DNA sequence.

mRNA Messenger RNA; produced in the nucleus from DNA to carry the genetic information to the ribosomes for the construction of a protein.

mutation A change in the DNA base sequence that may alter the formation of a protein in the cell.

nucleic acid A large molecule composed of nucleotides; found as a double helix in DNA and as the single strands of RNA.

nucleoside The combination of a pentose sugar and a base.

nucleotide Building block of a nucleic acid consisting of a base, a pentose sugar (ribose or deoxyribose), and a phosphate group.

phosphodiester linkage The phosphate link that joins the 3′ hydroxyl group in one nucleotide to the phosphate group on the 5′ carbon atom in the next nucleotide.

point mutation A mutation that replaces one base in a DNA with a different base.

polymerase chain reaction (PCR) A procedure in which a strand of DNA is copied many times by mixing it with DNA polymerase and a mixture of deoxyribonucleotides, and subjecting it to repeated cycles of heating and cooling.

primary structure The sequence of nucleotides in nucleic acids.

recombinant DNA DNA combined from different organisms to form new, synthetic DNA.

replication The process of duplicating DNA by pairing the bases on each parent strand with their complementary bases.

retrovirus A virus that contains RNA as its genetic material and that synthesizes a complementary DNA strand inside a cell.

RNA Ribonucleic acid; a type of nucleic acid that is a single strand of nucleotides containing ribose, phosphate, and the four bases: adenine, cytosine, guanine, and uracil.

rRNA Ribosomal RNA; the most prevalent type of RNA and a major component of the ribosomes.

transcription The transfer of genetic information from DNA by the formation of mRNA.

translation The interpretation of the codons in mRNA as amino acids in a peptide.

tRNA Transfer RNA; an RNA that places a specific amino acid into a peptide chain at the ribosome so that a protein can be made. There is one or more tRNA for each of the 20 different amino acids.

virus A small particle containing DNA or RNA in a protein coat that requires a host cell for replication.

CORE CHEMISTRY SKILLS

The chapter Section containing each Core Chemistry Skill is shown in parentheses at the end of each heading.

Writing the Complementary DNA Strand (17.3)

- During DNA replication, new DNA strands are made along each of the original DNA strands.
- The new strand of DNA is made by forming hydrogen bonds with the bases in the template strand: A with T and T with A; C with G and G with C.

Example: Write the complementary base sequence for the following DNA segment:
A T T C G G T A C

Answer: T A A G C C A T G

Writing the mRNA Segment for a DNA Template (17.4)

- Transcription is the process that produces mRNA from one strand of DNA.
- The bases in the mRNA are complementary to the DNA, except that A in DNA is paired with U in RNA.

Example: What is the mRNA produced from the DNA segment C G C A T G T C A?

Answer: G C G U A C A G U

Writing the Amino Acid for an mRNA Codon (17.5)

- The genetic code consists of a sequence of three bases (codons) that specifies the order for the amino acids in a protein.
- The codon AUG signals the start of transcription and codons UAG, UGA, and UAA signal it to stop.

Example: Use three-letter and one-letter abbreviations to write the amino acid sequence you would expect in a peptide for the mRNA sequence CCG UAU GGG.

Answer: Pro–Tyr–Gly, PYG

UNDERSTANDING THE CONCEPTS

The chapter Sections to review are shown in parentheses at the end of each problem.

17.89 Answer the following questions for the given section of DNA: (17.3, 17.4, 17.5)

a. Complete the bases in the parent and new strands.

Parent strand:	A		G	T				C	T
New strand:		C			G	C	G		

b. Using the new strand as a template, write the mRNA sequence.

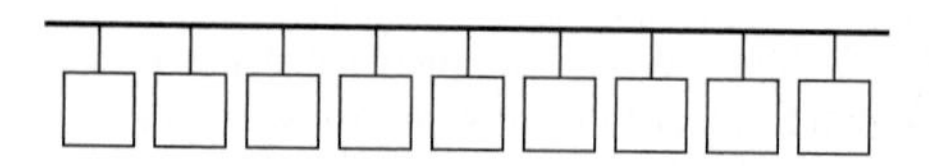

c. Write the three-letter symbols for the amino acids that would go into the peptide from the mRNA you wrote in part **b**.

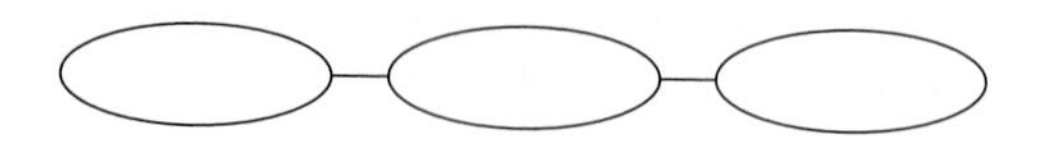

17.90 Suppose a mutation occurs in the DNA section in problem 17.89, and the first base in the parent chain, adenine, is replaced by guanine. (17.3, 17.4, 17.5, 17.6)

a. What type of mutation has occurred?

b. Using the new strand that results from this mutation, write the order of bases in the altered mRNA.

c. Write the three-letter symbols for the amino acids that would go into the peptide from the mRNA you wrote in part **b**.

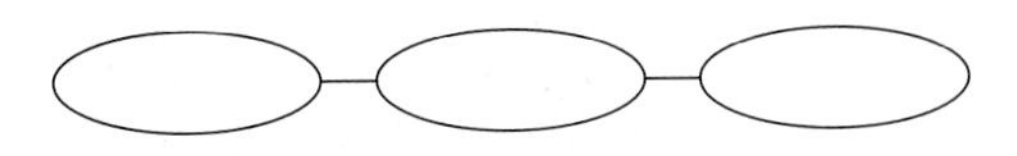

d. What effect, if any, might this mutation have on the structure and/or function of the resulting protein?

ADDITIONAL PRACTICE PROBLEMS

17.91 Identify each of the following bases as a pyrimidine or a purine: (17.1)

a. cytosine **b.** adenine
c. uracil **d.** thymine
e. guanine

17.92 Indicate if each of the bases in problem 17.91 is found in DNA only, RNA only, or both DNA and RNA. (17.1)

17.93 Identify the base and sugar in each of the following nucleosides: (17.1)

a. deoxythymidine **b.** adenosine
c. cytidine **d.** deoxyguanosine

17.94 Identify the base and sugar in each of the following nucleotides: (17.1)

a. CMP **b.** dAMP
c. dTMP **d.** UMP

17.95 How do the bases thymine and uracil differ? (17.1)

17.96 How do the bases cytosine and uracil differ? (17.1)

17.97 Draw the condensed structural formula for CMP. (17.1)

17.98 Draw the condensed structural formula for dGMP. (17.1)

17.99 What is similar about the primary structure of RNA and DNA? (17.2)

17.100 What is different about the primary structure of RNA and DNA? (17.2)

17.101 If the DNA double helix in salmon contains 28% adenine, what is the percentage of thymine, guanine, and cytosine? (17.3)

17.102 If the DNA double helix in humans contains 20% cytosine, what is the percentage of guanine, adenine, and thymine? (17.3)

17.103 In DNA, how many hydrogen bonds form between adenine and thymine? (17.3)

17.104 In DNA, how many hydrogen bonds form between guanine and cytosine? (17.3)

17.105 Write the complementary base sequence for each of the following DNA segments: (17.3)

a. G A C T T A G G C
b. T G C A A A C T A G C T
c. A T C G A T C G A T C G

17.106 Write the complementary base sequence for each of the following DNA segments: (17.3)

a. T T A C G G A C C G C
b. A T A G C C C T T A C T G G
c. G G C C T A C C T T A A C G A C G

17.107 Match the following statements with rRNA, mRNA, or tRNA: (17.4)

a. is the smallest type of RNA
b. makes up the highest percentage of RNA in the cell
c. carries genetic information from the nucleus to the ribosomes

17.108 Match the following statements with rRNA, mRNA, or tRNA: (17.4)

a. combines with proteins to form ribosomes
b. brings amino acids to the ribosomes for protein synthesis
c. acts as a template for protein synthesis

17.109 What are the possible codons for each of the following amino acids? (17.5)

a. threonine **b.** serine
c. cysteine

17.110 What are the possible codons for each of the following amino acids? (17.5)

a. valine **b.** arginine
c. histidine

17.111 What is the amino acid for each of the following codons? (17.5)

a. AAG **b.** AUU
c. CGA

17.112 What is the amino acid for each of the following codons? (17.5)

a. CAA **b.** GGC
c. UGG

17.113 What is the anticodon on tRNA for each of the following codons in an mRNA? (17.5)

a. AGC **b.** UAU
c. CCA

17.114 What is the anticodon on tRNA for each of the following codons in an mRNA? (17.5)

a. GUG **b.** CCC
c. GAA

Clinical Applications

17.115 Endorphins are polypeptides that reduce pain. What is the amino acid order for the endorphin leucine enkephalin (leu-enkephalin), which has the following mRNA? (17.5)

AUG UAC GGU GGA UUU CUA UAA

17.116 Endorphins are polypeptides that reduce pain. What is the amino acid order for the endorphin methionine enkephalin (met-enkephalin), which has the following mRNA? (17.5)

AUG UAC GGU GGA UUU AUG UAA

CHALLENGE PROBLEMS

The following problems are related to the topics in this chapter. However, they do not all follow the chapter order, and they require you to combine concepts and skills from several Sections. These problems will help you increase your critical thinking skills and prepare for your next exam.

17.117 Oxytocin is a peptide that contains nine amino acids. How many nucleotides would be found in the mRNA for this peptide? (17.5)

17.118 A polypeptide contains 36 amino acids. How many nucleotides would be found in the mRNA for this polypeptide? (17.5)

17.119 What is the difference between a DNA virus and a retrovirus? (17.8)

17.120 Why are there no base pairs in DNA between adenine and guanine or thymine and cytosine? (17.3)

17.121 Match each of the following processes (**1** to **5**) with one of the items (**a** to **e**): (17.3, 17.4, 17.5, 17.7, 17.8)

(1) replication of DNA (2) transcription
(3) translation (4) recombinant DNA
(5) reverse transcription

a. DNA polymerase
b. mRNA is synthesized from nuclear DNA
c. viruses
d. restriction enzymes
e. tRNA molecules bond to codons

17.122 Match each of the following processes (**1** to **5**) with one of the items (**a** to **e**): (17.3, 17.4, 17.5, 17.7, 17.8)

(1) replication of DNA (2) transcription
(3) translation (4) recombinant DNA
(5) reverse transcription

a. amino acids are linked together
b. RNA template is used to synthesize DNA
c. helicase unwinds DNA
d. genetic information is transferred from DNA
e. sticky ends join new DNA segment

ANSWERS

Answers to Selected Practice Problems

17.1 **a.** pyrimidine **b.** purine

17.3 **a.** DNA **b.** both DNA and RNA

17.5 deoxyadenosine monophosphate (dAMP), deoxythymidine monophosphate (dTMP), deoxycytidine monophosphate (dCMP), and deoxyguanosine monophosphate (dGMP)

17.7 **a.** nucleoside **b.** nucleoside
c. nucleoside **d.** nucleotide

17.9 **a.** both DNA and RNA **b.** RNA only
c. DNA only **d.** both DNA and RNA

17.11

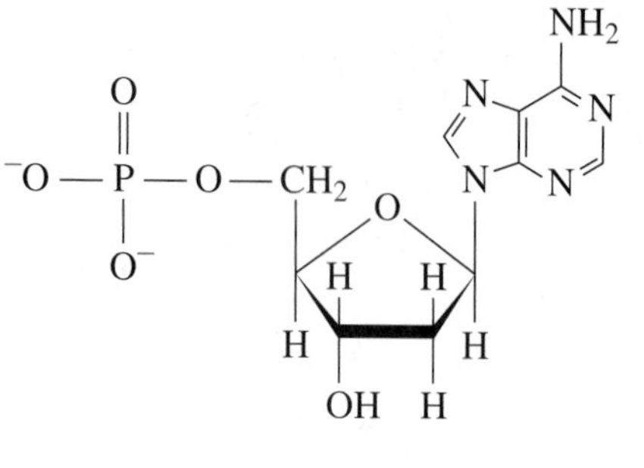

17.13

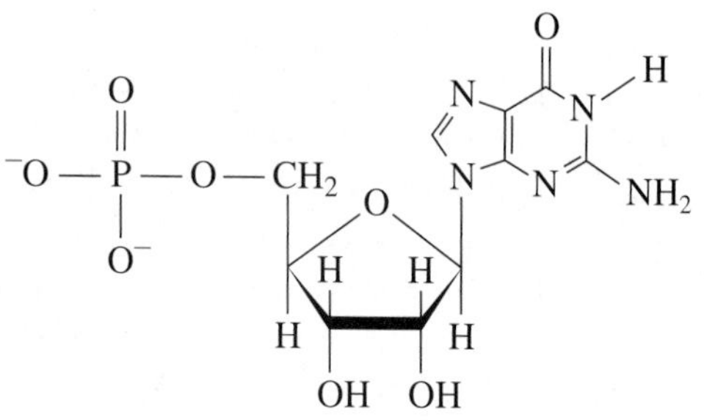

17.15 The nucleotides in nucleic acids are held together by phosphodiester linkages between the 3′ OH group of a sugar (ribose or deoxyribose) and a phosphate group on the 5′ carbon of another sugar.

17.17 –sugar–phosphate–sugar–phosphate–

17.19 the free 5′ phosphate on the 5′ carbon of ribose or deoxyribose of a nucleic acid

17.21

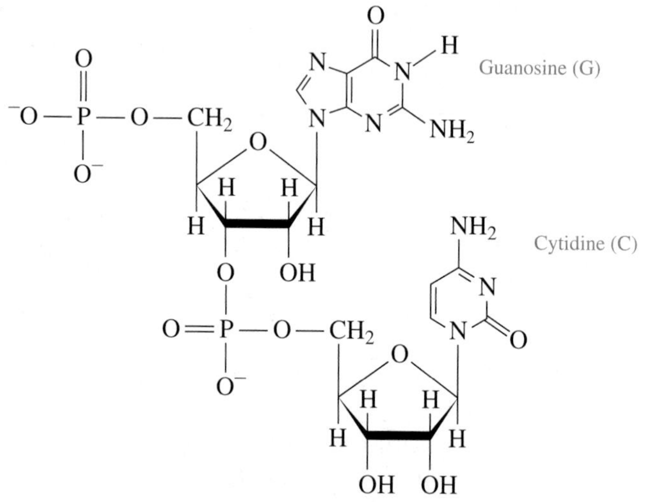

Guanosine (G)

Cytidine (C)

17.23 Structural features of DNA include that it is shaped like a double helix, it contains a sugar–phosphate backbone, the nitrogen-containing bases are hydrogen bonded between strands, and the strands run in opposite directions. A forms two hydrogen bonds to T, and G forms three hydrogen bonds to C.

17.25 The two DNA strands are held together by hydrogen bonds between the complementary bases in each strand.

17.27 a. T T T T T T **b.** C C C C C C
c. T C A G G T C C A **d.** G A C A T A T G C A A T

17.29 The enzyme helicase unwinds the DNA helix so that the parent DNA strands can be replicated into daughter DNA strands.

17.31 Once the DNA strands separate, the DNA polymerase pairs each of the bases with its complementary base and produces two exact copies of the original DNA.

17.33 ribosomal RNA, messenger RNA, and transfer RNA

17.35 A ribosome consists of a small subunit and a large subunit that contain rRNA combined with proteins.

17.37 In transcription, the sequence of nucleotides on a DNA template (one strand) is used to produce the base sequence of a messenger RNA.

17.39 GGC UUC CAA GUG

17.41 A codon is a three-base sequence in mRNA that codes for a specific amino acid in a protein.

17.43 a. proline (Pro) **b.** asparagine (Asn)
c. glycine (Gly) **d.** arginine (Arg)

17.45 When AUG is the first codon, it signals the start of protein synthesis. Thereafter, AUG codes for methionine.

17.47 A codon is a base triplet in the mRNA. An anticodon is the complementary triplet on a tRNA for a specific amino acid.

17.49 initiation, chain elongation, and termination

17.51 a. Thr–Thr–Thr, TTT
b. Phe–Pro–Phe–Pro, FPFP
c. Tyr–Gly–Arg–Cys, YGRC

17.53 The new amino acid is joined by a peptide bond to the growing peptide chain. The ribosome moves to the next codon, which attaches to a tRNA carrying the next amino acid.

17.55 a. CGA UAU GGU UUU
b. GCU, AUA, CCA, AAA
c. Arg–Tyr–Gly–Phe, RYGF

17.57 a. AAA CAC UUG GUU GUG GAC
b. Lys–His–Leu–Val–Val–Asp, KHLVVD

17.59 In a point mutation, a base in DNA is replaced by a different base.

17.61 In a mutation caused by the deletion of a base, all the codons from the mutation onward are changed, which changes the order of amino acids in the rest of the polypeptide chain.

17.63 The normal triplet TTT forms a codon AAA, which codes for lysine. The mutation TTC forms a codon AAG, which also codes for lysine. There is no effect on the amino acid sequence.

17.65 a. Thr–Ser–Arg–Val
b. Thr–Thr–Arg–Val
c. Thr–Ser–Gly–Val
d. Thr–STOP. Protein synthesis would terminate early. If this occurs early in the formation of the polypeptide, the resulting protein will probably be nonfunctional.
e. The new protein will contain the sequence Asp–Ile–Thr–Gly.
f. The new protein will contain the sequence His–His–Gly.

17.67 a. GCC and GCA both code for alanine.
b. A vital ionic interaction in the tertiary structure of hemoglobin cannot be formed when the polar acidic glutamate is replaced by valine, which is nonpolar. The resulting hemoglobin is malformed and less capable of carrying oxygen.

17.69 *E. coli* bacterial cells contain several small circular plasmids of DNA that can be isolated easily. After the recombinant DNA is formed, *E. coli* multiply rapidly, producing many copies of the recombinant DNA in a relatively short time.

17.71 *E. coli* can be soaked in a detergent solution that disrupts the plasma membrane and releases the cell contents, including the plasmids, which are collected.

17.73 When a gene has been obtained using a restriction enzyme, it is mixed with plasmids that have been cut by the same enzyme. When mixed together, the sticky ends of the DNA fragments bond with the sticky ends of the plasmid DNA to form a recombinant DNA.

17.75 In DNA fingerprinting, restriction enzymes cut a sample DNA into fragments, which are sorted by size using gel electrophoresis. A radioactive probe that adheres to specific DNA sequences exposes an X-ray film and creates a pattern of dark and light bands called a DNA fingerprint.

17.77 DNA or RNA, but not both

17.79 a. A viral RNA is used to synthesize a viral DNA to produce the proteins for the protein coat, which allows the virus to replicate and leave the cell.
b. retrovirus

17.81 Nucleoside analogs such as AZT and ddI are similar to the nucleosides required to make viral DNA in reverse transcription. However, they interfere with the ability of the DNA to form and thereby disrupt the life cycle of the HIV virus.

17.83 Estrogen receptors bind to DNA and increase the production of mammary cells, which can lead to mutations and cancer.

17.85 Tamoxifen blocks the binding of estrogen to the estrogen receptor, which prevents the activation of the gene.

17.87 a. ACC UUA AUA GAC GAG AAG CGC
b. Thr–Leu–Ile–Asp–Glu–Lys–Arg, TLIDEKR
c. There is no change; Asp is still the fourth amino acid.

17.89 a.

Parent strand: A G G T C G C C T

New strand: T C C A G C G G A

b. A G G U C G C C U

c. Arg–Ser–Pro

17.91 a. pyrimidine **b.** purine
c. pyrimidine **d.** pyrimidine
e. purine

17.93 a. thymine and deoxyribose
b. adenine and ribose
c. cytosine and ribose
d. guanine and deoxyribose

17.95 They are both pyrimidines, but thymine has a methyl group.

17.97

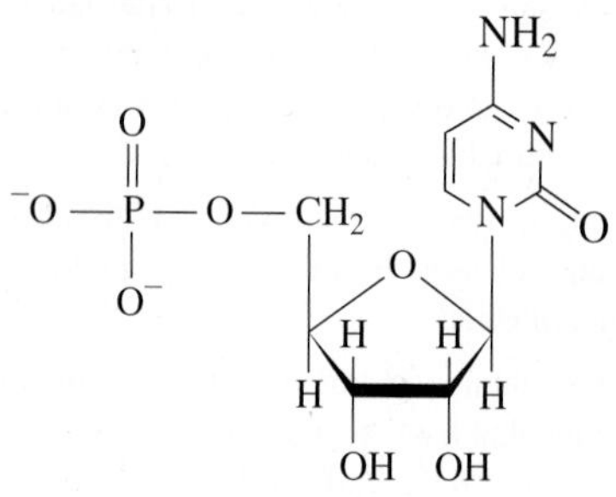

NH_2

O

$^{-}O-P-O-CH_2$

O^{-}

N N O

O

H H

H H

OH OH

17.99 They are both unbranched chains of nucleotides connected through phosphodiester linkages between alternating sugar and phosphate groups, with bases extending out from each sugar.

17.101 28% T, 22% G, and 22% C

17.103 two

17.105 **a.** C T G A A T C C G
b. A C G T T T G A T C G A
c. T A G C T A G C T A G C

17.107 **a.** tRNA **b.** rRNA **c.** mRNA

17.109 **a.** ACU, ACC, ACA, and ACG
b. UCU, UCC, UCA, UCG, AGU, and AGC
c. UGU and UGC

17.111 **a.** lysine **b.** isoleucine **c.** arginine

17.113 **a.** UCG **b.** AUA **c.** GGU

17.115 START–Tyr–Gly–Gly–Phe–Leu–STOP

17.117 Three nucleotides are needed to code for each amino acid, plus the start and stop codons consisting of three nucleotides each, which makes a minimum total of 33 nucleotides.

17.119 A DNA virus attaches to a cell and injects viral DNA that uses the host cell to produce copies of the DNA to make viral RNA. A retrovirus injects viral RNA from which complementary DNA is produced by reverse transcription.

17.121 **a.** (1) replication of DNA
b. (2) transcription
c. (5) reverse transcription
d. (4) recombinant DNA
e. (3) translation

18 Metabolic Pathways and ATP Production

LUKE IS 48 YEARS OLD AND WORKS AS A paramedic. Recently, bloodwork from his annual physical examination indicated a plasma cholesterol level of 256 mg/dL. Clinically, cholesterol levels are considered elevated if the total plasma cholesterol level exceeds 200 mg/dL. Luke's doctor ordered a liver profile that showed elevated liver enzymes: alanine transaminase (ALT) 282 Units/L (normal ALT 5 to 35 Units/L) and aspartate transaminase (AST) 226 Units/L (normal AST 5 to 50 Units/L). Luke's doctor ordered a medication to lower Luke's plasma cholesterol.

During Luke's career as a paramedic, he was exposed several times to blood and was accidently stuck by a needle containing infected blood. Luke takes 8 to 10 ibuprofen tablets per month for pain and uses herbs, garlic, ginkgo, and antioxidants. Although herbs, antioxidants, and ibuprofen can cause liver inflammation, they would not usually cause the elevation of liver enzymes reported in Luke's blood tests. A hepatitis profile showed that Luke was positive for antibodies to both hepatitis B and C. His doctor diagnosed Luke with chronic hepatitis C virus (HCV) infection. Hepatitis C is an infection caused by a virus that attacks the liver and leads to inflammation. Most people infected with the hepatitis C virus have no symptoms. Hepatitis C is usually passed by contact with contaminated blood or by needles shared during illegal drug use. As part of his treatment, Luke attended a class on living with hepatitis C given by Belinda, a public health nurse.

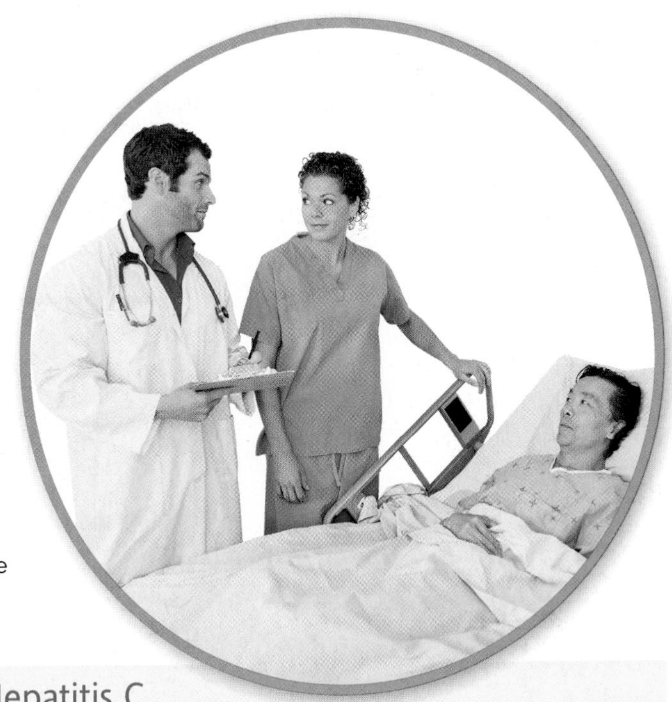

CAREER Public Health Nurse (PHN)

Hepatitis C virus is a common cause of liver disease and a major health problem worldwide. Patients with HCV require lifelong monitoring and are usually cared for by specialist teams including a public health nurse. A public health nurse works in public health departments, correctional facilities, occupational health facilities, schools, and organizations that aim to improve health at the community level. They often focus on high-risk populations such as the elderly, the homeless, teen mothers, and those at risk for a communicable disease such as hepatitis.

CLINICAL UPDATE Treatment of Luke's Hepatitis C

When the levels of Luke's liver enzymes remained elevated, his doctor prescribed a therapy of antiviral agents that inhibit the replication of the hepatitis C virus. Read more about this treatment and how it affected the levels of Luke's liver enzymes in the **CLINICAL UPDATE Treatment of Luke's Hepatitis C**, page 654.

LOOKING AHEAD

18.1 Metabolism and ATP Energy

LEARNING GOAL Describe the three stages of catabolism and the role of ATP.

When we eat food, the polysaccharides, fats, and proteins are digested to smaller molecules that can be absorbed into the cells of our body. As the glucose, fatty acids, and amino acids are broken down further, energy is released. Because we do not use all the energy from our foods at one time, we store energy in the cells as high-energy adenosine triphosphate (ATP). Our cells use the energy stored in ATP when they do work such as contracting muscles, synthesizing large molecules, sending nerve impulses, and moving substances across cell membranes.

The term **metabolism** refers to all the chemical reactions that provide energy and the substances required for continued cell growth. There are two types of metabolic reactions: *catabolic* and *anabolic*. In **catabolic reactions**, complex molecules are broken down to simpler ones with an accompanying release of energy. **Anabolic reactions** utilize energy available in the cell to build large molecules from simple ones.

This chapter focuses on the catabolic processes that happen in animal cells. We can think of these catabolic processes as occurring in three stages (see **FIGURE 18.1**).

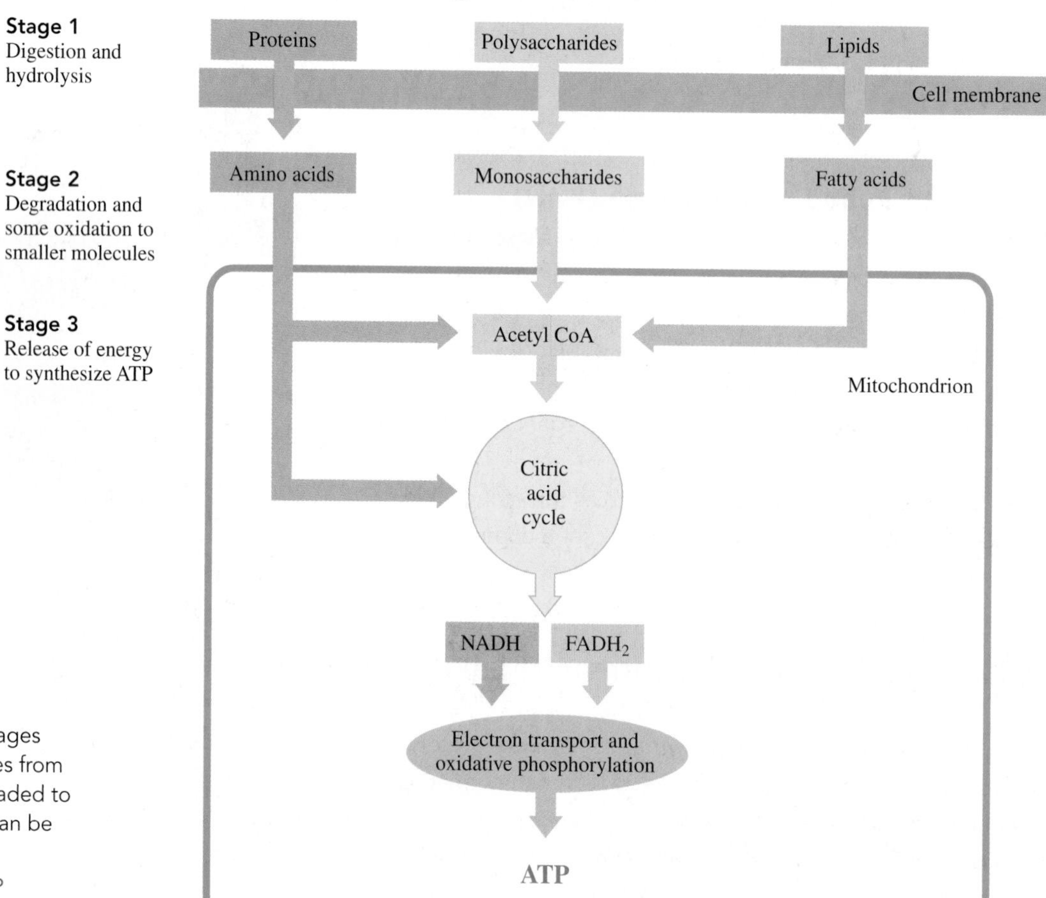

FIGURE 18.1 In the three stages of catabolism, large molecules from foods are digested and degraded to give smaller molecules that can be oxidized to produce ATP.

Q Where is most of the ATP produced in the cells?

Stage 1 Catabolism begins with the processes of **digestion** in which enzymes in the digestive tract break down large molecules into smaller ones. Polysaccharides break down to monosaccharides, fats break down to glycerol and fatty acids, and proteins yield amino acids. These digestion products diffuse into the bloodstream for transport to cells.

Stage 2 Within the cells, catabolic reactions continue as the digestion products are broken down further to yield smaller groups, such as the two-carbon acetyl group.

Stage 3 The major release of energy takes place in the mitochondria, where the two-carbon acetyl group is oxidized in the citric acid cycle and the reduced coenzymes NADH and $FADH_2$ are produced. As long as the cells have oxygen, NADH and $FADH_2$ can be reoxidized via electron transport to release the energy needed to synthesize ATP.

TEST

Try Practice Problems 18.1 to 18.4

SAMPLE PROBLEM 18.1 Metabolism

TRY IT FIRST

Identify each of the following as catabolic or anabolic:

a. digestion of polysaccharides
b. synthesis of proteins

SOLUTION

a. The breakdown of large molecules is catabolic.
b. The synthesis of large molecules is anabolic.

STUDY CHECK 18.1

Identify the oxidation of glucose to CO_2 and H_2O as catabolic or anabolic.

ANSWER

The oxidation of glucose to smaller molecules is catabolic.

Cell Structure for Metabolism

To understand metabolic reactions, we need to look at where these reactions take place in the cell (see **FIGURE 18.2**).

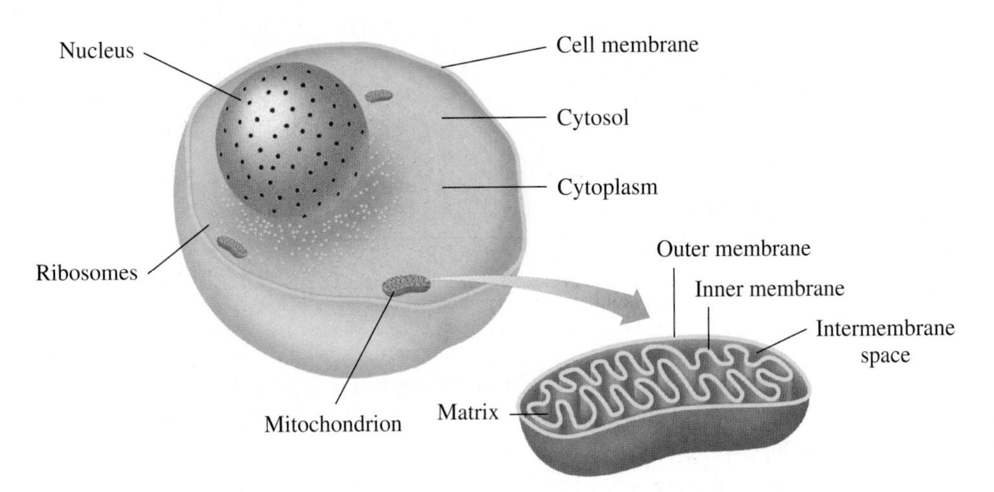

FIGURE 18.2 The diagram illustrates some of the major components of a typical animal cell.

Q What is the function of the mitochondria in a cell?

In animals, a *cell membrane* separates the materials inside the cell from the aqueous environment surrounding the cell. The *nucleus* contains the genes that control DNA replication and protein synthesis. The *cytoplasm* consists of all the materials between the nucleus and the cell membrane. The *cytosol*, the fluid part of the cytoplasm, is an aqueous solution of electrolytes and enzymes that catalyze many of the cell's chemical reactions.

Within the cytoplasm are specialized structures that carry out specific functions in the cell, some of which are labeled in Figure 18.2. The *ribosomes* are the sites of protein synthesis. The *mitochondria* are the energy-releasing factories of the cells. A mitochondrion has an *outer* and an *inner membrane*, with an *intermembrane space* between them. The fluid section surrounded by the inner membrane is called the *matrix*. Enzymes located in the matrix and along the inner membrane catalyze the oxidation of carbohydrates, fats, and amino acids. All these oxidation pathways eventually produce CO_2 and H_2O, and release energy, which is used to form energy-rich compounds. **TABLE 18.1** summarizes some of the functions of these components in animal cells.

TABLE 18.1 Functions of Some Major Components in Animal Cells

Component	Description and Function
Cell membrane	Separates the contents of a cell from the external environment and contains structures that communicate with other cells
Cytoplasm	Consists of the cellular contents between the cell membrane and nucleus
Cytosol	Fluid part of the cytoplasm that contains enzymes for many of the cell's chemical reactions
Mitochondrion	Contains the structures for the synthesis of ATP from energy-releasing reactions
Nucleus	Contains genetic information for the replication of DNA and the synthesis of protein
Ribosome	Site of protein synthesis using mRNA templates

ATP and Energy

In our cells, the energy released from the oxidation of the food we eat is stored in the form of a "high-energy" compound called *adenosine triphosphate* (ATP). The **ATP** molecule is composed of the base adenine, a ribose sugar, and three phosphate groups.

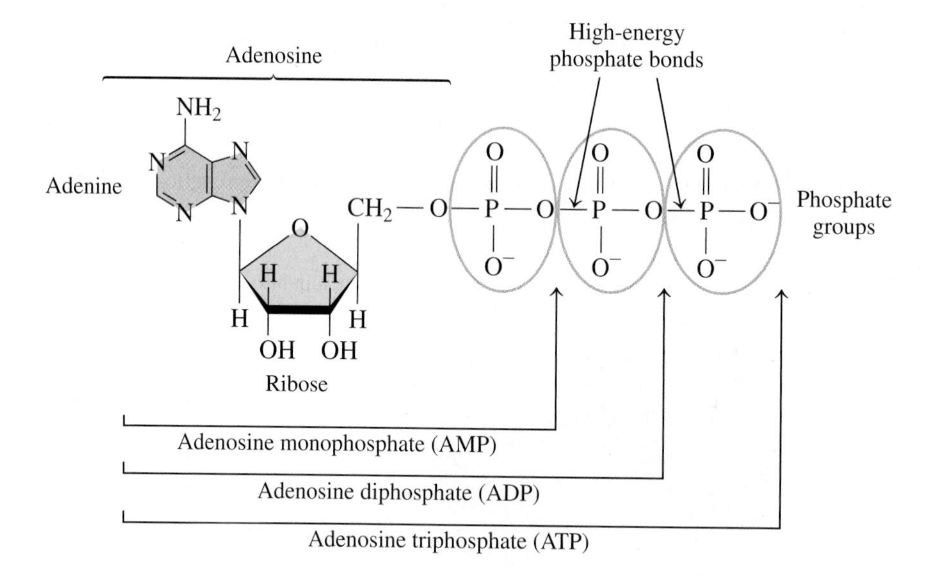

When ATP undergoes hydrolysis, the products are adenosine diphosphate (ADP), a phosphate group abbreviated as P_i and energy of 7.3 kcal per mole of ATP (31 kJ/mole). We can write this reaction as

$$\text{ATP} + H_2O \longrightarrow \text{ADP} + P_i + 7.3\ \text{kcal/mole (31 kJ/mole)}$$

The ADP can also hydrolyze to form adenosine monophosphate (AMP) and phosphate (P_i).

$$\text{ADP} + H_2O \longrightarrow \text{AMP} + P_i + 7.3\ \text{kcal/mole (31 kJ/mole)}$$

ENGAGE

Why are anabolic reactions that require energy always linked with the hydrolysis of a high-energy compound like ATP?

Every time we contract muscles, move substances across cellular membranes, send nerve signals, or synthesize an enzyme, we use energy from ATP hydrolysis. In a cell that is doing work (anabolic processes), 1 to 2 million ATP molecules may be hydrolyzed in one second. The amount of ATP hydrolyzed in one day can be as much as our body mass, even though only about 1 g of ATP is present in our cells at any given time.

When we take in food, the resulting catabolic reactions provide energy to regenerate ATP in our cells. Then 7.3 kcal/mole (31 kJ/mole) is used to make ATP from ADP and P_i (see **FIGURE 18.3**).

$$ADP + P_i + 7.3 \text{ kcal/mole (31 kJ/mole)} \longrightarrow ATP + H_2O$$

TEST

Try Practice Problems 18.5 and 18.6

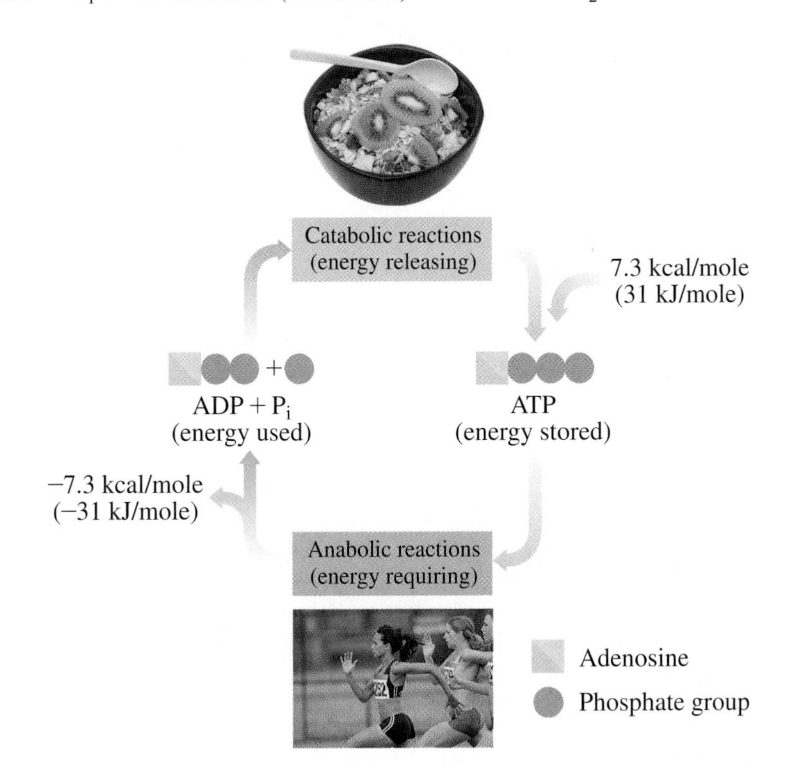

FIGURE 18.3 ATP, the energy-storage molecule, links energy-releasing reactions with energy-requiring reactions in the cells.

What type of reaction provides energy for ATP synthesis?

PRACTICE PROBLEMS

18.1 Metabolism and ATP Energy

LEARNING GOAL Describe the three stages of catabolism and the role of ATP.

18.1 What stage of catabolism involves the digestion of polysaccharides?

18.2 What stage of catabolism involves the conversion of small molecules to CO_2, H_2O, and energy for the synthesis of ATP?

18.3 Identify each of the following as catabolic or anabolic:
- **a.** synthesis of lipids from glycerol and fatty acids
- **b.** glucose adds P_i to form glucose-6-phosphate
- **c.** hydrolysis of ATP to ADP and P_i
- **d.** digestion of proteins in the stomach

18.4 Identify each of the following as catabolic or anabolic:
- **a.** digestion of fats to fatty acids and glycerol
- **b.** hydrolysis of proteins into amino acids
- **c.** synthesis of nucleic acids from nucleotides
- **d.** glucose and galactose form the disaccharide lactose

18.5 Why is ATP considered an energy-rich compound?

18.6 How much energy is obtained from the hydrolysis of ATP?

18.2 Digestion of Foods

LEARNING GOAL Identify the sites and products of digestion for carbohydrates, triacylglycerols, and proteins.

REVIEW

Identifying Fatty Acids (15.2)
Drawing Structures for Triacylglycerols (15.3)
Identifying the Primary, Secondary, Tertiary, and Quaternary Structures of Proteins (16.2, 16.3)
Describing Enzyme Action (16.4)

In stage 1 of catabolism, foods undergo digestion, a process that converts large molecules to smaller ones that can be absorbed by the body.

Digestion of Carbohydrates

We begin the digestion of carbohydrates as soon as we chew food. Enzymes produced in the salivary glands hydrolyze some of the α-glycosidic bonds in amylose and amylopectin, producing maltose, glucose, and smaller polysaccharides called dextrins, which contain

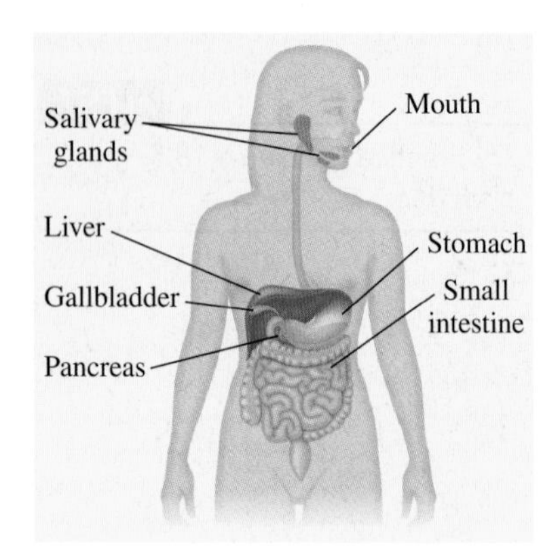

Carbohydrates begin digestion in the mouth, lipids in the small intestine, and proteins in the stomach and small intestine.

three to eight glucose units. After swallowing, the partially digested starches enter the acidic environment of the stomach, where the low pH stops carbohydrate digestion (see FIGURE 18.4).

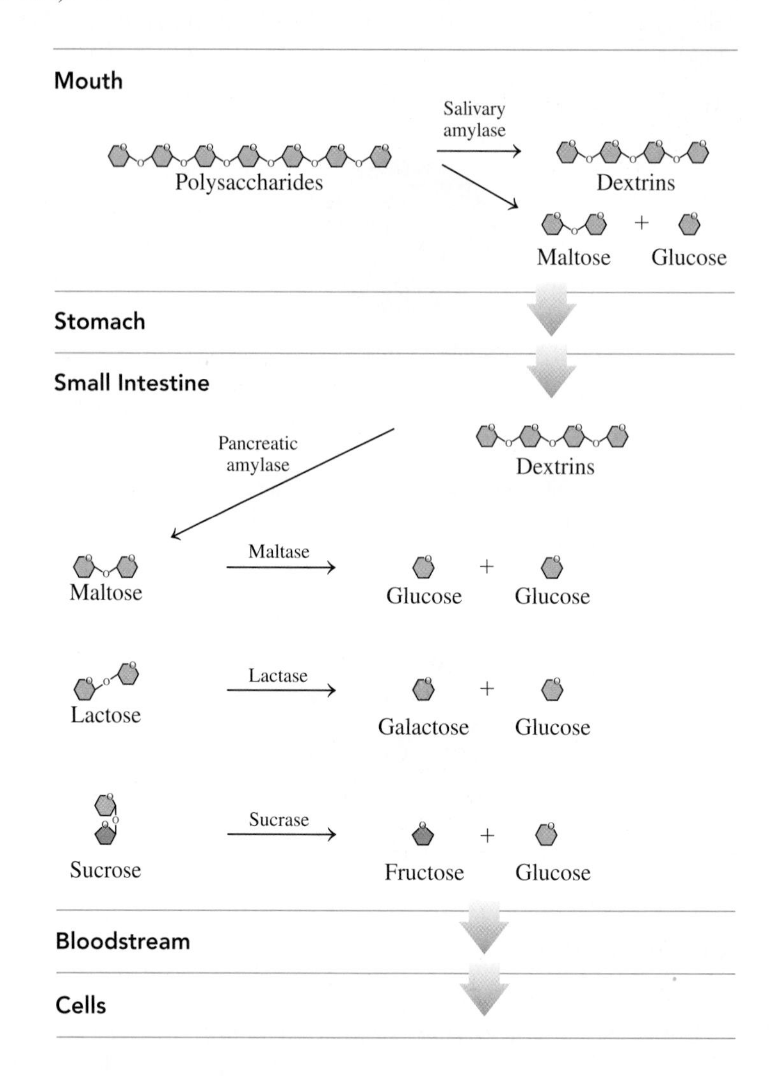

FIGURE 18.4 In stage 1 of catabolism, the digestion of carbohydrates begins in the mouth and is completed in the small intestine.

Q Why is there little or no digestion of carbohydrates in the stomach?

TEST

Try Practice Problems 18.7 and 18.8

In the small intestine, which has a pH of about 8, enzymes produced in the pancreas hydrolyze the remaining dextrins to maltose and glucose. Then enzymes produced in the mucosal cells that line the small intestine hydrolyze maltose as well as lactose and sucrose. The resulting monosaccharides are absorbed through the intestinal wall into the bloodstream, which carries them to the liver, where the hexoses fructose and galactose are converted to glucose. Glucose is the primary energy source for muscle contractions, red blood cells, and the brain.

CHEMISTRY LINK TO HEALTH

Lactose Intolerance

The disaccharide in milk is lactose, which is broken down by *lactase* in the intestinal tract to monosaccharides that are a source of energy. Infants and small children produce lactase to break down the lactose in milk. It is rare for an infant to lack the ability to produce lactase. However, the production of lactase decreases as many people age, which causes *lactose intolerance*. This condition affects approximately 25% of the people in the United States. A deficiency of lactase occurs in adults in many parts of the world, but in the United States it is prevalent among the African American, Hispanic, and Asian populations.

When lactose is not broken down into glucose and galactose, it cannot be absorbed through the intestinal wall and remains in the intestinal tract. Symptoms of lactose intolerance, which appear approximately $\frac{1}{2}$ to 1 h after ingesting milk or milk products, include nausea, abdominal cramps, and diarrhea. The severity of the symptoms depends on how much lactose is present in the food and how much lactase a person produces.

Treatment of Lactose Intolerance

One way to reduce the reaction to lactose is to avoid products that contain lactose. Another way is to ingest a product that contains lactase. The enzyme lactase is now available as tablets that are taken with meals, drops that are added to milk, or as additives in many dairy products such as milk. When lactase is added to milk that is left in the refrigerator for 24 h, the lactose level is reduced by 70 to 90%. Lactase pills or chewable tablets are taken when a person begins to eat a meal that contains dairy foods. If taken too far ahead of the meal, too much of the lactase will be degraded by stomach acid. If taken following a meal, the lactose will have already entered the lower intestine.

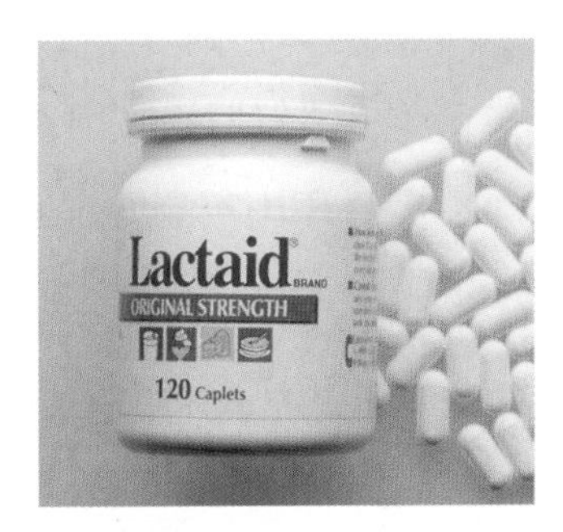

Lactaid contains an enzyme that aids the digestion of lactose.

SAMPLE PROBLEM 18.2 **Digestion of Carbohydrates**

TRY IT FIRST

Indicate the carbohydrates that undergo digestion in each of the following sites:

a. mouth **b.** stomach **c.** small intestine

SOLUTION

a. polysaccharides **b.** no carbohydrate digestion
c. dextrins, maltose, sucrose, and lactose

STUDY CHECK 18.2

Describe the digestion of amylose, a polysaccharide.

ANSWER

The digestion of amylose begins in the mouth where enzymes hydrolyze some of the glycosidic bonds. In the small intestine, enzymes hydrolyze more glycosidic bonds, and finally maltose is hydrolyzed to yield glucose.

Digestion of Fats

The digestion of dietary fats begins in the small intestine, when the hydrophobic fat globules mix with bile salts released from the gallbladder. In a process called *emulsification*, the bile salts break the fat globules into smaller droplets called *micelles*. Enzymes from the pancreas hydrolyze the triacylglycerols to yield monoacylglycerols and fatty acids, which are then absorbed into the intestinal lining where they recombine to form triacylglycerols. These nonpolar compounds are then coated with proteins to form lipoproteins called *chylomicrons*, which are more polar and soluble in the aqueous environment of the lymph and bloodstream (see **FIGURE 18.5**).

The chylomicrons transport the triacylglycerols to the cells of the heart, muscle, and adipose tissues. When energy is needed in the cells, enzymes hydrolyze the triacylglycerols to yield glycerol and fatty acids.

ENGAGE

Why are triacylglycerols insoluble in the lymph or bloodstream?

TEST

Try Practice Problems 18.9 to 18.12

$$\begin{array}{l} CH_2-O-\overset{\displaystyle O}{\overset{\|}{C}}-(CH_2)_{16}-CH_3 \\ | \\ CH-O-\overset{\displaystyle O}{\overset{\|}{C}}-(CH_2)_{16}-CH_3 \\ | \\ CH_2-O-\overset{\displaystyle O}{\overset{\|}{C}}-(CH_2)_{16}-CH_3 \end{array}$$

A triacylglycerol

The fat cells that make up adipose tissue are capable of storing unlimited quantities of triacylglycerols.

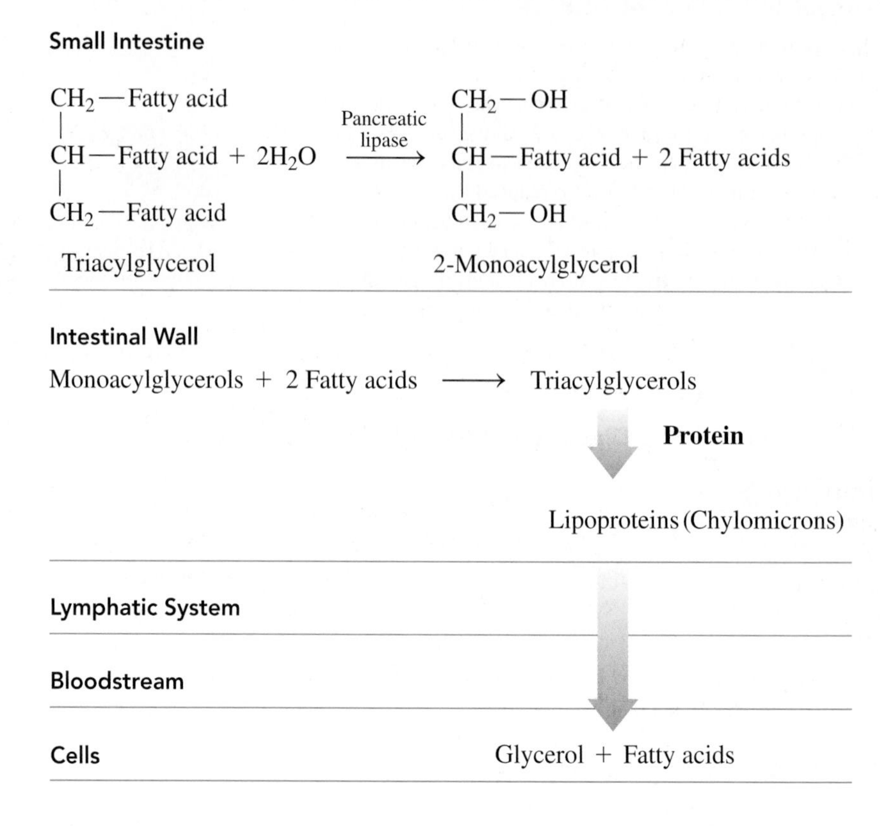

FIGURE 18.5 The triacylglycerols are hydrolyzed in the small intestine and re-formed in the intestinal wall where they bind to proteins for transport through the lymphatic system and bloodstream to the cells.

Why do chylomicrons form in the intestinal wall?

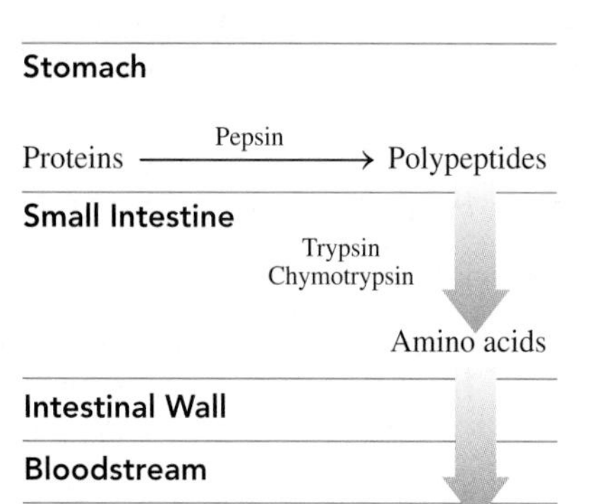

FIGURE 18.6 Proteins are hydrolyzed in the stomach and the small intestine.

What enzymes, secreted into the small intestine, hydrolyze peptides?

Digestion of Proteins

The major role of proteins is to provide amino acids for the synthesis of new proteins for the body and nitrogen atoms for the synthesis of compounds such as nucleotides. The digestion of proteins begins in the stomach, where hydrochloric acid (HCl) at pH 2 denatures the proteins and activates protease enzymes such as pepsin that begin to hydrolyze peptide bonds. Polypeptides move out of the stomach into the small intestine, where other proteases, such as trypsin and chymotrypsin, complete the hydrolysis of the peptides to amino acids. The amino acids are absorbed through the intestinal walls into the bloodstream for transport to the cells (see **FIGURE 18.6**).

PRACTICE PROBLEMS

18.2 Digestion of Foods

LEARNING GOAL Identify the sites and products of digestion for carbohydrates, triacylglycerols, and proteins.

18.7 What is the general type of reaction that occurs during the digestion of carbohydrates?

18.8 What is the purpose of digestion in stage 1?

18.9 What is the role of bile salts in lipid digestion?

18.10 How are insoluble triacylglycerols transported to the cells?

18.11 Where do dietary proteins undergo digestion in the body?

18.12 What are the end products of the digestion of proteins?

18.3 Coenzymes in Metabolic Pathways

LEARNING GOAL Describe the components and functions of the coenzymes NAD^+, FAD, and coenzyme A.

Several metabolic reactions that extract energy from our food involve oxidation and reduction reactions. In chemistry, oxidation is often associated with the loss of H atoms, whereas reduction is associated with the gain of H atoms. Often, we represent two H atoms as two hydrogen ions ($2H^+$) and two electrons ($2\,e^-$). In both oxidation and reduction, *coenzymes* are required to carry the hydrogen ions and electrons from or to the reacting substrate.

A coenzyme that gains hydrogen ions and electrons is reduced, whereas a coenzyme that loses hydrogen ions and electrons is oxidized.

In general, oxidation reactions release energy, and reduction reactions require energy. **TABLE 18.2** summarizes the characteristics of oxidation and reduction.

TABLE 18.2 Characteristics of Oxidation and Reduction in Metabolic Pathways

Oxidation	Reduction
Loss of electrons (e^-)	Gain of electrons (e^-)
Loss of hydrogen (H or H^+ and e^-)	Gain of hydrogen (H or H^+ and e^-)
Gain of oxygen	Loss of oxygen
Release of energy	Input of energy

NAD^+

NAD^+ (nicotinamide adenine dinucleotide) is an important coenzyme in which the vitamin *niacin* provides the *nicotinamide* group, which is bonded to ribose and the nucleotide ADP (see **FIGURE 18.7**). The oxidized form of NAD^+ undergoes reduction when a carbon atom in the nicotinamide ring reacts with two hydrogen atoms ($2H^+ + 2e^-$), leaving one H^+.

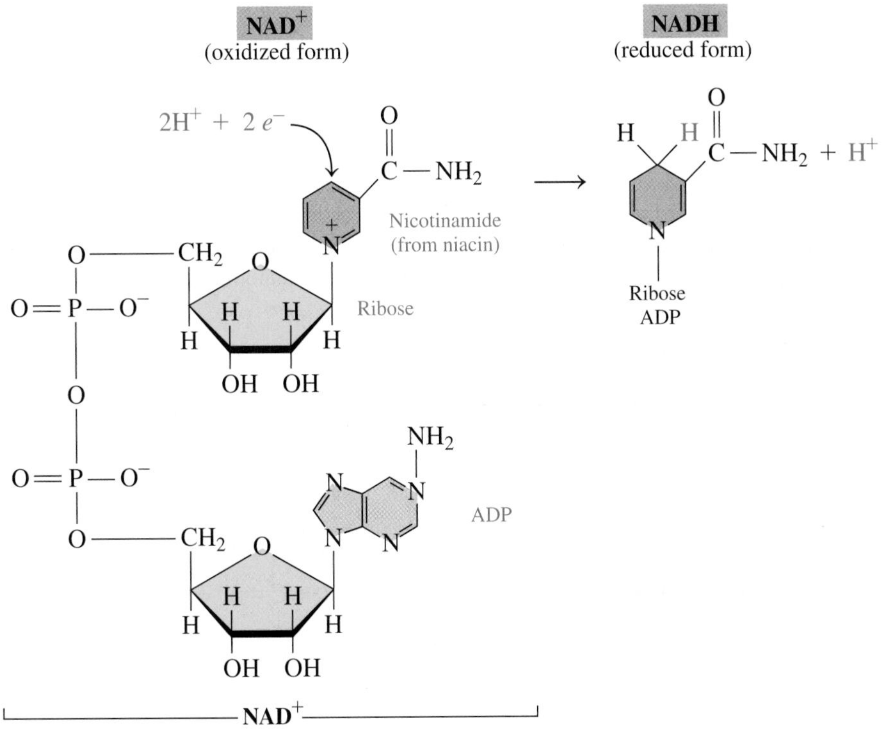

FIGURE 18.7 The coenzyme NAD^+ (nicotinamide adenine dinucleotide), which consists of a nicotinamide portion from the vitamin niacin, ribose, and adenosine diphosphate, is reduced to NADH + H^+ when a hydrogen ion and two electrons are added to the NAD^+.

Q Why is the conversion of NAD^+ to NADH and H^+ a reduction?

The NAD^+ coenzyme is required for metabolic reactions that produce carbon–oxygen double bonds (C=O), such as in the oxidation of alcohols to aldehydes and ketones. An example of such an oxidation–reduction reaction is the oxidation of ethanol in the liver to ethanal, with the corresponding reduction of NAD^+ to NADH and H^+.

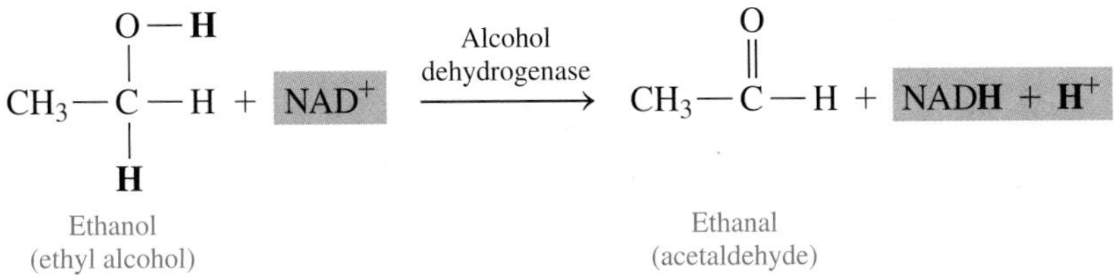

$$CH_3-CH_2-OH + NAD^+ \xrightarrow{\text{Alcohol dehydrogenase}} CH_3-C(=O)-H + NADH + H^+$$

Ethanol (ethyl alcohol) → Ethanal (acetaldehyde)

FAD

FAD (flavin adenine dinucleotide) is a coenzyme that contains the nucleotide ADP and riboflavin. Riboflavin, vitamin B_2, consists of ribitol (a sugar alcohol) and flavin. The oxidized form of FAD undergoes reduction when the two nitrogen atoms in the flavin part of the FAD coenzyme react with two hydrogen atoms ($2H^+ + 2e^-$), reducing FAD to $FADH_2$ (see **FIGURE 18.8**).

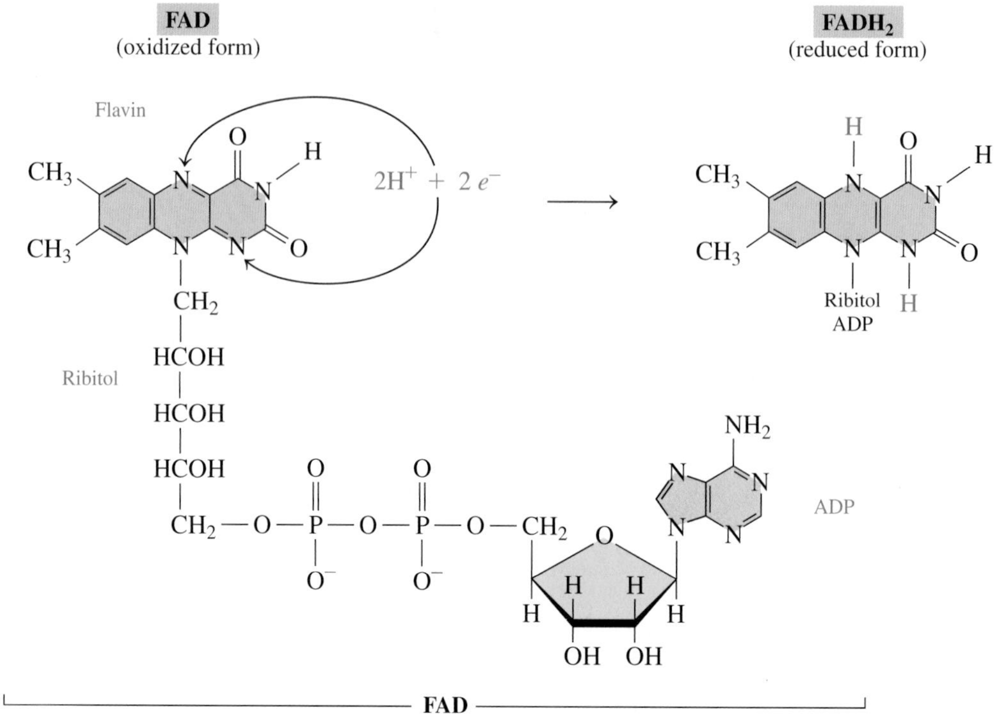

FAD (oxidized form)

Flavin

Ribitol

$2H^+ + 2e^-$ ⟶

$FADH_2$ (reduced form)

Ribitol ADP

ADP

FAD

FIGURE 18.8 The coenzyme FAD (flavin adenine dinucleotide), made from flavin, ribitol, and adenosine diphosphate, is reduced to $FADH_2$ when two hydrogen atoms are added to the FAD.

Q What is the type of reaction in which FAD accepts hydrogen?

FAD is used as a coenzyme when an oxidation reaction converts a carbon–carbon single bond to a carbon–carbon double bond (C=C). An example of such a reaction from the citric acid cycle is the conversion of the carbon–carbon single bond in succinate to a double bond in fumarate, with the corresponding reduction of FAD to $FADH_2$.

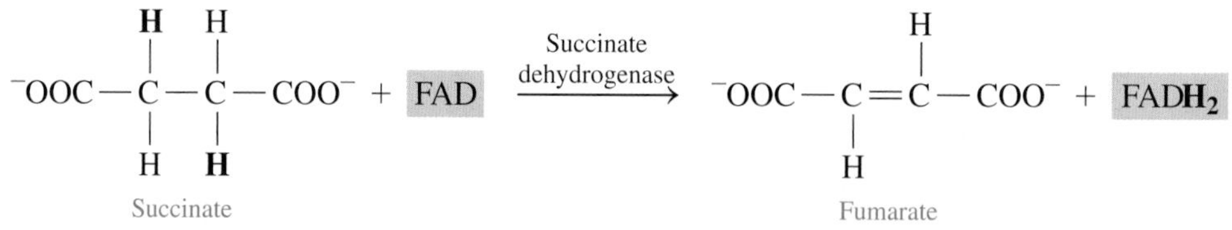

$$^-OOC-CH_2-CH_2-COO^- + FAD \xrightarrow{\text{Succinate dehydrogenase}} {}^-OOC-CH=CH-COO^- + FADH_2$$

Succinate → Fumarate

Coenzyme A

Coenzyme A (CoA) is made up of several components: pantothenic acid (vitamin B_5), phosphorylated ADP, and aminoethanethiol (see **FIGURE 18.9**).

An important function of coenzyme A is to prepare small acyl groups (represented by the letter A in the name), such as acetyl, for reactions with enzymes. The reactive feature of coenzyme A is the thiol group (—SH), which bonds to a two-carbon acetyl group to produce the energy-rich thioester **acetyl CoA**.

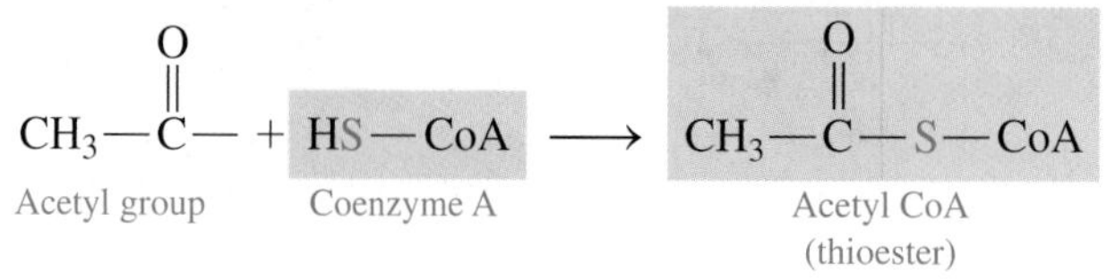

ENGAGE

Is coenzyme A oxidized or reduced when an acetyl group is added?

$$CH_3-\overset{O}{\overset{\|}{C}}- + HS-CoA \longrightarrow CH_3-\overset{O}{\overset{\|}{C}}-S-CoA$$

Acetyl group + Coenzyme A ⟶ Acetyl CoA (thioester)

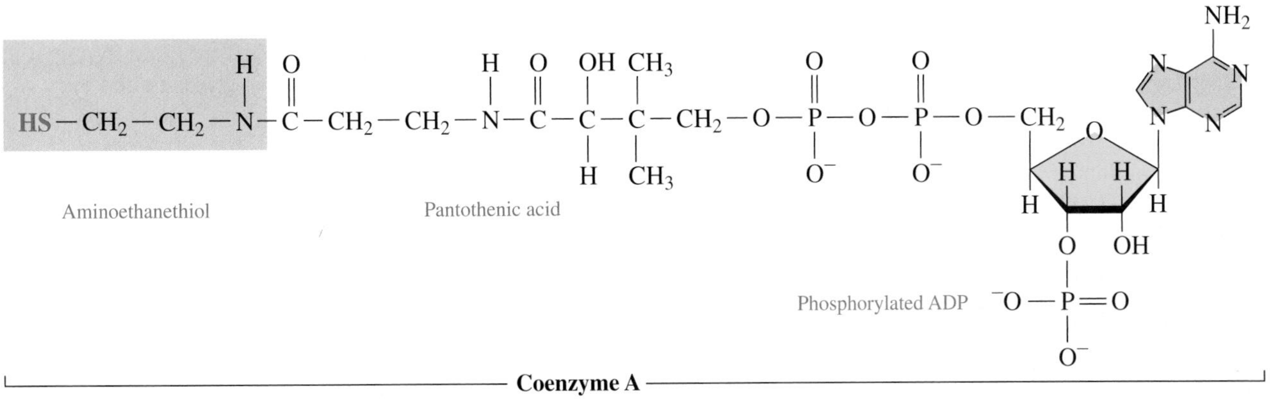

$HS-CH_2-CH_2-NH-C(=O)-CH_2-CH_2-NH-C(=O)-CH(OH)-C(CH_3)_2-CH_2-O-P(=O)(O^-)-O-P(=O)(O^-)-O-CH_2$–ribose(3′-$O-PO_3^{2-}$, 2′-OH)–adenine ($NH_2$)

Aminoethanethiol | Pantothenic acid | Phosphorylated ADP

Coenzyme A

FIGURE 18.9 Coenzyme A is derived from a phosphorylated adenosine diphosphate (ADP) and pantothenic acid bonded by an amide bond to aminoethanethiol, which contains the —SH reactive part of the molecule.

Q What part of coenzyme A reacts with a two-carbon acetyl group?

SAMPLE PROBLEM 18.3 Coenzymes

TRY IT FIRST

Describe the reactive part of each of the following coenzymes and the way each participates in metabolic pathways:

a. FAD **b.** NAD^+

SOLUTION

a. When two nitrogen atoms in the flavin accept 2H ($2H^+$ and $2\ e^-$), FAD is reduced to $FADH_2$. The FAD coenzyme participates in oxidation reactions that produce a carbon–carbon double bond (C=C).

b. When a carbon atom in the nicotinamide accepts 2H ($2H^+$ and $2\ e^-$), NAD^+ is reduced to $NADH + H^+$. The NAD^+ coenzyme participates in oxidation reactions that produce a carbon–oxygen double bond (C=O).

STUDY CHECK 18.3

Describe the reactive part of coenzyme A and the way it participates in metabolic pathways.

ANSWER

The thiol group (—SH) in coenzyme A combines with an acetyl group to form acetyl coenzyme A, which participates in the transfer of acetyl groups.

TEST

Try Practice Problems 18.13 to 18.18

PRACTICE PROBLEMS

18.3 Coenzymes in Metabolic Pathways

LEARNING GOAL Describe the components and functions of the coenzymes NAD^+, FAD, and coenzyme A.

18.13 Identify one or more coenzymes with each of the following components:
a. pantothenic acid **b.** niacin **c.** ribitol

18.14 Identify one or more coenzymes with each of the following components:
a. riboflavin **b.** adenine **c.** aminoethanethiol

18.15 Give the abbreviation for each of the following coenzymes:
a. reduced form of NAD^+ **b.** oxidized form of $FADH_2$

18.16 Give the abbreviation for each of the following coenzymes:
a. reduced form of FAD **b.** oxidized form of NADH

18.17 What coenzyme picks up hydrogen when a carbon–carbon double bond is formed?

18.18 What coenzyme picks up hydrogen when a carbon–oxygen double bond is formed?

18.4 Glycolysis: Oxidation of Glucose

LEARNING GOAL Describe the conversion of glucose to pyruvate in glycolysis and the subsequent conversion of pyruvate to acetyl CoA or lactate.

Glucose is the main energy source for the brain, skeletal muscles, and red blood cells.

The major source of energy for the body is the glucose produced when we digest the carbohydrates in our food, or from glycogen, a polysaccharide stored in the liver and skeletal muscle. Glucose in the bloodstream enters our cells where it undergoes degradation in a pathway called *glycolysis*. Early organisms used glycolysis to release energy from simple nutrients long before there was any oxygen in Earth's atmosphere. Glycolysis is an **anaerobic** process; no oxygen is required.

In **glycolysis**, a six-carbon glucose molecule is broken down to two molecules of three-carbon pyruvate (see **FIGURE 18.10**). All the reactions in glycolysis take place in the cytoplasm of the cell. In the first five reactions (1 to 5), called the *energy-investing phase*, energy from the hydrolysis of two ATP is used to form two three-carbon, energy-rich phosphate compounds. In the last five reactions (6 to 10), called the *energy-generating phase*, energy from the hydrolysis of the energy-rich phosphate compounds is used to synthesize four ATP.

CORE CHEMISTRY SKILL
Identifying the Compounds in Glycolysis

ENGAGE
In which steps of glycolysis is ADP phosphorylated to become ATP?

TEST
Try Practice Problems 18.19 to 18.22

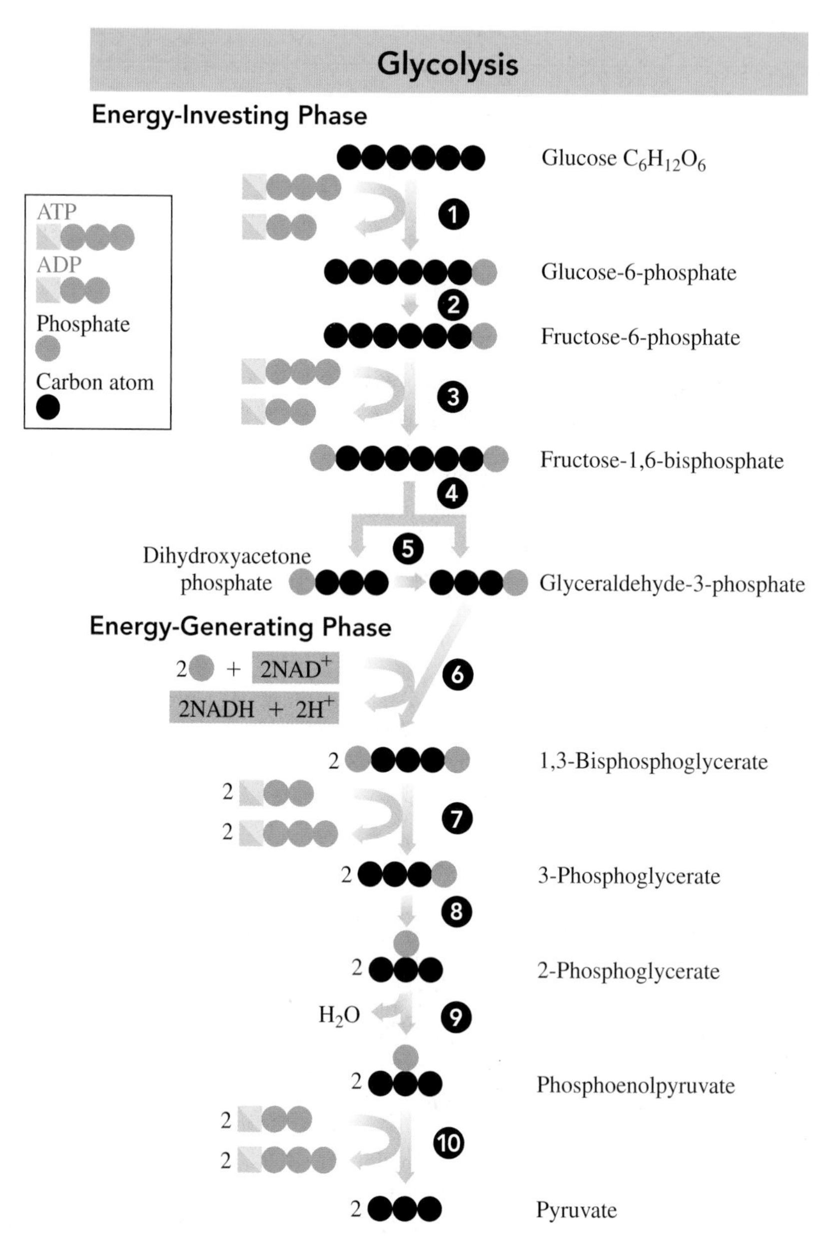

FIGURE 18.10 In glycolysis, the six-carbon glucose molecule is degraded to yield two three-carbon pyruvate molecules. A net of two ATP is produced along with two NADH.

Q Where in the glycolysis pathway is glucose cleaved to yield two three-carbon compounds?

Energy-Investing Reactions 1 to 5

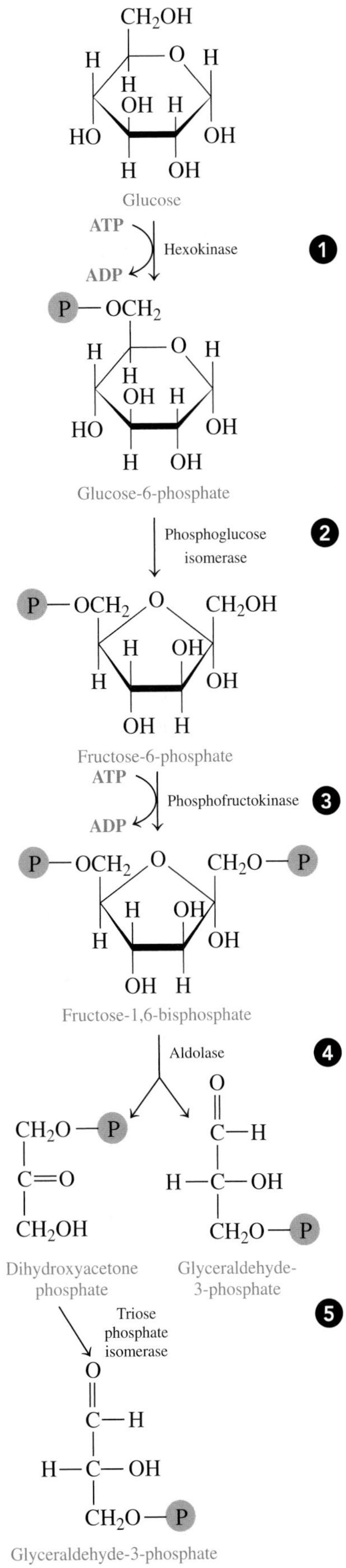

Reaction 1 Phosphorylation

In the initial reaction, a phosphate group from ATP is added to glucose to form glucose-6-phosphate and ADP.

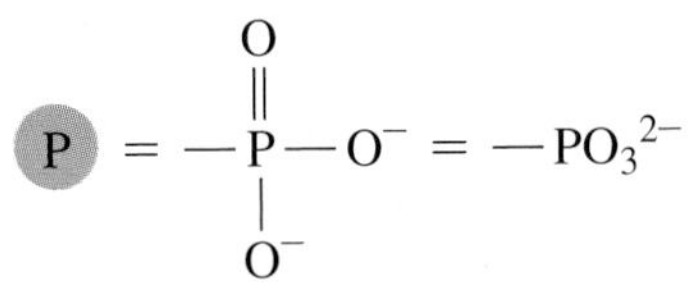

$$\text{(P)} = -\overset{\displaystyle O}{\overset{\|}{\underset{\underset{\displaystyle O^-}{|}}{P}}}-O^- = -PO_3^{2-}$$

Glucose

ATP → ADP (Hexokinase) 1

Glucose-6-phosphate

Reaction 2 Isomerization

The glucose-6-phosphate, the aldose from reaction 1, undergoes isomerization to fructose-6-phosphate, which is a ketose.

Phosphoglucose isomerase 2

Fructose-6-phosphate

Reaction 3 Phosphorylation

The hydrolysis of another ATP provides a second phosphate group, which converts fructose-6-phosphate to fructose-1,6-bisphosphate. The word *bisphosphate* is used to show that the two phosphate groups are on different carbons in fructose and not connected to each other.

ATP → ADP (Phosphofructokinase) 3

Fructose-1,6-bisphosphate

Reaction 4 Cleavage

Fructose-1,6-bisphosphate is split into two three-carbon phosphate isomers: dihydroxyacetone phosphate and glyceraldehyde-3-phosphate.

Aldolase 4

Dihydroxyacetone phosphate

Glyceraldehyde-3-phosphate

Reaction 5 Isomerization

Because dihydroxyacetone phosphate is a ketone, it cannot undergo further oxidation. However, it undergoes isomerization to provide a second molecule of glyceraldehyde-3-phosphate, which can be oxidized. Now all six carbon atoms from glucose are contained in two identical triose phosphates.

Triose phosphate isomerase 5

Glyceraldehyde-3-phosphate

Energy-Generating Reactions 6 to 10

In our discussion of glycolysis from this point, the two molecules of glyceraldehyde-3-phosphate produced in step 5 are undergoing the same reactions. For simplicity, we show the structures and reactions for only one three-carbon molecule for reactions 6 to 10.

Glyceraldehyde-3-phosphate

Reaction 6 Oxidation and Phosphorylation

The aldehyde group of each glyceraldehyde-3-phosphate is oxidized to a carboxyl group, while the coenzyme NAD^+ is reduced to NADH and H^+. A phosphate group (P_i) adds to each of the new carboxyl groups to form two molecules of the high-energy compound 1,3-bisphosphoglycerate.

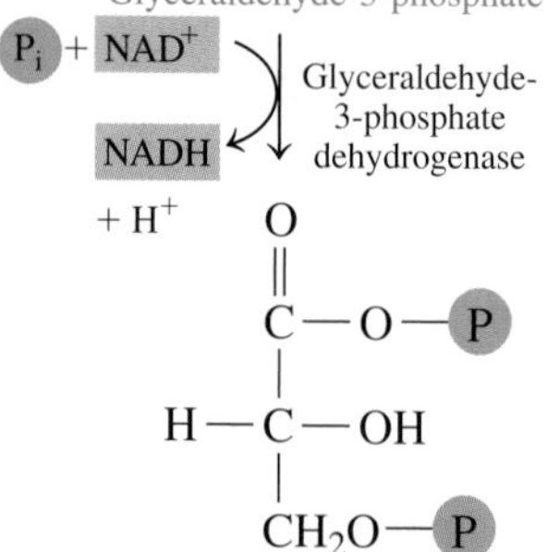

1,3-Bisphosphoglycerate

Reaction 7 Phosphate Transfer

A phosphate group from each 1,3-bisphosphoglycerate is transferred to two ADP molecules, yielding two molecules of the high-energy compound ATP. At this point in glycolysis, two ATP are produced, which balance the two ATP consumed in reactions 1 and 3.

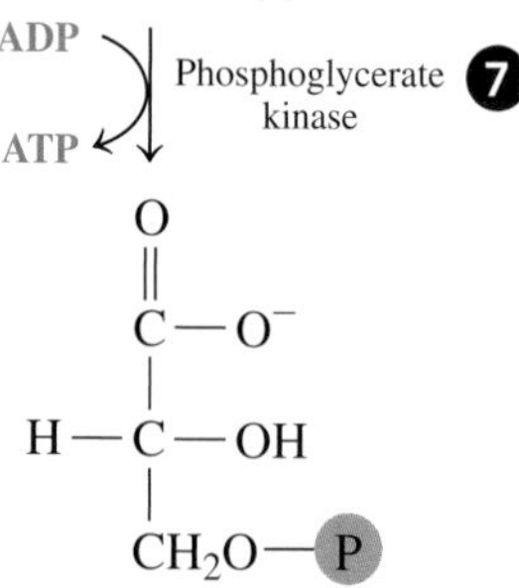

3-Phosphoglycerate

Reaction 8 Isomerization

Two 3-phosphoglycerate molecules undergo isomerization, which moves the phosphate group from carbon 3 to carbon 2, yielding two molecules of 2-phosphoglycerate.

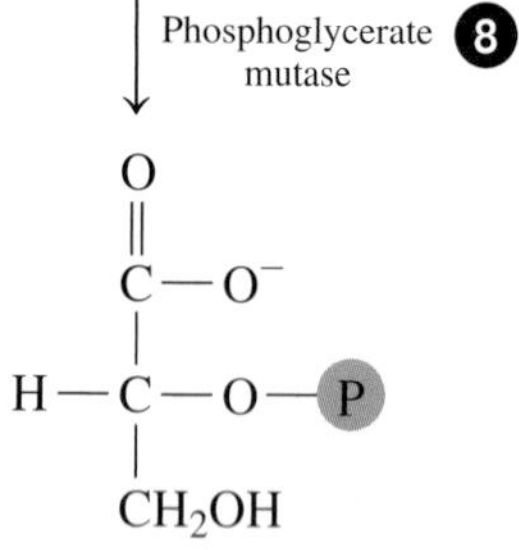

2-Phosphoglycerate

Reaction 9 Dehydration

Each of the phosphoglycerate molecules undergoes dehydration (loss of water), producing two molecules of phosphoenolpyruvate, a high-energy compound.

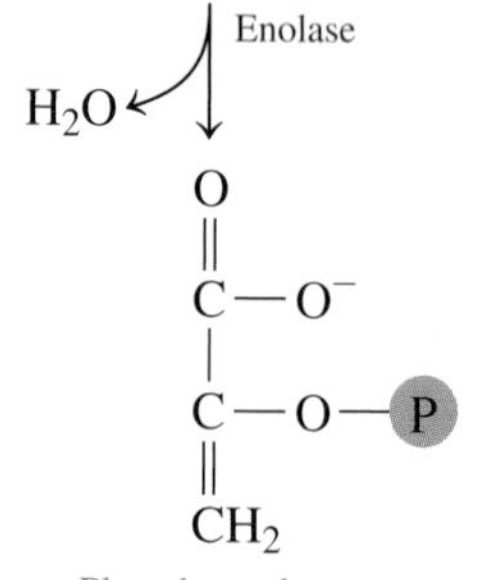

Phosphoenolpyruvate

Reaction 10 Phosphate Transfer

In a second direct phosphorylation, phosphate groups from two phosphoenolpyruvate molecules are transferred to two ADP to form two pyruvate and two ATP.

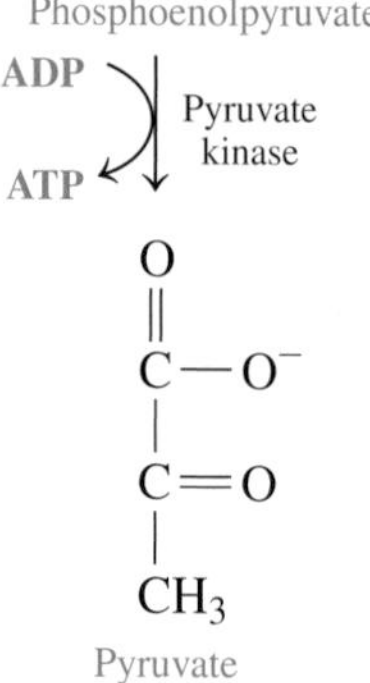

Pyruvate

SAMPLE PROBLEM 18.4 Reactions in Glycolysis

TRY IT FIRST

Identify each of the following reactions as an isomerization, phosphorylation, dehydration, or cleavage:

a. A phosphate group is transferred to ADP to form ATP.
b. 3-Phosphoglycerate is converted to 2-phosphoglycerate.
c. Water is removed from 2-phosphoglycerate.

SOLUTION

a. The transfer of a phosphate group to ADP to form ATP is phosphorylation.
b. The change in location of a phosphate group on a carbon chain is isomerization.
c. The loss of water is dehydration.

STUDY CHECK 18.4

Identify the reaction in which fructose-1,6-bisphosphate splits to form two three-carbon compounds as an isomerization, phosphorylation, dehydration, or cleavage.

ANSWER

The splitting of fructose-1,6-bisphosphate is cleavage.

TEST

Try Practice Problems 18.23 to 18.28

Summary of Glycolysis

In the glycolysis pathway, a six-carbon glucose molecule is converted to two three-carbon pyruvate molecules. Initially, two ATP are required to form fructose-1,6-bisphosphate. In later reactions, phosphate transfers produce a total of four ATP. Overall, glycolysis yields two ATP and two NADH when a glucose molecule is converted to two pyruvate.

$$\underset{\text{Glucose}}{C_6H_{12}O_6} + 2NAD^+ + 2ADP + 2P_i \longrightarrow \underset{\text{Pyruvate}}{2CH_3-\overset{\overset{\displaystyle O}{\|}}{C}-COO^-} + 2NADH + 2ATP + 4H^+ + 2H_2O$$

Pathways for Pyruvate

The pyruvate produced from glucose can now enter pathways that continue to extract energy. The available pathway depends on whether there is sufficient oxygen in the cell. Under **aerobic** conditions, oxygen is available to convert pyruvate to acetyl coenzyme A (acetyl CoA). When oxygen levels are low, pyruvate is reduced to lactate (see **FIGURE 18.11**).

Aerobic Conditions

In glycolysis, two ATP were generated when one glucose molecule was converted to two pyruvate. However, much more energy is obtained from glucose when oxygen levels are high in the cells. Under aerobic conditions, pyruvate moves from the cytoplasm into the mitochondria to be oxidized further. In a complex reaction, pyruvate is oxidized, and a carbon atom is removed from pyruvate as CO_2. The coenzyme NAD^+ is reduced during the oxidation. The resulting two-carbon acetyl compound is attached to CoA, producing acetyl CoA, an important intermediate in many metabolic pathways.

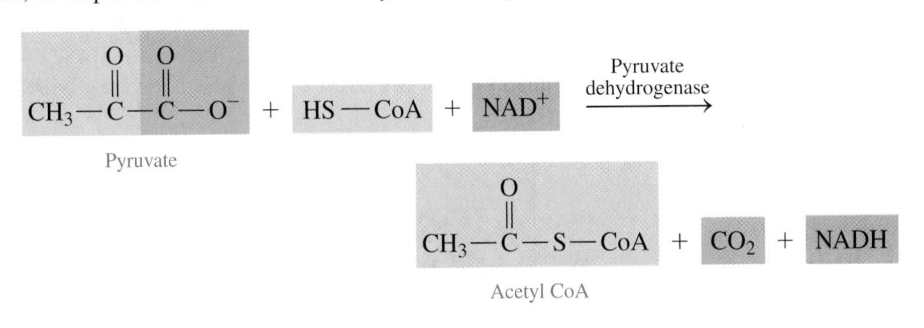

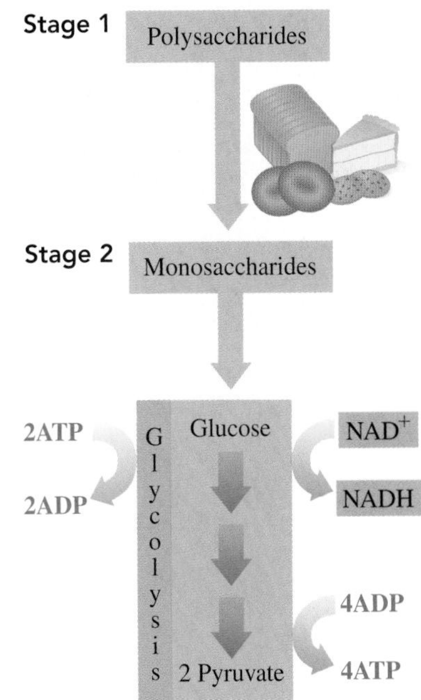

Glucose obtained from the digestion of polysaccharides is degraded in glycolysis to pyruvate.

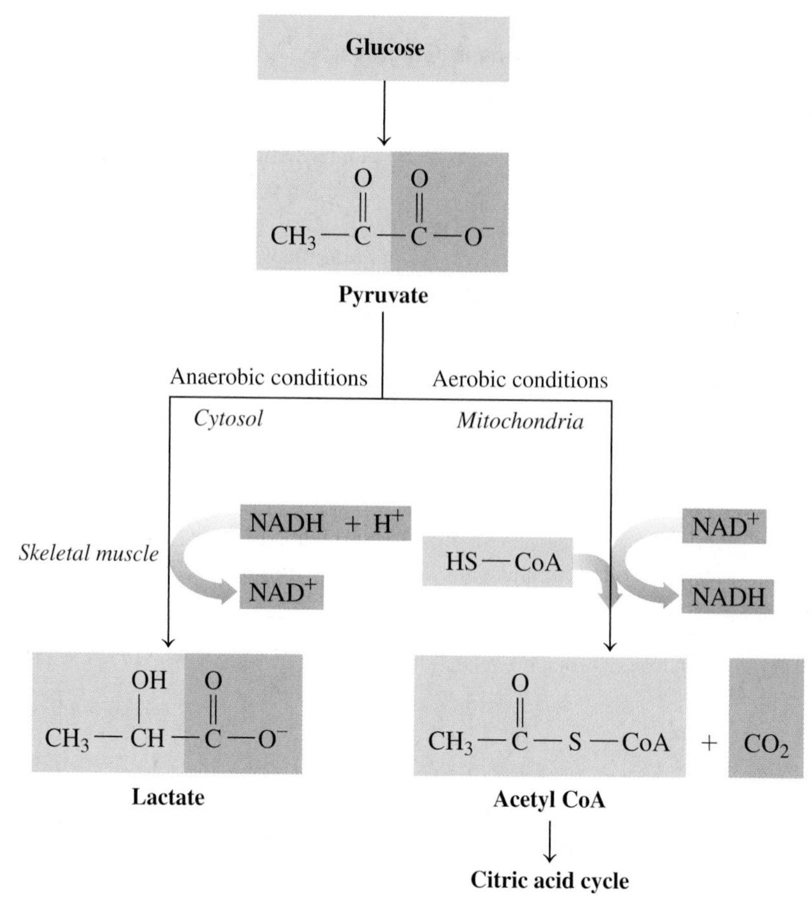

FIGURE 18.11 Pyruvate is converted to acetyl CoA under aerobic conditions and to lactate under anaerobic conditions.

Q During vigorous exercise, why does lactate accumulate in the muscles?

Anaerobic Conditions

After vigorous exercise, rapid breathing helps to repay the oxygen debt.

When we engage in strenuous exercise, the oxygen stored in our muscle cells is quickly depleted. Under anaerobic conditions, pyruvate remains in the cytoplasm where it is reduced to lactate. NAD^+ is produced and is used to oxidize more glyceraldehyde-3-phosphate in the glycolysis pathway, which produces a small but needed amount of ATP.

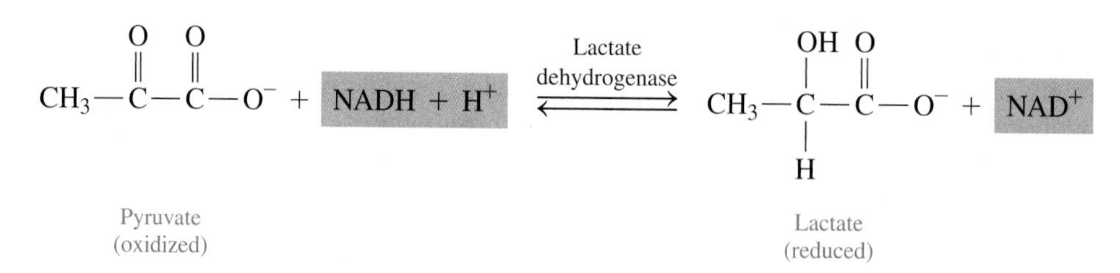

The accumulation of lactate causes the muscles to tire and become sore. After exercise, a person continues to breathe rapidly to repay the *oxygen debt* incurred during exercise. Most of the lactate is transported to the liver, where it is converted back into pyruvate.

TEST

Try Practice Problems 18.29 to 18.34

$$C_6H_{12}O_6 + 2ADP + 2P_i \longrightarrow 2CH_3{-}CH(OH){-}COO^- + 2ATP$$

Glucose → Lactate

PRACTICE PROBLEMS

18.4 Glycolysis: Oxidation of Glucose

LEARNING GOAL Describe the conversion of glucose to pyruvate in glycolysis and the subsequent conversion of pyruvate to acetyl CoA or lactate.

18.19 What is the starting compound of glycolysis?

18.20 What is the three-carbon end product of glycolysis?

18.21 How is ATP used in the initial steps of glycolysis?

18.22 How many ATP are used in the initial steps of glycolysis?

18.23 How does phosphorylation account for the production of ATP in glycolysis?

18.24 Why are there two ATP formed for one molecule of glucose?

18.25 What three-carbon intermediates are obtained when fructose-1,6-bisphosphate splits?

18.26 Why does one of the three-carbon intermediates undergo isomerization?

18.27 How many ATP or NADH are produced (or required) in each of the following steps in glycolysis?
a. glucose to glucose-6-phosphate
b. glyceraldehyde-3-phosphate to 1,3-bisphosphoglycerate
c. glucose to pyruvate

18.28 How many ATP or NADH are produced (or required) in each of the following steps in glycolysis?
a. 1,3-bisphosphoglycerate to 3-phosphoglycerate
b. fructose-6-phosphate to fructose-1,6-bisphosphate
c. phosphoenolpyruvate to pyruvate

18.29 What condition is needed in the cell to convert pyruvate to acetyl CoA?

18.30 What coenzymes are needed for the oxidation of pyruvate to acetyl CoA?

18.31 Write the overall equation for the conversion of pyruvate to acetyl CoA.

18.32 What is the product of pyruvate under anaerobic conditions?

18.33 How does the formation of lactate permit glycolysis to continue under anaerobic conditions?

18.34 After running a marathon, a runner has muscle pain and cramping. What might have occurred in the muscle cells to cause this?

18.5 The Citric Acid Cycle

LEARNING GOAL Describe the oxidation of acetyl CoA in the citric acid cycle.

The citric acid cycle is a series of reactions that connects the intermediate acetyl CoA from the catabolic pathways in stage 2 with electron transport and the synthesis of ATP in stage 3. As a central pathway in metabolism, the **citric acid cycle** uses the two-carbon acetyl group of acetyl CoA to produce CO_2 and the reduced coenzymes NADH and $FADH_2$. The citric acid cycle is named for the six-carbon citrate ion from citric acid ($C_6H_8O_7$), a tricarboxylic acid, which forms in the first reaction. The citric acid cycle is also known as the *tricarboxylic acid (TCA) cycle* or the *Krebs cycle*, named for H. A. Krebs, who received a Nobel Prize in 1953 for its discovery.

Overview of the Citric Acid Cycle

Six carbons move through the eight reactions of the citric acid cycle, with each cycle producing oxaloacetate (four carbons) and $2CO_2$ (see **FIGURE 18.12**). Each turn of the cycle

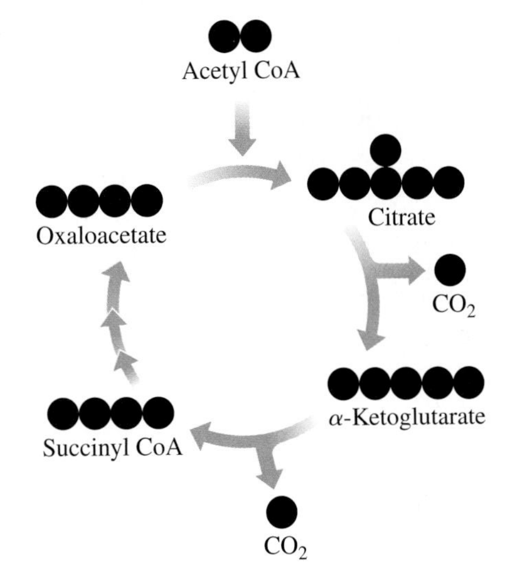

FIGURE 18.12 In the citric acid cycle, two carbon atoms are removed as CO_2 from six-carbon citrate to give four-carbon succinyl CoA, which is converted to four-carbon oxaloacetate.

Q How many carbon atoms are removed in one turn of the citric acid cycle?

includes four oxidation reactions, producing the reduced coenzymes NADH or $FADH_2$ from the energy released during the reactions. One GTP (converted to ATP in the cell) is also produced during the citric acid cycle. **FIGURE 18.13** shows the complete citric acid cycle.

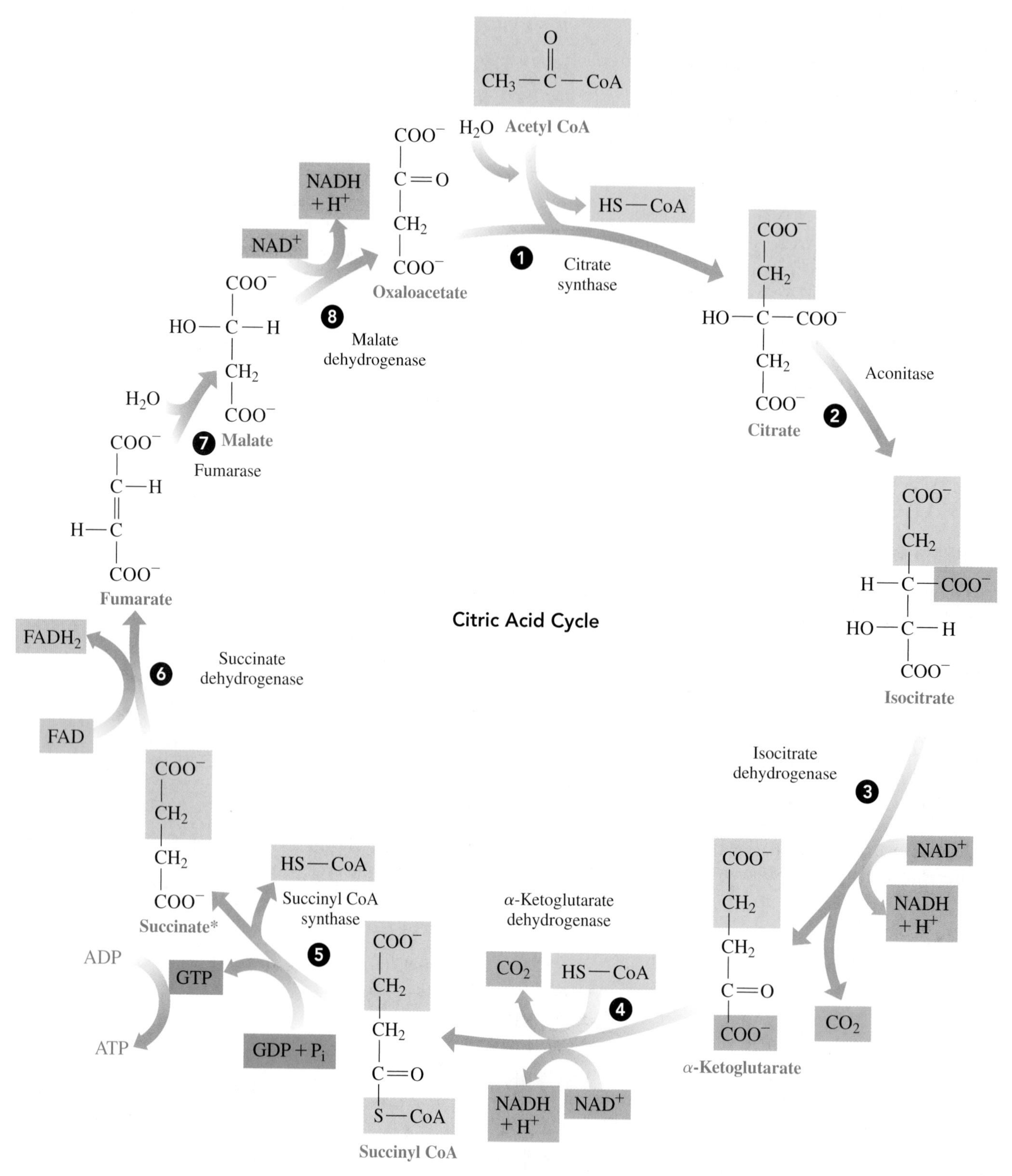

FIGURE 18.13 Each turn of the citric acid cycle regenerates oxaloacetate and produces 2 CO_2, 1 GTP, 3 NADH, and 1 $FADH_2$.

Q How many reactions in the citric acid cycle produce a reduced coenzyme?

Reaction 1 Formation of Citrate

In the first reaction of the citric acid cycle, the acetyl group (2C) from acetyl CoA bonds with oxaloacetate (4C) to yield citrate (6C).

CORE CHEMISTRY SKILL

Describing the Reactions in the Citric Acid Cycle

Reaction 2 Isomerization

The citrate produced in reaction 1 contains a tertiary alcohol group that cannot be oxidized further. In reaction 2, citrate undergoes isomerization to yield isocitrate, which provides a secondary alcohol group that can be oxidized in the next reaction.

Reaction 3 Oxidation and Decarboxylation

In reaction 3, an oxidation and a *decarboxylation* occur together. The secondary alcohol group in isocitrate is oxidized to a ketone. A **decarboxylation** converts a carboxylate group ($-COO^-$) to a CO_2 molecule producing α-ketoglutarate, which has five carbon atoms. The oxidation reaction also produces hydrogen ions and electrons that reduce NAD^+ to NADH and H^+. This reduced coenzyme NADH will be important in the energy-releasing reactions in electron transport.

Reaction 4 Oxidation and Decarboxylation

In reaction 4, α-ketoglutarate undergoes decarboxylation and oxidation to produce a four-carbon group that combines with CoA to form succinyl CoA (4C). As in reaction 3, this oxidation reaction also produces hydrogen ions and electrons that reduce NAD^+ to NADH and H^+. This forms another reduced NADH that will be important in the energy-releasing reactions in electron transport.

Reaction 5 Hydrolysis

In reaction 5, succinyl CoA undergoes hydrolysis to give succinate and CoA. The energy released is used to add a phosphate group (P_i) to GDP (guanosine diphosphate), which yields GTP (guanosine triphosphate), a high-energy compound similar to ATP.

Eventually, the GTP undergoes hydrolysis with a release of energy that is used to add a phosphate group to ADP to form ATP. This is the only time in the citric acid cycle that ATP is produced by a direct transfer of phosphate.

Reaction 6 Oxidation

In reaction 6, hydrogen is removed from each of two carbon atoms in succinate, which produces fumarate, a compound with a trans double bond. The formation of a carbon–carbon double bond ($C{=}C$) produces 2H that are used to reduce the coenzyme FAD to $FADH_2$. This reduced coenzyme $FADH_2$ is important in the energy-releasing reactions in electron transport.

Reaction 7 Hydration

In reaction 7, a hydration adds water to the double bond of fumarate to yield malate, which is a secondary alcohol.

Reaction 8 Oxidation

In reaction 8, the last step of the citric acid cycle, the secondary alcohol group in malate is oxidized to a carbonyl group ($C{=}O$), yielding oxaloacetate. For the third time in the citric acid cycle, an oxidation provides hydrogen ions and electrons for the reduction of NAD^+ to NADH and H^+.

ENGAGE

Which steps of the citric acid cycle involve an oxidation *without* a decarboxylation?

Summary of Products from the Citric Acid Cycle

We have seen that the citric acid cycle begins when a two-carbon acetyl group from acetyl CoA combines with a four-carbon oxaloacetate to form a six-carbon citrate. Through

Products from One Turn of the Citric Acid Cycle
2 CO_2
3 NADH and $3H^+$
1 $FADH_2$
1 GTP (1 ATP)
1 HS—CoA

oxidation, reduction, and decarboxylation, two carbon atoms are removed from citrate to yield two CO_2 and a four-carbon compound that undergoes reactions to regenerate oxaloacetate.

In the four oxidation reactions of one turn of the citric acid cycle, three NAD^+ are reduced to three NADH and one FAD is reduced to one $FADH_2$. One GDP is converted to GTP, which is used to convert one ADP to ATP. We can write an overall chemical equation for one complete turn of the citric acid cycle as follows:

$$\text{Acetyl CoA} + 3NAD^+ + FAD + GDP + P_i + 2H_2O \longrightarrow \text{HS—CoA} + 3NADH + 3H^+ + FADH_2 + GTP + 2CO_2$$

SAMPLE PROBLEM 18.5 Citric Acid Cycle

TRY IT FIRST

When one acetyl CoA completes the citric acid cycle, how many of each of the following is produced?

a. NADH **b.** ketone group **c.** CO_2

SOLUTION

a. One turn of the citric acid cycle produces three molecules of NADH.
b. Two ketone groups form when the secondary alcohol groups in isocitrate and malate are oxidized by NAD^+.
c. Two molecules of CO_2 are produced by the decarboxylation of isocitrate and α-ketoglutarate.

STUDY CHECK 18.5

What compound is a substrate in the first reaction of the citric acid cycle and a product in the last reaction?

ANSWER

oxaloacetate

TEST

Try Practice Problems 18.35 to 18.42

PRACTICE PROBLEMS

18.5 The Citric Acid Cycle

LEARNING GOAL Describe the oxidation of acetyl CoA in the citric acid cycle.

18.35 What are the products from one turn of the citric acid cycle?

18.36 What compounds are needed to start the citric acid cycle?

18.37 Identify the reaction(s) of the citric acid cycle that involve(s)
a. oxidation and decarboxylation
b. dehydration
c. reduction of NAD^+

18.38 Identify the reaction(s) of the citric acid cycle that involve(s)
a. reduction of FAD
b. direct phosphate transfer
c. hydration

18.39 Which reaction(s) of the citric acid cycle involve(s) the production of a carbon–carbon double bond?

18.40 What is the total NADH and total $FADH_2$ produced in one turn of the citric acid cycle?

18.41 Refer to the diagram of the citric acid cycle in Figure 18.13 to answer each of the following:
a. What are the six-carbon compounds?
b. How is the number of carbon atoms decreased?
c. What is the five-carbon compound?
d. What are the decarboxylation reactions?

18.42 Refer to the diagram of the citric acid cycle in Figure 18.13 to answer each of the following:
a. What is the yield of CO_2 molecules?
b. What are the four-carbon compounds?
c. What is the yield of GTP molecules?
d. In which reactions are secondary alcohols oxidized?

18.6 Electron Transport and Oxidative Phosphorylation

LEARNING GOAL Describe electron transport and the process of oxidative phosphorylation; calculate the ATP from the complete oxidation of glucose.

At this point in stage 3 of catabolism, for each glucose molecule that completes glycolysis, the oxidation of two pyruvate, and the citric acid cycle, four ATP along with 10 NADH and two $FADH_2$ are produced.

From One Glucose	ATP	Reduced Coenzymes	
Glycolysis	2	2 NADH	
Oxidation of 2 Pyruvate		2 NADH	
Citric Acid Cycle with 2 Acetyl CoA	2	6 NADH	2 $FADH_2$
Total for One Glucose	**4**	**10 NADH**	**2 $FADH_2$**

Electron Transport

In **electron transport**, hydrogen ions and electrons from NADH and $FADH_2$ are passed from one electron carrier to the next until they combine with oxygen to form H_2O. The energy released during electron transport is used to synthesize ATP from ADP and P_i, a process called *oxidative phosphorylation*. As long as oxygen is available for the mitochondria in the cell, electron transport and oxidative phosphorylation function to synthesize most of the ATP produced in the cell.

A mitochondrion consists of an outer membrane, an intermembrane space, and an inner membrane that surrounds the matrix. Along the highly folded inner membrane are the enzymes and electron carriers required for electron transport. Embedded within these membranes are four distinct protein complexes, labeled I, II, III, and IV. Two electron carriers, *coenzyme Q* and *cytochrome c*, are not firmly attached to the membrane. They function as mobile carriers shuttling electrons between the protein complexes that are bound to the inner membrane (see **FIGURE 18.14**).

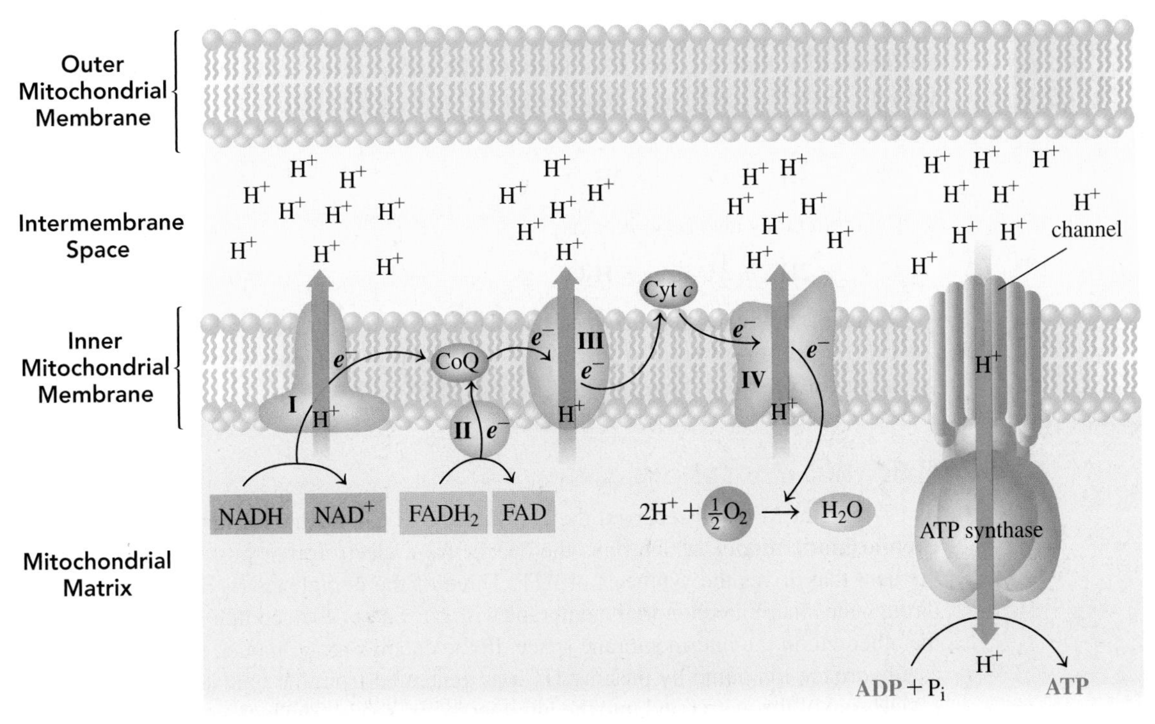

FIGURE 18.14 In electron transport, coenzymes NADH and $FADH_2$ are oxidized in enzyme complexes, providing electrons and hydrogen ions for ATP synthesis.

Q What pathway is the major source of NADH for electron transport?

ENGAGE

Which electron transport complex accepts electrons from NADH?

Complex I

Electron transport begins when hydrogen ions and electrons are transferred from NADH to complex I. The loss of hydrogen from NADH regenerates NAD^+, which becomes available to oxidize more substrates in oxidative pathways such as the citric acid cycle. The hydrogen ions and electrons are transferred to the mobile electron carrier coenzyme Q (CoQ), which carries electrons to complex II. The overall reaction in complex I is written as follows:

$$\mathrm{NADH + H^+} + \mathrm{CoQ} \longrightarrow \mathrm{NAD^+} + \mathrm{CoQ\mathbf{H_2}}$$

Complex II

In complex II, CoQ obtains hydrogen ions and electrons directly from $FADH_2$ that was generated by the conversion of succinate to fumarate in the citric acid cycle. This produces $CoQH_2$ and regenerates the oxidized coenzyme FAD, which becomes available again to oxidize more substrates in oxidative pathways.

ENGAGE

Which electron transport complex accepts electrons from $FADH_2$?

The overall reaction in complex II is written as follows:

$$\mathrm{FADH_2} + \mathrm{CoQ} \longrightarrow \mathrm{FAD} + \mathrm{CoQ\mathbf{H_2}}$$

Because complex II is at a lower energy level than complex I, the electrons from $FADH_2$ enter electron transport at a lower energy level than those from NADH.

Complex III

In complex III, two electrons are transferred from the mobile carrier $CoQH_2$ to a series of iron-containing proteins called *cytochromes* and eventually to cytochrome *c*, which is another mobile electron carrier. Cytochrome *c*, which contains Fe^{3+}/Fe^{2+}, is reduced when it gains an electron and oxidized when an electron is lost. As a mobile carrier, cytochrome *c* carries electrons to complex IV. The overall reaction in complex III is written as follows:

$$\mathrm{CoQH_2} + 2\mathrm{cyt}\ c\ (\mathrm{Fe^{3+}}) \longrightarrow \mathrm{CoQ} + \mathbf{2H^+} + 2\mathrm{cyt}\ c\ (\mathrm{Fe^{2+}})$$

(Oxidized) (Reduced)

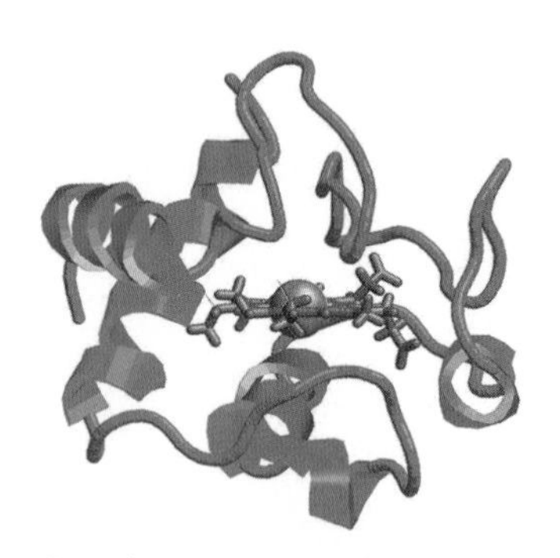

Cytochrome *c* carries one electron when Fe^{3+} is reduced to Fe^{2+} (orange sphere). The heme group (gray) holds the iron in place.

Complex IV

At complex IV, four electrons from four cytochrome *c* are passed to other electron carriers until the electrons combine with hydrogen ions and oxygen (O_2) to form two molecules of water. The overall reaction in complex IV is written as follows:

$$4\,e^- + \mathbf{4H^+} + \mathrm{O_2} \longrightarrow \mathbf{2H_2O}$$

This equation may also be written as:

$$2\,e^- + \mathbf{2H^+} + \tfrac{1}{2}\mathrm{O_2} \longrightarrow \mathbf{H_2O}$$

Overall, the reduced coenzymes NADH and $FADH_2$ from the citric acid cycle enter electron transport to provide hydrogen ions and electrons that react with oxygen, producing water and the oxidized coenzymes NAD^+ and FAD.

Oxidative Phosphorylation

In 1978, Peter Mitchell received the Nobel Prize in Chemistry for his theory called the **chemiosmotic model**, which links the energy from electron transport to a hydrogen ion gradient that drives the synthesis of ATP. Three of the complexes (I, III, and IV) extend through the inner mitochondrial membrane with one end of each complex in the matrix and the other end in the intermembrane space. In the chemiosmotic model, each complex acts as a hydrogen ion pump by pushing H^+ ions generated from the oxidation of NADH and $FADH_2$ out of the matrix and into the intermembrane space. The increase in H^+ concentration lowers the pH in the intermembrane space and creates a H^+ gradient. Because H^+ ions are positively charged, the lower pH and the electrical charge of the H^+ gradient make it an *electrochemical gradient*.

TEST

Try Practice Problems 18.43 to 18.50

To equalize the pH and the electrical charge between the intermembrane space and the matrix, the H^+ ions must return to the matrix. However, H^+ ions cannot move through the inner membrane. Instead, H^+ must return to the matrix by passing through a protein complex called **ATP synthase**. We might think of the flow of H^+ ions as a river that turns a waterwheel. The flow of H^+ ions from the intermembrane space through ATP synthase provides energy that is used to convert ADP to ATP. This process of **oxidative phosphorylation** couples the energy from electron transport to the synthesis of ATP from ADP and P_i.

$$\text{ADP} + \text{P}_i + \text{energy} \xrightarrow{\text{ATP synthase}} \text{ATP}$$

ATP Synthesis

When NADH enters electron transport at complex I, the energy transferred from its oxidation can be used to synthesize 2.5 ATP. When $FADH_2$ enters electron transport at complex II, which is at a lower energy level, its oxidation provides energy for the synthesis of 1.5 ATP. Older values of 3 ATP from NADH and 2 ATP from $FADH_2$ are often used in biology. However, current research indicates that the oxidation of one NADH yields 2.5 ATP and one $FADH_2$ yields 1.5 ATP.

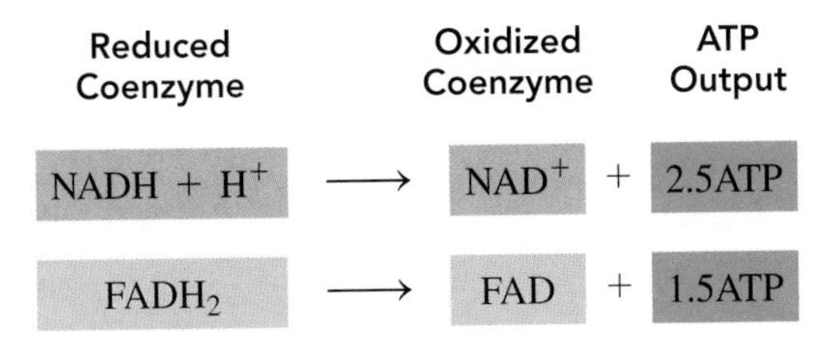

SAMPLE PROBLEM 18.6 ATP Synthesis

TRY IT FIRST

Why does the oxidation of NADH provide energy for the formation of 2.5 ATP whereas $FADH_2$ produces 1.5 ATP?

SOLUTION

Electrons from the oxidation of NADH enter electron transport at complex I. They pass through three complexes (I, III, and IV) pumping H^+ ions from the matrix into the intermembrane space. Electrons from $FADH_2$ enter electron transport at complex II, pumping H^+ ions through only two complexes (III and IV).

STUDY CHECK 18.6

Which complexes in electron transport act as hydrogen ion pumps?

ANSWER

The protein complexes I, III, and IV pump H^+ ions from the matrix to the intermembrane space.

TEST

Try Practice Problems 18.51 to 18.54

ATP from Glycolysis

In glycolysis, the oxidation of glucose stores energy in two NADH molecules as well as two ATP from direct phosphate transfer.

Therefore, in glycolysis, glucose yields a total of seven ATP: five ATP from two NADH, and two ATP from phosphorylation. In some cases, less than five ATP is produced from the two NADH from glycolysis because of the way these NADH molecules are transported from the cytosol into the mitochondrial matrix. To simplify our calculations of the ATP produced from glucose oxidation, we will not consider this potential reduction in ATP production.

$$\text{Glucose} \longrightarrow 2\ \text{pyruvate} + 2\text{ATP} + 2\text{NADH}$$

$$\text{Glucose} \longrightarrow 2\ \text{pyruvate} + 7\text{ATP}$$

ATP from the Oxidation of Two Pyruvate

Under aerobic conditions, pyruvate enters the mitochondria, where it is oxidized to give acetyl CoA, CO_2, and NADH. Because glucose yields two pyruvate, two NADH enter electron transport, where the oxidation of two pyruvate leads to the production of five ATP.

$$2\ \text{Pyruvate} \longrightarrow 2\ \text{acetyl CoA} + 2CO_2 + \boxed{5\text{ATP}}$$

ATP from the Citric Acid Cycle

One turn of the citric acid cycle produces two CO_2, three NADH, one $FADH_2$, and one ATP by direct phosphate transfer.

$$\begin{aligned} 3\ \text{NADH} \times 2.5\ \text{ATP/NADH} &= 7.5\ \text{ATP} \\ 1\ FADH_2 \times 1.5\ \text{ATP}/FADH_2 &= 1.5\ \text{ATP} \\ 1\ \text{GTP} \times 1\ \text{ATP/1 GTP} &= 1\ \text{ATP} \\ \hline \text{Total (one turn)} &= 10\ \text{ATP} \end{aligned}$$

Because one glucose produces two acetyl CoA, two turns of the citric acid cycle produce a total of 20 ATP.

$$\text{Acetyl CoA} \longrightarrow 2CO_2 + 10\text{ATP (one turn of citric acid cycle)}$$
$$2\ \text{Acetyl CoA} \longrightarrow 4CO_2 + 20\text{ATP (two turns of citric acid cycle)}$$

CHEMISTRY LINK TO HEALTH

ATP Synthase and Heating the Body

Some types of compounds, called *uncouplers*, separate the electron transport system from ATP synthase. They do this by providing an alternate route for hydrogen ions to return to the matrix. The electrons are transported to O_2 in electron transport, but ATP is not formed by ATP synthase.

Some uncouplers transport the H^+ ions through the inner mitochondrial membrane, which is normally impermeable to H^+ ions; others block the channel in ATP synthase. Compounds such as dicumarol and 2,4-dinitrophenol (DNP) are hydrophobic and bind with H^+ ions to carry them across the inner membrane. An antibiotic, oligomycin, blocks the channel, which does not allow any H^+ ions to return to the matrix. By removing H^+ ions or blocking the channel, there is no H^+ flow to drive ATP synthesis.

Dicumarol

2,4-Dinitrophenol (DNP)

When there is no mechanism for ATP synthesis, the energy of electron transport is released as heat. Certain animals that are adapted to cold climates have developed their own uncoupling system, which allows them to use electron transport energy for heat production. These animals have large amounts of a tissue called *brown fat*, which contains a high concentration of mitochondria. This tissue is brown because of the iron in the cytochromes of the mitochondria. The hydrogen ion pumps still operate in brown fat, but a protein embedded in the inner mitochondrial membrane allows the H^+ ions to bypass ATP synthase. The energy that would be used to synthesize ATP is released as heat.

In newborn babies, brown fat is used to generate heat because newborns have a small mass but a large surface area and need to produce more heat than adults. The brown fat deposits are located near major blood vessels, which carry the warmed blood to the body. Most adults have little or no brown fat, although someone who works outdoors in a cold climate will develop some brown fat deposits.

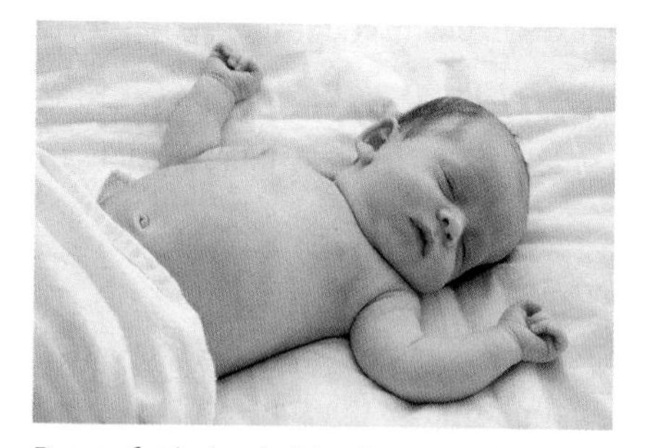

Brown fat helps babies keep warm.

CORE CHEMISTRY SKILL

Calculating the ATP Produced from Glucose

ATP from the Complete Oxidation of Glucose

The total ATP for the complete oxidation of glucose is calculated by combining the ATP produced from glycolysis, the oxidation of pyruvate, and the citric acid cycle (see **FIGURE 18.15**). The ATP produced for these reactions is given in **TABLE 18.3**.

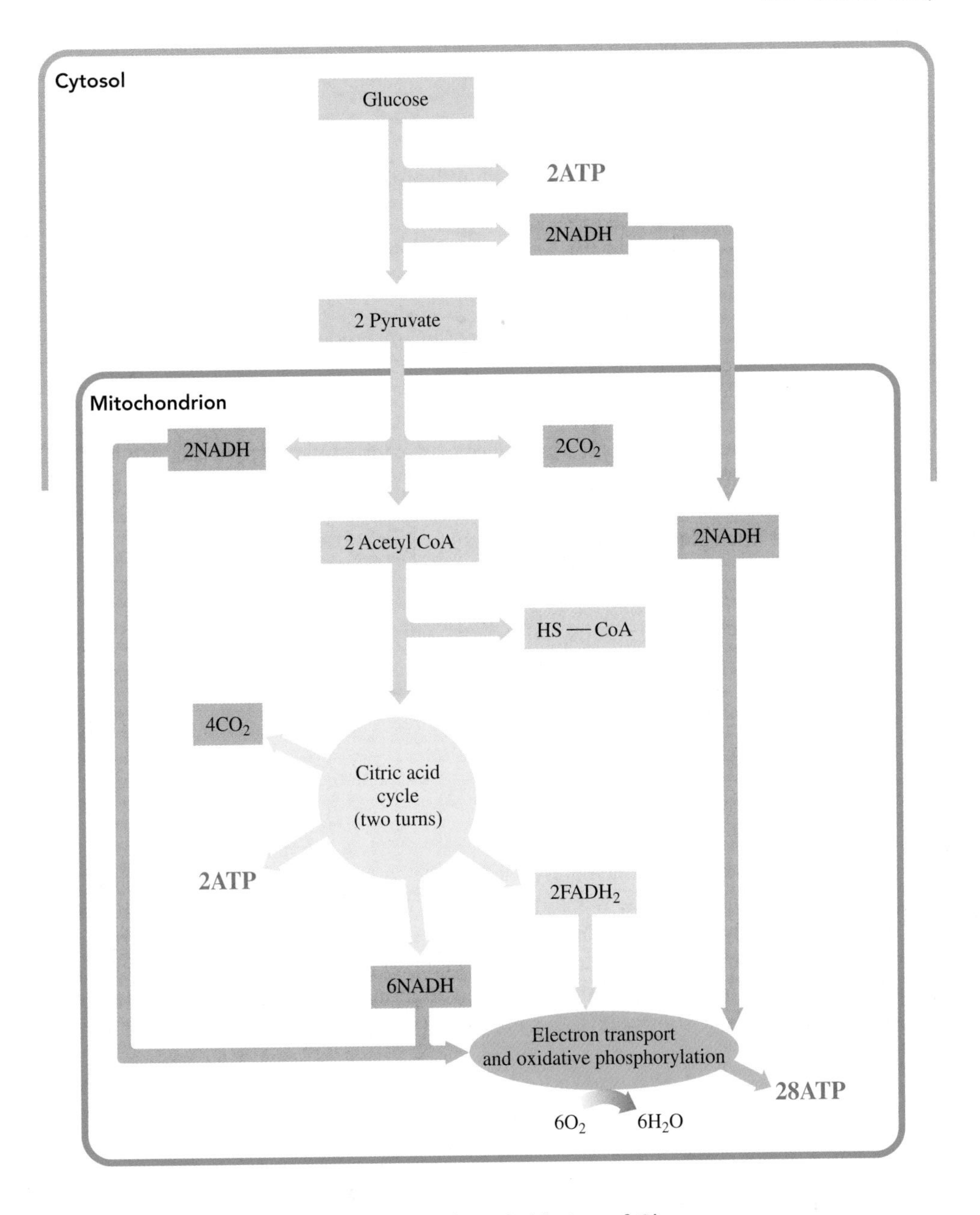

FIGURE 18.15 The complete oxidation of glucose to CO_2 and H_2O yields a total of 32 ATP.

Q What metabolic pathway produces most of the ATP from the oxidation of glucose?

TABLE 18.3 ATP Produced by the Complete Oxidation of Glucose

Pathway	Reaction	Coenzymes	ATP Yield
Glycolysis	Oxidation of glyceraldehyde-3-phosphate	2 NADH	5 ATP
	Direct phosphorylation (2 triose phosphate)		2 ATP
	Summary: $\underset{\text{Glucose}}{C_6H_{12}O_6} \longrightarrow 2\text{ pyruvate} + 2H_2O$		7 ATP
Oxidation and Decarboxylation	$2\text{ Pyruvate} \longrightarrow 2\text{ acetyl CoA} + 2CO_2$	2 NADH	5 ATP
Citric Acid Cycle (two turns)	Oxidation of 2 isocitrate	2 NADH	5 ATP
	Oxidation of 2 α-ketoglutarate	2 NADH	5 ATP
	2 Direct phosphate transfers (2 GTP)		2 ATP
	Oxidation of 2 succinate	2 $FADH_2$	3 ATP
	Oxidation of 2 malate	2 NADH	5 ATP
	Summary: $2\text{ Acetyl CoA} \longrightarrow 4CO_2 + 2H_2O$		20 ATP
Total Yield	$\underset{\text{Glucose}}{C_6H_{12}O_6} + 6O_2 \longrightarrow 6CO_2 + 6H_2O$		32 ATP

SAMPLE PROBLEM 18.7 ATP Production

TRY IT FIRST

How many ATP are produced during each of the following oxidations?

a. two pyruvate to two acetyl CoA
b. one glucose to two acetyl CoA

SOLUTION

a. The oxidation of two pyruvate to two acetyl CoA produces two NADH, which yields five ATP. We calculate this as:

2 NADH $\times$ 2.5 ATP/NADH = 5 ATP

b. Seven ATP are produced when one glucose is oxidized to two pyruvate, two by direct phosphorylation and five from the two NADH. Five ATP (from two NADH) are synthesized from the oxidation of two pyruvate to two acetyl CoA. Thus, a total of 12 ATP are produced when glucose is oxidized to yield two acetyl CoA.

STUDY CHECK 18.7

What are the sources of ATP for two turns of the citric acid cycle?

ANSWER

Six NADH provide 15 ATP, two $FADH_2$ provide three ATP, and two direct phosphate transfers provide two ATP.

TEST

Try Practice Problems 18.55 and 18.56

When glucose is not immediately used by the cells for energy, it is stored as glycogen in the liver and muscles. When the levels of glucose in the brain or blood become low, the glycogen reserves are hydrolyzed and glucose is released into the blood. If glycogen stores are depleted, some glucose can be synthesized from noncarbohydrate sources. It is the balance of all these reactions that maintains the necessary blood glucose level available to our cells and provides the necessary amount of ATP for our energy needs.

PRACTICE PROBLEMS

18.6 Electron Transport and Oxidative Phosphorylation

LEARNING GOAL Describe electron transport and the process of oxidative phosphorylation; calculate the ATP from the complete oxidation of glucose.

18.43 What reduced coenzymes provide the electrons for electron transport?

18.44 What happens to the energy level as electrons are passed along in electron transport?

18.45 How are electrons carried from complex I to complex III?

18.46 How are electrons carried from complex III to complex IV?

18.47 How is NADH oxidized in electron transport?

18.48 How is $FADH_2$ oxidized in electron transport?

18.49 What is meant by oxidative phosphorylation?

18.50 How is the H^+ gradient established?

18.51 According to the chemiosmotic theory, how does the H^+ gradient provide energy to synthesize ATP?

18.52 How does the phosphorylation of ADP occur?

18.53 How are glycolysis and the citric acid cycle linked to the production of ATP by electron transport?

18.54 Why does $FADH_2$ yield 1.5 ATP, using the electron transport system, but NADH yields 2.5 ATP?

18.55 What is the ATP energy yield associated with each of the following?
a. NADH $\longrightarrow$ NAD^+
b. glucose $\longrightarrow$ 2 pyruvate
c. 2 pyruvate $\longrightarrow$ 2 acetyl CoA + $2CO_2$
d. acetyl CoA $\longrightarrow$ $2CO_2$

18.56 What is the ATP energy yield associated with each of the following?
a. $FADH_2$ $\longrightarrow$ FAD
b. glucose + $6O_2$ $\longrightarrow$ $6CO_2$ + $6H_2O$
c. glucose $\longrightarrow$ 2 lactate
d. pyruvate $\longrightarrow$ lactate

18.7 Oxidation of Fatty Acids

LEARNING GOAL Describe the metabolic pathway of β oxidation; calculate the ATP from the complete oxidation of a fatty acid.

REVIEW

Writing Equations for Hydrogenation and Hydration (11.7)
Identifying Fatty Acids (15.2)

A large amount of energy is obtained when fatty acids undergo oxidation in the mitochondria to yield acetyl CoA. In stage 2 of fat catabolism, fatty acids undergo **beta oxidation (β oxidation)**, which removes two-carbon segments, one at a time, from the carboxyl end of the fatty acid.

β oxidation occurs here

$$CH_3-(CH_2)_{14}-\underset{\beta}{CH_2}-\underset{\alpha}{CH_2}-\overset{O}{\overset{\|}{C}}-OH$$

Stearic acid

Each cycle in β oxidation produces an acetyl CoA and a fatty acid that is shorter by two carbons. The cycle repeats until the original fatty acid is completely degraded to two-carbon acetyl CoA. Each acetyl CoA can then enter the citric acid cycle in the same way as the acetyl CoA derived from glucose.

Fatty Acid Activation

Fatty acids, which are produced in the cytosol, must be transported through the inner mitochondrial membrane before they can undergo β oxidation in the matrix. In an *activation* process, a fatty acid is combined with CoA to yield a fatty acyl CoA, which can be transported across the membrane. The energy for the activation is obtained from the hydrolysis of ATP to give AMP (adenosine monophosphate) and two phosphate ($2P_i$) groups. This is equivalent to the energy released from the hydrolysis of two ATP to two ADP.

$$\underset{\text{Fatty acid}}{R-\underset{\beta}{CH_2}-\underset{\alpha}{CH_2}-\overset{O}{\overset{\|}{C}}-OH} + ATP + HS-CoA \longrightarrow \underset{\text{Fatty acyl CoA}}{R-\underset{\beta}{CH_2}-\underset{\alpha}{CH_2}-\overset{O}{\overset{\|}{C}}-S-CoA} + AMP + 2P_i + H_2O$$

Reactions of the β Oxidation Cycle

In the mitochondrial matrix, fatty acyl CoA molecules undergo β oxidation, which is a cycle of four reactions that convert the $-CH_2-$ of the β-carbon (carbon 3) of a long-chain fatty acyl group into a keto group. Once the β-keto group is formed, a two-carbon acetyl group can be split from the carbon chain, which shortens the fatty acyl chain. The reaction for one cycle of β oxidation is written as follows:

$$\underset{\text{Fatty acyl CoA}}{R-\underset{\beta}{CH}-\underset{\alpha}{CH_2}-\overset{O}{\overset{\|}{C}}-S-CoA} + NAD^+ + FAD + H_2O + HS-CoA \longrightarrow$$

$$\underset{\text{New fatty acyl CoA (−2C)}}{R-\overset{O}{\overset{\|}{C}}-S-CoA} + \underset{\text{Acetyl CoA}}{CH_3-\overset{O}{\overset{\|}{C}}-S-CoA} + NADH + H^+ + FADH_2$$

The specific reactions in β oxidation are described for capric acid (C_{10}) (see **FIGURE 18.16**).

Reaction 1 Oxidation

In the first reaction of β oxidation, two hydrogen atoms are transferred from the α- and β-carbons of the activated fatty acid to FAD, yielding a trans carbon–carbon double bond and $FADH_2$.

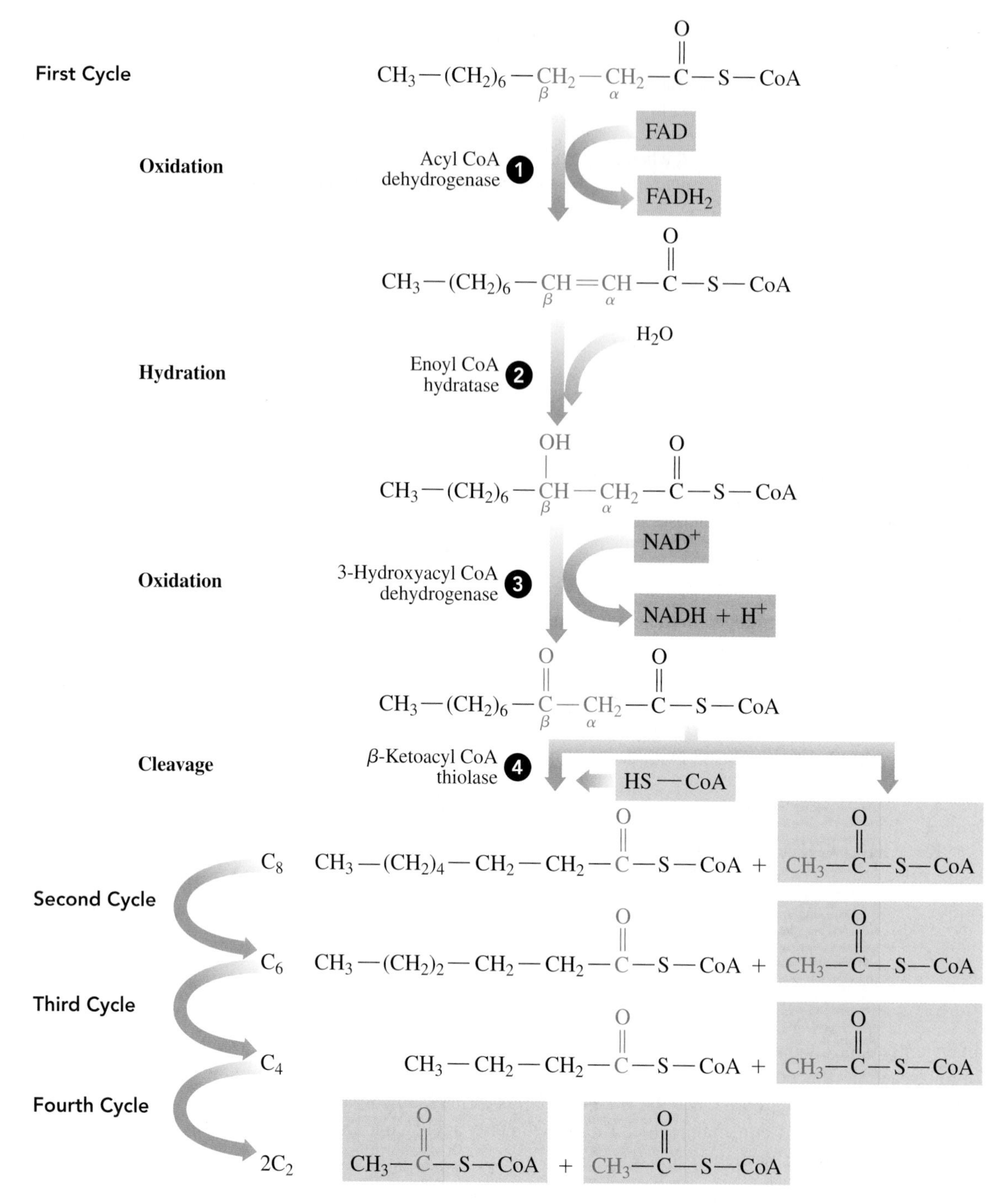

FIGURE 18.16 Capric acid (C_{10}) undergoes four oxidation cycles that repeat reactions 1 to 4 and yield five acetyl CoA molecules, four NADH, and four $FADH_2$.

Q How many NADH and $FADH_2$ molecules are produced in one cycle of β oxidation?

Reaction 2 Hydration

A hydration reaction adds the components of water to the trans double bond, which forms a secondary hydroxyl group (—OH) on the β-carbon.

Reaction 3 Oxidation

The secondary hydroxyl group (—OH) on the β-carbon is oxidized to yield a ketone. The hydrogen atoms removed in the oxidation are transferred to NAD^+ to yield the reduced coenzyme NADH and H^+.

Reaction 4 Cleavage

In the final step of β oxidation, the C_α—C_β bond is cleaved to yield a two-carbon acetyl CoA and a fatty acyl CoA that has been shortened by two carbons. The 8C fatty acyl CoA repeats the β oxidation cycle where it is cleaved to a 6C fatty acyl CoA, and then to a 4C fatty acyl CoA, which splits to give two acetyl CoA. All the acetyl CoA produced from the fatty acid can then be oxidized through the reactions of the citric acid cycle.

For capric acid, the overall equation for the four cycles of β oxidation is

$$\text{Caproyl } (C_{10}) \text{ CoA} + 4\text{HS—CoA} + 4\text{NAD}^+ + 4\text{FAD} \longrightarrow 5 \text{ acetyl CoA} + 4\text{NADH} + 4\text{H}^+ + 4\text{FADH}_2$$

SAMPLE PROBLEM 18.8 β Oxidation

TRY IT FIRST

Match each of the following (**a** to **d**) with one of the reactions (**1** to **4**) in the β oxidation cycle:

(1) first oxidation
(2) hydration
(3) second oxidation
(4) cleavage

a. Water is added to a trans double bond.
b. An acetyl CoA is removed.
c. FAD is reduced to $FADH_2$.
d. NAD^+ is reduced to NADH.

SOLUTION

a. (2) hydration
b. (4) cleavage
c. (1) first oxidation
d. (3) second oxidation

STUDY CHECK 18.8

Which coenzyme is needed in reaction 3 when a β-hydroxyl group is converted to a β-keto group?

ANSWER

NAD^+

Fatty Acid Length Determines Cycle Repeats

The number of carbon atoms in a fatty acid determines the number of times the cycle repeats and the number of acetyl CoA it produces. For example, the complete β oxidation of capric acid (C_{10}) produces five acetyl CoA, which is equal to one-half the number of carbon atoms in the fatty acid. Because the final turn of the cycle produces two acetyl CoA, the total number of times the cycle repeats is one less than the total number of acetyl groups it produces. Therefore, the C_{10} fatty acid goes through the cycle four times.

Fatty Acid	Number of Carbon Atoms	Number of Acetyl CoA	Number of β Oxidation Cycles
Capric acid	10	5	4
Myristic acid	14	7	6
Stearic acid	18	9	8

ENGAGE

Why would caproic acid, a C_6 saturated fatty acid, go through two β oxidation cycles?

ATP from Fatty Acid Oxidation

We can now determine the total energy yield from the oxidation of a particular fatty acid. In each β oxidation cycle, one NADH, one $FADH_2$, and one acetyl CoA are produced. The oxidation of each NADH leads to the synthesis of 2.5 ATP, whereas the oxidation of each $FADH_2$ leads to the synthesis of 1.5 ATP. Each time an acetyl CoA is oxidized through the citric acid cycle, sufficient energy is released to synthesize 10 ATP.

So far we know that the C_{10} acid goes through four turns of the cycle, which produces five acetyl CoA. We also need to remember that activation of the capric acid requires the equivalent of two ATP. We can set up the calculation as follows:

CORE CHEMISTRY SKILL

Calculating the ATP from Fatty Acid Oxidation (β Oxidation)

ATP Production from β Oxidation for Capric Acid (10C)	
Activation	−2 ATP
5 Acetyl CoA (10 C atoms × 1 acetyl CoA/2 C atoms)	
5 acetyl CoA × 10 ATP/acetyl CoA	50 ATP
4 β Oxidation Cycles	
4 NADH × 2.5 ATP/NADH	10 ATP
4 $FADH_2$ × 1.5 ATP/$FADH_2$	6 ATP
Total	64 ATP

SAMPLE PROBLEM 18.9 ATP Production from β Oxidation

TRY IT FIRST

How many ATP will be produced from the β oxidation of myristic acid, a C_{14} saturated fatty acid?

SOLUTION

	Given	Need	Connect
ANALYZE THE PROBLEM	β oxidation of myristic acid (C_{14})	total ATP produced	NADH, $FADH_2$, acetyl CoA

Myristic acid (C_{14}) requires six β oxidation cycles, which produce six NADH and six $FADH_2$. The total number of acetyl CoA is seven. The activation of myristic acid decreases the total ATP produced by two.

ATP Production from β Oxidation for Myristic Acid (C_{14})		
Activation of myristic acid:		= −2 ATP
Citric acid cycle:	$7\ \cancel{\text{acetyl CoA}} \times \frac{10\ \text{ATP}}{\cancel{\text{acetyl CoA}}}$	= 70 ATP
β oxidation:	$6\ \cancel{\text{NADH}} \times \frac{2.5\ \text{ATP}}{\cancel{\text{NADH}}}$	= 15 ATP
	$6\ \cancel{\text{FADH}_2} \times \frac{1.5\ \text{ATP}}{\cancel{\text{FADH}_2}}$	= 9 ATP
Total		= 92 ATP

STUDY CHECK 18.9

Calculate the total ATP produced from the β oxidation of cerotic acid (C_{26}), a saturated fatty acid found in beeswax and carnauba wax.

ANSWER

Cerotic acid (C_{26}) requires 12 β oxidation cycles (48 ATP) and produces 13 acetyl CoA molecules (130 ATP). Subtracting 2 ATP for activation produces a total of 176 ATP.

TEST

Try Practice Problems 18.57 to 18.60

CHEMISTRY LINK TO HEALTH

Stored Fat and Obesity

The storage of fat is an important survival feature in the lives of many animals. In hibernating animals, large amounts of stored fat provide the energy for the entire hibernation period, which could be several months.

In camels, large amounts of food are stored in the camel's hump, which is actually a huge fat deposit. When food resources are low, the camel can survive months without food or water by utilizing the fat reserves in the hump. Migratory birds preparing to fly long distances also store large amounts of fat. Whales are kept warm by a layer of body fat called "blubber" under their skin, which can be as thick as 30 cm. Blubber also provides energy when whales must survive long periods of starvation. Penguins also have blubber, which protects them from the cold and provides energy when they are incubating their eggs.

Humans also have the capability to store large amounts of fat, although they do not hibernate or usually have to survive for long periods of time without food. When early humans survived on sparse diets that were mostly vegetarian, about 20% of the dietary calories were from fats. Today, a typical diet includes more dairy products and foods with high fat levels, and as much as 60% of the calories are from fat. The U.S. Public Health Service now estimates that in the United States, more than one-third of adults are obese. Obesity is defined as a body weight that is more than 20% over an ideal weight. Obesity is a major factor in health problems such as diabetes, heart disease, high blood pressure, stroke, and gallstones, as well as some cancers and some forms of arthritis.

At one time, we thought that obesity was simply a problem of eating too much. However, research now indicates that certain pathways in lipid and carbohydrate metabolism may cause excessive weight gain in some people. In 1995, scientists discovered a hormone called *leptin* that is produced in fat cells. When fat stores are full, high levels of leptin signal the brain to limit the intake of food. When fat stores are low, leptin production decreases, which signals the brain to increase food intake. Leptin acts on the liver and skeletal muscles, where it stimulates fatty acid oxidation in the mitochondria, which decreases fat storage.

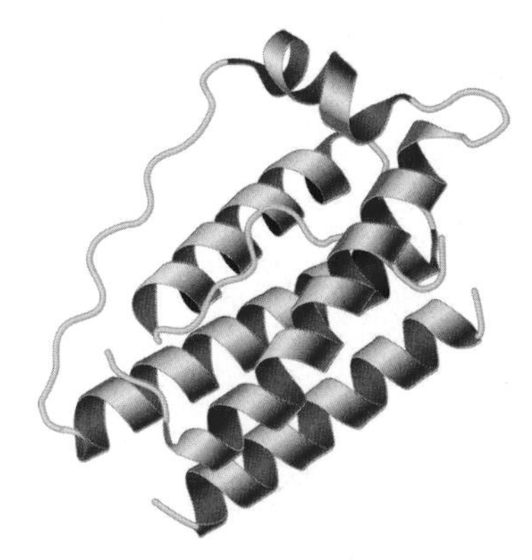

The leptin hormone helps regulate appetite and metabolism.

Major research is currently being done to find the causes of obesity. Scientists are studying differences in the rate of leptin production, degrees of resistance to leptin, and possible combinations of these factors. After a person has dieted and lost weight, the leptin level drops. This decrease in leptin may cause an increase in hunger and food intake while slowing metabolism, which starts the weight-gain cycle all over again. Currently, studies are being made to assess the safety of leptin therapy following weight loss.

Marine mammals have thick layers of blubber that serve as insulation as well as energy storage.

A camel stores large amounts of fat in its hump.

Oxidation of Unsaturated Fatty Acids

Many of the fats in our diets, especially the oils, contain unsaturated fatty acids that have one or more double bonds. Thus, in β oxidation, the fatty acid is ready for hydration. Since no $FADH_2$ is formed in this step, the energy from the β oxidation of unsaturated fatty acids is slightly less than the energy from saturated fatty acids. However, for simplicity, we will assume that the total ATP production is the same for both saturated and unsaturated fatty acids.

Ketone Bodies

When carbohydrates are not available to meet energy needs, the body breaks down fatty acids, which undergo β oxidation to acetyl CoA. Normally, acetyl CoA would enter the citric acid cycle for further oxidation. However, when large quantities of fatty acids are degraded, the citric acid cycle cannot utilize all of the acetyl CoA. As a result, acetyl CoA accumulates in the liver, where acetyl CoA molecules combine to form compounds called **ketone bodies** in a pathway known as *ketogenesis* (see **FIGURE 18.17**).

Ketone bodies are produced mostly in the liver and transported to cells in the heart, brain, and skeletal muscle, where small amounts of energy can be obtained by converting acetoacetate or β-hydroxybutyrate back to acetyl CoA.

$$\beta\text{-Hydroxybutyrate} \longrightarrow \text{acetoacetate} + 2\ \text{CoA} \longrightarrow 2\ \text{acetyl CoA}$$

FIGURE 18.17 In ketogenesis, acetyl CoA molecules combine to produce ketone bodies: acetoacetate, β-hydroxybutyrate, and acetone.

Q What condition in the body leads to the formation of ketone bodies?

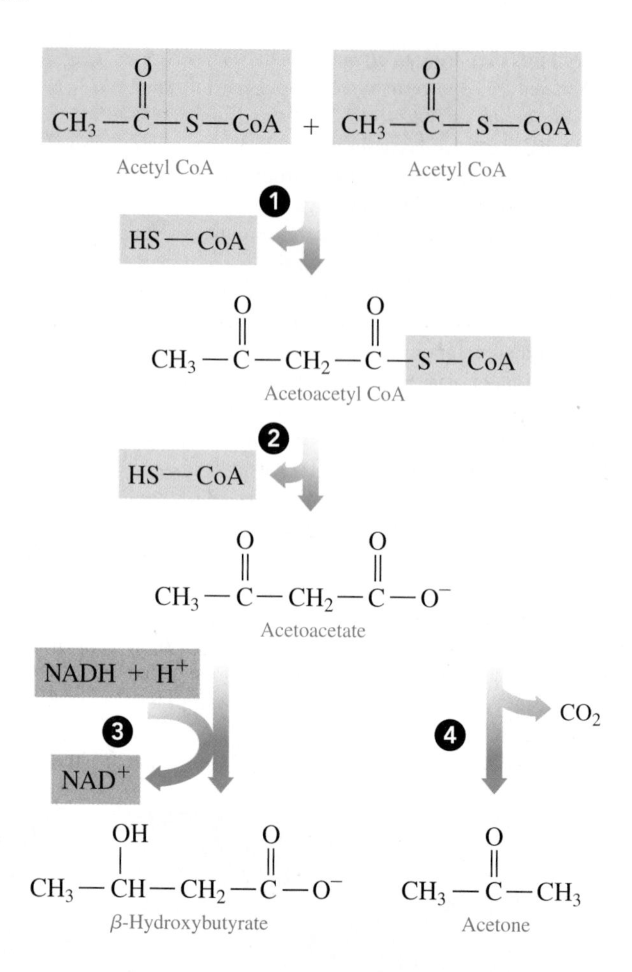

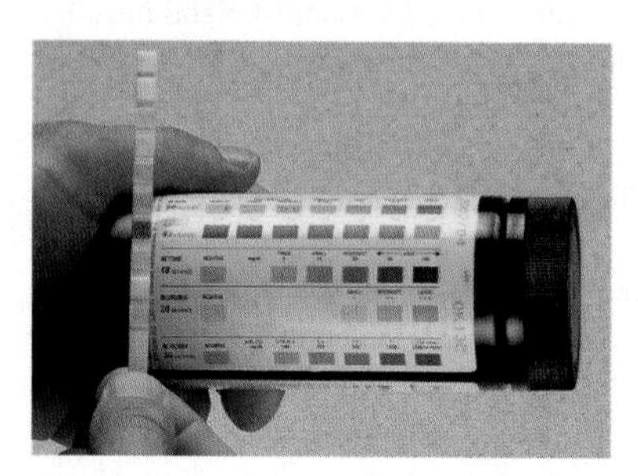

A test strip indicates the level of ketone bodies in a urine sample.

Ketosis

The accumulation of ketone bodies may lead to a condition called *ketosis*, which occurs in severe diabetes, diets high in fat and low in carbohydrates, alcoholism, and starvation. Because two of the ketone bodies are acids, they can lower the blood pH below 7.4, which is *acidosis*, a condition that often accompanies ketosis. A drop in blood pH can interfere with the ability of the blood to carry oxygen and cause breathing difficulties.

TEST

Try Practice Problems 18.61 to 18.64

PRACTICE PROBLEMS

18.7 Oxidation of Fatty Acids

LEARNING GOAL Describe the metabolic pathway of β oxidation; calculate the ATP from the complete oxidation of a fatty acid.

18.57 Caprylic acid, $CH_3-(CH_2)_6-COOH$, is a C_8 fatty acid found in milk.
a. State the number of β oxidation cycles for the complete oxidation of caprylic acid.
b. State the number of acetyl CoA from the complete oxidation of caprylic acid.
c. How many ATP are generated from the complete oxidation of caprylic acid?

18.58 Lignoceric acid, $CH_3-(CH_2)_{22}-COOH$, is a C_{24} fatty acid found in peanut oil in small amounts.
a. State the number of β oxidation cycles for the complete oxidation of lignoceric acid.
b. State the number of acetyl CoA from the complete oxidation of lignoceric acid.
c. How many ATP are generated from the complete oxidation of lignoceric acid?

18.59 Consider the complete oxidation of oleic acid, $CH_3-(CH_2)_7-CH=CH-(CH_2)_7-COOH$, which is a C_{18} monounsaturated fatty acid.
a. How many cycles of β oxidation are needed?
b. How many acetyl CoA are produced?
c. How many ATP are generated from the oxidation of oleic acid?

18.60 Consider the complete oxidation of palmitoleic acid, $CH_3-(CH_2)_5-CH=CH-(CH_2)_7-COOH$, which is a C_{16} monounsaturated fatty acid found in animal and vegetable oils.
a. How many cycles of β oxidation are needed?
b. How many acetyl CoA are produced?
c. How many ATP are generated from the oxidation of palmitoleic acid?

Clinical Applications

18.61 When are ketone bodies produced in the body?

18.62 Why would a person who is fasting have high levels of acetyl CoA?

18.63 What are some conditions that characterize ketosis?

18.64 Why do diabetics produce high levels of ketone bodies?

CHEMISTRY LINK TO HEALTH

Ketone Bodies and Diabetes

Blood glucose is elevated within 30 min following a meal containing carbohydrates. The elevated level of glucose stimulates the secretion of the hormone insulin from the pancreas, which increases the flow of glucose into muscle and adipose tissue for the synthesis of glycogen. As blood glucose levels drop, the secretion of insulin decreases. When blood glucose is low, another hormone, glucagon, is secreted by the pancreas, which stimulates the breakdown of glycogen in the liver to yield glucose.

In *diabetes mellitus*, glucose cannot be utilized or stored as glycogen because insulin is not secreted or does not function properly. In type 1, *insulin-dependent diabetes*, which often begins in childhood, the pancreas produces inadequate levels of insulin. This type of diabetes can result from damage to the pancreas by viral infections or from genetic mutations. In type 2, *insulin-resistant diabetes*, which usually occurs in adults, insulin is produced, but insulin receptors are not responsive. Thus a person with type 2 diabetes does not respond to insulin therapy. *Gestational diabetes* can occur during pregnancy, but blood glucose levels usually return to normal after the baby is born. Pregnant women with diabetes tend to gain weight and have large babies.

In all types of diabetes, insufficient amounts of glucose are available in the muscle, liver, and adipose tissue. As a result, liver cells synthesize glucose from noncarbohydrate sources and break down fat, elevating the acetyl CoA level. Excess acetyl CoA undergoes *ketogenesis*, and ketone bodies accumulate in the blood. The odor of acetone can be detected on the breath of a person with uncontrolled diabetes who is in ketosis.

In uncontrolled diabetes, the concentration of blood glucose exceeds the ability of the kidney to reabsorb glucose, and glucose appears in the urine. High levels of glucose increase the osmotic pressure in the blood, which leads to an increase in urine output. Symptoms of diabetes include frequent urination and excessive thirst. Treatment for diabetes includes diet changes to limit carbohydrate intake and may require medication such as a daily injection of insulin or pills taken by mouth.

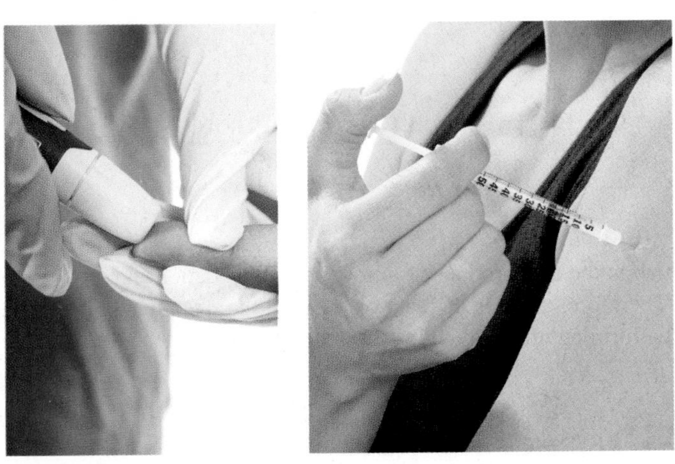

Type 1 diabetes can be treated with injections of insulin.

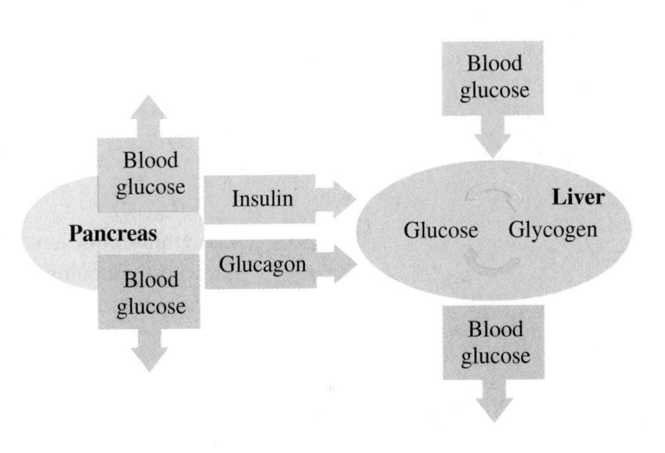

18.8 Degradation of Amino Acids

LEARNING GOAL Describe the reactions of transamination, oxidative deamination, and the entry of amino acid carbons into the citric acid cycle.

REVIEW

Drawing the Structure for an Amino Acid at Physiological pH (16.1)
Describing Enzyme Action (16.4)

The major role of protein in the diet is to provide amino acids for the synthesis of new proteins for the body and nitrogen atoms for the synthesis of compounds such as nucleotides. We have seen that carbohydrates and lipids are major sources of energy, but when they are not available, amino acids can be degraded to compounds that enter energy-releasing pathways.

Transamination

The degradation of amino acids occurs primarily in the liver. In a **transamination** reaction, an α-amino group is transferred from an amino acid to an α-keto acid, usually α-ketoglutarate. A new amino acid and a new α-keto acid are produced. The enzymes that catalyze transfer of amino groups are known as transaminases or aminotransferases.

We can write an equation to show the transfer of the amino group from alanine to α-ketoglutarate to yield glutamate, the new amino acid, and the α-keto acid pyruvate. Pyruvate can now react to form acetyl CoA, which can enter the citric acid cycle.

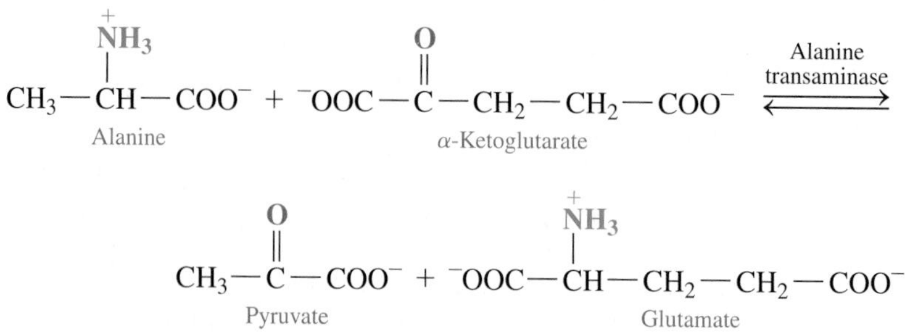

$$CH_3-\overset{\overset{+}{N}H_3}{\overset{|}{C}H}-COO^- + {}^-OOC-\overset{\overset{O}{\|}}{C}-CH_2-CH_2-COO^- \underset{\text{transaminase}}{\overset{\text{Alanine}}{\rightleftharpoons}}$$

Alanine α-Ketoglutarate

$$CH_3-\overset{\overset{O}{\|}}{C}-COO^- + {}^-OOC-\overset{\overset{+}{N}H_3}{\overset{|}{C}H}-CH_2-CH_2-COO^-$$

Pyruvate Glutamate

TEST

Try Practice Problems 18.65 and 18.66

Oxidative Deamination

In a process called **oxidative deamination**, the amino group ($-NH_3^+$) in glutamate is removed as an ammonium ion, NH_4^+. The reaction regenerates α-ketoglutarate, which can participate in transamination with an amino acid. The oxidation provides hydrogens for the NAD^+ coenzyme.

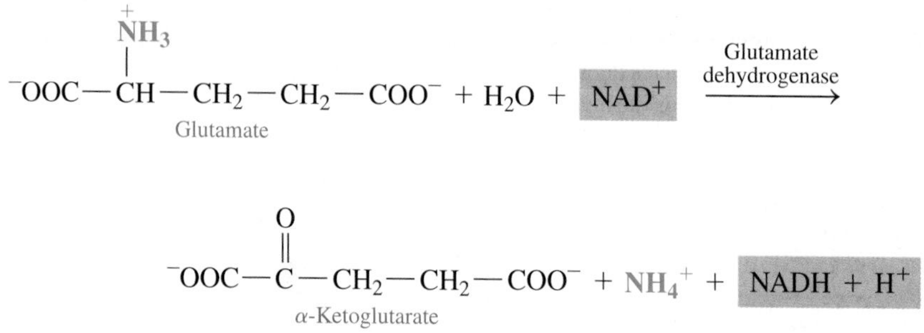

$${}^-OOC-\overset{\overset{+}{N}H_3}{\overset{|}{C}H}-CH_2-CH_2-COO^- + H_2O + NAD^+ \xrightarrow{\text{Glutamate dehydrogenase}}$$

Glutamate

$${}^-OOC-\overset{\overset{O}{\|}}{C}-CH_2-CH_2-COO^- + NH_4^+ + NADH + H^+$$

α-Ketoglutarate

Urea Cycle

The ammonium ion, which is the end product of amino acid degradation, is toxic if it is allowed to accumulate. Therefore, a series of reactions, called the **urea cycle**, detoxifies ammonium ion (NH_4^+) by forming urea, which is excreted in the urine.

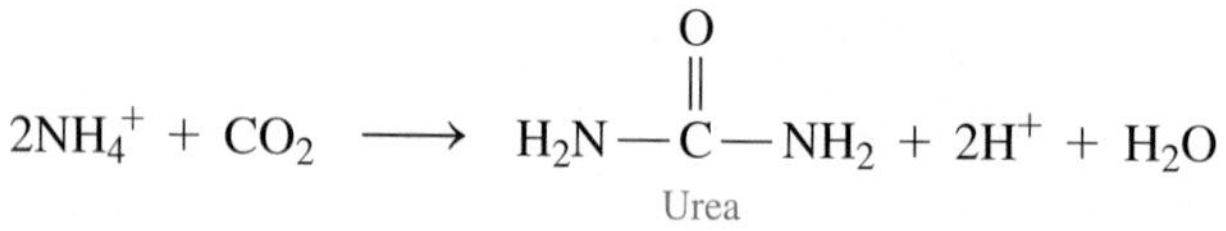

$$2NH_4^+ + CO_2 \longrightarrow H_2N-\overset{\overset{O}{\|}}{C}-NH_2 + 2H^+ + H_2O$$

Urea

In one day, a typical adult may excrete about 25 to 30 g of urea in the urine. This amount increases when a diet is high in protein. If urea is not properly excreted, as may be the case when the kidneys are not functioning properly, it builds up quickly to a toxic level. To detect renal disease, the *blood urea nitrogen* (BUN) level is measured. If the BUN is high, protein intake must be reduced, and hemodialysis may be needed to remove toxic nitrogen waste from the blood.

TEST

Try Practice Problems 18.67 and 18.68

SAMPLE PROBLEM 18.10 Transamination and Oxidative Deamination

TRY IT FIRST

Indicate whether each of the following represents a transamination or an oxidative deamination:

a. Glutamate is converted to α-ketoglutarate and NH_4^+.
b. Alanine and α-ketoglutarate react to form pyruvate and glutamate.

SOLUTION

a. oxidative deamination
b. transamination

STUDY CHECK 18.10

What is the source of the nitrogen atoms used to produce urea?

ANSWER

The process of oxidative deamination involves the removal of the amino group of glutamate as ammonium.

ATP Energy from Amino Acids

Normally, only a small amount (about 10%) of our energy needs is supplied by amino acids. However, more energy is extracted from amino acids in conditions such as fasting or starvation, when carbohydrate and fat stores are exhausted. If amino acids remain the only source of energy for a long period of time, the breakdown of body proteins eventually leads to a destruction of essential body tissues.

Once the carbon skeletons of amino acids are obtained from transamination, they can be used as intermediates of the citric acid cycle to provide reduced coenzymes for electron transport. Some amino acids can enter the citric acid cycle at more than one place (see **FIGURE 18.18**).

ENGAGE

Where can carbon atoms from isoleucine enter the citric acid cycle?

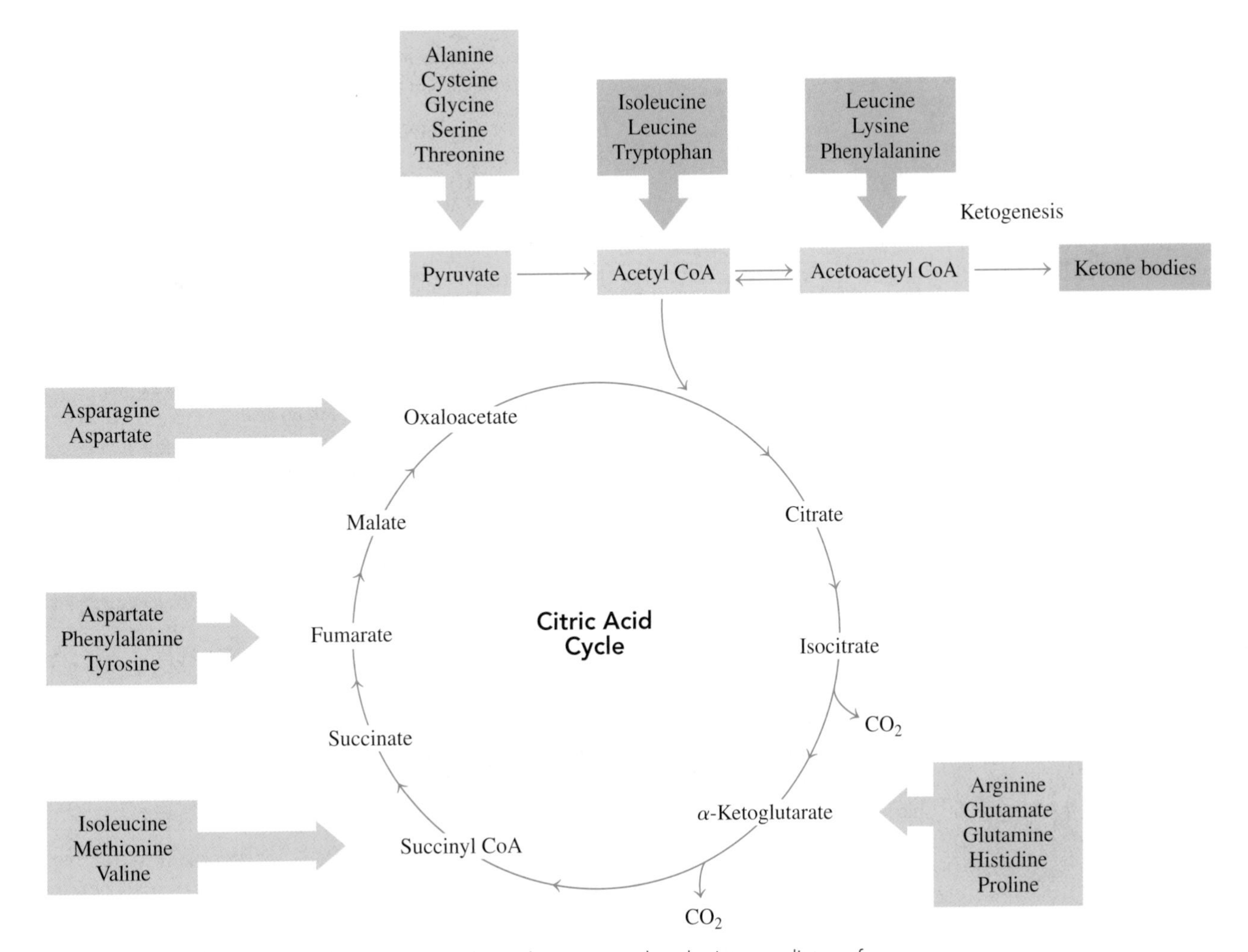

FIGURE 18.18 Carbon atoms from amino acids can be converted to the intermediates of the citric acid cycle. Carbon atoms from amino acids in gold boxes can also produce ketone bodies.

Q What compound in the citric acid cycle is obtained from the carbon atoms of alanine and glycine?

SAMPLE PROBLEM 18.11 Carbon Atoms from Amino Acids

TRY IT FIRST

At what point do the carbon atoms of each of the following amino acids enter catabolic pathways?

a. valine **b.** proline **c.** glycine

SOLUTION

a. Carbon atoms from valine enter the citric acid cycle as succinyl CoA.
b. Carbon atoms from proline enter the citric acid cycle as α-ketoglutarate.
c. Carbon atoms from glycine enter metabolic pathways as pyruvate.

STUDY CHECK 18.11

Which amino acids enter the citric acid cycle as α-ketoglutarate?

ANSWER

Arginine, glutamate, glutamine, histidine, and proline provide carbon atoms for α-ketoglutarate.

TEST

Try Practice Problems 18.69 and 18.70

PRACTICE PROBLEMS

18.8 Degradation of Amino Acids

LEARNING GOAL Describe the reactions of transamination, oxidative deamination, and the entry of amino acid carbons into the citric acid cycle.

18.65 Draw the condensed structural formula for the α-keto acid produced from each of the following in transamination:

a. $H-\underset{\underset{}{}}{\overset{\overset{+}{NH_3}}{|}}{CH}-COO^-$ Glycine

b. $CH_3-\overset{\overset{+}{NH_3}}{|}{CH}-COO^-$ Alanine

18.66 Draw the condensed structural formula for the α-keto acid produced from each of the following in transamination:

a. $^-OOC-CH_2-\overset{\overset{+}{NH_3}}{|}{CH}-COO^-$ Aspartate

b. $CH_3-CH_2-\overset{CH_3}{|}{CH}-\overset{\overset{+}{NH_3}}{|}{CH}-COO^-$ Isoleucine

18.67 Why does the body convert NH_4^+ to urea?

18.68 Draw the condensed structural formula for urea.

18.69 What metabolic substrate(s) are produced from the carbon atoms of each of the following amino acids?
a. alanine **b.** aspartate **c.** tyrosine **d.** glutamine

18.70 What metabolic substrate(s) are produced from the carbon atoms of each of the following amino acids?
a. leucine **b.** asparagine **c.** cysteine **d.** arginine

CLINICAL UPDATE Treatment of Luke's Hepatitis C

When Luke has a follow-up liver evaluation three months later, his liver enzymes are more elevated: ALT 356 Units/L and AST 418 Units/L. His doctor suggests that Luke begin interferon and ribavirin therapy for six months. Interferon and ribavirin work together as antiviral agents to inhibit the replication of the hepatitis C virus and strengthen the immune system. The interferon is injected subcutaneously or intramuscularly three

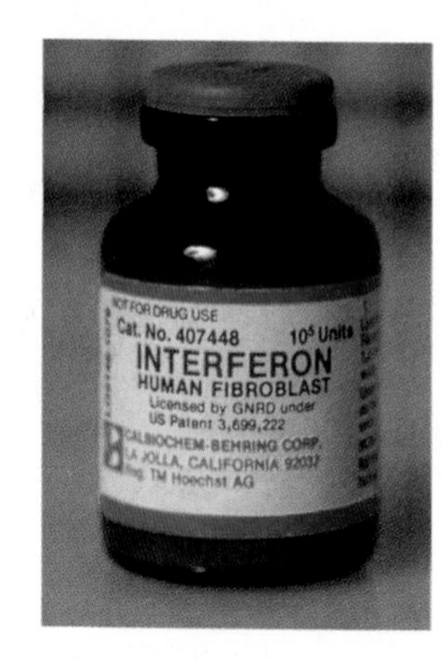

Interferon is one of the antiviral agents used to treat hepatitis C.

times a week for nine months. The ribavarin is taken orally as 400-mg tablets twice a day.

After eight weeks of treatment, Luke's liver enzymes have lowered to ALT 85 Units/L and AST 115 Units/L. After four months of therapy, his ALT and AST enzyme levels are each within the normal range. Luke will continue to have a liver profile every three months for the next year, and then every six months. Monitoring of his liver enzymes will continue throughout his life.

Clinical Applications

18.71 Draw the condensed structural formulas for the products of the reaction of alanine and α-ketoglutarate, which is catalyzed by alanine transaminase (ALT).

18.72 Draw the condensed structural formulas for the products of the reaction of aspartate and α-ketoglutarate, which is catalyzed by aspartate transaminase (AST).

CONCEPT MAP

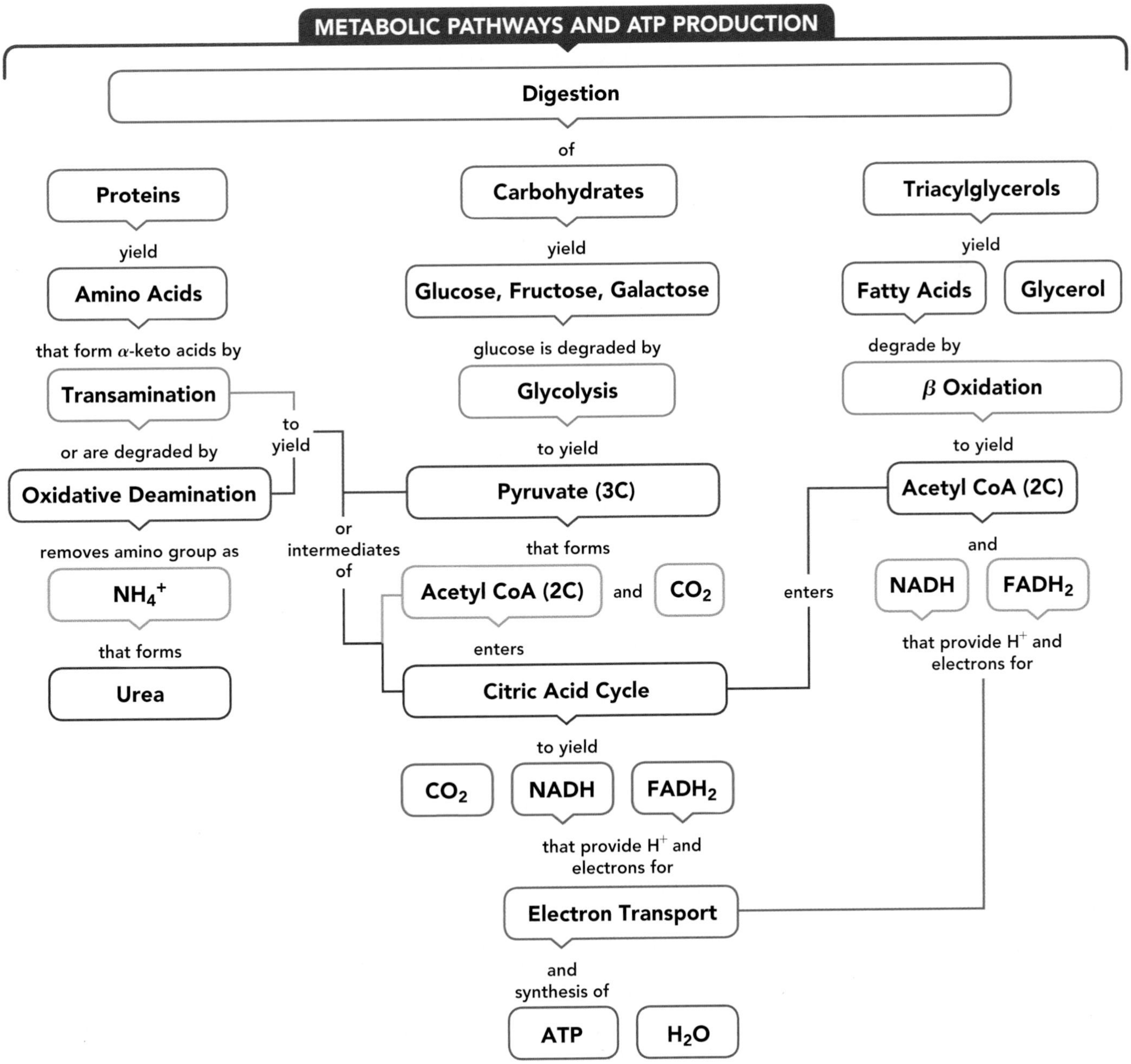

METABOLIC PATHWAYS AND ATP PRODUCTION

Digestion

of

Proteins

yield

Amino Acids

that form α-keto acids by

Transamination

or are degraded by

Oxidative Deamination

to yield

removes amino group as

NH_4^+

that forms

Urea

Carbohydrates

yield

Glucose, Fructose, Galactose

glucose is degraded by

Glycolysis

to yield

Pyruvate (3C)

or intermediates of

that forms

Acetyl CoA (2C) and **CO_2**

enters

Citric Acid Cycle

to yield

CO_2 **NADH** **$FADH_2$**

that provide H^+ and electrons for

Electron Transport

and synthesis of

ATP **H_2O**

Triacylglycerols

yield

Fatty Acids **Glycerol**

degrade by

β Oxidation

to yield

Acetyl CoA (2C)

enters

and

NADH **$FADH_2$**

that provide H^+ and electrons for

CHAPTER REVIEW

18.1 Metabolism and ATP Energy

LEARNING GOAL Describe the three stages of catabolism and the role of ATP.

- Metabolism includes all the catabolic and anabolic reactions that occur in the cells.
- Catabolic reactions degrade large molecules into smaller ones with an accompanying release of energy.
- Anabolic reactions require energy to synthesize larger molecules from smaller ones.
- The three stages of catabolism are digestion of food, degradation of larger molecules into smaller groups such as acetyl and pyruvate, and the oxidation of the acetyl and pyruvate groups to release energy for ATP synthesis.
- Energy released through catabolic reactions is stored as adenosine triphosphate (ATP), a high-energy compound that is hydrolyzed when energy is required by anabolic reactions.

18.2 Digestion of Foods

LEARNING GOAL Identify the sites and products of digestion for carbohydrates, triacylglycerols, and proteins.

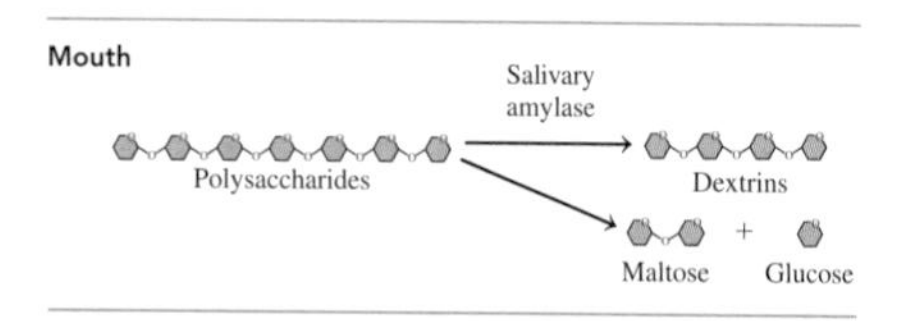

- The digestion of carbohydrates is a series of reactions that breaks down polysaccharides into smaller polysaccharides (dextrins), and eventually to monosaccharides glucose, galactose, and fructose.
- These monomers can be absorbed through the intestinal wall into the bloodstream to be carried to cells where they provide energy and carbon atoms for synthesis of new molecules.
- Triacylglycerols are hydrolyzed in the small intestine to monoacylglycerol and fatty acids, which enter the intestinal wall and form new triacylglycerols. They bind with proteins to form chylomicrons, which transport them through the lymphatic system and bloodstream to the tissues.
- The digestion of proteins, which begins in the stomach and continues in the small intestine, involves the hydrolysis of peptide bonds to yield amino acids that are absorbed through the intestinal wall and transported to the cells.

18.3 Coenzymes in Metabolic Pathways

LEARNING GOAL Describe the components and functions of the coenzymes NAD^+, FAD, and coenzyme A.

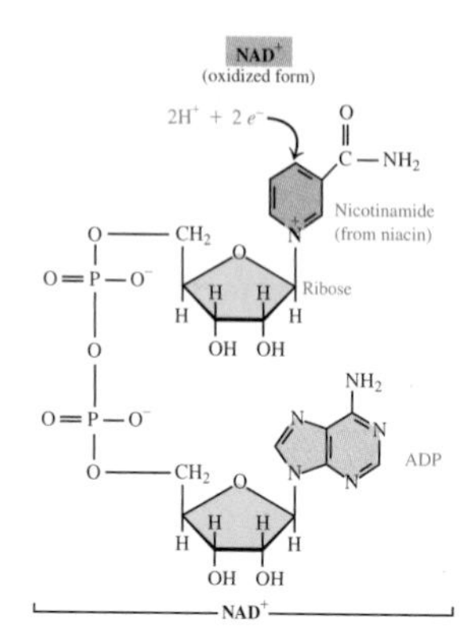

- FAD and NAD^+ are the oxidized forms of coenzymes that participate in oxidation–reduction reactions.
- When they pick up hydrogen ions and electrons, they are reduced to $FADH_2$ and NADH + H^+.
- Coenzyme A contains a thiol group that usually bonds with a two-carbon acetyl group (acetyl CoA).

18.4 Glycolysis: Oxidation of Glucose

LEARNING GOAL Describe the conversion of glucose to pyruvate in glycolysis and the subsequent conversion of pyruvate to acetyl CoA or lactate.

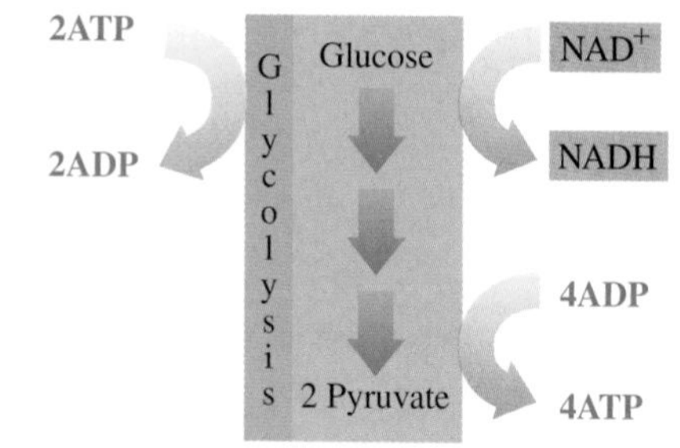

- Glycolysis, which occurs in the cytosol, consists of 10 reactions that degrade glucose (6C) to two pyruvate (3C).
- The overall series of reactions yields two NADH and two ATP.
- Under aerobic conditions, pyruvate is oxidized in the mitochondria to acetyl CoA.
- In the absence of oxygen, pyruvate is reduced to lactate and NAD^+ is regenerated for the continuation of glycolysis.

18.5 The Citric Acid Cycle

LEARNING GOAL Describe the oxidation of acetyl CoA in the citric acid cycle.

- In the first of a sequence of reactions called the citric acid cycle, an acetyl group is combined with oxaloacetate to yield citrate.
- Citrate undergoes oxidation and decarboxylation to yield two CO_2, one GTP, three NADH, and one $FADH_2$ with the regeneration of oxaloacetate.
- The direct phosphate transfer of ADP by GTP yields ATP.

18.6 Electron Transport and Oxidative Phosphorylation

LEARNING GOAL Describe electron transport and the process of oxidative phosphorylation; calculate the ATP from the complete oxidation of glucose.

- The reduced coenzymes NADH and $FADH_2$ from various metabolic pathways are oxidized to NAD^+ and FAD when their hydrogen ions and electrons are transferred to the electron transport system.
- The energy released is used to synthesize ATP from ADP and P_i.
- The final acceptor, O_2, combines with hydrogen ions and electrons to yield H_2O.
- The protein complexes in electron transport move hydrogen ions into the intermembrane space, which produces a H^+ gradient.
- As the H^+ ions return to the matrix by way of ATP synthase, energy is released.
- This energy is used to drive the synthesis of ATP in a process known as oxidative phosphorylation.
- The oxidation of NADH yields 2.5 ATP, and $FADH_2$ yields 1.5 ATP.
- Under aerobic conditions, the complete oxidation of glucose yields a total of 32 ATP.

18.7 Oxidation of Fatty Acids

LEARNING GOAL Describe the metabolic pathway of β oxidation; calculate the ATP from the complete oxidation of a fatty acid.

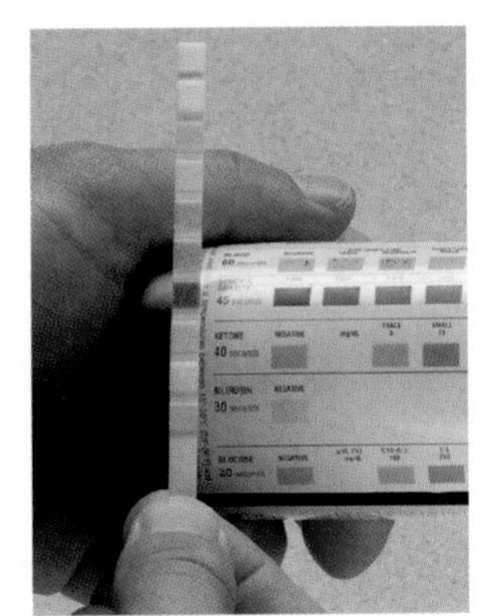

- When needed as an energy source, fatty acids are linked to coenzyme A and transported into the mitochondria where they undergo β oxidation.
- The fatty acyl CoA is oxidized to yield a shorter fatty acyl CoA, acetyl CoA, and reduced coenzymes NADH and $FADH_2$.
- Although the energy from a particular fatty acid depends on its length, each oxidation cycle yields 4 ATP with another 10 ATP from the acetyl CoA that enters the citric acid cycle.
- When high levels of acetyl CoA are present in the cell, they enter the ketogenesis pathway, forming ketone bodies such as acetoacetate, which cause ketosis and acidosis.

18.8 Degradation of Amino Acids

LEARNING GOAL Describe the reactions of transamination, oxidative deamination, and the entry of amino acid carbons into the citric acid cycle.

$$\underset{\text{Alanine}}{CH_3-CH(\overset{+}{N}H_3)-COO^-} + \underset{\alpha\text{-Ketoglutarate}}{^-OOC-C(=O)-CH_2-CH_2-COO^-} \xrightleftharpoons{\text{Alanine transaminase}} \underset{\text{Pyruvate}}{CH_3-C(=O)-COO^-} + \underset{\text{Glutamate}}{^-OOC-CH(\overset{+}{N}H_3)-CH_2-CH_2-COO^-}$$

- When the amount of amino acids in the cells exceeds that needed for synthesis of nitrogen compounds, the process of transamination converts them to α-keto acids and glutamate.
- Oxidative deamination of glutamate produces ammonium ions and α-ketoglutarate.
- Ammonium ions from oxidative deamination are converted to urea.
- The carbon atoms from the degradation of amino acids can enter the citric acid cycle or other metabolic pathways.

SUMMARY OF REACTIONS

The chapter Sections to review are shown after the name of each reaction.

Hydrolysis of ATP (18.1)

$$ATP + H_2O \longrightarrow ADP + P_i + 7.3\ \text{kcal/mole (31 kJ/mole)}$$

Formation of ATP (18.1)

$$ADP + P_i + 7.3\ \text{kcal/mole (31 kJ/mole)} \longrightarrow ATP + H_2O$$

Hydrolysis of Disaccharides (18.2)

$$\text{Maltose} + H_2O \xrightarrow{\text{Maltase}} \text{glucose} + \text{glucose}$$

$$\text{Lactose} + H_2O \xrightarrow{\text{Lactase}} \text{galactose} + \text{glucose}$$

$$\text{Sucrose} + H_2O \xrightarrow{\text{Sucrase}} \text{fructose} + \text{glucose}$$

Hydrolysis of Triacylglycerols (18.2)

$$\text{Triacylglycerols} + 3H_2O \xrightarrow{\text{Lipases}} \text{glycerol} + 3\ \text{fatty acids}$$

Hydrolysis of Proteins (18.2)

$$\text{Protein} + H_2O \xrightarrow{\text{Proteases}} \text{amino acids}$$

Reduction of NAD⁺ and FAD (18.3)

$$NAD^+ + 2H^+ + 2e^- \longrightarrow NADH + H^+$$

$$FAD + 2H^+ + 2e^- \longrightarrow FADH_2$$

Glycolysis (18.4)

$$\underset{\text{Glucose}}{C_6H_{12}O_6} + 2ADP + 2P_i + 2NAD^+ \longrightarrow$$

$$\underset{\text{Pyruvate}}{2CH_3-C(=O)-COO^-} + 2ATP + 2NADH + 4H^+ + 2H_2O$$

Oxidation of Pyruvate to Acetyl CoA (Aerobic) (18.4)

$$\underset{\text{Pyruvate}}{CH_3-\overset{O}{\overset{\|}{C}}-COO^-} + HS-CoA + NAD^+ \xrightarrow{\text{Pyruvate dehydrogenase}} \underset{\text{Acetyl CoA}}{CH_3-\overset{O}{\overset{\|}{C}}-S-CoA} + CO_2 + NADH$$

Reduction of Pyruvate to Lactate (Anaerobic) (18.4)

$$\underset{\text{Pyruvate}}{CH_3-\overset{O}{\overset{\|}{C}}-COO^-} + NADH + H^+ \underset{}{\overset{\text{Lactate dehydrogenase}}{\rightleftarrows}} \underset{\text{Lactate}}{CH_3-\overset{OH}{\overset{|}{C}H}-COO^-} + NAD^+$$

Citric Acid Cycle (18.5)

$$\text{Acetyl CoA} + 3NAD^+ + FAD + GDP + P_i + 2H_2O \longrightarrow HS-CoA + 3NADH + 3H^+ + FADH_2 + GTP + 2CO_2$$

Oxidation of Glucose (Complete) (18.6)

$$C_6H_{12}O_6 + 6O_2 + 32ADP + 32P_i \longrightarrow 6CO_2 + 6H_2O + 32ATP$$

β Oxidation of Fatty Acids (18.7)

$$\underset{\text{Fatty acyl CoA}}{R-\mathbf{CH_2}-CH_2-\overset{O}{\overset{\|}{C}}-S-CoA} + NAD^+ + FAD + H_2O + HS-CoA \longrightarrow$$

$$\underset{\text{New fatty acyl CoA (–2C)}}{R-\overset{\mathbf{O}}{\overset{\|}{\mathbf{C}}}-S-CoA} + \underset{\text{Acetyl CoA}}{CH_3-\overset{O}{\overset{\|}{C}}-S-CoA} + NADH + H^+ + FADH_2$$

Transamination (18.8)

$$\underset{\text{Alanine}}{CH_3-\overset{\overset{+}{N}H_3}{\overset{|}{C}H}-COO^-} + \underset{\alpha\text{-Ketoglutarate}}{^-OOC-\overset{O}{\overset{\|}{C}}-CH_2-CH_2-COO^-} \overset{\text{Alanine transaminase}}{\rightleftarrows}$$

$$\underset{\text{Pyruvate}}{CH_3-\overset{O}{\overset{\|}{C}}-COO^-} + \underset{\text{Glutamate}}{^-OOC-\overset{\overset{+}{N}H_3}{\overset{|}{C}H}-CH_2-CH_2-COO^-}$$

Oxidative Deamination (18.8)

$$\underset{\text{Glutamate}}{^-OOC-\overset{\overset{+}{N}H_3}{\overset{|}{C}H}-CH_2-CH_2-COO^-} + H_2O + NAD^+ \xrightarrow{\text{Glutamate dehydrogenase}}$$

$$\underset{\alpha\text{-Ketoglutarate}}{^-OOC-\overset{O}{\overset{\|}{C}}-CH_2-CH_2-COO^-} + NH_4^+ + NADH + H^+$$

Urea Cycle (18.8)

$$2NH_4^+ + CO_2 \longrightarrow \underset{\text{Urea}}{H_2N-\overset{O}{\overset{\|}{C}}-NH_2} + 2H^+ + H_2O$$

KEY TERMS

acetyl CoA The compound that forms when a two-carbon acetyl group bonds to coenzyme A.
aerobic An oxygen-containing environment in the cells.
anabolic reaction A metabolic reaction that requires energy to build large molecules from small molecules.
anaerobic A condition in cells when there is no oxygen.
ATP Adenosine triphosphate; a high-energy compound that stores energy in the cells. It consists of adenine, a ribose sugar, and three phosphate groups.
ATP synthase An enzyme complex that uses the energy released by H^+ ions returning to the matrix to synthesize ATP from ADP and P_i.
beta (β) oxidation The degradation of fatty acids that removes two-carbon segments from the fatty acid at the oxidized β-carbon.
catabolic reaction A metabolic reaction that releases energy for the cell through the degradation and oxidation of glucose and other molecules.
chemiosmotic model The conservation of energy from electron transport by pumping H^+ ions into the intermembrane space to produce a H^+ gradient that provides the energy to synthesize ATP.
citric acid cycle A series of oxidation reactions in the mitochondria that convert acetyl CoA to CO_2 and yield NADH and $FADH_2$. It is also called the Krebs cycle.
coenzyme A (CoA) A coenzyme that transports acyl and acetyl groups.
decarboxylation The loss of a carbon atom in the form of CO_2.
digestion The processes in the gastrointestinal tract that break down large food molecules to smaller ones that pass through the intestinal membrane into the bloodstream.
electron transport A series of reactions in the mitochondria that transfer electrons from NADH and $FADH_2$ to electron carriers and finally to O_2, which produces H_2O. The energy released is used to synthesize ATP from ADP and P_i.
FAD A coenzyme (flavin adenine dinucleotide) for oxidation reactions that form carbon–carbon double bonds.
glycolysis The 10 oxidation reactions of glucose that yield two pyruvate molecules.
ketone bodies The products of ketogenesis: acetoacetate, β-hydroxybutyrate, and acetone.
metabolism All the chemical reactions in living cells that carry out molecular and energy transformations.
NAD^+ A coenzyme (nicotinamide adenine dinucleotide) for oxidation reactions that form carbon–oxygen double bonds.
oxidative deamination The loss of ammonium ion when glutamate is degraded to α-ketoglutarate.
oxidative phosphorylation The synthesis of ATP from ADP and P_i, using energy generated by the oxidation reactions during electron transport.
transamination The transfer of an amino group from an amino acid to an α-keto acid.
urea cycle The process in which ammonium ions from the degradation of amino acids are converted to urea.

CORE CHEMISTRY SKILLS

The chapter Section containing each Core Chemistry Skill is shown in parentheses at the end of each heading.

Identifying the Compounds in Glycolysis (18.4)

- Glycolysis, which occurs in the cytosol, consists of 10 reactions that degrade glucose (6C) to two pyruvate (3C).
- The overall series of reactions in glycolysis yields two NADH and two ATP.

Example: Identify the reaction(s) and compounds in glycolysis that have each of the following:

a. requires ATP
b. converts a six-carbon compound to two three-carbon compounds
c. converts ADP to ATP

Answer: **a.** The phosphorylation of glucose in reaction 1 and fructose-6-phosphate in reaction 3 each require one ATP.
b. In reaction 4, the six-carbon fructose-1,6-bisphosphate is cleaved into two three-carbon compounds.
c. Phosphate transfer in reactions 7 and 10 converts ADP to ATP.

Describing the Reactions in the Citric Acid Cycle (18.5)

- In the initial reaction of the citric acid cycle, an acetyl group combines with oxaloacetate to yield citrate.
- Citrate undergoes several reactions including oxidation, decarboxylation, and hydration to yield two CO_2, one GTP, three NADH, and one $FADH_2$, and to regenerate oxaloacetate.

Example: Identify the compound(s) in the citric acid cycle that undergo each of the following changes:

a. loses a CO_2 molecule
b. adds water

Answer: **a.** Isocitrate (6C) undergoes oxidation and decarboxylation to form α-ketoglutarate (5C), and α-ketoglutarate (5C) undergoes oxidation and decarboxylation to form succinyl CoA (4C).
b. Fumarate, which has a double bond, adds an H_2O molecule to form malate.

Calculating the ATP Produced from Glucose (18.6)

- The reduced coenzymes NADH and $FADH_2$ from various metabolic pathways are oxidized to NAD^+ and FAD when their H^+ ions and electrons are transferred to the electron transport system.
- The energy released is used to synthesize ATP from ADP and P_i.
- The final acceptor, O_2, combines with H^+ ions and electrons to yield H_2O.
- The protein complexes I, III, and IV in electron transport move hydrogen ions into the intermembrane space, which produces a H^+ gradient.
- As the H^+ ions return to the matrix by way of ATP synthase, ATP energy is generated in a process known as oxidative phosphorylation.
- The oxidation of NADH yields 2.5 ATP, and the oxidation of $FADH_2$ yields 1.5 ATP.
- Under aerobic conditions, the complete oxidation of glucose yields a total of 32 ATP.

Example: Calculate the ATP produced from each of the following:

a. one glucose to two pyruvate
b. one NADH to one NAD^+
c. complete oxidation of one glucose to 6 CO_2 and 6 H_2O

Answer: **a.** Glucose is converted to two pyruvate during glycolysis, which produces 2 ATP and 2 NADH for a total of 7 ATP.
b. One NADH is converted to one NAD^+ in the electron transport system, which produces 2.5 ATP.
c. Glucose is completely oxidized to 6 CO_2 and 6 H_2O, which produces 32 ATP.

Calculating the ATP from Fatty Acid Oxidation (β Oxidation) (18.7)

- When needed as an energy source, fatty acids are linked to coenzyme A and transported into the mitochondria where they undergo β oxidation.
- The fatty acyl CoA is oxidized to yield a shorter fatty acyl CoA, acetyl CoA, and reduced coenzymes NADH and $FADH_2$.
- Although the energy from a particular fatty acid depends on its length, each oxidation cycle yields 4 ATP with another 10 ATP from each acetyl CoA that enters the citric acid cycle.
- When high levels of acetyl CoA are present in the cell, they enter the ketogenesis pathway, forming ketone bodies such as acetoacetate, which cause ketosis and acidosis.

Example: How many β oxidation cycles, acetyl CoA, and ATP are produced by the complete oxidation of montanic acid, the C_{28} saturated fatty acid?

Answer: A 28-carbon fatty acid will undergo 13 β oxidation cycles and produce 14 acetyl CoA. Each oxidation cycle produces 4 ATP (1 NADH and 1 $FADH_2$) or a total of 52 ATP. Each acetyl CoA produces 10 ATP in the citric acid cycle or a total of 140 ATP. The 192 ATP minus 2 ATP for activation gives a total of 190 ATP from montanic acid.

UNDERSTANDING THE CONCEPTS

The chapter Sections to review are shown in parentheses at the end of each problem.

18.73 Lauric acid is a C_{12} fatty acid found in coconut oil. (18.6, 18.7)

Coconut oil is high in lauric acid.

a. How many cycles of β oxidation are needed for the complete oxidation of lauric acid?
b. How many acetyl CoA are produced from the complete oxidation of lauric acid?
c. Calculate the total ATP yield from the complete β oxidation of lauric acid by completing the following:

Activation	$\longrightarrow$	−2 ATP
_____ acetyl CoA	$\longrightarrow$	_____ ATP
_____ NADH	$\longrightarrow$	_____ ATP
_____ $FADH_2$	$\longrightarrow$	_____ ATP
Total		_____ **ATP**

18.74 Arachidic acid is a C_{20} fatty acid found in peanut and fish oils. (18.6, 18.7)

Peanuts contain arachidic acid, a C_{20} saturated fatty acid.

a. How many cycles of β oxidation are needed for the complete oxidation of arachidic acid?
b. How many acetyl CoA are produced from the complete oxidation of arachidic acid?
c. Calculate the total ATP yield from the complete β oxidation of arachidic acid by completing the following:

Activation	$\longrightarrow$	−2 ATP
_____ acetyl CoA	$\longrightarrow$	_____ ATP
_____ NADH	$\longrightarrow$	_____ ATP
_____ $FADH_2$	$\longrightarrow$	_____ ATP
Total		_____ **ATP**

18.75 Identify the type of food as carbohydrate, fat, or protein that gives each of the following digestion products: (18.2)
a. glucose **b.** fatty acid **c.** maltose
d. glycerol **e.** amino acids **f.** dextrins

18.76 Identify each of the following as a six-carbon or a three-carbon compound and arrange them in the order in which they occur in glycolysis: (18.4)
a. 3-phosphoglycerate **b.** pyruvate
c. glucose-6-phosphate **d.** glucose
e. fructose-1,6-bisphosphate

Digestion is the first stage of catabolism.

ADDITIONAL PRACTICE PROBLEMS

18.77 Write an equation for the hydrolysis of ATP to ADP. (18.1)

18.78 At the gym, you expend 310 kcal riding the stationary bicycle for 1 h. How many moles of ATP will this require? (18.1)

18.79 How and where does lactose undergo digestion in the body? What are the products? (18.2)

18.80 How and where does sucrose undergo digestion in the body? What are the products? (18.2)

18.81 What are the reactant and product of glycolysis? (18.4)

18.82 What is the general type of reaction that takes place in the digestion of carbohydrates? (18.2)

18.83 When is pyruvate converted to lactate in the body? (18.4)

18.84 When pyruvate is used to form acetyl CoA, the product has only two carbon atoms. What happened to the third carbon? (18.4)

18.85 What is the main function of the citric acid cycle? (18.5)

18.86 Most metabolic pathways are not cycles. Why is the citric acid cycle considered to be a metabolic cycle? (18.5)

18.87 If there are no reactions in the citric acid cycle that use oxygen, O_2, why does the cycle operate only in aerobic conditions? (18.5)

18.88 What products of the citric acid cycle are needed for electron transport? (18.5, 18.6)

18.89 In the chemiosmotic model, how is energy provided to synthesize ATP? (18.6)

18.90 What is the effect of H^+ accumulation in the intermembrane space? (18.6)

18.91 How many ATP are produced when glucose is oxidized to pyruvate, compared to when glucose is oxidized to CO_2 and H_2O? (18.4, 18.5, 18.6)

18.92 What metabolic substrate(s) can be produced from the carbon atoms of each of the following amino acids? (18.7)
a. histidine **b.** isoleucine **c.** serine **d.** phenylalanine

CHALLENGE PROBLEMS

The following problems are related to the topics in this chapter. However, they do not all follow the chapter order, and they require you to combine concepts and skills from several Sections. These problems will help you increase your critical thinking skills and prepare for your next exam.

18.93 One cell at work may break down 2 million (2 000 000) ATP molecules in one second. Some researchers estimate that the human body has about 10^{13} cells. (18.1)
- **a.** How much energy, in kilocalories, would be used by the cells in the body in one day?
- **b.** If ATP has a molar mass of 507 g/mole, how many grams of ATP are hydrolyzed in one day?

18.94 State if each of the following processes release or require ATP: (18.4, 18.5, 18.6)
- **a.** citric acid cycle
- **b.** glucose forms two pyruvate
- **c.** pyruvate forms acetyl CoA
- **d.** glucose forms glucose-6-phosphate
- **e.** oxidation of α-ketoglutarate
- **f.** first six reactions of glycolysis
- **g.** activation of a fatty acid

18.95 Match the following ATP yields to reactions **a** to **g**: (18.4, 18.5, 18.6)

1.5 ATP	2.5 ATP	7 ATP	10 ATP
12 ATP	32 ATP	36 ATP	

- **a.** Glucose forms two pyruvate.
- **b.** Pyruvate forms acetyl CoA.
- **c.** Glucose forms two acetyl CoA.
- **d.** Acetyl CoA goes through one turn of the citric acid cycle.
- **e.** Caproic acid (C_6) is completely oxidized.
- **f.** NADH + H^+ is oxidized to NAD^+.
- **g.** $FADH_2$ is oxidized to FAD.

18.96 Identify each of the following reactions **a** to **e** in the β oxidation of palmitic acid, a C_{16} fatty acid, as (18.7)

(1) activation (2) first oxidation
(3) hydration (4) second oxidation
(5) cleavage

- **a.** Palmitoyl CoA and FAD form α,β-unsaturated palmitoyl CoA and $FADH_2$.
- **b.** β-Keto palmitoyl CoA forms myristoyl CoA and acetyl CoA.
- **c.** Palmitic acid, CoA, and ATP form palmitoyl CoA.
- **d.** α,β-Unsaturated palmitoyl CoA and H_2O form β-hydroxy palmitoyl CoA.
- **e.** β-Hydroxy palmitoyl CoA and NAD^+ form β-keto palmitoyl CoA and NADH + H^+.

18.97 Which of the following molecules will produce the most ATP per mole? (18.5, 18.6)
- **a.** glucose or maltose
- **b.** myristic acid, $CH_3—(CH_2)_{12}—COOH$, or stearic acid, $CH_3—(CH_2)_{16}—COOH$
- **c.** glucose or two acetyl CoA
- **d.** glucose or caprylic acid (C_8)
- **e.** citrate or succinate in one turn of the citric acid cycle

18.98 Which of the following molecules will produce the most ATP per mole? (18.5, 18.6)
- **a.** glucose or stearic acid (C_{18})
- **b.** glucose or two pyruvate
- **c.** two acetyl CoA or one palmitic acid (C_{16})
- **d.** lauric acid (C_{12}) or palmitic acid (C_{16})
- **e.** α-ketoglutarate or fumarate in one turn of the citric acid cycle

ANSWERS

Answers to Selected Practice Problems

18.1 The digestion of polysaccharides takes place in stage 1.

18.3 **a.** anabolic **b.** anabolic **c.** catabolic **d.** catabolic

18.5 The hydrolysis of the phosphodiester bond in ATP releases energy that is sufficient for energy-requiring processes in the cell.

18.7 Hydrolysis is the main reaction involved in the digestion of carbohydrates.

18.9 The bile salts emulsify fat to give small fat globules for lipase hydrolysis.

18.11 The digestion of proteins begins in the stomach and is completed in the small intestine.

18.13 a. coenzyme A **b.** NAD^+ **c.** FAD

18.15 a. NADH **b.** FAD

18.17 FAD

18.19 glucose

18.21 ATP is required in phosphorylation reactions.

18.23 ATP is produced in glycolysis by transferring a phosphate from 1,3-bisphosphoglycerate and from phosphoenolpyruvate directly to ADP.

18.25 glyceraldehyde-3-phosphate and dihydroxyacetone phosphate

18.27 a. 1 ATP is required.
b. 1 NADH is produced for each triose.
c. 2 ATP and 2 NADH are produced.

18.29 Aerobic (oxygen) conditions are needed.

18.31 The oxidation of pyruvate converts NAD^+ to NADH and H^+ and produces acetyl CoA and CO_2.

$$\text{Pyruvate} + NAD^+ + HS{-}CoA \longrightarrow \text{acetyl CoA} + CO_2 + NADH + H^+$$

18.33 When pyruvate is reduced to lactate, the NAD^+ is used to oxidize glyceraldehyde-3-phosphate, which recycles NADH.

18.35 $2CO_2$, $3NADH + 3H^+$, $FADH_2$, GTP (ATP), and HS—CoA

18.37 a. Two reactions, reactions 3 and 4, involve oxidation and decarboxylation.
b. Reaction 6 involves dehydration.
c. Reactions 3, 4, and 8 involve reduction of NAD^+.

18.39 In reaction 6, a carbon–carbon double bond (C=C) is formed.

18.41 a. citrate and isocitrate
b. In decarboxylation, a carbon atom is lost as CO_2.
c. α-ketoglutarate
d. Reactions 3 (isocitrate dehydrogenase) and 4 (α-ketoglutarate dehydrogenase) are decarboxylation reactions.

18.43 NADH and $FADH_2$

18.45 The mobile carrier coenzyme Q transfers electrons from complex I to III.

18.47 NADH transfers electrons to complex I to give NAD^+.

18.49 In oxidative phosphorylation, the energy from the oxidation reactions in electron transport is used to drive ATP synthesis.

18.51 Hydrogen ions return to a lower energy in the matrix by passing through ATP synthase, which releases energy to drive the synthesis of ATP.

18.53 The reduced coenzymes NADH and $FADH_2$ from glycolysis and the citric acid cycle transfer electrons to the electron transport system, which releases energy to drive the synthesis of ATP.

18.55 a. 2.5 ATP **b.** 7 ATP **c.** 5 ATP **d.** 10 ATP

18.57 a. 3 cycles of β oxidation **b.** 4 acetyl CoA
c. 50 ATP

18.59 a. 8 cycles of β oxidation **b.** 9 acetyl CoA
c. 90 ATP from 9 acetyl CoA (citric acid cycle) + 20 ATP from 8 NADH + 12 ATP from 8 $FADH_2$ − 2 ATP (activation) = 122 − 2 = 120 ATP

18.61 Ketone bodies form in the body when excess acetyl CoA results from the breakdown of large amounts of fat.

18.63 High levels of ketone bodies lead to ketosis, a condition characterized by acidosis (a drop in blood pH values), excessive urination, and strong thirst.

18.65 a. $CH_3{-}\overset{O}{\overset{\|}{C}}{-}COO^-$ **b.** $H{-}\overset{O}{\overset{\|}{C}}{-}COO^-$

18.67 NH_4^+ is toxic if allowed to accumulate in the body.

18.69 a. pyruvate **b.** oxaloacetate and fumarate
c. fumarate **d.** α-ketoglutarate

18.71 $\underset{\text{Pyruvate}}{CH_3{-}\overset{O}{\overset{\|}{C}}{-}COO^-} + \underset{\text{Glutamate}}{^-OOC{-}\overset{\overset{+}{N}H_3}{\overset{|}{C}H}{-}CH_2{-}CH_2{-}COO^-}$

18.73 a. Five cycles of β oxidation are needed.
b. Six acetyl CoA are produced.
c.

Activation		$\longrightarrow$	−2 ATP
6 acetyl CoA	× 10 ATP/acetyl CoA	$\longrightarrow$	60 ATP
5 NADH	× 2.5 ATP/NADH	$\longrightarrow$	12.5 ATP
5 $FADH_2$	× 1.5 ATP/$FADH_2$	$\longrightarrow$	7.5 ATP
Total			78 ATP

18.75 a. carbohydrate **b.** fat **c.** carbohydrate
d. fat **e.** protein **f.** carbohydrate

18.77 $ATP + H_2O \longrightarrow ADP + P_i + 7.3$ kcal/mole (31 kJ/mole)

18.79 Lactose undergoes digestion in the mucosal cells of the small intestine to yield galactose and glucose.

18.81 Glucose is the reactant, and pyruvate is the product of glycolysis.

18.83 Pyruvate is converted to lactate when oxygen is not present in the cell (anaerobic conditions) to regenerate NAD^+ for glycolysis.

18.85 The oxidation reactions of the citric acid cycle produce a source of reduced coenzymes for electron transport and ATP synthesis.

18.87 The oxidized coenzymes NAD^+ and FAD needed for the citric acid cycle are regenerated by electron transport, which requires oxygen.

18.89 Energy released as H^+ ions flow through ATP synthase and then back to the matrix is utilized for the synthesis of ATP.

18.91 The oxidation of glucose to pyruvate produces 7 ATP whereas the oxidation of glucose to CO_2 and H_2O produces 32 ATP.

18.93 a. 21 kcal **b.** 1500 g of ATP

18.95 a. 7 ATP/glucose **b.** 2.5 ATP/pyruvate
c. 12 ATP/glucose **d.** 10 ATP/acetyl CoA
e. 36 ATP/C_6 acid **f.** 2.5 ATP/NADH
g. 1.5 ATP/$FADH_2$

18.97 a. maltose **b.** stearic acid **c.** glucose
d. caprylic acid **e.** citrate

CI.33 Beano contains an enzyme that breaks down polysaccharides into mono- and disaccharides that are more digestible. It is used to diminish gas formation that can occur after eating foods such as beans or cruciferous vegetables like cabbage, Brussels sprouts, and broccoli. (16.4, 16.5)

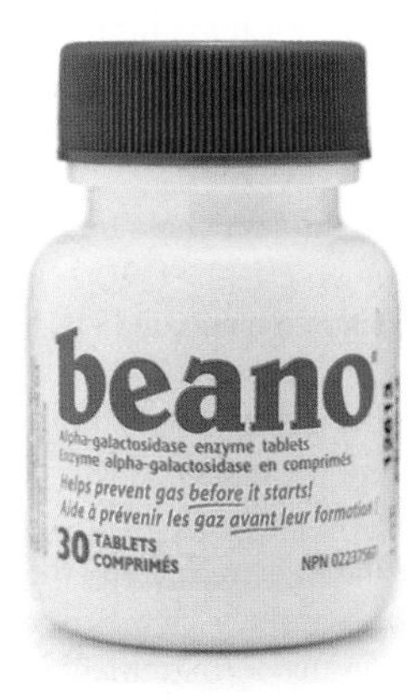

Beano is a dietary supplement that is used to reduce gas.

a. The label on beano says "contains alpha-galactosidase." What class of enzyme is present in beano?
b. What is the substrate for the enzyme?
c. The directions indicate you should not heat or cook beano. Why?

CI.34 In response to signals from the nervous system, the hypothalamus secretes a polypeptide hormone known as gonadotropin-releasing factor (GnRF), which stimulates the pituitary gland to release other hormones into the bloodstream. Two of these hormones are luteinizing hormone (LH) and follicle-stimulating hormone (FSH). GnRF is a decapeptide with the following primary structure: (16.1, 16.2)

Glu—His—Trp—Ser—Tyr—Gly—Leu—Arg—Pro—Gly

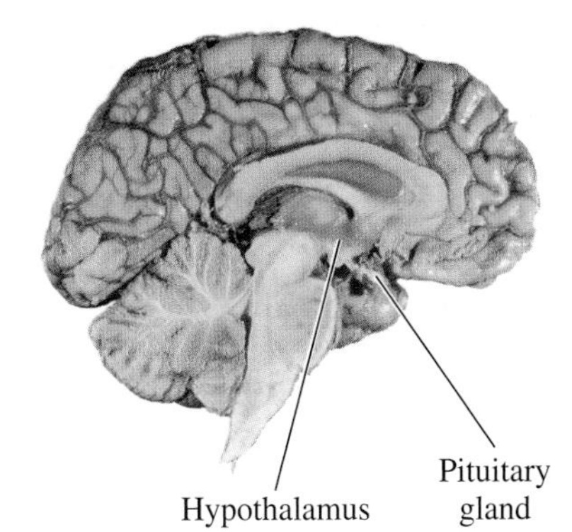

The hypothalamus secretes GnRF.

a. What is the amino acid at the N-terminus of GnRF?
b. What is the amino acid at the C-terminus of GnRF?
c. Which amino acids in GnRF are nonpolar or polar neutral?
d. Draw the condensed structural formulas for the acidic or basic amino acids in GnRF at physiological pH.

CI.35 Identify each of the following as a substance that is part of the citric acid cycle, electron transport, or both: (18.4, 18.5, 18.6)

a. succinate
b. $CoQH_2$
c. FAD
d. cyt *c*
e. H_2O
f. malate
g. NAD^+

CI.36 Use the energy value of 7.3 kcal per mole of ATP and determine the total energy, in kilocalories, stored as ATP from each of the following: (18.4, 18.5, 18.6)

a. the reactions of 1 mole of glucose in glycolysis
b. the oxidation of 2 moles of pyruvate to 2 moles of acetyl CoA
c. the complete oxidation of 1 mole of glucose to CO_2 and H_2O
d. the β oxidation of 1 mole of lauric acid, a C_{12} fatty acid, in β oxidation
e. the reaction of 1 mole of glutamate (from protein) in the citric acid cycle

CI.37 Acetyl CoA is the fuel for the citric acid cycle. It has the formula $C_{23}H_{38}N_7O_{17}P_3S$. (7.2, 18.3, 18.5)

a. What are the components of acetyl CoA?
b. What is the function of acetyl CoA?
c. Where does the acetyl group attach to CoA?
d. What is the molar mass, to three significant figures, of acetyl CoA?

CI.38 Behenic acid is a C_{22} saturated fatty acid found in peanut and canola oils. (18.7)

Behenic acid is one of the fatty acids found in peanut oil.

a. Draw the condensed structural formula for the activated form of behenic acid.
b. Indicate the α- and β-carbon atoms in behenoyl CoA.
c. How many cycles of β oxidation are needed for the complete oxidation of behenic acid?
d. How many acetyl CoA are produced from the complete oxidation of behenic acid?
e. Calculate the total ATP yield from the complete β oxidation of behenic acid by completing the following:

Activation	⟶	−2 ATP
______ acetyl CoA	⟶	______ ATP
______ NADH	⟶	______ ATP
______ $FADH_2$	⟶	______ ATP
Total		______ **ATP**

CI.39 Butter is a fat that contains 80.% by mass triacylglycerols; the rest is water. Assume the triacylglycerol in butter is glyceryl tripalmitate. (2.4, 2.5, 2.6, 7.2, 7.6, 14.4, 18.7)

Butter is high in triacylglycerols.

a. Write the balanced chemical equation for the hydrolysis of glyceryl tripalmitate.
b. What is the molar mass of glyceryl tripalmitate, $C_{51}H_{98}O_6$?
c. Calculate the ATP yield from the complete oxidation of palmitic acid, the C_{16} saturated fatty acid.
d. How many kilocalories are released from the palmitic acid in a 0.50-oz pat of butter?
e. If running for 1 h uses 750 kcal, how many pats of butter would provide the energy (kcal) for a 45-min run?

CI.40 Thalassemia is an inherited genetic mutation that limits the production of the beta chain needed for the formation of hemoglobin. If low levels of the beta chain are produced, there is a shortage of red blood cells (anemia); as a result, the body does not have sufficient amounts of oxygen. In one form of thalassemia, a single nucleotide is deleted in the DNA that codes for the beta chain. The mutation involves the deletion of thymine (T) from section 91 in the informational strand of normal DNA. (16.2, 17.3, 17.4, 17.5, 17.6)

89	90	91	92	93	94	95
—AGT	GAG	CTG	CAC	TGT	GAC	A. . . .

a. Write the complementary (template) strand for the normal DNA segment.
b. Write the mRNA sequence from normal DNA using the template strand in part **a**.
c. What amino acids are placed in the beta chain by this portion of mRNA?
d. What is the order of nucleotides in the mutated DNA segment?
e. Write the template strand for the mutated DNA segment.
f. Write the mRNA sequence from the mutated DNA segment, using the template strand in part **e**.
g. What amino acids are placed in the beta chain by the mutated DNA segment?
h. What type of mutation occurs in this form of thalassemia?
i. How might the properties of this section of the beta chain be different from the properties of the normal protein?
j. How might the level of structure in hemoglobin be affected if beta chains are not produced?

ANSWERS

CI.33 **a.** An alpha-galactosidase is a hydrolase.
b. The substrate is the $\alpha(1\rightarrow4)$-glycosidic bond of galactose.
c. High temperatures will denature the hydrolase enzyme so it no longer functions.

CI.35 **a.** citric acid cycle
b. electron transport
c. both
d. electron transport
e. both
f. citric acid cycle
g. both

CI.37 **a.** aminoethanethiol, pantothenic acid, phosphorylated adenosine diphosphate
b. Coenzyme A carries an acetyl group to the citric acid cycle for oxidation.
c. The acetyl group links to the S atom in the aminoethanethiol part of CoA.
d. 810. g/mole

CI.39 **a.**

$$\begin{array}{l} CH_2-O-\overset{O}{\overset{\|}{C}}-(CH_2)_{14}-CH_3 \\ | \\ CH-O-\overset{O}{\overset{\|}{C}}-(CH_2)_{14}-CH_3 \\ | \\ CH_2-O-\overset{O}{\overset{\|}{C}}-(CH_2)_{14}-CH_3 \end{array} + 3H_2O \longrightarrow \begin{array}{l} CH_2-OH \\ | \\ CH-OH \\ | \\ CH_2-OH \end{array} + 3HO-\overset{O}{\overset{\|}{C}}-(CH_2)_{14}-CH_3$$

b. glyceryl tripalmitate, 807 g/mole
c. 106 ATP
d. 33 kcal
e. 17 pats of butter

Credits

Cover Ralph Clevenger/Corbis Documentary/Getty Images
p. iv Nyaivanova/Shutterstock
p. iv AVAVA/Shutterstock
p. iv Network Photographer/Alamy Stock Photo
p. v Martin Harvey/Alamy Stock Photo
p. v Tyler Olson/Shutterstock
p. v Design Pics Inc/Alamy Stock Photo
p. vi Javier Larrea/AGE Fotostock
p. vi Adam Gault/Alamy
p. vi AJPhoto/Science Source
p. vii Lisa S./Shutterstock
p. vii Corepics/Fotolia
p. vii Viappy/Fotolia
p. viii Blend Images/Alamy Stock Photo
p. viii Bart Coenders/Coenders/E+/Getty Images
p. viii Michaeljung/Fotolia
p. ix Tetra Images/Alamy Stock Photo
p. ix RGB Ventures/SuperStock/Alamy Stock Photo
p. x KatarzynaBialasiewicz/iStock/Getty Images Plus
p. xiv Ladybug Photography by Bekky Murphy

Chapter 1

p. 1 Nyaivanova/Shutterstock
p. 2 *middle left:* Pearson Education, Inc.
p. 2 *bottom left:* Eric Schrader/Pearson Education, Inc.
p. 3 J. R. Eyerman/The LIFE Picture Collection/Getty Images
p. 4 *top right:* Peter van Evert/Alamy Stock Photo
p. 4 *middle left:* Slavko Sereda/Shutterstock
p. 4 *middle center:* Webphotographeer/Webphotographeer/E+/Getty Images
p. 4 *bottom left:* Pressmaster/Shutterstock
p. 5 Evgenyb/iStock/Getty Images
p. 8 *top left:* Russ Lappa/Pearson Education, Inc./Prentice Hall
p. 8 *bottom left:* Chris Schmidt/Schmidt/E+/Getty Images
p. 12 *top center:* Dmitry Rukhlenko/Fotolia
p. 12 *bottom left:* SpxChrome/Getty Images
p. 13 Kevin Peschke/Alamy Stock Photo
p. 17 *top left:* Jason Stitt/Shutterstock
p. 17 *bottom right:* Science Photo Library/Alamy Stock Photo
p. 19 Anyaivanova/Shutterstock
p. 20 *bottom left:* Pearson Education, Inc.
p. 20 *bottom right:* Pressmaster/Shutterstock
p. 21 *top left:* Chris Schmidt/Schmidt/E+/Getty Images
p. 21 *top right:* Kevin Peschke/Alamy Stock Photo
p. 21 *middle left:* Jason Stitt/Shutterstock
p. 22 OSTILL/iStock/Getty Images

Chapter 2

p. 25 AVAVA/Shutterstock
p. 26 *middle center:* Rod Ferris/Shutterstock
p. 26 *bottom left:* Andersen Ross/Blend Images/Getty Images
p. 26 *bottom center:* Pearson Education, Inc.
p. 27 *top right:* Richard Megna/Fundamental Photographs
p. 27 *middle center:* Pearson Education, Inc.
p. 27 *bottom right:* Pearson Education, Inc.
p. 44 sly/Fotolia
p. 30 Seth Christie/Getty Images
p. 32 Radu Razvan/Shutterstock
p. 33 Emin Kuliyev/Shutterstock
p. 36 *middle left:* Legolex/Fotolia
p. 36 *bottom left:* Monkey Business Images/Shutterstock
p. 37 *top left:* Eric Schrader/Pearson Education, Inc.
p. 37 *bottom right:* Pearson Education, Inc.
p. 40 *top left:* Eric Schrader/Pearson Education, Inc.
p. 40 *middle left:* Eric Schrader/Pearson Education, Inc.
p. 40 *bottom left:* Eric Schrader/Pearson Education, Inc.
p. 41 *middle right:* Julie Woodhouse f/Alamy Stock Photo
p. 41 *top right:* Peter Jordan/Alamy Stock Photo
p. 41 *bottom right:* ExQuisine/Fotolia
p. 44 Stephen Mcsweeny/Shutterstock
p. 45 WaveBreakMediaMicro/Fotolia
p. 46 Gina Sanders/Fotolia
p. 47 Pearson Education, Inc.
p. 48 *bottom left:* Pearson Education, Inc.
p. 48 *bottom right:* Pearson Education, Inc.
p. 49 *top left:* Pietro M. Motta/Science Source
p. 49 *top center:* Pietro M. Motta/Science Source
p. 49 *top right:* Voisin/Phanie/Science Source
p. 49 *bottom:* Syner-Comm/Alamy
p. 50 JPC-PROD/Fotolia
p. 51 Dallas Events Inc/Shutterstock
p. 52 *top left:* AVAVA/Shutterstock
p. 52 *top right:* Editorial Image, LLC/Alamy Stock Photo
p. 53 *2.3:* Emin Kuliyev/Shutterstock
p. 53 *2.5:* Eric Schrader/Pearson Education, Inc.
p. 53 *2.6:* Gina Sanders/Fotolia
p. 53 *2.7:* Pearson Education, Inc.
p. 55 Romans14/Fotolia
p. 56 Bugtiger/Fotolia
p. 57 Cristian Ciureanu/Alamy Stock Photo

Chapter 3

p. 60 Network Photographer/Alamy Stock Photo
p. 61 Pearson Education, Inc.
p. 61 Richard Megna/Fundamental Photographs
p. 61 Pearson Education, Inc.
p. 61 *top right:* Norman Chan/Fotolia
p. 61 *middle right:* Lenscap/Alamy Stock Photo
p. 61 *bottom right:* Steven May/Alamy Stock Photo
p. 62 Fotolia
p. 62 Pearson Education, Inc.
p. 62 Dvs71/Fotolia
p. 62 Pearson Education, Inc.
p. 62 *middle left:* Jules Selmes/David Murray/Dorling Kindersley Limited
p. 62 *bottom left:* Phawat/Fotolia
p. 63 *top left:* Richard Megna/Fundamental Photographs
p. 63 *top center:* Pearson Education, Inc.
p. 63 *middle right:* Charles Stirling (Diving)/Alamy Stock Photo
p. 64 *middle left:* Siede Preis/Getty Images
p. 64 *bottom left:* Justinb/Fotolia
p. 64 *bottom right:* Lisay/Getty Images
p. 65 *top right:* Elena Elisseeva/Shutterstock
p. 65 *bottom right:* Rafa Irusta/Shutterstock
p. 66 Westend61 GmbH/Creativ Studio Heinemann/Alamy Stock Photo
p. 68 Jordache/Shutterstock
p. 69 Phanie/Garo/Phanie/Alamy Stock Photo
p. 72 *top left:* Jfergusonphotos/Fotolia
p. 72 *bottom left:* Ian Hooton/SPL/Alamy Stock Photo
p. 74 NASA
p. 77 Stockbyte/Getty Images
p. 78 Celso Pupo/Fotolia
p. 79 *top right:* Alan Holden/Alamy Stock Photo
p. 79 *bottom right:* Joe/iStock/Getty Images
p. 80 John A. Rizzo/Getty Images
p. 80 John A. Rizzo/Getty Images
p. 80 Mark Downey/Getty Images
p. 81 Wsphotos/Getty Images
p. 82 John Scrivener/Getty Images
p. 83 *top center:* Pearson Education, Inc.
p. 84 John A. Rizzo/Getty Images
p. 84 John A. Rizzo/Getty Images
p. 84 Mark Downey/Getty Images
p. 86 Nikkytok/Fotolia
p. 87 Network Photographer/Alamy Stock Photo
p. 88 *bottom left:* Norman Chan/Fotolia
p. 88 *bottom right:* Justinb/Fotolia
p. 89 *middle left:* Jfergusonphotos/Fotolia
p. 89 *top right:* Celso Pupo/Fotolia
p. 91 *bottom left:* Pearson Education, Inc.
p. 91 *bottom center:* Spencer Jones/Getty Images
p. 91 *bottom right:* Teleginatania/Fotolia
p. 91 *middle right:* Starfotograf/iStock/Getty Images
p. 92 *middle left:* Fuat Kose/Getty Images
p. 93 Pearson Education, Inc.

p. 96 *top left:* Impala/Fotolia
p. 96 *middle left:* Shalom Ormsby/Getty Images
p. 96 *bottom right:* Pearson Education, Inc.
p. 97 *top left:* Aleksandr Volkov/Fotolia
p. 97 *top right:* Fuse/Getty Images

Chapter 4

p. 98 Martin Harvey/Alamy Stock Photo
p. 99 Pearson Education, Inc.
p. 100 Pearson Education, Inc.
p. 100 Pearson Education, Inc.
p. 100 Pearson Education, Inc.
p. 100 Pearson Education, Inc.
p. 100 Pearson Education, Inc./Pearson Science
p. 100 Marcel/Fotolia
p. 102 Pearson Education, Inc.
p. 102 Pearson Education, Inc.
p. 102 Pearson Education, Inc.
p. 102 Pearson Education, Inc.
p. 103 Pearson Education, Inc.
p. 104 Polushkin Ivan/Shutterstock
p. 106 *top left:* Pearson Education, Inc.
p. 106 *bottom left:* Graham J. Hills/Science Source
p. 108 Pearson Education, Inc.
p. 110 *top left:* Eric Schrader/Fundamental Photographs
p. 110 *top right:* Pearson Education, Inc.
p. 111 *bottom left:* Atiketta Sangasaeng/Shutterstock
p. 111 *bottom right:* Vtwinpixel/Getty Images
p. 113 Pearson Education, Inc.
p. 115 Okhotnikova/Fotolia
p. 119 *top:* Bruno Boissonnet/Science Source
p. 119 *bottom:* Cindy Minear/Shutterstock
p. 127 *bottom right:* Nigel Cattlin/Alamy Stock Photo
p. 128 Pearson Education, Inc.
p. 131 Oktay Ortakcioglu/Getty Images

Chapter 5

p. 136 Tyler Olson/Shutterstock
p. 138 Celig/Shutterstock
p. 139 Josh Sher/Science Source
p. 141 Uberphotos/Uberphotos/E+/Getty Images
p. 142 Pearson Education, Inc.
p. 143 BSIP SA/Alamy Stock Photo
p. 146 Pearson Education, Inc.
p. 147 *bottom left:* Pool for Yomiuri/Yomiuri Shimbun/AP Images
p. 147 *bottom center:* Don Farrall/Stockbyte/Getty Images
p. 148 U.S. Food & Drug Administration
p. 148 Pearson Education, Inc.
p. 149 Jürgen Schulzki/Alamy Stock Photo
p. 152 Library of Congress Prints and Photographs Division
p. 153 Jihad Siqlawi/AFP/Getty Images
p. 155 Editorial Image, LLC/Alamy Stock Photo
p. 156 Burger/Phanie/SuperStock
p. 156 Pasieka/Science Source
p. 156 Lawrence Berkeley National Library/Photodisc/Getty Images
p. 156 *bottom left:* National Cancer Institute/Photodisc/Getty Images
p. 156 *bottom right:* GJLP/Science Source
p. 157 *left:* Karen C. Timberlake
p. 157 *right:* Cytyc Hologic Corporation
p. 161 *middle left:* Foto-rolf/iStock/Getty Images
p. 161 *bottom left:* Tyler Olson/Shutterstock
p. 161 *bottom right:* CNRI/Science Source
p. 162 Pearson Education, Inc.
p. 163 Lawrence Berkeley National Library/Photodisc/Getty Images
p. 164 Pearson Education, Inc.

Chapter 6

p. 168 Don Hammond/Design Pics Inc/Alamy Stock Photo
p. 172 Pearson Education, Inc.
p. 174 *top left:* J. Palys/Shutterstock
p. 174 *middle left:* Richard Megna/Fundamental Photographs
p. 174 *middle right:* Pearson Education, Inc.
p. 174 *center:* Pearson Education, Inc.
p. 174 *bottom center:* Pearson Education, Inc.
p. 177 Pearson Education, Inc.
p. 179 Gary Blakeley/Shutterstock
p. 180 Mark Huls/Fotolia
p. 181 *middle left:* Pearson Education, Inc.
p. 181 *middle right:* Pearson Education, Inc.
p. 182 Smirnov Vladimir Itar-Tass Photos/Newscom
p. 183 Rachel Youdelman/Pearson Education, Inc.
p. 184 Studiomode/Alamy Stock Photo
p. 203 Don Hammond/Design Pics Inc/Alamy Stock Photo
p. 204 *bottom center:* Pearson Education, Inc.
p. 204 *bottom center:* Pearson Education, Inc.
p. 205 Kameel4u/Shutterstock
p. 214 *middle left:* Pearson Education, Inc.
p. 214 Augustin Ochsenreiter/Suedtiroler Archaeologiemuseum/dapd/AP Images
p. 214 *bottom right:* Editorial Image, LLC/Alamy
p. 215 Editorial Image, LLC/Alamy

Chapter 7

p. 216 Javier Larrea/AGE Fotostock
p. 217 *middle left:* Pearson Education, Inc.
p. 217 *middle right:* Lissart/Lissart/E+/Getty Images
p. 218 Reika/Shutterstock
p. 219 Pearson Education, Inc.
p. 221 *top left:* Helen Sessions/Alamy Stock Photo
p. 221 *bottom left:* Pearson Education, Inc.
p. 222 Pearson Education, Inc.
p. 222 Pearson Education, Inc.
p. 222 Pearson Education, Inc.
p. 222 Pearson Education, Inc.
p. 222 Pearson Education, Inc.
p. 222 *middle left:* Phil Degginger/Alamy Stock Photo
p. 227 *top left:* Pearson Education, Inc.
p. 224 Ruzanna/Shutterstock
p. 225 Pearson Education, Inc.
p. 226 Edward Westmacott/Shutterstock
p. 227 *middle left:* Pearson Education, Inc.
p. 227 *middle right:* Sciencephotos/Alamy Stock Photo
p. 227 *bottom left:* Lawrence Migdale/Science Source
p. 227 *bottom right:* Paul Michael Hughes/Shutterstock
p. 228 Pearson Education, Inc.
p. 230 Pearson Education, Inc.
p. 232 Richard Megna/Fundamental Photographs
p. 234 *middle left:* Pearson Education, Inc.
p. 234 *center:* Pearson Education, Inc.
p. 234 *middle right:* Pearson Education, Inc.
p. 235 *top center:* Pearson Education, Inc.
p. 235 *middle left:* Pearson Education, Inc.
p. 235 *middle right:* Pearson Education, Inc.
p. 236 *top center:* Pearson Education, Inc.
p. 236 *middle left:* Sergiy Zavgorodny/Shutterstock
p. 238 *top left:* Getty Images
p. 238 *bottom left:* Stinkyt/iStock/Getty Images
p. 239 *middle left:* Pearson Education, Inc.
p. 239 *middle center:* Pearson Education, Inc.
p. 239 *middle right:* Pearson Education, Inc.
p. 241 *bottom left:* Pearson Education, Inc.
p. 241 *bottom center:* Pearson Education, Inc.
p. 241 *bottom right:* Pearson Education, Inc.
p. 242 *middle left:* Pearson Education, Inc.
p. 242 *middle center:* Pearson Education, Inc.
p. 242 *middle right:* Pearson Education, Inc.
p. 244 Thinkstock/Getty Images
p. 246 Richard Megna/Fundamental Photographs
p. 247 Pearson Education, Inc.
p. 250 *top left:* Dario Secen/Lumi/Getty Images
p. 250 *bottom left:* Lissart/Lissart/E+/Getty Images
p. 250 *bottom right:* Pearson Education, Inc.
p. 251 *top:* Pearson Education, Inc.
p. 251 *top left:* Pearson Education, Inc.
p. 251 *top center:* Pearson Education, Inc.
p. 251 *top right:* Pearson Education, Inc.
p. 251 *middle right:* Thinkstock/Getty Images
p. 252 *bottom left:* Gabriele Rohde/Fotolia
p. 252 *middle right:* Ljupco Smokovski/Fotolia
p. 254 *top left:* Helen Sessions/Alamy Stock Photo
p. 254 *middle left:* Pearson Education, Inc.
p. 256 *bottom left:* Pearson Education, Inc.
p. 256 *bottom right:* Pearson Education, Inc.
p. 259 Adam Gault/Science Photo Library/Alamy Stock Photo

Chapter 8

p. 261 NASA
p. 262 Koszivu/Fotolia
p. 263 GybasDigiPhoto/Shutterstock
p. 264 Kenneth William Caleno/Shutterstock
p. 265 Levent Konuk/Shutterstock
p. 268 Larry Brownstein/Photodisc/Getty Images
p. 271 Prasit Rodphan/Shutterstock
p. 272 Andrey Nekrasov/Alamy Stock Photo
p. 275 Tomasz Wojnarowicz/Fotolia
p. 278 Mic Smith/Alamy Stock Photo
p. 279 Pearson Education, Inc.
p. 280 Andrey Nekrasov/Alamy Stock Photo
p. 282 Library of Congress

Chapter 9

p. 285 AJPhoto/Science Source
p. 286 *left:* Pearson Education, Inc.
p. 286 *right:* Pearson Education, Inc.
p. 287 Pearson Education, Inc.
p. 288 *top:* Thinkstock Images/Getty Images
p. 288 *bottom:* Pearson Education, Inc.
p. 289 Pearson Education, Inc.
p. 290 *top:* Pearson Education, Inc.
p. 290 *center:* Pearson Education, Inc.
p. 290 *bottom:* Pearson Education, Inc.
p. 293 Comstock Images/Getty Images
p. 294 *middle left:* Pearson Education, Inc.
p. 294 *middle right:* Pearson Education, Inc.
p. 295 *left:* Dr. P. Marazzi/Science Source
p. 295 *right:* remik44992/Shutterstock
p. 296 *bottom center:* Pearson Education, Inc.
p. 297 *left:* Pearson Education, Inc.
p. 297 *middle left:* Pearson Education, Inc.
p. 297 *middle right:* Pearson Education, Inc.
p. 297 *right:* Pearson Education, Inc.
p. 297 *bottom right:* CNRI/Science Source
p. 299 *middle right:* Pearson Education, Inc.
p. 299 *bottom right:* Pearson Education, Inc.
p. 301 *top:* Pearson Education, Inc.
p. 301 *bottom:* Pearson Education, Inc.
p. 302 Pearson Education, Inc.
p. 306 *middle left:* Pearson Education, Inc.
p. 306 *middle center:* Pearson Education, Inc.
p. 306 *middle right:* Pearson Education, Inc.
p. 306 *bottom left:* Pearson Education, Inc.
p. 306 *bottom right:* Pearson Education, Inc.
p. 311 *top right:* Thinkstock Images/ Getty Images
p. 311 *bottom left:* Pearson Education, Inc.
p. 311 *bottom center:* Pearson Education, Inc.
p. 311 *bottom right:* Pearson Education, Inc.
p. 313 Picsfive/Shutterstock
p. 314 *top:* AJPhoto/Science Source
p. 314 *bottom:* Pearson Education, Inc.
p. 315 *top left:* Pearson Education, Inc.
p. 315 *top right:* Pearson Education, Inc.
p. 315 *middle left:* Pearson Education, Inc.
p. 317 *middle left:* Stepan Popov/ Getty Images
p. 317 *bottom left:* Brand X Pictures/ Getty images
p. 321 *middle left:* Wrangler/Shutterstock
p. 321 *top right:* Eric Schrader/Pearson Education, Inc.
p. 321 *middle right:* IS293/Alamy Stock Photo
p. 322 *top left:* Donvictorio/Shutterstock
p. 322 *middle left:* Patsy Michaud/ Shutterstock
p. 322 *bottom left:* Pearson Education, Inc.
p. 322 *top right:* Photolibrary/Getty Images
p. 322 *middle right:* Mark sykes/Science Photo Library/Alamy Stock Photo
p. 322 *bottom right:* Rostislav Sedlacek/ Fotolia
p. 323 Isifa Image Service/Alamy Stock Photo

Chapter 10

p. 324 Lisa S./Shutterstock
p. 325 Kristina Afanasyeva/iStock/ Getty Images
p. 326 *top:* Kul Bhatia/Science Source
p. 326 *bottom:* Lukas Gojda/Shutterstock
p. 331 *bottom left:* Pearson Education, Inc.
p. 331 *bottom right:* Pearson Education, Inc.
p. 332 *middle left:* Richard Megna/ Fundamental Photographs
p. 333 *left:* Keith Homan/Alamy Stock Photo
p. 333 *right:* Eric Schrader/Fundamental Photographs
p. 336 Incamerastock/Alamy Stock Photo
p. 341 *top:* Pearson Education, Inc.
p. 341 *top center:* Pearson Education, Inc.
p. 341 *top right:* Pearson Education, Inc.
p. 341 *middle left:* Pearson Education, Inc.
p. 341 *middle center:* Pearson Education, Inc.
p. 341 *middle right:* Pearson Education, Inc.
p. 341 *bottom center:* Pearson Education, Inc.
p. 341 *bottom right:* Pearson Education, Inc.
p. 341 *bottom left:* Pearson Education, Inc.
p. 342 *top left:* Pearson Education, Inc.
p. 342 *top center:* Pearson Education, Inc.
p. 342 *top right:* Pearson Education, Inc.
p. 342 *middle left:* Alexander Gospodinov/ Fotolia
p. 343 Pearson Education, Inc.
p. 345 Charles D. Winters/Science Source
p. 346 Richard Megna/Fundamental Photographs
p. 347 Pearson Education, Inc.
p. 348 *bottom left:* Pearson Education, Inc.
p. 348 *bottom center:* Pearson Education, Inc.
p. 348 *bottom right:* Pearson Education, Inc.
p. 350 Pearson Education, Inc.
p. 355 Lisa S./Shutterstock
p. 356 *middle left:* Pearson Education, Inc.
p. 356 *top right:* Pearson Education, Inc.
p. 356 *middle right:* Pearson Education, Inc.
p. 359 Geotrac/iStock/Getty Images
p. 361 Universal Images Group/SuperStock

Chapter 11

p. 363 Corepics/Fotolia
p. 364 Pearson Education, Inc.
p. 365 *left:* lillisphotography/Getty Images
p. 365 *right:* Pearson Education, Inc.
p. 375 *top:* Pearson Education, Inc.
p. 375 *center:* Roza/Fotolia
p. 375 *bottom:* HansChris/HansChris/ E+/Getty Images
p. 376 *bottom left:* Tim Hall/Getty Images
p. 376 *bottom right:* Glen Jones/ Shutterstock
p. 380 *middle left:* Pearson Education, Inc.
p. 380 *bottom left:* Pearson Education, Inc.
p. 381 Nigel Cattlin/Alamy Stock Photo
p. 383 Pearson Education, Inc.
p. 385 *bottom left:* Alexandr Kan$\tilde{A}\mu$kin/ Shutterstock
p. 385 *bottom center:* Vblinov/Shutterstock
p. 385 *bottom right:* Dirkr Richter/ Getty Images
p. 388 Mediscan/Alamy Stock Photo
p. 389 Corepics/Fotolia
p. 390 *top left:* Pearson Education, Inc.
p. 390 *bottom left:* Roza/Fotolia
p. 390 *middle right:* Pearson Education, Inc.
p. 392 *bottom left:* Pearson Education, Inc.
p. 392 *bottom center:* Sean M. Carroll/ Fotolia
p. 394 Stockbyte/Thinkstock Images/ Getty Images
p. 395 *middle left:* Hemera Technologies/ PhotoObjects/Getty Images
p. 395 *middle right:* Igor kisselev/ Shutterstock

Chapter 12

p. 398 Viappy/Fotolia
p. 402 *middle left:* ASA studio/Shutterstock
p. 402 *bottom right:* Pearson Education, Inc.
p. 403 *top left:* Pearson Education, Inc.
p. 403 *middle right:* Ultrashock/Shutterstock
p. 404 *top center:* Pearson Education, Inc.
p. 404 *bottom right:* Renata Kazakova/ Shutterstock
p. 407 *top right:* Pictorial Press/Alamy Stock Photo
p. 407 *middle right:* Katharine Andriotis/ Alamy Stock Photo
p. 407 *bottom left:* Antagain/Antagain/ E+/Getty images
p. 412 Pearson Education, Inc.
p. 413 Pearson Education, Inc.
p. 414 David Murray/Dorling Kindersley, Ltd.
p. 417 *top right:* Schankz/Shutterstock
p. 417 *bottom right:* Marjan Laznik/ Laznik/E+/Getty Images
p. 418 *middle left:* Pearson Education, Inc.
p. 418 *middle right:* Pearson Education, Inc.
p. 418 *bottom left:* Pearson Education, Inc.
p. 421 *top left:* Viappy/Fotolia
p. 421 *top right:* Wavebreakmedia/ Shutterstock
p. 424 WavebreakmediaMicro/Fotolia
p. 425 *top:* Tim UR/Fotolia
p. 425 *bottom:* C Squared Studios/ Photodisc/Getty Images
p. 430 *top left:* Charles D. Winters/Science Source
p. 430 *middle left:* Eric Schrader/Pearson Education, Inc.
p. 430 *top right:* Gvictoria/IStock/ Getty Images
p. 430 *middle right:* A-wrangler/iStock/ Getty Image
p. 431 *top left:* Kakimage/Alamy Stock Photo
p. 431 *top right:* Mark Mirror/Shutterstock

Chapter 13

p. 432 Rolf Bruderer/Blend Images/Alamy Stock Photo
p. 433 *middle left:* AZP Worldwide/ Shutterstock
p. 433 *middle right:* Pearson Education, Inc.
p. 434 *top:* Pearson Education, Inc.
p. 434 *top left:* Paul Reid/Shutterstock
p. 434 *middle left:* Danny E Hooks/ Shutterstock
p. 434 *center:* Studioshots/Alamy Stock Photo
p. 436 *middle left:* Pearson Education, Inc.
p. 436 *center:* Pearson Education, Inc.
p. 436 *middle right:* Pearson Education, Inc.
p. 437 *top left:* Pearson Education, Inc.
p. 437 *top center:* Pearson Education, Inc.
p. 437 *top right:* Pearson Education, Inc.
p. 441 *middle left:* Q-Images/Alamy Stock Photo
p. 441 *middle right:* Maksim Striganov/ Shutterstock

p. 442 Kristoffer Tripplaar/Alamy Stock Photo
p. 445 *top right:* Pearson Education, Inc.
p. 445 *middle right:* David hancock/Alamy Stock Photo
p. 445 *bottom center:* Courtesy of Thermo Fisher Scientific Inc.
p. 450 Pearson Education, Inc.
p. 451 Pearson Education, Inc.
p. 452 Science Photo Library/Alamy Stock Photo
p. 454 *middle left:* Pearson Education, Inc.
p. 454 *bottom left:* Pearson Education, Inc.
p. 456 Pearson Education, Inc.
p. 457 Keith Brofsky/Photodisc/Getty Images
p. 459 Pearson Education, Inc.
p. 460 Danny E Hooks/Shutterstock
p. 461 Rolf Bruderer/Blend Images/Alamy Stock Photo
p. 463 *middle right:* Pearson Education, Inc.
p. 463 *bottom right:* Danny E Hooks/ Shutterstock
p. 466 *top left:* Pearson Education, Inc.
p. 466 *top right:* Danny E Hooks/ Shutterstock
p. 466 *middle left:* Pearson Education, Inc.
p. 466 *middle right:* Pearson Education, Inc.

Chapter 14

p. 470 BartCo/BartCo/E+/Getty Images
p. 471 *top:* Skynet/Fotolia
p. 471 *bottom:* Pearson Education, Inc.
p. 475 Pearson Education, Inc.
p. 476 *middle left:* Jose Manuel Ribeiro/ Reuters
p. 476 *bottom:* Pearson Education, Inc.
p. 478 *top:* Egorxfi/Fotolia
p. 478 *bottom:* Mates/Fotolia
p. 479 *top:* Digerati/Fotolia
p. 479 *bottom:* Pearson Education, Inc.
p. 480 *top:* Lynn Watson/Shutterstock
p. 480 *bottom:* Luisa Leal/Shutterstock
p. 481 *top:* Pearson Education, Inc.
p. 481 *bottom:* Pearson Education, Inc.
p. 482 Lee Brown/Alamy Stock Photo
p. 483 Hlphoto/Shutterstock
p. 486 *top left:* Koenig/Blickwinkel/Alamy Stock Photo
p. 486 *top right:* Imagestate Media Partners Limited - Impact Photos/Alamy Stock Photo
p. 488 *top:* Pearson Education, Inc.
p. 488 *bottom:* Pearson Education, Inc.
p. 489 Brasil2/iStock/Getty Images
p. 490 *middle left:* Pearson Education, Inc.
p. 490 *bottom left:* Pearson Education, Inc.
p. 490 *bottom right:* De Meester Johan/ Arterra Picture Library/Alamy Stock Photo
p. 494 Pearson Education, Inc.
p. 497 *middle left:* BartCo/BartCo/E+/ Getty Images
p. 497 *bottom left:* United States Department of Agriculture
p. 498 Pearson Education, Inc.
p. 501 DenisNata/Shutterstock
p. 502 *top left:* Monika Wisniewska/ Shutterstock
p. 502 *middle left:* Eric Schrader/Pearson Education, Inc.
p. 502 *top right:* Eric Schrader/Pearson Education, Inc.
p. 504 Oksanka8306/Fotolia
p. 505 Malyshchyts Viktar/Fotolia

Chapter 15

p. 513 *bottom left:* Chas53/Getty Images
p. 513 *bottom right:* Dr3amer/ Getty Images
p. 520 Pearson Education, Inc.
p. 522 Pearson Education, Inc.
p. 531 *top right:* National Heart, Lung, and Blood Institute
p. 531 *middle right:* National Heart, Lung, and Blood Institute
p. 532 Universal Images Group/BSIP/ Getty Images
p. 509 Michaeljung/Fotolia
p. 546 *top left:* David Zaitz/Alamy Stock Photo
p. 546 *top right:* Pearson Education, Inc.
p. 546 *bottom right:* Analia Valeria Urani/ Shutterstock
p. 547 Dustin Dennis/Shutterstock
p. 516 Jupiterimages/Getty Images
p. 517 *middle right:* Tischenko Irina/ Shutterstock
p. 517 *bottom right:* Blickwinkel/Alamy Stock Photo
p. 519 Dawn Wilson Photography/Moment/ Getty Images
p. 518 Kuleczka/Shutterstock
p. 523 Shelby Allison/Shutterstock
p. 527 Audrey Snider-Bell/Shutterstock
p. 529 Santibhavank P/Shutterstock
p. 534 CoverSpot Photography/Alamy Stock Photo
p. 538 *middle left:* Michaeljung/Fotolia
p. 538 *bottom left:* Robert Clare/Alamy Stock Photo
p. 538 *middle right:* Matt Rourke/AP Images
p. 539 *middle left:* Pearson Education, Inc.
p. 539 *middle right:* Pearson Education, Inc.
p. 541 *bottom left:* Ekkachai/Shutterstock
p. 541 *bottom right:* Sampsyseeds/ Getty Images
p. 542 *top right:* Pearson Education, Inc.

Chapter 16

p. 543 *top left:* Egal/iStock/Getty Images
p. 543 *top right:* Lauree Feldman/ Photolibrary/Getty Images
p. 543 *middle right:* Pearson Education, Inc.
p. 548 Tetra Images/Alamy Stock Photo
p. 549 *middle right:* Fuse/Corbis/ Getty Images
p. 549 *bottom left:* Center for Disease Control (CDC) - PHIL
p. 555 Pearson Education, Inc.
p. 557 APP Pharmaceuticals
p. 558 Brasil2/Getty Images
p. 560 Laurence Mouton/PhotoAlto sas/ Alamy Stock Photo
p. 565 Pearson Education, Inc.
p. 572 Halstenbach/Halstenbach/E+/ Getty Images
p. 575 Chris Howes/Wild Places Photography/Alamy Stock Photo
p. 577 Fuse/Corbis/Getty Images
p. 580 *top left:* Jane/Jane/E+/Getty Images
p. 580 *middle left:* Pearson Education, Inc.
p. 580 *bottom left:* Eric Schrader/Pearson Education, Inc.
p. 580 *top right:* Andy Crawford/Dorling Kindersley Limited
p. 580 *top center:* Dave Rudkin/Dorling Kindersley Limited
p. 581 Pearson Education, Inc.

Chapter 17

p. 584 Malcom Park/RGB Ventures/ SuperStock/Alamy Stock Photo
p. 603 khunaspix/123rf
p. 606 Tek Image/Science Source
p. 610 Alfred Pasieka/Science Source
p. 611 Malcom Park/RGB Ventures/ SuperStock/Alamy Stock Photo
p. 612 Mark Thomas/Alamy Stock Photo
p. 613 *top right:* Khunaspix/123rf
p. 613 *middle right:* Tek Image/Science Source

Chapter 18

p. 619 Katarzyna Bialasiewicz/iStock/ Getty Images
p. 620 Viafilms/iStock/Getty Images
p. 623 *top center:* Rukxstockphoto/ Fotolia
p. 623 *middle center:* Koji Aoki/ Getty Images
p. 625 *top right:* Wil & Deni McIntyre/ Getty Images
p. 625 *bottom left:* Science Photo Library/ Alamy Stock Photo
p. 630 Alexey Fursov/Shutterstock
p. 635 Andy Clark/Reuters
p. 643 Jill Lang/Shutterstock
p. 649 *bottom left:* Dpruter/Getty Images
p. 649 *bottom right:* Javier Gonz Alez/ Getty Images
p. 651 Pearson Education, Inc.
p. 652 *middle left:* Dewayne Flowers/ Shutterstock
p. 652 *middle center:* Dmitry Lobanov/ Shutterstock
p. 655 *bottom left:* Katarzyna Bialasiewicz/ iStock/Getty Images
p. 655 *bottom right:* Southern Illinois University/Science Source
p. 657 Viafilms/iStock/Getty Images
p. 658 Pearson Education, Inc.
p. 661 *middle left:* Floortje/Floortje/ E+/Getty Images
p. 661 *bottom left:* Pearson Education, Inc.
p. 661 *bottom right:* MBI/Alamy Stock Photo
p. 664 *top left:* Helen Sessions/Alamy Stock Photo
p. 664 *bottom left:* Martin M. Rotker/ Science Source
p. 664 *middle right:* Eric Schrader/Pearson Education, Inc.
p. 665 Pearson Education, Inc.

Glossary/Index

A

Absolute zero, 68, 261
Acceptor stem, 594
Acetaminophen, 203, 486, 494
Acetic acid, 351–352, 471, 473
Acetyl CoA The compound that forms when a two-carbon acetyl group bonds to coenzyme A, 628–629
Acetylcholinesterase, 574
Acetylene, 376–377
Achirality, 436, 438
Acid A substance that dissolves in water and produces hydrogen ions, according to the Arrhenius theory. All acids are hydrogen ion donors, according to the Brønsted–Lowry theory, 324–362
  Brønsted–Lowry, 327–330
  buffers and, 350–354
  carboxylic, 471–477
  characteristics of, 326
  conjugate acid–base pairs, 328–330
  dissociation of water, 338–340
  equilibrium with bases, 333–337
  naming, 325–326
  pH scale and, 340–346
  reactions of, 346–350
  strengths of, 330–333
  titration of, 348–349
Acid hydrolysis, 482–483, 495
Acid reflux disease, 324, 354
Acidic amino acid An amino acid that has an R group with a carboxylate group, 550
Acidic solutions, 338–339
Acidosis A physiological condition in which the blood pH is lower than 7.35, 352–353, 650
Acids and bases, 323–362
Acquired immune deficiency syndrome (AIDS), 608–609
Actinides, 102, 549
Activation, fatty acid, 645
Activation energy The energy needed upon collision to break apart the bonds of the reacting molecules, 246
Active site A pocket in a part of the tertiary enzyme structure that binds substrate and catalyzes a reaction, 567, 568
Active transport, 537
Activity The rate at which an enzyme catalyzes the reaction that converts substrate to product, 571–576
Acupuncture, 557
Addition
  with measured numbers, 34
  of positive and negative numbers, 11
Addition A reaction in which atoms or groups of atoms bond to a carbon–carbon double bond. Addition reactions include the addition of hydrogen (hydrogenation) and water (hydration), 382–385
Adenine, 586–587, 591–592
  in ATP, 622–623
Adenosine diphosphate (ADP), 622
Adenosine triphosphate. *See ATP*
Adrenal corticosteroids, 535
Advantame, 456
Aerobic An oxygen-containing environment in the cells, 633
AIDS (acquired immune deficiency syndrome), 608–609
Air
  as gas mixture, 277
  as solution, 286
Albinism, 603
Alchemists, 4
Alcohol An organic compound that contains the hydroxyl functional group (—OH) attached to a carbon chain, 384, 399
Aldehyde An organic compound with a carbonyl functional group bonded to at least one hydrogen, 408, 434
Alditols, 451
Aldose A monosaccharide that contains an aldehyde group, 434
Aldosterone, 535
Alkali metals Elements of Group 1A (1) except hydrogen; these are soft, shiny metals with one valence electron, 102
Alkaline earth metals Group 2A (2) elements, which have two valence electrons, 102
Alkaloids, 490
Alkalosis A physiological condition in which the blood pH is higher than 7.45, 353
Alkane A hydrocarbon that contains only single bonds between carbon atoms, 363, 366–376
  condensed structural and line-angle formulas for, 367–368
  cycloalkanes, 368–369
  names and formulas of, 367, 371–372
  properties of, 375–376
  structural formulas for, 373–374
  with substituents, 370–375
Alkene A hydrocarbon that contains one or more carbon–carbon double bonds (C=C), 376–378
  addition reactions for, 382–385
  cis–trans isomers, 379–382
  hydration of, 384
  hydrogenation of, 382–383
  identifying, 376–377
  naming, 377–378
Alkyl group An alkane minus one hydrogen atom. Alkyl groups are named like the alkanes except a *yl* ending replaces *ane*, 371–372
Alkyne A hydrocarbon that contains one or more carbon–carbon triple bonds (C≡C), 376–378
  identifying, 376–377
  naming, 377–378
Allergic reactions, 486
Alpha (α) helix A secondary level of protein structure, in which hydrogen bonds connect the N—H of one peptide bond with the C=O of a peptide bond farther along in the chain to form a coiled or corkscrew structure, 558, 559
Alpha particle A nuclear particle identical to a helium nucleus, symbol α or ${}^{4}_{2}He$, 137, 138
  decay of, 140–141
  shielding against, 139
Altitude, atmospheric pressure and, 260–261, 263–264
Aluminum
  specific heat of, 78
  symbol for, 99
Aluminum carbonate, 184
Aluminum hydroxide, 183
Aluminum oxide, 174, 177
Alveoli, 529
Alzheimer's disease, 560
Amidation, 492, 553
Amide An organic compound containing the carbonyl group attached to an amino group or a substituted nitrogen atom, 491–497
  in health and medicine, 494
  hydrolysis of, 495–496
  naming, 493
  preparation of, 492–493
  solubility of in water, 494
Amine An organic compound containing a nitrogen atom attached to one, two, or three hydrocarbon groups, 484–491
  aromatic, 485–486
  as bases in water, 487–488
  in health and medicine, 486–487
  line-angle formulas for, 485
  naming and classifying, 484–485
  in plants, 490
  solubility of in water, 487
Amino acid The building block of proteins, consisting of an ammonium group, a carboxylate group, and a unique R group attached to the α carbon, 549–553
  ATP energy from, 653–654
  carbon atoms from, 653–654
  classification of, 550
  degradation of in citric acid cycle, 651–653
  essential, 555
  oxidative deamination of, 652
  primary structure and, 555–558
  structure of, 551–552
  transamination of, 652
  urea cycle and, 652
Amino alcohols, 525–528
Aminoacyl–tRNA synthetase, 597
Aminotransferases, 651

B

Beta particle A particle identical to an electron, symbol $_{-1}^{0}e$ or β, that forms in the nucleus when a neutron changes to a proton and an electron, 137–138
decay of, 142–143
shielding against, 139
Beta (β)-pleated sheet A secondary level of protein structure that consists of hydrogen bonds between peptide links in parallel polypeptide chains, 558–559
Bicarbonate, 181
Bicarbonates, acid reactions with, 347, 349–350
Bile salts, 532
Biocatalysts, 248
Biogenic amines, 486
Bismuth, 178
Blood oxygen levels, 216, 249–250
Blood plasma
buffers in, 351, 352–353
electrolytes in, 293
lipoproteins, 532–534
Blood pressure, 25, 52, 262–263
Blood types, carbohydrates and, 456–457
Blood urea nitrogen (BUN), 652
Blubber, 649
Body fat measurement, 40
Body heat, 642
Body temperature variations, 71
Boiling The formation of bubbles of gas throughout a liquid, 82
Boiling point (bp) The temperature at which a liquid changes to gas (boils) and gas changes to liquid (condenses), 82
Bombardment, radioactive isotope production via, 145–146
Bonding pairs, 189–190
Bones
density of, 49
radiological dating of, 153
Boron
molecular compounds, 192
radioisotopes in medicine, 155
symbol for, 99
Boyle's law A gas law stating that the pressure of a gas is inversely related to the volume when temperature and moles of the gas do not change, 265–268
Brachytherapy, 157
Branched alkanes, 370–371
Breast cancer, 157, 584, 611
genetic testing for, 606
Breathing
gas mixture in air, 277
mixtures for, 63
pressure–volume relationship in, 266–267
Bromine, 102
Brønsted–Lowry acids and bases An acid is a hydrogen ion donor, and a base is a hydrogen ion acceptor, 327–330, 487–488
Brown fat, 642
Buckminsterfullerene, 111
Buckyball, 111
Buffer solution A solution of a weak acid and its conjugate base or a weak base and its conjugate acid that maintains the pH by neutralizing added acid or base, 350–354
Burns, 363, 389
Butane, 375
Butanoic acid, 471, 472, 474
Butter, 520
Butyric acid, 471, 472

C

Cadmium
symbol for, 99
Caffeine, 490
Calcium, 98
electron arrangement, 118
essential to health, 105
ions in the body, 172–173
symbol for, 99
Calcium carbonate, 349–350
Calculations, significant figures in, 31–35
Calculators
adding significant zeros with, 33
operations on, 11–12
scientific notation and, 19
Calorie (cal) The amount of heat energy that raises the temperature of exactly 1 g of water by exactly 1 °C, 72–73, 74–77
Calorimeter, 74–75
Camels, fat storage in humps of, 648–649
Cancer, 610
breast, 157, 584, 611
cholesterol and, 531
Kaposi's sarcoma, 608
radiation of, 139
brachytherapy, 157
radiotherapy and, 155
Canola oil, 520
Capric acid, 645–647
Caraway, 441
Carbohydrate A simple or complex sugar composed of carbon, hydrogen, and oxygen, 364, 432–469
blood types and, 456–457
chemical properties of, 450–453
chiral molecules, 436–443
digestion of, 623–625
disaccharides, 453–458
Fischer projections of, 439–440, 443–446
Haworth structures of, 446–450
monosaccharides, 433, 434–435, 443–453
polysaccharides, 459–461
types of, 433–434
Carbon
atomic mass of, 114
Avogadro's number, 217
chiral, 437
electron arrangement, 117
essential to health, 104
forms of, environment and, 111–112
naturally occurring radioisotopes, 149
radioisotopes, 137
symbol for, 99
Carbon cycle, 433
Carbon dating, 152–153
Carbon dioxide, 186
in the atmosphere, 260
climate change and, 73–74
as molecular compound, 185
molecule shape, 196
production of from methane and oxygen, 228–229
Carbon disulfide, 186
Carbon monoxide, 237–238
Carbon-14, 137
beta decay of, 142
half-life of, 152
in radiological dating, 152–153
Carbonates, acid reactions with, 347
Carbonyl group A functional group that contains a carbon–oxygen double bond (C=O), 408
Carboxyl group A functional group found in carboxylic acids composed of carbonyl and hydroxyl groups, 471
Carboxylate ion The anion produced when a carboxylic acid donates a hydrogen ion to water, 473–474
Carboxylate salt A carboxylate ion and the metal ion from the base that is the product of neutralization of a carboxylic acid, 475
Carboxylic acid An organic compound containing the carboxyl group, 471–477
acidity of, 473–474
amides and, 491–497
common names of, 471–472
IUPAC names of, 471, 472
in metabolism, 476
neutralization of, 475–476
properties of, 473–476
solubility of, 473
Carcinogens, 610
Cardiac imaging, 136, 161
Careers
clinical laboratory technician, 324
clinical lipid specialist, 509
diabetes nurse, 432
dialysis nurse, 285
dietitian, 60
emergency medical technician, 363
environmental health practitioner, 470
exercise physiologist, 216
farmer, 98, 127
firefighter, 363
forensic scientist, 1–2
histology technician, 586
pharmacy technician, 168
physician assistant, 548
public health nurse, 619
radiation technologist, 136
registered nurse, 25
respiratory therapist, 259
Carvone, 441
Casein, 549
Catabolic reaction A metabolic reaction that releases energy for the cell by the degradation and oxidation of glucose and other molecules, 620

D

F

I

J

K

L

M

Q

R

T

U

V

Metric and SI Units and Some Useful Conversion Factors

Length	SI Unit Meter (m)	Volume	SI Unit Cubic Meter (m^3)	Mass	SI Unit Kilogram (kg)
1 meter (m) = 100 centimeters (cm)		1 liter (L) = 1000 milliliters (mL)		1 kilogram (kg) = 1000 grams (g)	
1 meter (m) = 1000 millimeters (mm)		1 mL = 1 cm^3		1 g = 1000 milligrams (mg)	
1 cm = 10 mm		1 L = 1.06 quart (qt)		1 kg = 2.20 lb	
1 kilometer (km) = 0.621 mile (mi)		1 qt = 946 mL		1 lb = 454 g	
1 inch (in.) = 2.54 cm (exact)				1 mole = 6.02×10^{23} particles	
				Water density = 1.00 g/mL (at 4°C)	

Temperature	SI Unit Kelvin (K)	Pressure	SI Unit Pascal (Pa)	Energy	SI Unit Joule (J)
$T_F = 1.8(T_C) + 32$		1 atm = 760 mmHg		1 calorie (cal) = 4.184 J (exact)	
$T_C = \frac{T_F - 32}{1.8}$		1 atm = 101.325 kPa		1 kcal = 1000 cal	
$T_K = T_C + 273$		1 atm = 760 Torr			
		1 mole of gas = 22.4 L (STP)		**Water** Heat of fusion = 334 J/g; 80. cal/g Heat of vaporization = 2260 J/g; 540 cal/g Specific heat (*SH*) = 4.184 J/g °C; 1.00 cal/g °C	

Metric and SI Prefixes

Prefix	Symbol	Scientific Notation
Prefixes That Increase the Size of the Unit		
tera	T	10^{12}
giga	G	10^{9}
mega	M	10^{6}
kilo	k	10^{3}
Prefixes That Decrease the Size of the Unit		
deci	d	10^{-1}
centi	c	10^{-2}
milli	m	10^{-3}
micro	μ (mc)	10^{-6}
nano	n	10^{-9}
pico	p	10^{-12}

Formulas and Molar Masses of Some Typical Compounds

Name	Formula	Molar Mass (g/mole)	Name	Formula	Molar Mass (g/mole)
Ammonia	NH_3	17.03	Hydrogen chloride	HCl	36.46
Ammonium chloride	NH_4Cl	53.49	Iron(III) oxide	Fe_2O_3	159.70
Ammonium sulfate	$(NH_4)_2SO_4$	132.15	Magnesium oxide	MgO	40.31
Bromine	Br_2	159.80	Methane	CH_4	16.04
Butane	C_4H_{10}	58.12	Nitrogen	N_2	28.02
Calcium carbonate	$CaCO_3$	100.09	Oxygen	O_2	32.00
Calcium chloride	$CaCl_2$	110.98	Potassium carbonate	K_2CO_3	138.21
Calcium hydroxide	$Ca(OH)_2$	74.10	Potassium nitrate	KNO_3	101.11
Calcium oxide	CaO	56.08	Propane	C_3H_8	44.09
Carbon dioxide	CO_2	44.01	Sodium chloride	$NaCl$	58.44
Chlorine	Cl_2	70.90	Sodium hydroxide	$NaOH$	40.00
Copper(II) sulfide	CuS	95.62	Sulfur trioxide	SO_3	80.07
Hydrogen	H_2	2.016	Water	H_2O	18.02